Second Edition

CONCUSSIVE BRAIN TRAUMA

Neurobehavioral Impairment and Maladaptation

Second Edition

CONCUSSIVE BRAIN TRAUMA

Neurobehavioral Impairment and Maladaptation

Rolland S. Parker, Ph.D., F.A.P.A.
Consulting Neuropsychologist
Adjunct Professor of Clinical Neurology, NYU Medical School
New York, NY

CRC Press is an imprint of the
Taylor & Francis Group, an **informa** business

CRC Press
Taylor & Francis Group
6000 Broken Sound Parkway NW, Suite 300
Boca Raton, FL 33487-2742

Printed in the United States of America on acid-free paper
Version Date: 20110713

International Standard Book Number: 978-0-8493-8039-6 (Hardback)

Library of Congress Cataloging-in-Publication Data

Parker, Rolland S., author.
Concussive brain trauma : neurobehavioral impairment and maladaptation / Rolland S. Parker. -- Second edition.
p. ; cm.
Includes bibliographical references and index.
ISBN 978-0-8493-8039-6 (hardcover : alkaline paper)
1. Brain--Concussion. 2. Brain damage. I. Title.
[DNLM: 1. Brain Concussion--physiopathology. 2. Neurobehavioral Manifestations. WL 354]

RC394.C7P374 2012
616.8'047--dc22 2010043712

Visit the Taylor & Francis Web site at
http://www.taylorandfrancis.com

and the CRC Press Web site at
http://www.crcpress.com

This book is dedicated to my wife, Irmgard. It could be completed only with her emotional support, great effort, and patience with my involvement.

Contents

Preface

The frequently offered statement that a concussion reflects a "minor" head injury and usually "resolves" in a relatively brief interval (e.g., 90 days) is a misconception for both children and adults. Clinical experience has made it apparent that a sizable (although not precisely known) proportion of concussion victims remain impaired for very long periods of time with a substantial reduction in their quality of life. Consistent with lack of understanding about the potential neuropathology and the adaptive disorders following concussive accidents, adults with chronic symptoms are described disrespectfully as "the miserable minority" accused of malingering, seeking secondary gain, or burdened by an "emotional overlay." Frequent research design allows for error, reporting only on a very narrow range of cognitive measurements. A realistic estimate of the meaning of the findings is usually not performed through comparing the patient or group with an estimated preinjury performance baseline or qualitative estimate of functioning. A narrow sample of potential functions also contributes to error by not assessing a possible domain of impairment or distress.

To enable the researcher and the clinician to consider the documented outcome of concussion and other brain injury conditions, this book presents a comprehensive taxonomy of neurobehavioral disorders. This can guide the planning of an examination, referral to other specialists, and the integration and reporting of examination findings, as well as alert the practitioner to gaps in available data, leading to more effective treatment, diagnosis, and a better assessment of the outcome of an accident or other neurological condition. Several neglected areas of concern are addressed in detail: comorbid injuries from somatic areas during an accident creating damage to nervous system; the chronic somatic and pathophysiological effects that somatic injuries have upon performance and central nervous system (CNS) functioning (neurological, physiological, and stress/personality); late developing disorders; and subjective reactions to impairment and fear. Considerable attention is also paid to the chronic disorder (the "unhealed wound"). Adaptive capacity may be impaired by injury to the musculoskeletal system, other soft tissues, the spinal column, the spinal cord, unhealed tissue, pain, and other stressors. Injured neural pathways and centers do not account for all neurobehavioral impairment and distress. Unhealed tissue, pain, and other stressors cause physiological interference with adaptation by chronic dysregulation of important physiological systems, as well as cause a reduced enjoyment of life, capacity for social relations, and work. Particular concern arises for the immune, inflammatory, endocrine, circadian, and autonomic nervous systems. Dysregulation of these systems results in a bidirectional influence via the internal milieu between the soma and the CNS. Following trauma there are mutual system changes since neuroactive chemical substances ordinarily prevented from entering the brain can penetrate more easily due to disturbance of the various blood-brain barriers (BBB). After injury, dysregulation of production of neurobehavioral and other substances may create a neural "static" that reduces performance, alters the pattern of functioning of nervous and somatic tissues, and causes mood disturbances. These may be misattributed to traumatic brain injury or, if there is not documentation of a condition, the patient may be accused of malingering, exaggerating his or her symptoms, or having an "emotional overlay."

The complexity of neural support for adaptive functioning accounts for the impairing effects caused by the mechanical forces occurring in an accident. The detailed presentations of the neural basis for neurobehavioral functions help to account for these effects. This is a forensic application, helping to document the "credibility" of an accident involving compensation.

The chronic condition is given much weight. After a considerable interval, the association between the symptom and the accident may not be apparent. Prior practitioners may not have alerted the patient that there may be extended consequences of the injury. This denies both the patient and the current practitioner the information required to determine the correct etiology of a condition. Moreover, it is unusual for practitioners to routinely ask a patient about prior accidents or injuries that could lead to appropriate study and attribution.

The special nature of children's brain injuries and their physiological and neurobehavioral consequences has been studied extensively. There are problems in determining an accident's contribution to a disorder after a certain interval. Children's deficits of physiological, cognitive, and personality maturity may not be observable for years. The child may be nonverbal or deliberately evasive, the care provider may not reveal an accident to avoid responsibility, and subsequent disorders may become apparent only when the child does not manifest developmental milestones.

The reality of an accident (i.e., that both brain and body are injured) is expressed in these three neglected physiological mechanisms:

1. Breakdown of the allostatic system (far more significant than homeostasis for physiological adaptation during activities of daily living and stress). Chronic stress from impairment, pain, and injury leads to allostatic load (also known as "burnout"), which contributes to stress-related illness, vulnerability to contagious diseases, and the common traumatic brain injury (TBI) symptom of fatigability.
2. "Unhealed wounds" (i.e., chronic posttraumatic complaints) cause continued posttraumatic functioning (dysregulation) and fatigue of the immune, inflammatory, endocrine, circadian, and autonomic nervous systems.
3. Acute and chronic stress effects (the internal milieu) participate in bidirectional signals between the brain and the physiological systems (hormones, neurotransmitters, neuromodulators, cytokines, endogenous peptides, etc.):
   a. Identical biochemicals are secreted by neural and somatic organs.
   b. Particular biochemicals may activate multiple brain and somatic organs depending upon the location of receptors and their subtypes within particular tissues. There is a genetic variability for distribution.
   c. The widespread distribution of specific receptors in many types of organs, in addition to the "official" target, may account for drug side effects and genetic differences in medication effectiveness.
   d. After trauma, there is a change in the amount and distribution of neuroactive substances. These enter the brain's internal milieu through both the impaired BBB and other barriers, as well as the relatively unbarricaded circumventricular organs (CVO).
   e. The consequences for mood, arousal, cognition, and so on may be mistaken for the direct effects of TBI. However, if there is no neurological or radiological evidence for brain trauma, then the patient's complaints may be incorrectly assessed as faked.

Thus, the reader and the authors will collaborate in a study of an important public health problem, involving millions of persons ever year. This task is interesting, complex, sometimes ambiguous, but always challenging and important.

Acknowledgments

My husband Rolland did not live to see the publication of this book. He died on September 9, 2010 of cancer.

It was Rolland's lifelong goal to share his knowledge and research of traumatic brain injury and educate practitioners to help increase professional and public awareness of the so-called minor head injury. He devoted 10 years of his life to writing the second edition of *Concussive Brain Trauma*, completed it at the end of 2009, and received the page proofs in late spring 2010. At that time Rolland was already very ill. When he realized that he would be unable to undertake the overwhelming task of making the necessary corrections to the proofs, he was devastated.

In desperation I called the publisher CRC Press. It was my great fortune to reach Barbara Ellen Norwitz, Executive Editor, who listened compassionately to the sad news and immediately assured me that she and her colleague Jill J. Jurgensen, Senior Project Coordinator, would personally do all necessary work in preparation for the publication of the second edition of *Concussive Brain Trauma*. When I conveyed this message to him, Rolland became tranquil.

I cannot adequately express my thanks to Barbara Norwitz, Jill Jurgensen, Jessica Vakili, and Robert Sims for their commitment to Rolland's work. I am deeply grateful to them for making the posthumous publication possible.

Irmgard Parker

1 Concussion
Not Always "Minor" Head Injury

OVERVIEW: "MILD" TBI (MTBI) MAY NOT BE SO "MILD"

THE EXPERIENCE OF CONCUSSION

Clinical example of a car/truck collision: A 34-year-old man was driving a car on a highway. "All of a sudden I got hit by a huge trailer truck." He was going at 20 mph and was slowing down because of the traffic. His car was rear-ended by a truck; he did not know the speed of the truck. "I was hit several times." He was shaking all over. His head hit the corner of the headrest, which was not totally padded. "It was a sunny day. I felt so good." For hours he didn't know what was going on. "I was totally shocked. I can't describe how troubled I felt at the moment. I felt that this will never stop when this truck was hitting me. I felt like it was going to be forever." He didn't know how long it was before emergency assistance came. "I am totally different person. I can't think straight. I can't socialize with people. I am getting into arguments. I lost my friends. I broke up with my girl friend. I can't do anything. I can't go out to do sports."

The outcome of the accident was that he was affected by central visual scotoma due to parieto-occipital lesion. He never returned to work, abandoned plans for a graduate degree, suffered from chronic pain, required a cane, and so forth.

OVERVIEW OF COMORBID TBI AND SOMATIC INJURY

CONCUSSION

Concussion is the acute psychological experience of trauma incurred through head impact, acceleration, or both: an alteration or limited loss of consciousness (LOC). Generally, it is without sufficient neurotrauma to be detected by neuroimaging procedures. The limit of "lesser" LOC is about 20 min. The postconcussion syndrome (PCS) occasionally occurs without documented LOC. This may be due to the *geometry of the accident*: Impact and acceleration of the head and neck may not affect those central nervous system (CNS) centers and tracts maintaining awareness. Persistent postconcussion symptoms are caused directly by neural injury (CNS, cranial nerves, peripheral nerves, and brachial plexuses), unhealed somatic injuries, and physiological disruption.

The *postconcussion syndrome (PCS)* describes the multiple consequences of mechanical injuries to the brain that frequently result in comorbid traumatic brain injury (TBI) and somatic injuries. Its definition derives from common, but not universal, alterations of consciousness (AOC) after impact and/or acceleration and deceleration of the head and neck. Understanding TBI and AOC is enhanced by the examiner's attempted reconstruction of the accident: geometric configuration of the head and body; the physical environment; and the size, direction, and nature of the injuring physical force and surface. Characteristic AOC includes retrograde and anterograde amnesia, a limited interval of unconsciousness, and confusion or disorientation varying from minutes to years with unpredictable disappearance. Injuries are caused by physical forces (impact; acceleration/deceleration) from injuries in motor vehicle accidents (MVAs), assault, falls, falling objects, blast, and also electrical accidents.

Varied Outcomes

PCS (Parker, 2001), a so-called "mild" TBI, is a major *public health* problem. Trauma refers to the anatomical and emotional damage incurred by the person, and it is a multisystem reaction (Zellweger et al., 2001). Concurrent generalized brain effects occur due to mechanical forces, accompanied by brain and somatic physiological reactions to the trauma. These cause varying functional and injury-related behavioral changes. The nature of the event creates varied modal physical forces, and within each type of injury are different patterns of TBI.

The range of potential chronic disorders is frequently minimized in research and clinical practice. The range of potential neurobehavioral disorders after an accident is presented as a "taxonomy of neurobehavioral disorders." A cerebral disorder ("concussion") is frequently comorbid with significant somatic injuries. Thus, the clinician should consider a wide range of disorders, extending beyond the personal specialty. Concussion reflects a heterogeneous and complex spectrum of brain and somatic pathologies, clinical severity, and baseline prognostic risk (Maas et al., 2007). Neglect of particular domains in the acute injury period may miss significant signs of a potentially significant and chronic disorder. Thus, concussive-level injury may be both undiagnosed and untreated. Symptoms predicting a negative outcome have been described.

Concussive accidents result in varied comorbid psychological and anatomical conditions. Previously reported outcome markers have not been replicated in the literature. Further, typical outcome studies have concentrated upon a narrow range of cognitive, neurological, and physiological symptoms, without comparing posttraumatic functioning with an adaptive baseline. Therefore, there are varying patterns of neurobehavioral outcomes. A wide range of documented neurobehavioral disorders and symptoms is presented as a taxonomy of neurobehavioral disorders. This will be useful in planning an examination of a patient, preparing a report, and assessing the clinical outcome after an accident. Psychological and injury-related stress represent different, although overlapping, syndromes, whose effects are genetic, cellular, systemic, neurological, physiological, and mental.

"Head injuries" (without fracture or surgical involvement) have highly varied outcomes, from apparent loss of symptoms (so-called "minor TBI") to significant chronic impairment. Millions in the United States suffer "polytrauma" (comorbid brain and somatic injury), resulting in complex impairment interfering with safety, employment, family and social relationships, and quality of life. A high proportion is disabled and impoverished, and cannot obtain services due to inability to obtain transportation, lack of insurance, or denial of the necessity for treatment.

Errors of assessing the PCS are common: (1) No record of head examination after an accident. (2) The patient is not alerted to potential long-term consequences or the need for follow-up. (3) Negative radiological or neurological findings are misinterpreted to mean that no diffuse TBI has occurred. (4) Dysfunctioning is attributed to an "emotional overlay" without competent personality examination. (5) Practitioners may deny the attribution of persistent TBI symptoms based upon the stereotype that "minor" head injuries "resolve" in 3 months. Complainers are considered the "miserable minority," and are suspected of malingering, factitious disorders, or secondary gain. (6) Status is assessed prematurely, not considering interference with adaptive capacity or possible late-developing symptoms.

IGNORING THE RANGE OF POSSIBLE PCS DISORDERS

Clinical example: Emergency room (ER) personnel disregarding the actual accident: The patient was a man in his late 40s who was referred for study to confirm whether he was qualified for disability. Nine years earlier he had fallen from a height estimated at 40 to 45 feet while attempting to enter his own apartment when he was locked out. This vignette is based upon an intensive interview and review of the hospital records. The reader is directed to the height of the fall, the lack of close attention of the emergency squad, the patient's misleading description of his own condition, and the hospital's unawareness that his dysregulated behavior prognosticated a severe personality disorder. His discharge was described as "routine" a week after the accident. There had been an incorrect statement by his insurer that his policy had lapsed.

Observations: He was alert, cooperative, and did not appear to exaggerate or mislead concerning his condition. His affect was somewhat flat; despite a significant history of injury and emotional distress, he did not clearly exhibit anxiety, depression, or anger. The examiner had the impression that he was a reliable historian. Pain and orthopedic injury were apparent.

Record review: The patient was described as undomiciled, evicted from his apartment, and living alone. He was stated to withhold personal information, to display strange affect, and to be alert and oriented × 3 (time, place, person); memory appears to be intact. He had a skill in the construction industry, and was unemployed about 6 months.

His Glasgow Coma Score (GCS) was 8, which is usually categorized as a moderate brain injury. He denied LOC and was assessed as alert and oriented in three modalities. His statement at the accident scene that "I feel fine" is surprising in the context of the characteristics of the accident. It may indicate actual lack of orientation or denial. His examination revealed a bruise to his forehead, that is, a small abrasion with a hematoma on the left forehead.

Computed tomography (CT) of brain indicated mild enlargement of all the cerebrospinal fluid-containing spaces, consistent with cerebral atrophy.

Psychiatric consultation: Depression; alcohol-induced anxiety disorder; apparently he took the back brace off, which was dangerous, and could risk injury to his spine and spinal cord.

Velocity of his body upon striking the ground in feet per second was estimated: If the fall height was 40 feet, it was 50.8 feet/sec, and 53.8 if the height was 45 feet.

Abstract of interview: He stated that he graduated from a technical high school known for its high standards. He had an associate degree. He progressed to more skilled construction work over the years.

His description of the scene asserted that the emergency personnel did not recognize that he was injured. He asserted that he was dazed for about an hour. In the hospital he experienced pain, confusion, and fear. Self-described as disoriented and disorganized. (This self-description is consistent with hospital records, which indicated a level of cognitive or emotional disturbance anticipating later behavioral disorganization. The hospital record describes his discharge as "routine." This examiner's review suggested that apart from his documented injuries, there was significant evidence of impulsivity and possible disorientation, which indicated the need for postdischarge counseling and rehabilitation.)

He described his condition following his release from the hospital: "very little memory; began living on the street; I can't completely remember where I was or who I was talking to; began drinking because I was in pain. I was redeeming cans." He met his present lady companion 3 years after the accident.

- Preexisting conditions: Significant accidents or illness not reported.
- Subsequent conditions: Alcohol addiction, which he attributes to pain.

Complaints of impairment:

Sensory—close vision; dizziness; tinnitus; headaches; pain in wrists, ankles, elbow; disturbed sleep (nightmares; wakes up several times); oversensitivity to cold and loud sounds; partial-seizure like sensory phenomena (sound and somesthetic; electroencephalogram [EEG] recommended); physiological (fatigability; treated for a thyroid condition); affect (increased anxiety; more sensitive to anger; depressed but denies suicidal thoughts); memory and grasp of reality not completely recovered; could not work (fear of heights with use of a ladder; trouble steadying an electrical saw with his right hand); identity (asserts that he was a member of the high-IQ group Mensa). He wonders why he behaves the way that he does and is pessimistic about the future.

Status—Unemployed; is reliant upon his companion for some assistance. Asserts that their relationship is firm.

Diagnostic impression: PCS, severe, chronic.

Recommendations: Further neuropsychological study; internal medicine study; neurological consultation; orthopedic examination.

Comorbid Somatic and Neurological Trauma

PCS is the consequence of a mechanical injury, often ignored by clinicians, or treated without awareness of its wide range of disorders, discomforts and dysfunctions, or chronic outcome. A person with possible TBI usually has comorbid somatic trauma in other parts of the body (polytrauma) and frequently experiences acute or chronic stress, anxiety, or depressive disorder. "The unhealed wound" creates neurobehavioral effects in the postconcussive syndrome, overlapping with dysfunctions directly due to mechanical brain and head injuries, musculoskeletal trauma, and pain. Lateralized dysfunctions such as restriction of range of movement, adaptive difficulties, and interruption with nerve pathways including muscular, ligaments, and bone occur. Vascular Injury and stress results, creating vasospasm affecting the arteries of the neck and skull base leading to dizziness and faintness, and hypoperfusion of the brain parenchyma and basal ganglia and thalamus (lesions detected by single photon emission computerized tomography [SPECT] and other imaging procedures).

Consequently, the use of mTBI as a nomenclature frequently contributes to an inaccurate assessment of the actual traumatic basis of the patient's adaptive disorder. This leads to reduced quality of professional service. It is necessary in clinical and formal diagnosis to differentiate between anatomical findings, for example, without definite radiological or CNS neurological evidence, and the neurobehavioral consequences, which can be disabling. Accuracy is enhanced by giving extensive consideration to the noncerebral injuries and their disabling and physiological consequences. A useful clinical guideline is: "If the patient has an accident, and is examined for a head injury, closely examine the body. If the patient is examined for somatic injury, don't forget about the head." Examination of records frequently reveals that there is no evidence for injury victims that while in acute care any examination of the head was performed.

Posttraumatic Syndrome: Initial Examination

The posttraumatic syndrome may be seen when the posttraumatic examination of the patient as incomplete. Thus, ER physicians caring for patients with a head injury should conduct a thorough examination, including radiological examination if indicated, and reassure the patient that no serious damage has been caused (Boes et al., 2008). The present writer would modify this advice: since patients and family may fear a skull fracture or brain injury, the reassurance should be restricted to the current medical findings. If the history of the accident and/or head injury is sufficient to warrant an ER examination, the patient should be counseled concerning frequent symptoms of the PCS and advised to consult a neurologist or other practitioner with head injury experience for examination. Further, at any time in the future when presenting a history, the facts of the accident should be reported.

If in the acute or chronic phase the possible contribution of an accident to a given disorder or complaint is ignored, this leads to a dissociation between the symptoms of the chronic stage and their etiology in cerebral or somatic injury. Depending upon the professional specialty of the caregiver, one set of disorders may be neglected or misdiagnosed as to their actual trauma. Lack of initial correct diagnosis of cerebral injury, and its comorbid accompaniments, reduces the quantity of TBI in public health statistics, neglect of safety in public affairs, and nonattribution of later disorders to a single or multiple accidents. Thus, the differentiation between "head injury" and brain injury (Fearnside & Simson, 2005) is only the beginning of comprehensive assessment.

ADAPTATION

Adaptation is a biological concept that helps the neurobehavioral examiner to organize and assess current status. The clinician studies the effects of an accident or medical problem upon the patient's capacity and style of dealing with the daily requirements of life: the distinctive style of coping with the changing environment, including psychological changes and moods that develop, and the complex physiological, functional, and subjective reactions. Adaptation may also be maladaptive or self-destructive (solutions to problems are irrelevant, make the situation worse, or involve avoidable errors) (Parker, 1981). After accidents causing TBI, emotional disorders, persistent cognitive loss, and other neuropsychological symptoms impair greatly the capacity to adapt (Parker, 1996).

The components of adaptive success are *genetic* (hereditary), *phenotypic* (expression of genes in a particular personal history), and *stylistic* (learning and preferences). Adaptation is expressed as *mobility* (changing environment), *autoplastic* (personality changes), and *alloplastic* (changing some aspect of our environment) (Parker, 1981, pp. 43–44; 1990, p. 76). Advantageous adaptive efforts create a lifestyle that is safe, mature, independent, and enjoyable. This is characterized by productivity, pleasant moods, and self-esteem. Under demands, adaptation utilizes these functions to redirect behavior: increase arousal, alertness, cognition, vigilance, and focus attention. Adaptive success requires the integrity of most of the range of neurobehavioral functions and specifically the ability to be independent; mobility; good judgment (anticipating the consequences of one's actions and decisions); effective error prevention and monitoring of ongoing activities; freedom from dysphoric feelings, such as anxiety; lack of pain; adequate sensory abilities; strength; range of motion; lack of scarring or other embarrassing circumstances; and absence of impairment.

Effective adaptability leads to productivity, pleasant moods, and self-esteem. Yet, there are persons of historic accomplishments who had both significant illness and impairment of the organs necessary for their productivity (Beethoven's deafness and generalized organ disorder revealed at autopsy, for example).

THE POSTCONCUSSION SYNDROME

There are different levels of precision by which the numerous PCS symptoms may be arranged.

Factor analysis: Acute concussion evaluation (ACE).
Somatic (24% of variance accounted for): Sensitivity to light and noise; nausea; dizziness; headaches; drowsiness; balance; fatigue.
Emotional (8.5%): Fatigue; more emotional; sadness; irritability; nervousness; sleeping more than usual.
Cognitive (7%): Difficulty remembering; feeling mentally foggy; difficulty concentrating; feeling slowed down.
Sleep (6.8%): Sleeping less than usual; trouble falling asleep; sleeping more than usual.

Only 48% of the variance was accounted for. There was no systematic relationship between ACE total score and age or ethnic group membership. More symptoms were recorded by girls than boys. Higher scores were associated with LOC and anterograde amnesia, but not retrograde amnesia. Use of such an instrument presumes follow-up with referral to a specialist should there be little or no improvement (Gioia et al., 2008).

THE ACCIDENT VICTIM

Reasonably complete assessment of an injured person is a complex process. Somatic trauma is attributed to mechanical strains causing injuries to spinal muscles, ligaments, and stretching of spinal cord and brainstem structures, including the hypothalamus (Smith, 1989). In addition to the neurobehavioral effects of both nervous system and somatic injuries that may create disorders over an extended range of neurobehavioral domains, the examiner considers such factors as genetic, developmental, and preexisting pathological conditions (medical and psychological), cultural and socioeconomic aspects of the lifestyle that influence style and quality of observed behavior and formal performance, prior accidents and other medical conditions, the motivation of the patient within the social response to the alleged accident and impairment, recommendations for treatment (which may or may not be financially available), expected outcome of the injuries, possible late-developing symptoms, and prognosis. The medical examinations used for the accident victim (neurological, orthopedic, SPECT, magnetic resonance imaging [MRI], CT, EEG, X-ray) document different domains of trauma than the complex neurobehavioral functions studied by the neuropsychologist. The latter are more sensitive to neurobehavioral disorders consequent to TBI at the nonsurgical level of brain injury. The examiner will also be alert to the systemic consequences of chronic injuries, such as loss of stamina (fatigability), mood disorders, and vulnerability to diseases that are described as stress-related (Schnurr & Green, 2004).

Accidents are complex events. Their pathological consequences include the entire range of human performance and experience. The range of dysfunctions far exceeds those attributable primarily to the brain. Thus, the examiner needs to be aware of the direct and systemic effects of significant comorbid injury to any other part of the body. A variety of considerations of importance in claims of personal injury will be considered: (1) physical forces and the environment of the accident that create bodily injury; (2) trauma, or personal injury from an accident, and the phases that tissue goes through in a chronic condition; (3) the range of neurobehavioral disorders caused by an accident; (4) biomechanics, that is, the interaction between environmental objects, physical forces, and tissue that results in injury; (5) polytrauma, that is, multiple injuries that are comorbid with head injuries and the complex physiological interactions when such wounds remain unhealed; (6) the neurobehavioral, health, and stamina consequences of chronic conditions of injury and stress; and (7) various professional considerations in performing a forensic neuropsychological examination to assess claims of personal injury after an accident. An injury causing concussion or other TBI is usually accompanied by trauma to various other parts of the body, a condition termed "polytrauma" by the armed forces: CNS; cranial nerves; vascular; muscular; ligaments; bone; systemic (circadian, hormonal, inflammatory, immune); and psychodynamic.

Follow-up

Frequently, an accident victim is examined in the Emergency Department (ED) and released without any record to indicate that the head was examined, or whether the physician gave any counseling to the patient or parent. The need for a follow-up at the hospital or usual caregiver is neglected. The possibility of a later hemorrhage should be conveyed. The following procedure has been recommended to increase the likelihood of a follow-up: demonstrate to the patient the degree of neurological injury; educate the patient about the nature of mild head injury (MHI) and the postconcussive syndrome; arrange for an appropriate consultant in the ED for a more precise evaluation, further education, and follow-up; telephone high-risk patients after discharge to reinforce education and need for follow-up; and establish community-wide specialty clinics through a coordinated effort among primary care physicians, specialists, and insurance companies (Bazarian et al., 2000).

Risk-based Classification of "Mild" Head Injury

An approach to the nomenclature for "trivial, minor, or MHI" has been offered that recognizes that the initial presentation may discourage assessing the risk of short-term deterioration. It is reality-based, while noting that summarizing the previous adjectives into the term "MHI" violates two basic principles: a "mild" head injury is properly confined only to accidents occurring to someone else, and this term conflates behavioral disturbance with a statement about the lack of scan-detected trauma. The following is designed to detect possible deterioration due to intracranial hematoma of patients who initially appear to be at low risk. The basic "minor" patient had a GCS of 15 without LOC, amnesia, vomiting, or diffuse headache. Other groups are defined as low-, medium-, or high-risk MHI by the presence of particular criteria: GCS of 14–15, symptoms (LOC, amnesia, vomiting, diffuse headache), CT scan (e.g., fracture, hemorrhage, lucency within the hemorrhage), LOC (Servadel et al., 2001).

Our Guiding Theses

Concussive brain trauma initiates a multisystem dysregulation in addition to cerebral injury. The somatic and emotional pathology create dysfunctioning beyond their direct effects. TBI is a process initiated at a stage of a person's development, which proceeds through a series of phases. Its outcome is dependent upon the developmental stage, the functional baseline at the time of trauma, the persons stress resistance capacity, the nature of the brain injury, including polytrauma (i.e., both brain damage and accompanying somatic injury), and whether environmental interventions are benign and useful, or neglecting patient treatment, and motivated by other interests than patient treatment and eventual recovery.

- Long-lasting, chronic injury ("the unhealed wound") causes direct and indirect impairment and distress. Physiological substances are carried through the vascular and some nerve pathways to the brain, which further creates neurobehavioral impairment ("the internal milieu"). Some of the chronic neurobehavioral symptoms after head injury are due to somatic injury. They are not functions only of cerebral derangement; rather, they are reactions to dysregulation of a group of physiological systems.
- The same neuromodulatory pathways are used in trauma and in health. Endocrine and other systems have bidirectional signals to the brain, affecting both brain and peripheral responses in both stress and adaptive functions. Specific functions (e.g., hormonal effects upon defined tissues) actually have widespread expression in alternate tissues, which are not the familiar targets (including the brain). Examples are some drug side effects and cerebral neurobehavioral effects of substances released by the hormonal, immune, and inflammatory systems after injury.
- Physiological dysregulation: Systems (immune; inflammatory; hormonal; circadian); disruption of the barriers between the multiple liquid compartments surrounding the brain, causing infiltration of toxins, hormones, neurotransmitters, and immunoinflammatory signals generally excluded from the brain.

INTRODUCTION TO CONCUSSIVE BRAIN TRAUMA

Concussion is a complex phenomenon, often ignored by clinicians, or treated without awareness of the wide range of disorders and discomforts that are experienced by the patient, or its potential chronic outcome. It is the consequence of a mechanical injury, that is, the aftermath of various forces and directions applied to the head and body during MVAs, falling objects, falls, assaults, and so forth.

A historical review of the medical history of concussion from the Sumerians and Egyptians to date (McCrory & Berkovic, 2001) offers such topics as understanding the anatomy of the head, lateralizing neurological dysfunctions (eye deviations and aphasia), eventual differentiation of concussion from basal skull fracture and major brain injury, rapid disappearance of symptoms, developing awareness of the nature of trauma with the availability of microscopic study, distinct behavioral disorders following a blow, awareness of later-developing symptoms, and differentiation of functional from structural injury. This text elucidates the inclusion of modern concepts into understanding concussive brain trauma, namely that

- Head injury is usually associated with injury to other parts of the body. Therefore, a wide range of symptom presentation and etiology must be considered;
- Persistent physiological dysregulation affects outcome and health;
- Psychodynamic reactions occur as a result of the injury and impairment and are a part of the PCS;
- Persistent symptoms may be described as stressful, are components of the outcome, and influence health and stamina.

CONTROVERSIES CONCERNING CONCUSSIVE BRAIN TRAUMA

Weak Clinical Information Gathering

There are numerous reasons for the controversies concerning the nature of, and the outcome of, concussive-level accidents, including the complexity of comorbid trauma. Here we will consider deficiencies of clinical information gathering (Parker, 1995):

- *False information*: Questions may be asked while the patient is in a state of altered consciousness. An example is the assertion of "no LOC" by emergency personnel who arrived on the scene after the patient recovered consciousness. A patient may report later that they were not certain that correct information was given, or that they were in a state of confusion.
- *Inadequate examination*: Examination of the head after a somatic injury is often ignored according to the absence of data in the hospital record. Apparently all attention is paid to somatic injuries.
- *Ignoring the limitations of radiological scanning*: CT and MRI have their uses, but they will not detect diffuse axonal injury (DAI), which causes much of the mental impairment after a less serious head injury. Please note that 10–15% of patients with clinically severe head injury have a normal CT scan. In this circumstance, other vascular disruptions may exist, and angiography should be considered (Evans et al., 2007).
- *Ignoring the limitations of the neurological examination*: Sensorimotor procedures can be diagnostically reliable, but they will not detect impairment of complex functions. When the physician studies complex cognitive functions, what is lacking is the precision of the usual standardized psychological procedures and potential performance reduction from an estimated preinjury baseline.
- *Lack of inquiry concerning prior TBI*: This may result in ignoring the potential for greater current impairment consequent to the reduced ceiling of ability and occult injuries not detected by the examiner or reported by the patient.
- *Lack of reporting trauma to the patient or caregiver*: The existence of a TBI or of potential late-developing symptoms is not relayed to the patient. This does not encourage a wider examination than that which is customary for the clinician.
- *Not studying, or misattributing subjective symptoms*: The examiner may not observe, inquire into, or report emotional distress, or may assume incorrectly that these are not diagnostic. Thus, stress cerebral personality disorders are ignored.
- *Comorbid neural and somatic trauma*: Body injuries can create performance reduction that may be misattributed to a brain lesion. Similarly, the bidirectional signaling between the soma and brain creates cerebral dysfunctions consequent to systemic changes, which affect brain functioning due to reduced efficiency of the blood–brain and other barriers.
- *Lack of any examination*: After an accident, the patient may not seek an examination despite significant injuries.

- *Misattributing the effects of comorbid somatic and brain injuries*: Somatic injuries may interfere with neurological or other performance in a way that mimics cerebral disorder. Systemic dysregulation may create neuroactive substances that interfere with brain function, being permitted to enter the CNS due to reduced efficiency of the blood–brain and other barriers.

What Is the Impairing Effect of a Concussion?

Imprecise measurement: The definitions of TBI and PCS available to the clinician are imprecise. The GCS range of 13–15 in practice seems to be a broad range of trauma, augmented by varied comorbid intracranial injuries also having a relatively broad span. Further, there are cases in which after some mechanical injury in which there is no report of LOC, there is clinical evidence for impaired functioning. There is no agreed-upon definition differentiating "mild" and "moderate" TBI, with different intervals having been proposed. Part of the confusion stems from a lack of precise observation and recording at the scene of the accident, and also patient misrepresentation of posttraumatic amnesia (PTA) as LOC. Moreover, the alteration of consciousness depends in part upon the geometry of the patient's body and head within the field of force (which may be only accelerational or include a direct impact at different angles to the head).

Measurement imprecision: The outcome of patients with head impact followed by postconcussive symptoms was examined one month or longer after the accident. Varying levels of altered consciousness were studied by comparing those with a brief LOC ("concussion") with those who were disorientated, confused, with or without amnesia, but no LOC ("mild concussion"). Only procedures measuring short-term memory, abstraction, and concentration were utilized. Similar levels of dysfunction were obtained with injuries associated with a brief LOC or being dazed without LOC, and examination before or after 3 months. Dysfunction varied between procedures, leading to the conclusion that even the most sensitive neuropsychological tests do not tap some subtle processing deficits (Leininger et al., 1990).

What Proportion of PCS Victims Has Long-Lasting Conditions?

The assertion that concussion is a condition that "quickly resolves" is commonplace in the literature and among clinicians. Iverson (2005) summarizes the conclusion of a large group of students of PCS as asserting that for most people the injury is self-limiting and follows a generally predictable course, with permanent cognitive, psychological, or psychosocial problems due to biological effects of the injury as uncommon in trauma patients and rare in athletes. The statistics seem at variance with the previous impression: It is asserted that of approximately 230,000 persons hospitalized annually with TBI, an estimated 83,000 are discharged with TBI-related deficits, and 5.3 million Americans currently have TBI-related disability. There are individuals who are unaware of the TBI patients needs. Adding to the imprecision of this conclusion is the unavailability of information concerning the outcome of a major segment of the injured population: At least 25% of people with mTBI do not seek treatment; 14% are seen in private offices or clinics; and athletes with minor concussion conceal that event. Further, the trauma of the identified concussion victim varies: 7–20% of mTBIs have bleeding, bruising, or swelling (*complicated mTBI*) on the day-of-injury CT. While a GSC score of 15 is associated with 5% positive CTs, a score of 13 manifests 30% abnormality. Persistent symptoms may be accompanied by positive findings on positron emission tomography (PET) and SPECT. Complicated mTBI manifests poor neuropsychological performance at intervals up to 3–5 years. Reference is made to a well-designed study whose members (elderly and younger) had radically different outcomes. Referring to the group whose functional outcome (old and young) at 6 months was generally good to excellent, one wonders what is the range of disorders sampled so that potential disorders were not missed? The present writer often assesses research concerning outcome as sampling so narrow a group of potential disorders that statements concerning outcome are overly general.

What, then, is the significance of the group of patients who seek varied types of treatment for extended periods, do not return to work, are impaired on the job or in school, have documented musculoskeletal or neurological difficulties as well as impaired balance, gait, stance, and so forth? It is apparent that "concussion" is an abstraction representing a wide range of injury, and not specifying the nature or site(s) of trauma. Considering PCS as a condition with comorbid TBI and somatic disorders will provide insight into this paradox.

What Is the Nature of PCS?

PCS has been attributed to emotional reactions, physiological responses, symptom exaggeration, secondary gain, attentional disorder, dysfunction of information processing and mental speed, DAI, cranial nerve injury, emotional reaction to being injured, and so forth. PCS is actually a summary of a disparate group of dysfunctions consequent to injuries to the various components of the nervous system, the soma, and the physiological and psychological consequences of chronic unhealed injuries. Proper attention to the effects of somatic trauma and posttraumatic physiological and other hea[illegible] ıtifies conditions that respond to treatment, a[illegible] ·oach to the diagnosis of a person with a cred[illegible] racteristic symptoms, whether there are pre[illegible] ımented impairment, and noncerebral, soma[illegible] esented in this text is markedly different: If [illegible] fter an injury, a large proportion of individu[illegible] have not been previously identified.

How Permanent Is T[illegible]

There is a bitter partisa[illegible] ıl assertion is the issue of permanency of the s[illegible] is an "mild" traumatic brain injury (mTBI) th[illegible] ss. Earlier discussions have considered concu[illegible] t detectable pathology. Some clinicians believ[illegible] ad injury do not seek treatment. The specific [illegible] tes that the patient has recovered from traum[illegible] ıformation-processing speed (Mittenberg & S[illegible]).

On the other side, one review indicates that the prevalence of complaints 6 months after the trauma is 20–80%. Female sex, more advanced age, and prior mTBI are associated with poor outcome (De Krujik et al., 2002). Thirty-five percent of the 1.5 million TBIs recorded in 1991 received medical attention but were not admitted to the hospital. While these are considered "mild," they can have long-lasting sequelae. A small proportion of individuals hospitalized with mild TBI experience long-term disability (Guerrero et al., 2000).

Measurement suggests that litigating patients manifest reduced IQs, or Memory Index scores 7 points lower than those not seeking compensation. Further, the cognitive and emotional sequelae of MHI are considered distinct. mTBI is a faulty description that does not differentiate between behavioral disorders and brain trauma that cannot be documented with current medical procedures after an impact or hyperextension/hypoflexion injury. A team with a psychiatric affiliation (Hall et al., 2005) notes that PCS is difficult to define medically, since many symptoms are subjective. The diagnostic criteria are varied, and may be defined by the specialty of the physician who examines the patient (neurology, psychiatry, pain management), the clinical setting in which the patient is seen, and whether more rigorous research criteria are applied.

More recently it is acknowledged that there is some Permanent damage (Blumbergs, 2005). A substantial, yet unknown, proportion of concussive accident victims experiences significant chronic posttraumatic impairment and disturbed quality of life (Parker, 2001). The National Center for Injury Prevention and Control (2003, Preface), in a report to Congress, stated that "the consequences of mTBI may not, in fact, be mild" Clinical research has provided evidence that these injuries can cause serious lasting problems.

A particular definition of mTBI was the basis for a comparison of accident victims with and without head injury (Kraus et al., 2005): An uncomplicated head injury is generally characterized as a concussion. In such injuries there is evidence of impact and acceleration/deceleration and resulting in one or more of the following symptoms/signs: confusion or disorientation; observed or reported LOC for less than 30 min; PTA for less than 24 h; and a GCS of 13–15 on arrival at the ED. Review of findings illustrates the complexity of head injury outcome: After 6 months, the mTBI cohort was more likely to report headaches, double vision, memory or learning problems, and lower tolerance for alcohol. Yet, they were less likely to report a change of employment or falling, and were less likely to be taking prescribed medication, and exhibited no difference in use of medical service. In short, overall, the outcome was both benign and impaired upon standard clinical symptoms resolution (Thompson et al., 2005).

How Minor Is mTBI?

Clinical Example: "Minor" Head Injury: A woman with a responsible office job in which she had difficult contacts with the public was struck by a falling object while sitting at work.

"I was dizzy, headaches, nausea. I felt sick. I felt I was dying. I couldn't remember where I was sometimes." She has a severe hearing loss. In a quiet room she doesn't hear if one ear is covered by the telephone, and she can't hear a car or truck. She has motion sickness, and never returned to work because of headaches, dizziness, nausea, memory loss, slowness in thinking and moving, lack of motivation, feeling tired at all times, and becoming dizzy when bending. She could cope with situations, but now everything is big. She avoids taking care of the checkbook because she feels incompetent. She loses things and doesn't write letters to people. "I never was depressed before. Now, I'm always depressed." She self-described herself as angry, irritable, and very short-tempered. "I'm just angry that nobody understands me."

Concealment: She tries to hide such feelings as dizziness. She doesn't want people to see the way she is. They would say: "Why don't you see the doctor when I say I'll topple over. They don't understand me. "Maybe you have a brain tumor."

Personality change: Her life is very sedate. She does a little house cleaning and "really doesn't care. I watch TV. Others say I'm not like I was. I was active, I ran. Now I walk very slowly. That's not me. I see very few of my friends. I'm not very good company."

Her family's reaction: She was an all-around person, good housekeeper, good wife. Now, she is half the woman she used to be. She could out-think anybody, with or without a calculator. She was also good verbally at voicing her opinions. She also held down a full-time job. Her wits are slow, not totally blunt, but dull. She can't think as fast as she used to. She stumbles with mathematics, has trouble recalling words. Everything is on the tip of her tongue. Asked what's wrong, she replies: "Nothing." She doesn't clean the house. She doesn't remember what I said to her during a conversation. Within half an hour she says: "You didn't tell me that." Physically, she can't seem to do what she used to do, for example, waking up at dawn, rushing around the house to get everybody fed, get yetting herself dressed to go to work, cleaning the house, coming home, doing food shopping, Generally being a wife, mother, and worker at the same time.

Mood change: Crowded places and noise make her turn pale as a ghost, get an immediate headache, and become dizzy and nauseous. She can't take confusion.

She cries for no reason, and she's like somebody who's been in a war. She wakes up in the middle of the night from nightmares. She has temper outbursts, and fights over meaningless things that never bothered her before. During the working day she had patience to deal with irate members of the public. If they used profanity, she had the temperament to handle them in a professional manner. Now she loses her temper over something that was very minute. She fights with her children, getting into an argument for no reason at all. She constantly sleeps, and jumps as though something scared her.

Physical trauma: For just walking down stairs she has to be helped. Now she would have trouble walking to the bus without assistance, and she could not maintain her balance while the bus is moving. Bending over (which she would have to do at work to get to filing cabinets) also throws off her equilibrium. At work, she could not get on an elevator and not have her equilibrium thrown off. She could not lift 10 lbs nor bend. File drawers below waist level are difficult to use.

The writer avoids the use of "minor" traumatic brain injury as disparaging to the patient. While the term "minor" traumatic brain injury will be mentioned in citing published studies, the writer discourages its clinical and forensic use: first, as generally used, it erroneously combines the issues
of brain trauma and traumat individuals who are catego-
rized with minor TBI experie itions. The term
PCS is more precise. This r ral disturbance.
It is anatomically correct, s dents with more
extensive AOC and firm det f the assessment
involves inquiry into such is mpacting object,
or surrounding motor vehic mpacting object;
the physical qualities of the mber and nature
of impacts. A detailed inter , and perhaps the
location, of the trauma. In a ctim of a striking
automobile." Interview may person was again
knocked against the oncom impacts involving
different parts of the body

The nomenclature of "m ild" TBI illustrate
subdural hemorrhage, cont matoma (Samson,
2005). It may be inferred t h greater alertness
as to the frequent significa table disorders, we
might agree that the ultim

I offer the concept that a person with head injury, that is, possible TBI usually has comorbid somatic trauma in other parts of the body, and frequently experiences some time of acute or chronic stress, anxiety, or depressive disorder. Stress, anxiety, and chronic somatic conditions ("the unhealed wound) manifest deleterious neurobehavioral effects that overlap with the frequently detected dysfunctions after mechanical brain and head injuries, that is, PCS. Head injuries, from which we infer potential TBI, are accompanied by somatic injuries, which contribute to somatic disturbances and secondary brain damage. PCS symptoms are based upon multiple traumas: CNS and peripheral neurological; somatic trauma (muscular, ligaments, bone); vascular (injury and stress create vasospasm affecting the arteries of the neck and skull base). The result is interference with cerebral perfusion causing dizzy spells, hypoperfusion of the brain parenchyma and basal ganglia and thalamus (lesions detected by SPECT and other imaging procedures); psychodynamic, and so forth, damage to the very *organ of adaptation*, that is, the brain. The interaction of physiological systems, hormonal, inflammatory, and immune affects the CNS and is transmitted by systems affecting cognition, health, stamina, and mood.

Imprecise Definitions and Controversies with Multiple Definitions

The conceptual history of concussion has focused upon such topics as brain damage, whether structural disorders are neural or perhaps vascular and/or respiratory, and whether it was a structural or functional disturbance (McCrory & Berkovic, 2001). A major lack in these definitions and current clinical practice is ignoring the behavioral consequences of the comorbid, chronic bodily injury that commonly accompanies concussive events (*polytrauma*). There are differing criteria for concussive brain trauma: the length of time unconscious; referring to both a brain lesion and behavioral dysfunction, without acknowledging that there can be a paradoxical discrepancy disparity between lesser degrees of neurotrauma and the clinically established level of dysfunction; symptoms that overlap with posttraumatic stress disorder or psychiatric disorders; and referring to subjective complaints that occur only in a fraction of cases. The accidents reported as evidence of TBI vary with the population: TBI was reported by a greater proportion of psychiatric patients than staff and students and a higher proportion of psychiatric than medical patient males; the percentage of persons reporting two or more episodes of LOC ranged from 2% of staff and students, to 10% of medical patients, and between 1 and 17% of psychiatric patients (McGuire et al., 1998). These considerations have evolved into many definitions that describe concussive TBI: adding to the imprecision is the widely used nomenclature (so-called "minor" TBI), which discourages more through exploration.

What is the criterion of concussion? There are significant omissions in the conventional PCS list that are sufficiently frequent to be included as part of the PCS: seizure like phenomena, cerebral personality disorders, and altered sense of identity. Since they are not in any "official" list, regrettably, they are often not sought after an accident. These symptoms can be elicited by the clinician through a thorough examination that considers their possible expression. A narrow focus of the clinical examination should be avoided because it does not consider subjective symptoms, minimizes their importance, or attributes them to secondary gain for subjective or monetary gain.

Head injury: In general, these traumatic components are referred to: trauma to the brain or spinal cord; altered or LOC; scalp or forehead laceration, skull fracture or injury; unconsciousness; amnesia; neurological deficit; seizure; physical damage to the cranial contents; LOC; PTA; skull swelling, abrasion, contusion, or laceration; blow to the head; fracture at the base of the skull (Fearnside & Simpson, 2005). Various organizations have provided disparate criteria.

One of the definitions for "head injury" excludes patients with abrasions, lacerations, or fractures limited to the face or facial skeleton, as well as those with foreign bodies in the nose, ears, or eyes. Birth injuries also were excluded (Brookes et al., 1990). The present writer observes that exclusion of those with seemingly superficial injury or facial fractures ignores the possibility of a sufficient combination of impact and neck and head rotation to create neurobehaviorally significant TBI, probably *DAI*.

The American Academy of Pediatrics' definition (cited by Kirkwood et al., 2008) of "minor closed head injury" specifies normal mental status on initial examination; no abnormal or focal neurological findings; no skull fracture; LOC < 2 min; possible seizure, headache, or lethargy.

The American Congress of Rehabilitation Medicine's definition of "mild traumatic brain injury" only requires that there be focal neurological deficits that may be transient, and accepts any alteration of mental state, as well as a period of LOC up to 30 min.

Brain injury: One may differentiate between lesser levels of TBI in which there are documented brain lesions (*"complicated" brain injury*) and those with impaired functioning without scanning documentation of TBI (*postconcussive syndrome*). The National Center for Injury Prevention (2002) refers to brain injuries as follows: no medical care, nonhospital-based care, hospital ED care, and inpatient hospital care of more than 24 h that is classified as more severe than mTBI. For impairment in persons with a history of mTBI the following list of symptoms not present before injury or those made worse in severity or frequency is referred: problems with memory, concentration, and emotional control; headaches; fatigue; irritability; dizziness; blurred vision; and seizures. Current limitations in reported mTBI functional status are basic activities of daily living (e.g., personal care, ambulation, travel), major activities (e.g., work, school, homemaking), leisure and recreation, social integration, and financial independence.

Postconcussional disorder or syndrome: *DSM-IV-TR* (postconcussional disorder) and *ICD-10* have different defining symptoms. DSM-IV specifies interference with social role functioning and exclusion of dementia due to head trauma and other explanatory symptoms. (Neither of these criteria specifies the nature and intensity of mechanical forces or somatic injury.) In a particular sample of patients both with a range of intensity of TBI and without TBI, far more patients were included in the ICD criteria than DSM-IV, while both criteria had limited specificity to brain injury since PCS criteria could be met by patients with nonbrain trauma. The DSM-IV-TR criteria of clinical significance and cognitive deficit were not specific to TBI. It was inferred that these two PCS criteria encourage diagnostic disagreement and that complaints of PCS symptoms are insufficient to document the diagnosis of mTBI (Boake et al., 2005).

The First International Conference on Concussion in Sport offered this definition of concussion: a complex pathophysiological process affecting the brain induced by traumatic biomechanical forces:

- May be caused by a direct blow to the head, face, neck, or elsewhere on the body with an impulsive force transmitted to the head.
- Typically results in the rapid onset of short-lived impairment of neurological function that resolves spontaneously.
- May result in neuropathological changes, but the acute clinical symptoms largely reflect a functional rather than a structural injury.
- Results in a graded set of clinical syndromes that may or may not result in LOC. Resolution of the clinical and cognitive symptoms typically follows a sequential course.
- Concussion is typically associated with grossly normal structural imaging studies.

Additional concussion definitions by the American Academy of Neurology and the California Concussion Scale are available (Ruff, 2005). The advantage of using DSM-IV are as follows: (1) physical, cognitive, and emotional symptoms are incorporated; (2) the multiaxial diagnosis permits using Axis I for the PCS and other psychiatric diagnoses; Axis II permits use of premorbid emotional risks; Axis III permits current somatic injuries; and Axes IV and V express social and economic stressors. The following PCD modifiers were recommended for particular clinical features: objective neuropathological, neurocognitive, psychopathological, and mixed neurocognitive and psychopathological.

Assumption Concerning Basic Trauma

The differentiation has been made between *psychogenesis*_ (premorbid conditions, postinjury psychological factors or outright malingering, rather than an alteration in brain function) and *physiogenesis*_ (acute neuropathology and other abnormalities in brain function; Yeates & Taylor, 2005). Patients who meet the familiar characteristics of concussive brain injury, that is, GCS of 13–15; brief, retroactive, and/or PTA; or LOC less than 1h, may have TBI that is difficult to document.

Varying Procedural Sensitivity

- Changes distributed across many synapses and neurons can be difficult to measure and quantify (Kaas, 2002).
- Retrospective inquiry utilizing a few questions, for example, concerning intensity and interval of altered consciousness, are suspect for validity and reliability.
- It is useful to document who is the nonpatient responder describing altered consciousness, including professional qualifications (Corrigan & Bogner, 2007). Statements in the record about non-LOC come from persons who arrived on the accident scene too late to describe the injured person.
- Frequently used radiological scans are not sensitive to lesser TBI or DAI. *Computerized axonal tomography* is insensitive to axonal injury, and has a radioactive burden that limits the use of nonurgent, repeat imaging to assess neuropathology. *MRI*, particularly after

the acute period, may reveal lesions at the gray-white interface, in white matter such as the splenium of the corpus callosum consistent with gliosis, and punctate hemorrhagic lesions.

Diffusion tensor imaging (DTI) is a type of MRI that assesses myelin sheaths and cell membranes in vivo, which restrict the movement of water molecules. Water molecules move faster parallel to the major axes of nerve fibers rather than perpendicular to them. Another measure is *mean diffusivity (MD)*, which is the average diffusion in all directions.

Diffusion-weighted imaging (DWI) is also an indicator of water transport within cells or axons. It detects DAI, which is considered the primary cause of disability and poor outcome. It also detects acute ischemia, and differentiates cytotoxic and vasogenic ischemia. Its measurement is the ADC. Even excluding areas that appeared abnormal on T2 weighted MRI images, DWI was valuable in evaluating injury, particularly in brain imaging that appeared normal on conventional imaging. It detects abnormalities not revealed by conventional MRI, and reveals diffusion abnormalities hours or days before they are revealed by CT (see Galloway et al., 2008).

Single photon emitting computerized tomography (SPECT) reveals lesions (e.g., thalamic and basal ganglia) in patients reported negative with CT and MRI (Abu-Judah et al., 1999).

A particular finding may not be specific to TBI. While routine EEG is unlikely to reveal abnormalities, quantitative EEG is considered more sensitive. One study utilized mild head trauma patients (blunt object) with symptoms persisting >1 year, that is, psychiatric, somatic, and/or cognitive complaints that had developed within the first week following ED-documented head trauma. Any exclusive whiplash injury was excluded. Procedures vary in their sensitivity to TBI. Some microscopic lesions are not detected by presumably sensitive procedures such as MRI. Magnetoencephalography (MEG) was the most sensitive (19/30 patients), followed by SPECT (12/30), and least sensitive was MRI (4/30). MEG-measured slow waves are associated with cell loss and disrupted local interconnections, but areas that appear structurally intact but with decreased blood flow measured by SPECT. MRI, as routinely used clinically, is significantly less likely to be abnormal in these patients than functional methods. MEG and SPECT did not provide only redundant information (Lewine, 2007). SPECT was able to detect some basal ganglia lesions after concussions that were not detectable by MRI and CT (Abu-Judeh et al., 1999).

Narrow range of examination: The clinician's scope of practice may be so narrow that injuries and dysfunctions outside the examiner's specialty are not considered.

Prior accidents are not considered: Study of the patient's history does not include inquiry concerning possible prior accidents.

IGNORED TBI: "THE SILENT EPIDEMIC"

Data inadequacies concerning TBI in an individual incident or lifetime history are quite high, that is, paucity, ambiguity, or inaccuracy. While mild TBI represents an estimated 75–90% of all brain injuries, there is substantial underidentification of this condition, so the actual incidence is likely to be much higher. Neither the ED nor the primary care office specializes in its diagnosis and treatment. Rather than standardized procedures, assessment is often cursory, focusing upon readily identifiable or high-frequency symptoms (e.g., LOC, headache, memory loss). Consequently, treatment is often unsystematic, with limited focus, unmanaged for days or weeks due to lack of identification, and the possibility of exacerbation due to physical and cognitive exertion, possible reinjury (Gioia et al., 2008).

Many clinicians do not recognize potential TBI, or if in doubt, do not alert their patients that later discomforts should be further studied with an accident in mind. The public understanding of TBI is so poor that it is officially referred to by the U.S. National Center for Injury Prevention and Control as "the silent epidemic," since "the actual number of TBIs that occur in the U.S. is not known and much of the public is unaware of the impact of TBI ... and many of the people who are seen in outpatient clinics or doctors' offices and who do not receive medical care are not included" (Langlois et al., 2004; 2006). Various samples indicate that 25–60% of apparent TBI victims report no medical attention. This may be due to the *delay of initial symptoms*, that is, neck pain may be delayed for several days; headaches, back pain, and upper extremity symptoms may have their onset several months later (Nordhoff et al., 1996a). In the acute phase (ED), the usual radiological examinations are not sensitive to TBI and DAI. Yet, it is required to have timely and appropriate access to medical and nonmedical care and services to facilitate recovery, decrease the adverse outcomes of TBI, and promote general health. One study showed that 35% of people with TBI had at least one unmet need a year later. The most common barrier was lack of awareness of available services. Multiple forms of outreach are required since a single type of contact (e.g., a postcard) is ineffective (Seymour et al., 2008).

Problems of Data Collection

Many projects studying the etiology and outcome of TBI have flaws that either minimize the outcome of recognized TBI or do not identify victims of TBI:

- No use of matched control groups: Potential dysfunctions or deficits are not identified.
- Narrow range of cognitive functions: While some components of cognition can be measured with relatively time-economic procedures, loss of capacity is identified more precisely with complex procedures, perhaps with a higher ceiling.
- No premorbid baseline: Without a baseline, utilization of the imprecise assessment that the person has a "normal" or "average" score on neuropsychological tests essentially does not detect any possible loss from the prior level of performance. It invites the interpretation that "nothing happened!"
- Underestimating the outcome of so-called "minor" TBI: Trauma of a lesser intensity than maximum, accompanied by somatic injury, can impair a child so that physiological and mental development is below expectations, and an adult to the point that capacity to work, maintain independence and a family, and participate in the community is ruined.
- Disregarding later expression of TBI: Psychological, physiological, neurological, and chronic stress.
- Ignoring posttraumatic personality changes: dysregulated behavior, stress and anxiety, and other mood disorders such as anger and depression.
- Ignoring the comorbid effects of body trauma: Chronic physical injuries ("the unhealed wound") can create a chronic posttraumatic stress disorder (CPTSD) leading to "burnout" of systems (psychological and internal physiological) known as "allostatic load," creating vulnerability to contagious and stress-related medical conditions and the common postaccident loss of stamina.

A concussive brain injury merits a wide range of study in terms of both acute and later loss of function and outcome after a recovery period. Data on potential TBI victims will be missed when, initially, only transient confusion is experienced; in older persons, when the TBI occurred in childhood; when questionnaire self-report does not contain relevant details; when past injuries are forgotten; when the victim does not seek examination or treatment immediately after the trauma; when there are premorbid substance abuse problems or economic disadvantages (Corrigan & Bogner, 2007; Dettmer et al., 2007); when children's injuries are concealed by caregivers, or by the child, or the child cannot report it. Outcome data are lost for impoverished people who cannot pay for complex diagnostic procedures (refusal by defendants). Data completeness is influenced by the phase at which information was gathered.

Some groups are not identified or treated, or are selected by some other criterion (school children with behavior problems, nursing home residents, prisoners or other offenders, substance abusers, combat soldiers). It is difficult to obtain verifying information from records of persons attended by a school nurse, athletic trainer, or emergency medical technician in the field. "*Inactive diagnoses*" lead to an underestimate of prevalence in a group, for example, for nursing home residents, when a diagnosis such as TBI is excluded because it is considered irrelevant to the person's current needs. In some facilities, there is an issue whether TBI should be reported if there is recovery to the point that it no longer affects the patient's health, function, or behavior. It may be reported when it is considered a permanent condition whose consequences may change over time (Karon et al., 2007).

Some statistics have depended upon a few self-administered interview items, but the possibility of omission is high. The preferred procedure for detecting prior TBI is a face-to-face interview by an informed professional. Additional epidemiological issues include lack of TBI awareness by the attending professional, the limits of self-report (unawareness, poor recall, stigma), and unrecognized mild TBI (repetitive sports injuries, repetitive domestic violence) (Corrigan & Bogner, 2007).

TRAUMATIC BRAIN INJURY EPIDEMIOLOGY

Adults

A problem of definition: The older concept of TBI without detectable pathology being termed concussion has been replaced by terms such as MHI and minor traumatic brain injury (mTBI). Their criteria have not been universally accepted. The postconcussional syndrome (American Psychiatric Association, 2000, pp. 760–762) specifies altered consciousness, cognitive assessment, and the presence after 3 months of three or more of these symptoms (fatigability; sleep disorder; headache; vertigo or dizziness; irritability or poorly provoked aggression; anxiety, depression, affective lability; personality change; apathy or lack of spontaneity).

Overlapping symptoms: The onset, following head trauma, of a decline in impairment in social or occupational functioning not accounted for by other disorders. Limitations of this definition include the fact that these symptoms also occur in people without known TBI and a concussive head injury is frequently comorbid with impairing somatic trauma, which participates directly in a bidirectional physiological exchange with the CNS and creates symptoms that are considered to be neuropoychological. Preexisting personality variables, historical socioeconomic patterns, developmental history, and prior trauma influence the persistence of postconcussional symptoms.

Some population details: The distribution of TBI has been described as bimodal with peaks in young adulthood and old age. This has implications for the determination of outcome (Rapoport and Feinstein). U.S. government statistics include ED visits, hospitalizations, and deaths. They do not include persons treated for TBI in other settings, those that do not seek immediate treatment, or persons treated after an accident for other injuries, with no record of the condition of the head being studied. Average characteristics for the years 1995–2002 were 50,000 deaths, 235,000 hospitalizations, 1,111,000 ED visits, and an unknown number receiving other medical care or no care (Langlois et al., 2004). The leading cause of TBI (ED visits, hospitalization, and deaths) is falls (28%), followed by MVAs (20%), assault (11%), other causes (32%), and unknown (9%) (Langlois et al., 2004; 2006). A new category has been added for analysis. *Struck by/against* are those events in which a person was struck unintentionally by another person or object (e.g., falling debris or a ball in sports), or in which a person struck against an object, such as a wall or another person. It was the third leading cause of TBI (18%) after falls (32%) and MVA (19%). Many sports-related TBIs are in this group (Rutland-Brown et al., 2006). Males are about 1.5 times more likely than females to sustain a TBI. The two age groups at highest risk for TBI are 0- to 4-year-olds and 15- to 19-year-olds. Certain military duties (paratrooper) increase the risk of a TBI. African Americans have the highest death rate from

TBI (Centers for Disease Control, 2006). *Persons with disabilities* are four to ten times as likely to become victims of violence, abuse, or neglect as nondisabled persons (Reichard et al., 2007).

Substance abuse: In one sample (*adult substance abusers*), 29% had one TBI, almost 50% incurred *multiple TBIs,* and *17%* incurred their first injury before age 15 (Corrigan & Bogner, 2007). Alcohol has been stated to be the predominant risk factor for TBI. Post-TBI risk factors for heavy drinking include male gender, younger age, and history of substance abuse prior to TBI (Horner et al., 2005). Between 38% and 63% of substance abusers reported TBI. Those who drink *alcohol* have an increased risk of TBI than those who did not. In the sample, two or more TBIs with LOC reflected serious mental health problems (depression, anxiety, hallucinations, suicidal thoughts and attempts). When depression and anxiety were held constant, these patients had more months of marijuana and tranquillizer use (Walker et al., 2007 citing Corrigan et al., 1955). A Swedish study of inpatients and outpatients treated at an ED after a brain trauma causing any degree of altered consciousness revealed that at least 17% were under the influence of alcohol, particularly male bicyclists. The rate of intracerebral hemorrhage increased with increasing age and decreased with decreasing GCS scores (Stryke et al., 2007). Various studies of *offenders* indicate lifetime histories of TBI from over half to 87% and 88% (with a history of multiple injuries from 29% to 56.7%). Fifty percent suffered LOC, and 40% required medical care for a head injury (Diamond et al., 2007).

Nursing Home Residents

Residents of *nursing homes* had an incidence of 0.8% of TBI, which increased to 5.6% when the definition of TBI was enhanced by utilizing diagnostic codes that included the long-term consequences and complications of TBI (Karon et al., 2007). Severe cognitive impairment was more frequent in patients diagnosed with TBI (22%) than with those with neither TBI nor dementia (5%). They were more likely to be physically abusive and to behave inappropriately or disruptively (Gabella et al., 2007). Changes of behavior, function, or cognition may be mistakenly attributed to conditions common to nursing home residents, for example, dementia, infection, or medication (Karon et al., 2007).

Sports

There has been considerable public interest in the possibility that repetitive sports injuries ("*subconcussive impacts*") create traumatic encephalopathy, depression and suicidal thoughts, and other behavioral disturbances (Schwarz, 2007a). Some injuries are routine; others are violations of rules, that is, helmet-to-helmet impacts (Thamel, 2007), while others are directed with intent to disable (boxing). With the specific intent of producing brain injury, the uppercut is a rotating force greatly exceeding 600 kg that is delivered in less than one-tenth of a second. A jab (linear hit) is much less likely to cause a concussion than an uppercut of the same force. Brain protein markers of trauma are detected. Twenty percent of professional boxers develop *dementia pugilistica*, also known as *chronic traumatic encephalopathy* (slurring dysarthria, gait ataxia, disequilibrium, headache, memory, information-processing speed, finger-tapping speed, attention and concentration, sequencing judgment, abstraction, reasoning, planning, organization). Clinical symptoms may occur 10–20 years after retirement. Autopsy revealed brain damage (cortex, cerebellum, substantia nigra, subcortical) and neurofibrillary tangles) (Bazarian, 2009). The delayed pathophysiology of boxing may cause deficits months or years later, even though the acute injury is not apparent (Robinson, 2007). One group of college athletes was examined a mean of 89.4 days after a MHI. These individuals had been cleared for sport participation based upon standard clinical symptoms resolution (Thompson et al., 2005). The conclusions are based upon a combination of postural and EEG findings. There was decrease in EEG power in all bandwidths studied in concussed subjects, especially in standing postures. There was sustained postural instability, especially under a no-vision condition. It was suggested that long-term residual balance problems may be observed with appropriate research methodology. A change in EEG delta amplitude occurred between sitting and standing positions. The brains of college players (17–18 years old) are still in development, creating vulnerability for later impairment of higher-level cognitive functions. Development of symptoms of impairment can be concealed by the desire not to appear weak or to be removed from the team, and also a lack of awareness of the meaning of concussive symptoms by players and coaches. Information concerning the epidemiology of specific sports; falls; and the effect of alcohol, drugs, personal violence, and assault is offered by Fearnside and Simpson (2005).

Repeated concussions: Sports injuries represent a different sample of concussion than the general population: highly conditioned, with different psychological and motivational characteristics than most other groups. The present writer notes that sports injuries have far less impact and acceleration changes than such accidents as MVAs, falls from a height, and falling objects. When football players have been knocked unconscious, some have been returned to play the same day, reportedly consistent with the National Football Leagues Committee on Brain Injuries. It was believed that there was no risk for further concussions or prolonged neuropsychological symptoms. It was separately acknowledged that the data for some players' LOC was incomplete (Schwarz, 2007b). Thus, when football players have been knocked unconscious, the ease of return to play on the same day has proven fallacious. However, sports injuries are characterized by a several-day stage of acute physiological effects, followed by longer-lasting effects influenced by a range of psychological or motivational factors (Barr, 2007).

Sports injuries are sufficiently traumatic to cause permanent brain damage. Hockey and football have been described as having a culture of "playing hurt" (Zinzer, 2007). Coaches have been accused of premature pressure to return to play soon after head injuries. The possibility (or probability) of prior concussions may not be considered in a decision as to whether a player should continue immediately. The National Football League found that retired players under age 50 were almost 20 times more likely to report memory problems than would be expected in the general population (Talan, 2010). Hockey head injuries occur from checks to the head delivered at high speeds with players' shoulder pads. Please pay good attention: A general manager cautioned us as follows: "You've got to be careful what you do when you talk about rule changes. Hitting is part of our game, and you don't want to change the fundamental nature of the game" (Schwartz & Klein, 2009).

Second impact syndrome: This refers to the situation when an athlete sustains another head injury before the previous ones have cleared. (The frequency of danger is controversial.) It may be precipitated by a minor blow to the chest imparting motional force to the head. Although the player appears briefly dazed and completes the play, they may collapse, with coma, loss of eye movement, pupillary dilation, and respiratory failure. Mortality is 50%, and morbidity 100%. It may be consequent to a loss of cerebral autoregulation causing, after the second injury, vascular and cerebral compliance changes. It requires restriction of head-injured players' return to competition (Le Roux et al., 2000).

The National Health Interview Survey (NHIS, 1998) indicated that for the 12 months prior to 1991, of the estimated 1.54 million brain injuries occurring in the United States, 20%, or approximately 306,000, were attributable to sports or other physical activity, that is, an incidence of 124 per 100,000. It is noted that the milder and medically unattended brain injuries may be under-reported because of lack of awareness of their occurrence. The findings are divided between competitive sports (111,000, led by basketball, baseball, and football) and recreation (105,000, led by playground activities, swimming and water sports, skiing and other snow sports, skating [inline, roller, board]), horseback riding, exercise, and weightlifting). Approximately 100,000 of these injuries were concussive (Lovell & Collins, 1998). Estimations of hospitalizations or deaths vary between 7,000 and 35,000. During a 6-year period, there were 249,000 trampoline injuries in children 18 years old and younger treated in hospital EDs in the United States (Smith, 1998). A study of high school athletes revealed that 5.5% of reported injuries were mTBI (Powell & Barber-Foss, 1999). Of these, football injuries were the most numerous (63.4%), followed by wrestling (10.5%), girls' soccer (6.2%), and boys' soccer (5.7%).

CONCUSSION: AN OVERVIEW

Concussion is the acute psychological experience of mechanical trauma to the head. It is incurred through head impact and/or acceleration, usually accompanied by some alteration or limited LOC, and generally without neurotrauma detectable by neuroimaging procedures or requiring neurosurgery. Lack of significant alteration of consciousness is not needed for the diagnosis of concussion, since the condition is characterized by a mechanical brain injury. The range of altered consciousness described as lesser or minor TBI has been reported to be from 20 min up to 1 h. There may be transient neurological signs, including convulsions or other movements, blurred vision, confusion (mental state changes), feeling dazed, dizziness, focal neurologic symptoms, headache, and nausea. Radiological signs of lesions are usually absent, or if present, are assessed as minimal. If they are neurologically significant, although not a surgical event, the condition has been called a complicated mTBI.

The level of altered consciousness, or its lack, is somewhat determined by the geometry of the accident, that is, speed and the geometric relationship of the head, axis of the body, and the surrounding environment movement relative to each other, direction of movement of the involved person and objects, the developmental stage of the person, and the physical characteristics of any surrounding or impacted surface.

Concussive brain trauma results from a physical accident that creates impact to the head, and possibly to the neck and other parts of the body, with head rotation in varied planes. Impact and/or acceleration and deceleration cause movement of the brain within the skull in various planes. TBI occurs when there are brain structure changes due to impact and pressure waves, stretching, shearing with the skull and internally, rotating in various planes, cavitation, and cellular trauma. Some neurobehavioral dysfunctions consequent to somatic injury may be misattributed to nervous system disorder.

This writer regards the PCS as a polytraumatic neurobehavioral phenomenon involving comorbid brain, head, neck, and somatic injuries. It is expressed as a wide range of common symptoms: neurological, musculoskeletal, physiological, and psychological. Chronic symptoms may be consequent to unhealed injuries incurred at widely separated sites of injury: neurological (CNS, spinal cord, cranial, and peripheral nerves, brachial plexus); musculoskeletal trauma causing pain, restricted motion, and peripheral nerve injury; dysregulation of physiological systems (immune, inflammatory, hormonal, circadian, autonomic), which creates neurobehavioral disorders; breakdown of the multiple barriers separating the liquid compartments from the brain, permitting infiltration of toxins, hormones, neurotransmitters, and immunoinflammatory signals generally excluded; the allostatic state: a posttraumatic physiological dysregulation culminating in exhaustion as a result of a persistent unhealed injury (vulnerability to communicable and systemic disease and fatigability); disorders of identity (feeling impaired, unattractive, injured, threatened); and mood disorders (anxiety, depression, anger, or numbness).

Outcome depends upon the site and extent of the lesion, preexisting conditions and style of adaptation (baseline), symptom resolution or evolution into a chronic condition, patient's reaction to being injured and impaired, the availability of treatment, and social support or rejection. The practitioner must consider the possibility of both symptom remission and late-developing symptoms occurring months or years later: physiological dysregulation, neurological, and psychodynamic. Children may incur disorders of physiological maturity due to injury to the hypothalamic–pituitary axis, or deficits of cognitive or personality development. These may not be detected since they do not develop at the expected stage.

INTERVIEWING AN ASSAULT VICTIM

Introduction: The experience of this woman illustrates the detailed information that can be elicited through a structured interview. Critical data cannot be elicited through formal neurological and neuropsychological procedures. She experienced both neurobehavioral disturbance and somatic injuries.

Interview with translator: This is an example of the use of the interview in elucidating a large variety of symptoms.

Development: Born in a foreign country, middle-aged, in the United States for fewer than 10 years. At birth was living with her parents.

Education (age at graduation): She graduated college in her mid-twenties. There was no major; her career was nursing. Study required 4 years.

Military service: None.

How old were you when you received your first full time-job? Mid-teens. Sewing in a factory.

What kind of work do you do most of the time? Cook in a restaurant.

Are you working now? What is your job title? She was moved from cook to cashier after the accident because she was getting dizzy.

Occupation at the time of injury: Cook.

Did you ever return to work after you were injured or ill? She stayed out of work for two weeks.

Health history:

Early development: Knows of no constitutional problems

Serious illness: Denied

Accidents before the injury we are studying:

Head injury: Denied

Chemical exposure: Denied

Electrical injury: Denied

Anoxia/carbon monoxide: Denied

History of emotional problems: Treatment denied

Accidents after the injury we are studying: Denied

Describe the accident: She went down to do the laundry. She went down the stairs, and there was somebody hiding on the stairs. She didn't notice him until she was on the second set of stairs going down. She passed him by.

He followed her; she asked him if he wanted something. He said that he was waiting for the super of the building. The super owed him money (examiner brings interview back to injury). He followed her downstairs, stopped, and said, "Hey you." She turned, she was struck, and does not know whether this man had something in his hand. She lost consciousness. She tries to picture that she was struck, perhaps fell on a level, but rolled down an adjacent staircase. There were ten steps down, and she woke up at the base of the second set of stairs. She had a big red mark on her left hip and buttock; there was a red mark on her left shoulder; also bruise on left shin; she could not open up her mouth when she was in the hospital. Her face was swollen on the left side. There is intermittent clicking. (Examiner suggests that she mention this condition to her dentist.) She points out that three front right teeth are loose, and there is a space between two incisors. Since the accident she has lower back pain, and when she tries to extend her left knee there is pain in the back of the knee, and she has to help herself.

She has reported these pains, but some of them are becoming worse over time.

Did you have confusion, did you feel dazed, or lose consciousness? She was aware that she was struck, was unconscious, and then was disoriented. There was a pay phone right outside the building, and she called for help. On the basis of questioning the paramedic, she estimates that she was unconscious when the paramedic arrived.

What was your last memory before the accident?

Do you remember the moment in which you were injured?

Were you dazed or unconscious? For how long? First memory after the accident: When she came to, everything was blurry, and she felt heavy to stand up.

Did you go to the hospital? The ambulance took her. She was injured at 7:30 A.M. and was discharged from the hospital at 7 P.M. She wanted to go home because she thought that nobody knew what happened. Her nephews aged 18 and 14 were without an adult in the house.

Do you feel like the same or a different person since the accident? She feels very different. She forgets things more often. She doesn't feel safe crossing the street. She is scared of every man, but is looking at the face of every man to try to find the man that did it. She turns on the stove, leaves the kitchen, and completely forgets. The things she did before, she can do now, but it takes a lot more time. She is not completely sure but thinks that she has less mental ability. She is not sure if it's mental capacity or physical capacity, but she feels she is just not able to do it.

Were there any health or other bodily changes after the accident? Refers only to pain and forgetfulness.

Are you in any kind of medical treatment? Yes

Medication: For dizziness and pain

Current somatic concern:

Headache: Right after the event, the pain on the left side of her head and outside noise through the left ear would bother her.

Pain elsewhere: No additional pain to report.

Sleep problems: No trouble falling asleep. Wakes up due to nightmares, and without nightmares. She wakes up one to five times at night. Until 3 weeks ago, nightmares were almost every night; since then, less frequently. When she wakes up, she knows where she is; perhaps not nightmares.

Are you more sensitive after the accident to:

Heat: It seems too early to tell.

Cold: Also

Bright light: Her persistent reaction is as though one shone a flashlight directly into her eyes and then took it away. However, she does not complain about particular illumination as being too bright.

Loud sound: She is more sensitive than before to loud noises like trains.

Seizures at any time in your life: No

Dizziness: For about 5 days the dizziness even occurred when she was in bed. It was so strong that sometimes she would not know that she had urinated on herself. It happened on the 5–6th and 8–9th days after the accident. Sometimes she is dizzy, but was not before. When she lies down, it feels like everything around her is spinning. When she gets up, her head is too heavy to keep straight; she has to hold her head.

Balance: She feels as though she will fall to the back and right.

Range of motion: (See above with left knee.) None other.

Visual: Everything looks dim. This condition is not improving.

Hearing: She hears low sounds but sometimes she gets confused with words.

Movement: She can be lying down and her torso will jerk. No tremor.

Autonomic:

Sweating: Yes, but it might have been before she was injured.

Nausea: No

Frequent urination: No.

Bowels: No.

Appetite change, weight change, activity level change: She may have lost a few pounds. Less active, although she has lost weight.

Easily tired: More easily tired.

Do you get sick more often after the accident? No

Have others told you that you did something wrong and you did not know it? No, except for her forgetfulness.

Anxiety and hyperarousal:

Nightmares or bad dreams: Bad dreams. Her brother or one of her nephews is killed. Sometimes she is being followed and she has to hide.

Flashbacks: No

Avoidance due to anxiety or fear: No particular avoidance, but she doesn't want to be alone any more.

Rapid heartbeats: Yes. When she sees men of short stature and she can't see their face.

Deep breathing: No.

Jumpiness: She is jumpy, mainly when she is outside. When she sees somebody that looks like the guy, somewhat threatening.

Do you ever have strange experiences? Things happen and you can't explain it. (partial seizures screen)

Vision: Both sides. Can't identify.

Hearing: No

Feel something move on your skin: Both sides in the legs, and occasionally in the arms.

Tastes something others do not taste: No

Smells something others do not smell: (She works in a restaurant.) Sometimes the food smells rotten. She does not remember this happening elsewhere. (Advised to report this to her neurologist if the symptom becomes worse.)

Do you ever think that somebody is standing behind you (but there is nobody there)?: Yes (although this is characteristic of her injury)

Sudden mood change for no reason: Sometimes she gets aggressive. Somebody turns the music on loud, and she says if you don't turn it off I'll punch you. She has never struck anybody.

Does it ever happen to you that you lose time, you don't know what happens for a period: *Get to a place and don't know how you got there*: She gets on a train and she would miss her stop or she wouldn't know what the last stop was. She would go to the grocery store, and by the time she got back she would not know which way she went out, where she's been, or how she got there.

Is your memory normal or is there a change?

Change in level and control of anxiety: She panics when she is on the street or in her apartment by herself and she thinks somebody is turning the knob.

Change in level and control of anger: She was not aggressive before, but now she is easily annoyed.

Change in level and control of depression (dull and empty OR sad OR both): Feels depressed on occasion. Sadness and emptiness.

Change in level and control of sexuality: She has been having problems with her partner.

Dissociative experiences:

Does the outside world ever feel unreal? Some things she can't understand. She doesn't understand what happened to her. She can't understand the world, but it is not strange.

Does your body feel changed or unreal? No

Do you ever feel that something has happened before that you know has never really happened before? Not very often

What were you interested in or enjoyed doing before? Cooking and cleaning

What do you like to do now? Nothing

Do you get tired more easily? What makes you tired? Yes. She doesn't feel like doing anything; she doesn't even want to go out.

Any change in your social life? She used to go out with her nephews to the movies; now she does not because she doesn't feel like going out.

Do you need help at home or in getting around? She can help herself.

Is someone available to help you with personal needs or obtaining assistance because of health or other problems? She lived with her two nephews. If she doesn't feel like doing something, they help her.

Worst effect of the injury: Her emotional state. She doesn't feel secure. She even thought about moving.

Will your condition get better, get worse, or remain the same? She thinks that it will get better.

Are you able to handle your problems, or they are too much for you? There are too many problems.

What do you do to make yourself feel better? She tries to forget about the incident. Reads a book or a magazine to forget about it.

Is there anything else you want to tell me? No

2 Introduction to the Postconcussion Syndrome

OVERVIEW

Postconcussion syndrome (PCS) is a group of symptoms that commonly focuses on symptoms characteristic of lesser brain trauma, that is, excluding moderate or severe traumatic brain injury (TBI). A useful criterion may be for "moderate" TBI to involve more than 1 h of posttraumatic amnesia (PTA), perhaps with brain injury detected on radiological scans, for example, contusions and small hemorrhages. "Severe" TBI could be considered for surgical cases with extended unconsciousness. PCS is controversial insofar as it has been attributed to emotional reactions, physiological responses, symptom exaggeration, secondary gain, attentional disorder, dysfunction of informational processing and mental speed, diffuse axonal injury (DAI), cranial nerve injury, emotional reaction to being injured, and so forth. In fact, accidents are complex events whose pathological consequences include the entire range of human performance and experience. An accident causing brain injury has multiple psychological and significant noncerebral biological components. This extensive phenomenon is known as the postconcussive syndrome, that is, behavioral reactions to comorbid, chronic somatic, neural, and psychological trauma. The consequences of the original injury, the effects of having an "unhealed wound," and the psychological consequences of the initial and chronic trauma are a *disruption of the adaptation of the accident victim within the usual environment.*

A significant yet neglected component of PCS, particularly with chronic complaints, is somatic injury, which has a complex effect upon performance: The injured body's performance resembles in some ways performance after brain and spinal cord injury (SCI)—the psychological effects of being impaired after injury can lead to passivity and pessimism consistent with some brain injury lesions, or the numbing of the posttraumatic stress disorder (PTSD). Acute and chronic unhealed injuries cause various systems (immune, endocrine, inflammatory) to create chemical substances that influence neurobehavioral central nervous system (CNS) activity, consistent with the posttraumatic reduced efficiency of various barriers that separate the brain from the bloodstream and cerebrospinal fluid.

The site of noncerebral trauma has been described by the Abbreviated Injury Scale (AIS) (Fearnside & Simson, 2005) as head or neck, face, chest, abdomen or pelvic contents, extremities or pelvic girdle, external. Additional trauma sites are the spinal column and other bones, muscles, fascia, and viscera. One sample of 8-year-old children manifested 26% multiple injuries (Klonoff et al., 1995). The brain and soma interact with profound holistic behavioral and physiological changes. *Adaptation* is the integrated way in which a person or species copes with its environment: genetic (hereditary), phenotypic (expression of genes in a particular personal history), and stylistic (learning and preferences). It is assumed that TBI interferes with flexibility and capacity to deal with complex and difficult problems. Substantial inability to cope is indicated by later difficulties in school, employment, family relations, and community participation.

THE MAYO CLASSIFICATION SYSTEM: AN APPROACH TO THE SEVERITY OF TBI

A new approach to classifying the severity of TBI takes into account the frequent gaps in medical records, and the effects of nontraumatic factors upon widely used indicators as the Glasgow Coma Score (GCS) and loss of consciousness (LOC): the interval after the accident at which it is recorded, roadside sedation, intoxication, fractures, systemic or psychological shock, or organ system failure. A range of neuroimaging techniques is considered for differentiating grades of TBI severity. Classification is based upon multiple criteria, minimizing the consequences of missing information from single criteria. Each grade is based upon the maximally relevant documentation available for a sample of 1,678 cases (Malec, 2007a).

Moderate-severe (definite) TBI: death; LOC of 30 min or more; posttraumatic anterograde amnesia of 2 h or more; worst GCS in first 24 h of less than 13 (unless invalidated by intoxication, sedation, systemic shock); one or more of the following: intracerebral hematoma, subdural hematoma, epidural hematoma, cerebral contusion, penetration of the dura, subarachnoid hemorrhage, brainstem injury.

Mild: None of the previous criteria apply, but one of the following applies: LOC less than 20 min; posttraumatic anterograde amnesia momentary to less than 24 h; depressed, basilar, or linear skull fracture with dura intact.

Symptomatic (possible) TBI: if one of the following symptoms are present: blurred vision, confusion (mental state changes), feeling dazed, dizziness, focal neurologic symptoms, headache, nausea.

Pending further experience, the mild and symptomatic grades correspond to the usual category of concussion. However, the present writer notes that the clinician would recognize the complications ensuant to skull fracture and significant, chronic somatic injury.

PRINCIPLES OF POST CONCUSSIVE SYNDROME

Neurobehavioral and adaptive dysfunctions after an accident are diverse and complex. There are characteristic sensorimotor and cognitive symptoms, for example, dizziness, visual and balance symptoms, and so forth. These are attributable to injury to the cranial nerves, spinal cord and peripheral nerves, somatic structures (musculoskeletal system, internal organs and glands), and emotional reactions to injury and impairment.

These principles can aid the clinician's organization of findings and impressions for assessment, treatment, assessing outcome, and aiding re-entry into the community.

- Adaptation to the world, its demands, and characteristics are central regulators of behavior and quality of life.
- Although an alteration of consciousness is a defining characteristic of concussion, its quality in a specific case can be difficult to assess precisely.

Comorbid Conditions

The spectrum of injuries caused by a concussive event creates adaptive difficulties in coping with the social and physical environments. The phenomenon of concussion is usually co-morbid with somatic injury caused by mechanical forces that cause impact against nearby surfaces, disruption of tissues caused by rapid movement ("strain" exceeding elastic limits of tissue, i.e., "whiplash"), shearing forces separating tissue, and crush (external and internal). The clinician considers injury at various levels: from gross anatomical and systemic physiological reactions, down to tissue damage of different dimensions (microscopic, membrane, cellular, subcellular) and genetic.

- The PCS comprises a variety of disorders and complaints consequent to neurological, systemic, somatic, and psychological disruptions. These are caused by injury to the nervous system and also the chronic effects of somatic trauma ("the unhealed wound").
- The brain is in a bidirectional exchange with the physiological systems of the body via both blood and neurological pathways. These affect both brain and peripheral responses in injury-related, stress, and adaptive functions. Specific substances (e.g., hormones) actually have widespread expression in other tissues (e.g., the brain), which are not the familiar targets. The brain responds to signal substances best known for somatic activity (psychoneuroimmunology); drug side effects; and the hormonal, immune, and inflammatory systems.
- Chronic dysregulation of physiological and psychological systems leads to stress-related disorders, fatigability, and vulnerability to contagious diseases.
- The various "barriers" (e.g., blood-brain and several others) are not completely effective under normal circumstances (e.g., circumventricular organs have no blood-brain barrier [BBB]), and are even less effective after trauma or illness. This causes brain functioning to be affected by psychoactive substances that are present in small amounts or blocked from access. Thus, posttrauma, the liquid environment (internal milieu [IM]) conveys to it a wide variety of neuroactive substances, stemming the reaction of various systems (hormonal, inflammatory, and immune) to acute and chronic injury. The chronic functioning of these systems is described as "dysregulation."
- Somatic conditions affecting performance are caused by somatic injury to soft tissue, traumatic dysregulation, and breakdown of the brain barriers. Resultant disorders may be ignored or misattributed to brain injury. If so, the focus of treatment can be incorrect, or the patient can be described as malingering or somatizing.

The condition of stress, acute and chronic, is a significant component of PCS.

- The accurate assessment of a patient's status is extremely complex. A multidisciplinary approach requires study of a wide range of functions and considers that brain trauma proceeds through several phases that may evolve over years (trauma, physiological, neurological, chronic stress).
- The individual characteristics of the patient should be considered in determining status: Age, preexisting conditions (accidents, illness, poor education), baseline level of personality and performance, postaccident illnesses and further accidents, the interval since the injury (phases of trauma), wide-range sampling of neurobehavioral domains (the taxonomy of neurobehavioral disorders), psychological stress and other reactions to being in a frightening and/or impairing accident, the inability or unwillingness of a patient to completely disclose posttraumatic dysfunctions and discomforts ("expressive deficits"), exaggeration caused by angry and despairing reaction to denial of treatment and compensation, deliberate attempts to obtain undeserved compensation, the level of social support from friends, family, and agencies mandated to help injured people (governmental, insurance companies, religious organizations).
- The current generation of neuroimaging and clinical neurological procedures may not reveal brain trauma at the microscopic or cellular level (e.g., DAI and molecular changes). The comprehensive neuropsychological examination and single photon emission computed tomography (SPECT) provide some evidence for brain trauma of a diffuse nature not detected by other imaging procedures, for example, computed tomography (CT) and magnetic resonance imaging (MRI).

- Stress is a frequent comorbid condition whose neurobehavioral and physiological symptoms overlap considerably with TBI, and are frequently comorbid with TBI. Chronic posttraumatic stress (physiological and psychological) is caused by unhealed trauma (anatomic and psychological) injuries, and leads to physiological and neurobehavioral dysregulation. Therefore, accident-related stress differs considerably from psychological trauma (rape, assault, fear of death, harassment, or other fear-provoking events) or continued devastating environmental circumstances (combat, concentration and prisoner of war camps, extreme weather).

Dysregulation outside the nervous system is the source of many symptoms. Since these are not reflected in neurological and neuroimaging procedures, they can be *misattributed* to faking, somatization, symptom exaggeration, or other psychodynamic reasons. Consequently, the patient is not offered appropriate treatment or compensation, and the struggle for rehabilitation becomes more difficult. Several somatic systems will be described, including chemical-signaling substances that may affect adaptation.

COMPLEX FOUNDATIONS OF THE POSTCONCUSSION SYNDROME

The Brain and its Internal Milieu

The brain is an interactive site for receiving and responding to chemical and neural messages through the CNS, cranial nerves (e.g., the vagus, CN X), and the bloodstream. Normal functioning consists of a multisystem, bidirectional, and integrated signaling network between the brain, somatic structures, and the world. Physiological substances are carried through the vascular and some nerve pathways to the brain ("the IM"). This neurochemical matrix aids understanding of *PCS*, since a head injury may create physiological reactions to somatic and cerebral trauma, whose expression varies over time. Brain–body signaling affects condition and ultimate outcome, as well as mental and physiological reactions to being injured and unable to heal the body and spirit. Chronic complaints and symptoms suggest that there may be an "*unhealed wound*" contributing to neurobehavioral, psychological, and stress-related dysfunctioning. Focal cortical lesions, associated with *BBB* disturbance, and reduced blood flow, may be the underlying pathogenesis of some cases of PCS syndrome.

Dysregulation of the Internal Milieu

The IM is a basic concept introduced by the French physiologist Claude Bernard more than 100 years ago. It refers to the extracellular fluid, which circulates throughout the body, for example, in the blood supply and diffuses through the capillary walls (Guyton & Hall, 2000, pp. 2–3). In addition to nutrients and metabolites, the IM carries the products of physiological systems that respond to acute and chronic trauma. A long-lasting response, prolonged after the acute injury is being referred to here as dysregulation of the internal environment. These chemical and autonomic nervous system reactions have inappropriate neurobehavioral effects and lead to significant vulnerability to both disease and fatigability, a common complaint after TBI.

The true origin of some neurobehavioral disorders in somatic reaction to trauma is often ignored for various reasons:

- There may be a very narrow focus within the practitioner's specialty so that other disorders are not identified.
- The patient may not call the dysfunction to the examiner's attention.
- A long time interval between an injury and the examination misleads the practitioner so that the association between somatic injury and neurobehavioral disorder is ignored.

Neurobehavioral dysfunctions and discomforts (created by accidents, stress, and disease) can be organized into a neurobehavioral taxonomy. The organizing domains are neurological, cognitive, adaptive, personality and intrapersonal, and special problems of children. The organization of dysfunctions and discomforts, as well as the placement of a symptom within the taxonomy, enhances understanding of its etiology. When the clinician is aware of the fact that seemingly similar conditions may have widely different etiologies, selecting the correct domain increases the precision of the correct focus for treatment.

Examples of similar but unrelated symptoms:

- Language disorders can be separated into motor and cognitive disorders.
- Depression may be a cerebral personality disorder or a psychodynamic reaction to one's life situation.
- Fatigability with depression can be ignored or be misattributed to cerebral dysfunction or emotional reactions, when in fact their origin can include persistent stress, TBI, illness, sickness behavior, mood disturbance, memory loss, and hyperarousal.

Polytrauma: Comorbid Somatic and Brain Injury

Polytrauma refers to *multisystem trauma*, for example, comorbid brain and somatic trauma. Clinical assessment will be incomplete and perhaps significantly inaccurate when the examiner does not consider the multiple effects of trauma often accompanying a "head injury."

Clinical Example

The patient was a teacher struck on the head with a heavy object. They found blood in the ear chamber. The neurosurgeon exposed the left side of the head and the vestibular balance system. The patient was injured in June and operated on in October.

Health: "Devastating in the beginning to the extreme. I couldn't stand up. I lost my balance. They gave me medication that controlled the balance problem, but I was progressively using more of the medication to stabilize my balance. I couldn't work. I couldn't be up on my feet more than 2–3 h and then I had to lie down, cover my head with a black blanket or something dark. Elevating my legs helped. I would bring them up. There were no frequent colds and coughs. My problem with my bowels and bladder were troublesome. Before the operation I couldn't leave the house without being sure that I did not eat. I did not eat to be sure that I had nothing in my system. There was no control. I knew every bathroom on the way from home to work. They told me they wouldn't pay me if I didn't show up. If I had difficulties I could go into one of the empty rooms. My left leg used to go dead and I would walk into a room and sit down. When I walked in the hallway at school I would stagger. I used marked lines on the floor as a guide. I could walk pretty close to being straight so that I wouldn't fall or zig zag, look like a fool."

Stamina: "It slowly came back. Came back where I could work a full day." At one time he could be productive for 16 h, including work. Now when he goes home he has to lie down. "When I start to feel tired or exhausted, I go to sleep almost within 10 sec." He wonders if he is passing out. (Has been referred to Dr. L.) "I used to go bicycle riding with my wife. I haven't done that since the accident. Because of the balance problem. I am scared out of my mind about that. I used to go to gym and I had a chart of what weight I pressed. I noticed after the accident I could pick up less than 50% of the weight. It was something I couldn't understand." He has worked on that. He formerly pressed 150 and now he presses 80, He was hunching over, kids in school called his attention to it. He realized he was trying to watch his feet when he was walking to make sure everything was all right.

Spinal cord injury: Twenty percent of patients with *SCI* have isolated SCI, while others have other significant injuries: TBI, hemothorax or pneumothorax, extremity fractures, or internal injuries. Tissue perfusion is decreased by the systematic inflammatory and hemorrhagic hypotension. A young, strong person striking one's forehead protects the cervical cord from hyperextension through dispersion of translation forces and cushioning the spinal cord with cerebrospinal fluid. A weaker person, perhaps with an abnormality of the occipital–cervical junction, may suffer an SCI because of the associated ligamentous laxity and cervical stenosis (Santiago & Fessler, 2004).

Sexual disorder and spinal cord injury: *Libido* is often adversely affected by both complete and incomplete SCI. This is probably secondary to concerns about body image, poor sexual performance, and continence during sexual acts. Direct stimulation leading to *erection* involves the hypogastric plexus (T11–T12) and sacral parasympathetics. Men with complete SCI maintain reflex erection with perineal stimulation but not psychogenic input. Men with complete SCI below L2 often maintain psychogenic erections but lose reflexive erections. *Ejaculation* involves the coordination of sympathetic, parasympathetic, and somatic nervous systems; injury to any of these can result in ejaculatory difficulty (Santiago & Fessler, 2004).

Bladder dysfunction and spinal cord injury: Complications with bladder function are the second leading cause of death in patients with SCI social discomfort, poor hygiene, infection, sepsis, and a variety of other medical conditions. Normal bladder function depends upon coordination between the cerebral cortex, hypothalamus, brainstem, and spinal cord, and their influence upon the sympathetic, parasympathetic, and somatic nervous systems. (See details in Santiago & Fessler, 2004.) From the viewpoint of trauma, we are concerned also with micturition involving the pontine micturition center, the hypothalamic preoptic center, and contraction of the bladder's detrusor smooth muscle with input from S2–S4 via the pelvic nerves. Inhibition of micturition and bladder relaxation involves the cingulum and premotor areas, and levels T11–L2 via the hypogastric nerves. Contraction of the external sphincter striated muscle that comprises the bladder's external sphincter is mediated by efferent nerves originating in S1–S4 running in the pudendal nerves.

Bowel dysfunction and spinal cord injury: Bowel functioning is vulnerable to neurological injuries at various levels: the enteric, autonomic, and somatic nervous systems. Spinal disorders can occur at the upper and lower motor levels. Disorders include loss of voluntary inhibition, constipation with frequent episodes of incontinence, loss of sensation, and delay of gastric emptying creating abdominal distention that inhibits diaphragmatic excursion. Dietary, surgical, and pharmacological management may be needed, with the involvement of the patient, family, and caregivers (Santiago & Fessler, 2004).

Etiology of Symptom Expression

Although PCS symptoms can have an emotional component, this assertion should only follow formal study of the patient's personality. An accident creating TBI has a high probability of also creating an acute or persistent PTSD. Both PCS and PTSD present with an array of symptoms that interact with each other and whose etiology may be created by noncerebral, somatic injuries in skeleton and soft tissue.

A review (Gasquoine, 1997) suggests that persistent postconcussive symptoms are influenced by cognitive, emotional, and motivational factors. Demographic and personality factors were considered of lesser significance, although I consider this controversial.

The magnitude of the subjective complaint may be influenced by premorbid conditions and also psychosocial factors, injury severity, and financial concerns. In order to study the effect of certain variables upon subjective symptom endorsement, a subset of items from the self-report Minnesota Multiphasic Personality Inventory-2 (MMPI-2) is used. The Fake Bad Scale (FBS) consists of items that are believed to be sensitive to magnify somatic, emotional, and cognitive distress in individuals involved in personal injury litigation. However, prior psychiatric history could also influence the scores. It was determined that the highest scores were created by individuals with mild TBI in litigation, followed by mild TBI not in litigation, with the lowest scores from persons with moderate-to-severe TBI who were not seeking compensation. Thus, FBS index is elevated in some individuals who are not seeking compensation after an accident. This finding may be explained by psychosocial adversity that makes them more sensitive to normally transient symptoms of a physical injury. High responding individuals endorsed evidence for physical complaints (e.g., headaches, which may be exacerbated by stress) and cognitive inefficiency, emotional despair, and being misunderstood or maltreated by their environment. Many readers will identify with this writer's experience that family problems and difficulties of obtaining merited compensation occur frequently, and are consistent with this interpretation. It may occasionally be the equivalent of the psychotherapeutic event known as *"the cry for help"* (Miller & Donders, 2001).

The severity of postconcussion complaints in a series of mTBI is based upon the pattern of complaints in the emergency room (De Krujik, 2004):

- The presence of headache, dizziness, and nausea within 6 h of a "mild" TBI predicts the severity of posttraumatic complaints (PTC) after 6 months. Half of the patients with a combination of these three symptoms still had PTC after 6 months, and the other half were considered fully recovered.
- The reported incidence rate of PCS is variable, perhaps attributable to variations in assessment measures and sampling criteria.
- After 2 weeks, the severest complaints were forgetfulness, drowsiness, headaches, dizziness, trouble concentrating, and lightheadedness. After 6 months, the severity of most complaints had declined to pretrauma levels, but medians for headache, dizziness, and drowsiness were still increased.
- In a brief interval, measured in months, measurement differences with control groups tend to disappear, but differences in such functions as information processing, choice reaction time, verbal memory, visual–motor speed, Wechsler Adult Intelligence Scale-Revised IQ, remain. Apparent rapid recovery following "mild" head injury leaves open the reason for the persistence of postconcussion symptoms, even with the evidence for some cognitive deficit.
- Rapid recovery has been associated with stricter exclusion criteria (e.g., intracranial trauma) and the study of a wider range of potential impairment.
- Preinjury personal characteristics may play a role in symptom persistence.
- Mood disorders or emotional distress may prolong cognitive disturbance.
- Motivation affects performance level.

PCS is a potentially persistent reaction, often resistant to treatment, with controversial causes (Mittenberg et al., 1991) and controversial outcome. Controversy is caused by several findings: (1) Seemingly mild blows, even without LOC, can cause considerable neuropsychological dysfunctioning. (2) A wide variety of physiological and psychological systems that are impaired after an accident causing head and other somatic injury creates a question whether the symptoms are a unitary "syndrome" or variable. In one study, only loss of concentration and memory were related, without any other symptom cluster (Montgomery et al., 1977). It has been suggested that there are different postconcussive syndromes reflecting distinct patient groups.

There have been efforts to order the various PCS symptoms using factor analysis: emotional, organic, and psychosocial (Slagle, 1990); dysthymic, vegetative/bodily complaints, and cognitive (Bohnen et al., 1995); cognitive, emotional, and motivational, with demographic and personality factors of lesser significance (I consider this controversial) (Gasquoine, 1997). Cicerone & Kalmar (1995) detected four clusters, and several solitary symptoms were identified in one study of mTBI:

1. Affective (irritability, frustration, anxiety, depression)
2. Cognitive (concentration, memory, slowed thinking, decision making, and fatigue)
3. Sensorimotor (dizziness, imbalance, incoordination, nausea, visual dysfunction, appetite changes)

4. Sensory (sensitivity to light and noise).
5. Independent sensory symptoms were headache, sleep disturbance, numbness, hearing loss, change in taste, and change in smell.

It was observed that when similar symptoms were studied in a matched nonconcussed population, a rather different pattern was observed.

Other procedures have detected PCS being associated with psychiatric status, negative affectivity, or neuroticism (Fox et al., 1995a).

Fox et al. (1995a) assert that "suffering a bump on the head" without LOC is presumably without brain damage, and therefore discovered symptoms that do not suggest PCS. "Psychological trauma" leads to PCS complaints. This formulation is imprecise in the writer's clinical experience. While some complaints are not physiologically based (neurological or somatic), brief LOC and lack of LOC are not precise guides. To some extent, LOC depends upon how the impact affects motion of the skull and brain. In particular, the relative motion of the impacting object and head, when parallel to the body axis, is less likely to create LOC, contrasted with a head impact causing transaxial motion of the head within the skull, or angular motion (laterally or in the vertical plane).

Certain symptoms are associated with being knocked out, though their status in the usual PCS lists is denied (tremor, 44%; loss of interest, 68%; confusion, 61%; and broken bones, 32%) (Fox et al., 1995). However, the frequency of such symptoms is not surprising after an injury severe enough to cause LOC. It is noteworthy that in this sample of 50 consecutive patients (all involved in litigation or workers' compensation claims), LOC was assessed in only 52% of the cases, while 28% exhibited psychiatric symptoms severe enough to warrant psychiatric diagnosis and referral. The difficulty of correctly attributing these symptoms was pointed out: Measures of depression are influenced by endorsement of cognitive and physical symptoms, and various symptoms contributing to a diagnosis of major depression are a direct consequence of neurological disorder.

The approach to the recognition and classification of neurobehavioral disorders expressed here has a wide conceptual range than those utilized previously. The symptoms are assigned to the various domains of the taxonomy, which has a clinical and physiological, rather than psychometric, foundation. The list evolves from my clinical practice (Parker, 2001) and other sources: Bohnen et al., 1995; Mittenberg & Strauman, 2000; and Gasquoine, 1997. No attempt has been made to use any estimate of frequency; it is simply assumed that the inclusion in a research list, study of the symptom, or my own observations should be considered by the practitioner as a possibility in clinical study of an accident victim.

The population under study is primarily persons who have experienced mechanical accidents, with potential effects of unhealed wounds. From one point of view, the trauma is quite brief when compared to continued combat or incarceration in a prisoner of war or concentration camp. On the other hand, after an accident, people whose pain, humiliation, restricted range of motion, change of identity, and functional impairment extends over years may have psychological and physiological effects similar to those undergoing long intervals of gross stress. Perhaps, when the fear, injury, and feeling of impairment are persistent, the physiological and psychological conditions more closely approach the original description of PTSD. A gross or continuous emotional injury or environmental stressor may eventually injure the body or spirit. It may be a psychological trauma that is too intense for the body to reduce or encapsulate, or some physical or mental injury that is persistently unhealed. The consequence is dysregulation of physical or emotional systems, causing physical or mental disorder. Thus, PTSD, as usually described, is an incomplete statement at best in an injured person, that is, with people who are exposed to danger, humiliation, and helplessness, these reports do not emphasize either people with injury or the consequences of the injuries that induced the familiar PTSD. PTSD may involve an unhealed mental wound. However, an unhealed physical wound creates long-lasting and different psychological phenomena that are usually not attributed to an accident that occurred long before. Its results, while unrecognized or wrongly attributed, may interfere with the quality of life as well as being disabling. Vulnerability to a stressor varies among people. It is dependent on the personal reaction to being threatened, injured, or impaired; coping skills' social response to the injured person; prior experience; temperament; and genetic and experiential factors. The natural history of a trauma varies with the injury and the person.

Persistence of the Syndrome

Persistence of PCS is controversial. One review, while acknowledging that PCS may remit spontaneously, asserts that estimates of the prevalence of symptoms at 3 months varies between 75% and 80% in epidemiological studies. PCS symptoms and neuropsychological deficits have been observed 3 years postinjury (Mittenberg et al., 1997). This contrasts with another review asserting that cognitive complaints occur during acute hospitalization and for 1–3 months later (memory, attention, information processing), but generally resolved during the 3 months after hospitalization. Resolution of cognitive impairment is related to resolution of abnormalities in CT and MRI (Mittenberg & Strauman, 2000). Other studies report persistence of symptoms up to 6 months, with cognitive deficits found mainly in divided and selective attention (Bohnen et al., 1995). Persistence has been attributed to *symptom expectancies* after head injury, that is, a bias creating selective attention to internal states. Normal premorbid symptoms can also be selectively attributed to the accident. When symptom expectations are confirmed, anxiety about their significance maintains selective attention (Mittenberg & Strauman, 2000).

POSTCONCUSSIVE SYNDROME: A WORKING DEFINITION

Concussion is an ambiguous term with two components: alterations of consciousness and commonly amnesia, and brain trauma of a nonsurgical extent, for example, *DAI*, perhaps with contusions. The alteration of consciousness occurs on a spectrum, that is, from disorientation, confusion, amnesia (retrograde and anterograde, i.e., PTA), or actual LOC for a brief interval, for example, less than 1 h. Chronic alterations of consciousness may continue indefinitely only to clear up unpredictably. Where there is evidence for multiple contusions, hemorrhage, ischemia, brain swelling, and extended LOC, moderate or severe brain damage is the appropriate diagnosis.

Concussion describes the neurobehavioral consequence of an accident in which a brain injury is likely to have occurred due to mechanical forces. There may be a combination of direct head impact and/or rapid movement of the head on the neck ("whiplash") caused by hyperextension and hyperflexion after an accident accelerates the body, neck, and head, followed by another impact or restraints that cause the person's body to be propelled backwards. While there may not be gross trauma to the head or brain, this may be only a fraction of the traumatic event. The diagnosis of concussion should be limited to individuals with no or few minor radiologically detected brain injuries (e.g., contusions), with few or lacking minor neurological symptoms, without hospitalization for more than a few days for observation, lacking LOC for more than an hour.

Thus, concussion is an injury to the brain caused by a sudden acceleration or deceleration of the head and/or head impact that also results in immediate, but temporary, alteration in brain functions such as LOC, blurred vision, dizziness, amnesia, or memory impairment (Guskiewicz, 2001). While alteration or LOC is usual, in the writer's experience, the presence of altered consciousness seems to depend upon the overall geometry of the accident: direction of impact, position of the head and body, and position (and composition) of surfaces impacted.

EXAMINING THE PATIENT WITH PCS

An accident creating a head injury has multiple effects (cerebral, somatic, social, psychological, and biological) with disparate neurological, somatic, and psychological etiologies. To increase the completeness of neurobehavioral conditions considered in assessment, one may take into consideration the taxonomy of neurobehavioral disorders. It is recommended that the examiner determine the details of an injury or frightening event, overcome patient reluctance or inability to offer information, assesses alterations of consciousness and memory (Crovitz et al., 1992), estimate deviation from the preinjury baseline, consider the interval between injury and examination, and document a wide range of performance. It is important to avoid inappropriate diagnostic categories based upon behavioral and mood disorders whose symptoms superficially resemble TBI (see Table I) but are not trauma-related. TBI may be ignored in children due to the secrecy of the child, the defensiveness of the parent, and the ignorance of many pediatricians about diagnosis and outcome. The examiner may err by utilizing adjustment disorders with conduct disturbances. Accuracy affects incidence placed into public health statistics. Apart from malingering, some patients play the sick role (factitious disorders) to obtain support in a crisis or as a lifestyle. Recognizing the complexity of PTSD and TBI, eliciting information pertaining to diagnosis, need for treatment, and issues of compensation and disability will often be a multidisciplinary endeavor.

DIAGNOSIS OF THE POSTCONCUSSION SYNDROME

The moment of injury (primary phase) need be no more than a prologue; the outcome initiated by the pattern of injury can be severely disabling and uncomfortable. Later events in the drama of TBI will be modified by genetic characteristics, preexisting conditions, correct assessment and treatment of the symptom etiology, and community reaction to the patient's impairment. Initial assessment should consider:

- *Event*: Physical qualities of the mechanical forces and the surrounding environment
- *Mechanical forces and impact*: The moment of injury and reconstruction of the accident to assess trauma
- *Gross neurological injury*: Central, autonomic, and peripheral nervous systems
- *Cellular and subcellular neurological injury*: Stretching, impact (external, internal), membranes, intracellular neurotoxins, crushing, perhaps resulting in DAI
- *Somatic injury*: Musculoskeletal (head, neck, trunk and spinal column, limbs; internal organs and glands; cervical blood vessels; vocalization structures)
- *Basic trauma*: The differentiation has been made between *psychogenesis* (premorbid conditions, postinjury psychological factors, or outright malingering, rather than an alteration in brain function) and *physiogenesis* (acute neuropathology and other abnormalities in brain function (Yeates & Taylor, 2005). Patients who meet the familiar characteristics of concussive brain injury, that is, GCS of 13–15, brief, retroactive and/or PTA, or LOC less than 1 h, may have TBI that is difficult to document.

Noncerebral somatic trauma creates many PCS symptoms. An accident causing a head injury usually causes injuries to other parts of the body, that is, bone, soft tissue (muscles, ligaments, tendons, vasculature), internal organs, peripheral nerves, and somatic organs. These injuries often create chronic symptoms that later examiners associate with the concussive accident (see Figure 2.1). In the case of children, the possibility of avoidable developmental disorders (puberty, growth, maturation of personality) is ignored.

THE WIDE-RANGING EXAMINATION

Using a wide-ranging examination within one's specialty, as well as referral to other professions, reduces the likelihood of making diagnostic errors. Some member of the team should provide an intensive interview. The patient's assertion that they never had a significant accident must always be closely examined, with questions specifically addressed to motor vehicle accidents (MVAs), falling; falling objects; electrical, chemical, anoxic conditions; and assault. Since previous examiners

FIGURE 2.1 Figure drawings by an adult, illustrating regression of "mild" head injury after several years. (a) and (b) were early drawings, and (c) and (d) were made later.

(1) may not have examined the patient carefully for possible head injury or (2) may not be aware of potential chronic dysfunctions, the patient's lack of understanding could lead to false information being offered to the current clinician. The preexisting condition is not revealed to the health care provider. Interviews of MVA victims often reveal previously undocumented head impacts with significant neuropsychological deficits (Parker & Rosenblum, 1996).

The following taxonomy offers in a systematic way the very wide range of adaptive behaviors that are vulnerable to brain injury, somatic injury, and stress. The actual range of disorders provides a reference to assess research involving the area of epidemiology and the consequences of accidents causing concussion and more severe levels of TBI. It is apparent that much research evidence of the proportion of chronic dysfunctions after head injury ("resolution") and the applicability of measures of response bias or symptom validity are based upon an insufficient range of study of potential dysfunctions. The frequent assertion that only 15% of accident victims categorized as "mild" have persistent symptoms must be regarded skeptically. It was recently estimated that 15–30% of patients with minor trauma report postconcussive symptoms persisting for many years; thus, there are 200,000–400,000 new patients in the United States each year with a persistent postconcussive syndrome from mild injury (Lewine, 2007).

A reduced libido may result from cerebral damage, but also medical illness and such stress-related conditions as anxiety. It encourages a more accurate conception of the actual etiology of the consequences of an accident or other complex medical condition. Organizing the complex expression of symptoms with disparate traumatic bases into a formal taxon (domain) permits a more clearly focused treatment plan. The need for specialists from various disciplines becomes clear. For example, where Bohnen et al. (1995) assign sleep difficulties and heart palpitations to the "dysthymic factor," the offered taxonomy has biological components permitting assignment of symptoms such as the following to more precise domains: "neural-sleep disturbance" and "somatic-vegetative/physiological."

Use of the taxonomy has important advantages in planning an examination to include sampling in the various domains:

1. It reduces the possibility that significant impairment will be ignored.
2. Elicited information can be presented in an orderly way. Findings relevant to a particular taxon, elicited from multiple procedures, are presented together (although results of a specific test, e.g., standard scores, percentiles) are presented separately.
3. Redundancy is reduced since the examiner is alerted to the general function that is examined by a particular procedure.

Symptoms may be *positive*, for example, intrusive anxiety and partial seizures; *negative*, for example, avoidance; or *distracting, stress-related* (e.g., headaches, pain, intrusive anxiety).

Neurobehavioral dysfunctioning may be defined in part as a deviation from the baseline.

THE TAXONOMY OF NEUROBEHAVIORAL DISORDERS

I. Neurological

Arousal and self-awareness: Hypoarousal; hyperarousal

Focused attention and vigilance

Seizure activity: Posttraumatic seizures (grand mal; complex partial seizures; focal; simple partial seizures (hallucinations of light, smell, taste, touch, hearing)

Seizurelike activity of unknown etiology (SLAUE)

Sensorimotor integration (CNS injury at any level): Body schema; incoordination (reduced proprioceptive feedback); injury to spinal column, bones, and soft tissue; tremor; mobility; motor speed; weakness; lateralized differences

Vocalmotor communication: Dysarthria; dysprosodia

Sensory: Visual (scotoma; diplopia); hypersensitivity to light and noise; numbness; hearing (loss; tinnitus); vertigo; dizziness; imbalance; taste; smell

Somesthetic: Smooth pursuit; body schema (depersonalization due to disturbed body schema of the parietal lobes), cerebellum; loss of proprioceptive input

Oversensitivity (heat, cold, bright light, loud sound)

Impulse autoregulation (cerebral personality disorders): Reduced initiation (apathy); poorly controlled ("frontal lobe syndrome"); impulsive; poor judgment; inappropriate; childlike; inability to learn from experience

II. Dysregulation of the Brain's Internal Milieu: CNS–Systemic Interaction

Systemic and autonomic: Metabolism; sweating; nausea, frequent urination; bowel control; hyperventilation.

Disturbance of sleep-wake cycle: Bad dreams and nightmares; sleep initiation; interrupted sleep and inability to return to sleep; fatigue upon awakening; impaired daytime performance ("microsleeps"); fatigue; sexual (reduced or increased libido; impotence).
Structural damage to the hypothalamus and pituitary gland: Diabetes insipidus (polyuria); hypopituitarism; metabolic disorders.
Circadian: Secretion; physiological functions, for example, body temperature.
Stress (acute and chronic): Hyperarousal and hypoarousal (heart rate, blood pressure, endocrine); bidirectional neuroendocrine secretion (soma and CNS); nausea.
Systemic dysregulation (allostatic load, i.e., burnout consequent to unhealed somatic injuries), that is, hormonal, immunological, inflammatory, circadian, autonomic nervous systems; fatigability; autoimmune diseases; cardiac; cancer, and so forth.

III. Somatic Dysfunctioning: Tissue Damage and Pain

Headaches: Head injury; referred from trigger points of head, neck, limbs, torso (often unilateral); migraine; tension; vascular.
Pain: Alterations of gait and other movements.
Fibrosis: Interference with muscular and neural functioning.
Vascular (traumatic and stress interference with perfusion by the internal carotid, vertebral, and vertebrobasilar arteries into the cerebral parenchyma): Dizzy spells.
Spinal cord and ANS injury: Sexual (bidirectional changes of libido); bowel; urinary disorders.
Gait, stance.

IV. General Intelligence

Differential disorders of ability according to context: Structured, unstructured, and visuoconstructive; problem solving; comprehension; learning ability; effective decisions; concept formation, including abstractions; complex performance.

V. Cognition Loss

Concentration (mental control for a useful interval)
Change focus of attention and set
Reduced personal safety: Loss of comprehension, alertness, memory, judgment
Information processing/executive function: Processing speed; working memory; multitasking; foresight/judgment; error monitoring; planning; mature problem solving (preplanning vs. trial and error)
Memory and learning: Short term; long term; modality-specific (visual, auditory, somesthetic, etc.); PTA (anterograde; retrograde)
Insight: Inaccurate self-report
Communications (cognitive): Aphasia: fluent, nonfluent; reduced comprehension, receptive and expression

VI. Psychodynamic/Personality Regression

Impaired emotional regulation: Lability, irritability, aggression
Anxiety control: Psychological defenses
Identity/reduced self-esteem: Unattractive, damaged, incompetent, impoverished, victimized, rejected, vulnerable, unwilling to reveal distress

Role in life: Passive; dependent
Affect and mood: Enhanced and poorly controlled anger, depression, and anxiety

VII. Reduced Adaptive Capacity

Reduced competence: Ineffective in school, work, domestic responsibilities, family needs
Social change: Withdrawal and detachment due to shame; reduced social skills (cognitive disorder); inability to enjoy leisure activities; inability to love; loss of empathy; antagonistic; less attractive (reduced energy, money, mobility, and cognitive ability)

VIII. Stress Reactions: Acute and Chronic

Reduced resilience: Can't cope with stress
Hyperarousal: Heart rate; breathing; intrusive images and thoughts (anxiety); hyperalertness
Hypoarousal: Numbing; avoidance, libido
Dissociation (anxiety): Depersonalization (self has changed); derealization (outside world has changed
Event: Preoccupation; persistent re-experience
Fatigability after relatively minor effort
"Distracting" symptoms: Interfere with quality of life. Seizures, mobility, pain and headache, tinnitus, immobility, imbalance, and so forth
Weltanschauung (perception of the world): Bleak, unsupportive, hostile.
Poor morale: Does not expect improvement

IX. Developmental Problems of Children

Physical development: Puberty (premature, delayed, absent; reduced height and weight)
Motor skills
Cognition, learning, educability
Social rejection: Academic and mental ability; physical skills; social skills (immature, impulsive, affect control

MECHANICAL BRAIN TRAUMA—EXTENT

Biomechanical events create widespread neurological injury. The clinician considers the physical forces affecting the brain and their effects upon an organ surrounded by bones and meninges with sharp edges, penetrated by vessels, with shearing effects at the surface and internally. Integrate the mechanical forces operating upon the body, the characteristics of the surrounding physical environment, and the moment of inertia the accident imparted to the body with an understanding of the characteristics of tissue and the actual anatomy of head, neck, torso, and limbs (Parker, 2001, Chapter 5).

The injury may be widespread, focal, accompanied by skull fracture or infection, penetrating (knife or bullet), or mixed. There are levels of diffuse injury (Adams et al., 1989; Maxwell & Graham, 2003; Parker & Rosenblum, 1996). The shearing force may stretch but not sever the axon, though it initiates a pathological cascade of events. There are increasing levels of demonstrable patterns of damage to white matter, most properly referred to as DAI, from microscopic (scattered injured axons) to gross. TBI triggers ionic and metabolic events from which the cell may recover, or degenerate and die. These may be observable anatomic lesions, or only alterations in cellular metabolism, subtle morphologic abnormalities, or persistent derangements in neurotransmission. The relative paucity of overt histological damage in the presence of intracellular dysfunctioning (ionic and metabolic) is reflected in negative CT scans after concussion. Markers of injured nerve fibers are immunocytochemical

(ICC) labeling with antibodies to β-amyloid precursor protein, or silver-stained preparation. Brain trauma (except for the least intensity) passes through various phases.

PREEXISTING CONDITIONS

The concept of "programming" or "imprinting" describes a predisposition to later behavioral effects caused by the events of a sensitive period in prenatal or early development. The consequence may be lifelong structural and functional effects. The combination of the mother's stress during the prenatal period and the family ambience affects the level and nature of response to adversity and overt stress. It may predispose the fetus to later dysregulation (Seckl & Meaney, 2004). Among the events and conditions affecting the outcome of PCS will be significant or chronic illness; neurotrauma (including neurotoxins); the quality and length of education; the family structure, ambience, and socioeconomic level; and whether development took place in a different community and culture.

PHASES OF TRAUMATIC BRAIN INJURY

This section is an overview, designed to alert the clinician that TBI is a complex and changing reaction. The distinction between immediate (primary) and consequent (secondary) brain damage can lead to effective management, since it encourages prevention and treatment of such pathologies as brain swelling, hemorrhage, and infection, that is, reducing secondary brain damage (Mendelow & Crawford, 2005). Due to the variety of posttraumatic developments, assessment should include a prognosis (predicted outcome) and a statement concerning potential later symptoms, avoiding premature establishment of the "outcome" of an accident. A more complete presentation is offered in the chapters on acute and chronic brain injury. Trauma refers to the anatomical and emotional damage incurred by the person. The neuropsychological examiner's duty is to explain to the court the process of TBI and its potentially complex pathology and neurobehavioral outcome in the instance of a particular claim of injury. If review of the records and interview of the patient suggest that no significant injury has occurred, this should be stated. However, such a statement made by an examiner who is not proficient in neurotrauma and its consequences runs the risk of having a more learned adversary expert present evidence based upon superior knowledge that will question one's competence in this highly technical and complex area.

The outcome of brain injury is not fully described by the primary injury: Further trauma may develop when there is hemorrhage, gross contusions, skull fracture, and so forth. TBI is usually comorbid with polytrauma, that is, multiple injuries occur: neurological, somatic, mental, and social. Physiological dysregulation caused by unhealed chronic injuries in other parts of the body affect later brain functioning. TBI does not remain static. The brain undergoes a series of changes after the initial trauma, which will be described. Chronic conditions develop characteristic disorders of physiological dysregulation and health disorders.

The PCS does not remain static: After the initial trauma, neurotrauma and neurobehavioral functioning undergo a series of changes.

Primary phase: The immediate mechanical brain and somatic tissue damage: The application of physical force directly to the head and/or body creates neurological injury at the anatomical, cellular, subcellular, and genetic levels (cerebral, cerebellar, spinal, cranial and peripheral nerves) and somatic injury (fracture of the skull, damage to the spinal column, soft tissues (vascular, muscular, ligaments), bone, and systems (endocrine, inflammatory, immunological, circadian, and autonomic systems).

Secondary phase: Continuing injury: Primary neurotrauma may be greatly aggravated by secondary trauma (hemorrhage, swelling, ischemia, contusions, skull fracture, subcellular reactions, neuronal degeneration, etc.), occurring over a period of hours to days. Neural, biochemical, molecular, genetic, and systematic changes may be protective or grossly destructive and fatal.

Tertiary (late-developing physiological disorders: Physiological (systemic) reactions to the injury enter bidirectional signaling pathways (the brain's IM) between the body and the brain, and also pass the impaired BBB. Some neurobehavioral changes are of physiological origin, not TBI. These are derived from damage to the hypothalamic-pituitary target endocrine gland axis (HPA), the physiological reactions of chronic stress, and the "burnout" or allopathic load).

Quaternary (Late-Developing Neurological Disorders): Examples include neuronal degeneration consequent to pathway injury, posttraumatic epilepsy, movement disorders, hippocampal damage due to stress effects (catecholamines), and premature dementia.

Pentary phase: Chronic stress effects: These are consequent to the effects of (1) persistent unhealed somatic injuries upon physiological systems (hormonal, inflammatory, immune, circadian, and autonomic), leading to "burnout" or the allostatic load, and (2) chronic emotional distress (impairment, "distracting symptoms," for example, immobility, seizures, tinnitus, imbalance, social rejection, and so forth. Chronic stress consequences include vulnerability to cancer, cardiac disease, fatigability, and so forth.

Developmental problems of children: Injuries to the HPA can result in developmental delays, which may be neither recognized nor attributed to earlier trauma: Precocious, delayed, or absent puberty; growth deficit; sensorimotor performance; social incompetence; immature cognition and learning; inadequate academic achievement and vocational capacity; and personality maturity.

3 The Central Nervous System
Organization of Behavior

OVERVIEW

Adaptation involves the integration of neural and physiological systems. Adaptation involves the brain's interaction with the internal milieu and environmental reality, and then organizing and modifying ongoing physiological, musculoskeletal, mental, social, and emotional activities. We will focus on central nervous system (CNS) structures and pathways that are behaviorally significant and that participate in adaptive functioning with the purpose of clarifying posttraumatic and other impairments, thus enhancing examination precision and the focus of treatment. Rather than "centers" performing stated functions, there is a central role of widely distributed networks with bidirectional interaction, integrated sensorimotor responses, and complex mental and action functions supported by cortical-subcortical-brainstem circuits. This widespread locale of the neurological support of behavior renders the brain vulnerable to mechanical forces. Thus, seemingly trivial or occult head injuries are paradoxically impairing.

NEUROTRANSMISSION

Neurons' neurosecretory function involves release of neurotransmitters into synapses via axons. The neurotransmitter may act directly on the cell membrane or it may stimulate internal chemical reactions known as *second messengers*. These are large molecules within the cell membrane that modify and amplify signals that are brought by neurotransmitters and mediators. Neurotransmitter effects may be, directly or indirectly, excitatory or inhibitory, depending upon the characteristics of the neuron and its influence on the excitatory or inhibitory neurons in its region (Barabam, 2005). They act on specific membrane receptors to produce biochemical and excitability changes in the stimulated cell.

OTHER SIGNALING SYSTEMS

While neurotransmitters are the active agent in widespread systems (Dunn, 1996), there are numerous other signaling systems. Behavior control has varied sources. There is direct control of the cortex, by specific signals from the lower brain areas, and also more general control, by the secretion of excitatory or inhibitory neurotransmitter hormonal agents into the brain substance. Neurohormones may persist for minutes or hours, in contrast with some rapid action and disappearance of some neurotransmitters. *Neuroactive peptides* are so large that ordinarily they are restricted from passage into the brain.

NEURONS AND RELEASING HORMONE

Neurohypophyseal cells (stimulating secretion in the posterior pituitary) are found in the magnocellular nuclei of the hypothalamus and secrete vasopressin and oxytocin via the neural pathway into the posterior pituitary (neurohypophysis). *Hypophyseotropic neurons* secrete releasing hormones, which are released from the median eminence that are carried by the pituitary portal vessels to the anterior pituitary. *Hypothalamic projection neurons* target sympathetic preganglionic spinal neurons. Cell groups with long projections can be defined by their neurotransmitters (Guyton & Hall, 2005, pp. 730–731; Saper et al., 2000).

Major Modululatory Systems

Following are the major modulatory systems of the brain (Guyton & Hall, 2005, pp. 730–731; Marrocco & Field, 2002; Saper, 2000b):

Noradrenergic: Medulla—nucleus of the solitary tract and dorsal motor vagal nucleus project to hypothalamus, control cardiovascular and endocrine functions. Locus ceruleus maintains vigilance and responsiveness to unexpected stimuli, projecting to cerebral cortex and cerebellum, and descending projections to the brainstem and spinal cord.

Adrenergic: Medulla—project to spinal vasomotor neurons and to the hypothalamus, modulating cardiovascular and endocrine responses.

Dopaminergic: Midbrain and forebrain—Ascending input to the striatum initiating motor impulses; innervate frontal and temporal cortices, and the limbic structures of the basal forebrain. These pathways participate in emotion, thought, and memory storage. From the hypothalamus, descending pathways to the spinal cord regulate sympathetic preganglionic neurons. Wall of third ventricle—Participates in the tuberoinfundibular hypothalamic neuroendocrine system. Dopaminergic neurons are found in the olfactory system and the retina.

Serotonergic: Midline of the brainstem in the raphe nuclei—Descending projections to the spinal cord and motor and autonomic systems, as well as to the dorsal spinal horn, modulate perception of pain. From the pons and midbrain projects to virtually all of the forebrain. Serotonergic pathways regulate hypothalamic cardiovascular and thermoregulatory control and modulate the response of cortical neurons.

Cholinergic: Brainstem and forebrain—Used by somatic and autonomic motor neurons. Descending projection to the pontine and medullary reticular formation. Ascending projections to the thalamus. Participate in regulating wake-sleep cycles.

Histaminergic: Hypothalamic tuberomammillary nucleus—Project from the spinal cord to the entire cortical mantle. May maintain arousal in the forebrain.

Synapses

Neurobehavioral activity depends upon a wide variety of subordinate functions, including somatic, neurological, and mental. These closely regulate each other by numerous messengers, both chemical and neurological. *Synapses* between neurons are described as excitatory, inhibitory, and modulatory (Connors, 2005, chapter 12). The *modulatory synapse* has little effect of its own but modifies the effects of the other two types. Brainstem noradrenalin and serotonin neurons influence the forebrain structures that determine the arousal level. Cholinergic neurons in the basal nucleus of Meynert send impulses to almost all sections of the neocortex, where they participate in attentional mechanisms that sharpen cognitive or perceptual processes. Modulatory systems function in collaboration with perception. Using hunger and blood sugar level as an example, initially sensory and modulatory processes determine blood glucose levels, and later, sensory and modulatory systems focus the sensory apparatus on stimuli that are relevant to feeding (Amaral, 2000).

Chemical Receptors Determine Cell Responses

Cells communicate with each other to coordinate their functions. The neuron-receiving neurotransmitters are not passive recipients of this information. The message is determined by both the neurotransmitter receptors and the neurotransmitters themselves. In fact, there are more types of neurotransmitter receptors than neurotransmitters (Knapp et al., 2003). Receptors are proteins within the cell membrane to which substances bind; these are known generically as *ligands*. If neurotransmitters, hormones, and paracrine signals are considered the *first messengers* binding to a receptor, intracellular changes create *second messengers* that diffuse through the cell and change the activity of the intracellular effector system (Barabam, 2005; Changeux, 1993; Junqueira & Carneiro, 2005; Marks et al., 1996, pp. 679–685). Target cells are those that have specific receptors for a signal (e.g., neurotransmitter or hormone) and do not bind to other hormones. Different types of target cells contain different numbers of receptors for a particular hormone, varying from less than 100 to more than a million/cell. Receptor subtypes meditate different behaviors and response to particular medications (e.g., anticonvulsants and anxiolytics) (Zorumski et al., 2005). The receptor density determines the physiological response. Receptor characteristics influence physiological regulation and drug actions (intended and side effects).

Although receptors are given generic names, they respond to different substances than the traditional signal. *Acetylcholine* receptors, which stimulate muscle, also bind to *nicotine*, which mimics acetylcholine's stimulatory effects, as well as to the poison *curare*, which prevents acetylcholine and nicotine from stimulating a muscle. Receptors (e.g., acetyl choline and GABA) come in multiple varieties. A particular receptor in one part of the brain may have somewhat different properties than a variant elsewhere. This could lead to design of drugs that selectively impede or enhance the action of particular receptor types (e.g., treating brain-signaling disorders) (Changeux, 1993).

Although particular nuclei and tracts produce and convey particular neurotransmitters, these systems are not directly correlated with traditional functional systems or topographic subdivisions of the CNS. The neurotransmitter effects are actually determined by the nuclei targeted and the subtype of receptors stimulated. Five major neurotransmitters bind to at least 33 pre- and postsynaptic receptor types (Marrocco & Field, 2002). Thus, a drug that targets a particular system also affects other systems with similar receptors ("side effects") (Swanson, 2003).

The same messengers (e.g., hormones, neurotransmitters and neuromodulators, hypothalamic releasing hormones etc.) can serve as a (1) neurotransmitter; (2) hormone secreted by a neuron; and (3) a hormone secreted by an endocrine cell (Aron et al., 2004, including Table 5.1). Various tissues may share common ligands and receptors, so that the activation of one system may be accompanied by different changes in others (Marucha et al., 2001). Perhaps this is the basis of medication side effects. The complexity of the internal milieu may be illustrated by the chromaffin cells of the adrenal medulla (Kvetnansky & McCarty, 2000). Originally considered to be under the exclusive control of acetylcholine (ACH) and concerned with the secretion of the catecholamines norepinephrine and epinephrine, these cells are now known to release a great number of peptides. Various biologically active neuropeptides are co-localized with ACH and have a variety of behaviorally significant receptors.

Intracellular Signaling Pathways

Intracellular signaling pathways enable cells to maintain separate channels of information, which are integrated when appropriate. These have spatial and temporal features that integrate and code the signal, achieving optimal handling of different types of information. These signals bridge the spatial and temporal gap between sensory and motor output. Multiple intracellular signaling pathways allow neurons to detect coincident presentation of numerous stimuli and alter its response to current accordingly (Barabam, 2005).

Neurotrophic Factors: Development, Plasticity, Recovery

Neurotrophins are signaling molecules with multiple functions. They are proteins that regulate cell survival, nervous system development, neural regressive events (cell death, axon collateral elimination, dendritic pruning), activity-dependent and functional plasticity of the nervous system, repair of the damaged nervous system, and so on. In the mature brain, neurotrophins function in plasticity and maintain neuronal phenotype and function. Axonal injury leading to circuit disruption affects both function and trophic signaling between neuronal populations. Posttraumatic application of trophic factors is neuroprotective, maintaining neural survival and promoting circuit reorganization and functional recovery (learning and memory) (Dietrich & Bramlett, 2004). Neurotrophins protect neurons from endogenous toxic effects triggered by injury of disease (e.g., axotomy or glutamate excitotoxicity) (Bhattacharyya & Svendsen, 2003). Plasticity may be needed for adult nervous system function and for recovery following brain injury. *Nerve growth factor* (*NGF*) was the first of many *neurotrophins* (NT—neurotrophic factors) identified.

Neurotrophins regulate the response to neuronal injury (and development) including apoptosis. Neuron effects (including apoptosis) depend upon signaling. They can be bidirectional (i.e., to or from the target tissue [injured tissue] back to the cell body). In addition to cell survival, they also participate in learning, memory, and behavior, as well as neurotransmitter release, mechanosensation, pain, synaptic plasticity, depression, substance abuse, neurodegeneration, neuropathy, and cancer. They have been implicated in the pathophysiology of neurodegenerative and psychiatric disorders (Arvin & Holtzman, 2003; Bhattacharyya & Svendsen, 2003; Lee & Chao, 2005).

Apoptosis or Programmed Cell Death: Neuronal Injury

Development: Programmed cell death (PCD) is a normal cell response crucial to tissue remodeling during development and metamorphosis. Redundant neurons (50%) die in the process of apoptosis (programmed cell death) (i.e., regrowth or reorganization of central nervous connections). NTs maintain the neuronal phenotype (essentially the mature characteristics) and function. Necrotic cells have lost the competition for neurotrophins such as nerve growth factor (NGF), which blocks apoptosis.

Neuronal injury: When injury and disease compromise apoptosis, neurotrophins protect neurons from endogenous toxic events triggered by axotomy or glutamate excitotoxicity. NT expression is unregulated in both the peripheral and central nervous systems; axon regeneration and nerve repair is unregulated in the peripheral nervous system. It is found in a variety of tissues that die due to certain cellular mechanisms and are replaced by new cells following cell division and differentiation of stem cells. In this highly regulated process, in coordination with the generation of new cells, homeostatically maintains the number of tissue cells, organ size, and function. PCD is facilitated by cytokines, corticosteroids, and so on. During development it is observed in cortical, subcortical, ANS, and spinal regions. It serves the purpose of removal of neurons that have mis-migrated in the nervous system, or whose axons have gone astray or have innervated inappropriate targets. During adult neurogenesis PCD may be necessary for the efficient incorporation of newly generated neurons into functional circuits. Neurogenesis is restricted primarily in adults to the olfactory bulb and the dentate gyrus. Controlled cell disassembly on schedule occurs without adverse effect upon the organism (Buss et al., 2006).

VARIETIES OF CIRCUITRY

Brain Size and Connectivity

Large brains create a problem of connectivity (e.g., as axons lengthen to maintain speed their diameters must increase). As the brain size (number of neurons) increases, the number of interconnections increases much faster than the number of neurons. Large brains devote more of their mass to connections, with a decrease in cell body density and an increase in cortical thickness. However, an increase in connection does not compensate for need for increased ratio of connectivity or longer connection distances.

The brain becomes more modular: (1) Increases the number of processing areas so that the areas become smaller; (2) confines most connections within an area; and (3) subdivides areas into smaller areas such as columns (visual) or modules.

Connections that require long, thick axons are reduced: (1) Functionally related areas are grouped together; (2) cerebral hemispheres become specialized so that it is not necessary to send information from one hemisphere to another. The proportion of callosal axons are reduced, particularly those of the larger diameters needed for faster conduction time (Kaas & Preuss, 2003).

Vulnerability to DAI

DAI creates gross impairment and unpredictable consequences by interfering with *integration of complex circuits*, some of which are arranged in *networks* with cross-communicating pathways having a consensus output (Angevine, 2002; Bloom, 2003). Multiple sites of injury and disruption of axons that integrate multiple brain centers reduce performance efficiency without conspicuous damage to neural centers.

Local circuit neurons: Connections are primarily in the immediate vicinity, with the flow of information expanded and contracted by interneurons.

Parallel processing: The use of several different pathways or clusters to convey the same information. The advantages include increased processing speed, reliability, and processing capacity. It is considered more efficient than serial processing for complex tasks (Eriksson, 2002).

The linear model, or pathway specific: Long hierarchical neuronal connections make the transmission highly sequential (note the modification by the Recurrent model), for the primary sensory and motor systems.

Single source divergent circuitry (i.e., interregional interneurons such as the hypothalamus, pons, and medulla): Extends divergent connections to many target cells, almost all of which are outside the originating brain region.

Most long pathways cross at the midline: Most sensory and motor events have contrelateral cortical different systems that decussate at different levels. Exceptions include bilateral representation by certain cranial nerves such as vision (each eye is represented in both hemispheres; the temporal half of the retina remaining ipsilateral while the nasal half crosses at the optic chiasm), hearing, and innervation of the forehead by the facial nerve. Audition and olfaction also have some bilateral representation.

Multiple pathway or lens model: Particular stimuli utilize several pathways. While the convergence is obligatory, the outcome of trauma depends upon poorly understood variables.

Synchronization (Oscillation): A process of self-organization by which cell assemblies are synchronized and their activities oscillate.

Recurrent circuits (transmission is multiplex): The pathway between the cortex, the subcortex, and the periphery is simultaneously *bidirectional* (Nicolelis et al., 2002).

SYNCHRONIZATION (OSCILLATION)

EEG reflects signals generated by the cerebral cortex and represents the summation of excitatory and inhibitory potentials. *EEG frequency ranges* are represented by *alpha* (8–12 Hz), *beta* (more than 12 Hz), *delta* (1–3 Hz), and *theta* (4–7 Hz). In normal adults, the waking pattern is usually alpha prominent over the occipital area. The amount and amplitude of slow activity (delta and theta) is closely correlated with the depth of sleep. Slow frequencies are abundant in newborns and young children, but disappear progressively with maturation (Emerson & Pedley, 2004). Transitory generalized slowing is common after a concussion. A persistent, continuous area of slow wave activity suggests a cerebral contusion, even in the absence of clinical or CT abnormality. Unilateral voltage depression suggests a subdural hematoma. In the first 3 months after trauma, the EEG does not predict post-traumatic epilepsy.

There is a mechanism for integration within and between brain areas that is vulnerable to TBI. DAI interferes with higher-order brain functions (e.g., perception, action programming, and memory traces), which are subserved by dynamic assemblies of self-organized neurons. These are synchronized and oscillate in phase. Neural modulators alter the activity of particular neurons, which alter the activity of the remainder of the network that feeds back and modifies the initially affected neurons. Thus, the network response is modulated, rearranged as a whole, contributing to rhythmic oscillations, and the response is a *distributed* function of the network (Hooper, 1995). *Information processing* takes place in networks of neurons. Sustained, coherent activity, particularly in the thalamocortical circuits, supports *wakefulness*, *attention*, *mood*, and *sleep*. Here, changes of neuronal activity are associated with the state of behavioral arousal. Pacemaker activity also drives oscillatory firing in the thalamus (Zorumski et al., 2005). Synchronized oscillations in a neuronal network may be the basic unit forming attentional percepts requiring processing between different neuronal groups. Most *cognitive processes* involve the parallel activity of multiple cortical areas. In a network, neurons function in synchrony. Integrated groups of oscillating neurons interact with each other through excitatory and inhibitory local circuits, with feedforward and feedback elements. Neuronal circuit oscillation could link individual neurons to form a functional unit and an inhibitory functional unit (Da Silva, 2002). *EEG coherence* is a measure of the similarity of two EEG signals (i.e., phase consistency between pairs of signals in each frequency band after a time delay at two locations). It is a measure of functional interactions between oscillating brain systems. A related concept is *EEG synchrony* or *desynchronization*, which refers to sources oscillating approximately in phase so that their individual contributions to EEG add up. Desynchronization is often associated with EEG amplitude reduction (Nunez, 2002).

THE RECURRENT NEURAL NETWORK

There are several organizing principles of information flow in the brain.

Bidirectional

There is a two-way flow between the cerebral cortex and the periphery, interacting with off-axis nuclei (e.g., the cerebellum, basal ganglia). Sensory and motor activity transmission in any direction is affected by incoming stimuli from the opposite pole to relay stations (Livingston, 1985, p. 1157). Execution of any sequence or gestalt of behavior involves a flow of neural and environmental information in which sensory and motor hierarchies interact with one another. The interactions are bidirectional, including feedforward and feedback (Fuster, 1995). "Top-down" influences mediate "*somatosensory gating*" and precede tactile discrimination. Tactile neuronal responses are affected by the motor activity that underlies the behavior used by a particular species to explore its environment.

Corticocortical Feedback

The basic flow of information involves input from the whole cerebral cortex, which is processed and integrated before being fed back to frontal cortical areas via various parallel and closed basal ganglia-thalamocortical loops (Smith & Sidbe, 2003). The older concept of somatosensory functioning (i.e., stimuli affecting the periphery and proceeding to the cortical level [*feedforward*]), has been replaced by the more comprehensive *recurrent* model.

Network Influences

Anatomical: Between different levels of the neurological network. The latter takes into account interactions at and between all levels of sensory circuitry (cortical and subcortical—spinal cord, brainstem, thalamus).

Chemical: The effects upon the reaction pattern by neuroactive substances in the internal milieu. Neuromodulators alter both synaptic strength and inherent cellular properties. Some cellular properties are long lasting rather than ceasing after a stimulus. Long-lasting cellular properties are important (membrane changes; synaptic strength; inherent cellular properties). Changes in the network activity cannot be explained by solely considering the neurons directly affected by a neuromodulator. The networks are highly interconnected and the active responses of the neurons stimulated by the modulator alter the activity of the rest of the network's neurons. These in turn feed back onto, and modify the responses of, the directly affected neurons. The network functions as a biological response as a whole (i.e., both rhythmicity and responses to modulatory input are *distributed* functions) (Hooper, 1995).

Sensorimotor functioning requires the integrated and timely functioning of an extensive range of the CNS and the peripheral nervous system. The normal condition is neurological control over the level and pattern of afferent input from the environment or body that is transmitted centrally. Motor reflexes and control are influenced by the transmission from the receptors to the sensory cortex (i.e., interaction of centers for motor programming and control with sensory receptors). The thalamus is more than a relay station. Its reciprocal interaction with cortical areas summates tactile stimuli, which are compared with cortically stored templates of previous tactile experience (Nicolelis et al., 2002). Sensory information in the CNS is processed by stages in sequence, by relay nuclei of the spinal cord, the brainstem, the thalamus, and the cerebral cortex. At each of these processing stations, sensory input arrives from adjacent receptors. There are local feedforward and feedback circuits for inhibition in the relay nucleus. Inhibitory interneurons are activated by more distant sites (e.g., the cerebral cortex). Thus, higher centers control the flow of information through relay nuclei. Inhibition from distant regions of the brain need not be related to the intensity of the sensory-evoked responses. In the intermediate receptor fields there is enhanced contrast between stimuli, which gives the sensory systems higher resolution of detail. Also, networks of inhibitory neurons transform information to emphasize the strongest signals (Gardner & Martin, 2000).

Each of the major systems (sensory, motor, motivational, autonomic) is divided into components that function simultaneously. All processing levels receive the convergence of bidirectional *multiple parallel* ascending projections (spinocortical), and descending feedback pathways (corticofugal and corticocortical) (Nicolelis et al., 2002). *Multiple activities occur simultaneously.* Processing includes alternation of functioning with other areas, facilitation and inhibition, parallel or sequential transmission, synthesizing, analyzing, modification by information received through monitoring ongoing action and external events, modifying output to match an anticipated outcome, and so on. This applies to sensorimotor, higher cognitive functions, and autonomic activities. To use *vision* as an example, receptors, cortical nerve centers, the movement of eye muscles, and the movement of the limbs are served by different neural groups. Simultaneous activity is processed at relay stations. Neuronal pathways converge at particular nuclei where activity is modified by facilitation, inhibition, or information from other centers and retransmitted in both directions and off neuraxis. Ultimately, transmission proceeds via a "final common pathway" to a target organ. The advantage of this complex arrangement is multiple: (1) a unified sensorimotor motor system is established; and (2) past experiences enhance the perception of former tactile experiences and the learning of new ones. One categorization of data under study is whether the type of data under study is "*top-down*" (i.e., the nervous system uses past experiences), or "*bottom-up*" (i.e., expectations and predictions needed to interpret novel sensory experience). Another way of utilizing these rubrics is "top-down" from the real world of interest (e.g., stereotypes, attitudes, self-knowledge), or "bottom-up" (i.e., the neural system used to isolate and recognize details).

Two-way sensorimotor functions are one system. Adequate motor performance requires the complex integration of sensory information with programming; coordination centers such as the cerebellum; motor output; and feedback to neurological centers of the spinal column and brain. Gait, stance or posture, balance, and fine motor coordination require sensorimotor integration of the head, trunk, limbs, hands, and fingers. In addition, it has been asserted that maintaining complex motor activities requires conscious effort (Thompson, 2006).

AROUSAL

Whether arousal is a unitary function has been controversial. It was formerly defined as a level of nonspecific neuronal excitability derived from structures formerly known as the reticular formation but now generally referred to as specific chemically defined or thalamic systems that innervate the forebrain. A number of neuroanatomically and neurochemically distinct systems project to various parts of the cortex from subcortical nuclei: cholinergic basal forebrain, noradrenergic locus ceruleus, dopaminergic median forebrain bundle, and the serotonergic dorsal raphe nucleus. The anterior cingulate may play a role in regulating arousal in response to task demands, indicating how the vigilant attention system might interface with subcortical arousal mechanisms. The right frontoparietal system, which has extensive connections with the cingulate, seems to modulate arousal when it is not externally generated by task demand or external stimulus (Robertson & Garvan, 2004).

Arousal describes the degree to which the individual appears able to interact with the internal and external environments (Bleck, 2003). Arousal is differentiated from *attention*, which is the orientation of sensory receptors to stimuli within an already aroused organism. Reduced levels of arousal are characterized by lethargy. At high levels the patient is hyperalert, easily distracted by irrelevant stimuli, and therefore incapable of sustaining attention. In the awake state, alertness, attentiveness, memory, orientation, logical thought, and emotional stability are under brainstem diencephalon control: noradrenergic locus ceruleus (LC); serotonergic raphe nuclei of the pons; and histaminergic hypothalamic neurons (Hobson & Pace-Schott, 2003). Arousal is associated with consciousness and activation (Andreassi, 2000, pp. 401–405). It is a necessary condition for any kind of adaptive interaction with the environment. It varies from *tonic arousal* (slow fluctuations related to the circadian rhythm, food intake, drug effects, etc.), and *phasic arousal* or rapid fluctuations occurring in response to nonthreatening environmental stimuli. The latter condition slows the heart and other organs in contrast to fear-evoking stimuli. *Awareness* reflects the depth and content of the aroused condition; it is dependent on arousal otherwise one lacks awareness. *Attention* implies the ability to respond to particular types of stimuli or to some motivated, ongoing activity. Abnormal states of arousal dysregulate behavioral and physiological functioning.

Arousal is also a general description of the level of physiological activity, including the brain, the autonomic nervous system, and activated soma. It has also been defined as the ability to mobilize metabolic energy to meet environmental or internal demands on behavior. Arousal describes a complex system involved with such functions as maintaining homeostasis or governing levels of activation appropriate to an ongoing task (Heller, 1993). It is differentiable from emotional valence and includes multiple systems such as the autonomic, the electrocortical, and the behavioral. The right hemisphere is believed to modulate autonomic arousal in both emotional and nonemotional conditions. Nevertheless, personality factors (e.g., extraversion and optimism) may co-vary more robustly with right hemisphere activation than situation-induced emotions. Peripheral arousal stems from activation of the ANS, affecting internal organs directly, and skeletal musculature indirectly (Marrocco & Field, 2002).

Activation refers to the relationship between levels of physiological activity and changes in behavior. Performance effectiveness had been characterized by the familiar inverted "U"-shaped curve. Andreassi (2000) cites Malmo (1962) as indicting that the ascending reticular activating system (ARAS), by regulating the level of cortical excitability, regulates performance. However, subsequent research did not support this generalization. Perhaps the inverted "U" will occur when levels of activation are purposely manipulated to produce very high and low levels of physiological activity. Yet, cortical autonomic and somatic variables do not always change similarly. Activation and arousal (Lindsley, 1987) reflect central nervous system and autonomic nervous system activity (Porges, 1993) that influences levels and states of consciousness.

Wakefulness functions in the maintenance of sensory input from multiple receptors, capacity for accessing memory and directing attention, readjustments of posture, and motor output (Hobson & Pace-Schott, 2003). It also varies in its extremes from to hyperalertness, and is characterized by the intensity of stimulation needed to elicit a meaningful response. The interpretation performance on mental state examinations is influenced by abnormal states of arousal (Weintraub, 2000).

Arousal has a variety of components: an influence upon the level of consciousness, the functional level of numerous neurobehavioral domains, and various physiological functions. It has been described as (1) the effect of pathways from monaminergic cell groups in the brainstem and hypothalamus upon the cerebral cortex that increase wakefulness and vigilance; and (2) the responsiveness of cortical and thalamic neurons to sensory stimuli. Various pathways involve both arousal and functional modulation. The major developmental changes in arousal are the *growth of inhibitory processes* that control and shape the response, with an increase in the range of internal and environmental stimuli that activate these processes. There are varied *arousal indices*. The EEG voltage is inversely related to the degree of behavioral arousal. Changes in electrical activity follow an environmental stimulus, a shift of attention, or a change in the task. Other measures of arousal are physiological functions (electrodermal responses, changes in blood chemistry, early gene expression, heart rate, etc.).

Arousing structures occur in the brainstem and forebrain. Each of the six major brain modulatory systems has extensive connections to most areas of the brain and plays a major role in sensory, motor, and arousal tone (Saper, 2006). Some components of arousal occur in utero. Arousal in the neonate is frequent and is associated with hunger and discomfort. The direction of arousal change is towards *inhibition* (e.g., the soothing voice of the parent). Arousal is not a unitary condition. There are various arousal systems, including external stimuli and control systems of the brain. The *frontal cortex* (orbitofrontal) regulates the thalamic reticular nucleus and cholinergic basal forebrain structures, impulse activity of the locus ceruleus, and limbic system emotional arousal (Marrocco & Field, 2002). There are various measures of arousal (e.g., fear-evoking stimuli increases heart rate and other autonomic indices), while phasic alerting caused by orienting towards a nonthreatening stimulus slows the heart and other internal organs. Arousal is a function of different systems (peripheral, central, and nervous systems) interacting with other systems and projecting to different areas of the brain. Brainstem cells project rostrally to cortex and subcortical structures, and also project caudally to the spinal cord. Each system makes synaptic contacts with multiple cortical areas. A given cortical area is usually innervated by two or more systems. The arousal systems facilitate intrinsic subcortical and cortical activity and facilitate or inhibit activity in another system. Each system exerts effects on families of neurotransmitter receptors that produce a variety of pre- and postsynaptic effects ("*cross talk*"). Arousal systems can exert influence upon either narrow or broad areas of the cerebrum according to conditions. The broadest connections are noradrenergic and serotonergic. The transmitters may produce excitatory or inhibitory effects, with onset and duration varying widely (Marrocco & Field, 2002).

THE AROUSAL SYSTEM

The arousal system involves both brainstem and cerebrum, and therefore is vulnerable to trauma and medical disorders. Ascending pathways from the upper brainstem, hypothalamus, and basal forebrain diffusely innervate the thalamus and cerebral cortex and keep them in a state in which they can transmit and respond appropriately to incoming sensory information. One branch is formed by ascending pathways from monoaminergic cell groups in the brainstem and hypothalamus, which proceed to the cerebral cortex and thalamus, increasing wakefulness and vigilance. This is joined by various cholinergic inputs to form the *ascending arousal system*. This system branches at the junction of the midbrain and diencephalon. One branch enters the thalamus, activating and modulating thalamic relay nuclei, and also intralaminar and related nuclei with diffuse cortical projections. The other branch travels through the lateral hypothalamic area and is joined by ascending output from the hypothalamus and the *basal forebrain*. The basal forebrain contains the largest concentration of cholinergic neurons (*nucleus basalis of Meynert*). This branch projects from the brainstem diffusely to the cerebral cortex. Lesions disrupting either of these branches impair consciousness (Damasio & Anderson, 2003; Saper, 2002).

Cholinergic inhibitory areas (*CIA*): The CIA is a relatively unfamiliar mechanism involved in unconsciousness. It is hypothesized that muscarinic cholinergic systems facilitate unconscious processes and monaminergic systems facilitate conscious processes. There is a fine balance such that modification of activity in one system has effects upon the other. An overview suggests that suppression of dopaminergic and cholinergic systems and/or the activation of the serotonergic system contributes to concussive unconsciousness (Sasaki et al., 1987).

Muscle atonia: The linkage between activation of brainstem motor inhibitory systems and inactivation of brainstem facilitory systems may underlie reduced muscle tone. This system projects from the mesencephalon, pons, and medulla (Mileykovskiy et al., 2000). The corticotropin-releasing factor in pontine sites may interact with glutamate and acetylcholine to generate muscle atonia (Lai & Siegel, 1992). Injection of acetylcholine into the PIA elicited a significant increase in both glycine and GABA release in the hypoglossal nucleus and the lumbar ventral horn. This was associated with *atonia*. These neurotransmitters are believed to play an important role in the regulation of upper airway and postural muscle tone (Kodama et al., 2004).

Sleep: The dorsolateral PIA and medial medullary reticular formation mediates the atonia of REM sleep (Vakalopoulos, 2006). The dorsolateral *pontine inhibitory area* (*PIA*) and the medial medullary reticular formation mediate the muscle atonia of REM sleep. Corticotropin-releasing factor in pontine sites may interact with glutamate and acetylcholine to generate muscle atonia (Lai & Siegel, 1992). It is hypothesized that a glutamatergic pathway from the pontine inhibitory area to the medullary nucleus magnocellularis is responsible for the suppression of muscle tone in REM sleep (Kodama et al., 1998).

Loss of consciousness: A pattern has been suggested for a reversible comatose state following concussive head injury. Study of fluid percussion head injury in the cat offers information concerning coma (transient flaccidity of postural muscles and areflexia). During the initial period of generalized areflexia there was increase of excitability of motoneuronal pools in the spinal cord. During subsequent flaccidity, spinal cord somatomotor functions were depressed while afferent input transmission was recovering. This pattern was hypothesized to derive from a predominance of cholinergic activity within the PIA of the rostral pons. Local rates of glucose utilization increase here. There are descending inhibitory influences upon postural somatomotor functions, including the suppression of eye-opening responses without EEG slow waves (Katayama et al., 1988). A mesopontine tegmental anesthesia area has been described that is involved in loss of consciousness, slow-wave cortical EEG, motor suppression, and sensory suppression. There are four ascending pathways that may mediate loss of consciousness: intralaminar thalamic nuclei projecting to the cortex; pontomesencephalic tegmentum, diencephalic, and basal forebrain nuclei projecting cortical and part of the ascending arousal system; projections to subcortical forebrain (septal area, hypothalamus, zona inserta, and striato-pallidal system) that may indirectly affect cortical arousal and hippocampal theta rhythm; and projections to the frontal cortex. Several of these areas, including the frontal cortex, have reciprocal relations with the neural origin (Sukhotinski et al., 2007).

Muscarinic system: Noting the existence of a *muscarinic brainstem system* whose activation produces components of reflex inhibition and behavioral suppression, it is inferred that mechanisms mediating traumatic unconsciousness are likely to be distinct from those mediating enduring behavioral deficits (Hayes & Dixon, 1994). The pontomesencephalic brainstem center is a muscarinic brainstem system ventromedial to the locus ceruleus, as has been demonstrated in the rat and cat.

Cholinergic centers have a particular role in maintenance of consciousness. *Concussive unconsciousness* in the cat may be attributable in part to activation of cholinergic pontine sites. A low level of concussive injury was associated with increased local glucose utilization (Hayes et al., 1984). Enhancement of cholinergic transmission with physostigmine accelerates the recovery of global and hemispheric CBF and also the contrelateral CBF (Scremin et al., 1997). Injection of carbachol, a cholinergic agonist into this region results in pronounced suppression of postural somatomotor sympathetic visceromotor functions. This system can be organized to regulate reactions to events in the external environment, allowing expression of integrated behaviors (Katayama et al., 1984). Active inhibitory mechanisms in the brainstem modulate sensory input and/or motor output in response to changing environmental events or vegetative states, including noxious sensory input (Lyeth et al., 1988). Generalized cholinergic release contributes to convulsive seizures associated with death, at least in rats (Lyeth et al., 1988). There appears to be an initial nonspecific period of brain disorganization characterized by generalized areflexia with muscle hypertonia (Lyeth et al., 1984). Subsequently, there are active cholinergic inhibitory processes that create behavioral suppression and reversible LOC following low levels of concussive brain injury.

The Ascending Reticular Activating System (ARAS)

One function of the reticular formation is activation of the brain for behavioral arousal and different levels of awareness. The ARAS consists of the axons of a network extending from the medulla, the upper brainstem (including the midbrain), the hypothalamus, and basal forebrain. It includes the following systems: pedunculopontine, parabrachial, medial raphe, locus ceruleus, tuberomammillary, melanin-concentrating hormones and orexin, and the basal forebrain. Its function is increasing wakefulness and the responsiveness of cortical and thalamic neurons to sensory stimuli (arousal). In contrast to direct sensory pathways through the thalamus, their function is to diffusely innervate the thalamus and cerebral cortex and maintain them in a state in which they can transmit and respond appropriately to incoming sensory information. These pathways innervate the reticular nucleus and intralaminar nuclei of the thalamus. The ARAS divides into two branches at the junction of the midbrain and diencephalon. One branch (*direct*) enters the thalamus, where it activates and modulates thalamic relay nuclei and intralaminar nuclei with diffuse cortical projections. The other branch (*indirect and diffuse*) travels through the lateral hypothalamic area to be joined by ascending output from the hypothalamus and basal forebrains. These diffusely innervate the cerebral cortex (Saper, 2000).

The *locus ceruleus* (*LC*) of the pons is a noradrenergic (NE) nucleus in the floor of the cerebral aqueduct and the fourth ventricle. It is part of the brainstem reticular formation, and is associated with serotonergic centers (brainstem raphe nuclei lying from the caudal medulla to the midbrain), as well as pontine cholinergic centers. According to recent study, afferents are received from the medulla oblongata and the ventral tegmental area, prefrontal cortex, and raphe nucleus. The LC has widespread norepinephric projections. It plays a role in the fearful stress response, as well as depression, panic attacks, and anxiety disorder. It also activates the peripheral sympathetic nervous system (SNS), preparing for a physical response to urgent stimuli and increasing the attention and vigilance that prepares cognitively for adaptive responses to unexpected stimuli. As an organizer of incoming sensory information, it participates in mediating the stress response. Malfunctioning has consequences for the serotonergic and dopaminergic systems. The LC and the midbrain motor region facilitate muscle tone. Pathways curve around the hypothalamus, basal ganglia, and frontal cortex (Drouin & Tassin, 2002; Martin, 2002, pp. 31–32; Reynolds et al., 2002; Silver et al., 1991). There are ascending projection to cerebral cortex involved in integration, hypothalamus, projections to the cerebellar cortex, the dorsal thalamus, and the basal forebrain (including the hippocampus), and descending to sensory regions of the brainstem and spinal cord. Thus, projection patterns are primarily to sensory structures, as well as to cortical structures involved in integration (Hobson & Pace-Schott, 2003; Role & Kelly, 1991). Descending tracts project to the brainstem and spinal cord. These potentiate responses to sensory inputs while reducing spontaneous or low-level activity, improving the *signal-to-noise* ratio. LC discharge rate is associated with arousal and attention to environmental stimuli (Valentino et al., 1993). The responsiveness of target neurons to other inputs is enhanced. LC activation by stressors biases target neurons toward processing information about the stressor. This enhances processes that *counteract the stressor or maintain homeostasis*. Chronic CRH release would result in persistently elevated LC discharge and norepinephrine in targets, possibly resulting in hyperarousal and *loss of selective attention*.

Visceral adrenergic cell groups (epinephrine) project to the sympathetic preganglionic column of the spinal cord and are thought to provide tonic input to vasomotor neurons. Projections to the hypothalamus modulate cardiovascular and endocrine responses. Other neurons project to the nucleus of the solitary tract and to the ascending pathway to the parabrachial nucleus, transmitting gastrointestinal information. Dopaminergic cells groups of the midbrain and forebrain include the substantia nigra and midbrain tegmental regions. Ascending pathways input to the telencephalon, including the nigrostriatal pathway innervating the striatum and are believed to be involved in initiating motor responses.

Cerebral Cortical Arousal

Consciousness is considered to be a distributed system (see Chapter 4). The cerebral cortex exchanges stimulation with lower centers. Cortical arousal stimuli from the reticular formation via the thalamic diffuse-projection nuclei are relayed to the basal ganglia, the basal forebrain (including the hippocampus), the dorsal thalamus, the hypothalamus, the cerebellum, and the neocortex (Kelly & Dodd, 1991; Role & Kelly, 1991). The reticular core influences multiple brain areas and receives sensory information from the cortex via the thalamus. Arousal, awareness, and attention are dependent upon this system.

Cortical arousal (ARAS) is mediated by an *indirect system* and a *direct mechanism* (nonthalamic) via cholinergic, serotonergic, noradrenergic, and histaminergic arousal systems that originate in the brainstem, basal forebrain, or hypothalamus and do not pass through the thalamus. The *thalamus* plays a role in arousal, consciousness, affective behavior, memory, and integrates sensory and motor activities (Afifi & Bergman, 1998, pp. 258, 272).

Locus ceruleus (*LC*): LC activity is increased upon detection of a significant rewarding of threatening stimulus, as well as during certain phases of the sleep-wake cycle. LC inputs to the forebrain increase the efficiency with which the innervated neurons respond to particular environmental stimuli (Blessing, 2002). Neocortical EEG patterns associated with alertness are stimulated by noradrenergic neurons in the LC, serotonin neurons in the median and dorsal raphe nuclei, and acetylcholine neurons in the pedunculopontine tegmental nuclei of the brainstem. Robust LC activity during waking results in a heightened release of norepinephrine throughout the neuraxis that enhances the signal-to-noise ratio for external stimuli. Intermediate range activity maintains the ability to respond selectively to meaningful stimuli while suppressing inappropriate behaviors. Very low spontaneous LC activity reduces neural responsiveness to the external world and may facilitate sleep (Reynolds et al., 2003).

ORGANIZATION OF NEUROLOGICAL SYSTEMS

The nomenclature for areas of the cerebral cortex (*Brodmann's area*) and *long association bundles* (white matter) is found in Nolte (2002, pp. 531–537). Efferents to ipsilateral cortical areas vary from very short ones that never leave the cortex to shaped fibers that dip under one sulcus to reach the next gyrus, as well as longer association tracts that innervate a different lobe. Large areas of cortex may contribute to relatively narrow functions, but the performance of complex tasks may be impaired by widely separated cortical areas. Cortical activity represents the interaction of excitatory and inhibitory neurones. Circuits respond briefly to changes of input, limiting the duration of a cortical response. By selectively responding to what is new and changing with time, we respond to what is adaptively significant. Also, projection to a group of neurons from different groups of coactive neurons simultaneously creates "*perceptual binding*." Since it is activated by different aspects of the same stimulus, the object is seen as a whole rather than as fragments (Kaas, 2000b).

Varieties of Neurological Organization

Diffuse TBI creates a variety of disorganizations depending upon its location:

Circuitry integrating an identifiable cortical area with other areas
Complex units composed of multiple cortical areas
Cortical areas projecting to other levels of the nervous system

Levels of organizational complexity (Amaral, 2000b):

Multiple functionality: A particular structure may contain components of several different functional systems.

Stages: Projection in a system occurs in several stages and is modified at each step.

Multiple input: A single neuron receives input from thousands of presynaptic neurons that govern its output. Interneurons only contact local cells at the same processing stage.

Domains within a structure: An example of a single well-differentiated structure containing segments of essentially different functions is the cerebellar *dentate nucleus*. Dentate output comprises two separate output channels: The motor domain exclusively projects to the primary motor cortex and premotor areas of the cerebral cortex; the nonmotor domain projects to prefrontal and posterior parietal cortices. The output channels are part of closed-loop cerebrocerebellar circuits. Higher-order nonmotor areas (e.g., anterior and posterior cingulate cortex) project to the cerebellum. These areas are likely to be the target of cerebellar output as part of closed-loop circuits (Dum et al., 2002).

Modularity

There are several possible ways of conceiving the neural structure that performs a particular function.

- A relatively large, identifiable brain structure performs a relatively *specific* function (see the chapter on Cognition).
- A large structure is subdividable into *modules* (a small area with a unique function).
- Different configurations of structures, recycled as it were, can be considered as a *quasimodule* (a configuration performing a specific [disjunctive] function), defined as a specified degree of participation in a specified set of tasks (Lloyd, 2000).

Topographic Organization

The receptive surfaces are sometimes organized topographically through each processing stage (see Chapter 15). Neighboring groups of cells of the retinal, the cochlea, and the skin surface project to neighboring groups of cells in the thalamus, and then in the cortex. This maintains an orderly *topographic neural map* of information from the receptive surface to the cortex. These neural maps reflect not only the position of receptors but also their density, which determines the sensitivity to sensory stimuli.

Information is organized hierarchically. This determines sensitivity to specific characteristics of stimuli. An example of this is visual neurons projecting from retinal, to thalamus, to visual cortex, to association cortex. While the thalamus is sensitive to light in a particular part of the visual field, the convergence of axons in the primary visual cortex and then association cortex results in increasing specificity of response (e.g., firing only when a bar of light with a particular orientation, or a face).

Most neural pathways are bilaterally symmetrical, but cross to the opposite side (decussate) at different anatomical levels within the brain. Consequently, sensory and motor activities on one side of the body are mediated by the cerebral hemisphere on the opposite side.

Association Areas

Sensory information is processed in parallel relays. Information from peripheral receptors of a particular modality projects to a *primary sensory cortex*. This projects to a higher area of sensory cortex known as a *unimodal association area*, which integrates afferent information that may arrive on different pathways for that sensory modality. The unimodal association areas project to *multimodal sensory association areas* that integrate information about more than one sensory modality.

Primary sensory cortex: Each of the sensory cortices have a *unimodal sensory association area*. To illustrate, we will consider the distribution of the *optic nerve tract*. The retinal axons proceed to the optic chiasma, where some fibers decussate to the opposite side. The major visual pathway for conscious vision is to the dorsal lateral geniculate nucleus of the thalamus, which projects to the striate cortex of the occipital lobe. Spatial information is related from the striate cortex to the posterior parietal cortex. Form and color information is relayed to the inferotemporal cortex, which is involved in complex functioning, including visual recognition of objects and individuals. Visual projections to the midbrain roof (superior colliculus) are involved with spatial projections of the visual world. Similar associated maps involving auditory and somatosensory space are projected into the deeper layers. Superior colliculus visual input is related to the pulvinar (dorsal thalamus) and then to the extrastriate visual cortical areas bordering the primary visual cortex of the occipital lobe (Butler, 2002) (see Figures 3.1, 3.2, and 3.3).

Multimodal sensory association areas (Saper, 2000a):

- *Posterior* (margin of the parietal, temporal, and occipital lobes): Visualspatial localization, perception, language, attention
- *Anterior* (prefrontal cortex rostral to the postcentral gyrus): Motor planning, language production, judgment
- *Limbic* (medial edge of the cerebral hemispheres): Emotion and memory

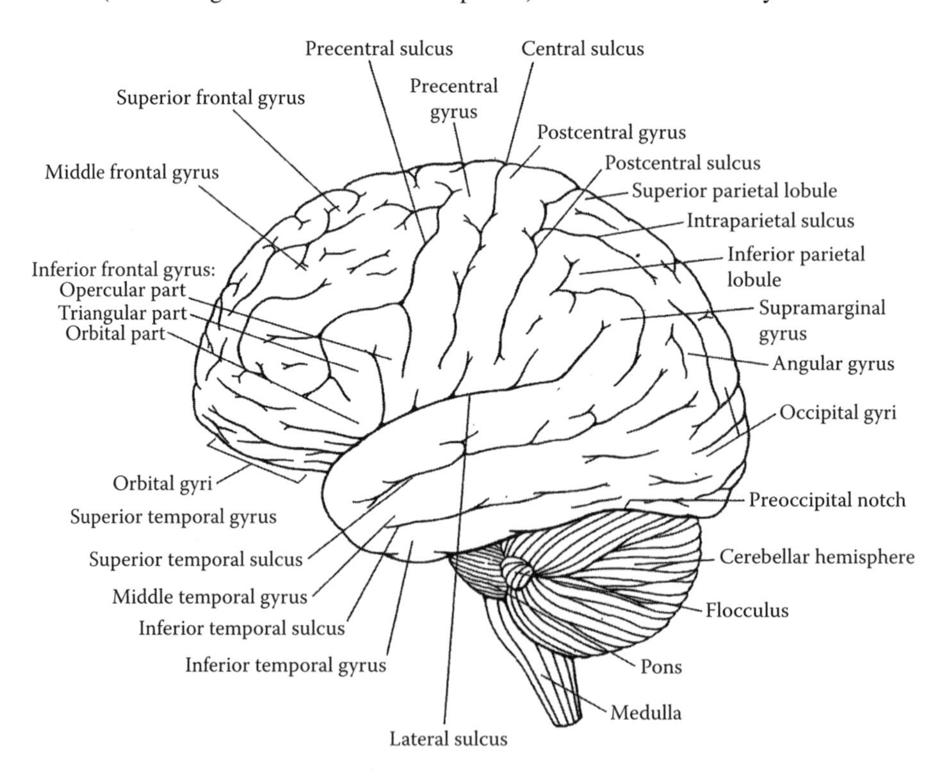

FIGURE 3.1 Brain—gross anatomy (lateral view). (From Martin, J.H., *Neuroanatomy*, McGraw Hill, 2003, p. 411, Figure 11.1. With permission.)

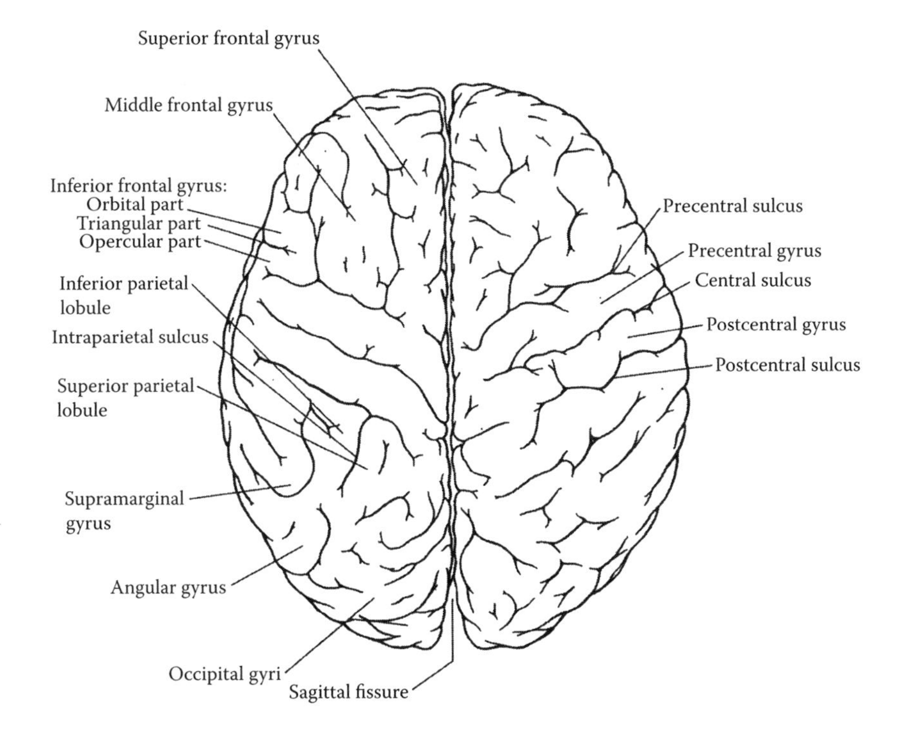

FIGURE 3.2 Brain cerebral hemispheres. (From Martin, J. H., *Neuroanatomy*, McGraw Hill, 2003, p. 411, Figure 11.1. With permission.)

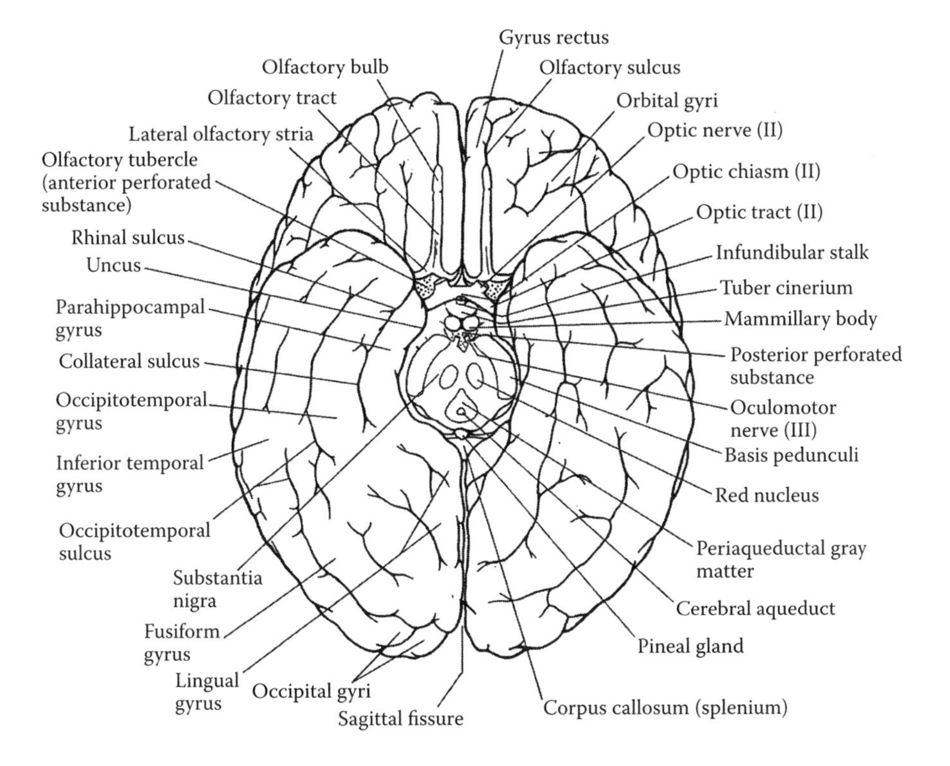

FIGURE 3.3 Inferior aspect of brain showing brain stem in relation to cranial nerves and associated structures. (From Martin, J. H., *Neuroanatomy*, McGraw Hill, 2003, p. 411, Figure 11.1. With permission.)

Motor association cortex (premotor [motor preparation and programs]): The most important functions of the prefrontal association area are to weigh the consequences of future actions and to plan and organize actions by integrating sensory information from the environment and the body. This prefrontal association area cortex comprises the dorsal prefrontal association area, the medial prefrontal cortex, and the ventralorbitofrontal cortex. These receive afferent input from the mediodorsal thalamic nucleus.

Primary motor cortex (frontal lobe within the precentral gyrus): The motor cortex regulates the movement of a joint along a vector (Saper et al., 2000).

Converging sensory information is integrated into a polysensory event. Posterior sensory association areas are highly interconnected with frontal association areas that convert plans about future behavior into concrete motor responses. The interactions between the posterior and anterior association areas are critical in guiding behavior. The dorsolateral prefrontal association cortex and parietal association cortex are densely interconnected and both project to numerous common cortical and subcortical areas (Figure 3.4). Neurons in the posterior association areas are most tightly linked to the sensory rather than the motor aspect of complex behavior. While neurons of the premotor cortex may selectively respond to certain sensory stimuli, they will fire only if motor output is required (Amaral, 2000b; Saper et al., 2000).

Sexual Dimorphism

Sexual dimorphism has been discovered in the hypothalamic *preoptic area*, which affects reproductive behavior and ovulation (Breedlove & Hampson, 2002, including Table 3.2 of anatomical studies showing sex differences in the human brain). Sex differences in cortical organization are *function specific*. There is evidence that sex differences exist in the degree that linguistic and nonverbal functions are lateralized in the two hemispheres. It appears that the two sides of the brain are more fully differentiated in men than in women, while women have more bilateral representation of functions in which the two hemispheres share processing capabilities. The degree of lateralization in women is variable: for a particular function it may be less or more than in men. Motor and acoustic functions related to speaking show sex differences in intra- but not interhemispheric organization, while higher-order, abstract, or complex verbal functions seem to show interhemispheric differences (Hampson, 2002).

Gender Considerations for Behavior

The developmental focus is upon *sex steroids*, which have an influence prenatally, and in males seems to be primarily prepuberty. Examples of a neurobehavioral influence upon *gender role behaviors* include childhood toy preferences; sex of preferred playmates; rehearsal through play of adult role as spouse or caregiver; preference for rough-and-tumble play; physical aggressiveness. Perceptual asymmetries (divided visual fields or dichotic listening) are smaller in women than men. *Abilities* are sexually differentiated. Males tend to excel in spatial abilities (mental rotation; route learning; visualization of spatial relationships; gross motor skills involving strength). Females excel in verbal skills (fluency; rate of speech acquisition; spelling; grammar), computational accuracy and procedural knowledge; fine motor skills and finger dexterity; short-term memory (including object locations). Females may program a sequence of movements more efficiently than males. Nevertheless, not all functions are less lateralized in females. Some may be more lateralized (Hampson, 2002).

MOTOR CONTROL AND PATHWAYS

I will use the term "motor control" in the generic sense of sensorimotor performance of all skeletal muscles for any purpose. "Locomotion" will refer to ability to move from one place to another.

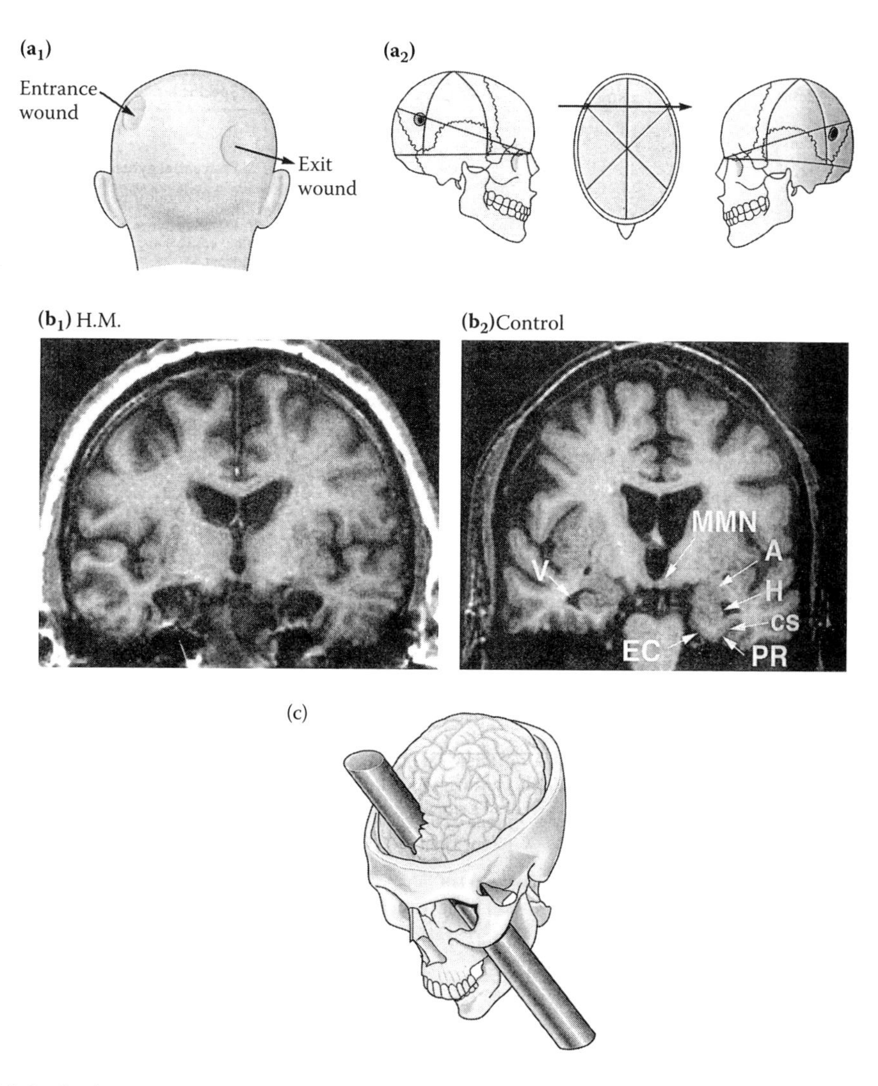

FIGURE 3.4 Brain—cerebral hemispheres. Important insights into the function of cortinal association areas has come from observations on patients with specific injuries of the cerebral cortex. (a_1) The drawing shows the path of a bullet in a soldier wounded in World War I. The bullet entered the skull over the dorsolateral parietal lobe on the left and exited through the ventrolateral parietal lobe on the right. This patient was studied by Gordon Holmes, an English neurologist, who derived the importance of the parietal lobe in visuospatial integration from his conservations. (a_2) drawing from the work of Aleksander Luria showing the path taken by a bullet through the parietal lobes of a Russian soldier in Worl War II. The soldier's visouspatial deficit was nearly identical to Holmes's patient. B. This magnetic resonance (MR) image shows the bilateral to moval of the medial temporal lobe including the hippocampus in patient H.M. (b_1) Scan of H.M.'s brain. (b_2) Scan of a control subject's brain. A = amygdala; H = hippocampus; EC = entorhinal cortex; CS = collateral sulcus; PR = perirhinal sulcus; MMN = medial mammillary nucleus. (Courtesy of D. Amaral). C. A drawing of a computer reconstruction of the passage of a tamping iron through the brain of Phineas Gage over a century ago. This injury resulted in severe personality changes that dominated our understanding of the function of the frontal lobes. (Adapted from Damasio et al., 1994). (From Kandel, E. R., et al., *Principles of Neural Science*, 4th ed., McGraw Hill, 2000. With permission.)

Motor coordination may be described as gross or fine, depending upon the size and number of muscles, limbs, and the trunk under control, as well as the precision and trajectory of the motion. *Voluntary movements* require that higher levels of our motor system dissociate the initiation of a movement from its informational content (Krakauer & Ghez, 2000).

Motor Cortex

Motor control cannot be understood as a one-way command system. The connections between motor cortex and muscles changes constantly on the basis of feedback from the periphery. This permits regulation of high-level and low-level movement parameters (Graziano, 2006). Afferent pathways from the reticular formation, the cerebellum, and the basal motor nuclei (corpus striatum, thalamus, subthalamus, midbrain) project to the thalamus and then to the motor cortex, modulating activity (DeMyer, 1988, pp. 239–244).

Major axonal control of motions descends from the motor cortex, the red nucleus, the pontomedullary reticular formation, and the vestibular nuclei. Two systems perform synergistically. The *medial system* provides postural control for stance, ambulation, and orientation of the head. Vestibular and reticular nuclei axons descend in ventromedial pathways in the spinal cord. The *lateral voluntary system* superimposes sophisticated, voluntary movements in response to stimulation from the motor cortex forming the medullary pyramids, decussating, and decussating to the lateral column with the exception of a small minority of uncrossed corticospinal axons (Schieber & Baker, 2000). *Premotor areas* control high-order aspects of movement. The *primary motor cortex* then decomposes movement into simple components in a body map that is communicated to the spinal cord for execution. The primary motor cortex (Brodmann area 4; M1) is made up of overlapping areas that control the contrelateral action of single muscles or muscle groups. It receives input from the primary somatosensory cortex (Brodmann areas 1, 2, 3) and from posterior area 5, which with area 7 is involved in integrating multiple sensory modalities for motor planning. Premotor damage causes more complex motor impairment than does damage to the primary motor cortex (Krakauer & Ghez, 2000). The *supplementary motor area* (premotor) instructs the motor area which neurons to fire and in what order they should be fired in order to achieve a spatial temporal trajectory. The motor areas of the cortex have reciprocal relations with these centers: directly with the thalamic; indirectly from ventral anterior and ventrolateral nuclei, which forward stimuli from the globus pallidus and the cerebellum (which receives information from the pons), and from projections to the supplementary motor cortex, premotor cortex (Brodmann area 6), and primary motor cortex; indirectly via the somatosensory cortex; the posterior parietal cortex relays visual and somatosensory information to the motor areas. Two large subcortical motor systems send output via the thalamus to the cortical motor systems (i.e., the basal ganglia [inhibitory] and the cerebellum [excitatory]) (Schieber & Baker, 2003).

Recent developments offer an ethological perspective. The primary motor cortex is not a map of muscles and joints, but rather represents complex actions. The premotor areas also play a higher hierarchical role, but emphasize different categories of complex movement (Graziano, 2006).

Widespread Connections

Motor cortex circuitry is very complex. In addition to the descending spinal pathway, motor cortex neurons have lateral connections to neighboring cortical neurons, other cortical areas, and a variety of subcortical structures. These may link different joints and body parts into more complex movements. Stimulation of one point in the motor cortex evokes excitation and inhibition of muscles actuating many joints. Each muscle also receives input from many cortical locations. *Gait* (walking) has multiple neurological components: *Supraspinal centers* signal starting, stopping, speed, and direction of stepping. The *basal ganglia* initiate walking and the quality of stepping, mediated through the cerebral cortex and brainstem structures. The *cerebellum* modulates the rate, rhythm, and force of voluntary stepping. *Gait* uses a repetitive pattern of limb motion that advances body weight over the supporting limb, then swings the unloaded limb forward for the next step. Normal function depends on free joint mobility and selective muscle action providing an efficient sequence of joint actions. Each lower limb is under the control of 32 major muscles (plus hip rotators and intrinsic muscles of the foot). Thus, lesions of the motor area deprive the patient of a normal walking pattern. Furthermore, accidents causing TBI create unpredictable errors in the patient's gait by introducing varying mixtures of spasticity, contractures, primitive flexion and extension synergies, and impaired selective control (Perry, 1999).

MOSAIC

There is an output of the cortical motor system that projects to the spinal cord, influencing each other through lateral connections.

- Stimulation of a descending pathway is accompanied by activation or diverse pathways through the motor network.
- Subregions of the motor cortex offer different categories of behaviorally useful actions, (e.g., interactions between the hand and the mouth, reaching, and defensive maneuvers).
- Single neurons may contribute to idiosyncratic, multijoint complex patterns reflecting the organism's behavioral repertoire.
- The output from cortex to muscles is not fixed; rather, it is fluid, changing on the basis of feedback from muscles and joints. Thus, it supports control of combinations of higher-order movements and low-level motor actions needed to produce diverse actions in the organism's repertoire (Graziano, 2006).

The *secondary (supplemental) motor cortex* (Brodmann areas 6,8) controls muscles on both sides. Stimulation can evoke complex postural movements. Cortical projections are both ipsilateral (areas 4,6 [lateral part],5,7 to contrelateral supplementary motor area, and bilaterally to the spinal cord) (Parent, 1996, p. 914). Movement programming prior to action is associated with negative potentials and increased blood flow in the premotor and supplementary cortices. While *corticoreticular fibers* originate in all areas of the cortex, the largest numbers arise from the motor and premotor areas and descend with fibers of the corticospinal tract to enter the brainstem reticular formation (locus ceruleus and raphe nuclei). Neuromodulatory projections alter the excitability of spinal neurons by descending noradrenergic and serotonergic projections from the locus ceruleus and raphe nuclei (Schieber & Baker, 2003).

LOCOMOTION

Locomotion depends upon rhythmic and repetitive movements that can be controlled by both relatively low levels of the nervous system and higher levels that direct locomotion in unpredictable or complex environments. In humans, spinal networks are more dependent upon supraspinal structures than quadrupeds for the basic motor pattern of stepping. While stepping rhythmicity can be produced by neuronal circuits entirely within the spinal cord, descending signals from the brainstem and motor cortex initiate locomotion and adjust stepping movements to immediate circumstances. The cerebellum receives signals from peripheral receptors and the *spinal central pattern generators* via *spinocerebellar pathways*. The spinal circuits can be activated by tonic descending signals from the brain. While these do not require sensory input, they are strongly regulated by limb proprioceptors. In contrast, *exteroceptors* are located in the skin. These adjust stepping to external stimuli and provide important feedback about body movements (Pearson & Gordon, 2000b).

SENSORY CONTRIBUTION TO MOTION CONTROL

Directed motions utilize registration of sensory information concerning the initial position of the target, location of the moving body structures, a model of the intended final position of the movement, monitoring of the motion to determine any error deviation from the intended position of the body relative to the target, and integration of the position of the eyes and ears relative to the limbs, trunk, and target. Senses involved are somesthesis, proprioception, vision, vestibular input, and hearing.

Movement of adaptive value, with reduced or absent sensory input, requires afferent guidance for task and environmental characteristics. In rapid-patterned movements (e.g., playing a musical instrument, sports motions), sensory feedback is too slow to modify movements already underway. Movement is preceded by sensory input that provided the CNS with information on the biomechanical state of the limbs and body in the preceding period. Probably an actual pattern describing the entire movement is prepared at some subcortical level as well.

SOMESTHESIS

The *stretch reflex* provides information as to the extent of muscle contraction and whether this is appropriate to the intended motion. It is a proprioceptive function that has a feedback function modifying central commands for voluntary motion. It keeps the muscle length close to a desired reference value. Hyperactive or hypoactive stretch reflexes can result from CNS lesions since motor neuron excitability is dependent upon both excitatory and inhibitory descending influences (Pearson & Gordon, 2000a). Motor control benefits from the fractionation of spatial information. Spatial guidance of motor behavior requires that the neurons controlling movement of the eyes, head, and arm also represent the location of visible targets. Nevertheless, despite occipital blindness ("*blindsight*"), such persons can locate a target (Colby & Olson, 2003). Loss of afferent input causes motion to be irregular and uncoordinated. When walking, sensory information is required about ground reaction forces and displacement of the support, internal forces and displacements, and the relative velocity of body segments. Spatial orientation is required for precise movements (Prochazka & Yakovenko, 2002).

VISUAL

Particular neurons in the retina respond specifically to moving images. Visual parameters include stereopsis (visual input be slightly different for each eye), fixation at varying distances, and form (significant for coordinated activity and gait). When head movement changes objects in the visual field, this information supplements that coming from the semicircular canals. Retinal cells synapse in the lateral geniculate nucleus, finally reaching the striate cortex. Separate pathways originating in the retina remain partially segregated in the striate cortex and proceed through the extrastriate occipital region on two routes, separate parallel systems:

Dorsal stream (posterior parietal lobe): The dorsal pathway is involved with *what* an object is, as well as conscious spatial awareness, with spatial guidance of actions such as reaching and grasping. Dorsal stream areas contain an extensive representation of the peripheral visual field. They are specialized for the detection and analysis of moving visual images.

Ventral stream (inferior temporal lobe): The ventral pathway is involved in recognizing visual patterns (i.e., *what* an object is, including *faces*) (Colby & Olson, 2002; Wurtz & Kandel, 2000).

PROPRIOCEPTIVE

Proprioception monitors limb position in space. It is essential for coordination and grading of muscle contraction and maintaining equilibrium. The primary receptors are the muscle spindles. The *conscious components* are the posterior column pathways; the *unconscious components* are the spinocerebellar pathways (Campbell, 2005, p. 434). Other peripheral sense organs dealing with proprioception are located in the muscles, tendons, and joints (Pacinian corpuscles). Proprioception refers to sensations responding to pressure, tension, stretching or contraction of muscle fibers, joint movement, changes in the position of the body or its parts, and related stimuli (e.g., cutaneous afferents). The integrity of the ascending spinal pathways is needed for skilled, purposeful movements. Proprioception contributes to reflex stabilization of the head and body. Signals from neck proprioception are a negative feedback compensatory system (i.e., they activate neck muscles to oppose their lengthening—*cervicocollic* or *neckstretch reflex*) (Schieber & Baker, 2003). When proprioception is decreased, the contribution of vision to controlling postural sway is increased. Proprioceptive reflexes, stimulated by input from the muscles, ligaments, bones, joints and tendons, adjust the motor output according to the biomechanical state of the body and limbs. Muscle tone (e.g., stretch reflexes) is dependent upon descending influences, resulting in hyperactive or hypoactive reflexes after CNS lesions (Pearson & Gordon, 2000a).

Rapid Reaction Time

The complexity of receiving information rapidly and adjusting limbs, hands, and torso with extreme rapidity may be imagined for any sport or musical instrument. Reaction time increases with complexity, as does requirement for choice. To achieve a precise and timely motor command, in addition to motor input and programming of preferred response, visual information about movement is projected from the parietal lobe to the frontal lobe (Grillner, 2003). The visuomotor system uses motion signals arising from eye and head movements to update the positions of targets rapidly and redirect the hand to compensate for body movements (Whitney et al., 2003). Slower reaction time is a well-established consequence of TBI. Reaction time has been established as a measure sensitive to injury to various cortical and subcortical regions. It is useful in quantifying the motor deficits of patients with movement disorders (Jahanshahi, 2003).

CENTRAL CONTROL OF AFFERENT INPUT

CNS Nuclei receive afferent input from distance and kinesthetic receptors in order to influence motor output to meet adaptive requirements. As just described, afferent input is modulated and integrated at all levels to create an integrated sensorimotor system. First-order receptors project to second-order and higher neurons. The relay neurons preprocess sensory information and determine whether it is to be transmitted to the cortex. They filter out noise and give preference to stronger sequences from single fibers or activity transmitted simultaneously by multiple receptors. Convergent reception from sensory receptors is interpreted by higher-order neurons in the context of activity from neighboring input channels (Gardner & Martin, 2000).

Corticofugal signals influence both sensory and motor functions. They are transmitted from the cerebral cortex to sensory relay stations of the thalamus, medulla, and spinal cord. There is CNS control over sensory coding and level at the periphery (Dobie & Rubel, 1989). Sensory input is potentiated by more central sites that determine what will be sensed and alter the evaluation and response to these sensations. There are active inhibitory mechanisms in the brainstem that modulate sensory input and/or motor output in response to changing environmental events or vegetative states (Lyeth et al., 1988). These signals serve as gatekeepers, regulating the quantity of mechanosensory information ascending through the CNS. For controlling the sensitivity of sensory input they are inhibitory. This decreases the lateral spread of sensory signals into adjacent neurons, while increasing the degree of sharpness in the signal pattern. It keeps the sensory system operating in a range of sensitivity that is not so low that the signals are ineffectual or so high that the system becomes swamped beyond its capacity to differentiate sensory patterns. This principle is used by all the sensory systems (Guyton & Hall, 2000, pp. 350–351; 130). Interneurons and motor neurons receive input from axons descending from higher centers. Facilitation of inhibition of different populations of interneurons modifies reflex responses and coordinates complex motions also through facilitation and inhibition of opposing muscles. Input from the cerebral cortex and subcortical nuclei projects to the spinal cord. The brainstem receives this input and projects to the spinal cord as two systems:

- *Medial descending* contributes to the control of posture by integrating visual, vestibular, and somatosensory information.
- *Lateral descending* controls more distal limb muscles (i.e., goal-directed movements of the arm and hand).
- Other brainstem circuits control movements of the eyes and head (Ghez & Krakauer, 2000).

Voluntary movements: Sensory processing generates an internal representation of the world, while motor processing begins with the internal representation of the desired result of movement. Goal-directed actions may be externally or internally initiated. With practice, the person anticipates and corrects for the geometric location of target and body. The desired state is represented by a reference signal . Direct moment-to-moment control of the limb is known as *feedback* or *servo-control.* The *error signal* adjusts the output, which can be changed by varying the reference signal. Feedback systems may be described by their *gain.* A high-gain system vigorously minimizes deviations from the target state. It will be unstable if the *phase lag* is long (e.g., interneurons creating long lags between input and output). Low-gain systems are slow because of repeated slow corrections. Feedback from muscle mechanoreceptors helps to maintain the position of the limbs or the force applied to objects. Lack of this information causes disruptions of posture and movement. One does not detect the position of limbs or detect objects through touch of the fingers, grasping an object, or maintaining a hand in one position. *Feedforward (anticipatory) control* relies upon experience as well as sensory events. It compensates for movements, changes in the center of mass when breathing, and catching a ball. When a ball hits the hand and displaces it, then feedback adjusts the hand's position (Ghez & Krakauer, 2000).

There is CNS control over sensory coding and level at the periphery (Sobie & Rubel, 1889). Sensory input is potentiated by more central sites that determine what will be sensed, and alter the evaluation and response to these sensations. There are active inhibitory mechanisms in the brainstem that modulate sensory input and/or motor output in response to changing environmental events or vegetative states (Lyeth et al., 1988). Corticofugal signals are transmitted from the cerebral cortex to relay stations of the thalamus, medulla, and spinal cord. They serve as gatekeepers, regulating the quantity of mechanosensory information ascending through the CNS. They are inhibitory, controlling the sensitivity of sensory input. This decreases lateral spread of sensory signals into adjacent neurons while increasing the degree of sharpness in the signal pattern. The sensory system is kept operating in a range of sensitivity that is not so low that the signals are ineffectual or so high that the system becomes swamped beyond its capacity to differentiate sensory patterns. This principle is used by all the sensory systems (Guyton & Hall, 2000, pp. 350–351; Martin, 2003, p. 130).

The primary sensory areas receive most of their information from the thalamus. The primary somatosensory cortex is caudal to the central sulcus on the postcentral gyrus of the parietal lobe. Touch provides a representation of the body surface and outer space, modifiable by experience. *Attentiveness* integrates the representation of the body with vision and movement. The conscious self functions together with the representation of the body in visual space, whether actual, imagined, or remembered (Kandel, 2000d).

Sensory input stems from the following:

- *Bodily senses* (cutaneous and subcutaneous mechanoreceptors, thermal receptors, nociceptors, muscle and skeletal mechanoreceptors) (Gardner et al., 2000).
- Head position and saccadic eye movements: These create successive frames of reference (retinotopic; head centered; body-centered or posture) (Kandel & Wurtz, 2000). Eye

movements are described as *saccadic*: movement of both eyes to change the point of fixation that are simultaneous, abrupt, rapid, small, or jerky.
- Orientation in space with movement: Vestibular apparatus; proprioceptive information (muscle and skeletal mechanoreceptors which are found in muscle spindles, Golgi tendon organs, joint capsules, and stretch sensitive free endings).

Motor systems produce movement by translating neural systems into *muscle contractile force* (Ghez & Krakauer, 2006). Several parameters describe forces involved in motor activities: magnitude and direction; force bias, referring to a constant force (e.g., gravity); static (isometric) force, referring to postural control; and dynamic force referring to changing force patterns (kinematic movement planning; a mass to be accelerated). Motor cortex activity refers to spatial representation of the motor trajectory. Visual spatial tasks include directional transformations and trajectory planning. Since a force bias is always present, the force exerted by a subject is made up of both dynamic and static components and is represented at the motoneuron pools by the confluence of dynamic and postural (static) inputs from supraspinal and spinal interneuronal systems. Convergence provides an integrated signal to the motoneuron pool (Georgopoulos et al., 2002).

Ballistic Movements

These are movements occurring so rapidly that it is impossible to receive feedback information from the periphery to the cerebellum or from the cerebellum back to the motor cortex before the movements are over. This first involves excitation and then delayed inhibition. The lateral cerebellar hemispheres are associated with planning and performing intricate sequential patterns of movement, especially of the hands, fingers, and the speaking apparatus. Communication is mostly between the lateral cerebellar areas and with the premotor area and primary and association somatosensory areas. Facial and vocal control is localized in the vermis (as are disturbances in the control of axial and trunk muscles during antigravity posture). Lesions may result in slurring and slowing of speech with a characteristic "one-word-at-a-time" quality known as *scanning speech* (Ghez & Thach, 2000). The cerebellar dentate nuclei display activity for the sequential activity that will happen a fraction of a second later (Guyton & Hall, 2000, p. 654). Assuming that projection of the right cerebellar hemisphere to the left precentral gyrus participates in motor control, it has been suggested that increased hemodynamic response is related to the articulatory level of speech production (Ackermann et al., 1998). The right posterolateral cerebellum may assist the left cerebral cortex in helping an individual to learn to generate specific types of constrained spoken two-word associations, not associated with overt or mental movement coupled to a verb. It may also help an individual to automate word associations during speech (Gebhart, Petersen, & Thach, 2002).

At higher levels of the CNS, sensorimotor integration involves multiple sensory functions: somatosensory, vestibular, vision, hearing, tactile, etc. This sensory information concerns the distant environment, our body (direction and speed of the torso and limbs); and close contacts (objects, people, the platform).

The lateral (voluntary) corticospinal tract superimposes upon the medial system (stance, orientation, position of the head) the ability to make sophisticated, voluntary movement responses to complex features of the external environment (perceived through the senses and stored memories, knowledge, and emotion). The lateral system includes fibers that originate primarily in the motor cortex Brodmann areas 4,6 (Heilman & Valenstein, 2003). Some form the cerebral peduncle of the midbrain. The remaining corticospinal fibers originate in the parietal somatosensory areas (1,2,3,5,7) and project to the dorsal horn of the spinal cord to control sensory input (Ghez & Krakauer, 2000; Schieber & Baker, 2003). Each primary motor cortical area receives inputs from the parietal somatosensory cortex. Some neurons receive proprioceptive input from the muscles to which they project. For example, neurons in the hand region respond to tactile stimulation from the digits and palms. The primary motor cortex receives input from posterior parietal area 5, which, along with area 7, is involved in integrating multiple sensory modalities for motor planning (Krakauer & Ghez, 2000).

- Movements of different complexity, reflexes, rhythmic, and voluntary movement, are differently organized in the CNS. The premotor cortex receives input from parietal area 7, the basal ganglia, and the cerebellum via the ventrolateral thalamus. Interconnections between premotor areas may allow working memory to influence aspects of motor planning. The brainstem integrates spinal reflexes into automated motions involving posture and locomotion. Areas of the cortex project to the brainstem and spinal cord initiate and control more complex voluntary motions. Graceful and effortless movement, carried out automatically, depends upon the continuous flow of visual, somatosensory and postural information. When trauma or disease interferes with vision, vestibular or proprioceptive input, compensation by substituting other sensory modalities result in motions that are conspicuous and disabling.
- The brain builds up an internal representation of the external world, and also generates a representation of movement plans. The brain state must correlate with sensory and motor variables (Cisek, 2002). Although the motor cortex is the first step of Sherrington's "final common path," motor programs utilizing preplanning, sequences, and patterns of motion are modified by the cerebellum, the basal ganglia, and brainstem motor nuclei. Individual muscles and joints are represented in a complex mosaic permitting the cortex to reorganize combinations of movements suitable for specific tasks.

SENSORIMOTOR INTEGRATION

Adequate motor performance is a complex action involving integrated sensorimotor functions, sensory and cognitive information processing, comprehension, judgment, and capacity to cope with multiple and conflicting sensory stimuli while the body is stable, moving, or unbalanced. Neurological localization is often difficult because of the distributed layout of the circuits over the entire nervous system, although some disorders are based upon well-defined sensory or motor pathways. In addition to relatively focused lesional damage, diffuse and/or widespread injury, including subcortical and peripheral injuries, result in loss of complex functions and programs.

MOTOR NETWORKS

The cortices of the motor strip and culmination of somatosensory, auditory, and visual afferent input are described as transmission platforms guiding motor action. Structures in the diencephalon, the mesencephalon, and probably the rhombencephalon have bilateral functional connections with the cerebral hemispheres (Penfield & Roberts, 1959, p. 21). Motor control is distributed over long sensory and motor circuits, as well as numerous cortical, subcortical, brainstem, and cerebellar nuclei, with central integration and CNS-peripheral modification in both directions. Thus, it is vulnerable to diffuse and focal brain injury and spinal cord , peripheral neurological lesions, and somatic damage.

Sensory and motor functions are processed by bilateral, complex, and interacting centers at all levels of the neuraxis. Well-defined pathways can be described as a ladder with struts (i.e., two-way influences at each rung for both afferent and efferent transmission [cortical, thalamic, basal ganglia, cerebellar, and spinal levels]). Impaired sensory information occurs when there is interference with the two-way neural traffic, modifying the level of afferent transmission between the receptors and

CNS centers. Motor influence may be indirect. Proprioception modulates motor activities in real time in contrast with some special senses of which the afferent pattern itself is modulated by cortical and intermediate centers. When information is disturbed due to injuries to the sensory organs, neural pathways, and cortical reception areas, there will be fine motor disorders or vision, sense of touch, or poor hearing interfering with verbal communication.

MULTILEVEL AND INTERHEMISPHERIC INTEGRATION

Awareness of the complex pathways between the levels of the CNS, and with such off-axis structures as the cerebellum, limbic area, basal ganglia, and the amygdala hippocampal complex, as well as the connection of the important multisensory association areas and other cortical structures, contributes to understanding the vulnerability of the brain to impact and shearing forces, as well as the basis for some impaired functions.

There are three types of integrative pathways in the cerebral hemispheres: commissural, association, and projection. They play a role in such domains as cognition, control over the internal environment, emotional reactions, sensorimotor programming and responses, body schema, information processing (verbal and nonverbal), and executive function.

COMMISSURAL

These connect fiber paths across the midline and connect homologous areas—and sometimes different structures—in the opposite cerebral hemisphere. They are located at various levels of the neuraxis. Some pathways connect nonadjacent areas (e.g., via the corpus callosum or anterior commissure), and others connect interacting nuclei (e.g., in the thalamus).

The corpus callosum: The anterior edge of this prominent landmark is called the *genu* and posterior edge is called the *splenium*. Premotor axons cross in the rostrum and genu, while motor and somatosensory axons are found primarily in the body. Inability to transfer *somesthetic* information is observed after callosal disconnection (Kakei et al., 1999). The corpus callosum plays an important role in interhemispheric transfer of information, learned experience, sensory discriminations, sensory experience, and memory. It probably integrates concepts and problem solving by providing different types of information to bear on a task. Transfer across the callosum includes learned discriminations (visual and tactual), sensory experience, and memory. It is believed to play an important inhibiting function as well (e.g., when we alternate between verbal and nonverbal functioning, or during the release represented by nighttime dreaming). It is easily damaged in traumatic injury. Severing the corpus callosum may not cause gross deficits of cognitive functioning, although on close examination deficits of communication can be detected.

The anterior commissure is a band of fibers that bifurcates. It connects the olfactory bulb with the contrelateral olfactory region, as well as the ipsilateral middle and inferior temporal gyri.

The *hippocampal commissure* is composed of efferent fiber tracts that are part of the fornix, which originate in the entorhinal cortex and hippocampus. The fibers proceed backward from each hippocampus, arch forward and join the fibers from the contrelateral hippocampus to form the fornix. Some fibers pass to the opposite entorhinal area (anterior medial portion of the temporal lobe) and hippocampus (Brodal, 1981, p. 681). The circuit is continued to the anterior insula and frontal cortex via the uncinate fasciculus (see below). Refer also to fornix (below) for subsequent portions of this circuit.

The *posterior commissure* lies rostral to the superior colliculus, where the cerebral aqueduct opens up into the third ventricle. It appears involved in eyelid and vertical eye movements (Carpenter & Sutin, 1983, p. 425).

Areas not having callosal connections: There are no direct commissural connections with the contrelateral side for the striate area (visual cortex), the hand, and, to a lesser region, the foot of the somatosensory area, the primary sensorimotor area, and the primary acoustic area (Brodal, 1981, pp. 803, 843). By inference, projection fibers communicate with nearby areas of the cortex (ipsilateral) and then newly formed patterns are transmitted contralaterally.

Association Tracts

These tracts connect the gyri, the lobes, or the widely separated areas within each cerebral hemisphere (Figure 3.5). Connections may be local or widely separated (Pansky et al., 1988, pp. 98–99).

- *Short*: Connect adjacent convolutions through the floors of the sulci.
- *Long*: Connect cortical regions in different lobes within the same hemisphere. They are behaviorally significant since they participate in patterns and sequences of complex "higher functions," and due to their length they are vulnerable to many kinds of impact and secondary damage.

Lateral Aspect of Hemisphere

- The *inferior longitudinal fasciculus* passes from the occipital lobe to the inferior and lateral temporal lobe.
- The *inferior occipitofrontal fasciculus* connects the cortex of the lateral or inferolateral portion of the frontal lobe with the cortex of the occipital lobe and inferior temporal lobe.
- The *uncinate fasciculus* crosses the lateral cerebral fissure, connecting inferior frontal lobe gyri with the uncus and other portions of the temporal lobe. Significant in verbal processing.
- The *superior longitudinal fasciculus* connects the frontal lobe's superior and middle gyri with the temporal lobe (more posteriorly than the uncinate fasciculus), and with the occipital lobe, passing over the lateral surface of the hemisphere (Ranson & Clark, 1959, pp. 325–326).
- The *arcuate fasciculus* starts as a portion of the superior longitudinal fasciculus, which passes over the insula into the temporal lobe.
- The *occipitofrontal fasciculus* radiates from the frontal lobes into the temporal and occipital lobes.

Medial Aspect of the Hemisphere

The *cingulum* is a band of tissues within the cingulate gyrus that connect the anterior perforated substance (which receives some olfactory input) with the parahippocampal gyrus. As an association pathway of the limbic system, it projects to the uncinate fasciculus in the temporal lobe tip, which projects to the inferofrontal region.

The *superior occipitofrontal fasciculus* is located between the corpus callosum and internal capsule. It connects the occipital and the temporal lobes with frontal lobes and insula.

The *stratum calcarium* is a sheet of fibers curving around the bottom of the calcarine fissures (which subdivides the visual area) from the cuneus (visual area above calcarine fissure) to the lingual gyrus (medial, inferior portion of the occipital pole).

Fornix

This band of fibers is the main efferent system of the hippocampus, projecting to the mammillary body and commissural stimulation of the contrelateral side (commissure of the fornix). The fornix divides into two bands:

Posterior commissural fibers traverse the thalamus en route to the mammillary body (hypothalamus), giving off fibers to the thalamus (anterior and intralaminar nuclei).

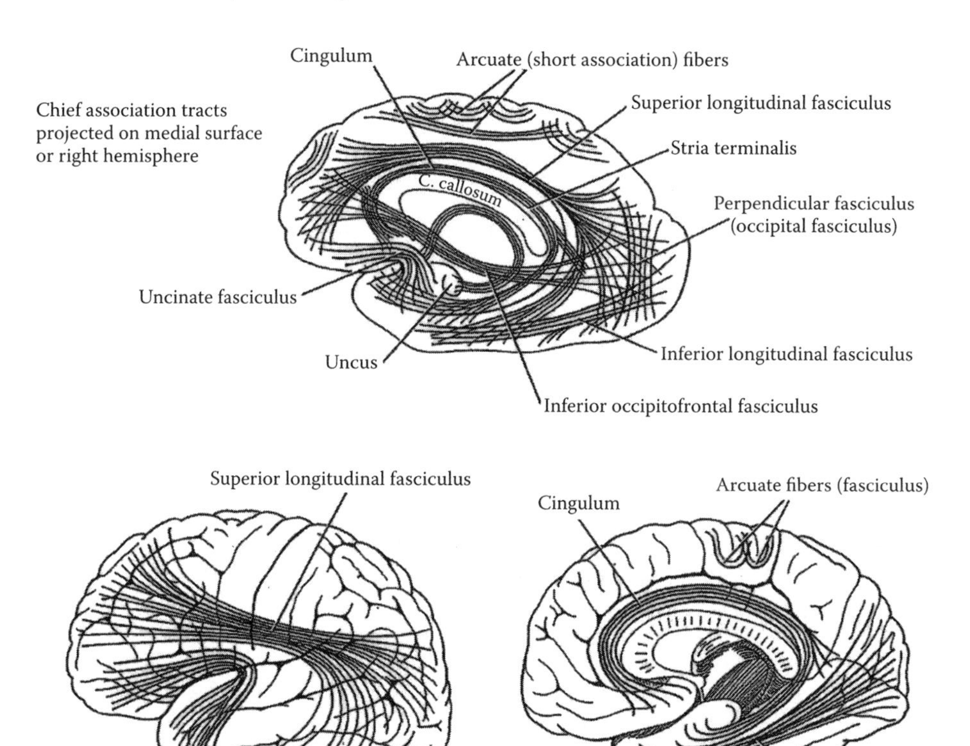

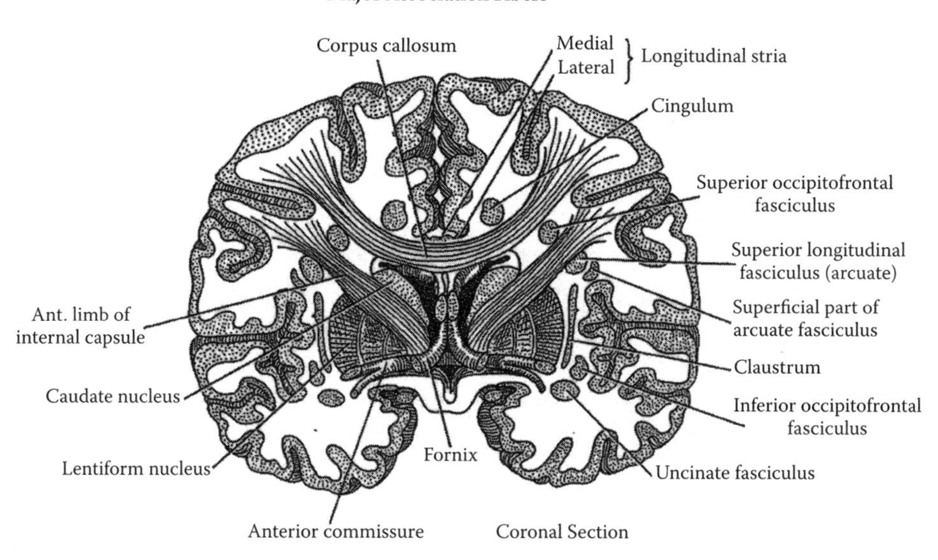

FIGURE 3.5 These tracts connect the gyri, the lobes, or the widely separated areas within each cerebral hemisphere. Connections may be local or widely separated. (From Pansky, B. *Review of Neuroscience*, 2nd ed., Macmillan, 1988. With permission.)

The *anterior band* (precommissural fornix fibers) is distributed to the septal nuclei, the hypothalamus, and the midbrain central gray (significant for arousal). From the anterior nucleus of the thalamus, the circuit projects to the cingulate gyrus via the cingulum fiber tract and entorhinal cortex, and then back to the hippocampus.

Projection

Projection neurons or pathways connect different levels of the neuraxis (e.g., corticospinal [from cortex to spine]). They include reciprocal connections between the cerebral cortex and the thalamus, brainstem, and spinal cord. Fibers converge toward the brainstem. These fibers form the *internal capsule*, which includes ascending and descending tracts between the cortex and the spinal cord, as well as to the motor nuclei of the brainstem and to the thalamus (Figure 3.6).

Corticospinal: The corticospinal tract originates in the motor cortex and the frontal and parietal lobes. Corticospinal motor information is modulated by sensory information and information from other motor regions. Accurate and properly sequenced voluntary movement is enhanced by tactile, visual, and proprioceptive stimuli (Amaral, 2000a). The corticospinal tract is the upper motor neurons (UMNs). These descend through the internal capsule to form the corticospinal tract (pyramids), with 90% decussating in the medulla. Two percent form the uncrossed lateral corticospinal tract, while 8% form the uncrossed anterior corticospinal tract (Parent, 1996, pp. 384–385). The uncrossed fibers descend as the anterior corticospinal tract and descend in the spinal cord. The remaining fibers descend in the ipsilateral ventral corticospinal tract via the internal capsule to synapse on motor neurons of the *nucleus ambiguus* (special visceral motor; medulla). Many of its motor functions can be taken over by the rubrospinal tract. The corticospinal tract, together with the rubrospinal tract controls fine, skilled manipulations of the extremities. The pyramidal

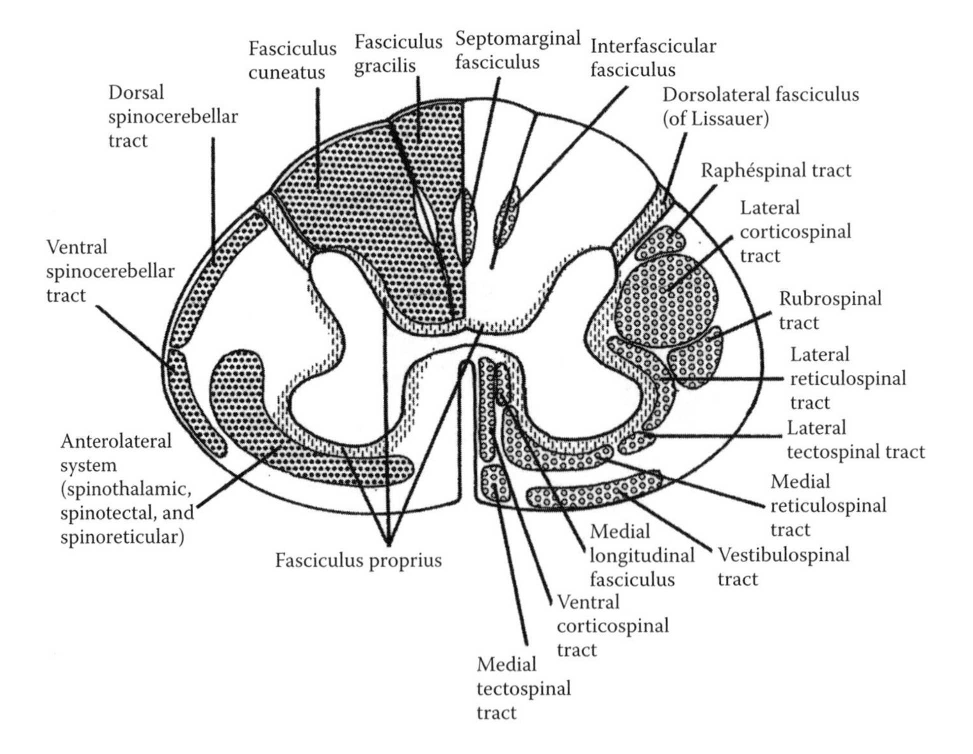

FIGURE 3.6 Some of the major fiber tracts of the spinal cord's white matter as represented at midcervical level. (From Burt, A., *Textbook of Neuroanatomy*, W.B. Saunders, 1993. With permission.)

tract contributes collaterals to extrapyramidal pathways. Thus, signals to the spinal cord to elicit a movement are accompanied via collateral signals via the extrapyramidal tract.

Pyramidal tract fibers originate in the motor cortex, premotor cortex, and in the first sensory area of the parietal lobe. The latter influences sensory input by modulating transmission of data through general sensory pathways. Fibers of the primary motor and premotor areas of the frontal lobe exercise muscular control, including fine coordination. The pyramidal tract gives off fibers to the basal ganglia, thalamus, red nucleus, reticular formation, pontine nuclei, and the inferior olivary complex. These fibers come together in the *internal capsule*, forming three-fifths of the *basis pedunculi* of the midbrain, and enter the pons. At the caudal end of the pons, the corticospinal tract forms the pyramids of the medulla, and most fibers of the corticospinal tracts decussate and enter the dorsolateral funiculus of the spinal cord to form the lateral corticospinal tract. Ninety percent of corticospinal fibers decussate in the pyramidal decussation in the lower medulla to form the lateral corticospinal tract. Two percent form the uncrossed lateral corticospinal tract, while 8% form the uncrossed anterior corticospinal tract (Parent, 1996, pp. 384–385). The remaining fibers descend in the ipsilateral ventral corticospinal tract via the internal capsule to synapse on motor neurons of the nucleus ambiguus (special visceral motor; medulla). Both arm and leg fibers appear to intermingle at the level of decussation, contrary to earlier beliefs. Perhaps only uncrossed upper-extremity fibers may travel in the uncrossed ventral corticospinal tracts (Santiago & Fessler, 2004).

Brainstem decussation of UMN: The remaining fibers descend in the ipsilateral ventral corticospinal tract via the internal capsule to synapse on motor neurons of the *nucleus ambiguus* (special visceral motor; medulla). Both arm and leg fibers appear to intermingle at the level of decussation, contrary to earlier beliefs. Perhaps only uncrossed upper-extremity fibers may travel in the uncrossed ventral corticospinal tracts (Santiago & Fessler, 2004). The neurons with which these descending fibers synapse are known as *lower motor neurons* (*LMN*). Clinical symptoms depend upon four characteristics of the lesion (Younger, 1999): (1) the level (the higher the lesion the greater the motor, sensory, and autonomic dysfunctions); (2) the extent in the transverse plane (complete or incomplete cord syndromes); (3) extent in the longitudinal plane (number of spinal segments involved); and (4) the duration and speed of injury.

Final common path: Contraction of skeletal muscle occurs when the nervous impulse is conducted down the axons of the motor neurons, which lie in the ventral horn of the spinal cord and exit the spinal cord via the ventral roots and continue into the ventral and dorsal rami of the spinal nerves. The alpha motor neurons innervate the extrafusal muscle fibers. The smaller gamma motor neurons innervate the intrafusal fibers of the muscle spindle. The dorsal rami innervate the muscles of the neck and trunk. The ventral rami innervate the rest of the trunk and the extremities.

Corticobulbar Tracts project to the motor nuclei of the cranial nerves: V, VII, IX, X, and XII.

The *extrapyramidal tract* motor pathways that influence the lower motor neurons of the spinal cord, but that do not send their axons into the pyramidal tract (the reticulospinal, rubrospinal, olivospinal, vestibulospinal, and tectospinal tracts). It is multisynaptic, and reaches motor ganglia of the brainstem and reticular formation. It forms the background of fine motion (i.e., posture, automatic movement, and unconscious adjustment of muscular tone). Actually, the corticospinal tract and the extrapyramidal system are extensively interconnected, and other parts of the brain participate in voluntary movement (Barr & Kiernan, 1988, pp. 334–337; DeLong, 2000; Myer, 1988, p. 243).

Extrapyramidal innervation influencing the axial muscles (neck, back, abdomen, pelvis) is largely crossed, while fibers influencing limb muscles are largely uncrossed (Goetz & Pappert, 1996, p. 172). Four areas give rise to descending tracts (i.e., vestibular formation, reticular formation, red nucleus, and the tectum). Tectospinal tract fibers integrate auditory and visual signals. After decussating, they proceed to cervical regions of the spinal cord and may play a role in orientation in space.

Ascending projection tracts include input from the various nuclei of the thalamus.

Peripheral Nervous System (PNS)

The motor units extend in a segmental pattern from cells lying in the anterior gray matter to target muscles in the limbs and trunk. The nerve is formed when motor (anterior) and sensory (posterior) roots combine just distal to the dorsal root ganglion. After a brief intraforaminal course, posterior branches (posterior rami) extend backwards to provide paravertebral muscles. Anterior branches (rami) extend forward, supplying the trunk and giving rise to the roots of the plexuses. Tissues surrounding larger structures and eventually the nerve trunk are the perineurium and epineurium. The latter melds into a loose layer of protective tissue called the mesoneurium, which allows some degree of passive movement in the transverse and particularly the longitudinal planes. The connective tissue elements and the specialized perineurium create a blood-nerve barrier, maintaining an immunologically privileged nerve microenvironment. The barrier also provides structure, tensile strength, and elasticity to the nerve trunk. After more severe grades of injury, the success or failure of neuroregeneration depends on the nature and degree of the injury to the connective tissue. Regeneration is interfered with by the formation of a neuroma filling the gap between nerve stumps and the replacement of a target muscle by fibrotic tissue (2–3 years). Sensory reinervation may restore useful function to a limb in up to 7 years. Grading of nerve trauma includes different levels of both nerve trunk and connective tissue. Metabolic disease, including diabetes, can interfere with regeneration.

MOTOR CONTROL AND PROGRAMMING

Motor programs are the product of premotor and primary motor operating in conjunction with sensory and association areas (Krakauer & Ghez, 2006).

Cerebellum

The cerebellum influences the motor system by evaluating disparities between intention and action, and also by adjusting the operation of motor centers in the cortex and brainstem while a movement is in progress. It is provided with extensive information about the goals, commands, and feedback signals associated with the programming and execution of movement. Its projections are mainly focused on the premotor and motor systems of the cerebral cortex and brainstem. Corrective signals are feedforward or anticipatory actions that operate on the descending motor systems of the brainstem and cerebral cortex (Ghez & Thach, 2000).

The sequences and intensities of muscle contraction are controlled by the cerebellum, basal ganglia, and sensory cortex, hence utilizing feedback from the basal ganglia and cerebellum (Guyton & Hall, 2000, pp. 669–670). The cerebellum refines the force and timing of locomotor and postural responses by affecting the activity of the reticulospinal, vestibulospinal, and rubrospinal pathways. These are phasically active during stepping and contribute to coordinated rhythmic stepping (Camicioli & Nutt, 2003). The cerebellum controls the ipsilateral limbs via connections with the spinal cord, brainstem, and contrelateral motor cortex via the thalamus (Brazis et al., 2001).

Basal Ganglia

Nomenclature of the basal ganglia (Parker, 1990, pp. 65–66).

- *Neostriatum* = striatum = putamen + caudate nucleus
- *Paleostriatum* = pallidum = globus pallidus
- *Lentiform nucleus* = globus pallidus + putamen

The basal ganglia and cerebellum closely modulate motor function. The thalamus, basal ganglia, and cerebellum may be involved in fluency, volume, articulation, and rhythm of speech. Forebrain-basal ganglia loops are active in the control of sequential movements (Doupe &

Kuhl, 1998). Basal ganglia output is inhibitory (Mink, 1999). The basal ganglia and cerebellum are related to the motor system, the limbic portions of the telencephalon, and to catecholamine-containing neurons in the midbrain, thus encompassing motor- and limbic-related forebrain systems (Butler & Hodos, 1996, p. 223). The basal ganglia is involved in the selection of movements, while the cerebellum processes sensory information and integrates it with movement (Crosson et al., 2002). The bulk of input from most of the cerebral cortex is to the striatum (putamen + caudate nucleus). There is no direct input from the motor systems. The striatum sends focused and convergent inhibitory projection to the output nuclei (the internal segment of the globus pallidus and subtantia nigra). In turn, these project inhibitory output to motor areas in the brainstem and thalamus, but not to the spinal motor circuits.

Cerebellum

The cerebellum participates in coordination of somatic motor activity (posture and rapid muscular activities), regulation of muscle tone, and equilibrium. Its function is to help combine the relative motions of different body parts into a single organized movement (Hore et al., 2003). The cerebellum plays an important role in adjusting for the dynamics of one's body movements. Cerebellar injury results in slowness or inability to adapt for different loads that the body is required to cope with in skilled movement (Bastian, 2002). The cerebellum influences posture and movement through the red nucleus, which modulates descending projections to the brainstem and the spinal cord. Its major influence on movement is via its connections to the ventral nuclear group of the thalamus, which connect directly to the motor cortex. Subsequently, the cerebellum compares information about the intention of an upcoming movement with what actually occurs; it offers control signals to correct for differences between intent and action. The medial lemniscus, the basal ganglia, and the cerebellum influence different portions of the somatosensory and motor regions of the cortex (Amaral, 2000a). Midline structures are involved in the control of motor execution, balance, and eye movement. Lesions result in disturbances of stance, gait, and ocular movement. The lateral cerebellar hemispheres are involved in motor planning, with lesions primarily affecting limb movements (Tinnmann & Diener, 2003). The lateral cerebellar hemispheres are associated with planning and performing intricate sequential patterns of movement, especially for the hands, the fingers, and the speaking apparatus. Communication is mostly between the lateral cerebellar areas and with the premotor area and primary and association somatosensory areas.

Afferents: The cerebellum has no direct connection with the cerebral cortex. The two halves of the cerebellum control and receive input from muscles of the ipsilateral side of the body. The cerebellar cortex receives afferent information from limb and extraocular motor systems, as well as varied sensory systems via the three cerebellar peduncles: spinocerebellar, vestibular, corticopontine, proprioceptive, cutaneous, vestibular, and visual. The cerebellum receives impulses primarily via the inferior and middle cerebellar peduncles (muscle spindles; touch receptors of the skin; joints). Afferent tracts are olivocerebellar, corticocerebellar, vestibulocerebellar, reticulocerebellar, spinocerebellar, and also spinocerebellar afferents via the superior cerebellar peduncle. The spinocerebellum is important in the control of body musculature. It receives somatic information from the trunk and the limbs. This includes internal feedback signals for correcting inaccurate movements (motor and sensory information). The cerebrocerebellum is involved with planning movements. Afferent and efferent connections are to the contrelateral cerebral cortex. Corticopontine fibers reach the cerebellum via the middle cerebellar peduncle, projecting to the contrelateral cerebellar hemisphere. Climbing fibers terminate in the contrelateral dendrites of the Purkinje cells (Goetz & Pappert, 2004, pp. 117–118; Hubbard & Workman, 1998, pp. 592–594). Nearly all have extracerebellar origin in the inferior olivary complex. Input is from the motor cortex, pretectal region (visual system), and spinoolivary tracts (cutaneous, muscle, joint proprioceptors). Major *excitatory* inputs are (1) the climbing fibers that originate in the inferior olivary nuclear complex and via the olivocerebellar tract project to the cortical Purkinje cells; (2) mossy fibers originating primarily in the spinal cord, and also the pontine nuclei, vestibular nuclei, and reticular formation nuclei (Martin, p. 312); (3) mossy fibers from all other cerebellar afferents (→ deep nuclear cells → granular layer of the cortex → Purkinje cells). All other cell types are *inhibitory.*

Efferents: The superior peduncles contain mostly efferent axons, the middle cerebellar peduncle contains only afferent axons, and the inferior cerebellar peduncle contains both afferent and efferent axons. The Purkinje cell is the only output from the cerebellar cortex. Axons pass through the granular layer and white matter to synapse with the deep cerebellar nuclei and, to a lesser extent, with the vestibular nuclei. They are *inhibitory* and use GABA as their neurotransmitter. Thus, efferent output is a balance between excitatory and inhibitory input to the cerebellar nuclei. The dentate nuclei project contralaterally to the thalamus, which projects to the premotor and primary motor cortices to influence planning and initiation of voluntary movements. The emboliform and globose nuclei project to the contrelateral red nucleus, which participates in the control of proximal limb muscles via the rubrospinal tract. The fastigial nucleus contributes to the vestibulospinal and reticulospinal tracts.

Error monitoring: The cerebellum is provided with extensive information about the goals, commands, and feedback signals associated with the programming and execution of movement. It receives input from the motor cortex, which informs the cerebellum of an intended movement before it is initiated. Its projections are mainly focused on the premotor and motor systems of the cerebral cortex and brainstem. It utilizes information from proprioceptors to monitor body position, muscular tension, and muscle length. The cerebellum influences the motor system by evaluating disparities between intention and action, and also by adjusting the operation of motor centers in the cortex and brainstem while a movement is in progress. In particular, by supplementing and correlating the activities of other motor areas, it plays a role in controlling posture and correcting rapid movements initiated by the cerebral cortex.

As a "*comparator*," it compensates for errors in movement by comparing intention with performance. It receives information about plans for movement and feedback about motor performance, and projects to descending motor systems (Goetz & Pappert, 1996, p. 179; Hubbard & Workman, 1998, p. 583). It generates an error signal to the cortex, which can adjust output to meet the demands of the situation. Corrective signals are feedforward or anticipatory actions that operate on the descending motor systems of the brainstem and cerebral cortex (Ghez & Thach, 2000). It may be noted that cerebellar influences on motor performance are indirect. Skilled movements involving multiple joints and muscles, including finger control, involve precise *timing* and *amplitude of movement.* When performing an overarm toss, persons with cerebellar injury display do poorly due to increased variability in both the timing and force of finger extension. There is variability according to chronicity and whether the lesion is lateral or medial (Hore et al., 2002).

Spinocerebellum (vermis; intermediate hemisphere; fastigial and interposed nuclei): Receives somatic sensory input from the spinal cord and other structures, and controls posture and movements of the trunk and limbs. The spinocerebellum projects to both deep nuclei and the thalamic projections to the motor cortex, thus controlling components of the descending motor pathways affecting the head, neck, proximal parts of the limb, and axial musculature (DeLong, 2000). The spinocerebellar tracts transmit internal feedback signals for correcting inaccurate movement (Martin, 2003, p. 305). The pathways originate from the ventral regions of spinal gray matter, receiving information from motor pathways as well as sensory information from afferent fibers. They enter the cerebellum from the inferior cerebellar peduncle. Information is recognized from different parts of the limbs, trunk, or head in order to coordinate movements. Two nuclei in the caudal medulla are noteworthy. The more rostral accessory cuneate nucleus relays somatic sensory information from the upper trunk and upper limb to the cerebellum for the control of movements. The more caudal cuneate nucleus transmits somatic sensory information to the thalamus for perception. The gracile and cuneate nuclei neurons decussate and ascend in the medial lemniscus to provide sensory information to the thalamus (ventral posterior lateral and medial nuclei), and thence to the somatic sensory cortex (Gardner & Martin, 2000).

The *vestibulocerebellum* receives afferents from the vestibular nuclei (see above). It coordinates an individual's ability to use vestibular information to control eye movements (Ghez & Thach, 2000). The *vestibulo-ocular reflex* involves the vestibular system's perception of head movement and causes the eyeball to move in the opposite direction.

The *vermis* receives direct projection from vestibular neurons, as well as visual and auditory stimuli relayed by brainstem nuclei. Deep nuclei influence motor neurons through projections into the descending pathways.

Cerebrocerebellum (lateral hemisphere; dentate nucleus): Receives input indirectly from the cerebral cortex and participates in the planning of movement. The major input is from the contrelateral cerebral cortex (motor sensory and association areas) via pontine nuclei. Efferent projection is via the dentate nucleus, which projects to (1) the thalamic ventrolateral nucleus, a motor relay nucleus to the cortex; and to (2) the parvocellular division of the red nucleus, which projects to the inferior olivary nucleus, which provides input to the cerebellum.

Vestibulocerebellum (flocculonodular lobe): Receives input from the vestibular labyrinth, the nuclei of which substitute for deep cerebellar nuclei. It controls gaze through the combined control of eye and head movements. Vestibular afferents are the only primary sensory neurons that project directly to the cerebellum. The vestibulocerebellum projects to the vestibular nuclei, which play a role in neck muscle and eye muscle control. It controls gaze through eye and neck muscle control over head movements (Martin, 2003, pp. 304–311).

Motor Skill Learning

Motor learning commences with all or many muscles of the hand or a limb, and proceeds toward the use of a minimum amount of activity necessary for the activity (Bach-Y-Rita, 2002). The potential for fatigue is thus reduced, as is greater motor precision. Willingham (1992) differentiates between motor and other forms of learning. Learning motor skills is a unitary faculty (i.e., all motor skills are handled by the same memory system). Motor skill learning processes reside in the same site as plasticity. This contrasts with other memory processes in which the components may be dissociable (only particular functions are affected by trauma or other events). There are no separate structures (e.g., the hippocampus), the damage of which interrupts memory but not other cognitive processes.

Perceptual motor integration: The integration of perceptual and motor integrations. This may be carried out by area 7a and the cerebellum, thalamus, and PMC.

Visuomotor association: Perceptual information determines which movement is made (e.g., depressing the brake with a red light). Lesions of the premotor cortex create difficulties in associating arbitrary visual cues with motor responses.

Visual guidance: Feedback about an ongoing movement (e.g., road conditions determining steering wheel movement). Cerebellar damage leads to difficulties in integration of visual and motor information.

Degrees of freedom: Even simple motions have a wide range of trajectories. The challenge of some motor skills (e.g., athletics) is the reduction of the spatial and temporal variability of a movement. Consider a ballerina running and leaping into the arms of a moving partner! If one considers degrees of freedom to be an aspect of precise motor control through reducing the range of trajectories, motor control benefits from minimizing jerks, considering the mechanical qualities of the limbs, and organizing actions in terms of muscle groups.

Serial order (sequencing): Typing and musical instrument playing are examples of this. The planning of a sequence of actions may occur prior to the actual availability of perceptual stimuli to guide or carry out the action. A loop between the supplementary motor area (SMA), the basal ganglia, and returning to the SMA participates in the circuit. The loop may be a high-level planner rather than a sequencer of motor actions.

Synergies or muscle groups: Degrees of freedom are reduced by organizing action in terms of muscle groups (rather than individual muscles) for some movements. Stimulation of a single corticospinal neuron leads to the activation of several muscles. Stimulation of a primary motor cortex neuron can lead to simultaneous flexion and extension of wrist muscles to fix its position in preparation for a finger movement. To maintain posture, humans make use of a very limited number of motor synergies in different combinations.

Praxis/programs (*how to*): Praxis involves the utilization and integration of various motor and mental units that utilize a variety of sensory and informational processes, such as imitating gestures, gesturing to command (pantomime), symbol recognition, sequencing, and spatial, temporal, and speech-language processing. Some cognitive processes have been described as motor plans that have been rehearsed and evaluated for their adaptational value. The Praxis system involves the premotor cortex, motor cortex, Wernicke's area, and visual association areas, and transmission via the corpus callosum and arcuate fasciculus (which connects the middle and superior frontal gyri with parts of the temporal lobe).

Cognition

Another view of the *posterior parietal cortex* (*PPC*) is emerging (i.e., that it serves both a sensory and motor role, and also serves higher-level cognitive functions related to action): it forms *motor intentions* (i.e., early plans for movement). Motor intentions are described as high level and abstract, such as decision-making. An initial vague intention becomes modified to specify the limb, the trajectory of the movement to avoid obstacles, the coordination of eye and hand movements, the avoidance of obstacles, and so on. Inability to plan movements is described as *apraxia* (Andersen & Buneo, 2002). There is a discontinuity between intent based upon accurate perceptions and action.

Program Storage and Representation

Liepmann (references given in Schilder, 1950) suggested that every action is based upon an anticipatory plan that has a specific structure: the final aim; a model of the body; the single actions necessary for the actualization of the plan; and a representation of the object (visual, tactile, acoustic, or images in readiness). The motor commands can be monitored ("*efference copies*") and compared with proprioceptive feedback from motor execution. We may become aware of motor commands when there is a discrepancy between the command and its execution (Goldenberg, 2002).

Movement representations seem to be stored in the left inferior parietal lobule and then forwarded to the motor or premotor areas. There, they program the premotor cortex, being transcoded into time-space or visuokinesthetic representations, which selectively activate the motor cortex to perform learned skilled movements. Performance may depend upon visual (imitation) or verbal (command) stimulation, or knowledge of specific procedures (how to hammer) versus association of information (using a hammer to bang a nail) (Heilman et al., 1997, 2004). Programs for action are assembled particularly within the basal ganglia, the brainstem, and reticular formation. Action programs require information about the initial and intended positions, as well as intermediate states. The body gestalt and relationship of parts, including posture, are modified by peripheral feedback from pressure receptors in the feet, the vestibular system, the eyes, and the proprioceptors in the neck and spinal column (Goldenberg, 2002; Parent, 1996, pp. 909–928; Pocock & Richards, 2004, pp. 173–182).

Memory functions of the frontal lobe provide an extended time frame during which more complex patterns may be organized (Fuster, cited by Tucker, 1995). One differentiates between reflexive motions and purposeful activities. Praxic programs provide these instructions of how to position one's limb when performing skilled movements, including use of tools and objects; the spatial trajectory and limb orientation of the skilled movement, including information, with regards to the object or to one's own body; how rapidly to move in space, including the timing of the limb's action; how to imitate a movement; how to solve mechanical problems; and ordering components of an act to achieve a goal (Heilman et al., 2004).

There are two types of neurons involved with hand motion: *Muscle-like* (force) and *movement* (independent of the pattern of muscular activity that generated the movement) (Kakei et al., 1999).

Basal Ganglia

The basal ganglia and cerebellum closely modulate motor function. The basal ganglia (including the substantia nigra and subthalamic nucleus) are involved in complex circuits among themselves, as well as functioning in feedback circuits with the cerebral cortex: the premotor cortex, the dorsolateral and lateral orbitofrontal cortex, and the anterior cingulate cortex. The basal ganglia are also associated with motivational, behavioral, and mood disorders. Since a feedback loop extends from all parts of the cerebral cortex to various basal ganglia continuing to the thalamus, and then back to primarily the motor cortex, it has been suggested that the basal ganglia, like the cerebellum, process sensory and memory information used to regulate motor function. They appear to be involved in basic, perhaps innate, movement patterns, possibly elaborating programs in response to cues from cortical association areas (Pocock & Richards, 2004, pp. 180–181). Forebrain-basal ganglia loops are active in the control of sequential movements (Doupe & Kuhl, 1999). Basal ganglia output is inhibitory (Mink, 1999). Basal ganglia and cerebellum are related to the motor system, the limbic portions of the telencephalon, and to catecholamine-containing neurons in the midbrain, thus encompassing motor and limbic-related forebrain systems (Butler & Hodos, 1996, p. 223). The basal ganglia are involved in the selection of movements, while the cerebellum processes sensory information and integrates it with movement (Crosson, 2002). The bulk of the input from most of the cerebral cortex is to the striatum (putamen + caudate nucleus). There is no direct input from the motor systems. The striatum sends focused, convergent inhibitory projection to the output nuclei (the internal segment of the globus pallidus and substantia nigra). In turn, these project inhibitory output to motor areas in the brainstem and the thalamus, but not to the spinal motor circuits.

Spinal Control of Gait

A basic mechanism is *central pattern generators* (CPGs), which in humans provide rhythmic motor patterns for walking, breathing, chewing, swimming, micturition, and sexual reflexes. The CPG is sited within the lumbosacral spinal cord, but is modified by supraspinal control and peripheral sensory information (mechanoreceptors and cutaneous receptors). It creates oscillatory motor output without any oscillatory input. Control of walking is generated by assemblies of CPGs (spinal interneurons), which are interlinked by propriospinal fibers to facilitate interlimb coordination (Thompson, 2004). By *spinal automaticity*, it is meant the concept of ability to carry out complex but routine tasks without "conscious thought" (walking across a room; shorter steps taken by the inside limb when walking in an arc; enhancing flexion of the ipsilateral leg when stepping over an object versus enhancing extension in the contrelateral limb, etc.). Proprioception is interpreted by the spinal cord analogously to the manner in which the visual system processes information. Based upon peripheral information (proprioceptive, cutaneous), the CPG receives, interprets, and predicts the appropriate sequences of action for the step cycle. The corticospinal tract, based upon animal study, seems to make fine adjustments rather than generating the basic locomotor pattern, while executing standing and stepping, as well as adapting to varying loads, speeds of stepping, turning, and stepping over objects (Edgerton et al., 2004).

THALAMUS

The thalamus is more than a relay station. Its reciprocal interaction with cortical areas summates tactile stimuli, which are compared with cortically stored templates of previous tactile experience (Nicolelis et al., 2002).

Sensory relay paths to the cortex and subcortical structures lie in the dorsal thalamus. The cortex sends topographically ordered projections to the respective thalamic relay centers. The thalamus then relays information between sensory, motor, and associative brain regions. Thalamic midline and intralaminar nuclei are a relay station for incoming impulses from the brainstem reticular formation, projecting to the ventral anterior nucleus and thence to the cortex (Ge & Grossman, 2008).

Intralaminar nuclei form a separate projecting system with a modulatory role over the cortex and the basal ganglia, representing the final stage of the ascending activating system for mental arousal.

The *motor control system* receives input from the globus pallidus and cerebellar dentate nucleus and sends efferents to the motor and premotor cerebral cortex.

The *limbic* thalamic group receives hypothalamic input (mammillary) and projects to the cingulate cortex, hippocampus, septum, and amygdala.

The *associative* thalamic group receives major inputs from the secondary sensorimotor, limbic, and associative cortex, and projects into higher-order cortical areas of the frontal, parietal, and temporal lobes (Puelles & Rubenstein, 2002).

FOREBRAIN

The forebrain is an embryological concept that illustrates the complex interconnections accounting for both brain function and its vulnerability to diffuse injury. It is an extremely complex assembly of functionally diverse structures regulating most aspects of cognition, homeostasis, and behavior. The most rostral portion of the CNS, it includes the most anterior region (telencephalon and the cerebral cortex) and the diencephalon, hypothalamus, thalamus, pineal gland, and so on. Its organization is similar for mammals, birds, reptiles, and amphibians (although we note that the cortex is relatively unique for mammals) (Puelles & Rubenstein, 2002).

The septal complex is a collection of forebrain cell groups lying medial to the lateral ventricles. Its bidirectional connections with the hippocampus permit integration of information from all sensory modalities. Efferent connections participate in endocrine stress reactions, including projections to the hypothalamus that influence the HPA axis (periventricular nucleus and preoptic area). It also seems to play a role in defensive activities (i.e., an important role in behavioral priorities during stress) (Watts, 2000).

FRONTAL LOBE

Neurobehavioral impairment is frequent after impact to the anterior and posterior portions of the skull hyperextension/hyperflexion injury. This reflects the multiple regulatory tasks occurring in the frontal lobe. Whereas the posterior cortex supports *representational operations*, the frontal cortex recruits the multiple levels of the nervous system for extended *goal-directed behavior.*

Brainstem and limbic functions are integrated with cognitive and motor planning. Significant varying temperamental disorders reflect the different motivational biases for motor control reflect input from various sections of the paralimbic cortex. Hemispheric specialization contributes to emotional self-regulation, but the alternative explanation is disinhibition of the contrelateral hemisphere versus disinhibition of ipsilateral subcortical structures. The frontal lobe participates in the arousal system, which explains deficits of motivation and initiative following lesions. Emotional disorders cannot be clearly attributed to a particular lobe (Tucker et al., 1995).

Orbitofrontal cortex: The orbitofrontal cortex receives projections from the visual, auditory, and somatosensory cortex, the amygdala, the entorhinal cortex, the cingulate gyrus, and the mediodorsal thalamic nucleus. Efferent pathways include brain regions regulating autonomic, neuroendocrine, and skeletal-motor function. The orbitofrontal and subcortical limbic areas may interact in learning and unlearning of the significance of fear-producing sensory events and choice and implementation of behavior needed for survival. This cortex may code the characteristics of the experience as rewarding or adverse since its sensory responses depend upon the meaning of the experience. This would facilitate behavioral effectiveness through changing behavior in the light of changing circumstances. Damage to it causes changes in mood, affect, and social and perseverative behavior (Charney et al., 1995). It appears that responses resembling *sympathetic nervous system* are elicited from subcallosal and orbital cortex, while those resembling parasympathetic nervous system (with respiratory inhibition) are elicited from the anterior cingulate cortex (Buchanan & Powell, 1993). The ventral *anterior cingulate* has been conceived of as a "*visceral motor cortex*." It is considered to contribute to emotion by activating visceral and somatic states that are important for the emotional experience (Neafsey et al., 1993).

THE AMYGDALA/HIPPOCAMPAL COMPLEX

This group is activated by ascending catecholaminergic neurons from the brainstem, and by conditioned *fear*, possibly from cortical association areas. The entorhinal cortex of the hippocampus is the main recipient of extrinsic sensory information arriving from polysensory neocortical regions (the temporal, insular, cingulate, and frontal lobes, which are relayed to other hippocampal fields via the perforant path). The amygdala projects to visually related areas of the temporal and occipital cortex, and with the polysensory regions of temporal, insular, cingulate, and frontal lobes. It projects to the prefrontal cortex directly and indirectly to the thalamus and the striatum (Charney et al., 1995). The entorhinal cortex receives projections from the lateral amygdala nucleus and numerous other regions. Perhaps the neuronal loop including cortex, hippocampus, and amygdala may attach cognitive significance to fear-inducing events, thus facilitating memory traces and enabling initiation of adaptive behavioral responses. Neocortical and subcortical projections to the lateral and basal lateral regions are relayed to the central nucleus, which reciprocally interacts with brainstem solitary and parabrachial nuclei that receive visceral information. Stimulation of the amygdala focuses attention upon fear-provoking events and activates the circuitry of the startle response. Other regions involved in organizing fear include components of the forebrain: the prefrontal cortex, the perirhinal cortex, and the bed nucleus of the stria terminalis. Infusions of CRH directly into this structure facilitate fear and anxiety-related responses (Schulkin, 2003, pp. 66–71, 88). CRH is a production is a significant response to stress.

HIPPOCAMPUS

The hypothalamus is the master controller of autonomic function, innervating all autonomic relay centers, including sympathetic neurons. It maintains homeostasis by integrating endocrine, autonomic, and behavioral responses (Cone et al., 2003). The hippocampus inhibits the amygdala, the PVN, and LC/sympathetic systems (Tsigos & Chrousos, 1996).

Complex functions enable the amygdaloid complex to participate in organisms' awareness of the significance of environmental stimuli. Single amygdalar cells may respond selectively or to various combinations of many different sensory modalities (somatosensory, visual, auditory, visceral) (Nolte, 2002, p. 579). A large number of neurotransmitters and other neuroactive substances are found here, reflecting the input from numerous sources (cortex, basal ganglia, forebrain, diencephalon, lower brainstem). Connections with the cognitive neocortex and visceral brainstem provide a link that is central to emotion. Reciprocal connection with the prefrontal cortex occurs via the thalamic dorsal medial nucleus. The human amygdala receives highly processed cognitive information. The amygdala integrates cognitive with biological motivations. Damage to parallel cognitive and/or limbic pathways that modulate behavior can have devastating effects on the subject's emotional life. Amygdala stimulation can evoke the various *subjective mental phenomena* of emotion (feelings, dream- or memory-like images, visceral sensations) and the *bodily signs of emotion* (visceromotor activation, general arousal, hormonal secretion, facial movements) (Halgren, 1992). Of special significance to neurotrauma and fear are projections from the suprachiasmic nucleus (vasopressin) and vasopressin and oxytocin (paraventricular nucleus). Environmental input is reflected in reciprocal innervations with the thalamus and also unidirectional input to the medial dorsal nucleus. The amygdala projects to frontal, insular, and cingulate cortices, and also to autonomic and visceral centers in the brainstem (and diencephalon), which mediate alterations of cardiac, respiratory, and visceral functions (Amaral et al., 1992). The amygdala (medial temporal) regulates emotional responses, projects to the PSNS via the vagus nerve, and has other autonomic regulatory functions via the hypothalamus (Martin, 1996, pp. 457–458). It processes anxiety-inducing stimuli and organizes adaptive behavior, autonomic, neuroendocrine, and skeletal-motor responses. Various nuclei project to the midbrain, pons, medulla (affecting autonomic functions), and to the locus ceruleus. Amygdaloid projections to the striatum (which also receives efferents from the prefrontal cortex) and ventral tegmentum may be involved in developing motor responses to threatening environmental stimuli that involve past experiences. Amygdalar stimulation often causes an animal to stop what it is doing and become very attentive, followed by defense, raging aggression, or fleeing. Bilateral amygdala destruction causes a great decrease of aggression, which seems to be part of a memory deficit that impairs the ability to learn or remember the appropriate emotional and autonomic responses to stimuli. In humans, the most common response is fear with autonomic reactions (pupillary dilation, release of adrenaline, increased heart rate) (Nolte, 2002, p. 580). Projections of the central nucleus of the amygdala to the lateral hypothalamus may be involved in the sympathetic autonomic nervous system reactions of traumatic stress (fear and anxiety). Projections to the dorsal motor nucleus of the vagus are involved in autonomic aspects of fear and anxiety. Projections to the parabrachial nucleus may be involved in respiratory changes during fear (Davis, 1992). The central nucleus of the amygdala projects directly to the PVN of the hypothalamus. Its indirect projections to the PVN may mediate the neuroendocrine responses to stressful or fearful stimuli.

Approach/avoidance: The amygdala plays a role in the retrieval and then emotional analysis of information for a given stressor. In response to an emotional stressor it can stimulate components of the stress and dopaminergic systems. CRH neurons of the amygdala respond to glucocorticoids by stimulating anxiety (Tsigos & Chrousos, 1996). The amygdala evaluates stimuli as to whether or when they may be harmful. It is activated by a thalamo-amygdala pathway (a rapid channel without a very accurate representation of the external events) and by representation from the auditory sensory cortex (slower, multisynaptic, but more accurate stimulus representation). This mechanism allows an initial protective response to be continued or aborted if inappropriate. Amygdala projections are involved in SNS activity, motor behavior, and control of pain.

Sensory information is transformed by the amygdala, thus contributing to *learning and conditioning*. The amygdala, perhaps modulated by CRH, participates in *conditioned fear responses* (Stout et al., 1995). Since conditioning refers to the pairing of an aversive stimulus with one that initially has no behavioral effect, it may be a model for the *avoidance reaction of trauma*. The amygdala is involved in expression of conditioned fear through its direct projection to the brainstem nuclei involved in the startle reflex; it may also be a site where neural activity that is produced by conditioned and unconditioned stimuli converges. Furthermore, animal studies indicate that lesions of the amygdala block the effect of a conditioned stimulus. The amygdala seems to contribute to plasticity through mediating the acquisition and extinction of conditioned fear (Davis, 1992). Since the amygdala also processes positive stimuli, it participates in determining whether a stimulus merits *approach* or *avoidance*, depending on the past history (LeDoux, 1995). A range of moods is elicited by amygdala stimulation in humans (fear, anxiety, pleasure, and anger). Additionally, stimulation results in autonomic, neuroendocrine, and reflex functions frequently accompanying fear and anxiety states (i.e., respiration, BP, and heart rate increases, as well as jaw movements and alterations of facial expression characteristic of fear) (Charney et al., 1995). The amygdala (central nucleus) plays a critical role in fear conditioning. It is a center with connections to all cortical sensory systems, the thalamus, and the hypothalamus. It appears to integrate the interactions of the sensory information, emotional tone, and neuromodulation needed for consolidation of long-term memory. (Discussion of the amygdala's role in memory is continued in the section on Memory and Stress.)

BRAINSTEM AND CERVICAL

Nuclei in the Midbrain

- Ventral tegmental area, projecting to the nucleus accumbens, prefrontal cortex, amygdala and hippocampus
- Substantia nigra, projecting to dorsal striatum (caudate and putamen of the basal ganglia)
- Medulla oblongata

CNS control over the SNS is expressed through centers in the *medulla oblongata* (reticular formation; raphe nucleus), pons, hypothalamus (paraventricular N.), and the spinal cord (intermediolateral cell column). The neurotransmitters regulating preganglionic neurons are epinephrine, dopamine, norepinephrine, serotonin, and oxytocin (paraventricular N.). CNS activity will also be affected by the characteristics of the extracellular fluid (electrolytes; hormones) and temperature.

The *solitary nucleus* is the first viscerosensory relay, and it initiates medullar reflexes that control cardiovascular, respiratory, and other autonomic functions (Benarroch, 1997, p. 10). It receives the following afferents: baroreceptors; cardiac; pulmonary; carotid and aortic chemoreceptors; facial N., VII, glossopharyngeal N., IX, and vagus N., X. The solitary N. relays viscerosensory information directly to the forebrain and, via the parabrachial N., to the hypothalamus, the amygdala, the septum, the cortex, and the periaqueductal gray. It also participates in circuits controlling respiration, circulation, vomiting, motor control of the stomach, heart rate, and micturition via the vagus N. and ventrolateral medullary reticular formation (Benarroch, 1997, pp. 10–12; Iversen et al., 2000). One *negative feedback* mechanism comprises afferent stimulation from the baroreceptor mechanism via the *tractus solitarius*. Axons leave the spinal cord between T-1 and L-2, synapse with postganglionic sympathetic neurons in the paravertebral sympathetic ganglia, or pass through the ganglia of T-5 through L-2, forming the splanchnic nerves innervating the adrenal medulla. Others synapse with postganglionic sympathetic neurons in various plexes to innervate the heart and other viscera (Landsberg & Young, 1992).

Ventrolateral medulla: Adrenergic projection to the brainstem and spinal autonomic nuclei, which play an important role in *regulating blood pressure.*

Pontomedullary reticular formation: Projects to autonomic preganglionic neurons in the brainstem and spinal cord. They also project to spinal motor and premotor neurons so that they may coordinate complex behavioral responses such as *defense reactions involving both visceral and somatic changes* (e.g., startled by an unexpected, loud noise, our BP rises).

SYMPATHOADRENOMEDULLARY (CATECHOLAMINE)

What is referred to here are two separate anatomical systems involved with catecholamine utilization, the SNS, and the adrenal medulla. The major catecholamines involved in stress are epinephrine and norepinephrine, which can act as neurotransmitters or hormones (Granner, 1993; Marks et al., 1996, pp. 385–386). There are two major classes or receptors (α-adrenergic and ß-adrenergic), each with two subclasses. The β receptors work directly on fuel metabolism in the liver, fat, muscle, and so on. Activity is increased when chronic stress of a single type is augmented by novel environmental stressors (Kvetnansky, 2004). Long-term repeated stress alters the sensitivity of neurons in the locus ceruleus and other brainstem nuclei through activation of gene expression in the adrenal medulla.

The *locus ceruleus* (LC) of the brainstem is the principle reticular site for noradrenergic neurons and the largest central noradrenergic nucleus. These neurons increase their firing rate from all stimuli that elicit attention, with stressful or noxious stimuli eliciting larger and more sustained increases in firing (Zigmond et al., 1995). CRH elevates LC firing, which in the face of danger interrupts lower-priority behavior such as sleep and upregulates CNS tone. It is essential for coordination of the CNS with the HPA, as well as the acquisition of fear-conditioned responses (Woodward, 1995). It has an extensive efferent projection system. The hypothalamus plays a regulatory role for autonomic function. Although afferent input is described as more restricted, it includes forebrain structures such as neocortex, amygdala, and hypothalamus, and brainstem monaminergic neurons from the raphe nuclei and a variety of sensory relay areas from the spinal cord and nucleus of the solitary tract.

Thus, the LC is responsive to external stress, noxious stimuli, and *alterations in homeostasis* and *trauma*, such as alterations in blood volume, BP, heat, hypoglycemia, blood loss, increase in pCO_2, and reduction in BP (Charney et al., 1995). An animal experiment indicates that the activation of catecholaminergic afferents to the prefrontal cortex is not specific to stress, but also occurs in response to nonstressors with positive motivational valance (Taber & Fibiger, 1997).

Cholinergic

The cholinergic system has multiple functions. It activates in parallel the adrenal medulla (catecholamines) and the SNS. Also, it participates with the LC-noradrenergic system in widespread arousal functions that interfere with sleep (Woodward, 1995).

Serotonergic System

Serotonergic projections are considered to be the most expansive neurotransmitter system. They influence stress-related activities (pain, vascular function, hormone secretion) and other functions (behavior, appetite, sexual activity, temperature). Most serotonin neurons have cell bodies along the midline of the brainstem (raphe) and send ascending and descending projections to the CNS. Serotonin influences CRF release in the paraventricular nucleus (PVN) (see below) and elsewhere to produce cardiovascular effects. Reduced serotonergic function may underlie *depression* and other behaviors (Fuller, 1996). Inescapable shock, but not escapable shock, decreases cerebral serotonin levels, while immobilization increases serotonin turnover (Southwick et al., 1992). Drugs that are effective in treating depression act on both the serotonergic and noradrenergic systems, particularly *selective serotonin reuptake inhibitors* (*SSRI*). However, depression is not single disorder. Disturbances in various transmitter systems can lead to depression. The noradrenergic and serotonergic systems overlap at both the systemic and cellular levels (Kandel, 2000d).

Raphe nuclei are part of the paramedian reticular formation. They lie on both sides of the midline (Connors, 2005; Chapter 12). Projections are part of the ARAS, innervating most of the CNS in the same diffuse way as the locus ceruleus. They use serotonin as their neurotransmitter, receiving strong inputs from the hypothalamus and projecting to spinal autonomic nuclei. The raphe nuclei are involved in the control of sleep-wake cycles, different stages of sleep, control of mood, and certain types of emotional behavior. Projections of the raphe spinal system are important in *suppressing dorsal horn pain transmission*. The raphe pontine micturition center projects to the parasympathetic cell column of the sacral spinal cord (Afifi & Bergman, 1998, p. 153), and serotonergic fibers project to the hippocampus (p. 432).

The raphe nuclei send descending input to the motor and autonomic systems of the spinal cord. These project to the spinal dorsal horn and modulate pain perception. *Serotonergic neurons* in the pons and midbrain project to virtually the whole of the forebrain, helping to regulate wake-sleep cycles, affective behavior, food intake, thermoregulation, and sexual behavior. Descending projections of the lower pons and medulla project to the brainstem and spinal cord, where they participate in regulating motor tone and pain perception. Serotonergic pathways play an important regulatory role in hypothalamic, cardiovascular, and thermoregulatory control. They also modulate the responsiveness of cortical neurons. Cholinergic cell groups are found in the brainstem, forebrain, and mesopontine tegmentum. They provide ascending cholinergic innervation of the thalamus and descending projection to the pontine and medullary reticular formation. These play an important role in regulating wake-sleep cycles. *Histaminergic* cell groups are located in the posterior lateral hypothalamus. Projections range from the spinal cord to the cortical mantle.

Histaminergic neurons of the tuberomammillary nucleus may help in maintaining arousal in the forebrain (Saper et al., 2000).

Dopaminergic

The dopaminergic system has cortical and limbic components that are activated by the LC-NE/sympathetic noradrenergic system. Projections to the frontal cortex are thought to be involved in anticipatory phenomena and cognitive functions, while mesolimbic neurons of the nucleus accumbens are believed to be involved in motivation, reinforcement, and reward (Tsigos & Chrousos, 1996). Dopaminergic neurons are preferentially activated in the mesocortex by cues conditioned to stressors. The dopaminergic innervation of the *prefrontal cortex* is very sensitive to stress. They arise from mesencephalic neurons (Deutch & Young, 1995). Chronic stress and repeated cocaine exposure increase dopamine transmission in response to acute stress (Southwick et al., 1992).

Benzodiazepines (BZDs) and Gamma-Aminobutyric Acid (GABA)

There is a dense network of GABA nerve fibers evenly distributed throughout the hypothalamus, and a plexus of GABA terminates in the median eminence. This system seems to participate in the inhibition of CRH release (Calogero, 1995). BZDs and GABA (the major inhibitory neurotransmitter in the CNS) are involved in the inhibition of seizures and anxiety reduction (Ballenger, 1995). Uncontrollable stress causes changes in their receptors, and seems related to alterations in the norepinephrine, dopamine, and endogenous steroid systems (Southwick et al., 1992).

The PAG integrates autonomic, somatic, antinociceptive, and defense reactions to stress (Benarroch, 1997, p. 397). Stimulation elicits behavioral reactions that contribute to survival: *arousal* (alertness, vigilance, cognition, attention); *enhanced analgesia*; *inhibition of vegetative functions* (feeding, reproduction); and *redirection of energy* (oxygen, nutrients) to the CNS and stressed body sites (Ader, 1996). *Cholecystokinin* (CCK), an anxiogenic peptide first found in the gut, occurs in fibers and terminals of afferents to the PAG. Receptors are found in areas associated with anxiety (amygdala, medial hypothalamus). The blockade of particular receptors reduced anxiety-like behavior in rats exposed to cats, presumably in limbic structures implicated in anxiety. Perhaps CCK blockers have a potential to provide poststressor pharmacological prophylaxis against anxiety disorder post stress (Ademec et al., 1997). Among the *endocrine axes that are inhibited* are reproductive; growth; thyroid; fat muscle and bone metabolism; gastrointestinal function. The immune/inflammatory reaction is also inhibited (Tsigos & Chrousos, 1996).

Brainstem Reticular Formation (RF)

Although it has become customary to consider the RF as regulating arousal in the rostral (cerebrum) and caudal (spinal cord) areas, it has been suggested that this is too general. Rather, the RF should be considered as groups of neurons controlling particular physiological functions. In fact, the cerebral hemispheres and cerebellum control less complex sets of interneurons and premotor neurons in the RF. This produces a finely coordinated discharge of brainstem motoneurons. The brainstem does participate in the awake state, since damage may create impaired alertness, including comatose state. Experimental activation of the cortex via the brainstem involves noradrenalin neurons of the LC, which has efferent connections to the thalamus, cerebral cortex, cerebellum, and spinal cord. Alertness is also associated with serotonin neurons in the raphe nuclei and acetylcholine neurons in the pedunculopontine tegmental nuclei. The LC is stimulated when the person has detected a significant potentially rewarding or threatening stimulus, as well as during some phases of the sleep-wake cycle (Blessing, 2002). Afferents from the cerebral cortex stimulate the hypothalamus, the efferents of which innervate the brainstem reticular formation. This area integrates barometric and chemoreflex controls of the circulation. Efferents are sent to the preganglionic nerve terminals (T1-L2). Preganglionic fibers (cholinergic) stimulate postganglionic fibers in a wide range of vascular territories (α and β receptors). It is presumed that high adrenal activity can initiate the aggravation of post-injury neurologic impairment and cause systemic metabolic derangements (Merisalu et al., 1996).

AMYGDALOID NUCLEAR COMPLEX

The amygdala (Parent, 1996, pp. 773–786) is situated in the dorsomedial portion of the temporal lobe in front of, and partly above the tip of, the inferior horn of the lateral ventricle. It is continuous with the uncus of the parahippocampal gyrus.

SUBCORTICAL AMYGDALOID CONNECTIONS

Olfactory system: Nearly all parts of the amygdaloid nuclear complex receive direct or indirect olfactory pathways, while neurons in the amygdala's nucleus of the olfactory tract and elsewhere send fibers to the olfactory bulb. The amygdala reciprocally connects with the *basal forebrain* (including the basal nucleus of Meynert, which projects cholinergic fibers back to the amygdala). Efferent fibers reach the thalamus, hypothalamus, and brainstem. There, projection to the ventral striatum is one of the most substantial amygdalal efferents. Fibers terminate in the caudate nucleus and putamen. *Amygdalohippocampal* projections are to the rostral entorhinal cortex, hippocampus, and subiculum. There is a less powerful influence by the hippocampus, since *hippocampo-amygdaloid* projections are less prominent. The amygdala projects to the *mediodorsal thalamic nucleus*. This nucleus projects to the region of the orbitofrontal cortex that receives direct amygdalocortical projections, but not back to the amygdala. There is also an amygdaloid projection to the midline thalamic nuclei that reciprocally innervates the amygdala. The bed nucleus of the stria terminalis is a relay for projections from the amygdala to the *hypothalamus*.

Amygdaloneocortical connections: The amygdala receives high stages of sensory processing from the visual, auditory, and somatosensory cortices. It receives an enormous array of convergent sensory information (Parent, 1996, p. 782). It does not receive information from the primary sensory cortices but rather, using vision as an example, input from the inferotemporal cortex. Reciprocal connections from the amygdala to unimodal sensory cortices are more widespread than corticoamygdaloid projections. The amygdala projects to virtually all visually related areas of the temporal and occipital cortex, suggesting that it modulates sensory processing at every stage in the cortical hierarchy. Thus, considering subcortical amygdaloid projections, the amygdala represents an important relay whereby external stimuli active and influence emotions.

Through its cortical projections, the amygdala modulates sensory processing according to affective states. In unanesthetized animals it has been established that stimulation of the amygdala creates an "*arrest*" *reaction* in which all spontaneous ongoing activities cease as the animal assumes an attitude of aroused attention. It appears as the initial phase of flight or defense reactions. Flight (fear) and defense (rage and aggression) have been elicited from different regions of the amygdaloid complex. The intensity of the stimulus determines the magnitude of the response, which guilds gradually and outlasts the period of stimulation. *Long-term potentiation* may be the reason for the long-lasting aspect of this response. Fear and rage reactions are signs of emotional involvement and activity of the autonomic nervous system (pupillary dilatation, piloerection, growing, hissing). Lesional studies suggest that destruction of the ventral amygdalofugal projections interfere with the defense reactions from amygdalal stimulation. In humans, *stimulation of the amygdaloid region* elicits fear, confused states, disturbances of awareness, and amnesia for events taking place during the stimulation. While rage is the most common response to animal amygdaloid stimulation, it is rarely associated with temporal lobe seizures. Bilateral lesions in humans cause a decrease of aggressive and assaultive behavior. Sometimes unilateral lesions are sufficient to bring about improvement.

Visceral and autonomic responses: *Responses may be sympathetic or parasympathetic.* Alterations of respiratory rate, rhythm and amplitude, and also inhibition of respiration. Cardiovascular response involves an increase and decrease in arterial BP and alterations in heart rate. Gastrointestinal motility and secretion may be inhibited or activated. Defecation or micturition may be produced. Observed are piloerection, salivation, pupillary changes, and alterations of body temperature.

Endocrine responses: The amygdala participates with the hypothalamus in controlling hypophysial secretions: release of adrenocorticotropin (ACTH); gonadotrophic hormone; lactogenic responses. Stimulation of arousal and emotion-eliciting areas produce increased adrenocortical output. *Bilateral lesions can cause elevated levels of serum ACTH, presumably due to release of an inhibitory influence.* The corticomedial amygdaloid nuclei, when stimulated, may induce ovulation. In the female, it appears to have estrogen-concentrating neurons that are part of a system extending into hypothalamic and limbic structures. The amygdala participates in secretion of luteinizing and follicle-stimulating hormones.

Integrator of autonomic and visceral functions: The amygdala integrates autonomic and visceral functions through reciprocal connections with the hypothalamus and brainstem visceral nuclei. It is also involved in complex cognitive functions that globally influence emotion and behavior. These functions involve virtually all regions of the cerebral cortex and all sensory modalities. The amygdala represents the interface between hypothalamus and brainstem visceral centers, as well as the cerebral cortex concerned with cognitive functions. Referring to Kluver-Bucy syndrome, bizarre behavior occurs because visual stimuli have no meaning. Impulses generated by sensory input and the cerebral cortex excite visceral and somatic systems that provide the physiologic expression of emotion and goal-directed behavior. Reciprocal connections function to correlate emotional expression and meaningful behavior.

The *ventral amygdalofugal pathway* enters the lateral preoptic area, the hypothalamus, the septal region, and magnocellular part of the dorsomedial nucleus of the thalamus. The *stria terminalis* terminates in the bed nucleus of the stria terminalis and the anterior hypothalamic area. The complexity of interactions of the amygdala is reflected in its multitude of neurotransmitters and neurotransmitter-related molecules (Parent, 1996, p. 775): the inhibitory amino acid transmitter GABA; the monamines serotonin (arising from midbrain raphe nucleus), the presence of which is more massive than noradrenaline (locus ceruleus), dopamine (arising in the ventral tegmentum, and substantia nigra; the excitatory transmitters glutamate and aspartate; acetyl choline (arising in the basal nucleus of Meynert); a variety of neuroactive peptides, including somatostatin, neuropeptide Y (which has an anxiolytic action at the amygdala level), opiate receptors, enkephalin), and vasopressin immunoreactive fibers (likely arising in the magnocellular neurosecretory nuclei of the hypothalamus). Ninety percent of the cholinergic fibers of the basal nucleus project to widespread cortical regions. It may be the single major source of cholinergic innervation of the entire cerebral cortex (Parent, 1996, p. 788).

Amygdaloid fibers originating in the corticomedial nuclear group terminate in the anterior *hypothalamus* (supraoptic and paraventricular nuclei). A more substantial amygdalohypothalamic projection terminates in the ventromedial hypothalamic nucleus. The hypothalamic ventromedial nucleus is innervated by fibers from the amygdalohippocampal transitional area and the subiculum, as well as the other amygdaloid nuclei. The central amygdaloid nucleus arborizes the rostrocaudal extent of the lateral hypothalamus, the tuberomammillary and supramammillary nuclei, and the midbrain tegmentum. On the other hand, hypothalamoamygdaloid projections are not prominent. The *brainstem* and *spinal cord* are innervated by fibers from the central amygdaloid nucleus and the bed nucleus of the stria terminalis (see Figures 3.7 and 3.8). *Efferent fibers* descend through the midbrain, pons, and medulla, with some extending to the spinal cord, distributing collaterals to structures implicated in autonomic control: periaqueductal gray;

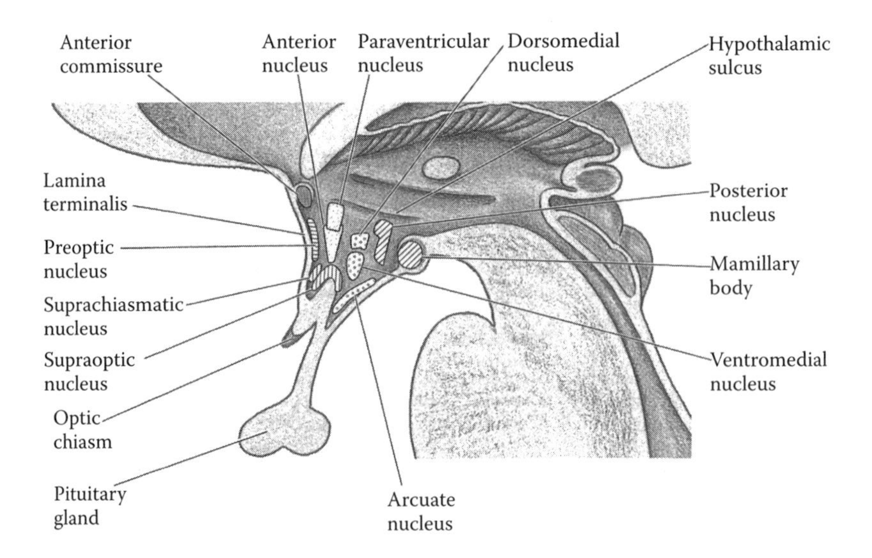

FIGURE 3.7 Inferior aspect of brain showing brain stem in relation to cranial nerves and associated structures. Schematic diagram showing nuclei within each of the regions of the medial hypothalamus. (From Afifi, A.K., & Bergman, R.A., *Functional Neuroanatomy*, McGraw Hill, New York, 1998. With permission.)

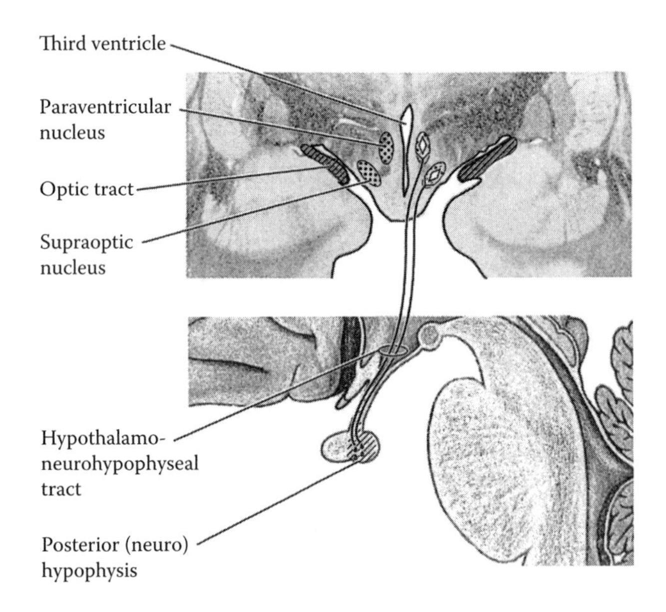

FIGURE 3.8 Schematic diagram showing the hypothalamo-neurohypophyseal system. (From Afifi, A.K., & Bergman, R.A., *Functional Neuroanatomy*, McGraw Hill, New York, 1998. With permission.)

parabrachial nucleus; dorsal nucleus of the vagus nerve; and the reticular formation. Many of these structures reciprocally project to the central nucleus. In the midbrain they reciprocally innervate dopaminergic amygdaloid projections. In the pons, fibers terminate on the parabrachial nuclei; in the medulla, they terminate in the nucleus of the solitary tract and dorsal motor nucleus of the vagus nerve. *Afferent fibers* are received by the central nucleus of the amygdala, and other fibers from adrenergic projections from the lower medulla, noradrenergic projections from the locus ceruleus, and serotonic projections from the midbrain raphe nuclei. These do not appear to have reciprocal amygdalofugal projections, in contrast to mesoamygdaloid dopaminergic projections. The presence of *benzodiazipine/GABA receptors* is consistent with the amygdala's control of emotional behavior (Parent, 1996, p. 788).

Locus ceruleus and norepinephrine: Signals spread throughout the brain and are primarily, but not exclusively, excitatory. Involved in dreaming and REM.

Substantia nigra and dopamine: Inhibitory transmitter to the basal ganglia; excitatory elsewhere.

Raphe nuclei and serotonin: Fibers secrete serotonin in the diencephalon cerebral cortex and spinal cord. Pain suppression at the cord level; an inhibitory role enabling normal sleep in the cerebrum and diencephalon.

Gigantocellular neurons of the reticular excitatory area and acetylcholine: Pass upward to the higher levels of the brain and downwards via the reticulospinal tract to the spinal cord. Activation leads to a wake, excited nervous system.

SPINAL CORD

SENSORY INPUT

Sensory information from the muscles, joints, and skin ensures the correct pattern of muscle activity (i.e., precise or finely coordinated). Sherrington's concept remains influential (i.e., that simple reflexes are the basic units for movement, and complex sequences are produced by combining them). However, reflexes can be modified to adapt to the task, and many coordinated movements can be initiated and maintained in the absence of patterned sensory input (Pearson & Gordon, 2000a).

MOTOR INPUT

The remaining fibers descend in the ipsilateral ventral corticospinal tract via the internal capsule to synapse on motor neurons of the nucleus ambiguus (special visceral motor; medulla). Both arm and leg fibers appear to intermingle at the level of decussation, contrary to earlier beliefs. Perhaps only uncrossed upper-extremity fibers may travel in the uncrossed ventral corticospinal tracts (Santiago & Fessler, 2004). The neurons with which descending fibers synapse are known as *lower motor neurons* (LMNs).

Spinal segments (neuromeres) preserve the primitive arrangement. They have an arrangement of sensory and motor columns similar to the brainstem, where there are motor and sensory roots represented by cranial nerve nuclei. The neuromeres are reflected in somatic segments (metameres), and are subdivided as myotomes, scleratomes, and dermatomes (Parent, 1996, pp. 264–268, 325).

Final common path: Contraction of the skeletal muscle occurs when the nervous impulse is conducted down the axons of the motor neurons. These lie in the ventral horn of the spinal cord and exit the spinal cord via the ventral roots, and continue into the ventral and dorsal rami of the spinal nerves that innervate the muscles of the neck, the trunk, and the extremities. The dorsal rami innervate the muscles of the neck and trunk. The ventral rami innervate the rest of the trunk and the extremities. Detailed descriptions and diagrams of a spinal nerve in its anatomical site are found in Netter (1983, p. 38) and of the thoracic nerves and spinal column within the chest are also found in Netter (1983, p. 140).

The neurons with which descending fibers synapse are LMN. Clinical symptoms depend upon four characteristics of the lesion (Younger, 1999): (1) the level (the higher the lesion the greater the motor, sensory, and autonomic dysfunctions); (2) the extent in the transverse plane (complete or incomplete cord syndromes); (3) the extent in the longitudinal plane (number of spinal segments involved); and (4) the duration and speed of injury.

PERIPHERAL NERVOUS SYSTEM

The motor units extend in a segmental pattern from the cells lying in the anterior gray matter to target muscles in the limbs and trunk. The nerve is formed when motor (anterior) and sensory (posterior) roots combine just distal to the dorsal root ganglion. After a brief intraforaminal course, posterior branches (posterior rami) extend backwards to provide paravertebral muscles. Anterior branches (rami) extend forward supplying the trunk and giving rise to the roots of the plexuses. Tissues surrounding larger structures and eventually the nerve trunk are the perineurium and epineurium. The latter melds into a loose layer of protective tissue called the mesoneurium, which allow some degree of passive movement in the transverse, particularly the longitudinal planes. The connective tissue elements and the specialized perineurium create a blood-nerve barrier maintaining an immunologically privileged nerve microenvironment. It also provides structure, tensile strength, and elasticity to the nerve trunk. After more severe grades of injury, the success or failure of neuroregeneration depends on the nature and degree of an injury to the connective tissue. Regeneration is interfered with by the formation of a neuroma, filling the gap between nerve stumps, as well as replacement of a target muscle by fibrotic tissue in 2–3 years. Sensory reinervation may restore useful function to a limb in up to 7 years. Grading of nerve trauma includes different levels of both nerve trunk and connective tissue. Metabolic disease, including diabetes, can interfere with regeneration.

SPECIAL SENSES

Acquisition of sensory information requires movements of the entire body or its sensory organs or appendages in a manner that orients the sensory organs to its target. The cortical maps devoted to vision, known as somatosensation, are plastic and subject to reorganization (Lackner & DiZio, 2002).

SOMESTHESIS

Dysfunctions include surface features such as texture (e.g., roughness) and spatial features (i.e., discriminating shapes or sizes). Skin receptors respond to the stretching of the skin and thus contribute to proprioception (kinesthesis). Sensation of touch, position, vibration, and so on, are projected to multiple cortical areas (i.e., the contrelateral parietal lobe, primarily the posterior depth of the central sulcus, the postcentral gyrus, the anterior wall of the postcentral gyrus, the supramarginal gyrus, and the Sylvian fissure, and the insula). There are multiple detailed somatotroph maps in the primary somatosensory cortex (SI), such as the postcentral gyrus (areas 3a, 3b, 1, 2 of Brodman; Parent, 1996, p. 882), the secondary somatosensory cortex (SII) in the inferior parietal lobule (Burton, 2002; Hsiao & Vega-Bermudez, 2002), and elsewhere. The intracortical sensory stream originating in SI and SII includes (1) a ventrally directed pathway that passes through the insular and entorhinal cortices before reaching the hippocampus and (2) a caudally directed pathway that projects to the posterior parietal cortex before reaching the prefrontal cortex. Information is received from mechanoreceptors (joint capsules, joint ligaments, and skin) detecting touch, pressure for surfaces, and objects. They signal limb position and movement in the full range of motion, particularly the extremes (Prochazka & Yakovenko, 2002). Peripheral receptors project to the spinal cord (dorsal horn) and brainstem. Uncrossed pathways stem from both limbs and the upper trunk via the dorsal spinocerebellar and cuneocerebellar tracts. The ventral spinocerebellar tract receives sensation from the lower limbs, decussates within the spinal cord, and projects via the superior peduncle to the cerebellum. Some sensation for limb position and touch proceeds in the spinal dorsal column to the cuneate and gracile fasciculae and nuclei, decussates in the medulla, ascends to the thalamus in the medial lemniscus, then projects to the somatosensory cortex, parietal and frontal association areas, the cerebellum, and the motor cortex (Martin, 2003, pp. 112, 308). The *basal ganglia*, which play a major role in control of *voluntary movement*, receive their major input from the cerebral cortex (sensory information and movement), and send their output to the brainstem and, via the thalamus, back to the prefrontal, premotor, and motor cortices (DeLong, 2000). *Epicritic sensations* are the fine aspects or touch, including gentle contact of the skin and localizing the position (*topognosis*) that is touched, vibration, spatial detail and texture of surfaces, and two-point discrimination (Gardner et al., 2000). Somatosensory information from the limbs and trunk is conveyed by the dorsal root ganglia (rendering it vulnerable to spinal column injuries). Somatosensory information from cranial structures (face, lips, oral cavity, conjunctiva, and dura mater) is transmitted by the trigeminal sensory neurons.

Proprioception

Proprioception refers to the sense of one's own body and limbs without using vision: (1) stationary position of the limbs and (2) sense of limb movement (kinesthesia). Information needed for motion is received from muscles, joints, and tendons, somatosensory receptors (touch, pressure, vibration, skin stretch, texture), the vestibular system, distance receptors (vision and hearing), and pain. Axons ascending in the dorsal column from the sacral region are placed at the midline and are placed more laterally at higher levels. These divide into the medial gracile and lateral cuneate fascicles, which terminate in the medullary gracile and cuneate nuclei. Second-order neurons decussate to form the medial lemniscus, shifting laterally to terminate in the ventral posterior lateral nucleus of the thalamus. The proprioceptive and touch neurons project to the primary sensory cortex of the postcentral gyrus (Gardner et al., 2000).

This reminds us of the familiar homunculi for the motor somesthetic strips and body representations in the cerebellum (Goetz & Pappert, 1956, p. 178; Martin, 2003, pp. 304–307). The spinocerebellum is important in the control of body musculature. It receives somatic information from the trunk (vermis) and limbs (intermediate hemisphere). This includes internal feedback signals for correcting inaccurate movements (motor and sensory information). The cerebrocerebellum (lateral hemisphere and dentate nucleus) is involved with planning movements. Afferent and efferent connections are to the contrelateral cerebral cortex. The vestibulocerebellum receives afferents from the vestibular nuclei.

4 Acute Alterations of Consciousness

OVERVIEW

Consciousness is a complex function that has evolved and been selected for its adaptive benefits. It mediates information gathering and responsiveness in both the internal milieu and the external world. With even slight dysfunctions of consciousness, a person's adaptive capacity will be impaired and their quality of life will be reduced. The person will have difficulties in attending to and processing information from both the external and internal world. Safety, decision making, employment, and community functions are inefficient.

WHAT IS CONSCIOUSNESS?

Consciousness is the mental condition of a normal person when awake, which implies responsiveness to stimuli and awareness of the self and environment, and involves interaction with multiple ongoing neurobehavioral activities. However, being awake (i.e., simply having one's eyes open) does not necessarily imply attention or awareness. Some sensorimotor processing that require judgment, input, and output occur without apparent awareness. The collective representation of mental operations, mental imagery, and self-signaling through language can be called consciousness. It has been considered to be the integration of qualitative information processing in human brains (Ommaya & Ommaya, 1997). Consciousness has been defined as

> that state of awareness in an organism that is characterized by maximum capacity to integrate and utilize sensory input and motor output to achieve accurate storage and retrieval of events and actions related to contemporary time and space, coupled with the ability to feel the quality of these events and recall ongoing actions and events, as well as reflecting upon them (Ommaya, 1996; Ommaya & Gennarelli, 1974).

Mood and feelings are part of consciousness; therefore the writer offers: "I feel. Therefore, I am conscious."

ADAPTIVE FUNCTIONS OF CONSCIOUSNESS

Lezak (1989) conceptualizes three aspects of self-awareness: (1) appreciation of one's physical status; (2) one's relationship with the physical environment; and (3) appreciation of oneself as a distinctive person in a social environment. To Kihlstrom and Tobias (1991), influenced by William James, consciousness is linked to the self. Awareness occurs only if there is a link between the mental representation of an event and some mental representation of the self as its agent or experiencer. The professional in the area of assessment and treatment of cerebral dysfunction *utilizes the capacity to have insight and to empathize as a valuable technical tool.* Consciousness and social comprehension participate in a wide range of behaviors that Bradshaw (1997, pp. 158–180) considers socially positive (altruism, empathy, reciprocity) or socially negative (exploitation, deceit, retaliation, manipulation, plot). Awareness of likeness to another contributes to imitative learning that is allied to social facilitation and stimulus enhancement. This increased the likelihood of performing actions already in the early hominid's repertory. Capacity for consciousness also creates social vulnerabilities. Without correctly monitored information acute awareness can create paranoia (i.e., the attribution of devious motives to others).

NEUROLOGICAL ASPECTS OF ALTERED CONSCIOUSNESS

Ascending Reticular Activating System (ARAS)

The *cerebral cortex* exchanges stimulation with the Ascending Reticular Activating System (ARAS), which is part of a network extending from the medulla to the midbrain. One function of the reticular formation (RF) is activation of the brain for behavioral arousal and different levels of awareness (Role & Kelly, 1991). Ascending projections terminate in the dorsal thalamus, hypothalamus, cerebellum, basal forebrain (including the hippocampus) and neocortex (Role & Kelly, 1991). Traumatic brain injury (TBI) causes loss of consciousness (LOC) through energy transmitted in the form of tissue movement to the ascending ARAS, or to projection tracts to thalamus and thence to cortex, which are damaged, destroyed, or temporarily impaired. Lesions of the hypothalamus may produce coma, perhaps by interrupting reticular axons (Martin et al., 1990). In addition, due to impact-caused brain movement, vascular compression and secondary ischemia impair brainstem reticular activating system structures, leading to changes in consciousness.

Arousal stimuli from the RF are relayed to the cortex, basal ganglia, basal forebrain, and other thalamic nuclei via the thalamic diffuse-projection nuclei (Kelly & Dodd, 1991). The reticular core influences multiple brain areas, and receives sensory information from the cortex via the thalamus. Arousal, awareness, and attention are dependent upon this system. Consciousness itself is considered to be a distributed system. Trauma affecting the cortex and subcortical white matter affects consciousness by suppressing reticular cells through lack of input (Goetz, 2004). The somatosensory-evoked response (SER) is divided into a P1 component (lemniscal elements) and a P2 component (cortically recorded signal). Abolition of P2 coincided with onset of paralytic coma, and its return with the restoration of animals' responsiveness and motor performance. P2 was always preserved in the translation injury, and no concussion was in evidence. Latency was increased for interhemispheric cortical-cortical transfer only in the rotated, but not in the translated group, persisting long after the return of P2-indicated adequate conduction through the RF (Ommaya & Gennarelli, 1974).

Shearing effects reaching the well-protected mesencephalic part of the brainstem is one source of traumatic unconsciousness. The effects are considered to begin at the brain surface in mild cases and extend inward to the diencephalic-mesencephalic core at the most severe levels of trauma (Ommaya & Gennarelli, 1974). This is the area where dysfunction is usually considered to be the prime contributor to LOC after brain impact. Cells responsible for cerebral activation are found only in the rostral portion of the brainstem. However, activity of the RF alone does not account for variations of consciousness. The rostral brainstem contains neurons required for wakefulness, while the caudal brainstem contains neurons necessary for sleep (Kelly, 1991a, 1991b).

Diffuse Injuries

Diffuse injuries cause widespread dysfunction of both cerebral hemispheres and disconnect the diencephalons- or brainstem-activating centers from hemispheric activity. Contusions are not a cause of LOC at the time of injury per se, although they can be linked to focal seizures or specific functional deficits when found adjacent to eloquent areas (Povlishock & Christman, 1994). Transient alterations in the neurotransmitter systems may contribute to transient coma and other reversible biochemical dysfunctions in the neuron (Miller, 1989).

Brainstem Movement, LOC, and Apnea

Impact can move the brain into the foramen magnum (Kuijpers et al., 1995). The brainstem can move as much as several centimeters, which is responsible for the LOC. The brain rotates around its own long axis (brainstem tethered to the spinal column as it exits the foramen magnum), causing compression of fibers, cells, and blood vessels, and leading to dysfunction of consciousness, including coma. Rotational injuries are more likely to cause suppression of behavioral responses to stimulation (so-called concussion) than translational acceleration. This accounts for LOC and severe intellectual deficits that are attributable to multiple diffuse shearing lesions deep within the brain set up by rotational mechanisms (Pang, 1989). Contralateral head rotation with hyperextension may not create symptoms until hours or days after the injury, including aphasia, altered consciousness, seizures, motor or sensory disturbance (monoparesis, hemiparesis), and sometimes slight drowsiness, retrograde amnesia (RA), confusion, and Horner syndrome (due to disruption of sympathetic fibers in the carotid wall). With no evidence of neck injury, carotid artery as a site of trauma may be ignored (Davis & Zimmerman, 1983).

A blow along the bodily axis (axial) causes mass motions of the brain toward the junction with the spinal cord, resulting in involvement of the midbrain and other subcortical structures.

- Lesions above the lower third of the pons must destroy the paramedial reticulum bilaterally to interrupt consciousness.
- Lesions located between the lower third of the pons and posterior diencephalon, and that are either acute or large, produce stupor or coma if bilateral. Length of coma is not statistically associated with lateralization of mass lesion (Levin & Eisenberg, 1984).
- Lesions below the lower third of the pons do not cause unconsciousness (Walton, 1985, citing Plum, 1972, 1980).

Cholinopontine Inhibitory Area (Cholinergic Pontine Sites): Cholinergic centers have a particular role in the maintenance of consciousness. Enhancement of cholinergic transmission with physostigmine accelerates the recovery of level global and hemispheric CBF (cerebral blood flow), and also the contralateral CBF (Scremin et al., 1997). A less familiar mechanism involved in unconsciousness is the pontomesencephalic brainstem lesion, a muscarinic brainstem system ventromedial to the locus ceruleus, demonstrated in the rat and cat. Microinjections of the cholinergic agonist, carbachol, into an area ventromedial to the principle nucleus of the locus coeruleus produces suppression of postural somatomotor and sympathetic visceromotor functions, as well as profound unresponsiveness to external stimuli (Hayes et al., 1984, 1992; Katayama et al., 1984, 1985; Lyeth et al., 1988). Noting the existence of *a muscarinic brainstem system* whose activation produces components of reflex inhibition and behavioral suppression, it is inferred that the mechanisms mediating traumatic unconsciousness are likely to be distinct from those mediating enduring behavioral deficits (Hayes & Dixon, 1994).

This system can be organized to regulate reactions to events in the external environment, allowing expression of integrated behaviors (Katayama et al., 1984). Active inhibitory mechanisms in the brainstem modulate sensory input and/or motor output in response to changing environmental events or vegetative states, including noxious sensory input (Lyeth et al., 1988). Generalized cholinergic release contributes to convulsive seizures associated with death, at least in rats (Lyeth et al., 1988). There appears to be an initial nonspecific period of brain disorganization characterized by generalized areflexia with muscle hypertonia (Lyeth et al., 1984). Subsequently, there are active cholinergic inhibitory processes that create behavioral suppression and reversible LOC following low levels of concussive brain injury. In the cat, it was demonstrated that a low level of concussive injury was associated with increased local glucose utilization (Hayes et al., 1984).

Reduced responsiveness is effected through (1) ascending (cranial nerve) and (2) descending (medullary brainstem spinal somatic and visceral motoneurons and dorsal horn cells).

- *Postural somatomotor*: complete loss of muscle tone; abolition of flexion; righting; and placing reflexes. Some cells may participate in normal postural atonia during desynchronized sleep.
- *Visceromotor functions*: reduced sympathetic tone (miotic pupils; reduced blood pressure and heart rate; failure of external stimuli to produce respiratory and heart rate changes). This may involve suppression of spinal cord sympathetic outflow independent of cranial nerve effects.
- Nociceptive somatosensory: heat and pressure.

Electrophysiological aspects: The electrophysiological characteristics of concussion are variable. Yet, it has been proposed that concussion is a consequence of such functions as paralysis of neuronal and reflex function, or traumatic depolarization of nerve membranes, producing a massive discharge or excitation analogous to an epileptic seizure (Shetter & Demakas, 1979). The initial 10 sec after concussion do not show excitatory or convulsive Electroencephalography (EEG) activity. Subsequently, there is a reversible depression in electrical amplitude and frequency, particularly in the medial RF. Sciatic stimuli that evoke responses in the medical RF are temporarily abolished, although not those in the medial lemniscus (Shetter & Demakas, 1979).

Somatosensory evoked potentials (SEP) in an experimental study that utilized monkeys suggested that unconsciousness occurred and disappeared inversely with conduction through the mesencephalic reticular (alerting) system (Ommaya & Gennarelli, 1974). More severe cases produced irreversible effects with neurological and behavioral deficits such as the persistent vegetative state (Ommaya & Gennarelli, 1974). That damage can be produced independently of LOC, and the association of deeper lesions with concussion caused only with rotation supports a hypothesis that a cerebral concussion that is sufficient to produce paralytic coma requires shear strains involving the cerebral cortex and deeper structures. Specifically, brainstem involvement seems to be required.

A significant disturbance of consciousness is associated with depressed cerebral metabolism, a loss of integration of responses in the high-frequency EEG range, partially functional sensory circuits even in the vegetative state, and preservation of language responsive networks in the context of preserved thalamocortical network preservation. Changes in the chemical properties of the cell may imply a change in frequency properties, accounting for a hypersynchrony that interferes with the cell assembly. This becomes a pacemaker, removing transmission units from the cell assembly, reducing responsiveness, and interfering with complex intellectual activities. Potential mechanisms include diaschisis (crossed synaptic downregulation), restricted hypersynchronous discharges, and damage to specific pathways due to alteration of subcellular processes without neuronal death. Although data was abstracted from patients with severe brain injury, it gives insight into some of the disruptions occurring in lesser injury (e.g., concussion) (Schiff, 2006, citing Hebb, 1949).

ALTERATIONS OF CONSCIOUSNESS

LOC-defining mild head injury has been variously asserted from merely being dazed to not exceeding 6 hours' LOC. Twenty minutes' maximum LOC seems preferable to this writer. In fact, *if there are no witnesses, and the patient is the source, it is impossible to differentiate posttraumatic amnesia (PTA) from actual LOC*. Durations of LOC as short as 3–5 min can lead to identifiable structural brain damage (Teasdale & Mathew, 1996). Furthermore, there may be a lucid interval until vascular or other complications cause the onset of PTA (Corkin et al., 1987).

LUCID INTERVAL

While LOC or its alterations are part of the definition of a concussion, it must be remembered that not all brain trauma is accompanied by LOC. It is noteworthy that some individuals who ultimately die may be able to talk until pathological processes become excessive (i.e., they experience a "*lucid interval*")

(Adams et al., 1985). This "*talk and die*" phenomenon may be due to an expanding intracranial hematoma following an absent or brief LOC. It may complicate a brain injury of any severity (Gennerelli & Graham, 2005). The lucid interval is usually absent in diffuse axonal injury (see below).

Confusion of LOC and PTA

Clouded consciousness and confusion can cause PTA to be commonly misinterpreted by the patient to be LOC. The accident victim may be responsive to questions (e.g., "does it hurt?"; "move your arm") without being sufficiently conscious to lay down memories. An altered state of consciousness (feeling dazed; "seeing stars") may be followed by an extended period of disorientation ("I don't feel like the same person. I don't feel oriented. I feel like I'm outside of myself, watching myself, dizzy, difficulties with direction [getting to places]").

A "mild" TBI causes acute disruption of brain functioning. Initially, the patient is dazed, confused, and temporarily disoriented, and often has memory gaps for the injury for a time afterwards of seconds to days (PTA, [i.e., posttraumatic stress disorder, PTSD]), or loss of memory for the period immediately preceding the impact (anterograde amnesia). The initial cognitive difficulties are common but generally short-lived, and include cognitive slowing, poor concentration, and attention difficulties. Decreased reaction time has been documented in individuals with a lesser grade of concussion (no LOC with symptoms lasting up to 24 h, and up to 35 days following emergency care for an uncomplicated mild traumatic brain injury [mTBI]). Another cognitive consequence is *proactive amnesia* (i.e., reduced ability to acquire new related information). This may be a consequence of defective inhibitory attentional mechanisms, perhaps cause by lesions of the frontopolar cortex. PI may be responsible for working memory difficulties of aging. When effect size is considered after mTBI is separated from moderate and severe cases, it is small and reduced with an increasing postinjury interval of up to more than 89 days (Vanderploeg et al., 2005).

ACUTE STATUS AFTER ACCIDENTS

Acute Neurological Assessment

This section focuses upon the conscious patient who does not require emergency care. A normal skull X-ray does not obviate the need for a computed tomography (CT) scan. Thus, a detailed, careful, and repeated neurological examination is needed for all patients with a minor head injury (Marion et al., 2004; Simpson, 2005; Wilberger, 2000). CT is utilized if there is reason to believe that there is a lesion of neurosurgical significance. While some assert that the level of consciousness is the most reliable empirical measure of impaired cerebral function after a closed head injury, the examiner also considers the interval after the impact, other potential causes of impaired consciousness (e.g., alcohol, hypoxia, drugs, hypotension), and the health before injury. Later, the duration of LOC is considered as a measure of cerebral injury.

Head palpation: Apart from emergency medical procedures, the head should be palpated for fractures, lacerations, or penetrating wounds (Marion et al., 2004). Orbital swelling may imply an anterior fossa fracture. Bruising behind the ear may indicate a fractured petrous bone. Blood or CSF (cerebrospinal fluid) leakage from the nose or ear should be noted as a sign of possible basal skull fracture, and a sample of fluid sent for immunological analysis for Beta-2 transferrin as a marker for CSF (Simpson, 2005). By head palpation, the writer has more than once detected tender spots and elevations in patients whose heads seem never to have been examined by any health care provider for years after an impact, according to the combined evidence of records and patient report.

Alterations of consciousness symptoms are physiological, psychological reactions, partial seizure phenomena and hyperarousal, and also variations in the intensity of experience, ranging from gaps in awareness, to total absorption in activities (one is not aware of what is happening) and imaginative involvement (uncertainty as to whether something has occurred in reality as opposed to having dreamed it). The latter are items from dissociative experiences scale (DES). The frequency of items on DES varies in the population (cited in Alper et al., 1997).

Neurological signs: LOC for only a few minutes, but without coma. Temporary disturbance of neurological function without LOC (e.g., loss of eye opening; unsteady gait; abolition of the corneal reflex [both pupils dilated and unresponsive to light, for varying periods]).

Altered consciousness: The accident victim may not be stunned or dazed, but later will complain of headaches or difficulty concentrating. Minimal PTA.

Clouded consciousness: Slow waking or drowsiness after LOC. Stunned, dazed; short-term confusion; dazed appearance. Athletes complain of "having their bell rung." Sensoriums clear quickly, usually in less than a minute.

Systemic

No physiological or behavioral abnormalities;

Immediate increase of systemic arterial pressure; bradycardia;

Respiratory irregularity; with LOC there is an apneic period, or from irregular gasping to apnea of increasing duration, to permanent respiratory arrest.

SOURCES OF DIAGNOSTIC AMBIGUITY WITH NONSURGICAL ("MINOR") TRAUMATIC BRAIN INJURY

Contrary to expectations, in an experimental study, the induction of cold pain led to a small but significantly greater report of pain and not greater pain tolerance. It is possible that the pain reduced the potency of the dissociative manipulation, which was not applied at the magnitude comparable to what occurs during a traumatic event (Horowitz & Telch, 2007). The level of consciousness may be reduced (hypoarousal or enhanced hyperarousal), and both of these conditions may be trauma related. They alert the examiner to trauma and to possible disease or disorder (neurological, physiological, and psychological). This finding may explain why other neurobehavioral functions are impaired. The issue of the association between lesser TBI and LOC is complex. Brain trauma can occur without LOC, and significant impairment may occur with an accident creating only brief LOC (penetrating injury, whiplash, shaken child, boxing and other sports injuries, etc.). Altered consciousness may occur for only a momentary interval after a head injury or there can be disorders that continue for extended periods, or even originate much later, such as posttraumatic epilepsy or enhanced risk for Alzheimer disease.

The Glasgow Coma Scale: Assessment of the patient's arousal and responsiveness is usually performed at the accident scene and emergency department (ED) with the Glasgow Coma Scale (GCS), a cumulated series of 15 responses in 3 areas: speech (oriented, conversing), eye opening (spontaneous), and motor (obeys verbal commands). The GCS is recommended as part of a neurological evaluation, but should be supplemented with a mental status examination, pupillary assessment, and observation for motor asymmetry. The range of 13–15 is described as "mild" TBI. Nevertheless, even patients with the maximum score of 15 can be lethargic or manifest memory loss—a sign of possible intracranial injury. These subtle abnormalities are a marker for severe TBI. With GCS down to 13, the focal neurological deficit or abnormal mental status was more significant as evidence of intracranial pathology than the presence or absence of a skull fracture. Such patients may display CT abnormalities and require operative intervention. It has been recommended that a GCS score of 13 should be considered a "moderate" rather than a "low" level of injury.

TRAUMATIC STRESS AFTER AMNESIA AND UNCONSCIOUSNESS

The question of diagnosis has been raised concerning amnesia for the event after a head injury (i.e., if there is amnesia then PTSD is by definition impossible). PTSD's "neuropsychological" symptoms have been identified even when a head injury is not involved. It can occur even with amnesia for the event (Knight, 1997). Baggaley and Rose (1990) offer the case history of a 22-year-old soldier who was injured in an explosion. The patient suffered significant closed head injury, PTA for 36 h and RA of 8 h, and developed PTSD.

The period of stress can begin after a trauma (e.g., when a person awakens after an accident in a hospital bed, not knowing who they are, or where they are, or what happened, or why the tubes are coming out of one's body, why there is pain, or even why there has been a loss of limb). There is a feeling of loss of control, identity and memory (Kohl, 1984). The events of the recovery period, as well as personal or social reconstruction of the event, seem to suffice to create the symptom pattern even in the absence of memory for the defining unusual life-threatening events. Referring to the controversy as to whether amnesia (or LOC) precludes the possibility of PTSD, Watson (1990) cites two cases with patchy awareness of the accident and in the ED. PTSD in the sense of reexperiencing the event was limited to reexperiencing specific pain and other symptoms. In addition, it can be the experience of altered consciousness prior to LOC. One patient was in an accident in which an elevator had a free fall for 50 feet, his legs were pushed into his abdomen, and he was in a coma 2 week for 2 weeks.

Clinical Examples: Waking Up Confused in a Hospital: The period of stress can begin when a person awakens from an accident in a hospital bed, not knowing who they are, where they are, what happened, why the tubes are coming out of one's body, why there is pain, or even why there has been a loss of limb. There is a feeling of loss of control, identity, and memory (Kohl, 1984). This is a significant issue in the controversy of whether LOC precludes the diagnosis of concussive brain injury.

A woman was unconscious for 3 days after a motor vehicle accident (MVA): "I kept hearing voices. I wanted to tell my mother that I was all right but I couldn't. I couldn't figure out where I was. I had the respirator on, and a neck brace. Scary and confusing. When I first woke up I didn't know what happened. She had no pain on awakening. My mother kept on repeat over and over that there was a car accident and I was in a coma." The woman remained in the hospital for 15 days.

A youth drifted in and out of consciousness. He woke up at 6–7 at night: "I saw my friends. I said my head hurts. I said 'hi' to my friends, and then I went back to sleep. It was as though I was dreaming. [After waking up] I didn't know what I was doing there. I felt scared. I felt bad. I didn't know what happened to me. I had a fever."

How long before you felt like yourself?: A week after he woke up he felt all right. "I was still in a daze. I felt weird when I left the hospital. The sun and the street felt as though I was never outside before."

Clinical example: Extended period of altered consciousness: This case illustrates an extended period of altered consciousness, following a high-speed MVA without LOC. It also illustrates an approach to interviewing to obtain information concerning the trauma.

The patient was very active at the accident scene and was unaware that she had a significant cognitive disorder until she returned to her job, which required high skills, attention to detail, and communication of technical information. The patient was a middle-aged woman who was in a MVA. Her consulting job in business development required high computer skills. She had a bachelor's degree with graduate credits. She had been in a prior MVA.

Describe the accident that we are studying: She was driving within the speed limit of 45 mph. A car turned in front of her and her vehicle struck the car from the side. It was a head-on collision. The impact was sufficient to make the other car roll over at least once. She was seatbelted: "I remember looking up at the traffic signal and asking 'what happened?' The light was yellow. The air bag did not go off. When I slammed on the brakes my body went forward. I was in a big car, an SUV. My head hit the top of the windshield." She struck the anterior vertex, and also injured her neck and left shoulder.

"2 weeks later I went to physical therapy. I told her about the pain in my lower back. She didn't report this to Dr. H. until weeks later. The PT refused to treat it without a prescription."

What was your last memory before the accident?: She was not unconscious. She remembers driving before the accident.

Do you remember the moment in which you were injured?: She thinks that she remembers the impact. "I jumped out of the car and ran to help the people in the car. They were moaning and groaning. I went back to my car."

Were you dazed or unconscious? For how long?: "I felt dazed, felt unreal, it was not happening. The paramedicals wanted me to go to the hospital by ambulance. I refused. I had a friend pick me up and I went to my friend's house."

If you were in a car, were you wearing a set belt?: "Yes."

Were there any bruises or cuts on your head?: **"**Yes. The center of my forehead, it was raised, and red. Where the hairline begins on the center of my head there was an abrasion that went back 3 inches on the top of my head."

Did you have pain anywhere on your body?: "At the time, my neck and shoulder were sore, and got progressively worse. Later sacroiliac pain, radiating down to her left foot."

Did you go to the hospital? If not, when did you get a medical examination?: "3 to 4 h later. I started feeling nauseated, dazed, dizzy."

Do you feel like the same or a different person since the accident?: "Different. I don't think my mind works the same. I have trouble retrieving the right word that I want to use."

Did you ever return to work after you were injured? How much time did you take off?: She immediately attempted to work at home, but had doubts about its value. The accident occurred on a Saturday and she attempted to work at home the following Monday.

Did you have any problems returning to school or work?: "I don't remember a lot about Monday. I don't remember whether I went into the company. Possibly over the telephone, a colleague became aware that I was forgetting repeatedly the name of somebody; I saw that I had a head injury over the weekend. I tried to work at home to Thursday, I realized that my mind wasn't working right, I wasn't thinking right, and I had an appointment with the neurologist the following Monday. I was very tired, I felt like I was dazed, like in a fog. I just knew that I wasn't thinking. I did go to a meeting on Friday, I was late because I couldn't find my car keys. I got to a meeting; I was there with a colleague. There was a Conference call with a client. I was taking notes on a computer. At the end, when I distributed the notes for everyone to review, I was told that I missed a major point. It was very rare that I would miss a point." (Another physician) diagnosed me with a postconcussive syndrome; we agreed that after 6 months I would have neuropsychological examination."

Lucid Interval

A lucid interval is defined as whether a patient talked between the time of injury and death (Graham et al., 1989). Absence of an immediate LOC, or when a patient is unconscious and then regains consciousness, is associated with hemorrhage. This may occur after severe injury even a week later. Another example offered is a child who has fallen or suffered a blow to the head and is only momentarily unconscious (Adams & Victor, 1989, pp. 704–707).

VARYING LEVELS OF LOC

The concussive occurs with varying levels of LOC. This may be consequent to the *geometry of the accident*. Whether impact and acceleration of the head and neck affect those central nervous system (CNS) centers and tracts that maintain awareness depends upon the geometry of the impacting structures, their speed, and mutual direction. The clinician will be aided by *reconstructing the accident* (i.e., interaction of head and body position with the physical environment): physical forces (the speed and acceleration of the person and injuring objects or surfaces); the impacts of the head and body (number, direction, and force involved); character of impacting surfaces (hardness, edges, points); speed of medical attention, etc.

Vignettes: Extended Alterations of Consciousness

A man was in an accident in which an elevator had a free fall for 50 feet. His legs were pushed into his abdomen and he was in a coma for 2 weeks. Interviewing indicated the experience of altered consciousness prior to LOC.

Memory for events after the accident: What do you remember about what happened when you woke up? It's very difficult for me to say what was the first thing that I remembered when I woke up because I do not know if it is real or not. Two weeks after the accident they transferred me from Brooklyn to Manhattan to a different hospital. I remember every detail. But in reality it didn't happen. They didn't have policemen in the front of the ambulance, it wasn't at night, it wasn't over the bridges. They stopped the traffic for me, I was on a stretcher. I had a lot of pain. There were details about his wife in another ambulance. (He created a story with a lot of "beautiful or ugly pictures"). In reality these things did not happen.

One month later he remembered:

"When I woke up I was covered and I did not feel my body. I just remember that my left hand was very heavy. I was looking at the nurse. How many pairs of hands and feet they had because this way I would understand what I have and what I am missing. Whether I have one hand or three. After I was watching the nurses I saw that they had two hands, two feet, one head, and five fingers on each hand. It was very important. The fingers were very important. I didn't know how many I had to have. I remember I was counting again and again the fingers. I couldn't concentrate. That was my biggest problem. It was such a big job to build a mountain to come to the conclusion how many fingers I had to have. For a long time I couldn't feel comfortable with the answer. The biggest problem was that I couldn't move my body at all. I was very happy that the nurses came into eye range so that I could reevaluate them again."

He did not lose his use of language. "People came to me, I don't remember that they came to me, but I spoke their languages. In my opinion I am just a good as before." Approximately a month later he felt very clear.

A 62-year-old man was struck by a car, suffered left frontal contusions and a skull fracture, and was comatose for 5 days. He doesn't remember being struck, but did not exhibit RA. When he woke up, he didn't know who the people were around him, or why he was in the hospital, although he knew where he was. Twelve days elapsed before he knew who he was and who all the people were. Two months later he tried to make sense of his life. He thought he was supposed to go on a trip and would say: "I'm leaving tomorrow." He mixed up his family name and first name.

The issue of LOC and memory for the accident is exceedingly complex.

- The level of consciousness in the acute state is not a precise guide for the severity of a brain injury.
- Brain trauma can occur without LOC (bullet penetration; body impact without pressure waves or brain movement along particular axes).
- Significant impairment may occur after an accident creating only brief LOC.
- Alterations of consciousness can persist for years, only to clear up unaccountably.
- Extradural hematoma may occur after low-energy injuries with no LOC or only transient alteration.

Phases of Recovery of Consciousness

Povlishock (2005) categorizes recovery of consciousness as follows: unconsciousness, occurring immediately postinjury without lucid interval; emerging consciousness and confusion with anterograde amnesia; postconfusional restoration of cognitive function. The progression is from deficits of arousal and consciousness, to increasing attention with anterograde amnesia, the increasing efficiency of attention, memory, executive functioning, processing speed, insight, and social awareness.

Behavioral components of altered consciousness include changed arousal, attention, motivation, sensory input, motor output, and cognitive and affective function (Harris & Berger, 1991; Hayes & Ellison, 1989).

POSTCONCUSSIVE ALTERATIONS OF CONSCIOUSNESS

When a mechanical head injury occurs with alterations of consciousness, though not necessarily unconsciousness, it is evidence for at least temporary disruption of brain functioning. Many individuals experience subsequent long-lasting neurobehavioral disruptions. Clinical experience suggests that many statements from the patient or the records that indicate an LOC are incorrect (i.e., it is likely that there was a profound PTA for a period). Alterations of consciousness may be accompanied by *social dysfunctioning*. Social behavior creates capacity for initiation of predictive strategies, detaching the observer from dependence on current inputs by linking them to events that are distant in time or space, allowing flexible rather than automatic processing, and facilitating social predictions through inferring the cognition of another being (Weiskrantz, 1992). Bisiach (1992) noted that our inferences concerning someone else (including a patient being studied) assume that we are ascribing behavior to someone that could be ascribed to oneself.

DESCRIPTORS OF CONSCIOUSNESS

Arousal describes the degree to which the individual appears able to interact with the internal and external environments (Bleck, 2003). *Awareness* reflects the depth and content of the aroused statement; it is dependent on arousal otherwise one lacks awareness. *Attention* implies the ability to respond to particular types of stimuli or to some motivated, ongoing activity. *Unconscious* refers to absence of awareness. *Coma* is a profound state of unconsciousness characterized by little or no response to stimuli, absence of reflexes, and suspension of voluntary activity.

POSTTRAUMATIC AMNESIA

AMNESIA AS A MARKER FOR ALTERED CONSCIOUSNESS

Amnesia is a general term referring to loss of memory. Deficits may occur in the process of sensory or imaginal acquisition, retention, or retrieval. While memory loss is considered to be a characteristic consequence of brain damage, the fact that brain damage is frequently consequent to head injury or other traumatic event opens the question of psychogenic origins of memory disorder ("functional"; "psychological"). The examiner has to consider whether such events as impaired long-term memory, RA, anterograde amnesia, and loss of working memory may be attributable to mental dissociative conditions rather than impaired brain structure or function. Other conditions to consider include repression; state dependent learning with retrieval of memory traces dependent upon limbic and amygdala circuits that add bodily information to incoming events; psychotic states; depression; the use of alcohol; and the use of minor tranquilizers. Even within PTSD, symptoms can vary between intrusive memories (recollections, dreams, flashbacks) and amnesia. Therefore, the separation of causative factors between "organic" and "functional" can be very difficult (Mace & Trimble, 1991).

POSTTRAUMATIC AMNESIA

Posttraumatic amnesia is a state of altered consciousness of varying intensity, with deficits of memory. It is frequently comorbid with disorientation or confusion, and dysfunction of arousal, attention, mood, and behavior. There may be "islands of memory" with or without intense TBI. The patient appears to be alert, but will remember little or nothing of his or her experiences after a blow to the head or other trauma. It may follow coma or unconsciousness, but in estimating the level of TBI, these conditions should be separately specified. PTA should be differentiated from disorientation: a person may correctly answer questions about people, place, and time without recollecting that he or she has been asked. For adults, the resolution of PTA may be considered to be accurately reporting what has happened, the circumstances of their hospitalization, and stating such items as month, year, and the day of the month and of the week (Arffa, 1998).

PTA is measured by the time between receiving a head injury and the resumption of normal, continuous memory. It is common for head injury victims to remember posttraumatic events (e.g., details of an accident scene, or transport to a hospital) followed by amnesia for an extended period after arriving at the hospital. PTA is a diagnostically confusing phenomenon since the patient appears normal, even during a screening examination, concealing that brain trauma has occurred and leading the clinician not to study the mental state more carefully. There may be a lucid interval until vascular or other complications cause the onset of PTA (Corkin et al., 1987). PTA does include any period of LOC or coma. There is a low correlation (+.3) between length of coma and PTA, with more variability concerning outcome when PTA is two weeks or less (Wilson, 1991).

Selhorst (1989) differentiates PTA from impaired consciousness, since amnesia is always longer, while the patients are fully conscious and responsive to the environment in a normal fashion. PTA may be contrasted with coma (i.e., LOC characterized by lack of behavioral function). It is a diagnostically misleading phenomenon since the patient appears alert, concealing the current brain dysfunctioning, but will remember nothing of his or her experiences. Indeed, there may be a lucid interval until vascular or other complications cause the onset of PTA (Corkin et al., 1987). PTA refers to the time from injury until the return of a full ongoing memory process. Length of PTA has been considered as a measure of severity of TBI. The original classification by Russell and Smith (cited by King et al., 1997) offered mild head injury as less than 1-h PTA, moderate head injury 1–24 h, severe head injury as 1–7 days, and very severe head injury as more than 7 days.

Anterograde amnesia (*AA*) refers to varied difficulties in laying down new memories after the accident. Some examples include rote memory (i.e., remembering something precisely) and procedural memory for learning and remembering a general function (i.e., how to do something). Here, storage and retrieval of data refers to memory. Learning refers to general skills, not, as sometimes used by other authors, as storage of information. Problem solving differs from memory insofar as it involves manipulation of information, not the retrieval of some data already in place.

Anterograde amnesia may be subdivided into two types: (1) posttraumatic (anterograde) amnesia (immediately subsequent to a blow or traumatic coma) and (2) chronic anterograde amnesia. After a period of recovery, and resolution of traumatic disorders of consciousness, continued (perhaps permanent) memory impairment is called anterograde amnesia. The length of AA is influenced by focal brain lesions, extracranial injuries, family briefing, and the age of the person (AA estimates in children under the age of 8–10 is unreliable) (Simpson, 2005).

Retrograde Amnesia (RA): RA refers to the loss of memory for personal experiences and other learned materials prior to LOC. It may have either a neurophysiological or psychogenic contribution. The hippocampus, which participates in locating experiences in space and time, can be suppressed after severe stress with its high corticosteroid levels.

It is my impression that RA will be observed in 10% of chronic, mild traumatic brain damaged patients. Ribot's Law summarizes retrieval by asserting that susceptibility of memories to disruption are inversely related to their age, and that prior repeated retrieval of memories (more for older ones) increases their resistance to decay. Partial RA may also occur, and its level is higher during the period of PTA (Levin et al., 1992). RA can also be subdivided into two types: (1) acute pretraumatic retrograde amnesia (loss of memory for events immediately prior to an accident) (this appears rather rarely) and (2) long-term retrograde amnesia (which accompanies various medical conditions as well as (though infrequently) trauma.

Example of combined retrograde and anterograde amnesia after being struck by automobile: Patient awake upon arrival, fully oriented.

Were you dazed or unconscious?: She asserts that she was unconscious about 15 min to one half hour. She still doesn't feel like herself. She believes that her looks are different. She is going through a lot of pain: "I used to be organized in my mind and outside. Now, my memory is so bad. I blank out and ask: 'What are you talking about.'" She remembers being struck, before that being in a shopping center (acute pretraumatic RA). Subsequent to the accident, she was told about certain events that she does not remember. Her memory returned in the operating room. She remembers somebody in blue woke her up, and then she was in the operating room: "They wanted to open a hole in my throat but I told them 'no' " (acute anterograde amnesia).

Acute Pretraumatic Retrograde Amnesia: When it does occur, recovery may progress with return of memory for all but the few minutes prior to injury. In a military sample of penetrating and non-penetrating injuries (excluding severe injuries), it lasted for seconds or minutes (Corkin et al., 1987). This is in accord with the writer's civilian experience (i.e., that RA is rarely extensive). Weinstein (1991) suggests that extensive RA may have a meaningful (psychological) component. Even though an accident victim does not remember the impact, the nonamnesiac patient can offer information based upon what he or she has been told. Denial and confabulation can play a role through symbolic representation of current deficits or misrepresentations of the last preinjury memory.

Clinical Example: PTA in an adult male: A man remembers being injured and is not sure if he was unconscious. Subsequently, there were gaps in his memory or perhaps intermittent periods of unconsciousness: "People say they saw me but I don't remember seeing them." He remembers being carried off, not the ambulance ride, then being examined in the hospital room. He didn't feel like himself when he woke up in the intensive care unit, and didn't realize what happened until his father arrived.

Problems of estimating length of PTA: While the length of PTA has been offered as a measure of cerebral damage, estimating it is very imprecise, since one must consider LOC, altered consciousness, PTA itself, and dissociation as a protective psychological mechanism. It is difficult to differentiate short traumatic impairment of consciousness from dissociative reactions or very short PTA (Radanov et al., 1992). Moreover, PTA can be overestimated due to periods of natural sleep or impaired consciousness due to medication, alcohol, or drugs. A patient may be oriented in time and place but not remember that questions were asked. Information from the patient is vital for estimating LOC or PTA. The accident victim may be responsive to questions ("does it hurt?"; "move your arm") without being sufficiently conscious to lay down memories. The observer and recorder at the accident scene or emergency room may be too preoccupied with somatic trauma to inquire into alterations of consciousness. Minor lapses of consciousness may not be attributed importance by the accident victim or emergency squad (e.g., feeling dazed; "seeing stars"). This may be followed by an extended period of disorientation ("I don't feel like the same person. I don't feel oriented. I feel like I'm outside of myself, watching myself, dizzy, difficulties with direction [getting to places]").

PTA can occur after head trauma without the patient being comatose; the patient may be responsive to the environment without remembering what is going on. While the patient seems to be totally awake and responsive, he or she is actually confused, disoriented, and lacking the ability to acquire and retrieve new information. Measurement or estimate of the interval of PTA is essentially vague, being dependent upon the patient's self-estimate of recovery of memory. For some individuals, PTA terminates with a memory event (e.g., ambulance ride; arrival at the hospital) during the period of injury or later. For others, recovery may be slow. Previously, the period of coma was included in PTA, but more recently, attention is directed specifically to confusion, disorientation, and amnesia. One may test capacity for recent memory through storage and retrieval of recently presented words or visual items (Selhorst, 1991). Concurrent measurement of the length of PTA may be measured more reliably than retrospectively, particularly with shorter intervals between testing (King et al., 1997).

While PTA is commonly considered to be a measure of the severity of brain injury, the length of PTA cannot be measured with precision. Different intervals could be determined by retrospective or prospective assessment (Forrester et al., 1994). Imprecision in using the retrospective technique is enhanced by unreliable recollections; confabulations; situational, cognitive, and emotional context at the time of recall; confusing events of a prior head injury with the current one; briefing by family or others; or inebriation at the time of injury (Forrester et al., 1994).

Selhorst (1989) differentiates PTA from impaired consciousness. Amnesia is always longer, occurring while the patients are apparently fully conscious and normally responsive to the environment. The following categorization of head injury (Selhorst, 1989, citing Russell et al., 1946, 1961) is offered since it is commonly used: mild, 1-h PTA; moderate, 1–24 h PTA; severe, 1–7 days PTA; very severe, 7 days or longer PTA. The present author observes that the actual neurobehavioral adaptive dysfunctions probably correlate poorly with the above categories. However, Guthkelch (1980) determined that the most common cause of very prolonged disability in patients whose PTAs exceeded 1 week was brainstem injury, some of whom remained in a vegetative state until they died.

POSTTRAUMATIC MEMORY DISTURBANCE

Encoding After Trauma

It is possible to consider some alterations of consciousness from the viewpoint of disturbance of memory function, as well as the active disturbance of the registration of the environment and the self. Perhaps the initial intense state of *arousal* initially imprints particular stimuli, which are associated with the ongoing systemic and psychological storm. In comparison with ordinary memory-storing events, the traumatic event may have been disorganizing or imprinted a partial or unusual configuration of stimuli. When particular components are confronted in imagination or in the environment by conditioned association, a state of combined neurological and physiological arousal may be elicited. Ongoing function involves multiple levels of representation between certain networks—primary sensory (neocortex) and paralimbic. There are reentrant connections linking each level, which must be functional to allow memory consolidation. Stronger memory representations (accurate or distorted) occur with *long-term potentiation* (i.e., increased activity of a neuron causing strengthened activation) or potentiation of an afferent synapse active at that time.

Also relevant to understanding dissociative alterations of consciousness is *kindling*, which characteristically involves the amygdala and hippocampus. Repeated stimulation of a cortical site results in increasing discharge with each event. The repeated stimulation has been attributed to conditioning, or the sensitivity of the person to being reminded of an anxiety-arousing event. Kindling predominates in the amygdala, which recruits secondary sites in other limbic cortices. Their after discharges are as great as that of the primary site. *Limbic centers* mediate both emotional reactions and memory consolidation. Extreme emotional stress may create emotional sensitization to particular stimuli, memory disorganization, and pathology of corticolimbic functioning. In daily experience, limbic networks respond in proportion to the personal significance of each event, leading to proportional consolidation pin which the event's significance is integral to its representation (Tucker & Luu, 1998).

Ordinarily, the cortex makes use of a very gradual learning strategy. New information is interleaved with ongoing content, which permits the gradual learning of the structure of informational domains. Ordinarily, the hippocampus permits the rapid acquisition of new information without catastrophic disruption of the existing structured system of knowledge gradually built up through experience. An attempt at more rapid consolidation than the system can cope with is expected to create catastrophic interference (McClellend, 1998). One can infer that a stressful event of the scope that we are considering actually overwhelms the system, perhaps disturbing current mental contents, as well as offering disorganized representation of the trauma.

Impaired consciousness may interfere with full encoding of the traumatic events. In addition, during the period of memory, consolidation adrenergic hormones modulate memory formation. This function depends upon the basolateral amygdala, and occurs when the learning conditions are sufficiently arousing, whether or not the emotions are positive or negative. Because epinephrine does not readily cross the blood brain barrier (BBB), memory modulation may take place through vagal input via ascending tracts to the *nucleus of the solitary tract*. While repeated activation of catecholamine memory modulation system creates PTSD through overconsolidation of trauma-related memory, serotonergic activation underlies PTSD in some patients (Cahill, 1999).

Stress (e.g., Vietnam War combat veterans) can be associated with memory deficits and poorer performance of a simple task requiring attention, visual scanning, and psychomotor speed (Trails A and B). Some dysfunction may be associated with antianxiety and cardiac *medications* as well (Beckham et al., 1998). The intensity of headache did not appear to impair memory test performance when no headache of mild posttraumatic headache patients were compared with those with moderate or severe pain; 55% of the subjects experienced intensified existing headaches, but headaches were triggered for only 5% of the sample. Headache victims report difficulties in handling workloads, particularly complicated ones; operating complex machinery; and experiencing interpersonal difficulties. Nevertheless, they may learn to pace themselves to reduce stress. However, a testing situation implies limited distractions, a supportive tester, and concern about the outcome of the results. Thus, maximum effort may encourage maximal effort leading to pain intensity amplification during complex cognitive tasks (Lake et al., 1999).

The Amnestic Syndrome

The amnestic syndrome refers to a severe and relatively pure impairment in new learning as a result of brain damage (i.e., inability to remember recent events in the context of relatively intact cognitive, linguistic, perceptual abilities, and retained fund of general information) (Albert & Lafleche, 1991; Brandt, 1992; Schachter et al., 1991). Amnestic patients perform poorly on tests for explicit memory, and almost normally on tests of skill acquisition, perceptual priming, and so forth. Recognition is more effective than recall (Brandt, 1992). Working (short-term) memory is preserved, although retention across delays is impaired. Amnestic patients have variable ability to retain information short term, perhaps varying with the procedure, and whether or not interference occurs between exposure and retrieval (Schachter et al., 1991).

Traumatic Stress and PTA

The question has been raised concerning amnesia for the event after a head injury. Nightmares and flashbacks may occur even in circumstances in which there is loss of memory for the accident, as well as extensive RA. One example was a young woman who was in a 7-month coma after an MVA, with both RA and long-term memory loss. The period of stress begins when a person awakens after an accident in a hospital bed, not knowing who they are, or where they are, what happened, why the tubes are coming out of his or her body, why there is pain, or even why there has been a loss of limb. The events of the recovery period, as well as personal or social reconstruction of the event, seem to suffice to create the symptom pattern even in the absence of memory for the defining unusual life-threatening event. PTSD's "neuropsychological" symptoms have been identified even when a head injury is not involved. It can occur even with amnesia for the event (Knight, 1997). Baggaley and Rose (1990) offer the case history of a 22-year-old soldier who was injured in an explosion. The patient had significant closed head injury, PTA for 36 h and RA of 8 h, and developed PTSD.

Some Representative Dysfunctions of Consciousness

Global imperception of stimulation from the environment or body; seizures; confusion; strange experiences; depersonalization and derealization (also anxiety-based); delirium; stupor; locked-in syndrome; sleep dysfunction; neglect of part or half of one's body (sensory and motor); changed sense of self (depersonalization; loss of detail). Dementia, delusions, confusion, and inattention do not affect the level of arousal, with the exception of advanced dementia. PTA, technically a memory problem, can also be considered to be an altered state of consciousness.

Wakefulness and the Unconscious

Consciousness implies the state of being awake (i.e., realistic, capable usually of movement, and with self-control). Awareness, in contrast, simply means that some type of stimulus (internal or external) is represented in the mind. Awareness involves perceptions similar to others (Prigatano & Schachter, 1991) and is supported by attention, concentration, and motivation. It has links with cognitive processes that are impacted after head trauma. A dream may reflect awareness of external light or noise or a vague bodily sensation while being asleep. While aware, one may experience fantasy, images, or an unconscious process. Wakeful consciousness fluctuates between keen alertness (external awareness), deep concentration (constricted field of attention), and general inattentiveness and drowsiness (Adams & Victor, 1989, p. 273). A decreased level of consciousness (including medical diseases) almost always indicates a neurological dysfunction (Black, 1988). The effect of psychodynamically active but repressed mental contents upon awareness and action without being at the center of attention is not discussed here.

Knowing

Cognition is defined as the process of knowing. Thus, consciousness involves the certainty that one is oriented in environment and body. The person knows and gives meaning to external and internal stimuli. That is why depersonalization and derealization are considered altered states, because of the lack of certainty that what one is experiencing is accurate or real. Experience and knowing (epistemology) are interacting areas. It is necessary to discriminate between some basic awareness we could call consciousness, and its registration in some system such as language related mechanisms of the left hemisphere (Farah & Feinberg, 1997). The belief that one knows is lost when the person has lost the ability to state what one knows. This is observed in word seeking, when the patient is convinced that a response would have been forthcoming if a brain injury had not occurred.

Alertness

Attention is a conscious effort and/or a mental set to detect one or more predetermined stimulus, task, or danger. It is required to absorb useful information, as well as to integrate thought, action, and internal experience with emotional responsiveness and external expression (Lezak, 1989). Attention may be general or focused, and may be directed to the environment or intrapersonal space. This includes orienting to sensory stimuli, such as locations in visual space, detecting target events (sensory or memory), maintaining an alert state, and shifting attention.

CONTRIBUTION OF MOOD AND FEELINGS TO AWARENESS

Affect, mood, and feelings have a great claim to be the central components of experience. Perhaps feelings, not cognition, are the core experience of consciousness: *"I feel therefore I am"* (Parker, 1981, frontispiece). Indeed, clinical experience suggests that anhedonia, the lack of an enlivening mood (regardless of valence), can be a marker of brain damage. In the locked-in state, with gross restriction upon the range of stimuli available to be reconstructed into new experiences, who would deny that the person's mood may be the core of consciousness?

Although alterations of consciousness may be attributed to temporary or persistent neurological damage, there is considerable similarity in the subjective experience and other symptoms of head-injured and emotionally traumatized persons (Parker, 2002). Comorbid diagnostic elements are cerebral trauma, psychodynamic effects of anxiety that disrupt one's security and lifestyle, and the persistent neurobehavioral effects of stress.

DISSOCIATION

Clinical Example: Dissociation After an MVA: A woman who had a graduate degree was in a two-car accident. There is evidence of considerable loss of mental ability and both short- and long-term memory. She was the driver of a car struck by another car. She was seatbelted. Her car was struck on the left front. Both cars were travelling at city speeds. She felt a jerk. She did not anticipate the accident because she claims she was going with the traffic light. She does not remember hearing or smelling anything. She saw the other car collide with her car:

> I tried to come out of my car. I remember telling her that she ran the lights. When the two cars collided, for a moment I went into shock. I don't know what happened. I wanted to know if it really happened or if it is a dream. I didn't believe that it actually happened. I was going through a period of denial. I didn't believe it actually happened because I was so sure that I had the light. There was a change in my awareness. I feel that I was spaced out.

She cannot state what she means by this. There is no claim of retroactive amnesia. She is uncertain whether there is a period that she does not remember.

What happened to your body?: "I felt a jerk in my neck." She does not remember hitting her head, nor does she remember any bruises the next day.

Clinical Examples: Depersonalization: A man was struck by a heavy falling object on his head, searing a construction hard hat, his neck, and spinal column. He saw it coming but it was so fast that he could not protect himself. He saw another man holding him. He had memory of this but, "everything was completely different. Like it wasn't me. It wasn't real. It was like a nightmare. Even today I don't feel that everything is real. The worse thing, when I am going for a walk to the park, I feel something strange. I feel that I am outside my body and the whole world is like a different planet. What am I doing here? Its very strange. It makes me think, What's going on? It makes me think about life. Maybe its true that human beings are body and the soul. I may be more soul now. Its a strange feeling."

"I've got the feeling I'm looking at myself. Its somebody else, a kind of movie. My mind is detached from my body."

Note: Depersonalization can be a rare manifestation of complex partial seizures, more often found in schizophrenia (DeLorenzo, 1991). One may also consider a differential diagnosis between epilepsy and a phobic-anxiety depersonalization syndrome (Trimble, 1991, citing Harper & Roth, 1962).

Clinical Example: Derealization:

Do you remember the accident (i.e., being injured)? He remembers hitting his head. Then somebody came over and asked if he was okay. He couldn't speak but took a card out of his pocket, which had his phone on it.

"I don't think I told the person anything, I just took the card out and gave it to them. There was a two-block stretch and he could see his wife running. (Derealization), the air looked like pieces of clear Jello that was soft and pieces moving around, that's sort of the way the air looked. She was running towards the car, two blocks, and to me it seemed like she was running in slow motion. It seemed like she was running through this jello-like substance in slow motion. It seemed like it took her an eternity to travel those two blocks. She asked me if I was okay and I had trouble talking; I couldn't get my mouth to move."

Clinical Example: RA; PTA, and Dissociation: A 16-year-old boy was hit by a motor vehicle. The hospital report stated incorrectly there was no LOC. He remembers that around 9 pm he was playing and he saw his friends running. The last thing he remembers was walking on 169th Street and the accident occurred on 168th St.

"I don"t remember getting hit. I never even felt it." He drifted in and out of consciousness. He woke up at 6–7 at night. "Before that I saw my friends. I said my head hurts. I said 'hi' to my friends, and then I went back to sleep. It was as though I was dreaming."

After waking up: "I didn't know what I was doing there. I felt scared. I felt bad. I didn't know what happened to me. I had a fever.

How long before you felt like yourself?: "I was still in a daze. I felt weird when I left the hospital. The Sun and the street felt as though I was never outside before."

Dissociation: It is concluded that he is suffering from both anterograde and RA.

Rorschach Examples:

- *Reenactment of Trauma ("Loss of Distance")*: *Rorschach sample*: A woman who watched while a car struck her own stopped car: "Tidal wave coming over an island."
- *Depersonalization: Changed or confused identity*: "caricature" (woman doesn't feel like the same person)

Clinical example: Depersonalization, the belief that one is dead: A woman was threatened by an oncoming car, extended an arm to protect herself, and suffered back injuries that created persistent pain. It took a long time after the accident to realize she was still alive. "For many weeks I thought I was dead. I was confused. Am I alive or dead? Till now I still dream about the car. I have been interactive with people, then I know I am not dead... They wanted me to go to the hospital. I didn't want to go because I didn't know if I was alive or not. I was in shock. I saw myself as flying."

Perhaps contributing to the depth of her experience was an event when she was 7–8 years old. She did not go to school in the usual car because her mother had just given birth. This car was in an accident, killing several people, and causing a man whom she knew well to lose an arm.

Etiology

Varied Etiology

Dissociation is a process that may have varied etiology. Experience is changed, because the normal integration of awareness with thoughts, feelings, memories, or actions is disturbed (i.e., lose their connection with each other; perceptions and memories may be compartmentalized). While mental components may be unavailable to consciousness, they exert control (Spiegel, 1991). The clinician should consider that altered states of consciousness may arise from cerebral dysfunction or as a psychological defense.

Dissociative phenomena such as amnesia and flashbacks are differentiated from intrusive disturbances such as nightmares and traumatic memories and from partial seizure phenomena and hyperarousal. The role of purely "psychological" etiology is given credence by many, although the author wonders about the state of physiological hyperarousal as the true cause of this reaction. After head trauma, hyperarousal is certain an internal environment event with widespread consequences.

PTSD is conceived of as disturbing the integration of consciousness, memory, identity and perception of the environment, while contrasting function dissociation helps the person to cope with reexperiencing traumatic memories. The development of dissociative symptoms in a group of policemen after a traumatic event was associated with the absence, or increasing complexity of the PTSD (partial or complete) (Carlier et al., 1996). The level of awareness of events after a frightening experience varies between patients. With lesser levels of dissociation, the individual can monitor one's relationship to the environment, which capacity is lost in more complete dissociation. Learning during dissociation may be state dependent (Krystal et al., 1995). Defenses fail when the victim is reminded of the event by internal or external stimuli (see Chapter 16). The consequences may be episodic rage attacks (Salley & Teiling, 1984) or fugue states (Spiegel, 1988). Traumatic events may be repressed or dissociated entirely or in part, or suppressed (Blank, 1994). Dissociation during combat trauma was associated with a greater risk for PTSD and dissociative symptomatology (Bremner & Brett, 1997).

Lesions of the cerebral cortex and thalamus can alter the content of consciousness without altering the state of consciousness (Walton, 1985, p. 640) (i.e., partial seizures) (see below). Dissociative phenomena (e.g., altered sense of self and amnesia) should be differentiated from reduced arousal, which can cause alterations of consciousness commencing with head trauma that can sometimes persist for years and then leave, permitting clear awareness without any apparent cause.

Dissociation and depersonalization seem to be protective psychological mechanisms. Dissociative memory phenomena, such as amnesia and flashbacks, are differentiated from intrusive disturbances such as nightmares and traumatic memories. It is difficult to differentiate short traumatic impairment of consciousness from dissociative reactions or very short PTA (Radanov et al., 1992).

Dissociative Phenomena

Dissociation is a process that may have varied etiology, in which experience is changed because the normal integration of awareness with thoughts, feelings, memory, or actions is disturbed. Dissociative phenomena such as amnesia and flashbacks are differentiated from intrusive disturbances such as nightmares and traumatic memories, physiological reactions, partial seizure phenomena, and hyperarousal. The complex comorbidity of dissociative symptoms is illustrated by group of policemen after a traumatic event. This symptom was associated with the absence or increasing complexity of the PTSD (partial or complete). PTSD is conceived of as disturbing the integration of consciousness, memory, identity, and perception of the environment; dissociation helps the person to cope with reexperiencing traumatic memories (Carlier et al., 1996).

Altered states of consciousness after head trauma may arise from cerebral dysfunction or as a psychological defense. The level of awareness of events after a frightening experience varies between patients. With lesser levels of dissociation, the individual can monitor one's relationship to the environment, in which capacity is lost in more complete dissociation. Learning during dissociation may be state dependent (Krystal et al., 1995). Defenses fail when the victim is reminded of the event by internal or external stimuli. The consequences may be episodic rage attacks (Salley & Teiling, 1984) or fugue states (Spiegel, 1988). Traumatic events may be repressed or dissociated entirely or in part, or suppressed (Blank, 1994). Dissociation during combat trauma was associated with a greater risk for PTSD and dissociative symptomatology (Bremner & Brett, 1997).

A patient in PTA might appear to be normal (i.e., clinically considered to be uninjured or slightly injured), so inaccurate statistics of non-TBI may be entered after an accident. The patient may have apparent short-term memory and seem lucid in the period after the accident, but subsequently be unable to recall either the accident or current conversations the next day. The failure was to consolidate the events. Moreover, "memories" of the events around the accident may be inaccurate.

The association of dissociative phenomena with trauma requires two cautions: (1) verification (the events have not always been verified) and (2) patients with dissociative tendencies have a strong propensity to have a firm conviction of the reality of ideas suggested to them by others (Nemiah, 1995). The development of dissociative symptoms in a group of policemen after a traumatic event was associated with the absence or increasing complexity of PTSD (partial or complete). The level of awareness of events after a frightening experience varies between patients. With lesser levels of dissociation, the individual can monitor one's relationship to the environment, which is lost in more complete dissociation. Learning during dissociation may be state dependent (Krystal et al., 1995). Defenses fail when the victim is reminded of the event by internal or external stimuli. The consequences may be episodic rage attacks (Salley & Teiling, 1984) or fugue states (Spiegel, 1988). Traumatic events may be repressed or dissociated entirely or in part, or suppressed (Blank, 1994). Dissociation during combat trauma was associated with a greater risk for PTSD and dissociative symptomatology (Bremner & Brett, 1997). Even when dissociation is addressed as a psychological stress phenomenon, neurobiological phenomena include memory and hyperarousal, so that the overlapping with TBI dysfunctions is considerable. For example, tunnel vision, which is sometimes considered to be a sign of hysterical sensory loss, has been obtained through administration of ketamine, an antagonist of one type of glutamate receptor. Dissociative states are also obtained through serotonergic hallucinogens, a neurotransmitter involved in stress-related functions (Krystal et al., 1995). Other examples of the difficulty of determining the etiology of dissociative-related phenomena are found. The association of phobic anxiety and panic attacks with depersonalization (Steinberg, 1991) is no doubt are more common after head injury. Furthermore, the characteristics and examples offered for psychogenic amnesia (PA, including inability to recall certain important information) overlap with cerebral trauma (retroactive amnesia and PTA). Yet, the criterion for PA that dysfunction of neurobiological systems is reversible (Loewenstein, 1991) would not be determined as incorrect for some time.

Anxiety Basis for Depersonalization or Derealization

The patient feels unreal, or sees the self outside of the self, experiencing emotional numbness and affective loss. It is represented in the DSM-IV-TR (2000, p. 532) as "dissociative disorder not otherwise specified." Depersonalization can be an individual symptom, is found in a variety of conditions (including other dissociative disorders), may or may not be a separate clinical entity in a particular patient, and has numerous etiologic theories. It is most frequently associated with panic, but also with anxiety, and/or agoraphobia, depression, seizures (preictal, and postictal), substance abuse, organic illness, depression, and as a medication side effect. It is characterized by an absence of feeling, a reduction of vividness, a sense of detachment, of being real or an automaton, and may be a state of low arousal. Sometimes the experience is pleasurable and can be followed by an appreciation of life. It may involve heightened arousal that is dissociated from consciousness (Steinberg, 1991). Psychogenic fugues can cause a prolonged altered responsiveness to the environment that mimics convulsive status epilepticus (DeLorenzo, 1991).

In 1986, Grigsby stated the difficulty of differentiating functional from neural etiology of depersonalization in cases of head trauma. More recently (personal communication), Grigsby stated that depersonalization reflects a *physiological* disturbance of the brain's dynamic equilibrium, whose mechanism is likely to be a deficit in the integration of perceptual, visceral, affective, and cognitive information. This is a conceptualization of the evidence that depersonalization can result from the high level of arousal accompanying panic. He offered a case in which there was mild neuropsychological disturbance and a depersonalization syndrome: "Neuropsychological assessment was largely within normal limits, and the patient's complaints were in large part a function of feelings of depersonalization and derealization.... [Depersonalization] appears to have been of psychogenic origin. Initially is probably served as a defensive response to a potentially life-threatening danger ... Her life was disrupted and she began to question her previous lifestyle."

SELF

The sense of self is a complex phenomenon. From the viewpoint of evolutionary biology, the self is the individual, a coherent unit selected to nourish, protect, and reproduce (Queller & Strassmann, 2002). Psychologically, it is the experience that participates in guiding action and reaction to events (Parker, 1983). Its components include consciousness or self-awareness, body image (sensorimotor), and identity (stress and psychodynamic). The key components of *self-awareness* are reality and integrity (i.e., "I am one person"); continuity ("I was the same person in the past as I will be in the future"); boundary of the self ("I can distinguish between myself and the rest of the world as not-self"); and activity of self ("I am thinking, doing, feeling") (Yager & Gitlin, 2005). Ordinarily, the sense of self is experienced as integrated. One can differentiate between the self as experiencer or observer, and an objectified self that is regarded as having feelings, qualities, and knowing; and includes the phases of "Me" as subject, or "I" as actor. Any aspect of experience that comes into awareness may have a meaning attributed to it. Its affect upon our self (identity) and its feelings can determine our mood and select and direct our actions. Thus our self, or whom we think we are, is part of our consciousness, and a determinant of feelings and activities that interact with our attitude to our self or identity (Parker, 1983). In turn, our identity is a determinant of action. Mental representation of the self resides in working memory along with a coexisting representation of the current external environment. Disruptions of the sense of self as integrated within one's person and with the environment are considered to be dissociative disorders (see below).

The self as a stable mental component represents a *memory system*. Reorganization may involve catastrophic destruction. A catastrophe (emotional stress) may create both disorganization of the earlier self and a different, highly cathected self. If the new experience is incompatible with the self, then the self must be reorganized.

Repression, understood as loss of access to the initial self, need not be defensive in the psychoanalytic concept. Repression occurs due to disorganization, or to the existence of fragmentary experiences that are not fully encoded in the self. Information that is not consistent with the existing structure (memory of the self) cannot be processed (Tucker & Luu, 1998).

5 The Internal Milieu
Brain and Body

OVERVIEW

Claude Bernard (1813–1878) was responsible for a major breakthrough in understanding the fundamental principles of organic life, one that is valid still today. It is his concept of "homeostasis," or controlled stability of the internal milieu, or internal environment, of cells and tissues. He proposed that "the fixity of the internal environment is the condition for free life," and offered the following explanation (Sabbatini, 1998): "The living body, though it has need of the surrounding environment, is nevertheless relatively independent of it. This independence which the organism has of its external environment, derives from the fact that in the living being, the tissues are in fact withdrawn from direct external influences and are protected by a veritable internal environment which is constituted, in particular, by the fluids circulating in the body."

The physiological dysregulation caused by physical and emotional trauma can be expected to have disruptive neurobehavioral consequences. Body functions influence the efficiency and pattern of the central nervous system (CNS). Successful, ecologically significant adaptation, and enjoyment of life occurs through intersystem, bidirectional regulation of ongoing activities while avoiding pathology and disease. This chapter is intended to give an extensive overview of physiological mechanisms that support brain functioning, and also adaptive responses to the external environment and internal experiences. Physical and emotional trauma can interfere with these functions acutely. If there is not substantial recovery, there is a possibility of interference with the quality of life, long-lasting complaints described as posttraumatic stress disorder (PTSD), fatigue and loss of stamina, physiological "burnout" (allostatic state), and ultimately, vulnerability to a range of stress-related medical disorders. This sequence is described in subsequent chapters.

The clinician and neuroscientist are concerned with more than afferent stimulation, reception, inhibition, and effector actions between different parts of the nervous system. Normal, posttraumatic, and diseased functioning have significant exchanges between physiological and neural systems. The immune, nervous, and endocrine systems are functionally interconnected (Levite, 2000). These function acutely to maintain the person's health; an unhealed wound, however, causes persistent functioning, of an interval uncharacteristic of the untraumatized person, and ultimately dysregulation due to physiological fatigue. Thus, attention is given to system interactions that are frequently ignored in both the acute and chronic states of concussive and more serious head and co-morbid somatic trauma: hormonal, immune, inflammatory, and autonomic. This reflects the neurobehavioral consequences of activity and dysregulation of these systems, which affect brain functioning via the *bidirectional signals* through the internal milieu.

Neurobehavioral success requires that physiological and neurobehavioral functions are coordinated with each other, that components of a system operate in sequence, and that tissues create and respond to appropriate signals, which are maintained within narrow bounds. Assessment, treatment, and outcome are enhanced by understanding the exchange between the physiological and neurological systems (Parker, 2005, 2008). Well-focused counseling and medical treatment are enhanced by understanding their contributions to the stamina and health of these functions and their disruption of homeostasis, allostasis, metabolism, the internal environment of the brain, and other components of health support.

It has been asserted that there are thousands of control systems, from the genetic and cellular levels, to macroregulation of parts of organs, up to controlling the inter-relations between organs. In uninjured and unstressed persons, performance and well-being depends upon adaptation to the outer and inner worlds. This is contingent upon adequate self-regulation of mental and physiological systems. Each system is in two-way signaling relationships with many others. This has been described as *integrated bidirectionally regulated neuroendocrine axes (NEA)* (Marchetti et al., 2001). This has been termed *the hard-wired neuroimmune network*. Homeostasis is maintained by an intensive infrastructure of vessels, circulating hormones, and direct fiber connections. These link the brain with all viscera, which provides adaptive reactions of cardiovascular, respiratory, metabolic, and immunological host defense functions (Downing & Miyan, 2000; Standing, 2000b). The neural and neuroendocrine systems cooperate by modulating the immunological reaction, that is, the system of vessels, circulating hormones, and nerve fibers linking the brain with all viscera. Systemic efficiency is coordinated by the process of *allostasis*, a physiological control mechanism operating at times of stress and higher demands than homeostasis. Mutually and simultaneously, the brain and physiological systems receive incoming stimuli and respond, thus integrating the organism's response to changes in its external and internal environments. Physiological systems express chemical signals that are often also expressed by, and responded to, in the CNS.

The internal milieu is served by a multidimensional, bidirectional, neurobehavioral signaling network.

THE INTERNAL MILIEU (LIQUID ENVIRONMENT)

The internal milieu is the fluid environment within and surrounding cells: intracellular fluid, extracellular fluid (ECF), the bloodstream, and cerebrospinal fluid. Survival requires that it remain quite constant so that the body's tissues can function effectively. Tissues must respond precisely and promptly to unusual internal and environmental events in order to maintain a constant physiological level of substances to meet physical and physiological conditions. Chemical messengers that cannot cross the blood-brain barrier (BBB) (e.g., large molecules such as peptides in the general circulation (BBB) have indirect effects, for example, insulin-mediated changes in the blood glucose concentration). Major systems, tissues, and individual cells require, maintain, and function within a healthy, reasonably steady state. While the reactions of numerous body systems to external changes in order to lessen their impact has been attributed to homeostasis, this serves to maintain particular physiological conditions precisely ("set points"). However, with major demands (nonsleep activity, stress, injury, illness), the process of integrated functioning is described by the different process of allostasis.

The internal milieu is the consequence of the interaction between numerous physiological systems. Proper neuronal function requires a regulated extracellular environment that maintains within a narrow range the concentrations of sodium, potassium, calcium ions, nutrients, and so forth. (homeostasis). Notably, this includes sensorimotor processing. These solutes are relatively small and are not excluded from passage into the brain cells by the various barriers.

Two characteristics of the liquid environment are essential to the functioning of the brain and all other tissues (Giebisch & Windhager, 2005):

1. The volume of ECF, that is, the water balance, which maintains function by regulating blood pressure, permitting tissue perfusion of dissolved substances.
2. Extracellular osmolality, or the levels of different substances, individually and cumulatively, in the ECF. Receptors in the hypothalamus detect changes in the plasma osmolality. Signals are sent to brain areas that control thirst, regulating water volume and water excretion antidiuretic hormone [ADH]). The Na^+ level determines the ECF volume, that

is, the amount of liquid within which nutrients, salts, gases, signals, etc., are dissolved. Abnormalities in a particular system can dysregulate a reciprocal system.

In addition, the liquid environment's *BP* is maintained on a short-term level, with signals from pressure receptors, by the autonomic nervous system (ANS), which modulates the heart and blood vessels. Long-term effects (hours to days) are maintained through kidney control of water excretion (nervous, humoral, and hemodynamic mechanisms).

Microenvironment

This refers to the *ECF* surrounding the neurons and neuroglia. It is important because it supplies the cells with oxygen and nutrients, and removes wastes. In addition, it supplements neural transmission by the synapses with the process known as *nonsynaptic diffusion neurotransmission (NDN)*. After traumatic brain injury (TBI), there may be an increase in the number of receptors (upregulation) on the neural membrane in the area of the impaired pathway. Synaptic transmission is used for precise, rapid initiation and ending of individual movements (musical instrument). NDN is used for mass-sustained functions (sleep, mood, hunger) and for adequate preparation at the brain level, perhaps mediated by noradrenaline (piano playing is an example). This utilization of diffusion may take part in vigilance, significant for a rehabilitation program (Bach-Y-Rita, 2002).

Jankovic (1994) uses the concept of microenvironment to describe the components that unify the immune, nervous, and endocrine systems: (1) *lymphoid cells* (T and B lymphocytes); (2) *nonlymphoid cells* (macrophages, epithelial cells, and dendritic cells that interact with lymphocytes and other cells of the microenvironment; (3) *visiting cells* (lymphoid and nonlymphoid cells, which enter the immune milieu and influence the immune network; (4) *neurons* (utilizing various neurotransmitters that influence the microenvironment by neurohumoral activity; (5) *hormones* (released from remote sites by endocrine glands, which enter the immune milieu via capillaries and influence the recognition processes; (6) *biologically active substances* (lymphokines, monokines, complement, immunoglobulins) produced by both lymphoid and nonlymphoid cells in situ or elsewhere; (7) *neurotransmitters* (of numerous types and structures, including the classical types, opioid, and other peptides); (8) *membranous and intracellular receptors* (these are molecules that maintain intercommunications and the microenvironmental circuits; (9) *ions* (sodium, potassium, magnesium, calcium) that maintain signals through cell membranes, affecting immune stimulation and inhibition; (10) *electrical, magnetic, and electromagnetic fields* created by differences in ionic concentrations and charged molecules in various compartments of the micromilieu; (11) *higher nervous activity* (mind and psyche) that influences the microenvironment through stress-related, emotional, aggressive, and other behaviors.

Water

Water represents 60% of body weight in men and 50% in women. Changes in whole-body water content lead to changes in osmolality to which the CNS is extremely sensitive. Disturbances of ±15% lead to severe disturbances of CNS function (Giebisch & Windhager, 2005).

Cerebrospinal Fluid (CSF)

Cerebrospinal fluid is secreted by the choroid plexuses, which are found in the walls of the lateral ventricles and the roofs of the third and fourth ventricles. The CSF's composition is regulated by the choroid plexus, whose function is similar to that of the kidney. It removes products of metabolism into the venous sinuses and transmits chemical signals (e.g., the neurotransmitters serotonin and dopamine) through the choroid plexus into the blood (Ransom, 2003). Seventy percent of the CSF is secreted by the choroid plexus, and 30% comes from the capillary bed of the brain and metabolic water production. The subarachnoid space is also filled with CSF.

These structures are the carriers of solutes that evade the BBB. CSF is filtered from the blood by the choroid plexus in the lateral, third, and fourth ventricles. It exits the fourth ventricle through the foramina of Lushka and Magendie to the arachnoid villae of the subarachnoid space, which absorbs the CSF back into the bloodstream (Blessing, 2002). CSF removes waste products of neuronal metabolism, drugs, and other substances diffusing into the brain from the blood. As it streams over the ventricular and pial surfaces of the brain, CSF drains solutes and carries them through the arachnoid villi into the venous blood. It integrates brain and peripheral endocrine functions. Hormones or hormone-releasing factors from the hypothalamus are secreted either into the extracellular space or directly into the CSF. They are carried to the median eminence and thence into the hypophyseal portal system (Parent, 1996, pp. 14–15). The CSF is also moved by active transport (not passive filtration, as in other tissues) through the BBB of the capillary walls into the interstitial space of CNS (Morganti-Kossman et al., 2005). Footlike processes from astrocytes form a continuous layer around the blood vessels participating in the BBB.

Circumventricular Organs

Some substances enter the brain through structures whose surface is on the cerebral ventricles, that is, the *circumventricular organs (CVO)*. The CVO are located adjacent to the third ventricle (subfornical system: subcommissural organ; organum vasculosum of the lamina terminalis [OVLT], pineal, and part of the median eminence; and at the roof of the fourth ventricle [area postrema]). Their vascular lining (endothelium) is permeable (Somjen, 2005, Chapter 1). The extracellular environments of these neurons and neuroglia are in a steady state with the CSF, an important factor in CNS homeostasis. This barrier is formed by the choroid epithelial cells, which creates active and passive transport for the major part of the CSF (Ronken & van Scharrenburg, 2005).

Cerebral Vasculature

The brain accounts for 2% of body weight, but it receives nearly 20% of the cardiac output. There are two control mechanisms: the larger cerebral vessels are influenced by a balance of sympathetic and parasympathetic tone; medium and small cerebral vessels are altered primarily by the separate mechanism of autoregulation (Jane et al., 2002). The blood supply of the brain is provided by two pairs of arterial trunks.

Anterior: The two *internal carotid arteries* supply the anterior and middle cerebral arteries, which perfuse the anterior half of the thalamus, the basal ganglia, the corpus callosum, most of the internal capsule, the medial and lateral surfaces of the frontal and parietal lobes, and the lateral surface of the temporal lobe.

Posterior: *Branches of the vertebral arteries* pass through the vertebral foramina to supply the meninges and portions of the cervical spinal cord. The vertebral arteries traverse the foramen of the sixth cervical vertebra, ascend through the transverse foramina of the remaining cervical vertebrae, and enter the foramen magnum and then the posterior fossa. The two vertebral arteries fuse at the junction of the medulla and pons to form the midline *basilar artery*, which proceeds rostrally along the anterior surface of the pons. Intracranial branches of the vertebral and basilar arteries supply the cervical spinal cord, the medulla, pons, midbrain cerebellum, posterior parts of the diencephalon, and parts of the occipital and temporal lobes of the brain. A branch of the basilar artery supplies the cochlea and vestibular apparatus.

Circle of Willis (arterial): A polygonal pathway on the base of the brain that connects the anterior with the posterior circulations.

Cerebral venous system: Scalp and facial veins may drain to the dural sinuses via *emissary (bridging) veins* passing through the skull and orbit (Bullock & Nathoo, 2005). Veins are found on the surface and within the brain parenchyma. Sinuses occur within the medial and external dura mater.

Cross-sectional diagrams of the brain with the arterial and venous territories are found in Kretschmann and Weinrich (2004, pp. 253–285) and Jane et al. (2002). Diagrams and text of the entry and distribution of the basilar (posterior) and internal carotid arteries (anterior) from the skull into the brain and its surface, the Circle of Willis, and the arterial supply of the basal ganglia and thalamus have been provided by Parent (1996, pp. 98–105) and Netter (1983, pp. 43–66).

HOMEOSTASIS

Homeostasis is a survival process functioning to sustain cell survival and function, for example, autonomic functions (BP; heart rate; breathing; temperature; and the level of salt, water, acid-base, and oxygen [Kaye & Lightman, 2005]). Homeostasis regulates the body for a narrow range of stress, assuming that organismic stability requires that all parameters of its internal milieu must remain constant, a classic principle of Claude Bernard and Walter Cannon. This holds in isolated organs and tissue. However, it poorly describes functioning in an intact organism. Autonomic functions are linked to the maintenance of homeostasis. They are modulated by sensory input arising from controlled organs, that is, mechanoreceptors (heart, lungs, gut, major thoracic arteries) and specialized sensors that detect circulatory oxygen levels or gut and hepatic concentrations of lipids or glucose (Verberne, 2003). An important homeostatic integrating function is the interaction between the immune system (IS) and CNS (Raison et al., 2005).

All physiological parameters vary with a behavioral state. The level of specified functions is restrained within a narrow range ("set points"). Comparators signal deviations from the set point. Homeostasis is maintained by a wide variety of physiological functions: input from cerebral, limbic, brainstem, and afferents; and effectors such as the ANS and several hormonal systems, including the hypothalamic-pituitary axis (HPA) target organs. The pituitary bridges and integrates the neural and endocrine mechanisms of homeostasis (Barrett, 2005a). Although metabolism is a homeostatic function, it is entrained to circadian rhythms.

Metabolic Homeostasis

A balance is required between need and availability, that is, between carbohydrate, fat, and protein intake; their storage when present in excess of immediate need; and their mobilization and synthesis when in demand. The body's control takes place at many levels: systemic level, (arterial pressure and blood volume), internal milieu (body core temperature and such chemicals as oxygen, glucose, and ions [potassium, calcium, and hydrogen]), intracellular, and energy. Homeostasis is maintained by *negative feedback* in which the controlled variable (e.g., blood level of a hormone) determines the rate of secretion of the hormone. The signal produces the effect that brings the condition back to the set point. In *positive feedback* the controlled variable increases the hormone secretion. In a *closed loop* regulation is entirely restricted to the interacting regulatory glands, while in an *open loop* the nervous system influences the feedback loop. The set point is maintained by balancing actions to keep the function constant, for example, the interaction of insulin and blood glucose. These functions are redundant, permitting some systems to fail while others maintain homeostasis. The "milieu intèrieur," that is, the sum of all the cells, is consequent to the priorities given for survival of one system over another. One system may be instructed to reduce its performance (e.g., the reduction of sweat to control temperature to maintain blood volume; if the physical demand does not stop, then heat stroke may occur).

The enteric nervous system (ENS) plays a crucial role in the maintenance of homeostasis. Its neurons record variations in the tension of the gut wall and the chemical environment. Its neurons also control the muscles of the gut wall, the vasculature, and the secretory activity of the mucosa. Its activity can be modified by the CNS, parasympathetic nervous system (PSNS), and somatic nervous system (SNS).

HOMEOSTATIC SET POINTS

Homeostasis implies that there is a *body set point*, that is, an optimal level for each physiological system, to which functions are returned through bodily mechanisms. This traditional view is incomplete: The *set point varies* with (a) circadian rhythms, (b) the interaction between the time of measurement for one system and another system (biological or environmental) that may have a different rhythm, whether shorter or longer, and (c) anticipation of future events (a short-lived process that may become self-sustaining). Examples of set points are osmoregulation under posterior pituitary control (Robertson, 2001) and heat regulation (hypothalamic control).

Homeostatic set points are a set of physiological functions occurring in different organs and using varied means of communication. They are designed to maintain an organ's output within a narrow range. This has been described as "time unqualified feedback" (Halberg et al., 2000). Temperature and oxygen level are examples.

THERMOREGULATION: AN EXAMPLE OF A SET POINT

One of the frequent postconcussive symptoms is paradoxical reactions to temperature, which is experienced as uncomfortably cold or warm although others find the ambience to be comfortable.

Dysthermia (hypothermia and hyperthermia) has multiple causes: congenital, drugs, metabolic, trauma, tumors, and miscellaneous. Disorders after brain trauma are more common than is generally recognized, and are considered to be unregulated changes in the hypothalamic set point. There are a variety of patterns: flattened curves, reversed cycles, sleep phase delay, irregular unexplained fluctuations, and exaggerated reactions (Chaney & Olmstead, 1994). Mammals and birds maintain a constant control of *core body temperature (CBT)* by neurologic, hormonal, cardiovascular, and metabolic systems, despite diurnal, geographical, seasonal, and temperature variations.

The homeostatic mechanisms maintaining a constant CBT are (1) hormonal control on basal metabolic rate (BMR) (short term) and body mass (long term); (2) hypothalamic/spinal neuronal network that sets and controls CBT; (3) SNS and PSNS, which govern effector mechanisms that warm or cool the body; (4) a steady-state bulk of energy expenditure represented by the activities of muscular and adipose tissues and neurochemical regulation (Jacobson & Abrams, 1999). Body heat is created through resting metabolic rate, a special oxidative process, and shivering. Resting BMR is under central control by thyroid hormones thyroxine (T_4) and triiodothyronine (T_3). Thyroid hormones have modulatory effects upon insulin, glucagon, growth hormone (GH), and adrenaline (Gatti et al., 2000).

The critical temperature set point is in the hypothalamus, which balances heat loss and production. Local warming or cooling of *thermosensitive neurons* in the hypothalamus and spinal regions triggers panting, autonomic responses, and shivering. Thermosensitive neurons in the preoptic nucleus of the anterior hypothalamus respond to warming by initiating sweating and vasodilation. Trauma results in interference with heat-dissipating mechanisms and thus creates hyperthermia. Thermal information is conveyed to the posterior hypothalamus's effectors for heat generation and dissipation. The hypothalamus generates the *circadian rhythm* of temperature regulation: a one-to-two-point degree difference in temperature between the 6 AM low point and the 6 PM high point. Damage to the posterior hypothalamus and mesencephalon usually causes hypothermia or poikilothermia (body temperature varies with the ambient temperature). In addition, thermoreceptors of the skin, viscera, and spinal cord provide input to the preoptic anterior hypothalamus to maintain temperature balance (Jacobson & Abrams, 1999).

Sympathetic activation can cause reduced skin blood flow, piloerection, heat production in brown adipose tissue, muscular shivering, and increased respiratory rate. Parasympathetic activation causes bradycardia and increased sweat gland secretion (Gatti et al., 2000). Above the set point temperature, sweating begins; below it, shivering begins. The degree of activity is determined mainly by the heat temperature receptors in the anterior hypothalamic-preoptic area of the hypothalamus. However, there is a slight peripheral contribution of peripheral signals to body temperature regulation through alteration of the set point of the hypothalamic temperature regulation center. The set point increases as the skin temperature decreases. When the skin temperature is high, sweating begins at a lower hypothalamic temperature than when the skin temperature is low. It is important for sweating to be inhibited when the skin temperature is low; otherwise, the combined effect of low skin temperature and sweating would cause too much body heat loss. When the skin temperature is cold, even when the hypothalamus temperature is on the hot side of normal, this drives the hypothalamic threshold (in the dorsomedial hypothalamus) to the point of shivering, which avoids deeply depressed body temperature through increasing heat production. This is a variable set point for physiological phenomena that varies with time of day and their adjustment in anticipation of future events.

When the body temperature falls even a fraction of a degree below the set point, shivering is initiated through bilateral signals of the lateral columns to anterior motor neurons, increasing their tone to the point of the shivering activity, which can increase body heat production to four to five times that of normal.

Moreover, an increase in sympathetic stimulation causes an immediate increase in the rate of cellular metabolism (*chemical thermogenesis*). In this abbreviated explication of temperature control, afferent and efferent tracts, the SNS, and catecholamine level were mentioned. This reminds us that victims of TBI may complain of the posttraumatic onset of temperature dysregulation (Guyton & Hall, 2000, pp. 828–830).

Endocrine Maintenance of Homeostasis

Many victims of TBI and stress manifest weight and stamina changes. Metabolism is influenced by numerous *hormones*. Hormones are divided into anabolic (insulin), which stores fuels; counter-regulatory (contrainsular, i.e., opposed to the actions of insulin: glucagon; EP; norepinephrine; cortisol; somatostatin; GH; thyroid hormone [Jameson, 2001; Marks et al., 1996, pp. 374–378, 703]). Hormones of metabolic homeostasis respond to changes in dietary intake and physiological state in order to adjust the availability of fuels (Marks et al., 1996, p. 376). Hormones act on the respective target organ to reestablish homeostasis. Homeostasis is controlled through feedback loops from the endocrine glands that maintain hormone levels rather precisely by feeding back to the hypothalamus and the pituitary, for example, inhibitive (negative) feedback from the target organ. Endocrine regulation is complex, and multiple glands may respond—as many as a dozen after a meal, for example (Habener, 2003). *Intertissue integration* is achieved by the concentration of nutrients or metabolites, which affects the rate of utilization and storage in different tissues; hormones carry messages to individual tissues about the physiological state of the body (nutrient supply or demand), and the CNS uses neural signals to control tissue metabolism directly or through the release of hormones.

The most important hormones maintaining homeostasis are (Jameson, 2001; Marks et al., 1996): *Thyroid* hormone, which controls about 25% of basal metabolism in most tissues; *glucagon*, secreted by the pancreas, which mobilizes fuels; and *cortisol*, which exerts a permissive action for many hormones in addition to its own direct effects. It adjusts for changing requirements by increasing amounts of amino acids, glucose, and fatty acids. Other important hormones include parathyroid hormone (PTH), which regulates calcium and phosphorus levels, stimulating amino acid mobilization; *vasopressin*, which regulates serum osmolality by controlling renal water clearance; *mineralocorticoids*, which control vascular volume and serum electrolyte (Na+ and K+) concentrations; and *EP and norepinephrine*, which stimulate the metabolic rate of virtually all tissues in acute stress by increasing the rate of glycogenolysis within cells. Maximal sympathetic stimulation may increase metabolic rate by 25–100% (Pocock & Richards, 2004, p. 574). *Insulin* maintains euglycemia in the fed and fasted states. Insulin and glucagon, which offer adversary roles, are synthesized and released in direct response to changing levels of circulating fuel in the blood. The release of cortisol,

EP, and norepinephrine is mediated by neuronal signals that indicate a demand for fuel. Interruption of circadian rhythms may disturb cortisol production (Merola et al., 1994).

Brain Homeostasis

Various processes maintain the stability of the brain's microenvironment. Functions that are controlled homeostatically are cerebral blood flow (CBF), BP, and cerebral metabolic rates of glucose, lactate, and oxygen ($CMRO_2$). *Homeostatic signaling* maintains the brain's set point through modulating synaptic efficiency and membrane excitability. Homeostatic signaling systems at the level of nerve and muscle cells interface with the mechanisms of neural plasticity to ensure that neural activity remains stable. Compensation is bidirectional, increasing or decreasing cellular excitability (Davis, 2006). The BBB actively regulates brain homeostasis, for example, the passage of solutes in both directions between brain and blood (De vries & Prat, 2005). There are no known sympathetic vasodilators to the cerebral vessels. The normal pH, or acid/base level is 7.4, that is, slightly alkaline.

The brain constitutes 2% of body weight, but receives 15% of resting blood flow, 20% of resting O^2, and 50% of glucose utilization. It has been described as *metabolically fragile* due to its high rate of energy consumption, absence of significant amounts of stored fuel (glycogen), and the rapid development of cellular damage when adenosine triphosphate (ATP) is depleted. Glucose and oxygen are derived momentarily from the blood. During excessive brain activity, neuronal metabolism can increase 100–150%. Sudden cessation of blood flow to the brain can cause loss of consciousness (LOC) in 5–10 sec (Guyton & Hall, 2006, p. 767; Ransom, 2003). ATP, among numerous other functions, provides the energy for maintaining the gradient in which the cell has a high concentration of K^+ and a low concentration of Na^+ relative to the external medium.

Nevertheless, during periods of activity, increases of blood flow and glucose consumption are much more than the increase of oxygen consumption. The vast majority of energy consumed is provided by the metabolism of glucose to carbon dioxide and water. The ratio of oxygen consumed by the brain to that delivered by flowing blood is constant across the brain. While oxygen consumption of gray matter is four times that of white matter, the ratio of oxygen delivered to that consumed is the same, that is, *oxygen extraction fraction (OEF)*. Even if blood flow and glucose consumption increase by 10%, oxygen consumption does not, since local energy consumption of a task-related response can be only 1%. Nevertheless, glucose utilization (*glycolosis*) can continue for 40 min following cessation of task performance, that is, Wisconsin Card Sorting Test. The major use of the brain's energy actually stems from neural signaling processes and other intrinsic neural activity. The brain is not primarily a system responding to changes; rather, sensory information interacts with the ongoing operation of the system (Raichle & Mintun, 2006).

CEREBRAL AUTOREGULATION

Cerebral autoregulation is defined as maintenance of constant CBF in spite of changes in cerebral perfusion pressure, that is, mean arterial blood pressure (MABP) of approximately 50–150 mm Hg (Junger et al., 1997; Low, 2003). Larger cerebral vessels are modulated by the balance of SNS and PSNS tone, while medium and small cerebral vessels are altered primarily by autoregulation. The anterior circulation is supplied by the two internal carotid arteries, which supply the telencephalon (cerebrum) and the diencephalon (anterior brainstem, i.e., thalamus and hypothalamus). Autoregulation refers to the function of the local vasculature to dilate and constrict to compensate for neuronal activity, metabolism, and BP. A constant blood flow is maintained over a wide range of BPs: among the blood factors influencing vasodilation and constriction, and the response of smooth muscle cells within the cerebral vessels that constrict or dilate in response to wall stress, shear, and arterial pressure are levels of carbon dioxide, hydrogen ions (pH), potassium ions, and oxygen (Berne & Levy, 1998, p. 492; Guyton & Hall, 2006, p. 762; Jane et al., 2002). Sympathetic control is provided to the internal carotid artery by postganglionic fibers that originate in the *superior cervical ganglia* and accompany cerebral vessels of the internal carotid territory. Sympathetic vasoconstriction is described as weak, and the role of probably parasympathetic vasodilator fibers is unclear (Pocock & Richards, 2004).

Blood flow is kept within closely defined limits, in contrast to other organs. The brain is highly sensitive to ischemia, and unconsciousness occurs with only a few seconds' interruption of blood flow (Pocock & Richards, 2004). Several functions maintain brain perfusion (Iadecola, 2004): (1) *Systemic circulation*: blood is redirected from other areas to the cerebral circulation by the brain through its humoral and neural influence over the cardiovascular system. (2) *Cerebrovascular autoregulation* counteracts the cerebrovascular effects of fluctuations in arterial pressure. (3) *Functional regulation of CBF*, that is, functional hyperemia occurs when the activity of a brain region increases and flow to that region also increases. In this regard, it is significant that increased CBF may occur in the absence of spiking activity (Iadecola, 2004). CBF increase is not unequivocally linked to a specific electrophysiological event; rather, it is partially dependent on local circuitry and on the balance between excitation and inhibition. (4) Local metabolism affects autoregulation. The brain initiates reflexes with local and/or systemic effects (Pocock & Richards, 2004).

Sympathetic Effects

Sympathetic control of smooth muscle contraction of cerebral vessels is described as both weak and strong, with stress hormones playing a role. Cerebral sympathetic stimulation can constrict the cerebral arteries markedly to prevent high pressure from reaching the smaller blood vessels and causing stroke (Guyton & Hall, 1996, p. 785). The stress response has been described as increased BP and circulatory mobilization in the absence of significant metabolic demands. Individual reactions are inconsistent, and may vary from no change in BP to an increase of 50 mmHg or more. There are individual differences in hemodynamic patterns. Stress may result in varied patterns causing the same outcome: higher BP. Some persons manifest an increase in CO and a decrease in systemic vascular resistance (SVR). Others may vasoconstrict, leading to increased BP resulting entirely from increased SVR. *Hypertension* is related to increase in BP and to vascular *smooth vessel hypertrophy*, which creates SVR. This is a physiological adaptation to increased BP, which allows the vessel to maintain its ability to regulate blood flow at higher pressures. There is also a condition of *hyperkinetic circulatory state* in which BP elevations are due to abnormally high (CO) levels. Stress modifies renal function (sodium and fluid retention in some persons, which may heighten and prolong stress-induced BP elevation) (Sherwood & Carels, 2000). One may raise the question whether individuals with hypertension prior to an accident may be more vulnerable to cerebral accidents at the time of trauma.

The vasomotor center (anterolateral portion of the medulla) has noradrenergic fibers that excite the vasoconstrictor neurons of the SNS, affecting the vasculature of the brain (Guyton & Hall, 1996, pp. 210–211). Sympathetic fibers reach cerebral vessels by these routes: (1) carotid territory via postganglionic fibers that originate in the superior cervical ganglia (Guyton & Hall, 1996, p. 285); and (2) innervation of the vertebrobasilar territory via fibers that arise from the stellate ganglion (Mathew, 1995) (fusion of first thoracic ganglion with the inferior cervical ganglion of the bilateral sympathetic trunks) (Parent, 1996, p. 299). The fibers follow the tunica adventitia of the common and internal carotid arteries, possibly innervating the rostral part of the circle of Willis. Intracerebral arterioles are supplied with perivascular sympathetic nerves, whereas cerebral microvessels, capillaries, and venules may be supplied with or closely associated with intraparenchymal adrenergic nerves. Cerebral blood vessels are also innervated by intraparenchymal fibers, which originate from the locus ceruleus (LC).

METABOLISM

Victims of accidents causing TBI frequently report changes of weight, fatigue, loss of stamina, depression, and oversensitivity to heat and/or cold. Dysregulation of metabolism may be considered a contributor. Metabolism is the transformation of substances to provide energy, the synthesis of proteins for growth and repair. The breakdown of large molecules is catabolism, and the synthesis of large molecules from smaller ones is anabolism. Since hormones alter the way foodstuffs are utilized, they may alter the metabolic rate. Participating are catecholamines (adrenal medulla and SNS), thyroid, growth and male sex hormones, and female gonadal steroids. Hypothyroidism produces reduced tolerance to cold; increased thyroid secretion may cause the patient to feel hot (Pocock & Richards, 2004, pp. 574–575). Involved are both intracellular relationships and transfer between organs (e.g., energy transfer between adipose tissue to liver to muscles and back).

Metabolism is influenced by numerous *hormones*: *anabolic (insulin)*, which stores fuels; and *counterregulatory (contrainsular)*, that is, opposed to the actions of insulin: glucagon and EP, norepinephrine, cortisol, somatostatin, GH, and thyroid. Hormones of metabolic homeostasis respond to changes in dietary intake and physiological state in order to adjust the availability of fuels. Cortisol mediates changes in fuel metabolism that occur over long periods, for example, starvation (Jameson, 2001; Marks et al., 1996, pp. 374–378, 385, 703). Thus, stress hormones are essential in regulating healthy metabolism. Cortisol and insulin store energy after eating, and glucagon and catecholamines liberate energy during fasting (Brindley & Rolland, 1989).

Trauma can cause hormonal changes in metabolism that create secondary influences on the CNS. Examples include depression associated with hypothyroidism, psychosis occurring with glucocorticoid excess, and coma occurring with insulin excess (Baxter et al., 2004). Coordinated control of energy metabolism and glucose homeostasis require communication between organs and tissues. The brain integrates information obtained from several tissues and organs via both humoral and neuronal pathways. A *neuronal pathway* participates in cross-talk between the liver and adipose tissue. Afferent stimuli from the vagus and efferent sympathetic nerves to adipose tissue regulate energy expenditure, systemic insulin sensitivity, glucose metabolism, and fat distribution between the liver and the periphery. *Leptin* and other hormones convey information about energy storage from adipose tissue to the hypothalamus (Uno et al., 2006).

THE AUTONOMIC NERVOUS SYSTEM: ADAPTATION AND STRESS

The ANS is the neural circuitry maintaining homeostasis and health. Its divisions are enteric, parasympathetic cholinergic, sympathetic cholinergic, sympathetic noradrenergic, and adrenomedulary. The IS is subject to neural and hormonal modulation, including that stimulated by stress. The ANS is not only neuronal; its components do not function autonomously from the NS. The ANS also includes various neuroendocrine systems: vasopressin, renin-angiotensin, and hypothalamo-pituitary-adrenocortical (Goldstein, 2003). It consists of both spinal and cranial nerves. The SNS facilitates the mobilization of energy, increases catabolism, and promotes those physiological responses supporting activity (including emergency responses). The PSNS (craniosacral) facilitates conservation of energy and supports physiological responses that conserve energy and increase anabolism, as well as physiological responses that promote rest, digestion, and restoration of body reserves (Powley, 2003).

The SNS and PSNS generally counterbalance these alternating functions, but each has independent activities. Sympathetic preganglionic neurons are primarily located in the intermediolateral cell column of the T2 to L1 segments of the spinal cord, controlling vasomotor, muscle vasomotor, sudomotor, pilomotor, and visceromotor effectors. Cranial parasympathetic neurons are located in the general visceral efferent column of the brainstem and sacral spinal cord.

The ANS contributes to adaptive physiological functioning during ordinary activities by altering organic performance during demands, danger, injury, and stress. It is sensitive to internal and external stimuli, and its effectors control organs with survival and reproductive importance. The ANS responds to external stimuli, for example, regulating pupil size to different intensities of light and changes of temperature.

The CNS component of the ANS, hypothalamus, and other portions of the limbic nervous system are involved with behaviors such as emotion and activities having survival value (visceral behaviors such as feeding, drinking, thermoregulation, reproduction, defense, and aggression). Afferent fibers trigger reflexes with some accompanying sensory experiences (pain, hunger, thirst, nausea, visceral distention, and chemical senses). The effectors of the ANS are smooth muscle, cardiac muscle, and glands. When internal stimulation indicates that a change in the body's internal milieu is required, the CNS, through its autonomic outflow, leads to compensatory functions.

Central Autonomic Network

In place of the initial concept of hypothalamic "centers" controlling autonomic and other homeostatic functions, the central autonomic network (CAN) is considered an effector structure involved with integrated autonomic, endocrine, and behavioral responses subserving homeostatic and social behaviors (emotional, sexual, etc.). Input to the CAN is both *humoral* and *visceral afferent*. The CAN (Bennarroch, 1997, pp. 1–28) involves multiple areas distributed throughout the neuraxis: insular, anterior cingulate, and ventromedial prefrontal cortices; central nucleus of the amygdala; bed nucleus of the stria terminalis; several nuclei of the hypothalamus; periaqueductal gray matter; parabrachial nucleus; nucleus of the solitary tract (NST); ventrolateral medulla; ventromedial medulla; and medullary lateral tegmental field (Benarroch, 1997a, pp. 3–4). These are reciprocally interconnected and constitute a functional unit. The insula is the primary viscerosensory cortex. The insular and medial prefrontal cortices are involved in the highest level of integration of viscerosensory and visceromotor responses (Benarroch, 1997a, p. 10). The anterior cingulate gyrus and the medial prefrontal cortex are described as visceral "premotor" and autonomic cortices (Benarroch, 1997a, pp. 4, 10). The CAN is involved in tonic, reflexive, and adaptive control over the ANS, and integrates autonomic, hormonal, immunomodulatory, and pain-controlling responses to internal or external stimuli. In patients with medial temporal lobe epilepsy, visceral symptoms are associated with hypometabolism of the posterior *insular cortex*, while emotional symptoms are associated with hypometabolism of the anterior portion of the ipsilateral insular cortex (Dupont et al., 2003). The anterior cingulate gyrus and insular cortex are intensively activated by pain, and the insular cortex has direct projections to the limbic system (Martin, 2003, p. 130). (The insular cortex is deep within the lateral sulcus; includes portions of the frontal, temporal, and parietal lobes; and participates in taste and internal bodily senses.)

Sympathetic preganglionic neurons are primarily located in the intermediolateral cell column of the T2 to L1 segments of the spinal cord, controlling vasomotor, muscle vasomotor, sudomotor, pilomotor, and visceromotor effectors (pp. 14–15). Cranial parasympathetic neurons are located in the general visceral efferent column of the brainstem and sacral spinal cord. Transmission of information within the CAN involves virtually all neuroactive chemical substances so far described.

Cervical Sympathetic Ganglia

Cerebral vasoreactivity is under the control of the SNS through complex CNS circuits (medulla oblongata; pons; hypothalamus), *feedback* through the ECF (electrolytes; hormones), *temperature*, and *negative feedback baroreceptor mechanism* of the *tractus solitarius* (Landsberg & Young, 1992). Cerebral vasospasm can be relieved through electrical stellate or cervical ganglia blocks (Jenkner, 1995, pp. 63–73). In a personal communication (Fritz Jenkner, MD, Vienna, Austria), it was reported that hemispheric flow measured by electrical resistance ("rheoencephalography") increased towards normal levels after stellate block in 11 patients (varied etiology, including head trauma).

Cardiovascular Regulation

The ANS mediates many of the *cardiovascular changes* following brain injury. Cardiac output was normal following a gunshot to the brain in experimental animals pretreated with atropine: The untreated animals' low cardiac output and bradycardia are consistent with a cholinergic effect, and the initial transient hypertension is consistent with a sympathetic effect. The cholinergic effect tends to predominate, producing a net reduction in heart rate and myocardial contractility. Cholinergic blockage is more successful than adrenergic blockage in improving the animals' cardiovascular status (Brown et al., 1982).

Cardiovascular regulation, particularly BP, is integrated by the medullary cardiovascular center. Baroreceptors (actually sensitive to stretch) are positive and negative feedback loops influencing vasodilation and bradycardia, versus vasoconstriction and tachycardia. Efferent pathways controlling the heart and dilation of the blood vessels originate in the medulla. Afferent centers (baroreceptors) are in the aortic arch (right and left *vagus nerves*) and the heart (vagus nerve, CN X) and the carotid body at the intersection of the external and internal carotid arteries (glossopharyngeal nerve, CN IX), both of which enter the medullary *nucleus of the tractus solitarius*, thence to the *dorsal motor nucleus of the vagus*, which provides efferent stimulation to the heart (Boulpaep, 2005). Vagal afferents provide the medulla information about hollow organs (distention of the heart, stomach, bronchioles, and blood vessels), and also chemical information (oxygen and carbon dioxide levels, blood pH, and glucose (Richerson & Boron, 2003).

Respiratory Regulation

Respiration maintains oxygen and carbon dioxide pressure within narrow limits. Neuronal pools subserving involuntary ventilation originate in the caudal medulla and give rise to descending pathways in the ventrolateral brainstem and spinal cord. The fastigial nuclei of the cerebellum are involved in ventilatory inhibition via the anterior and posterior lobes, including volitional breathing (Simon, 2001). Since aprosodia has been described as a lateralized phenomenon, it is noteworthy that, in fact, the fastigial nuclei give rise to both crossed and uncrossed axons.

Cholinergic nerves are the major neural bronchoconstrictor mechanism in determining *airway caliber*, that is, controlling smooth muscle tone, airway blood flow, and mucus secretion. They arise in the nucleus ambiguus, travel down the vagus nerve, and synapse in parasympathetic ganglia (Barnes, 2003). Airway constriction is enhanced by vagus nerve (parasympathetic) stimulation of the smooth muscle of the cartilaginous airways, membranous bronchioles (primarily smooth muscle), and alveolar ducts. Sympathetic stimulation inhibits airway constriction (Guyton & Hall, 2000, pp. 440–441; Staub, 1998b). Although sympathetic innervation of the central portions of the lung is weak, the bronchial tree is exposed to circulating norepinephrine and EP (beta receptors) released into the blood by SNS stimulation of the adrenal medulla. PSNS fibers (vagus nerves) release acetylcholine (ACH), causing mild-to-moderate constriction of the bronchiole.

Nucleus of the Tractus Solitarius

The solitary nucleus is placed dorsally and bilaterally in the medulla, between the dorsal afferent columns, that is, fasciculus gracilis (medial), fasciculus cuneatus (more lateral), and extends into the caudal portion of the pons. It is the first relay and integrative center for viscerosensory inputs: (1) It initiates medullary reflexes controlling cardiovascular, respiratory, and other autonomic functions; and (2) it relays viscerosensory information to all other regions of the autonomic core (Benarroch, 1997a, p. 10).

THE GASTROINTESTINAL TRACT

The effectors of the gastrointestinal (GI) tract control motility, which is required to ensure adequate absorption of nutrients, electrolytes, and fluid. The lamina propria of the intestines receives an extensive plexus of nerves that rise from the ENS, sympathetic, and dorsal root ganglia (Bellinger et al., 2001). The GI tract is under a diffuse and complex control of higher centers and the brainstem via the PSNS and SNS, which stimulate the ENS. This may account for the lack of close association between stress and gut dysfunctioning discussed later. The GI tract has direct SNS stimulation. PSNS control is via ganglia, which are part of the enteric NS. Within the gut are the mucosal, submucosal, and myenteric plexes. There is a feedback loop to the brainstem (NTS and *dorsal motor nucleus*, see diagrams [Parent, 1996, p. 303; Snape, 1996, p. 681]).

Several IS products (e.g., interleukins) impinge on the GI tract (Murison, 2000) where they diffuse to secretory and smooth muscle cells, affect their activities, and modulate the function of neurons in the GI tract (Berne & Levy, 1998, p. 591).

THE ENTERIC NERVOUS SYSTEM (ENS)

The ENS is a highly complex, semiautonomous network, extending over the full extent of the GI tract, including the pancreas and liver. Its approximately 80–100 million neurons are as many as the spinal cord. Although some peripheral ganglia are composed of postganglionic parasympathetic neurons, the ENS does consist of ganglions located in the wall of the GI tract. These contain reflex pathways controlling contraction of the muscular coats of the GI tract, secretion of gastric acid, intestinal transport of water and electrolytes, mucosal blood flow, and so forth. Separation from autonomic input may have no obvious effect upon some tissues (Standing, 2005, p. 238). Afferent neurons are sensitive to extension, chemistry (including specific nutrients), and mechanical stimulation. These activate interneurons that relay signals to secretomotor neurons. Among the neurotransmitters present in both the GI tract and the brain are ACH and vasoactive intestinal peptide (VIP) (Connors, 2005a). The ENS controls intestinal motility and secretion, although some controversy exists whether this is largely independent of CNS influences, since there is bidirectional communication between the GI tract and the high CNS. Also, cells of the IS in the intestine are sensitive to neurotransmitters. These mast cells process information from the brain to the ENS, respond to interneurones of the ENS, and affect the smooth muscle of the intestine. In short, GI function is under the control of the ENS, GI hormones, and the IS, a fail-safe mechanism in times of impaired function (Binder, 2005). Parasympathetic inputs (dorsal motor N.) of the vagus control the motility of the esophagus and stomach, while the sacral parasympathetic N. contributes to control of motility of the distal colon and rectum. Local reflexes control gut motility and secretion, triggered by such stimuli as local distension of the intestinal wall, distortion of the mucosa (lining), and chemical contents in the lumen (Benarroch, 2007). The number of neurons, the interconnections, and control of motility have suggested that it is a relatively independent "brain" in the gut. Nevertheless, extrinsic projections from the CNS are involved with motor responses of the GI tract other than peristalsis. The ENS also serves as an extensive postganglionic station for the vagus nerve (CN X) (Powley, 2003). It has more neurons than the ANS, and its anatomical and functional complexity is second only to the CNS. It is composed of sensory neurones responsive to tension of the gut wall and chemical environment, and also interneurons and motor neurons that control muscles of the gut wall, vasculature, and secretions. Its agonist and antagonist neuropeptide neurotransmitters are also utilized in the CNS. The complex interaction between excitatory and inhibitory neurotransmitters coordinates bowel activity through effects on nerves, muscle, muscosa, and blood vessels. It controls the motor and secretory activities of the digestive system, the tension of the gut wall and the vasculature, and records the chemical environment. Thus, it plays a crucial role in the maintenance of homeostasis (Parent, 1996, p. 302), and controls the function of the GI tract, pancreas, and gallbladder (Iverson et al., 2000).

Bidirectional communications occur between the ENS and CNS (using *irritable bowel syndrome* as a model). Stress reactions are modulated by neural pathways, and also by immunological and endocrinological mechanisms. Vagal efferents (PSNS) to the integrative program circuitry of the ENS minibrain (interneurons, sensory, and motor neurons). Central processing and autonomic regulation play a role in what has been called the *brain-gut* dialogue. Psychosocial factors strongly affect the digestive tract (Mulak & Bonar, 2004). Most of the neurotransmitters and modulators that function in the CNS also function in the GI tract. Enteric neurons often contain more than one neurotransmitter or neuromodulator, and may release more than one neuroactive substance. They are responsive to a variety of chemicals originating elsewhere, for example, hormones or medications The ENS shares some neuroactive peptides with the CNS.

A variety of responses involve projections from the CNS: Although the enteric system is regulated to some extent by parasympathetic and sympathetic inputs, interneurons between neural plexi in the wall of the GI tract connect afferent sensory fibers with efferent neurons to smooth muscles and secretory cells to form reflex arcs that are located wholly within the GI tract wall (Genuth, 1998c). The enteric NS is the primary source of gut motor control and secretion. However, in times of *emergency or stress*, the PSNS and ANS can override intrinsic enteric activity (Parent, 1996, pp. 302–303; Snape, 1996).

PSNS and SNS Innervation

Extrinsic innervation of the gut by the parasympathetic and sympathetic components of the ANS provides a second level of control of motility and secretion, but can override intrinsic enteric activity in cases of emergency or stress. Brain corticotropin-releasing factor (CRF) activates both sacral PSNS and SNS outflow to the GI tract. The PSNS innervating the gut originates in the medulla oblongata and sacral spinal cord. Vagal efferents synapse within the enteric NS, innervating the whole GI tract, except the rectum. Much of the control of the GI function can be achieved by the enteric NS independently of central influences. Its neurons greatly outnumber by far the rest of the autonomic neurons, that is, probably over 100 million. The CNS exerts powerful control over GI function, mainly through the vagus nerve. Parasympathetic innervation of the alimentary tract is provided mainly by the motor neuron of the vagus nerve, augmented by sacral PSNS preganglionic efferent neurons traveling within the pelvic nerve. From the perspective of stress and mood, it is significant that the vagus N. interfaces with the nucleus ambiguus (striated esophageal muscles) and the nucleus tractus solitarius (NTS), which receives afferent stimuli, visceral stimuli (chemical, mechanical, and temperature changes), and projects to the amygdala, the reticular formation, and esophageal motoneurons creating a vagovagal esophageal reflex (Taché, 2003).

HORMONAL SYSTEM

The classical definition of a hormone is a substance produced in a ductless gland, secreted directly into the blood, and delivered to a target gland, enhancing its growth and activity. Chemical factors affect strength, stamina, mood, cognitive efficiency, and so forth. A hormone functions in cell-cell signaling of various kinds: work in the same tissue (*autocrine*), adjacent tissues (*paracrine*), or distally (transported through the bloodstream as a *hormone*) (MacGillivary, 2001). *Autocrine* refers to cellular self-stimulation of a factor and a specific receptor for it. Cells of the IS have a substantial influence, serving as autocrine and paracrine factors exercising local control over a variety of immune functions (Weigent & Blalock, 1996). Some hormones are released into the circulation in a precursor form and acquire a biologically active form only after being modified by enzymes, for example, thyroid hormones $T_4 \rightarrow T_3$ (Pekary & Hershman, 2001). The hormones released to a distant target organ have a specific action, determined by binding to a *receptor* in target tissue. A signaling molecule binding to a receptor is known as a *ligand*. The hormone receptor binding is transduced into a *signal* within the cell, causing a physiological response such as enzyme activation or new protein synthesis for cell growth or differentiation. One type, *polypeptide hormones*, conveys information among cells and organs, controlling growth, development, reproduction, and maintenance of metabolic homeostasis. These hormones act on distant organs by being transported through the bloodstream as well as being cell-to-cell communicators. The latter function is exemplified by their secretion within neurons of the central, autonomic, and peripheral nervous systems, where they act as neurotransmitters (Habener, 2003).

As mentioned, the classical definition of a hormone is a substance produced in a gland, secreted into the blood, and delivered to a specific target gland. However, the same chemical may be delivered to nontraditional organs, where, with the appropriate receptors, a different response occurs. Further, varied receptors for the particular chemical exist, so that genetic differences occur between individuals according to their receptor pattern.

Multiple hormones cooperate in the coordination of development, reproduction, and homeostasis. Particular hormones act at multiple sites. There are a variety of receptors for a given hormone, located in different organs and tissues, whose response determines the hormone's biological effect. Each hormone can have multiple responses, combining to give the overall effect (Jameson, 1996).

Hormones and related substances are secreted all over the body, not merely in the classic endocrine glands. However, as a signaling tool responding to an integrated assessment of the system, designed for a long-distance or generalized systemic response by relatively specific tissues, one must consider the hypothalamus, which is high in the hierarchy of endocrine functions. This has been described as the efferent arm of the visceral brain, receiving information from the periphery, which it integrates with that of the internal milieu in order to adjust important functions such as sympathetic activity and endocrine glands (Brown & Zwilling, 1996).

Types of Hormones

Hormones, released by endocrines and transported through the bloodstream, bind to specific receptor molecules to regulate target tissue function. Just as in the case of drugs, hormones are classified according to the functions of the receptor that mediates the effect (Frohman & Felig, 2001). Hormones are derived from the varied major classes of biologic molecules (Baxter et al., 2004). They are divided into classes (Baxter et al., 2004; Jameson, 2001): *amino acid derivatives* (dopamine, catecholamine, thyroid hormones); *small neuropeptides* (gonadotropin-releasing hormone [GnRH]); *large proteins* (insulin, luteinizing hormone [LH], PTH); *steroid hormones* (cortisol, estrogen); *vitamin derivatives* (retinoids, e.g., vitamins A and D); and *cholesterol derivatives.* Almost all hormones influence brain activity. Neural and endocrine factors influence the IS, while cytokines (secretions of lymphocytes, monocytes, and vascular elements) modulate both the neural and endocrine functions. Both nerve and endocrine cells have secretory and electrical potential capacities (Reichlin, 1998). The immune, neuroendocrine, and CNS systems have reciprocal effects on paracrine, autocrine, and endocrine control of release of cytokines, hormones, and growth factors (Ferrari et al., 2000). These regulate the complex multidirectional communications among these systems. Such aspects of cell biology aid the understanding of the aging brain physiologically or during neurodegeneration, for example, dementia of the Alzheimer's type (*DAT*).

Hormonal Action

Hormone action is highly complex. Hormone secretions activate behavioral traits in three major ways (Wingfield et al., 1997): (1) Hormones that are secreted in a paracrine fashion can act as neurotransmitters or neuromodulators; (2) True endocrines are secreted into the bloodstream, enter the CNS, and organize or activate behavior; (3) In a combination of items 1 and 2, a blood-borne hormone can act on neurotransmitters or neuromodulators to regulate behavior. It appears hormonal signals are restricted at times during the developmental cycle. It may be inferred that disruption of endocrine secretion interferes with behavioral signaling, activation, and organization, as well as affects physiology and morphology. The body's more than 100 hormones are capable of interacting with more than one cell type distributed in various tissues (Kahn et al., 1998). Particular hormones with characteristic functions can also create effects similar to other hormones by binding to the latter's characteristic *receptors*, for example, glucocorticoids and insulin. They may antagonize or enhance the responses to other hormones similarly through effects upon their receptors. Tissue responses to hormones vary with the receptors they contain.

The substances are secreted by the CNS and by neural elements that develop with the primeval gut. Hormones are concentrated at particular tissues for specific purposes (e.g., testosterone's effect upon spermatogenesis in the testis). Control of these glands evolved in the CNS, that is, *releasing hormones* (RH), released in the hypothalamus that stimulate intermediate hormones in the anterior pituitary gland (Frohman & Felig, 2001).

The gonads and adrenal gland produce five major groups of *steroid hormones*: estrogens, progestogens, androgens, glucocorticoids, and mineralocorticoids (Beato & Klug, 2000). An example of the behavioral significance of the receptor is found in thyrotoxicosis, after trauma, surgery, or illness. The number of binding sites for catecholamines increases. Thus, the heart and nerve tissues have increased sensitivity to circulating catecholamines. Under stress, the outpouring of catecholamines, together with higher levels of free T_3 and T_4 (consequent to decreased binding to thyroxin-binding globulin), cause an acute crisis, including agitation, restlessness, delirium, and coma (Greenspan, 2004).

HYPOTHALAMIC HORMONAL CONTROL

The hypothalamus is now considered to be the primary hormonal regulator, replacing the pituitary as the "master gland." It controls major endocrine systems, and is also involved in nonendocrine activities (e.g., regulation of body temperature, thirst, and food intake). There is a sequence (axis) of *hypothalamic releasing hormones (RH)* stimulating hormones of the various pituitary divisions, which in turn stimulate the release of hormones in the various target glands secreting hormones into the blood circulation. An exception is the role of the pituitary gland as the regulator of growth (Reiter & Rosenfeld, 2003). Anterior pituitary hormone release and synthesis is controlled by a family of peptidic releasing and inhibiting hormones, which are synthesized in hypothalamic neurons and released into the hypophysial portal vessels (McCann, 2006).

Hypothalamic neurones project to many sites related to pituitary control and to other sites. Further, hypothalamic hormones have a widespread distribution. The presence of hypothalamic hormones at an anatomic region or projection suggests that the anatomic focus may be functionally related to the correlated pituitary hormone. For example, CRF, the major releasing factor for corticotropin, is found not only in the hypothalamic paraventricular nucleus (PVN), but also in hypothalamic centers projecting to brainstem autonomic centers. Since adrenocorticotropin is critical in the regulation of processes associated with stress regulation. CRF in an extrahypothalamic region may indicate that this region is important in modulating stress behavior, or at least adrenocorticotropin release (Jacobson & Abrams, 1999). See Figure 3.8.

Hypothalamic hormones are secreted episodically, in some instances with a circadian rhythm. Glandular secretion control includes CNS signals (stress, afferent signals, neuropeptides) affecting the secretion of hypothalamic hormones and neuropeptides, which have (1) *systemic effects* or (2) *control target endocrine glands* via trophic hormones. Other hypothalamic hormones suppress GH, prolactin (PRL), or thyroid-stimulating hormone (TSH) secretion. Control of hormone stimulation occurs in particular cases via both the hypothalamus and pituitary gland, which monitor circulating hormone concentration and secrete trophic RH that activate hormone synthesis and release LH, follicle-stimulating hormone (FSH), TSH, and adrenocorticotrophic hormone (ACTH). Target hormones provide negative feedback via the hypothalamus-releasing hormones (Kronenberg et al., 2003). Feedback loops maintain hormone levels within a narrow range. The system can be perturbed by changes in circadian periodicity and stress. Acute physical stress can result in an increase in

ACTH, GH, PRL, EP, and glucagon. GH and PRL have no target organs to supply feedback hormone (Molitch, 2001). ACTH and cortisol release are mediated by the limbic and reticular-activating systems, for example, temperature, burns, trauma, major surgery, hypoglycemia, and intensive exercise (Fuller & Fuller, 2002; Merola et al., 1994).

Somatostatin (GH inhibitory factor): This is found in the brain (primarily hypothalamus, but including the CVO), the spinal cord, and numerous somatic tissues. It inhibits GH and TSH secretion, and has a wide variety of endocrine and organ effects, that is, digestive, vascular, IS, and so forth (Ghosh & O'Dorisio, 2001; Molitch, 2001).

Hypophysiotropic Releasing Hormones

These neuropeptides are releasing and inhibiting hormones that are synthesized. This is one way that the brain controls the anterior pituitary gland. There are generally at least two hypothalamic hormones for each pituitary hormone. These are synthesized in hypothalamic neurones and released into the hypophysial portal vessels. Their cell bodies are in nuclei surrounding the third ventricle. Capillaries, outside the BBB coalesce into *portal veins*, which carry releasing and inhibiting factors to the *troph cells* that actually secrete the anterior pituitary hormones. In addition, short portal vessels carry blood from the neural lobe (posterior pituitary gland) across the intermediate lobe to the anterior pituitary lobe, thus delivering neural lobe hormones directly to the anterior lobe in high concentrations. These substances integrate hormonal and neural mechanisms by acting both as secretagogues for anterior pituitary hormones and as extrapituitary peptide neurotransmitters (Menzaghi et al., 1993). These are primarily secreted by the paraventricular hypothalamic (PVH) and arcuate (ARC) nuclei. RH are secreted by small-diameter neurons mainly in the periventricular portion of the hypothalamus and discharged into the *median eminence and neural stalk*. They enter leaky capillaries and travel via *pituitary portal veins* to the anterior pituitary. Since the main target of four of the anterior pituitary hormones (TSH, ACTH, LH, and FSH) is other endocrine tissue, that is, are themselves RH, the anterior pituitary has an integrative function (Barrett, 2005a; Netter, 1983, pp. 203–211 for text and diagrams of the hypothalamus and pituitary gland).

Secretions of the hypothalamic RH are influenced by multiple feedback from environmental and systematic reactions and levels. Of special concern is the HPA axis (comprehensive diagram, Cone et al., 2003) and its key limbic regulator (the hippocampus), which are sensitive to glucocorticoids and their perinatal programming actions (Seckl & Meaney, 2004). RH are secreted episodically, in some instances with a circadian rhythm. *Hypothalamic physiotropic* hormones are those that are secreted into the hypophysiotropic portal blood vessels and serve hormonal release of the anterior pituitary gland (growth hormone-releasing hormone [GHRH]), somatostatin, dopamine, thyrotropin-releasing hormone (TRH); corticotropin-releasing hormone (CRH), and GnRH. Most anterior pituitary hormones are controlled by stimulatory hormones but GH and PRL are also regulated by inhibitory hormones. Some RH control multiple hormones (TRH stimulates both TSH and PRL release). Some pituitary hormones are stimulated by more than one RH, for example, ACTH release is stimulated by both CRH and ADH (Barrett, 2005b; Webb & Baxter, 2007).

HYPOTHALAMIC AUTONOMIC CONTROL

The hypothalamus is the master controller of autonomic function, and innervates all autonomic relay centers, including sympathetic neurons. The hypothalamus has been described as the efferent arm of the visceral brain. It maintains homeostasis by integrating endocrine, autonomic, and behavioral responses (Cone et al., 2003). Information from the periphery is combined with that of the internal milieu, leading to the adjustment of sympathetic activity and endocrine secretions (Brown & Zwilling, 1996). Neurohormones with anterior pituitary-regulating functions (*hypophyseotropic hormones*) are located in the hypothalamus and the median eminence. If the substance is of unknown chemical nature, it is termed a *releasing factor*; if its identity is established, it is an *RH* (Reichlin, 1998).

Hypothalamic Neurotransmitters

These affect the CNS and the anterior and posterior pituitary glands (Webb & Baxter, 2007).

Bioactive amines: Dopamine, norepinephrine, EP, serotonin ACH, gamma-aminobutyric acid (GABA), and histamine
Neuropeptides: VIP, substance P (SP), neurotensin, components of the renin-angiotensin system, cholecystokinin (CCK), opioid peptides, neuropeptide Y, and so forth.
Amino acids: Glutamate and glycine

GnRH is the neuroendocrinological trigger and pacemaker of mammalian reproduction. It is released in the median eminence and stimulates the release of both FSH and LH. Libido involves the participation of both hypothalamic and extrahypothalamic sites. In cases of hypothalamic disease, loss of libido is associated with impaired release of GnRH, with a subsequent decrease in testicular testosterone in men. These levels can be low in women with ACTH deficiency and secondary adrenal insufficiency. In women, libido is more related to adrenal androgens (Cooper, 2004). *Luteal hormone-releasing hormone* (*LHRH)* neurons migrate from the medial olfactory epithelium into the basal forebrain. Failure of this migration results in suppression of the pituitary-gonadal axis. The physiological development of an operative hypothalamic-pituitary-gonadal (HPG) system requires the presence of an intact IS (Marchetti et al., 2001). LHRH functions in the brain-pituitary-gonadal axis; it also directly influences immune organs and cells. GnRH neurons migrate from the medial olfactory epithelium into the basal forebrain (septal-preoptic-hypothalamic), that is, medial basal hypothalamus, the infundibulum, and periventricular area of the third ventricle (Cone et al., 2003); from these regions, axons are sent to the median eminence. Failure of this migration results in suppression of the pituitary-gonadal axis (Cone et al., 2003; Marchetti et al., 2001). Thus, LHRH cell bodies are scattered through the base of the brain through various regions. The physiological development of an operative HPG system requires the presence of an intact IS (Marchetti et al., 2001). GnRH axons extend to other regions of the hypothalamus and to the cortex. These do not seem to be implicated in the control of sexual behavior in humans. GnRH is secreted in a pulsatile manner, stimulating the anterior pituitary secretion of LH and FSH, which stimulate the secretion of gonadal hormones in the ovary and testis (spermatogenesis) (Junqueira & Carneiro, 2005, pp. 397–399). Astroglia stimulate LHRH production via growth factors and prostaglandins (Wetsel et al., 1998).

Neurotransmitter systems from the brainstem, limbic system, and other areas of the hypothalamus convey information to GnRH neurons. Glutamate and norepinephrine provide stimulatory drive to the reproductive axis, while GABA and endogenous opioid peptides provide inhibitory drive to GnRH neurons. CRH, released from the hypothalamus during stress, in addition to releasing ACTH by the pituitary and corticosteroids by the adrenals, inhibits central gonadotropin release (Ferin, 2000). The endogenous opioid peptides function in times of stress in analgesia, neuroendocrine regulation (including suppression of reproductive function), autonomic regulation, modulation of cardiovascular responses, CBF, and regulation of immune function and behavior. Chronic stress results in the interruption of the normal reproductive cycle (Ferin, 2000).

- PVH: TRH, CRH, Somatostatin
- ARC: GHRH, GnRH, dopamine

The releasing factors stimulate the Anterior Pituitary (AP) to secrete trophic hormones that stimulate the target endocrine glands (ACTH, TSH, GH, LH, FSH, PRL). Some of these are multifunctional. Most anterior pituitary hormones are controlled by stimulating RH. However, GH and PRL are also regulated by inhibitory hormones. TRH is stored in the median eminence of the hypothalamus and transported via the pituitary portal venous system down the pituitary stalk to the anterior pituitary gland, where it controls the synthesis and release of TSH. Like CRH, it is found in other portions of the brain (Barrett, 2005b).

Magnocellular (Posterior Pituitary Target)

These are located in the supraoptic nucleus (SON) and PVH nuclei (medial basal hypothalamus) whose targets are the *kidney, uterus, and mammary glands*. Magnocellular neurons secrete arginine vasopressin (AVP) or oxytocin (OXY), which are carried by a neuronal process via the pituitary stalk to the posterior lobe of the pituitary. The *posterior pituitary* lobe, an extension of the brain, contains cell bodies whose origin is in the supraoptic and paraventricular nuclei of the hypothalamus. The posterior lobe is one of the CVO whose vessels breach the BBB.

HYPOTHALAMIC SECRETIONS

The neurosecreting nuclei are found in the PVH, ARC, and lateral hypothalamic area (LHA). They secrete AVP, OXY, melanin-concentrating hormone (MCH), orexin-hypocretin (ORX), and proopiomelanocretin (POMC). POMC is the precursor of both ACTH and the opioid β-endorphin, an endogenous opiate involved in stress and high exercise-related analgesia. The targets are neurones (spinal sympathetic preganglionic neurons; autonomic preganglionic neurones in the brainstem). This deficiency may reflect hypothalamic damage and be linked with posttraumatic sleep-wake disorders (Baumann et al., 2005). Endorphins are neuroactive peptides, including enkephalins, with morphine like actions. CRH activates β-endorphin, which inhibits both the adrenergic and HPA components of the stress response (Cummings, 1985; Friedman & McEwen, 2004; Money et al., 1996).

Corticotropin-Releasing Hormone/Factor (CRH/CRF)

There is a HPA under minute-to-minute control by hypothalamic releasing factors (CRF). The hypothalamic factor stimulates ACTH secretion in the corticotroph cells of the anterior pituitary. CRF was finally identified as a relatively large neuropeptide and later named CRH (Barrett, 2005a, p. 1011), with a comprehensive diagram of hypothalamic-pituitary interaction (Reichlin, 1992). CRH is stored by neurons of the median eminence and released into the venous portal system to be transported to the anterior pituitary. Although the term CRF is still currently used, I will use the term CRH to avoid the error that two separate substances are involved. CRH activates corticotrophs of the anterior pituitary to secrete ACTH, which in turn stimulates secretion of glucocorticoids, mineralocorticoids, and androgenic steroids from the adrenal cortex (see next).

CRH/CRF Stress Reaction

Corticotropin-releasing hormone, which is widely distributed through the brain, plays a complex role in integrating the *stress reaction* of endocrine, autonomic, immunological, and behavioral responses to stress. CRF is released after stress, and is believed to integrate the behavioral, neuroendocrine, and autonomic responses to stress. By activating the SNS, it reduces cellular immune function and specific antibody responses, with implications for the association between increased SNS outflow and reduced immune function in aging and stress. There are effects upon behavior, the HPA axis, and the ANS. The brain communicates with the IS via the ANS. Preganglionic spinal cord fibers synapse in the sympathetic chain or collateral ganglia and enter the lymphoid organs along with the vasculature (Irwin, 1993). CRH mediates the HPA axis, the SNS outflow in the stress response, enhances immunosuppression, cyktokine production, and local inflammation. A primary stress response is the release of CRH and *AVP* (also known as ADH) from the hypothalamus into the hypothalamic-hypophyseal portal circulation. This carries them to the anterior pituitary, stimulating the release of ACTH, which is transported through the blood supply to the adrenal cortex, increasing cortisol secretion into systemic blood, and then commencing a negative feedback loop. Glucocorticoids (cortisol) inhibit secretion of GRH, AVP, and ACTH (Orth et al., 1992). Since AVP may not be glucocorticoid suppressible, it may be important in maintaining chronic stress responses (Michelson et al., 1995). There does not seem to be a feedback system, but secretion may be inhibited by the release of dopamine from the hypothalamus, for example, during breast stimulation (Barrett, 2005a).

To illustrate the complex systemic interactions and effects, there are extrapituitary sources of ACTH, such as *CCK*, which is described as a more potent ACTH secretagogue than the far better sources CRH and vasopressin (Fehm & Born, 1989). Catecholamines and CRH neurones participate in reciprocal reverberatory innervation. Serotonergic and cholinergic systems stimulate these systems also (Tsigos & Chrousos, 1996). *Cortisol* activates the metabolic processes necessary for sustained physical demands and tissue repair.

CRH is a neuropeptide controlling release of the anterior pituitary hormone *ACTH*. CRH's primary location is the PVN of the hypothalamus. The PVN stimulates the pituitary-adrenal-cortical axis during stress. After stress, depletion of hypothalamic CRH is followed by a "compensatory" upregulation of CRH production (Baram et al., 1997). Fibers project to the median eminence of the hypothalamus, participating in control of the pituitary-adrenal axis. CRH is also found in the hypothalamus, cortex, limbic system, and spinal cord. These systems are major mediators of behavioral responses to stimuli producing fear and anxiety. The hypothalamic CRH system differs in function and regulation from the extrahypothalamic CRH system. CSF levels of CRH probably reflect the activity of an extrahypothalamic system. In the CNS, the highest density of CRH receptors are in the cerebral cortex and cerebellum (Grigoriadis et al., 1996). It is also found in the LC, dorsal motor nucleus of the vagus, amygdala, and certain portions of the basal ganglia. It likely functions as a CNS neurotransmitter in these regions. Plasma CRH represents secretion from peripheral sources and hypothalamic secretion in the median eminence, while CSF concentrations represent activity of extrahypothalamic and perhaps hypothalamic CRH neurons. CRH neurons of cortex, limbic region, and brainstem are in close proximity to the ventricles, and probably contribute to CSF levels (Hartline et al., 1996). CRH is also found in the lung, liver, and GI tract, with unknown functional significance (Reichlin, 1992). CRH has been detected in the placenta, testis, ovary, gut, IS, adrenals, and heart, where it may act as a paracrine regulator of neurotransmitters. Further, CRH dysregulation may be involved in the development of *immune dysfunction* (Stenzel-Poore et al., 1996). CRH hypersecretion is suspected to mediate endocrine changes and behavioral symptoms of major depressive disorder (Bisette, 1997). It is significant in *energy expenditure*. CRH containing cell bodies implicated in autonomic and behavioral regulation are found in the medial preoptic (MePO) area of the hypothalamus (Richard, 1993). CRH has an *anorectic effect* opposed by somatostatin and OXY. Paradoxically, CRH has been demonstrated to enhance cold-induced thermogenesis in brown adipose tissue. Cytokines, secreted by IS cells, induce fever and alter the regulation of energy balance. They increase energy expenditure and decrease energy intake.

In addition, CRH may function as a neurotransmitter (Valentino et al., 1993). It may have a global role in the stress response, serving as a neurotransmitter in regions outside of the HPA, mediating behavioral and autonomic responses to stress. Introcerebroventricular administration, in doses stimulating ACTH release, even in adrenalectomized animals, altered both behavior and cardiovascular and GI function. Acute and chronic stress increase CRH levels in the *LC*, acting as a neurotransmitter to enhance LC discharge rates. The LC's extensive system of projections enables the CNS to influence endocrine, autonomic, sensory, motor, and behavioral functions.

INTEGRATED CONTROL OF CORTICOTROPIN-RELEASING HORMONE SECRETION

Corticotropin-releasing hormone gene regulation is downregulated by glucocorticoids (Baram et al., 1997). CRH is concentrated in the *PVN* of the hypothalamus, central nucleus of the amygdala, bed nucleus of the stria terminalis, substantia innominata, and the locus cerulus. Different pathways stimulate the CRF neuron, invoked by different stimuli (Levine, 1993). CRH in the CNS has behavioral effects independent of the pituitary-adrenal axis. Thus, it acts both as a secretagogue for anterior pituitary hormones and as an extrapituitary peptide neurotransmitter, similar to other hypothalamic releasing factors. Stress causes its release into the pituitary portal system and increased synthesis within the PVN. Centrally administered CRF agonists create these *behavioral effects*: suppress exploration of an unfamiliar environment; increased activity in a familiar environment; decrease operant conflict responding; alter learning and memory performance; produce taste and place aversion; enhance behavioral sensitization; increase stress-induced freezing; enhance defensive burying; induce seizures in high dosage; facilitate acoustic startle; induce conditioned fear; disrupt sexual behavior; decrease food intake; and increase grooming (Heinrichs et al., 1995). The action of CRF seems mediated within the CNS independent of the pituitary-adrenal axis. It has activating properties and coordinates behavioral responses to stress (Menzaghi et al., 1993). The hippocampus inhibits the pituitary-adrenal-cortical axis. It inhibits release of CRH by PVN cells, and may be more sensitive to changes in steroid levels than the pituitary (Watson & Akil, 1991).

Depletion of hypothalamic CRH is followed by compensatory "upregulation" of CRH gene expression, with modulation by the hippocampus and the amygdala (primarily its central nucleus. *Extrathalamic CRH* has important actions: It may activate biological systems within the brain in response to stress (catecholamines and autonomic arousal), suppress release of LH and GH, and suppress sexual activity and feeding activity, while stimulating respiration and behavioral arousal (Molitch, 2001).

The CRH receptors and negative feedback: CRH receptors are enumerated as CRF. The number of receptors modulates neurotransmission in both the endocrine and neuronal effector components of the response to stress (*"upregulation" and "downregulation"*). It is hypothesized that there is a hypersecretion of neuronal CRH (CRF) in PTSD, derived from numerous areas outside of the median eminence (Bremner et al., 1997). Increased secretion of CRH induces ACTH release from the pituitary and also increases plasma corticosterone (CORT) in rats. Negative feedback by glucocorticoids terminates the response. There are reciprocal reverberatory circuits, including ultra-short negative feedback loops, between the CNS, CRH, and catecholaminergic systems. These systems receive stimulatory innervation from serotonergic and cholinergic systems, and inhibitory input from GABA and benzodiazepine, and opioid systems of the brain, as well as hormonal feedback from HPA axis glucocorticoids (Tsigos & Chrousos, 1996). ACTH and cortisol are secreted in a circadian and ultradian (more frequent) rhythm, with the highest levels in early morning and a nadir in late afternoon and evening (Michelson et al., 1995). *Negative feedback* of glucocorticoid secretion inhibits ACTH and CRH, and occurs in three phases: (1) *Fast* feedback occurs within seconds; (2) *delayed* 2–4 h, and persists 40–48 h; (3) *slow* occurs from prolonged exposure to corticosteroids for more than 24 h, and it affects ACTH release and synthesis (Wilkinson, 1989). Negative humoral feedback loops (Molitch, 1995) involve the adrenal gland, cerebrum (hypothalamus and elsewhere), and the pituitary gland. These maintain body homeostasis by regulating target hormone levels within a narrow range (see "Lack of a BBB," that is, "The Circumventricular Organs—CVO"). The concept of negative feedback has limitations, that is, independent of level of cortisol: Cortisol release is circadian (low in late afternoon and evening; during sleep); limbic and reticular-activating system release of ACTH increases in severe stress, which cannot be modified by glucocorticoids as in moderate stress (Merola et al., 1994).

ALLOSTASIS: MAJOR INTEGRATIVE CONTROL

Allostasis is a major, and still unfamiliar, expansion of the mechanism of physiological control. During activity, particularly stress, it is more active than homeostasis in integrating physiological systems under the conditions of great demand (McEwen, 1995, 2004).

DEFINITIONS

Allostasis: The systems that maintain physiological systems in balance despite circadian changes, as future events are considered and difficult events actually occur. In contrast to homeostasis (whose most characteristic state is perhaps sleep), in allostasis, there are variable set points with time of day, anticipation of difficulty, and actual difficult events. It achieves stability through functional change, in contrast to homeostasis, whose set points are static or if otherwise, fatal.

Homeostasis: The stability of physiological systems that maintain life, for example, pH, body temperature, glucose levels, and oxygen tension, which are maintained within a narrow range of values (set points).

Allostatic states: When allostatic systems are activated for a physiologically excessive interval, there is an imbalance of the primary mediators, that is, excessive reduction of some and inadequate production of others. Examples include hypertension, perturbed cortisol rhythm in major depression or after chronic sleep deprivation, chronic elevation of inflammatory cytokines, and low cortisol in CFS.

Allostatic load: This is a persistent stress reaction. There is a state of chronic deviation from the normal range of functioning. Eventually, the system regresses to a physiological inability to maintain regulation within useful limits: loss of stamina, nonimpairing bodily disturbances, frequent minor illnesses, and mood disturbance, creating a grave vulnerability to developing minor illness. Within limits, this is an adaptive response to demands, reflecting the wear and tear from the daily and seasonal routines used to obtain food, survive, and obtain extra energy for unusual events and demands.

Allostatic overload ("burnout"): This is the consequence of an additional load caused by chronic unpredictable and uncontrollable events in the environment, disease, disturbance, and social interactions. Allostatic load becomes allostatic overload, serving no useful purpose and predisposing the individual to disease. It reflects the cost to the brain and body of the accumulations of damage and pathological states that accumulate over time (Koob, 2003).

HIPPOCAMPUS AND AMYGDALA CONTROL

The HPA is modulated through activation by the amygdala via the central nucleus, and inhibited by the hippocampus via the ventral subiculum and the prefrontal cortex, thus shutting off the HPA stress response (McEwen, 2003). The failure to turn off the HPA axis and SNS efficiently has been documented to age-related decline in animals, but not definitively in humans (McEwen, 1998). A psychoendocrine model of the control and response of the HPA system has been based upon a *cognitive theory of arousal*, that is, more mental than traumatic (Levine, 1993, citing Berlyne). Comparison is the mechanism that elicits or reduces arousal: novelty, uncertainty, and conflict. In addition, whether or not there is an endocrine response to stress, as well as its magnitude, there appears to be the presence of familiar social partners. The defining variable of capacity to cope with stress is a reduction or elimination of the secretion of adrenal glucocorticoids. The hippocampus inhibits the amygdala, the PVN, and LC/sympathetic systems (Tsigos & Chrousos, 1996).

THE AMYGDALA

The amygdala regulates both ANS activity and ACTH and cortisol production through its central nucleus. Thus, structural remodeling, in the context of stress, is likely to contribute to impaired cognitive function and affect regulation. These structural changes may be reversible if the stress is terminated on time. Chronic stress is a predisposing and precipitating factor in psychiatric illness (McEwen, 2004).

THE PITUITARY GLAND

The pituitary gland is predominantly controlled by hypothalamic stimulation, although both stimulatory and inhibitory factors probably exist for all pituitary hormones (GH, thyrotropin, adrenocorticotropin, FSH, LH). In contrast, PRL is under tonic inhibition by the hypothalamus (Jacobson & Abrams, 1999). Pituitary secretion is controlled by both endocrine feedback by circulating hormones through the vasculature and neural feedback to the hypothalamus. This is followed by the release of hypothalamic RH, and paracrine and autocrine secretions of the pituitary itself. All pituitary glands either have nervous system inputs that alter the set point of the feedback control system or have open loop elements that influence or override the closed loop controls (Cone et al., 2003). The median eminence of the hypothalamus is covered by capillaries, which penetrate it and come in contact with nerve endings. This capillary plexus is gathered into *portal veins* that enter the anterior pituitary lobe. The neurohypophysis is composed of the terminals of the hypothalamic nerve tract and capillaries. These do not have the BBB characteristic of most brain capillaries; rather, they permit diffusion of secretions into the circulation. The nerve tracts entering the posterior pituitary arise in paired nuclei: the SON above the optic tract, and the PVN on each side of the third ventricle. These tracts descend through the infundibulum and neural stalk to terminate in the neural lobe. *Vasopressin* and *OXY* are produced by the hypothalamo-neurohypophysial tract. Axons terminate in the median eminence and release neuropeptides into the pituitary portal circulation (Reichlin, 1998).

THE POSTERIOR PITUITARY (NEUROHYPOPHYSIS)

The posterior pituitary gland is part of the brain, and receives secretions from the hypothalamic PVN and supraoptic nuclei (Reeves et al., 1998; Robertson, 2001). Unlike the anterior pituitary, it is not a gland; rather, it is the distal axon terminals of these hypothalamic nuclei, which make up the neurohypophysis (Robinson, 2007). It is considered to be a distinct entity for clinical purposes.

VASOPRESSIN

Vasopressin, also known as the ADH, regulates blood volume and pressure (Robinson & Verbalis, 2003), and is a mediator of the stress reaction. It is a water-retaining hormone, with the central sensing system (osmostat) located in the hypothalamus just anterior to the third ventricle. Magnocellular hormones are neurons clustered in the paired paraventricular nuclei and paired supraoptic nuclei. Descending fibers from the PVN contain *OXY* and *vasopressin* (Parent, 1996, p. 720). The synthesized hormones are packaged into granules and transported by hypothalamic axons along nerve pathways of the pituitary stalk to the posterior pituitary gland, where the neurohormones *AVP or ADH* and *OXY* are released (Barrett, 2005a) from nerve endings directly into the general circulation.

Vasopressin is involved in the regulation of body water, that is, altering the concentration of solids in the blood (*osmolality*). Disorders are expressed as abnormal retention or excretion of water (Robinson & Verbalis, 2002). Water volume interacts with cardiac output and arterial pressure. Secretion of AVP is stimulated by osmoreceptors, in particular CVOs, which, lacking a BBB would therefore be sensitive to the water balance of the CSF. The osmosensitive neurons project to the PVN and supraoptic nuclei of the anterior hypothalamus. These nuclei synthesize AVP and then transport it to the posterior pituitary, which also lacks a BBB, permitting the AVP to enter the general circulation (Giebisch & Windhager, 2005). It is the water-retaining hormone, and along with thirst, is the primary regulator of osmolality. Its release is controlled by blood volume and pressure

(Cone et al., 2003; Robinson, 2007), mediated by sympathetic input and changes in sodium balance (aldosterone, renin, angiotensin). Vasopressin contributes to increased BP by constricting blood vessels, increasing water reabsorption in the kidneys, and conserving free water (Jameson, 2001). This process conserves water and intravascular volume. The adrenal cortical hormone, the mineralocorticoid aldosterone, controls fluid volume and BP through a different process, that is, controlling sodium retention (Dluhy et al., 2003).

Oxytocin

Lactation: The neuroanatomy of OXY secretion is similar to that of vasopressin (Robinson, 2007). Another hypothalamic-pituitary hormone (*PRL*) participates in lactation. The stimulus of suckling, after a half-minute delay, initiates a reflex (spinal cord, lateral cervical nucleus, midbrain), eventually reaching the oxytocinergic magnocelluar neurones in the hypothalamic supraoptic and paraventricular nuclei. OXY is released in a pulsatile fashion producing, a pumping action in the breast alveoli and promoting maximum emptying of milk at the nipple.

Parturition: During pregnancy, high levels of *progesterone* suppress uterine muscle excitability, while *estrogen* activates many events that initiate *parturition*. Uterine myometrial cells' contraction is responsive to OXY. As parturition approaches, inhibitory factors decrease, with a dramatic increase in the responsiveness of the uterus to OXY, which may play a role in the onset and intensity of the expulsive phase (Reichlin, 1998). When labor begins, reflexes from the contracting uterus increase OXY secretion. It stimulates uterine muscle to contract maximally and to clamp blood vessels to decrease blood loss. In addition, OXY has been associated with feeding behavior and satiety, gonadotropin secretion, response to stress, decrease of stress, and so forth (Robinson & Verbalis, 2003).

THE ANTERIOR PITUITARY (ADENOHYPOPHYSIS)

The anterior pituitary synthesizes and secretes six peptide hormones—GH, TSH, ACTH, LH, FSH, and PRL—under the control of hypothalamic RH and *inhibitory factors* (Barrett, 2005a). Anterior pituitary cell function is influenced by hypothalamic and paracrine pituitary-derived factors, and factors reaching the adenohypophysis via the systemic circulation (adrenal steroid hormones and cytokines derived from the IS) (Antoni, 2000).

Adrenocorticotrophic Hormone

Adrenocorticotrophic hormone is a peptide hormone, secreted in the anterior pituitary gland, and carried by the peripheral circulation to its effector, the adrenal cortex. It stimulates the synthesis and secretion of glucocorticoids, mineralocorticoids, and adrenal androgens. The level of secretion represents a balance between *stimulation and negative feedback*. There are late negative feedback effects, that is, 3 days later, upon ACTH secretion by HPA axis hormones (CRH and ACTH, but not cortisol) at the hypothalamic and pituitary levels (Posener et al., 1998). Stimulation is by the hypothalamic hormones *CRH and AVP*. Higher levels influencing CRH secretion include hypoglycemia, circadian rhythms, stress, and neurotransmitters. ACTH levels are higher during pregnancy, some periods during the menstrual cycle related to estrogen, and psychiatric illnesses (major depression and schizophrenia) and tumors of the pituitary gland and adrenal cortex. Negative feedback regulates ACTH secretion via closed loops of various lengths, that is, glucocorticoids from the adrenal cortex feeding back to the hypothalamus, hippocampus, and anterior pituitary, and CRH acting directly on the hypothalamus to control its own secretion (Rhodes, 2000). *CRH* receptors are found throughout the brain and spinal cord. This peptide is synthesized in many peripheral cells, including immune cells. Thus, CRH has other important CNS functions related to, or independent of, stimulating ACTH release. It causes central arousal, increased SNS activity, and increased BP.

It decreases reproductive function by decreasing synthesis of GnRH, and also decreases feeding activity and growth. In immune cells, CRH stimulates release of cytokines and augments their activity on target cells (Berne & Levy, 1998, p. 882). Stress stimulation of ACTH is mediated by CRH (circadian) and vasopressin. The actual prior ACTH level does not blunt the ACTH response to stress (Aron et al., 2004).

Growth Hormone (Somatotropin); Growth Factors

The primary function of GH is the promotion of linear growth. Its secretion is mediated by two hypothalamic hormones: GHRH and *somatostatin* (GH-inhibiting hormone) (Barrett, 2005b). Both contribute to GH's episodic secretion, tightly regulated by neural, metabolic, and hormonal factors (Aron et al., 2007a). GHRH and its inhibitor, somatostatin, are released in a pulsatile pattern during stages III and IV of deep sleep (Hansen & Cook, 1993). It is a polypeptide synthesized and secreted by specialized cells of the lateral wings of the anterior pituitary gland. Secretion is neurologically controlled by the hypothalamus, and is released in a circadian rhythm, responds rapidly to stress, and is blocked by pituitary stalk section (Melmed & Kleinberg, 2003). It is released by GHRH and is inhibited by somatostatin. Its level received negative feedback at the GHRH of the hypothalamus, and also other neurotransmitters and neuroactive substances. It is involved with the neuroendocrine system, for example, thyroid hormone, glucocorticoid sex steroid hormones, and metabolic fuels. A target for GH is IGF-1, a growth factor that affects neural outgrowth and nervous system development. GH and IGF have similar effects on somatic growth and development, but act more efficiently in combination (Gore & Roberts, 2003). Secretion is mediated by GHRH and somatostatin, and affected by maturation, aging, sleep, nutritional status, fasting, exercise, stress, and gonadal steroids.

Growth occurs at different rates during each developmental phase. In infancy, growth is primarily dependent upon nutrition, whereas in childhood, GH is the major determinant of growth. Sex hormones and GH control the adolescent growth spurt (MacGillivary, 2001). There is an interplay of GH/insulinlike growth factor (IGF), gonadal steroids, and T_4. GH rises two- to threefold during puberty, with an earlier and greater rise in girls (Grumbach & Styne, 2003; Reiter & Rosenfeld, 2003). Its primary function is the promotion of linear growth, which is primarily mediated by IGF-I. The latter increases protein synthesis. Insulin and thyroid hormones also have growth-promoting effects, and are needed for optimal growth. GH decreases protein catabolism by mobilizing fat as a more efficient fuel source. By sparing proteins, it promotes growth and development.

GH functions include metabolism of carbohydrates, lipids, and protein; and its targets include blood cartilage and bone, visceral organs, kidney, and skin. GH release is increased by sleep, exercise stress, and postprandial hypoglycemia. GH decrease is consequent to hyperglycemia and elevated free fatty acids, hormones, somatostatin, progesterone, glucocorticoids, and various neurotransmitters and drugs (Aron et al., 2004). These suppress the secretion of TSH, and of insulin and glucagon from the pancreas, along with other functions that influence the utilization of fuels. Exercise, stress, high protein meals, and fasting cause a rise in the mean GH level in humans. This is due to the increase in the frequency of secretory pulses of somatotrophs. Head trauma may cause isolated GH deficiency or multiple anterior pituitary deficiencies (Barrett, 2005b).

GH deficiency has reversible adverse factors in adults: increased fat and decreased lean tissue, and lower bone density. GH deficiency appears to be associated with psychological abnormalities; treatment was associated with decreased mental distress and increased well-being (Snyder, 2001).

Growth factors and neurotrophic factors: Growth factors play a role in development, the function of the adult nervous system, maintenance of structural integrity, and regulation of synaptic plasticity. Since these processes are altered in the degenerative events following acute nervous system injury, it has been speculated that they are involved in the structural alterations in response to injury and disease. After trauma or loss of innervation related to inflammation or immune activation, changes in synthesis or release of growth factors may protect vulnerable neurons or may enhance the detrimental effects of the injury (Bellinger et al., 2001).

Thyroid-stimulating Hormone

Thyroid-stimulating hormone acts on the thyroid gland to induce thyroid hormone synthesis. It constitutes about 5% of functional anterior pituitary cells.

Prolactin

Prolactin is secreted from lactotrophs of the anterior pituitary gland. It is under the inhibitory control of dopamine, which is largely secreted by hypothalamic tuberoinfundibular cells, reaching the anterior pituitary lactotrophs through the hypothalamic-pituitary portal system. PRL stimulates postpartum lactation. It is a lymphocyte growth factor that stimulates immune responsiveness (Melmed & Kleinberg, 2003). During pregnancy, in concert with many other hormones, it promotes additional breast development in preparation for milk production (Aron et al., 2004).

INTERMEDIATE PITUITARY

Melanocyte-stimulating Hormone (MSH; Melanotropin Peptides)

The *melanocortin* refers to the various melanotropin peptides. MSH is formed in the intermediate pituitary gland (in fetal life, and in the anterior pituitary at other times) as part of a larger molecule that includes ACTH, β-lipotropin, and β-endorphin. Some of these have potent opioid actions (morphinelike) in the CNS (Barrett, 2005a; Guyton & Hall, 2000, p. 880). α-MSH is considered more important as a centrally active neuropeptide than as a pituitary hormone. Synthesized in the hypothalamus, MSH receptors are widely distributed in the brain (Reichlin, 1998). In some lower animals, MSH is controlled by the hypothalamus in response to light or other environmental factors. While it is best known for causing melanocytes in the skin to form the black pigment melanin, it has significant central actions, that is, inhibiting and stimulating secretion of hormones, thermoregulation, behavioral (sex, learning, attention, motivation, arousal), and cardiovascular (including increasing CBF, promoting functional recovery from brain lesions, inhibiting opiates, and modulating the release of dopamine and norepinephrine) (Reichlin, 1998). Its neurobehavioral effects include (Hruby et al., 2003): learning, enhancing feeding behavior, sexual response, addiction, stress response, pain, injury, immune response, cardiovascular function, temperature control, kidney function, pigmentation, and inhibiting feeding through the satiety center of the hypothalamus (Nadel, 2003). α-MSH antagonizes the effects of proinflammatory cytokines, and induces the production of the anti-inflammatory cytokine IL-10 (Skottner et al., 2003). In the hypothalamus *melanocyte-stimulating hormone-releasing factor (MRF)* stimulates the production of MSH and β-endorphin (Iverson et al., 2000).

Just as there are several types of MSH, there are numerous *melanocortin receptors (MC-R)* that bind to them. MC-Rs regulate such physiological processes as feeding behavior, energy metabolism, stress, pigmentation, and immune response (Reizes et al., 2003): MC1R (pigmentation); MC2R (regulator of corticosteroid production by the adrenals); MC3R and MC4R (function in the nervous system regulating feeding and sexual behavior through involvement with central reward and the central dopaminergic systems); MC5R (pigmentation) (Wikberg et al., 2003). MC1R seems involved in the anti-inflammatory effects of MSHα (Skottner, 2003). α-MSH induces penile erections when administered in the lateral ventricle, specific hypothalamic nuclei, and the spinal cod (Wessells et al., 2003). MC4R functions in the regulation of food intake and energy balance, and potentially in reproduction; NS is found in the cortex, thalamus, hypothalamus, brainstem, and spinal cord (Foster et al., 2003; Schioth et al., 2003).

Energy homeostasis utilizes a central serotonergic (5-HT) system that converges the lateral ARC nucleus of the hypothalamus. Hormones inhibit food intake and increase energy expenditure by activation of POMC neurons. These modulate the melanocortin pathway as a final common output mechanism to promote energy homeostasis. This may be involved as a target for the prevention and treatment of obesity and type II diabetes (Cowley et al., 2003; Heisler et al., 2003).

THYROID GLAND

The thyroid gland is the largest endocrine organ, secreting primarily the hormone thyroxine (T_4), which has little biological activity, and is a precursor of T_3. Thyroid hormones promote normal growth and development (Rekiter & Rosenfeld, 2003), and regulate homeostatic functions, including energy and heat production. Other cells secrete calcitonin, which is involved in calcium homeostasis. Deficiency of the thyroid gland in fetal life or at birth results in retention of the infantile characteristics of the brain. If not corrected in early postnatal life, the damage is irreversible: hypoplasia of cortical neurons, retarded myelination, and reduced vascularity. In adult life, CBF is reduced, while cerebral oxygen consumption is normal, except for severe cases, which may exhibit cerebral hypoxia (Larsen & Davies, 2003).

Control of Thyroid Function

1. *Hypothalamic-pituitary-thyroid axis:* Hypothalamic TRH (predominantly the PVN) stimulates the anterior pituitary to release TSH, stimulating hormone secretion by the thyroid gland. TRH is also produced in the hypothalamus and outside the nervous system. Somatostatin and dopamine reduce TRH secretion. Negative feedback is provided by thyroid hormone and cortisol in the PVN and hypothalamic ARC nucleus (Visser & Fliers, 2000). 2. *Autoregulation* of hormone synthesis in relationship to the gland's iodine supply, and so forth (Greenspan, 2004). 3. *Neural*: In addition to input from the *suprachiasmal nucleus (SCN)*, nerves originating from the cervical ganglia and the vagus nerve terminate within the thyroid gland (Young & Landsberg, 1998). 4. *Glucocorticoids* (cortisol) suppress the thyroid axis during stress at both central and peripheral levels (Visser & Fliers, 2000).

Hypothyroidism

Thyroid hormone is essential for the development of the CNS. As mentioned, deficiency in fetal life or at birth causes retention of the *brain's infantile characteristics*: hypoplasia of cortical neurons with poor development of cellular processes, retarded myelination, and reduced vascularity. In the *adult*, deficiency of thyroid hormones causes less severe manifestations that usually respond to hormonal treatment. Hypothyroidism may be accompanied by reduced CBF (with usually normal cerebral oxygen consumption), decreased cardiac output (secondary to a drop in heart rate and contractility), and decreased BMR. Symptoms include mild lassitude, fatigue, slight anemia, constipation, apathy, cold intolerance, menstrual irregularities, loss of hair, weight gain, lethargy, somnolence, headaches, and psychiatric disorders (paranoid, depressive, agitation). There are reduced intellectual functions (speech, memory, mental speed) (Larsen & Davies, 2003).

Stress Effects

Stress (injury) creates the effect known as the "low (T_3) syndrome," that is, thyroid suppression, which can be mediated by glucocorticoids (Visser & Fliers, 2000). There is downregulation of conversion of T_4 to T_3 in tissues such as the liver. The reduced energy expenditure and protein catabolism is considered to be a beneficial adaptation mechanism to sustain the stressful condition. This condition is associated with stimulation of the HPA axis. Further, inflammation is accompanied by induction of cytokines, for example, *tumor necrosis factor (TNFα)* and IL-1 and 6. IL-6 in particular is a mediator of alterations of thyroid hormone metabolism.

THE PARATHYROID GLAND

Parathyroid hormone level is controlled by the calcium ion level. In the kidneys, PTH enhances calcium resorption and phosphate loss. In bone, there is more resorption than synthesis. Related issues affect vulnerability to osteoporosis, that is, the sedentary lifestyle common to persons with TBI, and low calcium intake (Barrett & Barrett, 2005).

GROWTH AXIS

Glucocorticoids antagonize the effects of sex hormones and GH on fat tissue catabolism, and muscle and bone anabolism. The consequence is increased visceral adiposity and decreased lean body mass (bone and muscle). Trauma may provoke hyperglycemia due to hypersecretion of insulin antagonistic hormones (Sherwin, 1996). It is believed that when nutrients and other substances should be redirected to the brain, suppression of GH and increases of *somatostatin* secretion ensues (Tsigos & Chrousos, 1996). Somatostatin is found all over the brain and spinal cord, but has its highest concentration in the hypothalamus. It inhibits GH secretion, and less so TSH (Molitch, 1995; Utiger, 1995). Prolonged elevated concentrations of glucocorticoids are toxic, and are associated with neuronal death, medical diseases, impaired affect, and cognitive disorders (Yehuda et al., 1995). Initially, there may be acute elevations of GH in plasma. Later, glucocorticoids suppress GH and inhibit somatomedin C and other growth factor effects on their target tissue.

Insulin: The primary anabolic hormone. It maintains euglycemia in the fed and fasted states. It is the major anabolic hormone: It promotes storage of nutrients in liver, muscle, and adipose tissue; inhibits fuel mobilization; and stimulates synthesis in various tissues of over 50 proteins, some of which contribute to growth (functions similar to IGF).

Thyroid hormone: Controls about 25% of basal metabolism in most tissues. It increases the rate of fuel consumption, and increases the sensitivity of target cells to other insulin counterregulatory hormones.

Glucagon: A contrainsular hormone secreted by the islets of Langerhans of the pancreas. It acts on certain tissues to make energy stored as glycogen and fat available, increasing blood glucose content. Its target is the liver, where its functions include the formation of glucose and ketogenesis (a metabolic process found when fatty acid levels are elevated, e.g., fasting, starving, and high-fat, low-carbohydrate diet, of interest to those who consider this approach to poorly controlled seizures).

Cortisol: Exerts a permissive action for many hormones, in addition to its own direct effects. Cortisol is involved in negative feedback, limiting feedback of its own production at the hypothalamic (CRH) and anterior pituitary levels (ACTH).

Parathyroid hormone: Regulates calcium and phosphorus levels.

Vasopressin: Regulates serum osmolality by controlling renal water clearance.

Mineralocorticoids: Control vascular volume and serum electrolyte (Na+ and K+) concentrations.

THE ADRENAL CORTEX

Activation of the HPA axis is one of the two primary stress reactions (in addition to SNS). The cortex of the adrenal gland comprises the external 90% of the gland by weight. It secretes more than 50 steroids, whose synthetics are significant as pharmacotherapeutic agents. There are three principle categories of steroid hormones (Miller & Tyrrell, 1995; Ross et al., 1995, pp. 614–641; Stewart, 2003): *glucocorticoids*, which influence carbohydrate, fat, and protein metabolism (cortisol, CORT);

mineralocorticoids, which affect the regulation of electrolyte and water balance (aldosterone, deoxcorticosterone), and *sex steroids* (androstenedione; dehydroepiandrosterone [DHEA]).

MINERALOCORTICOIDS

These are adrenal cortex hormones that control vascular volume and serum electrolytes (sodium and potassium) concentrations (e.g., *aldosterone*). These act on the kidneys to increase the reabsorption of Na+ and the excretion of K+ and hydrogen (H+) in order to control blood volume and pressure.

ALDOSTERONE

Aldosterone controls *electrolyte homeostasis*, that is, blood and urine. Thus, they affect BP and intravascular volume through regulation of renal sodium retention. The brain is a target organ for *its role in sodium and potassium* balance. These are significant issues in *cerebral autoregulation*. Aldosterone secretion is regulated by the renin-angiotensin enzyme system (originating in the kidney and circulation), ACTH, sodium ion concentration, and glucocorticoids. Angiotensin II stimulates arteriolar vasoconstriction within a few seconds. One may speculate that poststress vasospasm (impairing the internal carotid artery) is a contributor to cerebral ischemia that is not much recognized. Early and late vasospasm is considered to be a significant entity in head trauma (Chestnut, 1996), occurring in up to 25% of patients with head injury (Batjer et al., 1993). Experimental studies indicate that stress, aldosterone, and deoxycorticosterone increase *salt appetite. This effect is potentiated* by glucocorticoid hormones (Weisinger & Denton, 2000). Excessive mineralocorticoids can result in excessive serum sodium concentration due to interference with kidney functioning (Reeves et al., 1998).

GLUCOCORTICOIDS

Glucocorticoid functions can be summarized as follows:

- *Permissive*: Occur in the resting state, that is, permitting other hormones or immunological factors to accomplish their function at a normal level, that is, maintaining homeostasis at a basal rate.
- *Regulatory*: At stress-induced levels they prevent overreaction of the components of the IS, which if unchecked would lead to tissue injury or such conditions as *Cushing's syndrome* (excess) or *Addison's disease* (deficiency), resulting in muscular weakness, low BP, depression, anorexia, loss of weight, and hypoglycemia (Yamamoto & Friedman, 1996).

PHYSIOLOGICAL FUNCTIONS

Energy and metabolic control: Glucocorticoids are named for their ability to increase the storage of glucose in the liver and to synthesize glycogen. They regulate energy expenditure during high levels of activity, transformation of substances into energy supplies, distribution of energy supplies, energy storage and conservation, control of normal metabolism and the various cellular elements that participate in cellular functioning, and influencing the effects of other hormonal systems upon energy supply and utilization. Glucocorticoids promote normal metabolism (carbohydrates, proteins, fats, lipids, nucleic acids, and bone). They increase the availability of glucose and fatty acids as sources of energy and in preparation for stress (*energy storage and conservation*). They maintain serum glucose, regulate carbohydrate metabolism, stimulate mobilization of fats from storage for energy utilization, recycle proteins and amino acids through the liver, and play a regulatory role in development, growth inhibition, and inhibition of the reproductive function.

Cellular: Cortisol and CORT also inhibit DNA synthesis, and in most tissues inhibit ribonucleic acid (RNA) and protein synthesis and accelerate protein catabolism. In the liver, RNA synthesis and protein synthesis are enhanced (Aron et al., 2004). Corticosteroids increase insulin resistance, decrease glucose utilization, increase hepatic glucose production, and impair insulin secretion. In the liver, corticosteroids enhance the conversion of amino acids (protein) to carbohydrate. In muscle and adipose tissue they inhibit glucose uptake. They also increase appetite, thus causing fat deposition in the face, cervical area, trunk, and abdomen. They regulate cardiovascular function and BP, water excretion, and electrolyte balance. In muscles they break down protein into amino acids. In adipose tissue they mobilize fat, which is an alternative fuel to glucose (Barrett, 2005c).

Functional Disorders

Trauma or other serious illness may provoke hyperglycemia due to hypersecretion of insulin antagonistic hormones (Sherwin, 1996). Carbohydrate effects may precipitate diabetes or worsen glycemic control in persons with diabetes mellitus. Yet, although secondary to insulin in controlling glucose metabolism, their influence upon blood sugar plays a protective role against glucose deprivation (Miller & Chrousos, 2001). In addition, exercise-stimulated catecholamines can increase plasma glucose excessively, promote ketone body formation, and possibly lead to ketoacidosis. Conversely, excessively circulating insulin may reduce glucose production and increase glucose entry into muscles, leading to hypoglycemia (Brown & Zwilling, 1996; Powers, 2001). Glucocorticoids antagonize the effects of sex hormones and GH on fat tissue catabolism, and muscle and bone anabolism. The consequence is increased visceral adiposity and decreased lean body mass (bone and muscle). Glucocorticoid effects upon *intermediary metabolism* create deleterious effects on muscle, bone, connective tissue, and lymphatic tissue.

Cortisol Release

Occurs during stress after the sequence of CRH hypothalamus—ACTH anterior pituitary gland—adrenal secretion (Brown & Zwilling, 1996). Cortisol secretion (Aron et al., 2007b) is under neuroendocrine control:

Circadian rhythm: The CNS regulates both the number and magnitude of CRH and ACTH secretory episodes. Cortisol secretion is low in the late evening, and continues to decline in the first hours of sleep. It increases in the third to fifth hours of sleep. There are major secretory episodes in the sixth to eighth hours. A decline begins with wakefulness. This last period represents about half of the daily cortisol secretion.

Stress responsiveness: Plasma ACTH and cortisol secretion are responsive to physical stress, which if prolonged abolish circadian periodicity. Regulation of the HPA axis is coordinated with the IS. For example, IL-I stimulates ACTH secretion, while cortisol inhibits IL-1 synthesis.

Feedback inhibition: ACTH and cortisol secretion is regulated by feedback inhibition by glucocorticoids of CRH, ACTH, and cortisol secretion at the pituitary and hypothalamic levels. Continued glucocorticoid presence results in reduced ACTH levels and ultimate responsiveness to stimulation. There is suppression of CRH and ACTH release, and atrophy of adrenal cortical areas involved with secretion. The suppressed HPA axis fails to respond to stress and stimulation.

Neuroimmunomodulation

The neuroimmunoregulatory process takes place in a *neuroendocrine milieu* sensitive to the influence of the individual's perception of, and response to, external events. This milieu is defined by the innervation of lymphoid organs, the availability of neurotransmitters for interactions with cells of the IS, and interaction between pituitary-endocrine organ and lymphocyte-derived hormones (Ader, 1996). The cells of the IS secrete more than 20 neuropeptides. These have a substantial influence, serving as autocrine and paracrine factors exercising local control over a variety of immune functions. IS cells contain neuropeptides and also are a source of pituitary, hypothalamic, and neural peptides. The IS manufactures neuroendocrine hormones, and it is affected by hypothalamic releasing hormones. IS cells produce GHRH and CRF. While plasma hormone contributions by lymphocytes do not reach pituitary gland levels, they are mobile and can deposit the hormone at the target site.

Two functions are performed by lymphocyte-secreted peptic hormones:

- Endogenous regulators of the IS
- Serving as a sensory system, conveying information from the IS to the neuroendocrine system

Some stimuli are not recognizable by the central and peripheral nervous systems (bacteria; tumors; viruses; antigens). Noncognitive stimuli are recognized by immunocytes, which are converted into information in the form of peptide hormones, neurotransmitters, and cytokines. These are conveyed to the neuroendocrine system, and a physiologic change occurs (Weigent & Blalock, 1996).

The IS is modulated by GH, gonadotropin, and PRL (*stress modulation*).

Glucocorticoids have a *stress-modulating effect* (Kaye & Lightman, 2005), playing the main role in the suppression of immune and inflammatory responses. Cortisol and CORT serve as both the final effectors of the HPA axis and control of the stress reaction of the HPA through negative feedback (anterior pituitary gland, hippocampus, hypothalamus [PVN], frontal cortex) (Chrousus, 1998). *Negative feedback* and *circadian rhythms* affect CRH and ACTH production, regulating glucocorticoid secretion. If stress overcomes these feedback controls, the result is elevated levels of glucocortocoids, leading to enhancement or suppression of defense mechanisms. High levels of glucocorticoids can be associated with severe infections due to the immunosuppressive effect (Yamamoto & Friedman, 1996).

Concerning the chronic stress reaction, *cortisol* is considered to be the main hormone responsible for *allostatic stress* responses. It is also an *immunoregulatory hormone*. Only 2–15% is free in the circulating blood, whereas the remainder is bound to cells and proteins. Chronic stress results in sustained increase of plasma glucocorticoid levels. Cortisol has these functions: suppresses immune and reproductive function, promotes analgesia, activates the ANS (Bremner et al., 1997), maintains blood glucose production from protein, facilitates fat metabolism, modulates CNS function, affects the IS, and maintains vascular responsiveness to EP and norepinephrine. It limits excessive inflammatory responses to prevent further volume loss and tissue damage. In the CNS it stimulates the peripheral sympathetic nervous system, which mediates adaptive cardiovascular responses, including increased BP and pulse rate. It induces appropriate behavioral responses. Physical stressors influencing adrenocortical function include pain, trauma, burns, bone fracture, shock, exercise, inflammatory cytokines, severe pain, anticipation of conflict, embarrassment, or failure (Genuth, 1998b; Reichlin, 1998). Only free cortisol appears in the saliva, where its level can be measured (Kirschbaum & Hellhammer, 2000). One must consider the difficulty of accurately measuring its secretion, since it is episodic (circadian), and 24-h collection is difficult (Miller & Chrousos, 2001).

ADRENOCORTICAL SEX STEROIDS

ADRENAL ANDROGENS/GONADOCORTICOIDS

Adrenal androgens (sex steroids) are principally regulated by ACTH, and include *DHEA*. These have little inherent androgen activity, but are converted to active androgens and estrogens in peripheral tissue. Normally, the conversion of adrenal androstenedione to testosterone accounts for a negligible 5% of the production rate. Excess secretion in boys causes premature penile enlargement and early development of secondary sexual characteristics. In females, the adrenal substantially contributes to total androgen production by conversion of androstenedione to testosterone. This process varies in amount with the menstrual cycle. In pathological conditions, for example, Cushing's syndrome, adrenal carcinoma, and congenital adrenal hyperplasia, virilization, acne, and hirsutism will be clinically observed (Aron et al., 2007). DHEA is a weak androgen, and the only sex hormone secreted in significant physiological amounts by the adrenal cortex (Junqueira & Carneiro, 2005, p. 405). It plays a role in mediating some female secondary sexual characteristics, and is involved in T-cell regulation and the functioning of glucocorticoids. Although DHEA has been described as high in cases of allostatic load (see earlier), in Vietnam veterans with long-term, co-morbid PTSD, there were abnormally low levels of DHEA-S (Boscarino, 2004). *Gonadocorticoids* have been described as being present in insignificant amounts in the adrenal cortex (Ross et al., 1995, p. 614). Adrenal androgens are >50% of circulating androgens in premenopausal females, but in males, the contribution is much smaller because of the testicular production of androgens (Stewart, 2003). However, precocious puberty has been observed after TBI in girls and boys, although increased levels of DHEA appear independent of ACTH or cortisol levels, and are not regulated by the HPG axis (Towbin et al., 1996).

THE ADRENAL MEDULLA (CHROMAFFIN CELLS)

The medulla's biologically active products are biogenetic amines (*EP*, *norepinephrine*, and *dopamine*), which are also CNS neurotransmitters. Their effects are similar to those induced by the SNS. *EP* also serves as a hormone in the periphery as a rapid communication system (instantaneous neural effects). Hormonal effects are detectable within minutes. EP functions during the stress of simple exercise: It increases blood flow to muscle, relaxes the smooth muscles of the bronchioles to increase lung ventilation, and degrades muscle glycogen to provide a ready fuel source; in adipose tissue it furnishes free fatty acids for sustained muscular activity; in the liver it activates glycogenesis to maintain the blood supply of glucose and it increases fuel supply by decreasing insulin levels (Barrett, 2005c). Catecholamines *regulate the secretion of hormones* via the sympathetic nerves and the adrenal medulla: pancreas; thyroid; parathyroid; ovary; testis; pineal; adrenal cortex; etc.). Further control is through *feedback loops*. EP is considered to be a major contributor to the regulation of memory formation. It does not enter the brain in significant amounts, but does contribute to the increase of *circulating glucose* levels. Glucose has a wide range of effects upon brain functions, interacting with several neurotransmitter systems, but varying with the brain site (Gold & McCarty, 1995).

The sympathoadrenal system is an efferent limb of the nervous system. Catecholamines influence virtually all tissues and many functions. Connections between the cerebral cortex and the sympathetic centers that regulate sympathoadrenal outflow are affected by conscious mental processes. Anticipation of a particular activity may activate the sympathoadrenal system before the activity begins, thereby stimulating catecholamine-responsive processes in advance. Directly mediated catecholamine events take place in seconds compared with the longer time course of action of most hormones.

The adrenal medulla is the sole source of circulating catecholamines norepinephrine and EP, that is, in distinction to other glandular tissue, control of catecholamine secretion is within the CNS and there is no endocrine feedback influencing its level (Barrett, 2005c). It is involved in stress responses, and since it is contiguous with the adrenal cortex, it is exposed to much higher cortisol concentrations than other tissues, which affect its functions (Miller & Chrousos, 2001). The adrenal medulla functions as an enlarged and specialized sympathetic ganglion without axons: The cells of the sympathetic adrenal medullary system (SAM) discharge catecholamines directly into the bloodstream and thus function as endocrine rather than nerve cells (Genuth, 1998a). Embryologically it is a portion of the sympathochromaffin system. The chromaffin cells constitute the adrenal medulla. It is innervated by preganglionic SNS neurons originating in the prevertebral ganglia that release ACH, SP, and other peptides (splanchnic nerve) and terminate on cells of the adrenal medulla. The secretory cells are modified neurons, analogous to postganglionic SNS neurons. Instead of releasing a neurotransmitter onto an effector organ, they release it into the bloodstream producing a widespread sympathetic response. There is no PSNS stimulation (Moore & Dalley, 1999, pp. 48–51). The blood supply for medullary norepinephrine-containing cells is derived primarily from capillaries emanating directly from the adrenal gland's external arterial supply. EP-containing cells are supplied by sinusoids that have passed through the adrenal cortex and thus contain secreted glucocorticoids, which catalyze norepinephrine to EP (Romrell, p. 618). The core of the adrenal gland secretes catecholamines (20% norepinephrine [NA] and 80% EP), enter the circulation, and act on distal tissues just like other hormones. These secretions are described as mimicking, all at once, increases in activity in most of the sympathetic pathways (Money et al., 1996).

THE GONADS

The Testis

The major functions of the adult testis are to produce the sex steroid hormone, testosterone, and mature spermatozoa. An *anti-Muellerian hormone* (a peptide) also drives masculine development. While a fetal structure in development in females forms the fallopian tubes, uterus, cervix, and inner vagina, in the male the Muellerian ducts regress. Testicular steroids masculinize behavior early in life, while their absence results in the feminization of sexual behavior (Breedlove & Hampson, 2002). At puberty, increased production of testosterone by the testis is necessary for normal male secondary sexual characteristics and function, and the initiation of spermatogenesis. In the adult male, continued production of testosterone and sperm are necessary for virilization, sexual function, and fertility. GnRH neurones receive neuromodulatory input from many neuronal hypothalamic and brainstem systems. There is a hormonal control sequence commencing with GnRH in the hypothalamus, which is secreted into the hypothalamic capillary bed, then transported through portal veins descending in the pituitary (infundibular) stalk to another capillary bed in the anterior pituitary, which causes stimulation of synthesis and release of LH and FSH from the gonadotropin-producing cells. The infundibular and hypophysial portal system are susceptible to pituitary stalk damage associated with basilar skull fracture, which may cause defects in GnRH secretion and ultimately secondary hypogonadism and other anterior pituitary loss. Negative feedback by testosterone, and its metabolites estradiol and dihydrotestosterone, on gonadotropin secretion occurs at both the levels of hypothalamic GnRH and pituitary gonadotropin secretion (Matsumoto, 2001).

The Ovary

Ovulatory competence is controlled by the hypothalamus. For ovulation to take place, the hypothalamus must release a surge of GnRH in response to the estrogen stimulation. This is a transient positive feedback, differing from the usual case where steroids inhibit GnRH release, that is, negative feedback (Breedlove & Hampson, 2002).

Puberty

During puberty there is a change in blood composition and energy requirement along with increased gonadal hormone levels and maturation of secondary sexual characteristics. Gonadal steroids do not contribute substantially to normal growth before puberty. However, along with GH they play a role in the pubertal growth spurt (rapid linear growth and increased skeletal maturation [Grumbach & Styne, 2003; Reiter & Rosenfeld, 2003]).

Male sexual function requires coordination of these activities: libido, erection, ejaculation, orgasm, and detumescence.

ENDOCRINE AND IMMUNE SYSTEMS' INTERACTIONS

The exchange between the immune and endocrine systems has been described as "extensive cross-talk" (Baxter et al., 2004). Luteal Hormone Releasing Hormone (LHRH) is a major channel of communication between the hypothalamus-hypophyseal-gonadal axis and the thymus gland (Marchetti et al., 2001). The interaction of sex steroids, aminergic and peptidergic signals, results in the appropriate coordinated release of pulses of LHRH. LHRH is the major channel of communication between the axis and the thymus. In the thalamus they regulate genes controlling T-cell responses. Hypothalamic LHRH controls the release of FSH and LH leading to gonadal production of sex steroids. Gonadal hormones feed information to the thymus and hypothalamus. Metabolic and interactive pathways related to hypothalamic control of appetite impinge on LHRH centers. The sex steroids control the production of LHRH and thymic peptides. There are also neural pathways innervating the immune and endocrine organs, in addition to modulation by glucocorticoids and catecholamines. Activation of the HPA axis in response to stressful stimuli is influenced by the sex steroid hormonal milieu. Steroid hormones are not merely immunosuppressive or immunoenhancing but also immunomodulatory. Their central role is in the interactive communication between the neuroendocrine and ISs. Glucocorticoids induce atrophy of the cortex of the thymus. The sex steroid hormones might have a role in immunomodulating the stress response. Estrogens inhibit IL-6 production, which enhances cortisol levels by activating the HPA axis. The severity of, or susceptibility to, inflammatory diseases in response to a given proinflammatory trigger may depend on sex steroid modulation of HPA activity, in addition to genetic factors.

THE IMMUNE SYSTEM (IS) AND THE HYPOTHALAMIC-PITUITARY AXIS (HPA)

The IS participates in a regulatory feedback loop with the HPA axis (Crofford, 2002). Rheumatoid arthritis is considered to be a dysregulation of the inflammatory process. Immune and inflammatory mediators stimulate HPA hormone secretion.

Interleukin-1 (IL-1) stimulates secretion of CRH and AVP from the hypothalamus. It is produced in the hippocampus and hypothalamus, coordinating the CNS response to stress. It synergizes with CRH to stimulate pituitary production of ACTH directly (Crofford, 2002). IL-1 is one of the major cytokines involved in initiation of the inflammation process. It is involved in regulating the release of pituitary hormones (HPA axis), and as a feedback regulatory mechanism between the nervous and ISs (Marchetti et al., 2001).

Interleukin-6 (IL-6) also stimulates ACTH and cortisol release, via CRH and AVP production by the hypothalamus, and directly by the pituitary.

Tumor necrosis factor-alpha (TNF-α), (a proinflammatory cytokine) acutely stimulates the HPA axis and influences the release and metabolism of thyroid hormones (Baxter et al., 2004). There is delayed activation by IL-2 and *interferon (IFN-ã)*. These have neuroregulatory activities, including cell viability and transmission functions. In addition to acting directly on neural cells, peripheral cytokines may trigger CNS effects indirectly by activating neuronal afferents (Hanisch, 2001). TNF-α, IL-1, and IL-6 all contribute to the suppression of TSH.

ADRENAL GLAND AND IMMUNE SYSTEM

The duration, quality, and direction of stress-induced alterations of immunity are influenced by the quality and quantity of stressful stimulation; the individual's capacity to cope effectively with stressful events; the quality and quantity of immunogenic stimulation; sampling times and the particular aspect of the immune function chosen for measurement; the experiential history of the individual and the existing social and environmental conditions upon which stressful and immunogenic stimulation are superimposed; a variety of host factors such as species, strain, age, sex, and nutritional state; and interactions among these variables (Ader, 1996). Stress has more than an immunosuppressive function; it can also keep under control infections such as colds until pressure is released. Mild acute stressors may enhance measures of immunity (Dunn, 1996). The adrenal gland products (glucocorticoids and catecholamines) switch the IS from specific reactivity to nonspecific defense. The IS is activated nonspecifically, leading to cytokine production similar to acute infections, and may culminate in endocrine and metabolic responses characteristic of the acute injury phase.

Both the brain and the IS are responsive to cytokines such as interferon, producing ACTH and endorphins, explaining IS stimulation of the brain. Since IS cells produce neuropeptides, the exchange of information between the brain and the IS is bidirectional. Activation of the IS is accompanied by changes in hypothalamic autonomic and endocrine processes, and changes in behavior. Both the SNS and PSNS have direct effects upon lymphocytes in IS tissues (thymus, bone marrow, spleen, lymph nodes). These fiber connections develop early in ontogeny. Handling in early life affects subsequent antibody responses in mice. Marked diminished fiber supply of IS tissues with age perhaps contribute to immunosenescence. Stressful experiences both reduce and increase the host's defense mechanisms (Ader, 1996; Dunn, 1996) thereby altering susceptibility to bacterial and viral infections. It has been hypothesized that increased SNS function (characteristic of stress) would be associated with a decline in immune function. Evidence includes the *immunosuppression* found with SNS agonist drugs (Arnason, 1997). Mice exposed to physical stressors along with associated environmental cues exhibited inhibition of the immune response when they received an immunological challenge by injection with another species' erythrocytes. Conditioned immunosuppression decreases in aged mice. It is noteworthy that immune organs can be directly stimulated, that there are neurohormonal receptors on lymphoid cells, CNS cells produce cytokines (intercellular messengers), and there is neurohormonal regulation of immune reactions. Mouse resistance to infection by parasites is reduced when they are re-exposed to cues signalling stress-induced immunosuppression (Gorczynski, 1996).

Thymus

The development and function of the thymus are regulated by the nervous and neuroendocrine systems. Developmental hormones determine whether it produces T lymphocytes or functions as an endocrine gland (Mocchegiani et al., 1994). The thymus gland, through the adrenergic system, plays an important role in maintaining the interactions among immune, endocrine, and nervous functions (Basso et al., 1994). Thymus development, in turn, is regulated by the nervous and endocrine systems (Mocchegiani et al., 1994). The activated IS (*thymus gland*) with its product *thymosin* stimulates the HPG and adrenal axes (Trainin et al., 1996).

The thymus seems to secrete hypophyseotropic factors that modulate pituitary hormone secretion, thus providing a link between the immune and the neuroendocrine systems (Goya et al., 1994). Another thymic target is β-adrenergic receptors of the cerebral cortex (Basso et al., 1994). *Noradrenergic sympathetic innervation of lymphoid organs is likely* (Machelska et al., 2001). The SNS and ISs overlap in their signaling systems (e.g., neurotransmitters and cytokines). The thymus and bone marrow are innervated by the SNS, that is, superior cervical ganglion and other upper cervical chain ganglia (Bellinger et al., 1992), which is believed to have a modulatory role on immunocyte maturation. SNS noradrenergic neurons are traced into the spleen, thymus, and lymphoid tissue of the gut. β-Adrenergic agonist drugs inhibit many immune responses, while sympathetic ablation augments immune responses. The possibility of the association of augmented immune functions is suggested by the association of progressive multiple sclerosis with impaired SNS function. There seems to be an increase in overall immune responsiveness (Arnason, 1997).

CIRCADIAN CLOCKS (BRAINSTEM)

Circadian rhythms are biological functions varying from seconds, to days, and to months. The brainstem has three clocks: (1) the hypothalamic suprachiasmic nucleus (SCN); (2) the nonrapid eye movement (NREM)–rapid eye movement (REM) *pontine sleep clock*; and (3) 90–100 min periodicity during the daily activity (nonsleep) phase. This refers to waxing and waning of attention and motor activity (Hobson & Pace-Schott, 2003). These cycles persist in the absence of time cues, that is, are not simply driven by the 24-h environmental cycle. There are three functional components to the SCN circadian pacemaker: a photoreceptive input to the SCN, a photoreceptive input from the eye to the SCN, and a myriad of rhythmic outputs contributing to the periodicity of the circadian pacemaker (Provencio, 2005). After an injury, the control and adaptive efficiency of basic body cycles, as well as widespread neural and hormonal systems, may be disrupted. These functions are vulnerable to diffuse brain injury.

CONTROLLED FUNCTIONS

- *Neural control of rest-activity cycles*: The sleep-wake cycle (including slow-wave sleep), rhythms in psychomotor performance, memory, sensory perception
- *Autonomic regulation*: Respiration, CBT, systolic and diastolic BP and heart rate (whose lowest point is between 2 and 4 AM and peak close to noon), skin temperature
- *Survival*: Feeding and drinking
- *Secretion*: Plasma cortisol, melatonin; patterns of endocrine secretion; plasma GH, calcium excretion; REM sleep; plasma corticosteroids, body core temperature, and potassium excretion (Kupferman, 1991). Sleep is associated with secretion of PRL, and slow-wave sleep with GH secretion. In puberty, LH secretion increases during sleep. In mature women, sleep inhibits LH secretion in the early follicular phase of menstruation. Sleep onset is associated with the inhibition of TSH and of the ACTH-cortisol axis. Peak of saliva melatonin in normal persons occurs at 3:30 AM, and oral temperature reaches its minimum around 5:00 AM.

SUPRACHIASMIC NUCLEUS

The SCN of the hypothalamus is the endogenous circadian pacemaker. The SCN responds to environmental cues such as light, social, and physical activities through input from the retinohypothalamic tract (RHT), melatonin from the pineal gland, and neuropeptide Y from the intergeniculate leaflet (IGL). It controls or influences attention, endocrine cycles, body temperature, melatonin secretion, and the sleep-wake cycle (Zee & Manthena, 2007). It projects primarily to other areas of the hypothalamus, and to several nonhypothalamic structures of the diencephalon and basal forebrain, controlling circadian rhythmicity (Moore, 2003). It plays a critical role in the sleep-wake cycle, diurnal variations in thermoregulation, feeding and drinking behaviors, and regulation of endocrine rhythms, such as secretion of melatonin by the pineal gland (Berne & Levy, 1998, p. 350). The circadian clock(s') *pacemakers* are located in the SCN of the hypothalamus (Fuller & Fuller, 2002). They interact with the REM and NREM sleep clock in the pons. The pontine reticular formation produces a state reminiscent of REM sleep, while lesions of the dorsal pons results in loss of the muscle inhibition characteristic of REM sleep (Czeisler et al., 2005).

SCN output: There are *descending nerve impulses, via the superior cervical ganglia*, to sympathetic postganglionic fibers (noradrenergic), which innervate the pineal gland. This completes the circuit by which light and dark affects levels of endocrine secretion (Reichlin, 1998, p. 216). Thus, pineal function could be vulnerable to *neck trauma*. The SCN projects to the medial preoptic hypothalamic region (MePO/AH), lateral hypothalamus, retrochiasmic region, and a region immediately ventral to the PVN. MePO neurons are involved in sleep induction through connections with the adjacent basal forebrain. Projections from the SCN to the PVN appear to be important for circadian regulation of pineal and autonomic function. The PVN in turn projects to the preganglionic sympathetic neurons innervating the superior cervical ganglion. Noradrenergic neurons of the superior cervical ganglion project to the pineal gland. Sympathetic inputs stimulate synthesis of melatonin by the pineal gland. This pathway may be responsible for the light suppression of melatonin. Both warm-sensitive neurons of the MePO/AH and "sleep active" neurons in the basal forebrain inhibit activity of neurons of the midbrain reticular formation and posterior hypothalamus that are part of the reticular arousal system.

SCN afferent input: While the pineal gland is light-sensitive in some species, in humans, direct afferent stimulation to the coordinator of the internal clock has varied sources. Circadian rhythms are considered part of the greater visual system, the *RHT* (Morin & Allen, 2006). Variations in exposure to light alter the daily melatonin signal, which then alters other functions, including reproduction and possibly the timing of puberty. Nonvisual retinohypothalamic fibers also innervate the SCN, offering information about light and dark states, serving as the biological clock that integrates the cyclical environment and the circadian rhythms (Parent, 1996, p. 712). Fibers receiving information concerning the external light-dark cycle arise from retinal ganglion cells, traverse the optic nerve and chiasm, and project bilaterally to the hypothalamic SCN. This is the pathway by which external light controls pineal gland activity. The *reproductive axis* is synchronized by the retinal-SCN-pineal link (Molitch, 2001; Mulcahy et al., 2003).

Thalamic: The SCN also receives input from thalamic nuclei, that is, the ventral lateral geniculate N. and the paraventricular N. (Parent, 1996, p. 718). The *IGL* conveys information about physical activity and general excitement (Waterhouse, 2007). Photic information from ganglion cells of the retina generates slow oscillations in the IGL, which participates in basic rhythms of the biological clock through the secretion of neuropeptides and neurohormones (Lewandowski & Blasiak, 2004).

Median raphe N.: There is an ascending afferent projection to the SCN from the serotonic median raphe N. (Morin & Allen, 2006). In the rat, it is involved in analgesia and sleep. The raphe fires regularly during waking (activity increased by visual stimulation), slowly during slow-wave sleep, and not at all during REM sleep (Moore, 2003). Lesions produce an increase in locomotor activity, insomnia, hyper-reactivity, and aggression (Brodal, 1981, pp. 415–416).

Principle Properties

Entrainment: This is the process by which an endogenous rhythm is regulated by a *zeitgeber* (an external cue, e.g., the light-dark cycle) to synchronize endogenous rhythms. Rhythms are normally entrained to a light-dark cycle with a period of 24 h and a stable phase relationship to the timing of the solar cycle. It is mediated by the photopigment melanopsin, which is located in retinal ganglion cells that project through the RHT to the SCN and the *intrageniculate leaflet (IGL)* independently of the familiar retinal circuits to the geniculate colliculus and accessory optic systems. The body's natural rhythm is now estimated at 24.2 h (Czeisler et al., 2005), which is periodically reset by light stimuli. For the body clock to be useful, it should be adjusted to the actual solar day. Rhythmic cues in the environment are known as *zeitgebers* (time-givers) whose effect on the body clock depends on its time of presentation (a phase advance, delay, or no phase shift). The primary zeitgebers are the light-dark cycle and rhythmic secretion of the pineal hormone *melatonin*, during nocturnal sleep in healthy people. Circadian rhythms persist in the absence of a light-dark cycle, with a free-running period that is slightly more than 24 h. The mammalian eye appears to have a circadian pacemaker maintaining the rhythm of visual sensitivity.

Generation and regulation: The output of molecular and synaptic processes creates the pacemaker that controls effector systems as part of a neural system. Since its principle output is the sleep-wake cycle, its major function is the temporal organization of the behavioral state to maximize the effectiveness of adaptive waking behavior. Activity and rest occur depending upon the particular mammal, that is, when the dominant senses can aid feeding, reproducing, and avoidance of predators. Rhythms are generated by (1) pacemakers, (2) photoreceptors and photoreceptor input to the pacemakers, and (3) pacemaker output to systems under circadian control. The rhythms are temperature-compensated, that is, the period length has a minor variation in response to external or internal temperature. The various mechanisms creating daily rhythms have an *endogenous* (the body clock) and an *exogenous* component (lifestyle) (Waterhouse, 2007). Sleep, hormonal secretion, alertness, and so forth, all respond to various rhythms controlled by biological clocks. One distinguishes between diurnal or circadian physiological rhythms and the sleep-wake cycle (Kelly, 1991). These rhythms manifest peaks and troughs at regular times during the 24-h day, although varying between functions. Feedback loops maintain hormonal levels within a narrow range, that is, *set points*. Circadian rhythms are quite regular and may be perturbed by stress, illness, nutritional status, and other factors (Molitch, 2001). Circadian rhythms are largely under the control of *pacemakers* located in the SCN of the hypothalamus. They interact with the REM and NREM sleep clock in the pons. The body clock brings about daily rhythms in core temperature, plasma hormone concentrations, and the sleep-wake cycle. The release of CRH, AVP, ACTH, and cortisol are controlled by pacemakers and perturbed by changes in lighting, feeding schedules, and activity. They are disrupted when a stressor is imposed (Chrousos, 1998).

Disturbance

Sleep deprivation rapidly resets behavioral rhythms, including simple arousal. Environmentally entrained rhythms have implications for shift workers and zone travelers and recipients of timed administration of drugs or radiation. Tissues might be entrained to a specific phase without disrupting SCN-driven rhythms such as sleep or body temperature (Stokkan et al., 2001). Disturbance of physiological rhythms should be considered one of the pathological effects of emotional stress and physical injury.

Disease

A variety of diseases manifest a circadian rhythm: allergies, symptoms of rheumatoid or osteoarthritis, and duodenal or peptic ulcers. Response to medications shows a circadian rhythm, since the body's responses vary with time of day. It is useful to prescribe medications in terms of the circadian rhythm of sleep-wake rather than clock time (Sothern & Roitman-Johnson, 2001).

THE SLEEP-WAKE CYCLE

Sleep is highly complex and incompletely understood. Although circadian clocks have received the highest attention, other biological functions contribute to sleep control and such conditions as narcolepsy. Regulating processes include the circadian and homeostatic (balancing the drive for sleep with actual time spent asleep).

Neurotransmitters in Sleep Regulation

Modulators of the sleep-wake and REM-NREM sleep-wake systems include many neurotransmitter systems: cholinergic, serotonergic, norepinephrine, and histamine. Dopamine does not vary with the sleep-wake cycle as do the others. It modulates the aminergic and cholinergic systems, and may induce or intensify nightmares. Histamine is the neurotransmitter of an arousal system originating in the posterior hypothalamus, which innervates the forebrain and brainstem. It is wake-promoting. Posterior hypothalamic lesions of its nuclei can result in hypersomnolence (Hobson & Pace-Schott, 2003).

Hypocretins: The hypocretin system and the circadian seem to exchange activation. The hypocretins are two peptides that are synthesized exclusively in the lateral, posterior, and perifornical hypothalamus. They project to such monoaminergic centers as the LC noradrenergic, raphe nucleus (serotonergic), and ventral tegmental (dopaminergic) areas. Hypocretin is an excitatory neuropeptide that activates multiple neuromodulatory systems involved in the regulation of sleep-wake behavior. It activates neuromodulatory cell groups during wakefulness, but less so during slow-wave sleep. Its role during REM sleep is uncertain. While low hypocretin levels have been observed after head trauma, and may result in sleepiness, disturbed capacity to sleep seems more characteristic. Hypocretins may be important in HPA regulation during arousal, and in vagally mediated gastric acid secretion, SNS activation, and cardiovascular function (increased BP and heart rate). This arousal effect may modulate autonomic and sensory functions of the spinal cord, promote energy consumption, produce hyperthermia, and simulate sympathetic tone (Taheri et al., 2006).

Hypocretin-1 is involved in regulation of the sleep-wake cycle. Its level is lower in 97% of TBI patients with computed tomography (CT) changes.

Sleep and waking are controlled by two opposing factors: homeostatic drive for sleep, and a circadian arousal stimulus. The cycle is modulated by circadian influences, thermoregulatory mechanisms, possibly immune function (Czeisler et al., 2005), and social cues (Moore, 2003). The medial preoptic/anterior hypothalamic (MePO/AH) region is involved in thermoregulation (Benarroch, 1997a, p. 348). Sleep disorders may have an intimate involvement with endocrine, thermoregulatory, and autonomic functions, and are influenced by the issue of whether the light/day is appropriately timed with the attempt to sleep. The sleep-wake rhythm is characteristically 24 h (diurnal), but can drift to 25 or more hours (circadian) with isolation from light, temperature, social cues, and knowledge of time (Kelly, 1991). The pacemakers interact with the REM and NREM sleep clock in the pons. The ease of sleep initiation and maintenance depends on both previous wake time and the circadian rhythm of *core temperature*. The minimum core temperature occurs between 3 and 7 AM. Sleep is easiest to initiate when core temperature is at its lowest or falling rapidly, and most difficult when body temperature is rising rapidly or is high. Waking occurs when core temperature is rising or high. Inappropriate phasing of the body clock makes sleeping difficult to initiate and maintain.

Sleep is affected by CBT, which appears to have a strong synchronizing effect on the cycle. The ability of most individuals to fall asleep is greatest at the nadir (4–5 AM), or as CBT declines. REM sleep is most likely to occur when temperature increases between 4 and 6 AM. More than the SCN is involved in modulating body temperature. The circadian timing system promotes waking and alertness, and also interacts with a sleep drive to promote consolidated sleep.

Neural Circuitry

The generation of *sleep* is associated with the medullary reticular formation, and that of *wakefulness* with the brainstem reticular formation, the midbrain, subthalamus, and the basal forebrain. Circuitry for sleep and wakefulness is distributed along an axial core of neurons from the brainstem to the rostral forebrain. The sleep-wake cycle is controlled by centers located in the pons, and their periodicity, or circadian rhythm, located in the *SCN* of the hypothalamus (Kupferman, 1991). Its pacemaker is in the thalamic SCN that is served by multioscillatory input. Sleep arises from complex interactions among brainstem and diencephalic structures that belong to, or are intimately interconnected with, the CAN. REM sleep is controlled by the pontine reticular formation. Neurons in the ventrolateral preoptic hypothalamus inhibit histaminic cell groups of the posterior tuberomammillary N. Their widespread projections are important in maintaining arousal (Czeisler et al., 2005).

THE PINEAL GLAND

The pineal gland is believed to coordinate the organism's response to environmental challenges. It is both a CVO (permeable to large molecules from the blood) and an endocrine organ secreting *melatonin*. It plays an important immunoregulatory role (Maestroni & Conti, 1992). It influences the circadian cycle, gonadal function, and development through the release of melatonin, TRH, somatostatin, GnRH, and norepinephrine into the general circulation and CSF (Aron et al., 2007a). It is involved in the chronobiologic regulation of immune, psychoendocrine, and metabolic functions. The pineal modulates brain neuropeptide secretion and activity, which in turn influence most biologic functions, including the inflammatory response and immune reactions (Lissoni et al., 1994). The pineal gland integrates information encoded by light into secretions that underlie biological rhythmicity. However, it is not directly light-sensitive and its rhythms are not entrained to the external light-dark cycle (Cone et al., 2003). It is involved with a variety of organs associated with the IS through feedback loops (Jankovic, 1994). The anatomical substrate for light to regulate the secretion of melatonin is via the circadian pacemaker to the SCN via the sympathetic superior cervical ganglion, which provides noradrenergic innervation to the pineal.

Melatonin is synthesized from serotonin and secreted from the pineal gland. Its secretion is controlled by the CNS via the *superior cervical sympathetic neurones*, whose activity increases at night. The RHT both entrains the SCN and mediates the effect of bright light in suppressing melatonin production at night. Melatonin secretion is entrained by the day-night cycle, particularly darkness. It plays a role in modulating many circadian rhythms, and its synthesis is closely controlled (Cone et al., 2003). Melatonin regulates the reproductive axis and the timing and onset of puberty (Low, 2003). The concentration of melatonin-synthesizing enzymes in the pineal is increased by activation of the SNS, that is, retina–suprachiasmic N. of the hypothalamus paraventricular N. of the hypothalamus–spinal cord to thalamic preganglionic synapse–postganglionic N. to superior cervical ganglion–pineal. Its level is increased in response to hypoglycemia and darkness. It can correct immunodeficiency states consequent to acute stress. In the presence of cytotoxic injury, it stimulates T-cells to produce cytokines (Maestroni, 1996). In neuromodulation of the IS, the effects of melatonin are mediated by opioid peptides (Lissoni et al., 1994). Melatonin rhythms are not affected by sleep deprivation. Melatonin has some effect upon suppressing human puberty by acting upon the hypothalamus and pituitary to inhibit gonadotropin secretion. The SCN is rich in melatonin receptors. Since it regulates pineal gland secretion of melatonin and other circadian rhythms, it appears to be a site of negative feedback regulation of the pineal as well as the intrinsic circadian pacemaker.

The concentration of melatonin-synthesizing enzymes in the pineal gland is increased by activation of the SNS. An effect of hypothalamic damage on melatonin effects is suggested by the following anatomical detail. There is an upward extension of anterior pituitary glandular tissue (pars tuberalis), which extends forward to envelop the base of the hypothalamus, including the endocrinologically significant median eminence. This tissue is primarily gonadotropes and thyrotropes, and melatonin could act on this area to release pituitary hormones through nerve endings or blood flow from the hypophyseal portal blood vessels.

MSH has been considered a mediator of immunity and inflammation. Through its effects upon *melanocortin-1 receptor* (MC1R), it downregulates the production of proinflammatory and immunomodulating cytokines. The cytokine-synthesis inhibitor IL-10 is upregulated (Luger et al., 2003).

INTERNAL MILIEU SENSORY INFORMATION

A sensory system providing information to the brain about the state of the internal organs: the *forebrain sensory representation* devoted to visceral sensation in the insular cortex.

PARASYMPATHETIC VISCERAL SENSORY: CRANIAL NERVES

Trigeminal (IV): Internal face and head visceral information
Facial (VII): Visceral sensory providing taste from the tongue
Glossopharyngeal (IX): Visceral sensory input (including taste) from the hard palate, upper part of the oropharynx, and carotid body

Vagus (X): Visceral sensation (including taste) from the lower part of the oropharynx, larynx, trachea, esophagus, and thoracic and abdominal organs, with the exception of the pelvic viscera. There is vagal sensory representation in the insula, where it is not identified as emotional experience.

S2–4: The pelvic viscera are innervated by nerves from the second through fourth spinal sacral spinal segments.

Visceral afferents from all four cranial nerves terminate in the *nucleus of the solitary tract (NTS).* This information is relayed by the ipsilateral *parabrachial nucleus* to a *contralateral thalamic ventrolateral* relay nucleus (ventroposterior parvicellular N.), thence to the forebrain, that is, the *insula*, acting as a sensory field. The NTS has other projections to the forebrain, which seem more important for *arousal* than sensory information.

Spinal (Sympathetic) Visceral Afferent System

This refers to information pathways of ascending spinal visceral afferent neurons. Spinal visceral afferents convey information about *temperature* and *nociception* (mechanical, chemical, thermal). General visceral afferent fibers accompany their efferent counterparts. Unicellular fibers are contained in the vagus, glossopharyngeal and possibly other cranial nerves; S2–4 spinal nerves; and upper lumbar and thoracic nerves. Some visceral afferents also have a motor function: vasodilation, venular permeability, smooth muscle contractility, etc. Neurogenic inflammation, an effect upon leukocytes and fibroblasts, may play a *trophic role* in the maintenance of tissue integrity and repair of trauma (Standing, 2005, pp. 238–239).

Visceral afferents are essential for circulatory, respiratory, cough, swallowing and autonomic reflexes, chemoreceptors, visceral pain, hunger, and vesical distention (Standing, 2005, pp. 238–239). Exercise information from muscle chemosensation may arise at all spinal levels. The sympathetic preganglionic neurons are located at thoracic levels. Visceral organs send afferents to dorsal horn neurons at different spinal levels than the axonal entrance to the spinal cord (e.g., cardiac pain from T2 to C5). Most ascending spinal visceral afferents are thought to converge with musculoskeletal and cutaneous afferents, and ascend via the spinothalamic and spinoreticular tracts. They provide collaterals that converge with the cranial nerve visceral afferent pathways at virtually every level including a continuation of the cranial nerve visceral sensory thalamus and cortex (Saper, 2002).

The Solitary Nucleus Complex

The solitary nucleus is the first relay and integrative center for viscerosensory inputs: (1) It initiates medullary reflexes controlling cardiovascular, respiratory, and other autonomic functions. and (2) It relays viscerosensory information to all other regions of the autonomic core (Benarroch, 1997a, p. 10). Local connections, ascending, and descending projections are involved in the integration of visceral afferent information and autonomic function. It is placed dorsally and bilaterally in the medulla (between the dorsal afferent columns, i.e., fasciculus gracilis [medial] fasciculus cuneatus, more lateral), and extends into the caudal portion of the pons. At the caudal portion of the complex, the cell mass is continuous with that of the contralateral side.

The solitary nucleus projects to the nucleus ambiguus, which is a column of cells in the reticular formation (Parent, 1996, p. 448); the parabrachial nuclei in the rostral pons, and the ventral posteromedial nucleus pars parvicellularis of the thalamus, involved with gustatory sensation; to the dorsal motor nucleus of spinal segments C3–5 and anterior horn cells of thoracic spinal segments involved in coughing and vomiting reflexes; hypothalamic PVN, which may play a role in regulation of cardiovascular function (either control over vasopressin release from the posterior pituitary, or activating spinal sympathetic preganglionic neurons through axons of the PVN that reach thoracic spinal segments).

The *fasciculus solitarius* is formed by visceral afferents contributed by the vagus, glossopharyngeal, and facial nerves. The rostral solitary nucleus is termed gustatory nucleus, and the caudal portion is the cardiorespiratory nucleus. The gustatory nucleus receives special afferent fibers from the facial and glossopharyngeal nerves. The remainder of the solitary nucleus (*cardiorespiratory nucleus*) receives viscerotropic fibers from the vagus, facial, and glossopharyngeal nerves (alimentary, pulmonary, carotid sinus). Ascending input from visceral receptors are chemoreceptors (blood carbon dioxide and oxygen) and mechanoreceptors (arterial pressure). Descending spinal projections contact sympathetic preganglionic neurons in the intermediolateral nucleus of the thoracic and lumbar spinal cord, mediating control of heart rate and BP. These projections are either direct or via adrenergic neurons in the ventrolateral medulla. Projections to the medullary preganglionic parasympathetic neurons are important in regulation of heart rate and motility and secretions of the gut.

Nucleus Ambiguus

The caudal solitary nucleus also projects to parasympathetic preganglionic neurons and somatic motor neurons in the *nucleus ambiguus*, which innervates pharyngeal and laryngeal muscles (in medulla, ventrolateral to the solitary nucleus and tract [Martin, 2003, p. 400, Fig. 13–10]). Bilateral corticobulbar fibers innervating cranial nerve nuclei are composed of axons from the premotor, motor, and other cortical areas, which descend through the internal capsule to synapse on motor neurons in the nucleus ambiguus (Wilson-Pauwels et al., 1988, p. 128). This branchial motor component (visceral efferent) of the glossopharyngeal nerve innervates the constriction muscles of the pharynx, intrinsic muscles of the larynx, and also the soft palate (Brazis et al., 2001, p. 333). It receives projections involved in the voluntary control of swallowing and phonation, and impulses from receptors in the pharyngeal and laryngeal muscles, etc., which convey impulses involved in reflexes such as coughing, vomiting, and pharyngeal and laryngeal reflexes. Ipsilaterally, secondary solitary nucleus fibers project to the nucleus ambiguus and the surrounding reticular formation, the parabrachial nucleus in the rostral pons, and the ventral posteromedial nucleus of the thalamus (concerned with gustatory sensation). The solitary nucleus also projects to the hypothalamic PVN, playing a role in regulation of cardiovascular function, either by exerting control over vasopressin release from the neural lobe of the pituitary or by activating spinal sympathetic preganglionic neurons via axons that reach thoracic spinal segments (Parent, 1996, p. 445). Another input to the N. ambiguus is the solitary N., the first viscerosensory relay that initiates medullar reflexes controlling cardiovascular, respiratory, and other autonomic functions (Benarroch, 1997a, p. 10) and relays viscerosensory information to the forebrain. It receives the following afferents: baroreceptors; cardiac; pulmonary (glossopharyngeal N.); carotid and aortic chemoreceptors (vagus N.); and participates in circuits controlling respiration, circulation, vomiting, and micturition (Benarroch, 1997a, pp. 10–12). Awareness of these functions aids our understanding of the influence of physiological and stress effects upon vocalization.

The pharynx serves both respiration and swallowing. Details of muscular control over movements of the pharynx and larynx during swallowing do not concern us directly. However, one may observe that sensory impulses are distributed via glossopharyngeal and trigeminal nerves to the medulla oblongata via the tractus solitarius. Thus, feedback becomes part of the background of homeostatic and stress reactions. Motor impulses from the swallowing center of the medulla and lower pons are transmitted by cranial nerves 5, 9, 10, and 12, and a few of the superior cervical nerves (Guyton & Hall, 2000, p. 729). The velopharyngeal valving mechanism regulates contraction of the muscles of the soft palate and pharynx, controlling force of contractions, range of movements, and coordination of muscle movement.

The nucleus of the solitary tract is coextensive with the physiologically defined dorsal medullary respiratory center (nucleus ambiguus and surrounding portions of the medullary reticular formation).

PARABRACHIAL NUCLEUS: VISCERAL AFFERENTS

The ascending projections of the cardiorespiratory nucleus are focused in the parabrachial nucleus, which is located in the pons adjacent to the principle output path of the cerebellum, the superior cerebellar peduncular. The parabrachial nucleus transmits visceral afferent information rostrally to the hypothalamus, amygdala, and the forebrain. It participates in circuits controlling respiration, circulation, vomiting, and micturition (Bennarroch, 1997, p. 10).

Visceral: The NTS (dorsal medulla) and the parabrachial nucleus (dorsolateral pons) function in both the reception of visceral stimuli and projection to the central nucleus of the amygdala. The visceral sensory component of the glossopharyngeal nerve monitors oxygen and carbon dioxide tension in circulating blood (Boulpaep, 2005; Wilson-Pauwels et al., 1988, p. 121). These sensations are transmitted via the carotid nerve to the NTS whence connections are made with the reticular formation, the hypothalamus, and the medullary cardiovascular center for the reflex control of the internal milieu. Information is integrated from other pathways, and control of arterial pressure, integrated with the respiration rate, via efferent impulses from both branches of the ANS, achieves homeostatic levels of oxygen and carbon dioxide (Boulpaep, 2005).

AUTONOMIC RESPONSE TO STRESS

Sensory innervation plays an important role in neural control of visceral function. These include *mechanosensory visceral afferents* that innervate the target tissues of the ANS. Motor neurons of the SNS receive afferent input from most of the autonomic targets. These visceral afferents convey pain. They are also capable of conveying an impulse, in the opposite direction of sensory stimulation, to the motor neuron (antidromic), which produces inflammatory and vascular responses. They are activated by distention and stretching of visceral muscles, which may reach the level of pain. *Chemosensory nerve endings* monitor visceral functions and are the afferent limb for many autonomic reflexes (Gardner et al., 2000; Powley, 2003). Noxious stimuli projected via the spinothalamic tract send collaterals to the medulla, stimulating sympathetic reflexes such as heart rate and BP. Visceral sensory information (taste) is from the anterior two-thirds of the tongue by the facial (CN VI) nerve, while afferents from the posterior oral cavity are conveyed by the glossopharyngeal (CN IX) nerve. Information about internal states from the thoracic and abdominal cavities is carried via the vagus nerve (CN X). All of these pathways synapse in the *nucleus of the solitary tract* (Iversen et al., 2000).

The integrated afferent information (spinal cord and cranial nerves) enables neurons at multiple levels of the CNS to generate specific response patterns along the lines of functionally related groups. These patterns are integrated at different brainstem levels with parasympathetic, endocrine, and behavioral components. Individual *pattern generators* at different levels of the neuraxis are hierarchically arranged, allowing individual response patterns to become part of larger responses. For example, pattern generators interact for vascular responses to emotional stress and fight-or-flight reactions. The infralimbic cortex serves as an emotional motor cortex. It receives input from prelimbic and cingular areas. It ensures that emotional reactions are linked to autonomic and endocrine responses by projections to the hypothalamus and ventrolateral medulla. The hypothalamic pattern generators activate a coordinated pattern of autonomic, endocrine, and behavioral responses. The hypothalamus activates a region of the periaqueductal gray (midbrain), which creates a pattern to different structures involving vasoconstriction, vasodilation, increased cardiac output, and adrenal secretion. This pattern is due to its integration with a medullary response controlling tissue perfusion in different areas and release of CRH. Functioning at different levels, different combinations of pattern generators produce the entire range of highly differentiated, complex patterns necessary to maintain homeostasis and defense against threat (Saper, 2002).

THE IMMUNE SYSTEM (IS)

The immune system is a complex system with significant neurobehavioral and health effects. It is comparable to the CNS in its complexity and specificity of its reactions. It is primarily autoregulated; therefore, it does not require a major input from the CNS, which nevertheless does play a modulatory role. It is made up of diverse yet interconnected components. Immune impairment could occur at various levels of this system (Vedhara & Wang, 2005). Components are found in the bone marrow, blood, lymph nodes, and spleen, and are dispersed through other tissues. It consists of interactive cells and soluble molecules whose function is to recognize foreign proteins, viruses, and bacteria as non-self (antigens), and then to send signals that lead other components (phagocytes, i.e., macrophages) to engulf and destroy wounded tissue, trauma-related substances, and microorganisms. Its cells are bathed in plasma, whose composition regulates the function of the immune cells (Rabin, 2005).

The IS is composed of primary lymphoid tissue (lymphocytes forming in the bone marrow [B] and the thymus [T]), different blood cells, and secondary lymphoid tissue (lymphocytes developing in the bone marrow and thymus). Its function is to patrol the blood and tissues to detect substances that are foreign to the body (*antigens*). Immune cells and tissues express *receptors* for transmitters and hormones emanating from the hormonal and ANSs. The immune response may be *cellular* or *humoral* (molecular, e.g., antibodies). *Antibodies* affect the infectious agents or their soluble products in an extracellular location (bacteria, viruses, parasites). The immune reaction may damage tissue or produce an allergy (Rabin, 2005).

Function: The primary function of the IS is the recognition and protection of the self from foreign invaders. In contrast to the CNS, which recognizes cognitive particular stimuli and responds cognitively (physical; emotional; chemical), the IS recognizes *noncognitive stimuli* (bacteria, viruses, tumors, antigens) that would go unnoticed if not recognized by the IS. When responded to by immunocytes, these structures become information in the form of peptide hormones, neurotransmitters, and cytokines that is conveyed to the neuroendocrine system. In an analogous fashion, the CNS and peripheral NS recognize cognitive stimuli, creating similar hormonal information that is conveyed to and recognized by hormonal and neurotransmitter receptors (Blalock, 1994). These are conveyed to the neuroendocrine system, and a physiological change occurs. Biological stress can originate in the IS. Immune cells can create a local antinociceptive effect in inflamed peripheral tissue. Cognitive recognition by the central and peripheral nervous systems results in hormonal information being conveyed to hormone and neurotransmitter receptors on immunocytes, with immunologic and neurologic changes resulting (Weigent & Blalock, 1996).

The function of the IS is multiple: to recognize foreign proteins, viruses, and bacteria as non-self; to set forth signals that lead other components of the IS to destroy nonself (phagocytes, macrophages) that engulf microorganisms and trauma-related substances. *Aging involves a gradual decline of all organ systems, including immune and neuroendocrine dysregulation. Elevation of a single component of the internal milieu may alter the function of the IS* (Pinedo & Dahn, 2005). *One function of the IS is to patrol the blood and tissues to detect substances that are foreign to the body (antigens).* When a foreign material is detected by the IS, the IS is activated and the foreign material is killed (or if an infectious substance, inactivated and removed). This process may damage tissue or produce an allergy (Bruce, 2005). The physiological development of an operative *HPG system* requires the presence of an intact IS (Marchetti et al., 2001). LHRH neurons migrate from the medial olfactory epithelium into the basal forebrain. Failure of this migration results in suppression of the pituitary-gonadal axis. Cytokines and TNF affect the nervous system by entering areas with a relatively weak BBB, for example, the *preoptic N.* of the hypothalamus.

Cellular Components of the Immune System

All the cellular components of the blood originate in the stem cells of the bone marrow, that is, lymphocytes and a variety of other types, including erythrocytes (red cells that are oxygen carrying), leukocytes (white blood cells), and megakaryocytes (produce platelets involved in blood clotting). When mature, they migrate to guard the peripheral tissues, circulating in the blood and in the vessels of the lymphatic system (Janeway et al., 2001, pp. 1–10). The lymphocytes are identified by their place of differentiation: T (thymus) and B (bone marrow). Another lymphocyte is the natural killer cells (see next). These blood cells may participate in both immune and inflammatory reactions. Leukocyte traffic is regulated by chemokines to sites of inflammation (Muller et al., 2003). Many of the cells in lymphoid organs are mobile. Nerves innervate lymphoid organ compartments where there is influence of specific subsets of cells by neurotransmitters. These neurotransmitters enter the circulation and provide feedback via the CNS (Barrett, 2005c).

Macrophages: Both glucocorticoids and catecholamine hormones affect the functional capacity of the macrophage. Catecholamines can suppress or stimulate macrophage function. Macrophages produce a variety of *inflammatory cytokines*, which serve as signals to the nervous system, as well as other functions (IL-1 and -6, and TNF-α). These substances, also known as *endogenous pyrogens*, elevate body temperature in two ways: (1) acting on the hypothalamus, altering the body's temperature regulator; and (2) acting on muscle and fat cells, altering energy mobilization to increase body temperature. At elevated temperatures, bacterial and viral replication is decreased, and the adaptive immune response operates more efficiently. *Catecholamines* alter many macrophage functions, including cytokine production. Glucocorticoids (1) enhance neutrophil survival by increasing their number in the circulation, and (2) inhibit apoptosis (programmed cell death) of neutrophils, thus extending their functional life span (Boomershine et al., 2001). Cytokine can be released from lymphoid cells to modulate nerve terminal activity viability and neurotransmitter release. Nerves can synthesize and release cytokines such as histamine, IL-1, and IL-6. Cells of the IS can synthesize and secrete neurotransmitters or neurohormones, including CRH, SP, and NE (Barrett, 2003c).

Cytokines

Cytokines are polypeptides that serve as regulatory or effector molecules of the immune system (Giulian, 1994). An increase of distress may influence cytokine dysregulation by way of neuroendocrine changes, which intensify physical symptoms. There is an increase in distress in reaction to mounting symptoms as a vicious cycle. This positive feedback loop has been related to the chronicity of CFS and its refractoriness to intervention based solely on symptom reduction (Patarca-Montero et al., 2001).

The Immunosuppressive Response

It is paradoxical that immune function is suppressed when an organism may be injured, with the possibility of infection from the attacker and enhanced vulnerability to inflammatory diseases. Glucocorticoids are the most potent endogenous inhibitors of immune and inflammatory processes (Crofford, 2002). Immunosuppression increases susceptibility to infections and cancer, and exacerbates inflammatory diseases like psoriasis, asthma, arthritis, and lupus erythematosus. However, it also ameliorates autoimmune and inflammatory disorders (Dhabar & McEwen, 2001). The *autoimmune response* is a failure of immunosuppression. Acute stress is not yet proven to be a cause of immunosuppression (La Via & Workman, 1998).

Cytokines, produced during injury, lead to neuroendocrine and metabolic changes, but must be kept under control because excess production has fatal consequences due to neurotoxicity. Endocrine response may be modulated by nerve endings in the area of insult, leading to endocrine alterations influencing cytokine production (Berczi & Szentivanhi, 1996). Cytokine release contributes to secondary neuronal trauma, with late neuronal injury mediated by IL-1, IL-6, and TNF (Rosman, 1999).

Monokines are soluble cytokines that mediate immune responses. They are produced by mononuclear phagocytes (monocytes or macrophages). Proinflammatory cytokines (IL-1, IL-6, TNF-α) increase ACTH secretion directly or by augmenting the effect of CRH (CRF). Yet, glucocorticoids themselves exert negative feedback and control their secretion2003). IL-1 (i.e., leukocyte-derived pyrogen) is the principle messenger activating the ACTH-adrenal axis, leading to glucocorticoid secretion. Of the cytokines, IL-1 is the principal messenger to stimulate synthesis and release of CRH from the hypothalamus. Other cytokines also stimulate the ACTH-adrenal axis and influence the release of other pituitary hormones (Berczi & Szentivanhi, 1996).

Lymphokines are soluble cytokines that mediate immune responses. They are released by sensitized lymphocytes on contact with an antigen.

Innate Immune Response

These responses are immediately available to combat a wide variety of pathogens without prior exposure. They are the first line of defense, but lack the ability to recognize certain pathogens and to provide specific protective immunity that prevents reinfection. The initial innate response to stimuli from the outside world is converted into the defense mechanisms of the adaptive response, ensuring that innate mechanisms function until the more finely tuned adaptive immunity develops (Zanetti et al., 2003). This function is carried out by *phagocytes (macrophages)* that cope with disease carriers or injured tissue from trauma. Autoimmune diseases are consequent to the immune tissues misreading ordinary tissues as foreign. The following quotation is from Janeway (2001, p. 13) but refers to disease control: Bacterial molecules binding to these receptors trigger the macrophage to engulf the bacterium and also induce the secretion of biologically active molecules. Activated macrophages secrete *cytokines* and *chemokines*. They also release proteins known as *chemokines* that attract cells with chemokine receptors such as neutrophils and monocytes from the bloodstream (chemotaxis) (Janeway et al., 2001, p. 71).

One component of the innate immunity response is mediated by a family of molecules called *Toll-like receptors (TLR)*, made by many defensive cells. When they detect an invader, they trigger the production of signaling proteins that induce inflammation and direct the body to mount the immune response (O'Neill, 2005). Another defense is carried out by *phagocytes (macrophages)* that cope with disease carriers, or injured tissue from trauma. Macrophages search for invaders, for example, blood cells leaking from a damaged vessel (bruise) or bacteria. When they detect a foreign protein (e.g., bacteria), they engulf and destroy it, and secrete cytokines, some of which recruit other cells to the site and put the IS on full alert.

Adaptive Immune Response

This specific immune response refers to the production of antibodies against a particular pathogen. The ability to distinguish between self and non-self is the *sine qua non* of the adaptive IS. Failure to follow this plan results in such *autoimmune diseases* as systemic lupus erythematosus (SLE) (Kong et al., 2003). It is essential that immune cells (T-cells) not attack the body. The IS develops in fetal life and is qualitatively quite complete at delivery, although certain cells are present in reduced numbers (Hanson et al., 2003). Self-tolerance (as opposed to autoimmunity) develops as a maturational process in the IS, that is, *dendritic cells* (Steinman et al., 2003). The adaptive immune response may confer protective immunity to reinfection with the same pathogen. Its mediator is the *T-cell (T lymphocyte)*, which is derived from the thymus. *Antibodies* are substances that are specific for an infecting pathogen. They are produced by plasma cells, bind specifically to its corresponding antigen, and neutralize them or prepare them for destruction by phagocytes. Phagocytic macrophages have surface receptors that recognize and bind to constituents of many bacterial surfaces. This triggers the macrophage to engulf the bacterium and to secrete cytokines, that is, proteins that affect the behavior of other cells that have receptors for them. In addition to pituitary cells, *PRL* is secreted by lymphocytes and T-cells, illustrating the multiple sources of biologically active substances (Dardenne & Savino, 1994). The SNS exerts inhibitory control over the IS (Arnason, 1997).

Bidirectional chemical sensory organ: The cells of the IS both contain receptors for neuropeptides and are a source of pituitary, hypothalamic, and neural peptides. Immune cells are mobile and deposit their hormones at the target site. These hormones are endogenous regulators of the IS, but also convey information to the neuroendocrine system. It is a sensory function insofar as leukocytes recognize stimuli that are not recognizable by the CNS and peripheral nervous systems, presumably trauma as well as disease entities and components) (Falaschi et al., 1994). The IS's cellular components express specific receptors for almost all types of signaling molecules, that is, neurotransmitters, steroid and peptide hormones, cytokines, and growth factors. The nervous system influences the immune response by three pathways: peripheral autonomic and sensory innervation of lymphoid organs; hypothalamo-hypophysial axis; diffuse neuroendocrine system (Geenen et al., 1994). The neuroendocrine system monitors and controls the physical and chemical variables of the internal milieu. Antigenic recognition creates an image of the macromolecular and cellular components of the body to which the IS responds, maintaining a kind of homeostasis (Goya et al., 1994).

IMMUNE SYSTEM AS A SENSORY ORGAN

The IS has been considered to be a sensory organ analogous to the nervous system, that is, it recognizes "noncognitive stimuli," such as bacteria, viruses, and tumors. Only molecules foreign to the host are immunogenic (Parslow, 2001). Carrying out a variety of immune responses is consequent to cross-talk between immunocompetent cells and the expression of different receptors. To regulate immunity and expression, a complex network of communicating signals is used, for example, cytokines, chemokines, and growth factors (Luger et al., 2003). The capacity to distinguish between self and non-self is the particular quality of the adaptive IS. Injured tissue from trauma as well as disease carriers are examples of non-self substances. Autoimmune diseases are consequent to the immune tissues misreading ordinary tissues as foreign.

MESSENGERS

Cytokines deliver a messenger to the CNS. This stimulates CRF release, and consequently cortisol release, which reduces immune activation and production of the increased cytokine levels. This is a regulatory mechanism that keeps the CNS IS in balance (Heninger, 1995). The stimuli responded to by the central and peripheral nervous systems have been described as *cognitive stimuli* (physical; emotional; chemical). In contrast, certain stimuli would go unnoticed if not recognized by the IS. These are termed *noncognitive stimuli*, for example, bacteria, viruses, tumors, and antigens. When responded to by immunocytes, these structures become information, in the form of peptide hormones, neurotransmitters, and cytokines, that is conveyed to the neuroendocrine system. In an analogous fashion, the CNS and peripheral NS recognize cognitive stimuli, creating similar hormonal information that is conveyed to and recognized by hormonal and neurotransmitter receptors (Blalock, 1994). Through *antigenic recognition*, the IS perceives an internal image of the macromolecular and cellular components of the body and reacts to alterations. This can be considered a "biological" homeostasis (Goya et al., 1994).

IMMUNE CELL REGULATORS

The cells of the immune system both contain receptors for neuropeptides and also are a source of pituitary, hypothalamic and neural peptides. Immune cells are mobile and deposit their hormones at the target site. These hormones are endogenous regulators of the immune system, but also convey information to the neuroendocrine system. It is a sensory function insofar as leukocytes recognize stimuli that are not recognizable by the CNS and peripheral nervous systems, presumably *traumatized tissues* as well as disease entities and components). These are termed "*noncognitive stimuli*" (Falaschi, 1994). Recognition is transformed into information in the form of peptide hormones, neurotransmitters, and cytokines. These are conveyed to the neuroendocrine system and a physiological change occurs. Biological stress can originate in the immune system. Immune cells can create a local antinociceptive effect in inflamed peripheral tissue. Cognitive recognition by the central and peripheral nervous system results in hormonal information being conveyed to hormone and neurotransmitter receptors on immunocytes, with immunologic and neurologic changes resulting. (Weigent & Blalock, 1996). The immune system's cellular components express specific receptors for almost all types of signaling molecules, i.e., neurotransmitters, steroid and peptide hormones, cytokines, and growth factors. The nervous system influences the immune response by three pathways: Peripheral autonomic and sensory innervation of lymphoid organs; hypothalamo-hypophysial axis; diffuse neuroendocrine system (Geenen et al., 1994). The neuroendocrine system monitors and controls the physical and chemical variables of the internal milieu. Antigenic recognition creates an image of the macromolecular and cellular components of the body to which the immune system responds maintaining a kind of homeostasis (Goya et al., 1994).

Detecting Danger

The usual description of the immune system is that it is stimulated by *foreignness*. By discriminating between self and non-self it detects and protects us against potentially dangerous material within our bodies. A more recent overview suggests that it is more concerned with entities that are *dangerous*, that is, do damage rather than are foreign. *It is brought into action by alarm signals from injured tissues, rather than the recognition of non-self* (Matzinger, 2002). It then directs the immune response against non-self rather than self components through T- and B-cells (Raison et al., 2005). It perceives an internal image of the macromolecular and cellular components of the body (antigenic recognition) and reacts to alterations in this image, thus participating in the "*biological*" *homeostasis* of the organism (Goya et al., 1994).

Immune System and CNS Interactions

The IS and the CNS are great memory and communication systems, which are in constant bidirectional communication through endocrine hormones, neurotransmitters, and cytokines. It has been said that the IS and the brain "talk to each other" (Besedovsky & Del Rey, 2001). The neural-immune circuitry permits behavioral influences to modulate the IS, including physical and psychological stressors. *Neuroimmunology* refers to immune reactions involving brain, nerves, and muscles, while neuroimmunomodulation refers to the bidirectional interaction of the nervous and ISs (Reichlin, 1998). Brain peptides and their receptors exist within the IS. The products of an activated IS function as neurotransmitters, while immune cells have receptors for classical neurotransmitters and neuropeptides. Leukocytic peptides include substances found in the neuroendocrine and nervous systems.

The CNS sends *neuroendocrine signals* to specific subsets of IS cells, and direct *neural connections* to specific cell subsets of lymphoid organs. Immune activities modulated by the CNS include innate immunity, cell-mediated immunity, autoimmunity, and reaction to infections. In return, IS release of cytokines signals the peripheral and central NS directly and indirectly. These signals act through the hypothalamus (visceral, neuroendocrine) and the *forebrain* (cognitive, affective). *Inflammatory cytokines* act through the hypothalamus to activate two significant components of the stress axes: the HPA and the sympathetic component of the ANS (Felten & Maida, 2000). Immune cells can also create a local antinociceptive effect in inflamed peripheral tissue.

Bidirectional Signaling

Psychoneuroimmunology (psychoimmunology) is the study of behaviorally associated immunological changes and immunologically associated behavioral changes resulting from reciprocal interactions among the neural, endocrine, and immune systems. It has a long evolutionary history extending back to the invertebrates. Immune cytokines influence complex mechanisms involving neuronal circuits, for example, thermoregulation, food intake, sleeping patterns, and behavior (Besedovsky & Del Rey, 2001). Both immunopharmacologic effects and immune responses themselves can be classically conditioned (Cohen & Kinney, 2001). There is a bidirectional signaling between the two great memory and communications systems, that is, the CNS and the cells and organs of the IS. Bidirectional signals utilize endocrine hormones, neurotransmitters, and cytokines. The CNS sends out neuroendocrine and direct neural connections to subsets of immunological cells. Both individual cellular functions and collective immune responses are influenced. Neural-immune circuitry influences visceral and neuroendocrine functions of the *hypothalamus*, and *forebrain* functions of affect and cognition. Inflammatory cytokines act through the hypothalamus to activate the HPA axis and the SNS. In addition, hormones influence immune function: TSH; thyroid hormones; gonadal steroids (estrogen and androgens); posterior pituitary hormones (vasopressin and OXY); and melatonin (pineal gland) (Felten & Maida, 2002).

Behavioral Factors

Psychological activity influences the IS, and therefore the individual's state of health. These mechanisms are largely obscure but represent a balance between reaction to an antigen and overreaction of the entire IS to this particular antigen (Schlesinger & Yodfat, 1996). Biological stress can originate in the IS. Stress-induced alterations of immunity are affected by the nature of the stressful stimulation; capacity of the individual to cope with stress; the nature of immunogenic stimulation; temporal relationship between stress and immunogenic stimulation; experiential history; and social and environmental conditions. Prolonged psychosocial stress associated with higher catecholamine levels may lead to hypertension, acute myocardial infarction, and sudden death (Schlesinger & Yodfat, 1996). Immune responses can be conditioned (*Pavlovian conditioning*), altering the outcome of immune-mediated diseases (Felten & Maida, 2002). The behavioral and emotional states that accompany the perception of, and the effort to adapt to, events in the real world can influence immune responses. Endocrine, autonomic, and neural activity changes during the course of immune responses indicate that the IS could convey information to the CNS.

ANS activity and neuroendocrine outflow via the pituitary can influence immune function. In addition, it is involved with catecholamine-mediated immunoregulation of cholinergic signals to the IS, modulation of the IS-brain feedback by autonomic pathways, and the role of nonadrenergic, and noncholinergic neurotransmitters involved in the central control of the ANS (Schauenstein et al., 2000).

Sympathetic Nervous System Regulation

Sympathetic noradrenergic nerve fibers signal cells of the IS and are capable of evoking major changes in their responsiveness. Yet, both SNS and PSNS innervate the spleen and thymus (Sternberg, 2003). Further, steroids in the blood affect lymphocytes more than sympathetic activity (Weiss et al., 1994). Sympathetic noradrenergic nerve fibers seem to come into apparent direct contact with lymphocytes and macrophages of lymphoid organs, both primary (e.g., thymus; bone marrow) and secondary (spleen; lymph nodes) (rodent). These nerve fibers form close contact with lymphocytes early in ontogeny, appear to influence early immunological development and compartmentation, and diminish markedly with age. Sympathetic noradrenergic nerve fibers are evidence for a connection between the CNS and the IS (Ader, 1996). The involvement of the SNS in control of the IS is indicated by alteration of immune function after *spinal cord injury* (Campagnolo et al., 1994).

NERVOUS AND IMMUNE SYSTEM INTERACTIONS

The IS is part of an integrated system of adaptive processes, affecting both the CNS and neuroendocrine system. The nervous system is complex and highly branched, making possible simultaneous signals, or gradually to one or several organs, contributing to an orchestrated operation of the IS (Levite, 2000). It is subject to some regulation by the brain through a bidirectional exchange of information, that is, in response to *internal milieu* changes (Dunn, 1996). The nervous system is linked with the IS through channels of biological signaling, that is, hormonal and neurotransmitter mediators. Some neural communication channels to the IS include hypothalamic-anterior pituitary-target organ axes, posterior pituitary hormones, and autonomic outflow to organs of the immunosensory mechanisms of neural-immune signaling (Felten & Maida, 2002). Since the nervous system seems to be able to directly and rapidly trigger or silence various T-cell functions, it may convey important messages that differ from familiar immunological signaling between exclusively IS substances. Information about local conditions is conveyed back to the nervous system by ascending sensory nerves.

Multiple pathways link the brain with the IS: peripheral sympathetic innervation of lymphoid organs; neuroendocrine outflow from the HPA (Geenen et al., 1994). Signals of the internal milieu provide molecules that are perceived by the IS via cell surfaces or internal receptors on the surface of lymphocytes, monocytes/macrophages, and granulocytes. Thus, all immunoregulatory processes take place within a neuroendocrine milieu that is sensitive to the individual's perception of, and response to, events of his external world. Activation of the IS is accompanied by changes in hypothalamic, autonomic, and endocrine processes, and by changes in behavior. Magnification of the potential interaction of neuroendocrine and immune processes occurs due to the fact that cells of the IS activated by immunogenic stimuli are capable of producing a variety of neuropeptides (Ader, 1996).

Immune function is influenced by the CNS through chemical messengers: pituitary (ACTH; β-endorphin; PRL; GH; TSH). The adrenal medulla secretes enkephalins and dynorphins. Endorphins and other neuropeptides are also secreted by the SNS (Dunn, 1996).

Peptide and Dopamine

One basis for neuroimmune interplay is likely to be innervation of lymphoid organs by peptide and dopamine release in areas of inflammation and foreign body invasion. There is also direct signaling to T-cells in the circulation. Nerves can synthesize and release immunelike substances such as cytokines (histamine, IL-1, and IL-6). Cells of the IS synthesize and secrete neurotransmitters or neurohormones (ACTH; CRH; β-endorphin NE, etc.) (Bellinger et al., 2001). Immune cell products (cytokines, lymphokines, lymphotoxins and chemokines released from lymphoid cells) modulate nerve terminal activity, viability, and neurotransmitter release. These products can also enter the circulation to provide feedback to the CNS.

Noradrenergic Sympathetic Terminals

Lymphoid organs are innervated with SNS fibers. Norepinephrine (NE) is released upon the administration of either antigen or cytokines. Sympathetic tone may be activated or suppressed during psychological and behavioral interventions for such clinical conditions as depression or human immunodeficiency virus (HIV) infection, and numerous other medical states (Sanders & Straub, 2001).

Bidirectional Central Nervous System and Immune System Interactions

The CNS regulates many aspects of immune function via the SNS and neuroendocrine pathways to organs and cells of the IS. The IS is controlled by an integrated circuitry of limbic cortex, limbic forebrain, hypothalamus, and brainstem autonomic nuclei. These centers have ascending and descending interconnections.

CNS and IS exchange is bidirectional through hormones, neurotransmitters, and cytokines interacting with receptors found on cells in both systems. There are bidirectional interactions between the immune and nervous systems at all levels, including the brain, pituitary, peripheral nervous system, neuroendocrine mechanisms, and the ANS. The interaction between pituitary, endocrine organ, and lymphocyte-derived hormones defines the *neuroendocrine milieu* in which immune responses occur. This adds another level of complexity to the cellular interactions that drive immune responses (Ader, 1996). Damaged or stimulated structures affect the immune reaction. There is asymmetrical brain control of the IS in rodents and higher incidence of autoimmune and allergic diseases in left handers. Positive and negative reinforcement processes exert modulatory activity on immune functions, involving ventral tegmental, lateral hypothalamic, and midbrain (periaqueductal central gray) (Jankovic, 1994).

The IS's products communicate with the CNS, altering neural activity influencing behavior, hormone release, and autonomic function. Thus, the immune cells function as a *diffuse sense organ*, informing the CNS about events in the periphery relating to infection and injury (Maier et al., 2001). The brain and IS form a *bidirectional communication network*. CNS circuitry involved in IS modulation is integrated with nuclei that regulate both autonomic and neuroendocrine outflow, and has ascending and descending connections. Included are brain areas mediating affective and cognitive processes, stress responses, aversive conditioning, and mood disorders such as depression. These regions, responsive to cytokines and immunization, have the highest concentration of glucocorticoid receptors (discussed under "Stress"), and link some endocrine signals with neuronal outflow to the autonomic and neuroendocrine systems. Stress–related neuroendocrine and neurochemical changes are part of larger system that modulates functional activities of lymphocytes and phagocytes. This aids the organism to maintain *homeostasis* by responding to external and internal changes subsequent to bidirectional signaling. This system seems to mediate such behavior as depression, response to stressors, aversive conditioning, and the emotional context of sensory input from the environment and the body.

Activation of the IS is accompanied by changes in hypothalamic, autonomic, and endocrine processes, and also behavior. Noradrenergic and serotonergic innervation from the brainstem regulates visceral, neuroendocrine, cognitive, and affective events. Organs of the IS appear to be target organs of direct autonomic stimulation (Felton et al., 1991). Products of the IS communicate directly with specific regions of the brain, directly or indirectly, with effects that include changes in neuronal activity and metabolism of neurotransmitter systems. In turn, these regulate visceral, autonomic, and neuroendocrine functions.

Immune System–Hypothalamic Signals

Cytokines may serve as neuromediators in the neuroimmune communication. Transport across the BBB may allow circulating cytokines (e.g., IL-2) to penetrate the BBB, thus gaining access to their receptors in neuronal and glial cell populations (Hanisch, 2001). Cytokines likely reach the hypothalamus by crossing at leaky areas in the BBB, that is, the CVO, active transport across the BBB, and through receptors expressed in cerebral blood vessels that activate secondary messengers such as nitrous oxide (NO) and prostaglandins (Molitch, 2001). Cytokines elicit a neuroendocrine response, and initiate major metabolic alterations, that is, protein loss from muscles and catabolism, on the one hand, and selected protein synthesis on the other (Berczi & Szentivanhi, 1996). Stimulation of the HPA axis, in turn, profoundly inhibits the inflammatory immune response and production of cytokines through cortisol. The endocrine response to injury may be modulated by nerve endings at the site of injury, which influences cytokine production. It may be inferred that *brain damage* itself would elicit a stress-system anti-inflammatory response. α-MSH is an inflammatory inhibitor. Further, cytokines affect both the hypothalamus and the pituitary to alter the release of hypothalamic releasing hormones and pituitary hormones. *Endocrines affected* by cytokines include ACTH, PRL, GH, TSH, LH, and LHRH (Geenen et al., 1994; McCann et al., 1994). Cytokine receptors may be present on the secreting or nearby cells, or a distant target (Hsu et al., 1996). The molecules that signal the brain that an immune event has occurred are such proinflammatory cytokines as IL-1 and TNF-α. They reach the brain via afferent pathways innervating the injured area, and a humoral pathway utilizing blood-borne cytokines and activated circulating immune cells type relayed by special parts of the brain with a deficient BBB (CVO) (Dantzer, 2000).

Psychosocial or physical stressors are communicated by varied endocrine outflow from the CNS via the HPA axis. Norepinephrine is the neurotransmitter from SNS to the IS (lymphoid organs). There is bidirectional signaling between the IS and the CNS by cytokines (including IL-1; IL-6; and TNF-α). Cytokines cross the BBB. A variety of neuropeptides located in nerve terminals innervate the lymphoid organs, for example, SP and neuropeptide Y (Kinney & Cohen, 2001).

The Brain's Immunoreactivity

The brain is immunologically reactive. It produces immunologically active factors, and is closely involved with the systemic immune response. CNS cells respond to inflammatory stimuli, and also secrete cytokines (Chitnis & Khoury, 2004). *Neuropeptides* have a substantial effect on IS functions and are also produced by the IS. Among the hormones produced by lymphocytes are TSH, PRL, GHRH, and hypothalamic releasing hormones (CRF). Some neuropeptides commonly found in the peripheral nervous system are produced by cells of the IS, and have profound effects upon it (Weigent & Blalock, 1996). α-MSH downregulates the production of proinflammatory and immunomodulating cytokines (Luger et al., 2003). Yet, the brain is considered to be an *immunologically privileged site* because tissue grafts placed there are not rejected or attacked, although antigens sequestered there may be the target of an *autoimmune attack, for example, multiple sclerosis.* Humoral factors (presumably cytokines) that affect the immune response are produced here. While there are protective qualities against immune processes, antigens sequestered here may be the target of autoimmune attack, for example, multiple sclerosis (Janeway et al., 2005, pp. 570–571).

Exchange Between the Immune System and Peripheral Nerves

Somatic nervous system *efferent* fibers innervate these IS organs: bone marrow, spleen, thymus, and lymph nodes. This effect seems modulated by *afferent* fibers of the vagus nerve (PSNS), which relay immune signals to the nucleus of the solitary tract, which has pathways to other brain regions influencing CNS response to IS activation. The effect upon the immune organ depends upon the local concentration of the neurotransmitter, and according to animal studies age, sex, and genetic character (Raison et al., 2005). Peripheral nerves are not completely stable: Their neurons depend on trophic factors produced by their target to maintain innervation. Thus, changes in homeostasis in lymphoid tissues often result in changes in neurotransmitter turnover, or remodeling of nerves that distribute to them. Neurotransmitters can influence specific functions of subsets of cells within lymphoid organs (specificity). Immune activation can induce changes in nerve fibers that innervate lymphoid organs, for example, altered neurotransmitter metabolism, induction of neurotransmitters in the nerves, and promotion of axonal sprouting or neuronal degeneration.

Neural-immune signaling: Cells of the IS possess elaborate systems of proteins that enable them to respond to signals that are (1) self-generated, (2) derived from other immunocytes, (3) conveyed by the neuroendocrine and ANSs, or (4) generated in response to foreign substances in their microenvironment. Neuroimmune signaling is a dynamic and interactive process, the understanding of which requires integrative investigation of neurotransmitter, hormonal, and cytokine influences on multiple target cells. It is clear even at this stage in studying neural-immune signaling TNT-Tunneling Nanotubes (1) there is extensive cross-talk between the nervous and ISs, (2) neurotransmitters and endocrine signals can profoundly alter immune function, (3) cytokines can profoundly influence nervous system function under normal conditions and in states of pathology, and (4) neural and immune mediators share common intercellular signaling pathways (Lorton et al., 2001).

Immunocytes can be exposed to specific neurotransmitters, to which they may respond. Nerves are distributed to sites where cells of the IS form the lymphoid tissue. Specific neurotransmitters convey messages. Changes in signaling pathways alter immune function under different physiological conditions and in disease states. Moreover, neurotransmitters may be produced and secreted by IS cells, acting in an autocrine and paracrine fashion. Since nerves have autoreceptors specific for their own neurotransmitters, secretion of neurotransmitters from immunocytes that reside adjacent to nerves is likely to regulate neurotransmitter release from nerves in lymphoid organs. Furthermore, signal molecules from the nervous system may arrive in lymphoid tissue via the circulation as hormones. Norepinephrine may reach the cells of the IS by various routes. There is also cholinergic innervation of lymphoid organs (Bellinger et al., 2001).

Immuno-CNS signals: There is bidirectional communication between the CNS and the IS, specifically the anatomical, physiological, and psychological domains (Kemeny et al., 1992), and between the immune and neuroendocrine systems. Signals pathways include the blood system, peripheral nerves, and cranial nerve X. The signals include psychoneuroimmune substances, the nervous system (traditional transmission), and hormonal transmission (to posterior pituitary gland; immunoinflammatory products, e.g., via vagus nerve as a response to trauma); humoral (hormonal; immunoinflammatory products secreted into the bloodstream, e.g., cytokines, etc.), local (paracrine) or used within the cell (autocrine). *Feedback ("cross-talk")* refers to chemical products of stress and trauma that regulate physiological processes through feedback (hypothalamic-pituitary target organ axes; various structures within the brain, including the hypothalamus, brainstem, etc.). Feedback is positive (stimulating) or negative (inhibitory). Hormones released by targets of the anterior pituitary gland feed back to the hypothalamus and anterior pituitary, with negative feedback bringing the stress reaction under control. Since various neurotransmitters affect pituitary hormone secretion, directly or indirectly, by modulating the activities of various hypothalamic hormones, pituitary hormone secretion reflects the integration of all these inputs (Molitch, 2001).

Neurobehavioral Interactions

The IS involved in behavioral interactions between the psychosocial and biological environments (Bohus & Koolhaas, 1996). There is a link between neural activity and altered immune responses, including psychosocial factors such as bereavement, marital separation, depression, and examination stress (Felton et al., 1991). The endocrine and ISs, which mutually regulate each other (Grossman, 1991), share the property of memory. There is direct CNS innervation of the IS, but also IS receptors for neurotransmitters, neurohormones, and neuropeptides. *Stress*-related neuroendocrine and neurochemical changes are part of a larger system that modulates functional activities of lymphocytes and phagocytes.

The IS can be conditioned bidirectionally between itself and the CNS. Signals travel via the hypothalamic-pituitary neuroendocrine axis and the ANS. The CNS detects alterations in immune reactivity. Immune responses can be altered by psychological stress, and by lesions in the limbic forebrain, hypothalamus, brainstem reticular formation, and autonomic regions (Bellinger et al., 1992). The IS seems to act as a *"receptor sensorial organ,"* emitting a response to which the CNS is capable of responding. One of the channels of communication between the neuroendocrine and IS is by receptors that exist on immune cells (Ader, 1996). They communicate information to the brain in diverse physiological conditions. For example, products of activated immune cells can increase the firing rate of noradrenergic neurons (Geenen et al., 1994). This would relate to an *"afferent humoral system"* that accounts for the fact that peripheral endocrine events are conveyed to the brain, and also, the effects of all hormones of the pituitary-adrenal system upon CNS functions (Fehm & Born, 1989). There are bidirectional effects between hypothalamic releasing hormones (having direct effects upon the IS) and the IS (which manufactures neuroendocrine hormones, e.g., CRF and GHRH). Neuropeptides have an important effect on the IS, including pituitary hormones, hypothalamic releasing hormones, and other substances (Weigent & Blalock, 1996). The IS contains receptors for neuropeptides and is a source of pituitary, hypothalamic, and neuropeptides. The secretion level of immune cells is not comparable to that of the pituitary, since they are mobile and can locally deposit the hormone at the target site. Thus, they serve as (1) endogenous regulators of the IS; and (2) relayers of information to the neuroendocrine system. There is an IS sensory function since leukocytes recognize stimuli not recognizable by the peripheral and CNSs. The recognition of *noncognitive stimuli* by immunocytes is converted into information, that is, peptide hormones, neurotransmitters, and cytokines, and conveyed to the neuroendocrine system leading to a physiological change (Weigent & Blalock, 1996).

THE INFLAMMATORY SYSTEM

Inflammation is characterized by pain, redness, heat, and swelling consequent to increase of vascular diameter, reduced velocity of blood flow (especially in smaller vessels), and changes in the endothelial cells that allow leukocytes to become attached and then migrate into the tissues (extravasation). The *febrile reaction* includes these neurobehavioral symptoms: shivering, malaise, and anorexia, and endocrine effects (increased secretion of AVP and corticosteroids). These are attributed to cytokine effects on neuronal targets in the CNS. Inflammation, as may accompany somatic injury, nociceptive, somatosensory, and visceral afferent stimulation, can stimulate catecholaminergic and CRH neuronal systems via ascending spinal pathways.

INFLAMMATION

Inflammation is a localized protective response, elicited by tissue injury or destruction, which destroys, dilutes, or sequesters the injurious agent and the injured tissue; protects against infection; and regulates the activities of tissue cells that repair the injured tissue. Since inflammation is triggered not only by microbial infection, but also by blunt trauma, radiation injury, lacerations, thermal or chemical burns, it is not considered an immune reaction. However, immune and inflammatory reactions are closely linked, and often promote and enhance one another. *Tissue injury*, as caused by trauma, bacteria, chemicals, heat, etc., is followed by the release of multiple substances that cause dramatic secondary changes in the tissues, that is, *inflammation* (Guyton & Hall, 2000, p. 397). Inflammation plays a role in the development of subarachnoid hemorrhage (SAH) (Provencio & Vora, 2005). CNS effects stemming from cytokines that penetrate the brain, or are produced within it, can promote peripheral inflammation (Lipton et al., 1994). The delicate balance between pro- and anti-inflammatory cytokines is easily disrupted by adverse conditions. The severity of trauma, and the duration of hypotension, will determine whether tissue repair will be initiated by a localized inflammation or whether there will be a systemic induction of a generalized inflammatory process. Massive stress also results in a significant activation of the anti-inflammatory cytokine TNF, and various interleukins, representing either immunosuppression or immunoparalysis. Immunodeficiency (depressed immune function) in trauma victims is associated with enhanced concentrations of inflammatory cytokines (Faist et al., 2004). Under physiological conditions, reactions are maintained by cells that circulate or are tissue-resident. Severe and protracted injury causes proliferation of fibroblasts and endothelial cells at the site, which form a permanent scar.

INFLAMMATORY MEDIATORS

Posttraumatic inflammation has a profound systemic and neurobehavioral effect.

Cytokines (IL-1; IL-6; TNF-α) are termed *endogenous pyrogens* because they cause fever and derive from an endogenous source. At higher temperatures, most pathogens are more vulnerable, adaptive immune responses are more intense, and host cells are also protected from the deleterious effects of TNF-α. These induce acute-phase proteins, including C-reactive protein (CRP) (Janeway et al., 2001, pp. 78–80). Most tissues of the body have protective protein coats that repel the phagocytes, whereas most dead tissues (wounds, presumably) and foreign particles have no protective coats, making them subject to phagocytosis (Guyton & Hall, 2000, p. 394). The intensity of the inflammation is proportional to the degree of tissue injury.

Inflammation causes the release of vasodilators (histamine and cytokines), which increases the number of open capillaries and the functional surface area. Cytokines cause widening of the inter-endothelial clefts, resulting in enhanced filtration of fluid from the capillaries, causing tissue swelling (Boron & Boulpaep, 2003, p. 473).

Nitrous oxide (NO) may mediate the proinflammatory actions of cytokines and bradykinin, and act synergistically with eicosanoids (Hsu et al., 1996).

α-MSH (secreted by the intermediate pituitary gland) modulates inflammation through both central influences (adrenergic pathways) and local sites of action, that is, has anti-inflammatory activity.

Neuroactive Substances

Cholinergic nerves may influence the inflammatory process, playing an integral role in host defense (Barnes, 2003). Many inflammatory events are controlled by cytokines and other small regulatory molecules called *inflammatory mediators*. Some products causing this reactions are known *neuroactive substances*: histamine, prostaglandins, leukotrienes, and cytokines are inflammatory mediators; bradykinin; serotonin; prostaglandins; products of the complement system; leukotrienes; reaction products of the blood-clotting system, and *lymphokines* released by sensitized T-cells, which also are part of the IS (Berne & Levy, 1998, p. 591; Boulpaep, 2003; Guyton & Hall, 2000, p. 397). Histamine, prostaglandins, leukotrienes, and cytokines are inflammatory mediators. Several of these inflammatory mediators activate the *macrophage system*, which within a few hours begins to devour the destroyed tissue, although at the time, the macrophages may also injure the still-living tissue cells.

Immune cytokines and other humoral mediators of inflammation are described as the "afferent limb" of the feedback loop through which the immune/inflammatory systems and the CNS communicate (Tsigos & Chrousos, 1996). Inflammation (as may accompany somatic injury), as well as nociceptive, somatosensory, and visceral afferent stimulation, can stimulate catecholaminergic and CRH neuronal systems via ascending spinal pathways.

SP is a member of the tachykinin family of hormones, which cause smooth muscle contraction and vasodilation. Its immunological functions include the release of monocyte-derived cytokines and of inflammatory mediators from lymphocytes and macrophages (Schwartzman, 1996). SP neurons in IS tissues modulate its functions. Activation by SP may facilitate neuronal regeneration and ward off infection or inflammation accompanying injury. Thus, SP could play a role as intermediary between the neuronal and ISs (Jonakait et al., 1991). SP has multiple inflammatory and immune modulator functions, interacting with some IS cells.

There is an overlap between the *symptoms of viral illness* caused by the effects of circulating cytokines on SNS nerves (e.g., lassitude, fatigue, myalgia, headache) and subjective reactions after trauma causing TBI (Arnason, 1997).

Neuroendocrine Effects

Control of the inflammatory system is performed by the HPA axis, the SNS and PSNS, and local release of neuropeptides, which tend to be proinflammatory (Sternberg, 2003). *Monokines*, released by inflammatory tissue, for example, IL-1, which mediates many of the inflammatory responses. It also stimulates synthesis and release of CRH from the hypothalamus and the release of ACTH from the pituitary. *Sarcoidosis* involves activation of inflammatory cells (e.g., T lymphocytes, which secrete various *cytokines* (IL-1 and -2, interferon-γ, and TNF. Other cytokines also elicit secretion of ACTH (Molitch, 2001).

Cytokines elicit a neuroendocrine response, and initiate major metabolic alterations, that is, protein loss from muscles and catabolism, on the one hand, and selected protein synthesis on the other (Berczi & Szentivanhi, 1996). An example of the neuroendocrine effects of cytokines on homeostasis is the fact that IL-1β is hypoglycemic despite its stimulation of glucocorticoid output and decrease of hepatic glucagon content. This stems from stimulation of CRF-producing neurons in the hypothalamus (Del Rey et al., 1998). Stimulation of the HPA axis, in turn, profoundly inhibits the inflammatory immune response and production of cytokines through cortisol. The endocrine response to injury may be modulated by nerve endings at the site of injury, which influences cytokine production.

Secondary Brain Injury

The inflammatory response is triggered by molecules that signify tissue damage, regardless of its specific cause. Many inflammatory events are controlled by cytokines and other small regulatory molecules. They are secreted by injured or distressed host cells, while others are by-products either of tissue injury per se or of the host's reaction to it. Acute inflammation may be an important determinant of the ultimate outcome.

Wound Repair

The cytokines and chemokines released by macrophages in response to bacterial constituents (trauma) initiate inflammation, which promotes the *repair of injured tissue*, a nonimmunological role (Janeway et al., 2001, p. 41). *Cholinergic nerves* may influence the inflammatory process, playing an integral role in host defense (Barnes, 2003).

Microglia

Microglial cells are derived from an immune cell lineage. They shape the brain's response to any insult: Rapidly activated by brain injury, they release neurotoxic substances, become phagocytic (Ransom, 2005), and release proinflammatory cytokines that regulate inflammatory response at the periphery. Brain cells themselves have receptors for these and other cytokines (Dantzer, 2000). Brain function and behavior are affected by lymphokines produced by an activated IS. Stress effects on the IS, particularly psychosocial stressors, depend on the nature of the environment, personality factors, and the IS location in particular compartments. Corticosteroids are mediators of stress effects. Inflammatory cells, particularly reactive microglia and invading macrophages, release both cell promoting and cell-destroying factors. Microglia are a distinctive class of mononuclear phagocytes that persist within the CNS after the perinatal period. They release growth factors, including cytokines, and destroy myelin and myelin-producing oligodendroglia. During CNS injury, mononuclear phagocytes engulf tissue debris. This stimulates the release of cytotoxic factors, which contributes to the ensuing loss of neurologic function after the perinatal period. Inflammatory responses lead to secondary loss of neurons and deterioration in neurologic function. Microglia are the only brain cells producing significant amounts of IL-1 and TNFα. Microglia and inflammatory cells release neuron-killing factors, including N-methyl-D-aspartate (NMDA)-receptor agonists (Giulian, 1994).

Signaling is carried out by the activation of microglia, a response common to all inflammatory and necrotizing processes in the brain. Immunomodulators (mitogens or activators of ameboid microglia) amplify inflammation by acting upon intrinsic brain mononuclear phagocytes. This is a regulatory system whereby secretion of brain-derived growth factors or systemic production of cytokines controls the microglial cell population, that is, suppressing the consequences of CNS inflammation.

Inflammatory processes occur within 24 h of acute brain trauma, with leukocytes accumulating and macrophages secreting cytokines. The latter create neuroendocrine and metabolic changes, and potentially neurotoxicity. Immunostaining of brains of individuals with mild head injury with recorded LOC as short as 60 sec, surviving up to 99 days after the injury, reveals multifocal axonal injury indicated by an antibody for *amyloid precursor protein (APP)* (Gennarelli & Graham, 1998). While all types of focal axonal injury immunostain for β-APP, it raises the question of the higher incidence of *DAT* after head injury.

There is rapid production of cytokines, chemokines, and prostoglandins that are responsible for autonomic changes of the acute phase. The proinflammatory cytokine interleukin-1 (IL-1B) is probably the most important molecule modulating cerebral functions during systemic and localized inflammatory insults (Rivest, 2003). The inflammatory response is stimulated by cytokines, which are produced by activated macrophages. Inflammatory mediators include prostaglandins, leukotrienes, and PAF. Following a wound, injury to blood vessels triggers enzyme cascades: The *kinin system* produces inflammatory mediators, including *bradykinin*, leading to increased vascular permeability and increase of proteins. Pain encourages immobilization. The *coagulation system* is another cascade of plasma proteins (Janeway et al., 2001, p. 42). They are secreted by injured or distressed host cells, while others are by-products either of tissue injury per se (e.g., fragments of collagen) or of the host's reaction to it. Thus, the inflammatory response is triggered by molecules that signify tissue damage, regardless of its specific cause. The changes in inflamed blood vessels attract cells of the IS into an injured or infected tissue. Injured skin, joints, or muscles have a reduced threshold for pain,that is, *hyperalgesia. Secondary hyperalgesia* results in inflammation involving processes near peripheral receptors and mechanisms in the CNS. Damaged tissue causes a variety of chemical substances to be released from itself, blood vessels, and nerve endings: bradykinin, prostaglandins, serotonin, SP, K+, H+, and others (Connors, 2005b). The dilatation and increased permeability of vessels near injured tissue results in part from a spinal reflex: Pain receptors stimulated by the injury transmit afferent signals to the spinal cord, where they act on autonomic motor neurons to cause relaxation of arteriolar smooth muscles at the injured site. A neural-independent component of the vascular response is triggered by substances produced at the injured site, which act directly on the local vessels. Among the *vasoactive mediators* are histamine, prostaglandins, and leukotrienes (Parslow & Bainton, 2001). The pathways and events vary according to the inciting stimulus, portal of entry, and characteristics of the host. Some outcomes are detrimental. While inflammatory reactions have evolved to inactivate or eliminate injurious substances or limit their spread, these same reactions can injure host tissues or interfere with normal functions (Terr, 2001).

Histamine, prostaglandins, leukotrienes, and cytokines are inflammatory mediators. Contributing to *secondary brain injury* is acute inflammation, which may be an important determinant of the ultimate outcome. NO may mediate the proinflammatory actions of cytokines and bradykinin, and act synergistically with eicosanoids (Hsu et al., 1996).

Components of inflammation: Although infection is a familiar cause of inflammation, we will emphasize the consequences of wounds. Inflammation is characterized by an increase of the number of open capillaries increasing the functional surface area; vasodilation of local blood vessels with excessive blood flow; increased permeability of the capillaries allowing leakage of large quantities of fluid into the interstitial spaces; perhaps clotting of this fluid because of leakage of excess amounts of fibrinogen and other proteins; migration into the tissue of large number of granulocytes and monocytes; and, swelling of the tissue cells. Enhanced vascular permeability and edema are mediated by prostaglandins, kinins, histamine, leukotrienes, and PAF (Miller & Chrousos, 2001). The fibrinogen clots block off the tissue spaces and lymphatics of the inflamed area, delaying the spread of toxic products.The role of the *sympathetic and peripheral nervous systems* in mediating swelling and hyperalgesia at remote sites from an injury is called *reflex neurogenic inflammation (RNI)* (this plays a role in *reflex sympathetic dystrophy [RSD]*). This process is mediated segmentally at the spinal cord level.

SP has multiple inflammatory and immune modulator functions, interacting with some IS cells. The consequence is the release of inflammatory mediators from lymphocytes and macrophages, and of monocyte-derived cytokines (Schwartzman, 1996). The *GI IS* has a mass of cells, with immune function approximately equal to the combined mass of immunocytes in the rest of the body. Reacting to food antigens and pathogenic microorganisms, it is involved in celiac disease, inflammatory bowel disease, and Crohn's disease (Kutchai, 1998).

Inflammatory mediators are tissue products causing these reactions. These include known *neuroactive substances*: histamine and cytokines; bradykinin; serotonin; prostaglandins; products of the complement system; leukotrienes; reaction products of the blood-clotting system, and *lymphokines* released by sensitized T-cells, which also are part of the IS (Berne & Levy, 1998, p. 591; Boulpaep, 2003; Guyton & Hall, 2000, p. 397). Inflammation causes the release of vasodilators (histamine and cytokines), which increases the number of open capillaries and the functional surface area. Cytokines cause widening of the interendothelial clefts, resulting in enhanced filtration of fluid from the capillaries, causing tissue swelling (Boron & Boulpaep, 2003, p. 473). *SP* is a member of the tachykinin family of hormones, which cause smooth muscle contraction and vasodilation. Its immunological functions include the release of monocyte-derived cytokines and of inflammatory mediators from lymphocytes and macrophages (Schwartzman, 1996). SP neurons in IS tissues modulate its functions. Activation by SP may facilitate neuronal regeneration and ward off infection or inflammation accompanying injury. Thus, SP could play a role as an intermediary between the neuronal and ISs (Jonakait et al., 1991).

Hypothalamic-Pituitary Axis

Trauma activates the HPA axis via "*inflammatory cytokines*." Control of the inflammatory system is performed by the HPA axis, the SNS and PSNS, and local release of neuropeptides, which tend to be proinflammatory (Sternberg, 2003).

Severe head injury can result in cerebral edema, a result of the breakdown of the normally tight endothelial barrier of the cerebral vessels. The enclosed skull prevents expansion of the brain, leading to occlusion of the cerebral microcirculation (Boulpaep, 2003).

The cytokines and chemokines released by macrophages in response to bacterial constituents (trauma) initiate inflammation, which promotes the *repair of injured tissue*, a nonimmunological role (Janeway et al., 2001, p. 41). *Cholinergic nerves* may influence the inflammatory process, paying an integral role in host defense (Barnes, 2003).

Another possible interfering effect upon memory is indicated by the finding that *cytokines* (released during enhanced immune response) inhibit long-term potentiation in hippocampal slices. The effects of the cytokines are so dire, that it is believed that they are normally modulated by cytokine antagonists. These become available during host challenges, such as inflammation, that is, a consequence of an accident-related injury. One of these is the neuropeptide *α-MSH*, important in communication among neuroendocrine, immune, and CNSs (Catania et al., 1994). *Melatonin* acts upon the hypothalamus and pituitary to inhibit gonadotropin secretion.

Cognitive Effect

Inflammation plays a role in Alzheimer's and other neurodegenerative diseases. A high level of serum markers of inflammation (IL-6; CRP) was associated with poor cognitive performance and greater risk of cognitive decline over 2 years of follow-up. The effects were cumulative. It is hypothesized that inflammation contributes to cognitive decline in the elderly. However, nonsteroidal anti-inflammatory drug (NSAID) use did not change the association between inflammation and cognitive decline (Yaffe et al., 2003).

Acute-phase response: The inflammatory component is a cytokine-mediated response to infection and some injuries (Parslow & Bainton, 2001). In addition to infection, it is triggered by trauma, burns, and tissue necrosis. It occurs after severe head injury, including fever without evidence of infection, protein changes, and enhanced levels of zinc and copper, with effects both beneficial and detrimental to recovery (Young & Ott, 1996). It is considered a primitive nonspecific defense reaction mediated by the liver that intensifies some aspects of innate immunity. When hepatocytes are exposed to particular cytokines (IL-1, IL-6, TNF-α), the innate response occurs. These cytokines, together with a poorly defined neural response, give rise to the fever, somnolence, and loss of appetite occurring as part of the acute phase response.

During injury, the IS is activated, leading to cytokine production similar to acute infections, and could culminate in endocrine and metabolic responses characteristic of the acute-phase response. The endocrine response to injury may be modulated by nerve endings located in the injured area, whose reactions influence cytokine production. The neuroendocrine system keeps cytokine production under tight control during the acute-phase response through increased glucocorticoid secretion. Excess cytokine production would be fatal due to neurotoxicity (Berczi & Szentivanhi, 1996). Trauma and immune-mediated inflammatory responses that compromise the BBB cause proinflammatory cytokines to be induced in the brain (Dantzer, 2000). Accumulation of vasoactive eicosanoids and inflammatory cytokines (TNF) cause a partial breakdown of the BBB (Laterra & Goldstein, 2000). Chronic systemic inflammation is accompanied by metabolic alterations, which promote *atherosclerosis* (Desiderio & Yoo, 2003). Inflammation and skeletal destruction in arthritis are induced and maintained by cooperative interaction between IL-1β and TNF-α cytokines, which act synergistically to promote inflammation (Campagnuolo et al., 2003). Damaged or stimulated brain structures (including the cortex) affect the immune reaction, including reduced immune function after ablation of the nucleus LC (Jankovic, 1994).

Fever

Fever as a reaction to injury is an adaptive mechanism of defense, mediated via neurons in the anterior hypothalamus (Maier et al., 2001). Inflammation, trauma, and tissue necrosis induce the production of pyrogenic cytokines (IL-1, TNF, and/or IL-6). These induce the production of prostaglandin E_2 (PGE_2). Peripherally, this has the effect of the myalgias and arthralgias that accompany fever. It is the induction of PGE_2 in the brain that starts the process of raising the hypothalamic set point for core temperature to febrile levels (Dinarello & Gelfand, 2001). Skin, joints, or muscles that are inflamed or damaged are *hyperalgesic*, that is, unusually sensitive with a reduced pain threshold or increased sensitivity to further stimuli. Damaged skin releases chemicals from itself, blood cells, and nerve endings: bradykinin, prostaglandins, serotonin, SP, K^+, H^+, and so forth, which trigger inflammation. When the blood vessels become leaky and cause edema (tissue swelling), nearby mast cells release histamine, which directly excites nociceptors. The spreading nociceptor axon branches release substances that sensitize nocireceptor terminals responsive to previously nonpainful stimuli (Aδ and C fibers) (Connors, 2005b).

Inflammation is a localized protective response, elicited by tissue injury or destruction. Its function is to destroy, dilute, or sequester the injurious agent and the injured tissue; protect against infection; and regulate the activities of cells that repair the injured tissue. As a response to infection, it provides macrophages (*microglial cells* in neural tissue) augmenting the killing of invading microorganisms, provides a physical barrier (microvascular coagulation) preventing the spread of the infection, and promotes the repair of injured tissue (described as a nonimmunological role. Vascular permeability is increased (Janeway et al., 2005, pp. 45–45). Inflammation is triggered by blunt trauma, radiation injury, lacerations, and thermal or chemical burns.

Immune and inflammatory reactions are closely linked, and often promote and enhance one another. Inflammation plays a role in the development of SAH (Provencio & Vora, 2005). CNS effects stemming from cytokines that penetrate the brain, or that are produced within it, can promote peripheral inflammation (Lipton et al., 1994). The delicate balance between pro- and anti-inflammatory cytokines is easily disrupted by adverse conditions. The anti-inflammatory stress response (glucocorticoids) interferes with wound healing by (1) reducing the ability of leukocytes to produce cytokine IL-1, and (2) suppression of sensory fibers, which release neuropeptides that induce the inflammatory response (Marucha et al., 2001). α-MSH (secreted by the intermediate pituitary gland) modulates inflammation through both central influences (adrenergic pathways) and local sites of action, that is, it has anti-inflammatory activity.

The severity of trauma and the duration of hypotension will determine whether tissue repair will be initiated by a localized inflammation or whether there will be a systemic induction of a generalized inflammatory process. Massive stress also results in a significant activation of the anti-inflammatory cytokine force of TNF, and various interleukins, representing either immunosuppression or immunoparalysis. Immunodeficiency (depressed immune function) in trauma victims is associated with enhanced concentrations of inflammatory cytokines (Faist et al., 2004). Under physiological conditions, reactions are maintained by cells that circulate or are tissue-resident. Severe and protracted injury causes proliferation of fibroblasts and endothelial cells at the site, which form a permanent scar. The changes in inflamed blood vessels attract cells of the IS into an injured or infected tissue. Injured skin, joints, or muscles have a reduced threshold for pain, that is, *hyperalgesia. Secondary hyperalgesia* results in inflammation involving processes near peripheral receptors and mechanisms in the CNS. Damaged tissue causes a variety of chemical substances to be released from itself, blood vessels, and nerve endings: bradykinin, prostaglandins, serotonin, SP, K+, H+, and others (Connors, 2005b). The dilatation and increased permeability of vessels near injured tissue results in part from a spinal reflex: Pain receptors stimulated by the injury transmit afferent signals to the spinal cord, where they act on autonomic motor neurons to cause relaxation of arteriolar smooth muscles at the injured site. A neural-independent component of the vascular response is triggered by substances produced at the injured site, which act directly on the local vessels. Among the *vasoactive mediators* are histamine, prostaglandins, and leukotrienes (Parslow & Bainton, 2001). The pathways and events vary according to the inciting stimulus, portal of entry, and characteristics of the host. Some outcomes are detrimental. While inflammatory reactions have evolved to inactivate or eliminate injurious substances or limit their spread, these same reactions can injure host tissues or interfere with normal functions (Terr, 2001).

Contributing to *secondary brain injury* is acute inflammation, which may be an important determinant of the ultimate outcome. NO may mediate the proinflammatory actions of cytokines and bradykinin, and act synergistically with eicosanoids (Hsu et al., 1996).

NEUROENDOCRINE–IMMUNOMODULATION

Hormones and CNS Effects

Receptors: The brain contains receptors for all six classes of steroid hormones (androgens, estrogens, progestins, glucocorticoids, mineralocorticoids, and vitamin D, as well as thyroid hormone. Each type of steroid receptor has a unique distribution in the brain, in neurons, and sometimes in glial cells. These receptors may interact, for example, hypothalamic neurons convert testosterone in male animals to estradiol, and then they occupy estradiol receptors.

Neurobehavioral: The steroid/thyroid hormone family coordinates behavioral states with other events in the body: Sexual behavior is coordinated with fertility. In the diurnal cycle of sleep and activity, metabolism is coordinated with food seeking and other active behaviors. Adrenal steroids participate in both diurnal rhythms of sleeping and waking, and the peripheral functions of glucocorticoids and mineralocorticoids (McEwen, 1992). *TSH* shortens the duration of pentobarbital anesthesia, while *somatostatin* depresses animals (McCann, 1992). *Vasopressin* acts centrally as a neurotransmitter, modulating the physiological processes that are integrated in the region to which its neurons project: pons, septum, amygdala, and hypothalamus (Le Moal et al., 1992).

Neurosecretory cells serve important integrative functions. Found in the hypothalamus, they propagate action potentials, and are sensitive to other neurons' neurotransmitters, hormones, and metabolites. They are involved in hypothalamic-pituitary regulation: neurohypophyseal and hypophysiotropic (these are discussed in the Chapter on endocrine functioning).

Histamine: A chemical found in all body tissues. Most relevant here is its capacity to dilate capillaries, which increases capillary permeability and results in a drop of BP. The H_3 receptor occurs in the CNS and peripheral nerves, and plays a role in the release of histamine and other neurotransmitters from neurons.

Secretions of somatic tissues may be identical to substances categorized as neuroendocrine or neurotransmitters. Thousands of different SE neurotransmitters, acting on more than 100 types of receptors, are used in the mammalian nervous system. Most of the neurotransmitters are similar or identical to substances that cells use for metabolism. The transmitter system involves the neuron, receptor, and second messenger systems, which regulate ion channels and enzymes. The effect of the transmitter depends on the nature of the components it interacts with, varying with different parts of the brain, and even of a neuron. The same molecule can be both a neurotransmitter and a hormone, for example, catecholamines are neurotransmitters when released by nerve terminals and hormones when released by the adrenal medulla. Furthermore, they use the same types of adrenergic receptors in the CNS and the peripheral tissue. TRH (hypothalamus) has neurotransmitter actions in the CNS (Baxter et al., 2004). Drugs affecting mental functions, and many brain disorders, involve neurotransmitter systems (Connors, 2005a).

Neuroendocrine–immunomodulation (NIM) is considered a single repertoire of neurotransmitters that are shared by the nervous, endocrine, and ISs, though not necessarily interacting within systems. Both neurons and endocrine cells can secrete into the bloodstream. Neurotransmitters can act as hormones, while hormones have neurotransmitter functions (Genuth, 1998a). Neuropeptides, hormones, and cytokines have a similar task that can be applied to different systems. Their role is defined according to the type of cell where they find their specific receptor. This is a contribution to the understanding of homeostasis (Panerai, 1994). The endocrine and nervous systems, and their interactions, underlie practically every regulator mechanism in the body. Neural regulation of endocrine function occurs through neurons that secrete hormones into the circulation (e.g., hypothalamus) and direct autonomic innervation of endocrine glands (Baxter et al., 2004). The intricate communication between the nervous, endocrine, and ISs is enhanced by the common set of hormones, neuropeptides, and cytokines shared by the immune and neuroendocrine systems. These systems share a similar group of receptors (Blalock, 1994).

A full understanding of the adaptational responses to stress involves the interaction between blood-borne molecules, that is, cytokines, adrenomedulary hormones (catecholamines), and the HPA—presumably including other target glands (Cullinan et al., 1995). Immune, central nervous, and neuroendocrine systems interact through cytokines and hormones. Immune and neuroendocrine systems interact and share common signal molecules and receptors. The IS and its products, for example, cytokines, can modulate neuroendocrine functions (Falaschi et al., 1994).

BODY–BRAIN SIGNALING SYSTEMS OF THE INTERNAL MILIEU

"The brain is the most prolific of all endocrine organs, producing scores of neurohormones within and beyond the boundaries of the endocrine hypothalamus" (Gold et al., 1987).

"The brain's earliest self-representational capacities arose as evolution found neural network solutions for coordinating and regulating inner-body signals, thereby improving behavioral strategies" (Churchland, 2002). Indeed, the integration of awareness of body states with plans for external action is attributed as a component of consciousness itself.

NEUROENDOCRINOLOGY

This refers to the interactions of the nervous system with the endocrine system (Jacobson & Abrams, 1999; Molitch, 2001; Reichlin, 1998) (see "Blood-Brain Barrier"). The hypothalamus and pituitary gland unit exercises control over both endocrine glands (thyroid, gonads, adrenals) and a wide range of physiological activities. The major regulatory system for virtually all physiologic activities is the mutual modulation of the endocrine and nervous systems. The IS interacts with the endocrine and nervous systems (Aron et al., 2007a). Neuroendocrine control of a wide range of immune responses occurs via neuropeptides and neurohormones (Brown & Zwilling, 1996). The CNS modulates the periphery through neurosecretory and synaptic connections, while the endocrine hormones modulate autonomic and metabolic processes, and provide direct feedback to the brain on the condition of the internal physiological environment. Both nerve and endocrine cells are involved in cell-to-cell communication, secrete chemical messengers, and create electrical activity.

The neuroendocrine milieu: The neuroendocrine system monitors and controls the physical and chemical variables of the internal milieu. Thus, the neuroimmunoregulatory process takes place in a *neuroendocrine milieu* that is sensitive to the influence of the individual's perception of, and response to, external events. This milieu is defined by the innervation of lymphoid organs, the availability of neurotransmitters for interactions with cells of the IS, interaction between pituitary-endocrine glands, and lymphocyte-derived hormones (Ader, 1996). More than 20 neuropeptides are secreted by cells of the IS. These have a substantial influence, serving as autocrine and paracrine factors exercising local control over a variety of immune functions. IS cells contain neuropeptides and also are a source of pituitary, hypothalamic, and neural peptides. The IS manufactures neuroendocrine hormones and may also produce hypothalamic releasing factors, which have direct effects upon it. Plasma hormone contributions by lymphocytes can be deposited at the target site. Two functions are, first, endogenous regulators of the IS, and second, recognition of stimuli not recognizable by the central and peripheral nervous systems (bacteria; tumors; viruses; antigens). Recognition of immunocytes is converted into information (peptide hormones; neurotransmitters; cytokines), which is conveyed to the neuroendocrine system.

Mechanisms of cell-to-cell communications: Autocrine messengers (secreted into the intercellular fluid and acting on the cells that secreted them; neural communication via synaptic junctions; paracrine messengers diffusing into the intercellular fluid and affecting adjacent cells without entering the bloodstream; and endocrine or circulating hormones). These secretions are multifunctional. Dopamine, norepinephrine, EP, somatostatin, GnRH, OXY, and vasopressin are neurotransmitters present in nerve endings, hormones secreted by neurons, and also hormones secreted by endocrine cells. Other chemical substances perform two of the three functions (Aron et al., 2007a, Figure 5–1). *Bidirectional signaling*: There is a bidirectional interaction between the immune and neuroendocrine systems. The molecular basis for communication is shared regulatory mediators: steroid hormones, neuropeptide, cytokines, and their receptors (Crofford, 2002). Neuropeptide transport systems can be unidirectional or bidirectional, that is, transporting peptides in and out of the brain. Penetration of the BBB depends upon their individual physiochemical properties (Neilan & Pasternak, 2003).

The brain is only one site for receipt and response to chemical and neural messages exchanged from distant locations. Other tissues receive stimuli, react to local conditions, and create chemical mediators that are expressed in different tissues, including the brain. The brain's reactions in turn influence the periphery for stress and adaptive functions. Thus, the far-flung exchange between the periphery and nervous system (CNS, autonomic, peripheral, and ENS) has a parallel in the combined chemical domains of the stress, inflammatory, and ISs. These super-systems influence each other. The various functions are influenced by the environment and occur in *real time* due to the influence of numerous circadian rhythmic cycles of greatly varying length.

Hypothalamic peptide hormones: These are released at hypothalamic and extrahypothalamic sites, and are believed to perform as neurotransmitters or neuromodulators. They subserve a synaptic communication system supplementing the classical neurotransmitters. This is performed by slower onset and longer neuropeptide effect on synaptic transmission. Examples include the generation of circadian rhythms associated with light-sensitive processes such as cortisol secretion and the estrus cycle; projections from the hypothalamus to the brainstem and spinal cord may have neuroregulatory influence upon responses to autonomic or sensory stimuli.

Immune–neuroendocrine interaction: The IS maintains homeostasis, being integrated with neural and endocrine factors. The IS's profound impact on the CNS is mediated via soluble molecules circulating in the bloodstream and acting on populations of supportive cells. The resultant neurophysiological responses play a role in restoring homeostasis and regulating innate immunity. This regulation involves the endocrine and autonomic NS, especially circuits controlling plasma release of glucocorticoids and control of increase in body temperature (fever). While it is believed that the brain itself has its own innate IS, functioning in viral and bacterial infections, this writer speculates that it will be active after brain and perhaps somatic trauma (Rivest, 2003).

Mutual messengers: Chemical signals influence the internal milieu (stress, response to injury, and effects of drugs and their side effects). The hormonal, inflammatory, and immune humoral systems are expressed and interact, frequently using the same chemical messengers and intracellular processes. Their complexity is comparable to that of the CNS. The same messengers (e.g., hormones, neurotransmitters and neuromodulators, hypothalamic releasing hormones, etc.) can serve as a (1) neurotransmitter, (2) hormone secreted by a neuron, and (3) a hormone secreted by an endocrine cell (Aron et al., 2007a, including Table 5–1). Various tissues may share common ligands and receptors, so that the activation of one system may be accompanied by different changes in others (Marucha et al., 2001). Perhaps this is the basis of medication side effects. The complexity of the internal milieu may be illustrated by the chromaffin cells of the adrenal medulla (Kvetnansky & McCarty, 2000). Originally considered to be under the exclusive control of ACH, and concerned with the secretion of the catecholamines norepinephrine and EP, these cells are now known to release a great number of peptides; various biologically active neuropeptides are co-localized with ACH, and have a variety of behaviorally significant receptors.

Multiple Pathways

In response to internal and environmental demands, signals travel via the internal milieu to offer negative and positive feedback to the brain and other organs. The *signal structures* of the internal environment are the blood system, peripheral nerves, and cranial nerve X. The molecular basis for communication are shared regulatory mediators (e.g., steroid hormones, neuropeptides, cytokines, neurotransmitters, hormones, and their receptors). Activated immune cells secrete cytokines, which serve as communicatory signals in cell-cell interactions (Pineda & Dahn, 2005). These multisystem messengers interact with components of the nervous, immune, and endocrine systems (Crofford, 2002; Panerai, 1994).

Multiple Cellular Stimulation

Central nervous system cells are continuously exposed to neurotransmitter, neuromodulator, and hormonal stimulations that are integrated spatially and temporally. *Neuromodulators* are chemical messengers that modify neuron sensitivity to synaptic stimulation or inhibition; some are produced in the nerve tissue, others are circulating steroids (Junqueira & Carneira, 2005, p. 159). Effector-signaling interactions occur within a receptor system, and between different receptors systems when responsiveness of one receptor system is regulated by activation of a different receptor system ("cross-talk") (Berg et al., 1998).

Complexity of Regulation

An integrated homeostatic network is composed of the immune, hormonal, nervous, and endocrine systems, requiring many brain sites to interpret, integrate, and organize the stress response (Goya et al., 1994; Kaye & Lightman, 2005). By regulation is meant maintaining functioning at an appropriate level: constant (homeostatic), adaptive to situational demands (allostatic), or stress responsive (overcoming immediate danger or situational demands) that may be momentary or persist for a long interval. Normal functioning can be defined as mutual regulation of multiple physiological systems to adapt to environmental necessities, and to avoid disease and injury, and maintain efficient adaptation in the person's ecological niche. The process of regulation involves the brain and other organ systems. Under adaptive conditions, each of these has both separate regulation and integration with the other systems.

Messenger/Receptor Significance

Particular messenger hormones, neurotransmitters, and neuropeptides have numerous effects upon the IS, depending upon the IS phase in which it is active (Raison et al., 2005, pp. 145–146). A particular signal can be created in numerous tissues and affect multiple organs, including those not ordinarily considered to be its targets. The signal and the receptor may have multiple subtypes. The response is not characteristic of the signal (e.g., a hormone or neurotransmitter), but depends upon which signal subtype attaches itself to the appropriate receptor subtype of the particular organ. Therefore, some messengers can stimulate a variety of tissues in the body if they have the appropriate receptors. Thus, formerly narrowly labelled substances are now known to be physiologically active in many tissues. The earlier concept of separate categories of chemical messengers (cytokines; hormones; neuropeptides) has been replaced with the concept that they are *ubiquitous messengers*, that is, their reaction is defined by the type of cell that possesses their characteristic receptor. Immune and neuroendocrine system cells use common signal molecules and receptors, while hormones and neuropeptides affect the activities of the IS. The immune and neuroendocrine systems share common signal molecules and receptors. IS products (e.g., cytokines) modulate neuroendocrine functions (Falaschi et al., 1994; Panerai, 1994). These immune-related chemicals are involved in brain function that includes processes such as behavior, neuroendocrine activity, sleep, neurodegeneration, fever, and depression. Particular substances (neurotransmitters; neuropeptides; hormones; cytokines) are each produced by and affect different systems. Immune and neuroendocrine system cells use common signal molecules and receptors, while hormones and neuropeptides affect the activities of the IS. IS products (e.g., cytokines) modulate neuroendocrine functions (Falaschi et al., 1994; Panerai, 1994).

Hormones

The familiar description of hormonal secretions described substances transported by blood to exert their effects upon distant tissues. The stress was upon the outflow from the HPA, in coordination with catecholamines secreted systemically into the blood from sympathetic NS axons. In fact, there is an inextricable structural and functional relationship between hormonal and CNSs (*psychoneuroendocrinology*). The brain is a target for hormonal regulation and also has secretory functions of its own. *Pleiotropy* refers to hormones serving a multiplicity of functions, that is, the same substance functions both within and outside the CNS. Almost all endocrine secretions are controlled by the brain, while virtually all hormones influence brain activity. Regulation of the body's interaction with the environment is influenced by direct and indirect neural control of hormonal secretions, hormonal effects on neural action, and interaction between the two systems. Hormones (secreted directly into the bloodstream) may serve as neurotransmitters, while classical neurotransmitters can function as hormones. Since neurones may have hormone receptors distinct from the area of the synapse, neuropeptide hormones may function as neuromodulators. Hormones and neurotransmitters are also secreted in unexpected tissues, that is, GI tract, thymus, and lung. The intricate communication between the nervous, endocrine, and ISs are enhanced by the common set of hormones, neuropeptides, and cytokines shared by the immune and neuroendocrine systems. These systems share a similar group of receptors (Blalock, 1994).

Hormones contribute to the plasticity of behavior, thus permitting adaptation to environmental changing demands, for example, stress. Behaviors both modulate and are derived from both.

The effect of a hormone depends upon the pattern of its release and the history of exposure during encoding periods (Harris et al., 2005). In addition, hormonal responses are elicited both by physiological and conditioned stimuli (Schulkin, 2003, p. 12).

The endocrine and nervous systems both utilize intercellular communication, but differ in mode, speed, and degree of localization of their effects. The ANS response to neurotransmitter release is rapid and localized. The *dispersed neuroendocrine system (DNS)* effects are slower and less localized in the sense that they may occur in contiguous cells, nearby cells reached by diffusion, or distant cells reached via the bloodstream. The endocrine system is slower and less localized, with effects that are stated to be specific and localized (Standing, 2005b), although this seems controversial, since specific endocrine receptors may be found in more than just the traditional target organs. Cells of the DNS (*paraneurons*) form a third division of the nervous system that supports, modifies, or amplifies the actions of neurons in the autonomic and somatic divisions.

Multiple hormones cooperate in the coordination of development, reproduction, and homeostasis. Particular hormones act at multiple sites. There are a variety of receptors for a given hormone, located in different organs and tissues, whose response determines the hormone's biological effect. Each hormone can have multiple responses, combining to give the overall effect (Jameson, 1996). Hormones and related substances are secreted all over the body, not merely in the classical endocrine glands: Alimentary tract, lung, heart, kidney, and so forth, are considered to be a DNS (Standing, 2005, Chapter 9). Particular hormones act at multiple sites. Hypophyseal hormones or part of the DNS can be detected widely throughout the CNS and PNS (Standing, 2005, chapter 4).

Other examples of functional complexity include (1) The multilocation production of CRH. It influences different brain centers and the hypothalamus. (2) While the stress system reduces immediate inflammation, chronic stress may aggravate it. Systemic integration involving physiological changes in one system affect the activation of the others by signals through chemical messengers, shared receptors and ligands (Marucha et al., 2001), and intracellular processes.

Hormone Receptor Type

The same chemical can have different effects, depending upon the organ of the receptor to which it binds. These signals lead to changes in the cell function (Kaye & Lightman, 2005). For example, reflecting the serotonin system's key modulatory role in the CNS in physiological and disease states, there are 14 different serotonin receptors found on a variety of neuronal types (Svenningsson et al., 2006). Naming a hormone for one effect or its endocrine gland can be confusing since two hormones can have the same effect through interaction with different receptors; the most familiar effect of a hormone can be only a subset of its actions; most glands produce multiple hormones; some hormones are made in more than one site; and different hormones can bind to the same receptor. The receptors describe the actual hormonal action, for example, α- and β-adrenergic receptors. The effects can facilitate (agonistic) or inhibit (antagonistic) organ activity.

Neuropeptides

Neuropeptides are inflammatory mediators released from neurons in response to local tissue injury that have a variety of neurobehavioral functions. They link the nervous system with the inflammatory response. Acting upon specific receptors, they produce modulation of neurotransmitter release, thermoregulation, regulation of emotionality and complex behaviors, physiological functions (regulation of reproduction; growth, food, and water intake; electrolyte balance; control of respiratory, cardiovascular, and GI functions; locomotion; modulation of emotions; memory, learning, responses to stress and pain). Involvement in such behavioral processes suggests that they may contribute to the symptoms and behaviors of major psychiatric illnesses as psychoses, mood disorders, and dementias (Neilan & Pasternak, 2003; Young et al., 2005). Some neuropeptides are implicated in psychiatric disorders, mood disturbances, and stresslike responses. These include somatostatin, CCK, SP, and neuropeptide Y (Young et al., 2003). Endogenous *opioid peptides* (e.g., endorphin, enkephalon, dynorphin) are widely distributed in the CNS and the periphery. Ordinarily, peptides are not expected to cross the BBB, but this restriction is less efficient after trauma. Endorphins and dynorphins (brain and spinal cord) are involved in pain perception, affective behavior, and motor control. They are upregulated following tissue damage and inflammation (Neilan & Pasternak, 2003). Immune cells have receptors for classical neurotransmitters and neuropeptides. Leukocytic peptides include substances found in the neuroendocrine and nervous systems.

System Interactions

Almost all hormones influence brain activity. Neural and endocrine factors influence the IS, while cytokines (secretions of lymphocytes, monocytes, and vascular elements) modulate both the neural and endocrine functions. The immune, neuroendocrine, and CNS systems have reciprocal effects on paracrine, autocrine, and endocrine control over the release of cytokines, hormones, and growth factors (Ferrari et al., 2000). These regulate the complex multidirectional communications among these systems. Both nerve and endocrine cells have secretory and electrical potential capacity. In addition to pituitary cells, PRL is secreted by lymphocytes and T-cells, illustrating the multiple sources of biologically active substances (Dardenne & Savino, 1994). The SNS exerts inhibitory control over the IS (Arnason, 1997; Reichlin, 1998).

Cytokines as Immune System Messengers

Cytokines modulate immune, neural, and endocrine functions, which transmit information from one part of the body to another. Their integration links aspects of cognitive and noncognitive neural activity with metabolic and hormonal homeostatic activity. The adaptational responses to stress involve the interaction between blood-borne molecules, that is, cytokines, adrenomedulary hormones (catecholamines), and the HPA (Cullinan et al., 1995). At the level of the CNS, pituitary, and adrenal glands, cytokines are involved in the regulation of sleep, feeding behavior, temperature, other homeostatic mechanisms, circadian influences upon these functions, the regulation of neurotransmission, and secretion of the most important peptides and hormones.

Cytokines communicate with the CNS to inform, protect, and destroy tissues behind the BBB. They cross the CVOs from the blood and through the BBB (Banks, 2001). Cytokine receptors have unique distributions in the brain. Since cytokines affect behavior via the CNS, as large molecules do not readily cross the BBB, it is useful to understand how they communicate with the brain:

Across the BBB
Porous areas of the BBB (CVO)
Conversion into secondary signals

Transmission of signals along the vagus nerve to the nucleus of the solitary tract, thence ascending catecholaminergic pathways to the hypothalamic PVN and other areas (Raison et al., 2005).

BARRIERS OF THE NERVOUS SYSTEM

Liquid Compartments and Barriers

The liquid compartments surrounding the brain are CSF, bloodstream, interstitial space (fluid), and intracellular fluid. The linings are composed of endothelial cells with varied characteristics. The efficiency of these barriers separating the nervous system from the liquid compartments is reduced by trauma and disease. The major compartments of the brain, that is, the CNS and CSF, are separated from a third compartment, the blood. The three fluid compartments of the brain are extracellular (interstitial), intracellular, and CSF. The brain's surrounding liquid, CSF, is secreted by the choroid plexus and also moved by active transport (not passive filtration, as in other tissues) through the BBB of the capillary walls into the interstitial space of CNS tissue. The product of endothelial secretion is the *interstitial fluid* of the CNS. The endothelium is leaky in the regions of the CVO (see later) (Somjen, Chapter 1).

Surfaces

In the brain there are two types of surfaces between liquids and the brain parenchyma: blood vessels and the ventricles. The capillary beds of the brain are nonfenestrated (tight junctions between the cells lining the endothelium—lining of the blood vessels and lymphatic vessels). This tight junction even sequesters blood glucose, which must enter the CSF via transporters in the endothelial cell membranes and then pass through the basement membrane. Since the maximum velocity is not much greater than the glucose utilization of the brain, decreased blood glucose renders the brain vulnerable to hypoglycemia. The choroid plexus is a capillary network surrounded by a lining or epithelium (Marks et al., 1996). The endothelium at the choroid plexus is somewhat structurally different, but some compounds may not penetrate. The arachnoid membrane (tight cellular junctions) isolates the CSF in the arachnoid space from blood vessels in the dura mater (*the arachnoid blood–CSF barrier* (Ransom, 2003). The *pia–glial membrane* is composed of astrocytes near the surface of the neuraxis. These send processes terminating as end feet to the pia mater, a thin vascular membrane adhering close to the surface of the CNS. This membrane thus faces the arachnoid space, which is filled with CSF (Angevine, 2002; Swanson, 2003). Thus, *blood-brain and blood-CSF barriers* have similar physiological effects (Greenberg & Atmar, 1996), although the BBB has 5,000 times the surface area of the blood-CSF barrier (Parent, 1996, p. 17). Blood vessels of the dura mater are outside the BBB.

In order to establish a distinct internal environment to maintain homeostasis, all internal and external surfaces of organs are covered with various kinds of epithelial tissues. The brain has a distinctive environment, with specific liquid compartments and barriers that are only partially effective screens against the influx of substances with maladaptive effects. The brain's barriers are composed of the endothelial linings of compartments, and have varied characteristics. To function as a barrier, the intercellular spaces of these epithelial sheets are sealed by what is termed a *tight junction*, including the microvascular brain endothelial cells forming the BBB (Kooij et al., 2005). Their efficiency is reduced by trauma and disease. The significant liquid compartments of the CNS are CSF, bloodstream, interstitial space (fluid), and intracellular fluid.

Blood-Brain Barrier (Vascular-Parenchyma; Blood-Retina)

The BBB is designed to create a stable environment for neurons to function effectively. It protects the microenvironment of the CNS from toxins, and also buffers fluctuations in blood composition (Fleegal et al., 2005; Wolburg & Warth, 2005). The BBB is effective in screening out many types of solutes, and permits the passage of other types. The size of a molecule, its lipid content, etc., are some of the determinants whether passage is permitted. Steroid and thyroid hormones cross the BBB, producing specific receptor-mediated actions. *Glucocorticoids* enter the brain through the BBB. They block stress-induced corticotropin release (Icone et al., 2003). *Glial cells* are important in inducing and maintaining the endothelial cell barrier of the cerebral microvessels. Astrocytes, in bidirectional communication with the CNS, receive signals from neighboring neurons and respond by releasing neuroactive substances. A complex neurovascular unit regulates CNS development, synaptic activity, permeability properties of the BBB, and the transport and diffusion properties of brain capillary endothelial cells composing the BBB. The BBB is also a facilitator of nutrient transport and a barrier to potentially harmful molecules. Enzymes in the endothelia or glial cells metabolize neurotransmitters, drugs, and toxins before they enter the brain and disrupt its functions (Banerjee & Bhat, 2007).

The *neuronal microenvironment* is protected from the blood by the BBB and choroid plexes. However, essential micronutrients, vitamins, and trace elements are selectively transported into the brain by the endothelial cells of the blood vessels. Both cytokines and the immune cells that produce them cross the BBB to release their mediators centrally (Cohen & Kinney, 2001). The BBB protects the brain's microenvironment from changes in the blood composition by serving as a restrictive diffusion barrier. Its physiological regulation involves a plastic, adaptive opening and closing of tight junctions according to local needs. The initial proof of such a barrier involved came from the blue coloring of all body tissues except the brain, eyes, and gonads after injection of a dye (Goldmann, 1909, cited by Somjen, 2004). That the BBB-like barrier is found in the gonads is more explicable by evolutionary psychoanalysts than neuroscientists. The BBB is permeable to certain substances, for example, lipids, in proportion to their solubility, and gases, such as O_2 and CO_2. There are other mechanisms as well that transport substances across the BBB. Most substances that must cross the BBB that are not lipid-soluble cross by specific carrier systems, including a glucose transporter to provide energy. Amino acids are transported across barrier endothelial cells by distinct systems (Laterra & Goldstein, 2000). The BBB excludes many toxic substances and prevents momentary interference with neuronal functioning by circulating transmitters such as norepinephrine and glutamate. Their blood level can increase greatly in response to stress or food. Cells lining the capillaries that create the BBB differ from ordinary capillaries: tight intercellular junctions and relative lack of transport of circulating chemicals. The BBB is more than a filter: It is an active regulator of the internal environment of the brain by effectively screening out many types of solutes. There is a balance between pro- and antipermeability trophic factors. The BBB is subjected to a bidirectional exchange of immune-mediated substances which are elicited by either cerebral or systemic inflammatory disorders (Morganti-Kossman et al., 2005).

Structure of the BBB: It is composed of endothelial cells, pericytes (which become part of the vascular barrier by increased phagocytosis after BBB injury), and astrocytes. The barrier is formed in the capillaries, that is, special endothelial that form tight junctions. The cells are further surrounded by a basement membrane, regulating the flow of macromolecules. The basement membrane itself is surrounded by astrocyte end feet providing the brain's epithelium with trophic factors that also modulate the permeability of the BBB. Since the astrocytes are in contact with neurons, they may receive input from the neurons and influence the local permeability of the capillaries in times of high local metabolic activity and energy demand (Ronken & van Scharrenburg, 2005). Yet, essential micronutrients, vitamins, and trace elements are selectively transported into the brain by the endothelial cells of the blood vessels. Their characteristics are determined by intrinsic factors of the brain and the brain's microenvironment, which includes neuronal and glial cells (Junqueira & Carneiro, 2005; Wolburg & Warth, 2005). Long processes of the astrocytes help the vessel lining endothelium to form tight junctions, formed by the capillary endothelial cells (Kandel, 2000b). Cells lining the capillaries, creating the BBB, differ from those of ordinary capillaries. There are tight intercellular junctions and relative lack of transport of circulating chemicals. Astrocytes help form the BBB, part of the lining of the brain's capillaries and venules. In balance, the BBB forms a metabolic and physiological barrier to separate the peripheral circulation from the brain extracellular fluid (BECF) (Fleegal et al., 2005).

Permeability of the BBB: Passage is permitted according to certain determinants, for example, the size of a molecule, its lipid content, etc. The BBB is not a complete barrier even during health. The olfactory nerve (I), whose neurons have a one-month life span, is not protected by the BBB; therefore, it offers access for viruses to enter the brain (Arbour, 2005). The BBB is permeable to certain substances, for example, lipids, in proportion to their solubility, and gases such as O_2 and CO_2). It plays an active role in regulating immune cell entry (Biernacki et al., 2005) along with other mechanisms that transport substances across the BBB (Laterra & Goldstein, 2000). The permeability of the BBB is affected by mediators of the *pain response*, including cytokines and chemokines

(Heurtult & Benoit, 2005). The BBB restricts many medications and diagnostic substances (Heurtult & Benoit, 2005; Zhang & Stanimiroic, 2005), and also many toxic substances. It prevents momentary interference with neuronal functioning by circulating transmitters such as norepinephrine and glutamate whose blood level can increase greatly in response to stress or food.

Blood-Nerve Barrier

The connective tissue elements of the nerves create an immunological barrier for the nerve microenvironment (Murray, 2004).

Cellular Barrier (Interstitial Space between Neurons)

This occurs within the brain parenchyma. It bars substances in the interstitial spaces between the neurons from entering them. The vascular system transports nutrients and molecules into the ECF surrounding neuronal cells. The product of endothelial secretion is the *interstitial fluid* of the CNS. The concentration of solutes in *BECF* fluctuates with neural activity. The composition of this compartment is controlled: The BBB and the CSF, synthesized by choroid-plexus epithelial cells, influence the composition of the BECF; the surrounding glial cells influence the BECF (Ransom, 2005). Neurons, glia, and capillaries, being packed tightly together, share a common extracellular space (Ransom, 2005).

Blood Cerebrospinal Fluid Barrier (Brain Ventricles Compartment)

This has been described as "leaky." These structures are the carriers of solutes that evade the BBB. CSF is filtered from the blood by the choroid plexus in the lateral, third, and fourth ventricles. CSF exits the fourth ventricle through the foramina of Lushka and Magendie to the arachnoid villi of the subarachnoid space, which absorb the CSF back into the bloodstream (Blessing, 2002). CSF removes waste products of neuronal metabolism, drugs, and other substances diffusing into the brain from the blood. As it streams over the ventricular and pial surfaces of the brain, it drains solutes and carries them through the arachnoid villi into the venous blood. It integrates brain and peripheral endocrine functions. Hormones or hormone-releasing factors from the hypothalamus are secreted into the extracellular space or directly into the CSF. They are carried to the median eminence and thence into the hypophyseal portal system (Parent, 1996, pp. 14–15).

One need for a BBB is based upon the assumption that the concentration of solutes in BECF fluctuates with neural activity, while changes can influence nerve cell behavior. The BECF compartment is protected from fluctuations by the BBB and the CSF, which influences the composition of the BECF, and the effects of the surrounding glial cells.

Limited Barrier: The Circumventricular Organs

These are secretory midline tissues in the third ventricle whose tissues and capillaries are permeable, therefore permitting the diffusion of large molecules from the general circulation (Aron et al., 2004). Some of these nuclei are major regulating centers: the subfornical organ; OVLT (supraoptic crest); the median eminence; the neurohypophysis; the pineal body; the subcommissural organ; the area postrema (group of cells on the dorsal surface of the medulla oblongata), and the posterior pituitary, thus permitting its hormones to enter the general circulation (Ransom, 2005). Not all tissue with inefficient BBB is part of the CVO system. The thalamic area in the region of the IGL cells (afferent input to the suprachiasmic N., i.e., biological clock) may have an attenuated BBB.

Neurons within the CVO are directly exposed to blood solutes and macromolecules, that is, humoral signals from endocrines, the autonomic NS, and behavioral centers within the CNS. CVO capillaries (and those of peripheral organs) have specialized functions for movement of substances into the cell, as well as permitting passive diffusion. Some substances that are excluded from the brain due to the BBB (e.g., peptides, proteins, cytokines, hormones, etc.) are admitted through the CVO, which lack the BBB (Parent, 1996, p. 23).

The CVO are permeable to proteins and peptides (Parent, 1996, p. 223). The BECF compartment is protected from fluctuations by the BBB. Its importance is illustrated by the fact that blood concentration of neurobehaviorally active solutes, for example, amino acids, solutes affecting neuronal activity (K^+ and H^+) vary with diet, hormones, inflammatory mediators such as cytokines, recency of eating, metabolism, athletic activity, illness, and age. If these substances directly entered the neuronal microenvironment, they would nonselectively activate receptors and disturb normal neurotransmission. Neurons within the CVO are directly exposed to blood solutes and macromolecules, that is, believed to be part of a neuroendocrine control system for maintaining levels of osmolity and hormone levels. Hormone signals received by CVO neurons, are conveyed to endocrine, autonomic, and behavioral centers within the CNS, The area postrema sends signals widely within the brainstem, particularly to the parabrachial nuclei which integrates visceral afferent information and autonomic function. In the median eminence, RHs are picked up by capillaries for carriage by the pituitary portal system to the anterior pituitary. The lack of a BBB in the posterior pituitary permits hormones that are released there to enter the general circulation. Cells lining the ventral surface of the medulla are especially sensitive to its hydrogen ion concentration (pH) (Blessing, 2002).

The Metabolic Blood Brain Barrier

The composition of the interstitial fluid surrounding neurons is controlled by enzyme systems specific to the BBB. Catecholamines and l-dopa are metabolized by endothelial enzymes. Although there is a transport system forl-dopa, it is rapidly metabolized in the vasculature's endothelial cells. Therefore, treatment of *Parkinson disease* requires that l-dopa be accompanied by a specific enzyme inhibitor (Laterra & Goldstein, 2000).

MICROENVIRONMENT OF THE BRAIN

Adaptation involves far more than an exchange of information and orders via the CNS. *Brain perfusion* may be controlled by neurons within the brain parenchyma. Neurally induced substances control microvascular tone, local CBF, and BBB function. Glial cells seem to be associated with *noradrenergic* nerve terminals in regulating local metabolic functions and BBB permeability. In contrast, brainstem raphe nuclei project *serotonergic* nerve fibers to specific brain areas throughout the brain, regulating CBFs in projection areas such as the cerebral cortex (Edvinsson, 2005).

Blood Brain (vascular—traditional blood-brain barrier (BBB). It is effective in screening out many types of solutes, and passes through other types. The size of a molecule, its lipid content, etc., are some determinants of passage. The information that the brain exchanges with the periphery via the blood and CSF leads to responses that contribute to adaptation or maladaptive (irrelevant to adaptive considerations) (see Kretschmann & Weinrich, 2004, pp. 239–252 for CSF containing spaces). The homeostasis of the cerebral microenvironment is affected by neurons, glia, and capillaries, packed tightly together, and acting as an integrated unit in cell-cell communication. The space surrounding individual neurons is part of the neuronal microenvironment (Iadecola, 2004; Ransom, 2005); this is composed of *glial cells* (Junqueira & Carneiro, 2005, p. 161).

The immune, nervous, and the endocrine systems constitute a multisystem *micromilieu.* Jankovic (1994) uses the term microenvironment to describe the components that unify the immune, nervous, and endocrine systems: (1) *lymphoid cells* (T and B lymphocytes); (2) *nonlymphoid cells* (macrophages, epithelial cells, and dendritic cells that interact with lymphocytes and other cells of the microenvironment); (3) *visiting cells* (lymphoid and nonlymphoid, which enter the immune milieu and influence the immune network); (4) *neurons* (utilizing various neurotransmitters that influence the microenvironment by neurohumoral activity); (5) *hormones* (released from remote sites by endocrine glands that enter the immune milieu via capillaries and influence the recognition processes); (6) *biologically active substances* (lymphokines, monokines, complement, immunoglobulins) produced by both lymphoid and nonlymphoid cells in situ or elsewhere; (7) *neurotransmitters* (of numerous types and structures, including the classical types, opioid, and other peptides); (8) *membranous and intracellular receptors* (these are molecules that maintain intercommunications and maintain the microenvironmental circuits; (9) *ions* (sodium, potassium, magnesium, calcium) that maintain signals through cell membranes, impacting on immune stimulation and inhibition; (10) *electrical, magnetic, and electromagnetic fields* created by differences in ionic concentrations and charged molecules in various compartments of the micromilieu; and (11) *higher nervous activity* (mind and psyche) that influences the microenvironment through stress-related, emotional, aggressive, and other behaviors (Wilbourne, 2003).

The functioning of the organism involves far more than an exchange of information and orders via the CNS. The brain receives information via the blood and CSF from the periphery that are both appropriate (contribute to adaptation) and maladaptive (irrelevant to adaptive considerations).

6 Biomechanics and Tissue Injuries

OVERVIEW

The mechanism by which the brain becomes injured in mechanical accidents is very complex, involving an interaction between physical objects, characteristics of the environment, mechanical forces, the anatomical structure of the brain, the nature of the brain's tissue, and its position within the complex structure of the skull, the skull's tethering to the neck and torso, and so on. Information is important for designing safe vehicles, roads, and buildings. The credibility that the accident actually caused the claimed impairment relies in part upon whether the interaction of structures and forces could have actually caused the claimed disorders that are a basis for compensation. Understanding physical principles leads to an improved capacity to reconstruct the accident, resulting in more complete and accurate interviewing of the patient and witnesses. This provides more information concerning the number and nature of impacts, which enhances diagnosis and referral for further study. Understanding some qualities of tissue and their reaction to physical forces and movement contributes to assessing the credibility of an accident claim, as well as to preventative measures (i.e., medical and engineering).

Biomechanics may be defined as the physical principles influencing the interaction between the person and the environment that causes injury. Formulas estimating the actual forces causing TBI (Traumatic brain injury) are available (Zhang et al., 2004). One differentiates between various injuries—contact (impact), inertial or nonimpact (acceleration/deceleration), and blast phenomena. Understanding some of the qualities of tissue and their reaction to physical forces and movement contributes to assessing the credibility of an accident claim.

PATHOMECHANICS: PHYSICAL PRINCIPLES INFLUENCING BODILY INJURY

Pathomechanics is the study of the physical concepts involved in an injury. The *mechanical dimensions affecting brain damage* are the *magnitude of the force* applied, the *velocity of the head* relative to surrounding body and environmental structures, the *direction* of force, the *point of contact*, the relative *mobility* of the skull, and the *angular and other directional components of the velocity of the brain* after trauma. Most TBI is the result of *contact* (impact) or *inertial* or *nonimpact* (acceleration/deceleration) of the brain, forces that tear apart the surface of the brain from the surrounding bony geometry and exiting vascular structures; brain movements relative to the enveloping dural membranes and venous sinuses; negative forces or *penetration*. The forces working on the skull (deformation and pressure waves) cause internal movement, deformation, rotation, and internal shearing injuries to brain tissue (Becker, 1989; Miller, 1977). The forces can be great enough to cause vascular injury (see Figures 6.1 through 6.4).

Dynamics or Mechanics (The Laws of Newton)

Mechanics describes the motion of bodies without regard to its causes (which includes acceleration, velocity, and vectors that combine the motion due to forces applied in different directions), but

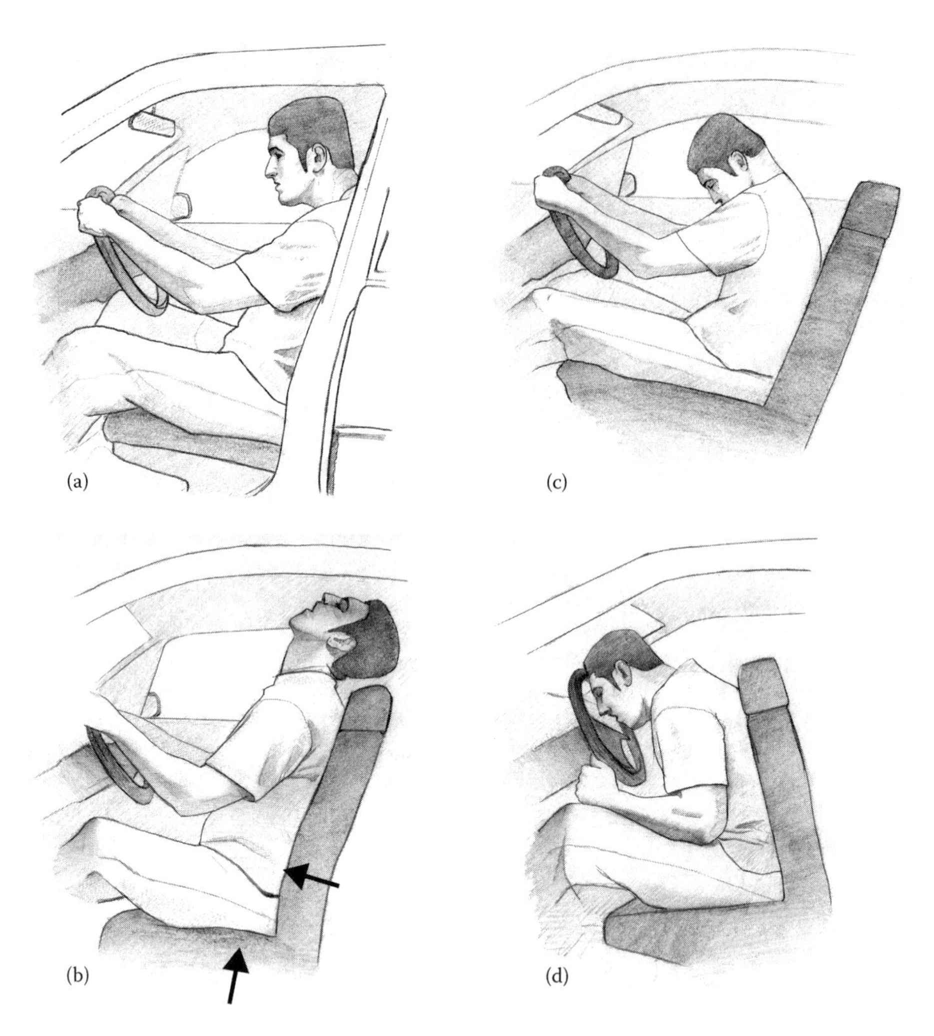

FIGURE 6.1 Head and neck motion when a vehicle is struck from the rear (whiplash). (Original illustrations by Chris McGrath.)

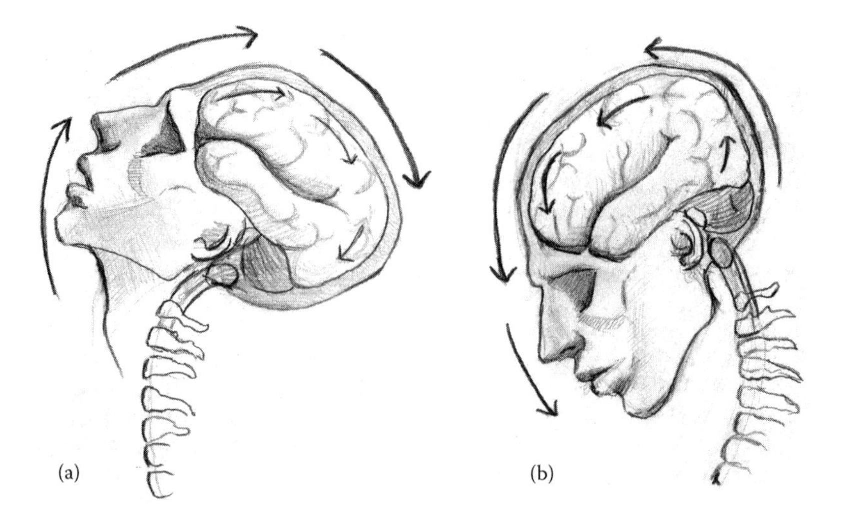

FIGURE 6.2 Brain motion during hyperflexion and hyperexension. (Original illustrations by Chris McGrath.)

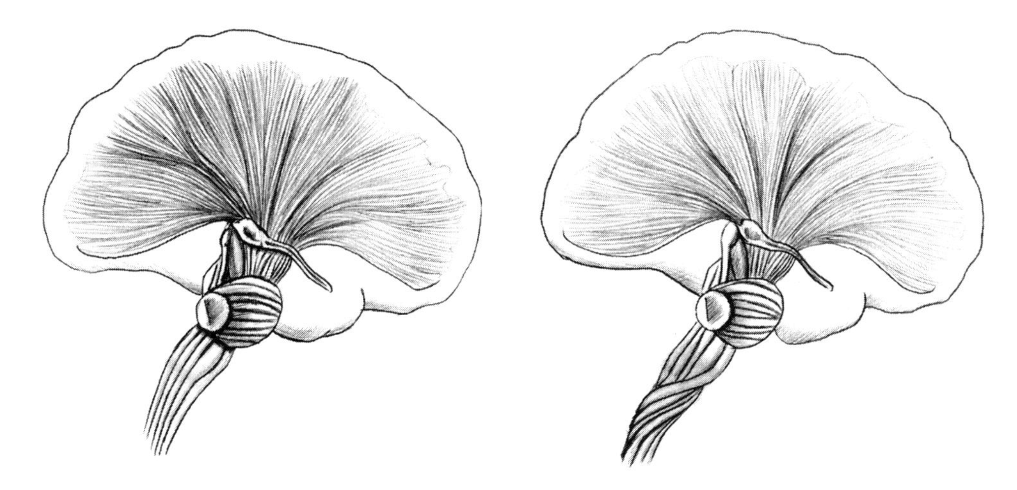

FIGURE 6.3 Rotation of the cerebral hemispheres around the brain stem. (Original illustration by Chris McGrath.)

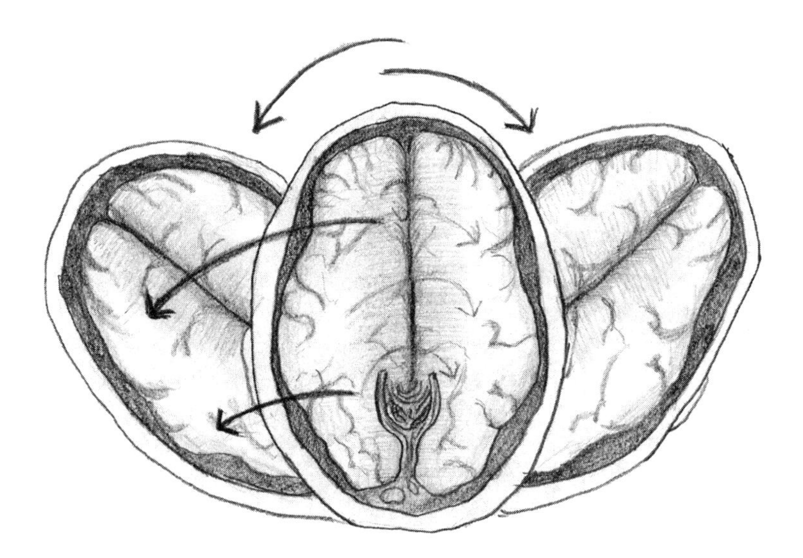

FIGURE 6.4 Lateral rotation around the brain stem. (Original illustration by Chris McGrath.)

also without reference to its mass or the forces acting on it. *Energy* is the capacity to do work (e.g., movement of the head and body with change of momentum of the surrounding vehicle or addition of energy to the person when struck by a vehicle). *Wave motion* is the propagation of energy or a disturbance.

Dynamics is the study of the forces and the changes in motion they cause (Newton's laws of motion). Mechanics is the study of the motion of material objects. Kinematics (Hunt et al., 2004) and kinetics refer to the interaction of force (impact) and mass (body structure and weight) and the description of motion of bodies (acceleration, velocity, and vectors that combine the motion due to forces applied in different directions). Force is defined as that which causes the acceleration of a material body. It has direction and duration. The direction of the force contributes to the location of TBI—the irregular shape of the brain and the skull, the brain's inhomogeneity of content, and firm structures that impact the brain when it is accelerated or decelerated (the ledges of the skull that support the brain, the meninges, and the penetrating blood vessels whose course is anatomically fixed and tears tissue moving past it). Motion of the body, vehicles, and brain are described as velocity, direction, acceleration, and deceleration.

Momentum: The mass of colliding objects and their velocity and their conservation in determining the postimpact movement of objects (e.g., vehicles) and the brain inside the skull.

Loading: The energy transmitted to the tissue. Strain deformations that damage tissues (impact, pressure waves, cavitation) cause a vacuum when the brain rebounds from impact with the skull, compression; shear; tension; torsion.

Movement

Velocity: Distance covered in a unit of time (i.e., speed = distance/time).

Acceleration: How quickly velocity changes (i.e., acceleration = change of velocity/time interval). Linear Acceleration: Direction: forward-backward; side to side; vertical. Rotational Acceleration: Direction: side to side; vertical.

Motion, Force, Mass, Kinetic Energy

To determine the likely area of injury and, by extension, the credibility of an accident where there are forensic or administrative considerations, the clinician is concerned with the interaction of physical forces within the structure of the torso, the neck, the head, and the brain tethered to the spinal cord. Motion of the body, the vehicles, and the brain is described in such terms as velocity, direction, acceleration, and deceleration.

Weight: Force of gravity (i.e., the force due to gravity that acts on an object's mass [how strongly that matter is attracted by gravity]).

Mass: Mass is the amount of matter and its inertia. More specifically, it is the measure of the inertia or sluggishness that an object exhibits in response to any effort made to start it, stop it, deflect it, or change its state of motion in any way. Mass and weight are proportional to each other.

KINETIC ENERGY

Physics of a Free Fall

Acceleration: Free fall is falling under the influence of gravity alone. What is the difference between a car hitting a haystack and a brick wall? In both cases there is a loss of momentum down to 0. According to the impulse-momentum relationship, the same change of momentum means the same impulse. By hitting a haystack instead of a wall, the time is extended but the force is reduced. Bending your knees when you hit the ground, or falling on several parts of your body in sequence, reduces the force upon the bones. For a ballet dancer, a wooden floor, which yields, reduces force and also possibility of injury. Another example is a boxer "rolling with the punches."

Calculating velocity: To calculate the velocity of a falling object when striking the ground, either a person striking the ground or an object falling upon someone:

> Velocity in feet per second when striking the ground after a free fall, if the height (h) in feet is known, and G is acceleration due to gravity (32.2. ft/sec sq).
>
> V = the square root of $2 \times 32.2 \times h$.
>
> (Formula provided by Prof. Mariusz Ziejewski.)

Kinetic energy: This may be described as the *conservation of energy*. The potential energy prior to a fall equals the kinetic energy as the object strikes the ground. The formula is $mgh = \frac{1}{2} mv^2$. Mass, gravitational acceleration, and height equals the kinetic energy as the object strikes the ground. Mass and gravitational acceleration remain constant. Thus, velocity and kinetic energy are directly related to height. The greater the change of momentum upon impact (varying with height), the larger the force or load applied. The injury pattern will depend upon which part of the body strikes the ground first and how the load is distributed. In an airplane crash, the helmeted head of a restrained pilot may decelerate from a velocity of several hundred miles per hour to zero in an interval of tenths of a second. The cranium does not contact a solid object but the brain is irreversibly damaged. In contrast, the stationary head of a machine operator may be slowly crushed by machinery producing massive fractures, extra-axial hematomas, and contusion, without loss of consciousness. A head striking a solid floor after a standing fall tears parasagittal bridging veins causing an acute subdural hemorrhage. Acceleration of a higher magnitude longer in duration (MVA) causes diffuse axonal injury (DAI) (Bullock & Nathoo, 2005).

Momentum, Force, Impulse, and Impact

The concept of momentum is significant in terms of movement of the brain within the skull, movement of a head and body in a car struck for the rear or striking another object. Momentum is resistance to acceleration and defined as Mass × Velocity. It is analogous to inertia in motion. *Inertia* is the tendency of a body to stay at rest, or if it is moving, the tendency to resist any change in the magnitude or the direction of the motion. Momentum has magnitude (how much) and direction (which direction) (i.e., it is a vector). As Newton's third law, it helps one understand the damage caused by impacts: whenever one object exerts a force on a second object, the second object exerts an equal and opposite force on the first. Momentum reflects size, weight and speed (i.e., a loaded tanker has more momentum than a soccer ball travelling at the same speed [velocity]).

Force: the rate of change of momentum.

Impulse: the *product* of force and time (e.g., lbs/sec). When a car crashes, the impact that represents force slows the car down and damages its parts. The KE of the moving car is turned into heat. KE is lost in an *inelastic collision*, in which bodies stick together or are deformed. As the bodies are permanently deformed, KE is turned into heat. In a completely inelastic, or partially elastic, impact, the KE of the system is transformed. A partially *elastic collision* occurs when two tennis balls collide or an orange bounces off the floor. In these cases, there is a loss of energy involved in a momentary deformation, with the resultant speed of the objects slower after rebound, reflecting loss of KE. Significance: Impact causing brain movement within the skull or body movement in a collision until the person crashes against an interior surface.

Conservation of momentum: In the absence of an external force, the momentum of a "system" remains unchanged. A collision is considered an internal system since the forces are "internal" in the sense that they act and react within the system itself. The effect of a collision (within or external to the body) only redistributes or shares whatever momentum exists before the collision. The mass of colliding objects and their velocity (momentum) and its conservation determines the postimpact movement of objects (e.g., vehicles; the brain inside the skull). This is described as *change of momentum*: the time by which the force acts affects change of momentum.

Acceleration and Deceleration

A general principle is offered: mild, moderate, and severe brain injuries created by both acceleration/deceleration and impact injuries are on a continuum. Direct impact to the head is not required for the creation of severe TBI (Gaetz, 2004).

Change of velocity is the best predictor of vehicular occupant injury or severity. Thus, the speed of the vehicle alone before the collision does not necessarily predict these factors. Injury can occur in some 5-mph crashes while others may have no significant injury at 45–55 mph. After the car has stopped, occupant movement may occur for about 200–400 msec.

Extremely short-lived acceleration, unless severe, is damped by the brain's structure and does not produce injury. Short acceleration duration shifts the focus of injury to the surface of the brain. Gradually, longer durations cause strains to propagate deeply into the brain's structural strength, resulting in concussion, DAI, and prolonged traumatic coma (Teasdale & Mathew, 1996). The amount of angular displacement and its acceleration and deceleration over time will depend on the impacting force's effect on the head or vehicle, the rate of acceleration and deceleration, and the angle through which the head, neck, and torso move (e.g., depending on the absence or presence of a seat belt, and its configuration, etc.).

Energy

This is the capacity to do work (e.g., movement of the head and body). There is a change of momentum with reference to the surrounding structure of the vehicle, or energy changing momentum when the person is struck by a vehicle. When a car hits and knocks a person through the air, the victim has gained KE through the application of *force*. Tissue injury is related to the energy of a projective: $k = ½ mv^2$ (k = kinetic energy, m =mass, v = velocity) (Britt, 2004). High-velocity injuries can have 60× more energy than handguns with lower missile velocities. If a person is on a stationary elevator that suddenly drops, *potential energy* (gravity) is released. KE of a vehicle is proportional to the square of its velocity. Thus, doubling the speed increases KE four times.

Loss of kinetic energy: An *elastic collision* occurs when objects collide without being permanently deformed. The collided object recovers the original configuration when the force is removed. An *inelastic collision* means that the struck object does not recover its original shape. The colliding objects become tangled or coupled together. Heat is generated. The total kinetic energy is less and KE lost is converted into heat. It is important to remember that *the brain is only partially elastic*. There may be a separate *neurotraumatic heat effect* (heat-shock effect).

Projectile Energy

High-velocity bullets have more destructive power (i.e., increased kinetic energy [v^2]). Injury is also related to penetration, the transfer of kinetic energy to the target (less if it exists in a straight path, the movement of tissue is out of the path of the bullet), a permanent cavity caused by the crushing effect of the projectile, and an increase of the surface area of the projectile-tissue interface (yaw, tumbling, projectile deformation, fragmentation). With reference to actual damage to the tissue, tissue elasticity is more significant than the energy transmitted to the tissue. Loss of kinetic energy is related to less elasticity of tissue with higher density, leading to retardation of the missile with energy or work transferred to the tissue. Destruction is related to the *expansion coefficient*. The equation E.C. = cm^3 indicates that for every erg of energy transmitted to the tissue there is a temporary cavity with a volume of 1 cm^3. A .25 caliber bullet from a handgun has a velocity of 810 ft/sec and amuzzle energy of 73 ft-lb. Larger weapons (.38–.45 caliber) have velocities from 855 to 1470 ft/sec (not directly proportional to the gun's caliber, but apparently varying with the bullet's design and charge) and muzzle energy from 255 to 1150 ft-lb. The writer has examined people with tangential bullet wounds to the head in which there was evidence of energy transmission causing TBI (Hunt et al., 2004). A case of hypoituitarism after a transfacial gunshot wound is reported, with intracranial damage due to energy transmitted during the trajectory of the bullet and bullet fragments. A bullet transmits more of its KE to surrounding tissues if it picks up bone during its trajectory. In addition, cavitation from the expanding property of some bullets and tumbling also explain injury to surrounding tissue (D'Angelica et al., 1995).

Force

Force is defined as that which causes the acceleration of a material body. It has direction and duration. The *direction* of the force contributes to the location of TBI, which is created by the irregular shape of the brain and the skull, the brain's inhomogeneity of content, and the firm structures that impact the brain when it is accelerated or decelerated. Blows to the head by a *fist* produce a violent acceleration in the same range as the unrestrained head striking the dashboard in low-speed car accidents. Trauma includes motion of the brain over the bony ridges of the skull, which produces contusions and lacerations of the inferior frontal and temporal lobes, or intracerebral hemorrhage due to gearing of small blood vessels within the brain parenchyma. The punchdrunk syndrome reported in 6% of professional *boxers* is consequent to trauma in multiple areas and is characterized by speech difficulties, clumsiness, and unsteadiness of gait that progress to disabling ataxia, dementia, spasticity, and extrapyramidal disturbances of Parkinson type. They exhibit βA4 protein deposition as diffuse plaques (Blumbergs, 2005).

The amount of mechanical stress depends upon the size of area to which the force is applied. Newton: a unit of force (i.e., the force required to impart to a kg of 1 m/sec). 1 Newton (N) = 0.2248 pounds force.

Impulse: equals Force × Time (that it acts). A force acting on an object changes its velocity. Therefore, since Momentum = Mass × Velocity, a force that changes its velocity changes its momentum. This is called the *impulse–momentum relationship.* The greater the impulse, the greater will be the change of momentum.

Modulus: defined as a quantity that expresses the degree to which substance possesses a property. It denotes that an applied stress causes permanent deformation (Hunt et al., 2004).

Torque: The tendency of a force to rotate the body to which it is applied. Torque changes rotational equilibrium and involves angular rotation. It involves magnitude, point of application, and direction. Torque is force times the length of the lever arm (the perpendicular distance from the axis of rotation to the point at which the force is applied). Force creates different torques at different lengths from the center of rotation.

Stress or pressure: the force/unit area (force divided by area of impact) that causes the displacement of one point relative to another. Since brain tissue stiffness is *age dependent*, the same brain indentation causes different stress at different ages (Levchakov et al., 2006).

Strain: the response of the material that is being stressed. One measure is the distance of the deformation divided by the total length of the material to which the stress was applied. It is measured as a proportional change in length in the direction of the compressive or tensile stress. If stress and strain are plotted on the same curve, the *elastic modulus* is that part of the curve in which the force does not cause permanent deformation. The remaining portion of the curve is the *plastic modulus.* Different depths of indentation produce different strain distributions at different ages, depending upon age-dependent elastic properties of the brain.

Tension: a condition of stretch, and the force is tensile force. *Ultimate tensile strength* is the force needed to break some material that is higher than the elastic limit.

Load

Load is the external force or stress to which a structure is subjected. *Loading* is energy transmitted to the tissue. How does the head or other part of the body react when a force or load is placed upon it? Loads are direct (e.g., impact of the head with another object) or indirect (e.g., acceleration of the head due to movement caused by change of speed or impact elsewhere on the body) (La Placa & Ziewjewski, 2004).

Duration of loading time affects trauma. It is the time that a force is applied. Duration partially determines the type of lesion. *Static loading* occurs when forces are applied to the head gradually and slowly (e.g., earthquakes and landslides, in which the head is squeezed slowly or is crushed), usually taking more than 200 msec. *Dynamic loading* occurs when the forces causing injury act in less than 200 msec. Dynamic loading (e.g., <50 msec) is the most common cause of TBI. *Impact loading* occurs when there is a collision between an object and the head. *Impulsive loading* is where there is no impact (e.g., whiplash, change of motion or impact elsewhere than the head). Brain maximum shear stress and coup/contrecoup pressures appear to occur in the duration of 15 msec to a standard head injury criterion (Ruan & Prasad, 1995). Brain–skull displacement is greater for brief (10 msec) rather than longer shocks (>20 msec). Short shocks release energy above 100 Hz, with the amount decreasing substantially with a duration as short as 10 msec. The relative displacement between brain and skull increases as shock duration decreases (Willinger et al., 1995).

Compression and Pressure Waves

Compressive force pushes two pieces together into less volume or space. However, the brain is incompressible: therefore, angular acceleration of the brain, accompanied by compression waves, sets up shearing strains (i.e., relative displacement of the different parts of the brain relative to each other). Rotational movement may contribute to compression of the frontal and temporal tips, yielding a contusion.

Clinical Example: Bullet Impact to Skull

The patient was a 9-year-old boy who was struck by a bullet that entered at the right of the cervical area and exited below the right ear. It was surmised by an examining MD and the hospital that the bullet could have impacted the skull and caused DAI through transmission of energy. No focal neurological signs were detected, although neuropsychological and quantified EEG dysfunctions were detected. The patient was stunned and thrown backwards. His mother reported no loss of consciousness, but subsequent severe and frequent headaches, difficulties with sleeping, and temperamental, memory, and concentration problems, with hypersensitivity to loud sounds. There was dysfunction in the brainstem and delay in the optic tracts bilaterally, attributed to regional DAI.

Pressure: Volume or bulk distortion stems from application of pressure to an enclosed structure. Pressure is one feature of *blunt injury*. Using a person knocked to the ground and striking his or her head as an example, it was observed that the brain and its coverings are vulnerable to degrees of blunt trauma that would rarely be lethal if applied to other body regions (Saukko & Knight, 2004, cited by Takamiya et al., 2007). Compressive strain occurs as a result of impact. Energy causes a crushing-type injury that results in deformation and interruption of the structural integrity of the organ. Volume or bulk distortion stems from application of pressure (external compression of the head and internal pressure caused by brain swelling). Since the brain is quite incompressible and since only a limited amount of pressure is absorbed when external forces applied, the brain can only move rather than be distorted inwardly. External application of pressure compresses the head. Internal pressure is caused by brain swelling.

Compressive wave strain: *Wave motion* is the propagation of energy or a disturbance after an impact load (*coup*) is applied to the head. It is the mechanism by which force radiates from the point of impact. *Pressure waves* occur after a skull impact. They also precede a projectile entering the brain and create *cavitation* (i.e., a cavity greater than the volume displaced by the missile) (Knightly & Pulliam, 1996).

A compression wave is propagated through molecules within a medium whenever a solid object is struck (e.g., head impact). The medium is cell walls, extracellular fluid, connective tissue, cell membranes and contents, and vessels and contents. The brain's *viscoelastic* qualities and varied structure makes it vulnerable to pressure waves that can induce shearing forces. Cerebral damage may occur due to *stress wave concentration due to contact forces* or *acceleration-induced brain damage resulting in tissue-tear hemorrhages* (Gennarelli & Graham, 1998). *Compression-rarefaction* is characterized by a change of volume without a change of internal shape. The skull is an inelastic container that contains three relatively *noncompressable substances*: brain, CSF, and blood. Since the brain is virtually incompressible, it has a lower tolerance to shear strains than compression strains (Adams et al., 1982). This mechanism may be involved in rotational movement's creation of *coup-contrecoup injury* (see contrecoup and cavitation).

Cavitation and *pressure waves*: *Cavitation* is defined as the creation of a cavity greater than the volume displaced. One example is the *shock wave* produced to the head and side of a projectile entering tissue (Knightly & Pulliam, 1996). If a pressure gradient exceeds atmospheric pressure at one pole, it is followed by movement away from the impact. The space drops below vapor pressure of the brain. Cavitation accounts for *contrecoup* lesions since the brain withstands positive pressure (compression) much better than negative pressure or cavitation (Gean, 1994, p. 163; Pang, 1989; Thibault & Gennarelli, 1985). Both whiplash and impact contribute to cavitation brain injury (Kuijpers et al., 1995; Nusholtz et al., 1996). (Note that doubt about this mechanism is expressed by Ommaya et al. [1971] and Povlishock [1989]). A compression wave is released with head impact. The contents of the skull and brain are compressed at the impact side, then a wave of tissue moves to the opposite pole (*contrecoup*) (Lockman, 1989; Pang, 1989; Unterharnscheidt, 1972). During impact the opposite side of the skull is moving faster than the brain and the brain lags behind. There is a pressure gradient that, when it exceeds atmospheric pressure at one pole, expands a space, and pressure then drops below vapor pressure of the brain. The liquid boils and changes rapidly to the gaseous state, causing instantaneous formation of gas bubbles with great violence, analogous to the sudden pressure changes in the center of an explosion (Chan & Liu, 1974). The space collapses violently to create brain trauma (Gurdjian, 1975, p. 175). Starting from the pathological observation that focal contusions may be quite small at the surface, involving only the cortex, or may extend into a large interior cavity, Ommaya (1990) assumes that the cavitation effect is most probably due to the bubbles caused by expansion of gases under low pressure in the blood of cerebrospinal fluid (CSF) rather than primarily in the brain. Gurdjian and Gurdjian (1975) describe the phenomenon as follows: "The brain is crowded at the impact site with high pressure development. [It] lags behind the faster moving skull...with development of underpressures...There may be tears in the interior of the brain in the direction of the force (and tears) of connecting veins over the convexity..." The high acceleration of the brain (during rebound) causes it to separate momentarily from the dura or causes the dura to separate from the skull. Uneven strength and density causes brain tissue to separate temporarily, creating contusions and intracerebral hematomas. The area under the contrecoup is damaged more than the area under the load (Thibault & Gennarelli, 1985). The interval during which impact, pressure waves, and brain inertia contributes to the formation of a cavity and violent collapse appears to be up to 5 msec (Ruan & Prasad, 1995).

Strain

Strain is the displacement of one point relative to another caused by stress (force). Different forces achieve a given type of deformation. The forms of strain are tensile strain, shear strain, compressive strain, and over-pressure (Hunt et al., 2004). Strain has been described as the "proximate cause" of tissue injury (Gennarelli & Graham, 1998). *Slow* application of strain is better tolerated than *rapid* strain, which leads to the brain becoming brittle and breaking (Teasdale & Mathew, 1996). In the spinal cord, tissue strain contributes to microvasculature pathology. Extravasation and physiological injury markers were more highly correlated to the rate of spinal cord compression than to the depth of compression. The microvasculature sensitivity to injury was affected by vessel caliber and the cells comprising the blood spinal cord barrier (Maikos & Schreiber, 2007).

Shearing and Shear Strain

Shearing refers to the differential speed of movement of adjacent tissues relevant to other brain planes (e.g., gray-white junction) or relative to the skull and blood vessels. Lateral rotation of the head on the neck, and axial rotation of the neck and skull, interacts with different distances of tissues from the center of rotation and the foramen magnum, and such factor as the weight of different tissues. Thus, adjacent tissues (brain planes, skull, vessels) move at different speeds. There may be no change in a particular plane, but there is a change in internal shape due to alteration of relative position. Shearing is characterized by a change of shape without a change of volume and is responsible for the vast majority of mechanically induced lesions (Gentry, 1989). Thus, shearing is the *deformation* of an elastic body caused by forces that produce an opposite but parallel sliding motion of the structure's planes. Shearing causes deformation (change of shape) when an external force causes different distances of movement of adjacent tissues. *Displacement* (the distance moved) of planes and structures contributes to shearing when rotational forces displace the parts of the brain at different speeds relative to their distance from the center of rotation. Shear strain has been described as "the most prominent mechanism of injury in minor head trauma" (Bailey & Gudeman, 1989).

Location of Shear Injuries

The most common location for shear injury is at the junction of tissues of differing density (e.g., the denser cortex shifts in relation to the less dense white matter) (Britt & Heiserman, 2000).

Scraping injuries of the brain against the inside of the cranium.

Separation of brain planes: gray-white junction.

Separations of structures adjacent to the skull: olfactory fibers projecting onto the cribriform plate; the hypothalamic-pituitary stalk extending to the pituitary gland embedded the sella turcica.

Brain propelled against fixed internal structures: blood vessels and brain tissue may be damaged by shearing forces since inertia and internal pressure waves move brain tissue against blood vessels tethered in place; the posterior corpus callosum scraping against the falx cerebri (dura mater).

In order to study the biomechanics of cerebral trauma, including brain injury mechanisms and tolerance to impact, three-dimensional models of human and porcine brains were constructed (King, 2004). They had sensors to monitor head kinematics and brain strains. The head was impacted frontally on the scalp in the mid-sagittal plane in an anterior-posterior direction. The findings were similar to some clinical evidence.

Shear strain can be a reasonable predictor of DAI. In the human model, but not the porcine one, there were high shear stresses at the coup site and the brainstem. On the surface of both hemispheres, as well as on the surface of the corpus callosum, large shear strains were observed, which can stretch the axon and cause diffuse DAI. Pressure was distributed uniformly side to side across the brain except at the partition between hemisphere and cerebellum, with compression at the impact point and tension at a point opposite to the impact. Negative pressure (tensile strain) at the contrecoup site can cause brain contusion. Thus, it can be a mechanism for contrecoup injury. Maximum tensile strain was also generated at the dorsolateral part of the rostral brainstem. Although diffuse brain injury is characteristic for concussive level trauma, with the frontal and temporal tips most vulnerable, the effect upon the brainstem, which is bent, compressed, and rotated, is often underestimated.

Sports injury as a shear model: Sports concussion is a sample of physical impact injury at its lower range. Using a brief neurological and neurocognitive screening measure for high school and college football players, significant impairment could be detected 15 min after an injury, even for players without LOC (loss of consciousness) or PTA (post traumatic amnesia). All groups returned to baseline functioning within 4 h (McCrea et al., 2002). This brings into perspective a study for which an anatomically correct model was created that modeled 24 head-to-head field collisions in a football game (Zhang et al., 2004). Shear stress was the injury predictor for concussion. Since peak pressure reflects the severity and extent of tissue response to an impact, intracranial pressure is a global response indicator for TBI. Physical formulas and experimental results describing the estimated acceleration and other physical forces that cause TBI were offered.

High shear stress concentrations were found in the central brain core (i.e., the upper brainstem and thalamus), which has special geometrical features and material composition. The induced shear may lead to a mild brain injury.

While translational head acceleration had a greater influence upon intracranial pressure, shear stress at brain center was more sensitive to rotational acceleration.

Elasticity

The varying ability of a material to recover its original length, shape, or volume after the stress is removed. It is a physical characteristic of the greatest importance, since force only creates trauma when the strain exceeds the elasticity of the particular tissue. If the amount of stretch exceeds the *elastic limit*, then the tissue need not break, but also does not return to its original length when the stress is removed. The brain and body may be torn by a single deformation of high pressure and rapid motion that exceeds the elastic limit. The *tensile strength* is the stress required to break a lengthy object by pulling on it. The *elastic limit* is the point at which the amount of stretch exceeds the capacity of the tissue to *return to its original length when the stress is removed*. The *tensile strength* is the stress required to break a lengthy object by pulling on it.

Pulse is a shock wave that is not repeated. Tissue displacement parallel to the direction of the motion of a pulse is described as *longitudinal*. Displacement at right angles to the direction of the pulse is *transverse*.

Rotational Injuries

Although doubt has been expressed (Teasell & Shapiro, 1998) as to the conclusiveness of the evidence that rotational injuries cause brain trauma, one may consider the anatomy of the brain within the head. The brainstem is tethered through the foramen magnum to the neck and spinal cord. There is particular vulnerability to movement of the frontal and temporal tips into the anterior and middle fossae. Parts of the brain may be compressed as the brainstem rotates while the spinal cord is tethered outside the foramen magnum. Rotational injuries (neurological and somatic) are referred to as "whiplash."

Knife or Glass Wounds

Piercing wounds have little energy, so that the damage depends upon striking vital structures. One is concerned only with the permanent cavity.

IMPACT DISTORTIONS OF THE SKULL

Clinical example: Low impact TBI: Assault: A female police officer was assaulted on the job.

Do you remember the accident (i.e., being injured)? While I was in the process of issuing a summons to someone who was smoking marijuana in the bathroom, his friend came up behind me and placed me in a military chokehold and starting squeezing my neck in a violent manner until I couldn't breathe. In the process of me trying to free myself I remember my head hitting the wall. I was knocked out. When I woke up, I was in the security room. I was told that I was unconscious for approximately 10–12 min. I remember not being able to speak, I remember holding my head because of the pain. I just kept holding my throat. They took me to the hospital after that. I remember having trouble breathing. Not being able to catch my breath, I can't talk about without ... it affects me.

Smell or taste anything? I tasted blood. I thought I bit my tongue but I didn't.

Memory for events after the accident: "There are definitely blank spots." She had bruises on ribs and arms and don't remember how that happened.

How did you feel when you woke up? "I was very dizzy. I didn't recognize where I was at first. They used ammonia or smelling salts." She was told she was kicked. That's where the bruises came from.

Were there any bruises, cuts, inflammation, or dirt on your head or face? Her right cheek had a bruise and under her left eye there was a swelling.

Neck, back, rib, arm or leg, shoulder pains? "My ribs were swollen."

Skull distortion follows from contact phenomena, including the pressure field generated within the cranium. This is accompanied by translation of the head (movement in a straight line), and rotation of the head (hyperflexion, hyperextension, lateral bending, and twisting of the head on the neck). Rotation and deformation of the skull contribute in about equal amounts to the injury potential. Doubt is expressed about pure translational effect as a brain-injuring factor (Ommaya et al., 1968, 1971). After impact, there may be a blow to a restricted, usually cortical, region with a diffuse pattern of functional disruption. The relative size of coup and contrecoup injuries depends on the nature of the impacting surface that creates acceleration or deceleration of the head and enclosed brain (Gennarelli & Graham, 1998). For example, an assault might involve a small hard impactor (e.g., a weapon), while a fall might be at a lower rate of speed against a broad padded surface (e.g., a carpeted floor).

If the blow is central and parallel to the long axis of a compliant skull, the shape of the skull changes from an ellipse to a circle (ellipsoidal deformation). Shortening of the axis of the brain causes shearing relative to the central region. Negative pressure zones relative to the perpendicular axis of the skull (now wider than the brain space) cause pressure at the center, resulting in extensive damage to periventricular and central structures. Trauma to the forehead can lead to increase of the transverse diameter of the skull at the level of the anterior fossa (wedging), causing deformation of the skull base and, consequently, basilar skull fracture (McElhaney et al., 1996).

PHYSICAL CHARACTERISTICS OF THE BRAIN

The brain is an incompressible substance enclosed in a rigid container with unyielding hard compartments. Brain tissue is inhomogeneous and anisotropic (i.e., its properties vary in different directions). It has fixed gray and white matter, ventricular spaces, meninges, and penetrating blood vessels attached outside the brain. For example, one may compare white and gray matter (i.e., the effects of an impact will vary according the direction of force). Regional differences after impact affect the stress and strain fields (Zhang et al., 2004). Ommaya (1990) described brain substance as a soft viscoelastic material. It is very easily deformed under shear or tension, but relatively resistant to compression strain because of its almost incompressible nature and its viscosity. There is maximal effect on the surface, diminishing toward the center of the spherical viscoelastic mass.

Stiffness is the opposite of compressibility (e.g., the deflection of a spring under a given load, as opposed to hardness).

Hardness is the unyielding quality of a surface, such as glass.

The brain is surrounded by the subarachnoid space and CSF. These do not provide much of a cushion in terms of significant acceleration or deceleration of the head. Brain and CSF have about the same specific gravity, which causes the brain to float. Contributions to mechanical damage after trauma are due to varying proportions of the brain's white matter, the subdivisions of the brain caused by great folds of the dura mater into compartments, and various shapes (smooth and rough) of the inner surface of the cranium. The different portions of the brain have varying water-tissue ratios—white matter, 60%; gray matter, 80% (Hayes & Ellison, 1989). The resulting structure is neither rigid nor relatively compressible. Cerebral injury depends on violation of the strength and adhesion of brain materials. Changes in age contribute to the relative vulnerability of the brain to damage. Injury would be intensified where there are junctional boundaries or a sudden transition between brain and hard tissues (e.g., dura, bone, meninges). Surface effects would be minimized where the skull interior is smooth and there are no venous attachments (e.g., occipital lobes).

SKULL–BRAIN INTERFACE

The skull-brain interface can be conceptualized as a coupled interface—the brain surface and inner surface of the closed skull being closely coupled. There is also the free interface (a free-slip condition permitting separation and only the transmission of compressive forces between brain and skull). The free-interface model agrees more closely with experimental test data performed on cadavers. The kinematic boundary of the head-neck junction can be visualized as a hinge and support. The presence of a kinematic restraint at the head-neck junction appears to displace peak pressure and shear stress to the skull base from the coup region. Specifically, force is applied centripetally (Gennarelli, 1986). Therefore, the relatively small attachment of the brainstem to the cerebral hemispheres, and the large mass ratio of the hemispheres and the brainstem, allow greater strain to occur in the brainstem. The mass ratio is important because the inertial effects of acceleration can produce a torque between the hemispheres and the brainstem, concentrating strain there. It is not known whether this effect is responsible for the effect on coma. The skull may hit a hard object or a hard object may strike the head, causing scalp laceration, skull fracture, extradural hematoma, some forms of cerebral contusion, and intracerebral hemorrhage. When the head is freely movable or strikes an object, there is a coup injury as the brain strikes the inner surface of the leading pole of the cranial dome (i.e., the calvarium). The head may then swing around until restricted by the neck to which it is tethered ("whiplash"), and then the brain may incur a second impact as it bounces and strikes the opposite surface (contrecoup). The neurotraumatic consequences (Becker, 1989) include brain movement, brain deformation, and brain rotation with accompanying internal shearing injuries to brain tissue. Adams et al. (1986) assert that acceleration/deceleration forces may cause, in addition to axonal damage, damage to blood vessels associated with deep intracerebral hematoma. Some hemorrhage is due to negative pressure forming a cavity as the skull moves faster than the brain following impact (Gurdjian, 1975, p. 175).

THE GEOMETRY OF HEAD/BRAIN MOTION

Planes of Brain Structure

Axial (parallel to spinal column); *coronal* (vertical slices)
Transaxial is horizontal along the base of the skull
Sagittal is the vertical midline
Parasagittal are slices off-center but in the same plane as sagittal.

Direction of Brain Movement

Translational Movement: Force applied through the center of gravity (mass) in the direction of the bodily axis or perpendicular to it in the anterior–posterior direction. It causes brain planes (including the cerebral cortex) to move parallel to each other, particularly in the forward direction, though not necessarily in straight lines. Pure translational acceleration creates pressure gradients, while pure rotational acceleration contributes to shear strain. Translational movement produces contusions and intracerebral hematoma, but not concussion.

Angular motion (up and down): The brain moves laterally, or up and down, relative to the body axis.

Oblique: Motions are in the combined sagittal and coronal planes.

Lateral: Sideways in the horizontal plane.

Determinants of Skull/Brain Motion

Due to the disorderly arrangement of forces, acceleration or deceleration, and restrictions upon movement, the following parameters cause head motion to occur simultaneously in various the axes and planes:

- The head's initial inertia may be followed by loading and transferred momentum due to impact or change of acceleration or direction due to the movement or restriction caused by adjacent surfaces, or due to being pulled by the torso, which differs from those of the torso, creating movement independent of the torso-neck tether

- Relative differences of velocity between head/brain and the torso due to differences of acceleration or deceleration
- The restraining effect upon the direction of movement of the head/enclosed brain unit, due to its being tethered by the neck to the torso (which may be stable or moving in a different direction). This creates a tendency to circular (angular) motion in various planes since the head is the end of a radius terminating in the torso
- Whether the skull is movable, *restrained*, or *rebounds*

Movement of Brain Components

- *Translational*: Moving as a unit in parallel planes
- *Shearing*: Moving in separate planes (*shearing*). After acceleration or deceleration in an accident, head motion is caused by its inertia
- *Angular motion* (both laterally and forward or backwards)
- *Oblique* motions are in the combined sagittal and coronal planes. After impact, the brain probably moves in complex directions (Gean, 1994, p. 149; Hayes & Ellison, 1989; Ommaya, 1990) creating a variety of lesions.

THE DIRECTION OF ENERGY AND BRAIN DEFORMATION

Brain movement occurs in various planes relative to supporting and penetrating surfaces—skull; meninges; blood vessels with fixed origins and insertions in brain, skull, and meninges; and exiting cranial nerves. If an impact is not on the head's anterior-posterior axis, then the skull receives an angular momentum (sagittal plane) and may rotate with the neck as the fixed end of the rotating radius). The CSF serves as a shock absorber only up to a certain level of force. The components of the body or of the brain may move as a unit (translational) or in separate planes (shearing). After acceleration or deceleration in an accident, head motion is caused by its inertia or *rotational* (angular [left-right], rotational [up-down]). After impact, the brain probably moves in complex directions, creating a variety of lesions (Gean, 1994, p. 149; Hayes & Ellison, 1989; Ommaya, 1990). Increasing head translational and rotational accelerations results in an increase of intracranial pressure (Zhang et al., 2004).

Association Between Lesion Type and Geometry of Movement

Angular acceleration (rotation) produces cerebral concussion. Rotational movement may also contribute to compression of the frontal and temporal tips yielding a contusion (Gennarelli, 1983). *Symmetrical damage* is expected only if the brain rotates in the horizontal or coronal plane. If the brain rotates so that one lobe moves up or down, opposite motion is expected in the contralateral lobe, resulting in asymmetrical degeneration (Stich, 1956). The frontal tips are susceptible to contusions, the base of the brain is susceptible to lacerations, and the corpus callosum, deep white matter, and brainstem incur DAI, perpendicularly to the bodily axis with anterior-posterior movement.

The Point of Impact and Site of Lesion

Force applied to the head perpendicular to the body axis produces primarily focal injuries. The skull and brain may also oscillate back and forth, causing more rubbing and blows. This accounts for the contusions and lacerations occurring at the orbital frontal region and frontal and temporal lobe tips. The point of impact is described as coup, the distant culmination of brain movement is contrecoup, and lesions in between are intermediate coup (Ommaya et al., 1971). There appears to be fewer contrecoup lesions in infants, increasing to a level in 4-year-old children that is slightly less than the proportion found in fatal injuries of adults (85–90%, McLaurin & Towbin, 1990). The issue of impact and associated lesion was studied excluding cases in which there was a skull fracture that tore the brain (Kirkpatrick, 1983).

1. *Frontal impact*: Contusions at frontal and temporal poles were more frequent and more severe than contrecoup.
2. *Lateral impact*: Contrecoup lesions of the temporal or frontal lobe were moderately more frequent and more severe. Impact site lesions were relatively infrequent.
3. *Posterior impact*: Posterior impacts usually damaged the cerebellum or occiput, but always caused contrecoup damage to the frontal and temporal lobes. Cavitation is not considered to be the sole force; rather shearing of the gyri over rough surfaces of the orbital plates and middle fossae creates most of the damage at these sites.

Shearing: Subdural Hemorrhage (SDH)

This is a lesion occurring after damage to the bridging veins that exit from the subarachnoidal space to the superior sagittal sinus (11 large pairs). Bridging veins are believed to usually stretch 30–35% before tearing. However, with abnormal body position, the veins may rupture with little force applied. Gender effect: For the same external force, a female's head exhibited significantly higher acceleration in comparison to males. Head acceleration for a female has been reported as 2 1/2 times greater than for a male in a collision. Physical dimensions interact with the difference of speed or direction of impact in determining the severity of injuries (Ziejewski, 2004).

The fact of less strain in a lateral impact than an occipital impact is attributed to the support that portion of the dura mater known as the falx cerebri (surrounds each cerebral hemisphere almost down to the corpus callosum). Strain on the bridging veins depends upon the amount of relative motion between the skull and the brain. These are oriented either forward or backward, with a measured length of 8.8–19.9 mm. In posterior–anterior motion, the maximal strain occurs in the shortest bridging veins, which are oriented in the plane of the motion and are angled in the direction of motion. SDH is more easily produced by an occipital than a frontal impact. In addition, anterior-posterior movement causes a higher strain than a lateral one. Loads in the lateral direction are more likely to cause DAI than impulses in the sagittal plane (Kleiven, 2003). Using *MRI* (*magnetic resonance imaging*) and *CT* (*computerized tomography*), the following generalizations concerning damage and outcome can be made: Diffuse injury has neuropsychological effects according to three main axes:

1. There is a strong anterior-posterior gradient, with most damage in frontal and temporal regions.
2. Depth of injury is related to overall severity of brain damage, although the relationship between depth of lesions and functional measures is mild.
3. There are often differences in lateralization of injury (Wilson & Wyper, 1992).

In a study that involved focal brain lesions, as well as considering the possible effects of diffuse brain trauma, it was determined that the depth of parenchymal lesion increased with traumatic force, producing more severe impairment of consciousness and worse outcome (Levin et al., 1997). Nevertheless, one study did not elicit a correlation between the apparent area in which the major force was applied and particular symptoms (Rutherford et al., 1977).

A glancing blow may create significant neurotrauma. It has two components of pressures in the fluid and stresses in the skull: (1) tangential traction with stress described as skew-symmetric radial; (2) axis-symmetric (Chan & Liu, 1974). It is modeled as abrupt surface traction applied to a small surface area of a spheroid. The tangential component of the load shifts the location of peak pressures in the fluid, producing stresses of higher frequency and magnitude since the fluid cannot transmit shear stress. Shear is manifested as relative motion of the scalp, skull, and intracranial contents (Thibault & Gennarelli, 1985). Reduced intellectual status associated with head injury was related to temporal horn size and third ventricle volume, but not to the degree of cerebral atrophy alone. Initial status of smaller brain size and reduced preinjury education can be associated with neuropathological vulnerability and intellectual impairment (Bigler et al., 1999). Tangential gunshot wounds have special mechanical considerations (Stone et al., 1996), while a missile that penetrates the body and exits the tissue retains kinetic energy and delivers less energy to the tissues. A missile that is retained (glancing) delivers its total energy to the tissue. A tangential injury is defined as a missile striking or grazing the cranium without penetrating the skull. The force may cause a linear skull fracture, a depressed fracture, or fragmentation of the inner table that leaves the outer table relatively or completely intact. It is believed that tangential gunshot injuries are likely to cause intracranial pathology, although many patients may have no neurological deficits.

The author has examined several patients in whom a falling object or structure propelled by a spring has struck the head with little alteration of consciousness or brief LOC, but with major dementia. One woman struck by a doorframe was stunned, but not hospitalized or treated at the scene. She suffered an estimated 18-point loss of WAIS-R Full Scale IQ. Another woman was struck by a falling bar, suffered an estimated 3-min LOC followed by an estimated loss of FSIQ of 16 points. A college student, struck by a piece of athletic equipment, scored an WAIS-R FSIQ of 86. On the Woodcock Johnson Tests of Cognitive Ability, his mean standard score was 62 (i.e., first percentile—generalized inefficiency of mental processing). He was not unconscious, but described himself as "out of it" for hours.

TRAUMA AND KEY ASPECTS OF HEAD ANATOMY

The Brain Is Tethered to the Spinal Column by the Brainstem

The foramen magnum is a large circular window through which the central nervous system exits the skull (see Figure 6.5). When impetus is transmitted to the body (e.g., in a fall or automobile accident), the head continues to move, perhaps to the front or rear, or to the side in an angular direction. It accelerates forward as far as the neck will permit and then swings back. Rotation of the cerebral hemispheres around the relatively fixed brainstem at the midbrain-subthalamic level is believed to create the maximal shearing stress (Adams & Victor, 1989, p. 697). This is part of the mechanism of "whiplash." The rapid to-and-fro motion at a high rate of speed stretches and tears brain tissue and brings the brain into contact with the inner surface of the skull. If the head is at an angle to the body when the motion in one direction stops, then rapid lateral motions

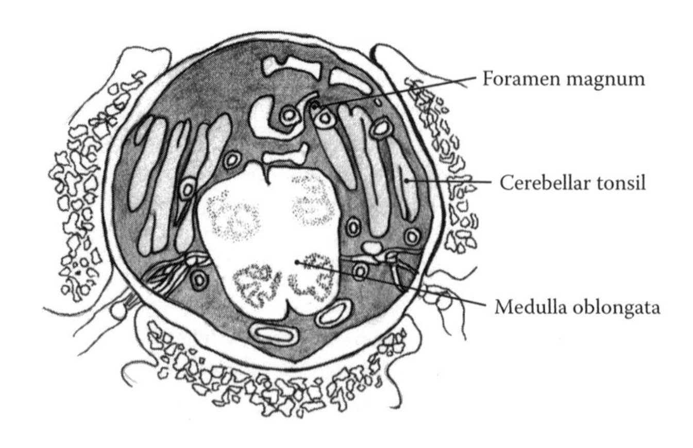

FIGURE 6.5 Foramen Magnum, brain stem, and cerebellar tonsils. (Kretschmann and Weinrich, *Cranial Neuroimaging and Clinical Neuroanatomy*, 1992, Figure 48b, p. 93, Thieme.)

(swirling) occur that can impact the inferior surface of the brain against the bottom of the skull (i.e., scraping it to cause lacerations). Animal models indicate *brainstem injuries* secondary to "whiplash" (Henry et al., 2000).

The Skull Consists of Flat and Sharp Surfaces

Sharp surfaces intrude between or are adjacent to brain areas. These include the frontal crest and crista galli (separating the frontal lobes). They are the anterior support for the dura mater (falx cerebri). When the brain hits a flat surface, the result may be a contusion or bruise. When the brain rotates on one of the sharp surfaces of the inside of the skull, scratches are called lacerations. This occurs characteristically to the frontal and temporal lobes (see diagrams in Adams & Victor, 1989, Chapter 35, for various types and locations of TBI). The frontal lobes (inferior orbital, polar surfaces), upper mesencephalon, and medial-inferior surface of the temporal lobes are particularly susceptible to damage (Stuss & Benson, 1986, p. 46).

CRANIAL FOSSAE

These are three partially enclosed sections supporting the brain and the middle and posterior fossae representing the base of the skull (Standing, 2005, pp. 461–463). During brain movement, there are structures that may lacerate it or may permit impact against its surfaces or enclosures (e.g., the temporal and frontal lobes are a common source of contusions).

The Anterior Fossa

This is the anterior third of the cranial floor, supporting the frontal lobes and roofing the orbits and the nasal fossae. On the floor of the fossa is an upward medial projection, the *crista galli*. The anterior fold of the dura mater (*falx cerebri*) is attached here, separating the frontal lobes. This structure ordinarily supports and protects the anterior brain, but when impact causes brain movement, the frontal lobes can crash into it, causing lacerations. Its free edge extends bilaterally and posteriorly over the corpus callosum until it fuses with the tentorium cerebelli more posteriorally. This arrangement (i.e., enclosure of the tip of the frontal lobes medially by the these midline structures, anteriorly by the frontal bone, and laterally by the temporal bone), accounts for *temporal tip contusions* after acceleration/deceleration and head impact injuries. The olfactory nerve roots exit through the medial cribriform plate.

The sphenoid bone and wing: The sphenoid bone supports the frontal lobe as the posterior base of the anterior fossa, the roof of the orbits, and nasal fossae. Its location makes it a candidate for the traumatic basis for memory impairment. The base of the frontal lobe rests upon the frontal bone anteriorly and rests upon the lesser wing of the sphenoid bone posteriorly, with a space below for the entrance of the tip of the temporal lobe that can be seen in the superior view (Standing, 2005, Figure 27.10, p. 462). A radiograph of the coronal plane illustrates the sharp edge of the lesser wing of the sphenoid bone, medially placed, bilaterally protruding into the space not supported by bone but occupied by the frontal lobe superiorly, with the entrance of the temporal tip within the middle fossa but coming to a dead end. Parasagittal sections illustrate the supported frontal lobe with the lesser wing terminating posteriorly as an edge, and the underlying temporal tip with the bony edge directed towards tissue connecting the temporal tip and the base of the frontal lobes (Kretschmann & Weinrich, 2004, p. 379, Figure 14, #4 & 5). A diagram of the passage of the uncinate fasciculus from the temporal tip to the ventral frontal lobe is offered by Pansky et al. (1988, p. 99). The frontal and temporal lobes are vulnerable to functional and structural disconnections caused by the intrusion of the sphenoid wing (Ommaya & Ommaya, 1997; MR diagram, Bigler et al., 1999) between them. The *lesser wing* extends from the midline laterally to the temporal bone. This is smooth and sharp edged, and overhangs the anterior portion of the middle fossa.

As quoted by Erin D. Bigler in the previous edition, "Oftentimes injury to the temporal lobe occurs because of its impact against the *lesser sphenoidal wing*. The sphenoid bone surrounds much of the medial undersurface and the anterior aspect of the temporal lobe. With high-velocity impact or rapid acceleration/deceleration the temporal lobe moves about in the middle cranial fossa. The movement of the temporal lobe impacting against the sphenoid is the basis of compression of temporal lobe structures against bone and also the temporal lobe may glide along or over the sphenoid, which may cause contusion and/or shearing effects. As illustrated in the CT imaging presented in this figure, this child ends up with significant temporal lobe contusion most probably related to the direct impact effect of the temporal lobe against the sphenoid (see Figures 6.6a and 6.6b). In clinical neuropsychology, it is likely that many of the memory and emotional changes that accompany TBI are related to damage produced by the temporal lobe coming into contact with the sphenoid and disrupting mesial

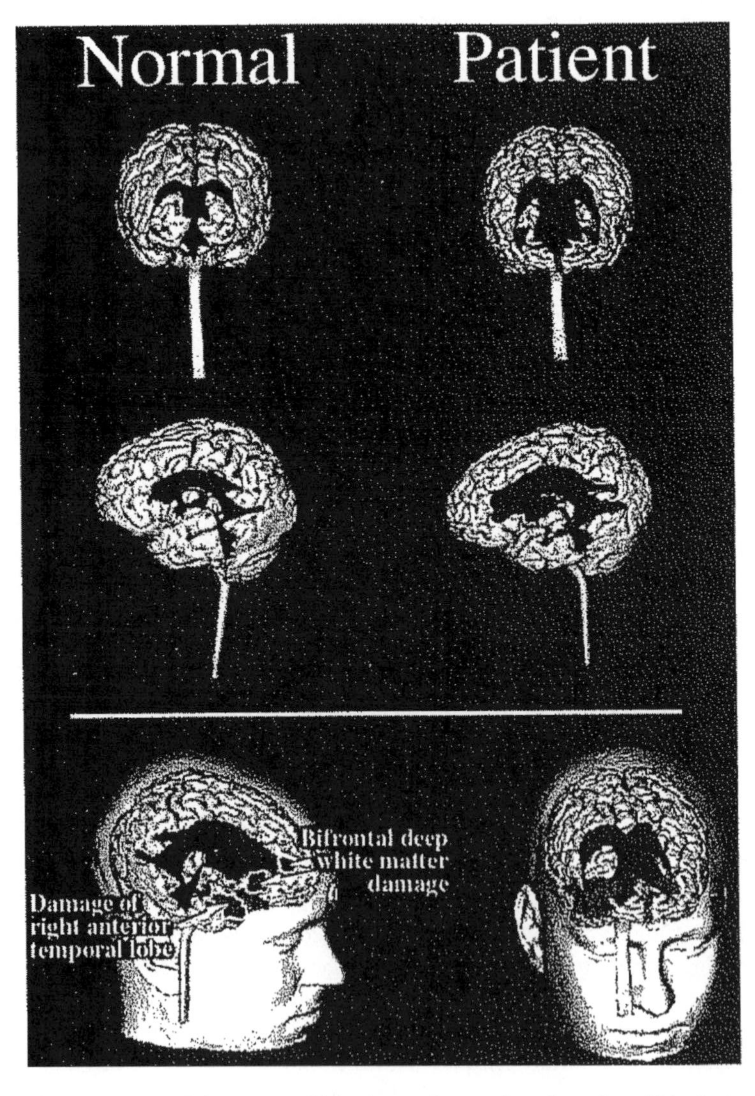

FIGURE 6.6a The movement of the temporal lobe impacting against the sphenoid is the basis of compression of temporal lobe structures against bone, and also, the temporal lobe may glide along or over the sphenoid, which may cause contusion and/or shearing effects. (Courtesy of Prof. Erin D. Bigler, Ph.D., Dept. of Psychology, Brigham Young University.)

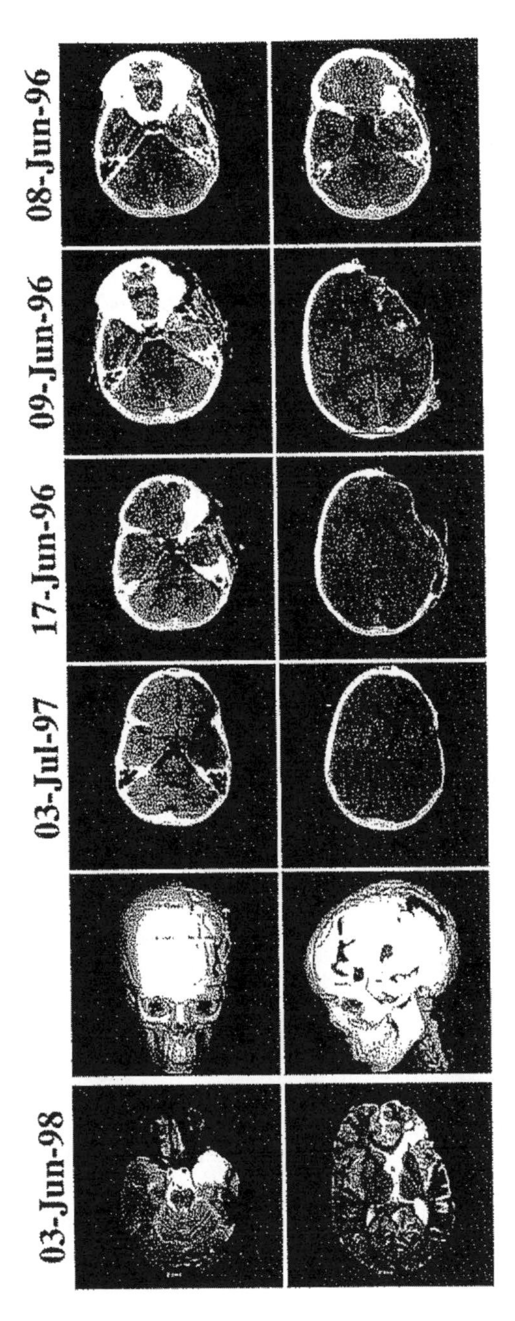

FIGURE 6.6b Progression of frontal and temporal lobe damage. (Courtesy of Prof. Erin D. Bigler, Ph.D., Dept. of Psychology, Brigham Young University.)

temporal lobe structures including subcortical structures such as the amygdala and hippocampus that sit just inside the parahippocampal gyrus and just above the fusiform gyrus."

The Middle Fossa

This supports the temporal lobe, which inserts under the lesser wing of the sphenoid bone. It is entered by the internal carotid artery (ICA) and existed by cranial nerves II–VIII.

High-speed impact and/or acceleration/deceleration injury render the medial temporal lobe (hippocampus and amygdala) vulnerable to mechanical deformation and contusion because of its location in the middle. TBI-related hippocampal damage may be a consequence of impact, excitotoxic, neuropathological changes, vascular damage, and transneuronal degeneration. There are implications from the fact that the hippocampus is predominantly a gray matter structure housed in the mesial temporal lobe, while its afferent pathway (the fornix) is suspended in the space of the third ventricle proceeding to the mammillary bodies. This pathway may be more susceptible to shear, strain, or tensile effects. This is consistent with the finding that there was a decrease in fornix area and hippocampal volume with increasing injury severity, although there was a minimal relationship between the volume and size of these structures after trauma (~11% of the variance). It is possible that different neuropathological processes reduce the size of both structures after trauma (Tate & Bigler, 2008).

The Posterior Fossa

This is the largest and deepest of the crania fossae and is the posterior part of the cranial base. It contains the cerebellum, pons, and medulla oblongata. Its floor in the occipita bone is the foramen magnum, through which the brain receives the spinal cord into the medulla. The Glossopharyngeal (IX), Vagus (X), and Accessory (XI) nerves exit the brain through the jugular foramen (lateral and above the foramen magnum). The Hypoglossal nerve (XII) and its recurrent branch (see Aprosodia) exits through the nearby medial hypoglossal canal.

CORPUS CALLOSUM AND FALX CEREBRI CUTS

The corpus callosum (integrating the two cerebral hemispheres) is vulnerable to impact injuries due to its proximity to the sharp edges of the dura mater, bilaterally extending down the internal surface of the cerebral hemispheres bilaterally (*falx cerebri*) (see diagrams in Kretschmann & Weinrich, 1992, pp. 26–43). Due to shearing injuries, axonal changes are found throughout the brain, most prominently in the corpus callosum and dorsolateral quadrants of the brainstem. Shearing lesions of the brainstem are associated with prolongation of impulses through the pontine-midbrain area (Gennarelli, 1987; Levin et al., 1987).

The dura mater supports and covers the brain and separates the two cerebral hemispheres. The falx cerebri acts as a dura partition and lies bilaterally in the midline and extends medially almost to the corpus callosum. This contributes to injury to the corpus callosum. The exposed edge extends posteriorly from the genu to the splenium, which is closest to the edge of the dura. Lateral motion can also scrape the top surface of the corpus callosum against the sharp edges of the falx cerebri and the incisura of the tentorium. These cuts may be considered lacerations (Rosenblum, 1989). It is an unyielding barrier and sharp edge during impact or acceleration that contributes to contusions. They can arise with high-velocity, rapid-deceleration head injuries, with sudden brain and arterial movement against the stationary edge of the falx creating a nidus for aneurysm development and present as delayed, acute intracerebral bleeding. After a high-velocity, rapid-acceleration/deceleration head injury, the edge of the dura mater extending down to the corpus callosum can create lacerations of the corpus callosum as the brain moves vertically and laterally (Rosenblum, 1989). It compresses and cuts the brain during brain swelling, causing hemorrhage, herniation, and other mass effects. In models of the human brain subject to shear the maximal tensile, larger shear strains were found (King, 2005).

Areas Vulnerable to TBI

Brainstem: Very vulnerable to twisting, pressure within the skull, etc. due to its location at the exit of the brain from the skull and continuation as the spinal cord. Damage here contributes to mortality of traumatic head injury. Injury above the corticospinal decussation can cause contralateral paralysis of the limbs.

Corpus callosum and the dura mater: The structure of the dura mater, a dense and inelastic membrane, creates a hazard when the lateral momentum of the brain causes it to be lacerated by the bilateral median edges of the dura (*falx cerebri*). Brain swelling also causes the medial temporal lobe under the meningeal *tentorium* (see Figures 6.7 through 6.10).

TISSUE INJURY

Biological Issues

Mechanisms of head injury, though complex, have been summarized as contact phenomena and acceleration (Gennarelli et al., 1982). Sufficiently intense mechanical forces interacting with

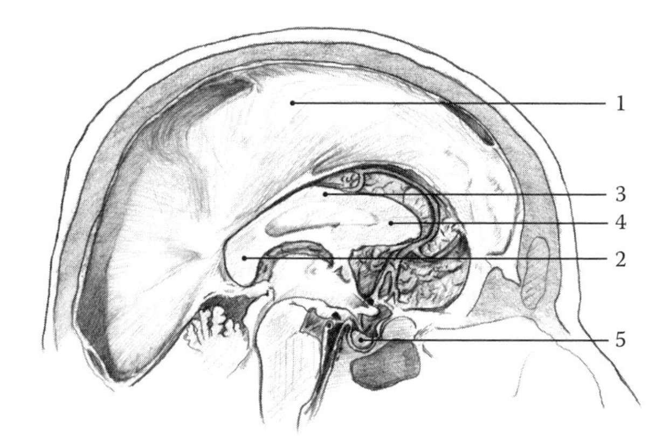

FIGURE 6.7 Sagittal section of the brain: (1) falx cerebri; (2) splenium; (3) corpus callosum; (4) ?; (5) pituitary gland. (From Rohan & Yokochi, *Color Atlas of the Anatomy*, Igaku-Shoin, 1993, p. 86. With permission.)

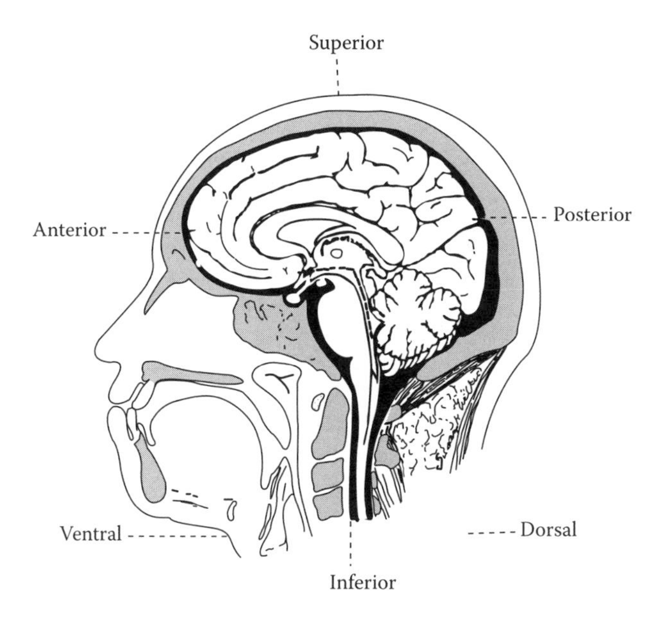

FIGURE 6.8 Sagittal section through head to show relationship of brain to anatomical terms of direction. Dotted line indicates bend in original axis of neural tube. Subarachnoid space is shown in black. (From Ranson & Clark, *Anatomy of the Nervous System*, l0th ed., Saunders, 1959. With permission.)

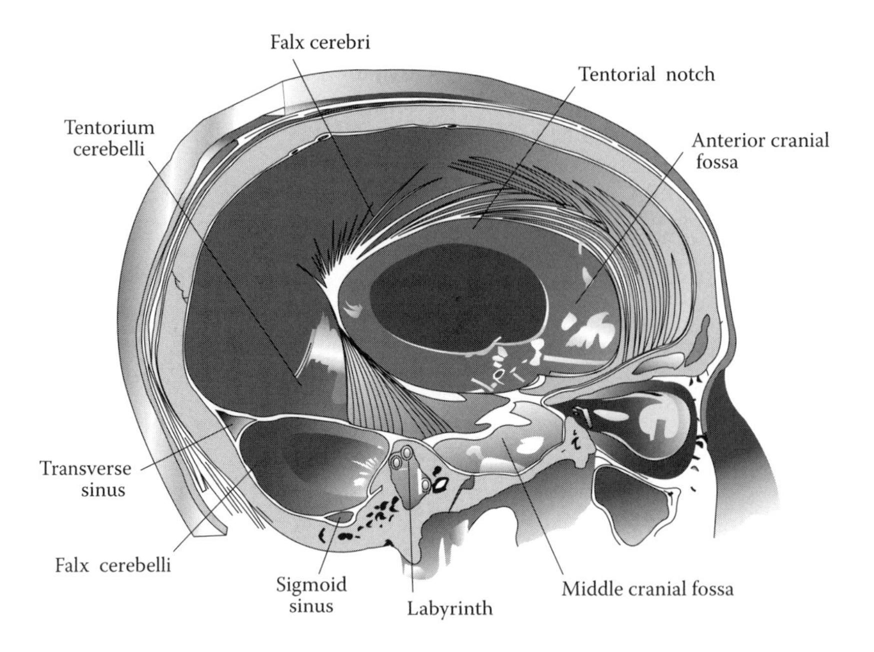

FIGURE 6.9 Sagittal section of the head showing the falx cerebri and the tentorium cerebelli. (Parent, *Carpenter's Neuroanatomy*, 9th ed., 1996, Williams and Wilkins.)

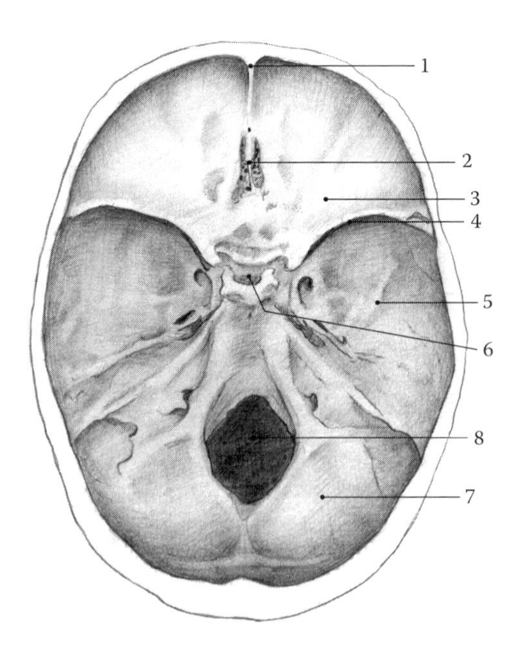

FIGURE 6.10 Base of the skull: (1) Frontal crest; (2) crista galli; (3) anterior fossa; (4) lesser wing of sphenoid bone; (5) middle fossa; (6) sella turcica; (7) posterior fossa; (8) foramen magnum. (Original illustration by Chris McGrath.)

the torso, neck, skull, and brain (Valadka, 2004) cause tissue changes that can may cause TBI. Different brain components are differentially deformable to the point where they permanently tear, move, or separate from the surrounding tissue. The brain is anchored within the cranial cavity by the parasagittal bridging veins, parasinusoidal granulations, cranial nerves, and the cerebellar tentorium. There is a significant tethering effect of the exiting blood vessels (Zhang et al., 2004) that may be expected to injure the brain as the vessels internal to it moves past them. When the head strikes the floor at the end of a fall, the brain movement tears parasagittal bridging veins, creating a subdural hemorrhage. Greater magnitudes of force with longer duration will result in DAI. Lesser magnitude and duration of the deceleration force can create transient unconsciousness (concussion) with few objective structural effects. After impact, deceleration forces result in reverberation and swirls within the gelatinous brain for many milliseconds. At the extremes of age, the brain is more vulnerable to vascular damage (Bullock & Nathoo, 2005). Sufficiently intense mechanical forces interacting with the torso, neck, skull and brain (Valadka, 2004) cause tissue changes that may result in TBI directly (impact, pressure waves). TBI may also directly result by acceleration/deceleration movements of the head, rotation, stretching, and shearing of the brain. The head injury is influenced by such factors as whether the head is free to move or is compressed and the shape of the head relative to the point of impact (Anderson & McLean, 2005). After head movement and impact and acceleration or deceleration, the brain, which is enclosed in the skull, moves against the skull's surfaces and compartments. These support the brain when all is well, but can scrape, cut, and impact the brain when movement and momentum brings the skull or body into contact with surfaces.

The brain is directly vulnerable to trauma because it is penetrable, soft, and not very elastic. The enclosing structure is firm (bone, blood vessels, and dura mater), which damage a moving and soft brain. Characteristics of the impacting surface (hardness [firm/yielding]; shape [sharp/dull]) affect the deformation of the skull and enclosed brain; penetration, range, or impact narrowing or dispersing the surface area to which energy is applied. The speed of the body and/or impacting surface affects the rate of deceleration or acceleration of the skull, creating brain shearing and tensile strain. With a sufficiently high rate of deceleration (e.g., from several hundred miles per hour to 0 over tenths of a second of a restrained aircraft pilot, or similarly in a MVA, with impact), while the cranium may not strike a solid object, pure deceleration injury damages axons (Bullock & Nathoo, 2005). The direction of the body/impacting object determine the plane(s) of rotation—axial, lateral, translational, or torque of brain structures such as fiber paths. It is frequently asserted that the CSF serves as a "cushion." One purpose is to keep the weight of the brain in the context of small gravity-induced movements from putting strain on pain-sensitive structures (Ransom, 2005). It seems likely that CSF is insufficient to prevent the brain from impacting the skull beyond a minor range of head impact or acceleration-deceleration.

It is frequently asserted that the *CSF* serves as a "cushion." One purpose is to keep the weight of the brain in the context of small gravity-induced movements from putting strain on pain-sensitive structures (Ransom, 2005). It seems likely that CSF is neither sufficient to prevent the brain from impacting the skull beyond a minor velocity of brain movement, nor would it be a protection in the rebound phenomenon described as cavitation.

Expansion: A bullet's passage causes a blunt trauma, creating a temporary cavity (i.e., a momentary stretch or movement of tissue away from the bullet).

Mechanical Forces and Trauma

The momentum of a physical accident directly and indirectly causes transmission of energy, momentum, and acceleration or deceleration of the brain that results in injury to the surface or interior of the brain confined inside the skull or tethered to the neck. There results hyperflexion, then hyperextension ("whiplash"). Moreover, the head may also contact hard surfaces due to body movement. The brain rotates rapidly around an axis centered in the cervical spine. Forces transmitted within the skull push, stretch, and compress the brain, creating movement that cuts tissues as they move against the meninges, exiting vessels, and surfaces of the brain. The result of application of force is affected by the fact that the brain is confined inside the skull, which is tethered to the neck.

Soft-Tissue Injury

Soft tissues refer to muscle, tendon, or ligament. Soft tissue internal injuries are difficult to localize. Their injuries reflect a range of dysfunctions, fibrositis, fibromyalgia, back or neck strain, sprain, etc. The head's soft tissues (muscle, ligaments, blood vessels, cartilage, bone, cervical disks) are stretched or otherwise forced *beyond the range of normal flexibility*. The regenerative capacity of skeletal muscle is limited after major muscle trauma or degeneration. Smooth muscle (e.g., blood vessels) is more easily replaced (Junqueira & Carneiro, 2005, p. 202).

Delayed symptoms: Neck pain may be delayed for several days (e.g., bleeding in the deep anterior and posterior cervical muscles). A delay of several months may occur for headaches, back pain, and upper extremity symptoms. While one study showed that 65% of radicular symptoms occur in the first 3 months, the remainder occurred later due to biomechanical changes within spinal joints and myofascial structures due to alterations of stability in ligaments and disks. Soft tissues are further chemically irritated by immobility, metabolite accumulation, and ischemia. Myofascial healing includes scar tissue and fibrocyte, which irritates peripheral nerve tissues and fibrosis at exiting nerve roots (Nordhoff et al., 1996a).

THE NECK: WHIPLASH INJURY

Whiplash is the most common residual problem after car accidents resulting in a large percentage of lost work and medical costs. Eighty percent of cases are asserted to resolve in 6 months. Risk factors for persistent pain include very severe initial neck pain, radicular arm pain, age above 60, and possibly female gender (Alexander, 2003).

- Upward thrust straightens the thoracic spine, generating axial compression and posterior shear forces that damage cervical facet joints.
- Stretching injury to cervical muscles and ligaments contribute to acute pain.

The neck is an extremely crowded area, vulnerable to blunt injury to the muscles and blood vessels, being pulled by the heavy, tethered head to stretching forces, and so on. Its tissues participate in PCS symptoms, affect outcome, and reduce quality of life—bone, nerves and nerve roots, muscles, blood vessels, fascia, blood vessels, cervical ganglia, cartilage and bone. Neck injuries are not noted to be life threatening but have achieved "great medicolegal notoriety" (Narayan, 1989). Evans (1997) observes that the pathology, psychological factors, prognostic studies, and persistent complaints after legal settlements all support an organic explanation for symptom persistence.

"Whiplash" (Hyperextension/Hyperflexion; Acceleration/Deceleration)

These are frequently considered exaggeration and self-limiting; there is ample evidence of associated chronic soft-tissue injury. Altered range of motion (reduced flexion and rotation, or increase of 30% of range of motion of the opposite side, accompanies soft-tissue injuries [alar ligament, which attaches the cranium to C1, and prevents excessive motion at the joints]) (Kaale et al., 2007; diagrams, Moore & Dalley, 2006, pp. 506–510). The patient may have a other comorbid trauma. Persistent posttraumatic stress reaction beyond the fear incurred from the two-car impact; TBI consequent to associated impact or shearing injuries. The somatic injury may result in headaches, fibromyalgia, inner ear dysfunction, and so on. The writer has examined several people who were totally impaired after whiplash, including a lawyer who mishandled his practice to the point he lost his license.

Acceleration and *Deceleration* forces are involved in the subdural hematomas that have been observed after roller coaster rides that were exceptionally high and fast. The ride's design created up-and-down, to-and-fro, and rotatory acceleration that produced tensile and shearing stress that apparently caused tearing of bridging veins, resulting in subdural hemorrhage (Fukutake et al., 2000).

Cervical Vasculature Dysfunctions

Fibers from the first five thoracic roots form the superior, middle, and stellate cervical ganglia. The clinician is concerned with trauma.

The ICA, which perforates the temporal to supply the forebrain (with the exception of the occipital lobe).

The bilateral vertebral arteries, which enter the brain through the foramen magnum at the anterolateral aspect of the medulla. They combine to form the unilateral basilar artery at the junction between the medulla and pons. In addition to somatic structures, they innervate the brainstem (medulla, pons, midbrain), cerebellum, and the occipital lobe.

There are symptoms of the PCS attributable to neck injury (vasculature and nervous system), as well as alterations of consciousness and balance. Separately, they perfuse the vertebrobasilar territory via fibers that arise from the *stellate ganglion*. This is formed from the lowest of the three cervical ganglia (frequent fusion of first thoracic ganglion with the inferior cervical ganglion of the bilateral sympathetic trunks) (Parent, 1996, p. 299) following the tunica adventitia of the common and ICAs, possibly innervating the rostral part of the circle of Willis (Mathew, 1995).

Shearing effects may injure the ICA as it emerges from the cavernous sinus and create diffuse brain injury as the brain moves past the arterial tree (Bandak, 1996). There may also be blunt injury to the carotid artery (Pretre et al., 1995). The combination of carotid occlusion and brain impact is more severe than carotid occlusion alone. A traumatized cerebral vasculature seems unable to respond to the reduced perfusion pressure associated with carotid artery occlusion (Cherian et al., 1996).

Stretching and flexion compression that cause vertebral artery injury create symptoms of cerebral ischemia. Flow is not reconstituted in the injured artery (Vaccaro et al., 1998). The vertebral artery is stretched in the region of the atlanto-occipital and atlanto-axial joints during head rotation. While the symptoms appear to be musculoskeletal in origin, the consequences can be vertebrobasilar ischemia, presented most commonly as a lateral medullary syndrome (Zafonte & Horn, 1999). Disruption of the trigeminal sensory vascular network and damage to the sympathetic cervical chain or the vertebral artery can be accompanied by numerous PCS symptoms (Jacome, 1986). Occult ligamentous injury to the cervical spine after trauma may contribute the pathogenesis of the vertebral artery by increasing the mobility of the neck.

Vertebral artery dissection, occlusion, or aneurysm may result from various forms of trauma to the head and neck, including excessive bending in chiropractic, yoga, calisthenics, archery, swimming, or ceiling painting (Teman et al., 1991). The vertebral artery can be damaged with minor head injury and even normal rotation. Lateral rotation of the head or hyperextension can cause vertebral artery obstruction with subsequent occult dissection of the artery with delayed symptoms and death (Auer et al., 1994).

Vertebrobasilar occlusion after minor head trauma, hyperextending or rotating neck injury, is most common in young people. The intensity of the trauma may be revealed by unilateral or bilateral facet joint dislocations (Hadani et al., 1997). The sharply turning vertebral arteries, as they emerge from the cervical atlas to the foramen magnum, are vulnerable to occlusion with head rotation. *Violent head motion* results in DAI and ruptured bridging veins resulting in subdural hematoma. Geometrically, the neck is a radius terminated by the head and tethered to the torso. Trauma is attributed to mechanical strains causing injuries to spinal muscles and ligaments and stretching of spinal cord and brainstem structures including the hypothalamus (Smith, 1989). Brainstem effects (torsion and pressure) may be a mechanical contribution to loss of consciousness (see Hohl [1974] for a review of the prognosis of "soft-tissue injuries of the neck" after automobile accidents, which were accompanied by unconsciousness in 10% of the patients). The muscle extension component is very significant in "whiplash" injury. The posttraumatic syndrome is manifested primarily as headaches of musculoskeletal or tension type. There can be damage to the anterior supporting muscles, the longus colli and the lateral and posterior elements (Fraser, 1994). Rapid acceleration or deceleration causes a stretching of the muscles, ligaments, and blood vessels of the neck. Myofascial injuries occur in the vast majority of neck injuries following whiplash, and may evolve into a myofascial pain syndrome. There is evidence of altered viscoelastic characteristics of muscle in patients with hypertonia following brain injury. Thus, velocity-dependent resistance to passive muscle lengthening in adults with brain injury is considered mechanical and unrelated to stretch induced reflex muscle contraction. This would contradict the belief that disturbance of components of gait is associated with stretch reflex hyperexcitability and inappropriate modulation of stretch reflexes. This is an example of a somatic injury that mimics neurological injury (Singer et al., 2003).

Internal Injuries

Soft-tissue internal injuries cause pain, interfere with systemic functioning, and affect outcome. We are particularly concerned with the inflammatory, immune, and hormonal systems. In addition to their traditional functions, they create changes in the interior milieu that affect the nervous system. The pain is difficult to localize (i.e., referred pain) due to imprecise signaling from internal organs. Different tissues have varying pain thresholds; radiation of pain is associated with experimental stimulation of the periosteum or tendinous attachments and with soreness and tenderness over bony prominences (Croft, 1995b).

A trigger point may radiate pain into the head or down the arm (Packard, 1999). The trauma can consist of edema, tendons and joints hemorrhage, and direct trauma to the nerve roots. The occipitocervical junction may undergo strain. Preexisting conditions can increase vulnerability (e.g., cervical spondylosis, osteophytes, degenerative joint changes, myofascial alterations from trauma, cervical stenosis, postsurgical conditions) (Nordhoff, 1996c). There may be injury to the intervertebral disk with narrowing of the foramen, and possible fibrosis and abnormal motility of the vertebral joints.

Local injury causes leakage of plasma contents from small blood vessels into the small blood vessels. This is an intermediate phase of inflammation (i.e., increased vascular permeability, with increases in tissue water, protein content, and morphology). Vascular leakage is decreased by CRH (Thomas et al., 1993). Within the context of the IS, there are two forms of cell death—*Apoptosis* and programmed cell death. Controlled disassembly on schedule occurs without adverse effect upon the organism.

Necrosis: After a severe cellular insult, the cell's contents are spilled into the surrounding environment. Their release induces inflammation. It has been suggested that the major physiological role of necrosis-induced inflammation is to induce a tissue repair response (i.e., injury). This is controlled by the innate immune system, detecting the molecular patterns of self and nonself. Immunologically, the recognition of self (which is characteristic of plants) is different than recognition of nonself found in all other recognition systems (Nasrallah, 2002). Necrotic cells express markers that induce tissue repair (i.e., cells of the innate immune system discriminate between normal cells and nonself/abnormal cells and to ignore, phagocytose, or kill them as appropriate) (Medzhitov & Janeway, 2002).

ANATOMICAL CHARACTERISTICS AFFECTING TRAUMA

Brain Anatomy

The brain's volume of tissue, blood, and CSF;
Presence of degenerative disease or previous injury (Liau et al., 1996; Miller, 1989; Ommaya & Hirsch, 1971)
Vasculature with particular exits and entrances into the cerebrum
Skull characteristics varying with age: shape; thickness; pliability
Scalp (thickness and mobility); thickness and adhesion of the dura to the skull; skin (has a protective factor in preventing skull fracture)
Ratio of brain mass (volume of tissue, blood, and CSF) to head weight
Size and shape of the tentorial hiatus (inner compartments with gaps for the brainstem formed by the dura mater)
Strength of the head–neck junction

Regional Structure of the Brain

Different regions of the brain have different cellular orientations. Therefore, there are different structural and functional tolerances. In addition, one considers the effect of brain movement when there are localized edges to the meninges, blood vessels that impede bran movement, irregular surfaces (base of skull), and blocking.

In addition to the high strain rates generated by angular acceleration, another injury mechanism is at work. It is important to consider the effect of *pressure waves passing through the brain tissue*. This orients us to consideration of experimental animal studies that deliver shocks to the brain with very little head motion (King, 2004).

Internal structures that create TBI are the ledges of the skull that support the brain (sphenoid wing), the meninges, and the penetrating blood vessels whose course is anatomically fixed and tears tissue moving past it.

Brain swelling (i.e., edema) is an increase in the volume of brain fluid. This causes the brain to be cut as it herniates through the dura, resulting in structural distortions as some portions move and others are stretched or pushed against hard surfaces.

Striking the skull has many pathological consequences, including pressure waves, skull fractures, and brain indentation. Mechanical forces damage nerve cell bodies and axons (axotomy), blood vessels, the meninges, soft tissues including muscle, tendons, ligaments, organs, as well as fracture the skull and other bones. Road accidents, for example, are claimed to transfer high kinetic energy to the nervous tissue, resulting in SDH and DAI (Kleiven, 2003). Contusions have several causes, such as brain movement that creates low-pressure area (i.e., *cavitation*, brain surface impact) (Gean, 1994, p. 163). *Intracerebral hemorrhage* is caused by cerebral contusion in the surface of the brain. In an acceleration-deceleration head injury (coup and contrecoup), the brain impacts its bony covering with the hemorrhages most likely in the basal frontal, anterior temporal, and occipital areas that frequently case multiple hemorrhages (see Figure 6.11) (Kase, 2004).

Trauma Location

Lesions directly underlying the skull where the blow was struck is called a *coup*. Damage to the opposite side is called a *contrecoup* and occurs primarily at the frontal and temporal poles (Cooper, 1987). Coup lesions are produced by deformation of the bone, with transient brain compression. The most significant lesions occur distant from the impact (i.e., when the head is free to move or is suddenly decelerated). Energy is transmitted from the skull through the brain, which may move as it receives it.

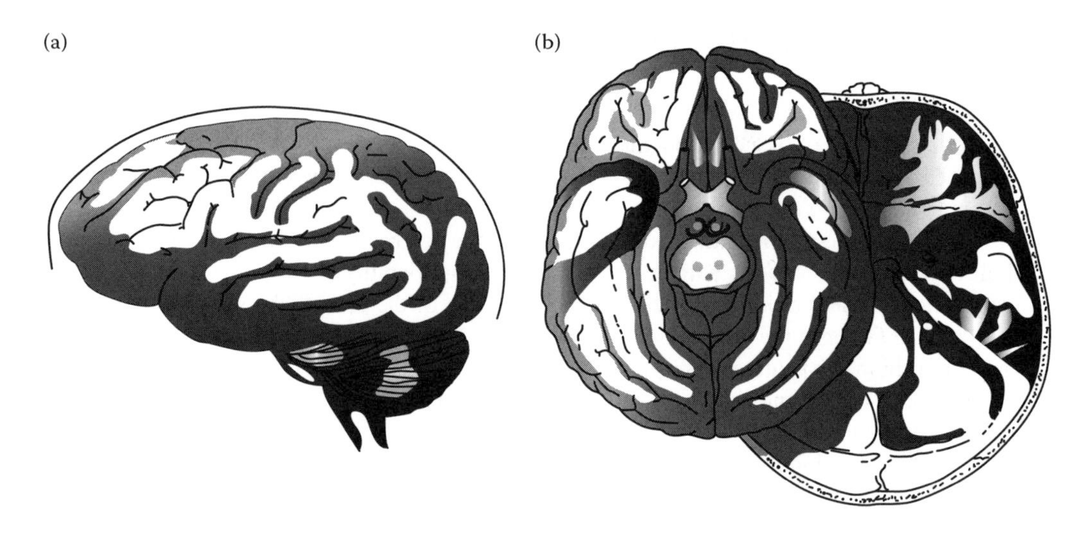

FIGURE 6.11 Cerebral contusion. A shows areas of the brain most commonly injured. B shows base of the brain and its relationship to the inner surface of the skull. Note how areas involved in this type of injury tend to correspond to bony prominences of the skull. (From Gean, A. D., *Imaging of Head Trauma,* Raven Press, New York, 1994. With permission.)

At impact, or with acceleration and deceleration, the brain is impelled forward to make violent contact with the skull. Shearing forces separate structures of the brain, cause pressure against the base of the skull, pull upon the brainstem, turn the entire brain around the brainstem, causing impaired circulation and injury or movement of the brain against sharp edges of the external covering of the brain described as leatherlike (i.e., the dura mater).

Tissue Deformation and TBI

Strain deformations damage tissues in a variety of ways (impact, pressure waves, a vacuum caused by cavitation when the brain rebounds from impact with the skull). The motion of the brain occurs in all directions: it may rotate laterally, radially along with the neck, move directly forward (translational), or its planes may move at different radii and speeds around an origin low in the skull (shear).

Torsion: This is defined as stress or deformation caused when one end of an object is twisted in one direction and the other end is held motionless or twisted in the opposite direction (see Torque). It is observed in "whiplash" injury (Pearce, 1992). The brain and neck are twisted in one direction while the spinal cord is motionless or twisted in the opposite direction. This contributes to tearing and pressure in the brainstem, crowded with nerve centers and blood vessels, and is a prime suspect in loss of consciousness.

Viscoelasticity and *brain deformation*: Brain deformation is a function of the *viscoelastic* nature of brain tissue. Time-dependent characteristics of brain tissue may indicate a significantly larger brain deformation due to the cumulative effect of consecutive multiple impacts (initial impact of the brain with the skull with a follow-up rebound effect). The tolerance level (or injury criterion level) is defined as the magnitude of loading indicated by the threshold of the injury criterion that produces a specific type of injury severity. There are large variations in tolerance levels between individuals (Ziejewski, 2004).

BIOMECHANICS

Biomechanics is concerned with how the anatomy of the skull and brain contributes to TBI. It describes the interaction of physical forces with the structure of the torso, neck, head, and brain as tethered to the spinal cord. The pattern of energy transfer between the patient and stationary of moving objects, with consideration to the interface with the person (e.g., nature of a projectile or the compartments of a vehicle), results in multiple injuries (Park et al., 2004).

Impacted environment: The trauma is influenced by the reaction of the body to the structure it hits or is impacted by. Less stiff impacting objects permit the head to decelerate over a larger distance, because of the effectiveness with which padding absorbs energy. If the head is free to move, the brain accelerates and decelerates and/or rotates in various planes leading to scraping, shearing, and impact trauma. If it is not free to move, the skull may be crushed. Head and brain injury are related to the location and extent of the skull deformation (Anderson & McLean, 2005).

Human body dynamics analysis refers to the mechanical understanding of what happens to the person in the course of an accident. Physical forces create bodily injury through motion, acceleration and deceleration, and usually some damaging contact with the environment. Movement within the body (i.e., shearing, crushing, and impact) participates in the primary injury. Reconstructing the accident can involve inquiring into such parameters as change of velocity, direction, duration of impact, body position, gender, height, weight, and vehicle interior design. The transmission of forces throughout the vehicle is influenced by the physical deformation of its structural components and their energy absorption capabilities. Body position at the moment of impact is very significant in determining the likelihood of injury. The distance between the driver, the driver's chest, and the steering wheel or instrument panel is protective, but the speed and angle of impact will determine how much the vehicle is pushed into this person's space or that of the passengers'. The seating position of the vehicle's occupant is a significant factor for injury production. The steering wheel becomes part of a hostile environment (and merely restraining one's body by grasping the wheel encourages tearing of soft tissue). Unrestrained passengers are thrown around, striking the deploying air bag or possibly hitting other passengers. For the driver or passenger in a rear end leaning forward, collision has a higher level of injury due to head displacement, seat mechanics, and deflection relative to the shoulder. The risk of sustaining a neck injury is increased by having one's neck turned at the moment of impact. Other issues include the extent of vehicular deformation, the speed of a collision, the elastic quality of the structure beyond which it is crushed, and deformation. It is implied that a vehicle designed to reduce damage, including at low speeds, produces higher dynamic loadings on their occupants than those that deform plastically. The relative weight, speed, and orientation of vehicles determine the dynamics of the vehicles and of the crash energy deformation. These will determine intrusion into the passenger space (Nordhoff & Emori, 1996).

Seat belt/head restraint/air bags: This section emphasizes findings for PCS level of injuries. In some models, the available headspace is inadequate. The value of seat belts varies at different vehicular speeds (see also below). One study indicated that the safety belt is effective as a head protector up to a certain speed but at higher speeds the head becomes one of the areas most exposed to injury. The frequency of injuries varies with the use or nonuse of seat belts, rear end, rollover crashes, or lateralization of the side of a crash if in the front seat, and so on. When the head strikes the steering wheel, combined facial and brain injury occurs in 26–32% of cases, with impact of the head against steering wheel occurring in 70% of the sample when speeds were above 40 km/h. The head of the unbelted driver strikes some part of the vehicle in 70% of cases. While air bags are described as reducing the impact force of the face and head into protruding objects within a vehicle, the inflation of an air bag is literally an explosion. The writer has examined several persons given consequential injuries by this device (McGehee, 1996). However, seat belts also have their risks. They are asserted to "increase the injurious forces delivered to the head and neck in low-speed collisions" (Croft, 1995a, p. 61). While seat belt use reduces the risk of head injury in low-to-moderate velocity collisions, it only mitigates the effect of higher-velocity crashes. Head impact risk varies with the height of the occupant and his positioning of the seat, as well as vehicle design. Increased belt length increases the risk of head injury due to head excursion. Moreover, in balance, while seat belts save lives (probably due in part to preventing ejection), there is an increased risk of whiplash and neck injuries. The seat belt causes the upper torso to move laterally and rotate. This increases the chance for rupture of the descending distal aorta with vertebral fracture, as well as an increased number of chest and abdominal injuries. The lap belt portion serves as a fulcrum for rapid-deceleration forces in a frontal or side collision, compressing the abdominal contents to the anterior portion of the vertebral bodies with ligament tears, disc injuries, or spinal fracture. Much more information is available in a review by Nordhoff and Emori (1996). A seat-belted person experiences faster acceleration of the neck when the body is seat-belted (all the momentum is concentrated on the tethered head) than he or she would experience head and brain momentum if the entire torso, neck, and head were free to move. Body proportion affects how a seat belt fits a person and how a person fits into the vehicle. Height affects how far a person's head is from the headrest and how far a person's knees are from the dashboard. Physical dimensions interact with the difference of speed or direction of impact in determining the severity of injuries (Ziejewski, 2004). With regards to head restraint, >10% disability was incurred by 10.5% of individuals with no head restraint, by 11.8% of those with adjustable head restraints, and by 6.1% with fixed head restraints. Thus, neck disability varies with the design of the head restraint (Nordhoff et al., 1996a).

It is useful to interview the patient precisely as to what to happened to his or her body. Sometimes information is elicited that is not on record (e.g., single or multiple impacts to the head, torso, limbs, and self-protective musculo-ligamentous soft tissues). Paradoxical experience of pain may suggest a lesion elsewhere than in the area of complaint. Referred pain (e.g., lateralized headaches or pain radiating to distant and apparently unrelated anatomical regions) is one example of this. If, for example, an unhealed trapezius lesion represents a trigger point radiating as a lateralized temporal headache, then an ordinarily effective headache medical approach might be ineffectual.

IMPACT, BRAIN MOTION AND TRAUMA

The direction of the body and impacting object determine the plane(s) of brain movement and rotation. The motion of the brain occurs in all directions: it may rotate laterally, radially along with the neck, move directly forward (translational), or its planes may move at different radii and speeds around an origin low in the skull (shear). Compression wave strain injures the brain's internal structure consequent to a skull impact that may not cause fracture or much contusion underneath the point of impact. A compression wave is propagated through molecules within a medium whenever a solid object is struck (e.g., head impact). The medium is cell walls, extracellular fluid, connective tissue, cell membranes and contents, or vessels and contents. Displacement parallel to the direction of motion of a pulse is called *longitudinal*. Transverse displacement at right angles to the direction of the pulse is the mechanism by which waves radiate from the point of impact. The brain's viscoelastic qualities and varied structure make it vulnerable to pressure waves that can induce shearing damage as different planes separate from each other. Volume or bulk distortion stems from application of pressure (external compression of the head and internal pressure caused by brain swelling). Since the brain is quite incompressible, when external forces are applied it can only move, rather than be distorted inwardly. External application of pressure compresses the head; internal pressure is caused by the brain's swelling. Cerebral damage can occur through stress wave concentration due to contact forces or acceleration-induced brain damage resulting in tissue-tear hemorrhages (Gennarelli & Graham, 1998). Compression-rarefaction is characterized by a change in volume without a change of internal shape. Since the brain is virtually incompressible, it has a lower tolerance to shear strains than to compression strains (Adams et al., 1982).

IMPACT

Impact creates brain movement and impact against other structures, pressure waves, skull fractures, and brain indentation. It also causes contusions due to brain movement that creates a low-pressure area (i.e., cavitation) (Gean, 1994, p. 163). Stretching and tearing of axons results in disturbed association between brain centers. Blood vessels and brain tissue may be damaged by shearing forces as inertia and internal pressure waves move brain tissue against blood vessels that are tethered in place. Petechial hemorrhage also occurs (Margoles, 1999).

Coup and contrecoup: Coup describes the place of impact and the brain lesion adjacent to the point of impact. Contrecoup refers to trauma occurring at the terminal impact of a pressure wave created by the coup impact. Coup creates pressure waves that move the brain and have the effects of contrecoup contusions, cavitation, DAI, and small hemorrhages from stretching blood vessels and contusions, as well as shearing. Contrecoup occurs in a different area of the brain, which is not necessarily precisely opposite the area of the coup injury. Since the brain is only loosely connected to the surrounding cranium, application of a load creates a compressive strain at the point of impact and sets the skull in motion along the line of force. Due to inertia, the motion of the brain lags behind the skull. When the skull comes to rest or recoils, the brain strikes the calvarium on the other side, generating another compressive strain.

Shearing Injuries

Neurons, axons, and capillaries are held together loosely. Shearing refers to the differential speed of movement of adjacent tissues (brain planes, inside of the skull, vessels) that results in scraping injuries of the brain against the inside of the cranium and the separation of brain planes (e.g., gray-white junction). Lateral rotation of the head on the neck and axial rotation of the neck and skull interact with different distances of tissues from the center of rotation and the foramen magnum, as well as such factors as the weight of different tissues. Thus, adjacent tissues (brain planes, skull, vessels) move at different speeds dependent upon their distance from the foramen magnum, lateral rotation of the head on the neck, and axial rotation of the neck and skull. The most common location for shear injury is at the junction of tissues of differing density (e.g., the denser cortex shifts in relation to the less dense white matter) (Britt & Heiserman, 2000).

Types of Shearing Injuries

Scraping injuries of the brain against the inside of the cranium.

Separation of brain planes: Gray-white junction.

Separations of structures adjacent to the skull: Olfactory fibers projecting onto the cribriform plate; the hypothalamic-pituitary stalk extending to the pituitary gland embedded the sella turcica.

Brain propelled against fixed internal structures: Blood vessels and brain tissue may be damaged by shearing forces since inertia and internal pressure waves move brain tissue against blood vessels tethered in place; the posterior corpus callosum scraping against the falx cerebri (the dura mater).

Diffuse Axonal Injury

Diffuse axonal injury occurs at sites with regional differences in compliance. DAI lesions usually occur at the gray-white matter interface when unequal rotational forces are applied during TBI, affecting areas with abrupt changes in tissue density and rigidity (Galloway et al., 2008). Levels of shear and tensile strains exceeding the brain's tolerance create various types of damage. Lesion visibility depends upon the paramagnetic qualities of varied blood products. Particularly vulnerable are interfaces between gray and white matter, brain and CSF, and brain and blood vessels (small capillary hemorrhages). This results in maximal shear forces during trauma (i.e., separation of tissues) (Garada et al., 1997). Different degrees of acceleration at different radii from the geometric center of the brain will separate these structures (Pang, 1989).

The lesions may or may not be hemorrhagic. Identification of small hemorrhages and their location offers information concerning the mechanism of injury. MRI detects injury patterns and secondary effects (edema, infarction, herniation, hemorrhage, etc.). About 50% of DAI lesions occur in the deep white matter or the corticomedullary junction (gray–white matter interface) of the frontal and temporal lobes. In addition, hemorrhagic lesions are observed in the brainstem and cerebellum. The presence of hemorrhage in DAI lesions suggests a poor prognosis. The precision of the detected lesion differs considerably according to the scanning procedure used. In adults, here are the predominant lesions of the frontal white matter, corpus callosum, brainstem, and diencephalon. Whether the same pattern occurs in children is uncertain (Tong et al., 2003).

Axonal changes are found throughout the brain, most prominently in the corpus callosum and dorsolateral quadrants of the brainstem. Shearing lesions of the brainstem is associated with prolongation of impulses through the pontine-midbrain area (Gennarelli, 1987; Levin et al., 1987). Lesions include parasagittal tissue-tear, superior medial frontoparietal white matter, corpus callosum, centrum semiovale, periventricular white and gray matter, internal capsule, basal ganglia, and brainstem (dorsal area of the midbrain and upper pons) (Gennarelli & Graham, 1998).

Rotation is the prime determinant of diffuse injuries. The further the blow is from the center of rotation, the greater its the acceleration and the greater the potential for shearing injury. The rotational trauma to the head produces a centripetal effect (i.e., progression of diffuse cortical-subcortical disconnection phenomena maximal at the periphery and enhanced at sites of structural inhomogeneity) (Ommaya & Gennarelli, 1974). Inertial loading must include a rotational component if a head injury is to produce diffuse brain injury and concussion (see diagrams of rotation and translation, p. 1878). The difference in torque between hemispheres and brainstem concentrates strain, which may be responsible for the occurrence of coma (Gennarelli, 1986). It is related to concussion and is caused only by inertial (angular or rotational acceleration) loading (20–25 msec), perhaps with impact against a soft surface, which is longer than that characteristic of subdural hemorrhage. Tissue deformation is greater at the gray–white junction than at the deeper white matter. Points near the center of rotation are likely to be damaged only if the acceleration is severe (Unterharnscheidt, 1972). Rotational injuries are more likely to cause suppression of behavioral responses to stimulation (so-called concussion). Angular rotation causes bilateral, although unequal, damage (Miller, 1989). There is a continuum of pathological effect from mild concussion to unconsciousness and severe neurotrauma that accompanies lengthening of the angular acceleration of the head (Gennarelli 1981, 1982, cited by Adams et al., 1982).

Surface shearing at the interface of the brain and skull. If there is insufficient shock absorption by the CSF, rotational gliding is hindered. Trauma occurs at rough surfaces where there is close contact between the brain and the skull and where dura mater-brain attachments impede brain motion (Cantu, 1997, 1998a). Other surface shearing forces affect confined tissues (pituitary gland), blood vessels entering and exiting the cranial cavity, and the rough edge of skull facing the lower surface of the cerebrum.

The smooth internal surface of the skull only rarely results in occipital and cerebellar concussions (Gean, 1994, p. 152). The moving brain is pushed against the enveloping surfaces and edges of the dura mater (i.e., the tentorium that separates the cerebral hemispheres from each other, as well as from the cerebellum). The parahippocampal gyri are vulnerable (Gennarelli & Graham, 1998). The corpus callosum (integrating the two cerebral hemispheres) is vulnerable to impact injuries due to the proximity of the sharp edges bilaterally of the dura mater extending down the internal surface of the cerebral hemispheres bilaterally (*falx cerebri*) (see diagrams in Kretschmann & Weinrich, 1992, pp. 26–43). Surface shearing forces occur at the rough anterior and middle fossa floors (basal frontal and temporal lobes), the crista galli (frontal poles), sphenoid ridge (temporal poles), and tentorial incisura (Gean, 1994, p. 152; Pang, 1989). Brain movement occurs against the knifelike lesser wing of the sphenoid bone (Gurdjian et al., 1968). The interposition of the lesser wing of the sphenoid bone with the frontal and temporal lobes explains various traumatic findings:

1. Lesions of the posterior orbitofrontal region immediately superior to the lesser wing of the sphenoid and of the anteroinferior temporal lobe adjacent to the greater wing of the sphenoid (MRI illustrations of Gean, 1994, p. 151).
2. The occurrence of memory disorder independent of the extent of concussive unconsciousness (Ommaya, 1996). Brainstem movement through the *foramen magnum* permits the

brain and fluids to be forced out and back, causing herniation contusion between the *cerebellar tonsils* and the foramen magnum (see LOC in Chapter 4; Gennarelli & Graham, 1998; diagrams in Kretschmann & Weinrich, 1992, pp. 92, 93; Parent, 1996, pp. 52, 53) (see Herniation).

Shearing of extensions into skull crevices:

1. Olfactory nerve fibers extending through the cribriform plate; other cranial nerves penetrate the skull.
2. Blood vessels penetrating from the brain through the dura to the skull are fixed at one end, and therefore subject to being torn by movement imparted to the brain.
3. The pituitary gland, within the sella turcica of the base of the skull, attached by the infundibulum to the hypothalamus, is subject to infundibular damage or having the stalk severed or injured.

Internal shearing between internal structures and planes may cause powerful tensile stresses. Rotation of the head is believed to cause a "swirling" of the brain that then creates shearing or tensile strains resulting in widespread damage to axons (tearing, twisting, and stretching of nerve fibers and vessels) (Adams et al., 1982; Bailey & Gudeman, 1989; Gennerelli, 1987). *Severe intellectual deficits* are attributable to multiple diffuse shearing lesions deep within the brain that are set up by rotational mechanisms (Pang, 1989). Deep rupture of a vein due to shearing creates a *gliding contusion* (Adams et al., 1986). Ischemic damage to the basal ganglia occurs from shearing effects upon the perforating branches of the middle cerebral artery (see Children's Brain Damage, Chapter 6).

Metabolic consequences of shearing injury: Immediately following impact, massive ion fluxes across the neuronal membranes cause widespread loss of resting potential and release of neurotransmitters into the extracerebral space. Within a few minutes, an intensely energy-dependent process occurs, resulting in an abrupt increase in glucose utilization. The attempt to restore ionic homeostasis involves reuptake of neurotransmitters and ion pumping. The increase of glucose metabolism is brief and maximally localized in those parts of the brain that are maximally deformed by the shearing force. Should there be contusions, hematoma, or focal infarction, glucose use increases for 5–7 days, but later there is decreased metabolism of oxygen and glucose for 1–4 weeks after impact. Glucose use is depressed in the event of coma (Bullock & Nathoo, 2005).

Brainstem Movement and Compression: Concussive Loss of Consciousness

Concussion has been attributed to stresses from pressure gradients and relative movements in the brainstem area. There is evidence that impact causes brain movement into the foramen magnum (Kuijpers et al., 1995). The brainstem can move as much as several centimeters, which is responsible for the *loss of consciousness. Vertex blunt impact* results in *brainstem movements* at the *craniosacral junction.* It has been demonstrated with rhesus monkey preparations that high levels of acceleration (up to 600 G) following occipital impact cause pressure gradients to the craniospinal junction. Pressure gradients occur either superioinferiorly, toward the spinal canal, or to a lesser degree extend inferosuperiorly, from the spinal canal toward the cranial cavity. In addition, relative movement of the *midbrain* and pressure waves in cases of closed-head injury are associated with concussion. These pressures are concentrated in the vicinity of the brainstem and craniospinal junction leading to concussion (Gurdjian et al., 1968, 1975, pp. 175, 179).

Cavitation

Cavitation accounts for contrecoup lesions since the brain withstands positive pressure (compression) much better than negative pressure or cavitation (Pang, 1989; Thibault & Gennarelli, 1985).

(It should be noted that Povlishock [1989] doubts that negative pressure waves can create contusions). Both whiplash and impact contribute to cavitation brain injury (Kuijpers et al., 1995; Nusholtz et al., 1996). (Note that doubt about this mechanism is expressed by Ommaya et al., 1971). During impact, the opposite side of the skull is moving faster than the brain and the brain lags behind. The contents of the skull and brain are compressed at the impact side before a wave of tissue moves to the opposite pole (*contrecoup*) (Lockman, 1989; Pang, 1989; Unterharnscheidt, 1972). Gurdjian and Gurdjian (1975) describe the phenomenon as follows: "The brain is crowded at the impact site with high pressure development. [It] lags behind the faster moving skull ... with development of underpressures ... There may be tears in the interior of the brain in the direction of the force (and tears) of connecting veins over the convexity." The high acceleration of the brain causes it to separate momentarily from the dura, or causes the dura to separate from the skull. The area under the contrecoup is damaged more than the area under the load (Thibault & Gennarelli, 1985). The interval during which impact, pressure waves, and brain inertia contributes to the formation of a cavity and violent collapse appears to be up to 5 msec (Ruan & Prasad, 1995).

As the brain rebounds from contrecoup, uneven strength and density cause brain tissue to separate temporarily, creating contusions and intracerebral hematomas. Starting from the pathological observation that focal contusions may be quite small at the surface, involving only the cortex, or may extend into a large interior cavity, Ommaya (1990) assumes that the cavitation effect is most probably caused by bubbles resulting from the expansion of gases under low pressure in the blood of CSF, rather than primarily in the brain.

Pressure Waves

Pressure waves move the brain and have the effects of contrecoup contusions, cavitation, DAI, and small hemorrhages from stretching blood vessels and contusions, as well as and shearing. Shearing refers to differential speed of movement of adjacent tissues (brain planes, skull, vessels) resulting in scraping injuries of the brain against the inside of the cranium, as well as separation of brain planes (e.g., gray-white junction). Rotation causes different parts of the brain to move at different speeds, dependent upon their distance from the foramen magnum—lateral rotation of the head on the neck and axial rotation of the neck and skull. The brainstem is tethered through the foramen magnum to the neck and spinal cord. A cumulative contusion diagram confirms that there is particular vulnerability to movement of the frontal and temporal tips into the anterior and middle fossae (Gean, 1994, p. 152). The inferior surfaces of the brain and the vertex, including the motor and somatosensory strips, are also vulnerable to contusion. Parts of the brain may be compressed as the brainstem rotates while the spinal cord is tethered outside the foramen magnum. Rotational injuries (neurological and somatic), are referred to as "whiplash".

The way the head moves determines which structures are important determinants of the structures that will suffer from acceleration. An unconstrained head undergoes numerous and unpredictable acceleration vectors (Gennerelli et al., 1982). The direction of the impacting object, in this writer's experience, can determine whether there is loss of consciousness. Impact in the axial plane (i.e., the spinal column, falling on one's feet or buttocks, being in an elevator that crashes, or a falling object striking the head), sometimes does not result in LOC, although a blow of equivalent force in a different plane or direction would be expected to do so. In a rear-end collision, the momentum of both cars is in the same direction. A frequent outcome is the head continuing in the same direction as the crashing car (whether or not the lead car is in motion), but force applied to the feet and knees by the floorboard and seat. The lower portion of the body moves forward, leaving the head trailing behind because more momentum is imparted to the lower body (hyperextension) followed by tensile forces bringing the head forward (hyperflexion). The entire pattern is the injurious "whiplash" that may result in documentable brain injury as well as soft tissue (neck, back, shoulder) and spinal column damage.

Tissue effects of acceleration and deceleration: Extremely short-lived acceleration, unless severe, is damped by the brain's structure and does not produce injury. Short acceleration duration shifts the focus of injury to the surface of the brain. Gradually longer durations cause strains to propagate deeply into the brain structural strength, resulting in concussion, DAI, and prolonged traumatic coma (Teasdale & Mathew, 1996). The amount of angular displacement and its acceleration and deceleration over time will depend upon the impacting force's effect on the head or vehicle, the rate of acceleration and deceleration, and the angle through which the head, neck, and torso moves (e.g., depending upon the absence or presence of a seat belt, and its configuration, etc.).

There is a continuum of pathological effects that accompanies the lengthening of the angular acceleration of the head according to studies on subhuman primates conducted by Gennarelli (1981, 1982, cited by Adams et al., 1982): Mild concussion, temporal and frontal contusions, and occasional intracerebral hematomas; torn bridging veins, resulting in the formation of rapidly fatal subdural hematoma; prolonged unconsciousness and DAI (i.e., hemorrhagic lesions of the corpus callosum and rostral brainstem), and axonal retraction balls in such white matter as the rostral brainstem and the parasagittal regions.

Types of strain and tissue effects: The brain and body may be torn by a single deformation of high pressure and rapid motion that exceeds the elastic limit. One considers the size, mass, and velocity at which contact occurs. Concentration of force (stress) affects the lesion—tensile (pulled apart), compressive (pushed together), and shear (adjacent parts are moved in different directions). Strain is consequent to force (i.e., the relationship of the deformed state of the body to its undeformed state). Brain components are differently deformable to the point of trauma where they permanently tear, crush, or separate from the surrounding tissue. *Compressive nerve injuries* most commonly occur after the nerves cross over bony protuberances at an exposed position (e.g., the ulnar nerve at the elbow and common peroneal nerve at the fibular head). After 4 years, motor recovery is unlikely because of extensive degeneration of muscle fibers (Murray, 2004).

ROSTRAL-CAUDAL BRAIN MOTION

Some cases cause up-and-down movement of the brainstem and posterior fossa in the direction of the foramen magnum. Depending on the amount and direction of the blow, impact causes mass motions with increased intracranial pressure, and pressure gradients and mass motions toward the foramen magnum. The size of the lesion (Pang, 1989) will depend on the degree of violence (with levels of pressure extending into the skull) and the shape of the gyrus (a flat gyrus with a large surface within the pressure region will be damaged more than a pointed gyrus exposing only its summit). In blunt head injury, damage occurs from the cortex inward toward the brainstem (Gurdjian & Gurdjian, 1975). In addition, with elastic deformation of the skull, the brain is depressed at the impact site, and bent outward at the other surface of the skull.

THE PRIMARY PHASE

The primary damage is the immediate anatomical lesion caused by mechanical forces. CNS injuries lead to an irreversible loss of loss of function due to lack of neurogenesis, poor regeneration, and the spread of degeneration (Schwartz, 2000). Also occurring are *loss of cerebral autoregulation, connection disruption* (tissue disruption interferes with nearby long axonal pathways); *diaschisis* or interference with the functioning of centers with which the damaged neural centers are integrated (e.g., cerebellum to contralateral thalamus); *cellular and molecular* biochemical responses that cause neurotoxic effects upon neural membranes, and cellular messenger systems causing necrosis.

Brain Trauma

In concussion, characteristic brain injury includes DAI, as well as cortical lacerations and contusions. Fifteen percent of temporal bone fractures are produced by blows to the occiput. A transverse fracture of the temporal bone may also cause a lesion to the brainstem (Heid et al., 2004). In one series, 5% of traffic accidents caused petrous bone (temporal) fractures. When a range of TBI injury is considered, diffuse neuronal and vascular damage occurs primarily in the area of corps callosum, midbrain-pons tegmentum, inferior colliculus, and the dorsolateral quadrant at the superior cerebellar peduncle (Zhang et al., 2004). *Explosive blast injury* creates some different and additional lesional disorders (Taber et al., 2006).

Elevated brain stresses, strains, intracranial pressures, and excessive motions between the brain and skull may cause contusions, concussions, DAI, and acute subdural hematomas. The fact that there are measured differences between children (peripheral white matter) and adults (deep gray and white matter) after TBI suggests that there are maturational, biomechanical, physiological, and metabolic differences in the response to injury (Galloway et al., 2008). It has been suggested that a TBI is caused by tissue strains above 10% applied at rates greater than 10 sec^{-1}. The brains of very young children and older adults brains are more vulnerable to vascular damage (*hemorrhage*) created by shearing forces (Bullock & Nathoo, 2005; Reilly & Bullock, 2005). Sheer stresses are a marker for subarachnoid hemorrhage and DAI, while around the brainstem region could be an injury predictor for concussion. Strains in the hippocampal region above a given rate create a larger number of cell deaths (Levchakov et al., 2006). In addition to neurological lesions, there occurs gross anatomical damage (e.g., torn blood vessels with hemorrhage and skull fracture). Using unembalmed fixed in place skulls, there was a failure of skull integrity at the range of 22–24 *Joules* (energy) with intermediate impact velocity creating single linear fracture and with only 33% higher velocity creating multiple fractures (Delye, et al., 2007). Varying levels of hemorrhage can occur even within the range of lesser TBI, including petechial hemorrhages (Margoles, 1999).

CONTUSIONS

Contusions are the most frequently encountered lesions after head injury (Cooper, 2000, diagram p. 12, p. 324). They usually occur at the apex of gyri and appear as multiple punctate hemorrhages, or streaks of hemorrhage progressing to bleeding into adjacent white matter. They are caused when direct impact to the skull deforms the bone, compressing the underlying parenchyma. Contusions with associated skull fracture are usually more significant. This may damage arterial or venous vessels. The severity of the contusion is proportional to the amount of energy transmitted through the bone to the brain, and inversely proportional to the size of the area of contact. Contusions are associated with other lesions, internal and external to the cerebral parenchyma, and are multiple in more than one-half of patients (see Figure 6.10). Larger contusions and those involving multiple areas often act as a mass lesion, threatening dangerous elevations of ICP. Nonsurgical treatment of larger contusions and mass effects, even with normal ICP and a relatively stable neurological condition, has the risk of sudden herniation 7–9 days after injury. Within the contusion blood leaks through the damaged Blood Brain Barrier (BBB). The combination of biochemical and physical changes leads to a mixed cytotoxic and vasogenic edema. Brain swelling adjacent to contusions is common. It is due to physical disruption of the tissue with damage to the BBB and loss of normal physiologic regulation of arterioles. The edema is considered vasogenic since water and electrolytes leak into the brain tissue and spread into the adjacent white matter (Graham & Gennarelli, 2000). There is evidence that the frequency of hematomas, contusions and contrecoup injury is less children than adults (Weiner & Weinberg, 2000).

Types of Contusions

Fracture: At the site of a fracture
Coup: At the site of an injury in the absence of a fracture
Contrecoup: Away from the point of injury

Herniation: An example of a mass effect, where the medial parts of the temporal lobes are impacted against the edge of the tentorium (meninges) or the cerebellar tonsils are impacted against the foramen magnum

Surface: Involves superficial gray matter

Gliding: These occur in the parasagittal regions and are attributed to rostrocaudal movements of the brain after impact (Blumbergs, 2005). They are hemorrhages in the cortex and subjacent white matter of the superior margins of the cerebral hemispheres.

Site of Contusions

The contusion pattern depends upon the direction and magnitude of the impact, whether the skull is fractured or penetrated, and whether the head is or is not accelerated by the impact, or is accelerated at the moment of impact or is accelerated and decelerated (Graham & Gennarelli, 2000).

- *Lateral fronto-temporal*: contralateral temporal lobe surface contusions and contusions of both uncinate gyri
- *Lateral temporoparietal*: contrecoup contusion of the temporal lobe
- *Midline occipital*: bilateral frontal and temporal lobe contusions
- *Lateral occipital*: contralateral contrecoup contusions of the frontal and temporal lobes
- *Frontal*: bilateral or unilateral contusions of the frontal and temporal lobes
- *Vertex*: contusions of the brainstem and tears of the corpus callosum and pituitary stalk
- *Surface contusions*: most severe at the frontal and temporal lobes irrespective of the impact site if the forces are sufficiently strong to impart movement of the brain over the irregular bony surfaces of the anterior and middle cranial fossae. This type of injury has been described to be accompanied by 40–80% of linear skull fractures.

Since both frontal and occipital impacts resulted in contusions that were most severe in the frontal lobes, it is inferred that the site of impact cannot be extrapolated as being diametrically opposite the area of the most severe contusion. After impact, the brain may be contused as it is propelled over the rough edges of the sphenoid wing or the floor of the frontal fossa. Veins at the site of contusions on the cortical surface may rupture into the subdural space. This has been observed in two-thirds of SDH surgical patients.

Contusion is the most frequently encountered head injury lesion of older persons (89% postmortem). Under the age of 1 year it occurs in less than one-half of cases. The usual hemorrhagic contusions are replaced by nonhemorrhagic tears of the white matter. A site of contusion can be the medial surfaces of the cerebral hemispheres as the brain is accelerated against the falx cerebri. Frontal and temporal lobes contusions are common. They are found only occasionally in the parietal lobe and least frequently in the occipital lobe. With regards to posterior contusions, SDH of the posterior fossa is rare. Cerebellar contusions usually occur directly underneath an occipital trauma, with the inferior surface of the cerebellum supposedly the most vulnerable to injury. A coexisting posterior fossa SDH is sometimes present. The occipital blow causing the brain to be propelled forward in the calvarium commonly results in contrecoup contusions and hematomas of the frontal and temporal poles (Blumsberg, 2005; Cooper, 2000).

Coup and Contrecoup

Clinical Example: Contrecoup impact; no LOC; positive SPECT; palpable head bruise; significant somatic injuries:

This study illustrates the impairing effect of an impact accident with comorbid brain and somatic injuries. It would meet the common but erroneous criterion of a "minor" TBI. A middle-aged person was struck by a metal object poorly fixed in place that was blown down by the wind.

There was nothing in the ambulance report or hospital "triage" that described the precise site of impact. The Glasgow Coma Score (GCS), a measure of altered consciousness, was 15 (the highest level). A nurse stated that the blow was in the "upper right" area of her head, presumably occipitoparietal. Palpation of her head by this examiner easily detected a bruise at the place of impact. According to the patient, no prior examiner (numerous medical and neuropsychological practitioners, in the acute and chronic phases) had examined her head despite her complaints of continued pain.

From Record Review: Chronic Posttraumatic Medical Symptoms

- Chronic pain: Cervical (muscle spasm altering normal lordosis), spine (sufficient to interfere with MRI examination), shoulder radiations to occipital region and neck, headaches, eyes
- Headaches
- Vomiting and nausea reflects systemic disorder
- Neurological: 2 × 3 mm cervical subligamentous disc herniation at the C4–5 level. Dizziness reflects apparent blow to the petrous area of the temporal bone, injuring the middle ear (audiological and balance studies are recommended—complaint of tinnitus); paresthesias; weakness
- Muscular weakness, difficulties of flexion and reaching

Spect Examination: A large area of decreased perfusion in the frontoparietal region of the left hemisphere (contrecoup) and a small area of decreased perfusion in the occipital region of the right hemisphere (coup)

Current Status: Unemployed. The patient was an honors college graduate and had been the self-employed proprietor of a service organization.

Major Findings of the Neuropsychological Examination: Almost perfect scores on two symptom validity tests (a previous examiner obtained suspicious results). Deficits of strength; sleep disorder; impairing pain; physiological distress (sleep; frequent urination; reduced mental speed; reduced capacity to avoid errors; significant loss of mental ability (particularly verbal) documented in both structured and unstructured situations; significant loss of both short term and long term memory; mood is depressed, angry, and anxious; reduced self-awareness and insight; pain; weakness; dissociative reactions.

When contusions occur at the site of impact (*coup*), the brain is transiently depressed by deformation and inbending of bone at the point of impact. They are produced by compressive forces caused by an area of skull inbending or tensile forces caused by negative pressure produced when an area of inbent skull suddenly snaps back into place. Contusions at the site of impact are usually minor. Those distant from the point of contact (*contrecoup*) occur when the stationary head is free to move at the time of impact, or the moving head is suddenly decelerated.

The most significant lesions are at a distance from the impact site (see Contrecoup) (Cooper, 2000). *Intermediate coup contusions* are intracerebral lesions occur deeply within the neural parenchyma between the impact site and the opposite side of the brain. *Contrecoup contusions* occur opposite the impact side.

Severe injury or a site confluent with other contused areas causes blood to displace the parenchyma, forming an intracerebral hematoma. Replacement of an entire lobe by blood is called a *burst lobe* (Britt & Heiserman, 2000). One may differentiate between coup, contrecoup, and herniation contusions (the medial parts of the temporal lobe impact against the edge of the tentorium; cerebellar tonsil impacts against the foramen magnum). Surface contusions are differentiated from gliding contusions (i.e., focal hemorrhages in the cortex and subjacent white matter of the superior margins of the cerebral hemispheres [common in DAI]). The pia-arachnoid is intact over contusions but torn in lacerations. Their hemorrhagic, swollen state evolves into shrunken brown scars.

Subdural Hemorrhage (SDH)/Bridging Veins

SDH is the most common traumatic mass lesion (Valadka, 2004). It originates in the space between the dura and arachnoidal meningeal layers and results in bleeding into the subdural space. It results from tearing of the bridging veins that drain blood from the surface of the hemisphere into the dural venous sinuses (Evans, 2001). SDH also arises from direct laceration of cortical arteries and veins from penetrating injuries that lacerate the brain and closed injuries causing large contusions of the temporal and frontal lobes. In one clinical group, 72% of acute subdural hemorrhage (ASDH) cases were due to fall or assault, with only 24% from motor vehicle accidents (MVA). Persons with SDH are less likely to have been in MVA than those without it (D. P. Becker cited by Thibault & Gennarelli, 1985). SDH is caused by acceleration induced by the impact, not the head contact per se. Thus, boxers hit on the chin or football players with heads set in motion by violent blows to the torso can receive ASDH. If the acceleration is reduced by padded or padded dashboards, deformable steering wheels, energy-absorbing assemblies, laminated windshields, there is a lower rate of deceleration, but increased potential for diffuse brain injury.

Late occurring SDH presents weeks or months after what appears to be a trivial head injury. It is particularly common in older individuals who have undergone some cerebral atrophy, infants with nonaccidental injury, and patients with shunts for hydrocephalus. The expansion is slow, with a long period of spatial compensation, so that there may be considerable distortion of the brain before it creates a significant rise in ICP. Rebleeding can occur in the chronic phase.

After impact, the brain is accelerated within the calvarium due to sagittal movement of the head. Even though frontal and temporal lobe contusions may not initially be associated with SDH, when intracranial pressure is released they can bleed at that time (Gudeman et al., 1989). Tearing of midline bridging veins in severe brain injury is associated with mass lesions requiring operation. Hemorrhage is often caused by tearing of the cortical veins, particularly where they enter the fixed portions of the dural sinuses. The three factors causing rupture of bridging veins (i.e., production of ASDH) are magnitude and rate of acceleration (strain rate), and the duration of acceleration. In an experimental procedure utilizing angular acceleration but not impact, increasing acceleration magnitude increased injury. Lesser levels were subconcussive. With increasing acceleration, mild or more severe concussion was produced. With more acceleration ASDH was produced. ASDH could be avoided by keeping the acceleration constant while increasing the pulse duration. Sagittal movement of the head imparted the acceleration causing rupture of parasagittal bridging veins. These have *viscoelastic* characteristics. As pulse duration increases, larger values of acceleration are required to cause failure of the bridging veins since the strain rate decreases. It was concluded that ASDH is associated with ruptured bridging veins because of rapid rate of acceleration onset (high strain rate) (Thibault & Gennarelli, 1982). They are usually located over the hemispheres, but can also be found between the occipital lobe and the tentorium cerebelli, as well as between the temporal lobe and the base of the skull.

About 72% of cases with acute SDH were produced by a fall or assault and only 24% were the result of MVA. SDH is the most common nonaccidental cranial injury of infants, often associated with retinal hemorrhages and skeletal injury. *Chronic SDH* can occur weeks or months after what appears to be a trivial head injury. It is particularly common in older individuals who have undergone some cerebral atrophy, infants with nonaccidental injury, and patients with shunts for hydrocephalus. The expansion is slow, with a long period of spatial compensation, so that there may be considerable distortion of the brain before it creates a significant rise in ICP.

Mechanism of SDH: Subdural veins are *strain-rate sensitive* (i.e., the rate at which they are deformed by acceleration causes rupture rather than head contact itself). This accounts for the fact that acute SDH is more common in falls, where the deceleration period is typically short, than in patients involved in MVA, where the rate of strain on the bridging veins is lower (Cooper, 2000; Graham & Gennarelli, 2000). Tearing caused by brain movement relative to the fixed dura mater, attached at one surface to the skull, and the other to the outer covering of the brain, results in SDH

(Bakay & Glasauer, 1980, p. 200). Rotational damage in particular tears midline-bridging veins creating SDH. Parasagittal bridging veins draining the surface of the hemisphere into the dural venous sinuses are stretched. If they are ruptured, venous blood escapes into the subdural space (Cooper, 2000). Tearing can occur with rapid cranial deceleration even with a relatively low magnitude of shearing force, such as a fall from a standing height to a solid surface or boxing injuries creating rotational acceleration of the head (Bullock & Nathoo, 2005).

The parasagittal bridging veins are ruptured when the cranium is rapidly decelerated with a relatively low magnitude of shear force (e.g., boxing injuries, a fall from a standing height to a solid surface). Avulsion of parasagittal and Sylvian bridging veins is usually accompanied by DAI and polar contusion (Bullock & Nathoo, 2005). Acute SDH is more common in falls, where deceleration period is typically shorter than in patients involved in MVA, where the rate of strain on the bridging veins is lower (Cooper, 2000; Graham & Gennarelli, 1998). After impact, the brain is accelerated within the calvarium due to sagittal movement of the head. Parasagittal bridging veins draining the surface of the hemisphere into the dural venous sinuses are stretched. If they are stretched so much that they are ruptured, venous blood escapes into the subdural space (Cooper, 2000). Tearing is caused by brain movement relative to the fixed dura mater. The latter is attached at one surface to the skull, and the other represents an outer covering of the brain. The result is SDH (Bakay & Glasauer, 1980, p. 200). Rotational damage in particular tears midline-bridging veins, leading to SDH). Even though frontal and temporal lobe contusions may not initially be associated with SDH, when intracranial pressure is released, they can bleed at that time (Gudeman et al., 1989). Hemorrhage may be caused by the tearing of midline bridging veins in severe brain injury. These are associated with mass lesions requiring operation: tearing of the cortical veins, particularly where they enter the fixed portions of the dural sinuses.

RECONSTRUCTING THE ACCIDENT

What the person saw or experienced is the key to reconstructing the accident—what happened to the person's body; mechanical forces, including motion of the whole body, head, torso, and limb impacts, as well as whiplash; and the nature of the impacting surfaces. Avoid factual statements that might be controverted in a law case by a witness. Examples might be precise descriptions of the speed or direction of another vehicle.

A key issue in many claims of TBI is the credibility of the accident: were the forces (e.g., of vehicular impact) sufficient to create injury? The physical principles explained above, as well as understanding some qualities of tissue and their reaction to physical forces and movement, all contribute to assessing the credibility of an accident claim. Awareness of the physical dimensions of accidents enhances the ability to gain clarification of what actually happened.

- Magnitude, duration, and speed of the force
- The stiffness of the struck object
- Direction of the force relative to the head and its point of contact
- Center of rotation of the brain and the head
- The velocity with which the body or head struck an object
- Relative mobility of the head, which swings variably after a blow or acceleration or deceleration of the body, as well as brain after impact
- Velocity and directional components of the striking object and of the brain after trauma
- Is the skull is penetrated (missile or other weapon)?
- Are the neck and skull accelerated then decelerated (whiplash or blunt trauma)?
- The distance through which a striking object falls (permitting statement concerning speed at the time of impact).

The reasonableness of a claim of personal injury is made more precise by considering the pattern of energy transfer between the patient and stationary or moving objects, and the interface with the person hardness of the struck surface, speed of a projectile, the enclosures of the surfaces of a vehicle, the presence of nearby occupants, whether the person is ejected from the vehicle or struck down, and how many further impacts occurred.

The interview is useful in reconstructing the details of the accident. Inquiry often reveals that when a car was struck from the rear, the head moved not only forward, where it was followed by reflexive elastic flexion, but also laterally. Additionally, there may be one or more significant impacts with the side windows, window frame, doorframe, head rest, front window, head of another passenger, and so on. Head impacts are often not stated in the emergency medical records. Evidence may be revealed by palpation of the scalp indicating tender spots. *How did the person become injured?* Possibly by impact to head; movement of neck and head; number of impacts; speed of body, vehicle, and any other impacting surfaces; or nature of impacting surface(s) (i.e., soft, hard). TBI has a variety of etiologies that may occur simultaneously (Parker, 2001, Chapter 5; Parker, 2005b). What were the characteristics of the *impacting surface* (hardness [firm/yielding]; shape [sharp/dull] affect; the deformation of the skull and enclosed brain; penetration, range or impact) that narrow or disperse the surface area when energy is applied? The speed of the body and/or impacting surface affects the rate of deceleration or acceleration of the skull, creating brain shearing, tensile strain, as well as the cause of injury (e.g., the number of head impacts, the hardness of the head-impacting surface, distance of falling objects, speed of vehicles, etc.).

Was there a serious impact by a hard object or surface, head motion (hyperflexion, hypoflexion, and rotation in various planes, including so-called “whiplash”)? The pattern of brain injuries will vary considerably, from DAI to contusions, mass effects, and hemorrhage (Reilly & Bullock, 2005). Also, was there evidence of injury beyond that to the head? Polytrauma is defined as multiple injuries in various sections of the body that are comorbid with TBI (Moore et al. 2004).

7 Acute Brain and Somatic Injury

OVERVIEW

Trauma has numerous dimensions, including tissue damage and loss of physiological autoregulation. Severe closed head injury produces a range of cerebral lesions, such as diffuse axonal injury; vascular lesions (including subdural hematoma); contusion; and neuronal degeneration in selectively vulnerable regions. Using animal experiments as models, the method of producing injury may be classified as *percussion concussion* (fluid percussion concussion and rigid indentation) or *acceleration concussion* (inertial injury and impact). Central *fluid percussion* (FP) injury tends to result in variable small injury in the region of the impact and scattered axonal damage mostly limited to the brainstem. Lateral and *parasagittal fluid percussion* (PFP) is characterized by lateral cortical contusion removed from the impact site. The PFP model produces widespread axonal perturbation in the forebrain, and tissue tears in white matter. *Traumatic axonal injury* (TBI) is characterized by rapid elongation of a nerve (e.g. the optic nerve) in situ. Cerebral contusion is the most common brain injury, and may be produced by both acceleration and percussion (Dietrich & Bramlett, 2004).

Comorbid somatic injuries accompanying TBI are established as increasing the rate of mortality in a variety of TBI populations (Baguley et al., 2000). Higher than expected *mortality rates* (i.e., reduced life expectancy [LE]) have been established in a variety of populations after post-traumatic intervals up to 15 years. The predicted mortality rate is 1.5%, with various studies of TBI populations, resembling the 5.7% of the report as a characteristic rate of 5 years postinjury. Lower-functioning individuals had a 100% mortality after 10–15 years, probably associated with immobility (e.g., *spinal cord injury* [SCI] and cerebral palsy). Patients with a SCI are diagnosed with TBI at a rate between 24% and 50%. TBI patients have a reported SCI between 1.2% and 6% (Elovic & Kirschblum, 1999).

Reduced LE is found in both high- and low-functioning TBI survivors, older as well as in the pediatric adolescent age range. Mortality rate is higher in males than in females. Higher mortality is associated with a previous psychiatric condition, but not alcohol or substance abuse. Risk-creating medical conditions were cardiorespiratory stress myocardial infarction, bronchopneumonia, and, for severe TBI, the presence of the APOE 4 allele (Baguley et al., 2000, see Mortality). Trauma is a multisystem reaction (Zellweger et al., 2001) with interaction of the immune, inflammatory, endocrine, stress, and circadian systems with the central nervous system. This outline will help the clinician assess the status of a patient in the acute or chronic phase to determine need for further assessment, treatment planning, and outcome. Trauma has numerous dimensions, including tissue damage and loss of physiological autoregulation. There are both local and systemic reactions (hormonal, inflammatory, immune) with later central neurobehavioral effects. Release of neuroactive and hormonal substances into the internal milieu, with reduced effect of the blood-brain and other barriers, as well as active signaling systems, creates a new environment in which the brain and physiological systems are in mutual exchange with neurobehavioral and physiological effects. I believe that parallel processing and extended circuits should be assumed at all times, together with diffuse axonal injury, which accounts for unlocalized dysfunctions. Thus, undetected injury might prevent the development of skills that have not yet been expressed, reflecting the frequent dysfunctioning without radiological evidence. The immediate physical injury (the primary phase) may be diffuse and microscopic, so symptoms or lesions are not detected by a neurological examination or radiological scans. After the primary TBI, the injury can evolve into further brain injury (the secondary phase), late-developing physiological disorders (the tertiary phase), later-developing neurological disorders (the quaternary phase), and chronic stress disorder (the pentary phase). Mild injury is usually manifested in axonal damage within brain parenchyma, while moderate-to-severe injuries result in diffuse injury with petechial hemorrhage and tearing of small vessels. Impact is not needed to create TBI, with mild acceleration/deceleration (A/D) forces injuring axons and dendrites in the presence of noninjured neural tissue (Gaetz, 2004).

An initial CNS reaction (prefrontal; limbic; brain stem) with a psychological component has the effector outcomes of (1) the hypothalamic-pituitary target endocrine axes; (2) the sympathoadrenomedullary (SAM) axes; alteration of stabilizing mechanisms (homeostatic and allostatic), which when chronic evolve into an allostatic overload characterized by loss of stamina and various diseases; and (3) tissue damage representing a breach of organismic integrity, as well as stimulating the innate immune system. Alertness to the systematic influences upon behavior after trauma, in addition to the neurological disorders, enhances the examiner's ability to perform a more comprehensive study of the injured person by considering a wide range of possible neurobehavioral dysfunctions and etiology. This also assists in determining the credibility of an injury or claim of psychological damage.

Posttraumatic disturbance of the internal milieu, with its effects upon the CNS, as well as the lesser efficiency of the blood-brain and other barriers in excluding neuroactive chemicals, creates multiple chemical products and consequent neurological activity that influences arousal, mood, synaptic characteristics, and bidirectional consequences of exchanges with hormonal, immune, and inflammatory systems. As we will see elsewhere, this activity is a response to "unhealed wounds" (i.e., the chronic effects of the accident), since the neurobehavioral reactions may be attributed directly to a disturbance of the nervous system. The physiologically based symptoms may be ignored, misattributed to brain injury, or, in the absence of physiological evidence, incorrectly concluded to have no basis for the disturbance. It is then incorrectly reported that the patient is malingering or experiencing somatization, or possibly experiencing an emotional "overlay" or other psychiatric disorder.

THE MAYO CLASSIFICATION SYSTEM: AN APPROACH TO SEVERITY OF TBI

A new approach to classifying the severity of TBI takes into account the frequent gaps in medical records and the effects of nontraumatic factors upon widely used indicators, such as the Glasgow Coma Score (GCS) and LOC (interval after the accident at which it is recorded; roadside sedation; intoxication; fractures; systemic or psychological shock; organ system failure). A range of neuroimaging techniques is considered for differentiating grades of TBI severity. Classification is based upon multiple criteria, minimizing the consequences of missing information from single criteria. Each grade is based upon the maximally relevant documentation available for a sample of 1678 cases (Malec, 2007a).

Moderate-Severe (*Definite*) TBI: Death; LOC of 30 min or more; posttraumatic anterograde amnesia of 2 h or more; worst GCS in first 24 h of less than 13 (unless invalidated by intoxication, sedation, systemic shock); one or more of the following: intracerebral hematoma; subdural hematoma; epidural hematoma; cerebral contusion; penetration of the dura; subarachnoid hemorrhage; and brainstem injury.

Mild TBI: None of the above apply, but one of the following apply: LOC less than 20 min; posttraumatic anterograde amnesia momentary to less than 24 h; and depressed, basilar, or linear skull fracture with dura intact.

Symptomatic (*Possible*) *TBI*: one of the following symptoms are present: blurred vision; confusion (mental state changes); feeling dazed; dizziness; focal neurologic symptoms; headache; or nausea.

Pending further experience, the Mild and Symptomatic grades correspond to the usual category of concussion. However, the writer notes that the clinician would recognize the complications ensuant to skull fracture and significant chronic somatic injury.

THE PRIMARY PHASE OF BRAIN INJURY

The primary damage is the immediate anatomical lesion caused by mechanical forces. CNS injuries lead to an irreversible loss of loss of function due to lack of neurogenesis, poor regeneration, and the spread of degeneration (Schwartz, 2000). In addition to neurological lesions, there occurs gross anatomical damage (e.g., torn blood vessels with hemorrhage and skull fracture). In unembalmed, fixed-in-place skulls, there is a failure of skull integrity at the range of 22–24 *Joules* (energy), with intermediate impact velocity creating single linear fracture and with only 33% of higher velocity creating multiple fractures (Delye et al., 2007). Varying levels of hemorrhage can occur even within the range of lesser TBI, including petechial hemorrhages (Margoles, 1999). In a concussion, characteristic brain injury includes diffuse axon injury, and cortical lacerations and contusions. Fifteen percent of temporal bone fractures are produced by blows to the occiput. A transverse fracture of the temporal bone may also cause a lesion to the brain stem (Heid et al., 2004). In one series, 5% of traffic accidents caused petrous bone (temporal) fractures. When a range of TBI injury is considered, diffuse neuronal and vascular damage occurs primarily in the areas of corpus callosum, midbrain-pons tegmentum, inferior colliculus, and the dorsolateral quadrant at the superior cerebellar peduncle (Zhang et al., 2004). *Explosive blast injury* creates some different and additional lesional disorders (Taber et al., 2006).

IMPACT

Striking the skull has many pathological consequences, such as pressure waves, skull fractures, and brain indentation. Mechanical forces damage nerve cell bodies and axons (axotomy); blood vessels; the meninges; soft tissues including muscle, tendons, ligaments, and organs; and fracture the skull and other bones. Road accidents, for example, are asserted to transfer high kinetic energy to the nervous tissue, resulting in SDH and diffuse axonal injury (Kleiven, 2003). Contusions have several causes, including brain movement that creates low-pressure area (i.e., *cavitation* and brain surface impact) (Gean, 1994, p. 163). *Intracerebral hemorrhage* is caused by cerebral contusion on the surface of the brain. In an acceleration–deceleration head injury (coup and contrecoup), the brain impacts its bony covering with the most likely hemorrhages in the basal frontal, the anterior temporal, and the occipital areas frequently casing multiple hemorrhages (Kase, 2004).

FALX CEREBRI CUTS

The dura mater contributes to injury to the corpus callosum. It supports and covers the brain, and separates the two cerebral hemispheres. It is an unyielding barrier and sharp edge during impact or acceleration contributing to contusions. Contusions can arise with high-velocity, rapid-deceleration head injuries, as well as with sudden brain and arterial movement against the stationary edge of the falx, creating a nidus for delayed traumatic *aneurysm* development and present as delayed, acute intracerebral bleeding. It compresses and cuts the brain during brain swelling, hemorrhage, herniation, and other mass effects. Among the signs are seizures and cranial nerve deficits (O'Brien et al., 1997). The falx cerebri bilaterally in the midline, and extends medially almost to the corpus callosum. The exposed edge extends from the genu anteriorly to the splenium, which is closest to the edge of the dura. Lateral motion can also scrape the top surface of the corpus callosum against the sharp edges of the falx cerebri and the incisura of the tentorium. These cuts may be considered lacerations (Rosenblum, 1989).

Structural Consequences

Impact: Structural damage and varied physiological dysfunctions occur after the immediate anatomical lesion caused by mechanical forces. Impact creates pressure waves, skull fractures, and brain indentation. It also causes contusions due to brain movement that creates low-pressure area (i.e., cavitation). The stretching and tearing of axons results in disturbed association between brain centers. The initial damage is caused by mechanical forces.

Pressure waves move the brain and have the effects of contrecoup contusions, cavitation, diffuse axonal injury (DAI), and small hemorrhages from stretching blood vessels and contusions, as well as shearing. Shearing refers to differential speed of movement of adjacent tissues (brain planes, skull, vessels), resulting in scraping injuries of the brain against the inside of the cranium, and separation of brain planes (e.g., grey–white junction). Rotation causes different parts of the brain to move at different speeds, dependent upon their distance from the foramen magnum (lateral rotation of the head on the neck; axial rotation of the neck and skull). The brain stem is tethered through the foramen magnum to the neck and spinal cord. A cumulative contusion diagram confirms that there is particular vulnerability to movement of the frontal and temporal tips into the anterior and middle fossae (Gean, 1994, p. 152). The inferior surface of the brain and the vertex, including the motor and somatosensory strips, are also vulnerable to contusion.

Risk Factors: Intracranial Injuries

There is a high incidence of risk factors for *intracranial injuries* (ICI). These were studied in a group of pediatric patients less than 16 years old). Of the patients, 16% with GSC score of 15 and (–) LOC had ICI manifested as subdural hematoma, epidural hematoma, subarachnoid hemorrhage, or brain contusion (Kushner, 2001). The GCS index was a good predictor for ICI with a total score of 12 or less, but not in the range 13–15 (mTBI). While unconsciousness did not correlate well with the risk of ICI, longer duration had a poor prognosis (Sharma et al., 2001). In another sample (Iversen et al., 2000), of those with GCS scores of 13–15 (mTBI) and a CT scan, 15.8% manifested *complicated* TBI (i.e., abnormal CT scans). There was a modest correlation between intracranial abnormalities and lower GCS scores, positive loss of consciousness, and greater frequency of skull fractures.

Whiplash

Parts of the brain may be compressed as the brain stem rotates while the spinal cord is tethered outside the foramen magnum. Rotational injuries (neurological and somatic), are referred to as "whiplash." TBI reports include *brain imaging* studies (hypometabolism in both temporal lobes, frontopolar, lateral temporal cortex, and putamen) and abnormal EEGs (a higher incidence 6 months postinjury in patients without closed head injury) (Henry et al., 2000).

Neurons

Impact, other forces, and brain and body movement operate upon tissues of different densities, resulting in displacement, stretching, shearing, pressure waves, brain rebound creating a low-pressure space (cavitation), skull fractures, and brain impact against internal structures and the skull. These create immediate contusions, lacerations, DAI, tearing of the vasculature, and injuries of the cell membrane (Gean, 1994, p. 163). Direct impact contusions are known as *coup*, while subsequent impact within the skull are known as *contrecoup*. Mechanical trauma subjects neurons to a heterogeneous field consisting of tensile, compressive, and shear strains. Cells may undergo immediate death if the structural threshold is surpassed. Death rate is increased with high-rate, high-magnitude loading. This is characteristic of *diffuse brain injury*, which seems to be associated with primary and secondary mechanisms of axonal damage rather than raised intracranial pressure (Wallesch et al., 2001). *Heterogeneity of cell death* has varied causes, including cell orientation with regards to the overall strain field; the viscoelastic characteristics of the cell and its matrix; morphology and cytostructure; and adhesions between cells and their matrix. Secondary alterations may be initiated in surviving cells due to altered signaling and the limited regenerative ability of the brain (Cullen & LaPlaca, 2006). *Stretching* and tearing of axons results in the disturbed association between brain centers. Stretch-induced axonal damage has been considered likely to be responsible for the postconcussive syndrome, cognitive deficits, and motor dysfunction (Bazarian et al., 2007). Severe stretching caused by blunt trauma can result in cranial nerves being torn loose from the brain stem at points of attachment or of angulation.

Electrophysiology

The acute behavioral symptoms of concussion have been explained electrophysiologically (impairment of consciousness; paralysis of reflexes; loss of memory):

1. The immediate postconcussive EEG is excitatory or epileptiform.
2. The cortical evoked potential waveform was totally lost in this period.

Mechanical loading of the head, through generation of turbulent rotatory and other movements of the cerebral hemispheres, increases the chance of a tissue-deforming collision or impact between the cortex and the bony walls of the skull (or barriers of the dura mater, RSP). Additionally, there is mechanically induced depolarization and synchronized discharge of cortical neurons (Shaw, 2002).

Membranal Damage

Mechanical trauma to the CNS (Anderson & Hall, 1994) has numerous consequences, including interference with transport of neurotransmitters caused by splitting of the axon. There is a *retraction ball* as a marker of the site. Potassium is released from cells *K+ efflux* (associated with axonal stretch and opening of voltage-dependent K+ channels). Trauma alters the microstructure of axons, which could affect water diffusion (see DTI/Anisotropy), causing a breakdown of myelin and downstream nerve terminals; neuronal swelling or shrinkage; and increase or decrease of extracellular space (Levin et al., 2008). Neuronal depolarization and disruption leads to release of excitatory amino acid (EAA). Increased vulnerability to a second injury (*second impact syndrome*) *ensues from the preliminary excessive loading with* Ca^{2+}, which may persist up to 2 days in experimental concussive brain injury (Lovell & Collins, 1998) (see also Macrophages; Apoptosis). *Mechanical strain* of axons is the primary mediator of axonal injury. The details are not neurobehaviorally illuminating, but the injury commences with defects in the cell membrane (*mechanoporation*), allowing for an influx of Ca^{+2}, sodium, chloride, and an efflux of potassium. There is a sequence of failure to propagate action potentials, production of oxygen-free radicals, glutamate neurotoxicity, and so on, before finally continuing to an irreversible CA^{+2} influx. An experimental study suggests that there is tolerance of tensile strain up to 65% of axonal length and delayed elasticity after deformation (Gaetz, 2004).

Labrynthine Concussion

A labrynthine concussion causing vertigo, hearing, and tinnitus may follow a blow to the head that does not cause a temporal bone fracture. The reason is that despite the protection of a body capsule, the labrynthine membranes are susceptible to blunt trauma. Occipital or mastoid region blows are most likely to produce labrynthine damage. Since forces are absorbed by an actual break in the bone, a blow that does not fracture the skull may cause more labrynthine damage than one in which the forces are absorbed. There may be loss of consciousness, or the onset may start several days later. Rapid improvement and the absence of brainstem symptoms suggest the peripheral localization of most lesions. The loss of hearing is unilateral; there is spontaneous nystagmus, the external canal and tympanic membrane may be normal, or there may be blood, CSF, or both in the external or middle ear if there is an associated temporal bone fracture. Sudden deafness is often partially or completely reversible. Intense acoustic stimulation is caused by pressure waves transmitted through the bones to the cochlea. Animal studies suggest that there is degeneration of hair cells and cochlea neurons in the middle turns of the cochlea, with pure tone hearing loss most pronounced at 4000–8000 Hz (Baloh, 1998, pp. 182–183).

PRIMARY NEUROLOGICAL INJURY

It has been stated that diffuse brain injuries result from the effects of head motion (i.e., without impact to the cranium, although usually this does occur). It might be considered a *head motion injury*, whose severity is determined by the direction, magnitude, and speed with which the head moves from rest or to rest. DAI may be observed after major trauma or may be detectable only microscopically, with autopsy. In cases of nonsurgical trauma, it can occur in the absence of intracranial hematoma, contusions, or increased intracranial pressure (Adams et al., 1989). DAI may take several hours to complete, with various changes to the cell's membrane and contents until the axon separates (Miller et al., 1996). Strich (1956), who primarily studied the most severe injuries, posited that white matter lesions were due to shearing injuries incurred at the moment of injury. Violent head motions produce strains and distortions within the brain resulting in shearing or stretching of nerve fibers and myelin sheaths with axonal damage. This is consequent to the white matter being exposed to a variety of forces. Depending upon the severity of injury, axonal damage may occur in isolation or in conjunction with actual tissue tears (i.e., small hemorrhages). The latter are usually in the central area of the brain (Gennarelli, 1986). DAI occurs as a primary brain damage at the moment of injury. Its etiology in the secondary brain has been rejected in favor of Strich's initial concept of primary brain damage (see Gennarelli & Graham, 1998, for extended discussion and imagery).

THALAMIC INJURY

Basal ganglia and thalamic lesions have also been detected by SPECT imaging in head injury without LOC, CT, or MRI findings (Abu-Judeh et al., 1999). mTBI is usually accompanied by discordance between clinical and radiographic findings, with a paucity of multiple resonance-documented DAI. Nevertheless, patients complain of stereotyped global symptoms. Reduced thalamic volume has been reported in the moderate-to-severe TBI levels (Ge & Grossman, 2008).

COUP-CONTRECOUP IMPACT

Direct *impact to the skull* deforms the bone, compressing the underlying parenchyma. This may damage arterial or venous vessels. Contusions with associated skull fracture are usually more significant. Within the contusion, blood leaks through the damaged blood-brain barrier. The combination of biochemical and physical changes leads to a mixed cytotoxic and vasogenic edema. Brain swelling is common adjacent to contusions, due to physical disruption of the tissue with damage to the BBB and loss of normal physiologic regulation of arterioles. The biochemical and physical changes in the region of the contusion lead to a cytotoxic and vasogenic edema. The edema is considered vasogenic since water and electrolytes leak into the brain tissue and spread into the adjacent white matter (Britt & Heiserman, 2000; Gennarelli & Graham, 1998). In a mouse study, *controlled cortical impact* (CCI) is followed by prolonged *acute edema* in older animals, which has a crucial impact on morbidity and mortality (Onszchuk et al., 2008).

Coup creates pressure waves that move the brain and have the effects of contrecoup contusions, cavitation, DAI, with small hemorrhages from stretching blood vessels and contusions, as well as shearing. Contrecoup occurs in a different area of the brain, not necessarily precisely opposite to the area of the coup injury. Since the brain is only loosely connected to the surrounding cranium, application of a load creates a compressive strain at the point of impact and sets the skull in motion along the line of force. Due to inertia, the motion of the brain lags behind the skull. When the skull comes to rest or recoils, the brain strikes the calvarium on the other side, and another compressive strain is generated.

Repeated blows to the head are a serious health problem in *sports*, notably boxing, American football, ice hockey, rugby, horse racing, and soccer. It is characterized by speech and gait disturbance, pyramidal and extrapyramidal features, personality changes, and psychiatric disease. In the early stages, symptoms are transient and reversible; in the later stages they are progressive. It is characterized by cerebral atrophy, septal fenestration, cerebellar tonsillar scaring, cavum septum pellucidum, loss of pigmented cells, and prominent neurofibrillary tangles. It has been considered to be consequent to an extra-long history of significant impacts (e.g., professional boxers and an association with the apolipoprotein E-4 gene). In the nonboxing population, this gene was associated with death and adverse outcome after TBI (McCrory, 2000).

Coup-Contrecoup Contusions

A contusion is a focal injury occurring when mechanical forces damage the small blood vessels and other tissue components of the neural parenchyma. Contusions at the impact site (coup) are considered minor compared to those at a distance from the impact site (contrecoup). The injuries may be to surface or within deeper structures. One differentiation between contusion and lacerations is that the pia-arachnoid is intact over contusions but torn in lacerations (Blumbergs, 2005; Cooper, 2000; Gennarelli & Graham, 1998) (see Figures 6.1 and 6.2).

Contusions occur at various sites: When contusions occur at the site of impact ("coup"), the brain is transiently depressed by deformation and inbending of bone at the point of impact. The severity of the contusion is proportional to the amount of energy transmitted through the bone to the brain, and inversely proportional to the size of the area of contact. For the most part, coup injuries are considered minor. Those distant from the point of contact ("contrecoup") occur when the stationary head is free to move at the time of impact, or the moving head is suddenly decelerated. After impact, the brain may be contused as it is propelled over the rough edges of the sphenoid wing or the floor of the frontal fossa. Veins at the site of contusions on the cortical surface may rupture into the subdural space. This has been observed in two-thirds of SDH surgical patients.

Cortical: These occur characteristically in the frontal and temporal poles, and also on the inferior surfaces of the frontal and temporal lobes, where brain tissue comes in contact with bony protuberances at the base of the skull. Contusions are found only occasionally in the parietal lobe, and least frequently in the occipital lobe.

Medial hemisphere surfaces: The brain is accelerated against the falx cerebri. While contusion is the most frequently encountered head injury lesion (89% postmortem), under the age of 1 year it occurs in less than one-half; the usual hemorrhagic contusions are replaced by nonhemorrhagic tears of the white matter. Patients with mTBI and persistent PCS have a high incidence of temporal lobe injuries, which are, according to the authors, attributable to the hippocampus and related structures. Abnormal temporal lobe findings on PET and SPECT in humans are considered analogous to the neuropathologic evidence of medial temporal injuries of animal studies after mTBI. This may explain the frequent findings of memory loss in this population (Umile et al., 2002). To contribute to our understanding of memory loss, the present author refers the reader to brain movement against the medial surface of the dura mater (see Falx Cerebri) and abrasion of the temporal tip by movement against the sphenoid wing during movement into the anterior fossa. The change in the percentage of the *volume of brain parenchyma* over time was significantly greater in patients than controls. After 11 months, whole brain atrophy was detected, and injury leading to LOC led to more atrophy (MacKenzie et al., 2002).

Posterior contusions: SDH of the posterior fossa is rare. Cerebellar contusions usually occur directly underneath an occipital trauma, with the inferior surface of the cerebellum supposedly the most vulnerable to injury. A coexisting posterior fossa SDH is sometimes present. The occipital blow causing the brain to be propelled forward in the calvarium commonly results in contrecoup contusions and hematomas of the frontal and temporal poles (Cooper, 2000).

Deep white and grey matter: "Intermediate" lesions are probably caused by shearing with rapid acceleration or deceleration of the brain at the time of impact.

Lacerations

A laceration is damage to the neural parenchyma occurring at the moment of injury, which may be limited to the cortex or extend into the deep white matter of the frontal and temporal lobes, or into the central grey matter. One differentiation between contusion and lacerations is that the pia-arachnoid is intact over contusions but torn in lacerations (Gennarelli & Graham, 1998). These occur characteristically in the frontal and temporal poles, and also on the inferior surfaces of the frontal and temporal lodes, where brain tissue comes in contact with bony protuberances at the base of the skull. They may be directly produced by penetration or depressed fracture or indirectly by tissue damage produced by mechanical forces, which are superficial or extending into deep structures (Blumbergs, 2005).

LESSER VASCULAR INJURY

Here we are concerned with the vascular after effects of an accident causing concussive brain trauma, excluding what may be called "surgical injury" (e.g., hemorrhage, edema, ischemia, reperfusion injury, etc.). The intensity of cerebral hemorrhage trauma may begin with *traumatic microbleeds* (TMB) whose presence (*hemorrhagic TBI*, as opposed to pure TBI) may be related to outcome. DAI may be diagnosed via detection of microbleeds by the specially sensitive 3T (Tesla) MRI systems. This may substitute for direct proof of DAI in the absence of reliable radiological tools (Scheid, 2007).

Cervical Vascular Trauma

Cervical trauma has been considered the cause of tinnitus, vertigo, and unsteadiness in a small proportion of cases. Whether neck injury can cause cervical vertigo is controversial, although this writer finds the purported causes reasonable—vertebral artery damage; multisynaptic pathways from the neck proprioceptors to the vestibular nuclei; somatosensory receptors from neck muscles, tendons, and joints affecting self-motion during locomotion; and labyrinthine concussion (Tusa & Brown, 1996).

Vasospasm

Early and late vasospasm is considered to be a significant entity in head trauma (Chestnut, 1996; Zubkov et al., 1999), occurring in up to 25% of patients (Batjer et al., 1993). Although there is evidence that concussion is associated with reduced or slowed cerebral circulation, increased level of vasospasm has been defined in terms of greater blood flow velocity in the middle cerebral artery. Severe vasospasm is associated with *subarachnoid hemorrhage* (SAH) after head trauma. SAH may be accompanied by a "*sympathetic storm*" (i.e., an acute disturbance of autonomic function including cardiovascular manifestations [arrhythmias, hypertension, and electrocardiographic alterations]). There is an adrenergic discharge, irritation of the hypothalamus by subarachnoid blood, and trauma to vasopressor centers. This creates cardiovascular risks and can influence outcome pose diagnostic problems (Ropper, 1997). That arterial narrowing can lead to impaired CBF is suggested by the correlation between the lowest CBF and the highest middle cerebral artery velocity during vasospasm (Martin et al., 1992). Injury severity is associated with the development of cerebral vasospasm. It has histological resemblance to aneurysmal vasospasm but is clinically milder (Zubkov et al., 1999).

Cerebro-arterial spasm is caused by sudden traction on the carotid artery sheath at the base of the brain, with symptoms including vascular-type headaches and a feeling of being dazed or stunned (Goldstein, 1991). Vasospasm is described as occurring after 48 h. Brain stem damage or subarachnoid hemorrhage is associated with vasospasm independently of dysregulation. Posttraumatic posterior cerebral vasospasm may be responsible for brain stem dysfunction Indeed, one-third of head-injured patients with anterior circulation vasospasm also had posterior circulation vasospasm, with unfavorable outcomes (citing Marshall et al., 1978, 1995). Even after MTBI, in the presence of injury to other systems, deficits of cerebral vascular autoregulation may cause cerebral ischemia in the event of decrease of blood pressure or cerebral edema consequent to increase in blood pressure (Labi & Horn, 1990). Vasospastic ischemia is most common after injury but can occur throughout the acute recovery period (Cherian et al., 1996). Ischemia (associated with vasospasm or mass effects) impairs the metabolic need of the brain, setting into motion multiple mechanisms of toxic metabolite formation and cell destruction. Blunt trauma can cause cerebral hypoperfusion (Dewitt et al., 1997).

The location of the hypoperfusion, as revealed by SPECT, in cases where reliable information was obtained concerning the mechanics of the trauma, does not correspond to the site of impact (coup) or contralateral site (contrecoup). Vasospasm involving the large basal intracranial arteries (ICA; middle cerebral; basilar) occurs in 25–40% of head-trauma patients. This association is statistically stronger for the most severely injured patients (Martin et al., 1995). The ICA may be damaged without a blow from stretching, tearing, or compression (Chandler, 1990). It can also be damaged by direct damage to neck structures (i.e., impact, stretching, tearing or compression of the ICA and other cervical vessels) (Chandler, 1990; Nordhoff et al., 1996a) caused by impact or hyperextension-hyperreflexion and rotation in various planes ("whiplash").

Reduced Cerebral Perfusion

When amateur sportsmen who did and did not box were compared with psychometric tests and SPECT, the nonboxers performed more efficiently on psychometric tests, and those with fewer bouts better than the more experienced boxers. The nonboxers had fewer regions of reduced cerebral perfusion (Kemp et al., 1995). Measurements taken 1 week to 3 years after injury of patient (those with lawsuit or unsettled insurance claim excluded) manifested reduced blood flow volume, as shown by increased circulation time and decreased amplitude. Initial lack of symptoms and normal circulation time may be followed 3 days later with complaints of postural dizziness and headache when circulation time is reduced. There may be a parallel symptom display and increased circulation time for several weeks. Symptoms mostly abate when circulation time returns to normal. It was speculated that the cause was increased arteriolar vasomotor tone (Taylor & Bell, 1966a). Referring to damage to the lowest four cranial nerves (Davies & King, 1997), vascular etiology has been proposed, suggesting brainstem ischemia secondary to vertebral artery spasm or compression. Unilateral dysfunction of nerves IX and X results in a unilateral paralysis of the larynx and soft palate, the aetiology usually being vascular, but also following trauma.

Watershed Injury (Ischemia)

After a period of ischemia or a hypotensive episode associated with an impaired level of consciousness, an infarct may be created at the boundary of a brain areas served by a single major artery (i.e., an area of reduced perfusion between major cerebral circulations). Watershed areas include visual, parietal, motor and sensory cortices, cerebellum, and basal ganglia. Symptoms include gait dysfunction, cortical blindness, and transcortical aphasia (with anosognosia) (Biller & Love, 2004; Snyder & Daroff, 2004).

PLEXOPATHIES

These conditions are relatively uncommon compared to peripheral nerve or nerve root diseases. Yet, they may cause major loss of neurologic function. The pattern of deficit does not match that of a root or peripheral nerve. Proximal muscles are affected, the sensory loss seems very dense, and sometimes SNS failure is indicated by dryness of the skin and lack of vasodilation (Dawson & Sabin, 1996). The various plexuses are vulnerable to traction (stretching), high-velocity closed injuries (high-velocity motor vehicle accidents, falls), root avulsion (separation of a nerve root from the spinal cord), sudden depression of the shoulder in football and similar sports, repetitive activities such as playing a keyboard or other musical instrument, whiplash injury, crush injuries, fall from a height, and various other medical conditions. Trauma results in injury to the myelin, axon, and connective tissue covering of plexus elements. Variations include focal demyelination with conduction slowing, demyelinating conduction block, and *axonotmesis* (loss of continuity of the axon and myelin with preservation of the connective tissue), leading to Wallerian degeneration (Evans et al., 2007).

The cervical plexus is most frequently injured by compression (e.g., American football). Weakness and paralysis of the diaphragm is consequent to bilateral phrenic nerve injury (Younger, 2005a) (see Plexuses).

Brachial plexus patients present with weakness, reflex change, and sensory loss, depending upon whether a part or the whole plexus is disturbed (Chad, 2004). The brachial plexus's location between trauma vulnerable structures (e.g., the highly mobile neck, shoulder, clavicle, first rib, and the arm) renders it vulnerable to trauma (impact, stretching, and traction, carrying heavy packs, firing weapons) (Evans et al., 2007). In the vast majority of patients there is blunt trauma striking the shoulder, compressing it between the clavicle and the first rib. Continued head movement increases the distance between them. It may result from injury to nearby bony structures. There is a loss of function in the shoulder and elbow joints, and the patient is unable to extend the wrist and fingers. Injury to the humeral head and axillary compression also occur. Unilateral muscular weakness may reflect brachial plexus injury of restricted range of motion due to fibrositis.

Neuropsychological Examination

Sensitive neuropsychological testing is indicated to detect the subtle deficits that could occur after ischemic damage (Junger et al., 1997). Thrombosis or embolism can occur due to trauma, resulting in ischemia or infarction (Hughes & Brownell, 1968; Teman et al., 1991). Injury to the carotid artery is indirect and may not be recognized when there is no penetrating wound of the neck. There is an effect on the corresponding cerebral hemisphere. The most common finding after nonimpact injury to the carotid is a thrombosis of the internal carotid artery (ICA), 2 cm distal to its origin with an associated intimal tear (Chandler, 1990). Carotid artery obstruction may be falsely attributed to direct injury to the brain or the spinal cord (left sided deafness, facial weakness, hemiplegia, hemianesthesia, left homonymous hemianopsia with defective conjugate movement of eyes to the left, and expressive aphasia). A traumatic thrombosis may not produce permanent neurological sequelae if the nondominant vertebral artery is involved or collateral circulation exists due to congenital defects of the circle of Willis (Teman et al., 1991).

SECONDARY PHASE (CONTINUING INJURY)

Secondary brain injury accounts for the majority of deaths after hospitalization for TBI. The primary phase of cellular injury is followed by secondary neuronal damage and chemical changes leading to cell death. These biochemical, molecular, and genetic reactions may be autodestructive or neuroprotective. Further traumatic changes after the moment of injury include anatomical and physiological reactions that increase gross TBI, extending the scope of brain damage and neurobehavioral disorders far beyond the primary, immediate injury (see Older Age at Injury). Secondary injury is a consequence to the structural and physiological disorders of the primary injury (e.g., overt tissue damage such as torn blood vessels, gross tissue destruction, inflammation and mass effects, etc.). Later neural, biochemical, molecular, genetic, and systematic changes that may be protective or grossly destructive. A series of events evolve, including continuing axonal injury and release of mediator substances, which cause injury from minutes to days after the impact. The axonal changes (regenerative and degenerative) (i.e., *secondary axotomy*) have been observed 59 days postinjury. It is noteworthy that diffuse traumatic axonal injury (TAI) is found in other conditions (i.e., *intoxication* and *chronic opiate abuse*) (Gaetz, 2004). Ultimately, the primary focal and diffuse damage evolves into a generalized atrophy with reduced overall *brain volume* (Merkley et al., 2008).

Secondary TBI creates cascades of physiological, biochemical, cellular, and subcellular (molecular and genetic) responses that expand the lesion. Paradoxically, the spread of injury-induced damage is an outcome of cytotoxicity and subsequent phagocytosis, both of which are parts of the healing process's response to injury (Schwartz, 2000). Blunt trauma creates pathophysiological effects common to the spinal cord and other portions of the CNS. Mechanical trauma results in a variety of biochemical processes that are prominent in the secondary autodestruction of neural tissue. The process known as *lipid peroxidation* creates *free radicals* or *reactive oxygen species* (ROS), which create further cell injury (Marks et al., 1996, pp. 327–339). Free radicals are the direct result of tissue damage, including iron compounds of blood leaving the vessels, which are countered by enzymes and vitamins C and E. ROS injures the neural membrane by interfering with *phospholipids*, which are a reservoir of intracellular messengers, and which anchor some proteins to cell membranes (Champ et al., 2005, p. 486). There is a release of other substances that participate in the secondary phase of CNS injury, including enhanced vascular permeability, produce edema, promote platelet aggregation, cause neutrophil infiltration, and so on, which worsen tissue injury (Anderson & Hall, 1994). Injury stimulates the upregulation of intracellular mechanisms that may lead to either regeneration or delayed cell death (Reilly, 2005). There is a cascade of cerebral compounds, metabolites occurring at different intervals after the injury. This is due to the combined effects of neuronal membrane damage and posttraumatic swelling of astrocytes. To prevent neuronal dysfunction, astrocytes absorb these metabolites and glutamate (Pascual et al., 2007) and absorb neurotransmitters from the synaptic cleft (Newell & Hedley-Whyte, 2003).

MASS EFFECTS AND CONTINUING TRAUMA

Among the consequences of physical trauma is the process by which initial focal brain injury progresses—the enlargement of contusions; hypotension; focal hypoxic ischemia; hemorrhage and hematoma; focal, unilateral, or bilateral edema and brain swelling; inflammation; brain swelling, and vascular stress effects (i.e., vasoconstriction); and expansion that presses parenchyma past the midline and, ultimately, herniation past the meninges or through the foramen magnum. Head trauma can result in a condition called *pituitary apoplexy*. It is an endocrine emergency with signs of intrapituitary damage, stalk deviation, and compressions of pituitary tissue. The neural and somatic disorders are gross, evolving over 1–2 days, with the conspicuous only stated here—severe headache, neck stiffness, progressive cranial nerve damage, cardiovascular collapse, and change in consciousness (Cone et al., 2003).

Postmortem studies indicate that secondary injury can continue for 100 days after RBI, although its nature is not well defined (Brooks et al., 2000; Chrousos, 1998). Further neuronal damage includes degeneration of central and peripheral nerve tissue; cellular and synaptic changes are believed to account for behavioral changes persisting weeks or months after TBI (see Cognition) (Bullock & Nathoo, 2005). Activation of a complex variety of physiological and molecular cascades initiates processes that persist for hours or days, aggravating the initial trauma. The pathophysiological sequelae consist of numerous molecular cascades that are set into motion immediately after primary injury. They are maintained for a long interval and compromise the brain's entire metabolism. Intracellular ionic and metabolic injury and changes in cellular homeostasis and metabolism after TBI may explain vulnerability. Neurochemical imbalance effects blood flow and BBB integrity, thus contributing to neuronal function (Morganti-Kossman et al., 2005).

Blood-Brain Barrier (BBB)

Breakdown of the *blood-brain barrier* (BBB) begins within 3 min of injury. Early BBB dysfunction results from mechanical damage to small venules in vulnerable regions (Dietrich & Bramlett, 2004). Focal axonal swelling and other changes occur within 15 min of TBI with further deterioration proceeding for a measured 6 h and ending in *axonal disconnection* (Bullock & Nathoo, 2005; Mendelow & Crawford, 2005). BBB breakdown influences the kind of materials that can perfuse into the brain. The uninjured BBB permits lipophilic substances to cross, with SPECT sensitivity to the substance's presence. However, a damaged BBB barrier will result in decreased uptake (i.e., a "cold" area) in a damaged brain. Other radioactive substances, ordinarily excluded, cross the damaged BBB. Their SPECT detection points to cortical dysfunction in conjunction with BBB disruption and hypoperfusion. EEG aberrations (contralateral parietal cortex), with reduced blood perfusion have been observed in patients presenting transient attacks of paresthesia suggestive of partial seizures. Focal cortical lesions, associated with BBB disturbance and reduced blood flow, may be the underlying pathogenesis of the PCS syndrome in some patients (Korn et al., 2005; also, communication by Hussein Abdel-Dayem, MD). A hyperpermeable BBB is considered a hallmark of TB pathophysiology, leading to increased access of ions, water, proteins, and free radicals to the brain parenchyma, loss of cerebral autoregulation, increased vasogenic edema, increased intracranial pressure, osmotic disruption, intraparenchymal hemorrhage, infiltration of proinflammatory mediators, systemic inflammation, and neuroinflammation. However, the BBB availability may be a window of opportunity for the delivery of therapeutic agents that do not traverse the normal, intact barrier (Onyszchuk et al., 2008).

SECONDARY PHASE: CELLULAR PATHOLOGY

Pathological cellular effects include damage to cellular membranes; ion channels of axons, neurons, and astrocytes; metabolic depression; neurochemical derangements; reactive glial changes; progressive degeneration of brain parenchyma; excitotoxicity; and free radical production. Cellular protein damage is highly toxic to cells, leading to the activation of stress genes (Truettner et al., 2007). Damage to the exoskeleton leads to blocks in axoplasmic transfer, accumulation of axonal materials, and ultimately delayed axonal disruption or secondary axotomy. After inflammation and trauma, astrocytes become hypertrophic, accompanied by these consequences following the release of chemical mediators of inflammation: an increase of blood flow and vascular permeability (edema or local swelling), chemotaxis (specific cell types are attracted by some molecules, and phagocytosis. A localized trauma can recruit astrocytes from as far as the contralateral side (Hof et al., 2003). Inflammatory cytokines contribute to tissue destruction (Marion et al., 2004). There is dysregulation of both inhibitory and excitatory neurotransmission following TBI. This is responsive to antioxidant treatment (Hi et al., 2006).

Chemical Changes

Pathophysiological damage (intracellular) adds to mechanical trauma (e.g., enhanced excitatory amino acid transmitters and oxygen-free radicals). Posttraumatic cellular changes include massive increases of extracellular potassium, extracellular calcium, decrease of intracellular magnesium, increase of glucose utilization, increased cerebral metabolic rate of oxygen ($CMRO_2$), and lactate accumulation. Reduced cerebral metabolism may reflect changes in cerebral perfusion, neuronal degeneration, anatomic reorganization, or imbalance of excitation and inhibition (Hovda, 1996).

Magnesium (Mg^{++}): Mg^{++} is involved with cellular energy metabolism, initiation of protein synthesis, maintenance of cellular membrane potential, and voltage gate control for NMDA receptors. Intracellular levels of Mg^{2+} may remain decreased for up to 4 days. It is associated with posttraumatic neurologic deficits in rats, including neurologic motor deficits. When the Mg^{++} block of NMDA receptors is relieved by depolarization, the response is to permit influx of Ca^{++} and further depolarization of the postsynaptic cell.

Calcium (Ca^{++}): Enhanced intracellular calcium mediates a variety of toxic effects. Influx of Ca^{2+} leads to cell damage, secondary cell death, or axonal malfunction and breakdown of neurofilaments and microtubule. *Lactate production* and accumulation is increased. Increased vulnerability to a second injury (*second impact syndrome* ensues from the preliminary excessive loading with Ca^{2+}, which may persist up to 2 days in experimental concussive brain injury.

Lactate: Elevated lactate has been implicated in neuronal dysfunction by inducing acidosis, membrane damage, altered blood-brain barrier permeability, and edema. It may leave affected cells more vulnerable to secondary ischemic damage.

There are alterations in response to signals caused by traumatic alteration of the *second messenger* phase of neurotrauma. Secondary posttraumatic cascades include excitotoxicity, free radical production, and activation of stress genes. Some secondary injuries include decreases in blood flow and metabolism, free radical formation, and perturbation of intracellular mechanisms that comprise a network of interacting functional, structural, cellular, and molecular mechanisms. These lead to *delayed cell death* and/or neurological dysfunctions that may be the cause of *epileptic seizures* hours or days after the impact. Other dysfunctions include free radical production; disruption of the membrane, disruption of the membrane's capacity to act as a controlling barrier between intracellular and extracellular solutes, inadequate oxygen (hypoxia) or ischemia (inadequate blood supply), release of toxins (external and traumatic excitotoxins and other substances), and change in cellular homeostasis and metabolism.

Excitotoxicity and Excitatory Amino Acids

Glutamate is the major neurotransmitter mediating CNS synaptic excitation. Two of its receptors, AMPA and NMDSA, are considered excitatory amino acid (EAA) receptors. They are widely distributed in the CNS and, together with Mg^{++}, control the influx of Ca^{++} into the neuron (Kutchai, 1998). The hippocampal and other limbic centers that play a role in *memory* disorders and spatial learning have a high density of EAA receptors. Compounds acting at EAA receptor sites have improved posttraumatic memory and neurological motor outcome (Smith & McIntosh, 1996).

Glutamate and aspartate increase after brain injury or secondary ischemia. Excitotoxicity contributes to secondary neuronal damage. Simple amino acids serve as neurotransmitters (glutamate, aspartate, glycine, etc.). They activate receptor proteins on release. If this results in electrical depolarization of the neurone they are termed *excitatory neurotransmitters*. This process is safe as long as it is transient; excitotoxicity refers to nerve cell impairment or death caused by prolonged or excessive receptor activation (Goetz, 2003). These neurotoxins create irreversible damage to neurons and glia. After concussion there ensues *K+ efflux* from cells owing to mechanical membrane disruption, axonal stretch, and opening of voltage-dependent K+ channels. Neuronal depolarization and disruption lead to the release of EAA. The first response to glutamate is depolarization of the postsynaptic cell due to influx Na^+ and K^+ consequent to glutamate stimulation. Excitotoxicity mediated through NMDA and muscarinic acetylcholine receptors may contribute to loss of CA1 pyramidal cells 1 week later.

Opioids

Dynorphins, endogenous opioid neuropeptides, are involved in antinociceptive and neuroendocrine signaling. At high concentrations or CNS pathology, they may be excitotoxic and contribute to secondary injury (Hauser et al., 2005).

Oxidative stress (*Peroxidation*) likely plays an important role in posttraumatic brain tissue damage and edema formation. Early after neurotrauma, there is *oxygen radical* (high valence unbound and highly reactive species of oxygen) formation. Reperfusion following resolution of ischemia or vasospasm leads to additional neurological injury, including phagocytic damage to the endothelium and surrounding tissues (Nemeth et al., 1997) and the release of oxygen-derived free radicals (Kirsch et al., 1992). This creates damage to vascular, neuronal, and glial membranes with excitotoxic, intracellular calcium overload and excitatory amino acid release (i.e., glutamate) (Hall, 1996). There is ischemia-related augmentation of active oxygen compounds that is accelerated by iron present in the brain tissue. Brain tissue is sensitive to peroxidation processes. Since there is a poor content of antioxidant agents, enhanced cell membrane lipid peroxidation is likely. Brain contusion is accompanied by enhanced lipid peroxidation persisting for at least 10 days postinjury. Enhanced lipid peroxidation processes seemed to correlate with the severity of head injury (contusion) (Kasprzak et al., 2001).

NEURODEGENERATION

Wallerian Degeneration

This refers to the posttraumatic effects upon postsynaptic neurons that undergo both degenerative and regenerative processes in the peripheral nervous system (PNS) and CNS. Axons in the PNS regenerate rapidly, while those in the CNS cannot. Interruption of the continuity of a nerve causes the peripheral stump to degenerate rapidly followed by regeneration and recovery of the organ if peripheral. Axonal sprouts grow from the proximal stump, enter the distal stump, and grow toward the nerve's end organs. Upon return to its target, the regenerated axon can form new neuromuscular junctions. Autonomic axons also reinnervate target organs, while sensory organs reinnervate muscle spindles (Sanes & Jessell, 2000; Vargas & Barres, 2007). If central, the pathway does not remain and a barrier to axonal growth develops due to astrocyte proliferation (*glial scarring*) (Krassioukov, 2002). This includes neurochemical and pathophysiological events, involving neurodegeneration, set into motion by the mechanical injury. Activation of complement pathways (immune) contributes to the demyelinization and neurodegeneration in TBI (Maikos & Schreiber, 2007). There is release of superoxidants that may be directly involved in the degeneration of spinal axons and BBB disruption after injury (Hall, 1996). Clinically, after concussion, some patients of middle age display enlarged sulci on brain scans. Some are assessed by radiologists as displaying premature dementia, although there was no evidence from preinjury performance. It is important that diffuse brain atrophy is displayed by widening of the ventricles and sulci. Nevertheless, the degree of cerebral atrophy correlating most with outcome parameters is the width of the third ventricle. A widened third ventricle 3 months after trauma portends a poorer neurobehavioral and vocational outcome (Britt & Heiserman, 2000).

White Matter (WM) Degeneration

When white matter (WM) loss is indicated by a scanning study, it may be attributed to normal aging or neurodegenerative disease, but rarely to trauma. A useful source of information is diffusion-weighted imaging (DTI), which studies disturbed axonal water transport from DAI occurs. DTI, which is white matter-specific, has been sensitive to white matter injury in the range from 24 h postinjury (normal appearance on conventional MRI) until 5 years after injury (see Brain Development). Patients whose MRI were consistent with diffuse axonal injury had FA and ADC scores consistent with severe disruption of white matter microstructure. The attribution of WM degeneration or loss is complex. WM integrity in patients with severe TBI from MVA averaging 7 years earlier had different areas of WM loss:

1. Associated with time since the injury but not age,
2. Associated with age, but not time since the injury.

This chronic sample probably involved multiple injury processes, such as primary damage (shearing, hemorrhages, contusions/contrecoup), secondary mechanisms (swelling), and tissue atrophy from Wallerian degeneration. There may be an interaction of multiple injury processes with age. The patients with TBI were all employed, either full or part-time, and living in the community. While the performance of those with TBI was lower on several neurocognitive measures, even when controlled for verbal IQ, their average performance was within the normal range and did not necessarily imply impairment. Thus, even persons with severe TBI might perform quite well over time (Kennedy et al., 2009; Levin et al., 2008).

SECONDARY PHASE: DISTURBED INTERNAL MILIEU

There are changes in whole brain systems affecting nutrient delivery, blood flow, brain metabolism, and neurological function (Bullock & Nathoo, 2005). Ischemia and hypoxia impair RNA transcription, which interferes with the synthesis of secretory products (Sarnat & Flores-Sarnet, 2004). There are indirect neuropathological consequences of trauma, as well as direct lesions—interruption of cerebral autoregulation, changed homeostasis (loss of cerebral and somatic autoregulation with raised cerebral blood pressure or hypotension), increased metabolic rate and changed organ functions, acute and chronic dysregulation of physiological systems (hormonal or stress, inflammatory, immune, and circadian), and vascular stress effects including vasoconstriction (Chrousos, 1998). Injury to the hypothalamus, by impairing central thermoregulatory mechanisms, can give rise to fever of central origin. Circadian changes are implied by loss of diurnal temperature variation. Dysregulation of the HPA axis (reduced reactivity and enhanced negative feedback suppression) may be the mechanism by which chronic "whiplash" symptoms are maintained (Gaab et al., 2005).

After injury, glucose production by the liver is increased and its oxidation is more than doubled, with an overall increase in glucose flow through the extracellular environment. With increasing severity of injury there may be hyperglycemia, which is associated with poor head injury outcome (Berger et al., 2002). Using FP TBI in rats, it was detected that 4 and 24 h posttrauma local cerebral metabolic rates of glucose (lCMRglu) were significantly depressed at the impact site. Moreover, milder abnormalities were also seen in noninjured areas. Metabolically compromised areas may or may not be subjected to histopathological damage. Ten days later, depressed lCMRglu was reported in the cerebral cortex, hippocampus, thalamus, and amygdala. Axonal damage leading to diffuse deafferentation is a consistent finding in all levels of TBI. This impairs exchange between cortical and subcortical regions and limits the responsiveness of remote circuit relay stations to peripheral activation. Examples were offered of brain stem dysfunctions, and functional and behavioral consequences of TBI (Dietrich et al., 1994).

IMMUNE AND INFLAMMATORY REACTIONS

Microglia secrete immunoregulatory cytokines, which respond to CNS damage including trauma (neuronal debris, blood clots, and dead cells) with microgliosis (cell proliferation), changes of structure, and immune cell location (trafficking). Inflammatory cells (microglia and invading macrophages) release cell-destroying inflammatory cytokines and cell-promoting factors. Cytokines are produced in both the IS and CNS. The inflammatory reaction (utilizing the chemical NO as well as other functions) involves different cell types—endogenous (microglia and astrocytes) and exogenous (blood-borne cells infiltrating areas where neuronal degeneration occurs) (Ziaja et al., 2007). IL-1 produces its effects in distant and specific brain regions by activation of cortical glutamergic pathways. By altering the structure of damaged neural tissue, these cells influence the pattern and degree of functional recovery (Allan, 2000; Giulian, 1994; Marion et al., 2004). Irreversible secondary damage develops postinjury over a period of hours to days.

TERTIARY PHASE: LATE-DEVELOPING PHYSIOLOGICAL DISORDERS

The direct physiological and medical consequences of damage to the neurological input and endocrine feedback of the hypothalamo-pituitary-target endocrine axes and other signaling or endocrine tissues.

Posttraumatic endocrine disorders include the anterior and posterior pituitary, thyroid, gonads, adrenal cortex and medulla, and other glands. Disorders include physiological developmental disorders of children, posterior pituitary disorders (diabetes insipidus; syndrome of inappropriate antidiuretic hormone [SIADH]) from head trauma, and hormonal disorders such as hypothyroidism, hypopituitarism, and growth hormone deficiency. For hormonal disorders consequent to chronic injury and stress effects, see Pentary Phase.

QUATERNARY PHASE: LATE-DEVELOPING NEUROLOGICAL CONDITIONS

TBI is strongly associated with some neurological disorders 6 or more months after injury. Clinicians should closely monitor their patients for development of these disorders. Late-developing neurological conditions include stress-related hippocampal damage, premature cerebral atrophy, enhanced incidence of Alzheimer disease with a genetic predisposition (Blumbergs, 2005), posttraumatic epilepsy, movement disorders, cochlear and vestibular dysfunctions, and Alzheimer and Parkinson disease with "mild" with LOC or moderate-to-severe TBI. Late-appearing movement disorders may occur after aberrant healing, ephaptic transmission after injury, remyelinization, or late inflammatory changes (Goetz & Pappert, 1996). The age of trauma may determine the interval before onset. Hemidystonia associated with brain damage has a longer latency in children injured before age 7 than in adults (Krauss & Jankovic, 2002). At present there does not seem to be sufficient evidence to link TBI with amyotrophic lateral sclerosis (ALS) or multiple sclerosis.

Neural plasticity is observed after deafferentation (experimentally produced in animal limbs). It is manifested as shifts in the representation of some other structures nearby on the cortical somatotopic map, replacing the lost input to the deprived territory. After time, the change can become a large magnitude of the sensory representation, with varying representation likelihood of substitution of the representation of particular structures (dominant or latent), some of which may unrelated. For example, with sufficient hand deafferentation, facial neurons may substitute for hand cortex (Florence, 2002).

PENTARY PHASE: CHRONIC STRESS EFFECTS UPON PERSONALITY AND HEALTH

The specific disorder known as posttraumatic stress disorder (PTSD) is only one of varied (and interactive) responses to stress (Yehuda et al., 1995). The physiological effects of TBI may elicit symptoms of PTSD, whose characteristics may be in the opposite direction to those of TBI (Edwards & Clark, 1997). The issue of the definition of PTSD requiring memory of the event is controversial. After TBI with LOC, the events of the accident may not be remembered, a distinction to the classical definition of PTSD.

Chronic stress reactions are consequent to persistent emotional distress, such as anxiety and depression, the psychological results of injury such as impairment, destruction of the quality of life, and the adaptive and neurobehavioral effects of unhealed injuries ("the unhealed wound"). There is evidence of *psychosis* appearing 2–3 years after the injury. Late physiological disorders and chronic physical injuries ("unhealed wounds") result in pain, impaired musculoskeletal performance, and dysregulation of the hormonal, inflammatory, and immune systems. The latter ultimately creates vulnerability to communicable disease, loss of stamina, and stress-related diseases, with neurobehavioral consequences. Late-stress and personality disorders includes such symptoms as change of identity centering on loss of self-esteem, vulnerability to further injury, mood disorder, reduced sexuality, poor morale, and emotional dysregulation (e.g., "frontal lobe syndrome.") Trauma is associated with altered host defense and susceptibility to infection (Buzdon et al., 1999). The trauma may persist for so long that its obscure origin is not recognized or considered. The patient is considered to be a malingerer or exaggerator, or treatment is based upon an incorrect diagnosis (see Credible Evidence).

DEVELOPMENTAL PROBLEMS OF CHILDREN

Injury can cause varied developmental delays that are commonly not attributed to the traumatic cause, such as disorders of physiological maturity; cognitive, personality, and sensorimotor functioning; social adjustment; educability; and capacity for employment.

TRAUMATIC BRAIN INJURY AND HOMEOSTATIC DYSREGULATION

Cerebral

Constant blood flow, locally maintained, ranges from approximately 70 to 150 mmHg. Cerebral perfusion pressure (CPP) is the difference between mean arterial pressure and intracranial pressure (ICP). Different measurements are taken for global and regional blood flow. The lower the CBF within the first 6 h, the greater the occurrence of global hypoperfusion; after 6 h global ischemia is uncommon except when it is associated with secondary injuries. When autoregulation is intact, decreases in BP cause a rise in ICP because of compensating vasodilation. In contrast, the opposite effect (ICP reduction with BP increase) is less likely because the induced vasoconstriction causes smaller CBV changes. The constant level of CBF in the presence of blood pressure fluctuation is achieved by the functioning of resistance-regulating arterioles of the cerebral circulation (Doberstein & Martin, 1996). Head injury results in dysregulation of cerebral homeostasis. The brain is more vulnerable to secondary phase injuries such as hypotension (i.e., less capable of maintaining an optimal CBF) (Engelborghs et al., 2000).

When cerebral perfusion is threatened, the brain's influence (humoral and neural) over the vascular system redirects flow from other circulatory districts to the cerebral circulation. With reference to local changes in CBF, activity-induced changes in hemodynamic responses require complex signaling mechanisms—*neurovascular signaling* depends on local circuitry, cellular composition, vascular architecture, the predominant signaling system, and the balance between excitation and inhibition. Specific activity-induced increases in CBF are the result of coordinated interactions between perivascular neurons, astrocytes, endothelial cells, and smooth muscle cells. This unit generates, coordinates, and transduces molecular signals that underlie changes in blood flow (Iadecola, 2004).

Cerebrovascular resistance (CVR) is expected to compensate for fall in BP. Normal autoregulation is represented by a rapid return to the baseline after an interventional reduction in arterial BP (Martin et al., 1995). *Normal CBF in adults* is stated to be 50 ± 10 mL/100/min. In children these values are elevated and may peak from 3 to 8 years in the 70 mL/g/min. After acceleration injury, cerebral metabolic rate of oxygen ($CMRO_2$) falls dramatically, more than the CBF. Cerebral perfusion is variable: normal, elevated, or reduced (Britt & Heiserman, 2000; Dewitt & Progh, 1997). *Diffusion* refers to the passage of water in tissue. In one model, CBF equals the ratio of cerebral blood volume (CBV) to mean transit time (MTT) (Wu et al., 2003).

Cellular: Glucose and Oxygen

Posttraumatic changes in cellular homeostasis and metabolism after TBI may explain vulnerability to cellular degeneration. A metabolic cascade affects the integrity of axons, from which they may recover or degenerate (Giza & Hovda, 2000). Experimental studies indicate that *glucose metabolism* may not stabilize for 10 days. This helps to understand the grave brain damage resulting from a second injury (e.g., in sports) following an initial trauma (i.e., incapacity for the hyperglycolysis necessary to adequately respond to the second ionic perturbation).

Immediately following impact that creates *shearing injury*, massive ion fluxes across the neuronal membranes cause widespread loss of resting potential and release of neurotransmitters into the extracerebral space. Within a few minutes, an intensely energy-dependent process occurs, resulting in an abrupt increase in glucose utilization. The attempt to restore ionic homeostasis involves reuptake of neurotransmitters and ion pumping. The increase of glucose metabolism is brief and maximally localized in those parts of the brain maximally deformed by the shearing force. Should there be contusions, hematoma, or focal infarction, glucose use increases for 5–7 days, but later decreased metabolism of oxygen and glucose for 1–4 weeks after impact. Glucose use is depressed in the event of coma (Bullock & Nathoo, 2005).

Hormonal Homeostasis

Hypothalamic damage interferes with maintenance of body homeostasis (Treip, 1970). Homeostatic regulation of hormone levels is vulnerable because it is controlled through feedback loops from the endocrine glands that maintain set points rather precisely by feeding back to the hypothalamus and the pituitary. Hypothalamic influence is positive in all instances except secretion of prolactin (PRL), so that damage causes a release of PRL (Molitch, 1995).

There may be prolonged posttraumatic *impairment of brain energy metabolism* (Geeraerts et al., 2006). Calorie and protein needs are elevated with acute TBI (Redmond & Lipp, 2006). The brains of very young children and older adults are more vulnerable to vascular damage (*hemorrhage*) created by shearing forces (see Physical Principles) (Bullock & Nathoo, 2005; Reilly & Bullock, 2005). Animal study of percussion TBI indicates regional changes (ipsilateral cerebral hemisphere) in glucose metabolism of acute (hours) increase and chronic (weeks) decrease in metabolic rates. Changes in the cell membrane and BBB were inferred, although the latter would not account for increased glucose metabolism across the entire hemisphere. The relatively short period of *metabolic depression* suggested neurological *sparing* when compared to older animals (Thomas et al., 2000).

Also occurring are *loss of cerebral autoregulation*, *connection disruption* (tissue disruption interferes with nearby long axonal pathways), *diaschisis* or interference with the functioning of centers with which the damaged neural centers are integrated (e.g., cerebellum to contralateral thalamus), *cellular and molecular* biochemical responses that cause neurotoxic effects upon neural membranes, and cellular messenger systems causing necrosis.

SPORTS INJURIES

Epidemiology

In 1998, The National Health Interview Survey (NHIS) estimated for the 12 months prior to 1991, approximately 306,000 (20%) of the estimated 1.54 million brain injuries occurring in the United States, 20% were attributable to sports or other physical activity (i.e., an incidence of 124 brain injuries per 100,000). It is noted that the milder and medically unattended brain injuries may be underreported because of lack of awareness about their occurrence. The findings are divided between competitive sports (111,000; led by basketball, baseball, and football) and recreation (105,000; led by playground activities, swimming and water sports, skiing and other snow sports, skating [in-line, roller, board]), horseback riding, exercise, and weight lifting). Of these injuries, 100,000 were concussive (Lovell & Collins, 1998). Estimations of hospitalizations or deaths vary between 7000 and 35000. During a 6-year period there were 249,000 trampoline injuries in children 18 years old and younger treated in hospital emergency departments in the United States (Smith, 1998). A study of high school athletes revealed that 5.5% of reported injuries were MTBI (Powell & Farber-Ross, 1999). Of these, football injuries were the most numerous (63.4%), followed by wrestling (10.5%), girls' soccer (6.2%), and boys' soccer (5.7%) (see Sports Injury and Return to Play). It is estimated that between 4% and 18% of sports injuries involve the maxillofacial region. Neck injuries resulting in quadriplegia remain a problem due to lack of equipment suitable to prevent it (Cantu, 1996). A football player's chance of experiencing concussion has been estimated to be as high as 19%, with cumulative concussions possible (i.e., the *second impact syndrome*) (Lovell & Collins, 1998). This concept that rapid occurrence of concussive accidents on the same day in children could cause catastrophic edema has come into doubt (Ropper & Gorson, 2007).

According to the Catastrophic Sports Industry Registry (CSIR), cited by Cantu (1998), the greatest incidence of catastrophic head injury occurs in football, gymnastics, ice hockey, and wrestling. Additional sports with a high risk of head injury are horseback riding, skydiving, martial arts, and rugby. The most common head injury victim in American sports is the teenage male football player. School sports with a significant incidence of head injury include pole vault and headfirst slides in baseball. Professional boxing has the highest number of deaths recorded (Cantu, 1998). Contact sports lead to TBI, with 250,000 concussion and eight deaths occurring every year in football. Other characteristic injuries involve boxing, martial arts competition, and high-velocity collisions in basketball, soccer, and ice hockey. Inherently dangerous sports are boxing, football, rugby, ice hockey, and mountain climbing. In gymnastics, most injuries occurred during training, with their severity strongly associated with the skill level of the gymnast. Advanced-level gymnasts suffered more serious injuries such as concussion and direct orofacial injuries. Experienced youthful skiers between 5 and 18 years old are noted for excessive speed. Loss of control was implicated since 58% involved collisions with stationary objects and helmet use was negligible. The average cost was $22,000. There were no deaths, but 26% of the victims had long-term sequelae (Shorter et al., 1996). Snowboarding has injury characteristics (including head, face, and spine) comparable to skiing accidents, although the overall incidence of injuries may be higher. The typical injured snowboarder is a young male, with drug or alcohol use in 6.8% of accidents and 18.6% injured participating for the first time (Chows et al., 1996).

Subconcussive Impacts

There has been considerable public interest in the possibility that repetitive sports injuries ("*subconcussive impacts*") create traumatic encephalopathy, depression and suicidal thoughts, and other behavioral disturbances (Schwarz, 2007a). Some injuries are routine, others are violations of rules (i.e., helmet-to-helmet impacts) (Thamel, 2007), while still others are directed with intent to disable (boxing). With the specific intent of producing brain injury, the uppercut is a rotating force greatly exceeding 600 kg that is delivered in less than one tenth of a second. A jab (linear hit) is much less likely to cause a concussion than an uppercut of the same force. Brain protein markers of trauma are detected. Twenty percent of professional boxers develop dementia pugilistica. The delayed pathophysiology of boxing may cause deficits months or years later, even though the acute injury is not apparent (Robinson, 2007). One group of college athletes was examined a mean of 89.4 days after an MHI. The conclusions are based upon a combination of postural and EEG findings. There was decrease in EEG power in all bandwidths studied in concussed subjects, especially in standing postures. There was sustained postural instability, especially under a no-vision condition. It was suggested that long-term residual balance problems may be observed with appropriate research methodology. There was a change in EEG delta amplitude between sitting and standing positions. These individuals had been cleared for sport participation based upon standard clinical symptoms resolution (Thompson et al., 2005). As will be detailed later (see Child TBI) the brains of college players (17–18 years old) are still in development, creating vulnerability for later impairment of higher-level cognitive functions. Development of symptoms of impairment can be concealed by the desire not to appear weak or to be removed from the team, and also lack of awareness by players and coaches.

Repeated Concussions

When football players have been knocked unconscious, some have returned to play on the same day, as is reportedly consistent with the National Football League's committee on brain injuries. Observations suggested that there was no risk of further concussion or prolonged neuropsychological symptoms. It was separately acknowledged that the data for some players' LOC was incomplete (Schwarz, 2007b). Hockey and football have been described as having a culture of "playing hurt" (Zinser, 2007). Coaches have been accused of putting premature pressure on players to return to play soon after head injuries. The possibility (or probability) of prior concussions may not be considered in a decision about whether a player should continue immediately. The writer notes that sports injuries have far less impact and acceleration changes than such accidents as MVA, falls from a height, and falling objects. Furthermore, when considering the following findings, the issue of whether there were any comorbid injuries is not specified. Sports injuries represent a different sample of concussion than that of the general population. It is highly conditioned, with different psychological and motivational characteristics than most other groups. A survey of findings suggests that mTBI is characterized by an initial physiological phase, followed by a several-day stage of acute physiological effects, followed by longer-lasting effects influenced by a range of psychological or motivational factors (Barr, 2007).

Repeated concussions within a short period (characteristic of athletes) can be fatal (*second impact syndrome*). These predispose the brain to vascular congestion from autoregulatory dysfunction. Experimental animal studies indicate that the axons swell after mild trauma (Kelly et al., 1991). With increasing numbers of bouts, boxers manifested both worse psychometric performance and more cerebral perfusion deficits as manifested by SPECT (Kemp et al., 1995). One study compared football players who had a baseline study with controls (Collins et al., 1999). On a symptom inventory, the group with no concussions expressed fewer symptoms than those who had single or multiple concussions. Those with two or more concussions performed worse on Trials B (attention and concentration) and the Symptom Digits Modalities Test (information processing speed). It was concluded that a history of concussion is significantly and independently associated with long-term deficits of executive functioning, speed of information processing, and an increase in self-reported symptoms. Furthermore, there was a significant interaction between the diagnosis of learning disability (LD) and having incurred two prior concussions, suggesting an additive effect of LD and multiple episodes of concussion on lowered functioning. Athletes with LD and multiple concussions performed in the brain-impairment range on the above measures. It was suggested that experiencing two or more prior concussions is associated with a lessening of cognitive skills, which, when combined with the deficits associated with LD, leads to even further compromised functioning. This would make academic achievement even more difficult for those athletes with multiple concussions who represented 20% of this sample.

Both "minor" and catastrophic brain injuries occur with significant cognitive dysfunctions caused by repeated minor head injuries (Warren & Bailes, 1998). The most serious injuries occur when the head is used as a battering ram with the neck flexed ("spearing position," as in football). Recognition of the danger has led to substantial injury reduction. In soccer, "heading" the ball, as well as the missile-like effect of a kicked ball, create a danger. A nonconcussive "heading" of the ball has been termed "microtrauma" or "subconcussive injury." The magnitude of this risk has not been ascertained because of insufficient studies of preinjury performance and uncertain comparability of control groups. The use of neuropsychological tests for athletes to measure recovery has not had conclusive results. Cognitive decline has been more firmly established for boxers who have sustained multiple knockouts (Ropper & Gorson, 2007).

Return to Play

Several consensus statements were presented by Cantu (2006), which are abstracted:

- Loss of consciousness cannot be relied upon as a measure of concussion severity.
- While an athlete is symptomatic following a concussion, scholastic activities and activities of daily living should be modified as to avoid intensifying or prolonging postconcussion symptoms.
- Pediatric cases could be managed using guidelines similar to those used in caring for adult patients.
- Severity of concussion is better estimated by number, duration, and severity of total postconcussion symptoms was more important than the single symptom of amnesia.
- Motor phenomena such as tonic posturing and convulsive movement, while dramatic, usually has a benign outcome and usually requires no specific treatment beyond the usual concussion management.
- The neuropsychological examination should never be a sole criterion to determine when an athlete should be allowed to return to play. It definitely has value but should not be performed until all signs and symptoms have resolved.
- Before any athlete is allowed to return to play, he or she should be asymptomatic at rest and exertion and must remain asymptomatic with exertion.
- Other considerations were the severity of the current injury, whether the injury appeared excessive for a minor blow, age, and possible implications for the developing brain, whether the athlete had any learning disabilities, the sport involved, and abnormal results on neurological exam. Concussive symptoms manifested on exertion and significant abnormalities on cognitive testing or imaging studies.

DAMAGED BLOOD-BRAIN BARRIER (BBB)

The BBB is involved in the status of the CNS's *immune privilege* (i.e., the restriction of the immune response). A second mechanism is the *immunosuppressive microenvironment*. The injured CNS does require immune intervention to limit damage and activate healing, but such intervention is limited because of the CNS's immune-privileged character. Opposing functions occur: inflammation interferes with CNS protection while actually participating in CNS repair. In fact, it is an autoimmune response (Schwartz, 2000). Glucocorticoids penetrate the BBB and also regulate it (Miller & Chrousos, 2001). Increased oxygen radicals contribute to damage to the microvasculature and the subsequent breakdown of the BBB. Cerebral contusion (Blumbergs, 2005) leads to breakdown of the BBB and also creates an inflammatory response, which contributes to edema (Tokutomi et al., 2003), including upregulation of proinflammatory cytokines within minutes, followed by induction of cerebral adhesion molecules and accumulation of neutrophils. Cerebral vessels dilate with an increase in the CSF (which essentially lacks the blood brain barrier) of hydrogen ion concentration and adenosine (which occurs with reduced oxygen supply, seizures, and increased carbon dioxide) (Berne & Levy, 1998). What is described as "mild" head trauma (i.e., coup and contrecoup injuries) and diffuse axonal injury from shear and tensile strain damage causes the release of excitatory neurotransmitters (e.g., acetylcholine, glutamate, and aspartate). This may result in impaired cerebrovascular autoregulation (Evans, 2001). Arterial hypotension or increased intracranial pressure can result in lowered cerebral perfusion pressure. Ischemia and luxury perfusion occur at different posttrauma periods (Nichols et al., 1996). Sudden increases in BP can be transmitted to the brain's microcirculation, contributing to secondary hemorrhage and edema. *Loss of autoregulation* prevents adjustment of cerebrovascular resistance. Cerebral blood flow (CBF) becomes entirely dependent on cerebral perfusion pressure (CPP). The rise in ICP with subsequent decrease in CPP can lead to ischemia (Miller et al., 1996). Loss of autoregulation may occur in some areas but not in others (Miller & Gudeman, 1986).

Astrocytes, glutamate, glutathione reduction: Astrocytes are responsible for neuronic, metabolic, and trophic support (Bambrick et al., 2004). They are the principle repository of brain glycogen while neurons have very little glycogen. Ca^{2+} influx into cells is the most significant event in the pathogenesis of ischemic brain damage. The principle route is through NMDA receptors. Mitochondrial dysfunction plays a central role in cerebral ischemia/reperfusion cell injury and death, including alteration of Ca^2 in various intracellular compartments. Astrocyte glutamate uptake and glutathione production are important for neuron survival:

(1) Massive release of glutamate occurs during ischemia. The consequent excitotoxicity plays a major role in neuronal death. Ischemic impairment (glutathione reduction, oxidative damage, impaired active uptake of glutamate) may promote neuron death, even in the absence of astrocyte cell death.
(2) Glutathione is an important brain antioxidant, as it acts as a free radical scavenger. It is important for maintaining other antioxidants in an active form.

Secondary Cell Injury, Dysfunction, and Death

Within the context of the IS, there are two forms of cell death—*apoptosis* and programmed cell death (PCD) (see Glossary).

Necrosis: After a severe cellular insult, the cell's contents are spilled into the surrounding environment. Their release induces inflammation. It has been suggested that the major physiological role of necrosis-induced inflammation is to induce a tissue repair response (i.e., injury). This is controlled by the innate immune system, which detects the molecular patterns of self and non-self. Necrotic cells express markers that induce tissue repair (i.e., cells of the innate immune system discriminate between normal cells and non-self/abnormal cells, and ignore, phagocytose, or kill them as appropriate) (Medzhitov & Janeway, 2002). However, the response of phagocytes is to release cytotoxic factors that contribute to further loss of neurologic function (Giulian, 2004). Necrosis and apoptosis are discussed in the inflammatory response.

Cellular injury may be caused by gross mechanical force, disruption of the membrane, or disruption of the membrane's capacity to act as a controlling barrier between intracellular and extracellular solutes, inadequate oxygen (hypoxia) or ischemia (inadequate blood supply), and toxins.

Oxidative stress (*peroxidation*) likely plays an important role in posttraumatic brain tissue damage and edema formation. Early after neurotrauma, there is oxygen radical (high valence unbound and highly reactive species of oxygen) formation. Reperfusion following resolution of ischemia or vasospasm leads to additional neurological injury, such as phagocytic damage to the endothelium and surrounding tissues (Nemeth et al., 1997) and the release of oxygen-derived free radicals (Kirsch et al., 1992). This creates damage to vascular, neuronal, and glial membranes, with excitotoxic, intracellular calcium overload, and excitatory amino acid release (i.e., glutamate) (Hall, 1996). There is ischemia-related augmentation of active oxygen compounds, which is accelerated by iron present in the brain tissue. Brain tissue is sensitive to peroxidation processes. Since there is a poor content of antioxidant agents, enhanced cell membrane lipid peroxidation is likely. Brain contusion is accompanied by enhanced lipid peroxidation persisting for at least 10 days postinjury. Enhanced lipid peroxidation processes seemed to correlate with the severity of head injury (contusion) (Kasprzak et al., 2001).

Enhanced intracellular calcium mediates a variety of toxic effects. Damage to the exoskeleton leads to blocks in axoplasmic transfer, accumulation of axonal materials and ultimately delayed axonal disruption or secondary axotomy.

Inflammatory processes occur within 24 h of acute brain trauma, with leukocytes accumulating and macrophages secreting cytokines that create neuroendocrine and metabolic changes, potentially neurotoxicity, and play an important role in the response of tissue to trauma and infection (Oppenheim & Johnson, 2003). Cytokines are released primarily in the cerebral white matter. Their analysis at a wound site (along with the dynamics of astroglia and inflammatory cells) may help in reconstructing the "crime scene" by offering information concerning the age of a wound and the case of death (Takamiya et al., 2007). Traumatized leukocytes produce interleukins and neurotrophic factors (Dietrich & Bramlett, 2004). The initial cell destruction and injury is continued with delayed secondary events that contribute substantially to neuronal cell loss and dysfunction.

Immune system: Immunostaining of brains of individuals with mild head injury with recorded loss of consciousness as short as 60 sec, surviving up to 99 days after the injury, reveals multifocal axonal injury indicated by an antibody for *amyloid precursor protein* (APP) (Gennarelli & Graham, 1998). This is a marker for impaired axonal transport (Gaetz, 2004). While all types of focal axonal injury immunostain for β-APP, it raises the question of the higher incidence of *Dementia of the Alzheimer's Type* (DAT) after head injury.

Excitotoxicity, mediated through NMDA and muscarinic acetylcholine receptors, may contribute to loss of CA1 pyramidal cells one week later. *Influx of* Ca^{2+} leads to cell damage, secondary cell death, or axonal malfunction and breakdown of neurofilaments and microtubules. *Lactate production* and accumulation is increased. Elevated lactate has been implicated in neuronal dysfunction by inducing acidosis, membrane damage, altered BBB permeability, and edema. It may leave affected cells more vulnerable to secondary ischemic damage. Pathophysiological damage (intracellular) adds to mechanical trauma (e.g., enhanced excitatory amino acid transmitters and oxygen-free radicals). Magnesium is involved with cellular energy metabolism, initiation of protein synthesis, maintenance of cellular membrane potential, and voltage gate control for NMDA receptors. Intracellular levels of Mg^{2+} may remain decreased for up to 4 days. It is associated with posttraumatic neurologic deficits in rats, including neurologic motor deficits.

Neuronal Cell Loss

Closed head injury can combine DAI, contusion, tissue tears at the grey–white matter junction, and neuronal degeneration. Chronic activity deprivation may lead to compensatory changes in postsynaptic receptor abundance (*quantal scaling*) (Davis, 2006). The initial cell destruction and injury is continued with delayed secondary events that contribute substantially to neuronal cell loss and dysfunction.

The process of diffuse axonal injury may take several hours to complete, with various changes to the cell's membrane and contents until the axon separates (Miller et al., 1996). This includes neurochemical and pathophysiological events, involving neurodegeneration, set into motion by the mechanical injury. There is release of superoxidants that may be directly involved in the degeneration of spinal axons and BBB disruption after injury (Hall, 1996). Traumatized leukocytes produce interleukins and neurotrophic factors (Dietrich & Bramlett, 2004). *Neurotoxicity* is mediated by acute release of EAAs, such as glutamate and aspartate, by initiating a cascade of metabolic disturbances. These produce cell swelling, vacuolization, and eventually cell death. EAA neurotransmitters can produce seizures and CNS neuronal death (i.e., *excitotoxicity*). Hippocampal and other limbic centers playing a role in memory disorders and spatial learning have a high density of EAA receptors. Compounds acting at EAA receptor have improved posttraumatic memory and neurological motor outcome (Smith & McIntosh, 1996). Inflammatory cytokines contribute to tissue destruction (Marion et al., 2004). Posttraumatic cellular changes include massive increases of extracellular potassium, extracellular calcium, decrease of intracellular magnesium, increase of glucose utilization, increased cerebral metabolic rate of oxygen ($CMRO_2$), and lactate accumulation. Reduced cerebral metabolism may reflect changes in cerebral perfusion, neuronal degeneration, anatomic reorganization, or imbalance of excitation and inhibition (Hovda, 1996).

Transneuronal degeneration: Neurons in a circuit interact in more ways than exchange of impulses; their metabolic equilibrium may derive from their interactions. Two types of degeneration are recognized. A *severed axon* may result in atrophy of skeletal muscle, though not usually postsynaptic neural effects. With *cellular* damage, transneuronal effects occur in the CNS and the PNS in a retrograde and orthograde direction. Neurons up to the tertiary level may become involved (Angevine, 2002).

CEREBRAL CIRCULATION DISORDERS

Cranial circulation and the content of the intra- and extracellular fluids are vulnerable to trauma of the neck and cranial contents. There are symptoms of the PCS and greater trauma attributable to neck injury (vasculature, nervous system, and vasospasm), including alterations of consciousness and balance. Cerebral circulation takes place within a rigid structure, the cranium, so that increase in arterial inflow must be associated with comparable increase in venous outflow. Interruption of the blood flow for as little as 5 sec can result in loss of consciousness, while ischemia for a few minutes results in irreversible tissue damage. Under circumstances of low pressure, there is vulnerability to infarction in the "*watershed areas*." These lie at the boundaries between major intracranial arteries, which are last in line for blood supply. The neurobehavioral disorders occurring will depend upon the injured under-perfused areas (i.e., motor, sensory, and association) with varied symptoms: motor, speech, cognitive, and sensory (Snyder & Daroff, 2004). The *ischemic penumbra* is the area of reduced perfusion surrounding the infarct (Santiago & Fessler, 2004; Wu et al., 2003).

CONCUSSION AND REDUCED CEREBRAL CIRCULATION

Early study of the association of postconcussion symptoms with slowed cerebral circulation was performed by Taylor and Bell (1966). Initial lack of symptoms and normal circulation time may be followed 3 days later by complaints of postural dizziness and headache, when circulation time is increased. This was attributed to increased arteriolar tone. Symptoms mostly abate when circulation time returns to normal. Cerebral autoregulation may remain intact for a few days and then deteriorate temporarily (Strebdel et al., 1997). When amateur sportsmen who did and did not box were compared with psychometric tests and SPECT, the nonboxers performed more efficiently on psychometric tests and those with fewer bouts performed better than the more experienced boxers. The nonboxers had fewer regions of reduced cerebral perfusion (Kemp et al., 1995). Measurements taken 1 week to 3 years after injury of patients (those with lawsuits or unsettled insurance claims excluded) manifested reduced blood flow volume, as shown by increased circulation time and decreased amplitude. There may be a parallel symptom display and increased circulation time for several weeks. Referring to damage to the lowest 4 cranial nerves (Davies, 1997), vascular aetiology has been proposed, suggesting brain stem ischaemia secondary to vertebral artery spasm or compression. Unilateral dysfunction of nerves IX and X results in a unilateral paralysis of the larynx and soft palate, the aetiology usually being vascular, but also following trauma.

NECK INJURY AND CONCUSSIVE SYMPTOMS

Varied symptoms ensuing are sometimes misattributed to cerebral damage when they are consequent to damage to neck and other somatic structures. Due to potential serious neurological sequelae, damage to extracranial arteries following closed head and neck injuries warrants consideration in differential diagnosis (Heilbrun & Ratcheson, 1972). A caution was offered that symptoms may appear several hours after a seemingly almost normal evaluation in the emergency department (Garrett & Hubbard, 1993). A neck injury, following blunt trauma, is usually followed by *posttraumatic dysautonomic cephalgia*—throbbing unilateral headache, ipsilateral mydriasis, and facial sweating. A "minor" head trauma (e.g., "heading" a soccer ball) may be rapidly followed by a headache indistinguishable from migraine. A head injury followed by recurring attacks of migraine can occur in the absence of history of similar headaches (Boes et al., 2008).

Mechanical Factors of Neck/Head Trauma

Shearing effects may injure the ICA as it emerges from the cavernous sinus and create diffuse brain injury as the brain moves past the arterial tree (Bandak, 1996). The combination of carotid occlusion and brain impact is more severe than carotid occlusion alone. Traumatized cerebral vasculature seems unable to respond to reduced perfusion pressure associated with carotid artery occlusion (Cherian et al., 1996).

Cervical Vasculature Trauma

One mechanism is a tear in the arterial intima, occlusion of the vessel, local thrombus formation, and then embolism followed by progressive thrombosis, and then ischemia resulting in neurological deficit (Heilbrun & Ratcheson, 1972). There are PCS symptoms attributable to neck injury (vasculature and nervous system), including alterations of consciousness and balance. Sympathetic fibers are vulnerable to trauma. They reach cerebral vessels by carotid territory via postganglionic fibers that originate in the superior cervical ganglia or by innervation of the vertebrobasilar territory via fibers that arise from the stellate ganglion (Parent, 1996, p. 299). This tract follows the tunica adventitia of the common and internal carotid arteries, possibly innervating the rostral part of the circle of Willis (Mathew, 1995).

The Vertebral Artery

The vertebral artery can be damaged with minor head injury and even normal head rotation. Vertebral artery dissection, occlusion, or *aneurysm* may result from various forms of trauma to the head and neck, including excessive bending in chiropractic, yoga, calisthenics, archery, swimming, ceiling painting, chiropractic, football, wrestling, and diving (Akiyama et al., 2001; Teman et al., 1991). Cervical traction and repeated neck trauma predisposes to vertebral artery dissection. The vertebral artery is relatively fixed between the transverse foramina of C1–2 (diagrams of cervical vertebrae, Moore & Dalley, 2006, p. 1047). Therefore it is easily stretched and narrowed during head rotation to the contralateral side. One case is reported leading to infarction bilaterally of the middle cerebellar peduncles (Akiyama, 2001). The sharply turning vertebral arteries, as they emerge from the cervical atlas to the foramen magnum, are vulnerable to occlusion with head rotation. Stretching and flexion compression that cause vertebral artery injury create symptoms of cerebral ischemia. Flow is not reconstituted in the injured artery (Vaccaro et al., 1998). The vertebral artery is stretched in the region of the atlanto-occipital and atlantoaxial joints during head rotation. While the symptoms appear to be musculoskeletal in origin, the consequences can be vertebrobasilar ischemia, presenting most commonly as a lateral medullary syndrome (Zafonte & Horn, 1999). Disruption of the trigeminal sensory vascular network and damage to the sympathetic cervical chain or to the vertebral artery can be accompanied by numerous PCS symptoms (Jacome, 1986). Occult ligamentous injury to the cervical spine after trauma may contribute the pathogenesis of the vertebral artery by increasing the mobility of the neck.

Vertebral artery obstruction with subsequent occult dissection of the artery can be followed by delayed symptoms, severe impairment, and death (Auer et al., 1994; Garrett & Hubbard, 1993). Vertebrobasilar occlusion after minor head trauma or hyperextending or rotating neck injury is most common in young people. It is accompanied by unilateral or bilateral facet joint dislocations (Hadani et al., 1997). A case described as "*amusement park stroke*" created vertebral artery dissection. This was attributed to neck hyperextension resulting from rotational forces applied to the vertebral artery at the C1–2 levels. This caused dissection of the artery, stenosis of the lumen, local thrombus formation, and then embolism, usually to the posterior cerebral artery (Burneo et al., 2000).

Carotid Artery

Injury to the carotid artery can occur without a blow, from stretching, tearing or compression, or minor trauma from a MVA. Presentation is delayed in onset and further delayed in recognition. Injury to the carotid may be a direct blow to the cervical carotid artery, or stretching caused by turning of the head to the opposite side, associated with hyperextension of the neck that causes it to be stretched and compressed against the lateral mass of the atlas (Chandler, 1990). Injuries of the arterial wall can occur after trivial trauma (Mokri, 2003) or after such sports as taekwondo (martial arts), softball (struck by ball), kickboxing, and French boxing (punches to the face) (Pary & Rodnitzki, 2003).

The internal carotid arteries are also vulnerable to extreme hyperextension or rotation of the head and neck, with fractures possible from the occiput to C6 or compression against the atlas (C1) (Chandler, 1990), particularly at bends when it passes the atlantoaxial joint or leaves the atlas to enter the carotid canal of the petrous bone (Davis & Zimmerman, 1983). Trauma to the internal carotid artery in the neck can cause thrombosis or embolism, resulting in ischaemia or infarction affecting the corresponding cerebral hemisphere. Dysfunctions may be falsely attributed to direct injury to the brain or spinal cord, such as left-sided deafness, facial weakness, hemiplegia, hemianesthesia, left homonymous hemianopia with defective conjugate movement of eyes to the left, and expressive aphasia (Hughes & Brownell, 1968).

Neurological Injury

Among the trauma incurred with rotational shear strain are disruption of the trigeminal sensory vascular network (involved in production of headaches), dysfunction of spinal afferents providing proprioceptive stimulation, and damage to sympathetic cervical chain and vertebral artery damage. Postganglionic sympathetic nerve supply stimulates the pineal gland (Reichlin, 1998). Traumatic lesions of sympathetic fibers can occur without obvious involvement of the adjacent vascular and neural structures, and it may correlate with the severity of "whiplash" injury (Jacome, 1986; Khurana & Nirankari, 1986). Interictal abnormalities include posteriorly predominant spike and wave or sharp and slow wave complex discharges. There are photoparoxysmal responses and, in children, posterior rhythmic slow waves or excessive beta activity. Posttraumatic syndrome is manifested primarily as headaches of the musculoskeletal or tension type. Disruption of the trigeminal sensory vascular network involved in production of headaches could follow brain concussion and rotational shear strain, dysfunction of spinal afferents providing proprioceptive stimulation, and damage to sympathetic cervical chain and vertebral artery damage:

1. Nerve damage: Near the spinal cord, damage can occur to the nerve rootlets (radiculopathy), or the cord itself may be damaged. The nerves can be compressed if tension pulls the nerve over bony encroachments of the cervical spine. In peripheral nerve injuries, adjacent veins can be injured and hemorrhage.
2. Spinal levels: C2-C3: Blurred vision (20/20 vision possible if each eye is tested separately); hyperacusis; tinnitus; otalgia; swelling of face and side of neck; dry eyes or mouth;

increased secretion of eyes or mouth; malfitting dentures; edema of salivary glands; dryness on swallowing; and altered mentation and asymmetrical opening of jaws (Fraser, 1994). Blurred vision is also due to transient disturbances in blood flow to the brain via the vertebral arteries, and also injury to the sympathetic nerves (Croft, 1995b).
C3–4: Increased tone of trapezius.
C7-T1 (Fraser, 1994): Dizziness, pain radiating to the upper dorsal area, heaviness of the head.

3. Vestibular: Cervical vertigo occurs with extension of the neck, rather than movement of the head into a position placing the posterior canal in the plane of gravity. Presumably, it is due to abnormal inputs from joint and muscle receptors of the neck (Herdman & Helminski, 1993).
4. Miscellaneous: Blurred vision, diplopia, horizontal diplopia, bilateral visual disturbances, photophobia, dysarthria, nausea, vomiting, vertigo, dizziness, syncope, confusion, numbness and dysesthesia (face, limb, body), hypoacusis, ear pain, bilateral facial dysesthesias, monoparesis, ataxia (Jacome, 1986). Dysphagia can be caused by damage to the esophagus or larynx, or by muscle spasms (Croft, 1995b). Speech deficits (hoarseness, nasal speech), which can be consequent to isolated injury to the hypoglossal nerve, seem to be associated with either hyperextension injuries to the cervical spine (i.e., atlanto-occipital junction) or with fractures of the occipital condyles. Damage may be unilateral or bilateral, and recovery is variable.
5. "Whiplash trauma" (flexion/extension neck injury) can result in smooth-pursuit abnormalities. Oculomotor dysfunction can be the result of involvement of the cervical proprioceptive system and possibly medullary lesions, that is, oculomotor dysfunctions may be consequent to dysfunction of the proprioceptive system (i.e., cervical afferent input disturbances of integration and tuning) (Heikkila & Wenngren, 1998).

SYSTEMIC EFFECTS OF TRAUMA

During and after illness and injury, important physiological systems function and continue to function. These send substances that are transported to the CNS by the blood stream and nerves (most importantly cranial nerve X, the vagus nerve).

1. Hormonal (stress) system
2. Inflammatory system
3. Immune system

After injury, the primary barrier between the BBB, which usually keeps out many substances, particularly large molecules, is less efficient. Therefore, those chemical products that attempt to heal the body enter the brain more easily. The BBB is not a complete barrier even during health; some substances enter the brain through the *circumventricular organs* (CVO). Not all tissue with inefficient BBB is part of the CVO system. The thalamic area in the region of the intergeniculate leaflet cells (afferent input to the suprachiasmic N. [i.e., biological clock] may have an attenuated BBB).

After a head injury, the acute reaction includes hypotension, hypoxia, hyperglycemia, anemia, sepsis and hyperthermia, increased systemic and ICP, and also hypoperfusion with hypoxemia, and hypercarbia (Dietrich & Bramlett, 2004; Marion et al., 2004). The systemic effect includes the acute-phase response, increased energy expenditure (hypermetabolism) in severely injured patients, hypometabolism for a period of up to 10 days in experimental animals (Hovda, 1996), hypercatabolism, increased protein turnover, altered vascular permeability, altered mineral metabolism, increased cytokine and hormone levels in the blood, urine, and cerebrospinal fluid (which peptides may be involved in secondary neuronal injury [e.g., IL-1, IL-6, and TNF]), hyperdynamic cardiovascular state, gastric ulceration, and altered glucose metabolism. Hyperglycemia may cause secondary injury to the brain. Hypoxia secondary to TBI and mild hypotension affects neurotrauma and outcome (Dietrich & Bramlett, 2004). *Average energy expenditure* increases 40% after brain injury; another estimate is 40–100% (Dwyer, 2001). The period of increased *metabolic rate* after brain injury may extend for as long as one year. Increased metabolic rate may be mediated by hormonal surge (serum insulin, cortisol, glucagons, catecholamines), cytokine increase, or brain inflammation. Severely brain injured persons lose weight (i.e., decreased muscle mass and visceral protein turnover). The *gut* expresses delayed gastric emptying, decreased pH, and increased gut permeability. The *liver* stimulates hepatic acute-phase responses (enhanced and decreased synthesis of particular proteins), elevated liver function tests (probably caused by inflammatory cytokine effects), and altered gene expression leading to altered rates of synthesis of various substances mediated by various cytokines (Young & Ott, 1996).

Posttraumatic Inflammatory Response—Acute Phase

The inflammatory component is a cytokine-mediated response (see Sickness Behavior) to infection and some injuries (Parslow & Bainton, 2001). In addition to infection, it is triggered by trauma, burns, and tissue necrosis. It occurs after severe head injury, including fever without evidence of infection, protein changes, and enhanced levels of zinc and copper with effects beneficial and detrimental to recovery (Young & Ott, 1996). It is considered as a primitive nonspecific defense reaction mediated by the liver, which intensifies some aspects of innate immunity. When hepatocytes are exposed to particular cytokines (IL-1, IL-6, TNF-α), the innate response occurs. These cytokines, together with a poorly defined neural response, give rise to the fever, somnolence, and loss of appetite that occur as part of the acute-phase response. Local injury causes leakage of plasma contents from small blood vessels into the small blood vessels. This is an intermediate phase of inflammation (i.e., increased vascular permeability with increases in tissue water, protein content, and morphology).

Posttraumatic Immune System Reaction

The immune system acute-phase response is designed to limit tissue damage. During injury, the immune system is activated, leading to cytokine production similar to acute infections, and could culminate in endocrine and metabolic responses characteristic of the acute-phase response. Inflammatory cytokines coordinate the systemic response with their effects upon the HPA axis and the CNS, mediating *symptoms of illness* including fever, loss of appetite, social withdrawal, and sleep changes. Varying IS molecular and cellular changes occur after *lesions* to anterior hypothalamus, hippocampus, amygdala, and left neocortex.

Stress: Due to the wide variety of external and internal experiences considered stressful, it is not possible to draw generalizations concerning outcome. However, acute and mild stress enhances cellular and humoral immunity while severe or chronic stressors suppress them. Immune-related disorders include infections (upper respiratory rhinovirus), cancer, and autoimmune diseases. Chronic stress correlates with decreased antibody responses to vaccines and delayed wound healing. If cancer develops (recognizing that there are various etiological vulnerabilities) individual differences of stress, coping style, and personality influence morbidity and survival (Raison et al., 2005). The endocrine response to injury may be modulated by nerve endings located in the injured area, whose reactions influence cytokine production. The neuroendocrine system keeps cytokine production under very tight control during the acute-phase response through increased glucocorticoid secretion. Excess cytokine production would be fatal due to neurotoxicity (Berczi & Szentivanyi, 1996). Trauma and immune-mediated inflammatory responses that compromise the BBB cause proinflammatory cytokines to be induced in the brain (Dantzer, 2000). Accumulation of vasoactive eicosanoides and inflammatory cytokines (TNF) cause a partial breakdown of the BBB (Laterra & Goldstein, 2000). Chronic systemic inflammation is accompanied by metabolic alterations that promote *atherosclerosis* (Desiderio & Yoo, 2003). Inflammation and skeletal destruction in arthritis are induced and maintained by cooperative interaction between IL-1β and TNF-α, cytokines that act synergistically to promote inflammation (Campagnuolo et al., 2003). Damaged or stimulated brain structure (including the cortex) affects the immune reaction, which includes reduced immune function after ablation of the nucleus locus ceruleus (Jankovic, 1994).

Pain

Musculoskeletal injury results in recruitment and sensitization of peripheral nerve endings, contributing to hyperalgia. In addition to the release of bradykinin and serotonin, *prostaglandin* is released. The latter stimulates the release of substance P, which sensitizes additional nociceptors adjacent to the injury. Nociceptive impulses create *neuroendocrine responses*. These alter the activity of the hypothalamus, the adrenal cortex, and the adrenal medulla. This *stress response to injury* is characterized by increased secretion of catabolic hormones (cortisol, glucagon, catacholamines), muscle wasting, impaired immunocompetence, and decreased resistance to infection. If the pain becomes chronic, humoral, sympathetic, and reflex tone alterations around the site of injury (e.g., an operation) increase discomfort and disability (Sinatra, 2003).

Fever

Fever as a reaction to injury is an adaptive mechanism of defense mediated via thermoregulatory centers in the anterior hypothalamus (Maier et al., 2001; Ross et al., 2003, p. 222). Inflammation, trauma, and tissue necrosis induce the production of pyrogenic cytokines (IL-1, TNF, and/or IL-6). These induce the production of prostaglandin E_2 (PGE_2). Peripherally, this has the effect of the myalgias and arthralgias that accompany fever. It is the induction of PGE_2 in the brain that starts the process of raising the hypothalamic set point for core temperature to febrile levels (Dinarello & Gelfand, 2001). Skin, joints, or muscles that are inflamed or damaged are *hyperalgesic* (i.e., unusually sensitive with a reduced pain threshold or increased sensitivity to further stimuli). Damaged skin releases chemicals from itself, blood cells, and nerve endings—bradykinin, prostaglandins, serotonin, substance P, K^+, H^+, and so on, which trigger inflammation. When the blood vessels become leaky and cause edema (tissue swelling), nearby mast cells release histamine, which directly excites nociceptors. The spreading nociceptor axon branches release substances that sensitize nocireceptor terminals responsive to previously nonpainful stimuli (Aδ and C fibers) (Connors, 2003).

Late-appearing movement disorders may occur after aberrant healing, ephaptic transmission after injury, remyelinization, or late inflammatory changes (Goetz & Pappert, 1996).

Tissue Damage

The inflammatory response is stimulated by cytokines that are produced by activated macrophages. Inflammatory mediators include prostaglandins, leukotrienes, and platelet activating factor (PAF). Following a wound, injury to blood vessels triggers enzyme cascades: the *kinin system* produces inflammatory mediators, including *bradykinin*, leading to increased vascular permeability and increase of proteins. After TBI, neuroinflammatory processes play dual opposing roles: they promote brain damage by releasing neurotoxins and promote repair of the damaged tissue (Bullock & Nathoo, 2005). The injury itself reduces the immune function, which may lead to delayed recovery (Yang & Glaser, 2005). *Melatonin* may reduce tissue destruction during inflammatory responses in various ways (i.e., scavenging free radicals and lowering the production of such agents as cytokines that contribute to cellular damage) (Reiter, 2000). The BBB plays an active role in the regulation of immune cell and molecule entry, playing a role in neuroinflammation (Biernacki et al., 2005). Although inflammation initiated by an injury is a defense, it is believed to contribute to neuropathology and secondary necrosis (Dietrich & Bramlett, 2004). Pain encourages immobilization. The *coagulation system* is another cascade of plasma proteins after injury to blood vessels (Janeway et al., 2005, p. 47). Thromboplastin is released locally or systematically after a severe injury, activates the intrinsic immune system, and leads to intracranial hematomas and systemic complications (Wilberger, 2000).

Connective tissue injury: Spaces left by tissues that do not continue to divide are filled by connective tissue, forming a *scar*. Skeletal muscle can undergo limited regeneration (Junqueira & Carneiro, 2005, pp. 91–92).

DYSREGULATION BY INJURY AND STRESS

Since physiological signaling and control pathways are multiply determined, an accident involving head and somatic injury and stress contributes to dysregulation (i.e., a variety of overlapping, frequently unrecognized or ignored neurobehavioral conditions). Injury initiates responses in the neuroendocrine control of a wide range of immune responses (Brown & Zwilling, 1996). Hormonal, inflammatory, immune, and psychological systems function continuously as long as there is an "unhealed wound" (tissue injury, dead tissue, interference with normal functioning by scar tissue, pain, inflammation, tissue debris, pain, etc.). Emotional stress directly effects somatic functioning, and indirectly CNS system functioning, through the secretion of neuroactive substances (see BBB, Neural Transport). Ordinarily, various physiological barriers (e.g., BBB) prevent inappropriate somatic substances in the internal milieu from affecting CNS functioning. After trauma, reduced efficiency permits neuroactive substances normally excluded to affect functioning. Circulating factors in the blood (e.g., hormones acting as neurotransmitters) interfere with synaptic communication (Hof et al., 2003). Dysregulated and otherwise disturbed physiological functions may be misattributed to cerebral damage or poor motivation. The physiological consequences are (1) PCS; (2) acute and persistent stress disorders; and (3) disorders of stamina and health. The initial hyperarousal and dysfunction (dysregulation) is ultimately succeeded by organ fatigue or burnout (see Allostatic Overload).

A high proportion of head injuries (including "whiplash") are accompanied by somatic and noncerebral injuries to the remaining nervous system. The immediate effect of the injury is an acute stress reaction: a profound disruption of the basic regularity of physiological functioning permitting survival. The initial stress reaction may condition various neurological systems to persistently affect mood (anxiety), memory (conditioned physiological responses creating hyperarousal), and altered pattern of physiological response. Many of these injuries create permanent tissue damage or are slow to heal. Furthermore, reduced effectiveness of the blood-brain and other barriers permits trauma-induced neuroactive substances to enter the brain, with potential effects upon mood and performance efficiency. If the injury is persistent, significant physiological systems are dysregulated from their basic allostatic condition (hormonal, inflammatory, and immune).

TRAUMATIC DISRUPTION OF THE INTERNAL ENVIRONMENT

Trauma creates considerable disruption of the internal environment. After a head injury, the acute reaction includes hypotension, hypoxia, hyperglycemia, anemia, sepsis and hyperthermia, increased systemic and intracranial pressure, hypoperfusion with hypoxemia, and hypercarbia (Dietrich & Bramlett, 2004; Marion et al., 2004).

METABOLISM

Victims of accidents causing TBI frequently report changes of weight, fatigue, loss of stamina, depression, and oversensitivity to heat and/or cold. Dysregulation of metabolism may be considered to be a contributor. Trauma can cause hormonal changes in metabolism that create secondary influences on the CNS. Examples include depression associated with hypothyroidism, psychosis occurring with glucocorticoid excess and coma occurring with insulin excess (Baxter et al., 2004).

Internal Milieu Signals

An accident with head injury may have comorbid long-lasting injury of multiple parts of the body ("*the unhealed wound*"). The neurobehavioral effects involve neuroactive chemicals influencing the CNS over various pathways in addition to the usual afferent system (e.g., vascular and cranial nerve transmission of psychoactive chemicals). Injured tissue and the long-lasting physiological efforts of the body to heal the wound have deleterious neurobehavioral affects, which are reflected as certain chronic PCS symptoms. Since the origin is somatic and problems exist long after the accident, there is often misattribution of the disorder to brain damage, faking or secondary gain, or the disturbance is completely ignored by a practitioner with a narrow focus of concern. The role of the hormonal, immune, and inflammatory systems will be emphasized. Particular physiological principles will be specified that are neurobehaviorally significant (i.e., homeostasis, allostasis, metabolism, the internal environment of the brain, and signaling systems).

Inflammation

Inflammation is a localized protective response, elicited by tissue injury or destruction, which destroys, dilutes, or sequesters the injurious agent and the injured tissue, protects against infection, and regulates the activities of tissue cells that repair the injured tissue. *Tissue injury*, as caused by trauma, bacteria, chemicals, heat, and so on, is followed by the release of multiple substances that cause dramatic secondary changes (i.e., *inflammation*) (Brooks et al., 2000; Guyton & Hall, 2000, p. 397). Following a wound, injury to blood vessels triggers enzyme cascades: the *kinin system* produces inflammatory mediators, including *bradykinin*, leading to increased vascular permeability and increase of proteins. Pain encourages immobilization. The *coagulation system* is another cascade of plasma proteins (see Trauma) (Janeway et al., 2005, p. 42). Severe and protracted injury causes proliferation of fibroblasts and endothelial cells at the site, which form a permanent scar. Inflammation plays a role in the development of subarachnoid hemorrhage (SAH) (Provencio & Vora, 2005). Since inflammation is triggered not only by microbial infection but also by blunt trauma, radiation injury, lacerations, and thermal or chemical burns, it is not considered to be an immune reaction. However, immune and inflammatory reactions are closely linked and often promote and enhance one another. Dysfunctioning based upon neuroinflammatory processes plays dual, opposing roles: it promotes brain damage by releasing neurotoxins and promotes repair of the damaged tissue (Bullock & Nathoo, 2005). Although inflammation initiated by an injury is a defense, it is believed to contribute to neuropathology and secondary necrosis (Dietrich & Bramlett, 2004). Tissue damage represents a breach of organismic integrity, stimulating the innate IS. There is rapid production of cytokines, chemokines, and prostoglandins that are responsible for autonomic changes of the acute phase. The proinflammatory cytokine Interleukin-1 (IL-1B) is probably the most important molecule modulating cerebral functions during systemic and localized inflammatory insults (Rivest, 2003). The inflammatory response is stimulated by cytokines that are produced by activated macrophages. Inflammatory mediators include prostaglandins, leukotrienes, and PAF.

The BBB plays an active role in the regulation of immune cell and molecule entry, as well as a role in neuroinflammation (Biernacki et al., 2005). CNS effects stemming from cytokines that penetrate the brain, or are produced within it, can promote peripheral inflammation (Lipton et al., 1994). The delicate balance between pro- and anti-inflammatory cytokines is easily disrupted by adverse conditions. The severity of trauma and the duration of hypotension will determine whether tissue repair will be initiated by a localized inflammation or whether there will be a systemic induction of a generalized inflammatory process. Massive stress also results in a significant activation of the anti-inflammatory cytokine TNF and various interleukins representing either immunosuppression or immunoparalysis. Immunodeficiency (depressed immune function) in trauma victims is associated with enhanced concentrations of inflammatory cytokines (Faist et al., 2004). Under physiological conditions, reactions are maintained by cells that circulate or are tissue resident.

Inflammatory Mediators

Inflammation is a localized protective response elicited by tissue injury or destruction. Its function is to destroy, dilute, or sequester the injurious agent and the injured tissue, protect against infection, and regulate the activities of tissue cells that repair the injured tissue. Inflammation causes the release of vasodilators (histamine and cytokines), which increase the number of open capillaries and the functional surface area. As a response to infection it provides macrophages (*microglial cells* in neural tissue) augmenting the killing of invading microorganisms, provides a physical barrier (microvascular coagulation) preventing the spread of the infection, and promotes the repair of injured tissue (described as a nonimmunological role). Vascular permeability is increased (see BBB) (Janeway et al., 2005, p. 45). Inflammation is triggered by blunt trauma, radiation injury, lacerations, or thermal or chemical burns.

The severity of trauma and the duration of hypotension will determine whether tissue repair will be initiated by a localized inflammation or whether there will be a systemic induction of a generalized inflammatory process. Massive stress also results in a significant activation of the anti-inflammatory cytokine force of TNF and various interleukins, representing either immunosuppression or immunoparalysis. Immunodeficiency (depressed immune function) in trauma victims is associated with enhanced concentrations of inflammatory cytokines (Faist et al., 2004). Under physiological conditions, reactions are maintained by cells that circulate or are tissue resident. Severe and protracted injury causes proliferation of fibroblasts and endothelial cells at the site, which form a permanent scar. The changes in inflamed blood vessels attract cells of the immune system into an injured or infected tissue. Injured skin, joints, or muscles have a reduced threshold for pain (i.e., *hyperalgesia*). *Secondary hyperalgesia* results in inflammation involving processes near peripheral receptors and mechanisms in the CNS. Damaged tissue causes a variety of chemical substances to be released from itself, blood vessels, and nerve endings such as bradykinin, prostaglandins, serotonin, substance P, K^+, H^+, and others (Connors, 2005). The dilatation and increased permeability of vessels near injured issue results in part from a spinal reflex: pain receptors stimulated by the injury transmit afferent signals to the spinal cord, where they act on autonomic motor neurons to cause relaxation of arteriolar smooth muscles at the injured site. A neural independent component of the vascular response is triggered by substances produced at the injured site that act directly on the local vessels (see inflammatory mediators). Among the *vasoactive mediators* are histamine, prostaglandins, and leukotrienes (Parslow & Bainton, 2001). The pathways and events vary according to the inciting stimulus, portal of entry, and characteristic of the host. Some outcomes are detrimental. While inflammatory reactions have evolved to inactivate or eliminate injurious substances or limit their spread, these same reactions can injure host tissues or interfere with normal functions (Terr, 2001).

Contributing to *secondary brain injury* is acute inflammation that may be an important determinant of the ultimate outcome. Nitric oxide (NO) may mediate the proinflammatory actions of cytokines and bradykinin and act synergistically with eicosanoids (Hsu et al., 1996).

Components of inflammation: Although infection is a familiar cause of inflammation, we will emphasize the consequences of wounds. Inflammation is characterized by increase of the number of open capillaries increasing the functional surface area; vasodilation of local blood vessels with excessive blood flow; increased permeability of the capillaries allowing leakage of large quantities of fluid into the interstitial spaces; perhaps clotting of this fluid because of leakage of excess amounts of fibrinogen and other proteins; migration into the tissue of large number of granulocytes and monocytes; and swelling of the tissue cells. Enhanced vascular permeability and edema are mediated by prostaglandins, kinins, histamine, leukotrienes, and platelet activating factor (Miller &

Chrousos, 2001). The fibrinogen clots block off the tissue spaces and lymphatics of the inflamed area, delaying the spread of toxic products. The role of the *sympathetic and peripheral nervous system* in mediating swelling and hyperalgesia at remote sites from an injury is called *reflex neurogenic inflammation* (RNI) (this plays a role in reflex sympathetic dystrophy). This process is mediated segmentally at the spinal cord level.

Substance P (SP) is a member of the tachykinin family of hormones, which cause smooth muscle contraction and vasodilation. Its immunological functions include the release of monocyte-derived cytokines and release of inflammatory mediators from lymphocytes and macrophages (Schwartzman, 1996). SP neurons in IS tissues modulate its functions. Activation by SP may facilitate neuronal regeneration or ward off infection or inflammation accompanying injury. Thus, SP could play a role as intermediary between the neuronal and immune systems (Jonakait et al., 1991). SP has multiple inflammatory and immune modulator functions, which interact with some IS cells. The consequence is the release of inflammatory mediators from lymphocytes and macrophages and of monocyte-derived cytokines (Schwartzman, 1996).

The *gastrointestinal immune system* has a mass of cells with immune function approximately equal to the combined mass of immunocytes in the rest of the body. Reacting to food antigens and pathogenic microorganisms, it is involved in celiac disease, inflammatory bowel disease, and Crohn disease (Kutchai, 1998).

Cytokines are released after injury and other challenges resulting in the inflammatory response. Their production by and influence upon leukocytes (white blood cells) led to the generic name *interleukin* (IL), but they are produced by a variety of cells. These and other humoral mediators of inflammation are described as the "afferent limb" of the feedback loop through which the immune/inflammatory systems and the CNS communicate (Tsigos & Chrousos, 1996). Inflammation, as may accompany somatic injury, nociceptive, somatosensory, and visceral afferent stimulation can stimulate catecholaminergic and CRH neuronal systems via ascending spinal pathways. Cytokines elicit a neuroendocrine response, and initiate major metabolic alterations (i.e., protein loss from muscles and catabolism) on the one hand, and selected protein synthesis on the other (Berczi & Szentivanyi, 1996). An example of the neuroendocrine effects of cytokines on homeostasis is the fact that IL-1β is hypoglycemic despite its stimulation of glucocorticoid output and decrease of hepatic glycogen content. This stems from stimulation of CRF-producing neurons in the hypothalamus (Del Rey et al., 1998). Stimulation of the HPA axis, in turn, profoundly inhibits the inflammatory immune response and production of cytokines through cortisol. The endocrine response to injury may be modulated by nerve endings at the site of injury, which influences cytokine production. There is an overlap between the *symptoms of viral illness* caused by the effects of circulating cytokines on SNS nerves (e.g., lassitude, fatigue, myalgia, headache) and subjective reactions after trauma causing TBI (Arnason, 1997). Another possible interfering effect upon memory is indicated by the finding that *cytokines* (released during enhanced immune response) inhibit long-term potentiation in hippocampal slices. The effects of the cytokines are so dire that it is believed that they are normally modulated by cytokine antagonists. These become available during host challenge such as inflammation (i.e., a consequence of an accident-related injury). One of these is the neuropeptide *á-melanotonin stimulating hormone*, which is important in communication among neuroendocrine, immune, and central nervous systems (Catania et al., 1994).

Immune System, Autoimmune Responses, and Other Medical Conditions

Autonomic nervous system activity and neuroendocrine outflow via the pituitary can influence immune function. Cytokines and hormones released by an activated IS can influence neural and endocrine processes. Regulatory peptides and receptors are expressed by both the nervous and the immune system, and each system is thereby capable of modulating the activities of the other. Immunologic reactivity can be modified by Pavlovian conditioning. The behavioral and emotional states that accompany the perception of, and the effort to adapt to, events in the real world can influence immune responses. The fact that there were endocrine, autonomic, and neural activity changes during the course of immune responses indicated that the IS could convey information to the CNS. Citing Besedovsky and Del Rey (2001) it is suggested that the IS acts as a receptor sensorial organ. It is accepted (Blalock, 1994) that brain peptides and their receptors exist within the IS and that the products of an activated IS function as neurotransmitters. Sympathetic noradrenergic nerve fibers signal cells of the IS and are capable of evoking major changes in their responsiveness (Felten et al., 1991). They seem to come into apparent direct contact with lymphocytes and macrophages of lymphoid organs (e.g., thymus; bone marrow) and secondary (spleen; lymph nodes) (rodent). These nerve fibers formed close contact with lymphocytes early in ontogeny, appear to influence early immunological development and compartmentation, and diminish markedly with age. Sympathetic noradrenergic nerve fibers are evidence for a connection between the CNS and the IS (Ader, 1996).

Microglia in the Inflammatory Process

Signaling is carried out by the activation of microglia, a response common to all inflammatory and necrotizing processes in the brain. Immunomodulators (mitogens or activators of ameboid microglia) amplify inflammation by action upon intrinsic brain mononuclear phagocytes. This is a regulatory system whereby secretion of brain-derived growth factors or systemic production of cytokines control the microglial cell population (i.e., suppressing the consequences of CNS inflammation).

Inflammatory cells, particularly reactive microglia and invading macrophages, release both cell-promoting and cell-destroying factors. Microglia are a distinctive class of mononuclear phagocytes that persist within the CNS after the perinatal period. They release growth factors, including cytokines, and destroy myelin and myelin-producing oligodendroglia. During CNS injury, mononuclear phagocytes engulf tissue debris. This stimulates the release of cytotoxic factors, which contributes to the ensuing loss of neurologic function releasing thin the CNS after the perinatal period. They release growth factors, including cytokines, destroy myelin and myelin-producing oligodendroglia. Inflammatory responses lead to secondary loss of neurons and deterioration in neurologic function. Microglia are the only brain cells producing significant amounts of IL-1 and TNF-α. Microglia and inflammatory cells release neuron-killing factors, including *N*-methyl-D-aspartate (NMDA)-receptor agonists (Giulian, 1994).

Only molecules foreign to the host are immunogenic (Parslow & Baintain, 2001). Carrying out a variety of immune responses is consequent to cross-talk between immunocompetent cells and the expression of different receptors. To regulate immunity and expression a complex network of communicating signals is used (e.g., cytokines, chemokines, and growth factors) (Luger et al., 2003). Capacity to distinguish between self and non-self is the particular quality of the adaptive IS. Failure to follow instructions taught to T cells produces the anti-self response (autoimmune diseases such as systemic lupus erythematosus) (Kong et al., 2003). Injured tissue from trauma as well as disease carriers are examples of non-self substances. Autoimmune diseases are consequent to the immune tissues misreading ordinary tissues as foreign. The IS is subject to neural and hormonal modulation, including that stimulated by stress. *Neuroimmunology* refers to immune reactions involving brain, nerves, and muscles, while *neuroimmunomodulation* refers to the interaction of the nervous and immune systems (Reichlin, 1998).

Microenvironment: This is Jankovic's term (1994) to describe the components that unify the immune, nervous, and endocrine systems: (1) *lymphoid cells* (T and B lymphocytes); (2) *nonlymphoid cells* (macrophages; epithelial cells, and dendritic cells that interact with lymphocytes and other cells of the microenvironment); (3) *visiting cells* (lymphoid and nonlymphoid which enter the immune milieu and influence the immune network); (4) *neurons* (utilizing various neurotransmitters that influence the microenvironment by neurohumoral activity); (5) *hormones* (released from remote sites by endocrine glands which enter the immune milieu via capillaries and influence the recognition processes); (6) *biologically active substances* (lymphokines, monokines, complement, immunoglobulins) produced by both lymphoid and nonlymphoid cells in situ or elsewhere; (7) *neurotransmitters* (of numerous types and structures, including the classical types, opioids, and other peptides); (8) *membranous and intracellular receptors* (these are molecules that maintain intercommunications and maintain the microenvironmental circuits); (9) *ions* (sodium, potassium, magnesium, calcium) that maintain signals through cell membranes, impacting on immune stimulation and inhibition; (10) *electrical*, *magnetic*, and *electromagnetic fields* created by differences in ionic concentrations and charged molecules in various compartments of the micromilieu; and (11) *higher nervous activity* (mind and psyche) that influences the microenvironment through stress-related, emotional, aggressive, and other behaviors.

The brain-immune relationship involves the cortex as well as the hypothalamus. There is asymmetrical brain control of the IS in rodents and higher incidence of autoimmune and allergic diseases in the left handers. Studies of rats indicate that both positive and negative reinforcement processes exert modulatory activity on the IS immune functions, including ventral tegmental, lateral hypothalamic, and periaqueductal central grey (Jankovic, 1994).

Innate Immune Response

This is immediately available to combat a wide variety of pathogens without prior exposure. It is the first line of defense, but lacks the ability to recognize certain pathogens or to provide specific protective immunity that prevents reinfection (see Adaptive Immune Response). The initial innate response to stimuli from the outside world is converted into the defense mechanisms of the adaptive response, ensuring that innate mechanisms function until the more finely tuned adaptive immunity develops (Zanetti et al., 2003). One component of the innate immunity response is mediated by a family of molecules called *toll-like receptors* (TLR), made by many defensive cells. When they detect an invader, they trigger the production of signaling proteins that induce inflammation and direct the body to mount the immune response (O'Neill, 2005). Another defense is carried out by *phagocytes* (*macrophages*), which cope with disease carriers, or injured tissue from trauma. Macrophages search for invaders (e.g., blood cells leaking from a damaged vessel [bruise] or bacteria). When they detect a foreign protein, they engulf and destroy it, and secrete cytokines, some of which recruit other cells to the site and put the IS on full alert. *Autoimmune diseases* are consequent to the immune tissues misreading ordinary tissues as foreign. The following quotation is from Janeway et al. (2005, p. 13), but refers to disease control: "Bacterial molecules binding to these receptors trigger the macrophage to engulf the bacterium and also induce the secretion of biologically active molecules. Activated macrophages secrete *cytokines* and *chemokines*, proteins using chemotaxis that attract cells with chemokine receptors such as neutrophils and monocytes from the bloodstream" (Janeway et al., 2005, p. 71).

The IS is part of an integrated system of adaptive processes and is subject to some regulation by the brain through a bidirectional exchange of information (Dunn, 1996). Activation of the IS is accompanied by changes in hypothalamic, autonomic, and endocrine processes, as well as changes in behavior. The interaction between the pituitary, endocrine organ, and lymphocyte-derived hormones that define the neuroendocrine milieu in which immune responses occur adds another level of complexity to the cellular interactions that drive immune responses (Ader, 1996). The IS is controlled by an integrated circuitry of limbic cortex, limbic forebrain, hypothalamus, and brainstem autonomic nuclei. These centers have ascending and descending connections and interconnect with the nervous system. CNS and IS interconnection is bidirectional through hormones, neurotransmitters, and cytokines interacting with receptors found on cells in both systems. This aids the organism to respond to external and internal changes that maintain homeostasis through bidirectional signaling. They seem to mediate such behavior as depression, response to stressors, aversive conditioning, and the emotional context of sensory input from the environment and the body. Noradrenergic and serotonergic innervation from the brain stem regulates visceral, neuroendocrine, cognitive, and affective events. Organs of the IS appear to be target organs of direct autonomic stimulation (Felton et al., 1991). Products of the IS communicate directly with specific regions of the brain, directly or indirectly, with effects that include changes in neuronal activity and metabolism of neurotransmitter systems. In turn, these regulate visceral, autonomic, and neuroendocrine functions.

Two pathways link the brain with the IS: (1) autonomic nervous system activity and (2) neuroendocrine outflow via the pituitary. Both routes provide biologically active molecules, which are perceived by the IS via cell surfaces or internal receptors on the surface of lymphocytes, monocytes/macrophages, and granulocytes. Thus, all immunoregulatory processes take place within a neuroendocrine milieu that is sensitive to the individual's perception of, and response to, events of his or her external world. Activation of the IS is accompanied by changes in hypothalamic, autonomic, and endocrine processes, as well as by changes in behavior. Magnification of the potential interaction of neuroendocrine and immune process occurs due to the fact that cells of the IS activated by immunogenic stimuli are capable of producing a variety of neuropeptides (Ader, 1996). Immune function is influenced by the CNS through chemical messengers: pituitary (ACTH, β-endorphin, prolactin, growth hormone, thyroid stimulating hormone). The adrenal medulla secretes enkephalins and dynorphins. Endorphins and other neuropeptides are also secreted by the SNS (Dunn, 1996).

The primary function of the IS is the recognition and protection of the self from foreign invaders. It is second only to the CNS in its complexity and the specificity of its reactions. It is primarily autoregulated, and therefore does not require a major input from the CNS, which nevertheless does play a modulatory role. Abnormalities in each system can cause dysfunctions in the reciprocal system. Cytokines and tumor necrosis factor affect the nervous system by entering areas with a relatively weak BBB (e.g., the preoptic N of the hypothalamus). CRF is released to reduce the production of cytokines (Heninger, 1995). The stimuli responding to be the central and peripheral nervous systems have been described as *cognitive* (physical, emotional, chemical). In contrast, certain stimuli will go unnoticed if not recognized by the IS. These are termed "*noncognitive stimuli*" (e.g., bacteria, viruses, tumors, antigens). When responded to by immunocytes, these structures become information in the form of peptide hormones, neurotransmitters, and cytokines, which are conveyed to the neuroendocrine system. In an analogous fashion, the CNS and the peripheral NS recognize cognitive stimuli, creating similar hormonal information that is conveyed to and recognized by hormonal and neurotransmitter receptors (Blalock, 1994). Through *antigenic recognition*, the IS perceives an internal image of the macromolecular and cellular components of the body and reacts to alterations. This can be considered a "biological" homeostasis (Goya et al., 1994).

Adaptive Immune Response

This specific immune response refers to the production of antibodies against a particular pathogen. The ability to distinguish between self and non-self is the sine qua non of the adaptive IS. Failure to follow this plan results in such *autoimmune diseases* as SLE (Kong et al., 2003). It is essential that immune cells (T cells) do not attack the body. The IS develops in fetal life and is qualitatively quite complete at delivery, although certain cells are present in reduced numbers (Hanson et al., 2003). Self-tolerance (as opposed to autoimmunity) develops as a maturational process in the IS (i.e., *dendritic cells*) (Steinman et al., 2003). The adaptive immune response may confer protective immunity to reinfection with the same pathogen. Its mediator is the *T cell* (*T lymphocyte*), which are derived from the thymus. *Antibodies* are substances that are specific to an infecting pathogen. They are produced by plasma cells, binding specifically to its corresponding antigen, and either neutralize them or prepare them for destruction by phagocytes. Phagocytic macrophages have surface receptors that recognize and bind to constituents of many bacterial surfaces. This triggers the macrophage to engulf the bacterium and to secrete cytokines (i.e., proteins that affect the behavior of other cells that have receptors for them).

In addition to pituitary cells, prolactin is secreted by lymphocytes and T cells, illustrating the multiple sources of biologically active substances (Dardenne & Savino, 1994). The SNS exerts inhibitory control over the IS (Arnason, 1997).

Chemical sensory organ: The cells of the IS both contain receptors for neuropeptides and also are a source of pituitary, hypothalamic, and neural peptides. Immune cells are mobile and deposit their hormones at the target site. These hormones are endogenous regulators of the IS, but also convey information to the neuroendocrine system. It is a sensory function insofar as leukocytes recognize stimuli that are not recognizable by the CNS and peripheral nervous systems (presumably trauma as well as disease entities and components). These are termed "*noncognitive stimuli*" (Falaschi, 1994). Recognition is transformed into information in the form of peptide hormones, neurotransmitters, and cytokines. These are conveyed to the neuroendocrine system and a physiological change occurs. Biological stress can originate in the IS. Immune cells can create a local antinociceptive effect in inflamed peripheral tissue. Cognitive recognition by the CNS and the PNS results in hormonal information being conveyed to hormone and neurotransmitter receptors on immunocytes, resulting in immunologic and neurologic changes (Weigent & Blalock, 1996). The IS's cellular components express specific receptors for almost all types of signaling molecules (i.e., neurotransmitters, steroid and peptide hormones, cytokines, and growth factors). The nervous system influences the immune response by three pathways: peripheral autonomic and sensory innervation of lymphoid organs, hypothalamo-hypophysial axis, and diffuse neuroendocrine system (Geenen et al., 1994). The neuroendocrine system monitors and controls the physical and chemical variables of the internal milieu. Antigenic recognition creates an image of the macromolecular and cellular components of the body to which the IS responds maintaining a kind of homeostasis (Goya et al., 1994).

The Immunosuppressive Response

It is paradoxical that immune function is suppressed when an organism may be injured with the possibility of infection from the attacker and enhances vulnerability to inflammatory diseases. Glucocorticoids are the most potent endogenous inhibitors of immune and inflammatory processes (Crofford, 2002). Immunosuppression increases susceptibility to infections and cancer, and exacerbates inflammatory diseases like psoriasis, asthma, arthritis, and lupus erythematosus. Chronic stress suppresses immune function and increases susceptibility to infections and cancer, but ameliorates autoimmune and inflammatory disorders (Dhabar & McEwen, 2001).

The immunosuppressive response may be based upon the need to avoid an autoimmune response after injury, when the organism might be exposed to fragments of its own cells. One may observe in advance that the immunosuppressive features of the stress response have side effects, including the negative consequence of increased susceptibility to infections (Yamamoto & Friedman, 1996). It can be conditioned, so that reexposure to the conditioned renders the organism more susceptible to environmental pathogens. The ANS plays a role in immunosuppression. Stress-induced immunosuppression might be mediated by depression of interferon and IL-2 (Weiss & Sundar, 1992). The sympathoadrenal system may be involved in ovarian and thyroid hypertrophy (Landsberg & Young, 1992). Stress-induced alterations of immunity, part of the adaptive process, reflect an integration between the CNS and the IS. An experimental study of rats indicates that after shock or in a conditioned fear-producing environment, a number of immune-related responses of the lymphocytes are greatly suppressed. The increased level of stress-related steroids only partially accounts for the immune effects. This role may be played by CRH (Weiss & Sundar, 1992). The immunosuppressive response may vary with dose (i.e., be stimulatory at lower doses).

WOUND HEALING: ACUTE

The purpose of wound healing is to reestablish the normal structure and function of a tissue after injury. Wound repair starts with inflammation (removal and repair of damaged tissue) and proceeds through a series of phases. There is a cascade of cell migration to the inflamed area (*chemotaxis*, which is stimulated by such locally secreted substances as cytokines). *Neutrophils* (a type of blood cell) are among the first cells to enter the connective tissue, aided by such substances as histamine. Neutrophils recognize and engulf the tissue damage. Most die and accumulate as pus. They also secrete pyrogens that stimulate the hypothalamic regulatory center, producing fever. Later, another type of blood cell (*monocytes*) also enters the connective tissue, transforming into macrophages that engulf debris, bacteria, and dead neutrophils (Ross et al., 2003, pp. 222–224). An immune response is generated against "transformed tissue," which is then considered to be an antigen. The effector cells, are *lymphocytes*, which are the effector cells of the IS responding to harmful substances. There are two components of immune response: (1) the antibody or humoral response marks invaders for destruction by other immune cells and (2) the more specific wound-healing cellular immune response, which targets transformed cells for destruction by lymphocytes (Ross et al., 2003, pp. 356–357).

The inflammatory cytokines TNF and IL not only damage the CNS through edema but also initiate wound healing through fibrosis and neovascularization (Bullock & Nathoo, 2005; Rabin, 2005). TNF plays a neuroprotective role in the immune response against excitotoxic, metabolic and oxidative damage. Noteworthy is production of growth factors by the inflammatory infiltrate and wound closure, which is orchestrated in part by cytokines (IL-1; TNF-α). Psychological stress-related *slowing of skin wound healing* is mediated by cytokine and the IS, cytokine, endocrine, and growth hormone dysregulation. *Relaxation/stress intervention methods* may modify the detrimental effects of psychological stress on physiological processes (Yang & Glaser, 2005). *Sensory neurons* are involved in neurogenic inflammation, releasing neuropeptides such as substance P, which induce inflammatory responses. Under the influence of nerve growth factor, neurons grow during wound healing. Cells in various tissues offer a localized protective response elicited by tissue injury or destruction, which destroys, dilutes, or sequesters the injurious agent and the injured tissue. The immune response seems somewhat limited in the brain due to the BBB and other conditions. However, brain cells, microglia, and astrocytes respond to inflammatory stimuli and also secrete cytokines (Chitnis & Khoury, 2004). Immune and inflammatory systems work together in the control and repair of wounds, with inflammation integrated into the nervous and endocrine systems. The wound-inflammatory response CNS effects, stemming from cytokines that penetrate the brain, or are produced within it, can promote peripheral inflammation. Cholinergic nerves may influence the inflammatory process, playing an integral role in host defense (Barnes, 2003).

Immunosuppression

Ordinarily damaged or stimulated structures affect the immune reaction that helps healing and prevents damage from dying cells. The neuroendocrine system affects the timing of the inflammatory process. Various substances may not be adequately expressed to perform their functions and healing will not proceed normally. Knowledge of psychoneuroimmunological processes can identify those at risk for a poor healing outcome. Individualized intervention involves the type of wound, the genetic makeup, and psychological factors. Therapeutic use of glucocorticoids conflicts with wound healing. We are concerned with increased risk of wound infection and 2–5 times the risk of wound complications (Marucha et al., 2001). Since glucocorticoids suppress sensory fibers in the periphery, *chronic stress* could reduce the inflammatory response mediated by these neurons and negatively influence wound healing. Depressed immune function is in conflict with the requirement for the body to defend itself against infection (Young & Ott, 1996). Since the proinflammatory filtrate is critical to the healing process, stress-induced anti-inflamation (glucocorticoids) conflicts with it.

Systemic factors have indirect effects on the wound. Since wound healing requires the expenditure of large amounts of energy, an adequate intake of nutrients is necessary. *Malnutrition* may cause delays in healing or result in chronic wounds. *Aging* has negative effects on wound healing, due in part to decrements of immune function. The SNS also plays a regulatory role in healing. Catecholamines alter blood flow and the movement of inflammatory cells, alter cell recruitment (epidermal cells), increase edema, and delay re-epithelialization. The rate of healing depends upon the location of the wound, the size and depth of the injured area, the presence of bacteria, and the age and overall health of the person.

Process

Wound healing reestablishes the normal structure and function of a tissue after injury. It is a sequential process: each component is dependent upon the preceding one. Since concussive brain injuries are frequently accompanied by somatic trauma, it is necessary to consider the relative rate of recovery for various tissues. Epithelial tissues of the skin and digestive tract regenerate most easily, while liver, bone, and skeletal tissue under repair to varying degrees (Yang & Glaser, 2005). Cells in various tissues offer a localized protective response elicited by tissue injury or destruction, which destroys, dilutes, or sequesters the injurious agent and the injured tissue. Immune and inflammatory systems work together in the control and repair of wounds, integrated with the nervous and endocrine systems. In contrast, stress hormones (CRH, CTH, cortisol) dysregulate immune responses, potentially having a detrimental effect on well-being, including delaying wound healing. Damaged or stimulated structures activate the inflammatory system, bringing inflammatory cells to the wound site. Later there is the arrival of macrophages (which debride wounds) and lymphocytes, both of which are sources of cytokines that are important for wound healing (Yang & Glaser, 2005). Inflammatory reactions may be blunted by increased cortisol concentrations to prevent autoimmune responses to tissue antigens released by cellular injury or excessive inflammatory responses that may cause tissue damage (Miller & Chrousos, 2001).

Immune Reaction: Injuries activate the immune reaction, whose function is wound identification, dead tissue removal, enhance healing and prevent damage from dying cells. Depressed immune function occurs despite requirement for the body to defend itself against infection (Young & Ott, 1996). The stress reaction of immunosuppression delays healing and renders the patient vulnerable to infection. Wound repair starts with inflammation and proceeds through a series of phases. Noteworthy is production or growth factors by the inflammatory infiltrate and wound closure that is orchestrated in part by cytokines (IL-1; TNF-α). Since the proinflammatory filtrate is critical to the healing process, stress-induced anti-inflammation (glucocorticoids) is in conflict. Therapeutic use of glucocorticoids conflicts with wound healing. We are concerned with increased risk of wound infection and a 2–5 times risk of wound complications. Individualized intervention involves the type of wound, the genetic makeup, and psychological factors (Marucha et al., 2001).

Damaged or stimulated structures affect the immune reaction that helps healing and prevents damage from dying cells. The purpose of wound healing is to reestablish the normal structure and function of a tissue after injury. Programmed cell death (apoptosis) is facilitated by cytokines, corticosteroids, and so on. The immune response seems somewhat limited in the brain due to the BBB and other conditions. However, brain cells, microglia, and astrocytes respond to inflammatory stimuli and also secrete cytokines (Chitnis & Khoury, 2004). Some glial cells are scavengers, removing debris after injury or neuronal death (Kandel, 2000b). Immune and inflammatory systems work together in the control and repair of wounds, with inflammation integrated into the nervous and endocrine systems. Depressed immune function occurs despite requirements by the body to defend itself against infection (Young & Ott, 1996). Inflammatory reactions may be blunted by increased cortisol concentrations to prevent autoimmune responses to tissue antigens released by cellular injury or excessive inflammatory responses that may cause tissue damage (Miller & Chrousos, 2001). The rate of healing depends upon the location of the wound, the size and depth of the injured area, the presence of bacteria, and the age and overall health of the person. The SNS also plays a regulatory role in healing. Catecholamines alter blood flow, the movement of inflammatory cells, and cell recruitment (epidermal cells), as well as increase edema and delay re-epithelialization. *Sensory neurons* are involved in neurogenic inflammation, releasing neuropeptides that induce inflammatory responses, such as substance B. Under the influence of nerve growth factor, neurons grow during wound healing. Since glucocorticoids suppress sensory fibers in the periphery, chronic stress could reduce the inflammatory response mediated by these neurons and influence wound healing. The neuroendocrine system affects the timing of the inflammatory process. Various substances may not be adequately expressed to perform their functions and healing will not proceed normally. Knowledge of psychoneuroimmunological processes can identify those at risk for a poor healing outcome. Individualized intervention involves the type of wound, genetic makeup, and psychological factors (Marucha et al., 2001).

Neurological Aspects

There is a bidirectional relationship between tissue cells and neurons at the local site. *Sensory neurons* are involved in neurogenic inflammation, releasing neuropeptides such as substance P, which induce inflammatory responses (Marucha et al., 2001). Since glucocorticoids suppress sensory fibers in the periphery, chronic stress could reduce the inflammatory response mediated by these neurons and influence wound healing.

When prolonged, central activity related to stress and unhappiness results in neuroendocrine imbalances having a disruptive effect upon the *adaptive immune response*. Nevertheless, under extreme conditions such as a concentration camp, a sense of moral commitment and acceptance of the reality of camp life increased the likelihood of survival (Totman, 1988). Knowledge of psychoneuroimmunological processes can identify those at risk for a poor healing outcome, yet some individuals are at risk for a disease that never materializes (Fisher, 1988). The *neuroendocrine system* recruits and activates inflammatory and tissue cells required for healing and affects the timing of the inflammatory process. Various substances may not be adequately expressed to perform their functions and healing will not proceed normally. The immune response seems somewhat limited in the brain due to the BBB and other conditions. However, brain cells, microglia and astrocytes respond to inflammatory stimuli and also secrete cytokines (Chitnis & Khoury, 2004).

Programmed cell death (apoptosis) is facilitated by cytokines, corticosteroids, and so on.

PHYSIOLOGICAL CONTROL AND DYSREGULATION

It is significant for the effects of chronic trauma and stress that the familiar concept of homeostasis is only a partial control system. The process of allostasis is more important in maintaining functioning by permitting wider physiological deviations at times of stress. However, after a chronic disorder dysregulation, physiological fatigue causes the body to succumb to "burnout," creating fatigability and susceptibility to a variety of stress-related diseases. Further, disruptions of the hormonal system in children by trauma to the hypothalamic-pituitary axis can create developmental disorders (see Children). Variation is achieved by multiple, mutually reinforcing neural and endocrine mechanisms that override homeostatic mechanisms. In fact, the brain has access through nerves to every tissue and internal signaling systems. Under allostatic control, the brain controls all mechanisms simultaneously, effecting the changes rapidly. While the homeostatic set point involves the negative feedback mechanisms unaware of need, allostasis adjusts physiological resources to needs or a situation rather than stabilizing it. Chronic arousal causes physiological and anatomical changes (e.g., thickened blood vessel muscle wall and downregulation of hormonal receptors). In a sense, the body becomes adapted to high levels of opiates, cortisol, ACTH, angiotensin, and so on. Entry into a relaxed condition is like withdrawal from catabolic hormones and might provide a physiological basis for conditions of high demand (e.g., "workaholism" or "Type-A" behavior). When chronic arousal is a form of adaptation, previously neutral signals anticipate arousing events and automatically reinforce the aroused endocrine and subjective states. Specific genes associated with arousal may be switched irreversibly into the active state (Sterling & Reason, 1988).

Dysregulation of homeostasis occurs in trauma and can result in a fatal outcome. In contrast, allostasis represents functions that change with adaptive needs, and dysregulation of these functions disrupts health and the quality of life. Homeostasis is a central requirement for survival, but its domain is less comprehensive than was believed in the classic views of Claude Bernard, Walter Cannon, and Hans Selye. When the capacity of homeostatic mechanisms is exceeded during high environmental or physical demands upon the body, the stress physiological control is known as allostasis. This involves accommodating a wide variety of environmental and systemic variations (e.g., circadian rhythms; environmental changes; physical demands such as exercise and threat; vocational, family, and community demands; emotional disturbances; injury and pain; illness, etc.).

The body responds to the external and internal environment by producing hormonal and neurotransmitter mediators that set into motion physiological responses that coordinate physiological responses to current situations. The stress mediators have protective, adaptive, and damaging effects. The brain is the site for integrating behavioral and neuroendocrine responses and functions influenced by diurnal rhythm (McEwen & Seeman, 2004; Schulkin, 2003, p. 11). Allostasis is a different process than homeostasis: it regulates the body during a wider range of stress through a coordinated wide range of functioning of particular systems. Allostasis is a adaptive mechanism for coping with temporary stress, a response shaped by genetic factors, early life experiences, and lifestyle factors that influence physiological responses to daily life events and to stressors. It describes the coping and adaptive processes that adjust a wider range of responses than homeostasis during adaptive necessity: A control mechanism or coordination of various physiological systems when confronting a wide range of change and stress, environmental, internal physiological status, and circadian rhythms. Examples include adjusting heart rate and blood pressure to sleeping, waking, and physical exertion. It corrects temporary dysfunction caused by high load. It differs from homeostasis, which brings functions back to a narrow range of performance ("set point"); rather, it permits varied changes between systems, coordinates output, and creates compensation, following greater deviation than homeostasis. In short, homeostasis continuously controls and restricts the range of performance of particular functions that are necessary for life, while allostasis keeps homeostatic functions in balance (McEwen, 2004), maintaining vigor, health, physical performance, and enjoyment of life. However, under conditions of injury or extended stress, the system enters into the condition of allostatic load, which reflects poor health and inability to meet ordinary demands of life. It provides physiological compensation for dysfunction caused by high load, utilizing the autonomic and endocrine systems, the HPA axis (and others), the cardiovascular, metabolic, and ISs (McEwen, 1998).

Allostasis is a constantly performing physiological process, the adaptive and survival value of which increases markedly after trauma. It regulates the body during a wide range of stress (i.e., the organism maintains internal stability through bodily variation). Examples include increasing sympathetic and HPA activity in order to promote adaptation and reestablish internal stability (Schulkin, 2003, p. 17). Allostasis maintains effective functioning by changing and balancing homeostatic domains in order to cope with environmental change and adaptive demands. Thus, it differs from homeostasis, whose function is to keep certain vital functions within a very narrow range (set point). In contrast, allostasis is the changing response to demands over a period of time, utilizing the autonomic and endocrine systems and influencing metabolic rate. It provides physiological compensation for dysfunction caused by high load. Stress requires both turning on physiological responses appropriate to the challenge and turning off the response when it is no longer needed (McEwen, 2000b).

Daily variability of physiological demands is sufficiently large to render adaptive capacity with set points of *homeostasis* too restrictive for healthy adaptation. This term means stability through internal change (e.g., secretion of mediators of allostasis such as cortisol and adrenaline) or lifestyle changes such as diet, sleep, and exercise.

Allostasis reflects the influence of the CNS upon systemic physiological regulation (Schulkin, 2003), utilizing the autonomic and endocrine systems and influencing the metabolic rate. Allostasis is the process for regulating the body for a wider range of stress. It comprises both behavioral and physiological processes that maintain internal parameters within the limits essential for life. Mediators of allostasis are varied (McEwen, 2004): They include chemicals such as cortisol and adrenaline, as well as lifestyle factors that are released in response to stressors. It appears that there are two major subtypes of the HPA axis stress response: overactivity and mild hypocortisolism (Hellhammer et al., 2004).

Current thinking suggests that the range of functioning in stress and circadian rhythms occurs in a wider range than the traditional homeostatic set points (i.e., allostasis). Allostasis is the maintenance of a wider range of functions that occur during the circadian rhythms and demanding activities. It is a regulating function as the body reacts to conditions that range from demanding, to acute, to persistent stress—sleep/wake cycle, the posture of the body, level of exercise, and usual environmental stressors (noise, crowding, isolation, hunger, extremes of temperature, physical danger, psychosocial stress, various infections). Yehuda and McEwen (2004) observe that the effects of stress are not always detrimental. Allostasis has been described as the way in which people cope with challenges (McEwen, 1995, 2000a).

Allostasis utilizes a variable set point for physiological phenomena that varies with time of day, as well as the adjustment of the set point in anticipation of future events. The brain and soma interact with profound holistic behavioral and physiological changes. The multiple traumata accompanying concussive TBI involve fright, injury, and a sense of impairment. These create persistent stress effects of both a psychological and physiological origin, including effects that are mental (e.g., fright), holistic (systemic and behavioral), and cellular and molecular (the expression of given genes). Allostatic autonomic changes represent multiple interacting systems (Uchino et al., 2001). The stress response serves the biological function of coping with threat to bodily or personal integrity per se, and integrates sensory, physiological, and subjective information. Persistent stress affects the stress victim's adaptation (including stamina and health). Alterations of allostasis caused by continued stress are described as allostatic load.

8 The Acute Stress Response

OVERVIEW

After traumatic brain injury (TBI), there is an evolution of the complex traumatic event, possibly accompanied by the posttraumatic stress disorder (PTSD), a partially emotional reaction to injury. The acute and chronic experience of stress that is comorbid with TBI after an accident is psychologically, physiologically, and neurologically different from the condition of the person experiencing stress under other circumstances (environment, psychological abuse, acute or chronic threat, rape, etc.). While a concussive accident may be momentary, or a period of stress extends through hospitalization and a difficult rehabilitation, other forms of stress can be far more prolonged: combat, incarceration, harassment, living conditions, and so on. Prolonged chronic stress does occur as an extended event with unhealed injuries. The examiner is alerted that using in all cases the same stress definition and reporting category (Posttraumatic Stress Disorder, American Psychiatric Association, DSM-IV, 309.81) invites imprecision, error, and lack of consideration of significant comorbid symptoms.

The condition usually referred to as "Concussion" frequently has a variety of comorbid psychological and anatomical conditions; therefore, there is a varying pattern of neurobehavioral outcomes. Brain injuries are often accompanied by slow-healing or permanent injuries to other parts of the body. Psychological and injury-related stress represent different—although overlapping—syndromes whose effects are genetic, cellular, systemic, neurological, physiological, and mental. Neglect of somatic effects contributes to errors: misattribution of etiology of particular symptoms, inviting inefficient or incorrect treatment modalities and lack of attention to particular trauma, as being outside the specialty of the practitioner.

The "Event" refers to a momentary injurious condition, such as an accident, abuse, or assault. Subsequent neurobehavioral and somatic reactions can be categorized as Acute (perhaps up to 3 months) or Chronic (continuous). Stress has such varied etiologies that the writer assumes that there are meaningful differences in the actual mental and physical trauma included in the same category of "Stress." Some events are brief (e.g., rape and physical injuries) while others are continuous (e.g., combat, imprisonment, harassment, pain, impairment, and the persistent secondary effects of an unhealed physical or mental injury). Stress symptoms vary from time to time. Its symptoms overlap the postconcussive syndrome (PCS).

DEFINING CHARACTERISTICS OF STRESS

Stress may be defined as an experience or injuring event (environmental, physical, or psychological) that the person cannot cope with and has long-lasting or permanent consequences. It is a threat that may be real or interpreted, resulting in physiological and/or behavioral responses. It also may be physiologically defined as a situation that elevates levels of adrenal glucocorticoids (GC) and catecholamines. The response is phasic: its initiation is hopefully adequate, protective to cells, and the response to stressors may either increase or reduce exposure to the threat (McEwen, 2000c). The stressor may create reactions that overwhelm physiological and psychological defenses, and sometimes create long-lasting symptoms as well. These are frequently not self-healing after a brief interval, and express effects that are psychologically or physically injurious, impairing, or distressing.

Stressors

These may be characterized in terms of *interval*, as discrete or encapsulated, or chronic and pervasive (Perry & Pollard, 1998). It may also be characterized in terms of *injury* (i.e. fright, health hazards, and/or physical injury). The posttraumatic stress reaction or other anxiety-generated condition can be initiated acutely or after recovery of consciousness. Stress reactions refer to the consequences of an event, either momentary or extended, and which may create a different condition (disease; nonhealing mental or physical injury).

Short-term memory losses: Patients with PTSD manifest deficits of short-term, though not long-term, memory (Everly & Horton, 1988). Whether this is due to interference by ongoing neurological hyper- or hypoarousal, or merely more routine concentration problems, remains to be determined.

Information Processing: Reduced concentration due to anxiety and avoidance of a variety of activities impairs new learning and problem solving.

Stress ordinarily is considered to be an event or stimulus that causes an abrupt and large change in autonomic activity and hormonal secretion. Prior experience and expectation modify the reaction. The writer has noticed that the events that create the PTSD in persons whose levels of daring and strength are commonplace do not lead to persistent stress in those who knowingly undertake activities that may be hazardous (e.g., policemen and athletes). In the stressed individual, the initially anxious condition alters perceptions so that *anticipation*, or expectation of danger, becomes unrealistic (McEwen, 1995). However, an accident creating comorbid head and body injury is a very complex phenomenon. This is reflected in the multiple components that are sometimes not considered in etiology and outcome: genetic and intracellular processes where intensity and extensity vary with the extent, persistence, pain, unhealed wounds, and impairment of *somatic injury*.

- Misattribution of neurobehavioral, health, and performance effects of ongoing unhealed wounds to cerebral disorders.
- Since concussive level trauma is frequently hard to document with radiological scans, the claim of particular deficits and dysfunctions can be misattributed to malingering or psychiatric disorder.
- The chronic stress condition (see below) may be ignored or not attributed to the concussive event.

Emphasizing the numerous and significant effects of comorbid physical trauma implies that physical accidents creating PTSD may have consequences that significantly differ from events with primarily brief psychological trauma. Chronic trauma (combat, incarceration, prisoner-of-war status; environmental hazards) might represent a third meaningfully different syndrome. However, information from non-TBI of trauma is included. Thus, many symptoms and traumatic etiologies are present that are not implied by the "official" description of PTSD (American Psychiatric Association, DSM-IV-TR, 2000, pp. 463–468).

The actual range of conditions described as "stress" is so complex that it is impossible to bring it under a single rubric. A possible explanation for persistent fear responses is an evolutionary, adaptive, and protective response that remains after the initial fear-stimulus vanishes as a real threat. By inference, this long-term memory is protective (Gardner, 2001). While PTSD was first studied after *extended exposure* (wartime exposure as combatant or civilian; concentration camp inmate), our focus here is based upon a *momentary overwhelming experience* (i.e., an accident) that is often followed by a chronic stressor (recovery period). In an analogy to "depression," as a diagnosis assigned to many essentially different etiologies, PTSD is really a variety of different disorders determined by the different traumata. Stress is an extremely complex neurobehavioral domain, involving comorbid yet disparate etiologies, multiple and changing symptoms, and complex outcomes. There are significantly different conditions labeled "stress," from the brief traumatic event of an accident causing TBI or another injury, to long-term experiences and events (e.g., environmental, hostile confinement with abuse and exhaustion, chronic unhealed injuries, etc.). Stress is both a direct outcome of trauma and a chronic consequence of injuries and unhealed somatic disturbances, the origins of which might be traumatic or medical. In particular, the diagnosis of PTSD is imprecise since there are substantially different traumata, which are, in this writer's opinion, wrongfully included in that single category. While a generic description is a person's inability to reduce or cope with distress, this writer believes that "stress" will ultimately be separated into numerous discriminable patterns of cause and expression. The emotional consequences of trauma are also diverse and comorbid. Perhaps they are part of a *trauma spectrum* sharing a stress-induced alteration in brain circuits and systems (Bremner, 2005).

The incidence of PTSD in the general population is estimated at 3% for 1 year and 8% for lifetime prevalence (Stein et al., 2008). In populations of TBI, it has been estimated as 20–33% (Turnbull et al., 2001). The stress effects of the initiating accident overlap with those effects caused by chronic symptoms, but the specific differences await further research—the possible exception being allostatic load. We will refer to PTSD as the *primary stressor*. PTSD and its variants may persist indefinitely even without somatic injury. Often, additional significant stress may be initiated by the consequences of the accident (emotional, adaptive, or medical), so that a *secondary stressor* exists that is persistent, increases discomfort, and presents a health hazard. The combined concussive stress response can persist for years, and involves a variety of physiological functions, subjective reactions, and overt behavioral responses and restitutive maneuvers.

The emphasis here is on the initial response to injury. The postconcussive stress disorder (with its overlapping symptoms the postconcussion syndrome [PCS]) and the chronic stress syndrome are considered in later chapters. The definition of PTSD (whose specifications are, unfortunately, usually incomplete) emphasizes persistence (acute if assessed less than 3 months after the event; chronic if the duration is 3 months or more). Consistent with the discussion of the acute anatomic injury, the emphasis is upon complex disturbances that hamper the person's adaptation to environmental requirements.

Complexity

The stress disorder *in toto* is a far more complex reaction than as provided by the formal definition of PTSD. A tissue injury (in numerous portions of the body), and reaction of various systems (autonomic, inflammatory, immune, hormonal) can be comorbid. These are concerned with stress control and wound repair. Thus, current understanding encompasses the *initial stress of fright* (awareness of danger to the body or one's life) and *neurobehavioral impairment*, *pain*, *loss of function*, and *mobility*. The stress system integrates neurosensory, endocrine, neuroendocrine, and humoral signals to alert the person that a certain site or event was life-threatening and therefore should be remembered and avoided. It also represents a wound control and danger-adaptive function that confronts healing and restorative procedures that often does not resolve in an adaptive and healthy body or state of mind for the injured person. The various stress components (that oppose the immune system) represent an interaction with systemic and psychological processes at all levels, from genetic and cellular to psychological: mood; cognitive; subjective and experiential awareness of being an injured and impaired person; dissociative experiences; traumatic memories; and associated intense emotions. These are interrelated and affect quality of life and the capacity for adaptive success.

THE STRESS WARNING SYSTEM

Hyperarousal

The stress-response system is an early warning system and a response system that drives escape and confrontation. The system initially attempts to cope with the acute reactions by the processes of homeostasis (maintaining precise levels of function for certain basic processes) and allostasis (recognizing the need for altered level of function, yet under integrated control) within a wider range of physiological mechanisms. In the face of threat, the response involves coordinated hormonal, autonomic, immune, and behavioral responses that allow for escape or adaptation. Within minutes, hypothalamic releasing hormones stimulate the release of pituitary hormones, with adrenocorticotropic hormone (ACTH) signaling the recruitment of the hypothalamic-pituitary adrenal (HPA) axis into the process. The actual cortisol response is much slower, with peak levels not seen for 15–20 min after the onset of the stress. The actual range of conditions described as stress is so complex that it is impossible to bring the primary reactive systems involved under a single rubric. A possible explanation for persistent fear responses is an evolutionary, adaptive, and protective response that remains after the initial fear-stimulus vanishes as a real threat. By inference, this long-term memory is protective (Gardner, 2001). While PTSD was first studied after *extended exposure* (wartime exposure as combattant or civilian; concentration camp inmate), our focus here is based upon a *momentary overwhelming experience* (i.e., an accident), which is often followed by a chronic stressor (recovery period). In an analogy to "depression," as a diagnosis assigned to many essentially different etiologies, PTSD is really a variety of different disorders determined by the different traumata. Stress is an extremely complex neurobehavioral domain, involving comorbid yet disparate etiologies, multiple and changing symptoms, and complex outcomes. There are significantly different conditions labeled "stress," from the brief traumatic event of an accident causing TBI or some other injury, to long-term experiences and events (environmental, hostile confinement with abuse and exhaustion, chronic unhealed injuries, etc.). Stress is both a direct outcome of trauma and a chronic consequence of injuries and unhealed somatic disturbances, the origins of which might be traumatic or medical. In particular, the diagnosis of PTSD is imprecise, since there are substantially different traumata that are, in this writer's opinion, wrongfully included in that single category. While a generic description is a person's inability to reduce or cope with distress, this writer believes that "stress" will ultimately be separated into numerous discriminable patterns of cause and expression. The emotional consequences of trauma are also diverse and comorbid. Perhaps they are part of a *trauma spectrum*, sharing a stress-induced alteration in brain circuits and systems (Bremner, 2005).

The incidence of PTSD in the general population is estimated at 3% for 1 year and 8% for lifetime prevalence (Stein et al., 2008). In populations of TBI it has been estimated as 20–33% (Turnbull et al., 2001). The stress effects of the initiating accident overlap with those effects caused by chronic symptoms, but the specific differences await further research—the possible exception being allostatic load. We will refer to PTSD as the *primary stressor*. PTSD and its variants may persist indefinitely even without somatic injury. Often, additional significant stress may be initiated by the consequences of the accident (emotional, adaptive, or medical), so that a *secondary stressor* exists that is persistent, increases discomfort, and presents a health hazard. The combined concussive stress response can persist for years, and involves a variety of physiological functions, subjective reactions, and overt behaviorial responses and restitutive maneuvers. The hypothalamic-pituitary-target gland axes, also known as the HPA axis: (1) the sympathetic nervous system's autonomic effects; (2) the sympathoadrenal system (SAS), leading to the release of ACTH, GC, catecholamines, and co-stored neuropeptides into the circulation (Boscarino, 2004; Sabban et al., 2004); and (3) the adrenal-medullary system releases catecholamines into the bloodstream.

SAS leads to the release of ACTH, GC, catecholamines, and co-stored neuropeptides into the circulation (Boscarino, 2004; Sabban et al., 2004). The autonomic nervous system (ANS) responds within seconds, mediated by catecholamines from sympathetic nerves and the adrenal medulla, with withdrawal of parasympathetic activity. Successful coping requires that the characteristics and intensity of the response must match that of the threat. To avoid psychological or physical pathologies, the duration of the response should be no longer than is necessary for a successful outcome (Kaye & Lightman, 2005).

THE RANGE OF "STRESS" REACTIONS

The range of acute dysfunctioning goes beyond PTSD as formally defined by DSM-4. It includes numerous other physiological processes and diagnoses. Dysfunctioning in our context is based upon at least two considerations: *tissue damage* and *loss of autoregulation*.

PTSD is common after an accident (Hickling et al., 1992; Parker & Rosenblum, 1996). There is individual variability in response to stress. Vulnerable individuals usually experience reduction of symptoms over time, with a fraction of the victims having chronic PTSD symptoms (Cohen & Zohar, 2004). Emotional consequences of a frightening event may be comorbid with chronic physical injuries, cognitive dysfunctions and substance abuse, depression, suicide, family disruption, and so on. Thus, the diagnostic outcome may be varied and even numerous, and require consideration of a wide range of data. Symptoms of PTSD and other trauma-related disorders may be ignored, or be misattributed to cerebral disorder, emotional factors, and other medical conditions, but not to physiological processes that have been dysregulated by the trauma. Stress phenomena overlap, and can be comorbid with, the PCS (Parker, 2002, 2005a, 2008). The requirement that the initiating trauma be created at the occasion of the accident is incomplete and often incorrect. PTSD can be elicited by later events.

The multiple traumata that accompany concussive TBI involve fright, injury, and a sense of impairment. These create persistent stress effects of both a psychological and physiological origin that include effects that are mental (e.g., fright), holistic (systemic and behavioral), and cellular and molecular (the expression of given genes). The stress response per se serves the biological function of coping with threat to bodily or personal integrity, and integrates sensory, physiological, and subjective information. The persistent dysfunctions and damage that create a chronic stress reaction have multiple effects upon the quality of life, including mood, health, and overt behavioral adaptations to environmental requirements. (The constellation of stress-related dysfunctions, including PTSD, is persistent, hard to treat, and creates significant maladaptation.)

SOME IMPLICATIONS OF STRESS

The descriptor "stress" is somewhat ambiguous, having been used to describe both the vivid, circumscribed event (e.g., an accident), and a long-lasting, demanding, or dangerous context such as job harassment or incarceration in a concentration or prisoner of war camp. An accident that has created a concussion can cause a wide range of emotional, physiological, and somatic injuries and disorders, leading to neurobehavioral impairments, discomforts, and maladaptation. When there are chronic consequences in addition to the immediate traumatic reaction to the event (acute stress), this additional burden may be considered to be persistent stress. To clarify, this writer suggests that the momentary or extended injurious event be termed the "stressor," and any mental or biological consequences be known as the "trauma." Further modifiers would be tissue (destruction or damage; completely or incompletely healed); function (response of such tissues as muscles, organs, tissues, and glands, including systems responding to injury [e.g., hormonal; inflammatory; immune]); psychological (refer to the Taxonomy of Adaptive Functions), including the issues of motivation, reaction to impairment, Identity, social reaction (support; neglect; antagonism); neuropsychological (direct effect upon behavior of injured central nervous system [CNS] function); and indirect reaction of the CNS to dysregulation of the internal milieu (IM). Persistent stress is due to failure to adapt to continuous physiological disorder caused by unhealed wounds, and is reflected in chronic dysfunctions and pain. Our consideration of dysfunctioning goes beyond PTSD as formally defined by DSM-4. It includes numerous other physiological processes and diagnoses. Dysfunctioning in our context is based upon at least two considerations: *tissue damage* and *loss of autoregulation*.

Stress has been considered to be a *general alarm response* of a self-regulating system. The general alarm is identical to physiological activation or *arousal*. One recognizes that there are tremendous *differences between people in the capacity and styles for coping with daily challenges*: (1) *how the situation is interpreted* (i.e., the perceived level of threat), and (2) the *physical capacity* of the person (i.e., physical condition and susceptibility to disease). After dysregulation (failure of homeostasis), the brain resets the system. While it has been claimed that what constitutes a "stressor" may depend upon how the brain filters and interprets it, it is assumed here that there are injuries and events so intense that they would be always dysregulating (Ursin, 1998). Current thinking observes that the biological and behavioral responses to stress are not uniform (Yehuda & McEwen, 2004). Stressors, individual vulnerability, and protective factors result in a complex set of cognitive, systemic, and behavioral responses.

THE COMPREHENSIVE ADAPTIVE RESPONSE

Any stressor (i.e., threat to the survival, adaptation, stability [homeostasis] of the IM) is counteracted by an adaptive response. The main component of the CNS reaction are hypothalamic and brainstem (corticotropin-releasing hormone, CRH)—specifically, the arginine vasopressin (AVPO) and locus ceruleus-noradrenalin (LC/NA)/sympathetic neurons of the hypothalamus and brainstem. These regulate the hypothalamic adrenal axis and other hormonal targets, as well as the systemic and adrenomedullary systems. The CRH and LC/NA systems participate in a reverberatory feedback loop, so that activation of one system tends to activate the other as well. Activation of the HPA axis and the LC/NA-ANS systems result in systemic elevation of GC (adrenal cortex) and catecholamines, which act to maintain homeostasis. These also influence immune system responses that have significant neurobehavioral, health, and other systemic effects. Stress products influence the immune system at baseline and elevated levels effecting infections, autoimmune/inflammatory, allergic, and neoplastic diseases (Chrousos, 2000).

PHYSIOLOGICAL

Physiological variables include heart rate (HR), skin conductance (SC), electromyogram (EMG), systolic blood pressure (SBP), and diastolic blood pressure (DBP). In one study of PTSD patients without report of physical injuries (compared with Vietnam War combat veterans with current PTSD, some with lifelong PTSD, and never having PTSD), at baseline HR and SC differentiated the groups; in response to an arithmetic stressor, HR, SBP, DBP; exposed to audiovisual and imagery scripts, all physiological measures and subjective ratings discriminated between the groups. Responders showed more severe PTSD symptoms and poorer functioning. However, not all individuals exposed to traumatic events qualify for a PTSD diagnosis. Self-protective or self-regulatory maneuvers can be used to disengage from trauma-related challenges or otherwise dampen physiological responses. Veterans with current PTSD experienced a broad range of psychological symptoms, multiple comorbid conditions, marital and familial dysfunction, vocational impairment, financial instability, and punitive involvement with the law (Keane et al., 1998).

THALAMUS

Thalamic nuclei collaborate with other brain regions involved in arousal, fear, and stress to heighten the salience of information elicited in moderate or controllable stress. The dissociative state may represent the intrusion of sleep-related disturbance due to inability of the thalamus to prevent sleep-related disturbances (e.g., sensory-like phenomena from reaching the cortex (Krystal et al., 1995). The thalamus, as well as the cortex, relays threatening stimuli to the amygdala, as well as relaying sensory information to the cortex and limbic forebrain (Charney et al., 1995).

HYPOTHALAMUS

The hypothalamus has multiple CNS connections with *descending autonomic pathways*, which originate both from the hypothalamus and various brain stem nuclei. Hypothalamic function is controlled by hormonal negative feedback and neural input from a wide variety of sources. It participates in endocrine functions, regulation of body temperature, food intake, and so on. The hypothalamus regulates the ANS, although many effects are due to ascending or descending pathways of the cerebral cortex or basal forebrain passing through it. It regulates five physiological functions: blood pressure (BP) and electrolyte composition; body temperature; energy metabolism; reproduction; and emergency responses to stress. It receives information from virtually the entire body and the retinal. Then it compares sensory information with biological set points. When it detects a deviation from set points it adjusts autonomic, endocrine, and behavioral responses to restore homeostasis (Iversen et al., 2000).

In trauma or shock, multiple stress-induced pathways converge on the hypothalamus, stimulating vasopressin and CRH. The Sympathetic Nervous System (SNS), multiple endocrine, and cytokine pathways are activated, releasing catecholamines, increased CO, and a primed musculoskeletal system. The increased CO is distributed to the skeletal musculature vasculature (Sherwood & Carels, 2000). Hormones and cytokines (IL-2, IL-6, and tumor necrosis factor α) increase ACTH and growth hormone (GH) production. ACTH stimulates the adrenal cortex to secrete cortisol, sustaining BP and dampening the inflammatory response.

Paraventricular nucleus (*PVN*): The PVN is an integrative system for neuroendocrine and autonomic responses to stress, receiving humoral, viscerosensory, and external information. The major hypothalamic nucleus for controlling sympathetic and parasympathetic functions is the PVN, using the neurotransmitters vasopressin and oxytocin, which are similar to those released by the magnocellular neurosecretory system. It is an integrative center that coordinates neuroendocrine and autonomic responses to stress by receiving humor and viscerosensory information (Benarroch, 1997a, p. 414). The PVN is a regulator of neuroendocrine and other autonomic stress responses.

PVN sensory input: The PVN receives input from the telencephalon; hypothalamus; amygdala; brainstem; suprachiasmic nucleus, vagus and glossopharyngeal nerves (via the nucleus of the solitary tract); and some vascular circumventricular organs (CVOs). The latter input stimulates patterns of ACTH secretion in response to homeostatic challenges, and provides control signals for neuroendocrine CRH neurons (Watts, 2000). Sensory input involves two tracts: (1) information from the thoracic and abdominal viscera via projections through relays in the ventrolateral medulla to the solitary tract; (2) blood borne substances signal via CVOs adjacent to the third ventricle. *Psychogenic*, *neurogenic*, or *emotional* input involves nociceptive and somatosensory pathways, and also cognitive and affective brain centers. Its *effector neurons* are (1) oxytocin and vasopressin neurons projecting to the posterior pituitary influencing BP, fluid homeostasis, lactation, and, parturition; (2) neurones projecting to the brainstem, which regulate autonomic responses including sympathoadrenal activation; and (3) neurons projecting to the median eminence that control ACTH synthesis and release (Cone et al., 2003).

PVN descending pathway: The PVN is described as the *origin* of the final common pathway controlling the HPA axis (Cullinan et al., 1995). The PVN projects to the median eminence of the hypothalamus (i.e., response to stressful cues by a transient CRH-induced elevation of plasma and ACTH) (Baram et al., 1997). Stimulation of the PVN of the hypothalamus causes it to release CRH into portal vessels of the median eminence, which, upon reaching the anterior pituitary, causes release of ACTH. ACTH, α-melanocyte stimulating hormone (α-MSH), and λMSH are endogenous peptide agonists derived from a precursor *proopiomelanocortin* (*POMC*) *peptide*.

The PVN influences neuroendocrine behavior and autonomic output. The descending path descends laterally through the hypothalamus and the brainstem. These pathways are distinct from those descending to the posterior pituitary. Axons enter the median forebrain bundle, leave the bundle, and run in the dorsolateral tegmentum in the midbrain, the pons, and the medulla. The descending autonomic pathway synapses on brainstem parasympathetic nuclei such as the dorsal motor nucleus of the vagus and spinal sympathetic neurons in the intermediolateral nucleus of the thoracic and lumbar segments, and also on the spinal parasympathetic neurons in the sacral cord. Other hypothalamic sites contribute axons to the descending visceromotor pathways (details not included). Hematopoiesis (blood forming system) is under sympathetic neural control, directed at bone marrow via adrenergic receptors (Maestroni, 1996).

Habituation and Desensitization

Acute stress may require up to several hours for the HPA hormones to almost completely recover resting levels, depending upon the intensity of the stressors and on whether ACTH or corticosterone (animals) is being measured. In contrast, a single exposure to a severe stressor can cause a long-term desensitization of the HPA-axis to similar types of stresses (Armario et al., 2004). Chronic intermittent exposure to a stressor of low or moderate intensity, within the context of familiarity, leads to minimal activation of cardiovascular and metabolic homeostasis, providing for significant conservation of energy expenditure.

Dishabituation: This is the enhancement of a physiological response to a novel stressor in animals that have been exposed repeatedly or continuously to an unrelated stressor. If presented with an unfamiliar stressor, there is a much greater behavioral and physiological challenge than presentation of the same stressor to a naive (unstressed) control. The plasma catecholamine response is amplified compared to naive controls exposed to a similar stress.

Prior exposure to a stressor affects the subsequent secretory response (e.g., the adrenal medulla). After several weeks of exposure to the same stressor, the response is reduced significantly. If the same animals are exposed to a novel stressor, the adrenal medullary response is significantly greater than that of a first-time stressed control group. Stressed rats, at least, express this pattern: the adrenal medulla exhibits a significant, long-lasting enhancement of catecholamine synthetic capacity. However, habituation results in significantly less epinephrine (EPI) release with daily stress exposure. However, when a repeatedly stressed animal is exposed to a novel stressor, EPI release is higher than that of a first-time exposed stressed control. *Other locals* for stress-induced increases in catecholamine sympathetic enzymes occur in various sympathetic ganglia, several hypothalamic nuclei, and noradrenergic cell bodies of the N. LC (Kvetnansky & McCarty, 2000). Several hormonal systems follow a pattern of sensitization and tolerance. Sensitization follows exposure to high intensity stimuli, which create recurring interruptions in cardiovascular and metabolic processes. The organism has a higher plasma level of norepinephrine (NE) and EPI than that in animals exposed to the stress for the first time. In contrast is *desensitization* (*tachyphylaxis*), or the diminished response to adrenergic agonists engendered by prior exposure to catecholamines (Landsberg & Young, 1992). GC also participate in adaptation to stress (Sorg & Kalivas, 1995). A transient sensitization response of GC (corticosterone in the rat) may initially occur. *Tolerance* appears after a few days, perhaps due to familiarity with the same stressful environment. Tolerance would not be observed if the animals are reexposed to a novel stimulus. Chronically stressed persons differ in their physiological patterns when not challenged. Adults with and without chronic life stress (assessed by questionnaires) were exposed to either 12-min mental arithmetic or nature video. Although the groups were almost identical at baseline in psychological, sympathetic, neuroendocrine and immunological domains, they reacted differently to acute stress: greater subjective distress, higher peak levels of EPI, lower peak levels of β-endorphin, and differences in Natural Killer (NK) cells cytotoxicity and distribution, and delayed recovery. It was concluded that exaggerated psychological distress and sympathomedullary peak reactivity occurs in persons with antecedent life stress (Pike et al., 1997).

THE PHYSIOLOGICAL STRESS REACTION

Cerebral

Since the adrenal cortex and the secretion of NE and EPI are key players in the stress reaction, CRH production is positively regulated by stressors and negatively related by GC (e.g., cortisol). It is also found in the raphe nucleus and the LC. These are the origin of major serotonergic and adrenergic projections to the forebrain, which are postulated to play a role in depression and anxiety. *Anxious behavior* follows CRH injection into the LC, while *panic disorder* may be associated with a blunted GH response (Harris et al., 2005). Emotional, physical, and immunological stressors activate hypothalamic control of neuroendocrine and autonomic responses. This centralized stress-response apparatus is critical to the adaptive changes in all categories of bodily function. Behavioral stressors are linked with immunological changes. Superimposed upon immune regulation by cytokine cascades, a strategic control of immune activity by the nervous system is achieved through a hierarchy of neural and endocrine contacts (Downing & Miyan, 2000). Stress may stimulate CRH neurons terminating on LC noradrenergic neurons to stimulate CRH release from the hypothalamic paraventricular N. (Young et al., 2005).

Sympathoadrenal system: The ANS is comprised of the parasympathetic nervous system (acetylcholine) and sympathetic nervous system (NE) in the CNS and sympathetic nerve endings at the periphery; EPI, which is secreted by the adrenal medulla into the systemic circulation; and dopamine, which is secreted by the peripheral sympathetic nerve endings). The SNS, controlled by the CNS, has rapid onset of actions of short duration as a result of the abbreviated half-lives of catecholamines (Dluhy et al., 2003).

Acute physiological function: There are at least two components: the response of *injured tissues* (neural, muscle, bone, etc.) and *disturbance of homeostasis.*

CHRONIC STRESS: PERSISTENT POSTTRAUMATIC STRESS REACTION (PPTSR)

An accepted scientific definition of stress still eludes us, interfering with defining and measuring its psychological, physiological, immune or endocrinal consequences (Kaye & Lightman, 2005). Nevertheless, stress contributes to dysfunctional performance and numerous diseases. The consequences of persistent stress disrupt the victim's *adaptation* (including stamina and health). Chronic stress implies that the body is paying a cost for maintaining its functioning in addition to compensating for normal range external or internal events manifesting diurnal variations (*allostasis*) (i.e., stability through change) (McEwen, 1995). Stress changes the patient's state of health in a morbid direction (i.e., *suboptimum health*). The characteristic malaise and fatigue of so many TBI victims is consistent with the concept that "health" is a continuum between complete well-being and death. A factor study of health is consistent with the clinical picture of some concussion patients as "*worn out*" (tiredness, emotional lability, and cognitive confusion) and "uptight and tense" (worry, tension, physical signs of anxiety) (Cox, 1988).

Chronic stress refers to a continuous condition (combat, imprisonment, harassment, pain, impairment), or the persistent secondary psychological effects of an unhealed physical or mental injury ("Distracting Symptoms") that hamper peace of mind, work, and social relationships. These are often consequent to an unhealed wound. Their secondary effect includes interference with work and other activities of daily living, such as through seizures, headaches, other pain, balance problems, reduced range of motion, and so on (Parker, 1995). Chronic stress has a much greater symptom range than the narrowly defined chronic PTSD. Examples of chronic trauma include an accident, concentration camps, combat, harassment, and unhealed wounds. Stress symptoms vary from time to time and may lead to physiological exhaustion (e.g., headaches, pain, balance and vertigo, reduced range of motion, etc.). Pre-existing coping difficulties include pragmatic communication (verbal and nonverbal); comprehension and expression; written and oral communications; capacity for self-report; productive efforts (e.g., school, work, family and social life, and community activities); and motivation for useful or pleasurable activities. Can the person lead a safe life or is some kind of supervision needed? How does the person cope with or improve the current condition? Is presentation to the provider motivated by occult need for financial or secondary gain? What are the retained or dysfunctional strengths? Can the person utilize or have access to social and community resources?

Stress: Immune Reactions

Immune reactions may be classified as:

- *Functional*: There is a change in the actual functioning of the immune cells.
- *Enumerative*: Changes in the number of particular types of immune cells
- *Redistributative*: There are changes in the distribution of immune cells in the different part of the body changing the local immune response. Chronic stress (caring for a spouse with dementia) causes increase of the proinflammatory response via increased cytokine levels of IL-6, resulting in increased medical morbidity. Activation of proinflammatory elements contributes to immunosuppression (Raison et al., 2005).

Disruption of equilibrium due to chronic, repeated, or physiologically exhausting stress may result in immunosuppressive conditions. There is reduced leukocyte redeployment, innate immunity, effector cell function, cell-mediated immunity, humoral immunity, and resistance to infections and cancer (Dhabar & McEwen, 2001) The inadequate response to behavioral and physical stressors includes infections and immune challenges (Dunn, 1996). *Enhancement of immunosuppression* was created under stressful, chronic circumstances, milder in intensity than those creating immunosuppression (Weiss & Sundar, 1992). Frequent consequences of TBI (chronic stress, depression, and social deprivation) activate the HPA axis, resulting in immunosuppression. Different emotional reactions may have different immunologic consequences (O'Leary, 1990). Depression of immune functions may persist after the stressor (psychiatric illness; caring for Alzheimer patients) has been removed, suggesting that some events may *damage immunocompetence irreversibly* (La Via & Workman, 1998). *Dysregulation of CRH* in the brain (i.e., decreases in its central secretion) is associated with pathological increases of immune function. Immune function is altered in other conditions that also have elevated sympathetic tone (i.e., depression and aging) (Friedman & Irwin, 1995).

Acute Reaction Pattern

The familiar PTSD, related to fright and occurring after injury, has well-characterized physiological processes, notably hormonal (hypothalamic-pituitary-target organ axes; sympatho-adrenal medullary reactions; cortical arousal). The reaction to a physical or psychological stress has been conceptualized as follows (Koetke & Doyle, 2005):

A. *Alarm reaction* (fight, flight, and, less frequently, recognized freezing and ambivalence).
B. "*Brake*" or "*Freeze/Immobility*": This writer offers the descriptor of "*Brake*" as a response to the acute stress. Under some circumstances the victim neither flees, attacks, or actively copes (i.e., acts as though locked in with the motor creating force to propel a vehicle, but the brakes are locked); the stress reaction is potent, but the physiological consequence is turmoil and inability to cope with the situation in any meaningful way. This may be analogous to the concept of *tonic immobility* (TI). This is considered to be part of a *dissociative reaction*. The PTSD model is linked to cyclical autonomic dysfunction, *triggered and maintained by kindling, and perpetuated by vagal tone and endogenous reward systems.*

C. *Resistance*: Coping in an attempt to return to homeostasis.
D. *Recovery*: Adequate adaption with return to a normal state.
E. *Exhaustion*: A failure to adapt, leading to inadequate hormonal production, organ damage, and other diseases.

A *defense cascade* may precede the stress reaction if the traumatic event is anticipated. The pre-encounter stage precedes awareness of danger. In the encounter stage, when approached by the threat (predator), the initial response may be to cease all movement ("freeze"). This is accompanied by focused attention, sustained cardiac deceleration, defensive analgesia, and potentiated startle. Assuming defense response (resistance, fight, or flight), the postencounter stage is characterized by the startle reflex, rapid acceleration of HR, and electrodermal activity. These are associated with fear. Unsuccessful escape or resistance results in TI (see below) (Marx et al., 2008). The physiological dysregulation caused by severe trauma can be accompanied by sufficient external and internal sensations to create an imprint, even if an integrated image of the actual moment of impact is not created (Parker, 2002). A clear event memory may not have been imprinted due to varied causes: dissociation, altered consciousness, confusion and fugue-like states, the neurobehavioral consequences of neuroactive substances in the IM (e.g., the hormones CRH and cortisol, and the neurotransmitters EPI and NE). These may contribute to imprinting salient but disorganized images.

A persistent stress disorder stemming from pain, slow healing of somatic injury, and psychological reaction to impairment is referred to as the *persistent postconcussion syndrome* (*PPCS*). Those with PPCS have multiple subjective and objective dysfunctions and complaints. This completely contradicts the older neurological textbooks that described concussion as a self-limiting phenomenon that resolved after 6 months. One cannot agree with so respected a scientist as Gennarelli (1993), who states that the mildest forms of head injury, characterized by confusion and disorientation, but unaccompanied by amnesia and lasting only momentarily, is "completely reversible and is associated with no sequelae." This perpetuating error is still common. There are several factors contributing to its perpetuation: (1) lack of follow-up on patients to determine the outcome; (2) ignoring late developing neurological (e.g., posttraumatic epilepsy, Parker 2001, pp. 163–165) and physiological conditions (Parker, 2001, pp. 283–286); (3) conducting a narrow range of examination, thus not sampling areas of disorder; and (4) ignoring expressive deficits (i.e., a patient's inability or unwillingness to communicate disorders) (Parker, 2001, pp. 263–268); as well as numerous other errors in gathering information or using a less than wide range of ecologically valid procedures (Parker, 1995).

The proportion of patients who complain of posttraumatic symptoms and recover without further medical attention is frequently stated to be about 85%, but here the question will be studied concerning the appropriate range of functions to be examined the length of the postinjury interval to be considered, and the criterion of "recovery." The condition of an unknown proportion of accident victims evolves into the postconcussional syndrome (PCS). The writer has examined several people who were not even knocked to the ground, having walked away from the scene slightly dazed, but were grossly impaired and no longer the same afterwards (Parker, 1990).

The "Unhealed Wound"

The posttraumatic concept of "*unhealed wounds*" refers to injuries that are not attributed to a prior accident since they persist so long that this obscure attribution is not considered. Thus, the traumatic basis for a current sign or complaint is unrecognized and the patient is considered to be a malingerer or exaggerator, or treatment is based upon an incorrect diagnosis. Some somatic reactions include (Friedman, 1991) SNS hyperarousal; hypofunction of HPA axis (absolute reduced cortisol levels, and related to catecholamines); abnormalities of the endogenous opioid system (stress-related analgesia; lowering of pain threshold at rest); and sleep abnormalities (traumatic nightmares, increased latency, decreased sleep time, increased awakenings and bodily movement).

Elevated glucocorticoids: If the stress is persistent, then elevated GC can cause insulin hypersecretion and insulin resistance. Elevated GC and insulin promotes obesity and facilitates atherosclerotic plaques. Insulin resistance and elevated cortisol are characteristics of endogenous depression, which individuals have an increased risk for cardiovascular disease. Chronic PTSD is also associated with hypervigilance and elevated SNS activity (McEwen, 1995). Physical and psychological stressors (trauma, exercise, hypoglycemia, fever, anxiety, exposure to novelty, and depression) increase cortisol levels. While acute elevations in cortisol have an adaptive function, excess over long periods have adverse physiological effects: elevations in BP, atherosclerosis, diabetes, immunosuppression, bone resorption, and muscle wasting (Hansen-Grant et al., 1998).

Connective tissue (poor wound healing, thinning of the skin, easy bruising); *bone* (inhibit bone formation and contribute to net bone resorption); calcium metabolism (decreased calcium absorption and increased urinary calcium excretion resulting in disabling osteoporosis); growth and development (inhibit growth in children by a direct effect on bone cells, although GC do accelerate the development of a number of systems and organs in fetal tissues); *blood cells and immunological function* (decrease the migration of inflammatory cells to sites of injury, a major mechanism of its anti-inflammatory actions and increased susceptibility to infection following chronic administration); *immunologic and inflammatory responsiveness*, which affect the hypothalamic-pituitary-adrenal axis. They inhibit release of effector substances such as interleukin-1 (IL-1), which presumably reduces secretion of CRH and ACTH.

Chronic Physiological Dysregulation

Functioning occurs in a wider range than the traditional homeostatic set points, and for an interval not closely related to the disruption. The response of *systems* to an unhealed injury (hormonal, inflammatory, immune, circadian) has a deleterious effect upon mental functioning, stamina, and health. Endocrine glands may be depleted according to chronicity and extent of hyperarousal and hypoarousal. Repeated activation of the *SAS* and HPA axis provokes responses that are both adaptive and maladaptive (Sabban et al., 2004). The magnitude of *SNS responses* to stress is variable. Some have a robust SNS response ("overreact"). People with unchecked persistent SNS hyperresponsiveness may be vulnerable to chronic anxiety, hypervigilance, fear, intrusive memories, and increased risk for hypertension and cardiovascular disease. Susceptibility to stress is affected by neurochemicals that help to maintain SNS secretion of NE within an optimal range (Southwick et al., 2003).

Vasopressin Interaction

Stress interacts with vasopressin functioning (e.g., vasopressin facilitates CRH-stimulated ACTH secretion). Vasopressin secretion is inhibited by GC. If there is reduced blood volume (*hypovolemia*), the renin-angiotensin-aldosterone system is activated to enhance vasoconstriction and to conserve sodium. Vasopressin promotes conversion of fat or protein to glycogen in the liver (*glyconeogenesis*). CRH cells also secrete vasopressin, which is a potent stress mediator (Watson & Akil, 1991). Vasopressin controls ACTH, since it causes corticotrophe cells to become relatively resistant to negative feedback action. Higher levels occur in chronic stress.

The Immune System and Stress

Stress reactions lead to an enhanced secretion of GC from the adrenal cortex (e.g., cortisol). Excess secretion (Styne, 1991a) can result in Cushing's syndrome (decreased growth rate followed by obesity). GC are potent immunosuppressive agents. Cortisol effectively depresses both the cellular and humoral immune system (Grossman, 1991). In addition to the glucocorticoid stress effect, the ANS alters the molecular composition of the microenvironment for lymphocytes (Dantzer, 2000). Vagal afferents seem to transmit peripheral inflammatory signals to the brain. Stress decreases the absolute number of leukocytes in the blood and their distribution throughout the body. SNS activation (with high levels of NE) may cause an increase of circulating leukocytes. In addition, stress and GC also cause decrease in monocytes and NK cell numbers. NK level is the most sensitive marker of stress-induced suppression of cellular immunity.

Activation of the HPA axis induces a decrease in circulating leukocytes. Leukocyte numbers return to the baseline 3 h after the cessation of stress (Dhabar & McEwen, 2001). Vagal afferents seem to transmit peripheral inflammatory signals to the brain. Furthermore, both sleep and sleep deprivation affect immune function (Krueger et al., 2000). GC can be immunosuppressive or immunoenhancing depending upon their concentration (pharmacological or natural), source (natural or synthetic), the effects of other physiological factors (hormones, cytokines, neurotransmitters), and other parameters. Part of the effect appears to be inhibition of proinflammatory molecules. *Acute stress* may enhance immune function (facilitating injury recovery and preventing infection and increase resistance to infections and cancer), but may also predispose one toward developing autoimmune or inflammatory disorders. During an immune response, cytokines signal the CNS via CRH pathways, activating the HPA axis and the SNS. There is a feedback loop involving the immune system, HPA axis, SNS, and the CNS. The afferent limb operates by blood-borne cytokines, which, via the circulation or afferents of the vagus nerve, activate the central components of the stress system. Inflammation may also stimulate the central NA stress system through cytokines and other mediators, which act on stress system neurons of the area postrema outside the blood-brain barrier (BBB), or reach neurons inside the barrier (Chrousos, 2000). Infection causes a stress response (i.e., of the HPA axis with increase of the catecholamine turnover in the brain and periphery) (Dantzer, 2000; Penedo & Dahn, 2005). *Chronic* stress may suppress the innate and adaptive immune systems, increasing susceptibility to infections (Ni & Redmond, 2006) and cancer, but ameliorate autoimmune and inflammatory disorders. Hormones released under conditions of stress are believed to induce changes in the absolute numbers, proportions, and distribution of leukocytes, illustrating *two types of stress reactions*. Psychosocial stress also makes a target of the immune system (McEwen, 2000).

The *genetic* component to the response of the immunoinflammatory responses to such stresses as injury contributes to the body's capacity to compensate (recovery rate response to medication and injury) (Johnson, 2004). Predisposition to a negative outcome (i.e., risk) reflects both inherited and acquired factors, including age, chronic health status, and injury severity. The particular physiological outcome (e.g., inflammation) reflects multigenic contributions that vary between individuals (Freeman et al., 2004). There are different stress-induced changes in the blood and spleen, after foot-shock was administered to rats (Cunnik et al., 1992). Stress elevation of GC and catecholamines results in binding these hormones to lymphocyte receptors, causing altered immune function. Altered production by stress also alters the production of cytokines, and ultimately decreases the ability of the innate defense system to develop a localized response needed to combat invading bacteria (Rabin, 2005).

Stressors activate the CNS and HPA and SAM axes, releasing hormones that mediate changes in the function and trafficking of immune cells. Stress-induced activation of the HPA axis influences the immune system. Lymphoid and myeloid (bone marrow) cells have receptors for neuroendocrine hormones, neuropeptides, EPI, and NE. Lymphoid organs (immune tissues, i.e., bone marrow, thymus spleen, and lymph nodes) are innervated by noradrenergic and peptidergic fibers of the ANS.

Immune cells are stimulated to release cytokines, which stimulate the hypothalamus, inducing the secretion of ACTH and cortisol by the pituitary gland and adrenal cortex. These "stress" hormones mediate changes in the function and trafficking of immune cells. Stress-associated hormones dysregulate immune responses and delay cutaneous wound healing. They are mediated by activation of the HPA axis and dysregulation of cytokine and growth factor expression (Yang & Glaser, 2005). The immune and neuroendocrine system share common signal molecules and receptors. These immune-related chemicals are involved in brain function (behavior, neuroendocrine activity, sleep, neurodegeneration, fever, and depression). *Psychoimmunology* is the study of the bidirectional communication between the CNS and the immune system, specifically anatomical, physiological, and psychological communication (Kemeny et al., 1992).

Loss of function: The patient's inability to function at the baseline level has multiple meanings. These are an inability to meet personal and community demands, and feelings of social unattractiveness, gross discomfort, and low self-esteem. Thus, incapacity is a source of stress in the person with chronic difficulties.

Short-term memory loss: Patients with PTSD manifest deficits in short-term, though not long-term, memory (Everly & Horton, 1988). Whether this is due to interference by ongoing neurological hyper- or hypoarousal, or merely more routine concentration problems, remains to be determined.

Information processing: Reduced concentration, due to anxiety and avoidance of a variety of activities, impairs new learning and problem solving.

Altered psychological functioning with time: Over an interval, adaptive capacity and pattern will be affected by unhealed injuries; natural healing; availability of treatment; regression; adaptive consequences of impairment and its effect upon well-being; burnout of stress-related tissues and morale; social and institutional support, neglect, or opposition; reaction of the CNS to dysregulation of the IM; and continued consequences of bodily reaction to unhealed tissue.

Symptomatic: Numbness alternating with hyperarousal. Allostasis (adaptive regulation to demand and change) yields to allostatic load, reflecting physiological exhaustion consequent to chronic dysregulation of bodily functions. The characteristic symptoms of intrusive anxiety (flashbacks and nightmares) may in turn create persistent physiological and psychological processes that maintain ongoing stressful functions or create new ones.

Clinical Examples: Flashbacks

- A youth who had a severe frontal impact in a two-car collision: "Sometimes I'm sitting in the front seat of my friend's car, I will just picture it happening over again. I get very nervous. I want to stop and get out of the car. I want him to just get to the destination. A couple of times a month."
- A man fell 30 feet from a collapsing scaffold. He thinks about the accident every day. He dreams about falling from a height, although this content is not related to the accident. It is not a nightmare, but he is relieved to find himself in his bed. He had a flashback when he was near an accident: "I was told I was as white as a ghost. My heart was pumping and all I could think of was the scaffold that collapsed. I would rather not go into high or dangerous conditions."
- A man who was rendered unconscious in a fatal automobile accident: "I see the shape of the oncoming car. I can imagine the sound of crushed metal and glass. Then the words of my wife mumbling. I feel very exhausted for a while, particularly during the flashback. I'm not involved. I'm looking from outside."
- A man who was close to the collapse of the World Trade Center, and was knocked unconscious by a falling object, was given a computerized alertness test (Vigil) with flashing letters on the computer screen. He stated that, "The flash bothered me. It startled me, It put a chill in. I see the flash of light in the building I was in."

Clinical Example: Bad Dreams

Dreams abut car accident: "Sometimes I have dreams about the car accident. I go to sleep, let's say its 10 o'clock. I wake up being startled, I jump up, and then I look at the clock and its a few minutes after I lay down." He is oriented where he is, "but its so realistic so realistic its like its happening again. *The accident is always in my head; I'm always thinking about it. All the time.* I've been in nonsense accidents, but this was a crazy accident. The doctor tells me 'how lucky you are'; he never did a surgery like that; he never saw that happen." Either it would have been a fatal accident or brain damage. (He describes the procedure of the bone flap, 40 staples, etc.)

Comparing PTSD and PPTSR

When the person experiences a traumatic event, the outcome is variable, both as to the variety of symptoms and their chronicity or permanence. The DSM-IV-TR description of PTSD is inadequate to describe the persistent stress reaction after a TBI involving an accident. It has a limited range of dysfunctions, does not specify their varying expression from time to time, inadequately describes the range of identity and affect disorders (Parker, 1996), and does not address the significant overlapping with the PCS (Parker, 2002; Parker & Rosenblum, 1996). PPTSR's etiology is far more extensive than PTSD, and includes chronic injury and environmental extremes. PPTSR may be classified as *positive neurobehavioral* ("*distracting*") *symptoms*: headaches; pain; restricted range of motion; mood (anxiety, psychodynamic depression); panic; anger; oversensitivity to sound and light; gait and range of motion; sleep disorder; and *negative neurobehavioral stress symptoms* (reduced motivation, social interest, initiative, activity).

STRESS RESPONSES OF THE NERVOUS SYSTEM

Central Nervous System

Laboratory-induced stress in humans indicates activation of specific striatal-limbic-pre-frontal circuits, specifically the medial prefrontal, anterior cingulate, caudate, putamen, thalamus, hippocampus, parahippocampal gyrus, and posterior cingulate regions, as well as midbrain. With the increased intensity and duration of the stressor, there enhanced utilization of dopamine in mesolimbic regions as the striatum occurs (Sinha et al., 2004). Patients with PTSD manifest volume reductions in the hippocampus (emotional memory) and medial prefrontal cortex (fear extinction) that appear to be correlated with the level of cognitive deficit. The prefrontal cortex functions to curb the fear response once danger has passed by modulating fear conditioning and fear extinction after the trauma. Thus, hypofunction results in the consolidation of fear, which becomes intractable after time. This may be related to increased affinity of a serotonin receptor. Cellular changes in serotonin (5-HT) receptors are consistent with the hypothesis that PTSD-associated behavioral and neurodegenerative changes are driven by excess serotonin. These receptors may affect fear extinction through mediation of the HPA axis glucocorticoid release (Harvey et al., 2004).

Depression, common in persons with TBI, is a stressful event. There is reduction in the volume of gray and white matter of the prefrontal cortex, reduction in hippocampal volume, decreased volume of the amygdala, and blood flow changes (McEwen & Chattarji, 2004). Stress-released corticosterone causes attenuation of response to 5-HT in the hippocampal CA1 region. Gradual attenuation of 5-HT responsiveness due to hypercortisolism may explain why genetically predisposed individuals can develop major depression after chronic stress or traumatic early life experience (Joëls & van Riel 2004).

Sympathetic Nervous System Hyperarousal-Release of Physiological Functions

Clinical Example: Hyperarousal

A woman whose car was struck from the rear as she watched: "In discussing the accident, I could feel that my pulse rate had quickened, my heart was beating more rapidly. 'Pins and needles' recurred." Victim felt a feeling of unease that preceded this arousal.

A 35-year-old woman pedestrian was hit on the left side while crossing the street. She was unconscious for an unknown period. When she regained consciousness, she experienced shortness of breath, palpitation, lost of sensation to entire left side of her body, headaches, dizziness, and left knee pain.

She was provided with pain medication and discharged home. Because of the pain in her left knee joint, she was hardly able to walk. Due to persistence and worsening of symptoms, she came to the physician's office for reevaluation and treatment. From the examination: The telephone ringing, banging of the door pounding on the floor, makes her heart beat and makes her jump. When she hears a lot of noise it feels as though her head wants to explode. When she sees a bright light, she cries and needs to close her eyes and close down her head.

The acute stress response, with SNS hyperactivity, incurs a high-energy cost with a problem of thermoregulation (Woodward, 1995). Symptoms include palpitations, headache, panic, nausea, and diarrhea. Acute gastritis is recognized as a stress reaction (erosive gastritis) (Merck Manual, 2006, p. 117). Ordinarily, gastric acid secretion is inhibited by parasympathetic fibers of the vagus nerve (X), operating over the cholinergic pre- and postganglionic pathways that are stimulated by the thought or smell of food, or by the chemical constitution of a meal. Stress, however, can induce both stimulation and inhibition of gastric secretion by SNS pathways in an unknown manner (Buchan, 1989). The SNS signals the *immune system* directly, as does the adrenal gland, which secretes the immune regulator EPI. NPY inhibits the continued release of NE so that the SNS does not "overshoot."

CNS, Catecholamines, SNS, Stress, and Anxiety

Catecholamines (dopamine; EPI; and NE) are associated with the pathophysiology of many stress-related disorders, including PTSD and major depression. Reduced cortisol and elevated catecholamine levels characterize PTSD, with an increased NE level correlating with PTSD severity. The major sources of plasma NE and EPI are the LC and the adrenal medulla. NE system of the brain is rapidly activated by stress and is believed to mediate increased alertness, vigilance, and other cognitive responses to stress. Stress-related elevation of NE may be related to reexperiencing, intrusive memories, and nightmares. Elevated plasma catecholamine levels are associated with SNS stimulation of the adrenal medulla and postganglionic activity during mental and physical stress: vascular pressor (vasodilation or constriction), myocardial, and BP effects (56) (Catt, 1995; Cryer, 1995). Stress also stimulates norepinephric neurons centrally (LC and SNS centers) (Zigmond et al., 1995). Oversensitivity to sympathetic stimulation has been implicated in the cerebral vasospasm associated with subarachnoid hemorrhage. Panic and anxiety result in symptoms suggesting cerebral ischemia (dizziness; unsteadiness; fainting).

Counter-Stress: Opiates (Alkaloids); Opioids (Peptides)

Maintenance of homeostasis during stress is a vital demand (Drolet et al., 2001). Specific stress-related neurotransmitter systems initiate changes at multiple levels of the brain. The endogenous opiates defensively terminate the stress response. They balance the stress demands with which the brain must cope with the detrimental effect of a sustained stress response. They are located (with varying densities) at all levels of the CNS, and in the peripheral and ANSs, and also in several endocrine tissues and target organs. The response of enkephalon to chronic stress appears to be both region-specific and, fortunately, resistant to habituation (Drolet et al., 2001). Endogenous opioids (*enkephalin* and *endorphin*) may be one of the major counter-stressor systems. They modulate the release of adrenaline (phasically) and cortisol (tonically), and also mediate *stress-induced analgesia*. Thus, they may act as a "brake" to inhibit the excessive stress response. By reducing pain threshold, particularly after injury, increased opioid levels allow the organism to focus on survival behaviors (Abbadie & Pasternak, 2002).

Opioid peptides comprise three sets of peptide classes, with a numerous classes of receptors in the brain and in the dorsal horn of the spinal cord, an important pain-processing region. The opioids are expressed throughout the CNS with a variety of functions: β-endorphins, enkephalons, and dynorphins, including stress-induced analgesia and pain perception (Harris et al., 2005; Stout et al., 1995). They are secreted with ACTH in response to CRH release in the pituitary. Behaviorally, they influence cognitive and affective responses, and also play a role in modifying human mental stress (Lombardi et al., 1994). *Endorphins* are morphine-like substances that are manufactured by the brain and pituitary gland (Connors, 2005b; Puder & Wardlaw, 2000). They are released both from the anterior pituitary into the peripheral circulation in parallel with ACTH during stress, and are also synthesized in the hypothalamic arcuate N. of the brain, as well as in the N. of the solitary tract of the brainstem, and projected widely to many regions of the brain. The circulating peptide has restricted access to the brain through the BBB.

Endogenous opiates: Opiates have multiple functions. They are involved in pain control and are powerful suppressants of central and peripheral noradrenergic activity. They are believed to act as neurohormones, neuromodulators, classical neurotransmitters, and as co-transmitters with classical neurotransmitters. In the context of the stress response, they are found in the hypothalamus, the brainstem, and the peripheral ANS. The CNS and ANS areas concerned with cardiovascular homeostasis contain opioids, actions of which are excitatory and inhibitory. Exogenous opiates and opioids depress cardiovascular function and circulating catecholamines, both in the basal state and in response to stress. Active opiate withdrawal increases noradrenergic activity. PTSD symptoms become exacerbated, a phenomenon similar to opiate withdrawal (Southwick et al., 1995). Opiates are powerful suppressants of central and peripheral noradrenergic activity, with opiates "treating" or dampening hypersensitivity and the accompanying symptoms. *Chronic numbing and blunting* of emotional responses may be due to stress-reduced analgesia, or dysregulation of opioid systems in PTSD. Active opiate withdrawal increases noradrenergic activity.

The neuroendocrine response to emotional stress (without tissue injury) resembles its response to injury, but it is not associated with a cytokine response (Berczi & Szentivanhi, 1996). Some consequences are corticosteroid resistance and susceptibility of the receptors to influence, which may be related to stress-related intracellular responses of the *genetic apparatus*. Stress can bias signaling in a receptor-specific fashion, creating *imbalance*. Mineralocorticoid receptor blockage enhances HPA activity in humans. Glucocorticoid receptor blockage attenuates and prolongs the response to a novel stressor. Inability to maintain homeostasis during an adverse situation may lead to neuroendocrine dysregulation with metabolic, cardiovascular, and behavioral pathology. When corticosteroids fail to optimize stress reactions, organs are exposed to improper corticosteroid concentrations for long periods. Coping with stress fails. Stress may bias receptor-signaling pathways, changing "good" corticosteroid actions into "bad" ones. If there are sustained positive reverberating loops, this would further the imbalance.

A multiple physiological response is expected in stress, with human studies indicating three stable factors: catecholamine, cortisol, and testosterone (Ursin, 1998). After a head injury, there is a hormonal surge that includes increased levels of serum insulin, cortisol, glucagon, and catecholamines, which are speculated to increase the metabolic rate. Higher concentrations of cytokines and hormones in the blood, urine, and cerebroventricular spinal fluid mediate some of the systemic effects and may be involved with secondary neuronal injury (Young & Ott, 1996). Maximal levels of adrenaline and cortisol after a head injury in nonsurvivors are assumed to have a toxic effect on vital and functionally active tissues. The release of *NE* does not occur in all panic-eliciting circumstances, but it may be associated with either panic onset or a startle response elicited by sudden symptom induction. GH release may not be associated with anxiety attacks. Laboratory studies that utilized *doxapram* and *lactate* indicated that panic and *phobic anxiety* might not be associated with HPA activation. In a study stimulating panic by doxapram, it was determined that fairly intense panic, accompanied by strong physiological reactions, can occur without clearly measurable activation of the HPA axis, and that cognitive manipulation (expectancy, perceived control, and anticipatory anxiety) could modulate neuroendocrine activity (ACTH, cortisol, and EPI) (Abelson et al., 1996).

The neuroendocrine response to emotional stress (without tissue injury) resembles response to injury, but is not associated with a cytokine response (Berczi & Szentivanhi, 1996). This differs from an event including a TBI, with a combination of initial fright; trauma to the brain and other components of the central, peripheral, and ANSs; sometimes slow-to-heal medical consequences of the injury; long-lasting humoral and other physiological dysfunctions consequent to continued stress and damage to the brain and or the pituitary gland; and secondary stress from adaptive problems caused by impairment, pain, dysphoria, loss of social support, and so on. The subjective, nervous, endocrine, immune, cerebrovascular and somatic systems interact at a period of stress to modulate homeostasis and defense responses. PTSD is a phenomenon that involves a high proportion of all major brain areas, and this inferred involvement is likely to increase with further knowledge. It has been speculated that the extreme situation involved in the induction of PTSD may involve "virtually every neuronal system in the brain" (Post et al., 1995).

Familiar brain functions participate in an integrated fashion in the syndrome of PTSD: Afferent sensory input permits assessment of the anxiety-provoking nature of the event. Prior experience is integrated into the event's cognitive appraisal. These interactions participate in the association of affective significance to specific stimuli, as well as the mobilization of adaptive responses. Following mobilization, efferent projections mediate neuroendocrine, autonomic, and skeletal responses to the original trauma, and account for the pathological reactions of subsequent traumatic cues resulting in anxiety-related signs and symptoms (Charney et al., 1995).

PTSD is a phenomenon that involves a high proportion of all major brain areas, and this inferred involvement is likely to increase with further knowledge. The unified nature of PTSD has been questioned (i.e., whether it is a "final common pathway" reached through a wide variety of relatively severe stressors). Different symptoms may vary with stressors, age, and the kind of trauma. Not only is the concept of "outside usual human experience" subjective, but even common events such as bereavement may be followed by PTSD (Sparr, 1995). In fact, since the symptoms and symptom clusters respond to different psychopharmacologic agents with separate modes or action, from a neurobiological viewpoint, stress may represent a *multisystem disorder.* It has been speculated that the extreme situation involved in the induction of PTSD may involve "virtually every neuronal system in the brain" (Post et al., 1995). Note, however, that our area of concern is far more extensive than PTSD as considered as a defined entity. After a head injury, one may expect in various combinations neurological trauma (potentially to central nervous, cranial nerves, peripheral and ANSs); fright; perhaps slow-to-heal tissue injuries; long-lasting endocrine other physiological dysfunctions consequent to continued stress and inflammatory system activity; and secondary stress from adaptive problems caused by impairment, pain, dysphoria, loss of social support, and so on. The subjective, nervous, endocrine, immune, and cerebrovascular systems and the somatic systems interact to modulate homeostasis and defense responses. Afferent sensory input permits assessment of the anxiety-provoking nature of the event. Prior experience is integrated into the event's cognitive appraisal. These interactions associate affective significance to specific stimuli and mobilize adaptive responses. Then, efferent projections mediate neuroendocrine, autonomic, and skeletal responses to the trauma, and later account for the pathological reactions and subsequent traumatic cues, which result in anxiety-related signs and symptoms (Charney et al., 1995).

Stress can serve as an *alarm reaction* when there is a profound disturbance of *homeostasis.* Stress has been conceptualized as a significant change in emotional and physiological homeostasis, in a situation in which the person is unable to cope and restore the changed equilibrium. Although homeostasis is a term used by physiologists to mean maintenance of static or constant conditions in the internal environment (Guyton & Hall, 2000, pp. 3–4), we shall see that this incompletely describes even normal functioning. A more comprehensive view of orderly physiological functioning during demand or stress is allostasis (see below). With continued demand or hardship, adaptability requires an alteration of the response, using flexible feedback loops whose performance may depend upon the prior history of the person. Homeostasis has an honorable history, but it is now viewed as reflecting a very narrow, although survival-necessary, range of functions. It is but a fraction of a more comprehensive concept that reflects diurnal and stress factors. This is continued in the concept of allostasis and allostatic load, below (Boulpaep & Boron, 2003).

Acute Stress Response of the Hypothalamic-Pituitary Gland Axes

The hypothalamic-pituitary-target gland endocrine axes, most usually identified by the HPA axis, links chronic and traumatic stress to mental and bodily disorders (Hellhammer et al., 2004). Its endocrine control is influenced by both input from other brain centers and feedback from the periphery. A stress threshold level is needed to elicit a particular endocrine response (Armario et al., 1996, 2004). Uncontrollable stress and fear, as well as LC stimulation causes increased NE turnover in the PVN, leading to increased release of CRH (Charney et al., 1995).

Stress is associated with HPA axis hyperactivity (depression; anorexia nervosa; cardiovascular and metabolic disorders). PTSD is associated with dysfunction of the HPA (i.e., mild hypocortisolism). Stress-related disorders are associated with mild hypocortosolism (fatigue and pain). Paradoxically, hypocortisolemia had a lower allostatic load but scored higher on depression, perceived stress, and physical complaints (Hellhammer et al., 2004).

Acute activation of the HPA axis is intended to be of limited duration to avoid adverse effects from catabolic and immunosuppressive activities. The PTSD syndrome evolves from the "acute" stage (the attempt to return to homeostasis) to "chronic" stages (sustained activation → exhaustion). Changes in stress and psychosocial variables can affect neurochemical responses that determine the biological response, specifically concerning the type of consequent symptoms associated with depression.

Major stress (e.g., trauma with pain and hypovolemia), elicits responses from the hypothalamus, the pituitary (anterior and posterior), and the adrenal cortex to facilitate survival. The CNS response includes synthesis and secretion of CRH and antidiuretic hormone (ADH) (Jameson, 1996). Intraventricular administration of CRH produces behavioral and neuroendocrine effects similar to those seen during fear and anxiety (Ladd et al., 2000). The psychiatric, circulatory, metabolic, and immune components of chronic stress are attributed to increased and prolonged production of CRH (Tsigos & Chrousos, 1996). When hypothalamic CRH is depleted, there is a "compensatory" upregulation of CRH gene expression. Limbic input from the *hippocampus* and *amygdala* further modulates stress-induced alteration of CRH gene expression. The magnitude of the effect is also determined by the abundance of receptors mediating CRH's effects (e.g., CRF1 in the hippocampus, amygdala, and cerebral cortex) (Baram et al., 1997). CRH is transported by portal blood flow to the anterior pituitary, stimulating the secretion of ACTH, which in turn stimulates adrenal-cortical secretion of GC (cortisol; corticosterone).

HPA axis and the immune system are generally in balance. Excessive HPA axis response increases susceptibility to infections or neoplasias, while defective HPA axis response leads to increased susceptibility to autoimmune/inflammatory diseases. The *HPA axis* has both neuronal and vascular input, including positive and negative feedback effects. Stimulation affects adrenal-cortical secretion and includes feedback of adrenal-cortical hormonal levels. Control of adrenal cortex functioning (steroids such as mineralocorticoids and GC) has been conceptualized as the *limbic-hypothalamic-pituitary-adrenal stress axis* (*LHPA*). Its activation promotes behavior directed at increasing the energy available for immediate response. This has been categorized as *adaptive redirection of energy* (e.g., increase in the concentration of oxygen and nutrients in the CNS and stressed sites; cardiovascular tone; respiratory rate; gluconeogenesis; and lipolysis) (Michelson et al., 1995). Dysregulation of the HPA (e.g., hypocortosolism) is associated with lack of drive and/or loss of feedback making a balanced homeostasis impossible (Willenberg et al., 2000).

ENDOCRINE REACTIONS AFTER TRAUMA

As long as people are coping psychologically, they do not show increased pituitary-adrenal function (Martin & Reichlin, 1989, p. 670). When a stress disorder appears, it may be extremely long lasting. The gross outflow of various neurotransmitters, internally manufactured opiates (endorphins), and other neuroactive substances cause permanent or long-lasting changes in neural circuits that account for the long-lasting effects of stress. The hormonal stress response is complex (i.e., multiple chemical responses at each phase, and evolution from the acute to the chronic phase) (De Kloet & Deruk, 2004).

Experimental study of rats indicates that after shock or a conditioned fear-producing environment, a number of immune-related responses of the lymphocytes were greatly suppressed. The increased level of stress-related steroids only partially accounts for the immune effects. This role may be played by CRH (Weiss & Sundar, 1992). The immunosuppressive response may vary with dose (i.e., be stimulatory at lower doses). CRH neurons are stimulated by multiple afferents conveying stress information: from the brainstem (pain, blood loss, infection, inflammation, metabolic demands); a complex multisynaptic pathway to the PVN while processing sensory information with cognitive and emotional input. Ascending stimuli arise in the brain stem or peri- and CVOs. Input may include psychosocial stimuli (real, anticipated, or imagined) that modulate emotional and cognitive processes. Processing of psychosocial information may occur in the amygdala, the hippocampus, and the frontal cortex, as well as in other limbic structures. This process, via a GABA network, modulates CRH release by the PVN, which integrates inhibitory and excitatory signals to initiate ACTH secretion. The afferent input determines the pattern of secretory substances released, which stimulate ACTH secretion in the anterior pituitary gland.

Pituitary secretion is regulated by complex, multilayered controls. This complicates assessment of hormonal events at hypothalamic and higher centers through the assay of pituitary hormone blood levels. Furthermore, there may be loss of the normal circadian rhythmic secretion of corticotropin (ACTH). Hypophysiotropic levels cannot be measured directly since pituitary hormone secretion is regulated by multilevel controls. The assay of pituitary hormones in the blood does not necessarily lead to information concerning the condition at the hypothalamus and higher levels. Responses to physiological stimuli can be paradoxical. Hypothalamic injury does not cause effects identical to primary pituitary insufficiency. The results are decreased secretion of most pituitary hormones, but also hypersecretion of hormones normally under inhibitory control. Examples are PRL and precocious puberty caused by the loss of normal restraint over gonadotropin maturation (Cone et al., 2003). Pituitary trauma yields decreased secretion of most pituitary hormones, with the exception of hypersecretion of hormones normally under inhibitory control by the hypothalamus (prolactin [PRL]; precocious puberty caused by loss of restraint over gonadotropic maturation).

The Sympathetic/Adrenomedullary System

Plasma levels increase rapidly in response to even minor stressors. When stress activated, the sympathetic/adrenomedullary system has immunosuppressive effects. There is a monotonic relationship between the intensity of a stressor and the increment in plasma levels of EPI (Kvetnansky & McCarty, 2000). Ordinarily, plasma (systemic) NE is an index of sympathetic neuronal activity, although changes in SNS activity may occur with no change in plasma NE concentration. Elevated plasma catecholamine levels occur during acute mental and physical stress. The adrenal medullae contribute 30–45% of the circulating NE in stressed animals, in contrast to 2–8% contributed by basal conditions. Only some of the numerous biological effects of the catecholamines are listed here: hemodynamic (vasodilation or constriction depending upon the receptor and catecholamine types); increased rate and force of myocardial contraction; increased systolic and DBP, controlled by parasympathetic nervous system; metabolic effects; bronchodilation; and hyperglycemia (EPI) (Cryer, 1995). NE is a major stress-response neurotransmitter: EPI and NE adrenal secretion propels the immediate stress response to a frightening stimulus; NE modulates the encoding of emotion to the stress reaction through its role in the amygdala; and in the frontal cortex, NE acts in the frontal cortex to inhibit novel over well-versed activities, activating the HPA axis and mediating SNS input to the adrenal gland. The stress response sets into motion more than SNS responses: anabolism of somatic tissue; changes in immunological functions sustained by hypercortisolemia;

neuroendocrine changes; stereotyped behaviors and affects associated with stress. *Melancholia* is associated with more than pessimism and hopelessness, it characterized by fear or hyperarousal (increased heart and respiration rates, more frequent eye blinking, abnormal skin responses). It is suggested that melancholia is in fact associated with dysregulation of the HPA axis and the NE system (Rinetti & Wong, 2003).

Elevated plasma catecholamine levels are associated with SNS stimulation of the adrenal medulla and postganglionic activity during mental and physical stress (vascular pressor [vasodilation or constriction]; myocardial and BP effects) (56) (Catt 1995; Cryer, 1995). Stress also stimulates norepinephric neurons centrally (LC and SNS centers) (55) (Zigmond et al., 1995). Oversensitivity to sympathetic stimulation has been implicated in the cerebral vasospasm associated with subarachnoid hemorrhage. Catacholamines may have the paradoxical effect of either vasospasm and high intracranial pressure (ICP) or vasodilatation with increase of blood supply. They also have a protective effect on ischemic brain tissues after TBI and contribute to recovery (Merisalu et al., 1996). In addition to the hormonal and neurotransmitter levels discussed above, stress (in this case motion sickness) decreases plasma TSH and increases plasma GH, plasma PRL, and ß-endorphin (Money et al., 1996).

In anticipation of an anxiety-provoking situation (medical school examination), or sometimes only after the stress experience, levels of plasma cortisol, PRL, and glucose are increased. *Hyperglycemia* and PRL levels were related to the intensity of the stress (Armario et al., 1996).

A common endocrine feature of the stress response is the activation of the HPA axis with increase of *cortisol* levels, the most important GC in humans. Cortisol affects a multitude of systems, including every aspect of stress-related homeostasis. There is a significant impact of genetic factors in HPA axis and related functions. This will eventually elucidate the mechanisms linking stress, the HPA axis, and HPA-related clinical states (Wüst et al., 2004).

Behavioral: A proposed role for GC is to inhibit catecholaminergic systems during resting conditions and under stress, which turns off NE defense reactions (Kvetnansky et al., 1995). While some GC are necessary for learning and adaptation, an excess amount may result in neurotoxicity (Southwick et al., 1992).

Within 2–3 min after ACTH administration, adrenal blood flow increases and cortisol is released (Miller & Tyrrell, 1995). The initial cortisol response to stress is the most brisk, with subsequent adaptation. Cortisol (also catecholamines, GH, and glucagon) oppose(s) the metabolic actions of insulin. The result resembles type II diabetes mellitus, and there may be altered distribution of body fat (Sherwin, 1996). The initial cortisol response to stress is the most brisk, with subsequent adaptation.

Cortisol variability: GC also have a role here. Patients experiencing posttraumatic stress (i.e., "burnout" and physical health problems) typically have a blunted cortisol awakening response (*hypocortisolism*), including fatigue, irritable bowel disease, low back pain, chronic pelvic pain, and diverse health problems. *Hypercortisolism* is associated with hypertension, abdominal obesity, type II diabetes, and enhanced stress sensitivity (Hellhammer et al., 2004).

Prolonged elevated concentrations of GC are toxic, and are associated with neuronal death, medical diseases, impaired affect, and cognitive disorders (Yehuda et al., 1995). Glucocorticoid steroids both suppress the immune response and maintain it (recovery from infections), depending upon the dose and bodily condition (Dunn, 1996). They may play a role in detoxification (Michelson, et al., 1995). An excess of adrenal GC inhibits growth, impairs tissue metabolism, and may be present as the wasting of bone, muscle, and subcutaneous connective tissue rather than as growth failure. This accounts for some of the major clinical morbidity of GC (osteoporosis, muscle weakness, and bruising) (Barrett, 2005c). *Corticoid withdrawal syndrome* is characterized by headache, lethargy, nausea, vomiting, anorexia, abdominal pain, fever, myalgia and arthralgia, postural hypotension, and so on (DeAngeles et al., 2001). Accident victims often exhibit some of these symptoms.

Stress as a complex function is illustrated by the simultaneous activation of CRH and *antidiuretic hormone* (vasopressin or ADH) in the PVN and adrenergic neurons elsewhere in the hypothalamus.

This is mutually reinforcing, since NE increases CRH release while CRH increases adrenergic discharge. ADH release also stimulates ACTH release, which elevates plasma cortisol levels, while adrenergic stimulation elevates plasma EPI and NE. The adaptive outcome includes the activation of the adrenal cortex and adrenal medulla, which interact with the immune system. The adrenergic stimulation increases plasma EPI and NE levels, which increase glucose production together with the similar action of cortisols. The combined presence shifts glucose utilization toward the CNS and away from peripheral tissues. Additionally, there is added availability of fatty acid to the heart and muscles, improved delivery of materials needed for immediate defense, and increased BP and cardiac output (Berne & Levy, 1998, p. 961). PTSD appears to have HPA alterations opposite to those of depression: PTSD patients have lower levels of cortisol and their hyperresponsivity to dexamethasone administration suggests enhanced negative feedback. Patients with PTSD may have a blunted ACTH response to CRH and an increased CSF CRH level (Hansen-Grant et al., 1998).

Mood

High glucocorticoid levels often result in symptoms of euphoria and apathy and lethargy in accident victims. With prolonged exposure, a variety of psychological abnormalities may also occur, including irritability, emotional lability, and depression. Psychosis, hyperkinetic, or manic behavior are less common, although overt psychoses occur in a small number of patients. Other effects are claimed impairment in cognitive functions (most commonly memory and concentration), increased appetite, decreased libido, and insomnia with decreased REM sleep and increased stage II sleep (Aron et al., 2004; Stewart, 2003).

Panic Attacks, Depression, and Symptom Misinterpretation

The overlap of panic attack and stress is somewhat ambiguous. Panic attacks are not accompanied by a concomitant rise in stress hormones (Goldstein & Halbreich, 1987). Studies of cortisol levels indicate inconsistent findings of increased cortisol levels in the afternoon and evening, with a blunted ACTH response to CRH. Panic attacks offer neuroendocrine findings similar to *depression*. The initial findings of increased cortisol levels in major depression appear to be confirmed, although there are both suppressors and nonsuppressors to the dexamethasone suppression test (Merola et al., 1994). There is also a cognitive component. Panic is largely influenced by the patient's interpretation of somatic symptoms (e.g., catastrophically interpreting tachycardia as a sign of impending heart attack or shortness of breath as a sign of suffering) (Bryant, 2001). Nevertheless, this writer cautions against too high a reliance upon patient interpretation as the basis for symptoms. While panic and postconcussive symptoms are described as involving cognitive and emotional response to somatic or perceptual sensations, this book amply documents the directly injurious effects of accidents comorbid to concussion and PTSD. However, it is acknowledged that anxiety results in pain-enhancing influences (fatigue, muscle tension, headaches, irritability, etc.).

Glucocorticoid Dysfunction

Thyroid gland: GC suppress TSH function (Melmed & Kleinberg, 2003). An excess of neurotransmitters such as NE can stimulate TSH, as happens during trauma (Williams, 1998). GC stimulate TRH directly, but their overall inhibitory effect on the thyroid axis results from negative glucocorticoid feedback in such structures as the hippocampus (Cone et al., 2003).

Stress-Injury-Related Inflammatory Dysfunction

Immune modulatory molecules can also stimulate the brain's hormonal stress response and start a cascade of hormones that ultimately result in the adrenal gland's release of anti-inflammatory corticosteroid hormones, which in excess (chronic stress) predispose the distressed host to more infection. On the other hand, a blunted immune response (genetic factors, drug therapy, surgery) can predispose the host to autoimmune disease (arthritis, systemic lupus erythematosus, allergic asthma, atopic dermatitis, etc.) (Sternberg, 2000).

TRAUMATIC DISTURBANCE OF CEREBRAL AUTOREGULATION

After brain injury, the "steady state" of close association between cerebral blood flow (CBF) and metabolism can be altered. There is a marked reduction in *cerebral metabolism* in regions of the brain remote from the site of injury (Hovda, 1996). In addition to catecholamines, *neuropeptide Y* is released from sympathetic neurones. It is elevated in stressed, depressed, and aged individuals (Friedman & Irwin, 1995). Neuropeptide Y may act as an endogenous anxiolytic, possibly by diminishing the effects of other stress-related peptides (Stout et al., 1995). The changes of ICP and cerebral perfusion pressure (CPP) after severe brain injury are not very relevant to concussive brain trauma (Reilly, 2005); only selected issues will be addressed. Autoregulation is sensitive to both minor and severe TBI (Strebel et al., 1997). Lack of cerebral autoregulation has been established after minor head injury, which may increase risk for secondary ischemic neuronal damage. After head injury, autoregulation is absent, reduced, or delayed, leading to moderate or transient hypotension, causing ischemia. After impact injury (in a rat model) there is transient hypertension and increased blood flow, followed by blood flow reduction below control values within minutes (Lam et al., 1997; Muir et al., 1992). Impaired cerebral autoregulation of vasomotor control occurs after a percussion injury (Junger et al., 1997). It is associated with poor outcome, even after a mild head injury (Zubkov et al., 1999). This may be a controversial point since it has been asserted that the relationship between impaired autoregulation and outcome after a head injury remains unclear (Martin & Doberstein, 1995). Post-TBI cardiovascular sequelae include hypertension and numerous other dysfunctions (Labi & Horn 1990). SPECT signs of hypoperfusion have been attributed to loss of cerebral autoregulation after a head trauma, including minor head injury (Lam et al., 1997). TBI results in reduced CBF and hypermetabolism, usually associated with the uncoupling of CBF and metabolism in the acute stage. There is then a potential risk of the brain developing ischemia with a reduction in CPP or, alternatively, hyperemia. Trauma-induced cerebrovascular dysregulation is a risk factor for diabetes and hypertension (Iadecola, 2004).

Neurological outcome in head-injured patients is inversely related to the amount of time that ICP remains greater than 20 mg Hg. The use of hypertonic substances to reduce ICP includes the hazard of mechanical shearing or tearing of the bridging vessels, which results in subarachnoid hemorrhage and other types of lesions (Khanna et al., 2000). Normal mental status implies adequate cerebral perfusion and probable absence of significant injury to the brain. Diminished mentation in the presence of tachycardia and hypotension is associated with shock with or without injury to the brain (Parks, 2004).

CBF is low in the newborn and increases to adulthood. After brain injury, there are species and age differences in the direction of change of CBF (Armstead & Kurth, 1994). In the adult cat, sustained vasodilation and increased CBF, as well as blunted cerebrovascular responsiveness to variations in systemic BP and CO_2 tension occur. In the adult rat, global suppression of CBF occurs. Generation of free radicals may have created changes in pial circulation and rise in systemic arterial pressure. Using fluid percussion upon newborn and juvenile pigs, it was determined that age is an important factor in the pathophysiology of TBI. The newborn was exquisitely sensitive to TBI, in terms of alteration of cerebral hemodynamic parameters. Brain injury resulted in marked hypoperfusion in the newborn; the pial arterioles are more constricted. In the juvenile pig, the circulation was slightly hyperreactive after TBI.

Disturbances of human cerebral circulation can be created by head trauma, stress reactions, and mechanical injury to the cerebral and neck vasculature. Taylor and Bell (1966), using Oldendorf's 1962 technique of recording γ-ray emission from injected ^{131}-I-labeled Hippuran, determined that the mean cerebral circulation time in 70 patients with postconcussional symptoms was 15% greater than the mean of 70 control patients. It was hypothesized that there was increased cerebral vasomotor resistance at the arterial level, and that CBF was reduced. There was a relationship between the improvement of symptoms and the normalization of circulation time.

Disturbed Cellular Homeostasis

Posttraumatic changes in cellular homeostasis and metabolism after TBI may explain vulnerability to cellular degeneration. A metabolic cascade affects the integrity of axons, from which they may either recover or degenerate (Giza, 2000). Experimental studies indicate that *glucose metabolism* may not stabilize for 10 days. This helps to understand the grave brain damage resulting from a second injury (e.g., in sports) following an initial trauma (i.e., incapacity for the hyperglycolysis necessary to adequately respond to the second ionic perturbation).

Dysregulation of Cerebral Metabolism

There is a *triphasic pattern of cerebral metabolism* of brief hypermetabolism, depressed metabolism with reduction of glucose metabolism not correlating with level of consciousness, and subsequent recovery of metabolic activity. Coping with the acute and chronic effects of injury and stress involves a complex interaction between somatic structures (e.g., injured tissue, glands, organs), the central, autonomic, and enteric nervous systems, and brain-interactive glands such as the pineal and pituitary. Cerebral metabolic requirements are reduced after a head injury. However, hemoglobin concentration may be reduced so that actual oxygen delivery may be lower for a given CBF value.

Cerebrovascular dysfunction due to TBI contributes to the increased sensitivity of the injured brain to secondary hypoxia and hypotension, with consequent increased mortality and morbidity occurring during the phase of secondary injury. The vascular injury renders the brain more susceptible to secondary injury due to impaired protective compensatory mechanisms (i.e., autoregulation). Reductions in CBF appear due to increased production of cerebral vasoconstrictors, or reduced synthesis or increased destruction of cerebral vasodilators (Dewitt & Prough, 2003). Reduced oxygen extraction from the microcirculation, out of proportion from the flow changes, suggests the consequences of more diffuse impairment to brain cells than after a gunshot injury. Neuronal integrity is maintained through control of brain metabolism and blood gas level, and is affected by a variety of metabolic and neurogenic effects (Reis & Golanov, 1996). A lesion in a pathway to a cortical region can create *hypometabolism* (i.e., diaschisis) (Caselli et al., 1991). Normal cerebral autoregulation prevents major changes from sympathetic stimulation. Decreases occur over the first 1–5 days, and later there may be increases in patients who improve neurologically. Increased extraction of oxygen compensates for reduced CBF resulting in maintained $CMRO_2$ (*hypoperfusion pattern*). When further decreases in CBF cannot be compensated for by increased oxygen extraction, the patient has exhausted the brain's compensatory mechanism for maintaining cerebral metabolism. The reduced $CMRO_2$ and increased lactate production produce the *ischemia pattern*. After a while, irreversible ischemia or infarction develops (*infarction pattern*) (Robertson, 1996).

For a group of severely brain injured children, $CMRO_2$, mean CBF, and arteriojugular venous oxygen difference ($AJDVO_2$) were at the highest levels early after the injury (e.g., in the normal range for 81–95%, respectively) (Sharples et al., 1995). CBF tended to be initially low (77% with only 6% above the upper limit of the normal range). The mean $AJVDO_2$ was in the normal range. The children who died had very low cerebral metabolic rates and very low cerebra oxygen extraction. It has been hypothesized that cerebral trauma results in *impaired brain mitochondrial respiratory function*; thus, O_2 is not used for aerobic metabolism. Normal oxygen extraction and metabolism do not reflect neural metabolic demands. Cerebral ischemic damage could occur despite an abundant supply of oxygen to the brain.

Glucose metabolism: Cerebral metabolism is almost entirely dependent on the oxidation of glucose, and requires a constant supply of oxygen by CBF [$CMR_{O2} = CBF \times AVDO_2$ (arteriovenous difference of oxygen)]. CBF is positively related to the diameter of the blood vessels and inversely related to the viscosity of the blood. While vessel diameter is a compensatory response for both metabolic and pressure autoregulation, in CO_2 response, the situation is different: the changes in diameter are primary; CBF and $AVDO_2$ follow passively. The purpose of *CO_2 reactivity* is unclear (Bouma & Muizelaar, 1991). CBF may be depressed for several days. This is associated with mismatch between *glucose delivery* and consumption that may predispose to secondary TBI. In fact, cerebral oxidative metabolism, which normally runs near maximal capacity, now requires an increase in glycolysis that is impaired by TBI. It is needed to rapidly restore cerebral ionic balance. Postinjury hypoxia, on the other hand, implies impaired substrate delivery, resulting in energy failure, inability to restore membrane potential, and cell death. Experimental studies indicate that *glucose metabolism* may not stabilize for 10 days. This helps to understand the grave brain damage resulting from a second injury (e.g., in sports) following an initial trauma (i.e., incapacity for the hyperglycolysis necessary to adequately respond to the second ionic perturbation). There is increased vulnerability to a second injury (*second impact syndrome*) ensuing from the preliminary excessive loading with Ca^{2+}, which may persist up to 2 days in experimental concussive brain injury.

ACUTE DYSREGULATION OF THE INTERNAL MILIEU AFTER TBI

Autonomic Nervous System Dysregulation

Some ANS functions occur as integrated rhythms (i.e., oscillations in HR), which are a function of both SNS and PSNS stimulation of the cardiovascular systems that control BP. These exist in a dynamic, coupled, nonlinear environment, serving as a communication pathway for biological systems (Goldstein et al., 1994). In a sample of patients with different levels of brain injury, it was determined that there is uncoupling of cardiovascular and autonomic activity in acute brain injury. This occurs at these levels: the brain, sinoatrial node, peripheral vasculature, and arterial baroreceptor. The severity of neurological injury and outcome are inversely associated with HR and BP variability. *Cervical cord injury* results in the uncoupling of the spinal pathways that link the supraspinal cardiovascular centers with peripheral sympathetic outflow to the sinoatrial node and peripheral vasculature. After severe brain injury, *dysautonomia* has been reported: Hypersudation, tachycardia, arterial hypertension, muscle hypertonia, hyperthermia and increased respiration rate (Cuny et al., 2001). A recommended battery of clinical tests of autonomic function include sudomotor function, cardiovagal function, adrenergic function, and plasma catecholamines (Low, 2003).

Hormonal Dysregulation

Homeostatic regulation of hormone levels is vulnerable because it is controlled through feedback loops from the endocrine glands that maintain set points rather precisely by feeding back to the hypothalamus and the pituitary. Hypothalamic influence is positive in all instances except in the secretion of PRL, so that damage causes a release of PRL (Molitch, 1996). Hypothalamic damage interferes with the maintenance of body homeostasis (Treip, 1970). A blunted hormonal stress response in animals and humans, whether present on a genetic basis, because of drug therapy, or because of surgical intervention, can all lead to increased susceptibility to inflammatory disease (arthritis, systemic lupus erythematosus, allergic asthma, atopic dermatitis, etc.) (Sternberg, 2000).

TRAUMA AND THE BLOOD BRAIN BARRIER

Genetic factors can predispose one to TBI-induced BBB dysfunction (Morganti-Kossman et al., 2005). The primary barrier between the BBB ordinarily keeps out many substances, particularly large molecules. The efficiency of the barrier is reduced by trauma and disease. The BBB changes during various CNS diseases. After CNS injury that leads to neuronal necrosis with breakdown of the BBB, microglia are promptly activated and aided by infiltrating macrophages from the circulation. Necrotic brain lesions induce prompt inflammation (Stoll et al., 2002). With BBB breakdown, chemical products involved in wound healing enter the brain more easily. Under pathophysiological conditions, trophic factors generate, maintain, and regulate BBB functions. This might make them a target of therapeutic intervention (e.g., stabilizing tight junctions in the BBB). After concussion, there may be vasogenic edema related to the acute rise in a trophic factor vascular endothelial growth factor (VEGF), which increases vascular permeability (i.e., the opening of tight junctions contributing to angiogenic edema).

After an injury, the primary BBB is less efficient. Therefore, those chemical products that attempt to heal the body enter the brain more easily—not necessarily a benign process! The BBB is not a complete barrier, even during health. Some substances enter the brain through structures whose surface is on the cerebral ventricles: these are known as the CVOs.

Animal models of TBI (impact and acceleration) have demonstrated perturbation of the BBB. Disruption leads to exudation of substances from the blood into the brain. Experimental fluid percussion of animals creates bilateral permeability changes in the neocortex, the dorsal hippocampus, subcortical white matter, and throughout the brainstem. Lateral percussion appears to evoke more unilateral change involving the same anatomic sites. The perturbation of the BBB appears to be transient, resolving in the first days after injury (Povlishock, 1996). Lactate accumulation at high concentrations has a deleterious effect on brain cells, including the breakdown of the BBB (Hovda, 1996).

Individual genetics may predispose the BBB to TBI-induced dysfunction. In severe TBI, at least, the BBB is impaired by the initial physical rupture due to tearing mechanisms, and also the later release of multiple soluble factors that attack the membrane of endothelial cells and cause the opening of tight junctions. Focal brain damage creates a more profound and prolonged BBB damage than does diffuse brain injury. Local hemorrhages and/or hematomas eventually release soluble factors that attack the membrane of endothelial cells, causing the tight junctions to open. This allows diffusion of the serum components in the interstitial space, increases water content, contributing to vasodilation and increased CBF, edema, and ICP. After TBI cytokines produced in the brain mediate the activation, migration, adhesion, and extravasation of immune cells across the BBB into the lesioned brain area (Morganti-Kossman et al., 2005).

Damaged BBB: GC penetrate the BBB and also regulate it (Miller & Chrousos, 2001). Increased oxygen radicals contribute to damage to the microvasculature and the subsequent breakdown of the BBB. Cerebral contusion creates an inflammatory response in the injured brain, including up-regulation of proinflammatory cytokines within minutes, followed by induction of cerebral adhesion molecules and accumulation of neutrophils. The acute inflammatory response may be associated with BBB damage, which contributes to edema (Tokutomi, 2003). Cerebral vessels dilate with an increase in the CSF (which essentially lacks the BBB) of hydrogen ion concentration adenosine (which occurs with reduced oxygen supply, seizures, and increased carbon dioxide) (Berne & Levy, 1998). What is described as "mild" head trauma (i.e., coup and contrecoup injuries), and diffuse axonal injury from shear and tensile strain damage, cause the release of excitatory neurotransmitters (e.g., acetylcholine, glutamate, and aspartate). This may result in impaired cerebrovascular autoregulation (Evans, 2001). Arterial hypotension or increased ICP can result in lowered CPP. Ischemia and luxury perfusion occur at different posttrauma periods (Nichols et al., 1996). Sudden increases in BP can be transmitted to the brain's microcirculation, contributing to secondary hemorrhage and edema. *Loss of autoregulation* prevents adjustment of cerebrovascular resistance. CBF becomes entirely dependent on CPP. The rise in ICP with subsequent decrease in CPP can lead to ischemia (21). Loss of autoregulation may occur in some areas but not others (50) (Miller & Gudeman, 1986).

Immunoinflammatory reactions: The brain is an immunologically active organ. Inflammation is initiated directly by brain cells after an injury (Morganti-Kossman et al., 2005). Under pathological conditions, disruption of BBB integrity might be followed by migration of activated leukocytes.

When leukocytes leave the circulation in parenchyma, a neuroinflammatory condition develops (Kooij et al., 2005).

Animal models of TBI (impact and acceleration) have demonstrated the perturbation of the BBB. Disruption leads to exudation of substances from the blood into the brain. The experimental fluid percussion of animals creates bilateral permeability change in the neocortex, the dorsal hippocampus, subcortical white matter, and throughout the brainstem. Lateral percussion appears to evoke more unilateral change involving the same anatomic sites. The perturbation of the BBB appears to be transient, resolving in the first days after injury (Povlishock, 1996). Lactate accumulation at high concentrations has a deleterious effect on brain cells, including a breakdown of the BBB (Hovda, 1996).

Peripherally generated cytokines reach the hypothalamus by crossing leaky areas in the BBB (CVOs), active transport across the BBB, and through receptors in cerebral blood vessels, activating second cellular messengers such as NO and prostaglandins (Molitch, 2001). In a series of head injury patients admitted within 1–9 h after closed head injury, at all levels of trauma there were increased levels of EPI, NE, ACTH, and cortisol that persisted during the entire first week after the trauma. In general, the levels of these chemicals were inversely correlated with the Glasgow Coma Scale (i.e., with severity of head injury). The pattern of change with time varied according to severity and whether or not the patient survived. The effect of the catecholamines upon brain ischemia and vasospasm is controversial (Koiv et al., 1997).

THE INFLAMMATORY SYSTEM

Inflammation

Inflammation is a generalized and localized protective response. It is elicited by *tissue injury* or destruction, which destroy, dilute or sequester the injurious agent and the injured tissue. Inflammation protects against infection, and also regulates the activities of tissue cells that repair the injured tissue. It is caused by blunt trauma, radiation injury, lacerations, thermal, or chemical burns, bacteria, and so on, and is followed by the release of multiple substances that cause inflammation in the tissues (Guyton & Hall, 2000, p. 397). The variety of causes results in inflammation not being considered as an immune reaction. However, immune and inflammatory reactions are closely linked, often promoting and enhancing one another. Postganglionic sympathetic fibers release CRH and NA. Sensory fibers respond to the inflamed condition, send signals to the CNS, and also secrete substances that are proinflammatory (substance P) and anti-inflammatory (somatostatin). *Polytrauma* (multiple injuries) can be accompanied by cerebral hypoperfusion and hypoxia, leading to a poor outcome. *The systemic inflammatory response syndrome* makes it difficult to maintain tissue perfusion (Santiago & Fessler, 2004). Thus, CNS damage may exceed the initial trauma.

CNS inflammatory effects: It is noteworthy that *the brain* is an immunologically active organ, with inflammation initiated by damaged cells, particularly microglia and astrocytes. When brain cells have been primed (exposed to injured cells), they secrete cytokines and chemokines that act upon the BBB, mediating the accumulation of peripheral leukocytes into the lesion cite (Morganti-Kossman et al., 2005). CNS effects, stemming from cytokines that penetrate the brain, or are produced within it, can promote peripheral inflammation (Lipton et al., 1994). The delicate balance between pro- and anti-inflammatory cytokines is easily disrupted by adverse conditions. The severity of trauma and the duration of hypotension will determine whether tissue repair will be initiated by a localized inflammation or whether there will be a systemic induction of a generalized inflammatory process. Trauma results in the three "*inflammatory cytokines*" (TNF-α, IL-1, IL-6) to activate the HPA axis (i.e., inducing CRH leading to vagus activity) (gastric motility) and stimulation of the LC and the norepinephric sympathetic NS. The stress-activated ANS induces secretion of Il-6, which causes glucocorticoid secretion and, by suppressing TNF-α and IL-1, plays a major role in controlling inflammation (Chrousos, 1998). On the other hand, *immunodeficiency* in trauma victims is associated with enhanced concentrations of inflammatory cytokines (Faist et al., 2004).

Severe and protracted injury causes proliferation of fibroblasts and endothelial cells at the site, which form a permanent scar. Cytokines and other small regulatory molecules, called *inflammatory mediators*, control many inflammatory events. They are secreted by injured or distressed host cells while others are byproducts either of tissue injury per se (e.g., fragments of collagen) or of the host's reaction to it. Thus, the inflammatory response is triggered by molecules that signify tissue damage, regardless of its specific cause. The changes in inflamed blood vessels attract cells of the immune system into an injured or infected tissue. Injured skin, joints, or muscles have a reduced threshold for pain (i.e., *hyperalgesia*). *Secondary hyperalgesia* results in inflammation, involving processes near peripheral receptors and mechanisms in the CNS. Damaged tissue causes a variety of chemical substances to be released from itself, blood vessels, and nerve endings (bradykinin, prostaglandins, serotonin, substance P, K+ H+, and others) (Connors, 2005). The dilatation and increased permeability of vessels near injured tissue results in part from a spinal reflex: pain receptors stimulated by the injury transmit afferent signals to the spinal cord, where they act on autonomic motor neurons to cause relaxation of arteriolar smooth muscles at the injured site. A neural independent component of the vascular response is triggered by substances produced at the injured site, which act directly on the local vessels. Among the *vasoactive mediators* are histamine, prostaglandins, and leukotrienes (Parslow & Bainton, 2001). The pathways and events vary according to the inciting stimulus, the portal of entry, and the characteristics of the host. Some outcomes are detrimental. While inflammatory reactions have evolved to inactivate or eliminate injurious substances or limit their spread, these same reactions can injure host tissues or interfere with normal functions (Terr, 2001).

Cognitive Effect

Inflammation plays a role in Alzheimer and other neurodegenerative diseases. A high level of serum markers of inflammation (IL-6; CRP) was associated with poor cognitive performance and greater risk of cognitive decline over 2 years of follow-up. The effects were cumulative. It is hypothesized that inflammation contributes to cognitive decline in the elderly. However, nonsteroid anti-inflammatory drug use (NSAID) did not change the association between inflammation and cognitive decline (Yaffe et al., 2003).

POSTTRAUMATIC CNS-IMMUNE SYSTEM SIGNALS

Injury, infection, or ischemia activates the immune system to produce cytokines, which coordinate the immunological and metabolic response. The cytokine response is regulated closely. While low levels of cytokine activity are beneficial, an over- or under-abundance of cytokine activity can impair organ function and cause shock or tissue injury. Protection of the host from cytokine excess is provided by pituitary-adrenal glucocorticoid system (cortisol), the anti-inflammatory cytokine system, and a recently discovered vagus nerve, the *cholinergic anti-inflammatory pathway*, which prevents tissue injury through inhibiting the release of TNF and other proinflammatory cytokines (Gallowitch-Puerta & Tracey, 2005). After trauma, the immune system interacts with the nervous, stress, inflammatory, autonomic, and other systems. Its messengers create neuropsychological effects. The immune system participates in a regulatory feedback loop with the HPA axis, with immune and inflammatory mediators stimulating HPA hormone synthesis and secretion (Crofford, 2002). CRF (also known as CRH), when injected into the lateral cerebral ventricles of rats, appeared to have anxiogenic effects, increased hypothalamic biogenic amines, and changed immune function to reflect its stress-consequent properties (Song et al., 1995). Stress-induced enhancement of immune function may be an adaptive response preparing for immunologic challenges (e.g., a wound or infection inflicted by an attacker). The brain's stress perception and stress hormone and neurotransmitter release serve as an early warning (Dhabar & McEwen, 2001).

Integration is reflected in the manipulation of neural and endocrine functions that alter immune responses; the antigenic stimulation that induces an immune response, resulting in changes in neural and endocrine function; and behavioral processes that are capable of influencing immunologic reactivity, while the immune status of an organism has consequences for behavior. Stressful stimuli and conditioning alter the development of immunologically mediated pathophysiological processes. These effects appear to be based upon regulatory feedback loops between the brain and the immune system, involving both neural and endocrine signals to the immune system, as well as signals from the immune system that are received by the nervous system and provoke further neural and endocrine adjustments.

BRAIN TRAUMA AND THE IMMUNE SYSTEM

Lesional studies indicate either enhanced or suppressed immune function in the hypothalamus, the limbic forebrain, the brainstem, the autonomic or reticular regions, and the cerebral cortex. The CNS circuitry involved in immune system modulation is integrated with nuclei that regulate both autonomic and neuroendocrine outflow and have ascending and descending connections. Included are brain areas that mediate affective and cognitive processes, stress responses, aversive conditioning, and mood disorders, such as depression. These regions, responsive to cytokines and immunization, have the highest concentration of glucocorticoid receptors, and link some endocrine signals with neuronal outflow to the autonomic and neuroendocrine systems. There is a link between neural activity and altered immune responses, including psychosocial factors such as bereavement, marital separation, depression, and examination stress (Felten et al., 1991). The endocrine and immune systems, which mutually regulate each other (Grossman, 1991), share the property of memory. There is direct CNS innervation of the immune system, but also immune system receptors for neurotransmitters, neurohormones, and neuropeptides. Stress (and depression) may lead to the impairment of immune function through hypersecretion of hypothalamic CRF, or through turning on a regulatory portion of the brain-immune system feedback loop (Gorman & Kertzner, 1990).

Tissue damage represents a breach of organismic integrity, stimulating the innate immune system. There is rapid production of cytokines, chemokines, and prostoglandins that are responsible for autonomic changes of the acute phase. The proinflammatory cytokine IL-1B is probably the most important molecule modulating cerebral functions during systemic and localized inflammatory insults (Rivest, 2003). Immune and inflammatory reactions are closely linked, and often promote and enhance one another. Inflammation plays a role in the development of subarachnoid hemorrhage (SAH) (Provencio & Vora, 2005). CNS effects stemming from cytokines that penetrate the brain, or are produced within it, can promote peripheral inflammation (Lipton et al., 1994). The delicate balance between pro- and anti-inflammatory cytokines is easily disrupted by adverse conditions. The anti-inflammatory stress response (GC) interferes with wound healing by reducing the ability of leukocytes to produce the cytokine IL-1, and also by suppressing sensory fibers, which release neuropeptides that induce the inflammatory response (Marucha et al., 2001). *Alpha-MSH*, secreted by the intermediate pituitary gland, modulates inflammation through both central influences (adrenergic pathways) and local sites of action (i.e., it has anti-inflammatory activity).

TRAUMA: THE IMMUNOSUPPRESSIVE RESPONSE

It is paradoxical that immune function is suppressed when an organism may be injured, with the possibility of infection from the attacker and enhanced vulnerability to inflammatory diseases. GC are the most potent endogenous inhibitors of immune and inflammatory processes (Crofford, 2002). Immunosuppression increases susceptibility to infections and cancer, and also exacerbates inflammatory diseases like psoriasis, asthma, arthritis, and lupus erythematosus. It ameliorates autoimmune and inflammatory disorders (Dhabar & McEwen, 2001). The *autoimmune response* is a failure of immunosuppression. Acute stress is not yet proven to be a cause of immunosuppression (La Via &

Workman, 1988). The stress response restrains an overresponse (e.g., excessive autoimmune and lymphoproliferative processes) and potentially harmful products of activated lymphocytes (Ader, 1996). Immune mechanisms may be associated with pathogenesis of *seizure disorders*. This may evolve from autoantibodies reacting with neurotissue antigens, influencing the function of brain neurons; a finding in traumatic brain lesions among may other disturbances (Jankovic, 1994). *Behavioral factors* influence the immune system, and therefore the individual's state of health. These mechanisms are largely obscure but represent a balance between reaction to an antigen and overreaction of the entire immune system to this particular antigen (Schlesinger & Yodfat, 1996). Stress-induced alterations of immunity are affected by the nature of the stressful stimulation; the capacity of the individual to cope with stress; the nature of immunogenic stimulation; the emporal relationship between stress and immunogenic stimulation; the experiential history; and social and environmental conditions. Prolonged psychosocial stress associated with higher catecholamine levels may lead to hypertension, acute myocardial infarction, and sudden death (Schlesinger & Yodfat, 1996).

TRAUMA: IMMUNOINFLAMMATORY RESPONSE

An effective immune response involves the synergy of many immune system (IS) components, which augment the overall response. Unguided amplification of all branches of the IS would be both inefficient and dangerous (e.g., leading to an autoimmune and inflammatory response). Intrinsic regulation of the IS mediated by cytokines and specialized T-cell reactions (Raison et al., 2005). Under stress, if adrenal steroid secretion does not increase appropriately, then the secretion of inflammatory cytokines increases, which ordinarily would be counter-regulated by these adrenal steroids. In rats, it has been demonstrated that inadequate HPA response results in increased vulnerability to autoimmune and inflammatory disturbances. Analogous human examples of HPA hyporesponsiveness may be fibromyalgia and chronic fatigue syndrome. Further clinical examples of allostatic load are offered:

(1) Elevated GC may damage hippocampal neurons, causing insulin hypersecretion and then insulin resistance, and leading to obesity and atherosclerotic plaque formation. The hippocampal damage interferes with the hippocampal ability to suppress negative associations and match expected outcome with reality. This further increases the likelihood of self-sustaining fears and anxiety that do not match with real events.
(2) PTSD is associated with hypervigilance and elevated SNS activity. While cortisol and insulin levels are not as high, chronically high sympathetic tone does have adverse consequences.
(3) Extreme physical exercise is associated with elevated cortisol and adrenal cortex and medulla size.

Although inflammation that is initiated by an injury is a defense, it is believed to contribute to neuropathology and secondary necrosis (Dietrich & Bramlett, 2004). *Inflammation* is characterized by pain, redness, heat and swelling consequent to an increase of vascular diameter, reduced velocity of blood flow (especially in smaller vessels), and changes in the endothelial cells that allow leukocytes to become attached and then migrate into the tissues (extravasation). Tissue damage represents a breach of organismic integrity, stimulating the innate IS. There is rapid production of cytokines, chemokines, and prostoglandins, which are responsible for autonomic changes of the acute phase. The proinflammatory cytokine IL-1B is probably the most important molecule modulating cerebral functions during systemic and localized inflammatory insults (Rivest, 2003). The inflammatory response is stimulated by cytokines, which are produced by activated macrophages. Inflammatory mediators include prostaglandins, leukotrienes, and platelet activating factor (PAF). Following a wound, injury to blood vessels triggers enzyme cascades. The *kinin system* produces inflammatory mediators, including *bradykinin*, leading to increased vascular permeability and increase of proteins. Pain encourages immobilization. In the *coagulation system* there is another cascade of plasma proteins (Janeway et al., 2005, p. 42).

Trauma activates the HPA axis via "*inflammatory cytokines*." *Severe head injury* can result in cerebral edema, a result of the breakdown of the normally tight endothelial barrier of the cerebral vessels. The enclosed skull prevents expansion of the brain, leading to occlusion of the cerebral microcirculation (Boulpaep & Boron, 2003). The *febrile reaction* includes neurobehavioral symptoms of shivering, malaise, and anorexia, as well as endocrine effects (increased secretion of arginine vasopressin and corticosteroids). These are attributed to cytokine effects on neuronal targets in the CNS. Inflammation, as may accompany somatic injury, nociceptive, somatosensory, and visceral afferent stimulation, can stimulate catecholaminergic and CRH neuronal systems via ascending spinal pathways.

Cytokines and trauma: Cytokines elicit a neuroendocrine response, and initiate major metabolic alterations (i.e., protein loss from muscles and catabolism on the one hand, and selected protein synthesis on the other) (Berczi & Szentivanhi, 1996). Stimulation of the HPA axis in turn profoundly inhibits the inflammatory immune response and production of cytokines through cortisol. The endocrine response to injury may be modulated by nerve endings at the site of injury, which influence cytokine production. There is an overlap between the symptoms of viral illness caused by the effects of circulating cytokines on SNS nerves (e.g., lassitude, fatigue, myalgia, headache) and subjective reactions after trauma causing TBI (Arnason, 1997). Another possible interfering effect upon memory is indicated by the finding that *cytokines* (released during enhanced immune response) inhibit long-term potentiation in hippocampal slices. The effects of the cytokines are so dire that it is believed that cytokine antagonists normally modulate them. These become available during host challenge, such as inflammation (i.e., a consequence of an accident-related injury). One of these is the neuropeptide, *alpha-melanotonin stimulating hormone*, which is important in communication among neuroendocrine, immune, and central nervous systems (Catania et al., 1994).

The *immunosuppressive response* may be based upon the need to avoid an autoimmune response after injury, when the organism might be exposed to fragments of its own cells. One may observe in advance that the immunosuppressive features of the stress response have side effects, including the negative consequence of increased susceptibility to infections (Yamamoto & Friedman, 1996). It can be conditioned so that reexposure renders the organism more susceptible to environmental pathogens. The ANS plays a role in immunosuppression. Stress-induced immunosuppression might be mediated by depression of interferon and IL-2 (Weiss & Sundar, 1992). The SAS may be involved in ovarian and thyroid hypertrophy (Landsberg & Young, 1992).

The Thymus Gland

The thymus gland's key role is the selection and differentiation of T lymphocytes that populate the peripheral immune system as part of acquired immunity. It is critical for normal immune function, secreting, in addition to thymic hormones (thymulin and others involved in T-cell maturation), cytokines (IL-1,2,4,6,7; TNF-α; IFN-γ) and neuroendocrine hormones (ACTH; CRH; GR [glucocorticoid receptor]; PRL; somatostatin; arginine vasopressin; corticosterone). Locally produced GC seem critical for normal T-cell differentiation during the neonatal period when corticosteroid synthesis by the adrenals is low. Locally produced CRH inducing ACTH modulates glucocorticoid synthesis (Vacchio, 2000).

The thymus is highly susceptible to stress. It plays a key role in the development of the IS during embryonic and early life, and also plays an active role in its maintenance via thymic hormones (Trainin et al., 1996). Stress leads to the pathway CRH to ACTH to corticosterone secretion by the adrenals, causing involution of particular thymocytes. Receptors for ACTH, GH and PRL are expressed in the thymus. These hormones augment thymocyte development and release of thymulin. This peptide enhances immune responses in the periphery (cytokines) and provides feedback to the HPA axis modulating pituitary function.

Sympathetic Nervous System

The SNS plays a role in the modulation of the IS. Lymphoid tissues are stimulated by numerous neurohormones, neuropeptides, and catecholamines (NE and EP) secreted by the SNS as a stress response. The consequences are a large number of alterations in the immune system. Receptors on a variety of immune system blood cells also respond to stress secreted NE and EP (Brown & Zwilling, 1996). The immune system in turn creates neurobehavioral effects and interacts with the stress system. SNS activity generally *inhibits* the immune system.

Neuroendocrine Responses

The integrated reaction to stress has been described as *neuroendocrinimmunomodulation.* Impaired neuroendocrine-immune counter-regulation might increase susceptibility to, or severity of, *auto-immune diseases.* It is suggested that the neuroendocrine response pattern is dependent upon the perception of stimuli and the behavioral response. In rodents, dominant and defense reactions, and submissive or defeat reactions are associated with either a sympathetic adrenal-medullary or a pituitary adrenal-cortical response pattern respectively (Ely, 1995). College music and biology majors reacted differently to music insofar as their cortisol and NE levels were concerned (Ely, 1995).

Adrenal Gland and Immune System

The duration, quality, and direction of stress-induced alterations of immunity are influenced by the quality and quantity of stressful stimulation; the individual's capacity to cope effectively with stressful events; the quality and quantity of immunogenic stimulation; sampling times and the particular aspect of immune function chosen for measurement; the experiential history of the individual and the existing social and environmental conditions upon which stressful and immunogenic stimulation are superimposed; a variety of host factors such as species, strain, age, sex, and nutritional state; and interactions among these variables (Ader, 1996). Stress has more than an immunosuppressive function; it can also keep infections such as colds under control until pressure is released. Mild acute stressors may enhance measures of immunity (Dunn, 1996). The adrenal gland products (GC and catecholamines) switch the immune system from specific reactivity to nonspecific defense. The IS is activated nonspecifically, leading to cytokine production similar to acute infections, and may culminate in endocrine and metabolic responses characteristic of the acute injury phase.

Both the brain and the immune system are responsive to cytokines such as interferon, producing ACTH and endorphins, which explains the IS stimulation of the brain. Since immune system cells produce neuropeptides, the exchange of information between the brain and the IS is bidirectional. Activation of the IS is accompanied by changes in hypothalamic, autonomic, and endocrine processes, as well as by changes in behavior. Both the SNS and PSNS have direct effects upon lymphocytes in immune system tissues (thymus, bone marrow, spleen, lymph nodes). These fiber connections develop early in ontogeny. Handling in early life affects subsequent antibody responses in mice. Marked diminished fiber supply of IS tissues with age perhaps contributes to immunosenescence. Stressful experiences both reduce and increase the host's defense mechanisms (Ader, 1996; Dunn, 1996), thereby altering susceptibility to bacterial and viral infections. It has been hypothesized that increased SNS function (characteristic of stress) would be associated with a decline in immune function. Evidence includes the *immunosuppression* found with SNS agonist drugs (Arnason, 1997). Mice exposed to physical stressors along with associated environmental cues exhibited inhibition of the immune response when they received an immunological challenge by injection with another species' erythrocytes. Conditioned immunosuppression decreases in aged mice. It is noteworthy that immune organs can be directly stimulated, that there are neurohormonal receptors on lymphoid cells, that CNS cells produce cytokines (intercellular messengers), and that there is neurohormonal regulation of immune reactions. Mouse resistance to infection by parasites is reduced when they are reexposed to cues that signal stress-induced immunosuppression (Gorczynski, 1996).

Immunoregulation

The neuroimmunoregulatory process takes place in a *neuroendocrine milieu* that is sensitive to the influence of the individual's perception of, and response to, external events. This milieu is defined by the innervation of lymphoid organs, the availability of neurotransmitters for interactions with cells of the immune system, interaction between pituitary-endocrine organ- and lymphocyte-derived hormones (Ader, 1996). There are over 20 neuropeptides secreted by cells of the immune system. These have a substantial influence, serving as autocrine and paracrine factors exercising local control of a variety of immune functions. IS cells contain neuropeptides and also are a source of pituitary, hypothalamic, and neural peptides. The IS manufactures neuroendocrine hormones and may also produce hypothalamic releasing factors, which have direct effects upon it. While plasma hormone contributions by lymphocytes do not reach pituitary gland levels, they are mobile and can deposit the hormone at the target site. Its two functions are, first, endogenous regulators of the immune system, and second, recognition of stimuli not recognizable by the central and peripheral nervous systems (bacteria; tumors; viruses; antigens). Recognition of immunocytes is converted into information (peptide hormones; neurotransmitters; cytokines), which is then conveyed to the neuroendocrine system. *MSH* has been considered to be a mediator of immunity and inflammation. Through its effects upon *melanocortin-1 receptor* (MC1R), it downregulates the production of proinflammatory and immunomodulating cytokines. The cytokine-synthesis inhibitor IL-10 is upregulated (Luger et al., 2003).

The *pineal gland* is a CVO, and therefore permeable to large molecules from the blood. It is innervated by the sympathetic superior cervical ganglion. It is involved through feedback loops with a variety of organs associated with the immune system (Jankovic, 1994). Melatonin is secreted directly into the general circulation. Its secretion is under sympathetic control, mediated by a multisynaptic pathway originating in the retina. Melatonin is involved in the photoperiodic control of circadian rhythms (Fink, 2000).

CNS-PERIPHERAL IMMUNE SYSTEM SIGNALS

The IS is involved in behavioral interactions between the psychosocial and biological environments (Bohus & Koolhaas, 1996). The brain and immune system form a bidirectional communication network. Stress-induced alterations of immunity, part of the adaptive process, reflect the integration between the CNS and IS. The CNS regulates many aspects of immune function via the SNS and neuroendocrine pathways to organs and cells of the immune system. There are bidirectional interactions between the immune and nervous systems at all levels, including the brain, pituitary, peripheral nervous system, neuroendocrine mechanisms, and the ANS. The immune system's products communicate with the CNS, altering neural activity to influence behavior, hormone release, and autonomic function. Thus, the immune cells function as a *diffuse sense organ*, informing the CNS about events in the periphery relating to infection and injury (Maier et al., 2001). Stress-related neuroendocrine and neurochemical changes are part of a larger system that modulates functional activities of lymphocytes and phagocytes. The molecules that signal to the brain that an immune event has occurred are such proinflammatory cytokines as IL-1 and TNF-α. They reach the brain via afferent pathways that innervate the injured area, and a humoral pathway utilizing blood-borne cytokines and activated circulating immune cells type relayed by special parts of the brain with a deficient BBB (CVO) (Dantzer, 2000).

Immune cells have receptors for classical neurotransmitters and neuropeptides. Leukocytic peptides include substances found in the neuroendocrine and nervous system. Microglial cells of the CNS, equivalents of macrophage, release proinflammatory cytokines that regulate inflammatory response at the periphery. Brain cells themselves have receptors for these and other cytokines (Dantzer, 2000). Brain function and behavior are affected by the lymphokines produced by an activated immune system. Stress effects on the IS, particularly psychosocial stressors, depend on the nature of the environment, personality factors, and the immune system location in particular compartments. Corticosteroids are mediators of stress effects.

The brain is immunologically reactive; it produces immunologically active factors and is closely involved with the systemic immune response. CNS cells respond to inflammatory stimuli and also secrete cytokines (Chitnis & Khoury, 2004). *Neuropeptides* have a substantial effect on IS functions and are also produced by the immune system. Among the hormones produced by lymphocytes are TSH, prolactin, GH-releasing hormones, and hypothalamic-releasing hormones (CRF and GHRH). Some neuropeptides commonly found in the peripheral nervous system are produced by cells of the IS and have profound effects upon it (Weigent & Blalock, 1996). MSH downregulates the production of proinflammatory and immunomodulating cytokines (Luger et al., 2003). Yet, the brain is considered to be an *immunologically privileged site* because the tissue grafts placed there are not rejected or attacked, although antigens sequestered there may be the target of an *autoimmune attack* (e.g., multiple sclerosis). Humoral factors (presumably cytokines) that affect the immune response are produced here. While there are protective qualities against immune processes, antigens sequestered here may be the target of autoimmune attack (e.g., multiple sclerosis) (Janeway et al., 2005, pp. 570–571).

Neural-immune signaling: Cells of the IS possess elaborate systems of proteins that enable them to respond to signals that are (1) self-generated, (2) derived from other immunocytes, (3) conveyed by the neuroendocrine and ANSs, or (4) generated in response to foreign substances in their microenvironment. Neuroimmune signaling is a dynamic and interactive process, the understanding of which requires integrative investigation of neurotransmitter, hormonal, and cytokine influences on multiple target cells. It is clear even at this stage in studying neural-immune-signaling TNT that (1) there is extensive cross-talk between the nervous and immune systems, (2) neurotransmitters and endocrine signals can profoundly alter immune function, (3) cytokines can profoundly influence nervous system function under normal condition sand in states of pathology, and (4) neural and immune mediators share common intercellular signaling pathways (Lorton et al., 2001).

Immunocytes can be exposed to specific neurotransmitters, to which they may respond. Nerves are distributed to sites where cells of the immune system form the lymphoid tissue. Specific neurotransmitters convey messages. Changes in signing pathways alter immune function under different physiological conditions and in disease sites. Moreover, neurotransmitters may be produced and secreted by immune system cells, acting in an autocrine and paracrine fashion. Since nerves have autoreceptors specific to their own neurotransmitters, secretion of neurotransmitters from immunocytes that reside adjacent to nerves is likely to regulate neurotransmitter release from nerves in lymphoid organs. Furthermore, signal molecules from the nervous system may arrive in lymphoid tissue via the circulation as hormones. NE may reach the cells of the immune system by various routes. There is also cholinergic innervation of lymphoid organs (Bellinger et al., 2001).

The IS can be conditioned bidirectionally between itself and the CNS. Signals travel via the hypothalamic-pituitary neuroendocrine axis and the ANS. The CNS detects alterations in immune reactivity. Immune responses can be altered by psychological stress and by lesions in the limbic forebrain, hypothalamus, brainstem reticular formation, and autonomic regions (Bellinger et al., 1992). The IS seems to act as a "*receptor sensorial organ*," emitting a response to which the CNS is capable of responding. One of the channels of communication between the neuroendocrine and immune system is by receptors that exist on immune cells (Ader, 1996). It communicates information to the brain in diverse physiological conditions. For example, products of activated immune cells can increase the firing rate of noradrenergic neurons (Geenen et al., 1994). This would relate to an "*afferent humoral system*" that accounts for the fact that peripheral endocrine events are conveyed to the brain, and also accounts for the effects of all hormones of the pituitary-adrenal system upon CNS functions (Fehm & Born, 1989). There are bidirectional effects between hypothalamic releasing hormones (having direct effects upon the immune system) and the immune system (which manufactures neuroendocrine hormones [e.g., CRF and GHRH]). Neuropeptides have an important effect on the immune system, including pituitary hormones, hypothalamic releasing hormones, and other substances (Weigent & Blalock, 1996). The immune system contains receptors for neuropeptides and is a source of pituitary peptides, hypothalamic peptides, and neuropeptides. While secretion level of immune cells is not comparable to that of the pituitary, since are mobile they can locally deposit the hormone at the target site. Thus, they serve as endogenous regulators of the IS and relayers of information to the neuroendocrine system. There is an IS sensory function since leukocytes recognize stimuli not recognizable by the peripheral and central nervous systems. *Noncognitive stimuli*, recognized by immunocytes, is converted into information (peptide hormones, neurotransmitters, and cytokines), which is conveyed to the neuroendocrine system leading to a physiological change (Weigent & Blalock, 1996).

Peripheral Nerves and the Immune System

Peripheral nerves are not completely stable: their neurons depend on trophic factors produced by their target to maintain innervation. Thus, changes in homeostasis in lymphoid tissues often result in changes in neurotransmitter turnover, or the remodeling of nerves that distribute to them. Neurotransmitters can influence specific functions of specific subsets of cells within lymphoid organs (specificity). Immune activation can induce changes in nerve fibers that innervate lymphoid organs (e.g., altered neurotransmitter metabolism), induction of neurotransmitters in the nerves, and promotion of axonal sprouting or neuronal degeneration. Attention may be focused on the spleen and lymph nodes, which are innervated by sympathetic branches of the SNS. Immune cell products (cytokines, lymphokines, lymphotoxins, and chemokines), can be released from lymphoid cells and modulate nerve terminal activity, viability, and neurotransmitter release. These products can also enter the circulation to provide feedback to the CNS. Nerves can synthesize and release such cytokines as histamine, IL-1, and IL-6, while cells of the immune system can synthesize and secrete neurotransmitters or neurohormones (ACTH; CRH; beta-endorphin NE, etc.) (Bellinger et al., 2001).

Circumventricular Organs (CVO) Entry

The cytokines IL-2 or IL-1 may enter the brain through the organum vasculosum lateral terminalis (which lacks a BBB) increasing production of CRH (or CRF, see below), stimulating ACTH release from the pituitary, leading to cortisol release from the adrenal cortex, inhibition of some aspects of the immune response, and increase of SNS innervation of the spleen (also inhibiting immune responses) (Arnason, 1997). Many of the CVOs are heavily innervated by the serotonin-containing fibers originating in the midbrain raphe nucleus. Hypothalamic hormones have been detected in high concentrations in the CVOs: TRH, LHRH; somatostatin; angiotensin II; substance P; adenohypophysial hormones (posterior pituitary); and neurohypophysial hormones (anterior pituitary) (Parent, 1996, p. 23).

Peripherally generated cytokines seem to reach the hypothalamus via several mechanisms: crossing at leaky portions of the BBB; active transport across the BBB; and through receptors in cerebral blood vessels with activation of second messengers such as nitrous oxide and prostaglandins (Molitch, 1995).

After trauma, the immune system interacts with nervous, stress, inflammatory, autonomic, and other systems. Its messengers create neuropsychological effects. The immune system participates in a regulatory feedback loop with the HPA axis, with immune and inflammatory mediators stimulating HPA hormone synthesis and secretion (Crofford, 2002). CRF, when injected into the lateral cerebral ventricles of rats, appeared to have anxiogenic effects, increased hypothalamic biogenic amines, and changed immune function reflective of its stress-consequent properties (Song et al., 1995). Stress-induced enhancement of immune function may be an adaptive response preparing for immunologic challenges (e.g., a wound or infection inflicted by an attacker). The brain's stress perception and stress hormone and neurotransmitter release serve as an early warning (Dhabar & McEwen, 2001).

Integration is reflected in the manipulation of neural and endocrine functions that alter immune responses; antigenic stimulation that induces an immune response, resulting in changes in neural and endocrine function; and behavioral processes that are capable of influencing immunologic reactivity, while the immune status of an organism has consequences for behavior. Stressful stimuli and conditioning alter the development of immunologically mediated pathophysiological processes. These effects appear to be based upon regulatory feedback loops between the brain and the immune system, involving both neural and endocrine signals to the immune system and signals from the immune system that are received by the nervous system, provoking further neural and endocrine adjustments. The local nervous system production of IL-1 is a growth factor for glial cells in traumatized rat brain and human multiple sclerosis brain (Ferrero et al., 1992). *Cachexia* is associated with elevated cytokines (IL-1, IL-6, TNF-α). Loss of skeletal muscle is a major source of morbidity and mortality in serious illness (e.g., cancer, AIDS, and organ failure) (Foster, et al., 2003). IL-1, a proinflammatory polypeptide cytokine produced by monocytes and other cell types, integrates the broad cellular reactions of the acute immune response. Since it also stimulates CRF and ACTH release to increase glucocorticoid output, it plays a role in the simultaneous regulation of the immune and stress systems. The CRF peptide has been described as regulating and coordinating the body's endocrine, autonomic, metabolic, behavioral, and emotional responses to stressful stimuli (Petrusz & Merchenthaler, 1992).

9 Chronic Posttraumatic Stress
Injury, Disease, and Burnout

OVERVIEW

The behavioral and adaptive status of an accident victim who does not make a quick and thorough recovery is subject to subtle impairment that is usually not recognized. A genetic vulnerability to a common but infrequently recognized dysfunction—hypopituitarism—has been discovered. From a biological viewpoint, chronic fear and activation of glucocorticoids are metabolically costly. Any unhealed injury to the body or nervous system is coped with by continuing reactions of the immune, inflammatory, hormonal, and autonomic nervous systems. From a biological viewpoint, chronic fear and the activation of glucocorticoids are *metabolically costly*. Although the acute stress response is considered adaptive, the chronic stress response contributes to disease. Threats to adaptation of long duration, and that are more injurious than the usual level and range of physiological functions, have multiple physiological consequences. Endocrine glands may be depleted according to chronicity and extent of hyperarousal and hypoarousal. The physiological effects of traumatic brain injury (TBI) (including damage to the hypothalamus, the hypophysis, and the pituitary gland) may elicit posttraumatic stress disorder (PTSD), the physiological characteristics of which may be in the opposite direction to those of TBI (Edwards & Clark, 1997). The characteristic symptoms of intrusive anxiety (flashbacks and nightmares) may in turn create persistent physiological and psychological processes that either maintain ongoing stressful functions or create new ones. Among the systems involved are the hormonal, the immune, and the inflammatory systems. Continued posttraumatic systemic functioning beyond the uninjured condition of the patient may be considered to be *dysregulation*. The dysregulated state is enhanced by lifestyle changes attributed to anxiety alternating with numbness, as well as by interference with physical health as the result of an injury (pain, restricted range of motion, loss of stamina, etc.). Consequently, the clinician is concerned with more than the obvious neurological, neuropsychological, psychodynamic, and injury effects. The brain's internal milieu is modified by the chronic effects of the injury that affect the need for treatment and outcome. Thus, the initial and chronic stress reaction to an accident that causes bodily injury are far more extensive than the familiar definition of PTSD. The chronic stress reaction is initially an expression of the systematic dysregulation and impairment referred to above. It evolves into the "burnout" condition known as "allostatic load."

Clinical example: Postconcussive stressors: A woman with a responsible office job in which she had difficult contacts with the public, was struck by a falling object while sitting at work.

"I was dizzy; headaches, nausea. I felt sick. I felt I was dying. I couldn't remember where I was sometimes." She has motion sickness, and never returned to work because of headaches, dizziness, nausea, memory loss, slowness in thinking and moving, lack of motivation, always tired, becoming dizzy when she bends. She avoids taking care of the checkbook due to feeling of *loss of competence*. She loses things. She doesn't write letters to people. "I never was depressed before. Now, I'm always depressed." She used to be able to cope with situations; now everything is a big deal. She describes herself as angry, irritable, and very short-tempered. *I'm just angry that nobody understands me*. As for sex: "I don't feel like doing anything."

She tries to hide her feelings (e.g., her dizziness). She doesn't want people to see the way she is, because she feels they would say: "Why don't you see the doctor?" when she says she might topple over. Or they might say, "Maybe you have a brain tumor." She feels no one understands her. Her life is very sedate. She does a little housecleaning, but "I really don't care. I watch TV. Others say

I'm not like I was. I was active, I ran. Now I walk very slowly. That's not me. I see very few of my friends. I'm not very good company."

Her family's reaction: "She was an all around person, good housekeeper, good wife. Now, she is half the woman she used to be. She could out-think anybody with or without a calculator. She was also good verbally at voicing her opinions. She also held down a full-time job. Her wits are slow—not totally blunt but dull. She can't think as fast as she used to. She stumbles with mathematics, has trouble recalling words. Everything is on the tip of her tongue. Physically she can't seem to do what she used to do, e.g., waking up at dawn, rush around house to get everybody fed, get herself dressed to go to work, cleaning the house, coming home, doing food shopping, generally being a wife, mother and worker at the same time.

"She cries for no reason, and she's like somebody who's been in a war. She wakes up in the middle of the night from nightmares. She has temper outbursts and fights over meaningless things that never bothered here before.

"Now she has to be helped just walking down the stairs. She has trouble walking to the bus without assistance, and she cannot maintain her balance while the bus is moving. Bending over [which she would have to do at work to get to filing cabinets] also throws off her equilibrium. At work, she could not get on an elevator and not have her equilibrium thrown off. She could not lift ten pounds nor bend over. And file drawers are below waist level.

"During the working day, she used to have the patience to deal with irate members of the public. If they used profanity, she had the temperament to handle them in a professional manner. Now she loses her temper over little things. She'll fight with her children, getting into arguments for no reason at all. She constantly sleeps. She randomly jumps, as though something scared her. I'll say, 'What's wrong?' She says, 'Nothing.' She doesn't clean the house. She doesn't remember what I said to her during a conversation; within half an hour she'll say, 'You didn't tell me that.' "

Crowded places and noise make her turn pale as a ghost. They give her an immediate headache, and make her feel dizzy and nauseous. She can't take confusion. She has a severe hearing loss. In a quiet room she can't hear if one ear is covered by the telephone, and she can't hear a car or truck.

General Adaptation Syndrome (Selye)

Hans Selye proposed a generalized persistent stress reaction, described as *general adaptation syndrome*: it is alert or alarm stage (sympathomedullary discharge and homeostatic disturbance), reacted to by a counter-shock phase (responses of the adrenal cortex and adrenal medulla), followed by a counter-shock reaction in which there are alterations in homeostatic processes. The stage of resistance involves adaptation to the effects of the stressor with increased susceptibility to the deleterious effects of other homeostatic challenges. Continued stress would lead to the third and final stage (i.e., exhaustion and death) (Felker & Hubbard, 1998; Hubbard & Workman, 1998). Selye's concept of the nonspecificity of the stress reaction and the centrality of the adrenal cortical response has been disputed (the current emphasis is the participation of multiple neural and neuroendocrine pathways with no single preeminent system). Prior stress history appears to influence the pattern of response for subsequent exposure to a stressor. Animal research suggests that exposing a stressed animal to an unfamiliar stressor is a greater challenge to behavioral and physiological homeostasis than exposing the same stressor to naive animals (McCarty & Pacak, 2000). In fact, an accident-creating head injury has multiple effects (cerebral, somatic, social, psychological, and biological) that may represent multiple diagnostic entities.

Somatic Injuries Comorbid with the Post Concussive Syndrome (PCS)

Somatic injury occurring in the concussive accident may be the most important consideration affecting current status and outcome. Even though the TBI represented by concussion may be relatively slight, body trauma by itself can be impairing. To illustrate this paradox, a sample of four patients in the writer's practice was selected using the criterion of comorbid diagnosis of postconcussive syndrome, nonhospitalization for injury, and extended psychotherapy to permit close observation of the patient's condition and also documentation of the chronicity of the somatic injury.

Patients: Four men examined 1 month to 6 years after their injury, then in psychotherapy for many years. Education: 7 years to college degree. Age at injury: 34–52 years old. All patients had postconcussive syndrome (head impact without hospitalization; posttraumatic altered consciousness ranging brief to long). Other diagnoses were adjustment disorder (depression and anxiety), PTSD, and dementia due to head trauma. Trauma: Two patients had rear end collisions; one patient had three separate rear end collisions over a period of years; one patient had his head and back struck by two falling objects from 9 to 10 feet, each approximately 20 lbs.

Primary symptoms: spinal column; general medical disorders (including cardiac); dental Temporomandibular joint disorder (TMJ); mobility and gait; dizziness and imbalance; partial seizures; sleep disorder (bad dreams); loss of stamina; muscle; dysarthria; pain; headache (lateralized and bilateral, traumatic and referred pain); visual (bilateral central scotoma, diplopia); tinnitus; autonomic (sweating, nausea, frequent urination, bowel control, hyperventilation); sexual (loss of libido, impotence); oversensitivity (heat, cold, bright light, loud sound).

The extended range of disorders in chronically symptomatic and disabled patients alerts the examining and treating clinicians that attention can be too highly focused upon, for example, neurological or neurobehavioral disorders. This can result in an examination that is incomplete and misrepresents the actual adaptive condition of the patient.

Initial Dysregulation

If not correctable by normal control processes (*allostasis*), the initial dysregulation by trauma, or persistent and excessive environmental demands eventually create health problems or permanent system damage (*allostatic load*). Wounds that remain unhealed contribute to chronic complaints and varied neurobehavioral dysfunctioning. Wound healing is often incomplete after mechanical trauma, including injured and dead tissue, interference with normal functioning by scar tissue, pain, inflammation, tissue debris, pain, and so on. Persistent *pain* creates adverse adaptive efforts. The response to injury interacts with the CNS. Sympathetic neurons in the anterior lateral horn system initiate increased cardiac activity, pronounced increase in peripheral vascular resistance, and redistribution of blood flow away from the heart and brain. These responses may initiate many *pathophysiological changes*: altered regional perfusion, compromised function of key target organs, activation of the rennin-angiotension system, increased platelet activation, and reflex efferent hyperactivity (Walker, 2004). The response to injury interacts with the CNS to create diagnostic confusion and problems of appropriate treatment. The physiological effect of trauma-related injuries is a frequently neglected issue after head injury.

INJURY, STRESS, AND WOUND HEALING

Chronic Stress and Wound Healing

Chronic stress impairs the process of wound healing, attributable to PIC (Biondi, 2001). *Stress reduces the immune response* (i.e., multifactor suppression of the production of various cytokines [IL-1; Tumor necrosis factor (TNF)], and also reduces the proliferation and activity of various types of cells (Song et al., 1995). Inflammatory cytokines are active in the immune and in the central nervous systems. They also signal the hypothalamus to release CRH, which activates pituitary-adrenal counter-regulation of inflammation through glucocorticoids. Thus, stress affects the outcome of wound healing, which is regulated by inflammatory mediators. When wound healing is delayed, procedures such as periodontal surgery may not have the therapeutic effect (Marucha et al., 2001). While various stressors suppress the immune response, pain enhances immune function provided that the thymus is intact. *Pain stimulation enhances immune cell production* through the effects of catecholamines released by the adrenal gland upon T cells (Fujiwara & Yokoyama, 1994).

Many *chronic symptoms* after an injury (e.g., limited range of motion, pain and headaches, reduced stamina, vulnerability to infectious disease, etc.), are consequent to damaged tissue (i.e., physiological reaction of the body to the intrusion of scars, fibrous and dead tissue, coping with injured tissue, etc.). The chronic reaction of the hormonal, inflammatory, and immune systems to an injury has neurobehavioral effects that can be mistaken for TBI. Among the consequences are tissue dysregulation and exhaustion (*allostatic load*), with varied symptoms, including *somatic* (tissue exhaustion, fatigue, burnout, organ damage, sexual dysfunction); *metabolic* (hormonal exhaustion, loss of stamina, fatigability); *neurobehavioral* (reduced social interest, reduced motivation); diseases (stress and communicable diseases); *illness behavior* (consequent to the CNS effects of cytokines, etc.) such as malaise, fatigue, loss of appetite, apathy, social withdrawal, inactivity. *Cytokines* alter CNS function, invoking fever, sleep, and illness behavior; activating the stress axes; causing affective, cognitive, and behavioral changes; and direct the immune cell invasion of, and interactions in, the CNS (Felten & Maida, 2002).

Wound healing is impaired by *chronic stress*: the effects of *PIC* (Biondi, 2001), foreign materials, bacterial infection, poor nutrition, *smoking*, *alcohol*, and *advancing age* (Yang & Glaser, 2005). *Inflammatory cytokines* activate the immune and the central nervous systems; they also signal the hypothalamus to release CRH to activate pituitary-adrenal counter-regulation of inflammation through glucocorticoids. Thus, stress impairs the outcome of wound healing by which is regulated by inflammatory mediators. While various stressors suppress the immune response, pain enhances immune function provided that the thymus is intact. *Pain stimulation enhances immune cell production* through the effects of catecholamines released by the adrenal gland upon T cells (Fujiwara & Yokoyama, 1994). *Stress reduces the immune response* (i.e., multifactor suppression of the production of various cytokines [IL-1α; IL-8; INF]), and also reduces the proliferation and activity of various types of cells. Psychological stress associated with surgery impaired the inflammatory response, leading to more painful, poorer, and slower surgical recovery. Relaxation and stress intervention procedures modify the detrimental effects of psychological stress on physiological processes (Song et al., 1995; Yang & Glaser, 2005). Inflammatory cytokines activate the immune and CNSs. They also signal the hypothalamus to release CRH, which activates pituitary-adrenal counter-regulation of inflammation through glucocorticoids. Thus, stress affects the outcome of wound healing, which is regulated by inflammatory mediators. When wound healing is delayed, such procedures as periodontal surgery may not have the therapeutic effect (Marucha et al., 2001). While various stressors suppress the immune response, pain enhances immune function provided that the thymus is intact. Pain enhances immune cell production through the effects of adrenal catecholamines upon T cells (Fujiwara & Yokoyama, 1994).

Body functions influence the efficiency and pattern of the CNS. Thus, brain control is not only consequent to afferent input, stimulation, and inhibition between different parts of the nervous system, but is also subsequent to efferent output. During and after illness and injury, important physiological systems function and may be actually dysregulated (the hormonal [stress] system, the inflammatory system, and the immune system). These send substances that are transported to the CNS by the blood stream and nerves (most importantly cranial nerve X, the vagus nerve). The same messengers (hormones, immune system products, neurotransmitters and neuromodulators, etc.) are expressed and function in many different parts of the body. The blood-brain barrier (BBB) is not a complete barrier even during health. Some substances enter the brain from the cerebral ventricles through the circumventricular organs (CVO). This creates the potential for neurobehavioral disorder. As discussed before, after trauma various internal milieu barriers that usually keep extraneous substances from the CNS are less efficient. It has been suggested that breakdown of the BBB permits autoantibodies to enter the brain with potential cognitive decline associated with *systemic lupus erythematosus* (*SLE*) (Carroll, 2004).

TRAUMATIC DYSREGULATION

Chronic complaints invite exploration of traumatic physiological dysregulation (loss of autoregulation). Prolonged somatic activity related to injury, stress, pain; stress reactions to chronic pain and impairment; and unhappiness all result in neuroendocrine imbalances. Chronic brain and somatic trauma lead to dysfunctions or loss of *autoregulation* of the brain and soma. These have a disruptive effect upon the adaptive immune response and others.

The persistent disturbance of the internal milieu contrasts with the control of systems during normal temporary demands (i.e., allostasis). This function is mediated in part by physiological systems that release neuroactive substances into the internal milieu (i.e., hormonal, inflammatory, and immune). If the injury is healed, and other stressors cease, the physiological systems resume their normal controls and limited deviation from some norm that is integrated with other tissues. With chronic injury, their persistent activity is considered to be "dysregulated." The ultimate ill health effects are termed "burnout" or "allostatic load."

SOME TRAUMATIC DYSREGULATION EFFECTS

Chronic somatic injury ("unhealed wounds") has direct and indirect consequences for the status of the internal milieu, leading to impaired adaptive functioning of the supportive physiological systems (hormonal, immune inflammatory, and autonomic). "Unhealed wounds" cause a persistent, ceaseless reaction designed to reduce the effects of the somatic and cerebral lesions (i.e., internal organs, soft tissue of the musculoskeletal system, pain, and reaction to lesions). Chronic neurobehavioral dysfunctioning has a significant effect upon the outcome of injury, as well as on posttraumatic health and stamina. The consequence is allostatic load (see below). After a lengthy period of reaction to injury and impairment (chronic stress reaction), physiological systems become exhausted ("burnout"). Consequently, the person is vulnerable to stress-related diseases, loss of stamina, and contagious diseases (allostatic load).

Several major and disparate neurobehavioral and health mechanisms mediate neurobehavioral consequences of chronic injuries:

- Chronic discharge of neuroactive chemicals into the brain's internal milieu; and,
- Transmission of psychoactive chemicals from the vegetative system via the Vagus N.

Initial altered physiological response: Altered output of the hypothalamic pituitary axis and other endocrinological functions of the pituitary gland, and persistent responses also of the immune, inflammatory, and autonomic systems.

DIAGNOSTIC ISSUES

Physiological dysfunctions contribute to neurobehavioral dysfunctioning that can be misattributed to cerebral injury, faking or secondary gain, if the practitioner does not consider the effects of a current or earlier injury. The chronic stress reaction also includes the subjective and realistic experience of impairment.

HOMEOSTATIC DYSREGULATION

One of the frequent postconcussive symptoms is the loss of control over body temperature. The patients frequently complain of feeling cold or warm in ambient temperatures that seem comfortable to others.

When one or more functional systems can no longer contribute, all cells of the body suffer (Guyton & Hall, 2000, p. 8). Trauma, whether mental or somatic, initiates multiple physiological functions, some of which restrain the stress function, heal some tissues, and control other tissues. Paradoxical consequences may have deleterious effects upon both soma and the CNS. The reactions of the hypothalamic-pituitary adrenal (HPA) axis and the sympathoadrenal medullary axis are primary when there is no accompanying injury. When the stressful event causes a wound, there are complex reactions of the hormonal, inflammatory, and immune systems. Trauma may cause loss of function due to tissue destruction, disturbed function of damaged tissue, effects of the healing or damaged tissue removal process, and stress reactions intended to reestablish adaptive regulation. Some effects exceed normal physiological or homeostatic limits or set points, or may create remote neurological and physiological consequences of signals derived from the above, or may manifest dysfunctions in the future as well. Thus, pathology substitutes for adaptation.

In the event of an accident or a disease process, there is a reciprocal relationship between the immune and inflammatory systems and the brain. These organs not only affect tissues and infiltrate the brain, but their characteristic chemical products are also formed in the brain with neurobehavioral effects. Trauma and stress may damage tissue required for a system to function, or may create loads that exceed the body's capacity to function within adaptively useful limits. Thus, the brain is merely a portion of an entire-body network that functions through multiple pathways, influencing healing and disease processes, somatic efficiency, moods, and probably cognitive efficiency as well.

While the stress response is adaptive for overcoming acute physical stressors, it may be more damaging than the stressor itself, resulting in the emergence of many stress-related diseases (Lambert et al., 1999). The effectiveness of the stress response is judged both by the rapidity and efficiency with which it mobilizes physiological mechanisms, and also by how quickly these physiological mechanisms return to prestress homeostatic and allostatic levels (Friedman & McEwen, 2004). The clinician's examination should explore multiple potential comorbid conditions, such as cerebral trauma; somatic injury (accompanied by acute and chronic dysregulation of various physiological systems); and posttraumatic stress (considered as a physiological and psychological condition that may be both acute and chronic). The consequences of stress must be considered in the treatment and assessment of any accident victim, including specifically the patient with assumed cerebral injury. Parallel to nervous system trauma is a set of impairing and discomfort-producing physiological conditions, which create separate conditions for maladaptation and impairment. The concept has already been offered that persistent symptoms and discomforts after an accident are consequent to a wide range of sites of bodily injury and dysregulation of internal tissue systems (inflammatory; immune; hormonal). Furthermore, one must consider disorders of identity, mood, and arousal. The mechanisms of the PTSD reflect and create varied mood disorders (e.g., anxiety, withdrawal, numbness), and disturbing experiences create hyperarousal (e.g., "jumpiness," nightmares, bad dreams, flashbacks). Dysregulation of the internal milieu occurs as a function of coping with body damage. Since there are defective barriers to the CNS for fluids entering the liquid environment of the nervous system (they are normally mostly excluded), the varieties of abnormal physiological functioning may account for such neurobehavioral effects as loss of stamina (fatigability), mood disorders, cognitive inefficiency, vulnerability to a variety of diseases, and so on. The cumulative effects over time of the parallel physiological and psychological dysfunctioning may be conceptualized as "stress."

Physiological dysregulation aids our understanding of *chronic complaints*: an unrecognized unhealed wound is not only uncomfortable and reduces performance efficiency, but may also stimulate the physiological systems mentioned above. Unawareness of the complaint's true etiology leads to neglected or ineffective treatment, or the complaints being seen as malingered, exaggerated, factitious, or having an "emotional overlay."

SLEEP–WAKE DISTURBANCE OF ENDOCRINE RHYTHM

Sleep disturbance may interfere with circadian control over hormonal function (Grumbach & Styne, 1998). Disruptions are common after concussive head injury. Sleep disturbance has profound effects upon adaptive efficiency, including disturbances of endocrine function, stress reactions to fatigue, enhanced danger in the use of power tools, and so on. Disturbances of circadian rhythm are discussed in Chapter 7 on Physiological Disturbances. An increase of CNS aminergic activity (e.g., norepinephrine [NE], serotonin, epinephrine [EPI]) results in hyperarousal with insomnia and perhaps stress. With sleepiness, attention and cognition decline and motor activity is impaired. Insomnia prevents the cholinergic system from exerting a restorative effect. Prolongation of sympathetic activity can be harmful to cardiovascular, cognitive, and behavioral functions. Decrease in CNS aminergic activity, such as in depression, has the opposite effect (Hobson, 1999). After severe TBI there may be considerable variability in intracranial pressure and cerebral perfusion pressure, which has implications for pharmacotherapy (Kropyvnytskyy et al., 1999).

A variety of behavioral disturbances (e.g., sleep and appetite) may be attributable to disorganization of the circadian rhythm. When neural control of gonadotropin regulation is perturbed (e.g., after stress), the amplitude of pulses is initially reduced and, if severe, ultradian cyclic rhythm may be lost completely.

Puberty: Maximum secretion of luteinizing hormone-releasing hormone (LHRH) occurs at night. If stress occurs prior to puberty, puberty may be delayed indefinitely. The fundamental change during puberty has long been thought to be due to a reduction in tonic hypothalamic inhibition of LHRH release.

Symptoms: Disruption of social patterns (i.e., connection between events or between particular responses and outcomes) may remove the social cues to which circadian rhythms are attuned (Healy & Williams, 1988). In women, an approximately 90-min ultradian rhythm is present, with larger bursts of gonadotropin secretion during sleep than during the day (Reichlin, 1998). There can be a *delayed sleep-phase syndrome* with chronic inability to fall asleep at a desired clock time to meet required work or study schedule (Quinto et al., 2000).

Hormones

It may be inferred that hormonal functions are altered by sleep-wake cycle interference, which is frequent after TBI: gonadal, cortisol (adrenal), thyroid, growth hormone (GH) (Akil, et al., 1999), and melatonin. The cortisol circadian rhythm is stronger in PTSD than normal, although there is reduced 24-h urinary excretion (Yehuda et al., 1996). Patients with *major depression* demonstrate disruption of the circadian rhythmicity of the pituitary-adrenal axis: elevation of adrenal glucocorticoid output (earlier morning surge) and enhanced cortisol secretion in the late afternoon. Reduced diurnal variation of cortisol secretion is associated with sleep deprivation (Crofford, 2002). Synchronized episodic increased gonadotrophin stimulation (on a circadian basis) sets into motion full development, ovulation, and spermatogenesis (Lee, 1996). The present writer speculates whether a child's sleep disorder post-TBI may thus interfere with optimal physiological development if the hypothalamic-pituitary-gonadal axes, dependent upon episodic release in the neonatal period and childhood, is perturbed. The dysregulating effects of sleep disturbances includes: interruption of endocrine secretion; disturbance of alertness and arousal at work, other safety considerations, and the effect of examination during a particular point in the circadian cycle (Williams et al., 2001).

The normal secretion of numerous hormones is under the control of circadian rhythms (i.e., secreted in a pulsatile manner during sleep at night) (Berne & Levy, 1998, p. 1007). If stress occurs prior to puberty, puberty may be delayed indefinitely. The fundamental change during puberty has long been thought to be due to a reduction in tonic hypothalamic inhibition of LHRH release. Circadian (daily) rhythms are the most common adaptation of living organisms. They affect the rest-activity cycle, variations in psychomotor performance, sensory perception, secretion of hormones, and regulation of core body temperature (Moore, 1999). Normally, circadian rhythms are entrained to the light-dark cycle with a stable phase relationship. The light-dark cycle is entrained with endogenous circadian rhythms (i.e., sleep-wake, behavioral, and hormonal). Within the overall wake-sleep cycle, there are sub-cycles in the sleep phase. A variety of hypothalamic and pituitary secretions are controlled on circadian (24-h) and ultradian (one pulse every 60–80 min) rhythms by the hypothalamus, which secretes in a pulsatile fashion. By inference, hormonal functions are be altered by sleep-wake cycle interference, which is frequent after TBI: gonadal, cortisol (adrenal), thyroid, GH (Akil et al., 1999), and melatonin. Synchronized episodic increased gonadotrophin stimulation (on a circadian basis) sets into motion full development and ovulation and spermatogenesis (Lee, 1996). The author speculates whether a child's sleep disorder post-TBI may thus interfere with optimal physiological development if the hypothalamic-pituitary-gonadal axis, dependent on episodic release in the neonatal period and childhood, is perturbed.

One of the frequent postconcussive symptoms is loss of control over body temperature. The patients frequently complain of feeling cold or warm in ambient temperatures that seem comfortable to others. In one study of patients hospitalized with severe head injury, only 1 of 10 patients manifested circadian rhythms for body temperature and heart rate with the peaks occurring at the appropriate times. The low (but significant) correlations between body temperature and heart rate suggests desynchronization of rhythms, and may have clinical significance (Lanuza et al., 1989).

NEUROLOGICAL ISSUES: STRESS

Hypoarousal

Animals subjected to continued stress (analogous to the experience of patients) have reduced NE concentrations in the hypothalamus, locus ceruleus (LC), and anterior cortex (Ricci & Wellman, 1990, citing Mayeux, 1986; Weiss et al., 1985). An alternate explanation might be the human equivalent of animal-model *stress-induced depression following inescapable shock*. This might correspond to the phase of hypoarousal.

Wound Healing

The process of healing or removing injured or dead tissue has significant neurobehavioral effects (i.e., a variety of biological products are discharged which are part of a bidirectional signaling system with the CNS). The intensity of the neurological effect is enhanced since in posttrauma the effectiveness of various barriers to systemic substances entering the brain is reduced. Consequently, here are alterations of neural functioning both within the brain and in the pattern of discharges of the hypothalamic-pituitary-target organ axes and cranial nerve and spinal cord regulation.

Cerebral Volume

The following data refers primarily to the chronic effects of noninjury stress. However, they may reveal vulnerability to the greater effects of TBI and/or other components of stress. Adult studies have determined in PTSD coexisting, smaller hippocampal volume and smaller cerebral tissue volumes. Combat veterans manifested smaller sulcal cerebral spinal fluid (CSF) and cranial volumes (corrected for body size, with TBI excluded), but not smaller cerebral tissue volume. There was a larger CSF volume in those with a prior history of *alcohol abuse* or dependence. It was surmised that the difference of cranial volume was related to a developmental divergence predating adult trauma, the cause of which may be environmental adversity (mothers with poor health; poor nutrition with neglect; HPA axis and mood disorder) (Woodward et al., 2007).

MECHANICAL TRAUMA TO THE HYPOTHALAMIC-PITUITARY AXIS

Head trauma can cause defects ranging from isolated adrenocorticotropic hormone (ACTH) deficiency to panhypopituitarism with diabetes insipidus (Molitch, 1995). Any suggestion of damage to the sella turcica should initiate a study of pituitary function, which may take years to develop. Symptoms of hypopituitarism may be total or partial (D'Angelica et al., 1995; Grossman & Sanfield, 1994). Total loss of pituitary secretion requires 90% glandular destruction and may accompany a prolonged coma. Recovery of pituitary function usually parallels neurological improvement. Residual impairment is not uncommon (Frohman, 1995). Prompt treatment is essential after suspicion of a pituitary injury. *Hypothalamic anovulation* results from the disorder of LHRH production in the arcuate nucleus of the hypothalamus and its secretion into the portal vessels from the median eminence. It can be caused by head trauma or by any other disorder of the CNS that interferes with this process (and also feedback by gonadal steroids mediated by brain neuromodulators such as NE, dopamine, and β-endorphin). It is also related to stressful events and vigorous exercise (Bulun & Adashi, 2003).

POSTTRAUMATIC DISORDERS OF THE HYPOTHALAMIC-PITUITARY AXES

The pituitary gland is vulnerable to trauma involving changes in momentum to the brain relative to the skull, or direct blows, since the gland is attached to a stalk at the base of the brain (infundibulum) and actually located within of the skull at the base of the brain (sella turcica). When the brain moves, this stalk and/or its attachment within the hypothalamus may be torn, stretched, or rotated. Head trauma later in life may cause hemorrhage in the area of the hypothalamus or pituitary gland, resulting in hypopituitarism with ACTH deficiency (Migeon & Lanes, 1996). Hypothalamic-pituitary trauma sites include the brain (e.g., vascular injury to the hypothalamus), pituitary stalk shearing and laceration, anterior pituitary lobe necrosis, and posterior lobe hemorrhage. Lesions are caused by blunt trauma; mass effects causing compression of the hypothalamo-hypophyseal portal vessels (Graham, 1996); pressure waves shearing movements of the brain tearing the pituitary stalk (Molitch, 1995); a bullet wound outside the brain in the temporomandibular area of the face, creating hypophyseal damage secondary to transmission of energy and bullet fragments (Molitch, 1995); and transfacial gunshot wounds (Carlier et al., 1966). Hematoma can damage the hypothalamus, the pituitary stalk, the blood vessels, or pituitary tissue directly. Inflammation around the area of the partially enclosed pituitary gland can injure its tissue. Pituitary damage from head injury may be due to sella turcica fracture, pituitary stalk section, trauma-induced vasospasm, or ischemic infarction affecting the hypophyseal-portal system (Melmed & Kleinberg, 2003). Basilar skull fractures that tear the pituitary stalk rupture the neural connections to neurohypophysis and the vascular connections to the adenohypophysis. This disrupts delivery of releasing or inhibiting factors to the pituitary. Cooper (1987) asserts that most of the damage to the hypothalamus and pituitary gland is secondary to raised intracranial pressure, brain shift, and the distortion of the brain.

The supply of pituitary hormones is controlled by numerous neural inputs and endocrine feedback to the hypothalamus. In turn, the hypothalamus provides stimulatory neuropeptides (NP) to the anterior pituitary via the pituitary portal circulation, and also to the posterior pituitary via long axonal projections. Early endocrine evaluation is essential when there is strong suspicion of a pituitary injury (e.g., any suggestion of injury to the sella turcica) (D'Angelica et al., 1995). A severe enough lesion on any part of the HPA leads to loss of endocrine homeostasis by disturbing excitatory and inhibitory stages of endocrine signaling to the hypothalamus and pituitary gland (Molitch, 1995).

HYPOTHALAMIC DISEASE

Signs and symptoms of hypothalamic disease include sexual dysfunction (hypogonadism or precocious puberty), diabetes insipidus, psychological disturbance, obesity or hyperphagia, somnolence, emaciation or anorexia, thermodysregulation, and sphincter disturbance (the following information is largely based upon Jacobson & Abrams, 1999). The signs are selected according to whether the may be directly observed or interviewed by the nonmedical practitioner (e.g., not requiring laboratory study or medical procedures not within the scope of practice of the nonmedical practitioner). Concern should stimulate medical consultation.

Signs/symptoms of hypothalamic disease: sexual development (hypogonadism or precocious puberty); diabetes insipidus; psychological disturbance; obesity or hyperphagia; somnolence; emaciation or anorexia; thermodysregulation; sphincter disturbance.

Hypothalamic-Pituitary Endocrinological Dysfunction (Selected)

Hypothyroidism: Deficiency results in a generalized slowing of metabolic processes. In children it results in a marked and permanent slowing of growth and development, including mental retardation when it occurs in infancy. It also results in placid behavior, cold intolerance, and hoarse voice.

Hypogonadism: Results in delayed puberty; amenorrhea; micropenis; infertility; diminished libido

Hypoadrenalism: Results in lethargy; postural hypotension

Hyperadrenalism (Cushing's syndrome): Results in obesity; growth delay; hirsutism; acne; striae; weakness

GH deficiency: Results in short stature; hypoglycemia; reduced muscle mass; increased abdominal fat mass; reduced vigor and concentration

GH excess: Results in gigantism; acromegaly

Hyperprolactinemia: Results in delayed puberty; galactorrhea; impotence; amenorrhea

Hypopituitarism symptoms (Schneider et al., 2007)

Hormone deficiency: GH

Symptoms: Anergia; poor quality of life

Clinical findings: Osteoporosis; visceral obesity; reduced lean muscle mass; dyslipidemia

Luteinizing hormone/follicular stimulating hormone (*LH/FSH*) (Sex steroid symptoms): loss of secondary hair

ENDOCRINE DYSFUNCTIONS

Endocrine disorders are a grossly unappreciated consequence of head injury due to the significant risk of hypothalamic and pituitary injury (somatic, sexual, and developmental problems). Decreased pituitary function can occur in children and adults from injuries that do not cause a loss of consciousness (LOC) and that may remain unrecognized for a lengthy period. Head trauma can result in hypothalamic failure (e.g., *acquired gonadotropin deficiency*). Head trauma and *child abuse* are clearly stated to be etiologic considerations for hypopituitarism (Findling & Tyrell, 1986). The degree of hormonal reduction reflects the severity of the trauma. An estimated 30.5% of patients with moderate-to-severe TBI incur pituitary dysfunction (Bondanelli et al., 2007).

With increased sensitivity to the existence of post-TBI endocrine disorders, the recently reported incidence is higher than previously thought (i.e., 35–59% with 17–22% having disorders in 2 axes). Anterior pituitary deficiencies have been diagnosed more than 5 years postinjury (Elovic, 2003). Head trauma can cause defects ranging from isolated ACTH deficiency to panhypopituitarism with diabetes insipidus (Molitch, 1995). Any suggestion of damage to the sella turcica should initiate study of pituitary function, which may take years to develop.

Panhypopituitarism

Panhypopituitarism has been defined as generalized or particularly severe hypopituitarism. In its complete form, it leads to absence of gonadal function and insufficiency of thyroid and adrenal cortical function. Some characteristic dysfunctions are dwarfism, regression of secondary sex characteristics, loss of libido, weight loss, fatigability, bradycardia, hypotension, pallor depression, and cachexia (Saunders Elsevier, 2007). Other symptoms described as being associated with panhypopituitarism are integrated into the general symptom list. Panhypopituitarism is rare, and is associated with alterations in arousal and awareness (Rosen & Cedars, 2004).

Symptoms: Excessive and inappropriate secretion of antidiuretic hormone (SIADH); thyroid dysfunction (Brontke et al., 1996); galactorrhea, amenorrhea, and hirsutism (Rebar, 1996). Inhibition of gonadotropin secretion resulting in secondary hypogonadism in boys (Matsumoto, 1996) may indicate hypothalamic damage with hyperprolactinemia. Loss of axillary and pubic hair (Woolf, 1992); growth failure; acquired gonadotropin deficiency (Ferrari & Crosignani, 1986); hypothyroidism; ACTH deficiency (D'Angelica et al., 1995); transient hypogonadotrophic hypogonadism (Woolf et al., 1986) or permanent hypogonadotrophic hypogonadism (Ferrari & Crosgnani, 1986); reduced thyroid function, possibly associated with a stress-related sympathetic discharge of catecholamines (Woolf et al.,1988); hyponatremia (cerebral salt wasting, or natriuresis); SIADH (vasopressin or antidiuretic hormone [ADH]) (Kokko, 1996); myxedema; dwarfism; obesity; hypothermia; bradycardia (Treip, 1970); dermatological (waxy or dry skin; wrinkling around eyes and ears); anemia; psychiatric disturbance (mental slowing; apathy; delusions; paranoid psychosis); hypoglycemia; weakness; lethargy; cold intolerance; decreased libido; constipation; bradycardia; hoarseness; myxedema; increased, decreased menstrual flow and amenorrhea; delayed resumption of menstruation (Brontke et al., 1996); hyperprolactinemia (persistently elevated prolactin is almost always associated with a hypothalamic-pituitary disorder), with such symptoms as amenorrhea, headaches, visual field defects, infertility, galactorrhea, and osteopenia (Attie et al., 1990; Ferrari & Crosignani, 1986; Grossman & Sanfield, 1994; Liau et al., 1996; Reichlin, 1992; Rosen & Cedars, 2004; Shaul et al., 1985).

Neurobehavioral Deficits

Hypopituitarism can exacerbate existing impairments of physical, cognitive, and psychosocial functioning caused by a TBI and may be the major impediment to successful rehabilitation. The examiner should consider referral to assess ACTH, GH, gonadotroph, thyroid study (TSH; Free T4), and basal morning cortisol. In women, cognitive function and memory appear to be impaired by low estrogen levels (Kelly et al., 2000; Lieberman et al., 2001; Schneider et al., 2005). The symptoms of anterior hypopituitarism (strength, aerobic capacity, sense of well-being, general health, vitality, and mental health, with commonly experienced depression and anxiety) may be overlooked or attributed to other causes.

Neurobehavioral, cognitive, and psychosocial dysfunctions of patients with moderate or more severe TBI are similar to those of *hypopituitarism*. Disorders after TBI may be mimicked by hypopituitarism since the pituitary regulates hormone production from the thyroid, gonads, and adrenals, which in turn regulate processes with physiological and/or psychological consequences. Evidence is cited that the height of patients with TBI is significantly shorter than that of patients with GH deficiency. Therefore, pituitary hormone deficiencies could result in suboptimal rehabilitation for patients with TBI-induced hypopituitarism (Ghigo et al., 2005): Fatigue, decreased lean body mass with increased body fat and dyslipidemia, reduced exercise tolerance and muscle strength, diabetes insipidus, decreased TSH and T4 levels, adrenal insufficiency, amenorrhea/infertility, erectile dysfunction, hyperprolactinemia, diminished cardiovascular function, impaired cognitive function, memory loss, decreased concentration, mood disturbances (anxiety and depression), irritability, insomnia, and a sense of social isolation.

Apolipoprotein E: Pituitary Dysfunction

Pituitary dysfunction can occur after blunt head injury in a high-speed impact (motor vehicle accidents [MVA]), lesser impact sports injury (e.g., boxing and kickboxing). Genetic polymorphisms play an important role in the susceptibility to, and the outcome from, CNS disorders. Apolipoprotein E genotype (APO E) has a role in neuronal repair and maintenance. It has been implicated in Alzheimer disease and TBI outcome. The eta 4 allele (APO E4) is related to chronic traumatic encephalopathy in boxing (Tarnriverdi et al., 2008).

There is evidence that particular alleles (genotypes) of the *APO E* influence both hypopituitary risk and TBI outcome. The E4 allele has a role in neuronal metabolism, maintenance, and repair. It is also a risk factor associated with the development of Alzheimer disease. Prior studies indicated that pituitary hormone deficiencies are reported more frequently in severe TBI; that pituitary hormone deficiencies, particularly GH disorder, could not be correlated to TBI severity; that pituitary hormone deficiencies could occur even after mild TBI; that genetic polymorphisms play an essential role in susceptibility to CNS disorders; and that the APO eta 4 allele was found in 3.7% of patients with a good TBI recovery and in 31% of patients without it.

In a sample of TBI patients examined for hypopituitarism (traffic accident, boxing, kick boxing) 25.8% of 93 SS had pituitary dysfunction. The ratio of pituitary dysfunction was significantly lower in SS with the APO *E3/E3 genotype* (17%) than those without this genotype (41.9%). The E3/E3 genotype decreases the risk of hypopituitarism after TBI. APO E is the primary apolipoprotein synthesized within the CNS, including the hypothalamic-pituitary region. It is upregulated after trauma and reduces the neuroinflammatory response (release of reactive oxygen species and inflammatory cytokines such as TNF alpha and IL-6). It is speculated that trauma-induced neuroinflammatory responses and variations in APO E neuronal repair mechanisms influence the pathogenesis of TBI-induced hypopituitarism (Tarnriverdi et al., 2008).

Late Endocrine Related Dysfunctions

Late *endocrine dysfunctions* are related to persistent demands upon the hypothalamic-pituitary-endocrine axes, as well as stress-related health disorders including immunosuppression, dysfunctions of various systems, and endocrine exhaustion. Common symptoms are accounted for by this model: fatigue, exhaustion, and burnout; illness behavior; loss of appetite; apathy, social withdrawal, inactivity, and sleepiness; sexual dysfunction and dysfunction of pediatric sexual development (see Allostasis and Allostatic Load). These persistent symptoms may be misattributed to symptom exaggeration or emotional factors (e.g., cytokines). This secretion can be reflected in mood disorders, cognitive dysfunctions, and presumably other disrupted neural patterns causing impairment.

The onset may be slow and insidious, depending on the magnitude of hypothalamic-pituitary damage, destruction of the anterior pituitary, or secondarily based upon loss of hypothalamic stimulating factors normally acting on the pituitary. Trauma may occur to the hypothalamus, pituitary stalk, or anterior pituitary (Aron et al., 2007a). Hypopituitarism after head trauma is usually expressed within a year, although some patients may overtly manifest signs of pituitary failure only after several decades. It is not known whether early onset hypopituitarism may be transitory. These authors disagree with others by asserting that the severity of symptoms need not be related to the severity of the injury. Virtually all patients with pituitary failure have a history of LOC after trauma. GH plays an important role in the recovery of the CNS and is an independent predictor of outcome. This indicates that patients undergoing a TBI-intensive recovery program may be positively influenced by normal GH secretion (Bondanelli et al., 2007). GH deficiency was reported in 2.4% of one series of hypopituitary adult patients (Abs et al., 1999, reported by Melmed & Kleinberg, 2003). When found after TBI, it may be due to its secretion in the wings of the pituitary gland, where the blood and oxygen supply is anatomically vulnerable. Patients with TBI may benefit from hormone replacement therapy since hypopituitarism may mimic the symptoms of TBI. Whether the degree of pituitary failure correlates with the severity of head trauma is doubted by Ghigo et al. (2005), but affirmed by others. Half of the patients have skull fracture. One-third have magnetic resource imaging (MRI) signs of hypothalamic or posterior pituitary hemorrhage or anterior lobe infarction, with diabetes insipidus the most common endocrine disorder (30% of patients). Gonadotropin failure, amenorrhea, and hyperprolactinemia can occur after months or years after date of trauma (Melmed & Kleinberg, 2003). Although the stress reaction is characteristically hyperpituitary, trauma may cause hypopituitarism (Cooper, 2004; Reiter & Rosenfeld, 2003) due to damage to either the anterior and posterior lobes of the pituitary gland or to both (Edwards & Clark, 1997), or to nonresponse by the HPA axis to stressors. Head trauma can cause hypopituitarism due to direct pituitary damage by a sella turcica fracture, pituitary stalk section, trauma-induced vasospasm, or ischemic infarction after blunt trauma. Hypothalamic stalk trauma consequences differ from pituitary insufficiency (Reichlin, 1998).

ALLOSTASIS AND STRESS

Although antireproductive, antigrowth, catabolic, and immunosuppressive, the initial stress response is temporarily beneficial. Chronic stress may lead to a pathological syndromal state with psychiatric, neuroendocrine, cardiovascular, metabolic, and immune components (Chrousos, 1998). The HPA becomes hyporesponsive, affecting stamina, health, and quality of life. Stress disrupts energy metabolism and caloric balance (Middleboe et al., 1992).

ALLOSTATIC LOAD

There are different types of allostatic load (McEwen, 2000a):

(1) Frequent stress results in chronic overexposure to stress hormones. Blood pressure surges may trigger myocardial infarction or accelerate atherosclerosis, which increases the risk of myocardial infarction.
(2) Failure to habituate to repeated challenges, as the individual who remains stressed by the necessity for public speaking.
(3) Inability to turn off allostatic responses (i.e., the termination of stress with the decline of these hormones during the diurnal cycle). Some people's blood pressure fails to recover after mental stress, with hypertension leading to atherosclerosis. Intense athletic training induces allostatic load [elevated sympathetic nervous system (SNS) and HPA activity that causes reduced body weight, amenorrhea, and anorexia nervosa].
(4) Inadequate allostatic responses triggering compensatory increases in other allostatic systems.

A high incidence of ACTH and adrenal insufficiencies (HPA axis) are reported after TBI. The pattern is consistent with *allostasis* (i.e., alterations in baseline function without responsiveness to and recovery from stressful stimuli). These findings were extended using mice as subjects of *controlled cortical impact* (*CCI*) (left parietal cortex). There are time-dependent effects of CCI upon the stress response, varying with the severity of the injury. Mild and moderate CCI attenuated the stress response within the first 3 weeks postinjury. Mild CCI enhanced and prolonged the HPA stress response for at least 10 weeks postinjury. The stress response has mixed adaptive effects. It is essential for maintaining homeostasis, but in a dysregulated state after TBI it creates a vulnerability for major depression and PTSD. Understanding the consequences associated with injury severity upon short- and long-term neuroendocrine allostasis will contribute to TBI diagnosis related mental health disturbances (Taylor et al., 2008a).

ALLOSTATIC STATE

Allostatic states can be sustained, for a limited amount of energy is available to fuel homeostatic mechanisms. When an imbalance continues for a long period and becomes independent of adequate energy reserves, then symptoms of allostatic overload appear. Abdominal obesity is an example. Allostatic states produce "wear and tear" on the regulatory systems in the brain and body. Allostatic load and allostatic overload refer to the cumulative result of an allostatic state that may be an adaptive response to demands. However, a superimposed load of unpredictable events in the environment, disease, human disturbance, and social interaction can increase allostatic load dramatically to become allostatic overload, which serves no useful purpose and predisposing one to disease. The sustained activity of mediators of allostasis is known as the *allostatic state*, which can lead to medical illness: elevated blood pressure; elevated inflammatory cytokine production when glucocorticoid levels are inadequate, or elevated diurnal production of glucocorticoids in major depressive illness contributing to mineral loss; abdominal obesity; and atrophy of brain structures. Allostatic states produce "wear and tear" on the regulatory systems of the brain and body (i.e., the cumulative result of an allostatic state, which is referred to as allostatic load and allostatic overload). Overload causes decline of end organ systems. Examples include both normal environmental and adaptive variations (fat deposition before animals migrate or fish spawn). Allostatic overload occurs when there is the superimposition of an additional load of environmental events (e.g., disease, human disturbance, and social interactions). These serve no useful purpose and predispose one to *disease*. Psychological stress, sleep deprivation, or a rich diet, in the context of inactivity, lead to chronically elevated glucocorticoids, and increased insulin level and resistance, promoting the deposition of body fat and the formation of atherosclerotic plaques in the coronary arteries. Long-term imbalances in hormonal and other regulatory systems, as well as major depression, are associated with changes in size and function of the hippocampus, the prefrontal cortex, and the amygdala.

Allostatic load was earlier defined as a persistent stress reaction. There is a state of chronic deviation from the normal range of functioning. Eventually, the system regresses to a physiological inability to maintain regulation within useful limits. Among the *measures of allostatic load* are systolic and diastolic blood pressure, waist-hip ratio, serum high-density lipoproteins (HDL) and total cholesterol, cholesterol/HDL ratio, blood plasma levels of glcosylated hemoglobin, serum dehydroepiandrosterone sulfate (DHEA-S), fibrinogen, overnight urinary cortisol excretion, and so on. Some paradoxical results are reported. Allostatic load need not be associated with life events and depression; it did not predict subjective physical complaints; and it was not found in hypocortisolemic elderly subjects; subjective physical complaints were higher in older hypocortisolemic subjects, but not explainable by allostatic load measures (Hellhammer et al., 2004; McEwen & Seeman, 1999).

ALLOSTATIC OVERLOAD

Allostatic overload has been defined as the excessive level, up to years, of mediators of allostasis, resulting from too much release or inefficient operations of the allostatic systems, including their failure to shut off mediators when not needed (McEwen, 2000a). It is the hidden price paid for functioning under continued stress (i.e., the need to maintain function in spite of an external load) (McEwen, 1995). It is characterized by excessive levels of the mediators of allostasis (i.e., too much release of these mediators, or the inefficient operation of the allostatic systems that produce the mediators and fail to shut off their release when not needed) (McEwen, 2000a).

Allostatic overload is the physiological and anatomical state after extended trauma/stress. It is the response to a persistent demand for functioning beyond a normal range. It is characterized by inability to adapt physiologically, with potential for disease and tissue damage. Persistent stress leads to allostatic load (i.e., a status requiring a high level of functions mediating homeostasis) (McEwen, 1995, 2000b). Four situations lead to an excessive allostatic load: frequent stress; failure to habituate; inability to shut off allostatic responses; and inadequate allostatic response, which triggers compensatory increases in other allostatic systems. The analogous syndrome for chronic job stress is the "*burnout*" syndrome of emotional exhaustion, depersonalization, and reduced personal accomplishment (Burke & Richardson, 1996; Chiu & Lilly, 2000).

This condition involves the expression of pathophysiology after chronic arousal and chronic fatigue, resulting in overactivation of regulating systems and the alterations of body set points (Schulkin, 2003, pp. 21, 161–162). Allostatic overload is consequent to repeated challenges, or misdirection of the physiological responses that constitute allostasis. Examples include failure to shut off mediators such as cortisol and catecholamines, failure to habituate to repeated challenges of the same kind, and failure to mount an adequate response to a challenge (inadequate secretion of glucocorticoids, leading to overproduction of inflammatory cytokines). The complexity of the stress physiological response is illustrated by the retest after 10 years of Holocaust survivors. There is an age-related flattening of the circadian rhythm. The age of two samples during the second examination after 10 years was for PTSD+ 75.0, and for PTSD− 78.7. Survivors with PTSD tend to have paradoxically lower cortisol levels. Those whose PTSD remitted manifested increased cortisol level, but this declined in those who developed PTSD or whose PTSD status did not change. Cortisol status predicted diagnostic status better than psychological variables. This may be due to problems of assessing PTSD in aged participants. The authors do not consider possibility of "burnout" of cortisol secretion (e.g., with relief of PTSD, the stress-related requirement for cortisol declined, and eventually the adrenal cortex recovered and its output reached a more optimal level). Also, it was suggested that symptom improvement reflects the difficulty of differentiating symptoms that became internalized and less recognizable after a lifetime with PTSD (Yehuda et al., 2007).

FATIGUABILITY, EXHAUSTION, AND BURNOUT

Clinical example: Loss of stamina after TBI: A teacher reported, "My energy slowly came back. Came back where I could work a full day." At one time he could be productive for 16 h, including work. Now when he goes home he has to lie down. "When I start to feel tired or exhausted, I go to sleep almost within 10 seconds." He wonders if he is passing out. "I used to go bicycle riding with my wife. I haven't done that since the accident. Because of the balance problem. I am scared out of my mind about that. I used to go to gym and had a chart of what weight I pressed. I noticed after the accident I could pick up less than 50% of the weight. It was something I couldn't understand." He has worked on that. He formerly pressed 150 lbs; now he presses 80. He was hunching over, kids in school called his attention to it. He realized he was trying to watch his feet when he was walking to make sure everything was all right.

Fatigue Complexity

Fatigue requires consideration from many perspectives. From a biological viewpoint, chronic fear and activation of glucocorticoids are metabolically costly. Sustained secretion causes brain and bone tissue to deteriorate, increasing the likelihood of neurotoxicity and endangerment through the loss of glucocorticoid receptors. High corticosterone and low testosterone levels have been demonstrated in a number of species (i.e., decrease of reproductive fitness), and with subordination and further decrease in the likelihood of successful reproduction in macaques (Schulkin, 2003, pp. 74–75). Pollution of the interior milieu, energy problems, breakdown of the nervous system, and motivational self-assessment regarding the demands of a task. It was related to neurasthenia, which is characterized by the fatigability of body and mind. It was organic in origin, resulting from environmental factors, and treated by rest. It is common to the educated and professional classes (Beard, cited by Appels, 2000). It is asserted that this state is currently described as fibromyalgia or chronic fatigue syndrome, although this writer is more sympathetic to a physiological mechanism (Appels, 2002). A major depressive episode may be accompanied by decreased energy, tiredness, and fatigue incurred without physical exertion (American Psychiatric Association, DSM-IV-TR, 2000, p. 350). Fatigue is common in persons with head and somatic injury after mechanical accidents.

The nature of the fatigue complaint should guide treatment due to the variety of components, the availability of different treatments, measurement of treatment efficacy, and the experience of different types of fatigue over the long period of difficulties experienced by some patients. In the sample reported, there was a decrease in fatigue in the first year postinjury, and then no further significant change. Individuals could be grouped as improved, stable, or declining, with those reporting increased fatigue demonstrating decline in functional outcome (Bushnik et al., 2007b).

When considering comorbidity of somatic trauma, it is essential to differentiate between *weakness* (reduction of normal power of one or more muscles) and *increased fatigability* (inability to sustain performance of an activity that should be normal for a person of the same age, gender, and size). Fatigue is differentiated from sleepiness. It is a sustained lack of energy, associated with loss of motivation and drive, but without behavioral criteria of sleepiness such as heaviness or drooping of the eyelids and sagging or nodding of the head (Chokroverty, 2004). Fatigue's descriptors include:

- An overwhelming sense of exhaustion with decreased capacity for physical and mental work regardless of adequate sleep.
- A state of weariness following exertion, mental or physical, characterized by a decreased capacity for work and reduced efficiency to respond to stimuli.
- Diminished motivation, interest, or task aversion.
- Observable signs or somatic symptoms.
- Reduced activity.
- Mental fatigue.
- Difficulty in staying awake.
- Reduced activity.
- Difficulty in completing tasks (Borgaro et al., 2005; Jha et al., 2008).

Major Dimensions of Fatigue

Physical fatigue is the end result of excessive energy consumption, depleted hormones or neurotransmitters, or diminished ability of muscle cells to contract. In one study (Borgaro et al., 2005), using questionnaires, fatigue was found to be unrelated to injury severity (unspecified whether TBI or somatic), the number of days from injury to assessment, cognitive impairment, and gender. It is not made clear whether severity refers to brain or comorbid somatic injury. There is no reference to the concept of allostasis (i.e., physiological "burnout" consequent to chronic stress of which a pain-provoking injury would be a suitable example).

Psychological fatigue is a subjective state of weariness related to reduced motivation, prolonged mental ability, or boredom (Jha et al., 2008). Fatigue has been defined in the medical subject headings (MeSH) of PubMed as the state of weariness following exertion, mental or physical, characterized by a decreased capacity for work and reduced efficiency to respond to stimuli. Weariness implies that the fatigue appears to be out of proportion to exertion, and may exist without exertion. Because of reduced capacity for work, daily activities, and enjoyment of social and recreational activities, fatigue engenders negative feelings, perhaps even upon one's entire daily life. *Endocrinological disturbance* contributes to weakness, tiredness, vague gastrointestinal discomfort, and weight loss or gain (Frohman & Felig, 2001). The patient characteristically expresses *loss of stamina* (i.e., *fatigability*).

Epidemiology

Fatigue worsens other symptoms such as depression or weakness in individuals with stroke. It is a prominent disabling symptom in chronic fatigue syndrome, multiple sclerosis, cancer, and the PCS. During the first PCS month, it was reported by 29–47% of patients, at 3 months by 22–37%, and after one year, 21% still experienced abnormal levels of fatigue. It has objective (muscular fatigue) and subjective elements (self-assessment of cognitive, physical and social functioning) (LaChapelle & Finlayson, 1998). In one study the general population had a fatigue prevalence of 6–7% and a lifetime prevalence of 24%. Elsewhere, its extent in the general population was estimated at 10–20%, with the range for TBI been estimated at 30–70%. Fatigue is found in a wide variety of medical conditions (Jha et al., 2008).

For patients with the PCS, fatigue is listed as the second or third most common complaint after headaches, dizziness, and memory. Fatigue is variably associated with cognitive performance.

While fatigue may be secondary to a sleep disturbance, not all fatigued patients report a sleep disturbance.

Injury

The range of fatigue complaints in persons with TBI has been estimated as 50–80%. Levels compare to those found with significant medical disorders. Fatigue worsens other symptoms such as depression or weakness in individuals with stroke. For individuals with TBI, fatigue is associated with depression, sleep disorders, pain, and possibly hypopituitarism. Persons with TBI reported more severe fatigue than those in the control group, and females more fatigue than males (Bushnik et al., 2007a; Dijkers & Bushnik, 2007). In the reported sample, the authors report relatively low measures of global fatigue: 16–32% at year one, and 21–34% at year 2, assessed as no change. Sleep quality, depression, and pain were the most prevalent disturbances. There were paradoxical findings insofar as hours of activity (household, school, and work did not relate to global fatigue, probably attributable to low levels of fatigue permitting more activity). Measures highly associated with fatigue were sleep quality, depression, pain, social integration, somatic, motor, and memory-attention complaints, consistent with prior findings (Bushnik et al., 2007a).

Assessment of Fatigue

It is necessary to differentiate between objective measures of fatigue and its subjective quality. One may quantify the extent of reduced performance after sustained effort, which will not specify the self-perceived state of fatigue. Some report tiredness or exhaustion after even minimal effort. Persons with TBI report both objective and subjective measures of fatigue, while describing it as their worst symptom. As has been discussed here, post-TBI fatigue is associated with neuroendocrine dysfunction and also pain, sleep problems, cognitive deficits, and depression. Their effect is to reduce energy reserves, with reductions in levels of activity and participation. Cognitive effort causes mental fatigue, which may result from an inability to expend sufficient mental effort considering the resources available and the amount of compensation needed to overcome impairment. Mental fatigue is characterized by the perception that thinking is increasingly effortful and ineffective. Research findings are mixed, with more errors having been reported at the end of an attention task. While self-reported fatigue was relatively independent of cognitive impairment and performance decrements on challenging cognitive tasks, persons with TBI experience greater distress and effort, as indicated by self-report and increased blood pressure. Fatigue-related performance changes are characteristic of divided attention tasks over an extended period in which controlled cognitive functioning exceed the available cognitive resources. Original findings were that performance on speed was associated with situational and day-to-day fatigue, but performance in other cognitive domains was unrelated to subjective fatigue. Objective changes in performance (accuracy) were not related to subjectively experienced fatigue. The most self-reported fatigued persons performed worse than those reporting less fatigue but their performance decline did not vary from the less fatigued persons (Ashman et al., 2007).

It has been proposed that the injured brain has to work harder due to the necessity of compensating for such impairments as attention and processing speed. This is reflected in increased blood pressure during performance. It is controversial as to whether mood and sleep disturbances are directly consequent to TBI or to fatigue itself. Research on the association between chronic pain and headaches and TBI fatigue is sparse. Furthermore, there is a measurement error insofar as measures of fatigue overlap with those of depression, pain, and other health problems. An empirical study suggested that a large proportion of the fatigue was more likely associated with TBI than with the comorbid conditions of pain, depression, and disrupted sleep. These symptoms accounted for twice the variance in the control group than TBI group. While stress and personality accounted for some of the variance, some etiology is attributable directly to TBI. Fatigue did not seem to limit participation in life activities. However, fatigue was associated with physical and mental health issues in performing activities of daily living. There may be a bidirectional relationship between health problems and fatigue, leading to an impaired quality of life. Although the intensity of fatigue interferes with the quality of functioning, it did not limit participation in either the TBI or control groups (i.e., frequency).

Loss of Stamina

The TBI victim characteristically expresses *loss of stamina* (i.e., fatigability). Impaired persons characteristically lose capacity to exert effort over a useful period of time at its former level. The variables are appetite, fuel metabolism, and absorption, transport, and oxidation of foodstuffs (Marks et al., 1996). Fatigue is differentiated from sleepiness. It is a sustained lack of energy, associated with loss of motivation and drive, but without behavioral criteria of sleepiness such as heaviness or drooping of the eyelids, sagging or nodding of the head (Chokroverty, 2004). Considering comorbidity of somatic trauma it is essential to differentiate between *weakness* (reduction of normal power of one or more muscles) and *increased fatigability* (inability to sustain performance of an activity that should be normal for a person of the same age, gender, and size (Olney & Aminoff, 2001). *Change of weight after injury or illness is common, occurring primarily in either direction or fluctuating around the initial value.*

Burnout and Stress

A syndrome descriptive of chronic job stress is "burnout" (i.e., emotional exhaustion, depersonalization, and reduced personal accomplishment) (Burke & Richardson, 1996) (see Allostasis and Stress and the following discussion). Stress in elderly caregivers was associated with poorer antibody response to influenza vaccine (Kiecolt-Glaser, cited by Vedhara & Wang, 2005). The initial stress response is temporarily beneficial, although antireproductive, antigrowth, catabolic, and immunosuppressive. Chronic stress may lead to a pathological syndromal state with psychiatric, neuroendocrine, cardiovascular, metabolic, and immune components (Chrousos, 1998). The amplitude of the release of pituitary hormones can be altered by prior stimulation that might have depleted a releasable pool of hormone (Molitch, 2001). The HPA becomes hyporesponsive, affecting stamina, health, and quality of life. Stress disrupts energy metabolism and caloric balance (Middleboe et al., 1992). A study compared two groups, mTBI and minor injury. Severe fatigue was associated with other symptoms (e.g., nausea and headache) (Stulemeijer et al., 2006a). Perhaps *loss of stamina* is related to the fact that catecholamines and cortisol shift.

FATIGUABILITY AND LOSS OF STAMINA

It has objective (muscular fatigue) and subjective elements (self-assessment of cognitive, physical and social functioning) (LaChapelle & Finlayson, 1998). From a biological viewpoint, chronic fear and activation of glucocorticoids are metabolically costly. Sustained secretion causes brain and bond tissue to deteriorate, increasing the likelihood of neurotoxicity and endangerment through the loss of glucocorticoid receptors (Schulkin, 2003, p. 75). Pollution of the interior milieu, an energy problem, breakdown of the nervous system, motivational self-assessment regarding the demands of a task. It was related to neurasthenia, which is characterized by the fatigability of body and mind, organic in origin, resulting from environmental factors, treated by rest, and common to the educated and professional classes (Beard, cited by Appels, 2000). It is asserted that this state is currently described as fibromyalgia or chronic fatigue syndrome although the present writer is more sympathetic to a physiological mechanism (Appels, 2000). A major depressive episode may be accompanied by decreased energy, tiredness, and fatigue incurred without physical exertion (American Psychiatric Association, 2000, p. 350). Fatigue is common in persons with head and somatic injury after mechanical accidents. *Hypothyroidism* may be accompanied by decreased cardiac output (secondary to a drop in heart rate and contractility) and by decreased basal metabolic rate. Symptoms include mild lassitude, fatigue, slight anemia, constipation, apathy, cold intolerance, menstrual irregularities, loss of hair, and weight gain (Larsen & Reed, 2003).

Daily life: It has been proposed that the injured brain has to work harder in order to compensate for such impairments as attention and processing speed. This is reflected in increased blood pressure during performance. It is controversial as to whether mood and sleep disturbances are directly consequent to TBI or to fatigue itself. Research on the association between chronic pain/headaches and TBI fatigue is sparse. Furthermore, there is a measurement error insofar as measures of fatigue overlap with those of depression, pain, and other health problems. An empirical study suggested that a large proportion of the fatigue was more likely associated with TBI than with the comorbid conditions of pain, depression, and disrupted sleep. These symptoms accounted for twice the variance in the control group than TBI group. While stress and personality accounted for some of the variance, some etiology is attributable directly to TBI. Fatigue did not seem to limit participation in life activities. However, fatigue was associated with physical and mental health issues in performing activities of daily living. There may be a bidirectional relationship between health problems and fatigue, leading to an impaired quality of life. Although the intensity of fatigue interferes with the quality of functioning, it did not limit participation in either the TBI or control groups (i.e., frequency). Although there may be individual differences, fatigue was not related to participation in activities rated as having high or low importance (Cantor et al., 2007). This writer has reservations about some of the inferences presented here. Perhaps there was a sampling bias in this study, but the writer's own observations indicate that when a patient has disabling injuries (e.g., radicular damage, herniated discs, peripheral nerve damage, etc.), that is, interferences with mobility, perhaps frontal lobe damage consequent to repeated falls, then the level of participation is significantly reduced, and these patients are considerably confined to their homes. They may report feelings of illness or not being well rather than fatigue per se.

Allostatic load is a cumulative measure of the cost of repetitive cycles of adaptive allostasis (Friedman & McEwen, 2004) and physiological dysregulation in multiple systems. Dysregulation of body systems is due to the continued effects of unhealed injuries after trauma, such as the hormonal, immune, inflammatory, and circadian systems creating chemical signals to the brain and other structures. If not correctable by normal control processes (*allostasis*), the initial dysregulation by trauma as well as persistent and excessive environmental demands eventually become physiologically costly, creating health problems or permanent system damage (*allostatic load*). "*Burnout*" is a concept that evolved from occupational and social psychology. It is the endpoint of a long-lasting stress process resulting in a negative state of physical, mental, and emotional exhaustion (i.e., disillusionment). When there is PTSD in the context of the person's inability to obtain a respite from the job, the condition may be magnified and complicated by "*burnout*" (Burke & Richardson, 1996).

Daily Life

Allostatic load reflects the pathophysiology of the physiological and anatomical states following the events of daily life that elevate activity of physiological systems and extend trauma/stress. Social disruption, which is common after impairing TBI, contributes to stress and disease. Changes in the social context lead to, and maintain, anxiety (Fisher, 1996). Allostatic load results from the impact of life experiences, genetic load, individual habits (diet, exercise, substance abuse, smoking, drinking), developmental experiences that set lifelong patterns of behavior, and physiological reactivity. This condition is important in the diagnosis and treatment of many illnesses (McEwen, 1998). Animal studies suggest that psychosocial stress impairs cognitive function and promotes disease (issues of competition and dominance) (McEwen & Seeman, 1999).

Persons with performance inefficiency (e.g., less than the estimated baseline or paradoxically low by other criteria) should be considered from the viewpoints of mental and physical trauma-created dysregulation and possibly the consequences of allostatic load.

Markers of Allostatic Load

The markers of allostatic load include serum DHEA, average systolic blood pressure, and 12-h overnight urinary excretion of EPI (Southwick et al., 2003). It occurs consequent to repeated activation of stress-responsive biological mediators such as glucocorticoids and catecholamines. Chronic overactivation of regulating systems causes *alterations of body set points* (Schulkin, 2003, p. 21). "*Wear and tear*" has been attributed to repeated cycles of allostasis and also the inefficient "turning on" or "shutting off" of these responses. Structural changes to cells occur in the hippocampus and amygdala as a result of allostatic load (McEwen, 2003a; McEwen & Seeman, 1999). There may be partial failure of the regulatory apparatus caused by trauma or persistent stress beyond a period in which recovery is possible. Stress hormones mobilize and activate bodily resources. An uninformed person may interpret their side effects (*hyperarousal*) as a disease (dizziness, tension, stomach cramps, palpitations and respiratory problems). Chronic elevated hormone levels lead to structural disorders. Reexperiencing unpleasant events or anticipating threats maintain pathological arousal (Fisher, 1996). The next phase is allostatic overload (see below).

Allostatic load reflects inability to turn the systems off or down even when they are *dangerous to health* (McEwen, 1998). Markers of allostatic load include (1) *frequent stress* (e.g., leading to myocardial infarction); (2) *failed shut down* (high blood pressure and glucocorticoids; obesity and diabetes); and (3) *inadequate response* (failure to challenge autoimmune responses and inflammation). When certain systems do not respond adequately, other systems overreact because they are not counter-regulated (e.g., inflammatory cytokines by adrenal steroids). One factor in allostatic load is the *immunologic memory* (i.e., a *delayed type hypersensitivity* [*DTH*] or response to a prior challenge). DTH is enhanced by acute stress, but is ultimately reduced by chronic stress, even when there is an acute stress challenge.

Physiological Mechanisms

Altered immune functioning is associated with depression, suppression of affectivity, a perceived sense of helplessness, and intrusive thoughts associated with an adverse event. A high stress level is also associated with increased depressive symptoms, dissatisfaction with social support, and limited uses of adaptive coping strategies (Fletcher et al., 1998).

Exhaustion of physiological systems due to persistent dysregulation after trauma: The latter can follow persistent effects of somatic injuries from the same accident that has caused TBI (cars, fibrous tissue, injured tissue, skeletal damage, peripheral nerve damage, internal injuries, limited range of motion, pain, and headaches). The dysregulated systems involved are the circadian, the hormonal, the inflammatory, and the immune systems. The consequences of chronic dysregulation may be summarized by the concept of allostatic load (Dhabar & McEwen 2001; McEwen, 2000a, 2005). This signifies systemic exhaustion, the neurobehavioral consequences of which are stress-related health disorders. Behavioral disorders, in addition to chronic PTSD symptoms, include illness behavior (malaise, fatigue, loss of appetite, apathy, social withdrawal, inactivity), reduced stamina (fatigability), vulnerability to infectious disease, tissue exhaustion, burnout, organ damage, sexual dysfunction, and stress related health disorders (Curle & Williams, 1996; Lombardi et al., 1994; Schnurr & Green, 2004). This constellation is consistent with the frequent complaints of patients concerning loss of energy, resulting in inability to fulfill the responsibilities of domestic and professional life. Furthermore, the clinician's conclusions should consider that the neurobehavioral effects of stress overlap those of TBI (Parker, 2002).

(1) *Overactivity of the HPA axis* (Hellhammer et al., 2004). Elevations in the stress hormones cortisol, NE, and EPI may be accompanied by decrements in immune function. Altered immune functioning is associated with depression, suppression of affectivity, perceived helplessness, and intrusive thoughts that are associated with the trauma. A high stress level is also associated with increased depressive symptoms, dissatisfaction with social support, and limited uses of adaptive coping strategies (Fletcher et al., 1998).
(2) Trauma and stress may damage or exhaust the tissue required for a system to function, or may create demands that exceed the body's capacity to function indefinitely at this level. Exhaustion of various physiological systems is due to persistent dysregulation after chronic injury (scars, fibrous and otherwise injured tissue, skeletal damage, peripheral nerve damage, internal injuries, limited range of motion, pain and headaches, concern with seizures, imbalance, vertigo, etc). The dysregulated systems are the circadian, the hormonal, the inflammatory, and the immune systems. The consequences of chronic dysregulation may be summarized by the concept of *allostatic load*, which signifies systemic exhaustion or "burnout" (Dhabar & McEwen 2001; McEwen, 2000a, 2005). Its neurobehavioral consequences are stress-related health disorders, chronic PTSD symptoms, illness behavior (malaise, fatigue, loss of appetite, apathy, social withdrawal, inactivity), reduced stamina (fatigability), vulnerability to infectious disease, tissue exhaustion, organ damage, sexual dysfunction (Curle & Williams, 1996; Lombardi et al., 1994; Schnurr & Green, 2004). This constellation is consistent with the frequent complaints of patients concerning loss of energy, resulting in inability to fulfil the responsibilities of domestic and professional life. PTSD is associated with high cortisol levels in acute stages and low cortisol in chronic PTSD. Another consideration is dysregulation of neurotransmitter systems that augment and attenuate threat responses. Sympathetic dysregulation in PTSD may be related to deficits in the *neuropeptide Y (NPY)* response to stress. NPY deficits may promote anxiety and distress, while higher levels can buffer these effects (Nijenhuis et al., 2002). NPY is extensively localized with NE in sympathetic nerve terminals. Its measurement in peripheral blood is a more dependable estimate of sympathetic activity than is NE measurement (Felten & Maida, 2002).

What is the cost of a response *in terms of stress or harmful interactions*? An example is the effect upon the hippocampus of the interactions between excitatory amino acids, serotonin, and glucocorticoids. Thus, allostasis refers to systems responding to the body state (sleep-wake cycle, physical activity, the external environment, circadian cycle, and threats to safety). Increased CNS activity (arousal) results in insomnia and stress, which contribute to harmful CNS effects on cardiovascular, cognitive, and behavioral effects. Prolonged sleep deprivation leads to immunodeficiency, from which can be inferred sleep's restorative function (Hobson & Pace-Schott, 2003).

Key players in stress responsivity, and functionally interrelated, are the *LC* (which regulates the autonomic nervous system (ANS)) and the hypothalamic paraventricular nucleus (PVN) (which regulates corticotropin releasing factor (CRF)/CRH release), which partially mediates the HPA axis.

The *LC*, when exposed to chronic stress, may have altered firing and NE release in target brain regions. NE also plays a role in fear conditioning and sensitization, and may be critical for the amygdala's activation of neuromodulatory influence on memory storage.

The *PVN* receives input from brainstem aminergic and peptidergic afferents, blood-borne information, and the limbic system and its associated regions (prefrontal cortex, hippocampus, amygdala, and local circuits) (Nijenhuis et al., 2002).

The PVN stimulates the pituitary-adrenal cortical axis during stress. After stress, depletion of hypothalamic CRH is followed by a "compensatory" upregulation of CRH production (Baram et al., 1997).

In a state of *exhaustion*, somatic and mental processes reinforce each other and may precede myocardial infarction. There is decreased cortisol excretion, while coronary patients have elevated titers of antibodies against microorganisms that are involved in the pathogenesis of coronary heart disease. They also have elevated levels of cytokines, which are markers of inflammation (IL-1, IL-6, and TNF-α). If prolonged stress causes or aggravates inflammation, the inflammation itself may amplify feelings of fatigue through the release of cytokines (Appels, 2000). There is some fluctuation of symptoms and cortisol levels in survivors of extreme stress (e.g., The Holocaust). *Cortisol* levels were measured. They were low in survivors with PTSD. Studied 7–8 years later, cortisol level was even lower, and there was a flat circadian rhythm due to reduced peak cortisol level. In a separate sample (measured after 10 years), the level increased when PTSD remitted and declined in participants who developed PTSD or whose PTSD status had not changed. Cortisol was a better predictor of changes in PTSD than psychometric measures of PTSD symptomatology (Yehuda et al., 2007). One may offer a different explanation: assuming that symptoms of PTSD reflect the experience of stress, after an extended time the secretion by the adrenal cortex becomes exhausted. Should the stress be relieved, secretion will approach a more normal level.

POSTERIOR PITUITARY DISORDERS

Overview

Polyuria, polydipsia, and thirst (diabetes insipidus) are associated with reduced levels of vasopressin after severe head trauma (Ganong & Kappy, 1993), but also may occur after minor head trauma (hypothalamus; hypophysectomy; retrograde degeneration of axons into the supraoptic and paraventricular hypothalamic nuclei) (Kern & Meislin, 1984; Robertson, 1996). Permanent ADH insufficiency is expected with delayed onset (Hadani et al., 1985).

Head Trauma

Impact can lead to transection of the pituitary stalk. Lesions of the hypothalamic supraoptic nuclei, the infundibulum, and the upper half of the pituitary stalk can denervate the posterior lobe of the pituitary (Crompton, 1971). After head trauma, there is sudden appearance of hypotonic polyuria consequent to interference with the transport of vasopressin (a water retaining, or ADH), which is synthesized in the paraventricular and supraoptic nuclei of the hypothalamus, and is transported down the long axons that comprise the supraopticohypophyseal tract to terminate in the posterior pituitary (Robinson, 1996). It is observed in children and adults, may occur within 24 h after injury, and resolves in only half of cases; the remaining half suffer from permanent vasopressin deficiency (Bode et al., 1996; Findling & Tyrell, 1986). Interruption of the blood supply to the hypothalamus and pituitary leads to neurogenic diabetes insipidus (Ramsey, 1986). Accidental head injuries may result in diabetes insipidus within 24 h after injury (Bode et al., 1996). In about half of these patients the disease resolves spontaneously. Some children recovering from traumatic stalk section and transient diabetes insipidus form an ectopic posterior pituitary.

Diabetes Insipidus

Diabetes insipidus (*vasopressin deficiency syndrome*) is a complication of closed head injury in children and adults due to posterior pituitary stalk lesion or hypothalamic damage with loss of vasopressin (Molitch, 2001; Robinson, 2004). The sensing system for water level (*osmostat*) is in a small area of the hypothalamus just anterior to the third ventricle. Hormone synthesis and regulation is high in the hypothalamus; thus, local trauma (e.g., severing the pituitary stalk) or surgery traumatizes only the axon terminals (Ramsey, 1986). Survival of even 10% of vasopressin neurons may be sufficient to maintain homeostasis without symptoms.

Alerting symptoms are polyuria, polydipsia, and thirst (Ganong & Kappy, 1993). These can develop from hypothalamic lesions or the destruction of vasopressin-containing neural fibers that terminate in the posterior pituitary. Diabetes insipidus (DI) is associated with reduced levels of vasopressin after severe head trauma, but may also occur after minor head trauma (hypothalamus;

hypophysectomy; retrograde degeneration of axons into the supraoptic and paraventricular hypothalamic nuclei) (Kern & Meislin, 1984; Robertson, 1996). Permanent ADH insufficiency is expected with delayed onset (Hadani et al., 1985).

Diabetes insipidus is a condition characterized by excretion of large volumes of urine that is hypotonic, dilute, and tasteless (insipid). This is a common disorder after closed head trauma (CHI), but does not seem to be recognized in later clinical contacts. This writer remembers one patient who, during a long examination, had to excuse himself repeatedly to urinate. This condition was posttraumatic and had never been studied by his physicians. It is associated with head injury (e.g., MVA) and panhypopituitarism (hypothyroidism and adrenal insufficiency). The trauma involved are hypothalamic or posterior pituitary (neurohypophysis) injury (Robinson & Verbalis, 2003). Dehydration creates thirst, and perhaps a desire for cold liquids as well. The writer has seen some patients who did not suffer such severe head trauma as to be unaware that their posttraumatic water ingestion and frequent urination followed their head injury.

Syndrome of Inappropriate Antidiuretic Hormone

This endocrine disorder is characterized by inability to excrete water so that the urine is hypertonic (highly concentrated) with potential fatigue, headache, nausea, and anorexia, which can progress to altered mental status, seizures, coma, and death. It results from the disruption of hypothalamic pituitary pathways caused by closed or penetrating head trauma, surgery, tumors, or numerous CNS disorders. This can trigger the unregulated increase of vasopressin (*ADH*), resulting in water retention. SIADH has been reported in 0.6% of patients with mild head injury, 10.6% of those with moderate head injury, and 4.7% of those with severe head injury (Kirby et al., 2007; Kokko, 1996). The clinical picture may be a marker for trauma (Kokko, 2004). Plasma levels of ADH (AVP) are elevated when secretion would ordinarily be suppressed (Robinson, 2007; Robinson & Verbalis, 2003). Water retention results in the production of decreased volumes of highly concentrated urine. There is a decrease in the osmotic pressure of body fluids (i.e., *hypoosmolality*). If the condition eventually results in the death of the last vasopressin neurons to reduce the vasopressin level, there might be a great increase in urine volume (Robinson & Verbalis, 2003). The cause is impairment of inhibitory control over the neurohypophysis (i.e., excessive secretion of AVP). Clinically it appears as water retention with dilution of body fluids, such as hyponatremia or deficiency of blood sodium (reduced sodium ions in the blood, high urine sodium, with potential increase of body water content by 10% without edema), or water intoxication that may include mild headache, confusion, anorexia, nausea, vomiting, coma, and convulsions (Robertson, 2005).

Primary Polydipsia

Primary polydipsia may follow acute trauma to the head, and represents a disorder of thirst stimulation (Robinson, 1996). Ingested water produces a reduction of osmolality (concentration of brain fluids), which turns off the secretion of vasopressin. Urine is not concentrated and liquid excretion is higher. Primary polydipsia is characterized by drinking even greater amounts of fluid than in diabetes insipidus, perhaps more than 20 L/day.

CHRONIC STRESS-RELATED ENDOCRINOLOGICAL DYSFUNCTION

Stress was described as manifesting as the type of metabolism that occurs when the action of the stress hormones (glucocorticoids, corticotropin, catecholamines, GH, glucagon) predominates over insulin. This pattern increases the risk for atherosclerosis (Brindley & Rolland, 1989). The secretion of the known pituitary hormones is affected by physical and psychologic stress (Martin & Reichlin, 1989, p. 669). While peripheral catecholamines (EPI and NE), SNS, and adrenal medullary response increase rapidly, cortisol (HPA axis response) decreases. In the hypothalamus, initial catecholamine level is decreased and eventually increases (Goldstein & Halbreich, 1987). Hyperprolactinemia occurs after stress and trauma. Signs include galactorrhea, sexual dysfunction, and suppressed gonadal function (short luteal phase, reduced central FSH and LH, decreased estradiol and amenorrhea) (Melmed & Kleinberg, 2003). Osteoporosis can be an endocrine disorder secondary to hypopituitarism with GH deficiency (Castels, 1996). From a behavioral point of view, one may wonder if lesser injuries are associated with reduced levels of these endocrine substances with obscure emotional consequences. Symptoms of reduced LH and FSH include male infertility, loss of libido, and erectile dysfunction. Symptoms of TSH include cold intolerance and weight gain, depression, fatigue, and cognitive impairment. Symptoms of ACTH include weakness, depression, anxiety, fatigue, and apathy (Schneider et al., 2005). In women, cognitive function and memory appear to be impaired by low estrogen levels (Kelly et al., 2000).

The HPA axis stress response is less well understood for chronic stress than acute stress, but it seems to be more driven by AVP than by CRH (Michelson et al., 1995). At the pituitary level, the main hormones released by stress are ACTH and prolactin (PRL). The role of PRL is not well understood (Antoni, 2000). ACTH secretion is enhanced by PIC (IL-1, IL-6; TNF-α), which also augment the effect of CRH. An increase of cortisol following a variety of stressors is a normal counter-regulatory response. Acute psychological stress raises cortisol levels, but secretion appears normal in patients with chronic anxiety and underlying psychotic illness.

Anterior pituitary: Abnormal activation of the stress system during *critical periods* (intrauterine, infancy, childhood, adolescence) may affect this system throughout life, causing a predisposition to pathologic states (Chrousos, 1998). ACTH (cortisol secretion) has a diurnal rhythm with early morning secretion exceeding evening secretion by at least twofold (Gill, 1996). It may be vulnerable to the sleep disorders that are common after TBI and stress. Victims of PTSD have been demonstrated to have enhanced negative feedback sensitivity of ACTH production to circulating cortisol (Kanter et al., 1998).

There are multiple interactions between the HPA axis, the gonads, the growth axis, and metabolism. The reproductive axis is inhibited by the HPA axis, while gonadal steroids reduce cortisol and circulating ACTH (Roca et al., 1998). The inhibition of the growth axis by the reproductive axis antagonizes fat tissue catabolism (lipolysis), as well as muscle and bone anabolism. This adds fat tissue and results in the loss of lean body mass. Stress system-related mood disorders (chronic anxiety or melancholic depression) are associated with GH level reduction and inhibition of thyroid axis function. Stress also affects the metabolic axis, gastrointestinal function, and is the immune system.

Sexual disorder: In men, stress, including surgical stress, induces hormonal changes (including LH/FSH secretion) that may lead to reduced *testosterone* levels. Stress-increased concentrations of NE, EPI, and serotonin may induce *altered testicular metabolism* or *testicular vasoconstriction*. *Anger* and *depression* are associated with enhanced stress, and therefore higher sympathetic tone and smooth muscle contraction, resulting in reduced capacity for erection. Related to this concept, abnormal levels of circulating NE and EPI contract penile cavernous tissue, reducing the capacity for arterial inflow, which may underlie *erectile failure during stress* (Lemack et al., 1988).

Stress and Depression

The level and nature of physiological activity during chronic stress may vary considerably. Behavioral sensitization is considered to be an enhanced response modification that follows repeated presentation of stimuli (Southwick et al., 1994, 1995).

Depression is associated with high-circulating cortisol concentrations (Stewart, 2003). *Melancholic depression* is characterized by CRH hypersecretion, impaired cortisol feedback, and hypercortisolaemia. Depression seems to have a role in the regulation of immunity, with widespread changes in the number of circulating immune cells in the blood (Irwin & Cole, 2005).

In contrast, *PTSD* is also characterized by central CRH overactivity, but relative hypocortisolaemia (Kaye & Lightman, 2005). Prolonged and increased production of CRH, which coordinates behavioral, neuroendocrine, autonomic, and immunologic adaption, could explain the pathogenesis and manifestations of chronic stress syndrome (psychiatric, circulatory, metabolic, and immune). There are numerous mood and endocrine disorders associated with enhanced (hyperarousal) and decreased (hypoarousal) functioning of the HPA axis (Tsigos & Chrousos, 1996). Experimental administration of CRF into the brains of animals seems to make the animals agitated and "on edge" (Weisinger & Denton, 2000). Chronic stress elicits a *vigilance reaction* involving the activation of the HPA (Benarroch, 1997). Chronic stress results in a sustained increase of plasma glucocorticoid levels, with potentiation of glucocorticoid responsiveness to subsequent stressors and neuronal hypersecretion of CRH (CRF) (Bremner et al., 1997). Prolonged ACTH exposure causes hypertrophy of the adrenal gland. Increasing glucocorticoid exposure in turn causes the pituitary to become unresponsive even to extreme stimulation, with slow recovery that may require months to return to normal function (Miller & Tyrrell, 1995). High physiological levels of glucocorticoids compromise the capacity of neurons to survive a variety of metabolic insults, including ischemia and excitotoxins (Sapolsky & Pulsinelli, 1985). The HPA axis is expected to shut down at the end of stress. Stressful affective responses that influence immune function increase susceptibility to a variety of diseases. Behavioral interventions that reduce anxiety or distress may decrease the intensity or duration of autonomic and endocrine responses, changing immune function and promoting wellness and/or recovery from disease (Ader, 2005). Emotional stress and psychological disorders activate the pituitary-adrenal stress response and suppress gonadotropin secretion, causing psychogenic amenorrhea, or inhibit GH secretion, causing psychosocial dwarfism (Cone et al., 2003).

VARIED ANTERIOR PITUITARY DISORDERS

Symptoms

Hypothalamic damage with hyperprolactinemia may cause galactorrhea, amenorrhea, and hirsutism (Rebar, 1996). Inhibition of gonadotropin secretion may result in secondary hypogonadism in boys (Matsumoto, 1996). Symptoms of anterior pituitary disorders include loss of axillary and pubic hair (Woolf, 1992); growth failure; acquired gonadotropin deficiency (Ferrari & Crosignani, 1986); hypothyroidism; ACTH deficiency (D'Angelica et al., 1995); transient hypogonadotrophic hypogonadism (Woolf et al., 1986) or permanent hypogonadotrophic hypogonadism (Ferrari & Crosgnani, 1986); reduced thyroid function, possibly associated with a stress-related sympathetic discharge of catecholamines (Woolf et al., 1988); hyponatremia (loss of sodium ions into the urine or salt wasting); natriuresis; myoedema; dwarfism; obesity; hypothermia; bradycardia (Treip, 1970); dermatological (waxy or dry skin; wrinkling around eyes and ears); anemia; psychiatric disturbance (mental slowing; apathy; delusions; paranoid psychosis); hypoglycemia; weakness; lethargy; cold intolerance; decreased libido; constipation; bradycardia; hoarseness; myxedema; increased and decreased menstrual flow and amenorrhea; and delayed resumption of menstruation (Brontke et al., 1996).

After brain trauma, both the *sympathoadrenal* and *HPA* axes are stimulated, and the localization and dimensions of the brain lesion influence the nature and pathological concentration of blood hormones. Brain injury may affect ACTH secretion in various ways: (1) stimulating cortical or subcortical areas that stimulate or inhibit the hypothalamus; (2) severing communications between the hypothalamus and other areas (e.g., inactivating integrating centers); or (3) stimulating, inhibiting, or destroying hypothalamic or pituitary centers. ACTH and cortisol levels rise within 15 min of surgery, returning to normal 8–10 h after surgery. CRH production has two effects, increasing the output of ACTH and cortisol. Also, the secretion of NE, which by feedback increases the secretion of CRH itself.

THYROID

Abnormal thyroid levels, derived from trauma, are characterized by thymus deficiency (i.e., *thymulin*) (Mocchegiani et al., 1994). Severe TBI initiates the impairment of thyroid function, manifesting mainly as low blood triiodothyronine (T3) and thyroid-stimulating hormone (Cernak et al., 1999). Following MTBI, there is a significant increase in serum thyroid-stimulating hormone levels.

ADRENAL CORTEX

An accident can create primary adrenal insufficiency (i.e., gradual loss of both glucocorticoid and mineralocorticoid activity). Some of the symptoms may be incorrectly ascribed to TBI directly (e.g., generalized weakness, fatigue, psychiatric symptoms, depression, apathy, and confusion). The loss of cortisol production interferes with feedback control, resulting in the overproduction of CRH by the hypothalamus and ACTH production by the anterior pituitary (Webster & Bell, 1997). After mild TBI, enhanced serum cortisol is normalized by the second day.

GLUCOCORTICOIDS

Stress functions: These were originally designated as those steroids that have glucose-regulating properties (e.g., cortisol, corticosterone, aldosterone). These are secreted by the adrenal cortex. Glucocorticoids affect behavior, emotional state, brain fluid compartmentalization, and the aging of the brain. Other glucocorticoid functions that participate in overcoming stress are carbohydrate metabolism, glycogen metabolism, lipid metabolism, protein and nucleic acid metabolism, inhibition of vasoactive and other inflammatory agents, muscle glucose and protein metabolism, leukocyte movement and function, the cardiovascular system and fluid and electrolyte balance, and bone and calcium metabolism (Miller & Tyrrell, 1995). *Cortisol* enters the blood and binds to receptors in the hypothalamus and pituitary, inhibiting the release of CRH system. Cushing syndrome is a chronic increase in glucocorticoids that may stem from a variety is sources within or without the adrenal cortex. A related complication of glucocorticoid excess is bone loss due to suppression of bone formation (Finkelstein, 1996).

Stress modulation: Glucocorticoids have a *stress-modulating effect* (Kaye & Lightman, 2005). They play the main role in suppression of immune and inflammatory responses, the SAM system. Functions include cardiovascular function and blood pressure, water excretion, electrolyte balance, and metabolism (carbohydrates, lipids, proteins, nucleic acids, bone), growth inhibition, and inhibit reproductive function.

Glucocorticoid excess: These are powerful anti-inflammatory agents. The distinction between immunosuppressive drugs (cell- or humoral-mediated immune responses) and anti-inflammatory drugs (nonspecific inflammatory cells) is not always clear, since drugs affect more than one cell type as well as the extensive interactions between the immune and inflammatory responses (Imboden et al., 2001). The immune modulatory molecules can stimulate the brain's hormonal stress response. A cascade of hormones may finally result in the adrenal gland's release of anti-inflammatory corticosteroid hormones. An excess of these (chronic stress) predispose a stressed host to more infection. Too small a response can predispose the person to autoimmune disease such as arthritis, since the immune response is not shut off and can go on unchecked. A blunted hormonal stress response in animals and humans (genetic, drug therapy, surgical intervention) can lead to increased susceptibility to inflammatory disease (arthritis, SLE, allergic asthma, atopic dermatitis) (Sternberg, 2000).

Sustained exposure to glucocorticoids also compromises the capacity of neurons to survive metabolic insults, and it appears to play a role in neuron loss during aging (Sapolsky & Pulsinelli, 1985). Brain damage induced by ischemic trauma, glutamate toxicity, and axonal transaction can be exacerbated by elevation of circulating glucocorticoids (Shohami et al., 1995). Excess can induce hyperphagia, pathologic insomnia, depression, and hallucinations. Hippocampal atrophy, as in Alzheimer disease, is associated with a high blood cortisol level and an elevated set point of feedback control. Hippocampal neurons are uniquely sensitive to glucocorticoids and can be damaged by associated stress. Severe psychological stress and pain may cause cerebral cortical atrophy, possibly related to damaging effects of glucocorticoids and excitotoxins) (Reichlin, 1998).

Cortisol and stress: It is a common misconception that high levels of adrenal cortical secretion always accompany stress, when in fact there are various patterns. Higher cortisol levels have been associated with the presence of PTSD despite similar experience (combat) or lower (compared to normals and major depression) (Cohen & Zohar, 2004). Cortisol levels measured after the acute stress period are inconsistent due to the different intervals after stress, different stressors, and so on. Acute PTSD may be characterized by high cortisol, and chronic PTSD may be characterized by low cortisol, a function of negative feedback inhibition in the HPA axis (Nijenhuis et al., 2002). Low cortisol patients experiencing posttraumatic stress ("burnout" and physical health problems) typically have a blunted cortisol awakening response (*hypocortisolism*). Symptoms include fatigue, irritable bowel disease, low back pain, chronic pelvic pain, and diverse health problems. *Hypercortisolism* is associated with hypertension, abdominal obesity, type II diabetes, and enhanced stress sensitivity (Hellhammer et al., 2004). Should there be intense noradrenergic activation in the presence of low cortisol levels at the time of trauma, it may result in a failure to contain the biological stress response with persistence of the stress reaction. High-circulating glucocorticoid levels facilitate the extinction of fear behaviors in rats, while a blockade of glucocorticoid synthesis prevents the extinction of conditioned fear (Schelling et al., 2004).

Groups with PTSD may have higher or lower cortisol than the control group. The longest post-stress interval was apparently a group of Holocaust survivors with PTSD, who had lower levels than survivors without PTSD as well as unexposed normals. In fact, low urinary cortisol levels were a function of the current severity of PTSD rather than exposure to trauma per se. PTSD victims may also have more intense circadian cortisol fluctuations (Yehuda et al., 1995).

Opioids: Termination of the stress response: Stress stimulates the hypothalamic β-endorphin, which inhibits the activity of the stress system while producing *analgesia*, with a possible influence upon the emotional tone. Inhibitory glucocorticoid feedback on the ACTH secretory response limits the duration of total tissue exposure to glucocorticoids, minimizing their catabolic, lipogenic, antireproductive, and immunosuppressive effects (Tsigos & Chrousos, 1996).

Mineralocorticoids

Aldosterone affects salt balance, as do glucocorticoids. They regulate renal sodium retention (i.e., are key components in sodium, potassium, blood pressure, and intravascular volume). Cortisol also plays a mineralocorticoid role. The brain, the mammary gland, and the pituitary gland appear to responsd to mineralocorticoids. Excess leads to a variety of disorders, including shock (Miller & Tyrrell, 1995). Aldosterone secretion is largely independent of corticotropin.

Adrenal insufficiency (Addison disease) (Hasinski, 1998) has symptoms are similar to those observed after head trauma (weakness, weight loss, loss of body hair in women, loss of libido, psychiatric symptoms). Hypoadrenocorticism (adrenal insufficiency) is secondary to CRH deficiency and/or ACTH secretion (e.g., hypopituitarism: fatigue, weakness, reduced libido are consistent with TBI). This can be related to negative-feedback-controlling cortisol secretion (Migeon & Lanes, 1996). Glucocorticoid deficiency leads to anorexia, apathy, cognitive disorder, stupor, and coma.

ADRENAL MEDULLA

Pancreas

Diabetes 2

I referred to a patient of mine who had diabetes mellitus type II, which went out of control briefly after a third whiplash in a brief interval.

The mechanism of stress disorder leading to diabetes 2 complications is noted below, and reflects the cognitive concerns noted by the above colleagues.

The description was offered of type I diabetes mellitus as insulin dependent, although this was also true to some extent for type II diabetes mellitus.

There has been some recent change in the differentiation of these conditions, according to the 15th edition of Harrison's *Principles of Internal Medicine.* For type I diabetes, the etiology is destruction of beta cell in the pancreas that leads to absolute insulin deficiency. Type IA DM results from autoimmune beta cell destruction, which usually leads to insulin deficiency. Type 1B DM is also characterized by insulin deficiency, as well as a tendency to develop ketosis. However, such individuals lack immunologic markers indicative of an autoimmune destructive process of the beta cells (Powers, 2001).

Contributors to type I DM are suspected to include medications, poisons, the ingestion by the mother of curing agents (*n*-nitroso compounds), antibodies to bovine albumen, and viruses. There is a genetic susceptibility (Unger & Foster, 1998).

Contributors to type II DM range from predominantly insulin resistance with relative insulin deficiency to a predominantly secretory defect with insulin resistance.

An example of disease consequent to stress was a man who had *type II diabetes mellitus* under control. This condition is characterized by variable degrees of insulin resistance, impaired insulin secretion, and increased glucose production. After a third whiplash in a relatively brief interval, the man's blood sugar rose considerably. Glucocorticoids increase insulin resistance and hepatic glucose production, decrease glucose utilization, and impair insulin secretion. These changes worsen glycemic control in diabetics and may precipitate diabetes in others. In addition, exercise-stimulated catecholamines can increase plasma glucose excessively, promote ketone body formation, and possibly lead to ketoacidosis. Conversely, excessively circulating insulin may reduce glucose production and increase glucose entry into muscles, leading to hypoglycemia (Powers, 2001).

Trauma or other serious illness may provoke hyperglycemia due to hypersecretion of insulin antagonistic hormones (Sherwin, 1996).

REPRODUCTIVE AXIS DISORDERS

The sexual experience and physiological response is maintained by a variety of systems that are vulnerable to trauma (CNS, somatic, and emotional). Somatic or pelvic injury has to be considered in sexual problems. This section will explicate some of the neurological and endocrinological bases for sexual response (for the somatic component, see Matsumoto, 2001). A high proportion of victims of head injury have reduced sexual performance and interest. Male sexual functions require coordinated regulation of libido, erection, ejaculation, orgasm, and detumescence. Psychodynamic components require a detailed study of clinical issues such as personality dynamics and psychopathology. This writer considers psychometric procedures as screening devices. It is not easy to differentiate between the cause of sexual dysfunction and the psychodynamic and social effects of an accident (e.g., attitude to being impaired, scarring, fear, etc.). This is similar to the consideration of the effects of any accident that involve potential neurological injury.

NEUROLOGICAL

Libido is generated by visual, tactile, imaginative, auditory, and gustatory erotic stimulation of the CNS. Erotic stimuli are received in cortical and subcortical brain regions (limbic system) and relayed via medial preoptic and anterior hypothalamic areas to the centers that regulate penile vascular smooth muscle and erection (the sympathetic thoracolumbar [T12–L2] and parasympathetic sacral spinal cord centers). *Psychogenic erections* are mediated by the latter structures, though induced by psychogenic stimuli. The neurological components of the sexual response are explicated in Matsumoto (2001): cortical and subcortical brain regions (primarily limbic system), brainstem, spinal cord pathways, and the autonomic innervation of pelvic organs. *Erection* is primarily a vascular event, regulated by both the peripheral and CNSs. *Reflexogenic erection* is mediated by parasympathetic innervation of the pelvic (pudendal) nerve (S2–S4), which supplies various nerve plexes of the internal structures (autonomic nuclei). Reflex erections are preserved with lesions above T11. Afferent impulses are conveyed from the pudendal nerve (S2–S4 roots) to the *sacral spinal erection center*. Efferent impulses travel through the same sacral routes (Fowler, 2004). The SNS controls contraction of the ductus deferens, seminal vesicle, prostatic and uterine muscle, and vasoconstriction. Preganglionic neurones are found in T10–L2. The PNS controls vasodilation and erection (Parent, 1996, Chapter 9, The Autonomic Nervous System).

The bladder, bowel, and organs of sexual function are neurologically similar. They share the same nerve roots with common peripheral nerves within the pelvis, and each is controlled by a unique set of CNS reflexes.

Brain nuclei sexual dimorphism has been demonstrated for a variety of brain regions, including vocal control mechanisms in birds. These are subject to steroid effects (including androgens) during development (Breedlove & Hampson, 2002). In frogs, at least, female sexual receptiveness is revealed by the length of the intervals between trill units.

Hormonal

The reproductive axis is inhibited at all levels by various components of the HPA axis (e.g., glucocorticoids and inflammatory cytokines). The effect is bidirectional. Sexual dysfunction after an accident or medical illness requires consideration of multiple systems: (1) psychodynamic, (2) neurological; and (3) psychoneuroendocrine. This has at least two aspects: TBI interference with homeostasis and neuroendocrine secretion; the neuroendocrine and allostatic effects of chronic stress stemming from unhealed wounds (participating in postconcussive syndrome) and persistent feelings of impairment (physiological and anatomical considerations will be emphasized here).

Posttraumatic epilepsy involves the consequences of neurological phenomena, endocrinological consequences, and the effects of anti-epileptic drugs (AED) (Harden, 2006). *Disruptions of reproductive function* include anovulatory cycles (increased risk for infertility, migraine, emotional disorders, reproductive cancers), altered hormonal levels, promotion of reproductive endocrine disorders (e.g., polycystic ovarian syndrome), and premature menopause. The possibility of seizures is an emotional burden, restricting mobility within the community, and raising issues of safety in swimming, driving, use of tools, and so on. There is a strong correlation between epilepsy and depression and anxiety. The impact of depression on adolescents is noteworthy (Baker, 2006). Children with epilepsy have a high proportion of psychiatric diagnoses, with only half of them receiving psychiatric treatment. Depression is associated with epilepsy, but antidepressants raise concern that they interfere with AEDs.

Neuroendocrine and Persistent Stress Effects

Hypothalamic anovulation is a result of disorder in the production of LHRH in the arcuate nucleus of the hypothalamus and its secretion into the portal vessels from the median eminence. It can be caused by head trauma or any other disorder of the CNS that interferes with this process (as well as feedback by gonadal steroids that are mediated by brain neuromodulators such as NE, dopamine, and β-endorphin). It is also related to stressful events and vigorous exercise (Bulun & Adashi, 2003). Prompt treatment is essential after suspicion of a pituitary injury.

Infertility in the absence of anatomical abnormalities was associated with psychological stress profiles (Lemack et al., 1988). Various components of the HPA axis and cytokines inhibit the reproductive axis. CRH and glucocorticoids inhibit gonadotropin-releasing hormone (GnRH), pituitary gonadotrophs, the gonads, and target tissues. Ovarian steroids in particular target the CRH gene, leading to potential *gender differences* in the stress response. In women, physical, psychological, metabolic, and pharmacological stresses may result in reversible reproductive axis suppression. These stresses are believed to act by inhibiting the GnRH pulse generator output as a final common mechanism. After cues are processed by the brainstem and higher CNS pathways are integrated and interpreted, output information is directed by hypothalamic nuclei. Anxiety or perceived threat can result in reversible suppression of the hypothalamic GnRH network, which directs gonadotropin secretion and mediates feedback control via ovarian steroid and glucoprotein signals. Various stressors affect GnRH release, including surgical trauma, psychiatric disorders, infectious disease, and strenuous exercise. Low levels of plasma estrogen cause or accelerate health problems (e.g., osteoporosis and subsequent fractures after menopause) (Rivest & Rivier, 1993). Suppression of the LH and the FSH causes relative quiescence of ovarian function with reduction of sex hormone steroid secretion, diminished gonadal steroid feedback control of the hypothalamo-pituitary unit, and *hypothalamic amenorrhea* (Rivest & Rivier, 1993; Veldhuis et al., 1998). Functional hypothalamic amenorrhea may also be preceded by psychologic stress, poor nutrition, or strenuous exercise (Rosen & Cedars, 2004). *Endometriosis* is associated with impairment of the immune system (Di Stefano et al., 1994).

IMMUNE SYSTEM CONSIDERATIONS

Cytokines: Behavioral Consequences

Regulation: Cytokines play a role in the whole body response to stressors (injury and illness) and the etiology of depressive symptoms. They signal the brain to produce varied effects: neurochemical, neuroimmune, neuroendocrine, and neurobehavioral (including "sickness behavior"; (Slimmer et al., 2001). The intensity and duration of immune responses are regulated through cell-to-cell communication (e.g., TNF, IL-1, and IL-6). These cytokines are products of the immune system (i.e., secreted by monocytes or macrophages in response to infection or other antigenic challenge) (Orth et al., 1992). Cytokines cross the BBB, and probably the choroid plexus of the brain and spinal cord. Their alteration in the injured or diseased CNS contributes to a deranged neuroimmune axis (Banks, 2001).

IL-1, which elicits behavioral responses characteristic of sickness (fever, anorexia, decreased locomotor activity, increased sleep) seems to be mediated by CRF, which in turn activates the HPA axis response. This would be stimulated by noradrenergic projections from the brainstem (Dunn, 1993). There are various mechanisms of immune system response. Vagal sensory mechanisms may contribute to CNS responses to immune stimuli from the abdominal or thoracic cavities. Blood-borne immune challenge may activate the CNS from other routes as well (Cone et al., 2003).

Chronic stress reduces the immune response (i.e., multifactor suppression of the production of various cytokines [IL-1; INF], and reduced proliferation and activity of various types of cells) (Song et al., 1995). *Cytokine release* during response to injury and chronic stress creates responses frequently observed in TBI, but the etiology may be primarily described as the behavioral sequelae of the physiological reaction to primarily noncerebral injury (Kop & Cohen, 2001; Raison et al., 2005). One mechanism by which they induce depression is the stimulation of CRH with HPA axis activation. Depression merits study of cortisol and thyroid hormone levels (Booth, 2005; Irwin & Cole, 2005; Kop & Cohen, 2001; Marshall & Rossio, 2000; Raison et al., 2005). The molecular effectors are inflammatory cytokines and mediators of inflammation (Chrousos, 2000).

Proinflammatory cytokines: PIC (TNF, IL-1, and IL-6) are secreted by cells in the periphery of the brain, as well as and by the neurons and glia of the brain. They activate or inhibit the release of hypothalamic peptides, altering the pattern of pituitary hormone secretion. They activate the SNS and HPA axis via the release of CRF (Slimmer et al., 2001). They also mediate neurotransmitter changes in the brain, as well as behaviors known as "sickness behavior." Additionally, the cytokines alter the responses of the hypothalamic peptides and other transmitters that act directly on the pituitary gland. They alter the response of the various pituitary cell types to the hypothalamic peptides and other transmitters that act directly on the gland. The pituitary hormones then modulate the responses of the immune cells, either directly or via secretions of their target glands in a medical condition or treatment (McCann, 2003). They induce the pattern of behavioral changes described as "sickness behavior" (see below). PIC induce *glucocorticoid resistance* to the negative feedback effects of circulating glucocorticoids in nervous, endocrine, and immune tissues. This would reduce control of the stress reaction mediated by the HPA axis. In animals, CRH induces symptoms associated with cytokine-induced sickness behavior (fever and sleep disturbance). It also produces behavior that resembles major depression. INFα induces inflammatory cytokines and symptoms of sickness behavior (i.e., *fatigue*) sufficient to interfere with daily functioning (i.e., *fatigue*). Agents that diminish cytokine production may be of value in stress-related conditions, including depression, and to psychological stressors (Raison et al., 2005). *IL-1* appears in the brain when stimuli activate astrocytes and microglia. It modifies EEG sleep patterns, activates the pituitary-adrenal axis, and elevates body temperature. IL-1 is produced by stimulation of astrocytes and microglia, with behavioral, neurochemical, and physiological effects and reduction of brain immune responses. Infusion of IL-1 into the brain suppresses the immune response with the contribution of adrenal factors. It also activates the pituitary-adrenal axis. The effect of IL-1 in the brain seems to activate the sympathetic ganglia, which in turn activate peripheral cellular immune reactions (Weiss et al., 1994). IL-1β activates the HPA axis in many species, with systematic elevation of corticosteroids acting as a brake on the immune system. Negative feedback limits immune hyperactivity, which would limit autoimmune hyperactivity and perhaps other physiological actions (Dunn, 2000). Inflammatory cytokines such as TNF-α are involved in physiological sleep regulation and in the hypnotic effects of sedating drugs. IL-6 suppresses REM sleep and produces a short period of subjective sleepiness (Pollmacher et al., 2000).

CHRONIC CONDITION: MULTIPLE TRAUMA

The brain and soma interact with profound holistic behavioral and physiological changes. The *multiple traumata* that accompany concussive TBI involve fright, comorbid injuries, and a sense of impairment. These create persistent stress effects of both a psychological and physiological origin, which include effects that are mental (e.g., fright), holistic (systemic and behavioral), and cellular and molecular (the expression of given genes). The stress response serves the biological function of coping with threat to bodily or personal integrity, and integrates sensory, physiological, and subjective information. Persistent stress affects the stress victim's adaptation (including his or her stamina and health).

Chronic dysfunctioning leads to the issues of allostasis, allostatic load and overload, and systemic dysregulation after chronic injury. The ultimate consequence of persistent dysfunctioning, beyond the capacity of the tissues to cope so long, is "*burnout*." This writer considers the physiological state of continuing injury and its neurobehavioral effects to be chronic stress. Stress directly effects somatic functioning and indirectly effects CNS functioning through the secretion of neuroactive substances. The latter enter the CNS through damage to the BBB, neural transport, and so on. Social factors (e.g., lack of support, feelings of loneliness, hostility) influence physiological factors in the immune system, with implications for wound repair, inflammation, natural killer (NK) cell response, tumor metastases, and immune response to infectious challenges (Felten & Maida, 2000).

After an injury, the inflammatory and immune humoral systems are expressed and interact, frequently using the same chemical messengers and intracellular processes. Their complexity is comparable to that of the CNS. In addition, there is a bidirectional interaction between the immune and neuroendocrine systems. The molecular basis for communication is shared regulatory mediators (steroid hormones, NP, and cytokines) and their receptors (Crofford, 2002). Neuroendocrine control of a wide range of immune responses occurs via NP and neurohormones (Brown & Zwilling, 1996). The inflammatory and immune systems function together.

PHYSIOLOGICAL DYSFUNCTIONS IN THE CHRONIC STAGE

The HPA axis stress response is less well understood for chronic stress than acute stress, but seems to be driven more by AVP than by CRH (Michelson et al., 1995). It is expected to shut down at the end of stress. In addition to considerable chemical changes during the acute and chronic phases of psychological stress, persons who have been injured in accidents may also have incurred damage to the CNS and peripheral structures that contribute directly and neurobehaviorally to persistent stress. Many somatic symptoms may be attributed to TBI and persistent stress. Their actual mechanism could be endocrinological (e.g., weakness, tiredness, vague, gastrointestinal discomfort, and weight loss or gain) (Frohman & Felig, 2001). Overactivity of adrenal steroids leads to insulin resistance, which accelerates progression towards *type II diabetes* (including abdominal obesity, atherosclerosis, and hypertension). Failure to turn off the hormonal stress response or to display a normal trough in the diurnal cortisol pattern leads to blood pressure elevations in work-related stress, turning off slowly in some persons with a family history of hypertension. *Sleep deprivation* leads to elevated evening cortisol and hyperglycemia within five days. Depressive illness leads to chronically elevated cortisol and loss of bone mineral mass (McEwen & Seeman, 1999).

HYPOTHALAMIC DYSTHERMIA

Temperature regulation may vary by day, week, or month. Instability of hypothalamic regulation is a concomitant of aging, failing health, and is also a predictor of shortened life span. Dysthermia has been observed in a group with brain damage and mental retardation, many with an IQ under 30 (Chaney & Olmstead, 1994). There were varied etiologies, including trauma, infection, and prenatal metabolic and chromosomal abnormalities. Symptoms included hypothermia, hypothermia, flat or exaggerated rhythms, reversed day–night cycles, unusual fluctuations, or sleep-phase delay. Prenatal brain injuries tended to have hyperthermia; perinatal, postnatal, and unknown etiologies were associated with hypothermia.

CARDIOVASCULAR

Mild social stressors can lead to bursts of increased blood pressure, causing structural thickening of the blood vessel wall and neuroendocrine changes that lead to a further blood pressure increase (Ely, 1995). Repeated surges of blood pressure accelerate atherosclerosis and synergize with metabolic hormones to produce type II diabetes. Activation of the immune and inflammatory systems are involved in the development of atherosclerosis (Steptoe & Brydon, 2005). Hormones associated with the stress response (CRH, ACTH, and adrenal cortical) can induce *salt appetite*, probably through effects on the central nucleus of the amygdala and the bed nucleus of the stria terminalis. *Mental stress* (Felker & Hubbard, 1998) is associated with sudden cardiac death that involves both SNS and PSNS mechanisms, as well as thyroxin, glucocorticoids, insulin, and CNS state. Chronic mental stress shifts the parasympathetic-sympathetic control of heart rate towards *sympathetic predominance*. It is recognized that it is difficult to clearly measure chronic mental stress. An increase of SNS activity predisposes one to dysrhythmias. However, it is not established that patients with anxiety-related psychiatric illness have a higher prevalence of cardiovascular disease than less anxious individuals. Mental stress may induce cardiovascular changes analogous to isotonic (aerobic) and isometric (anaerobic exercise) ones (i.e., changes that promote health or disease). *Chronic mental stress* has been related to cardiovascular disease, with symptoms such as bereavement, personality patterns, anxiety and depression, and occupational stress. Pathological components include atherogenesis, ischemia, hypertension, sudden cardiac death, and cardiac risk factors.

In a state of exhaustion, somatic and mental processes reinforce each other and may precede myocardial infarction. There is decreased cortisol excretion, while coronary patients have elevated antibody titers against the microorganisms involved in the pathogenesis of coronary heart disease. They also have elevated levels of cytokines, which are markers of inflammation (IL-1, IL-6, and TNF-α). If prolonged stress causes or aggravates inflammation, the inflammation itself may amplify feelings of fatigue through the release of cytokines (Appels, 2000).

Peripheral catecholamines, released as a result of central lesions, may generate cardiac arrhythmias. These have also been attributed to CNS lesions that involve irritation (e.g., epileptogenic lesions). However, arrhythmias can be caused by alteration of the symmetry of sympathetic activity, as well as plasma catecholamine levels. These changes are more likely to occur following lesions of the right hemisphere rather than the left hemisphere (Talman, 1997). Sympathetic preganglionic neurons that have a low level of activity at rest are modulated by the interactions with CNS influences and segmental influences of somatic and visceral afferent origin (Benarroch, 1997). Patients with a panic disorder have *sympathetic overactivity* and cholinergic underactivity in the wakeful period before sleep. The higher sympathetic tone is probably dependent on cognitive activity, and might play a role in fatal cardiac arrhythmia in cardiac disorder (Ferini-Strambi et al., 1996).

Respiratory System

Stress makes a nonspecific contribution to the development of respiratory disorders. Respiratory symptoms are a component of panic attacks (i.e., shortness of breath or feeling of choking or smothering) (American Psychiatric Association, 1994, p. 394). Psychological stress can affect upper respiratory illness (Bovbjerg & Stone, 1996) and may increase risk for developing illness (e.g., herpes and AIDS) in those who are immunocompromised, including the elderly and individuals receiving chemotherapy (Kennedy, 1996). There is a bidimensional relationship between emotions and respiration. Manipulation of respiration affects mood (i.e., slower rate and larger tidal volume is associated with reduction of subjective and physiological indices of anxiety). Voluntary hyperventilation produces symptoms of anxiety. Emotional arousal is associated with increased rate and irregularity of breathing. The most consistent findings for the role of stress in respiratory disorders have been found for naturally occurring and experimentally induced infections (Sriram & Silverman, 1998).

Gastrointestinal

Physiological functioning of this system is complicated by the addition of the separate nervous network called the "enteric nervous system" (see above), which functions semi-independently of the ANS and CNS.

Stress effects on this system may be the earliest medical study in this area (Beaumont's study of the gastrointestinal functions of an open gunshot wound to the abdomen of his patient in 1933). However, the intensive psychosomatic study of the Franz Alexander psychosomatic school, and its specific association of GI tract difficulties with personality types and social events, is not well established. In fact, medical conditions as ordinarily understood were then labeled personality disorders, which have been described as a double burden of physical illness and inability to adapt to an environment. Nevertheless, it has been determined that the development of GI symptoms are associated with childhood and adult physical and sexual abuse, which were reported in less than half of cases (Olden, 1998).

Somatostatin suppresses secretion of insulin, glucagon, various gastrointestinal hormones, and such functions as gastric acid secretion, gastric emptying, and splanchnic blood flow (Molitch, 1995). Gastrointestinal functioning is altered by the stress system, possibly with hypothalamic CRH acting as a mechanism for delayed gastric emptying with colonic motor activity increase (Stratakis & Chrousos, 1995). CRH acts in the PVN to induce stress-related inhibition of gastric emptying and the stimulation of colonic motor function (emptying) in response to stress, and acts on the LC to stimulate colonic transit alone (Tachè et al., 1993).

The behaviors associated with this list are consistent with those of hypoactivation after persistent PTSD. Although panic attacks have been associated with *irritable bowel syndrome* (IBS), the description of IBS as a disorder of gastrointestinal motility, and its treatment by changes of diet and pharmacological agents, imply a physical origin (Snape, 1996). The complexity of physiological control contributes to incomplete knowledge of brain-gut interactions. For each section of the GI tract, stress-related GI disorders exist in multiple subsets of patients (Olden, 1998).

Reproductive Axis

Infertility in the absence of anatomical abnormalities was associated with psychological stress profiles (Lemack et al., 1988). Various components of the HPA axis and cytokines inhibit the reproductive axis. CRH and glucocorticoids inhibit GnRH, pituitary gonadotrophs, the gonads, and target tissues. Ovarian steroids in particular target the CRH gene, leading to potential *gender differences* in the stress response. In women, physical, psychological, metabolic, and pharmacological stresses may result in reversible reproductive axis suppression (amenorrhea and functional chronic anovulatory syndrome) (Ferin, 2001). They are believed to act by inhibiting the GnRH pulse generator output as a final common mechanism. After cues processed by the brainstem and the higher CNS pathways are integrated and interpreted, output information is directed by hypothalamic nuclei. Anxiety or perceived threats can result in reversible suppression of the hypothalamic GnRH network, which directs gonadotropin secretion and mediates feedback control via ovarian steroid and glucoprotein signals. Suppression of LH and FSH causes relative quiescence of ovarian function with reduction of sex hormone steroid secretion, diminished gonadal steroid feedback control of the hypothalamo-pituitary unit, and *hypothalamic amenorrhea* (Veldhuis et al., 1998). Disorder of the endocrine or other systems can result in *anovulation* or *irregular menses* (Bulun & Adashi, 2003). *Endometriosis* is associated with impairment of the IS (Di Stefano et al., 1994). In men, stress, including surgical stress, induces hormonal changes (including LH/FSH secretion), leading to a reduction of *testosterone* levels. Stress-increased concentrations of NE, EPI, and serotonin may induce *altered testicular metabolism* or *testicular vasoconstriction. Anger and depression* are associated with enhanced stress, and therefore higher sympathetic tone and smooth muscle contraction, resulting in reduced capacity for erection. Related to this concept, abnormal levels of circulating NE and EPI contract penile cavernous tissue, reducing the capacity for arterial inflow, which may underlie *erectile failure during stress* (Lemack et al., 1988).

Cognitive Disorders and Stress

Chronic stress has a variety of cognitive effects depending upon the various characteristics of the stressor: length of exposure, intensity, gender and hormone, age, and the brain target organ of stress-related hormones (Sapolsky, 2004).

DISEASE AND STRESS

Among the mediating factors contributing to the relationship between stressors and disease are coping mechanisms, aspects of disease (weight loss), nutrition, circadian rhythms, psychosocial (housing; life changes; power; loss of affiliation or social support), genetic and environmental history, and such physiological mechanisms as hormones (Plaut & Friedman, 1981, cited by Schlesinger & Yodfat, 1996). Health effects of trauma and other childhood adversities are broad and not related to any particular psychiatric or other disorder (McEwen, 2003b). Older TBI patients were more likely than nondisabled persons to report metabolic, endocrine, and neurological problems. Younger TBI patients resembled older persons with TBI in their report of difficulty falling asleep (Breed et al., 2004). Sedentary *lifestyle*, a frequent outcome of TBI, is a risk factor for *osteoporosis* (Barrett & Barrett, 2005).

Mood

Affective responses to stressful circumstances are accompanied by autonomic and neuroendocrine changes that influence immune function and increase susceptibility to a variety of diseases. The common nonspecific effects of stress alone do not explain disease processes, since different stressors have variable effects to the same immune response or disease process. There are *individual differences* in immunological reactivity, hormonal and autonomic responses, and personality (perception and appraisal of situations). *Temporal considerations* affect the consequences. Exposure to a stressor may precede exposure to a pathogen. Acute stress may be superimposed on a chronic disease. Stress influences the expression or progression of disease, not its induction or cause. Factors that increase vulnerability include a genetic predisposition to particular diseases (autoimmune diseases, diabetes, some cancers); remission from chronic disease; and aging (reduced immunocompetence) (Ader, 2005).

Stress-related disease processes are associated with altered function of (1) *hypothalamic pituitary target endocrine axes* (the hormonal system), and (2) *autonomic nervous system functions* (implying involvement of the CNS and spinal cord), the *immune system*, and the *inflammatory system*. It is assumed that multiple physiological activities are associated with any disorder.

Autonomic reactivity (psychophysiological): Hyperresponsivity to stress; increased cardiovascular morbidity. Psychophysiological studies (heart rate; blood pressure; galvanic skin response) indicate that there may be no baseline differences between combat veterans (with or without anxiety but without PTSD) and the normal control group. However, psychophysiologic reactivity occurred in both veterans and civilians who had PTSD (Southwick et al., 1995).

Inflammatory system: In healthy person, tissue injury is resolved by the inflammatory response before the IS activated (*acute inflammation*). The chronically ill person or a person experiencing a high level of stress may have an inefficient acute inflammatory response, thus the IS becomes activated and participates in the inflammatory response (*chronic inflammation*). Trauma survivors have higher than average rates of many chronic diseases and often die prematurely. The human stress response activates inflammation, including heart disease, diabetes, multiple sclerosis, and Alzheimer disease. *PTSD* patients have higher levels of cardiovascular, disease, (hypertension and heart disease, respiratory disease (asthma and chronic obstructive pulmonary disease), chronic pain syndromes (fibromyalgia, arthritis, migraine), gastrointestinal diseases (ulcerative colitis and ulcers), cancer, chronic fatigue syndrome, multiple chemical-sensitivity, suicidal attempts, poor quality of life, and long- and short-term disability (see Allostatic Load) (Kendall-Tackett, 2008). The clinician may be alerted that there is an overlap between the symptoms of PTSD and concussion (Parker, 2002), and that a stress symptom pattern may be consequent to the illness and not the reverse.

The inflammatory system function is augmented by psychological factors to increase the risk of *coronary artery disease* (Kop & Cohen, 2001). Psychological risk factors include hostility, depression, acute mental stress, such as outbursts of anger (although the association is indirect), anxiety, and low socioeconomic status. This pattern is consistent with frequent clinical findings in PCS. The inflammatory cytokines IL-1β and TNF-α play a role in the induction of the osseous lesions of *arthritis*.

Extremes of HPA axis activity (Appels, 2000): The increased activity seen in chronic stress is associated with melancholic depression, panic disorder, diabetes mellitus, central obesity, Cushing syndrome, elevated excretion of cortisol, and immunosuppression. *Decreased activity* occurs after prolonged exposure to stress and is associated with adrenal insufficiency, atypical depression, chronic fatigue syndrome, fibromyalgia, nicotine withdrawal, decreased free cortisol secretion, and increased activation of inflammation.

Some symptoms of HPA axis activity are *connective tissue* (poor wound healing, thinning of the skin, easy bruising); *bone* (inhibit bone formation and contribute to net bone resorption); calcium metabolism (decreased calcium absorption and increased urinary calcium excretion resulting in disabling osteoporosis); growth and development (inhibit growth in children by a direct effect on bone cells, although glucocorticoids do accelerate the development of a number of systems and organs in fetal tissues); *blood cells and immunological function* (decrease the migration of inflammatory cells to sites of injury, a major mechanism of its anti-inflammatory actions and increased susceptibility to infection following chronic administration); and *immunologic and inflammatory responsiveness*, which affect the HPA axis. They inhibit release of effector substances such as IL-1, which presumably reduces secretion of CRH and ACTH.

Cardiac Disorders

Peripheral catecholamines, released as a result of central lesions, may generate cardiac arrhythmias. These have also been attributed to CNS lesions that involve irritation (e.g., epileptogenic lesions). However, arrhythmias can be caused by an alteration of the symmetry of sympathetic activity, and by plasma catecholamine levels as well. These changes are more likely to occur following lesions of the right hemisphere rather than the left hemisphere (Talman, 1997). Sympathetic preganglionic neurons that have a low level of activity at rest are modulated by the interactions with CNS influences and segmental influences of somatic and visceral afferent origin (Benarroch, 1997). Patients with a panic disorder have *sympathetic overactivity* and cholinergic underactivity in the wakeful period before sleep. The higher sympathetic tone is probably dependent on cognitive activity, and might play a role in fatal cardiac arrhythmia in cardiac disorder (Ferini-Strambi et al., 1996).

Somatic reactions (see Friedman, 1990, for a summary): SNS hyperarousal; hypofunction of HPA axis (absolute reduced cortisol levels, and related to catecholamines); abnormalities of the endogenous opioid system (stress-related analgesia; lowering of pain threshold at rest); sleep abnormalities (traumatic nightmares, increased latency, decreased sleep time, increased awakenings, and bodily movement). Various traumata (combat, transportation, natural disasters) create a syndrome of hyperarousal. Its change to *reexperiencing symptom cluster* is consequent to the nature of psychological distress following trauma-related cues (i.e., incorporating the meaning of stressor into the process of mediating psychophysiological reactivity) (Prins et al., 1995).

Somatic forms of anxiety: Exaggerated startle response; diarrhea; frequent urinating; sweating; tremor; rapid heart beat; frequent urination.

Sleep disturbances: Difficulty in falling asleep or in staying awake; change of diurnal sleep patterns; bad dreams; nightmares. Interruption of circadian rhythms may disturb cortisol production (Merola et al., 1994).

Reexperiencing Symptom Cluster

Hyperarousal disturbances include physiological reactions (tachycardia; breathing) and neural reactions (dissociation; nightmares); neurological changes (e.g., memory loss associated with reduction of hippocampal neurons); and repeated intrusive memories that are probably associated with hormonal and hyperarousal changes at the moment of stress. Various traumata (combat, transportation, natural disasters) create a syndrome of hyperarousal. Its change to a reexperiencing symptom cluster is consequent to trauma-related cues. This manifests the importance of *stressor meaning* in mediating psychophysiological reactivity (Prins et al., 1995). A case is reported where the reexperiencing of the traumatic event became *perseverated*, occurring without remission for 7–10 days (King, 2000). While initial reexperiencing and hyperarousal predicted PTSD symptom level after 6 months, peritraumatic dissociation had little predictive power (Wittmer et al., 2006).

SICKNESS (ILLNESS) BEHAVIOR

PIC, CNS and PNS cells, and glia (microglia, astrocytes) mediate a set of sickness behaviors (feeding, activity level, sleep and social interactions), as well as neurotransmitter changes in the brain. Immune molecules (the interleukins) signal the brain through many routes (through the bloodstream and through nerve pathways). When the brain receives such signals, one experiences a set of feelings and behaviors that, lumped together, are called sickness behavior (Sternberg, 2000). Clinically, this condition is manifested as fever, slow-wave sleep, malaise, fatigue, loss of appetite, diminished sexuality, apathy, social withdrawal, inactivity, and sleepiness (Berczi & Szentivanhi, 1996; Dantzer, 2000; Felten & Maida, 2000). Clinical administration of cytokines (particularly proinflammatory cytokines such as IL-1, IL-2 Il-6, TNF, and interferon-α/β) develops the behavioral symptoms called *sickness behavior* (Hansen-Grant et al., 1998). The intensity and duration of immune responses are regulated through cell-to-cell communication (e.g., TNF, IL-1, and IL-6). These cytokines are products of the immune system (i.e., secreted by monocytes or macrophages in response to infection or other antigenic challenge) (Orth et al., 1992). IL-1, which elicits behavioral responses that are characteristic of sickness (fever, anorexia, decreased locomotor activity, increased sleep), seems to be mediated by CRF, which in turn activates the HPA response. This would be stimulated by noradrenergic projections from the brainstem (Dunn, 1993). Vagal sensory mechanisms may contribute to CNS responses to immune stimuli from the abdominal or thoracic cavities; blood-borne immune challenge may activate the CNS from other routes as well (Cone et al., 2003).

In a state of exhaustion, somatic and mental processes reinforce each other and may precede myocardial infarction. There is decreased cortisol excretion, while coronary patients have elevated antibody titers against the microorganisms involved in the pathogenesis of coronary heart disease. They also have elevated levels of cytokines, which are markers of inflammation (IL-1, IL-6, and TNF-α). If prolonged stress causes or aggravates inflammation, the inflammation itself may amplify feelings of fatigue through the release of cytokines (Appels, 2000).

One may differentiate between illness behavior (e.g., sensitivity to physical sensations, labeling these as symptoms or as a disease, and then seeking medical care) and some objective evidence of illness. Stress is associated with increased illness behavior, but the association with verified objective disease is complex and less firm (Cohen & Williamson, 1991). Illness behavior does have *adaptive value*, in part from the organism-fighting function of fever, sleep hypomotility, anorexia, and reduced libido. It may have the consequence of hiding and escaping predators (Dunn, 1996).

As long as people are coping psychologically, they do not show increased pituitary-adrenal function (Martin & Reichlin, 1989, p. 670). When a stress disorder appears, it may be extremely long lasting. The gross outflow of various neurotransmitters, internally manufactured opiates (endorphins) and other neuroactive substances, cause permanent or long-lasting changes in neural circuits that account for the long-lasting effects of stress.

It may be a mistake to consider reduced output at home and elsewhere as a sign of secondary gain, discouragement, malingering, or the like. Fatigability and reduced stamina are common after chronic concussion and other levels of TBI. These are associated with reduced adaptive capacity and decreased cardiac reserve. The practitioner should consider chronicity, since such conditions may express allostatic load or system "burnout." Walking as an exercise may be encouraged, albeit with caution by the practitioner. Apart from systemic health problems, one must consider dizziness, pain, imbalance, and so on. Collaboration with a physiatrist or physical therapist familiar with the problems of the TBI patient may be necessary.

PSYCHONEUROIMMUNOLOGY–HEALTH PSYCHOLOGY

The expression of emotions affects the immune system, including mimicking various affective states (i.e., method actors). Expressing interpersonal emotion creates changes in physical appearance, with concomitant changes in nervous, hormonal, and immune system activity. Hostile behavior is associated with immunological downregulation, increased blood pressure, higher cortisol, NK cell levels, and cytotoxicity (Booth, 2005).

Behavioral interventions improve health outcomes and are assumed to impact the immune system. These applications evolve from various nontraumatic medical conditions and are applicable to trauma victims. They permit behavioral treatment (cognition, affect, insight, social relations) in a medical approach to trauma and may be useful in normalizing stress reactions (Antoni, 2005).

Relaxation and imagery techniques: Procedures are taught to reduce anxiety, tension, and other stress responses that are mediated by elevations or dysregulation of neuroendocrine systems. These include progressive muscular relaxation and imagery.

Massage: Sessions with varying strokes affect levels of numerous immune cell markers and reduced cortisol and catecholamine levels.

Biofeedback, imagery, and hypnosis: The patient visualizes himself as becoming more relaxed. In a state of relaxation, the writer uses visualization with imagination of visual, olfactory, and aural stimuli, creating images of very calm, comfortable, and beautiful conditions. Create a text according to your own preference.

Physical exercise: This a stress management technique can reduce depression and anxiety, and alter immune system markers.

Cognitive behavioral stress management techniques: This has been applied to numerous medical conditions, including postoperative pain.

The experience of a threat is determined by the appraisal of the situation and the perception of one's resources. There is an interaction between cognitions and feelings that affect emotions, tension level, and behavior, as well as behavioral responses in a social context.

Since an individual's appraisals, beliefs, and expectations mediate aspects of health, disease, and response to treatment, one is taught new patterns of thinking, feeling, and behavior in order to achieve control over emotional states and maladaptive behaviors. Stressor appraisals are modified; different coping responses are suggested integrating social support.

Cognitive restructuring: Homework and in-session activities lead to insight concerning the self-defeating appraisals of stressors. This reduces extreme reactions and premature giving up. Efficient initial appraisal may affect the magnitude and duration of neuroendocrine responses, possibly affecting health outcome.

Coping skills training: Varying levels of coping reaction and resources moderate environmental burdens. The pretraumatic baseline capacities of impaired individuals may be impoverished. In addition to the role of advocate, the practitioner teaches skills of daily living. The patient is trained to choose problem-focused coping strategies for the controllable aspects of stress (planning), and emotion-focused strategies (venting, relaxation, seeking emotional support) to deal with uncontrollable aspects of life.

Two main dimensions of coping strategies to stressful events and trauma have been identified (Littleton et al., 2007).

Problem Focused Or Emotion Focused

Problem-focused coping strategies directly address the problem by seeking information about the stressor, making a plan of action, and then planning to manage or resolve the stressor. Emotion-focused strategies manage the emotional stress by disengaging from stress-related emotions, seeking emotional support, and venting emotions. This style is not one of emotional regulation since it concentrates on conscious processes that address a stressor. Problem-focused strategies are more effective in controllable situations, whereas emotion-focused processes are more effective in uncontrollable situations.

Approach-Focused Or Avoidance Focused

Approach strategies are focused on the stressor or one's reaction to it, by seeking emotional support, planning to resolve the stressor, and seeking information about it. Avoidance strategies focus on avoiding the stressor or one's reaction to it by withdrawing from others, denying that the stressor exists, and disengaging one's thoughts and feelings from it. Reliance on avoidance strategies seems maladaptive and is associated with emotional distress. Distress is related to age, consistent with the view that some individuals are ineffective in developing coping strategies and thus over time experience diminished energy and cognitive changes.

Anger management: Collaterals who do not accept the diagnosis of TBI (or other condition), nor the impairment of the victim, create anger. Patients can become aware of the situations in which they experience anger, their style of expressing anger, and strategies to change so that their needs are met. Insight and resolution of stress may lead to improved social support.

Assertiveness training: Insight into one's own style and teaching the advantages of different styles of communicating is followed by an examination of whether one's style is appropriate to one's mood and likely to meet one's needs. Avoiding conflict may have beneficial stress effects.

Social support building: Social support can moderate stress effects on health, neuroendocrine parameters, and immune system functioning. Loneliness itself has stress manifestations. Improved social support of patients with medical conditions improves mood and the levels of the immune system.

EMOTIONAL DYSFUNCTIONS INTERACTING WITH STRESS

Changes in *lifestyle* can be stressful, and can interact with *aging*. Stress-induced decrements in immunity mimic age-related immunosuppression. Older adults facing a chronic stressor may face a physiological disadvantage. Progressive decrease in thymus size is correlated with increased glucocorticoid levels (Penedo & Dahn, 2005). Chronic stress can lead to *habituation*, although this need not prevent long-term detrimental changes of a physical condition, such as reduced resistance to infectious agents, hypertension, and behavioral disturbances (Antoni, 2000). When one considers the enormous complexity of both psychosocial conditions and physiological processes related to the IS, and the effects of health behaviors, health, medications, and so on, one may understand the uncertainty as to the clinical impact upon immunologic changes resulting from stress or conditioning have on health or illness (Schlesinger & Yodfat, 1996). Immunization leads to parallel production of antibodies and increase in firing rate of neurons within the ventromedial hypothalamus.

PSYCHOSOCIAL PARAMETERS

Stress components include *mood* (feelings of depression, loneliness, and hopelessness); *personality qualities* such as "hardiness" (see above), need for power, isolation with lack of social support, and hostility; coping power; family support; *proneness to various conditions* (cancer; coronary; autoimmune). *Depression* is associated with increased presynaptic release of CRH followed by downregulation of postsynaptic receptors (Hansen-Grant et al., 1998). Treatment with prednisone, or withdrawal from it, is associated with depression and also euphoria, irritability, insomnia, difficulty concentrating, and so on (DeAngelis et al., 2001). CRH (hypothalamus) increases the pituitary secretion of ACTH, which increases adrenal cortisol production. It is believed that depression and exhaustion have different neurohormonal concomitants since some forms of depression (hyperphagia and hypersomnia) are characterized by inactivation of the CRH system (Kop & Cohen, 2001). *Social support* acts as a buffer against the psychobiological effects of stress in disease. *Psychotherapy* alters the level of neuroendocrine hormones, especially cortisol (Walker et al., 2005).

STRESSFUL LIFE EVENTS AND VULNERABILITY TO ILLNESS

Emotional stress and psychological disorders activate the pituitary-adrenal stress response and suppress gonadotropin secretion, causing psychogenic amenorrhea or inhibiting GH secretion, which results in psychosocial dwarfism (Cone et al., 2003). Stress may increase susceptibility to illnesses and mortality, although only 5% of its variance is accounted for by life event scores. (An exception may be Graves disease insofar as the association of it with life events seems strong.) Reduced effectiveness of the immune response is associated with bereavement, separation, and divorce. The implication for the TBI victim is a common impairment of social relationships and networks (Kiecolt-Glaser & Glaser, 1992). Infectious disease is influenced by ethnic, environmental, and stress timing considerations (Brown & Zwilling, 1996). Chronic stress may reduce the capacity to terminate glucocorticoid secretion, implying reduced immune capacity.

Differentiating "Organic" From "Functional" Emotional Distress: "Organic" in this context means attributable directly to some kind of somatic damage. One popular psychometric tool, MMPI, was only marginally successful in identifying patients with psychogenic PTSD in a general psychiatric population (Silver & Salamone-Genovese, 1991), creating doubt about its use in instances of simultaneous PTSD and TBI. Psychogenic manifestations may closely resemble or may coexist with organic ones, and it may be difficult to discriminate between the two (Haerer, 1992, p. 733). Physical symptoms created by trauma when reported could elevate scores having a different psychiatric implication.

There are personality differences between patient groups according to type of injury (back vs. extremity), and with time since an accident (Beals & Hickman, 1972). Using the MMPI as a criterion, patients with recent physical injuries tended to display moderately severe depressive reactions, mild hypochondriacal, and mild hysterical symptoms, while those with long-term physical impairment showed moderately severe hypochondriacal and hysterical reactions and only mild depression. Injured patients' psychopathology was significantly greater in magnitude and complexity than that of extremity-injured workers. Physicians and psychologists were effective in assessing psychopathological contribution to return to work of extremity-injured patients, but physicians' ratings did not have prognostic value in assessing back-injured patients. While this specifically illustrates the difficulty of assessing "low back pain," it reflects the complexity in cases of TBI when determining the psychological contribution to disability as opposed to the "objective" cause of impairment. Beal and Hickman (1972) concluded that it is important to document the importance of psychological evaluation for optimum rehabilitation effort, and suggest that the "whole man concept" is a useful, and perhaps necessary, consideration in the rehabilitation of the industrially injured workman. This writer would extend the concept to the brain-damaged victim.

The findings were interpreted to mean that, with constant physical disability over time, patients increasingly elaborate on and exaggerate their symptoms, are less depressed by their predicament, and the likelihood of return-to-work changes. The greater the severity of the physical disability, the more deceit was measured, perhaps related to dependency on compensation and need to have their physician recognize their disability. Yet, malingering was considered rare, in the sense of describing symptoms not truly present. This writer notes that if reasonable settlement and medical treatment are denied, the patient becomes increasingly desperate and angry, which is expected to influence the need to focus upon and present symptoms to a clinician or examiner.

An additive rating was used to assess the workers' physical impairment, vocational skill, and psychological factors regarding their return to work. When the cumulative ratings reached a certain level, no workers returned to work. Yet, while proportion returning to work fell with increased physical and psychological handicaps, it did not do so with a decreased level of vocational skill. This finding is discrepant with other work (Rimel et al., 1981), which has suggested that the motivation to return is reduced with lesser vocational achievement. Study of minor head injury 3 months after an accident determined that executives and business managers returned to work 100% of the time, minor professionals 83% of the time, clerical sales 79% of the time, machine operators 63% of the time, and unskilled laborers 57% of the time. These results are attributable to motivation and retained resources. The earlier the application of management of the condition, the greater the likelihood of returning to work was.

Premature Return-to-Work As an Ecological Demand

Premature return to work or school may create unexpected and atypical difficulty, with failure leading to additional problems. Unexpected failure at previously well-managed tasks results in anxiety, loss of self-esteem, and depression. When environmental demands exceed cognitive capacities, nonspecific symptoms such as headaches and dizziness may occur (Gasquoine, 1997). During the phase of reduced cognitive function, there is frustration and anxiety, which perpetuate post-concussional symptoms. One study suggested that for patients still in the hospital, stress does not play a significant role in worsening postconcussive symptoms. However, a small subset (29%) had worsening symptoms associated with an apparently premature return to work (Moss et al., 1994). The significance of a circumscribed deficit depends on whether the worker meets the requirements of a particular position, noting that the range of opportunities may be restricted (Goldstein, 1942, p. 218). Unexplained reduction in the quality of performance may be falsely assumed to be due to a defect in motivation rather than to a lack of ability. The interaction of increasing anxiety, self-doubt, and the desire to avoid threatening situations, coupled with the dissatisfaction or anger of family or colleagues, can be misinterpreted as an "emotional" disorder or "personality change" (Boll, 1982).

10 Comorbid Posttraumatic Stress Disorder and Concussion

OVERVIEW

This chapter is concerned with *complex comorbid diagnostic entities. Concussive brain trauma* (CBT), also known as minor traumatic brain injury (mTBI), and posttraumatic stress disorder (*PTSD*), a poorly defined and complex condition (see DSM-IV-TR, American Psychiatric Association, 2000). The APA is inefficient in answering many questions, including (1) how to identify and diagnosis so-called "minor" head injury; (2) whether PTSD can occur in the presence of traumatic brain injury (TBI) associated loss of consciousness (LOC) or posttraumatic amnesia (PTA); and (3) whether or not dissociation has the same meaning in PTSD and CBT. There is a significant problem in generalizing between the meaning of PTSD in individuals who have and have not incurred TBI. TBI trauma would seem to be significantly different than the reaction of a person without brain injury.

The symptom complex of concussive injury is due to comorbid neuropsychological and somatic injuries, PTSD, and other emotional problems. These considerably overlap, although it may be surmised that symptoms with the same label, following different etiologies, might in fact be substantially different. For example, since PTSD and simple phobia appear to support fear networks that are highly specific to individuals and easily activated, a particular patient might have a unique pattern of disorders.

Victims of CBT commonly experience persistent pain, unhealed tissue, neurobehavioral impairment, impaired range of function, and so on. Similarly, PTSD is a syndrome or cluster of symptoms with numbers and severity varying along a continuum. Patients may display several signs of PTSD, yet fall short of the strict formal diagnosis. This may be described as *partial PTSD*, with PTSD-like symptoms (Harvey et al., 2003). It should be considered in reported complaints that the change in postconcussional symptoms after an accident did not differ when concussive and nonconcussed back pain victims were compared. When the two groups were combined, there was a correlation between symptom change and measures of emotional distress. The highest correlation was with the MMPI Depression Scale (Gasquoine, 2000).

Preventing PTSD and other responses to physical injury has a high public health value that is incompletely addressed by the mental health system. Mental health sequelae accentuate disability and worsen functional outcome. A clinical goal would be to integrate physiological reactivity with trauma-related cues into the diagnosis. One assessment is that cognitive-behavioral-oriented approaches to preventing PTSD are very effective. Critical incident stress debriefing is not useful, and may even be harmful, and there is limited efficacy of currently available psychotherapeutic and pharmacotherapeutic approaches to treating chronic PTSD. An attempt to obtain the cooperation of acutely injured patients in the emergency room to participate in a psychopharmacotherapeutic study of severely injured trauma victims recruited fewer than 10% of eligible patients. Possible reasons were desire to leave the hospital, or participate in a procedure that would delay them, denial of possible adverse mental health outcomes, the belief that not thinking about the outcome might be protective, and concern for the possible medical side effects (Stein et al., 2008).

THE CHRONIC STRESS SYSTEM

INTEGRATION

The stress system integrates neurosensory, neuroendocrine, endocrine, and humoral signals to alert a person that a certain site or event was life-threatening and therefore should be remembered and avoided. It also represents a wound control and danger adaptive function that confronts healing and restorative procedures that often does not resolve in an adaptive and healthy body or state of mind of the injured person. The various stress components (opposing the immune system) represent an interaction with systemic and psychological processes at all levels, from genetic and cellular to psychological (mood; cognitive; subjective and experiential awareness of being an injured and impaired person; dissociative experiences; traumatic memories; and associated intense emotions). These are interrelated and affect a patient's quality of life and capacity for adaptive success.

NEUROBEHAVIORAL DYSREGULATION

The direct disturbance of brain functions and the systemic reaction to unhealed wounds cause the body to respond in a complex yet integrated physiological response, which extends far beyond that of the usual definitions of PTSD and CBT (Parker, 2005). Emotions and injuries mobilize a large variety of physiological responses to maintain adaptation. The body's complex signaling capacity is created by the response to distant cellular receptors to substances designed for other purposes. This creates a potential for maladaptive reactions in, as well as exhaustion of, particular tissues. Furthermore, hyperarousal of organs is consequent to conditioned traumatic cues that result in anxiety-related signs and symptoms. An injury creates multisystem disruption, which should be considered in order to obtain a reasonably complete understanding of the patient's condition, a comprehensive treatment plan, and an estimation of that patient's outcome. Rasmusson and Charney (2000) describe a variety of options under the rubrics of cognitive-behavioral and psychodynamic therapies for PTSD.

Acute stress has been defined as an increase in stress-responsive physiological parameters for a period from minutes to hours, while *chronic stress* persists for several hours a day for a number of days or even months (Dhabar & McEwen, 2001). It is useful to differentiate between the experience of fright (PTSD), and the adaptive response to injury and impairment (stress). While PTSD was first studied after *extended exposure* (wartime exposure for both combatant or civilians; concentration camp inmates), our focus here is based upon a *momentary overwhelming experience* (i.e., an accident) that is often followed by a chronic stressor (recovery period). The stress effects of an accident overlap with the more chronic exposure. PTSD or its variants may persist indefinitely even without somatic injury. Additional significant stress (emotional, adaptive, or medical) is often initiated by the consequences of the *primary stressor* (i.e., an accident), so that a *secondary stressor* exists that is persistent, increases discomfort, and presents a health hazard to the patient. The combined concussive stress response can persist for years and involves a variety of physiological functions, subjective reactions, and overt behavioral responses and restitutive maneuvers.

The *initial, immediate response* is mediated by corticotrophin-releasing hormone-1 (CRH-1), which organizes the behavioral, sympathetic nervous system (SNS), and HPA responses to a stressor. The immediate stress reaction is counterbalanced by hypothalamic adrenal-axis (HPA) activity and stress-induced elevation in the circulating levels of glucocorticoids secreted by the adrenal cortex. Corticosteroids are conceptualized as *glucocorticoids* (cortisol and corticosterone) and *mineralocorticoids* (aldosterone). In PTSD there are increased cerebral spinal fluid (CSF) concentrations of corticotropin releasing factor (CRF) and somatostatin (Bremner et al., 1997). A *slower mode* facilitates behavioral adaptation, promotes recovery, and reestablishes homeostasis. It is coordinated by a CRH-2 receptor system, which facilitates the recovery of homeostasis. Cortisol and corticosterone serve in both systems. They determine the *threshold* or sensitivity of the fast-responding mode. In high concentration, these same hormones facilitate *termination of the stress response*. Their reactions are mediated in the brain by two receptor types. The first, mineralocorticoid *receptors* (*MR*), maintain neuronal homeostasis and limit the disturbance by stress. MR appear to activate signaling pathways that prevent the disturbance of homeostasis by facilitating the selection of an appropriate coping strategy. The second receptor type, *glucocorticoid receptors* (*GR*), help a victim to recover from a stress challenge and store the experience for future situations. These receptors activate signaling pathways aimed at facilitating behavioral adaption, with an adequate coping strategy stored for use at the next encounter. MR and GR are co-localized, particularly in limbic neurons. It is possible to block each type of receptor, differentially affecting HPA axis activity. Response to a novel situation was used to indicate stress. Limbic-cortical brain circuits that respond to a novel situation (attention, appraisal, fear, reward) abundantly express MR and GR.

POSTTRAUMATIC DISSOCIATION

Clinical example: Dissociation after motor vehicle accident (*MVA*): A woman with a graduate degree was in a two-car accident. There is evidence of considerable loss of mental ability and both short-term and long-term memory. The woman was the driver of a car that was struck by another car. She was seat-belted. Her car was struck in the left front. Both cars were travelling at city speeds. She felt a jerk. She did not anticipate the accident because she claims she was going with the traffic light. She does not remember hearing or smelling anything, but she saw the other car collide with her car. "I tried to come out of my car. I remember telling her that she ran the lights. When the two cars collided, for a moment I went into shock. I don't know what happened. I wanted to know if it really happened or if it is a dream. I didn't believe that it actually happened. I was going through a period of denial. I didn't believe it actually happened because I was so sure that I had the light. There was a change in my awareness. I feel that I was spaced out." She cannot state what she means by this. There is no claim of retroactive amnesia. She is uncertain as to whether there is a period that she does not remember. **What happened to your body?**: "I felt a jerk in my neck." She does not remember hitting her head, nor does she remember any of the bruises she had the next day.

Defining Characteristics

Dissociation is described as a disruption of the usually integrated functions of consciousness, memory, identity, or perception of the environment. The disturbance may be sudden or gradual, transient or chronic. Its presence worsens the patient's quality of life, in addition to the other effects of trauma. Similarly to the subjective and social consequences of partial seizures, the reaction by the person may raise questions in the subject's mind or collaterals as to his or her mental competence. The existence of this phenomenon will usually not be detected during a formal examination, but may be considered from an interview (e.g., after the question): "Are you a changed person after the accident (or other trauma)?"

Complexity

Dissociation is a complex and controversial phenomenon with implications involving the meaning of the Freudian entity of hysteria, more narrowly defined conditions (somatoform disorders, conversion, somatization), and dissociative phenomena such as amnesia, fugue, stupor, trance, possession disorder, movement and sensation disorders, and so forth. There are conceptual conflicts such as disagreements between the DSM-IV and the international classification of diseases (ICD) diagnostic nomenclatures, and psychodynamically and biologically oriented and also "mentalist" vs. behavioristically oriented clinicians. Dissociation includes autonomic, energy, and sensory phenomena, and can be elicited by medical, illicit drug use, and alcohol, as well as stress-related and other neurotransmitter phenomena (Lowenstein & Putnam, 2005).

Dissociation may occur as a component of immediate PTSD or as an aspect of partial seizures (Paraiso & Devinsky, 1997). It is more common during trauma than subsequently. There may be a sudden or gradual alteration in mental integration (i.e., identity, memory, or consciousness). Amnesia about the recollection of events can occur during the dissociative state, as can depersonalization, derealization, autoscopy, or personality changes. Capacity for dissociation occurs along a continuum and is enhanced by childhood trauma.

Dissociation as a Defensive Reaction to Trauma

There is an issue as to whether altered states of consciousness that follow concussions represent a psychogenic phenomenon with alternative discrete states of consciousness, or are consequent to cerebral dysfunction. Dissociative phenomena (e.g., altered sense of self and amnesia) should be differentiated from traumatically reduced arousal, which commences with head trauma and can sometimes persist for years. Sudden trauma violates basic assumptions and the sense of control, representing a sudden extreme discontinuity in a person's experience. Dissociation wards off ongoing trauma, making it a defense that insulates the person from overwhelming experiences that may generalize to maladaptive stress reactions. Defenses ward off wishes, fears, and memories, permitting the compartmentalization of perceptions and memories. Avoiding the memory of an event could avoid reexperiencing pain, fear and helplessness, although dissociation prevents working through of the memory later (Spiegel, 1991).

Numbing since the trauma is a dissociative symptom that overlaps with TBI. A posttraumatic study of earthquake victims indicated that more victims with TBI reported numbing since the trauma (Jones et al., 2005; citing Kato et al., 1996). The dissociative reaction may be described as more of a dreamy state than one of confusion, loss of comprehension about one's situation, and expressed fear. People with multiple personality disorder, a dissociative phenomenon, may have episodes of "time loss" or blackouts for complex activities, analogous to absence seizures.

When only mental symptoms are involved, the addition of somatic phenomena such as conversion and hysteria is controversial. Conversion phenomena may mimic somatic illness. Childhood trauma, psychological needs, conflicts, or an inability to cope can be central ("splitting of the ego"), although neurological and psychiatric pathology is also present. While dissociated affects, fantasies, memories, and so on are removed from consciousness, they may be later recalled or manifest ongoing effects, such as an ego-alien disturbance of sensorimotor function (Loewenstein, 1991; Nemiah, 1991). However, the association of *phobic anxiety* and *panic attacks* with *depersonalization* (Steinberg, 1991) is more common after head injury. Furthermore, the characteristics offered for psychogenic amnesia (PA), which include the inability to recall certain important information, overlaps with cerebral trauma (retroactive amnesia and PTA).

POSTTRAUMATIC DEPERSONALIZATION AND DEREALIZATION

These phenomena may be considered to be *disturbances of the self.* Defense against trauma is described as allowing the victim to avoid the reality of one's situation, or watching it as an observer by leaving one's body (Cozolino & Siegel, 2005). The fear that one is going crazy causes concealment for long periods before revealing it to a health care professional.

Depersonalization Definition

A feeling of detachment from, and being an outside observer, of one's mental processes or body, or feeling like an automaton or as if in a dream. The person feels detached or estranged from himself or herself., that is, an automaton or observer of their own mental processes, life, or body. The body is experienced as changed (e.g., larger or smaller).

Derealization Definition

The world appears strange, foreign, or dreamlike. The person is both observer and actor. This condition also occurs in schizophrenia and panic attacks, as well as dissociation.

The literature infers both an organic and psychodynamic origin. Depersonalization can be a rare manifestation of complex partial seizures (CPS), which are more often found in schizophrenia (DeLorenzo, 1991). Features that differentiate temporal lobe epilepsy (TLE) and a phobic-anxiety depersonalization syndrome are found in Trimble (1991), citing Harper & Roth (1962). The patient feels unreal, or sees the self outside of the self, experiencing emotional numbness and affective loss. It is characterized by an absence of feelings, a reduction of vividness, a sense of detachment, of being real or an automaton, and may be a state of low arousal. Sometimes the experience is pleasurable and can be followed by an appreciation of life. It may involve heightened arousal that is dissociated from consciousness (Steinberg, 1991). Psychogenic fugues can cause a prolonged altered responsiveness to the environment that mimics convulsive status epilepticus (DeLorenzo, 1991). Depersonalization can be found in a variety of conditions (including other dissociative disorders), and may or may not be a separate clinical entity in a particular patient. There are numerous theories concerning its etiology. It is most frequently associated with panic, but also with anxiety and/or agorophobia, depression, seizures (preictal, ictal, and postictal), substance abuse, organic illness, depression, and as a side effect of medication.

In 1986, Grigsby stated the difficulty of differentiating functional from neural etiology of depersonalization in cases of head trauma. More recently (personal communication), Grigsby stated that depersonalization reflects a physiological disturbance of the brain's dynamic equilibrium, the mechanism of which is likely to be a deficit in the integration of perceptual, visceral, affective, and cognitive information. This is a conceptualization of the evidence that depersonalization can result from the high level of arousal accompanying panic.

DEFINING QUALITIES OF PTSD

The acute or initial response to a trauma is characterized by a psychological reaction, frequently fear of death or injury, helplessness, humiliation, anxiety, dissociation, disorientation, confusion and usually an alteration of consciousness. PTSD may also be considered from a psychophysiological viewpoint (Orr & Kalouupek, 1997). However, psychophysiological measures, although they provide unique information, are not inherently more valid or more objective than a self-report during interviews. In fact, PTSD diagnosis is based upon subjective information that is not comparable to information recorded directly from physiological systems.

The definition of PTSD has been criticized for putting more emphasis upon the event than the vulnerability of the individual who seeks treatment. Distress is more associated with individual differences (e.g., trait emotionality, general negative affectivity [neuroticism], emotional expressiveness) than with event characteristics (Bowman, 1999). There are many practitioners who assert that the definition of PTSD should include remembering the trauma, which is an inconsistency when PTSD and concussion are stated to be comorbid. The nature of the sample shapes one's impressions and conclusions. The writer's own practice is focused upon lesser injuries, although the discussion represents a wide survey of students of this issue. In this writer's experience they represent entities that frequently coexist. Combat veterans who had incurred a *blast* with comorbid PTSD had more persistent symptoms than those with different types of trauma (Trudeau et al., 1998).

Both CBT and PTSD involve an event that is likely to be unusual and unexpected, and certainly threatening to bodily integrity, sense of identity, and life. Thus, the diagnostic process after an accident creating a head injury is complex, since one of the defining characteristics of CBT (PTA) is considered to be a negative marker for the alternative condition (i.e., PTSD [exposure and reactions to the traumatic event, and reexperiencing symptoms]).

"Mild" TBI seems to alter the course and nature of PTSD. Patients experience fewer posttraumatic stress symptoms after a MVA than non–head-injured patients. They displayed after 6 months more intrusive memories than the uninjured group (i.e., a course of posttraumatic adjustment different from other forms of PTSD) (Bryant et al., 2000). The pattern of posttraumatic adjustment differs: patients without TBI initially report more fear and intrusive memories than those with mild brain injury (MBI), with the difference disappearing 6 months later. Intrusive memories decreased in those with PTSD, while those with MBI displayed increased intrusive memories (Bryant et al., 2000).

Commonly associated with PTSD are depression, panic disorder, substance abuse, somatization disorder, and dissociative disorders. However, depression and panic disorder with PTSD may be neurobiologically distinct. Drug abuse or alcohol, if used as an avoidance reaction, can have an adverse adaptive effect (Rasmusson & Charney, 2000). Thus, when there is a complex presentation, the clinician must decide whether to assign one diagnosis or the other, or decide that they are comorbid with overlapping and superficially similar symptoms, thus comprising two diagnoses. Many treatment options should be considered for the chronic PTSD patient who may have complex comorbidity (Schnurr et al., 2007). The diagnostic decision is complex since neurological injury may cause ambiguous symptoms, patients underreport some symptoms, in part to an avoidant coping style, and do not attribute some PTSD symptoms to the trauma.

Incidence of PTSD

Civilian: One sample suggested that the overall risk of PTSD following exposure to civilian stress was 9.2%, about one-third less than other studies that had focused on the worst traumas. A serious car crash was experienced by 8.2% of the sample, with the risk of PTSD being 2.3% and that for "other serious accident" being 16.8%. Being badly beaten up (presumably with risk of TBI) had an incidence of 3.9% and PTSD of 31.9% expression. Consistent with other reports, women had a twofold higher risk of incurring PTSD than men after a similar event (Blanchard et al., 1997). Within 1–4 months after an MVA, 39% met the DSM-III-R criteria for PTSD (Breslau et al., 1998). A study was undertaken of 119 adolescents, ages 8–16, who were involved in MVA (Mirza et al., 1998). The involvement of a parent in the same car accident was associated with severe scores on the Frederick's Reaction Index at 6 weeks but not at 6 months. Forty-five percent had PTSD symptoms 6 weeks following a relatively minor injury. Fourteen of 19 categorized with a severe level of PTSD symptoms at 6 weeks were categorized as having severe symptoms at 6 months. There were a variety of comorbid psychiatric disorders.

Military: In Iraq, a Department of Defense screening identified 20.3–42.4% of soldiers requiring mental health treatment. The active duty component (as opposed to the Reserve and National Guard) who were assessed with mental health risk were nearly 3 times as likely to use services within 90 days than those without a mental health risk. Nine percent of this sample were women. In comparing two PTSD questionnaire assessments of the same troops, half of those reporting symptoms on the first survey reported symptomatic improvement on the second, while twice as many were identified on the second who did not have an initial high PTSD score (Milliken et al., 2007).

THE RANGE OF PTSD SYMPTOMS

PTSD and simple phobia appear to have fear networks that are highly specific to individuals and easily activated.

The Stressful Event: Criterion A

What constitutes a "stressor" may depend upon how the brain filters and interprets it (Ursin, 1998). This is an issue of vulnerability to stress. The stressor may be a physical or an emotional trauma.

The effects are so extreme that the victim cannot cope with them, nor resume the normal level and quality of adaptation. One is unable repair the tissue damage, gain control over physiological disorder, or relieve the behavioral impairment and discomfort. There are a wide variety of symptoms that fluctuate.

The examiner differentiates between the environmental or bodily event that is a trauma and the stress reaction that is its result (DSM-IV-TR, p. 467):

1. The person experienced, witnessed, or was confronted with an event or events that involved actual or threatened death or serious injury or a threat to the physical integrity of self or others.
2. The person's response involved intense fear, helplessness, or horror. (In children, this may be expressed instead by disorganized or agitated.)

The diagnosis of PTSD is only made when the stressor satisfies criterion A, although symptoms may follow that are relatively low magnitude stressors. It is cautioned that relatively minor trauma should not be classified as stressors. The dose-response relationship between trauma and PTSD is still imprecise. The examiner considers the impact of trauma over the lifespan (Weathers & Keane, 2007).

"Distracting" symptoms: The symptomatic consequences of unhealed somatic injury can contribute to *persistent posttraumatic stress reactions* (*PPSR*): hyperarousal, emotional distress, pain, depression, headaches, balance problems, restricted range of motion, changes of identity, partial and generalized seizures, and signs of hyperarousal (Parker, 1995; Steinberg, 2000). These serve as "*distractors*" that affect adaptive efficiency and quality of life. In a study of motor vehicle accidents (Parker & Rosenblum, 1996), in addition to TBI, the following symptoms and comorbid diagnoses were present, most of which correspond to DSM-3R: depression (12); PTSD (16); conversion reaction (1); affective disorder (2); mixed neurotic reaction (1); anxiety reaction (1); and cerebral personality disorders, persistent altered consciousness, psychodynamic reactions to impairment, complex reactions expressing neurological, somatic, and psychological dysfunctions (sexuality and somatization). Thirty of thirty-three subjects reported headaches at some time (one migraine). Of the three who did not report headaches, one claimed neck and back pain, and the other two seemed to be pain free. Pain affects attentional capacity, processing speed, and psychomotor speed. It is also associated with such stressors as mood change, somatic awareness, sleep disturbance, and fatigue (Machulda et al., 1998). Yet, a rating of PTSD positive or negative was not associated with the status of chronic pain as present or absent (Coffey et al., 2006).

HYPERAROUSAL

Autonomic Responses

Autonomic responses to repeated presentations of intense stimuli are greater in individuals with PTSD. In chronic cases, SNS activity such as blood pressure and heart rate present persistent increases, with an increased risk of hypertension and cardiovascular disease. Skin conductance is a useful measure for emotional arousal. Heart rate seems to be a better indicator of active avoidance behavior, whereas skin conductance points to active inhibition. Physiological reactivity is affected by race, sex, age, menstrual cycle, and physical fitness. One-third to one-quarter of individuals diagnosed with PTSD are physiologically nonreactive. Patients frequently report rapid heartbeats. *Cardiovascular reactivity* displays higher or lower values, compared to baseline, when encountering stimuli or situations experienced as engaging, challenging, or aversive. The possible etiological reaction is that repeated elicitation of the defense reaction (enhanced blood pressure and heart rate) translates into vascular hypertrophy, causing hypertension in genetically susceptible individuals. Risk factors include male gender, family history of coronary heart disease, borderline hypertension, and type A behavior (Georgiades & Fredrikson, 2000). Veterans with PTSD show heightened noradrenergic functioning. Medications that decrease noradrenergic transmission (e.g., propranolol, clonidine) greatly ameliorate PTSD symptoms (Kandel, 2000e). The etiology of hypertension is usually multifactorial. It is associated with repeated or sustained stress, not acute stressors (Steptoe, 2000). Heart rates of 80–90 beats/min or higher are sometimes used as a criteria for hyperarousal.

Sleep Disturbances

Comorbid PTSD will augment sleep disorders (Friedman & McEwen, 2004) such as traumatic nightmares, insomnia, increased awakenings, reduced sleep time, and increased motor activity. Traumatic nightmares differ from stage IV night terrors and rapid eye movement (REM) dream anxiety attacks. Persistent insomnia after an emotionally stressful event may persist due to the patient's hyperarousal caused by persistent efforts to sleep (i.e., the sleep environment becomes a condition). Persistent insomnia can lead to impaired daytime function, accidents, major depression, and increased utilization of health care resources (Czeisler et al., 2005). Sleep disturbances (SD) in older individuals are associated with stress-related HPA axis activation (Penedo & Dahn, 2005).

Measures of sleep initiation and maintenance do not distinguish from those who are developing PTSD soon after trauma. In fact, expectation of disturbed sleeps affects sleep phenomena. Chronic PTSD is associated with the disruption of REM sleep continuity (awakening and arousal). One study did not show a relationship between PTSD symptoms and the timing of onset of or the amount of REM sleep. On the other hand, stress and sleep deprivation seem to exacerbate sleep paralysis, night terrors, and sleep-related panic attacks (Hirschkowitz & Moore, 2000). Cognitive activation during REM sleep may be adaptive during the aftermath of trauma. Thus, targeting REM sleep disruption may be an important goal for treating sleep disturbance after trauma (Mellman et al., 2007).

SDs are a significant component of PTSD. Increased PTSD severity is associated with more severe SD. Sleep quality may be influenced by such patient- and disorder-related characteristics as age, gender, type of trauma, PTSD chronicity and severity, and psychiatric comorbidity. More severe core clinical features of PTSD often complicate its outcome. Increased attention to this symptom may facilitate efficient intervention strategies (Germaine et al., 2004). However, in one sample of traumatized individuals (noninjury), sleep disorder severity did not differ according to gender, age groups, type of trauma, PTSD chronicity, or psychiatric comorbidity.

Localization of dream components has been suggested by functional neuroimaging, such as perceptual aspects (occipital and temporal cortices). Patients with occipitotemporal lesions may report cessation of visual dreams; emotional features (activation of amydalar complexes, orbitofrontal cortex, anterior cingulate cortex); memory content (mesio-temporal areas); alteration of logical reasoning, working memory, episodic memory, and executive functions (relative hypoactivation of the prefrontal cortex) (Autret et al., 2001).

Night Terrors (Sleep Terrors), Pavor Nocturnus

These parasomnias are brief nondream, non-REM experiences usually arising during slow-wave sleep (SWS) and involve awakening with an intense fear, a sharp piercing scream, tachycardia, tachypnea, sweating, and an increased muscle tone. The person is usually unresponsive, confused, and disoriented during the episode, and may be amnesic about it the next morning (Hirshkowitz & Moore, 2000). Sleep terrors are associated with daytime stress, the use of sedatives and hypnotics (and other predisposing factors), and with the greatest change of arousal state (i.e., from SWS to awakening). Thus, they are change of state dependent . They also occur during nonrapid eye movement (NREM) sleep. The sleeper awakes with a scream and appearance of terror, cannot be consoled for some time, and recollects difficulties in breathing and moving (feels "locked in"), but has none or minimal recollection of mental activity. Sleep terrors decline with age (along with delta sleep, and associated delta sleep conditions such as enuresis and somnambulism) and can be suppressed by benzodiazepines, which suppress delta sleep. In contrast, REM and associated dreams neither decline nor are suppressed by the same drugs (Broughton, 1989; Hirshkowitz & Moore, 2000; Kelly, 1991a).

Nightmares and Bad Dreams

Nightmares may be defined as long, frightening dreams that awaken a sleeper from REM sleep, or at least awaken the sleeper. They have been observed 40 years after trauma in a population of Dutch war veterans. Nightmares are distinguished from night terrors and sleep paralysis with hypnagogic (transition from wakefulness to sleep) or hypnopompic (transition from sleep to wakefulness) hallucinations (Hirshkowitz & Moore, 2000). These awakenings are accompanied by panic, dysphoria, or anxiety. They seem to be disorders of arousal from SWS, and are similar to sleepwalking and confusional arousals. Subjectively, these can be differentiated by a question as to whether a dream has occurred. Other types of awakening are sleep-related panic attacks and sleep paralysis associated with hypnagogic or hypnopompic hallucinations. A dream's structure implies imagery, narrative, and plot. The nightmare may be functional insofar as it is the representation of emotion with the gradual making of connections with other life events. The frequency of psychopathology is greater in individuals suffering from chronic nightmares. Nightmares usually resolve, but persistent nightmares with recurrent dream content can be a marker for significant psychopathology. Treatment is indicated when they produce serious distress, insomnia, or interference with daily activities. Sleep deprivation exacerbates nightmares, night terrors, and sleep paralysis attacks, while insomnia resulting from fear of sleep can pose a serious obstacle to treatment. In addition to trauma, it may be elicited by specific medications (e.g., those that potentiate REM sleep and others that produce REM sleep on withdrawal) (reserpine, thiothixene, alpha methyldopa, propranolol, and levo). Drug abuse and drug or alcohol withdrawal also enhance nightmares. Nightmares are more common with certain traits, such as thin personality boundaries, genetic factors, and early childhood experiences.

It has been asserted that the nature of *PTSD nightmares* is poorly understood. They are reported in 92% of PTSD patients with flashbacks, but only 57% of PTSD patients without flashbacks. While they resemble REM sleep nightmares (have a high REM density [eye movements per minute of REM sleep]), they occur in all stages of sleep, including sleep onset. They contain imagery, narrative, and plot, but are unusual for their repetitive, recurring nature, often without elaboration or symbolism. REM sleep disturbance may contribute to the pathogenesis of PTSD. Combat veterans with PTSD appear to have elevated REM sleep percentage, longer average REM cycle duration, more leg muscle activity during REM sleep (Hartmann, 2003; Hirshkowitz & Moore, 2000).

HYPERAROUSAL: REEXPERIENCING AND INTRUSIVE IMAGES

Some alterations of consciousness are common in PCS and in PTSD.

Flashbacks

A flashback is defined as the involuntary recurrence of some aspect of a hallucinatory experience or a perceptual distortion some time after taking a hallucinogen (Fischer, 1986). When they occur after a TBI, one cannot be certain whether this is the consequence of a temporary high level of neurotransmitters related to the TBI or part of a different process (i.e., PTSD). The formal criteria for a reexperienced event are recurrent, intrusive recollections, distressing dreams, flashbacks/dissociative experiences, distress at exposure to symbolic cues, physiological reactivity to symbolic cues.

Failure to develop a coping strategy is expected to result in stimulus generalization and consequently susceptibility to "flashbacks" (Deutch & Young, 1995). Flashbacks are common occurrences after trauma but they also occur as a normal experience. In addition, they are common after accidents, perhaps being caused by the profound neurotransmitter effects. They are also state-related (i.e., evoked by imagery, melodies, and symbols of the content of an experience). A woman who was struck by a car offered this example of a flashback: "I can see the car coming and myself flying."

One characteristic of flashbacks is considered to be reexperiencing one of more sensory modalities (Krystal et al., 1995). Repeated conditioned or triggered flashbacks may acquire a kindling capacity, leading to their occurrence in the absence of environmental triggering. Dissociation may be comorbid with concussion. The correct identification of dissociation is significant since the presence and length of the PTA interval is considered to be a measure of TBI intensity. Therefore, confusion of a different phenomena for a symptom based upon neurotrauma would lead to incorrect diagnosis, assessment of performance, and possibly incorrect treatment. Dissociative amnesia (DA) may occur in three forms: generalized (patients have no recall of any events in their lives); systematized (certain kinds of events are forgotten); and circumscribed (an island of amnesia creates a gap in the patient's historical memory). DA can be differentiated from CPS since the episode is triggered by a particular emotional stress. The memory of the episode may also be enhanced as opposed to permanent loss of the episode.

Some flashbacks are state related (i.e., evoked by imagery, melodies and symbols of the content of an experience, or emotional needs and available information). State-relatedness accounts for the amnesia between different states of arousal. These phenomena are accompanied by expected levels of anxiety. They may be subdivided into types: involuntary recollections of a recalled event; pseudomemories of events; and imagery that involves reexperiencing an event (Bryant, 1996). Flashbacks are not necessarily the reestablishment of a single arousing experience; they are considered by some to be a product of repetition of a particular mental state. Repeated conditioned or triggered flashbacks may acquire a kindling capacity, which can lead to their occurrence in the absence of environmental triggering. Failure to develop a coping strategy is expected to result in stimulus generalization and consequently susceptibility to flashbacks (Deutch & Young, 1995). One characteristic of flashbacks is considered to be reexperiencing one or more sensory modalities (Krystal et al., 1995). Thus, flashbacks have a varied etiology; they need not be an actual memory of the accident.

- A normal experience.
- The involuntary recurrence of some aspect of a hallucinatory experience, or a perceptual distortion some time after taking a hallucinogen (Fischer, 1986).
- Flashbacks may occur as a *dissociative state* (America Psychiatric Association, 2000, p. 464).
- Persistent intrusive anxiety is associated with physiological imprinting at the time of injury or postinjury state (nightmares and flashbacks).
- Reexperiencing a trauma ("loss of distance") is associated with dissociative disorders and multiple personality disorder. The intrusive effects of anxiety, perhaps controlled by dissociative mechanisms, can interfere with cognitive function (Armstrong & Loewenstein, 1990).
- Directly reexperiencing a trauma, a very complex condition that is common after accident. A woman who was struck by a car offered: "I can see the car coming and myself flying."
- A personally meaningful postaccident reconstruction, perhaps from photographs or the events as described by others.
- An image partially shaped by an ongoing mental state.

Arousal and Affect

It is subject to question whether reexperiencing the trauma (recollections, dreams, flashbacks, illusions, hallucinations) should be considered to be memories or affective disorders. Physiological reactivity on being exposed to internal or external cues that resemble or symbolize the event would be better classified under *increased arousal.* In any event, the repetition of a flashback, with its affective and behavioral disturbance, can progress in a kindling-like phenomenon to autonomous occurrences (Post et al., 1999). Flashbacks are state related (i.e., evoked by imagery, melodies, and symbols of the content of an experience). Psychoactive substances or hypnosis induce the level of arousal that prevails during the initial experience. Psychoactive substances (e.g., alcohol and LSD) can initiate flashback experiences, which can be stimulated by psychoactive substances different from the one that imprinted them. State-relatedness accounts for the amnesia between different states of arousal.

Using 123 patients who contacted a posttraumatic stress unit after an MVA (but excluding those with head injury), it was determined that patients with low anxiety (not meeting PTSD criteria) had more vivid visual imagery than those patients with PTSD and specific phobia subjects (i.e., ability to experience visual imagery diminished as anxiety increased). However, in the separate PTSD group, visual imagery ability correlated with flashback and nightmare frequency. Thus, these intrusive symptoms might reflect a preexisting skill rather than be reflective primarily of fear (Bryant & Harvey, 1996).

Physiological Contribution to Flashbacks

After a "biologically important event... all recently active circuits may be 'printed'" by neurotransmitters and allied substances (Livingston, 1985, p. 1270). This creates sensitization to dangerous stimuli or generalizations that are reminiscent of them, repeating the trauma and creating new synaptic connections that fixate it (van der Kolk, 1988). Intrusive memories have the characteristics of early memories (i.e., without symbolic and linguistic representations or an autobiographical context). They may be considered to be primitive thinking. Psychoactive substances or hypnosis induce the level of arousal prevailing during the initial experience. Psychoactive substances (e.g., alcohol and LSD) can initiate flashback experiences. These can be stimulated by different psychoactive substances than the ones that imprinted a flashback experience. Fischer (1986) suggests that flashbacks are based on these neuroanatomical considerations, such as linkage between a sensory modality and the limbic system and cross-modal associations occurring in the inferior parietal lobe (junction of the angular and supramarginal gyri) that enable and enhance cross-modal associations in perception, thinking, and language.

Traumatic Memories

Traumatic memories may be relatively indelible (overconsolidation of memory) due to the overstimulation of stress-response hormones and neuromodulators of thalamo-amygdala pathways. Glucocorticoids enhance memory consolidation for emotionally aversive experiences in animals and healthy humans (Schelling et al., 2004). However, increased serum levels of cortisol during the acute stress experience may be protective with regards to later development of PTSD. Stress levels of hydrocortisone in patients with traumatic memories may prevent chronic stress symptoms by interfering with other aspects of memory. Stress is associated with the inhibition of new neuronal growth in the hippocampus, which is reversible with *selective serotonin reuptake inhibitors* (*SSRI*) and other medications. PTSD of various etiologies and gender differences are accompanied by reduced hippocampal volume. Treatment was associated with improved declarative memory and hippocampal volume (Bremner & Vermetten, 2004).

ADDITIONAL STRESS COMPONENTS

Identity

The personality dynamic has been well studied by the *Rorschach inkblot procedure* for this and other stress-related characteristics. Symbolic response groups and single examples are offered: under the control of greater forces; ongoing victimization or personal destruction; vulnerable to further injury; trapped; feels helpless or powerless; under the control of mysterious forces; defeated ("the battle is lost"); loss of self-control and sense of integration; bodily damage; mutilated; internal anatomy unattractive; reduced self-esteem; still struggling; hampered due to injuries; loss of vitality including reference to death; cry for help.

Weltanschauung (World View)

The meaning of life; feels one's environment is unstable or destroyed; loss of stability; manmade dangerous events; overwhelming natural forces; feels the world is unsupportive or unfriendly.

Motor

Brake or tension: The familiar canon that the stressed individual responds with a "*fight or flight*" is incomplete. It was observed decades ago that there is a more common alternate response for civilized life (i.e., "brake!"). When we are hyperaroused, our "motor" may be running but our foot is on the "brake," since we do not know how to cope with the stressor. Clinically, we would expect the condition of *tension* (Parker, 1982, pp. 55–56).

Tonic immobility: Inability to move is an unlearned response observed in animals (regardless of their opportunity to do so), although this has been reported in rape, plane crashes, and so on ("scared stiff"; "frozen with fear"; "limp"; "faint, trembling and cold"). It is differentiated from *learned helplessness*, a learned response that usually requires several trials of inescapable aversive stimulation. From an evolutionary perspective, when confronted with danger, all resources are concentrated on the immediate danger, and the repertoire of reactions is limited to that which has evolved for protection. The cognitive component is not considered to be a form of peritraumatic dissociation, although both are considered to be a result of extreme fear. It may be associated with undisrupted or enhanced event-related memory, consciousness, and learning. Some degree of perceived or real restraint is consistent with anecdotal reports (e.g., sexual assault and plane crashes) (Marx et al., 2008; Zoellner, 2008).

Health Disorders

TBI and bodily injury create persistent stress that may have secondary health and behavioral consequences. Physiological dysregulation of the autonomic and other organic systems, disturbed sleep, anxiety, "distracting" symptoms, and so on interfere with physiological functions (e.g., the circadian rhythms) and emotional well-being. Stressful experiences affect a broad range of physiological functions while enhancing the likelihood of many diseases (Cooper, 1996; Friedman et al., 1996). In the domain of medical conditions involved with stress, as in psychiatric conditions, study is rendered difficult by questions of definition, comorbidity, and inadequate tools for mental and physiological measurement. Stressful life events create a risk for a variety of illnesses, including infectious disease, although the correlations account for only about 10% of the variance (Kiecolt-Glaser & Glaser, 1991). The variety of neurobehavioral disorders after stress can be understood in part as a consequence of the wide range of physiological dysfunctions occurring in the PTSD (Southwick et al., 1992). Negative health effects of TBI have multiple origins, including stress-related damage to the immune system, other CNS autonomic effects, direct consequences of trauma, and deteriorating health due to reduced nutrition and access to health care. The *stress process* has been described as the stressors (environmental events), the physiologic response to the stressors, and the health consequences. A wide variety of health conditions are associated with family life events (children and adults).

Reaction to impairment (including the social context): motivation to struggle, to give up, to cheat; to play the sick role; to recover;

"Distracting" symptoms: pain; restricted movement; seizures; imbalance; anxiety and depression; weakness and loss of stamina; illness;

Reaction to posttraumatic lifestyle: motivation to struggle, to give up, to cheat; to play the sick role; to recover.

PTSD AFTER LOSS OF CONSCIOUSNESS

Since PTSD is defined as involving an unusual event, lack of memory would appear to presuppose that there cannot be PTSD by definition, as though the event were not noteworthy. Some authors assert that PTSD is "rarely diagnosed in patients with head injuries who do not recall their trauma." The controversy as to whether the PCS and PTSD can be comorbid arises from the observation that CBT is usually associated with PTA or at least clouded consciousness concerning the period of the accident, while the PTSD definition specifies retention of the memory of the trauma (American Psychiatric Association, 2000, pp. 463–468). The conclusion of various studies that PTSD does not occur after minor TBI was questioned on procedural grounds by Bryant (2001).

CLINICAL EXAMPLES OF POST-TBI PTSD

Vignettes: Waking Up Confused in a Hospital

These posttraumatic experiences are vivid enough to be reasonable precipitating events.

Didn't know what happened: A woman was unconscious for 3 days after an MVA. "I kept hearing voices. I wanted to tell my mother that I was all right but I couldn't. I couldn't figure out where I was. I had the respirator on, and a neck brace. It was scary and confusing. When I first woke up I didn't know what happened." She had no pain on awakening. "My mother kept on repeating over and over that there was a car accident and I was in a coma. She remained in the hospital for 15 days."

Patchy awareness in a hospital: A youth was hospitalized after a head injury. He drifted in and out of consciousness. He woke up at 6–7 one night. "Before that I saw my friends. I said my head hurt. I said 'hi' to my friends, and then I went back to sleep. It was as though I was dreaming." **After waking up**: "I didn't know what I was doing there. I felt scared. I felt bad. I didn't know what happened to me. I had a fever." **How long before you felt like yourself?**: A week after he woke up he felt all right. "I was still in a daze. I felt weird when I left the hospital. The sun and the street felt as though I was never outside before."

DYNAMICS OF DELAYED POSTTRAUMATIC ACCIDENT MEMORY

If one extends the range of frightening experiences to those that occur shortly after the immediate accident, a subset of persons experience PTSD in the context of PTA. While the definition of PTSD specifies memory of a traumatic event, the perceived traumatic event can be after the actual accident. The emotional trauma may occur after the mechanical event of the accident. After TBI, the frightening experiences of the accident may or may not be remembered—a significant exception to the classical definition of PTSD.

- The earliest posttraumatic memory upon awaking from a coma can be frightening (waking up on a road after a car crash; waking up confused and in pain in a hospital with physicians around and tubes coming out of one's body). Imagination or gory stories from collaterals about the accident may create new images.
- Terror may occur in an "*island of memory*," intruding between amnestic periods due to hyperarousal, creating altered consciousness or coma (King, 1997).

- Reconstruction of the event may be consciously or unconsciously determined by what the patient has been told as to what happened, photographs, police reports, and the patient's information on whether a person or other condition was responsible for the accident.
- Emergency personnel at the scene of the accident or at the hospital can erroneously classify a patient: they were not present immediately after the accident when the most intense altered consciousness may have been presented, or a patient in PTA might appear to be normally responsive, and thus falsely classified as not manifesting altered consciousness.

Epidemiology of Post-TBI PTSD

In the writer's experience, PTSD is common after an accident. Of 33 consecutive victims in a private office after a motor vehicle accident (most had cases for litigation, workers compensation, or no fault pending), 16 were comorbid for PCS and PTSD (Parker & Rosenblum, 1996). Other surveys are consistent with this view. Hickling et al. (1992) and Harvey et al. (2003) cite studies in which the coexistence of TBI and PTSD ranged from 0% to 48%; Bryant cites studies in which 33% and 40% of patients experienced PTSD after MVA. Fewer posttraumatic stress symptoms occur in mild head-injured patients than in non–head-injured patients. The prevalence of more fear and intrusive memories in PTSD alone in the first month decreased, while MBI displayed increased intrusive memories. This suggests that the course of MBI adjustment differs from other forms of PTSD.

Loss of Memory for the Accident

There are several possible attributions, including LOC, altered consciousness, PTA, or dissociation as a protective psychological mechanism. In a series of patients rated as having serious brain injury (coma lasting more than 6 h; PTA of one day or more), there was no relationship between length of disturbance of consciousness and PTSD score (Williams et al., 2001). There is a controversy as to whether PTSD can occur after an accident in which there is PTA (i.e., lack of memory for the TBI-causing event). Both entities involve an accident initiating a *complex physiological acute stress reaction*, *altered cerebral arousal* (perhaps lesser and greater than the baseline), and a *traumatic affective state.* There are complexities and perplexities involving the effects of stress upon consciousness (e.g., issues involving amnesia and dissociation). It is difficult to differentiate short traumatic impairment of consciousness from dissociative reactions or very short PTA (Radanov et al., 1992).

Limited Memory of the Event

This writer raises the question as to which experiences are contributory to the PTSD syndrome. Are they the memory of the fear and injury of the accident, or are they the stress of the recovery period, including perplexity as to what happened? Or imagination or gory stories about the accident that create new images? It is the writer's experience that PTSD (which is defined to included an event that is either experienced/witnessed and involving death or serious injury, or a response of fear, helplessness, or horror) can occur without any memory of the actual event. When a person awakens from accident in a hospital bed, not knowing who they are, where they are, what happened, why the tubes are coming out of their body, why there is pain, or even why there has been a loss of limb, there is a feeling of loss of control, as well as disorders of identity and memory (Kohl, 1984). Referring to the controversy as to whether amnesia (or LOC) precludes the possibility of PTSD, Watson (1990) cites two cases featuring patchy awareness of the accident and being in the emergency room. PTSD in the sense of reexperiencing the event was limited to reexperiencing specific pain and other symptoms.

Patients who had suffered a TBI were studied according to whether they manifested PTSD and the extent of their memories. PTSD was observed in 27% of the subsample who were not unconscious for an extended period, but was observed in only 3% of those unconscious for more than 12 h. Another study concluded that PTSD from a head impact was unlikely after an extended period of unconsciousness. It was recorded as to whether there were traumatic or bad memories, or no memories. Having a head injury with some amnesia about the event is not completely protective against development of PTSD. Only 15% of identified TBI victims in an emergency room sample responded to a questionnaire as having a high proportion of severe or very severe TBI. Those with *traumatic memories* had a higher ratings of PTSD symptoms. Lower levels of psychological distress were reported by those with *no memories* of the accident. The lowest reported symptoms were those experiencing distressing dreams, flashbacks, avoidance, and hypervigilance. Those with *undramatic memories* reported the least psychological distress and did not meet the criteria for PTSD. Comorbid traumatic stress symptoms occurred with both anxiety and depression. Higher ratings of the effect of *physical injuries* on functioning were related to more psychological distress (Turnbull et al., 2001).

Implicit Memory (PTSD with No Images)

Reexperiencing symptoms (flashbacks; feelings of distress) can occur without verbal or visual memories of the event. Fear conditioning centered in the limbic regions may be consequent to excessive noradrenergic activity, resulting in long-term enhancement of arousal. A soldier's experiences after being knocked unconscious by an improvised explosive device (IED) were described as patched together from his own memories and from what others told him later (Okie, 2005). Krikorian and Layton (1998) discuss a worker who was buried completely in sand in a construction accident. He was unconscious on admission to the hospital and comatose for 2 days. The period of exposure to grievous trauma before succumbing to anoxia exceeded the interval usually occurring in mechanical injuries. Later he developed cognitive problems and a variety of symptoms that fulfilled the criteria for PTSD. This was explained as being dependent upon implicit memory (i.e., mental representation of trauma can be encoded without consciousness). Perhaps there was participation of a *nondeclarative memory system* independent of episodic memory that was rendered dysfunctional temporarily as a result of anoxia but functioned subsequently.

Event Memory Cues

- Information received from family and professionals may be the frightening stimuli initiating PTSD.
- *Priming* makes the retrieval of explicit memory for episodes unnecessary. Later, a cue facilitates the subsequent identification of the previously unsalient stimuli that occurred after the initial trauma (Schachter et al., 1993).
- *Conditioning* can trigger stimuli that arouse a complex fear network, leading to intrusive reexperiencing of the trauma (Joseph & Masterson, 1999).
- *Reduced intellectual effectiveness*: reduced concentration due to anxiety and avoidance of a variety of activities impairs new learning and problem solving.

EMOTIONAL EXPERIENCES OF TRAUMA

Meaning of the Event

The patient experiences a profound ultimate stress (fear of damage to the brain and loss of competence, bodily and social attractiveness). The clinician may ask whether the personality consequences of a trauma derive from earlier meanings. To some extent this will depend upon the trauma and the personal history. Victims of *rape*, particularly those who reacted with *TI* and who received poor social support, are at greater risk for more severe PTSD symptomatology (Marx et al., 2008). However, on the basis of my clinical experience with *head injury victims*, utilizing intensive interviews and studying Rorschach symbolism, earlier psychodynamically significant experiences seem not to shape the meaning of the trauma. The experience of an accident possibly arises anew. The range of stressors is so great that one may not assume much common etiology in varied traumatic experiences, leading to the somewhat imprecise syndrome of PTSD. Even the universal assumptions of intense emotion as etiology is not established and may be related to events with prolonged trauma (Harvey et al., 2003).

Avoidance

Avoiding memory of a traumatic event could avoid reexperiencing pain, fear, and helplessness, although *dissociation* prevents working through that memory later (Spiegel, 1991). Parts of the past, present, and future, as well as parts of the self become disconnected. Fear-conditioned behavior (avoidance and arousal) may become uncoupled from the conscious memory, permitting the development of aversive emotional memories from circuits independent of those that involve explicit, declarative memory (Warden et al., 1997). Experience becomes fragmented, with deleterious effects upon the capacity to trust oneself and others (Hegeman, 1998). Nonassociation of symptoms with a TBI is supported by (1) the normal tendency not to connect terror with ongoing disruptions, and (2) professional failure to identify TBI acutely or later (Parker, 1995).

The long-term persistence of avoidance may be explained by a two-factor learning theory (Kolb, 1988):

1. *Conditioning*: Situations reminiscent of the original trauma result in a startle response (i.e., hypersensitivity of the sympathetic nervous system). It is acknowledged that learning theory is insufficient to account for all of the physiological disturbances.
2. *Instrumental learning*: The individual learns to avoid cues that arouse anxiety.

Dissociation

An TBI-causing accident is a shocking or overwhelming event. Sudden trauma violates basic assumptions and one's sense of control, creating a sudden extreme discontinuity in experience. The overall effect involves compartmentalization of perceptions and memories. The discontinuity of experience involves not only the extraordinary event, but also the psychodynamic mechanism of dissociation. Moreover, symptoms characteristic of concussion are also associated with dissociation found during PTSD in the emergency, acute, and chronic periods. Dissociation wards off ongoing trauma; it changes experience by disturbing the normal integration of awareness when thoughts, feelings, memory, or actions may be too painful and affect out of awareness. Coping may benefit from minimizing the trauma's importance; the disconnection between the event and the person's feeling of well-being. Nonassociation of symptoms with a TBI is enhanced by two factors: (1) the normal tendency not to connect terror with ongoing disruptions, and (2) professional failure to identify TBI acutely or later (Parker, 1995).

Dissociation has been described as the separation of a group of "personalities." There is an alternating pattern between the mental units described below:

- The *emotional personality* (*EP*) is fixated on the trauma.
- The *apparently normal personality* (*ANP*) is detached from trauma, experiences some amnesia or depersonalization from some components of the experience, and is dedicated to fulfilling functions of daily life.

An individual can have more than one EP and ANP. The extent of fragmentation depends upon the severity of traumatization in terms of developmental age, the chronicity and intensity of the traumatization, the relationship to the perpetrator, and the amount of support for and social recognition of the trauma. EP and ANP are considered to be separate systems, each with its own sense of self (Nijenhuis et al., 2002).

Dissociation associated with PTSD may be protective, perhaps related to the retroactive amnesia, confusion, and PTA that occur after head trauma. The nature of one's experience is changed because the normal integration of awareness with thoughts, feelings, memory, or actions is disturbed. Here, it serves a psychodynamic defensive purpose by keeping painful affects out of awareness. To the extent that coping involves minimization of the trauma's importance, it leads to a disconnection between an event and disruption of the person's feeling of well-being. Sudden trauma violates one's basic assumptions and the sense of control, creating sudden and extreme discontinuity in one's experience. Dissociation wards off ongoing trauma. Defenses ward off wishes, fears, and memories. The overall effect permits the compartmentalization of perceptions and memories.

Although by avoiding the memory of an event, one can avoid reexperiencing its pain, fear and sense of helplessness, dissociation also prevents working through of the memory later on (Spiegel, 1991). Parts of the past, present, and future, and also parts of the self, become disconnected. Fear-conditioned behavior (avoidance and arousal) may uncouple from the conscious memory, allowing for the development of aversive emotional memories from circuits independent of those involving explicit, declarative memory (Warden et al., 1997). Experience becomes fragmented, with deleterious effects upon the capacity to trust oneself and others (Hegeman, 1998).

Response Styles

Accident and stress victims have various coping styles, which may be taken into account in assessing outcome and planning treatment. These may be characterized through direct observation, intensive interviewing of the patient and collaterals, use of projective and psychometric personality tests, employment records, and so on. It is presumed that these develop psychodynamically before an accident and influence adaptive capacity after injury: withdrawing/detachment; conspicuous risk taking; engagement (fight/approach); disengagement (flight/peacemaking); brake (ambivalence with combined strategies of approach and anxious retreat); immobilization (reduced activity to avoid threat and conserve resources, including slow heart rate); coping to the extent possible; motivation to be high achiever; minimization of one's expression of complaints.

In addition, cognitive and coping styles have been described as enhancing some of the personality stress characteristics, such as exaggerating the adversity of events; memory bias and impairing the positive retrieval of positive experiences; attentional bias to negative experiences; and focusing excessive attention on sensations perceived to be aversive (Bryant, 2001). This writer suggests that these cognitive biases are also found in many individuals with depressive and masochistic qualities. The underreporting of symptoms also occurs.

TRAUMATIC CHANGE OF IDENTITY: RORSCHACH PERCEPTS

The following themes, with examples of responses to the Rorschach inkblot procedure, are all persons, children and adults, who have been involved in accidents creating head and somatic injuries, or severe threat. The stressed and injured person frequently perceives himself or herself as unattractive, damaged, impaired, victimized, and vulnerable to further injury, and also socially rejected. There are environmental reactions to losing social support, having insufficient money to continue former activities and social connections, lack of fair treatment with regards to compensation, and so on. Concern with disorders may vary between preoccupation with impairment and injustice, to lack of insight into significant dysfunctions.

Under the control of greater forces: "pinned"; "glued to the wall." The possibility of suicidal thoughts cannot be discounted, and is consistent with the pattern of great distress and poor emotional control.

Ongoing victimization or destruction: (A 12-year-old victim of an auto accident): "Two birds smashing into each other"; "falling out of the sky, alive at the moment, when he hits the ground he's going to be dead"; "a bat got beat up, trying to do something with his wings and he don't have the strength";

Vulnerable: "Orchid, delicate and fragile, cut off; Animals, the shell is not there... vulnerable to further damage, monkeys hanging on to a ledge";

Trapped: "A fly caught in a web"; "Dragon fly, tail caught on something";

Helpless, powerless: "I'm standing at the giant's feet. I can barely see his head. It gives me the feeling I"m small, he's big"; "Another man in command, he wants my head"; "fetus"; "suspended in the air; a rat, cut open, dead, used for an experiment";

Under the control of mysterious forces: "Witch doctor trying to show his power to the tribe";

Defeated—the battle is lost: "Vulture waiting for death so he can eat; "fly that's been dissected, that's why there's blood, I'm being examined"; "a bird that's flying down in speed and crashing; suffocating"; "Siamese twins trying to tear themselves apart, bleeding all over, split themselves".

Loss of self-control and sense of integration: "Swinging; balancing; whirling; spinning; pirouetting; off balance; slipped, falling backwards"; "someone burning, falling; assaulted" (KDabd); A boy with life-long seizures: "2 people falling down from a hill, frightened."

Bodily damage, mutilated: "Sheep hanging by their feet, and the head is cut off"; "a bat, with his head split open"; (A victim of overwhelming force): "squashed"; "tree split in half";

Deteriorated (This occurs in medical conditions and also persistent stress): "Petals falling off a flower"; "picking up the body, a pair of hands, pieces laying around"; "lying in a hospital; chair made of human arms and legs"; "wheat hanging off a stalk"; "worn out heels"; "burnt paper"; "stain";

Internal anatomy: "Tissue"; "an organ spattered with blood"; "X-ray"; "accident victim, who is damaged, broken, dead"; "kidney, blood coming out of the cut";

Unattractive (A man who required rhinoplasty after an auto accident): "Ugly face, disturbing, even my face";

Reduced self-esteem (A teacher who could no longer work): "Two kings kneeling down, scraped knee, blood"; (A woman who was molested): "Not sure if she is an angel" (I); (Seen by a 12-year-old girl): "Old people"; "dog smelling the dirt on the ground" (a girl with chronic foul-smelling vaginal discharge after severe vaginal-anal injuries);

Still struggling: "Holding on for dear life; struggling; wants to get the hell out of there; Crabs trying to survive."

Hampered due to injuries: "Women pulling on something real hard, and they can't pull it"; "Siamese Twins (limping due to broken leg); (A 20-year-old man, with diminution of cognitive functioning: "Man after a long day's work, sad, drawn-out face, panting like a dog, tongue hanging out, weary."

Loss of vitality including reference to death: (EAabd, his HTP): "A tree cut off at the roots"; "flower eaten by ants"; "a lady, no head, alarm or surprise, dead or killed"; "two lambs, hanging by the tongue, dead"; "mice are crawling on him to see if they could get meat"; "leaf falling, dead'; "someone gluing pieces of something together"; (Plate X) "something with heads and a worm body. An unborn baby, sucking on the blue thing coming out of their mouth, drained dead, the bone structure." This is consistent with his description of his injured arm as too skinny, his loss of activity and strength, and his figure drawings (simple, empty, lifeless, and suggesting a wary attitude).

Cry for help (A policewoman who was almost strangled on the job): "Two hands reaching out for help" (EHDabd); "supernatural being, arms raised to Heaven" (ws2ABD).

Depersonalization—changed or confused identity: "Humanoid, cave dweller"; Initial human figures changed into "apes" by a man who was perplexed by the "weird" or unaccountable mistakes he made; (man who described his body as not working as well as it once did) "deformed monster, made of garbage"; (a woman who didn't feel like the same person) "birds, I don't know what they are, they are not people"; "caricature" (woman doesn't feel like the same person) (SG); "an alien".

Derealization

"Flowers from 'Alice in Wonderland'"; "Fellini movie."

How can one differentiate neurotrauma from psychologically protective dissociation and anxiety-associated hyperarousal? One notes in advance that some characteristic symptoms of concussion are not specific to it (i.e., may accompany emotional and adjustment problems) (Reitan & Wolfson, 1994). Furthermore, particular symptoms of PTSD are frequently observed in both the acute and chronic phases of concussive brain injury (listed with diagnostic example coding). Differentiating symptoms include diplopia, photophobia, alcohol intolerance, fatigue, headache, and dizziness (Blank, 1994). PTSD and TBI have similar symptoms to disease of the hypothalamus, which is vulnerable to accidents. These symptoms might be otherwise attributable to PTSD or to psychodynamic reactions to impairment (epilepsy; sweating; cardiac arrhythmias; somnolence; paroxysmal rage, laughing, and crying; hallucinations; loss of sphincter control, etc.) (Reichlin, 1992). Many dissociative phenomena are more common in phobic-anxiety syndrome than in CPS, in which they may be a consequence of concussion (Theodore & Porter, 1995, p. 34): depersonalization; derealization; loss of feeling of familiarity (jamais vu; deja vu; illusions and distortions of perception, including body image changes; idea of a "presence." Differentiating characteristics of epileptic phenomena include symptoms that are more vivid, stable, and stereotyped; that are shorter (i.e., not for many minutes or hours); and with frequent association of psychic phenomena with other epileptic phenomena such as strange feelings in the stomach.

Hyperarousal and anxiety are characteristic symptoms of PTSD after accidents. Their primary biologic processes may contribute to the utility of dissociation as a defense: trauma-induced dysregulation of multiple neurobiologic systems, including the sympathetic nervous system (hyperarousal, intrusive memories, impulsivity, numbing, and substance abuse); fear conditioning (to the stimuli present at the time of trauma); behavioral sensitization (enhanced response magnitude after repeated presentations of stimuli (not characteristic of the one-time accident trauma); and neural mechanisms of learning and memory (accounting for the persistence of such symptoms as flashbacks). Released during stress, norepinephrine, epinephrine, and opioid peptides are neuromodulators that influence memory encoding and consolidation, including the apparently indelible thalamo-amygdala pathways (Southwick et al., 1994).

COGNITIVE PROCESSING

A recent review illustrates the great range and complexity of cognitive processing in individuals with PTSD. In addition to functions that are not considered in the familiar DSM-IV-TR description of PTSD, the question is explored as to whether particular disorders are significant, and whether dysfunctioning is a cause and a risk factor of a consequence of PTSD. One consequence is reduced verbal and, to a lesser extent, visual memory among those with PTSD, when compared to control groups. Stressed groups may show reduced verbal memory or reduced general intellectual performance. Decrements of verbal memory in PTSD are accounted for by impaired encoding rather than attention, storage, or retrieval. The overlapping response trends for patients with a variety of diagnostic entities. PTSD memory difficulties or inconsistencies could be inappropriately attributed to symptom malingering (Moore, 2009).

Cognition and emotion are intertwined. Fear contains representations of the feared stimuli, fear responses, and the meanings associated with these stimuli and responses. In female survivors of rape, TI fear for one's life, circumstances that prevent escape, and depersonalization are associated. Thus, with reduced motor response, cognitive-behavioral PTSD treatments designed to reduce physiological arousal may be inappropriate (Marx et al., 2008). PTSD is unique among mental disorders insofar as it requires specification of the experience that is presumed to have caused the symptoms to develop (Orr & Kalouupek, 1997). This is a controversial point (i.e., whether apparent LOC precludes the careful examiner from using this diagnosis). These authors assert that diagnostic accuracy is best served by a combination of procedures beyond self-reported symptoms: the clinical interview, psychological testing and psychophysiological assessment. While this diagnosis may be considered from a psychophysiological viewpoint, these measures, although they provide unique information, are not inherently more valid or more objective than the self-report of interviews. PTSD diagnosis is based upon subjective information that is not comparable to information recorded directly from physiological systems. A goal would be to include into the diagnosis objective physiological reactivity to trauma-related cues.

Environmentally produced alteration of physiological responsiveness to stressors is an interface between the cognitive regulation of behavior and the more basic response regulation, which may be genetic (Kraemer, 1997). *Processing* of the accident representations may continue (cognitive, behavioral, and physiological), changing what is experienced or recollected. Traumatic experiences after being encoded exist outside of awareness, yet influence ongoing emotions and behavior. Some severe TBI patients report a larger amount of situationally accessible memories (SAM) that have been elicited by trauma reminders. Reexperiencing symptoms (or perhaps the belief of full recall) may utilize events during the period of partial recall ("islands of memory"), perhaps *fear conditioning* made possible by a heightened physiological arousal response to an overwhelming trauma. Associations may be formed between trauma reminders and anxiety responses. *Noradrenergic activation* has been implicated—stress-induced reactivation of the locus ceruleus-hippocampus pathway, perpetuated by repetitive stress levels associated with reexperiencing symptoms of PTSD. Limbic damage also may result in altered noradrenergic activation that predisposes the TBI person to limbic-mediated processes that are associated with PTSD (Bryant et al., 2000). Included into the schema can be imagination or gory stories about the accident, as well as subsequent hyperarousal cues to pain and disability and perplexity and fear.

While it has been asserted that there can be full recall of the experience, one may experience reservations. Recall may be a reconstruction with others' information or confabulation, , without actual eyewitness reports verifying the accuracy of the report. Later symptoms can represent aspects of the trauma without conscious recall of the accident. A psychodynamic defense against anxiety can cause oscillation between intrusive images and numbing. Alteration of ongoing mental and physiological activities can create and imprint the false imagery or pseudomemories of the event, which initiate the syndrome: information received by family or friends, or the *implicit memory* of events, physical injury, or bodily trauma that occurred during a period of altered consciousness (Krikorian & Layton, 1998).

Cognitive Theories of PTSD

- The *fear network storage of information* (social-cognitive) about what should be escaped or avoided. Activation of this network causes intrusive recollections that arouse physiological reactions.
- *Wider meaning of the trauma and its associated content.* Oscillation occurs between intrusive images and nightmares about the trauma and numbing and denial. Intrusions represent the incompatibility between pretraumatic memories and new memories that are related to

the trauma. This represents an opportunity to integrate the traumatic memories; if this is performed inadequately, the memories result in PTSD. The *experience of trauma* shatters the individual's assumptions about personal invulnerabity, the meaningfulness of the world, and the self as positive or worthy.

- *Dual representation* of the network and social-cognitive theories. *Verbally accessible memories (VAM)* are verbal or visual memories that were attended to in full consciousness, and were encoded to permit deliberate search and retrieval strategies (i.e., can be consciously retrieved). Structured diagnostic interviews may elicit intrusive memories or conditioned emotional responses in the absence of explicit memory of the trauma. *SAMs* are subconsciously generated memories that may be experienced as flashbacks, a sense of reliving a trauma, or somatic (sensory, physiological, motor) sensations reminiscent of a traumatic experience. Peripherally perceived cues may not interact with the autobiographical knowledge base (Harvey et al., 2003).

STRESS-RELATED DIAGNOSES

Stressful experiences often do not evolve into the excessively specific characteristics of PTSD (American Psychiatric Association, 2000, pp. 467–468). Many people who ultimately manifest PTSD do not initially meet the acute stress disorder (ASD) criteria for three or more specific dissociative symptoms (Bryant et al., 2000, 2007). Persons with similar symptoms may have an etiology that does not match the high level of trauma of PTSD criterion A (Event).

It is noteworthy that TBI, with the likely alterations of consciousness and impaired memory, does not markedly influence the diagnostic symptoms. Ten symptom patterns do vary with the posttrauma interval: Shortly after the accident, TBI victims (compared to non-TBI victims) reported less fear and helplessness and more dissociation. Six weeks later, fewer TBI victims than non-TBI victims reported reliving and physiological reactions to trauma reminders (the issue of recall is introduced [i.e., the reported experience at the time of trauma may depend upon memory both short-term and long-term memory, RSP]). At 3 months posttrauma, there was no difference in PTSD symptom profile between non-TBI groups and TBI groups. The lack of difference between ASD and PTSD incidence between TBI and non-TBI victims after a 3-month interval was consistent with prior research. Other reports explicate the issue of symptom expression changing with time. There were comparable rates of ASD and PTSD in research comparing TBI and non-TBI victims. Persistent dissociation at 4 weeks posttrauma was a better predictor of PTSD symptom severity at 6 months than was dissociation experienced during the trauma. The concepts of dissociation and PTA overlap, although this writer suggests that the cerebral and physiological disruption of a TBI-causing accident create different mechanisms of memory loss (Jones et al., 2005).

Acute Stress Disorder

Acute stress disorder (308.3; DSM-IV-TR, pp. 471–472) is a syndrome evolving after an accident that differs from PTSD in terms of length of experience, numerical requirement of particular symptoms, and the emphasis upon dissociative reactions. Approximately three-quarters of trauma survivors who display ASD subsequently develop PTSD. The probability of ASD evolving into PTSD after 6 months was highest for those who showed symptoms of depersonalization, a sense of reliving the experience, and recurrent images of the trauma (Bryant et al., 2000). The following diagnostic descriptions are abstractions from the DSM-IV-TR:

A. Event:
1. Experienced/witnessed event involving death or serious injury, or a threat of serious injury, to self or others.
2. Response (fear, helplessness, horror).

B. *Dissociative symptoms* (three or more during or after the event): Numbing, reduced awareness ("dazed"), derealization, depersonalization, dissociative amnesia.
C. *Reexperiencing*: Images, thoughts, flashbacks, reminders, and so forth.
D. *Anxiety or increased arousal*: Sleep disorders, irritability, poor concentration, hypervigilance, exaggerated startle, motor restlessness.

PTSD: Criteria

A. Event:
   1. Experienced/witnessed event involving death or serious injury, or a threat of serious injury, to self or others.
   2. Response (fear, helplessness, horror).
B. *Event persistently reexperienced*: Recurrent, intrusive recollections; distressing dreams; flashbacks/dissociative experiences; distress at exposure to symbolic cues; physiological reactivity to symbolic cues. Rarely the person experiences dissociative states (a few seconds to hours or days), in which the event is relived and the person behaves as though reexperiencing the criterion again.
C. *Avoidance* (any three of the following): Thoughts, feelings, conversations, activities, places, people, inability to recall an important aspect of the event, diminished interest or participation, detachment or estrangement, restricted range of affect, foreshortened future.
D. *Persistent increased arousal* (*somatic forms of anxiety*): Sleep disorder; irritability/anger, concentration, hypervigilance, exaggerated startle response; diarrhea; frequent urinating; sweating; tremor; rapid heart beat.
E. Duration of the disturbance more than 1 month.
F. *Clinically significant distress or impairment*: social, occupational, or other important areas of functioning or pursuit of tasks of daily living.

Modifiers: Acute (less than 3 months); chronic (3 months or more); delayed onset (at least 6 months after the stressor. *Adjustment Disorder* with Mixed Anxiety and Depressed Mood, Acute or Chronic (309.28).

Subsyndromal PTSD: A syndrome involving only portions of the more extended type (Blanchard & Hickling, 1997, p. 38), which includes reexperiencing, avoidance, and hyperarousal (Bryant et al., 2007).

Adjustment Disorder with Mixed Anxiety and Depressed Mood (309.28)

The writer has observed that this posttraumatic stress reaction, which is of a different configuration than the above, may be the primary emotional disorder or comorbid with ASB or PTSD (DSM-IV-TR, pp. 679–683). The DSM-IV-TR description offers such stressors as termination of a romantic relationship, marked business difficulties, continuous living in a crime-ridden neighborhood, and so forth. Regrettably, having one's brain and body traumatized may result in an emotional disorder qualifying for this diagnosis.

Definition of Adjustment Disorder, Mixed Anxiety, and Depression

A. The symptoms must appear within 3 months after the onset of the stressor(s).
B. Clinically significant: Distress in excess of what would be expected (or) significant impairment of social, occupational, or academic functioning
C. Do not meet the criteria for another specific AXIS I disorder or represent an exacerbation of a preexisting AXIS I or II disorder.
D. The symptoms do not represent bereavement.
E. Once the stressor or its consequences are terminated, the symptoms do not persist for more than an additional 6 months. The stressor is classified as *acute* if it exists for less than 6 months, or *chronic* if it exists for more than 6 months.

PTSD IS MORE COMPLEX THAN ITS DEFINITION

The stress reaction is multisystem: *neural integration of autonomic functions* (cerebral and brainstem); *awareness and body schema*; *hypothalamic-adenohypophysis adrenal cortical* (HPA); *hypothalamic-anterior pituitary and other target endocrine, pineal gland, locus ceruleus-sympatho-adrenomedullary axis, immune system, cerebral circulation*, and *metabolic functions*. The stress system integrates a great variety of neurosensory and bloodborne signals that arrive through distinct pathways. It is believed to be influenced by a genetic component, and early and later life experiences. The acute phase is a period of extreme emotional stress with emotional sensitization, memory disorganization, and pathology of corticolimbic function (Tucker & Luu, 1998). Acute homeostatic responses to trauma involve multiple systems: the nervous system and hormonal system (adrenocortical, neuroendocrine stemming from the adrenal medulla and postganglionic neurones overflowing into the plasma), immune, and metabolic.

ENDOCRINOLOGICAL COMPONENTS OF PTSD

The arousal-related release of adrenal stress hormones (epinephrine and glucocorticoids) and neurotransmitters influences the long-term storage of memories of emotional significance. These hormones modulate the neuroadrenergic system within the amygdala. The amygdala, in turn, influences memory by regulating consolidation in other brain regions. This *memory modulation system* may be involved in the formation of traumatic memories and PTSD (McIntyre et al., 2003). Activation of the limbic system influences mood (chronic anxiety, melancholic depression), motivation, and reinforcement phenomena (Chrousos, 1998). Endocrine secretions are controlled both directly and indirectly by the brain and have feedback interaction between the brain and the individual glands (Wilson et al., 1998). An increase in adrenal mineralocorticoids during stress (e.g., aldosterone) results in increased potassium excretion (Whitehouse, 2000), which is associated with a general decrease in strength and possibly severe depression and age-related memory impairment (Frohman, 1995). Since hormonal effects are part of the individual's experience (Lombardi et al., 1994) and influence brain activity, stress initiates an interaction between experience and hormonal reactions. The acute and chronic phases of stress affect hormonal levels not related to specific organ dysfunctions. In many systematic illnesses, changes in pituitary function cannot be distinguished from "stress" due to the illness (Molitch, 1995, pp. 226, 271).

NEUROLOGICAL COMPONENTS OF PTSD

Various imaging studies of stress victims, primarily of those without comorbid TBI, have offered leads concerning the brain areas participating in PTSD response. The neocortex and the limbic systems interact to shape cognitive and emotional reactions. Cross-sectional volumetric studies report decreased hippocampal volumes but inconsistent localization. It is uncertain whether these findings are predispositional or consequent to trauma. Additional areas that have been implicated are the anterior cingulate, the medial temporal lobe, the medial prefrontal lobe, the neocortex and caudate, and so on. Since the *medial prefrontal cortex* (*MPFC*) has an inhibitory effect upon the *amygdala*, reduced activation may result in failure of *extinction* to fear, (i.e., potentially contributing to chronic anxiety). This is consistent with the vulnerability of the frontal lobes to impact trauma in dependent areas of the brain upon force direction and head orientation in the accident. It is speculated that the amygdala may only be activated by externally generated emotions. The *anterior cingulate* is described as having a role in attention and stimulus discrimination (i.e., participating in inhibiting acquisition and promoting extinction of fear conditioning). Position emission tomography (PET) studies indicate that in stress there is either decreased activation of or failure to activate the anterior cingulate in response to traumatic stimuli. Thus, impairment of its gating function contributes to PTSD. The *orbitofrontal cortex* is activated in anxiety disorders and activation of images of trauma. It has an inhibitory influence upon fear responses that are mediated by the amygdala. In rats, lesions prolong the extinction phase of fear conditioning. An overview of PTSD has suggested that a lower threshold of activation of the amygdala, perhaps with a dysfunction MPFC (anterior cingulate), creates an inability to inhibit the amygdalar response, leading to failure of extinction of fear responses. It appears that the hippocampus has a limited role in symptom mediation. Reduced hippocampal volume may be a downstream effect of chronic hyperarousal of the amygdala and/or MPFC (Villarreal & King, 2001).

Chronic Stress: The Persistent PTSD

Chronic complaints imply that there is an unhealed wound that is uncomfortable, reduces performance efficiency, and may stimulate the numerous physiological systems. The writer regards a long-standing postinjury condition as a *persistent PTSD*. Its range of disorders is far more extensive than that of the familiar *PTSD* (American Psychiatric Association, 2000; Parker, 2002). It is so multidimensional (experiences, neuropsychological reactions, physiological) that the "official" definition of PTSD may be considered as far too restrictive. Consider accident victims with combat trauma in addition to comorbid psychiatric conditions (Parker & Rosenblum, 1996). Their psychotherapists in the current Iraq War point out that current combat conditions vary considerably from the early traumatic environment, leading to the current definition (Carey, 2005).

COMORBID CONCUSSION AND PTSD

Concerning the metamorphosis from acute to chronic trauma responses, several possibilities have been offered. Animal study suggests that *extinction* (turning off the fear function) seems more easily enhanced within the first 30 days. Perhaps dysfunction of the *MPFC* causes failure to inhibit amygdala function. Repeated stress may cause hippocampal changes that are analogous to "*kindling*," a potential factor in the development of epilepsy. Subsequently, there is a long-term transfer of memories to the cerebral cortex. With chronic and repeated stress, *sensitization* may occur. This involves apparently a potentiated release of substances like norepinephrine following exposure to subsequent stressors (Bremner, 2005). Chronic neurobehavioral dysfunctions include mood, cognition, health, and loss of stamina. This author's focus will now be on tissue damage ("*unhealed wounds*") as a source of dysregulation and persistent stress. Dysregulation of bodily functions creates widespread disorders of the body's systems, behavioral functioning, stamina, and health. Chronic somatic complaints are a marker for continuing dysfunction of particular systems (hormonal, inflammatory, and immune). Since it usually is not recognized that the dysregulated internal environment disturbs a variety of neurobehavioral functions, chronic complaints (mood and cognition) may be ignored or be misattributed to TBI, malingering, emotional reasons, and so on.

The symptom patterns for PCS (Parker, 2001, Chapters 2 and 3) and PTSD highly overlap (Parker, 2002). A contributor to this ambiguity is comorbid stress-related hyperarousal. Heart rate is increased at 1 week and 1 month following severe TBI. In PTSD heart rate is increased at 6 months postinjury. In PTSD patients, heart rate declined in the interval from 1 week to 1 month (Jorge, 2006). The examiner of a highly anxious victim of an accident should also be alert to common subjective disturbances with varied etiologies, including reaction to the event, cerebral trauma, and reaction to chronic and significant impairment (reduced motivation, mood changes, and poor morale). Cytokine release during response to injury and chronic stress creates responses frequently observed in TBI, but the etiology may be described as primarily the behavioral sequelae of the physiological reaction to primarily noncerebral injury: sickness behavior (malaise, social withdrawal, somnolence, hyperesthesia) and depression (somnolence, anorexia, diminished libido, malaise, fatigue, SWS, apathy, and irritability (Kop & Cohen, 2001; Marshall & Rossio, 2000).

A subjective reaction to the impairment incurred in both PTSD and TBI elicits the experience of incapacity, reduced self-confidence, and can lead to the belief that one is socially unattractive. The sense of foreshortened future of PTSD resembles the discouragement from persistent impairment and discomfort of a TBI victim. Depression is common in both TBI and PTSD. Organic mood disorder (depressed type consequent to TBI) and/or organic anxiety disorder (Van Reekum et al., 2000) can be confused with the dysphoric moods of PTSD. Head injury predicts the experience of depression more than the severity of PTSD (Vasterling et al., 2000).

Both TBI and PTSD represent controversial topics in health care. The actual symptoms of PTSD overlap with TBI (Parker, 2002), although they may have different origins. For example, it is necessary to differentiate organic from dissociative amnesias. Organic amnesia (Steinberg, 2000) tends not to involve identity alteration, is permanent and does not lend itself to psychotherapeutic relief, and is more often anterograde than retrograde. Amnesia interacts with depression, which is very common after TBI in numerous forms of expression. Head injury predicts depression more than PTSD severity (Vasterling et al., 2000). Amnesia associated with depression is characterized by replies that are more frequently "I don't know" than near misses and confabulation.

Campbell (1999) notes that diagnostic decision-making is undermined by injuries that lack inter-rater reliability data for many classifications, and also by an atheoretical approach to definitions. One source of conceptual difficulty is the inadequate definition in the DSM-IV, which is not based upon the event of a concussive or other brain trauma. The definition of PTSD does not include a full range of symptoms, variability with the passage or time, nor cautions about symptoms overlapping with concussive brain injury. DSM-IV-TR ignores concussion (whiplash, mild TBI) as a significant entity, merely referring to it as "research criteria." Thus, it does not realistically describe the stress experiences of many people following an accident, nor does it identify or relate symptoms that fluctuate (e.g., numbing and hyperarousal) to physiological "burnout" (i.e., allostatic load) or other conditions that overlap with postconcussive syndrome.

Etiology

Initially, there is afferent sensory input that creates an assessment of the anxiety-provoking nature of the event. Prior experience is integrated into the event's cognitive appraisal. The emotional experience of the trauma becomes associated with specific stimuli. PTSD as a syndrome disturbs the integration of consciousness, memory, identity, and perception of the environment. Nevertheless, one component, dissociation, helps a person to cope with reexperiencing traumatic memories (Carlier et al., 1996). Another possible etiology is TBI. Frontal lobe dysfunction may result in imagery that is phenomenologically similar to PTSD reexperiencing symptoms (Bryant et al., 2000). Dissociative phenomena are associated with psychopathology, including child abuse, combat, and other trauma, and also with conditions that can be described as hysteric. Some dissociative symptoms can be elicited through hypnosis.

Memory, Cognition, and the Comorbidity of Concussion and PTSD

Priming is preserved in amnestic patients (i.e., retrieval need not depend upon explicit memory for episodes). Later, a cue facilitates the subsequent identification of the previously unsalient stimuli that occurred after the initial trauma. Priming is associated with the hippocampus and other limbic structures (Schachter et al., 1993). *Conditioning* can trigger stimuli that arouse a complex fear network, leading to the intrusive reexperiencing of a trauma (Joseph & Masterson, 1999). In addition, intrusive symptomatology is assumed to be associated with separate memory systems, implicit and explicit, which mediate the encoding and retrieval of information. Pseudomemories that are phenomenologically similar to flashbacks may be created (Bryant, 1996). A recollection contributing to traumatic memories and other symptoms of the PTSD may accompany a high level of arousal later on. With lower arousal, perhaps the subsequent PTSD would not develop.

Since concussion by definition refers to the alteration or loss of consciousness and deficits of memory, some assert that PTSD is "rarely diagnosed in patients with head injuries who do not recall their trauma."

Later Images as a Contributing Event to PTSD

Several mechanisms may account for PTSD in the absence of images of the accident, including the concept that the emotional trauma may occur after the mechanical event of the accident (i.e., waking up at the scene or in the hospital). The traumatic event can occur after recovery of consciousness, e.g., waking up in a hospital or at the scene of an accident. During this acute period, fear conditioning may occur due to overwhelming stress that leads to high noradrenergic activity involving long-term enhancement of arousal in the limbic system:

- Terror may occur in an "island of memory" intruding between amnestic periods due to hyperarousal creating altered consciousness or coma (King, 1997).
- Fear can begin when the patient wakes up in the trauma of the accident scene, emergency room, intensive care unit, and so forth. There can be perplexity as to what happened. The patient's imagination or gory stories about the accident may create new images.
- Reconstruction of the event, consciously or unconsciously, may be determined by what the patient has been told as to what happened, and whether he or she or some other condition was responsible for the accident. Processing may continue of the accident representations (cognitive, behavioral, and physiological), changing what is experienced or recollected. Included into the schema can be imagination or gory stories about the accident, and subsequent hyperarousal due to pain and disability, perplexity and fear. Later symptoms can represented aspects of the trauma without conscious recall of the accident. A psychodynamic defense against anxiety can cause oscillation between intrusive images and numbing.

Physiological Components of PTSD

Since repeated stimulation of the amygdala produces fear conditioning, it seems to suggest that, despite damage to cortical functions, fear conditioning may occur at subcortical levels. Persons with posttraumatic stress have attentional bias to negative experiences. They may focus on what they deem to be aversive sensations and misinterpret the meaning of these sensations, leading to elevated anxiety responses (Bryant et al., 2000).

Since PTSD is defined as involving an unusual event, lack of memory would appear to presuppose that there cannot be PTSD by definition. Experiences suggest that, if one extends the range of frightening experiences beyond the immediate accident, a subset of persons will experience PTSD in the context of PTA. Intensive care treatment can be an extremely stressful experience, with trauma (nightmares, respiratory distress, anxiety/panic, or pain) vividly present up to 10 years after discharge from ICU (Schelling et al., 2004). Dissociation associated with PTSD may be protective, perhaps related to the retroactive amnesia, confusion, and PTA occurring after head trauma. This writer raises the question as to which experiences contribute to PTSD syndrome. Are they the memory of the fear and injury of the accident, or are they the stress of the recovery period, including perplexity as to what happened, or imaginatory or gory stories about the accident that create new images?

How can one differentiate neurotrauma from a psychologically protective dissociation and anxiety-associated hyperarousal? One notes in advance that some characteristic symptoms of concussion are not specific to it (i.e., may accompany emotional and adjustment problems) (Reitan & Wolfson, 1994). Furthermore, particular symptoms of PTSD are frequently observed in both the acute and chronic phases of concussive brain injury (listed with diagnostic example coding). Differentiating symptoms include diplopia, photophobia, alcohol intolerance, fatigue, headache, and dizziness (Blank, 1994).

PTSD and TBI have similar symptoms to disease of the hypothalamus, which is vulnerable to accidents. These symptoms might be otherwise attributable to PTSD or psychodynamic reactions to impairment (Reichlin, 1992): epilepsy; sweating; cardiac arrhythmias; somnolence; paroxysmal rage, laughing, and crying; hallucinations; loss of sphincter control; and so forth. Many dissociative phenomena are more common in the phobic-anxiety syndrome than in CPS, which may be a consequence of concussion (Theodore & Porter, 1995, p. 34): depersonalization; derealization; loss of feeling of familiarity (*jamais vu*, déjà vu, illusions and distortions of perception including body image changes, idea of a "presence"). Differentiating characteristics of epileptic phenomena include symptoms that are more vivid, stable, and stereotyped, that are shorter (i.e., not for many minutes or hours), and, often, the association of psychic phenomena with other epileptic phenomena such as strange feelings in the stomach.

Although altered consciousness occurs after a mechanical brain trauma, quite similar symptoms occur after "psychological" events. At present, it is difficult to distinguish such symptoms from each other. LOC, altered consciousness, and PTA appear to be neurological disruptions. However, LOC may occur at other times than trauma and for other reasons than seizures (Remler & Daroff, 1991). Dissociation seems to be a protective psychological mechanism, yet it causes symptoms such as derealization and depersonalization, which may be consequent to acute or chronic brain injury. This increases the complexity of diagnosing TBI and its outcome. The outcome following interaction of TBI, posttraumatic stress hormones, and psychopathology is not yet clear (Harvey et al., 2003). This writer takes the view that dissociation may not directly represent permanent neurotrauma. Rather, the complexity of neurological and neurochemical effects that accompany a frightening experience or a sudden somatic injury can be conceptualized as a direct physiological cause of dissociation. Furthermore, one cannot assume that a "psychological" reaction does not have a physiological substrate.

Hyperarousal

Hyperarousal and anxiety are characteristic symptoms of PTSD after accidents and their primary biologic processes may contribute to the utility of dissociation as a defense: trauma-induced dysregulation of multiple neurobiologic systems, including the sympathetic nervous system (hyperarousal, intrusive memories, impulsivity, numbing, and substance abuse); fear conditioning (to the stimuli present at the time of trauma); behavioral sensitization (enhanced response magnitude after repeated presentations of stimuli (not characteristic of the one-time accident trauma); and neural mechanisms of learning and memory (accounting for the persistence of such symptoms as flashbacks). Released during stress, norepinephrine, epinephrine, and opioid peptides are neuromodulators that influence memory encoding and consolidation, including the apparently indelible thalamo-amygdala pathways (Southwick et al., 1994).

PTSD AND ALTERED CONSCIOUSNESS

Posttraumatic stress disorder as a syndrome creates a maladaptive integration of consciousness, memory, identity, and environmental perception. Dissociation helps a PTSD victim to cope with reexperiencing traumatic memories (Carlier et al., 1996). Dissociative phenomena are associated with psychopathology, including child abuse, combat, and other trauma, and with conditions that can be described as hysteric. Some dissociative symptoms can be elicited through hypnosis.

Diagnostic Stress Categories

The ASD (308.3, American Psychiatric Association, 1994, pp. 424–432) (DSM-IV) has certain markers that are remarkably similar to the experiences of many victims of head injury:

> (B). Either while experiencing or after experiencing the distressing event, the individual has three (or more) of the following dissociative symptoms: (1) a subjective sense of numbing, detachment, or

absence of emotional responsiveness; (2) a reduction in awareness of his or her surroundings (e.g., "being in a daze"); (3) derealization; (4) depersonalization; (5) dissociative amnesia.

(E). Marked symptoms of anxiety or increased arousal (e.g., difficulty sleeping; irritability; poor concentration; hypervigilance; exaggerated startle response; motor restlessness).

PTSD (309.81). The major criteria for PTSD are: (a) exposure to a traumatic event in which both confrontation with actual death or serious injury (as witness or participant) is accompanied by intense feelings of fear, helplessness, or horror; (b) the traumatic event is persistently reexperienced; (c) persistent avoidance of stimuli associated with the trauma and numbing of general responsiveness; (d) persistent symptoms of increased arousal.

Subsyndromal PTSD is a syndrome involving only portions of the more extended type (Blanchard & Hickling, 1997, p. 38).

ISSUES CONCERNING THE DEFINITION OF PTSD

The diagnosis of PTSD when there has been an accident creating head injury and potential brain trauma has special difficulties.

The stress definition does not reflect differences in events and persons: (1) the influence of the nature, severity, and meaning of the event, or (2) individual factors (personality, previous history; experience, support, and subsequent experience) (O'Brien, 1998, p. 2).

Symptoms of PTSD overlap with TBI: Both included create the ultimate stress (fear of damage to the brain and loss of bodily and social attractiveness). These clinical entities include similar symptoms with different origins. It is necessary to differentiate organic from dissociative amnesias. Organic amnesia (Steinberg, 2000) tends not to involve identity alteration; is permanent and does not lend itself to psychotherapeutic relief; and is more often anterograde than retrograde. One interaction is with depression, which is (in its numerous forms of expression) very common after TBI. Head injury predicts depression more than PTSD severity (Vasterling et al., 2000). Amnesia associated with depression is characterized by replies that are more frequently "I don't know" than near misses and confabulation.

VULNERABILITY

There are *individual differences* in vulnerability and outcome. Not all persons are *vulnerable* to a given level of serious trauma. For example, in a series of burn injuries, including of the face and hands, only about one-fourth of victims developed severe stress reactions within 2–3 weeks of the event. Only half of those with high initial stress scores developed a chronic stress reaction. Initial anxiety and *dissociation* predicated the development of posttraumatic stress symptoms. Three percent of victims showed a delayed stress reaction, suggesting that there are different pathways to chronic stress (Van Loey et al., 2003).

After the initial stress of fright (awareness of danger to the body or one's life), the stress victim copes with *neurobehavioral impairment*, pain , *loss of function*, *mobility problems*, and a belief that he or she is *socially undesirable*. Among the symptom determinants are the *meaning* of the experience, the *risk factors* that increase the likelihood of PTSD, *vulnerability* (development of PTSD without a risk factor), and *resilience* or experiencing adversity with a satisfactory outcome).

INDIVIDUAL DIFFERENCES IN REACTIONS

Postconcussive symptoms even occur in individuals without head injuries with an intensity suggesting the impact of *daily stress*. This suggests the contribution of stress to symptom maintenance (Machulda et al., 1998). Differences exist between persons suffering from PTSD (Alarcon et al., 1997), such as stage of symptom development; number and intensity of traumatic events; a neurotic or somatizing response; style of information processing (relative attention to threatening and neutral stimuli, distractibility, range of attended stimuli); onset, duration, and clinical course, evolving into the acute, chronic, and delayed types; personal premorbid factors (personality, psychophysiological responses to stress, family experiences); type of emotional conflicts; adaptational models for coping; and response to treatment. *Clinical subtypes* have been proposed: depressive; dissociative; somatomorphic (somatizing); psychotomorphic (psychotic like, though non-schizophrenic); and organomorphic (cognitive deficits).

Personal qualities that enhance stress resistance have been described (Parker, 1990, p. 78). *Coping* refers to managing excessive external and emotional demands. It is enhanced by optimism (belief in the possibility of a positive outcome) and reduced by hopelessness. *Hardiness* counteracts the adverse effects of stressful events. Hardy individuals have lower level of symptoms of physical illness and depression, believe that they have control over external events, and are committed to and flexible in adapting to change. *Stamina* is enhanced by self-esteem; a warm relationship to parents; an open, flexible approach to life; and minimal nervous tension, anxiety, depression or anger under stress (Thomas, 1982). It is affected by depression-like reduction of activity caused by stress-related endocrine changes (Williams, 1998). *Constitution* (i.e., health and strength) contributes to stress resistance. Stable, athletic, and tough-fibered individuals take a concussive injury in stride, while sensitive types may be so overwhelmed that they cannot expel the incident from their minds (Adams & Victor, 1985, p. 659). *The will to survive.* Dangerous and degrading circumstances bring forth individuals whose desire for self-preservation aids survival when others succumb to stress (e.g., concentration and prisoner of war camps).

There are multiple psychiatric consequences of head injury. PTSD considered alone has an accompanying 50–90% psychiatric comorbidity (Yehuda & Wong, 2000). PTSD or other anxiety-related diagnoses are common after head injury (Blanchard & Hickling, 1997; Parker & Rosenblum, 1996): depression (Busch & Alpern, 1998); major depression, bipolar affective disorder and anxiety disorders (Levin et al., 1997a; Van Reekum et al., 2000; and frontal lobe disorders Parker, 1990, Chapters 13 and 15). In one study (Parker & Rosenblum, 1966), 30 of 33 subjects reported headaches at some time (one migraine). Of the three who did not report headaches, one claimed neck and back pain, and the other two seemed to be pain-free. Thirty-one of 33 patients had a dual diagnosis of a psychiatric disorder. Two subjects had two additional diagnoses: one was dementia consequent to head injury; one was diffuse brain damage. The additional diagnoses (mostly corresponding to DSM-3R) were depression (12), PTSD (16), conversion reaction (1), affective disorder (2), mixed neurotic reaction (1), and anxiety reaction (1). There were, additionally, cerebral personality disorder, persistent altered consciousness, psychodynamic reactions to impairment, complex reactions expressing neurological, somatic, and psychological dysfunctions (sexuality and somatization) (Parker & Rosenblum, 1996).

Injuries and the persistent posttraumatic stress syndrome (*PPSS*): Contributors to PPSS include unhealed somatic injury or illness and "distracting" symptoms such as headache, pain, hyperarousal, depression, headaches (including migraine), somatic pain, balance problems, restricted range of motion, and changes of identity after TBI such as impairment (which reduce one's capacity to function), partial and generalized seizures, and hyperarousal (Parker, 1995). The PPSS has a subjective element: chronic pain does not produce the objective signs of acute pain (distorted facies; muscle tightening, elevation of pulse and blood pressure). Nevertheless, refractory pain, which has been thought to have a strong psychogenic component, in almost every case has a treatable but missed neurological or orthopedic condition (Guggenheim, 2000). The PPSS includes health and stamina problems, fear and hyperarousal, disability and impairment, persistent pain, subjective reactions, neurobehavioral impairment, and so on. Psychological response alone to severe stress without specific trauma has been associated with health problems in Vietnam War (Beckham et al., 1998) and Gulf War (Wagner et al., 2000) veterans. Allostasis is the ability to achieve stability or homeostasis. Persistent stress leads to allostatic load, an excessive level of mediators of allostasis (McEwen, 2000a). The analogous syndrome to accident-related stress described for chronic job stress is the "burnout" syndrome of emotional exhaustion, depersonalization, and reduced personal accomplishment (Burke & Richardson, 1996). Four situations lead to an allostatic load: frequent stress; failure to habituate; inability to shut off allostatic responses; and, inadequate allostatic response, which trigger compensatory increases in other allostatic systems. The initial stress response is temporarily beneficial, although antireproductive, antigrowth, catabolic and immunosuppressive. Still, chronic stress may lead to a pathological syndromal state with psychiatric, neuroendocrine, cardiovascular, metabolic and immune components (Chrousos, 1988). The HPA becomes hyporesponsive, affecting stamina, health, and quality of life. Disability is associated with a wide range of neurologic or neuropsychological symptoms (Middleboe et al., 1992).

Deficits in the Definition of PTSD

The definition is incomplete, disregards conditions overlapping with TBI, and ignores the fact that symptom intensity, direction and presence varies over time. Although the defining characteristic of PTSD is the intense dysphoric affect of the threatening event (fear; helplessness; horror), this DSM-IV criterion (A-2) does not occur for all accident victims (Brewin et al., 2000). Instead, what commonly occur are the experiences of numbing, dissociation, depression, loss of interest, detachment, and restricted range of affect (Feeny et al., 2000). Some of these symptoms are common to concussion. Dissociative symptoms, which may originate in anxiety or cerebral trauma, attenuate the victim's experience. Some victims report being in a daze, feeling numb, or not remembering the event. The finding that high initial levels of numbing predicted PTSD severity 3 months later appears to be inconsistent with its definition in terms of hyperarousal of affect. In addition, numbing added predictive value independently of depression and appears to be clinically distinct from dissociative symptoms. Furthermore, shame and anger affect subsequent PTSD developments, although they are not part of the definition. The addition of these affects would lead to more accurate diagnoses and treatments. The present definition also does not consider the potential chronic health and cerebral dysfunctions. Children's stress reactions are reviewed in Chapter 17.

AN APPROACH TO ASSESSING PTSD COMORBID WITH TBI

1. The postinjury interval should be specified, taking into consideration functions that vary with time (levels of arousal, and endocrine, autonomic, etc.) that may be enhanced or exhausted.
2. Comorbid somatic and head injuries should be specified since unhealed tissues and pain adds to psychological trauma and the PPSS.
3. The range of stress-related possible dysfunctions should be specified: physiological (particularly autonomic, sympathetic nervous system; adrenal cortex and medulla; immune system; arousal level and neurological disorders; and psychological reactions. O'Brien (1998, p. 47) offers the concept of posttraumatic illness (PTI).
4. For both PTSD and TBI the examiner must consider overlapping neurological, neurobehavioral, autonomic, and endocrinological symptoms, dissociative reactions, and personality reaction to chronic impairment.
5. Memory for the actual accident should not be specified for the definition of PTSD. Later stress or implicit memory may create an accurate or false memory contributing to this disorder.
6. The description of PTSD should not require simultaneous manifestation of all criteria (A-F): (a) the DSM-IV-TR (p. 466) recognizes that particular symptoms wax and wane, and may be reactivated for various reasons; (b) the definition should permit certain criteria appear opposed (i.e., avoidance); (c) and persistent arousal; (d) avoidance is a dynamic that many individuals employ, while in PTSD it suggests reduced resources and maladaptive management of anxiety (Bryant et al., 2000).

7. Numbness merits recognition as a separate criterion (currently in 'C'). Stress reactions can commence with numbing, dissociation and depression, loss of interest, detachment, and restricted range of affect (Feeny et al., 2000), in addition to the specified feelings of fear, helplessness, and horror. Numbness (anhedonia; various forms of depression) is common in concussion as well.
8. In PTSD there may be no disability, though DSM-IV-TR specifies impairment in social and occupational functioning (O'Brien, 1998, p. 47).
9. For assessment in general, and diagnosis in particular, the range of traumatic conditions described as "stress" is so broad that essentially different factors are included in a kind of generic definition.

11 Disorders of Sensation, Motion, and Body Schema

OVERVIEW

Neurological and somatic injuries create sensorimotor disorders that interfere with safety, quality of life, employment, household tasks, and mobility This chapter offers an overview of many sensory and motor functions and their organization in body schemata. Because of their widespread organization and projections, sensory and motor functions are highly vulnerable to mechanical head trauma. While some disorders (e.g., gait and posture) are not ordinarily considered to be postconcussion symptoms, they are presented in detail to sharpen the clinician's observations, increase the range of examination of concussed patients, and enhance treatment and assessment of outcome. Some disorders (e.g., spinal column lesions causing proprioceptive loss) are comorbid in the accident that has caused the head injury. Consequent motion disorders should be accurately assessed, since loss of coordination could be misattributed to cerebral damage. Observation of movement, stance, gait, and posture, as well as sensation, aids in diagnosis and thus a more sharply focused treatment of the injured person.

Some screening evaluation procedures use observational skills and easily obtained equipment. Screen for possible nervous system damage (e.g., lateralization, strength, fine motor coordination) and gross motor disturbance (movement, balance, gait, posture). Also screen for sensory disorders to study how these contribute to movement (including proprioception and vision) and body schema interviewing techniques should be used to assess for obtain from outcome, need for rehabilitation, employability, participation in sports, safety, and so on.

When environmental demands exceed a person's sensorimotor and cognitive capacities, the result may be inappropriate behavior with inadequate motor control, or behavioral deterioration (Weber & Verbanets, 1986). While it is claimed that the cortex's influence via the descending systems over spinal motor neurons allows for significant recovery in case of injury, an exception would be the projection from the motor cortex to the fingers and hands. Injury to these fibers results in the permanent loss of skilled movements such as the manipulation of small objects (Ghez & Krakauer, 2000).

PRINCIPLES OF SENSORIMOTOR ASSESSMENT

Diagnosis of impairment includes two principles:

(1) That movement is based upon a unified sensorimotor system. Trauma to motor or sensory circuits or centers will impair the supposedly alternative function.
(2) Sensorimotor disorders may be due to either neurological injury, somatic injury, or both. Body injury interferes with movement, creates pain, and may damage or impinge upon the nerve tracts.

Recently, assessment has stressed outcome measures that reflect functional impairment more than the measurement and treatment of specific impairments such as range of motion or resistance to stretch. Rehabilitation methods should involve more than motor impairment, also improvement in activities of daily living. It is necessary to assess higher level outcomes, not merely the properties of the muscle itself. Retraining includes weight-bearing posture, balance, endurance, and coordination of all limbs.

One offers the precept: the observation of complex rather than simple functions is more likely to detect subtle dysfunctions. The high level of competition required for athletes requires them to utilize simultaneously numerous cognitive resources. Thus, when faced with the question of returning to play, simple tests or balance testing used alone can cause residual impairments to go undetected (Thompson, 2006). For a detailed explication of sensorimotor examinations in adults, see Camicioli and Nutt (2003); for gait, see Ochi et al. (1999); and for children, see Gagnon et al. (2001).

POSTCONCUSSIVE CRANIAL NERVE SYMPTOMS

The cranial nerves (*CN*) leave the brain at various levels of the forebrain and brainstem and exit the skull through complex pathways to cranial, cervical, thoracic, and abdominal locations. Trauma to the skull and CN caused as they traverse and leave the brain, and bodily structures will affect functioning. CN have both sensory and motor components that innervate the head and viscera of the thorax and abdomen. Thus, they participate in the reactions to an injury of both the head and the soma. Some of these are motor reactions, some are sensory, and others are mixed (Butler, 2002). Particular functions and their disorders may be associated with multiple CN, with a nontraumatic etiology that includes aging, drug effects, and medical illness.

Location of Cranial Nerve Nuclei

- Forebrain (CN I Olfactory; CN II Optic).
- Midbrain (CN III Oculomotor; CN IV).
- Pons (CN V Trigeminal; CN VI Abducens; CN VII Facial; CN VIII Vestibulocochlear).
- Medullary (CN IX Glossopharyngeal; CN X Vagus; CN XI Spinal Accessory; CN XII Hypoglossal).

CHEMOSENSORY

Chemosensory systems allow organisms to orient themselves to their chemical environments and to evaluate and distinguish food resources from dangerous substances.

Olfaction–Olfactory nerve (I)

Olfaction is a sensation taken for granted; its loss creates both danger and reduced quality of life. It was among the first brain structures developed in primitive animals, and much of the remainder of the brain has been organized around it. Nevertheless, olfactory structures evolved in several different stages. The very old olfactory system subserves basic olfactory reflexes. The less old system provides automatic but partially learned control of food intake and aversion to toxic and unhealthy foods. The newer system is comparable to most other cortical sensory systems and is used for the conscious perception and analysis of olfaction (Guyton & Hall, 2006, pp. 668–670). The olfactory epithelium is located in the roof of the nasal cavity and extends onto the superior nasal conchae and the nasal septum. Neurosensory cells assemble into about 20 bundles, which traverse the cribriform plate of the ethmoid bone to synapse on the olfactory bulb located on the ventral surface of the frontal lobes. This proceeds posteriorly and then divides, with one branch decussating contralaterally via the anterior commissure and the other branch proceeding to the medial surface of the frontal lobes (stirring emotional responses of the limbic system), and to the temporal lobe (uncus, entorhinal area, the insular junction with the frontal lobe, and the amygdaloid body). The central mechanisms by which the brain recognizes and discriminates attractive and repulsive odorants and tastants, and thus makes behavioral decisions, are not well understood. Selective frontal lobe damage can cause defective odor quality discrimination without a decrease in odor detection (Damasio & Anderson, 2003). Aversive stimuli can create a false impression of olfactory capacity through the stimulation of the trigeminal nerve in the nasal mucosa.

Etiology of Olfactory Disorder

- *Medical*: Many medical illnesses and partial seizures, as well as traumatic brain injury (TBI), can create disturbances
- Regional growths and disease (viral, bacterial; allergic)
- Toxins (tobacco; benzine; cement; ammonia; cocaine; medications)
- Expansive processes to the anterior and middle fossae
- Endocrine disorders (Cushing disease; hypothyroidism; diabetes mellitus)
- Neurological disorders (Alzheimer; Parkinson; Korsakoff psychosis; Down syndrome; multiple sclerosis; epilepsy (including partial seizures); Schizophrenia
- Congenital Kallman de Morsier syndrome (hypogonadotropic hypogonadism; small size; delayed puberty; often transmitted by the X chromosome)
- Other possible causes include alcoholic cirrhosis and hepatic failure, renal failure, and acquired immune deficiency syndrome (AIDS)
- Advanced age (Doyon et al., 2002, pp. 71–73)

Trauma

A previous trauma or surgery should be considered as part of the history. An anteroposterior skull fracture parallel to the sagittal suture can tear the olfactory fibers traversing the cribriform plate (ipsilateral loss of olfaction). Sufficient anteroposterior movement of the brain caused by an impact with a hard surface may pull out or shear olfactory fibers at the cribriform plate. Such fractures may result in leakage of cerebrospinal fluid from the subarachnoid space into the nasal cavity, permitting the passage of air and infectious agents into the cranial cavity (Wilson-Pauwels et al., 1988, p. 6).

Smell affects the palatability of food and drinks and alerts the person to some dangers. Thus, impairment increases vulnerability to danger. A head injury can create olfactory loss due to movement of the ventral surface of the brain after a head acceleration or impact, which shears olfactory fibers originating in the epithelium, then piercing the cribriform plateand terminating in the olfactory bulb. Head trauma results in unilateral or bilateral smell impairment. Standardized assessments indicate that the incidence is about 50–60% after TBI, particularly after orbital-frontal injury. Occipital blows are more likely to cause shearing injury than frontal blows. Forty percent of impaired patients are unaware of their olfactory loss. Such anosmics have a great incidence of frontal lobe-mediated executive skill deficits, which creates a risk for poor vocational and community integration outcomes (Gelber & Callahan, 2004). Olfactory loss is more common when trauma is associated with loss of consciousness (LOC) or more severe head injury. In 50% of cases anosmia is only temporary, though the return of olfaction is not well understood. Recovery can be expected from within a few days to up to 5 years, with a sharp increase being observed in about 10 weeks after injury (Mutali & Rovit, 2000).

Taste–CN VII, IX, and X

The anterior 2/3 of the tongue projects by these routes: Lingual nerve, chorda tympani, facial nerve, tractus solitarius of the brain stem; back of the tongue and posterior regions of the mouth glossopharyngeal nerve VII to tractus solitarius (solitary N nucleus); the base of the tongue, pharynx and laryngeal regions (vagus N). The solitary N. projects to the thalamus (uncrossed) and then to the somatosensorial cortex (mouth) and the brain stem vomit centers.

The elementary taste sensations are described as sour, salty, sweet, bitter, and *umami* (i.e., food containing L-glutamate such as meat extracts and aging cheese) (Guyton & Hall, 2006, pp. 663–666). The receptor cells classify input as either attractive or aversive, with repulsive receptors being predominant. Taste is considered to be a central nervous system (CNS) phenomenon since previous experience with pleasant or unpleasant taste determines taste preferences. Multiple receptors are utilized for bitter substances (Voshall & Stocker, 2007).

Gustatory disorders are less frequent than olfactory ones. Primary sensory neurons are unique among sensory systems in that they are regularly replaced and regenerate. It is not possible to differentiate olfactory losses due to interference with the access of an odorant to the olfactory neuroepithelium, receptor region or central olfactory pathways (Lalwani & Snow, 2005). Examination of taste is recommended when there is a facial paralysis by dropping strong syrup or salt solution on each side of the tongue (Simpson, 2005). Sensations related to olfaction are mediated by the trigeminal, glossopharyngeal, and vagal afferents of the nose, oral cavity, tongue, pharynx, and larynx.

Clinical example: Olfactory and taste loss I: A middle-aged woman was knocked down by a car while riding her bicycle, and suffered brief post traumatic amnesia (PTA) or LOC. She had subarachnoid hemorrhage in the orbitofrontal and left parietooccipital region. She had a skilled public relations position to which she returned successful. She asserts almost complete loss of olfaction. In addition to sensory loss (of herself, her children, their dog, etc.), the injury has removed feelings of human warmth: "I can't smell my daughter's neck and hair. The worst is getting close to my children and not smelling their scent. I don't smell soap or Lysol when cleaning the house. People are more like robots. Now there is a barrier to everyone." Since she cannot determine the progress of her cooking; she cannot smell something burning. With reference to her lack of taste: "There is some loss, but it doesn't bother me as much as the loss of smell." She can still taste lemon, salt, and sugar, but not onion or garlic.

Vignette: Olfactory nerve loss: A medical student who suffered a head injury in a motor vehicle accident (MVA) with brief LOC or PTA was examined more than a year later: "During the last few weeks, during anatomy labs, my dissection group has been commenting on the odor of our cadaver. Initially, I brushed aside the fact that I did not have a similar complaint. I believed either that the others were very sensitive or that I was acclimated to the smell. Many students and professors commented on our cadaver's particularly foul stench. My face was leaning over the body and I could not smell what a student several feet away had responded to very strongly. My anatomy group began responding to my inability to smell. What they perceived as a 'rotting stench' was very much separate from the normal smells of preservation that I could perceive at other anatomy tables. Other students began having difficulty proceeding... I was perfectly comfortable, even with my face practically in the cadaver."

OPTIC PATHWAY AND VISUAL FIELD DYSFUNCTIONS

Overview

Visual pathways: Retina to lateral geniculate body to occipital cortex; tectum to participate in pupillary function; the superior colliculus to participate in eye movements and multisensory integration.

Post-TBI visual dysfunctions: Visual field defects; reduced visual acuity; reduced contrast sensitivity; unilateral or bilateral blurred vision (eyestrain with sustained visual tasks); binocular vision dysfunction (diplopia in some or all positions of gaze, reduced depth perception); nystagmus (abnormal eye oscillations [oscillopsia, nausea, blurred vision, visual confusion]); Deficits of pursuit (difficulty tracking in any plane); deficits of saccades (difficulty in rapid location of objects in space, difficulty with reading).

Optic Nerve (II)

Traumatic neuropathy can occur in the globular, the retinal, and the optic nerve (including swelling within the optic canal) (Kapoor & Ciuffreda, 2005; LeBlanc, 1995, p. 23); the optic chiasma; and can occur posterioraly through the various visual pathways. Optic nerve injury can occur with little or no evidence of significant head trauma, or with multisystem trauma or serious brain injury. Motor vehicle and motorcycle accidents are the most frequent cause, followed by falls, falling debris, assault, stab wounds, gunshot wounds, skateboarding, and seemingly trivial head trauma. The consequences are variable. There may be visual recovery, or there may be delayed visual loss that occurs hours to days after the injury. The victim is concerned with loss of light perception and field defects (Miller & Newman, 1999, pp. 278–290).

Visual field defects (areas of scotoma or lack of vision but without patient awareness) may occur with scattered defects or with the experience of visual inattention. This writer had a patient whose car was rear-ended by a truck, resulting in head impact to the rear with bilateral occipital lesions. He had unusual *central field loss* that required him to read with his finger, line by line. Some patients are aware of their visual field loss. Others may not be aware that part of their visual field is missing and have to be told of their visual inattention. *Hemianopia* alters one's basic activities of daily living, preventing driving and unaccompanied ambulation. Those with visual field defects may bump into objects on one side of their body, miss food on one side of the plate, have difficulty dressing one side of their body, and have difficulty navigating streets and buildings. Left hemianopia creates difficulty finding the beginning of the next line of print when reading. With right hemianopia, reading is slow and laborious and the saccade cautiously takes small steps to move into the right blind field. (Kapoor & Ciuffreda, 2005).

Indirect Trauma

This is caused by forces that are transmitted at a distance from the optic nerve (e.g., blunt trauma to the forehead that results in a transmission of force through the cranium to the confined intracanalicular portion of the nerve). One cause is a deceleration injury of sufficient momentum, with the force of the impact directed to the ipsilateral forehead or mid-face region. Shearing injures the optic nerve axons and its sheaths, particularly within the optic canal, where it is tightly bound. Although evidence is obtained from serious trauma, optic neuropathy is associated with sphenoid bone fracture; forces applied to the frontal bone in a deceleration injury are transmitted to and concentrated in the region of the optic canal. Secondary trauma involves vasoconstriction and swelling of the optic nerve within the optic canal (leaving the orbit). The optic nerve is subject to secondary chemical traumata (Miller & Newman, 1999, p. 284).

Direct Trauma

Direct trauma results from orbital or cerebral trauma, or a penetrating object that transgresses normal tissue planes and thereby disrupts the optic nerve's anatomic and functional integrity. These tend to produce immediate and severe visual loss with little likelihood of recovery. Assessment utilizes the history of the accident and medical, drug and drug allergy, and visual deficits before the injury. Examination includes visual field, visual acuity, and color vision.

Photosensitivity

This is a classic example of postconcussive syndrome (PCS) and occurs with or without pain caused by the contraction and relaxation of inflamed ocular tissue. It can be alleviated by using lenses of varying levels of tint indoors and outdoors. Objective findings include damage to rod-mediated photo mechanisms (without necessarily loss of night vision), which may disinhibit the bright light sensing pathway, as well as pupil dilation in one or both eyes (which particularly requires ultraviolet protection) (Kapoor & Ciuffreda, 2005; Suter, 2004).

OCULOMOTOR DYSFUNCTIONS

This refers to the functions of these CN (Miller & Newman, 1999, pp. 524–561; Saper, 2000b). The most common cause of IVth nerve palsy is head trauma (Liu, 1996). Eye movement control involves the cortex and the brainstem. Proprioceptive stimuli from the extrinsic eye muscles is received by the mesencephalic N of the trigeminal nerve.

Extrinsic Eye Muscles: CN

Cranial nerve III (*Oculomotor*): This nerve innervates all external eye muscles except the superior oblique and lateral rectus muscles. CN III damage is associated with ptosis, a dilated pupil and eyeball displacement that depend upon which muscles are injured or not innervated. CN III is particularly vulnerable to stretching and contusion caused by frontal head trauma, including skull fracture that causes trauma to the cavernous sinus and superior orbital fissure. Recovery can take as long as 3 years. *Synkinesis* may occur (i.e., misdirection of regenerating fibers of CN III to other extrinsic eye muscles and the pupil). Symptoms associated with CN III are diplopia, large pupil, the uneven dilation of pupils, ptosis, and outward deviation of the eye.

Cranial nerve IV (*Trochlear*): Superior oblique muscle (blunt head trauma from a orbital, frontal, basal, or oblique blow). The most common cause of CN IV palsy is head trauma (Liu, 1996). It is particularly apt to suffer an injury after closed head trauma due to impingement by the free edge of the tentorium cerebelli during a concussive blow (i.e., diplopia) (Parent, 1996, p. 5). Recovery from trochlear nerve palsy varies from complete to no recovery. One complaint is vertical diplopia when reading or looking down. Synkinesis can occur between the superior oblique muscle and one of the muscles elevating the tongue and hyoid (CN V,VII,XII), leading to double vision caused by swallowing (Miller & Newman, 1999, p. 561).

Cranial nerve VI (*Abducens*): Lateral rectus muscle. Symptoms associated with cranial nerve VI (CN VI) damage include diplopia, the inward deviation of the eye, and horizontal gaze paralysis towards the side of the lesion.

Bilateral Fixation Disorders

This is the direction of the gaze so that the image falls on the fovea centralis. Binocular fixation adequately trains both eyes on the same object. In addition to the pursuit system, there may be an independent visual fixation system, perhaps in the parietal lobe (Suter, 2004).

Convergence Insufficiency

Common in TBI, convergence insufficiency is a binocular condition in which the eyes cannot rotate inwardly and maintain single vision at close distances. It is associated with eyestrain, the intermittent closing of one eye, diplopia, abnormal sensitivity to visual motion, and illusions of apparent movement of a line of print.

Vertical oculomotor deviations vary in magnitude by gaze position and time of day. In addition, patients may report impaired binocular depth perception and headaches (Kapoor & Ciuffreda, 2005). For other deviations of maintaining fixation (heterotropia, heterophoria), see Miller and Newman (1999, pp. 501–504).

Version insufficiency: Version refers to the conjugate rotation of both eyes in the same direction. Deficits affect the ability to track objects smoothly as they move from one point to another. Versional oculomotor deficits create *reading problems* (reading slowly; loss of place while reading; misreading or rereading words and paragraphs; text that appears unsteady; and apparent visual motion).

Accommodation

This refers to the focusing of the lens of the eye at different distances, a function of the circular *ciliary* muscle that relaxes or extends the lens. Accommodation disorders are affected by CN III, central neurological disorders, or result as a side effect of medication. *Presbyopia* is the diminution of accommodation so that the near point of distinct vision moves farther from the eye. Its neural input is complex and includes parasympathetic stimulation of the ciliary muscle and the sphincter muscle of the iris, which constricts the pupil and increases the convexity of the lens for near vision (accommodation). Difficulty with near focus following concussion or craniocervical injury are commonly associated with headache. This is described as most common in patients with persistent complaints who are seeking compensation. However, an organic basis is difficult to establish. Still, the impairment of the highly complex neurophysiological control of the near focus response could be associated with any cerebral injury. Abnormal input from the upper posterior cervical roots or contusion to the side of the cervical cord could also affect the ascending pathways that influence parasympathetic outflow from the Edinger-Westphal nuclei (oculomotor nuclear complex of the midbrain). Presbyopia impairs the ability to sustain near vision for long periods without ocular fatigue, decreasing visual efficiency and reading ability. Symptoms include intermittent blurred vision, inability to sustain prolonged near vision, tearing, and headaches (Kapoor & Ciuffreda, 2005). Patients complaining of difficulties with near vision after head trauma are described as having a true PCS, with symptoms caused by disturbed comprehension and reduced accommodation from asthenopia as an indirect complication of the injury (weakness or easy fatigue of the visual organs, eye pain, dimness of vision, etc.). They are also described as attempting to gain psychologic or material compensation (Miller & Newman, 1999, pp. 475–476).

Nystagmus

Nystagmus is a nonvoluntary, rhythmic oscillation of the eyes, usually with clearly defined fast and slow components. Contributions come from the vestibular, optokinetic, and pursuit systems, as well as the brainstem (Baloh, 1998, pp. 63–70).

Gaze

Gaze is the act of looking steadily at an object; one is concerned with the ability to retain focus on an object when the body, or both, are steady or moving. During most natural head movements, visual, proprioceptive, and vestibular signals act synergistically to stabilize gaze (Baloh, 1998, p. 38). Deficits of gaze may occur with damage to the frontal eyefields, causing parietal lesions that disrupt the pursuit of targets moving toward the side of a lesion. Lateral gaze palsy, which is worse with gazing to the side of the lesion, suggests abducens CN VI trauma (Horton, 2005). Eye movement disorder can be due to mechanical restriction of an extraocular muscle from trauma to the head (Kapoor & Ciuffreda 2005).

Screening: Conjugate vision (bilateral) can be tested for detecting uneven extrinsic eye muscle function. The patient is seated facing the examiner.

Instruction is given to track a finger moving from the center of the vertical plane to 45° deviations around the circle.

With instruction to track the finger, the finger is rotated clockwise and counter-clockwise.

Attention is paid to the pathways from in which the two eyes do not move synchronously. This reflects a lack of synchronization or dysfunction of one or more extrinsic muscles. With the circular motion, an inquiry is made as to whether it causes the sensation of dizziness. An association has been observed between reports of dizziness and imbalance complaints.

Diplopia

A blurry or double image is caused by a lack of ocular fusion since the perceived object is projected to noncorresponding points of the retina. It may be monocular or binocular, and the determination is aided by covering one eye. It is more frequent with lesions on the extraocular muscles or the ocular motor nerves rather than supranuclear brain stem lesions (structures other than the motor neurones), which often result in gaze palsy. Multiple images in grid-like patterns suggest parietooccipital region damage with field and other complex defects.

Synkinesis

This is a simultaneous movement or coordinated set of movements supplied by different nerves or different branches of the same nerve. Although this is a normal phenomenon (sucking, chewing, conjugate eye movements), it occurs with CN IV trauma (see CN III, CN IV).

LABRYNTHINE (VESTIBULAR) DYSFUNCTION

The vestibule is essentially a space in the bony labyrinthe adjacent to the malleus, which transmits vibrations, and between the semicircular canals and the spiral cochlea. The function of the vestibular system is to transform the forces associated with head acceleration and gravity into a signal that creates a subjective orientation concerning head position in space, and that produces motor reflexes for postural and ocular stability (Baloh, 1998, p. 55). Vestibular loss refers to a reduction of signals from the inner ear balance organ (vestibular labirynthe). Labirynthine fractures are caused by a violent impact producing a shock wave that begins at the occipital bone and spreads to the petrosal mass of the temporal bone. Labrynthine fractures and dysfunctions usually described as vestibular frequently occur after an impact to the base of the skull or behind the ear.

Symptoms associated with fracture of the labrynth include facial paralysis; perilymphatic fistula with pneumolabrynth; a meningeal breach with otorrhea or rhinorrhea, indicating leakage of cerebral spinal fluid (CSF), or meningitis. *Somatic complaints* associated with TBI dysequilibrium include dizziness, vertigo, imbalance (veering or falls), visual blurring, fatigue and difficulty reading, tinnitus, difficulty distinguishing speech from background noise, difficulty hearing, and sensitivity to noise. Even recent minor recent injuries might indicate a balance or coordination problem. Difficulties may be observed in a busy environment or in an environment with visual or auditory distractions. Assessment is affected by musculoskeletal factors: direct injury, leg-length discrepancies, skeletal deformity, and fatigue (Doyon, 2002, pp. 160–161, 176; Richter, 2005).

Labirynthine Concussion

A labirynthine concussion causing vertigo, hearing, and tinnitus may follow a blow to the head that does not cause a temporal bone fracture. The reason is that despite the protection of a body capsule, the labirynthine membranes are susceptible to blunt trauma. Blows to the occipital or mastoid regions are most likely to produce labirynthine damage. Since forces are absorbed by an actual break in the bone, a blow that does not fracture the skull may cause more labirynthine damage than one in which the forces are absorbed. There may be LOC or the onset may occur several days later. Rapid improvement and the absence of brainstem symptoms suggest the peripheral localization of most lesions. The loss of hearing loss is unilateral, there is spontaneous nystagmus, and the external canal and tympanic membrane may be normal. There may also be blood, CSF, or both in the external or middle ear if there is an associated temporal bone fracture. Sudden deafness is often partially or completely reversible. Intense acoustic stimulation is caused by pressure waves that are transmitted through the bones to the cochlea. Animal studies suggest that there is degeneration of hair cells and cochlea neurons in the middle turns of the cochlea, with pure tone hearing loss being most pronounced at 4000–8000 Hz (Baloh, 1998, pp. 182–183).

Injury may stem from injury to the vestibular end organ, the vestibular nerve, or the root entry zone of the cranial nerve VIII (CN VIII). In vestibular deficits, the inner ear's sense of balance is essential when visual and somatosensory inputs are disrupted or provide conflicting information (Guskiweciz, 2001). A person without severe brain injury may be able to utilize sensory compensation and alternative balance strategies. Acute unilateral vestibular loss affects the vestibulo-ocular and vestibulospinal systems, causing autonomic symptoms and spatial disorientation. Acute vestibular loss causes symptoms such as vertigo, dysequilibrium, nausea, nystagmus that increases when visual fixation is removed, gait instability, pallor, and other signs of autonomic dysfunction. Compensation begins in the CNS for the remaining labirynthe to substitute as the alternate signal (Furmam, 2003).

We are concerned with the functions of structures included within the bony labirynthe, which is a series of hollow channels within the petrous portion of the temporal bone. It consists of the vestibule, semicircular canals, and the cochlea. The vestibule and labirynthe have overlapping structures and functions. The most important of these are the semicircular canals, the cochlea (concerned with hearing), and the utricle and saccule. The vestibular nerve senses linear and angular acceleration, which is used to maintain balance. This function is vital for precise, complex movement. This involves sensing the initial position of the external or bodily target, the location of the moving body structures, a model of the intended final position of the movement, the monitoring of the motion with determination of any error deviation, and the integration of information concerning the position of the eyes and ears as relative to the limbs and trunk.

The Central Nervous System and Vestibulocochlear N

Vestibular afferents project to the four vestibular nuclei that occupy the floor of the fourth ventricle of the medulla and pons. There are ipsilateral and bilateral interconnections with the vestibular complex. There are ascending thalamic connections, connections within the brain stem, and descending spinal connections. Some vestibular fibers project directly to the cerebellum. Labirynthine dysfunction may be categorized as peripheral (clear labirynthine dysfunction) or central (cerebellar or lateral medullary lesions). Apparent vestibular dysfunction can occur with and without documented CNS lesions. A syndrome includes hearing loss (which may be transient), hyperacusis, tinnitus, vertigo, dizziness, and loss of stapedial reflex responses. Blunt trauma to the head can lead to auditory dysfunction, probably due to diffuse axonal injury of the central auditory pathway (Nolle et al., 2004). One report asserts that a punch that turns the head can create sensory disorders and memory disturbance (Ohhashi et al., 2002). Transverse fractures of the temporal bone cause transient auditory-vestibular symptoms for milder injuries, but severe vertigo or total auditory loss for severe injuries. Longitudinal temporal fractures cause injury to the middle ear, with prominent conductive hearing loss and, less frequently, vestibular dysfunction. Unilateral injuries may include acute spontaneous nystagmus, impaired balance, and provoked vertigo (a spinning sensation elicited by head turning, sudden eye movements, and other challenging stimuli) (Heid et al., 2004; Richter, 2005).

Comorbidity: Vestibular disease can be comorbid with brain trauma, musculoskeletal disease, head and neck injury, and headache. Symptoms include LOC, emotional distress (anxiety without panic disorder; frustration, despair, depression), and cognitive complaints. In one study, active psychiatric morbidity, somatic concerns, and affective symptoms were more frequent and severe in patients with persistent vestibular symptoms, although these were unrelated to the specific character of the vestibular symptom. Psychiatric patients with complaints of dizziness but no evidence of peripheral vestibular disorder had a greater prevalence of unexplained medical symptoms, and self-reported somatic, depressive, and anxiety symptoms. While vestibular pathology improves during the first 4 weeks, mood disorders may not improve at the same rate (depression, anxiety, and interpersonal relationships). These may occur post-trauma due to central damage, CN damage, skull fracture, trauma to nuclei and CN, or injury to the external ocular muscles (Brazis et al., 2001, pp. 171–172).

Middle ear: The auditory ossicles transmit auditory vibrations. The tensor tympani muscles (dampens loud noises). Conductive hearing loss is mechanical (e.g., interference of movement of the oscicles, or transmitting this movement to the cochlea). Bony dislocations dominate middle ear sequelae and can cause labirynthine fistulae with impairing dysequilibrium. One patient in a low-energy MVA experienced head impact with a persistent impairing tinnitus (Granier, 2006). Neurobehavioral structures are listed (Moore & Dalley, 2006, pp. 1026–1037).

Inner ear: The cochlea (auditory branch) and the semicircular canals (vestibular branch). Mechanotransduction affects the sensory hair cells of the inner ear, which are vibrated by sound or head movements, to stimulate the vestibulocochlear nerve (VIII), which projects to the vestibular and cochlear nuclei in the brainstem (Standing, 2005, p. 663).

Hearing Loss (Cochlear Nerve)

There are multiple medical reasons for hearing loss, in addition to trauma.

Sensorineural hearing loss: Sensorineural hearing loss results from defects in the cochlea or cochlear nerve brainstem, and also in cortical connections. High-frequency sensorineural hearing loss refers to a deficit in perceiving tones or speech due to a lesion involving the cochlea (central) cochlear nerve and nuclei or the central auditory pathways. Injury of the inner ear causes unilateral hearing loss specific to the area of injury; consequently, there is varying loss of frequencies. Sound distortion is common and the two ears may be differently affected. A pure tone may be heard as noisy, rough, buzzing or a complex mixture of tones. The patient has difficulty hearing speech mixed in with background noise and may be annoyed by loud speech. Since auditory information is distributed bilaterally, damage to the auditory neurons, brainstem, or cortex causes more subtle problems, such as inability to localize sounds (Hudspeth, 2000; Vernick, 1996).

Conductive hearing loss: This disturbance results from lesions involving the external or middle ear. The tympanic membrane and ossicles act as a transformer, amplifying airborne vibrations to the inner ear fluid and the spiral organ of Corti. Trauma is only one cause of conductive hearing loss. The perceptual loss is greater for quiet speech than for loud speech.

Central hearing loss: This refers to lesions of the central interconnecting auditory pathways and the temporal lobes, medial geniculate bodies, inferior colliculi, and the cochlear nuclei. A patient usually does not have impaired hearing levels for pure tones. He or she can understand speech clearly spoken in a quiet environment. Background noise or competing messages cause more marked performance deterioration than in patients without impaired hearing. Since half of auditory fibers cross central to the cochlear nucleus, this is the most central structure in which a lesion can result in a unilateral hearing loss.

Tinnitus

Tinnitus is the perception of a sound, localized in one or both ears or in the skull, without outside stimulation. The sensation of hearing a noise or a tone may be subjective (i.e., heard only by the patient), which is a component of the PCS. It can also be due to lesions situated at any level of the auditory pathway—in fact, it can persist even after destruction of the inner ear. The sound is usually high pitched, not very intense, and its cause remains unknown. Traumatic causes include blast trauma, barometric trauma, and presbyacousia trauma. Objective sounds are audible by the examiner, that is, a vibration located in the craniocervical region, within or outside the ear. Pulsating tinnitus sounds are synchronous with the cardiac rhythm. These may be subjective or objective. Tinnitus may be caused by muscle spasms (Baloh, 1998, Chapter 9; Doyon et al., 2002, 178–186). Traumatic tinnitus is described as being more intense and bothersome than a nontraumatic condition, but does not vary with respect to pitch or masking or general hearing levels. It can be exacerbated by stress and anxiety (Tusa & Brown, 1996). Tinnitus may be caused by lesions of the external or middle ear, cochlear or auditory nerves, or the CNS. Pulsatile tinnitus may develop after head trauma if there is an arteriovenous malformation, particularly one involving the cavernous sinus. A tinnitus screening should include hearing loss from damage to the inner ear, noise-induced hearing loss, and hearing loss due to ototoxic drugs. Screening for a severe pathological process is indicated with unilateral, pulsating, or fluctuating tinnitus or vertigo. Many other medical disorders may be implicated (Brazis et al., 2001, p. 315). Tinnitus caused by lesions of the external or middle ear is usually accompanied by a conductive hearing loss with complaints that one's voice sounds hollow and that other sounds are muffled. Tinnitus caused by lesions of the cochlea or auditory nerve is usually associated with sensorineural hearing, with or without distortion of sounds. Tinnitus resulting from CNS lesions is usually not associated with hearing loss but with other neurological conditions.

Exploration is initiated by asking the patient about the characteristics of the noise (pulsatile vs. constant; unilateral vs. bilateral; intermittent vs. continuous). The noise may be subjective (heard only by the patient) or objective (heard from the physician) (Baloh, 1998, Chapter 9; Brazis et al., 2001, pp. 315–316; Vernick, 1996). Tinnitus may be traumatic in origin, although there are a large number of nontraumatic etiologies, lesions of the external or middle ear, cochlear or auditory nerves, the CNS, muscle spasms, vascular tumors, and so on. Pulsatile tinnitus may develop after head trauma if there is an arteriovenous malformation, particularly one involving the cavernous sinus. Tinnitus screening should include hearing loss from damage to the inner ear, noise-induced hearing loss, and hearing loss due to ototoxic drugs.

Imbalance

Imbalance is an objective sign insofar as it can be measured, although visual observations are useful. It refers to postural responses that interfere with an individual's ability to rise and remain erect during standing and locomotion. It is characterized by feelings of unsteadiness, an inability to walk a straight line or sudden turn and change in direction without veering or staggering, uncertainty of the exact position of the feet when walking, and an inability to appreciate the texture of the ground beneath one's feet (or the impression of walking on a spongy surface). Most patients with brainstem lesions have some balance problems, while balance problems from cerebral lesions are rare (Halmaji, 2003). Tumors of the nervous system have a slightly increased incidence in the first year after an injury. It is hypothesized that early tumors were already present at the time of injury but that a fall may have resulted from associated seizures, ataxia, or aphasia (Inskip et al., 1998).

Balance has been defined as a condition in which all forces and torques acting on the body are in equilibrium so that the person's center of mass (COM) is within his or her limits of stability. Balance requires that the COM of the upright body be maintained within the area of foot contact on the ground. The *otolith organs* of the vestibular system provide signals of the direction of gravity, which are related to head orientation but not the relative position of the COM. Balance can be impaired by somatic injury since tissue viscosity and elasticity, as well as inertia, act at each of the joints. The *index of stability* is the area within which a person will move as a function of his or her base of support (Thompson, 2006). Balance is the process of maintaining the *center of gravity* (*COG*) within the body's base of support. Its sensory component involves vision, bilateral labirynthine function, and proprioceptive information from the periphery (see Vestibular Contribution to Motion Control). If one considers a batter striking a rapidly approaching ball, a football player leaping to make a catch, a dancer en pointe, and so on, the combination of somatic, visual sensation, vestibular input (and a bit of practice), will illustrate a combined support for balance. Balance should be explored particularly after impact to the temporal region.

Somatosensory contact force indicates a downward direction. Contact forces (touch and pressure) are very important in the control of balance and locomotion. Touch stabilizes balance: standing persons in the dark are as stable with finger contact as they would be with full sight of their surroundings without finger contact. In the recovery period, patients learn to use tactile cues on the soles when walking to control balance. This becomes easier with practice but they will continue to experience difficulties on uneven or compliant surfaces under conditions of reduced illumination.

Vestibular dysfunction: Individuals who lose vestibular function have great difficulties with *balance and locomotion*. Lesions to the *proprioceptive system*, cerebellum, thalamus, or basal ganglia are considered. Asymmetry of tonic vestibulospinal activity leads to postural and gait imbalance, with the patient tending to fall toward the side of the lesion. Falls may be due to foot drop, shallow steps, proximal muscle weakness, or the impairment of postural reflexes (Thompson, 2004). It is stated that this unsteadiness in gait is rapidly compensated for, usually lasting less than a week. In a group of military personnel with concussion, balance problems were reduced in the first 4 weeks after injury, but did not return to control level (Gottshall et al., 2003).

Dizziness

Dizziness is a subjective symptom that can be experienced at rest or when in motion. It is a generic term for a variety of disorders, and may not include the vestibular system. It may be associated with visual disorders that are consequent to eye movement disorders (ophthalmoparetic), imbalance or unsteadiness (contributing to falls), and vertigo (an illusion of movement, most commonly rotation, although patients may report a sense of linear displacement or "tilt"). In its use it overlaps somewhat with vertigo, which refers to the sensation of rotation of the subject, or of objects around the subject. Dizziness is related to injury of the inner ear, peripheral neuropathies, the consequences of whiplash, musculoskeletal injuries, vascular disorders, and impairments of visuomotor control. Dizziness caused by vestibular lesions is usually worsened by rapid head motion. Central neurological signs of dizziness are almost always accompanied by other signs of CNS dysfunction (gaze-evoked nystagmus, facial weakness, other CN abnormalities, ataxia, hemisensory loss, or paralysis). Patients with dizziness have a significant risk of psychiatric dysfunction, perhaps as high as 50% in cases of panic disorder or depression. A significant proportion of patients with balance disorders experience anxiety. If disequilibrium or dizziness is of long duration it may be difficult to determine whether the symptom complex is the result of anxiety or depression or whether this anxiety or depression is secondary to the illness. Anxiety reduces the prognosis for resolution of dizziness and increases the handicap. However, "psychogenic" dizziness is not merely an emotional disorder (Richter, 2005). It is a sensation of unsteadiness (perhaps with a feeling of movement within the head), presyncope, unsteadiness, giddiness, faintness, imbalance, lightheadedness, and so on.

Posttraumatic Vertigo

Vertigo is a subjective symptom (i.e., the false sensation of objects moving with respect to the subject, or of the subject with respect to objects). Vegetative symptoms (nausea, vomiting, pallor, sweating) are frequently present (Younger, 2005e). Vertigo should be differentiated from faintness, lighted headedness, presyncope (decreased blood flow to the brain); disequilibrium, which is an imbalance or unsteadiness while standing or walking (loss of vestibular, proprioceptive, visual, or motor functioning); nonspecific or vague symptoms of floating, swimming, giddiness, rocking, falling, and spinning inside the head (associated with psychogenic dizziness, and found in anxiety, agorophobia, obsessive-compulsive disorder, and somatoform disorders including conversion and depression); and hyperventilation as a cause of dizziness (Brazis et al., 2001, p. 315). There are numerous causes of vertigo. Here we will focus upon trauma. Vertigo is described as a hallucination of movement. Episodic positional vertigo after head trauma suggests cupulolithiasis. Calcareous deposits of the labirynthe (otoconia) are displaced to a sensitive region of the posterior canal, making it more susceptible to stimulation in certain head positions. A spinning sensation suggests the peripheral vestibular apparatus. A complaint of incoordination or clumsiness makes one consider cerebellar dysfunction or peripheral neuropathy (Younger, 2005b). Lightheadedness or feelings of a swimming head suggest presyncope or syncope with systemic factors (vasodepressor syncope, postural hypotension, cardiac dysrhythmia). A unilateral injury to one peripheral end organ may result in asymmetrical input to the CNS. This may be interpreted as a sensation of turning or vertigo. CNS plasticity may compensate for an injured peripheral vestibular system. Despite differences in afferent input, the patient no longer experiences a sensation of vertigo and no nystagmus is present (Troost, 2004). Vertigo is a disorder of the vestibular system with the concomitant objective sign of *vestibulo-ocular nystagmus*. Afferent nerve impulses from the otoliths within the saccule and utricle, as well as the semicircular canals of each labirynthe, maintain a balanced tonic firing rate into the vestibular nuclei. Asymmetry of the baseline activity creates a sensation of movement. Vertigo may be associated with unilateral vestibular dysfunction. When patients with unilateral lesions fixate on an object, it appears to be blurred and moving away from the side of the lesion (slow phase of their nystagmus). Since the brain lacks eye proprioceptive information, it interprets target displacement on the retina as object movement rather than eye movement. In addition to head impact, vertigo may be involved with disorders of the *cervico-occipital junction* (trauma involving the upper cervical vertebrae) (Doyon et al., 2002, p. 176). The disorder involves the posterior labirynthe, the vestibular nerve, and centers involving integration of sensation. The consequence is *sensory conflict*. False information transmitted to the centers involved with equilibrium contradicts other sources of information that deny any movement.

Here, we are primarily concerned with head trauma, one form of which may include an ischemic stroke of the vertebral artery or the posteroinferior cerebellar artery. Other sources of vertigo, including tumors, are not considered here. Acute unilateral vestibular lesions cause postural and gait imbalance with a tendency to fall towards the side of the lesion. Imbalance usually lasts less than one week. Bilateral symmetrical vestibular loss results in a more pronounced and persistent unsteadiness, which can be incapacitating in older patients, particularly those with peripheral neuropathy, and impaired vision, which is worse and night and creates difficulty in compensating for the vestibular loss (Baloh, 2003; Doyon et al., 2002; Drislane, 1996).

Oscillopsia

This is an optical illusion where stationary objects move back and forth or up and down. When a patient with acute unilateral and peripheral vestibular damage attempts to fixate on an object, it will appear blurred and seem to be moving in the opposite direction of the spontaneous nystagmus. With bilateral loss of vestibular function, oscillopsia occurs with any head movement. When walking, the surroundings seem to be bouncing up and down leading to inability to fixate on objects. Visual tasks, including reading, requires holding the head still.

Vestibular Contribution to Motion Control

Maintaining balance requires visual, somatosensory (proprioceptive), and vestibular sensations. The vestibular system controls both balance and eye movements. The sensory organ detects head velocity, acceleration, and orientation to gravity, creating a signal that the brain uses to develop a subjective awareness of head position in space and producing motor reflexes for balance and equilibrium. Displacement is monitored by sensors for linear and angular acceleration (Baloh, 2003). The output is carried by the vestibulocochlear N. (VIII) to the lateral medulla, and thence to a central processor (vestibulonuclear complex and vestibulocerebellum). Vestibular nuclei integrate vestibular, proprioceptive, and somatosensory information (Hain & Helminski, 2003). Central processor outflow is sent to:

The oculomotor nuclei, to generate eye movements that compensate for the motion of the head (*vestibulo-ocular reflex*, or *VOR*). Movement in each semicircular canal stimulates particular external eye muscles while inhibiting others, resulting in eye movement in the plane of that canal (Baloh, 1998, p. 33). Impulses from the semicircular canals serve to move the eyes in the opposite direction of rotations of the head. Relays from the vestibular nuclei to the vestibulocerebellum adjust the VOR responses.

The ventral horn of the spinal cord, which generates compensatory head and body movement effectively stabilizing eye movements (*VOR*, *VCR*, and *vestibulospinal reflex* or *VSR*).

The cerebral cortex, which contributes to subjective awareness (head and body position and motion in the environment) (Hain & Helminski, 2003).

The vestibular system maintains the eyes' fixation on a stationary target during head and body movement by converting information about angular acceleration of the head into velocity information and sending it through the VOR pathways to the extraocular muscles. Since these are arranged in planes that are closely oriented to those of the canals, a pair of canals is connected to a pair of eye muscles, resulting in approximately parallel movement. It also maintains balance by combining additional information from visual and somatosensory inputs, with angular and linear acceleration information (including gravity) from the utricles and saccules of the inner ear, and transmitting it via the vestibulospinal tract to the spinal and lower extremity muscles.

PROPRIOCEPTION: IMBALANCE AND MOTOR CONTROL

Balance dysfunction may be difficult to detect when vision compensates for loss of proprioceptive input, or when performance is observed upon even surfaces. Touch cues (e.g., from the soles) can override proprioceptive misinformation about body orientation.

Screening Procedures

Walking with eyes directed forward: Cautiously ask the patient to walk with eyes directed forward. The examiner or another person should be close by in case of a possible fall. With a heavy person, this procedure may be disregarded if there is no suitable assistance. The writer remembers one man who almost collapsed at the moment of walking with his eyes directed forward instead of downward. One may immediately observe indecision, tilting, or imbalance.

Walking with eyes closed: Cautiously, when observation suggests a balance problem (deficiency in standing on one leg; noting that the patient looks down for guidance when walking towards the examiner), ask the patient to walk with eyes closed. It may be necessary to have the examiner or a collateral stand by the patient during this procedure. Sometimes a single step with eyes closed will reveal that visual orientation is required, since the patient immediately tilts. By implication, axial and limb proprioceptive impulses are impaired (Guskiewicz, 2001; Lackner & DiZio, 2002; Schieber & Baker, 2003).

Closed eyes complex proprioceptive maneuver: After simple motions (e.g., having the seated patient extend each arm to touch the examiner's finger), the following maneuver is performed with eyes closed. The examiner sits in front of the patient. At a comfortable distance, the patient is then told to look at the examiner's finger, which will not be moved. The patient would be asked to do certain motions with his or her eyes closed. First, the examiner determines a comfortable distance for the patient to extend an extended forefinger to touch one of the examiner's forefingers (with the four other fingers retracted). This avoids the possibility that contact would be made by other fingers.

This requires extending the arm beyond the legs. The instruction is given: "Look at the position of my finger. It will not move. Close your fist except for the second finger pointing forward. With your eyes closed, touch your nose (then when this is accomplished) touch your knee (then) touch my finger in space, and touch your nose."

If the patient opens his or her eyes, then the entire maneuver has to be repeated. The maneuver is performed bilaterally, starting with the dominant hand, then with the alternate hand, and is repeated at least once. The examiner observes the distance between the patient's forefinger and his own, whether the approach is direct or with circling motions, and whether the deviation is characteristically high or low or left or right. Thirty seconds is a reasonable time to determine whether the patient can direct a finger to the examiner's. When a second trial with the same arm is given, it is observed whether the direction of movement is better directed.

Lower limbs: These procedures assess proprioceptive functioning of the legs, in contrast to the upper limb concern of the eyes closed proprioceptive maneuver described above. The procedures include *heel-to-shin* maneuver (i.e., bringing the heel to the opposite knee, and sliding it down the anterior aspect of the tibia to the ankle) and *toe-to-finger* maneuver, where the examiner moves a finger to a new position with the patient *standing unassisted from a chair.*

Romberg test: Procedures for this test vary between examiners. The patient stands with feet together, first with eyes open and then with eyes closed (e.g., for 6 sec). A standard procedure is standing in the heel-to-toe position for 6–30 sec with eyes closed, though length decreases with age in adults. The *Romberg sign* is the patient exhibiting slight unsteadiness while standing with feet together and eyes closed.

POSTURE/STANCE

Stance is defined as the posture and general orientation of a person standing. The writer defines posture as the position of the body when standing as modified by muscular effort to offer a certain appearance (or lacking effort to stand erect), injuries (bone, soft tissues, nerves) pain, fatigue, and compensation for a balance disorder. Receptive and perceptual difficulties can cause adoption of head tilts or turns and shifts in posture. This creates or complicates problems of balance when standing and walking, of which patients are often unaware (Suter, 2004). Inappropriate interactions among the sensory inputs can cause a balance deficiency (e.g., excessive dependency upon one sense can cause intersensory conflict). Although posture is a sensorimotor activity, when instructed to stand "as still as possible," persons with lesser TBI showed electroencephalogram (EEG) changes, from which it was inferred that this instruction is a *cognitive load* (Thompson, 2006). A general muscular concept important in all activities is *tone*, which is defined as the unconscious and automatic induction of muscle power in response to an applied load. *Power* is the maximum force that can be generated by any stimulus or provocation, whether voluntary or reflexive, or normal or pathological. Diminished muscle power is more evidence for involvement of the motor unit rather than centrally. *Fatigue* has several causes: central perception of mood (weariness or sleepiness) and progressive loss of contraction from tetanic muscle stimulation (high-frequency stimulation causing contraction fusion) (Crawford, 2004).

MAINTAINING UPRIGHT POSTURE

Sensory organization involves timing, direction, and amplitude of corrective information from vestibular, visual, and proprioceptive (somatosensory) inputs, including muscles of the neck, head, calves, thighs, and back. Postural responses are also under *feed-forward control.* Some movements are so rapid that control is impossible if it depends upon afferent stimulation to make a round trip from the periphery to the brain and back. Feed-forward control utilizes anticipation of the outcome of a movement in coordination with the overall motor program and whether the movement has been performed correctly. If not, the brain corrects its signals each time the movement is performed.

This process has been described as adaptive control and delayed negative feedback (Guyton & Hall, 2006, p. 9).

Balance: The preferred sense for balance control comes from somatosensory information. Vestibular signals from the labirynthe (inner ear) provide information about the rotation, its orientation with respect to gravity, and the acceleration of the head in space (Schieber & Baker, 2003). Pressure receptors in the feet relay information about the distribution of weight relative to the COG. Contact forces supporting the body indicate the downward direction, which indicates the direction of gravity and the location of the body COM with respect to the body surfaces in contact with the ground. The otolith organs also provide signals relating to the direction of gravity. These signals emphasize head orientation. Unlike somatosensory contact cues do not convey information about the relative position of the center-of-mass (Prochazka & Yakovenko, 2002). The vestibular system, eyes, and neck and spinal column proprioceptors give information concerning the position of the head as relative to the environment (rotation and orientation with respect to gravity). Unilateral damage to the labirynthine section of CN VIII causes leaning or falling towards the side of the lesion. Patients with cerebellar lesions were confronted with perturbations in their visual surroundings, or of the platform on which they stood (disorder of *sensorimotor integration*). They were impaired in maintaining their stance because of a disruption to the temporal and spatial pattern of activation of muscle contractions (Willingham, 1992).

Reflexes

Vestibulocervical (neck) or vestibulospinal (lumbar) reflexes stabilize the posture of the head and body. They are negative feedback systems, exciting pathways that contract neck and limb muscles that (1) oppose the undesired movement, and (2) reduce the characteristic resonant frequency of the head (203 Hz) that occurs after bilateral semicircular canal damage.

Observation

Posture is vulnerable to trauma. An examiner will observe lateralized changes in shoulder height and spinal column orientation. Posture may be upright or stooped. The COG may be displaced either forward or backward. With the patient standing, the examiner will observe the position of the shoulders. Slight deviations from a horizontal position may indicate known or unreported damage to pectoral girdle and the associated soft tissue. The examiner should note wasting or hypertrophy of muscles (fat replacement of muscles or chronic contraction, including of the limbs and the face). Is the base wide or narrow? Is the neck vertical or tilted; mobile or stiff? Neck position, in addition the effect of trauma, may be a result of compensation for visual or auditory deficits. The examiner should continue the observation on through the waist, hips, legs, and feet. Ask the patient to turn slowly around. Is the body axis erect or tilted? Are the expected curvatures of the neck and spine normal, excessive, or straightened? The examiner will consider spasm of the paraspinal muscles, as well as cervical or other spinal trauma. Is the torso erect or tilted? The examiner should observe the feet: Are they splayed outwardly bilaterally or unilaterally, or adducted.

Arms: Tremor may be slight but detectable by requesting the patient to relax each hand, and then holding them to sense inconspicuous movements. There is generally an agreement between the examiner's impression of slight or absent tremor and the patient's awareness. Slight spasticity can be detected by the patient gently extending and flexing his or her wrist. When using a dynamometer to measure grip strength, even an obviously muscular person asserting that he or she is gripping very hard may manifest a low level or absent measured muscular pressure on the dynamometer. If muscle bulk indicates normal tone, and palpating the forearm does not suggest more than trivial tension, then faking may be cautiously considered. Electrophysiological study can aid in the differential diagnosis to consider incomplete voluntary effort, hysteria conversion, malingering, and combined functional and lesional disorder (69). *Dysdiadochokinesis* is the decomposition of alternating or fine repetitive movements. Deficits occur in the rate of alternation, the completeness of the sequence, and errors in the sequence and speed of the component parts of a multijoint movement. It may be screened by asking the patient to extend both arms and and then slowly and later rapidly to rotate the forearms in opposite and then parallel directions. At a certain speed the synchronization may break up (Timmann & Diener, 2003).

Reaching and grasp (goal-directed movements) require the sensorimotor transformation of sensory environmental representations into muscle control signals. Reaching depends upon the location of the target relative to the body, shoulders, or hand. Grasping is governed by the shape and dimensions of the object (Krakauer & Ghez, 2000). This visuomotor transformation is rendered more precise by proprioceptive input.

Lateralized differences: Numerical measurements of muscular strength (dynamometer) and psychomotor speed (grooved and ordinary pegboard) are recommended. In addition to normal variability, the examiner should consider the effects of pain, reduced range of motion caused by skeletal and soft tissue damage, as well as CNS lesions at different levels (resulting in ipsilateral or contralateral dysfunction). The frequent assumption that the expected dominant/nondominant measurement ratio is expected to be 10% faster or stronger is open to question. Bornstein (1985) studied this issue with these frequently used sensorimotor procedures: finger tapping; smedley dynamometer; and grooved pegboard test. Age and years of education were not related to intermanual differences. In a nonclinical population, a significant proportion of cases exhibited "atypical patterns of performance." About 30% of males and 20% of females offered the nonpreferred hand as superior to the preferred (writing) hand. An analogy was drawn with *IQ discrepancies* (e.g., performance and verbal IQ differences on the Wechsler Adult Intelligence Scale-Revised). Deviations of 30 IQ points in both directions exceeded the statistically derived expectations. A deviation of 10 points was required to be statistically significant at the .05 level. In actuality, in the normative sample, a deviation of 23 points was required for the .05 level.

Body Position

Information arrives from several sources. Pressure receptors in the feet relay information about the distribution of weight as relative to the COG. The proprioceptors in the neck inform the head and body about their relative positions. Gait and limb motions are dependent upon the integration of positional information (awareness of movement) position of limbs and joints, weight and tension (kinesthesis; proprioception; somesthesia) from muscles (muscle spindles of extensor muscles), joints, tendons (Golgi tendon organs) and hair cells with visual information.

The *vestibular system*, the eyes, and neck and spinal column proprioceptors give information concerning the position of the head as relative to the environment. Particular neurons in the retina respond specifically to moving images. When head movement changes objects in the visual field, this information supplements that coming from the semicircular canals. Postural control is the means by which balance is achieved in order to control the relation between the COM and the base of support (Gagnon et al., 2004). Visual, vestibular, and somatosensory sensors send commands to the muscles of the extremities, which generate contraction to maintain postural stability.

Performance After an Injury

Two components of walking are *multijointed limb coordination* and *postural control*, which are both affected by the cerebellum. Impairment can occur separately or for both. After an injury, visual and somatic proprioceptive systems compensate for postural and movement deficits stemming from vestibular (labrynthine) dysfunction. Postural control deficits are accompanied by only more abnormalities than a leg coordination deficit. Most impaired walking occurs in those with both leg incoordination and impaired postural control. When impairment of postural control and leg coordination during walking are compared, postural control is more impairing. It is suggested that on level surfaces locomotion is more dependent on cerebellar output to vestibular and the brain-stem motor regions than to the cortical motor regions (Morton & Bastian, 2002). Concussed athletes manifested a *postural stability deficit* (i.e., a sensory interaction problem preventing accurate use and exchange of sensory information from the visual, vestibular, and somatosensory systems) (Guskiewicz, 2001).

Force

Motor systems produce movement by translating neural systems into muscle contractile force (Ghez & Krakauer, 2000). Several parameters describe forces involved in motor activities: magnitude and direction; force bias, referring to a constant force (e.g., gravity); ostatic (isometric) force, referring to postural control; and dynamic force to changing force patterns (kinematic movement planning; a mass to be accelerated). Motor cortex activity refers to spatial representation of the motor trajectory. Visual-spatial tasks include directional transformations and trajectory planning. Since a force bias is always present, the force exerted by a subject is made up of both dynamic and static components and is represented at the motoneuron pools by the confluence of dynamic and postural (static) inputs from supraspinal and spinal interneuronal systems. Convergence of paths provides an integrated signal to the motoneuron pool (Georgopoulos et al., 2002).

PLEXUSES

Plexuses are intermediary neural elements situated between the emerging spinal nerves as they emerge from the intervertebral foramina and the main nerve trunks. Because of the intermingling of sensory, motor, and sympathetic nervous system (SNS) fibers, after an injury, the dysfunctions may be difficult to assign to a particular spinal segment. Details of the distribution of motor and sensory roots and neurological signs after trauma are offered by Brazis et al. (2001, pp. 75–90). Many thousands of axons traverse them while following intermingling roots. These are rearranged as distal nerve tracts. Plexus disorders tend to involve motor nerve axons of the arms and legs (see Younger, 2005a, for diagrams).

Cervical Plexus

The cervical plexus arises from the ventral rami of C1–C4 or C5 and branches from the *superior cervical sympathetic ganglion* (Moore & Dalley, 1999, pp. 294, 1010, with diagram; Parent, 1996, p. 275). Nerve elements include the hypoglossal, phrenic, and accessory nerves. Ascending fibers of the accessory nerve through the foramen magnum join bulbar rootlets of the vagus nerve, and the two leave the skull through the jugular foramen. Accessory nerve filaments receive branches from C2 to C4 and innervate the sternocleidomastoid muscle and the upper portion of the trapezius muscle. The cervical plexus is most frequently injured by compression. Weakness and paralysis of the diaphragm is consequent to bilateral phrenic nerve injury (Younger, 2005a).

Brachial Plexus

Fibers stem from cervical segments C5–C8 and T1, which unite above the level of the clavicle to form three trunks that carry motor, sensory, and postganglionic sympathetic fibers to the upper limb. Patients may present weakness, reflex change, and sensory loss, depending upon whether a part or the whole plexus is disturbed (Chad, 2004). The brachial plexus' location between trauma-vulnerable structures (e.g., the highly mobile neck, shoulder, clavicle, first rib, and the arm) renders it vulnerable to trauma (impact, stretching, and traction). In the vast majority of patients there is blunt trauma striking the shoulder, compressing it between the clavicle and the first rib. Continued head movement increases the distance between them. It may result from injury to nearby bony structures. There is a loss of function in the shoulder and elbow joints and the patient is unable to extend the wrist and fingers. Injury to the humeral head and axillary compression also occur. Unilateral muscular weakness may reflect brachial plexus injury of restricted range of motion due to fibrositis.

The pattern of motor and sensory dysfunction is related to the differential vulnerability of the roots of the nerves at different levels and whether they are anterior or posterior (Younger, 2005e). A complete lesion of the spinal nerves can result in paralysis of the arm (flail arm). Lesser injury results in loss of function of the elbow and shoulder joints, with inability to extend the wrist and fingers. The prognosis is variable (Ferrante & Wilbourne, 2000; Millesi, 1997; Wilbourne, 2003). Injuries affect the trunks and terminal nerves of the upper limb. Its location between the highly mobile neck and the arm, as well as other trauma-vulnerable structures, increases its vulnerability to traction injuries. Injury to the humeral head and axillary compression also occur. Accidental traction injury is illustrated by forcible separation of the arm or neck from the trunk. Traumatic plexography is due to direct trauma (stretch or fraction) or secondary injury from damage to structures around the shoulder and neck (fractures of the clavicle and first rib) (Chad, 2004). Lesions occur after high velocity trauma, falling from a height, contact sports ("burners" or "stingers"), and penetrating injury (gunshot and stab wounds). High-energy injuries cause substantial axon loss, while the "burner" syndrome has mild axon loss.

Lumbosacral Plexus

In contrast to the brachial plexus, trauma is responsible for only a small fraction of lumbosacral plexopathies. The lumbar plexus (L1 to a portion of L4) is formed within the psoas major muscle. Its motor innervation includes quadriceps and thigh adductors, and iliacus and psoas. The sacral plexus receives axons from part of L4, and L5–S3. It is formed on the anterior aspect of the piriformis muscle within the posterior aspect of the pelvis. Motor innervation includes pelvic muscles, glutei, hamstrings, and the leg and foot muscles.

MOTOR SKILL AND REGULATION

Here, we are concerned with motor performance as considered as a combined pattern of skill and regulation.

Muscular coordination refers to the processes that determine the sequencing and distribution of muscular contractions of the legs and trunk, which support balance. When active balancing is required (e.g., sports or bicycling), vision plays a significant role, but simultaneous input is received from proprioceptive functioning and motor programming (see Proprioception).

The physical characteristics of movement may be summarized as *kinetics* (the forces or torques that generate a movement) and *kinematics* (the description of the motion in terms of position, velocity, and acceleration). The writer takes into consideration Newton's second law of motion (force is required to change a body's velocity, whose acceleration is proportional to the net force acting on it); the constant effect of gravity, including its effect on the posture of a limb; the complications of multijoint movement; and the force of rotational motions described as torque. The spinal cord contains the motor programs for protective reflexes and locomotion. Systemic programs (swallowing, chewing, breathing, and fast saccadic eye movements) are located in the mesencephalon and medulla oblongata, with the spinal cord involved. Cortical control of movement is executed by direct corticospinal neurons and cortical fibers that project to brainstem nuclei, which in turn project to the spinal motor centers. Motion is influenced by sensory signals that provide information about the position of different parts of the body in space (see Body Schema). To obtain an effective neural command, information about the initial position of the body (e.g., the arm), ongoing movement, and the target must be integrated (Grillner, 2003).

In planning and regulating movement, the CNS takes into account posture and its rate of change Peripherally, for each muscle, both a *tonic* pattern of activity related to static postures and counteracting gravity and a *dynamic* component scaled with the speed of the movement is needed. Errors of movement can be due to errors in the transformation of extrinsic eye-centered coordinates into shoulder- or hand-centered coordinates. Forward regulation refers to muscular commands. Feedback provides a comparison of the actual motion with the expected motion. This error signal is used to update the motor commands. If errors persist over many trials, the models used to correct motion errors are modified so that their predictions of appropriate commands are brought into accord with the actual performance. The physical details need not concern most clinicians, but specialists in physical medicine and rehabilitation, including those concerned with the design and performance of prosthetic devices, will use this data. Learning appears to be modular. A trained movement in a modified force field adapts to this changed direction, but movements in directions far from that in which the training takes place are unaffected (Soechting & Flanders, 2002).

Motor skills are learned behavior that requires patterns of activity across sets of muscles. Skilled behavior requires the integration of a series of multi-joint movements into a coordinated action. *Abstract motor programs* do not specify the commands sent to the muscles or the particular muscles to be activated. This information is implemented by downstream systems that implement motor programs. Training with one set of muscles may not transfer well to another set, although transfer of knowledge may not be complete. *Sex differences* need to be considered. While males greatly exceed in strength after the early teenage years, females exceed in speed and ability of fine motor skills (manual and articulatory). Females may have an advantage in the central processing of motor commands since they may process sequences of movements more efficiently than males (Hampson, 2002).

The motor cortices: Movement planning involves components of the *posterior parietal cortex (PRC)*: The *parietal reach region*, an anterior intraparietal area specialized for grasping, and a *lateral intraparietal area* for saccadic eye movements (see diagram in Andersen & Buneo, 2002, p. 99). These regions are specialized for multisensory integration and transformation of coordinates from one domain of function into another. Attention and learning are distributed functions in the posterior parietal cortex (PPC), where lesions can lead to apraxia. The *motor cortex* reorganizes its topographical map as a function of practice. The cortical region becomes larger, paralleling learning-related changes in sensory cortices. The *prefrontal cortex* (PFC) functions are more general than those of motor learning, being in executive control rather than only in encoding motor skills for long-term storage. The *premotor cortex* fires when the movement is signaled by an external cue, whereas *supplementary motor cortex* fires when a sequence is produced from memory.

MOTOR LEARNING AND PROGRAMS

The Cerebellum

The cerebellum constitutes 15% of the brain's mas, and contains more than half of the neurons in the brain functions with both coordination and motor learning. It is concerned with the control of posture and multiple joint movements. It modulates muscle activity across multiple joints in anticipation of the mechanical forces that are generated by one's own movements and by forces in the environment (Bastian, 2002). Major movements also involve eye, trunk, and head movements, as well as with proprioceptive input to the cerebellum. The cerebellum also regulates the *timing* of movements (i.e., the precise representation of the temporal relationship between successive events). Above the brainstem-hypothalamic level of signal convergence is the cortical model of the body representation in relation to the environment. An *inverse model* calculates the initial path for a muscular action. It is followed by an error-predicting *forward model* that runs the command on a neuronal emulator (i.e., it anticipates the sensory consequences of motor commands). This is faster and safer than real world feedback. Acquisition of sensorimotor skills and the solution of complex motor problems occur with the access to background knowledge, goal priorities, and current sensory information. The inverse model responds to the error signal with an upgraded command. The difference between the expected and the actual sensory feedback seems to be an *error signal*. *Consciousness* may follow if the brain's wiring permits distinguishing between inner-world and outer-world representations and building a representational model of their relationship (Churchland, 2002; Fuchs, 2002). The present writer speculates that this function may be related to the cerebellar components of the body schema.

The Basal Ganglia

The basal ganglia are involved with scheduling movement elements in the appropriate order. It is possible that, as a skill develops, there is a shift in the locus of control from the cortex to the subcortex. This may be enabled by a representation in the basal ganglia of the sequence (i.e., while the representation of the sequence remains cortical, the basal ganglia provide a mechanism by which there is rapid progression of the overt movements as the sequence unfolds). The basal ganglia seem to operate on a higher level when selecting a particular goal. The cerebellum fine-tunes the motor program used in accomplishing a set goal. Both the basal ganglia and the cerebellum form loop-like circuits in which cortical inputs are processed and then relayed back to the cortex via the thalamus. They both have inhibitory projections to their output targets: the globus pallidus inhibits the thalamic nuclei and the Purkinje cells of the cerebellar cortex inhibit the cerebellar nuclei. They both utilize a divergent-convergent model: input signals are distributed across as range of neural networks and are then recombined into a more compact, topographic organization. In the basal ganglia, this output is directed to the globus pallidus. In the cerebellum, input from mossy fibers is reintegrated in the Purkinje cells (Hazeltine & Ivry, 2002).

GAIT

Gait Descriptors

The gait cycle refers to the sequence of events that begins with floor contact of one lower extremity and continues until that event is repeated with the same extremity: stance time (the time elapsed during which the limb is in contact with the walking surface); swing time (the remainder of the time of the gait cycle when the lower extremity is in the air); double support time (when the stance phase of one limb overlaps the stance phase of the contralateral limb); stride length (time required for one cycle, such as heel strike to successive ipsilateral foot contact); step length (the linear distance between two consecutive lower limb contralateral contacts); step time (time of step length); cadence (number of steps taken/unit of time, such as steps/minute); and base of support (distance between the center of the ankle joints).

Neurological Disorders of Gait

Neuroanatomical associations (Camicioli & Nutt, 2003) include the cerebral cortex (disequilibrium, freezing, falls, hyperreflexia); the brainstem (disequilibrium, astasia [incoordination with inability to stand], hyperreflexia); the basal ganglia (freezing, astasia, hyper- and hypokinetic gait, rigidity, tremor, bradykinesis, chorea, dystonia); the cerebellum (ataxia, dysmetria); the vestibular system (ataxia), difficulty with tandem walk; the spinal cord (spastic gait, loss of sensation, sensory ataxia, increased tone); the peripheral nerves (foot drop, peripheral weakness, sensory ataxia, sensory loss, reduced or absent DTRs); and muscle and neuromuscular junction (waddle, hip and shoulder girdle weakness). Gait disorders may result from intoxications, medication side effects, neurological disease, and conversion disorders such as astasia-abasia (i.e., inability to walk or stand normally, although normal leg movements can be performed while lying down or sitting) (Yager & Gitlin, 2005).

Ataxia is defined as inability to coordinate muscles during voluntary movement of the limbs, trunk, and eyes. *Cerebellar ataxia* is a disruption of timing of successive movements (Spencer et al., 2003). A symptom of cerebellar vermis disorder is gait ataxia, with widening of the base on walking. Tandem gait is difficult, and there are difficulties on turning. Continuous movements require noncerebellar control. The Romberg test is usually positive with cerebellar ataxia. There is an increased prominence in patients with proprioceptive or vestibular lesions (Subramony, 2004). *Sensory ataxia* arises from proprioceptive sensory loss in the lower extremities (large myelinated peripheral nerves, dorsal roots, or posterior columns). This illustrates the point that a particular symptom may evolve from different levels of the CNS. The patient may adapt to mild sensory abnormalities by walking slower or taking smaller steps. With more intensive sensory loss, the patient exhibits more sway when standing (a wider base, feet raised too high, thrown forward, and brought down too quickly). There may be inappropriate synergy of balance and gait if body schema maps are distorted (see Body Schema) (Camicioli & Nutt, 2003). *Vestibular ataxia* exhibits dysfunction depending upon the speed with which it develops. The deficit may emerge only in an environment in which visual or proprioceptive cues are reduced or deceptive. Patients with deficient VORs have difficulty in differentiating between movement of the self and environmental movement. Their gait is staggering and veering in response to erroneous self-motion. In the chronic state they describe unsteadiness but not vertigo (Camicioli & Nutt, 2003). Patients with a bilateral vestibular loss are ataxic (make extensive use of vision and are unsteady when their eyes are closed in the tandem position). Patients with chronic unilateral vestibular loss show little ataxia. Patients with superimposed posterior column sense deficit or cerebellar dysfunction are unsteady even with their eyes open (Hain & Micco, 2003).

Examination of Gait

Leg pain and weakness may have a neurological or musculoskeletal origin. The patient may engage in strategies to minimize pain by *antalgic gait* (avoiding bearing the full weight on the altered limb and limiting its range of motion) (Thompson, 2004). The examiner should note whether speed affects position of the COM, which in turn affects instability; speed of unobstructed and obstructed walking; foot (toe) clearance above the floor with and without obstructions, which varies with leading and trailing limbs; stepping times which varies with the height of the obstacle; relative joint angles of the ankle, knee and hip; trajectories of the heel and toe of each foot; cadence (steps per minute); gait speed (stride length divided by stride time; bilateral stride length (between consecutive heel contacts).

Gait and active postural disturbance are observed when the patient is asked to walk (e.g., 10–20 ft towards and away from the examiner) (see also, Stance). The clinician will observe (1) whether the patient's eyes are directed ahead or to the ground; (2) whether the base is narrow or broad; (3) whether the arms swing symmetrically and freely; (4) whether the torso midline is vertical or tilted (examiner should observe leg and foot trajectory [shallow or high stepping], then ask the patient to pretend to walk over an obstacle while walking); and (5) whether initiation is hesitant, shuffling, or with magnetic feet (seemingly glued to the floor). Is the pace regular or irregular, and of regular or irregular length and trajectory? Are the legs raised easily and with even trajectory and speed? This is an opportunity to observe balance, comparative strength, range of motion, and motor control.

Observe arm and shoulder movements. Irregularity may indicate compensation for a feeling of unsteadiness on one leg or poor timing associated with basal ganglia dysfunction. The clinician should observe hopping, standing on each leg, tandem walk, and rising from a chair with arms folded. Jerky steps (irregular rhythm, length, and trajectory) suggest ataxia. Patients with cerebellar disorder keep their feet close together and exhibit sideways trunk movement (lateropulsion). With more severe disorders they learn to stand with feet apart.

The clinician should inquire into speed of walking, the length of the steps taken, the width between the feet, the difficulties experienced initiating walking, the ability to turn, the tendency to trip or bump into things, and the appearance of being drunk. Walking confronts variable pathways: level, uneven, and oriented upward or downward. Obstacles are of various heights and involve knee flexion. Greater initial distance from the obstacle can be used to obtain greater clearance. The trailing foot passes closer to an obstacle than the leading foot. Ability to modify gait speed, obstacle avoidance, and head movement are predictors of falls with vestibular deficits (McFadyen et al., 2003). A reliable and valid clinical measure is available to assess step length and width (Van Loo et al., 2003).

Proprioception and Gait

If the patient walks with eyes down, the issue of visual-proprioceptive integration is raised. Since the base of support need not be widened after TBI (Ochi et al., 1999), it may be inferred that this symptom is more likely to be based upon a particular afferent neurological disorder (e.g., cerebellar or proprioceptive). Following the classic contribution of Sherrington, Pearson and Gordon (2000a), the clinician should note that stepping is adjusted by proprioceptors in the muscles, joints, and other body movements, as well as exteroceptors in the skin, which adjust stepping to external stimuli. *The step cycle* may be divided in two phases: *swing* occurs when the leg is off the ground (flexion and extension; *stance* is the interval during which the foot is in contact with the ground. Regarding neural control for stepping, supraspinal structures are not necessary for the basic motor pattern; rhythmicity is produced by neuronal circuits entirely within the spinal cord; spinal circuits can be activated by descending signals from the brain; and pattern generating networks are strongly regulated by input from limb proprioceptors, although this is not required.

Clinical Vignette: Loss of visual-proprioceptive integration: The patient was a right-handed 30-year-old woman who was knocked down by car and experienced several hours LOC. Preexisting conditions were denied.

Record Summary: Struck and injured her head (laceration), chest, right shoulder, both elbows, left knee, and right ankle, and in the process jolted her neck and back. LOC of several hours at Bellevue Hospital's emergency department. Experienced subarachnoid hemorrhage and a laceration; daily headaches; positional vertigo; blurred vision and nausea; bilateral ear pain; bilateral temporomandibular joint disorder (TMJ); impairment of memory and concentration; pain cervical, lumbosacral right anterior chest wall; right shoulder; right elbow, left arm, left knee (clicking and buckling), right ankle. Her head was lacerated in the right parietal region. Her neck was jolted.

Her gait was slow and based wide. She attributed this to a pain in her left knee. She was reluctant to walk. She felt dizzy if she didn't look down. She thought that she might fall and feared falling. She has fallen after losing her balance, and becomes dizzy if she gets up too fast. Bilateral reduction of grip strength was detected. Using the proprioceptive maneuver, her ability to locate an object in space from nonvisual proprioceptive cues was bilaterally poor as observed in several trials.

Assessment of the contribution of vision to movement: She did not exhibit hand tremor or spasm. Making a thumb to finger circle was slow bilaterally. She was given 16 children's blocks, and given two trials with each hand, then one trial with both hands, and instructed to make a as high a tower as possible. Her success with every trial, bilaterally, and with faster speed on the second trial, suggests good eye-arm coordination and ability to improve a motor task from experience. Motor planning and control seem spared. Thus, with visual information, she performed a difficult maneuver well. Her fine hand-eye coordination was assessed through the study of long lines and junctures in drawing tasks and handwriting. Her fine psychomotor performance did not seem overtly impaired.

Gait: Her gait was slow and based wide. She attributed this to a pain in her left knee. She was reluctant to walk. She feels dizzy if she doesn't look don. She thought that she might fall and fears falling. She was cautioned by the examiner about the danger of falling. She has fallen after losing her balance. She tried to get up fast and then became dizzy.

Grip strength (Smedley Hand Dynamometer, average of three trials): Preferred using her right hand.

Norms utilized: age and sex (Yeudall et al., 1987).

Kg		Norm	% of Norm
Right Hand:	14.2	33.9	42
Left Hand:	12.3	30.3	41

Ratio of average preferred to nonpreferred handgrip strength was 115%. A 10–20% preferred hand advantage is expected. Using appropriate sex standards and taking into account her age, stature (short), and occupation (hotel work), it appears that there is a *bilateral reduction of grip strength.*

Kinesthesis (awareness of the position of limbs and joints in space): With each hand separately, and eyes closed, the patient was asked to place a finger to her nose, to her knee, to the examiner's finger in space, and then back to her own nose. *Her ability to locate an object in space from nonvisual proprioceptive cues was bilaterally poor as observed in several trials.*

Tremor Observation

Intention Tremor

Fine tremor (*hands*): The patient is asked to extend arms with the fingers slightly open. Tremor may be visually observed, or may be detected by holding the fingers with the examiner's fingers or having patients stretch out their fingers and feeling the tips with the palm.

TRAUMA AND MOTION

Motor behavior is vulnerable to dysfunctioning after diffuse and more extensive brain damage to the circuits that integrate afferent and motor impulses. Confirmed trauma-induced or exacerbated movement disorders show scan evidence of basal ganglia or cerebral structures. However, traumatic lesions of the peripheral nervous system (PNS) or cortical regions might alter basal ganglia functioning indirectly. Posttraumatic movement disorders include *tremor* (*kinetic* carries out a movement; *postural* maintains a position) of the legs, hands, head, trunk, and tongue, commencing immediately to four weeks after an injury. Parkinson disease can follow severe head trauma or multiple episodes of mild head injury. *Dystonia* (slow muscle contortions causing an abnormal body posture, frequently with a twisting character) may occur in an accident without LOC and be expressed months or years later. Also reported are *tics*, *Tourette syndrome*, and *myoclonus* (brief, lightning-like involuntary jerks) (Evans, 1996; Goetz & Pappert, 1996). Some sensations modify motor activities directly through a kind of interneuronal modification of motor output between the receptor and the spinal cord motor control. Some examples include proprioceptive stimulation from the joints and muscles permitting skilled, goal-directed movements and loss of balance following vestibular injury that results from misinformation by the injured sensory receptors, leading to loss of limb and torso control.

Motor control is dysfunctional after particular frontal lobe injuries (Nunez, 2002): *primary motor cortex* (weakness and paralysis of the contralateral muscles); *premotor cortex* (producing movements with complex stimulation [e.g., miming use of a tool or arbitrary associations between stimulus and response]); *PFC* (lack of drive and impaired ability to execute plans).

Muscular weakness is not only due to damage to the motor cortex. It may be associated with "*positive signs-release phenomenon*" such as spasticity, or to a "*negative sign*" (e.g., weakness associated with fibrous tissue or other damage to muscle, or reduced neural input to the muscle). Thus, diseases affecting the descending pathways yield spasticity and cause disorders that are distributed diffusely in the limbs, the face, muscles, or large groups of muscles (e.g., flexors). Disorders of motor neurons result in denervation atrophy and reduced muscle volume, or affect muscles in a patchy way and may be limited to individual muscles. Nerve lesions result in weakness, reflecting the distribution of individual nerves (Ghez & Krakauer, 2000).

Locomotor performance after an injury or stroke has a spatiotemporal context with implications for the occurrence or avoidance of falls. The examiner should be alert to the contributions of foot, knee, and hip function to gait; the differences between functions of single muscles and groups; the difference between passive stretch and strength testing; the supine and upright positions; poor judgment, attentional demands (e.g., when using a walker), drug effects, dystonia, and psychogenic disorder; deficits of proprioception, vestibular function, visual disturbance including instability under poor lighting, clumsiness, pain, deformity of the bones and joints, cautious gait pattern (width of the base of the gait and length of step, including turns, weakness sufficient to disturb balance and gait, requiring the use of arms in order to rise from a chair, spasticity, and peripheral disorders of muscles and senses). Proximal weakness can be explored by inquiry concerning difficulty in going up stairs, rising from a chair, and reaching above one's head.

Chronicity of Motor Disorders

Clinical conditions after TBI-causing accidents include gait, posture, balance, both gross and fine movements of the extremities, changes of muscle tone of the torso and extremities, and such symptoms as tremor. The neural components of posttraumatic motion disorders are organized as follows: upper motor neuron lesions (UMN)—tone disorders; involuntary motion disorders—dystonia; intentional (when) disorders; praxic (how) disorders; body scheme disorders. Comorbid disorders with motor dysfunctions are somatic damage (torso and limbs, [i.e., skeleton, soft tissues]); the effector organs of the internal environment (hormone, inflammatory and immune systems); autonomic insufficiency; whether the onset is gradual or immediate; whether complaints refer to weakness or incoordination; the distribution of weakness (Younger, 1999), and genetic and psychosocial issues (Jankovic & Lang, 2004).

Motion disorders due to TBI can appear years after the injury. There is some ambiguity concerning the etiology of movement disorders. Many cases of trauma are not reported or recognized. Therefore, later symptoms cannot be related to a particular event (Bower et al., 2003; Parker, 1995). Pretrauma accidents and symptoms may not be reported or queried, or may be suppressed in the event of litigation. Recognition of later dysfunctions reduces misattribution and misdiagnosis. The multisystemic effects and the comorbid conditions of accidents are often ignored, with attention directed only to neurological or somatic injuries. Multidiscipline referrals may be needed for comprehensive assessment (Murray, 2004; Parker et al., 1997; Santiago & Fessler, 2004). Motor rehabilitation benefits from being varied. Changes in the motor cortex are driven by the acquisition of motor skills. Repetitive activity is useless. What is needed is an enriched environment that avoids repetition, an attractive environment for rehabilitation, family support, and a positive attitude by the rehabilitation professionals (Bach-Y-Rita, 2002).

Origins of major cortical inputs to the motor cortex: in the prefrontal, parietal, and temporal cortices and innervate the premotor and supplementary motor areas; primary sensory to primary motor cortex; corticocortical input arises in the opposite hemisphere, courses through the corpus callosum to contralateral homologous areas. In contrast, *fingers* do not receive callosal fibers, and are thus functionally independent of the other side.

Lesions of the Cerebellum

These disrupt the temporal properties of voluntary movements (i.e., reacting to the precise timing between events).

The *vermis* controls structures that are axial or that are bilaterally innervated (i.e., station and gait)—walking and coordination of the head and trunk. A patient with a mild vermian lesion has *gait ataxia*, where the base is widened, tandem gait is very difficult, and there may be decompensation on turning. With *truncal ataxia*, there is swaying and unsteadiness when standing and an inability to maintain upright posture. There is little or no abnormality of the extremities, although all coordinated movements may be poorly performed. Patients are more variable on motor tasks that require precise temporal representation. A difference between *sensory ataxia* (loss of proprioception) and cerebellar ataxia is that while in sensory ataxia, performance is not normal with eyes open, it worsens markedly with eyes closed. Tremor is due to voluntary visually guided corrections to deviations from the intended track. With cerebellar ataxia, it makes little difference whether or not there is vision to guide movement (Campbell, 2005, pp. 522–524). Patients with varied cerebellar degeneration have difficulty in the expression and timing of a conditioned eyeblink response (Topka et al., 1993).

Unicerebellar lesions: In the hemispheric syndrome, the manifestations are appendicular, not axial. Deficits are ipsilateral since the pathways are double-crossed. There is a disturbance of skilled movements of the extremities: ataxia; dysmetria; dyssynergy; dysdiadochokinesia; and hypotonicity affecting the arm and hand more than the leg and foot. Abnormalities of posture and gait are less than those of the vermis syndrome: swaying and falling toward the side of the lesion; inability to stand on one leg using the contralateral foot; and inability to bend the body towards the involved side without falling. The abnormalities often resemble a unilateral vestibular lesion. Walking may be unsteady, with deviation towards the involved side (Campbell, 2005, pp. 511–524).

A hierarchy of neural structures associated with movement has been described, extending from the spinal cord—including the hypothalamus, basal ganglia, cerebellum, and other subcortical structures—to the frontal cortex. The corresponding functions range from automatic, reflexive, and instinctual to schemes, programs, and plans of action. It has been suggested that the premotor region and supplementary motor area participate in the representation of actions that are defined by goals and trajectory, while the motor cortex represents the most concrete movements in terms of the direction and muscle groups required to accomplish them. Sequential behavior requires a flow of environmental and neural information involving the interaction of sensory and motor hierarchies. At all hierarchical levels, these interactions are bidirectional and include feed-forward and feedback. This is the neural apparatus by which perception translates into action. For an organized reaction, sensory and motor integration require the binding of the unit through time by working memory and preparatory motor set. This process is supported by the dorsolateral PFC. Working memory contributes to preparation for action by retaining information on which the action is contingent. It is likely that there is interaction between the PFC, the areas representing the information upon which the action is based, the inferotemporal cortex (visual short term memory), and the posterior areas. It is asserted that the dorsolateral PFC accomplishes syntax of an enormous range of actions (orderly arrangement), from sequences of skeletal movement to sequences of logical reasoning (Fuster, 1995).

DISORDERS OF MOTOR PROGRAMMING

Temporal Processing of Motor Programs

A program is a long, perhaps complex routine, involving goal-directed activity, including several components that are performed in a particular sequence and thus have a temporal dimension. The program may be so long that one part commences while the previous one is still proceeding. There is cooperation between the frontal and posterior cortex. A hierarchy of motor structures extends from the spinal cord to intermediate levels (basal ganglia and cerebellum) to the dorsolateral PFC, where temporally extended schemes, programs, and plans of action occur (novel, voluntary). The PFC represents the highest stage of the hierarchy of motor memory. In the dorsolateral frontal cortex there are interlinked stages of motor memory: prefrontal, premotor, and motor cortex. Linguistic structures are represented here: propositional speech (PFC), syntactic speech (Broca's area), and the pronunciation of words (dependent on the oropharyngeal musculature [primary motor cortex)]). In the PFC there is syntax of the temporally extended schemes, programs, and plans of action. Beginning with information on which action is based, preparation for prospective action is combined with temporary retention (working memory) and preparatory set. This bridges time in the perception-action cycle. Probably there is feedback from the PFC to areas representing information. The PFC coordinates the posterior perceptual network with the motor network. Patients with large dorsolateral lesions have planning difficulties, although people with PFC lesions may execute long and complex routines (Fuster, 1995).

Lesions do not clearly impair functions that fit into familiar syndromes. Unilateral dysfunctions may reflect disconnection of a limb from its programming representations. Some movement disorders reflect white matter disconnection (callosal and cerebral). Not all functions that are lateralized control contralateral performance. One must consider possible ipsilateral feedback via the corpus callosum. Some right-handed patients may have movement representations that are bilateral, which accounts for the paradoxical sparing of movement after massive unilateral cerebral lesions. Unilateral disorders may occur with lesions of a single hemisphere. Aphasia and apraxia may each occur independently of the other, negating the possibility that there is a unitary hemispheric function accounts for both.

Motor unawareness. Heilman (1991) contrasts the usual theory of neglect (sensory defects, sensory disconnection, and inattention) with his "feed-forward" or "intentional" theory of anosognosia. He posits an intention system (premotor cortex) that stimulates motor systems and forwards expectations to a monitor. If the intention and motor areas were destroyed, no information concerning either lack of kinesthesis or plan to move would reach the "monitor-comparator." The monitor would not know that the arm was supposed to move—that is, there is no mismatch between intention and sensory input—and one would not consider oneself paralyzed. If there is a lesion in the motor system, but the monitor-comparator is intact, then there is no input effector that is aware of one's intention.

APRAXIA

Apraxia is an inability to perform purposeful movements (Zoltan, 1992) and a cognitive motor disorder involving the loss or impairment of the ability to program motor systems despite intact motor function, sensation, task comprehension, and attention to command. There is a discontinuity between intent based upon accurate perceptions and action. Dyspraxias are not a disorder of primary sensory and motor functions. Probably they represent a disorder of motor programming and execution, specifically manifesting in visual-perceptual, dysphasic, attentional, or intellectual dysfunction (Miller, 1989). Lesions of either hemisphere can lead to apraxia, but the deficits are most severe following left hemisphere (LH) lesions. These patients exhibit the disorder when using either the contra- or ipsilesional hand. Within each hemisphere, apraxia occurs with both frontal and parietal lesions (Hazeltine & Ivry, 2002).

Apraxia has been described as a cognitive motor disorder that entails the loss or impairment of the ability to program motor systems to perform purposeful skilled movements (Heilman & Rothi, 2003). It has been described as inability to execute a program due to the prevention or disorganization of its realization. Related to executive function disorder, it may be characterized by an inability to initiate a concrete plan of action, lack of orientation in the environment, an inability to resynthesize after breaking a complex task into small bits, and losing some details and proceeding sequentially with the remainder (Prochalska, 2002). A combined dysarthria–dyspraxia may ensue from the compromise of primary and secondary motor cortex and their interconnections. Difficulty increases with fine motor commands rather than psychological complexity, as in true dyspraxias. Even in persons with lesser TBI, apraxia may be obvious or subtle, and influence physical performance.

Initiation and sequencing of functional motor acts should be observed to detect potential disorganization (Sutin, 2004). *Gesture agnosia* (*deficits*) are noteworthy, both meaningless and meaningful. Meaningful gestures include transitive (actual manipulation and use of tools), intransitive (symbolic [e.g., a salute]), and pantomime (the miming of tool use).

Lesions of the PPC can lead to apraxia, that is, inability to follow verbal commands for simple movements, difficulty in following a sequence of movements, difficulty in correctly shaping their hands in preparing to grasp objects. They may also suffer from *optic ataxia*. Without primary sensory or motor defects there is difficulty in estimating the location of stimuli in three-dimensional space. This causes pronounced errors in reaching movements (Andersen & Buneo, 2002). Apraxic deficits can result in the loss of independence and may increase the potential for injury (Cubo et al., 2003; Hammerstad, 2003; Heilman & Rothi, 2003; Heilman et al., 1997, 2004; Zoltan, 1992). Classes of action are *transitive* (gestures or actions demonstrating the use of a missing object), *conventional*, (gestures representative of culturally specific ideas), *natural*, or *nonrepresentative* (actions conveying no message). A motor disorder is not apraxic if it can be ascribed to other neurological conditions, to sensory defects, or to nonmotor cognitive disorders, for example, poor comprehension, agnosia, or poor attention. Strength is normal, but a movement cannot be performed if the cortical motor centers, which plan and provide the proper commands to execute the movement, fail.

Apraxia is more likely to be detected when sought directly rather than as a by-product of an examination focused upon specific neurological deficits. Here, dyspraxia is considered to be a lesser functional disorder. The literature does not encourage finding a pure type, which is a firm union of lesions, pathways, and behavior to create definite syndromes. Often, the examiner will detect a complex motor disorder with cognitive elements without identifying a classical syndrome.

Types of Apraxia

Body part affected: Buccofacial, unilateral limb, bilateral limb, and/or total body apraxia. The patients cannot articulate, or carry out facial movements to command (lick the lips, blow out a match, etc.).

Constructional apraxia: The inability to draw, copy, or construct a design, for example, geometric designs using pencil and paper, produce clear handwriting (drawings; Bender Gestalt), or three-dimensional part objects (block design; object assembly), whether on command or spontaneously. It is associated with lesions of the posterior parietal lobe, or of the junction between the occipital, parietal, and temporal lobes. It is correlated with dysfunctions of body schema, dressing (mistakes of orientation in putting on clothing or neglect), ability to use objects, and other activities of daily living. Constructional apraxia is associated with *clumsiness,* along with the inability to guide the hand toward an object (*optic ataxia*).

Ideomotor apraxia: Inability to perform a purposeful motor task to command, even when the idea of the task is fully understood; impaired imitation and production on command of meaningless gestures (Bartolo et al., 2008). This seems to be the most common type. According to the Liepmann's concept, in some cases apraxia may in some cases involve capacity to form a general plan but inability to transform it into action. There is inability to use objects, although knowledge of these objects can be expressed in words or through actions of the unaffected limbs. For example, being able to describe the use of a pencil, but not being able to write, to misuse a match and match box, or to find one's own nose. Loss of representations account for the spatial and temporal errors made after verbal commands. Errors are characterized by incorrect joint movements or stabilized joints, by incorrect orientation of tools when pantomiming, inability to coordinate multiple joint movements to obtain a desired spatial trajectory. Language and motor programs are separate from the motor strip. Thus, a callosal lesion can disconnect a contralateral motion program from the actual motor activity. Movement takes place without organization, that is, with spatial and temporal errors.

Ideational apraxia: Semantic (impaired gesture comprehension and production of familiar gestures), and procedural (impaired production of familiar gestures). This is a failure of the sequence, creating inability to carry out a task automatically to achieve a goal. For example, sealing an envelope before inserting the letter.

Conceptual apraxia: The patient makes content and tool-selection errors: A tool may be used as though it were a different tool; there is inability to recall which tool is associated with a particular object (selecting an incorrect tool to finish a task); inability to associate a function with a particular tool (selecting an inappropriate substitute when the correct tool is available).

Dressing apraxia: This is associated with deficits in spatial orientation and may include inability to initiate the task or errors in orienting the clothes to the body.

Dissociation apraxia: Patients can imitate or use objects, but cannot pantomime when instructed verbally or visually.

Conduction apraxia: Impairment when imitating, rather than pantomiming to command.

ATAXIA

Ataxia: Lack of accuracy or coordination of movement that is not due to paresis, alteration in tone, sensory loss, or the presence of involuntary movements. It is considered a cardinal sign of cerebellar disease. It relates to motor dysfunctions of the limbs, trunk, eyes, and medullary musculature. Among the disorders observable in nonsurgical level of TBI are ataxia of gait, limb ataxia, and dysarthria (ataxia of speech). Some muscular symptoms are *dysmetria*, (disturbance of the trajectory of placement of a body part during active movement), *hypometria* (undershooting a target), and *hypermetria* (overshooting the limb's goal). This dysfunction is attributed to the inability to generate early antagonistic muscle activity (Hore et al., 2002). Decomposition of movement is errors in the sequence and speed of the component parts of a movement, including a multijoint movement (Timmann & Diener, 2003).

SPINAL CORD TRAUMA

Spinal injury is common in accidents causing head injury. It has been estimated that there are 10,000 hospital admissions each year secondary to spinal cord injury, with 50% of the patients being between 16 and 30 years of age. There are 250,000 hospital admissions each year secondary to TBI. In patients with spinal cord trauma (SCI), 1.2%–6% have been diagnosed with TBI (Elovic & Kirschblum, 1999). Although motor function can be severely disrupted with gross spinal cord injury, the extensive degree of automaticity and plasticity has implications for rehabilitation.

Clinical symptoms depend upon four characteristics of the lesion: The level (the higher the lesion the greater the motor, sensory, and autonomic dysfunctions); the extent in the transverse plane (complete or incomplete cord syndromes); the extent in the longitudinal plane (number of spinal segments involved); and the duration and speed of injury. Rehabilitation of the patient with SCI may have the advantage of intact cognitive processing to learn new skills and adjust to disability. Thus, it is important to recognize whether TBI complicates the process (Watanabe et al., 1999). When SCI and TBI are comorbid, rehabilitation approaches, goals, and expectations require modification (Ricker & Regan, 1999).

PERIPHERAL NERVOUS SYSTEM

The motor units extend in a segmental pattern from cells lying in the anterior gray matter, extending to target muscles in the limbs and trunk. The nerve is formed when motor (anterior) and sensory (posterior) roots combine just distal to the dorsal root ganglion. After a brief intraforaminal course, posterior branches (posterior rami) extend backward to provide paravertebral muscles. Anterior branches (rami) extend forward, supplying the trunk and giving rise to the roots of the plexuses. Tissues surrounding the larger structures and, eventually, the nerve trunk are the perineurium and epineurium. The latter melds into a loose layer of protective tissue called the *mesoneurium*, which allows some degree of passive movement in the transverse, and particularly the longitudinal planes. The connective tissue elements and the specialized perineurium create a blood-nerve barrier that maintains an immunologically privileged nerve microenvironment. It also provides structure, tensile strength, and elasticity to the nerve trunk. After more severe grades of injury, success or failure of neuroregeneration depends on the nature and the degree of injury to the connective tissue. Regeneration is interfered by the formation of a neuroma that fills the gap between nerve stumps, and replaces a target muscle with fibrotic tissue every 2–3 years. Sensory reinnervation may restore useful function to a limb for up to 7 years. Grading of nerve trauma includes different levels of both nerve trunk and connective tissue. Metabolic disease, including diabetes, can interfere with regeneration.

BODY SCHEMA AND SELF-REPRESENTATION

The differentiation between body schema and self-representation is both conceptual and is concerned with the range of neurobehavioral functions encompassed. Perhaps the most significant and complex neurobehavioral extensions are complex planning, involving the environment and consciousness itself. Thus, body schema is more explicitly a kind of a map organized in the brain (specifically cortex and cerebellum), spinal cord, and perhaps elsewhere. It's organized information includes current status and change in, both, the internal milieu and the various body tissues. The spinal cord receives sensory input, integrates it and produces motor output, totally independent of the brain.

BODY SCHEMA

An intact body schema serves as a postural model of one's body parts, and their interrelationship, vital to both the service of movement and also to the creation of a body identity. Despite normal sensory input from the proprioceptive, visual, and vestibular systems, the patient may be unable to integrate the information because of abnormalities in higher sensory processing. Failure to integrate sensory information may be related to distortion of spatial maps: parietal cortex, putamen, ventral premotor cortex, superior colliculus, and frontal eye fields (Camicioli & Nutt, 2003). The brain also receives sensory information used to control the motor activity of the spinal reflexes and the central pattern generators. The sensory maps may be planar, representing a spatial map of the sensory environment it encodes. For example, visual representation or the somatic sensory receptors that form a map of the body surface. These are projected to the cortex as topographic representations of the visual world and the body surface (*somatotopy*). The sensory and motor maps are adjacent and are similar in basic layout. The scaling of the visual, somatosensory, and motor fields is not constant (*the magnification factor*). The cortical space allocated is determined by the behavioral importance, for example, center of gaze (visual), lips (somatosensory), and fingers (motor). Other sensory inputs are organized by qualities, rather than a geometric analogy of what is represented (taste, smell, and hearing) (Connors, 2005b).

BODY IMAGE AND SELF-REPRESENTATION

Body image is differentiated from the body schema, terms used by Schilder (1950, 1964). It is the experience of feelings, ideas (Zoltan, 1990), and values that guide action (Parker, 1983), that is, the *psychodynamic sense of identity*. The body image (subjective) is created by neurological stimulation and emotional conditioning (Parker, 1983); visceral and vestibular stimulation; moods, appetites, satisfaction of internal motives, and so on. The *interoceptive context* may provide the basis for the subjective awareness of the self as a feeling entity (Craig, 2003). Self-representation evolved as a means of coordinating inner-body signals to generate survival regulation, and later, action. It participates in the decision as to "what to do next." Pain, thirst, and threat reflect its functioning. Signals converge into the brain stem-hypothalamic axis and are integrated. This level is the nonconscious neurobiological platform for higher levels of self-representation. Varied types of networks participate: Representations of the external world, the internal environment and its needs, its posture, and so on. (Churchland, 2002). A different body schema occurs in varying body positions to predict and guide movements. Spatial perceptual representation is needed for complex movements. *Spatial perception* is achieved by integration of multiple sensory information, including somatosensory. Body representation, especially that of the *hand,* has a close relationship to spatial perception (Sumitani et al., 2007a). Some *deficits of body image* include feeling of bodily asymmetry (accompanying unilateral sensory or motor deficits), loss of the detailed awareness of one's body, and imbalance (accompanies seizures or vestibular dysfunctions) (see Neglect).

Neurological Basis of the Body Schema

There is not a single master map of one's body in the brain. The body has a cortical representation for each submodality of sensation (Kandel, 2000c). This differs from a unified "map" of the entire body. One may infer that lesions of afferent input (somesthesis and proprioception) will interfere with (1) the ongoing body model, and (2) the appropriate modification of the body schema with experience. The nervous system employs various sensory channels and representations. A particular body state may be represented incompletely in a given representation, and also differently and with contradictory results at different levels of the CNS, that is, *fractured somatotopy* (Goldenberg, 2002).

Somatosensory Cortex (Parietal Lobe)

Body schema is represented in the parietal cortex (see somesthesis, above) (Benton & Sivan, 1993). The parietal lobe receives afferents from the somatosensory and visual cortices, and has reciprocal connections with the premotor cortex. Thus it forms coherent images of the entire body, including motor commands, visual and somatosensory information. Each parietal lobe creates a representation of the *opposite hemibody*, which has significance for phantoms, for hemineglect, for denial of ownership, and for the supernumerary limbs. The splitting of the neural basis of conscious body representation contrasts with the feeling of a seamless body (Goldenberg, 2002). Symptoms are more characteristic of right parietal lesions, suggesting that *visuospatial impairment* is more severe after right brain damage. Note, however, that body information reaches the parietal lobe from both contralateral and ipsilateral sources. This internal map-like organization of sensory cortical areas shapes the way the sensory information is learned. This may involve a module involving modules, that is, areas in which particular stimulus features are represented, use of cortical areas in which the information is stored, and learning of a behavioral strategy (Diamond et al., 2002). Somatosensory integrated information may be significant in forming the body schema. Each of the four regions of the primary somatosensory cortex contains a complete map of the body surface (Kandel, 2000c). In this light, the cerebellum (primarily fastigial and dentate nuclei) projects via the intralaminar nuclei to the parietal cortex (superior parietal lobule, area PE, or 5 and 7 of Brodman) and to the frontal cortex and striatum (Nieuwenhuys et al., 1988, p. 233; Zilles, 1990). There is sensory integration in the multimodal association areas of the PRC, which receives input from the primary somatosensory cortex, the visual and auditory systems, and the hippocampus. The primary somatosensory cortex projects to higher order somatosensory areas of the anterior parietal lobe. The posterior parietal association areas receive input from the primary somatosensory cortex. The parietal lobe contributes to orientation in space and to awareness of bodily sensation.

Cortical maps of the body surface are not hard wired; they change with the use of afferent pathways. Each of the four areas of the primarily somatic sensory cortex (Brodmann's areas 31, 3b, l and 2) has its own complete representation of the body surface (Kandel, 2000c). The spatial coordinates of the body are organized in a parietooccipital temporal association area. A section of the PRC, extending into the superior occipital cortex, provides continuous analysis of the spatial coordinates of all parts of the body, as well as the surroundings of the body. This area receives visual sensory information from the posterior occipital cortex and simultaneously, somatosensory information from the anterior parietal cortex. This permits computation of the coordinates of the visual, auditory, and body surroundings. Damage to this part of the brain can lead to loss of recognition of the opposite side of the body, and failure to consider the opposite side either for receiving sensory experiences or for planning voluntary movement (Guyton & Hall, 2000, p. 665). *Depersonalization* is associated with areas responsible for an integrated body schema, that is, parietal somatosensory association and multimodal association areas. An integrational failure with respect to bodily cues, an aspect of consciousness, is related to dysfunctioning of the temporal, the parietal, and the occipital association areas. By inference, this is a disturbance of the hierarchy of sensory processing from primary sensory areas to unimodal then to the polymodal association areas, and finally to the PFC. Depersonalization and other negative psychoform symptoms (amnesia and derealization) are related to disturbance of the posterior association area's input into the PFC (Nijenhuis et al., 2002).

Somatotopic Organization

Somatotopic refers to the organization of a neurological structure according to major body parts (Schieber & Baker, 2003). Body schema has multiple components in numerous brain areas. Knowledge of our body is necessary for motor control, makes available conscious awareness and evaluation of our body's integrity, and serves social communication and learning. Stimuli from the joints and skin participate in creating a self-image (Schilder, 1950, p. 27). A somatotopic map of the body evolves from peripheral *sensory information,* affecting each level of motor control. A *motor hierarchy* is based upon somatotopic considerations. Each area of motor control receives input (premotor cortex) representing a rough somatotopic map of the body, and its motor output preserves this organization in the descending brain stem pathways. Each level of motor control receives peripheral sensory information that modifies its motor output. At each level of motor control, neurons project in parallel to sensory relay nuclei, for example, the thalamus, cerebellum, and so on. These recurrent pathways provide sensory and other processing systems with information about ongoing motor commands. This allows higher motor centers to control the information reaching them, that is, restricted to a particular task (Ghez & Krakauer, 2000).

Accurate purposeful movements (motor control) cannot be accomplished without proprioception: (1) the *limb-position sense*, that is, the sense of the stationary position of the limbs and (2) *inesthesia*, that is, the sense of limb movement. Perception of body configuration for motor control relies upon somatosensory input and functions outside of attention. The CNS uses spatial configuration for motor control at multiple levels from the spinal cord to the parietal lobe. Even at the highest level there is no master map of the entire body; rather there are multiple task- and body–part-specific representations. A new perception of the postural model of the body (gestalt) develops through experience of an active mental process and actions. One may infer that lesions of afferent input or integration will interfere with, both, the ongoing body model and also appropriate modification of the body schema with experience. Since the nervous system registers both sensory and motor centers, there are numerous mental models in different brain structures. These register different degrees of detail, integration of bodily functions and structure, and topographical accuracy. A particular body state may be represented incompletely in a given representation, and differently with contradictory information at different levels of the CNS (*fractured somatotopy*) (Goldenberg, 2002).

Cerebellum: There are ipsilateral sensory representations in the cerebellar cortex. Two inverted somatotopic maps represent the leg anteriorly within the anterior lobe; the face represented more posteriorly. In the posteriorly lobe the arrangement is reversed, with the face represented anteriorly (Burt, 1993, p. 360; Thimman & Dienst, 2003, p. 304, Figure 17–3). Auditory and visual afferents project to the central portion of the vermis. Somatosensory stimulation reaches the anterior and posterior lobes of the cerebellum. The anterior portion is unified, including the vermal and paravermal regions. The posterior portion is bilateral, in the paravermal areas (Burt, 1993, diagram, p. 360). There is a somatotopic representation of the body surface maintained by climbing fibers. Represented is a complete map of the body, with contiguous parts represented contiguously, with the extremities and face represented with higher resolution than the other areas of the body (Provini et al. 1998). Ability to experience identity is disturbed by damage to the various homunculi that support the body schema. It is vulnerable to head injury and emotional regression. This reminds us of the familiar homunculi for the motor somesthetic strips. There are body representations in the cerebellum (Martin, 2003, pp. 305–307; Goetz & Pappert, 1996, p. 178; Timmann & Diener, 2003).

Subcortical: According to the *recurrent model* (Nicolelis et al., 2002), a topographic representation of the (animal's) body surface occurs in the *subcortical nuclei* (spinal cord, brain stem, thalamus). Also, paralimbic networks represent the visceral and kinesthetic information integral to the representation of the bodily self (Tucker et al., 1995).

The insular cortex represents a topographic ordering of visceral sensation maintained all the way from the nucleus of the solitary tract (Saper, 2002). It may be an *internal alarm center* that evaluates distressing thoughts and body sensations with negative emotional significance. Since the insula has afferent and efferent connections with the amygdala, the alarm function may function in conjunction with it (Nijenhuis et al. 2002).

The motor cortex is arranged somatotopically, with a "*motor homunculus*" in the motor cortex, and a similar pattern of localization with respect to the somesthetic sense in the postcentral gyrus (Parent, 1996, p. 890). There are four representations of the complete body surface in the primary somatosensory cortex (Brodmann's areas 3a, 3b, 1, 2), not a single integrated somatosensory map in a limited section of the postcentral gyrus (Kandel, 2000c, Figure 20–6, p. 389). This sensory map roughly matches the motor map and receives input from somatosensory areas that may be reorganized after amputation, also causing reorganization of the motor cortex (Kaas, 2002a). Organs that require the greatest precision and control (lips, tongue fingers, hand, face) have greater cortical representation than body parts used in gross movements such as ambulation. Hands, face, and feet (sensors of objects) have the highest density of touch receptors. A cortical territory (e.g., the arm) overlaps considerably with the territory representing nearby parts. This evolves from the motor cortex's organization: convergence, divergence, and horizontal interconnection. Lesions in arm representation lead to degeneration of myelinated fibers in the cervical cord, while lesions of leg representation lead to degeneration extending to the lumbar cord (Krakauer & Ghez, 2000; Schieber & Baker, 2003).

Components of body scheme include *personal space* (the neural representation of the body surface involving touch and proprioception), *peripersonal space* (the space within arms reach), *extrapersonal space* (the larger environment around the body), and *imagined and remembered space* (the representation of spacial relations in the association cortex of the posterior parietal lobe) (Kandel, 2000c, p. 381). Body schema is represented in the parietal cortex (Benton & Sivan, 1993) and the cerebellum (Burt, 1993, diagram, p. 360). In addition, disorganization of the temporoparietal junction (e.g., by electric stimulation) creates an illusion of a shadow person, a phenomenon also observed in psychiatric and neurological patients. Since the illusory person closely mimicked the patient's body posture and position it appeared that the patient experienced a perception of her own body. The phenomenon is based upon multimodal mechanisms at the temporoparietal junction in self-attribution, and may be the basis for paranoia, persecution, and alien control (Arzyk et al. 2006).

Cerebellum: There are ipsilateral sensory representations in the cerebellar cortex. The cerebellar surface, as a site of sensory input, may play a role in mapping the external environment. It receives a wide range of input: visual coordinated with vestibular, a motor homunculus, proprioceptive

(Braitenberg, 2002). Somatosensory stimulation reaches the anterior and posterior lobes of the cerebellum. The anterior portion is unified, including the vermal and paravermal regions. The posterior portion is bilateral in the paravermal areas (Burt, 1993, diagram, p. 360). The cerebellum (primarily fastigial and dentate nuclei) connects via the intralaminar nuclei to the parietal cortex (superior parietal lobule, area PE, or 5, 7 of Brodman). It also projects to the frontal cortex and striatum (Nieuwenhuys et al. 1988; Zilles, 1990). Auditory and visual afferents project to the central portion of the vermis.

DISORDERS OF BODY SCHEMA

Somatosensory Loss

Although loss of sensation at the midline is often considered to be the criterion for a psychogenic disorder, as opposed to neurological disorders (due to sensations, the anatomical consideration that enter the nervous system (NS) from overlapping nerve endings), Rolak (1988) determined that, for both cortical and psychogenic deficits, sensory loss could stop at the midline. Similarly, loss of vibration over a long bone is considered psychogenic, since this sensation is widely transmitted. Patients with hemifacial numbness were tested at the midline with vibration and pinprick. There was no difference in splitting of sensation between patients with organic or psychogenic symptoms. These findings indicate caution in using the midline theory to identify hysterics or malingerers.

Neglect

Unilateral brain lesions may cause *spatial neglect,* which is considered to be a disorder of spatial attention, that is, profound inability to attend to contralateral personal and extrapersonal space, or hesitancy to initiate movement in contralesional space with or without stimulation. The most profound deficits are observed in right hemisphere lesions in right-handed persons (Nunez, 2002; Reynolds et al. 2003). Neglect of sensory stimulation arising from the limbs or from one side of the body can be attributable to right parietal damage (somesthetic dysfunctioning), although it has also been attributed to the nearby temporal gyrus (Andersen & Buneo, 2002). It is likely to be accompanied by inappropriate euphoria or indifference, patients may fail to perceive left-sided stimuli, and may dress, wash, or groom only the right side of the body, misperceive the side that is being stimulated, and so on (Joseph, 1988).

Distortions of the Body Schema

A dysfunctional body image is probably due to loss of somatosensory input or integrational cortex, that is, loss of detail of the body schema. Impaired body image comprises impaired conceptualization of one's own body or others' bodies; inability to identify body parts, phantom limbs, and left-right dysfunction (Benton, 1985a; Schilder, 1950). Deficits of the body schema contribute indirectly to motion disorders, somatagnosia, right-left discrimination, finger agnosia, unilateral neglect. When body information is incomplete or faulty, actions, particularly those directed toward our own body will be faulty (Schilder, 1950, p. 45).

Agnosia is the inability to obtain useful information from acknowledged, familiar stimuli. Lesions of the PRC produce inability to perceive objects through sensory channels, including *astereognosis*, inability to recognize the form of objects through touch.

Anosognosia is unawareness of the defect per se. Further deficits of body image include feeling of bodily asymmetry (accompanying unilateral sensory or motor deficits), loss of detailed awareness of one's own body, and imbalance (accompanying seizures or vestibular dysfunctions). Unawareness of hemiplegia (motor unawareness) or hemianopia has been termed anosognosia. It is likely to be accompanied by inappropriate euphoria or indifference. Anosognosia is characteristic of the acute stage of neurological disorder and tends to remit in chronic cases. While the ratio of right- to left-hemisphere damage accompanying anosognosia varies from 2:1 to 8:1, bihemispheric damage outnumbered lesions confined to a single hemisphere. It was associated with acute onset, secondary damage, edema or hemorrhage, bilateral location, and corticolimbic connections, which produced dysfunction of both hemispheres, but not focal cortical lesions (Weinstein, 1991).

Neglect

A state where the patient is unaware that one side of space is ignored in a sensorimotor sense is termed *neglect*, and it is treated as a spatial disorder and disorder of consciousness. Sensory unawareness, that is, neglect of sensory stimulation coming from the limbs, or one side of the body, can be attributable to right parietal damage (somesthetic dysfunctioning). Neglect is the polar opposite of body awareness, and may be expressed through unawareness of sensory or motor functions. Unawareness may occur in virtually all of the major neuropsychological syndromes. It is useful to differentiate between unawareness of the deficit and unawareness of the consequences of the deficit (Schachter & Prigatano, 1991). Patients may fail to perceive left-sided stimuli, and may dress, wash, or groom only the right side of the body, misperceiving the side that is being stimulated, and so on. The "alien limb" may be experienced as irritating or may be actually hated (Bisiach & Geminiani, 1991; Joseph, 1988).

Under normal waking conditions, sensory-driven neuronal groupings inhibit internally activated patterns (Bisiach & Geminiani, 1991). In their absence, fantastic images occur, or the person neglects part of external space (e.g., figure drawings representing only half of the body). The patient may understand that there is a sensory lack. However, when the central portion of the network is lesioned, the person cannot even conceive of the hemispace or body side, which is true neglect.

REORGANIZATION (PLASTICITY)

Cortical reorganization (*plasticity*) affects multiple levels of the somatosensory pathway. The loss of one link disturbs the functional system as a whole. Patterns of recovery have been described: *compensation and substitution of function* (new responses to solve tasks); *equipotentiality and vicariation* (if specific lesions do not cause a specific deficit, other areas are said to "take over" the function); *diaschisis* (Bach-Y-Rita, 2002) (at sites distant to the injury there is depressed metabolic activity or reduced neuronal activity). Examples include crossed-cerebellar inhibition, and interaction between the locus ceruleus and the ipsilateral sensorimotor cortex. During attempted repair, the undamaged terminals elsewhere may not function normally; *reorganization or unmasking of "latent synapses* (expansion of cortical receptive fields; synesthesias, phantom limb reactions, and visual stimulation activating the auditory cortex, and vice versa).

Not all reorganization is adaptive or beneficial. Spasticity and neural kindling that cause epilepsy are examples of plasticity. Consider also an area subserving input from the legs now responding to input from the arm due to the unmasking of previously silent pathways. Increased activity is necessary for neural reorganization to take place or to be maintained. However, forced use of environmental stimulation early after injury can worsen the extent of brain damage or increase the severity of behavioral deficits. Excessive secretion of excitatory neurotransmitters (glutamate) will kill vulnerable neurons (Stein & Hoffman, 2003).

The final outcome reflects network-wide and local-circuit modifications. In the uninjured person practice improves sensory and motor representation, that is, performance. Training in one task may alter circuits so that a given neuron is less effective in another. By implication there is a "zero sum game," indicating that training in one skill should not be hampered by practice in another. This has implications for TBI: New circuits have to be reformed through practice to be helpful and compensate for loss of input. After restricted motor cortex lesions, training on finger movement skills helps reform the cortex so that more neurons are devoted to finger movements. Adjacent cortical neurons extend themselves into cortical area deprived area deprived of sensory input. Reorganization of sensory representation occurs after a limited loss of sensory inputs. Postlesion reorganization manifests new cortical fields representing body parts that were not previously responsive to its stimulation, for example, sensory representation of the fingers (Kaas, 2002b; Stein & Hoffman, 2003). In the somatosensory system input change occurs in the parietal cortex and the somatosensory thalamus (ventroposterior N).

Deprived portions of the primary visual cortex are activated by intact surrounding portions of the retina, concealing the hole in the visual scene. It may be more useful to be aware of what is missing. After partial hearing loss, the cortex responds to different tones than those of the damaged part of the cochlea. The motor system is also reorganized: After amputations (monkeys), stimulation of the motor cortex that normally produces movements of the hand and forelimb evokes movement of the stump of the limb and the shoulder. The result of reorganization may not be benign: Inputs and outputs may be misaligned with misperception and error rather than compensation.

Phantom limb experience has been explained by cortical reorganization, that is, adjacent cortical neurons extending themselves into the deprived area. Sensory reactivations after peripheral sensory deprivation are initiated subcortically. In humans treated for pain after limb amputation, there can be reorganization of the somatosensory thalamus (Florence, 2002; Kaas et al. 2002, diagram p. 9) and of cortical circuits (Kandel, 2000c). In one case, the trigger zone was the skin of the stump. Deafferenting injuries produce reorganization of the somatosensory thalamus. New sensory input to the denervated cortex arises from a different peripheral structure for the lost input. In one case, the trigger zone was the skin of the stump.

12 Vocal Motor Disorders*

OVERVIEW

Vocalization is an extremely complex activity, requiring cognitive, affective, and motor planning, sequencing, modulation, social and somatic feedback, and control of a rapid sequence of nonrepetitive muscular movements through the segmental and branchiomeric musculature (see below), which requires spinal motor impulses, the autonomic nervous system, and cranial nerve (CN) input. It is influenced by developmental, hormonal, emotional, somatic, and neurological functions. Vocal motor disorders may be associated with swallowing disorders; at minimum, the rehabilitation team includes a speech-language pathologist, an otolaryngologist, and a neurologist (Stewart et al., 1995). The neuropsychologist and other examiners can perform a brief screening procedure (see Dysarthria, Screening). Neurological control includes speech centers in the cerebral cortex (Guyton & Hall, 2006, pp. 720–722), the basal ganglia, the cerebellum, the respiratory control centers of the brainstem, spinal cord reflexes (Guyton & Hall, 2000, p. 521), and CN V, VII, IX, X. Motor vocalization disorders have been described as an inability to sequence sound systems to form words instead of noises (Guyton & Hall, 2000, pp. 669–670), formulating the thoughts and motor control of vocalization. Motor prosody refers to the melodic and other communicative qualities of the voice. Prosody carries information about the speaker's feelings and attitudes. The nonverbal part of the speaker's affect is estimated at 90% (Van Lancker & Breitenstein, 2000). Prosody's somatic basis is expressed through organs that serve multiple functions: respiration (ventilation), phonation (voice content and quality), ingestion, and swallowing. This is vulnerable to a wide range of neurological and somatic injury, as are the somatic structures that support breathing and vocalization. Thus, the actual site associated with a patient's posttraumatic aprosodia may be difficult to localize. Since comorbidity occurs between dysprosody, dysarthria, dyspraxia, and dysphasia, the clinician's diagnostic skills are challenged. There is an issue as to whether apraxia is an articulatory (production by the vocal apparatus) disorder or an aphasic disorder (i.e., whether errors are articulatory or linguistic). The patient may not have lost the ability to produce particular phonemes, but the problem is with consistently encoding the correct phonemic representation of the word. The utterance is expressed with correct articulation, but deviates phonologically from the target word (Blumstein, 1991). While Benson (1993) classifies dysprosody as an example of aphasia, here we consider aprosodia to be a motor deficit. The appearance of apathy and indifference can alternate with feelings of anger, facetiousness, or boastfulness. Mutism, associated with frontal lobe dysfunction, is associated with well-articulated but hypophonic speech (Damasio & Anderson, 1993). In addition, a pattern has been reported of feeling normal emotions but with abnormal facial expressions (e.g., pseudobulbar palsy). The reflex mechanism for facial expression is released bilaterally from cortical control, resulting in involuntary laughing or crying. However, normal emotions are experienced (Heilman et al., 1993).

Speech is multidimensional behavior, involving an extremely complex neuromuscular anatomical basis, linguistic rules and conventions, cognition, learning, physiological and anatomical development, social conventions, social and vocational communication, the expression of feelings, and so on. Disordered vocalization is a varied group of disorders consequent to injuries at different levels of the central and peripheral nervous systems, the torso, the neck, the throat, and the face. After an

* With contributions from Sidney I. Silverman, D.D.S., Eleanor T. Silverman, M.A., CCC, SLP and Cynthia Braslau, M.A., CCC/SLP.

accident, vocalization problems may be caused by damage to the head, the trunk, the neck, CN X (vagus nerve), and the brain. Associated syndromes, including aphasia, are not clearly defined since they do not specify operations within the language-processing system. Symptoms also overlap with those of other conditions, and classical aphasic syndromes do not correlate well with described lesion sites (Caplan, 2003). The examiner should be observant of language disorders and screen for referral to speech pathologists.

Neurological voice disorders are the result of weakness, lack of coordination, involuntary movements or postures of the vocal folds, or some combination of these. It is possible to differentiate between *dysarthrias* (abnormal muscle strength for articulation of sounds or phonemes); *dysphonias* (disorders of voicing), which would be a deficit in execution of an intonation contour, rather than its planning (Van Lancker & Breitenstein, 2000); *apraxia* (inability to program sequences of phonemes, with consonant substitution rather than distortion as in dysarthria); *stuttering* (initial pauses and dysfluency of speech production without articulator or language disorder); and *aprosodia* (loss of vocal expression caused by injury somewhere in the motor vocalization apparatus, neural or somatic, see below). Treatment involves a multidisciplinary team (e.g., a speech-language pathologist, an otolanryngologist, and a neurologist) (Stewart et al., 2005). Vocalization is differentiated from emotional expression and language comprehension.

COMPONENTS OF VOCALIZATION

The range of communication channels or levels has been described (Borod, 1993a, 1993b; Doupe & Kuhl, 1999): facial, prosodic, lexical, gestural, postural, with an emotional dimension (i.e., pleasant-unpleasant and approach-avoidance, semantics), lexicon, and phonology. There are two components of communication that must be differentiated: *emotional* (which includes the voice quality referred to as prosody) and *propositional* (a complex code requiring semantic, lexical, syntactic, and phonemic decoding).

Phonemes

A phoneme is a single distinct sound, the minimal sound unit that contrasts meaning and defines a word in a language (e.g., /p/ and /b/). Phonemes consist of distinctive features of sound production (i.e., voicing, aspiration, roundedness, and the location and degree of maximal constriction of the vocal tract creating pitch). In a given language, some sequences are permitted while others are forbidden (Blumstein, 1991; Caplan et al., 1999). Air has to go through the larynx either whispered or voiced. *Voiced* is defined as sounds produced with vocal cord vibrations (/b/), as contrasted with *voiceless* air (t/p/,/s/) (i.e., without vibration of the vocal cords). For voiceless consonants, the vocal cords vibrate 30 msec after the stop consonant is released (s/). Phonemes are formed by the location and the maximal constriction of the vocal tract, as well as voicing (glottal or laryngeal vibrations) (i.e., speech sounds produced by vibration of the vocal cords with the opening between them, as b/d/c). A *glottal stop* is a speech sound made by the closure and then explosive release of the glottis. Opening and closing of the *velopharyngeal port* is required to produce appropriate nasal and oral resonance of speech and the intraoral pressures necessary for the articulation of phonemes, as well as to affect prosody and articulation in dysarthritic speakers. Dysfunction may result after lesions to the upper motor neurons (UMNs) that supply the bulbar region of the brainstem, and the lower motor neurons (LMNs) that supply muscles of the soft palate and pharynx, and subcortical structures such as the basal ganglia and cerebellum (Theodoros & Murdoch, 2001a). *Voice-onset time* is timing between the release of a stop consonant and the onset of glottal pulsing. Anterior patients have difficulty with phonetic dimensions requiring the timing of two independent articulators (Blumstein, 1991). The brain processes complex acoustic information and identifies a phoneme based on known categories of speech signals (Fitch et al., 1997). Contrast between related sounds involves both voicing and the place of articulation.

Phonetics

Phonetic units are the smallest elements to alter the meaning of a word. One may differentiate between phonetic and phonological characteristics of speech (Ladefoged, 1975, pp. 1, 24; Stoel-Gammon & Dunn, 1985, p. 52). Phonetics refers to the description of speech sounds, how they fall into patterns, and how they change. It has various components: *articulation*, or how speech is produced by the vocal apparatus (i.e., the place of articulation [structures such as bilabial, labiodental, etc.] and the manner of articulation [fricative or forcing air between two narrow surfaces to produce sounds such as /f or /s; nasal; voiced or accompanied by vibration of the vocal cords, such as /b/ /d/ /c/ /sh/ etc.]; *acoustic*, or the physical properties of sounds; and *psychoacoustics*, or the way in which speech sounds are perceived. Plosive refers to consonants produced by closing off the oral cavity and then being released with a burst of air (e.g., /p/ in stop).

Phonology

Phonology includes speech perception, production, cognition, and the motor aspects of speech: (1) The rules involving the structure and systematic pattern of sounds in language (i.e., a way of taking a small set of basic articulatory gestures and using them to form a vast set of possible words, with arbitrary meanings requiring memorization, like a consistent pattern in the language) (Dronkers et al., 2000; Nadeau, 2003, citing Akmajian et al., 1984); and (2) an organized classification system of speech sounds that occur as constructive units of the spoken language (Indefrey & Develt, 2000). While phonetic analysis is purely descriptive, phonology refers to the classification and organization of the units of speech sounds (phonemes). Together, phonology and articulation involve rapid articulatory precision of consonants and vowels in a variety of lengths and sound combinations. This characterizes the distinctive sounds of a language and the rules that describe the changes occurring in different relationships with other sounds. Phonological processing has been associated with a center for auditory images (Nadeau, 2003) in area 22 (superior temporal gyrus), apparently linked by the arcuate fasciculus to area 40 (postcentral gyrus, Parent, 1996, pp. 882–883) and to Broca's area. Phonological disorder is characterized by the failure to use developmentally expected speech sounds that are appropriate for an individual's age and dialect. This means errors in sound production and the use, representation, organization (e.g., substitutions of one sound for another), or omission of sounds (American Psychiatric Association, 2000, pp. 65–66). The consequence is interference with occupational, academic, and/or social functioning.

Prosody

Prosody is a major nonverbal component to speech meaning, social communication, and the expression of feelings. Its phonological quality includes the placement of stress, the intonation, and the intensity and duration within a sequence of phonemes.

Some parameters of prosody: Prosody refers to major properties of the speech wave (Benson, 1993; Ross et al., 2001; Shapiro & Danly, 1985; Theodoros et al., 2001a; Wilson, 1996). A prosodic profile would include monopitch, range, envelope, and intravowel variation (Leuschel & Docherty, 1996). *Pitch* or tone (i.e., pitch variation or steadiness); *fundamental frequency* (F_0) (the number of vocal fold vibrations/unit of time); *range* (maximum and minimum for each vowel); envelope (midpoint between minimum and maximum frequency of a vowel); *mean* (average of maximum and minimum of each vowel in the signal); *intonation* (change of pitch over time, including reduced frequency at the end of a sentence, particularly the last word); *duration* (word duration and pausing); *amplitude and stress* (rate of speech and rate fluctuations; loudness and its variability; *pattern* (phrase length, rhythm, melody, rate, prolonged intervals, short rushes, inflection, and timbre, lengthening the final word in the sentence). (Blumstein, 1988; Leuschel & Docherty, 1996).

Study of prosody: Ross et al. (2001) assert that detailed acoustic measures are not necessary when analyzing affective prosody in English since English speakers impart affect in their speech predominantly through changes of pitch over time (intonation). Their measure of vocal prosody was a *coefficient of variation* (*CV*) for each utterance, with a mean CV for 12 sentence sets and for 10 sec of spontaneous speech. After the stimuli were passed through a 70–300 Hz bandpass filter, there were measurements of affective-prosodic comprehension and word comprehension with judgment of affect. The clinician may observe whether the patient uses gestures and prosody, and may also observe the patient's style of repeating a variety or prosodic utterances, to identify the affect represented by gestures of the face and limbs (Ross et al., 2003).

The formal study of prosody may use these frequency characteristics: (1) average level at peaks; (2) average level at valleys; (3) average level at all peaks and valleys per sentence; (4) variability per sentence defined as the sum of all peak-to-valley and valley-to-peak changes in level; (5) variability per variation (i.e., average change in level peak-to-valley and valley-to-peak changes for each sentence); (6) excursion or range (i.e., highest peak minus lowest valley); (7) rise (i.e., valley-to-peak rise for interrogative sentences, taken on last word of the sentence) (Shapiro & Danly, 1985).

Affective components of language: Spontaneous prosody and gesturing; capacity for prosodic-affective repetition; prosodic-affective comprehension and comprehension of emotional gesturing. Change of pitch seems to be the best vocal cue for the perception of spontaneous affective, expressive prosody (attitude and affect) (Indefrey & Develt, 2000; Ross et al., 1981).

Linguistic contribution: Prosody conveys more information than merely arousal or mood valence (i.e., discrete emotions and linguistic information) (lexical, syntactic, semantic, and discourse) (Blumstein, 1988; Etcoff, 1991). Prosodic competence expresses subtle shades of meaning, while frequent lapses of expressive and receptive prosody may signify intellectual inferiority (Monrad-Krohn, 1947). The intonation or fundamental frequency varies with the grammatical mode. It rises during the last portion of an interrogative sentence and falls in declarative sentences. Other intonational parameters are the terminal falling of the fundamental frequency in the utterance's final position; the fundamental frequency declining over the full range of a declarative sentence; the fundamental frequency resetting in a complex sentence between clauses; continuation rises (frequency rises in the last syllable preceding a syntactic boundary); an incomplete sentence or utterance; and lengthening of the last word in a sentence.

Cognitive contribution: Prosodic and facial cues permit inference about the speaker's emotional state and intentions. Thus, motor aprosodia would impair the listener's capacity to understand some of the emotional communication. The voice enables identification of people even when the listener is suffering from prosopagnosia including facial recognition (De Gelder, 2000). Word production is an extensive sequence of processes starting with conceptual preparation (including articulation) and concluding with the spoken word (Indefrey & Develt, 2000). Speech perception and cognitive and motor planning skills also influence effective speech production. Ross and Rush (1981) have suggested that comprehension of emotional gesturing and prosadic comprehension vary between types of aprosodias. Affective comprehension is attributed to the right posterior temporoparietal operculum (Ross et al., 2003). In *schizophrenia*, the inability to modulate affective prosody and the difficulty in processing facial affect impoverishes interpersonal relationships, leading to social isolation (Ross et al., 2001). *Fetal alcohol exposure* and/or excessive alcohol ingestion affect affective-prosodic comprehension (Monnot et al., 2002).

Resonance

The original vocal energy (Caplan et al., 1999) depends upon an unimpeded flow of air through the larynx, either whispered or voiced. This sound is bounced around in the resonating chamber of the pharynx, which gives the essential personal character of the person's voice (i.e., the overtones, melody, prosody, and harmonics, and prosody). Without resonance, a voice has no affect. It is created by a group of tissue spaces that encompass the oropharyngeal and nasal recesses of the maxillofacial and neck areas. Resonance is dependent upon a velopharyngeal valving mechanism. Efficient opening and closing of this port produces nasal and oral resonance in speech and the intraoral pressure needed for the articulation of various phonemes (Theodoros & Murdoch, 2001a). Pitched sound is bounced around in the resonating chamber of the pharynx and augmented with additional acoustic properties like amplification and attenuation of the tone of the voiced speech. This creates the essential personal character of the person's voice (overtones, prosody, harmonics).

Phonation

This is the process that creates sound of varying pitch and intensity by vibrating the air passing through the larynx during exhalation under lung pressure. Without this element in speech one's voice has no effect, which is suggestive of brain injury and depression. Phonation is achieved by the larynx, the articulation by the structures of the mouth (lips, tongue, soft palate), and the resonance by structures of the mouth, nose, nasal sinuses, pharynx, and chest cavity. The laryngeal biomechanism is adapted to act as a vibrator, its vibrating element being the vocal folds or cords. These protrude from the lateral walls of the larynx toward the center of the glottis and are stretched and positioned by several muscles of the larynx (Guyton & Hall, 2006, p. 481). This mechanism is dependent upon respiration, physical effort, and phonation (Guyton & Hall, 2000, p. 442; Moore & Dalley, 1999, pp. 1038–1050; Williams, 1995, pp. 1644–1646). It involves the rapid, periodic opening and closing of the glottis through the separation and apposition of the vocal cords, which are paired elevations on the lateral wall of the larynx, extending from the tongue to the laryngeal inlet down to the trachea (Williams, 1995, Figure 12.63, p. 1726). Vocal folds are abducted, adducted (approximated), and rotated partially or completely by movements of the laryngeal muscles and cartilages. The degree of tension may be increased by 50% to create the highest tones.

Articulation

Articulation is the movement of speech organs involved in the pronunciation of a particular sound. The *formation of words* depends upon the rapid, orderly succession of individual muscle movements of the larynx, mouth, respiratory system, tongue, and vocal cords. The *articulators* have been described as the most sophisticated motor system. In addition to coordinating these organs, there must be *sequencing* (i.e., adjustment, in advance, of the intensity and duration of each sound). Articulation is activated by the facial and laryngeal regions of the motor cortex. Sequences and the intensity of muscular contractions are influenced by the cerebellum, the basal ganglia, and the SC (Guyton & Hall, 2000, p. 655, pp. 669–670).

Articulation commences with retrieval of a lemma (syntactic word), which has a semantic function in determining a word's place in the syntax of the sentence and thus its meaning. The lemma activates the phonological code of each of its morphemes (indivisible units of meaning). Anomia represents the blockade of phonological access with preserved access to syntactic information. Phonolological words include a syllabic structure, which is not identical with the actual separate words (i.e., a syllable can straddle the boundary between separate words). As we consider the building up of phonological phrases and larger units, we become involved with intonation. Sentences have an intonational contour. The pitch accent may be on a word within the sentence, followed by a falling boundary tone at the end indicating completion. A rising boundary tone invites continuation on the part of one or the participants in the conversation, including possibly the speaker (Indefrey & Develt. 2000).

Articulators include the mouth, lips, larynx, vocal cords, tongue, and jaw. These impede the expiratory breath stream in different positions to produce specific phonemes. The articulators of the mouth and nasal cavities create diversity (i.e., adding to the number of phonemes beyond vowel sounds). The integrity of the *tongue* is essential for the precise and rapid production of most speech sounds (Theodoros et al., 2001b). Articulation is under the control of the larynx and face area of the somatosensory cortex, the caudal midbrain, and the cerebellum. Articulation occurs when the voiced stream of air is modified by the resonating chambers. It is still not language; it has only the potential for vowel sound production. The vowels reach their final sound effect by the modifications of the shape and contours of the oral spaces. For example, there are the front vowel(s) when the mouth is opened widely between the lips as for "ah" and "e," and there are the back vowels like the "o" and "u" when the lips approximate each other as the air passes through the lip aperture. Whereas the vowels are formed by an uninterrupted flow of air from the lungs through the lips, the consonants are formed by interrupting the stream of air, either abruptly (as in the "k" "t" or "d" sounds) or slowly (as a continuous slow leaking of air to create a sibilant like the "s" or "th" sounds). Control and coordination of breath during phonation permits a person to articulate by interrupting the stream of air with the tongue, like the stops on a clarinet. This leads to about 25 consonants. Thirty phonemes permit the large vocabulary constituting a language.

Speech

Speech has been regarded as a tertiary structure that encompasses frequency and amplitude cues within a temporal envelope (Fitch et al., 1997). The association between speech signal and phonatory/articulatory actions is complex, based upon the organs participating in respiration, phonation, and articulation, and permitting similar acoustic results to be produced by a variety of mechanisms (Scering, 1986, cited by Etcoff, 1991). Acoustic processing has been defined by characteristics such as loudness and frequency, which are not of direct linguistic significance. Phonetic processing is concerned with extraction of linguistic features (e.g., place and manner of articulation of consonants) (Norris & Wise, 2000). Speech is divisible into two components: somatosensory feedback that is independent of acoustic output (Tremblay et al., 2003).

Vocalization Structures

Speech involves the skull cavities, the respiratory system, the respiratory control centers of the brain, and the articulation and resonance structures of the mouth and nasal cavities (Guyton & Hall, 2000, p. 442), including vocal cords with neural control.

Skull: Humans' unique craniofacial system and vocal tract results from an increasing curvature of the basicranium with descent of the tongue to form the superior-anterior phyaryngeal space. These changes are essentially complete by age 4 (Kent, 1998). The skull base, with its articulating mandible, must be coordinated with other facial structures for comprehensible speech. One notes that the length of the posterior cranial base to the anterior skull base grows until age 11 or 12 (Hilloowala et al., 1998).

Air flow and breath control: The active force in the production of the voice is the air flow (Van Leden, 1961). Voice production originates from the vibration emanating from the vocal folds (Razak et al., 1983) and then is processed by the posterior oral-pharyngeal and nasopharyngeal ports (Merson, 1967). Intelligible speech is dependent upon the muscular coordination expressed as breath control during phonation, tone (source of the vibration), strength, and the flexibility of the tongue (transverse, longitudinal, and vertical muscles). The tongue's greatest role is in eating functions (grasping, chewing, swallowing). Its integrity is necessary to distribute the airflow through the mouth (Danlioff et al., 1980) and for adequate precision and rate of speech utterances that depend on the acoustic and perceptual components of intelligibility (Theodoros et al., 2001b). The pharynx serves both respiration (significant in prosody) and swallowing.

Articulatory structures: Articulators include the mouth, lips, nasal cavities, tongue, jaw, soft palate, larynx, vocal cords, and tongue. These impede the expiratory breath stream in different positions to produce specific phonemes. The lips, jaw, velum, nasal cavity, and larynx also contribute to *speech intelligibility* (Razak et al., 1983). The articulators of the mouth and nasal cavities create diversity (i.e., adding to the number of phonemes beyond vowel sounds) (Guyton & Hall, 2000, pp. 669–670). This function is under the control of the larynx and the face area of the somatosensory cortex, caudal midbrain, and cerebellum. Articulation occurs when the voiced stream of air is modified by the resonating chambers. It is still not language; it has only the potential for vowel sound production. The vowels reach their final sound effect by the modifications in the shape and contours of the oral spaces. For example, there is the front vowel (s) when the mouth is opened widely between the lips, as for "ah" and "e," and back vowels like the "o" and "u" when the lips approximate each other as the air passes through the lip aperture. Control and coordination of breath during phonation permits the person to articulate by interrupting the stream of air with the tongue, like the stops on a clarinet. This leads to about 25 consonants. Now there are 30 phonemes permitting a big vocabulary constituting a language. It is necessary to add 20 or more consonants to the five or six vowels to construct a language. Whereas the vowels are formed by an uninterrupted flow of air from the lungs through the lips, the consonants are formed by interrupting the stream of air—either abruptly, as in the "k" "t" or "d" sounds, or as a continuous slow leaking of air to create a sibilant, like the "s" or "th" sounds.

Prosody, pitch, and articulation are dependent upon precise coordination of the laryngeal, pharyngeal, and respiratory subsystems. Phonation utilizes the larynx with its vocal folds or cords. Resonance utilizes the mouth, the nose and associated nasal sinuses, the pharynx, and the chest cavity (Guyton & Hall, 2000, p. 442; Murdoch & Theodoros, 2001). Articulation involves the muscular movements of the mouth, tongue, larynx, and vocal cords, which are responsible for the intonation, timing, and rapid changes in the intensity of the sequential sounds.

Larynx and vocal cords: The laryngeal biomechanism is dependent upon respiration, physical effort, and phonation (Guyton & Hall, 2000, p. 442; Moore & Dalley, 1999, pp. 1038–1050; Williams, 1995, pp. 1644–1646). It involves rapid, periodic opening and closing of the glottis through separation and apposition of the vocal cords, which are paired elevations on the lateral wall of the larynx that extend from the tongue to the laryngeal inlet and down to the trachea (diagram, Williams, 1995, Figure 12–63, p. 1726). Vocal folds are abducted, adducted (approximated), and rotated, either partially or completely, by movements of the laryngeal muscles and cartilages. The degree of tension may be increased by 50% to create the highest tones.

NERVOUS SYSTEM CONTROL OF VOCALIZATION

Cerebral Development

At 3 months, the dendritic density is greater in the RH than in the left, and is greater in the oral motor area of the cortex than in Broca's area. At 6 months, there is a peak in the development of the inner layers of the cortex in the language areas. By 15 months, the hippocampus is fully mature, with implications for memory. At 24 months, the dendritic density increases in Broca's area to catch up with that in the oral motor cortex. Left hemisphere (LH) dendritic density catches up with that in the RH. At 72 months, the dendritic density of Broca's area is now greater than that in the oral motor cortex (Kent, 1999).

Cerebral Cortex

The usual description of Broca's and Wernicke's area as expressive and receptive may be oversimplified. *Sensory information* is initially processed in the posterior area of the brain with the frontal region then playing a role. Pitch is processed by the RH and phonetic information by the left (Doupe & Kuhl, 1996). *Speech and language* are gated by the thalamic way station (Fitch et al., 1997) and proceed to the primary speech area, to Wernicke's area, to Broca's area via the arcuate fasciculus, and then to the motor cortex (Guyton & Hall, 2000, pp. 669–670), thus coordinating hearing and speech. Perception of species-specific vocalizations by macaques is mediated in the superior temporal gyrus, with the left temporal lobe apparently playing a predominant role in this perception (Heffner &

Heffner, 1984). It appears that Japanese macaques possess an area analogous to Wernicke's area. Feedback from the speaker's own voice stabilizes speech production. Vocal learning depends upon the modification of vocal output using auditory feedback as a guide (i.e., in speech sensory and motor processes are virtually inseparable) (Doupe & Kuhl, 1999). Internal feedback loops, which act in parallel with external feedback loops, detect mismatch between the anticipated and actual outcome of speech (personal communication, Rodney M. J. Cotterill, undated, Biophysics Faculty, Danish Technical University, Lyngby), function to maintain accurate vocalization.

Prosodic processing: The neural pathway for prosodic processing initially shares the *auditory channels* with environmental sounds and language (Voeller, 1998). The neural pathway for prosodic processing includes the superior temporal gyrus for judgment about tones contralateral to Wernicke's area (Brust, 2001; Zatorre, 2001). These pass to homotypical areas contralaterally via the rostral half of the corpus callosum (CC), travelling posteriorly to the parietal lobe fibers and anteriorly to the occipital lobe fibers. Referring to the pathways that convey receptive verbal meaning and their transformation into expressive prosody (i.e. pitch), it has been hypothesized that one may distinguish the brain organization of musical and linguistic structures. Grouping, meter, duration, contour, and timbral similarity are brain and mind systems shared by music and language, whereas linguistic syntax and semantic, and musical pitch relations are systems not shared (Lerdahl, 2001).

Motor output: Speech is a multidimensional phenomenon with control distributed throughout the central nervous system (CNS) (Miller, 1989). Stimulation of the secondary motor cortex can evoke vocalization (Pocock & Richards, 2004, p. 175). Prosody, pitch, and articulation are dependent upon precise coordination of the laryngeal, pharyngeal, and respiratory subsystems. Kent (1999) summarizes the CN innervation for the muscle systems used in speech production: nasopharyngeal (V,IX,X,XI); pharyngeal (IX,X); tongue (XI,XII); lips and circumoral muscles (VII); mandible (V); larynx, intrinsic muscles (X); larynx, extrinsic muscles (V,VII, spinal nerves); and the respiratory system (XI, spinal nerves). Oral sensorimotor systems develop throughout lifespan in concert with complex motor skills.

Motor influence and the prosodic and emotional information processing of the *inferior frontal gyrus* would traverse the rostral half of the CC (Brodal, 1981, pp. 804–805). The lips, jaw, and tongue participate in vocalization, swallowing, and chewing. Broca's area (Brodmann areas 44,45,46,47) or the prefrontal and premotor facial region of the cortex (Crank & Fox, 2002) control skilled motor patterns of the larynx, lips, mouth, respiratory system, and other accessory muscles of speech. The primary motor cortex for the face and mouth is in Brodmann area 6. The SC (occipital, parietal, temporal) and the prefrontal cortex (PFC) project to the premotor and supplementary motor cortices. These pass signals to the motor cortex. The PFC also receives signals from the SC. There is an asymmetrical reciprocity between these areas: interaction is not a closed loop; rather, the areas higher in the circuit receive a wider range of input than those lower in the hierarchy. It may be inferred that input to the motor areas is closely monitored by the cortices receiving peripheral sensory information.

UMN axons of CN arise in the primary motor cortex (Brodmann area 4) and the supplementary motor and lateral premotor cortices (Brodmann area 6). There is bilateral UMN innervation of the muscles of the pharynx, larynx, and jaw (Brookshire, 1997, p. 412). The upper facial muscles play a crucial role in the definition of some common facial expressions of emotion, such as happiness (Davidson, 1993). Preservation of emotionally motivated facial movements leads to the inference that input to the facial nucleus follows a different pathway than corticospinal input (Wilson-Pauwels et al., 1988, p. 88). Contralateral innervation is received by muscles of the lower face and tongue so that damage to the facial nerve results in contralateral loss of lower facial expression. Higher muscles of facial expression (frontalis and orbicularis oculi) continue to function after unilateral cortical injury since their nucleus receives bilateral input, thus compensating for lateralized cortical damage. The UMN of the tongue is in the precentral motor cortex (Guyton & Hall, 1996, pp. 635–636; Wilson-Pauwels et al., 1988). Disconnection of the arcuate fasciculus destroys feedforward and feedback projections that interconnect temporal, parietal, insular, and frontal cortical cortices, which are part of a network required for assembling phonemes into words and coordinating speech articulation (Dronkers et al., 2000). Articulation (muscular movements of the mouth, tongue, larynx, vocal cords, etc. responsible for intonation, timing, and rapid changes in intensities of sequential sounds) originates in the facial and laryngeal regions of the motor cortex (Burt, 1993, p. 449, for full map; Guyton & Hall, 2000, pp. 669–670; Williams, 1995, p. 1158; described as the inferior frontal gyrus, Bookheimer, 2002). Sensory impulses (swallowing) are distributed via glossopharyngeal and trigeminal nerves to the medulla oblongata via the tractus solitarius. Thus, feedback from vocalization structures is part of the background of homeostatic and stress reactions.

Feedback: *Auditory input and feedback* are used as a guide, unifying sensory and motor processes. *Kinesthetic* input arrives from the muscles of speech (tongue, lips, soft palate) to the postcentral region, contributing to articulation (Glezerman & Bolkoski, 1999, p. 86). It has been hypothesized that speech production is organized by control signals for movements and is associated with vocal-tract configurations. Thus, *somatosensory information* would be fundamental to the achievement of speech motions and there is dissociation between it and auditory feedback during speech production (Tremblay et al., 2003).

Cingulate stimulation (anterior limbic cortex) induces vocalization and makes direct connection to the midbrain *periaqueductal gray* (*PAG*) and the adjacent parabrachial tegmentum, from whence full calls are also stimulated (Doupe & Kuhl, 1998). According to Llinas (2001, pp. 116–117), in primates, particular PAG cells are active only when preceding vocalizations of a very specific pitch. This activity reaches a peak at the onset of the monkey's vocalization. Some of these cells respond to general auditory stimuli and are correlated with eye movements in certain directions. We infer that a quality of external auditory stimulation stimulates a call of a specific pitch.

Corticobulbar fibers (*UMN*) project via the internal capsule and descend in the corticospinal and corticobulbar tracts (accompanied by fibers from the somatic SC and adjacent temporal lobe region), terminating in the lower brainstem. They pass caudally in the internal capsule to (1) the reticular formation of the medulla (indirect system as internuncials), which project to CN nuclei (and other destinations), and (2) to the motor neurons of the nucleus ambiguus of the medulla (Parent, 1996, pp. 452–453; Wilson-Pauwels et al., 1988, pp. 128–129). Direct corticobulbar projections are received by nerves V, VII, XII and supraspinal nerves (Parent, 1996, p. 452). Bilateral corticobulbar innervation is projected to CN nuclei innervating muscle groups that cannot be contracted voluntarily on one side: laryngeal, pharyngeal, palatal, and upper facial. Direct innervation of pharynx and larynx occurs via nerves of the branchial arches, and nerves V, IX, and X (Wilson-Pauwels et al., 1988, pp. 148, 160). These have a very high component of autonomic nervous system (ANS) activity. Corticobulbar tracts project bilaterally to motor neurons of nerves V, VII, IX, X, and XII. Afferents to the nucleus ambiguus are received from the vagal, glossopharyngeal, and trigeminal nerves, which convey reflex impulses from the oral, pharyngeal, and respiratory mucosa (Mitsumoto, 2000; Parent, 1996, p. 449). The facial N. projects to the upper and lower face, while the rostral ambiguus N. projects to nerves IX, X, and XI. Several brainstem nuclei can be considered part of the UMN system (including the brainstem reticular formation) since they modulate spinal motor neurones (presumably including those that control breathing).

Corticospinal tract (*voluntary and volition ventilation*) (Simon, 2001): The major effect of cortical stimulation on ventilation is inhibitory, although stimulation of the motor and premotor cortex may modestly increase ventilation. Cortical "readiness potentials," originating from the supplementary motor and primary motor cortex, can be recorded prior to volitional inspiration or expiration. Volitional ventilation pathways descend from the cortex through the brainstem and spinal cord in the region of the corticospinal tract.

Basal Ganglia

The thalamus, the basal ganglia, and the cerebellum may be involved in fluency, volume, articulation, and rhythm of speech. The basal ganglia (and temporoparietal lesions) have been associated with emotional comprehension aprosody (Starkstein et al., 1994). The caudate participates in vocalization through auditory temporal processing (Fitch et al., 1997). Expressive aprosodic localization is assigned to the putamen, or laterally to the adjacent anterior insula, and subjacent white matter (Tranel, 2002). It has been also been stated to be dysregulation of the autonomic nervous system (Lane et al., 1997). The insula has also been credited with a major role in speech articulation (Cereda et al., 2002).

Cerebellum

The cerebellum receives tactile and proprioceptive impulses from the face and jaw. Input through the trigeminocerebellar tract (Bastian et al., 1999) via the dentate nucleus proceeds to area 46 (Bastian et al., 1999). Facial and vocal control is localized in the vermis (as well as disturbances in the control of axial and trunk muscles during antigravity posture); lesions may result in slurring and slowing of speech with a characteristic one-word-at-a-time quality known as *scanning speech* (Ghez & Thach, 2000). The cerebellar dentate nuclei display activity for the sequential activity that will happen a fraction of a second later (Guyton & Hall, 2000, p. 654). Assuming that projection of the right cerebellar hemisphere to the left precentral gyrus participates in motor control, it has been suggested that increased hemodynamic response is related to the articulatory level of speech production (Ackermann et al., 1998). The right posterolateral cerebellum may assist the left cerebral cortex in helping an individual to learn to generate specific types of constrained spoken two-word associations that are not associated with overt or mental movement coupled to a verb. It may also help an individual to automate word associations during speech (Gebhart et al., 2002).

Brainstem Nuclei for Vocalization and Respiration

Medullary respiratory center: The site of the automatic CNS rhythmic out put to muscles of ventilation (central pattern generator) is not known. Respiratory-related neurons are located primarily in the medulla, but also in the pons and other parts of the brainstem (Richerson & Boron, 2003). Varied pathways carry cortical and thalamic descending axons to the medullary respiratory area and the pontine pneumotaxic center. Separation of the medulla from the rest of the brain permits an essentially normal pattern of breathing (Staub, 1998b). Breathing control is maintained by overlapping functions—metabolic (automatic) and behavioral (voluntary). The respiration muscles are under both automatic and voluntary control and react to speech, locomotion, swallowing, coughing, and vomiting. These movements influence the medullary respiratory center, which controls the separate inspiratory and expiratory mechanisms. During normal respiration, there is interaction between upper airway muscles (motoneurons in the medulla) and pump muscles (motoneurons in the spinal cord) (Ezure, 1996). Breathing control is maintained by overlapping functions: metabolic (automatic) and behavioral (voluntary). There are three groups of speech and respiratory centers in the medulla and pons. The first is the dorsal respiratory group. Its function is inspiration, and its neurons are located in the nucleus of the tractus solitarius. Its afferents are the vagus and glossopharyngeal nerves. The second is the ventral respiratory group, which is involved in either expiration or inspiration, depending upon which neurons are stimulated. Its neurons are found in the nucleus ambiguus and nucleus retroambiguus. It is not involved in normal quiet respiration. The third group is the pneumotaxic center, which is located in the nucleus parabrachialis of the dorsal upper pons. It helps control the rate and pattern of breathing (i.e., the duration of the filling stage of the lung cycle) (Guyton & Hall, 2000, pp. 474–475).

The *reticular formation* controls swallowing and respiration through the integration of the tongue and the extensor and flexor state of the jaw muscles (Butler & Hodos, 1996, p. 164). Reticulobulbar and reticulospinal pathways coordinate with direct corticobulbar and corticospinal tracts.

The *nucleus ambiguus* receives rostral and caudal afferent impulses. It participates in the voluntary control of swallowing and phonation by innervating the striated muscles of the tongue, the intrinsic muscles of the larynx (constrictor), most of the striated muscles of the soft palate and pharynx, and the constrictor muscles of the pharynx (Brazis et al., 2001, p. 320; Martin, 1996, p. 400,

Figure 13–10; p. 45). Cortical input arrives from *cortical motor areas*, particularly the precentral gyrus, and is bilateral and indirect (Hermanowicz & Truong, 1999). Bilateral corticobulbar fibers, which innervate CN nuclei, have axons that descend through the internal capsule to synapse on motor neurons in the nucleus ambiguus (Wilson-Pauwels et al., 1988, pp. 128–129). The nucleus ambiguus is in the special visceral motor column of the medulla and projects to the muscles that control facial expression, the jaw, the pharynx, and the larynx (i.e., nerves V,VII,IX,X,XI) (Martin, 1996, p. 45). *Afferent input* is received from receptors in the pharyngeal and laryngeal muscles, which convey reflexive impulses for coughing, swallowing, and vomiting.

Lower motor neuron: The *somatic efferent column* (like its spinal counterpart) terminates on the skeletal muscles (e.g., the head, neck, jaw, and salivary glands) (Butler & Hodos, 1996, pp. 95, 102). Axons exit the medulla and travel briefly with the abducent nerve (XI) to reach the constrictor muscles of the pharynx and the intrinsic muscles of the larynx. LMN diseases include those of the muscles or of the myoneural junction (botulism; myasthenia gravis). The articulatory disturbance depends upon the nerves or muscles involved (Haerer, 1992, p. 272).

Cranial Nerves

Motor: Axons from LMN units in the brain stem, originating in the CN nuclei innervate muscles of the tongue, pharynx, larynx, palate, face, and extraocular muscles. CN transmit the responses organized by cortical activity, which are supported by the reflex and autonomic regulatory mechanisms. The pharynx and the soft palate are derived from branchial structures and are therefore supplied by nerves of the branchial arches (V,VII,IX,X, and X). The hypoglossus nerve (XII) is purely motor (the tongue). The vagus nerve (X) gives rise to the bilateral recurrent laryngeal nerves, which innervate the larynx. Their course varies bilaterally. At the root of the neck, the recurrent laryngeal nerves branch from the vagus nerve and are directed upwards, close to the larynx (Brazis et al., 2001, pp. 334–335; Hermanowicz & Truong, 1999, diagram p. 205; Williams, 1995, pp. 1648, 1649). Before entering the larynx, this nerve divides into motor and sensory rami. Since it serves such fine control as vocal pitch, the motor units are small in number, with an estimated ratio of 30 muscle fibers to each motor neuron (Williams, 1995, p. 1647). The superior laryngeal nerve is the sensory nerve for the cricothyroid muscle. Injury may result in mild hoarseness with some decrease in voice strength (Brazis et al., 2001, p. 334). The accessory nerve, along with the CN, supply the secondary respiratory muscles (sternocleidomastoid, trapezius). These become active during forced expiration (Richerson & Boron, 2003).

Phrenic nerves: Motor innervation of the diaphragm occurs via the *phrenic nerves* arising from the ventral rami of C3–C5 (Moore & Dalley, 1999, p. 294). Each phrenic nerve supplies motor, sensory, and sympathetic nerve fibers to half of the diaphragm. It rises primarily from C4, but also C3 and C5, and possibly from the subclavian nerve of the brachial plexus. Sensation is served by the phrenic, intercostal, and subcostal nerves (Moore & Dalley, 1999, p. 294). Each phrenic nerve within the neck runs caudally close to the body of the cervical vertebrae and passes underneath the clavicle before entering the thoracic cavity.

Afferent Innervation of Vocal Structures

The Vagus Nerve

1. *Visceral sensory afferents* (the mucous membrane of the epiglottis, the base of the tongue, and the majority of the larynx) travel in the internal laryngeal nerve. This arises in the inferior vagal ganglion of the carotid triangle and is continued as the internal laryngeal nerve, supplying afferents to the laryngeal mucous membrane of the vocal folds. Visceral stimuli are received from baroreceptor and chemoreceptor stimuli, as well as from the internal

organs. The axons transmit to the medulla and also to the tractus solitarius and participate in the reflex control of cardiovascular respiratory and gastrointestinal functions, innervating the reticular formation and hypothalamus.

2. *General sensory*: the larynx, pharynx, and so on, including sensation from vocal folds and the larynx below it. These sensations are carried with the visceral sensory fibers of the recurrent laryngeal nerve. Sensory fibers transmit to the superior laryngeal branch of the vagus nerve, to the inferior vagal ganglion, enter the medulla, project to the spinal trigeminal tract, the trigeminal thalamic tract, the contralateral ventral posterior nucleus of the thalamus, finally through the internal capsule to the SC. Nevertheless, it is not clear how afferent information is used to regulate speech movements (Kent, 1999).
3. *Visceral sensory*: These are fibers from the larynx, trachea, thoracic, abdominal viscera, etc. (Wilson-Pauwels et al., 1988, pp. 134–136). The medullary caudal *solitary nucleus* receives input from the viscerosensory receptors of the IXth (including taste) and Xth nerves, which project to the cardiorespiratory nucleus (Martin, 1996, p. 226), the first viscerosensory relay for mechanoreceptors and chemoreceptors (Martin, 1996). This initiates the medullary reflexes that control cardiovascular, respiratory, and other autonomic functions (Benarroch, 1997a, pp. 10–12; Martin 1996, p. 230). The solitary N. relays viscerosensory information to the forebrain and projects to parasympathetic preganglionic neurons and somatic motor neurons in the *nucleus ambiguus* (see Motor, above). Airway afferents are primarily vagal, projecting to the nucleus solitarius, with some fibers entering the spinal cord in the upper thoracic sympathetic trunks (Barnes, 2003).

The Spinal Cord

Spinal segments (neuromeres) preserve the primitive arrangement. They have an arrangement of sensory and motor columns similar to the brainstem, where there are motor and sensory roots represented by CN nuclei. The neuromeres are reflected in somatic segments (metameres), subdivided as *myotomes* (striated muscle), *scleratomes* (the area of bone innervated by a spinal segment), and *dermatomes* (the skin supplied by a single spinal root) (Parent, 1996, pp. 264–268, 325). The *cervical plexus* arises from the ventral rami of C1–4 or C5 and branches from the *superior cervical sympathetic ganglion* (Moore & Dalley, 1999, pp. 294, 1010, with diagram; Parent, 1996, p. 275).

Autonomic

Respiration maintains oxygen and carbon dioxide pressure within narrow limits. Neuronal pools that subserve involuntary ventilation originate in the caudal medulla and give rise to descending pathways in the ventrolateral brainstem and spinal cord. The fastigial nuclei of the cerebellum are involved in ventilatory inhibition via the anterior and the posterior lobes, including volitional breathing (Simon, 2001). Since aprosodia has been described as a lateralized phenomenon, it is noteworthy that it is in fact the fastigial nuclei that give rise to both crossed and uncrossed axons.

Cholinergic nerves are the major neural bronchoconstrictor mechanism in determining *airway caliber* (i.e., controlling smooth muscle tone, airway blood flow, and mucus secretion). They arise in the nucleus ambiguus, travel down the vagus nerve, and synapse in the parasympathetic ganglia (Barnes, 2003). Airway constriction is enhanced by vagus nerve (parasympathetic) stimulation of the smooth muscle of the cartilaginous airways, membranous bronchioles (primarily smooth muscle), and alveolar ducts. Sympathetic stimulation inhibits airway constriction (Guyton & Hall, 2000, pp. 440–441; Staub, 1998b). Although sympathetic innervation of the central portions of the lung is weak, the bronchial tree is exposed to the circulating norepinephrine and epinephrine (beta receptors) released into the blood by SNS stimulation of the adrenal medulla. PSNS fibers (vagus nerves) release acetylcholine, causing mild-to-moderate constriction of the bronchiole.

SOMATIC TRAUMA AND VOCALIZATION DISORDERS

Direct trauma to the neck and whiplash (hyperextension, hyperflexion and rotation of the neck) create symptoms that are sometimes misattributed to cerebral damage when actually resulting from damage to the neck and other somatic structures (Parker, 2001, pp. 121–127, with numerous references). The neck is an extremely crowded structure (Rohen & Yokochi, 1993, pp. 150–180). This creates vulnerability to injury through torsion, tension, or crushing of neck structures, including the recurrent laryngeal nerves (a branch of the vagus, which controls the vocal cords. The trunk of the vagus nerve itself is vulnerable to trauma (Davies & King, 1997). The recurrent laryngeal nerves ascend close to the notch between the trachea and the esophagus. Components of speech subject to traumatic brain injury (TBI) disruption include overall intelligibility, prosody, respiration, phonation, resonance, and articulation (Theodoros et al., 2001a, 2001b). The *phrenic nerves*, involved in diaphragmatic control, are vulnerable to trauma at multiple sites from the neck to the lower thorax (Lacomis, 1996). They are in close association with various veins and muscles, suggesting that it is possibly vulnerable to direct trauma and/or torsion or extension during hyperextension—hyperflexion injury (whiplash).

It is useful to distinguish a paralyzed vocal fold from a fixed one. Etiology includes mechanical trauma to the skeleton of the larynx (cricoarytenoid joint), collagen vascular disorders (e.g., rheumatoid arthritis, Lyme disease, and lupus erythematosus), damage to the vagus nerve (CNS disease, stroke, head trauma), vagus nerve damage in association with chest trauma, malignant lesions that invade the recurrent nerve, and virus infection of the recurrent nerve (Stewart et al., 1999). Since the diaphragm lacks bilateral cortical representation, hemispheric stroke results in attenuation of diaphragmatic excursion on the hemiplegic side during volitional breathing. There is also impairment of contraction of the intercostal muscles. Contralateral diaphragmatic contraction can be elicited with cortical stimulation that was localized by a PET study of increased cerebral blood flow in the motor strip, the right premotor cortex, the supplementary motor cortex, and the cerebellum (Simon, 2001). Hypokinetic dysarthria develops with worsening Parkinson disease (impairment in voice, articulation, and fluency, particularly decreased voice loudness with a monopitch and monoloud inflection and prosodic insufficiency). Hypokinetic dysarthria has been associated with damage to the caudate nucleus (Theodoros et al., 1995). Dysarthria with dysprosody was observed after a small unilateral putamenal lesion. Basal ganglia lesions creating aphasia and dysarthria are associated with a contralateral hemiplegia. They are usually caused by large lesions (infarct) that involve the caudate and lentiform nuclei as well as the internal capsule and the adjacent white matter (Bhatia & Marsden, 1994).

BREATH SUPPORT

Vital capacity is affected by the effort required for inspiration and expiration, muscle weaknesses, reduced postural control, and a trend to higher WAIS-R scores (McHenry, 2001). There is no one-to-one correspondence between a specific deviant speech dimension and the impairment of a particular component of the speech production apparatus. In addition, speech breathing is vulnerable to severe brain injury, which may include damage to the cerebral cortex, subcortical structures, and the cerebellum and/or brainstem, disrupting normal speech breathing process (Murdoch & Theodoros, 2001). Most pulmonary problems are directly related to the trauma (pneumothorax and flail chest) or are consequent to neurological deficits. TBI can cause abnormal chest wall movements during speech production, possibly from breakdown in the coordination of the movements of the chest wall, ribcage, abdomen, and diaphragm. Reduced breath support may evolve from impairment of other speech systems (i.e., laryngeal, velopharyngeal, and articulatory). Alterations of vital capacity and chest wall volumes affect speech breathing and the number of syllables produced per breath. Dysfunctions of the *velopharynx* contribute to impaired speech breathing by wasting expired air through the velopharyngeal port. Inefficient valving or an impaired respiratory system gives the impression that speakers with TBI "run out of air" (Netsell, 2002). Inadequate airflow for the duration of speech segments results in rate fluctuations and decreased rate of speech, as well as inadequate intraoral pressure for plosive sounds or for the intensity of airflow required for fricative consonant production (Theodoros et al., 1995).

Larynx

Since phonation is a highly integrated neurophysiological coordination, laryngeal dysfunction may result from impairment at many levels of the CNS: bilateral lesions of the UMN, damage to LMN supplying laryngeal musculature, lesions of the extrapyramidal system, and other locations centrally and peripherally. In a group of severe closed head injury (CHI) subjects, 75% displayed laryngeal dysfunction, with implications for intelligibility of speech that includes articulation (harshness, strained-strangled, hoarseness, intermittent breathiness). There is increased tone or spasticity in the laryngeal musculature, resulting in hyperadduction of the vocal folds, reduction in the size of the laryngeal aperture, and an increase in resistance to airflow during phonation. Some vocal effects are due to compensation for laryngeal impairment, wastage of expiratory airflow from other valving mechanisms, or deficits in other focal systems (Theodoros & Murdoch, 1996).

Dysphonia and laryngeal dyskinesia: Hyperfunctional, hypofunctional, or incoordinated laryngeal function may result from bilateral damage to the UMNs, lesions in specific LMNs, damage to the extrapyramidal system and its connections, or lesions of the cerebellum and its pathways (Cahill et al., 2003). It is an effect upon striated muscles, causing tongue protrusion and/or harsh, strained, or breathy voice (Jankovic & Lang, 2004). *Paradoxical vocal cord motion* (*PVCM*) is a form of laryngeal dyskinesia characterized by inappropriate adduction of the vocal cords during inspiration. It results in inspiratory stridor. It has varied etiology: cortical or UMN injury (e.g., severe static encephalopathy; cerebral vascular accidents); movement disorder (adductor laryngeal breathing dystonia; myoclonic disorder with palatal movements); "parkinsonism-plus" syndromes; drug-induced brain stem dysfunction; nuclear or LMN injury (abnormality of the vagal nuclei or recurrent laryngeal nerves usually with brainstem compression); and brainstem compression. Conditions less related to the scope of this chapter include factitious or malingering disorder and somatization/conversion disorder (Maschka et al., 1997). Another study, while confirming that laryngeal dysfunction is common after severe TBI, did not confirm preponderance of hyperfunctional voice production (Cahill et al., 2003). Low laryngeal airway resistance is associated with breathy and/or soft voice (McHenry, 1996).

Trachea: Trauma can occur to the trachea (stridor). Deficits are caused by tracheostomy devices and secondary hypoxic injury. Vocal cord paralysis is due to direct trauma or damage to the recurrent laryngeal nerve (Wiercisiewski & McDeavitt, 1998).

NERVOUS SYSTEM TRAUMA AND VOCALIZATION DISORDERS

CNS Disorders

While a speech disorder may be accompanied by the localization of a lesion, disordered performance reflects the performance after the brain has adjusted to a loss and alteration of function. Significant disorder of a particular aspect of vocalization is frequently accompanied by a variety of vocalization disorders. Abnormal articulation is based upon differential impairment of motor subsystems. Dysarthria constitutes one-third of communication impairments observed in *head-injured persons*. It is associated with severe brain injury. Its most common etiology is spasticity resulting from bilateral UMN damage, affecting pyramidal and extrapyramidal motor tracts (Theodoros et al., 2001b). Dysarthria is found in *amyotrophic lateral sclerosis associated with* degeneration of corticobulbar projections (Brown, 2001). *Postcentral* (*kinesthetic*) *articulatory dyspraxia* (Glezerman & Balkoski, 1999, p. 98) refers to substitution of particular articulatory components by similar ones, resulting in mispronunciation of consonants more than vowels. Slurred speech and speech arrest are associated with a lesion of the *supplementary motor area*. Dysarthria with inarticulate, intermittent explosive speech occurs commonly in *bilateral cerebral hemisphere*.

UPPER MOTOR NEURON (UMN) DISORDERS

UMN Diseases

Respiration: UMN damage may be comorbid with respiratory and limb weakness. UMN damage can affect automatic respiration through alveolar hypoventilation. The most common site of damage affecting central respiration is the medulla (Lacomis, 1996). *Limbic* stimulation causes the inhibition of ventilation, resulting in apnea and hypoxia, with cardiac slowing and hypotension.

Laryngeal dysfunction: Laryngeal dysfunction is vulnerable to bilateral lesions of the UMNs; LMNs that supply the laryngeal musculature; the extrapyramidal system; the cerebellum; and the vagus nerve (the superior laryngeal and recurrent laryngeal nerves). UMN weakness can cause laryngeal spasticity or weakness sufficient to create perceptible effects on the voice. This may be associated with less-bilateral UMN supply to the vagus nerve so that unilateral lesions make it more susceptible to the effects of weakness and spasticity on laryngeal function (Duffy & Folger, 1996). When these systems are not coordinated and adjusted in advance, the intensity of the duration for each successive sound causes dysarthria (see below). Some syllables are loud, some are weak, and some are held for too long or too short intervals, resulting in almost unintelligible speech (Guyton & Hall, 2000, p. 655).

Dysphonia refers to hyperfunctional, hypofunctional, or incoordinated laryngeal function due to abnormal, involuntary muscle contraction. The voice is interrupted, rough, and breathy, with increased voice arrests and decreased loudness (Younger, 2005c). It may result from bilateral damage to the UMN, lesions in specific LMN, and damage to the extrapyramidal system or the cerebellum and their pathways (Cahill et al., 2003). *Extrapyramidal disorders* are involved in sighing, forced expiration, inability to alter the respiratory rhythm voluntarily, and so on. Another study, while confirming that laryngeal dysfunction is common after severe TBI, did not confirm a preponderance of hyperfunctional voice production. Low laryngeal airway resistance is associated with breathy and/or soft voices (McHenry, 1996). The *tongue* has sufficient ipsilateral innervation to restrict the effects of UMN unilateral damage.

UMN damage can affect automatic respiration through alveolar hypoventilation. Dysarthria may originate from unilateral disorders of the UMN. In one sample, the most frequent abnormality was harshness, while 23% manifested abnormality of rate (usually slow), stress, or prosody (Duffy & Folger, 1996). In a group of dysarthrics ranging from 15 to 32 years of age, there were discrete, atrophic, and afferent fiber lesions in the descending motor fibers of the periventricular white matter. Eighty percent of these lesions were subcortical, which resulted in moderate-to-severe dysarthria. This has been attributed to weakness because of the involvement of the distributed neocortical system. With predominantly cortical lesions, mild-to-moderate dysarthria was found. Supratentorial disorders of the UMN of either hemisphere can create dysarthria (Duffy & Folger, 1996). The neocortical motor system can be involved in TBI. UMN weakness can cause laryngeal spasticity or weakness sufficient to create perceptible effects on voice. This may be associated with less-bilateral UMN supply to the vagus nerve, so that unilateral lesions make them more susceptible to the effects of weakness and spasticity on laryngeal function (Duffy & Folger, 1996). In one study of stroke, most patients had a contralateral lower facial weakness and more than half had clinically detectable tongue weakness. Jaw or palatal asymmetry or weakness was rare. This is consistent with the greater bilateral UMN supply to the trigeminal and vagus nerves, compared to the facial and hypoglossal nerves. *Bilateral UMN diseases* may cause spastic dysarthria, affecting phonation, articulation, resonation, and prosody: strained-strangled-harsh voice; imprecise consonant articulation; hypernasality; and low vocal pitch, with diminishing variations in pitch and loudness contributing to monotonous vocal quality.

Lower Moter Neuron Disorders

LMN disease (nerve fibers or neuromuscular junction) is associated with respiratory support problems. Weak muscles do not fully inflate the lungs, which reduces breath pressure at the vocal folds.

Furthermore, weak laryngeal muscles fail to fully adduct the vocal folds, creating speech with short utterances, frequent pauses for breath, weak vocal intensity, voice quality breathy, and vocal pitch low and monotonous. Damage to the *pons and medulla*, where the CN nuclei are located, creates dysarthria when the damaged nerves supply muscles producing speech. Localization of the injury (i.e., whether nerve trunk or nuclei) is aided by determining whether there is comorbid hemiplegia or hemiparesis of either the arm or the leg, since corticospinal tracts pass through the brainstem near the CN nuclei decussating below (Brookshire, 1997).

Dysphonia: Dysphonia can result from damage to the LMN (spinal and cranial) that supply the muscles of respiration, causing a change in speech described as flaccid dysphonia. In this condition, unilateral recurrent vagal or laryngeal nerve injury results in a voice quality that is described as hoarse, harsh, breathy with short phrases, reduced loudness and mild inhalator stridor. There is unilateral paralysis of almost all laryngeal muscles (except cricothyroid) (Brazis et al., 2001, p. 335; Parent, 1996, p. 449). Lesions of the glossopharyngeal, vagus, and accessory nerves associated with basal skull fractures may result in dysphonia (Brazis et al., 2001, p. 331). Bilateral vagus nerve lesions cause drooping of the palate, losing movement during phonation. When speaking, air escapes from the oral to the nasal cavity, giving a "nasal" quality to the voice (Brazis et al., 2001, p. 333). Compression of the recurrent laryngeal nerve with hoarseness occurs with thyroid carcinoma and Hashimoto disease (autoimmune thyroiditis type 2A) (Larsen et al., 1998, p. 479). Dysphonia may be described as flaccid (X nerve lesions, with vocal effects dependent upon the location of the lesion along the nerve pathway from brainstem to muscle); spastic (pseudobulbar, with a harsh strained-strangle quality due to vocal fold hyperadduction); hypokinetic (*Parkinsonian*); and monopitch, with reduced loudness due to muscle rigidity. Other dysphonias have been described as consequent to uncontrolled muscular movements that are jerking, slower, or tremulous due to reduced feedback of muscular performance (ataxic, choreic, dystonic).

Extrapyramidal Disorders

Extrapyramidal disorders are involved in sighing, forced expiration, inability to alter the respiratory rhythm voluntarily, and so on. *Limbic* stimulation causes inhibition of ventilation, resulting in apnea and hypoxia, with cardiac slowing and hypotension. Laryngeal dysfunction is vulnerable to bilateral lesions of the UMNs, LMNs that supply laryngeal musculature, the extrapyramidal system, the cerebellum and their connections, and the vagus nerve, that is, the superior laryngeal and recurrent laryngeal nerves. Damage to LMNs (spinal and cranial) that supply the muscles of respiration causes a change in speech described as flaccid dysarthria. Unilateral or bilateral damage to the vagus nerve alters displacement of the vocal folds (adduction of the glottis), depending upon whether the lesion is in the brainstem (aphonia due to soft palate paralysis), vagus nerve (aphonia), or its recurrent laryngeal branch (Rosenfield & Barroso, 2000). Abduction or adduction displacement in the vocal folds and supraglottic musculature changes the quality of the voice.

Basal Ganglia Disorders

The basal ganglia (and temporoparietal areas) play a role in emotional prosodic function (comprehension and production). Putamen lesions have been described with defects of articulation and prosody. Disorders do not seem to be lateralized. Some imprecision is indicated by the inclusion of patients with "an indifferent attitude" and "apparent affective indifference" (i.e., lack of spontaneous affective expression of the affects). Furthermore, depression and other affective disturbances have been described in basal ganglia disorders and Parkinson disease. Therefore, with these lesions, aprosodia may be comorbid for motorvocal and mood disorders (Van Lancker & Breitenstein, 2000). This motor influence may account for the fact that disorders may express poorly modulated, monotonous, and halting speech (Gilman et al., 1981). Resemblance between aprosodia and dysarthria occurs in hypokinetic dysarthria, the prosodic deficits of which stem from the effects of muscular rigidity, reduced force and range of movement, and variable speed of movement associated with basal ganglia pathology (Theodoros et al., 2001b).

Cerebellum

Damage to the cerebellum, brainstem, or their connections causes dysphonia through interference with laryngeal muscle coordination and tone, affecting vocal fold movements (Theodoros & Murdoch, 2001b). Dysrhythmias are associated with cerebellar damage (Theodoros et al., 1995). Dysarthria has been associated with hemispheric infarction on the left (Gilman et al., 1981, p. 226) and also in patients with pathology that is limited to the right cerebellar hemisphere (see Dysarthria) (Topka & Massaquoi, 2002). Disturbance of speech can result from damage to the ventrolateral nerve of the thalamus (Gilman et al., 1981, p. 225). The cerebellum projects to it from the dentate nucleus, contralaterally decussating to the red nucleus, and then projecting to the thalamus (Parent, 1996, p. 618). Tumors of the posterior fossa may manifest as ataxic dysarthria. The ten most deviant speech disorders are: excess and equal stress; irregular articulatory breakdown; distorted vowels; harsh voice; phonemes prolonged; intervals prolonged; monopitch; monoloudness; and slow rate (Hudson, 1990, citing Darley, 1969). A case is reported of a 5-year-old boy, right handed, with a midline medulloblastoma (considered to be an acquired cerebellar lesion), who, in addition to impaired spontaneous language, had dyspraxic, dysarthric, and agrammatic articulation (Levisohn et al., 2000). Degeneration of the cerebellum is associated with impairment of the discrimination of pitch (Parsons, 2001). Cerebellar dysarthrias may result in part from generalized hypotonia. In advanced *multiple sclerosis*, scanning speech (cerebellar dysarthria) is common (Hauser & Goodkin, 2001).

Motor Speech Disorders

Neurological nonverbal communication disorders include verbal expression as well as vocalization disorders, although both impair social development. In the clinical study of communications, one should consider that brain injury inhibits the production of some aspects of spontaneous facial expression as well as the complex neurological and somatic components of the vocalization apparatus. Children with acquired brain injury display communicative facial expressions for a smaller proportion of the time, are less likely to shift from one expression to another, and some may express reduced intensity of expression (Kupferberg et al., 2001).

Disorders occur in the CNS [integration, programming], the PNS, and somatic structures. Vocal assessment of speech disorder explores the parameters of respiration, phonation, resonance, articulation, prosody, facial musculature, diadochokinesis, reflexes, and intelligibility. Additional features include delayed initiation and impaired rate control (Jaegar et al., 2000). A case is reported of bilateral hypoglossal nerve paralysis following head trauma without skull base fracture, possibly due to traction injury (Noppeney & Nacimiento, 1999). Neuromuscular function includes the strength of muscular contraction; the speed, accuracy, and range of excursion of the muscle; the steadiness of the contraction; and the tone of the muscle (Murdoch et al., 1990). It is possible to differentiate between dysarthrias (abnormal muscle strength for articulation of sounds or phonemes), dysphonias (disorders of voicing), apraxia (inability to program sequences of phonemes, with consonant substitution rather than distortion, as in dysarthria), and stuttering (initial pauses and dysfluency of speech production without articulator or language disorder) (Theodoros et al., 2001b).

DYSPROSODIA

Clinical understanding of communications is hampered by the use of the term dysprosodia for manifestly different functions. It is used for (1) the comprehension of others' emotional communications; (2) the modulation of communication functions such as timing and linguistic stress to communicate meaning; and (3) the modulation of the motor elements of vocalization, including pitch control by the vocal cords, breathing, articulation, shape of the resonance chambers of the mouth and throat, and so forth. The writer defines prosody and dysprosody exclusively as a motor vocalization function, in which after trauma (neurological or somatic) or neurological or medical disease there is a change of audible emotional expression. Aprosodia is a variant in which the quality of vocalization is flat. Hyperprosody is an exaggerated prosody observed in a manic state or in individuals with aphasia who can access very few words but use exaggerated prosody to convey their feelings as much as possible (Wymer et al., 2002).

The anatomical basis for dysfunction can be neurological (CNS or PNS) or somatic (anatomical structures supporting vocalization and respiration). While there is input from CNS centers that mediate emotional comprehension, it is considered to be a separate function. It has been detected after concussive brain injury (Parker, 2001), severe CHI (Samuel et al., 1998), and a variety of neurological disorders.

Etiology

Aprosodia may result from congenital, neonatal, developmental, aging and senescent pathology, and/or major TBI (Theodoros et al., 2001a, 2001b). After mechanical accidents, aprosodia can be consequent to an extremely wide range of neurological and somatic injury, including innervation and damage to the larynx, face, jaw, and breathing apparatus. It may be comorbid with dyspraxia, medical conditions, and aphasia. Acquired aprosodia in children can be associated with right or nondominant cortical lesions (Nass, 2000). Various neurological structures to which speech disturbances have been related include the reticular system (initiation of speech), the limbic system (particularly RH for emotional aspects of communication), and the neocortical system (sensorimotor control of speech movements involving the entire vocal tract), as well as the supramarginal gyrus, basal ganglia, and neocerebellum (Theodoros et al., 2001a).

Comorbidity and differential diagnosis: Aprosodia is comorbid with severe dysarthria (Patel, 2002) and dysphoric mood (depressed, anxious, angry, emotionally deprived, etc.), (Cereda et al., 2002; Tabuchi et al., 2000). Apathy, blunted affect, and alexythymia require differentiation from aprosodia. While aprosodia has been described as unique to TBI (Jorge et al., 2000), it has also been associated with neurological problems such as autism and nonverbal learning disorders (Asperger disorder) (Davidovicz, 1996; Duane, 1996). In a study of stroke victims, it was found difficult to study aprosodia due to concurrent dysarthria, hypophonia, and shyness (Starkstein et al., 1994) Aprosodia is found in Wernicke's aphasia (but not expected in delirium, schizophrenia and mania) (Mendez & Cummings, 2002) and is also associated with alexithymia (Fricchione & Howanitz, 1985). Aprosodia (used in its generic sense) has been detected in alcoholism, schizophrenia, Alzheimer disease, Parkinson disease, autistic spectrum disorders, Broca's and Wernicke's aphasias, and alcohol exposure (detoxification and fetal alcohol exposure). In schizophrenia, it could be considered a negative symptom. Comorbidity with depression and apathy creates diagnostic difficulty. Aprosodia and apathy may display occasional inappropriate cheerfulness (Wymer et al., 2002).

Alexythymia has the following characteristics: deficient ability to express emotions caused by a lack of understanding of the emotions; concrete, bland, and tedious communication style; a tendency to be externally oriented in thinking; and poor ability to use imagery. One may speculate that a bland manner—confusable with aprosodia—could result from a lack of appropriate emotional expression due to a lack of understanding of one's emotions (Williams et al., 2001). Association has been demonstrated between self-reported head injury and levels of alexythymia (Williams et al., 2001).

Clinical Implications

Expressive dysprosodia interferes with communications in both the clinical and daily living domains. The injured person with aprosodia specifically describes experiences of distress or is assumed to be socially participating in a painful situation. There are changes in emotional quality, as created by variations of pitch and volume. The aprosodic patient has deep feelings that are well described verbally but that lack the expected nonverbal quality of distress. The monotonous vocal expression impairs social communication of emotion and meaning as well as vocational functioning (e.g., vocal communication) (Stringer, 1996). It can cause the intensity of one's distress to be underestimated. This is in contrast with a patient who may not be able to fantasize, describe his or her feelings, and communicate (Fricchione & Howanitz, 1985). The accident victim describes a frightening accident, dysphoric moods, and a bleak life, without the expected prosodic quality. The impression is bland, emotionally flat, or indifferent, leading to a misperception of the injured person's feelings. Normally, concurrent auditory and visual stimuli are congruent. Incongruent facial and vocal information is associated with changes in event related potential (ERP), known as *mismatch negativity* (De Gelder, 2000).

In daily living, incongruent information is reacted to with confusion and may be disregarded. There are job-related communication disturbances with the public and one's co-workers. As defined, the writers are concerned here primarily with the difficulties of communication due to an impaired rendering of the quality of voice by the speaker and as understood by the hearer.

Aprosodia can be considered a *deficit of self-monitoring*. The external loop utilizes the same superior temporal structures whether one is hearing one's own voice or that of another. The feedback permits some degree of output control (i.e., loudness and error correcting). The internal loop comprises internal modulation of the motor area (Indefrey & Develt, 2000). A prosodic output is not phonetic processing (specifically, linguistic features of speech including place and manner of articulation of consonants), but acoustic processing, which is loudness and frequency that is not of direct linguistic significance (Norris & Wise, 2000). Thus, aprosodia may be a deficiency of self-monitoring of vocal output.

In one study after CHI patients with major depression were excluded, there was no significant correlation between prosodic measurements, depression scale, or neurobehavioral assessment. Depression itself alters pitch and loudness (Samuel et al., 1998). A salient characteristic of aprosodia is loss of vocal pitch, which is considered to be the central aspect of all music (Zatorre, 2001). Patients with severe CHI were not able to change pitch and volume according to prosodic content, particularly for anger (Samuel et al., 1998). Aprosodia has long-term persistence, with devastating effects upon communications. It occurs in a range of loss that is very responsive to the clinical level of mood elevation, other manifestations of a patient's emotional state (Damasio, 1999), and psychopharmacologic effects. Speech therapy that focuses on loudness of speech production may create an elevation of prosodic outcome measures (Patel, 2002).

APROSODIA

Aprosodia may be defined as a motor vocalization dysfunction in which there is loss of audible emotional expression. It refers to inability to express moods nonverbally (i.e., with a musical, expressive quality), lacking the musical quality (pitch variations, volume, emphasis, pacing changes) that usually communicates mood, reaction, social expectation, desire to influence others, and so forth. Thus, aprosodia is a mismatch between verbal expressions of experienced emotional distress,and the intensity and quality of overt expression of feelings. It is differentiable from indifference, the inability to experience feelings deeply, or anhedonia.

Flat affect creates a false impression of indifference or lack of distress. This writer considers aprosodia to be a motor vocal disorder. It has been described as an impairment of the *pragmatics of nonverbal communication*, resulting in impoverished language and reduced communication proficiency (Jorge et al., 2000). This contrasts with cognitive and aphasic deficits that include incapacity for communicating inner states, inability to comprehend the emotional communications of others, deficit of symbolization, and comprehension of the complexity of emotions (Heilman et al., 1993; Lane et al., 1997).

The injured person describes feelings of distress in a flat voice without the emotionality that is expected with long suffering (i.e., physical and emotional pain). The anatomical basis for dysfunction can be neurological (central or peripheral nervous systems) or somatic (anatomical structures supporting vocalization and respiration). Aprosodia has been detected after concussive brain injury, severe CHI, and a variety of neurological disorders (Parker, 2001; Samuel et al., 1998). The author humbly rejects his former attribution of aprosodia as a cerebral personality disorder (Parker, 2001).

Specifying the Neural Mechanism of Aprosodia

The literature uses the term aprosodia to refer to both motor activity (e.g., pitch and intonation) and the comprehension of vocal qualities of the speaker's and affect (sensory prosody). Grossly different conditions are included as aprosodia, including vocal disorder, incomprehension of others' affect, motor expression of feelings, and the inability to understand one's own feelings (alexithymia). Some use the term as though aprosodia is a single mechanism, inclusive of both expressive and receptive symptoms. The term aprosodia has been used for highly varied behaviors, such as sensory input (visual, pitch, linguistic involving emotional comprehension) and motor or expressive vocal output (Ross et al., 2001; Starkstein et al., 1994). The claim has not been substantiated that RH prosody could be detected in cerebral areas homologous to those of aphasic syndromes (LH). This assertion is based upon studies primarily of cerebrovascular accidents (Ross & Rush, 1981). The author considers that the loss of recognition of others' moods, facial expressions, and gestures require different neuroanatomical mechanisms than motor prosodic vocalization. This is in contrast with the assumption of Ross et al. (2001) that schizophrenia is a RH disorder, since the symptoms they considered to be aprosodic (loss of affective comprehension) were detected. Furthermore, it has been largely based upon cerebrovascular accidents, neglecting mechanical accidents (i.e., concussion with brief alterations or LOC) (Parker, 2001).

Etiology and Neural Basis

Aprosodia may result from congenital, neonatal, developmental, aging and senescent pathology, and/or major TBI (Theodoros et al., 2001a, 2001c). Acquired aprosodia in children can be associated with right or nondominant cortical lesions (Nass, 2000). Various neurological structures to which speech disturbances have been related (Theodoros et al., 2001a) include the reticular system (initiation of speech), the limbic system (particularly the RH for emotional aspects of communication), and the neocortical system (sensorimotor control of speech motions involving the entire vocal tract), as well as the supramarginal gyrus, the basal ganglia, and the neocerebellum.

Extensive neurological and somatic structures participate in vocalization, lateralized to LH or RH, bilateral, and diaschisis effects contralateral structures. The basal ganglia (and temporoparietal lesions) have been associated with emotional comprehension aprosody (Starkstein et al., 1994). The caudate participates in vocalization through auditory temporal processing (Fitch et al., 1997). Expressive aprosodic localization has been attributed to the putamen, laterally adjacent to the anterior insula, and subjacent white matter (Tranel, 2002), as well as the dysregulation of the autonomic nervous system (Lane et al., 1997). The insula has also been credited with a major role in speech articulation (Cereda et al, 2002). Cerebellar tract damage contributes to *aprosodia* by impairing vocal timing, force, range, and direction of muscle movement. Vocalization deficits are consequent to disordered airflow, breath, rhythm, pitch, loudness, rate, resonance, articulatory precision, rate, and stress (Darley et al., 1975). Damage to the cerebellum or brain stem, or their connections, causes dysphonia through interference with laryngeal muscle coordination and tone, affecting vocal fold movements (Theodoros & Murdoch, 2001b).

Comorbidity and Differential Diagnosis

While aprosodia has been described as unique to TBI (Jorge et al., 2000), it has been associated with autism (Davidovicz, 1996) and nonverbal learning disorder/Asperger disorder (Duane, 1996). Aprosodia is comorbid with severe dysarthria (Patel, 2002), dysphoric mood (depressed, anxious, angry, emotionally deprived, etc.), and dysarthria (Cereda et al., 2002; Tabuchi et al., 2000). Apathy, blunted affect, and alexythymia require differentiation from aprosodia. In a study of stroke victims, it was found difficult to study aprosodia due to concurrent dysarthria, hypophonia, and shyness (Starkstein et al., 1994). Aprosodia is found in Wernicke's aphasia (but not expected in delirium, schizophrenia and mania) (Mendez & Cummings, 2000) and is associated with alexithymia (Fricchione & Howanitz, 1985).

Observational Examples of Aprosodia

Aprosodia is a mismatch between the verbal expression of emotional distress and the intensity and quality of overt expression of feelings. It is an inability to convincingly express feelings nonverbally or with the usual intonations and phrasings characteristic of emotions. It is differentiable from indifference, the inability to experience feelings deeply, or anhedonia. *Rorschach cues of dysphoria*: A rough guide for selecting Rorschach content was a bland exterior with report of intense distress and consistent responses. Informal analysis indicated that only about half had Rorschach content consistent with severe distress.

The writer reviewed cases of persons who had significant injuries and whose expression of affect was bland compared to their reported levels of distress. Selection was made on the basis of the use of the term "aprosodia" in the report or in review of cases for various reasons in which the clinical observations displayed dull affect. Thus, the clinical unit was complaints, clinical observation, and supporting Rorschach content. This discussion does not include possible alexythymia (i.e., inability to recognize one's own feelings). Attention was directed to persons who reported their feelings, but whose motor expression was reduced.

A 5-year-old girl's Rorschach: She was scarred on the face when struck by a projecting portion of a car. These percepts were offered calmly and quietly: "It looks like a monster, eyes, scratches on his face, walking around scaring everybody"; "Dracula killing a person"; "Dinosaur's body inside"; "Two people carrying a bag." The writer asked her whether it was easy or hard: "It's hard." The writer asked how they looked: "They don't look good." The writer asked about her mood: "They're in pain."

A high proportion of human-like figures were expressed (H), suggesting a feeling of detachment or unfamiliarity with one's new identity (injured, vulnerable, impaired, unattractive).

These were guided by an *emotional behavior checklist* and the notes made during the examination:

Expressed concern about problems: high; average; little; indifferent
Reaction to event: maximizes ( ) appropriate ( ) minimizes ( ) none( )
Level of overt affect: Generally high/normal/low/variable
Demeanor: Pleasant/resentful/unhappy/dull/cries/smiles or laughs
Range of affect: Reduced ( ) Normal ( ) Wide ( )
Depressed (sad): Y/N ... Anxious/startle: Y/N ... Angry: Y/N
Exaggerated personal difficulties: Yes/No

Observation that signs were lacking nonverbal overt expressions of anger, depression, or anxiety; average concern and appropriate reaction to her problems and the event; dull affect; restricted range of affect; demeanor may be pleasant, range of affect reduced; expression dull; did not appear to exaggerate his personal difficulties; voice was low and he tended to be somewhat monosyllabic;

affect dull, acknowledges depression when asked; affect rated as flat, although he was appropriate, related, cooperative, and well motivated.

Rorschach Cues of Aprosodia

In the author's outpatient practice—primarily for accidents creating head or other physical trauma and emotionally stressful experiences—a guide for selecting Rorschach content was utilized (see Definition, above): a bland voice and facial expression; a self-report of intense distress; and Rorschach content that was symbolically or specifically consistent with the reported feeling of distress or stressful events or injury. Informal analysis indicated that only about half of the patient sample had Rorschach content suggestive with severe distress. *Concept:* The clinical observations suggested dull affect, while the Rorschach offered evidence for feelings not clinically observed. One consideration might be alexythymia (i.e., inability to recognize one's own feelings).

Clinical example: A calm girl: An 11-year-old girl appeared to be clinically very pleasant and without apparent distress. She did not appear to be clinically depressed, angry, or anxious. Nevertheless, a wide range of distress was documented. The girl remembered being hit by the car and what she was doing at the time of the accident. Her cheek was scraped. When the Emergency Medical Services (EMS) arrived, they picked her up and taped something to her face and chin to stop the bleeding. She was frightened. She did not claim to have bad dreams at the time of the accident or afterwards.

Rorschach responses: There was a discrepancy between the girl's demeanor and experiences. Overtly, she was a pleasant and cooperative child. She denied any emotional distress. This is reflected in a Rorschach projection of two dancers, poorly constructed, who are "happy." Poorly perceived human figures are suggestive of discomfort with one's self-image. The girl's fantasies reveal that she would like to be assertive and outgoing. Her tendency to conceal her feelings is revealed by a "mask of a cat" and by the denial of anxiety with "a costume … [representing] a bat." There were specific signs of anxiety: "snake"; "bat"; "crawling bug"; "spiders"; "monster." The girl tried very hard to exercise emotional control (all of her Rorschach responses were precise, with no vagueness). This is unusual for a child, and represents a denial of normal emotionality and spontaneity. Moreover, the girl projected no color responses, which is noteworthy at any age, but here was indicative of depression.

Clinical example: Aprosodia—facial scarring: A 5-year-old girl's face was scarred when she was struck by the projecting portion of a car. The following Rorschach responses by the patient were offered calmly and quietly: "it looks like a monster, eyes, scratches on his face, walking around scaring everybody"; "Dracula killing a person"; "dinosaur's body inside"; "two people carrying a bag." When asked whether it was easy or hard: "Its hard." When asked how they looked: "They don't look good." When asked about mood: "They're in pain." A high proportion of human-like figures were expressed suggesting a feeling of detachment or possibly unfamiliarity with her new identity (injured, vulnerable, impaired, unattractive).

Clinical example: Aprosodia—repeated industrial impacts to head: A man suffered repeated blows from an industrial object swinging from a 600-lbs hook, which may have been glancing blows, as well as one direct blow from a 25 lbs falling object. His affect was very dull, without the characteristic sadness of depression. The man reported himself as disabled. His Rorschach responses were "two people holding a snowman's head" and a "Mad Hatter rabbit."

Clinical example: Aprosodia—deceleration impact: Patient was a 44-year-old taxi driver in a MVA. He was seat-belted. His nose and jaw struck the steering wheel and his head swung back and hit the partition. He showed an average concern and appropriate reaction to his problems and to the event. His demeanor was pleasant, but the level of overt affect was low, with a restricted range that was mostly dull. His lack of clinically apparent anxiety, depression, and/or anger was consistent with aprosodia. He stated that he had no hope for the future. He saw himself as injured. His Rorschach responses were "a chicken, no head, cut, dead" and an "injured monkey, bleeding" (in the accident, his nose was bleeding).

Clinical example: Aprosodia—3 industrial concussions: Patient had suffered three concussions in industrial accidents, two of which were incurred when he fell backward and hit his head. He saw stars once and blacked out twice. He was uncertain whether he incurred LOC or PTA. Patient had a pleasant demeanor but there was an apparent absence of depression, anxiety, or anger. There were anomia and dysarthria. His Rorschach response was of "two bears fighting, blood."

Dysarthria

Dysarthria may be defined as lack of motor control over peripheral speech organs, causing a disorder of articulation of single sounds that results in an inability to form or produce understandable speech. Word formation depends upon the rapid and orderly succession of individual muscle movements in the larynx, the mouth, and the respiratory system. Jumbled vocalization occurs when there is lack of coordination between these organs and an inability to adjust in advance the intensity of the sound or the duration of each successive sound (Guyton & Hall, 2000, p. 655). Dysarthria as a symptom has been aggravated by left-sided surgery to relieve posttraumatic tremor more so than right-sided surgery (Krauss & Jankovic, 2002).

Differential impairment of systems can occur. Dysfunction extends from slight distortion of produced consonants and vowels to unintelligible speech—or even anarthria (Jaeger et al., 2000). After TBI, there is a range of articulatory disturbances. Slurred, indistinct speech impedes communication with family, friends, and the general community, which in turn affects psychological, social, academic, and occupational aspects of life (Theodoros et al., 2001a). Weak, slurred, or indistinct speech should be more precisely described by clinicians (Rosenfield & Barroso, 2000). Even with severe dysarthria, pitch control for question/statement contrast is retained with a reduced range of frequency (Patel, 2002). Some brain-injured persons manifest reduced syllable repetition rates (*oral diadochokinesis*) and lengthened syllable duration. In the early period after onset, swallowing disorders are more common than speech disorders, often occurring in isolation from each other.

Dysarthria is consequent to mechanical disturbance of the tongue or larynx, neurological disorders at various levels (central or peripheral) (Kirshner, 2000), or physiological deficits of the speech-production mechanism (respiratory, velopharyngeal, and articulatory) (Guyton & Hall, 2000, p. 655; Swanberg et al., 2003). TBI forces patients to function near the limits of their orofacial motor system, with differential effects upon movements and strength of different portions of the tongue and lips. Reports vary as to the movements of articulatory structures (i.e., from large displacements to hypometria), including articulatory overshoot of the lower lip. Since the lips, the tongue, and the jaw have different biomechanical properties, the various articulators can be affected by hypotonia, spasticity, rigidity, and/or dystonia, resulting in hypometria, hypermetria, dysmetria, speed reduction, or acceleration (Jaeger et al., 2000). Disturbed muscular control alters speech due to compromise in the sound output of the supralaryngeal articulators (lips, tongue, jaw, palate) (Netsell, 1998).

Velopharyngeal: This refers to the soft palate and pharynx. The most characteristic velopharyngeal disorder is hyponasality, but in dysarthria it may affect prosody, articulation, and respiratory aspects of speech production. It is necessary to differentiate between dysprosody caused by dysarthria and dysprosody stemming from the RH (Wertz et al., 1998).

The velopharyngeal valving mechanism is vulnerable to damage to:

- The UMN supplying the nuclei of the bulbar region of the brain stem;
- The LMNs supplying the muscles of the soft palate and the pharynx;
- Subcortical structures (the basal ganglia, the cerebellum, and the connections of each);
- Cerebellar damage causes hypotonicity and impaired coordination of muscles of the soft palate and pharynx, leading to hypernasality and/or hyponasality (Theodoros & Murdoch, 2001a).

Inability to close the velopharynx "drains" subglottal air pressure, reduces loudness, and causes hypernasality (nasal emission) that distorts oral consonant production. When the speaker runs out of air, it results in an increased respiratory effort, which compounds the problems, especially when vital capacity is also reduced (Netsell, 1998). There can be changes in the force that are expressed by different portions of the tongue or lips (Jaeger et al., 2000). One half of a group of TBI patients had *velopharyngeal airway resistance* (*VAR*). Those with mild or absent dysarthria typically had no velopharyngeal deficits, while those with severe dysarthria had very low velopharyngeal resistance. VAR deficit was associated with perceived hypernasality. Discrepancies between VAR and perceived hypernasality may be caused by intelligibility or speaking style (McHenry, 1998). VAR and orifice area are associated with the perception of nasality. False positives were associated with a slow speaking rate (McHenry, 1999).

The tongue: Inability to move the tongue discretely interferes with the ability to interrupt the stream of air to create consonants and vowels or to make discrete sounds in an organized and repeatable fashion. There is evidence for reduction in the rate of speech production, as well as evidence of exaggerated protrusions and retractions of the tongue. Lip and tongue movements were performed with impaired speed and accuracy, distorting consonants and vowels. In a sample of dysarthric individuals with severe TBI, although there were reductions in tongue endurance and rate of repetitive movement, neither tongue strength nor fine pressure control differed significantly from that of the control group. Since there were no clear relationships between the physiological and perceptual parameters, it was suggested that the TBI subjects were compensating in different ways for the physiological impairments (Goozee et al., 2001).

The neuropathology of lip and tongue functions is likely to be multifactorial. One group of individuals with severe CHI manifested *orofacial force control* deficits, particularly strength, endurance, and rate of lip and tongue function. The various articulators could be affected by spasticity, flaccidity, rigidity, hypokinesis, and/or incoordination of movement. Tongue strength and endurance are associated with speech ability. Fatigue could account for impaired lip and tongue endurance. Tongue function was more severely compromised than lip function (Theodoros et al., 1995). Subjects with TBI were not significantly different in tongue strength from normal controls (children and adolescents). There were significant negative correlations between measures of tongue strength and endurance, and perception of articulatory precision and overall speech deficiency (Stierwalt et al., 1996). After TBI there may be inaccuracy in decelerating tongue movements on approach to the palate, resulting in an overshooting of the target and reducing the length of time that the tongue remained at the palate (Goozee et al., 2000). Inability to move the tongue discretely interferes with the ability to interrupt the stream of air in order to create consonants and vowels or to make discrete sounds in an organized and repeatable fashion. There is evidence for reduction in the rate of speech production, as well as evidence of exaggerated protrusions and retractions of the tongue. Lip and tongue movements were performed with impaired speed and accuracy, distorting consonants and vowels. In a sample of dysarthric individuals with severe TBI, although there were reductions in tongue endurance and rate of repetitive movement, neither tongue strength nor fine pressure control differed significantly from the control group. Since there were no clear relationships between the physiological and perceptual parameters, it was suggested that the TBI subjects were compensating in different ways for physiological impairments (Goozee et al., 2001).

The larynx: Phonatory disturbances can result from laryngeal dysfunction associated with respiratory, articulatory, and resonatory deficits, which impairs the intelligibility of speech (Goozee et al., 2001). Monotony of pitch and equalized stress in speech can reflect adverse effects of hypokinesia and rigidity of the laryngeal muscles with reduced abduction of the vocal folds. This symptom could also be due to accelerated speech (Fahn, 2002).

Dysarthria has been described as a group of speech disorders attributable to disturbance in muscular control of the speech mechanism. This results from impairment of the basic motor processes involved in the execution of speech. It is due to damage to the central or peripheral nervous system. After TBI, there is a range of articulatory disturbances. There is evidence of for reduction in the rate of speech production, as well as evidence of exaggerated protrusions and retractions of the tongue. Lip and tongue movements were performed with impaired speed and accuracy, distorting consonants and vowels (Theodoros et al., 2001b). Dyspraxia, on the other hand, is a motor speech programming disorder characterized by articulation errors and, secondarily, alterations of prosody. The individual has difficulty speaking because of a cerebral lesion that prevents—voluntarily or on command—the complex sequence of muscle contractions used in speaking (Murdoch et al., 1990, citing Darley et al., 1975).

Differential Diagnosis

It is possible to differentiate between dysarthrias (abnormal muscle strength for articulation of sounds or phonemes), dysphonias (disorders of voicing), apraxia (inability to program sequences of phonemes, with consonant substitution rather than distortion, as in dysarthria), and stuttering (initial pauses and dysfluency of speech production without articulator or language disorder). Aprosodia seems to differ from dysarthria through a different pattern of dysfunctioning of structures of the speech, swallowing, and breathing systems (see below). Dysarthria has been described as a group of speech disorders attributable to disturbance in the muscular control of the speech mechanism resulting from impairment of the basic motor processes involved in the execution of speech. It results from damage to the central or peripheral nervous systems. After TBI, there is a range of articulatory disturbances. There is evidence for reduction in the rate of speech production, as well as evidence of exaggerated protrusions and retractions of the tongue. Lip and tongue movements were performed with impaired speed and accuracy, which distorted consonants and vowels (Theodoros et al., 2001b). Dyspraxia, on the other hand, is a motor speech programming disorder characterized by errors of articulation and secondarily alterations of prosody. The individual has difficulty speaking because of a cerebral lesion that prevents—voluntarily or on command—the complex sequence of muscle contractions used in speaking (Murdoch et al., 1990, citing Darley et al., 1975).

Types: Seven types of dysarthria are described and characterized by different patterns of auditory-perceptual characteristics (Zeplin & Kent, 1996): flaccid, ataxic, hyperkinetic with chorea, mixed, spastic, hypokinetic, and hypokinetic with dystonia. The dimensions have been rated by judges, and deviate for each dysarthritic type depending upon the task (i.e., passage reading and syllable repetition). Discriminable qualities include pitch level; monopitch; monoloudness; inappropriate silences; audible inspiration; bizarreness; imprecise consonants, hypernasality; breathy continuous voice; phrases short; inappropriate silences; level of intelligibility; nasal emission; forced inspiration; inappropriate silences; stress level; fast or slow rate; strained-strangled; harsh voice; voice stoppage; pitch breaks; irregular articulatory breakdown; voice tremor; short phrases; and slow rate (Zeplin & Kent, 1996). Spastic dysarthria is characterized by a strained/strangled vocal quality and slow but regular speech rate.

Neurotrauma

There can be accrual of degeneration of brain tissues from experimentally induced minor repetitive trauma to the brain (Dixon et al., 1991; Uryu et al., 2002). There is also evidence of tissue atrophy to vocal structures from repetitive trauma to the teeth and jaws, observed when the nervous system causes tremor of the masticatory muscles and or subluxation of the temporomandibular joints via the CN (Silverman, 2002). Destruction of the somatic or neurological components of articulation can cause total or partial inability to speak distinctly (Guyton & Hall, 2000, p. 655). Loss of muscular control causes increased or decreased muscle tone as well as poor motor control with imprecise movement and slurred or indistinct speech. Intelligibility is a consequence of muscular paralysis, weakness, abnormal tone, or incoordination. These result in deficits of direction, range, force, endurance, timing, and regulation of movements (i.e., reduced intelligibility) (Carter et al., 1996). Dysarthrics have shorter groups of words per breath, which contributes to the reduction of the naturalness of speech. Dysarthrics also take breath at a location not syntactically correct, which misleads the speaker as to the meaning (Hammen & Yorkston, 1996).

The velopharyngeal valving mechanism is vulnerable to damage to the UMNs that supply the nuclei of the bulbar region of the brain stem, or LMNs that supply the muscles of the soft palate and pharynx and subcortical structures (the basal ganglia, the cerebellum, and the connections of each). Cerebellar damage causes hypotonicity and the impaired coordination of the muscles of the soft palate and pharynx, leading to hypernasality and/or hyponasality (Theodoros & Murdoch, 2001a). The most characteristic velopharyngeal disorder is hyponasality, but in dysarthria it may affect prosody, articulation, and the respiratory aspects of speech production. It is necessary to differentiate between dysprosody caused by dysarthria and dysprosody stemming from the RH (Wertz et al., 1998).

Dysarthria after TBI is found in about 30% of both children and adults (Stierwalt et al., 1996). *Military injuries* producing dysarthria (primarily missile wounds) primarily involved the peripheral nerves and most frequently trauma to the facial nerve. All dysarthric patients after CHI are reported to have a subclinical language deficit (Levin, 1991). Dysarthria can be complicated by apraxia of speech (a problem with sequencing (McHenry & Wilson, 1994) and the paralysis, incoordination, or weakness of the muscles of the speech mechanism (Murdoch et al., 1990).

One group of individuals with severe CHI manifested deficits of orofacial force control, particularly strength, endurance, and rate of lip and tongue functions. The neuropathology of lip and tongue functions is likely to be multifactorial. The various articulators could be affected by spasticity, flaccidity, rigidity, hypokinesis, and/or incoordination of movement. Tongue strength and endurance are associated with speech ability. Fatigue could account for impaired lip and tongue endurance. Tongue function was more severely compromised than lip function (Theodoros et al., 1995).

Medical Conditions

Dysarthria is found in *multisystem atrophy*, which is to be distinguished from pure autonomic failure. Its most common cause is weakness or paralysis of muscles required for speech, but exists in a variety of medical conditions affecting nerve-muscle junctions (e.g., myasthenia gravis) (Drachman, 2001) and the muscles themselves, damage to nerve fiber tracts connecting motor nerves to higher centers, compression or irritation of motor nerves, and psychosomatic conditions. These include vascular disorders, inflammation, demyelinating disease, immunological disorders, toxins, and extrinsic compression (Horner & Massey, 1996). Dysarthria may be the leading symptom of *hypothyroidism* and can be promptly resolved after hormone substitution (Stollberger et al., 2001). Dysarthria may be associated with dysphagia due to oropharyngeal dysfunction, and their onset may be paroxysm, progressive, and part of a multifaceted neurological syndrome. Dysarthria itself can be associated with disorders of every level of the nervous system (Horner & Massey, 1996).

Neurological disease can affect the palatopharyngeal component of the motor speech system, creating articulatory defects (Aronson, 1980, p. 122). Often after a stroke, patients vocalize with vowel sounds because they have a limited range of lingual motion and can only move their tongue around inside their mouth. Post-TBI dysarthria is characterized by the impairment of articulatory accuracy and the rate, timing, strength, endurance, and range of movement of the vocal speech organs. Dysphonias may be consequent to *myasthenia gravis* (breathy voice or nasal timbre [weakness of the palate] or a dysarthric "mushy" quality [tongue weakness]), or *Parkinson disease* (whispered voice) (Aronson, 1980; Drachman, 2001; Kirshner, 1996). Parkinson disease and dysarthria are characterized by reduced acoustic contrast, diminished energy in the higher frequencies, and mask-like faces accompanied by aprosodic, indistinct, or hypernasal speech. This may give the incorrect impression of depressed affect (Code, 1987, p. 98; Heilman et al., 1993). There is a similarity between the monotonous and fading manner with which a depressed patient speaks and that of an aprosodic patient with stroke of the nondominant hemisphere. However, the depressed patients also manifested lack of drive, boredom, and dullness (Naarding et al., 2003).

Flaccid is caused by Xth nerve lesions, with vocal effects dependent upon the location of the lesion along the nerve pathway from brainstem to muscle. *Spastic* (pseudobulbar) has a harsh strained/strangled quality due to vocal fold hyperaduction; hypokinetic (*Parkinsonian*) is monopitch with reduced loudness due to muscle rigidity. Other dysphonias have been described as consequent to uncontrolled muscular movements that are jerking, slower, or tremulous due to reduced feedback of muscular performance (ataxic, choreic, dystonic).

Neuromuscular: Changes in the frequency and amplitude of the voice can be caused by tremor in the intrinsic and extrinsic laryngeal muscles, the pharyngeal and expiratory muscles, and a variety of motor conditions, including laryngeal dystonic tremor (frequency oscillations more than amplitude oscillations) and focal laryngeal dystonia (when speech is initiated, laryngeal muscles contract inappropriately). A weak and breathy voice results from vocal cord immobility. Some deficits are due to the impairment of lip and tongue strength and endurance during speech production, affecting fine force control and the rate of repetitive movements (Theodoros et al., 2001b). Physiological measurements (Theodoros et al., 1995) have been made of vocal reflexes, respiration, lips, jaw, palate, laryngeal function, and the tongue. These have documented impaired lung capacity, reduced breath support, disturbance of vocal fold vibrations and laryngeal aerodynamics, reduction in loudness variation consequent to reduced lung capacity, incoordination of the force speed, range, and timing of the respiratory muscles, and breathiness consequent to incoordination of the laryngeal muscles involving intersubsystem incoordination. Audible inspiration may reflect incoordination of the respiratory and laryngeal subsystems, possibly due to cerebellar dysfunction, retracted tongue position, or deficit in timing due to lip and tongue movement. Accelerated speech rate can be due to failure of the articulator to reach his or her target position in the expected time due to muscle rigidity ("articulatory undershoot"). Reduced breath capacity and incoordination of the respiratory musculature accounts for apparent reduced breath support for speech, loudness decay, shortened phrases, and short rushes of speech. Weakness, inaccuracy, and reduced speed of the movements of the lips and tongue seemingly contribute to impaired articulation, rate, and fluency of speech. Increased tone and activity of the laryngeal musculature are consistent with hyperfunctional vocal fold activity. Voice tremor can result from vocal cord tremor, adductor dystonia, or neck tremor (Fahn, 2003).

APHASIA

Language disorders categorized as aphasia occur as disturbances of language or as a motor disorder. The classical locations to which disorders described as Broca's or Wernicke's aphasia are attributed have been demonstrated to be unreliable. In fact, the areas contributing to language processing in addition to Wernicke's and Broca's areas, and the arcuate fasciculus include the insula, anterior superior temporal gyrus, and a large swath of the temporal cortex. Each syndrome may be absent in the expected area or exist outside of that area. Both Broca's and Wernicke's aphasias can evolve into less-disabling conditions when the initial damage occurs on neighboring or connected brain areas. Thus, the lesion deficit assessment should await stabilization (Dronkers et al., 2000). Among the issues to be considered are motor and association pathways, such distinctions as word recognition and speech, planning of speech versus actual articulation, the difference between major linguistic tasks (speaking, comprehension) and specific operations, the location and extent of lesion site, and the interval since the cerebral damage. Classical syndromes are only related to lesion sites in cases of rapidly developing lesions such as stroke. In practice, a large proportion of cases are "mixed" or "unclassifiable" (Caplan, 2003). It is useful to differentiate phonemes (the minimal sound units) and phonology (their combination in the language system) from articulatory problems and phonological errors (substitution, simplification, addition, and transposition of phonemes).

Speech production of anterior- and posterior-injured aphasics have similar patterns. The pattern of errors—not their presence—characterizes speech-production deficits in aphasia. There is an issue as to whether apraxia is an articulatory or aphasic disorder (i.e., whether errors are linguistic or articulatory). The patient may not have lost the ability to produce particular phonemes, but the problem is consistently encoding the correct phonemic representation of the word. The utterance is articulatorily correct but deviates phonologically from the target word (Blumstein, 1991).

Motor, Broca's, and anterior aphasia: This condition is characterized by slow and effortful speed and lack of grammatic markers in language production (use of high frequency content words while omitting smaller function words that convey grammatic information). Damage to Broca's area makes it impossible to speak whole words, in contrast to producing uncoordinated utterances or single monosyllabic words. There can be an underlying comprehension disorder revealed by examination. Dysarthria and apraxia of speech frequently accompany Broca's aphasia. The latter has been attributed to a discrete lesion at the tip of the precentral gyrus within the insula (Dronkers et al., 2000). Closely associated cortical area controls appropriate respiratory function: respiratory activities of the vocal cords occur simultaneously with movements of the mouth and the tongue during speech. Motor aphasia is attributed to damage to Broca's speech area in 95% of individuals it is associated with LH damage (Guyton & Hall, 2000, p. 669). Broca's aphasia (originally described by Broca as aphemia) may be accompanied by right-sided motor weakness and buccofacial apraxia. Recovery can be accompanied by dysprosody characterized by a "foreign accent" and poorly articulated, hypophonic, slow, breath speech (Stuss & Benson, 1986, p. 164).

Anterior patients have difficulty producing phonetics requiring the timing of two articulators (voicing [vocal cords] and nasality [velum, releasing air from oral an nasal cavities]). They do not integrate articulatory movements from one segment to another (Blumstein, 1991). Broca's aphasia (nonfluent) manifests phonologic-prosodic deficits: reduced rate; reduced phrase length; excessive initial and interstitial pause; aberrations in syllabic, phrasal, and sentence prosody (stress, pause, duration, tone) (Aronson, 1978). While Benson (1993) classifies dysprosody as an example of aphasia, the authors consider aprosodia to be a motor deficit. The appearance of apathy and indifference can alternate with anger, facetiousness, or boastfulness. Mutism (associated with frontal lobe dysfunction) is associated with well-articulated but hypophonic speech (Damasio & Anderson, 1993). In addition, a pattern has been reported for feeling normal emotions with abnormal facial expressions (e.g., pseudobulbar palsy). The reflex mechanism for facial expression is released bilaterally from cortical control, resulting in involuntary laughing or crying. However, normal emotions are experienced (Heilman et al., 1993). The symptom of *witzelsucht* has been described as the behavior of a patient that alternates between apparent flat affect with facetiousness, boastfulness, and aggression (Damasio & Anderson, 1993).

Broca aphasics show several motor deficits. Posterior aphasics do not seem characterized by timing deficits or impairment of laryngeal control and have only a subtle phonetic impairment (Blumstein, 1988). There is a profound expressive speed deficit and also impairment of the fine control of the timing of phonetic units. Speech output is typically nonfluent, that is, produced slowly, often with seeming great effort, and with much groping and hesitation not only between words, but also in the production of a single word. The speech is dysarthric (i.e., seems slurred, consonants are not produced with precision, and, seem distorted to the examiner). The melody of speech is often flat and seems to lack full intonational structure. Broca patients are described as displaying distinct limitations in control of patterns of speech prosody. Expected frequency deviations do not appear in long sentence strings. Their prosodic disorder appears to be a manifestation of a higher-level linguistic impairment that affects the elaboration of the syntactic structure of an utterance. Prosodic deviations vary (i.e., flattened intonation or exaggerated range) depending upon the severity of the injury and the linguistic characteristics of the utterance.

Wernicke's aphasia: Wernicke's area is not clearly circumscribed (i.e., Brodman's area 22 or a larger temporoparietal area) (Kirshner, 2000). It is considered to be part of a processor of speech sounds that associates sounds with concepts, also involving the parts of the brain that subserve grammar, attention, social knowledge, and knowledge of the concepts corresponding to meanings of words (Dronkers et al., 2002). The speech of patients with Wernicke's aphasia is effortless, melodic, and produced at a normal rate. However, they have difficulty selecting words that accurately represent their intended meaning (verbal or semantic paraphasia), and also have difficulty comprehending sentences uttered by others (Dronkers et al., 2000). In contrast to Broca's aphasia, the speech of

Wernicke aphasics is fluent and well articulated (Blumstein, 1988). It is produced with facility and ease by the patient. Whatever hesitation occurs, it seems to reflect the failure to access a particular word rather than a difficulty in implementing its sound structure. Consonants are clearly articulated, as are vowels. Speech melody seems quite normal. Although patients with severe Wernicke's aphasia produce jargon or neologisms, they exhibit relatively normal speech prosody. In contrast with patients with Broca's asphasia, they retain significant abilities to plan an utterance, but fail to signal between syntactic boundaries of constituent linguistic structures. There is a question as to whether prosodic patterns reveal impairment of planning the target utterance or impairment of decoding and ultimately recoding the target utterance articulatorily.

VOCAL DYSPRAXIA AND APRAXIA

Motor vocalization disorders have been described as an inability to volitionally sequence sound systems to form words instead of noises (Guyton & Hall, 2000, pp. 669–670). Indicators of dyspraxia include involuntary movement of the articulators appearing superior to volitional movement; inconsistent errors during repetitive oral movements and speech production of phonemes, words, and phrases; deterioration of performance as the word length increases; automatic speech being more intelligible than propositional speech; searching behaviors are present in the child's speech; facial expression suggesting puzzlement about how to use the muscles of articulation to make the required sound; and altered prosody (i.e., rate of speech) (Murdoch et al., 1990). *Buccofacial apraxia* may be expressed as making incorrect movements when asked to pantomime various oral activities, substituting verbal descriptions for the movement. Apraxic errors of content occur (e.g., knowledge of tool action [use of a hammer] or tool-object knowledge [hammer is associated with a nail]). Spatial and temporal errors interfere with pantomime, imitation, and actual usage (Heilman et al., 1997). Such errors, including difficulty with imitation of meaningless movements, indicate that oral apraxia is not a form of asymbolia (Heilman & Rothi, 2003).

Dyspraxias are not a disorder of primary sensory and motor functions. Probably dyspraxias represent disorders of motor programming and execution, specifically manifesting in visual-perceptual, dysphasic, attentional, or intellectual dysfunction (Miller, 1989). There has been a controversy concerning the neurological dysfunction of apraxia (e.g., a disorder of central centers) disconnection between the areas for comprehending language and the area for outputting motor commands for speech sounds, and a dysfunction similar to limb-kinetic apraxia in which there is a loss of kinesthetic memories of a particular part of the body (Buckingham, 1991). A combined dysarthria-dyspraxia might ensue from the compromise of primary and secondary motor cortex and their interconnections. Difficulty increases with fine motor commands rather than psychological complexity, as in true dyspraxias. It is controversial as to whether dyspraxic disorders are motor, and therefore differ from dysphasia (phonological). The latter implies selection of improper sounds from a semantic lexicon rather than a motor disorder. If neurological, what is their neurological locale? Apraxic speech substitutes one sound for another and omits, adds, transposes, and distorts sounds. It includes equal and even stress, the insertion of inappropriate intersyllabic pauses, and the restriction and alteration of normal intonational contours (Wertz, 1985). The normal durational relationships of vowels and consonants are distorted. There is effortful grouping and repetitive attempts to produce a sound accurately. The rate of production is slowed. Prosodic disturbances probably reflect both the primary motor deficit as well as the patient's efforts to compensate. Speech-apraxic patients produce inconsistent errors of phoneme substitution and omission. This is commonly involved in speech-production difficulty in the aphasias. In contrast, dyspraxia is characterized by consistent abnormal articulation of phonemes (Kirshner, 2000). In children, one may differentiate verbal dyspraxia (short and laboriously produced short utterances) from buccal-oral dyspraxia (positioning muscles of articulation) and generalized dyspraxia (clumsiness) (Nass, 2000). Voice apraxia (hesitation) can result from corticobasal degeneration (Fahn, 2002).

OTHER MOTOR VOCAL DISORDERS

LMN disorder: The tongue has sufficient ipsilateral innervation to restrict the effects of *UMN unilateral damage. LMN disease* (nerve fibers or neuromuscular junction) is associated with problems of respiratory support. The weak muscles do not fully inflate the lungs, which reduce breath pressure at the vocal folds. Furthermore, weak laryngeal muscles fail to fully adduct the vocal folds and create speech with short utterances, frequent pauses for breath, weak vocal intensity, voice quality breathy, and vocal pitch low and monotonous. Damage to the *pons and medulla*, where CN nuclei are located, creates dysarthria when the damaged nerves supply muscles producing speech. Localization of the injury (i.e., whether nerve trunk or nuclei) is aided by determining whether there is comorbid hemiplegia or hemiparesis of either arm or leg since the corticospinal tracts pass through the brainstem near the CN nuclei decussating below (Brookshire, 1997).

Extrapyramidal disease includes *Parkinson disease*, where muscular rigidity causes a failure of full adduction of the vocal folds, reducing the loudness of the voice—sometimes to inaudibility. Articulation is also imprecise and indistinct since rigid muscles do not reach full excursion (speech rate is variable and speech is indistinct, with inappropriately placed pauses). *Dyskinesia* interrupts respiration, causing involuntary changes in breath pressure at the glottis. This leads to abrupt changes in vocal intensity and reduced articulatory accuracy that alternates with normal articulation. Speech is slow, with exaggerated variability in loudness, pitch, and inappropriate pauses, with periodic explosive articulations interrupting speech flow (Brookshire, 1997, pp. 416–417).

Stuttering is a disorder of the rhythm of speech (i.e., the speaker knows what is to be said, but is unable to say it because of an involuntary, repetitive prolongation or cessation of a sound). In children, it is often an inherited linguistic disorder (errors occur at grammatically important points in the sentence) (Nass, 2000). There is a repetition, lengthening, and inappropriate pausing in the generation of consonants, vowels, and words, which is related to control over the laryngeal sound source (Kirshner, 2000). Dysfluencies of stuttering occur at the beginning of sentences and phrases, not randomly. Auditory feedback effects fluency, for example, loud, broadband noise when speaking renders most stutterers fluent. Perhaps stuttering is a disorder of motor movement, interacting with language. During dysfluent output, the stutterer, in attempting to reach the target word, fluently engages in varied maneuvers such as holding or repeating the sound, facial distortions, and other struggling behaviors. A possible etiology is the underdamping of an oscillation system (i.e., a kind of tremor). Another possibility is unspecified or variable location of instability of a negative speech sensorimotor feedback loop (Rosenfield, 1997).

Stuttering may be associated with a variety of cortical locations (seemingly not occipital) and other speech disorders or no other speech disorder (Market et al., 1990; Rosenbek et al., 1978). "Acquired" stuttering can create a problem of differentiating between brain trauma and stress, even after documented TBI (Helm-Estabrooks & Hotz, 1998; Market et al., 1990) or as a reaction to language loss (Rosenbek et al., 1978). In one sample (Market et al., 1990), 38% were apparently due to head trauma, followed by 37% due to ischemia, and a wide variety of other potential causes. Fifty-six percent manifested stuttering less than one month after the apparent insult. Stuttering was found to be associated with a substantial reduction in the quality of employment or unemployment. Persons with developmental onset stuttering may manifest fear and avoidance of words and situations, but not those with sudden late onset symptoms.

DYSPHONIA

Unilateral recurrent vagal or laryngeal nerve injury results in flaccid dysphonia: voice quality is described as hoarse, harsh, and breathy with short phrases, reduced loudness and mild inhalator stridor. There is unilateral paralysis of almost all laryngeal muscles (except cricothyroid) (Brazis et al., 2001, p. 335; Parent, 1996, p. 449). Lesions of the glossopharyngeal, vagus, and accessory nerves associated with basal skull fractures may result in dysphonia (Brazis et al., 2001, p. 331). Bilateral vagus nerve lesions cause drooping of the palate, losing movement during phonation. When speaking, air escapes from the oral to the nasal cavity, giving a "nasal" quality to the voice (Brazis et al., 2001, p. 333). Compression of the recurrent laryngeal nerve with hoarseness occurs with thyroid carcinoma and Hashimoto disease (autoimmune thyroiditis Type 2A) (Larsen et al., 1998, p. 479).

Childrens' Vocalization Problems

In children with CHI, motor speech performance such as respiration, phonation, resonance, prosody and articulation should be studied simultaneously with speech and nonspeech activities. However, a majority of children with CHI will not present with significant speech disorders (Jordan, 1990). In children, articulatory delay, rather than dyspraxia, is more likely with left hemispheric lesions than right hemispheric lesions. Language disorders in children with LH vascular disorders is accompanied by dysarthria, oral dyspraxia, and a wide variety of additional linguistic disorders. There may be good recovery of linguistic skills if there are no seizures but slow linguistic progress (Ozanne & Murdoch, 1990). Children with expressive and receptive dysprosodia may have reduced development of social competence, shyness and isolation, and depression (Wymer et al., 2002).

Cerebellar disorders: Tumors of the posterior fossa may manifest as ataxic dysarthria. Their 10 most deviant speech disorders are excess and equal stress; irregular articulatory breakdown; distorted vowels; harsh voice; prolonged phonemes; prolonged intervals; monopitch; monoloudness; and slow rate (Hudson, 1990, citing Darley, 1969).

CLINICAL IMPLICATIONS

Social Impression

Communication can be impaired due to the loss of spontaneous facial and body gestures. Emotional expression has diagnostic significance for cerebral personality syndromes and chronic stress disorder (Parker, 2001, see Chapters 12, 13, 17). Misinterpretation of aprosodia could lead to the incorrect assessment as apathetic, indifferent, a misrepresentation of one's condition, malingering, or manifesting the dull affect of schizophrenia. Monotonous verbal output with decreased facial grimacing and sparse use of gestures could also be misinterpreted as an emotionless response signifying depression (Mendez & Cummings, 2002). It is often ignored by the clinician (Kiss & Ennis, 2001; Messner & Messner, 1988; Theodoros et al., 2001a). While emotional gesturing (and prosody) were described as features of the RH (Fricchione & Howanitz, 1985), the focus and controlled environment of the neuropsychological examination restricts the opportunity to make detailed observation. Upon observing aprosodia, further exploration is useful. The clinician may explore potential depression with the patient (i.e., whether or not there is an experience of sadness or emptiness). Should inquiry suggest a lack of experienced affect, study is invited for issues of endogenous depression (neurochemical or neurotraumatic) or endocrine dysfunction. Inability to express affect may cause one to consider the link between depression and chronic pain.

Treatment

A model for treatment has been proposed utilizing voice pitch biofeedback and modeling of affective communication. Alerting the family that the prosodic dysfunction may be traumatic and not psychodynamic can reduce guilt, improve interpersonal relations, and refocus to more appropriate problems (Heilman et al., 1993). Unintelligible speech has been treated by optimizing the physiological systems through palatal lift prostheses, reducing the speaking rate through pacing, phonetic transcription of words to minimize speech apraxia and improve word stress, visual cues from therapist's person as a model, and self-awareness of proprioceptive cues (McHenry & Wilson, 1994). For articulation, a treatment program should be designed taking into account individual differences of dysfunction (Theodoros et al., 1995).

DYSREGULATION OF THE INTERNAL MILIEU

Discussion

Respiration can be affected by vestibular dysfunction, a common disorder after head injury.

Aprosodia can lead to the misperception of a patient's emotional condition and mental level. The question has been raised as to whether flat, dysprosodic speech reflects a true paucity in the emotional realm or whether is it an inability to control prosodic elements of speech—perhaps only accidentally linked to the aberrant emotional qualities of a clinical population (Shapiro & Danly, 1985).

Receptive and Expressive Processing Is Distinguishable

Sensory prosody is processed earlier (attributed to the right hemisphere), as part of a distributed network. It contributes to the subsequent linguistic and vocalization network, utilizing all levels of the nervous system, with a final common pathway in the recurrent laryngeal nerves. Several functions have been described as aprosodic but are considered by this writer to be independent of vocal or motor expression until research demonstrates that they are part of the same symptom. These include loss of affective comprehension, aphasia, and processing facial affect (Ross et al., 2001). The somatic basis for prosody and aprosodia are expressed through organs that serve multiple functions: respiration (ventilation), phonation (voice content and quality, ingestion, and swallowing.

Sensory prosody (comprehension) has been described as a function bound to motor prosody (Chow & Cummings, 2000; Robinson & Starkstein, 2002). This is not supported by the findings of a sample of patients with Alzheimer disease (Testa et al., 2001). As dementia severity increased, performance on comprehension and repetition tasks became more impaired, while spontaneous affective prosody remained normal. In *multi-infarct dementia*, communication disorders that eventually evolve into dysarthria and kinetic mutism may at an earlier stage display deficient pitch, tone, and melodic qualities, while the content and organization of speech remain unaffected (Lezak, 1995, p. 201). A study of children who sustained severe head injury and developed posttraumatic mutism offered evidence for the involvement of mesencephalic structures (Dayer et al., 1998). Reduced comprehension of stimuli with aprosodic implications can lead to loss of comprehension of the person's own mood, resulting in denial of the person's depression. Thus, affective motor aprosodia (flat, monotonous verbal output) can be misinterpreted as apathy or depression. It is therefore more conservative to consider motor aprosodia to be a distinct entity when compared to other conditions in which there is a deficit of repetition or comprehension of emotionally intoned speech or loss of the propositional type. Propositional content is expression, repetition, or comprehension of sentences such as questions, statements, or commands (Ramasubbu & Kennedy, 1994). Cummings (1985), noting the significance of the bilateral limbic contribution to the experience of emotions, points out that right-sided lesions may impair prosody without altering the propositional component of verbal output. One team (Denes et al., 1984) studied cognitive assessment involving the discrimination of stimuli and judgment of emotional content. The sporadic analysis of vocalization to determine patients' prosodic characteristics is our main subject of interest.

SUMMARY

Vocalization is an extremely complex activity requiring cognitive, affective, and motor planning; sequencing; modulation; social and somatic feedback; and control of a rapid sequence of nonrepetitive muscular movements from the segmental and branchiomeric musculature via autonomic and CN input. It is influenced by developmental, hormonal, emotional, somatic, and neurological functions. Aprosodia has been used to describe a disparate group of symptoms and traits, yet motor or expressive prosody is differentiable from emotional (receptive) prosody, which provides input to prosodic expression. Motor prosody and aprosodia refer to melodic and other communicative qualities of the voice. Prosody is expressed through somatic structures supporting breathing and vocalization. Neurological control includes speech centers in the cerebral cortex, basal ganglia, cerebellum, respiratory control centers of the brainstem and spinal cord reflexes (Guyton & Hall, 2000, p. 521), and CN (V, VII, IX, X). Feedback is to brainstem autonomic centers. It appears that prosodic control is by a distributed network. Thus, the actual site associated with a patient's post-traumatic aprosodia may be difficult to localize. Since comorbidity occurs between dysprosody, dysarthria, dyspraxia, and dysphasia, a common neurological substrate can be inferred.

The issue of hormonal effects during development or at any time from stress or hormonal disorders was raised. Prosodic expression is related to LH Broca's area. The right prefrontal and temporal areas preferentially process prosodic information as part of the more general process of language comprehension (Bookheimer, 2002). Motor aprosodia is defined as a speech disorder with dysfunctions of pitch, rate, and rhythm by which different shades of meaning are conveyed. The characteristic monotone creates apparent blandness. Thus, persons who have incurred TBI, who express distress concerning the quality of their life, can appear to the clinician to be indifferent, faking, or as having a dysphoric mood disorder. Disorders of gesture, facial expression, expressive posture, and the comprehension of the emotions and moods of other people are presumed to be different functions.

It is inaccurate to ascribe this condition to RH damage, and this statement has been based upon studies focusing primarily upon cerebrovascular accidents. Extensive neurological and somatic structures participate in vocalization. Some are lateralized to the LH; some are bilateral. Diaschisis effects contralateral structures and the association of aprosodia with other disorders.

Motor aprosodia had been described as a cerebral mood disorder (Parker, 1990, 2001). This review suggests an alternate interpretation (i.e., a somatic/neurological disorder). The initial statement may prove to be correct in an unknown proportion of instances. There is a loss of vocal qualities that communicate and reflect emotion, particularly distress, in persons who report dysphoria or other distress because of their neurotraumatic condition. If one utilizes this definition of aprosodia involving vocalization, then one is required to study the functioning, neurological control, and status of the anatomical basis for vocalization (i.e., diaphragm, ribs, lungs, larynx with its vocal folds, and oropharyngeal area).

Since prosody offers motivation, mood, and linguistic information, aprosodia creates loss of communication capacity. The person is subject to being misunderstood both in terms of content, and in the other person's appraisal of the internal and realistic adaptation and level.

Research on aprosodia has studied the sensory (acoustic and visual) input of varying levels of stimulus organization, as well as vocal and cognitive output. Vocal output is studied through the analysis of the frequencies, their amplitude, the difference from highest pitch to lowest pitch, stresses, occasional changes within a sentence, and so forth. The emotional issues referred to include the loss of affect in one's own voice (aprosodia as defined here), the inability to recognize the feelings and affect of others, the inability to recognize one's own affect (alexithymia), and the loss of the ability to experience affect and feelings (blunting).

Damage to a variety of neurological and somatic structures can contribute to motor aprosodia. When a vocal disorder is clinically detected, it will be useful to review the medical records to assess whether there is evidence for a somatic or neurological etiology. Lack of information could invite referral for further study. While the right cerebral hemisphere has usually been implicated in aprosodia (as loosely defined), when one considers this as an expressive linguistic phenomenon, then damage to the basal ganglia and cerebellum must also be considered for their contribution to speech. Furthermore, the purported controlling connection between any of the various hemispheric components that have been implicated in aprosodia and control over the vocal apparatus is, in the writer's opinion, tenuous (Lovallo, Nixon and Ross, 2002).

Cold cognitions of emotion (Bowers et al., 1993) involve affect, judgment, and knowledge detached from the personal experience of an emotional state. The function is attributed to the RH. Knowledge of the emotional meaning of species-specific nonverbal communications, including tone of voice, is referred to as the nonverbal affect lexicon.

ANATOMY AND PHYSIOLOGY OF RESPIRATION AND VOCALIZATION

Speech involves the respiratory system, the respiratory control centers of the brain, the articulation and resonance structures of the mouth and nasal cavities (Guyton & Hall, 2000, p. 442), as well as the vocal cords with neural control and the space provided by the skull. Humans' unique *craniofacial system* and vocal tract result from an increasing curvature of the basicranium with descent of the tongue to form the superior-anterior pharyngeal space. These changes are essentially complete by age 4 (Kent, 1998). The skull base, with the articulating mandible, must be coordinated with other facial structures for comprehensible speech. Thus, one may note that the length from the posterior cranial base to the anterior skull base grows until age 11–12 (Hilloowala et al., 1998). *Articulation* involves the mouth (lips, tongue, jaw, and soft palate) and nasal cavities. Movements of the mouth, the tongue, the larynx, and the vocal cords are responsible for the intonation, timing, and rapid changes in intensities of the sequential sounds. Prosody, pitch, and articulation are dependent upon the precise coordination of the laryngeal, pharyngeal, and respiratory subsystems. *Phonation* utilizes the larynx, with its vocal folds or cords. *Resonance* utilizes the mouth, nose, associated nasal sinuses, pharynx, and chest cavity (Guyton & Hall, 2000, p. 442; Murdoch & Theodoros, 2001). The pharynx serves both respiration (significant in prosody) and swallowing. Sensory impulses for swallowing are distributed via glossopharyngeal and trigeminal nerves to the medulla oblongata via the tractus solitarius. Thus, feedback from vocalization structures is part of the background of homeostatic and stress reactions. People with TBI have a high incidence of impaired *respiratory support* for a range of speech deviations (Murdoch & Theodoros, 2001).

ANATOMY

Speech production is the product of a complex laryngeal apparatus that includes the muscles of the vocal folds, various cartilages, ligaments, and the mouth and throat. The structures of the laryngeal and other vocal apparatus, as well as their neural control, differ significantly from those used for the trunk, limbs, and the respiratory apparatus. They evolved through aeons from the bilateral spaces and cartilages, forming the gill arches of primitive fish, rather than the transaxial segments that form the limbs, ribs, and muscles utilized for movement (Black, 1970, pp. 240–242; Butler & Hodos, 1996, pp. 156, 256–258; Kent, 1978, pp. 146–158, 165, 361, 376). The five branchial nerves that evolved from the gill arches of primitive fish (V, VII, IX, X, XI) are both sensory and motor nerves. Detailed text and diagrams of the human branchial (pharyngeal) arches and their muscular and neural derivatives are found in Williams (1995, pp. 274–287). Thus, structures and neural control of vocalization are located around the body axis rather than as part of the axial and bilateral structures (spinal segments).

Vocalization structures: The active force in the production of the voice is the airflow (Van Leden, 1961). Voice production originates from the vibration emanating from the *vocal folds* (Razak et al., 1983). It is then processed by the posterior oral-pharyngeal and nasopharyngeal ports (Merson, 1967). The space between the vocal folds is the *rima glottidis*. Its shape is determined by the position, tension, and length of the vocal cords, as well as the intensity of expiration changes in vocal pitch (Moore & Dalley, 2006, with diagrams, pp. 1089–1095). Cartilages of the larynx, trachea, and bronchia maintain the larger respiratory tubes; branchial motor axons of glossopharyngeal nerve serve the stylopharyngeus muscle that elevates the pharynx during swallowing and speech (Wilson-Pauwels et al., 1988, p. 117). The fleshy tongue manipulates the food particles to and from moving teeth on route to the mouth and throat for swallowing. Complex musculature raises and lowers the tongue, pushes it forward, and changes its shape (Butler & Hodos, 1996, pp. 56–157). The basicranial region in human infants is similar to that of monkeys and apes (see Vocalization Structures, below).

Intelligible speech is dependent upon the muscular coordination expressed as breath control during phonation, tone (source of the vibration), strength, and flexibility of the tongue (transverse, longitudinal, and vertical muscles). The tongue's greatest role is in eating functions (grasping, chewing, swallowing). Its integrity is necessary to distribute the airflow through the mouth (Danlioff et al., 1980) and for adequate precision and rate of speech utterances that depend on the acoustic and perceptual components of intelligibility (Theodoros et al., 2001b). The lips, jaw, velum, nasal cavity, and larynx also contribute to *speech intelligibility* (Razak et al., 1983).

Speech production is acquired and refined when the respiratory, laryngeal, and supralaryngeal subsystems are growing. The structure is shaped by the forces of mastication and swallowing. The neuromotor regulation of the speech system reflects the synchronous growth and maturation of the peripheral organs. Mandibular control develops very early, reflecting the embryological fact that the mandibular and maxillary divisions of the trigeminal nerve are the earliest functioning peripheral afferents (see Kent, 1999, for development and use of the speech structures). Speech production requires delicate timing and the coordination of numerous muscles throughout the respiratory system and vocal tract (McHenry & Wilson, 1994): abdominal muscles, diaphragm, rib cage (Ezure, 1996), lungs, larynx, pharynx, tongue, mouth, and nasal cavities.

The tongue and jaw also transform the airflow of the voice into speech and alter vocal output (Theodoros & Murdoch, 2001b). The *velopharyngeal valving mechanism* regulates contraction of the muscles of the soft palate and pharynx, controlling the force of contractions, the range of movements, and the coordination of muscle movement. *Vocalization* utilizes the larynx, velopharynx, and orofacial structures, which move in and out of the air stream to produce the precise movements that result in individual speech sounds (Netsell, 2001).

Physiology

The forebrain influences ventilatory rate and rhythm (e.g., overbreathing, breath-holding, speaking, singing, laughing, crying, exercise, sleep, hypoxia, sighing, and gasping (Feldman et al., 2003; Parent, 1996, pp. 603–604).

Rhythmicity (Feldman et al., 2003): Respiratory rhythm is generated by a ventrolateral column extending from the facial nucleus to the spinal cord, which contains respiratory premotor neurons that project to spinal and cranial respiratory motoneurons and proprioceptive neurons. However, it is not certain that pacemaker neurons drive the normal rhythm. A significant rhythm-generating area occurs at the rostral end of this column (*pre-Bötzinger complex* [*pre-BötC*]). This group may be inspiratory (i.e., the diaphragmatic breathing function). A subordinate respiratory center functions in the medulla (*preinspiratory discharge* [*pre-I*]), which drives expiration. It also has pacemaker functions.

Plasticity: The respiratory apparatus is required to maintain gas homeostasis (O_2 and CO_2) and, when confronted by respiratory infection, environmental change, aging and weight change, pregnancy and injury. While respiratory control is somewhat *activity-dependent, neuromodulators* induce or maintain respiratory neuroplasticity (serotonergic, dopaminergic, or noradrenergic receptor activation).

Chemosensitivity: Sensory receptors (e.g., stretch, baroreceptors, spindle, etc.) respond to changes in homeostasis. Chemoreception is a widely distributed system. Different combinations of receptors provide adequate chemoreception in changing arousal state. *Chemoreceptors* respond to changed gas levels and pH (acid–base balance). Receptors to these changes are found in the carotid bodies, and also as central chemoreceptors in the brain. These sites may be involved in rhythm and pattern generation.

The vestibular system avoids reduced perfusion to the brain with changes of posture. Vestibular receptors provide input to the respiratory control center, cerebellum, locus coerulus, and so on. This system is coordinated with components of the speech-production process (laryngeal, velopharyngeal, and orofacial). It affects the accessory respiratory muscles of the upper airway, maintaining airway patency and acting as "valves" to regulate air flow. Vestibular input affects the discharge of various respiratory-related motor neurons: phrenic; intercostal; abdominal; recurrent laryngeal; and the hypoglossal nucleus and tongue protrusive muscles. It also leads to early onset of phrenic stimulation of inspiration and along with vestibular lesions modifies the discharge of both expiratory and inspiratory spinal nerves (Miller & Yates, 1996).

Respiration

Speech production is an aeromechanical event of aerodynamics (pressure, flow, volume) that maintains steady subglottal air pressure throughout an utterance. If one runs out of breath, one can only speak several words at a time. *Respiratory support* contributes to intelligible speech, particularly prosodic characteristics and speech naturalness. Air pressure is provided by (1) the downward and upward movement of the diaphragm muscles, which lengthens or shortens the chest cavity, and (2) the elevation and depression of the ribs to increase and decrease the anterior-posterior diameter of the chest cavity (Guyton & Hall, 2000, p. 432; Staub, 1998a, diagram, p. 526). This maintains the subglottal pressures necessary for normal speech production. Expiration is maintained by the internal intercostal muscles (intercostal nerves/thoracic spinal cord) and abdominal muscles (spinal nerves/ventral horn of lumbar spinal cord) (Richerson & Boron, 2003). Neurological impairment disturbs the contributions of the rib cage and the abdomen to change in lung volume level during conversation and reading. Among the vocal parameters are the number of syllables per breath group and the location and consistency of breaths (Hammen & Yorkston, 1996).

Respiration has been described as a pumping action that increases and decreases the volume of the thoracic cavity. The *thoracic wall* (Moore & Dalley, 1999, pp. 59–73), in concert with the diaphragm and abdominal wall, varies the volume of the thoracic cavity. Structures involved in "speech breathing" are the contents of the abdomen, the anterolateral abdominal wall and its muscles (which extend from the upper thorax to the pelvis), the rib cage (external and internal intercostal muscles), and the diaphragm (Moore & Dalley, 1999, pp. 175–188, pp. 291–294); Staub, 1998a). *Inspiration* creates negative pressure to bring fresh air into the lungs. It occurs when the diaphragm's muscular contraction causes the abdominal viscera to descend, the abdomen to protrude, and the chest cavity to expand, accomplished by the elevation of the lower ribs and the forward displacement of the sternum and upper ribs, with the increase in intrathoracic volume causing air to rush into the lungs from the upper air passages. Muscles of the ribcage and upper airways also function to increase thoracic volume (Lacomis, 1996). Inspiratory muscles are innervated by somatic motor nerves (including the phrenic nerve), the cells bodies of which are located at the C3 and C4 levels (Blessing, 2002). *Expiration* begins passively since lung tissue is elastic. The final part of expiration depends on the forced contraction of the chest and abdominal muscles. These are innervated by spinal motoneurones activated by expiratory motoneurones in the medulla, the Botzinger group, and the caudal ventral respiratory group (Blessing, 2002, Figure 5). The contraction and relaxation of the abdominal muscles push the contents against the diaphragm and compress the lungs, producing expiration (Guyton & Hall, 2000, pp. 432–433). Also, the elevation and depression of the ribs is controlled by the contraction and relaxation of the external intercostal muscles (Guyton & Hall, 2000, pp. 432–433). The *intercostal muscles* are considered to have both an expiratory and inspiratory function (Miller & Yates, 1996). Motor innervation of the diaphragm is via phrenic nerves arising from the ventral rami of C3–C5 (Moore & Dalley, 1999, p. 294). Stretching of intercostal muscles (dorsal roots T9–T12) excites intercostal and phrenic (diaphragm) motor neurons, causing the thoracic cavity to expand. Stimuli to segments T1–T8 inhibits phrenic motor neuron activity and terminates inspiration (Staub, 1998b). Expansion of the capacity of the cavity allows the lungs to expand, and through subsequent relaxation decreases the volume of the cavity causing the lungs to expel air. Air is distributed to the lungs via the larynx, trachea, bronchi, and bronchiole, as well as the alveoli, the passage of which must be kept open.

Pitch

Pitch control contributes to perception as a question or statement (Patel, 2002). Speakers vary widely in the size and shape of their vocal tracts, and thus the fundamental pitch of their speech. However, the relative combinations of frequency required to produce speech signals are consistent and replicable across speakers. Humans can control the primary vocal pitch (F_0), thus altering the distribution of energy in spectral bands by adjusting the filter properties of the vocal tract (Brauth et al., 1997).

Pitch is a complex outcome of the movement of the vocal folds (cords) as a consequence of actions of muscles, ligaments, and cartilage within the larynx (Guyton & Hall, 2000, p. 442). During normal breathing, the folds are wide open to allow easy air passage. During phonation, *pitch* is determined by the degree of stretch of the cords, which are moved together so that passage of air between them causes vibration, and by how tightly the cords are approximated to each other and by the mass of their edges. When exhalation is initiated, the vocal folds are stiffened and adducted to a position close enough to vibrate but not to touch (Guyton & Hall, 2006, p. 481). Stewart et al., Brin, 1999). The elastic quality of the folds, and the suction created by the air moving between them, brings the vocal folds back to the midline, while their oscillation results in phonation. Deviations from this alignment can lead to voice disorders. Pitch is determined by the degree of stretch of the folds, how tightly they are approximated to one another, and the mass of their edges. A sharp edge of the folds creates a high-pitched sound and blunt edges create a more bass pitch.

Formants

Speech has been described as a temporal complex acoustic signal composed of multiple frequencies (formants) (Doupe & Kuhl, 1998; Tremblay et al., 2003). These are encoded in the auditory cortex, reflecting the complexity of language, during the time constraints of ongoing speech. A formant is the consequence of the vocal tract's concentration of energy at particular frequencies. These change over times as short as 10 msec and may extend to 40 msec. Vowel sounds are temporally static, steady-state frequencies. Formant frequencies decrease with the growth of the vocal tract (Kwent, 1999). Consonants contain variable onset times and rapid transitions of frequencies that change within syllables (Fitch et al., 1997).

Speech

Speech has been regarded as a tertiary structure that encompasses frequency and amplitude cues within a temporal envelope (Fitch et al., 1997). The association between speech signal and phonatory/articulatory actions is complex, based upon organs participating in respiration, phonation, and articulation, and permitting similar acoustic results to be produced by a variety of mechanisms (Etcoff, 1991). Acoustic processing has been defined by characteristics such as loudness and frequency, which are not of direct linguistic significance. Phonetic processing is concerned with extraction of linguistic features (e.g., place and manner of articulation of consonants) (Norris & Wise, 2000). Somatosensory feedback is independent of acoustic output. The goal of speech movements is to achieve a position of the speech articulators and associated somatosensory input, not the sounds produced (Tremblay et al., 2003).

CEREBRAL LATERALIZATION AND VOCALIZATION

IS MOTOR APROSODIA A RIGHT HEMISPHERE (RH) DISORDER?

Aprosodia is, in this author's opinion, a motor vocal disorder in which the expressive qualities of speech are so lacking that the speaker offers a false impression of indifference. It has been described as a group of disparate phenomena attributed to RH control. The author will cheerfully reveal the end of the story (No! Aprosodia as an expression of the PCS is usually *not* a RH disorder!). Explicating the controversy helps one understand the complexity of the integration of both cerebral hemispheres in the motor expression of speech. The initial studies of prosody were performed on people with RH damage, with LH patients excluded because aphasia was thought to complicate the test procedure. Damage to either hemisphere can affect the processing of both emotional and linguistic prosody (Van Lancker & Breitenstein, 2000). Kirshner (2004) asserts that RH patients understand what is said, but not how it is said, since they lose their sense of metaphor, humor, sarcasm, irony, and related constituents of language that transcend literal meaning. There is an impairment of the pragmatics of language.

The assertion that aprosodia is a RH disorder conflates emotional expression, sensory reception, and other emotional processing (Borod, 1993a, 1993b; Bell et al., 1990; De Gelder, 2000; Lezak, 1995, p. 66). Etcoff's summary (1991) asserts that both hemispheres can decode prosodic cues. This hypothesis seems to be consequent to study of RH patients with stroke or embolism (Denes et al., 1984; Shapiro & Danly, 1985; Ross et al., 2003; Starkstein et al., 1994; Wertz et al., 1998), that is, focal neurotrauma. Motor aprosodia after mechanical TBI (Parker, 2001) has not been intensively studied. It has a more diffuse neurotraumatic basis, and is associated with a different outcome than vascular brain damage, suggesting that it is a distinct condition.

CONTRARY EVIDENCE CONCERNING RIGHT HEMISPHERE (RH) BASIS FOR APROSODIA

1. Neurological control of vocalization is widely distributed: LH; RH; brainstem; CN and spinal cord; subcortical nuclei (basal ganglia bilaterally); medial frontal; Broca's area; anterior temporal lobe, anterior cingulate cortex; the supplementary motor area of either hemisphere; and the corona radiata (Cancelliere & Kertesz, 1990; Nass & Stiles, 1996; Theodoros et al., 2002a).
2. Partial LH lateralization: The motor area for the mouth is lateralized to the LH, while Broca's area itself is bilaterally activated but shows greater activation in the LH (Crank & Fox, 2002). However, right-handed persons with LH dominance can develop motor aprosodia after RH damage (Kirshner, 2000b).
3. Somatic structures are non-neurological: Injured tissue contributing to aprosodia is vulnerable to illness and trauma. Pitch and articulation are laryngeal and oropharyngeal phenomena.
4. Commissural connections minimize lateralization: Callosal fibers connect most of the corresponding areas of the hemispheres, although cortical-subcortical circuits also transfer information (Brodal, 1981, pp. 804–805; Parent, 1996; Zilles, 1990). The CC is vulnerable to mechanical head injury (see CC and Falx Cerebri Cuts). The superior surface of the CC is vulnerable to being sheared by the inferior edges of the falx cerebri (Kretschmann & Weinrich, 1992, pp. 26–43; Parent, 1996, pp. 4–5). Complete commissurotomy is followed by short-term mutism with retained comprehension and writing, and may produce a diaschisis (loss of function far from the lesion, although neurally connected) in the speaking hemisphere (Trevarthen, 1990). Lateralized damage could create a respiratory or vocalization disorder through diaschisis as well. Matching of the affective and propositional components of language is attributed to transcallosal interaction between the "sensory" language areas of each hemisphere. It is assumed that the actual motor integration of these components of

language is is accomplished in the brainstem (Ross et al., 1981). Retention of capacity to curse and express other nonpropositional vocal behavior after cerebrovascular accidents suggests dissociation for propositional and nonpropositional vocal behavior (Brown & Cannito, 1996). Inability to transfer affective information from the RH via the CC to the propositional language areas of the LH can cause the voice to be monotonous, flat, and devoid of emotion. Spontaneous gesturing may be absent. Ross et al. (2003) observes that considerable interhemispheric interaction is needed to ensure temporal and behavioral integration of articulatory-verbal and affective prosidic elements. Affective-prosodic deficits after LH brain damage are attributed to loss of callosal integration of the language functions.

5. Some vocalization functions are sexually dimorphic. There are function-specific sex differences, which are more or less lateralized in women. Basic motor or acoustic functions related to speaking show sex differences in intrahemispheric but not interhemispheric functions. Speed and accuracy of articulation is improved at phases of the menstrual cycle characterized by high concentrations of estrogen (Hampson, 2002).

Variable Lateralization of Emotional Vocal Expression

The RH has an advantage for recognition of emotion while the LH participates in linguistic decisions. There is evidence that functions described as prosodic or emotional are vulnerable to cerebral damage from either side: cortical (frontal) or subcortical regions such as the amygdala or hippocampus (Heilman et al., 2003; Kolb & Taylor, 2000). Prosodic disturbance can occur with either right or LH damage; prosodic impairment following a RH lesion could be linguistic and not secondary to an affective disorder; lateralization of ability to perceive prosody may depend upon the difficulty of the task, etc. (Wertz et al., 1998). In fact, LH lesions impair mechanisms that process information within a time frame of tens of milliseconds. This leads to phonological disorders of acquired and developmental aphasias (Fitch et al., 1997). Variable expression of emotional behavior may reflect asymmetrical control of processes such as movement, language, or processing complex sensory information. Consistent with the lack of lateralization were the findings of Cancelliere and Kertesz (1990) that there were no hemispheric differences in a cerebral infarct group for varied aprosodia measures.

Right CHI impairs comprehension of the affective-prosodic elements of language (Ross, 1993). Reduction of affective prosody is correlated with negative behavioral symptoms and regarded as similar to the deficits found after right brain damage. Nevertheless, it may not be as lateralized as propositional language, and it is recognized that speech production entails interhemispheric interaction to ensure articulatory-verbal and affective-prosodic unification and temporal coherence. For example, surprise, sarcasm, or emphasis involve prolonging certain phonetic units are consistent with informing the RH of the words that will be articulated and their cadence. This creates the intended affective-prosodic intonation. One theory posits that left brain damage is associated with the loss of callosal integration of bilateral language functions and that right brain damage is associated with the loss of affective communication representation and loss of ability to modulate affective prosody. A limitation in applying this approach should be considered. Mildly concussed patients characteristically suffer from diffuse, bilateral brain injury. Yet the authors excluded this group from a recently sampled study (Ross et al., 2001).

More precise localization of affect processing: There are reports of both lateralization and anterior-posterior effects. Different findings, and exceptions within a particular structure, have been reported. In one case, motor aprosodia was described after bilateral infarction, with an emphasis upon subcortical damage destroying the motor neurons of the cortex. Other cases had right frontoparietal opercular damage (Ross et al., 1981). Localized RH lesions may not explain vocal aprosodia per se when the accident victim is experiencing appropriate distress following an injury. There is evidence that while the RH identifies the presence of emotion (e.g., emotional intonation), it does not differentiate different kinds of affect. Using identification of affect and discrimination tasks, the posterior RH contributes to the understanding of the relationships among acoustically transmitted emotions, while the LH does not contribute to sensitivity to the interrelations among emotional concepts (Denes et al., 1984).

The processing of music is controversial with reference to cerebral lateralization. Prosodic information and tone discrimination appear to be processed preferentially in the right temporal lobe. The right posterior temporal region overlaps with areas involved in judgment about tones, that is, nonlanguage interpretation. Emotional face processing is associated with the right PFC. The RH creates a complete representation of meaning and intent by integrating prosodic information with inferences while keeping track of the topic (Bookheimer, 2002). It has been stated that the only localization consensus has been the processing of pitch contour information in the RH superior temporal gyrus and frontal regions (Brust, 2001; Zatorre, 2001). In contrast, musical processing is described as multimodal, and cannot be sharply localized to one part of the brain or even to one hemisphere. The psychological whole is even more widely distributed. It is observed that for melody, pitch intervals are local whereas melodic contour is global; for temporal patterns, rhythm is local but meter is global. There are differences between musicians and nonmusicians during the discriminations of pitch and musical rhythm (Parsons, 2001). Lateralized brain damage does not produce uniform dysfunctions (Brust, 2001). In contradistinction to the emphasis upon cerebral lateralization is the finding of cerebellar processing of pitch or melody, bilaterally for musicians and with left-sided activation for nonmusicians. The left cerebellar projection to the right cerebral cortex supports the right temporal activation for pitch. This is believed to support not motor but sensory or cognitive processing. There is also concurrent activation during a rhythm task of the cingulate cortex and the basal ganglia for musicians and nonmusicians (Parsons, 2001).

Right Hemisphere (RH) Affect Processing

There are sex differences in cortical organization within the two hemispheres, creating differential vulnerability of men and women to reduced function: apraxia (considered here as a motor disorder) and aphasia occurring post-anterior or posterior brain damage. Sex differences are function-specific and caution is needed against drawing conclusions about lateralized verbal functions. Basic motor functions show sex differences in intrahemisphere but not interhemisphere organization, while higher-order abstract or more complex verbal functions seem to show interhemispheric differences. In addition, the RH is believed to modulate autonomic arousal in both emotional and nonemotional conditions (Hampson, 2002). Nevertheless, personality factors (e.g., extraversion and optimism) may co-vary more robustly with RH activation than situationally induced emotions (Heller, 1993). Bogen (1993) speculates that the RH contains representations of both species-typical facial expressions and affective-prosodic expression. Destruction of these representations or inability to access them would impair comprehension and discrimination of emotional prosody. While the RH is needed for emotional prosody (determining whether prosodically intoned sentences are statements, commands, or questions), both hemispheres participate in comprehending syntactic-propositional prosody. RH-damaged patients also have *distraction* defects that contribute to poor performance on emotional prosody tasks (Heilman et al., 2003). Furthermore, assuming that each hemisphere is more distracted by the stimuli it normally processes, Weintraub (2000) notes the complex components of communication (motor and prosody). Thus, right-sided cerebral lesions can interfere with the comprehension and production of affective and emphatic prosody, the nonverbal identification of facial affect, and the ability to adjust interpersonal behavior to the social context.

Disorders: Children with RH developmental disorders have difficulty recognizing and producing emotional prosody (Voeller, 1998). It has been hypothesized that the functional and anatomical organization of the RH's affective components of language resembles that of propositional language in the LH (Ross et al., 1981). However, this has not been well supported (see Aprosodia). Ross assumed that the aprosodias were a group of six disorders of language that were anatomically and functionally organized in the right cerebral hemisphere analogous to propositional language in the

LH: motor, sensory, global, transcortical motor, transcortical sensory, and mixed (Ross et al., 1981). Frontal opercular lesions are likely to impair production of affective prosody, whereas posterior temporoparietal lesions are likely to impair comprehension of affective prosody. Numerous exceptions have been reported (Ross et al., 2003).

The thesis of RH parallel functions for LH prosody received "mixed support" in one study. Seventeen of twenty patients with RH stroke displayed both dysprosody and dysarthria (Wertz et al., 1998). Support was not offered for the hypothesis that the organization of prosody in the RH mirrors that of propositional speech on the left side. Similar dysprosodic impairment was determined for infarcts in different regions of the RH. Furthermore, some patients may have suffered from unnoticed ("subclinical") small bilateral lesions.

RH damage has been associated with depression and deficits of comprehending emotional prosody (Etcoff, 1991; Messner & Messner, 1988) and indifference with pharmacological inactivation, contrasting with the catastrophic reaction with LH inactivation (Denes et al., 1984). The appearance of indifference has been associated with RH lesions (Heilman et al., 1993). That lack of emotional expression may not reflect the inner experience of distress is precisely the authors' thesis. Ovsiew (2002) has suggested, "Should the clinician consider receptor prosody within the same category as expressive prosody, the left hemisphere plays a role." Both normals and LH patients seem to rely on conceptual and acoustic cues in identifying and discriminating emotional sounds (Denes et al., 1984). Affective motor prosody is disrupted by lesions of the frontal opercular area, described as the RH equivalent of Broca's area (Mendez & Cummings, 2002). Also implicated have been the right premotor cortex or the basal ganglia, while receptive prosody is impaired by lesions of the posterior superior right temporal lobe (Wernicke's area) (Chow & Cummings, 2000). One might consider this to be an aspect of music agnosia (Peretz, 2001).

Left Hemisphere (LH) Speech Processing

The LH seems specialized for rapid acoustic change that enhances speech perception, consistent with its complex auditory discrimination in nonhuman species (Fitch et al., 1997). LH disturbance with demonstrated RH lesions contributed to disturbed prosody (Bradvik et al., 1991). fMRI indicators suggest that receptive prosody seems to be a RH function, serving the task of representing meaning and intent critical to social communication. Acquired aprosodia in children has been observed in congenital and in right or nondominant cortical and subcortical lesions (Nass, 1996).

The posterior temporal region, roughly contralateral to Wernicke's area and involved in making judgment about tones, is also involved in language tasks relevant to figurative, contextual, or connotative meaning. Emotional and prosodic information processing in the inferior frontal gyrus occurs in Brodmann's area 45, homologous to Broca's area activity in traditional language tasks. Prosodic expression is related to LH Broca's area. The right prefrontal and temporal areas preferentially process prosodic information as part of the more general process of language comprehension (Bookheimer, 2002).

Bilateral Vocalization Functioning

Some structures have bilateral representation of UMN (facial nerve fibers innervating the forehead; branchial motor component of the vagus nerve supplying the pharynx, soft palate, and one tongue muscle). Others have unilateral representation (e.g., the remaining facial muscles). There is a neurobehavioral distinction between hearing and expressing pitch since musically tone-deaf people do not speak in a monotone (Altenmuller, 2001). Assuming that "processing" refers to incoming stimuli, pitch modules appear to function in the posterior right supratemporal lobe. Professional training causes a shift from right to left hemisphere predominance, whereas temporal structures (meter and rhythm) function in the left temporal lobe (Altenmuller, 2001).

PROSODY AS COMMUNICATION

Prosody is discussed here as a component of the normal communications matrix existing in an unimpaired person without trauma or disease. Its disturbance after TBI is discussed under Aprosodia. Prosody contributes to the characteristic intonation of a language and pragmatics, that is, communicative intent (Wilson, 1996) and affective signaling (Ross et al., 2001). Yet, not all verbal communication is prosodic. The *propositional format* infers a conceptual structure that is expressible in language (refers to events and the states of the world), as opposed to the representational format of visual images, musical patterns, or motor images (Indefrey & Develt, 2000). One may differentiate between the intrinsic prosodic quality of a language or dialect and a later intellectual form of prosody that bears propositional values in subtle shades. Prosodic features aid word recognition by giving prominence to the key words, identifying word boundaries, and distinguishing between new and given information. Prosody was a literary reference dating back to the earliest years when poetry was first recorded. It was defined as "the science of metrical versification," when a sequence of coupled syllables have a stress on every second or third syllable, as, for example, in a Shakespeare sonnet (Random House Dictionary). It was not until the late 1940s and the early 1950s that "prosody" was embraced by the newly emerging speech sciences, establishing the concept that speech and language were truly biologic functions (Silverman, 1953). Monrad-Krohn (1947) described prosody as the affective components of language. Its original clinical definition had been restricted to vocal expression (i.e., the faculty of speech that conveys different shades of meaning irrespective of the words and the grammatical construction). Shades of meaning are expressed through stress on syllables and words, correct spacing (prolongation and shortening), rhythm, pauses, and correct placing of pitch on syllables and words. Tomkins (1995, cited by Wertz et al., 1998) added rate, juncture, and duration.

Social Impression

Communication can be impaired due to loss of spontaneous facial and bodily gestures. Emotional expression has diagnostic significance for cerebral personality syndromes and chronic stress disorder (Parker, 2001, Chapters 12, 13, 17). Misinterpretation of aprosodia could lead to the incorrect assessment as apathetic, indifferent, a misrepresentation of one's condition, malingering, or manifesting the dull affect of schizophrenia. Monotonous verbal output with decreased facial grimacing and sparse use of gestures could also be misinterpreted as an emotionless response signifying depression (Mendez & Cummings, 2002). Aprosodia is often ignored by the clinician (Kiss & Ennis, 2001; Messner & Messner, 1988; Theodoros et al., 2001a). While emotional gesturing and prosody were described as features of the RH (Fricchione & Howanitz, 1985), the focus and controlled environment of the neuropsychological examination restricts the opportunity to make detailed observation. Further exploration is useful upon observing aprosodia. The clinician may explore potential depression with the patient (i.e., whether there is an experience of sadness or emptiness). Should inquiry suggest reduced experienced affect, study is invited for issues of endogenous depression (neurochemical or neurotraumatic) or endocrine dysfunction. Inability to express affect may suggest a link between depression and chronic pain. It may be useful to utilize the healing effect of increasing arousal and thus mobilizing positive or negative emotions, while encouraging their release, that is, catharsis (Beutler et al., 1986).

Somatic Signaling

Prosody might have evolved as a "*somatic marker*," (i.e., a self-signaling [authors' term] primitive neurophysiological system). It appears to inform the frontal cortex of the state of the soma and interact with cognitive strategies under the influence of viscera, internal milieu, and skeletal musculature (Damasio & Anderson, 1993). Loss of the signal could lead to maladaptive outcomes at both conscious and unconscious levels due to interference with identification of both danger and positive opportunities. The evolutionary pressure to communicate verbally has contributed to the specialization of the left auditory cortex and the complementary structural and functional changes of the two hemispheres (Zatorre, 2001). It is speculated that phonemes may have evolved from the innate calls typical of some primates. Their acoustic features provide the basis for vocalizing information such as the presence of specific predators (signs) (Caplan et al., 1999). At birth, babies can discriminate the phonetic units of all the world's languages, a talent lost by the end of the first year of age (Kuhl, 2000). Heilman et al. (1993) assert that there is a species-typical affective expression that destruction of would impair both the comprehension and discrimination of emotional prosody. An alternate view is that vocal indicators of emotion are likely to be phylogenetically continuous, since the social functions they serve and the physiological mechanisms that produced them are similar in humans and primates (Etcoff, 1991).

Prosody and Learning of Speech

It has been asserted that the prosodic elements of speech are individually learned (Ross et al., 2001). *Linguistic prosody* meanings include statement and question, with emphasis meaning a noun or a verb (IMport vs. imPORT). *Emotional prosody* includes attitudes and emotions (e.g., emotional meanings such as happiness, anger, sadness, and fear). Speech learning depends upon *hearing others* and *hearing one's own vocalizations*. Fetuses learn prosodic cues while still in the *womb*; they hear their mother's voice through the abdominal wall and bone conduction. For newborns, the mean fundamental frequency is about 400 Hz, declining to about 250 Hz between the ages 4–7 for both boys and girls, with change in mean frequency to 130 Hz observed in boys at age 12, compared to 250 Hz for young women. This octave difference is maintained until old age (Kent, 1999). Children learn the emotional meaning of facial gestures significantly earlier than they learn to correctly identify vocal emotions. As children learn speech, they learn movement sequence (articulation) and its sensory consequences. Speech has been described as one of the gestural patterns but one that results in sounds (Kent, 1999). *Deaf infants* show deficits in babbling, an important milestone in early language acquisition. *Parentese speech* emphasizes prosody and the hyperarticulation of phonetic units. From birth, prosody plays an important role in social life. Maternal speech patterns are measurably distinct from frequencies used by these women in speech directed at adults. The prosodic speech structure of mothers is characteristically higher pitched, with a slower tempo, exaggerated intonation contours, and syntactically and semantically simplified with hyperarticulated vowel sounds. Prosody is an important component of infant language learning and maternal communication. At birth, infants discern differences between all phonetic units, which is considered to be an evolutionary heritage (Kuhl, 2000).

Cognitive Component

In learning a language, articulatory praxis and sound code (symbolic, language) are developed sequentially, consequent to the development of different cortical areas that form at different stages of brain phylogenesis (Glezerman & Bolkoski, 1999, p. 89). Children learn to correctly articulate sounds and sound sequences (phonetic mastery) but also to use these sounds in accordance with the phonological patterns of the language. A child may be aware of the phonemic difference of two sounds but be unable to produce that contrast (e.g., differentially producing /s/ and /sh/). Lack of prosodic development contributes to the failure to develop competent social skills. This alerts us to childhood aprosodia, which has been considered analogous to childhood aphasia (Bell et al., 1990).

Experience establishes *memory* representations for speech, specifying the phonetic units of the language. This is not a simple sensory memory; rather, it is a complex mapping in which the perception of the acoustic dimensions of speech are warped to create a *recognition network* that emphasizes phonetic differences and minimizes those not used in the language. There is a *perceptual magnetic effect* that enhances perception of that language. The Japanese do not discriminate American sounds. The subset of all possible vocalizations of a language that are perceptually learned by listening to others is known as an *acoustic target*. Perceptual learning constrains and guides what can be produced, creating an *auditory-articulatory map* (Doupe & Kuhl, 1998).

Steroid Effects

Experience influences the fine but not gross structure of song-control nuclei in birds. Steroid hormones play a crucial role in the development of sexual dimorphisms and also in the organization and activation of neural circuits of communication systems during development (Breedlove & Hampson, 2002; Kelley & Brenowitz, 2002). Neurotrophins sculpt neural circuits while estrogens more than androgens masculinize song-control nuclei (Bottjer & Arnold, 1997). Vocal courtship signals are an indicator of *vigor* and are subject to hormonal effects upon somatic and neurological sexual dimorphism (Kelley & Brenowitz, 2002). One may infer stress effects upon vocalization, whether directly experienced or as a consequence of an injury. Infection-related stress can lead to adrenal insufficiency or make apparent subtle adrenal dysfunctions (New & Rapaport, 1996). In women, a variety of *hormone-related language functions* change according to the *menstrual cycle*, including speed and accuracy of articulation (Hampson, 2002).

Musical Processing

Prosody is the musical quality of emotional expression. It includes pitch, interval, contour, rhythm, meter, timbre, and emotional response. Most processing components underlying language and music are not shared and are neuroanatomically separate (Peretz, 2001). *Vocal music* has been speculated to have an evolutionary origin, with language perhaps evolving as an alternative for primate physical grooming (Huron, 2001). It may have functioned in mate selection, conflict reduction to enhance social cohesion, group effort, communication, mood regulation for greater comfort, and so on. An alternate view is that music (vocal), rather than being a trait selected by evolution, is a cognitive capacity to process sound patterns discriminable by pitch and rhythm variation (Cross, 2001). Music as a fitness indicator can represent a unique display in sexual competition when one is below average in other characteristics (Miller, 2000).

Musical processing is detected in children as early as four months (i.e., sensitivity to temporal isochrony) (Peretz, 2001). The musical characteristics of caregiver/infant interaction (by this the authors infer prosody) lead to a capacity for emotional bonding and social regulation. By the end of the first year, perceptual abilities become constrained and infants cannot discriminate formerly accessible foreign language-contrasting sounds. An example is a Japanese infant's inability to discriminate American English /r/ from /l/ at 12 months, which they could do at 6 months. Auditory and visual (facial) speech representation in infants may have a LH involvement, noting that the language-processing areas of the brain include many more areas than Broca's and Wernicke's areas (Kuhl, 2000).

Peripheral and CN

A case is reported of bilateral *hypoglossal nerve* paralysis following head trauma without skull base fracture, possibly due to traction injury (Noppeney & Nacimiento, 1999). Unilateral or bilateral damage to the *vagus nerve* alters displacement of the vocal folds (adduction of the glottis), depending upon whether the lesion is in the brainstem (aphonia due to soft palate paralysis), the vagus nerve (aphonia), or its recurrent laryngeal branch (Rosenfield & Barroso, 2000). Abduction or adduction displacement in the vocal folds and supraglottic musculature changes the quality of the voice. The *recurrent laryngeal nerves* ascend close to the notch between the trachea and the esophagus, which may perhaps be crushed or sheared by internal movement. The *phrenic nerves* (diaphragm control) are vulnerable to trauma at multiple sites, from the neck to the lower thorax (Lacomis, 1996). These nerves, closely associated with various veins and muscles, are possibly vulnerable to direct trauma and/or torsion or extension during hyperextension (hyperflexion injury). Since the diaphragm lacks bilateral cortical representation, hemispheric stroke results in attenuation of diaphragmatic excursion on the hemiplegic side during volitional breathing, and also in impairment of the contraction of the intercostal muscles. Contralateral diaphragmatic contraction can be elicited with cortical stimulation, increasing cerebral blood flow in the motor strip, the right premotor cortex, the supplementary motor cortex, and the cerebellum (Simon, 2001).

Larynx: Since phonation requires highly integrated neurophysiological coordination, laryngeal dysfunction may result from impairment at the central and peripheral levels of the CNS (bilateral lesions of the UMN; damage to LMN supplying laryngeal musculature; lesions of the extrapyramidal system, etc.). Seventy-five percent of a group of severe CHI subjects displayed laryngeal dysfunction with some intelligibility of speech (articulation [harshness, strained-strangled, hoarseness, and intermittent breathiness]). The increased tone or spasticity in the laryngeal musculature results in hyperadduction of the vocal folds, reduction in the size of the laryngeal aperture, and an increase in resistance to airflow during phonation. Some vocal effects are due to compensation for laryngeal impairment, wastage of expiratory airflow from other valving mechanisms, or deficits in other focal systems (Theodoros & Murdoch, 1995).

Receptive and Expressive Processing

Sensory prosody is processed early in verbal comprehension and is attributed to the RH as part of a distributed network. It contributes to the subsequent linguistic and vocalization network, utilizing all levels of the nervous system, with a final common pathway in the recurrent laryngeal nerves. Cummings (1985), noting bilateral limbic contribution to the experience of emotions, points out that right-sided lesions may impair prosody without altering the propositional component of verbal output. *Propositional content* is the expression, repetition, or comprehension of sentences such as questions, statements, or commands (Ramasubbu & Kennedy, 1994). Sensory prosody (comprehension) has been described as a function bound to motor prosody (Chow & Cummings, 2000; Robinson & Starkstein, 2002). This was not supported by the findings of a sample of patients with Alzheimer disease (Testa et al., 2001). As dementia severity increased, performance on comprehension and repetition tasks became more impaired while spontaneous affective prosody remained normal. In *multi-infarct dementia*, communication disorders eventually evolve into dysarthria and kinetic mutism. At an earlier stage they may display deficient pitch, tone, and melodic qualities while the content and organization of speech remain unaffected (Lezak, 1995, p. 201). Several functions have been described as aprosodic but seem to be independent of vocal or motor expression: affective comprehension, aphasia, and processing facial affect (Ross et al., 2001). Reduced comprehension of stimuli with aprosodic implications can lead to reduced social comprehension (e.g., denial of the person's depression). Thus, motor aprosodia (flat, monotonous verbal output) can be misinterpreted as apathy or depression.

Motor Aprosodia

Motor aprosodia seems to be a distinct entity when compared to other conditions such as the comprehension of emotionally intoned speech. While there is input for vocalization from the CNS centers that mediate emotional comprehension, the author considers this to be a different function. Motor aprosodia had been described as a cerebral mood disorder (Parker, 1990, 2001). The author's current understanding is that this interpretation was incorrect. It is rather a somatic/neurological vocalization disorder, frequently observed in patients with concussion and related to neurological and/or somatic trauma.

Clinical Significance

The author's clinical experience suggests that aprosodia is among the most common symptoms of a TBI. Prosodic impairment can be linguistic and not secondary to affective disorder. Nevertheless, *depression* can be accompanied by increased pauses before answering and speech that is decreased in volume, inflection, amount or variety of content, or muteness (American Psychiatric Association, 2000).

After mechanical accidents, aprosodia can be consequent to an extremely wide range of neurological and somatic injury, including innervation and damage to the larynx, face, jaw, and breathing apparatus, as well as CNS and peripheral nerve injuries. Aprosodia may be comorbid with dyspraxia, medical conditions, and aphasia.

Since prosody offers motivation, mood, and linguistic information, aprosodia reduces the accuracy and range of communication. Loss of the voice's expressive quality can cause the patient's condition to be significantly underestimated.

Reduction of the credibility of the patient's self-description can cause an examiner to believe that the patient is faking or exaggerating (see Expressive Deficits).

Aprosodia leads to social loss and incompetence, affecting the person's role as parent or child, social companion, employee, and so on. For many employment conditions, it represents a lesser level of competence (staff and customer relations; public affairs) (Stringer, 1996). Maternal aprosodia may have developmental effects upon the child.

Aprosodia is differentiated from a patient's inability to fantasize or describe one's own feelings and communicate them (Fricchione & Howanitz, 1985). Depression itself alters pitch and loudness (Samuel et al., 1998). In one study after CHI, in which patients with major depression were excluded, there was no significant correlation between prosodic measurements, depression scale, or neurobehavioral assessment.

Communications

Aprosodia has long-term persistence with devastating effects upon communications. It occurs in a range of loss that is very responsive to the clinical level of mood elevation, other manifestations of a patient's emotional state (Damasio, 1999), and psychopharmacologic effects. Speech therapy, which focuses on loudness of speech production, may create an elevation of prosodic outcome measures (Patel, 2002). Aprosodia can lead to a misperception of the person's emotional condition and mental level. The question has been raised as to whether flat and dysprosodic speech reflects a true paucity in the emotional realm or whether it is an inability to control prosodic elements of speech perhaps only accidentally linked to the aberrant emotional qualities of a clinical population (Shapiro & Danly, 1985). A salient characteristic of aprosodia is the loss of vocal pitch, which is considered to be the central aspect of all music (Zatorre, 2001). Patients with severe CHI were not able to change pitch and volume according to prosodic content, particularly for anger (Samuel et al, 1998).

Aprosodia can be considered as a *deficit of self-monitorin* . An external vocalization loop utilizes the same superior temporal structures whether one is hearing one's own voice or another's. The feedback permits some degree of output control (i.e., loudness and error correcting). The internal loop comprises the internal modulation of the motor area (Indefrey & Develt, 2000). The prosodic output is not phonetic processing (linguistic features of speech, including place and manner of articulation of consonants) but acoustic processing (loudness and frequency) that is not of direct linguistic significance (Norris & Wise, 2000).

Comorbid Disorders

The *cerebellar cognitive affective syndrome* (Schmahmann & Sherman, 1997) is a syndrome with language, personality, executive, visual-spatial, and sequencing disorders, reminiscent of the frontal lobe syndrome and other conditions. It has been attributed to cerebellar degeneration, particularly of the posterior lobe and the vermis. Adult subjects with cerebellar disorders display unusual prosody and dysarthria, with tone or voice characterized by a high-pitched, whining, childish, and hypophonic quality.

Alexythymia has these characteristics: deficient ability to express emotions caused by a lack of understanding; a concrete, bland, and tedious communication style; a tendency to be externally oriented in thinking; and poor ability to use imagery. A bland manner, confusable with aprosodia, could result from a lack of appropriate emotional expression due to a lack of understanding of one's emotions (Williams et al., 2001). There is an association between self-reported head injury and levels of alexythymia (Williams et al., 2001).

Amusia is a variety of aprosodia that may occur in the absence of apraxia and with the preservation of cognitive functioning. One case is described as having monotonous speech without linguistic prosodic cues (interrogative, exclamative, or declarative), an inability to produce or repeat accent or intonation without emotional expression or facial expression, and an inability to sing notes or familiar songs or to recognize common melodies. Most reported cases involve the left dominant hemisphere and language areas (Confavreux et al., 1992).

Somatic Trauma

There may be overlap between the mechanisms of aprosodia and dysarthria. Direct somatic trauma resulting in dysarthria is attributed to motor and/or sensory impairments of neural mechanisms that subserve respiration, phonation, resonance, articulation, and/or prosody. The formation of words depends upon the rapid and orderly succession of individual muscle movements in the larynx, mouth, and respiratory system. Jumbled vocalization occurs when there is lack of coordination between these organs and an inability to adjust in advance the intensity of the sound or the duration of each successive sound. Some brain-injured persons manifest reduced syllable repetition rates and lengthened syllable duration. There is evidence of atrophy to vocal structures from repetitive trauma to the teeth and jaws, observed when the nervous system causes tremor of the masticatory muscles and/or subluxation of the temporomandibular joints via the CN (Silverman, 2002). Destruction of the somatic or neurological components of articulation can cause a total or partial inability to speak distinctly (Guyton & Hall, 2000, p. 655). Loss of muscular control causes increased or decreased muscle tone and poor motor control with imprecise movement, as well as slurred or indistinct speech. Intelligibility is a consequence of muscular paralysis, weakness, abnormal tone, or incoordination. These result in deficits of direction, range, force, endurance, timing, and regulation of movements (i.e., reduced intelligibility) (Carter et al., 1996).

13 Chronic Pain and Posttraumatic Headaches

OVERVIEW

Pain is an important dimension of health status and a marker of disabilities caused by neurologic or neuropsychological symptoms (Middleboe et al., 1992). This chapter reviews the neurological, somatic, and emotional components of pain, primarily in persons who suffer comorbid head and body injuries on a nonsurgical level. Those concerned with more serious injuries may refer to Rivera et al. (2008). The reader will be better prepared to help patients cope with the neurobehavioral consequences of chronic pain, to make referrals to specialized pain-management practitioners, and to understand medical and psychological treatment procedures. A review of medical approaches to pain management is found in Zasler et al. (2007), and in Martelli et al. (2007) for psychological approaches to pain management. Because of pain's widespread neurological distribution and bidirectional signaling, it is vulnerable to the diffuse injuries sustained in mechanical accidents. Pain disappears as tissues are repaired and inflammation is reduced. If healing is incomplete, acute pain evolves into the chronic state, with changes in neurological functioning and multiple maladaptive effects upon level of performance and quality of life.

Both neurologically and psychologically, pain is a complex phenomenon that is often difficult to understand. Pain's etiology may be difficult to determine precisely. A painful experience has multiple neural and sensory components that range from being entirely neurophysiological to entirely psychiatric (Hendler, 1990). Neural and somatic trauma is augmented by complicating emotional stress and medical conditions. It is intertwined with tissue damage, anxiety, and depression. After an injury pain may be referred from a trigger point to distant cranial and somatic targets. As a result of this complexity, its etiology may be mistaken and treatment misdirected.

While pain is a major stressor and is augmented when problems (internal or external) prevent the patient from directly communicating distress, it also serves multiple adaptive purposes. It alerts us to escape danger, discourages the use of an injured area, and allows it to heal. It alerts the injured person and the clinician to that person's emotional condition, the need for further study and treatment, and the outcome of an accident. It impairs many components of cognition, affects adaptive responses, and shapes mood and lifestyle. Discomfort reduces the victim's quality of life by interfering with activity, increasing fatigability, and reducing stamina. An excessive delay in the resolution of insurance claims and/or inappropriate denial, interacts with economic problems, and disturbed self-esteem from inability to work, creating further emotional distress.

NEUROLOGICAL ASPECTS OF PAIN

Circuits

Pain is usually, but not universally, considered to be a separate but complex sensory modality, peripherally and centrally with central circuits dispersed and chemically unique (Connors, 2005c). Actually, several neurotransmitters may be released from a single neuron or neuronal system. The classic pathways are nociceptors; the dorsal horn; spinothalamic, spinoreticular, and spinomesencephalic tracts; and the somatosensory cortex.

Direct pathway: This serves the fine discrimination of nociceptive input. Commencing in the dorsal horn, second-order neurons decussate one-to-three levels above the input of the afferent spinal neurons, then enter the lateral funiculus to form the spinothalamic tract. It ascends to the ventropostero-lateral thalamic nucleus, and then projects to the primary sensorimotor cortex of the parietal lobe.

Indirect pathway: This comprises a medial pain system mediating the autonomic, visceral, endocrine, arousal, and emotional aspects of pain perception. It ascends bilaterally, with poor somatotopic organization. It leaves the dorsal horn as the *paleospinothalamic tract* and ascends bilaterally in the ventrolateral spinal cord to synapse in the reticular formation, the midline and intralaminar nuclei of the thalamus (dorsal medial nucleus), and then projects diffusely to limbic structures, including the anterior cingulate gyrus.

Integrative Components

The *dorsal horn* receives local facilitory and inhibitory inputs and also descending influences from the brain. Ascending neurons are nociceptive specific and also wide-dynamic range (WDR) neurons that receive both nociceptive and non-nociceptive afferents. The distributed centers and circuits of pain are far more complex: there are multiple redundancies and feedback loops for ascending and descending inhibition and facilitation. Additional centers include the midbrain periacqueductal gray (PAG); pontine and mesencephalic reticular centers; and paralimbic and limbic structures. The PAG is a powerful antinociceptive system. The anterior cingulate gyrus participates in affective responses to pain and endogenous pain control circuitry (Zasler et al., 2007). There are central nervous system (CNS) modulatory circuits that *regulate the perception of pain* (i.e., interactions between nociceptive, nociceptive afferent and other afferent fibers). That varied classes of nociceptive neurons can be integrated in a pain response suggests that it is one aspect of a *homeostatic afferent pathway. Pain inhibition* circuits originate in the PAG via serotonergic neurons of the raphe nucleus and the nuclei of the *medulla* and pons (locus ceruleus and others). Nociceptive stimulation elicits cardiovascular responses and the activation of vagal responses that suppress pain that are projected to the raphe nucleus and the rostral ventrolateral medulla, and then to the dorsal horn, which receives nociceptive stimuli (Benarroch, 1997a, p. 406). Descending fibers of the *medullary raphe nuclei* modulate nociception and also thermoregulation, sympathetic vasomotor control, and some aspects of sexual organ function (Verberne, 2003). Chronic daily headache (CDH) may be the consequence of disturbance of the pain-reducing capacity of the PAG (Couch et al., 2007).

The dependency on experience and clinical variability makes pain difficult to treat clinically (Basbaum & Jessell, 2000). Actually, the cognitive sensation of pain is under higher control by the brain than other sensations: nociceptors may fire wildly in the absence of perception of pain, or pain may be crippling even though nociceptors are silent. Pain stems from homeostatic disturbance of tissue metabolic needs (e.g., excess lactic acid) during regional and whole-body homeostatic adjustments to muscular work. Different neurons respond to contraction or serve as sensors of tissue metabolic needs (e.g., lactic acid) to drive regional and whole-body homeostatic adjustments to muscular work. While muscles produce ongoing homeostatic adjustments without the signal of pain, a large increase in activity causes muscles to ache or burn, and synchronous activation causes a painful cramping sensation (Craig, 2003). Pain may be modified by nonpainful sensory input and by neural activity within various brain nuclei.

Referred, Projected, Radiating Pain

Referred pain may be defined as pain that arises from one anatomical site but is perceived as originating from a distant receptive field ("*convergence-projection*"):

- Visceral organs to areas of the body wall
- Intracranial structures

- Deep somatic structures (e.g., muscles and joints) to areas of the body usually included in the same dermatome or myotome as the structure originating the painful input.

Whether originating in a visceral organ or a deep somatic structure, referred pain is very often accompanied by secondary hyperalgesia and by trophic changes to the three tissues of the body wall (skin, subcutaneous tissue, and muscle). These are alterations of skin reactivity, increased thickness and consistency of the subcutaneous tissue, and reduced thickness of the muscle. Pain is normally never referred from the skin; it is characteristically localized and circumscribed at the injury site (Byrne & Waxman, 2004; Vecchiet & Giamberardino, 2003).

Pain Sensitization and Transmission

Pain is a very complex sensation, modified by anatomical and chemical aspects of transmission and modulation. It is both a stimulus and a response, with sensory, affective, and evaluative aspects (Fields, 1991). Pain can be produced by a widely distributed network in the sensitization of the peripheral or central components of the CNS (medial pain system and related structures could mediate the pain response in widely distributed structures in the brain). Peripheral nociceptors may be *sensitized* by local trauma. *Dynorphins and endogenous opioid neuropeptides* are involved in antinociceptive and neuroendocrine signaling. At high concentrations or CNS pathology, they may be excitotoxic and contribute to secondary injury. A *membrane polarization disorder* (voltage-gated sodium channels) contributes to these pain syndromes: inflammatory, neuropathic, and central pain associated with spinal cord injury (Benarroch, 2007a). Spontaneous *ectopic discharges* may appear near the dorsal root ganglia. Pain is enhanced by mechanical stimulation, endogenous algesic substances, inflammation accompanied by pressure and joint movement, hypoxia, and increased catecholamine concentration (Mense & Schaible, 1993). Abnormal electrical or chemical communication ("ephaptic crosstalk") between nerve fibers (e.g., afferents or the peripheral and sympathetic nervous systems) may occur (Galer, 1994).

"*Wind-up*" refers to progressive increase of dorsal horn neurons provoked by repetitive firing of *C fibers*: small, unmyelinated slow fibers whose transmission is diffuse, poorly localized, and burning or throbbing. *A delta fibers* have faster velocities, with pain that is sharp, localized, and well defined. *Central sensitization* is a response to afferent input (increased excitability of dorsal horn neurons, increased spontaneous activity, and reduced firing threshold in response to afferent input).

Stimulus Component

Pain thresholds fluctuate through the day and over longer intervals. One reason for the poor relationship between tissue damage and pain is shift in the *focus of attention* (Svebak, 2000). An unemployed person's attention to body states is facilitated by lack of external attentional demands. Skeletal muscles are the only case where the brain exerts some degree of volitional control of the activity level of an organ. With *emotional load* there is activation of the extrapyramidal pathways from the brain, inducing tension not intrinsic to the activity, which creates circulation side effects that results in local ischemia. Yet, discomfort is seen as originating in the skeletal muscles. Pain has a *discriminative* component, which signals the location and the intensity of the nociceptive stimulus, and an *affective* component with arousal, autonomic, endocrine, and motor manifestations. These two elements involve separate and parallel pathways through nociceptive dorsal horn neurons, thalamus, and somatosensory cortex. Spinal nociceptive neurons send direct projections to the *hypothalamus* (spinohypothalamic tract), terminating bilaterally in the paraventricular ventromedial, posterior, and lateral hypothalamus. Consequently, the hypothalamus reacts to nociception with sympathoexcitation and the release of cortieotropin releasing hormone (CRH), and may also contribute to emotional responses via its connections with the amygdala, septal region, and the prefrontal and cingulate cortices. The anterior cingulate gyrus participates in multiple aspects of pain, though not sensory coding of intensity: its affective or motivational significance; cognitive components such as decision-making to select competing actions such as avoidance or tolerance. Nociceptive information is integrated centrally with information previously processed (Casey et al., 2000). With association and comparison of painful and nonpainful stimuli, adaptive processes occur (i.e., problem solving, skill development, and arousal reduction) (Hamburger & Lohr, 1984, pp. 107–108).

Specificity: The specificity of pain as a neurological phenomenon is controversial (Craig, 2003). Pain is a sensation represented by specialized elements both peripherally and centrally. It is only one aspect of the representation of the physiological condition of the body (interoception) as distinguished from fine touch (exteroceptive). It is both a feeling and a motivation, like temperature, itch, thirst, and hunger. This system provides a direct cortical image of the state of the body (i.e., a subjective representation of the feelings from the body associated with emotion). There is integration of specific pain labeled lines and convergent somatic activity in the forebrain. There is no single pain center in the brain. Spinal and trigeminal afferents project to the brainstem and thalamus, and the latter projects to many areas of the brain. These are the substrates for the motor, hormonal, emotional, and memory responses that accompany pain (Oaklander, 2003).

Homeostasis: Somatic activity creates a well-organized hierarchical system that reflects the sense of the physiological condition of the body and awareness of feelings and emotion. Since homeostasis is a process that continuously integrates all aspects of the body's condition, one has a new prospect for understanding mysterious pain conditions such as *fibromyalgia* (*FM*) (deep aches and pain) that may ensue from homeostatic dysfunction instead of direct tissue damage (Younger, 2005). The current physiological condition (e.g., after vigorous muscular activity) is transmitted via lamina 1 of the spinal dorsal horn and decussating to the contralateral lateral spinothalamic tract, then to the major homeostatic integration sites in the brain stem. The range of receptors of this information is suggested by additional projections to the ventromedial nucleus of the thalamus, the interoceptive cortex, and the anterior cingulate gyrus. In turn, lamina 1 functioning is modulated by descending impulses from the brainstem, preautonomic sources, and the hypothalamus (Craig, 2003).

Nociceptors: Nociceptors are classified as mechanical, thermal, chemical, and polymodal. Nociceptors are tooth pulp and corneal neurons, as well as virtually all sensory neurons that penetrate the epidermis. They are differentiated by the level of myelinization. The slow-conducting unmyelinated C fibers produce the diffuse, persistent sensation called "second pain" and tend to be the predominant source of *chronic and visceral pain*. Spinal pain neurons are inhibited by primary afferents and also by low-threshold mechanoreceptors. Inflammatory cells release endogenous opiates that diminish nociceptive transmission in inflamed tissues. Nociceptors appear to project ipsilaterally within the spinal cord, but there may be indirect bilateral pathways as well. After integration and processing, most pain signals cross the spinal midline in the anterior commissure to travel in the anterolateral quadrant to the reticular formation or to the thalamus.

Nociceptors have efferent activity. They release substance P and the calcitonin gene-related peptide, which contribute to vasodilation and to the sensitization of nearby nerve endings. Damaged or dead nociceptors lose contact with their postsynaptic target cells within the dorsal horn. When second-order dorsal horn neurons are deprived of their peripheral input, they increase their gain, maximizing input from the remaining peripheral fibers. In extreme conditions, this can lead to spontaneous activity or long bursts of firing in the dorsal horn neurons after minimal input from the periphery. This contributes to *hyperalgesia* and *neuropathic pain*. Skin, joints, or muscles that become damaged or inflamed are unusually sensitive to further stimuli (*hyperalgesia*). Primary hyperalgesia occurs within the area of damaged tissue. Secondary hyperalgesia occurs within 20 min and involves processes both near peripheral receptors and mechanisms in the CNS. *Acute pain* may be transmitted by the lateral spinothalamic tract to the thalamus and somatosensory cortex. *Chronic pain* may be a transmission of the medial thalamic nuclei to area 24 of the anterior cingulate cortex (ACC) and other forebrain areas. The ACC is involved in the integration of cognition, affect, and response selection. It projects to the medial thalamic nuclei and may be involved in the modulation of reflex responses to noxious stimuli through its projection to the brainstem PAG (Zasler et al., 2005). Nociceptors may fire despite sensations of pain being absent, while pain may be crippling even though nociceptors are silent. They are chemoreceptors that respond to irritant chemicals released in the surrounding tissue by noxious thermal or mechanical stimuli, or to exogenous chemicals that penetrate the skin and bind to sensory endings (Gardner et al., 2000). Characteristic sharp, well-localized pain stimuli are mediated by both small myelinated and unmyelinated nociceptive fibers. Delayed and more diffuse burning pain is mediated by different polymodal receptors (Bosch & Smith, 2004).

Visceral pain: Most peripheral nerves contain sympathetic afferents and efferents, as well as somatic axons; segmental reflexes between the two systems cause the autonomic components of pain. This empathizes *visceral pain*, including muscle guarding, changes in vascular tone, and sweating. Touch is rudimentary in many organs, while deep pain is diffuse and poorly localized. Internal tissues are sensitive to hypoxia, ischemia, and the distention of hollow organs. Acute visceral pain produces widespread *autonomic effects* (seating, pallor, tachycardia and hypotension). Pain referral from deep to superficial structures is due to the convergence of somatic and visceral neurons in the spinal cord. Referral is also based upon the anatomical consideration that incoming afferents send collateral axons several segments up and down the spinal cord in Lissauer's tract. This leads to pain in adjacent, seemingly uninvolved areas (Oaklander, 2003). Referred visceral pain is classified as with and without hyperalgesia. Hyperalgesia appears soon after the first visceral episode. In visceral tissue, *referred hyperalgia* involves muscle tissues and tends to be accentuated by the repetition of painful visceral attacks. It occurs in the damaged area, near undamaged tissue, and in other undamaged viscera (Vecchiet & Giamberardino, 2003). Sensitization of injured tissue visceral nociceptors occurs due to spontaneous activity and the lowering of the threshold so that they are excited by the organ's normal activity. Referred hyperalgesia occurs when the somatic area is stimulated (i.e., there is *viscerosomatic convergence*): there is no separate visceral sensory pathway; there is a low proportion of visceral afferent fibers compared to those of somatic origin; there are visceral afferents that are not sensory receptors, rather they are concerned with the homeostatic regulation of the internal environment (mainly activated by chemical and mechanical signals that do not reach consciousness); single visceral afferent fibers; and there is viscerovisceral convergence onto spinal cord neurons. Afferent pathways include the spinothalamic pathway and several other pathways to the hypothalamus and amygdala. *Visceral referred pain* occurs when the somatic area of referral is stimulated, somatic input summates with the ongoing visceral activity, and the threshold for pain sensations is reached more rapidly (Cervero & Laird, 2003). The nociceptors of internal organs are sensitive to stretch and noxious chemical irritants. It is mapped viscerotopically to the brainstem but not to the cortex. Nociceptive fibers travel with the sympathetic fibers and enter the spinal core at a specific segmental level along with a spinal nerve. Since the pain is associated with a particular spinal nerve, it is referred to the dermatome associated with that spinal nerve. One does not know whether the pain is coming from the skin or from a visceral organ (Richerson, 2005).

Convergence: An integrated and plastic state represented by a pattern of convergent somatosensory activity within a distributed network. It is proposed that deep in the dorsal horn (lamina V) there are convergent *WDR* cells that project in the spinothalamic tract to the somatosensory thalamus and then to the primary somatosensory cortex. Here the discriminative components of pain are processed: localization, intensity, temporal profile, quality. There is also a widespread forebrain network by way of a multisynaptic pathway through the reticular formation and medial thalamus to generate the affective/motivational aspect of pain.

The reaction of these cells (spinal dorsal horn, lamina 5) is graded mechanically (brushing hair, light touch, pressure, pinch and squeeze; noxious heat, cold, inputs from visceral, and muscle and joint tissues). The "gate control" theory refers to inhibition of pain cells through rubbing or vibration by large-diameter afferents.

TRAUMATIC BASIS OF PAIN

Trauma is accompanied by hemorrhage, coagulation, and possibly necrosis, followed by clotting in the surrounding injured tissues, then lesser degrees of injury of necrosis. Finally, there is swelling, vascular disruption causing hypoxia or anoxia, inflammation, pain, and restricted range of motion. Chronic inflammation causes proliferation of fibrous tissue to confine the injured area and provide increasing strength. In turn, this causes contraction distortion of the tissues with loss of normal function. The inflammatory response prepares for the repair process by protecting uninvolved tissues and removing the debris from cellular necrosis. There is an increase in blood flow and vessel permeability, causing swelling. This results in pressure on the nerves causing pain, and secondarily limitation of motion consequent to the pain, limited function, and edema (Fisher, 1996; Nolan & Nordhoff, 1996). However, a differential diagnosis is needed to determine whether weakness and restricted range of motion are consequent to injuries that are peripheral, spinal, or in the brain.

INTERNAL MILIEU AND PAIN

Individuals who tend to complain about pain or symptoms in their head-neck region (common after the kinds of trauma discussed in this book) have higher daily cortisol levels than others (Kugler & Kalveram, 1989). Enhanced pain is associated with pro-inflammatory cytokines, and an analgesic effect with anti-inflammatory cytokines (Uceyler, 2007).

The interaction between the nervous and immune systems (Machelska et al., 2001) has implications for the level of experienced pain. These effects are *immunomodulatory* (pro- or anti-inflammatory), *hyperalgesic* (enhancing pain), or *analgesic* (inhibiting pain). Opioids are released from inflamed tissue which attenuates clinical pain. *Nerve growth factor* (*NGF*) and *brain-derived neurotrophic factor* (*BDNF*) are proteins that mediate pain experience. NGF accumulates in the inflammatory sites caused by noxious stimuli. It is a peripheral pain mediator, upregulated in a wide variety of inflammatory conditions. NGF increases sensitivity to mechanical stimuli of viscera and muscle that even undergo spontaneous activity. BDNF has neuroprotective, pro-nociceptive, and antinociceptive properties. It is expressed in the peripheral, spinal, brain, and visceral tissues. Patients suffering from CDH show increased levels of BDNF and NGF in their cerebral spinal fluid, which is correlated with the duration but not the intensity of pain (Pezet & McMahon, 2006). Since these markers of injury and pain are neuroactive, they illustrate the internal milieu's potential to affect cerebral functioning. When *opioid-containing immune cells* reach the site of inflammation, they secrete opioids to produce pain relief (Machelska et al., 2001).

DIAGNOSTIC PROBLEMS

Pain is not always a result of direct tissue damage. It may be referred from tissues whose afferents converge on an overlapping group of dorsal horn cells that subserve other locations. Pain may arise from *sympathetically maintained pain* of sensitized mechanoreceptors. Pain may arise centrally through abnormal activity in nociceptive and non-nociceptive pathways, alteration in the spontaneous electrical activity of central pathways, changes in neurotransmitter function and receptor activity, and alterations of descending modulatory pathways. Furthermore, with time the affective consequences of pain become important in the character of the pain syndrome: depression; sleep disorders; sequelae of disuse (muscle atrophy; limited joint mobility).

Strain refers to overwork or overstretch of tissue due to excessive effort or undue force. There may be localized pain and swelling. *Sprain* refers to a partial tear with localized pain and swelling, or to an injury to the tissues about a joint in which some of the fibers of a supporting ligament are torn away from the bone (ruptured) but the continuity of the ligament remains intact (Margoles, 1999).

Repair phase: Damaged cells are replaced with either the same type of tissue or scar tissue, depending on the degree of injury. The spinal cord and nerve rootlets can be injured in whiplash injuries of the neck: a damaged disc puts compression on the cord or nerve rootlets before passing into the intervertebral foramen. Central nervous tissue becomes fibrotic or scarred. Further damage may be caused by veins leaking into the intervertebral foramen. Ligaments and tendon repairs tend to be slower and are accompanied by more scarring than muscle tissue repair (Nolan & Nordhoff, 1996).

Additional metabolic causes of pain (common to soft tissue and the brain) include the slowing of cerebral circulation (Lishman, 1987; Taylor & Bell, 1966), which results in posttraumatic symptoms. Traumatic brain injury (TBI) results in a cascade of destructive events leading to impaired cellular function and destruction, including posttraumatic migraine headaches. This is a neurovascular condition involving vasoactive peptides (neurotransmitters) and the trigeminovascular system (Packard & Ham 1999). These neurotransmitters cause vasoconstriction, vasodilation, and transmission of nociceptor stimuli to the CNS.

Neck injuries and chronic pain reflect more severe collisions and are associated with the rupture of muscles and ligaments, tearing of cervical disks, disturbance of the labyrinth, and afferent input to the hindbrain (Merskey, 1993). Pain contributes to the flexion-withdrawal reflex, which minimizes further tissue damage. It also reinforces learned avoidance, thus minimizing the probability of future encounters with the same noxious stimulus. Pain in joints, muscles, soft tissues, and the like are a signal that the area should be rested, and also a subliminal stimulus to avoid chronic postural positions leading to joint destruction (Brown, 1989). The effects of chronic pain on sensitivity to experimental pain vary with different pain syndromes, though this may be an artifact of induction technique (Lee et al., 1993). Pain can be elicited by non-noxious stimuli and be present in the absence of noxious stimuli or nerve damage (psychogenic pain). Categories of pain are associated with different fiber types and neurotransmitters (Terenius, 1987). Individual differences in response are considerable in the verbal description of the pain experience, autonomic responses, and in local segmental reflexes such as flexion (Wall, 1987).

DESCRIPTORS OF PAIN

Pain as a Sensation

Different tissues have varying pain thresholds. Katz and Melzak (2003) have suggested three psychological dimensions subserved by physiologically specialized brain systems: *sensory-discriminative* (rapidly conducting spinal); *motivational-affective* (slow-conducting spinal tracts innervating reticular and limbic structures); and *cognitive-evaluative* (CNS or higher CNS processes that evaluate input in terms of past experiences, exerting control over the discriminative and motivational systems). These systems provide perceptual information, lead to motivation to escape or attack, and cognitive experiential information leading to assessment of the outcome of different response strategies. It is sometimes difficult to describe one's pain, although the terms may suggest diagnosis and the course of therapy. Katz and Melzak (2003) describe pain rating scales, questionnaires, and a scale that differentiates critical pain qualities.

Suffering: Traumatic pain has a highly varied expression, from exaggeration or unusual complaints to being a significant stressor, or even unexpressed, such as "*belle indifference*." However, the extent of suffering (affective-motivational) is dependent on tissue damage, mood (anxiety and depression), previous emotional responses to pain, sociocultural factors, and social support. Pain is enhanced when there are problems (internal or external) preventing the patient from directly communicating distress.

Perception of pain (cognitive-evaluative) is dependent on arousal, attention, and past pain experience. Svebak (2000) emphasizes the subjective nature of pain, since it can be perceived with no definable cause, and large tissue injury may not cause severe pain.

Dorsal root dysfunction (pain and sensory impairment, local or projected in a radicular or non-radicular (referred) distribution.

Radicular pain: This has great localizing value and arises from irritation of the dorsal roots. The pain is projected to the dermatome of the nerve root (e.g., cervical and lumbosacral nerve roots) (Byrne & Waxman, 2004).

Nociceptive Pain

Nociceptive pain results from ongoing injury to skin, bone, deep somatic tissues (viscera), meninges, cerebral and cranial vessels, muscles, fasciae, and temporomandibular and other joints. Pain is enhanced by mechanical stimulation, endogenous algesic substances, inflammation accompanied by pressure and joint movement, hypoxia, and increased catecholamine concentration (Mense & Schaible, 1993). Abnormal electrical or chemical communication (ephaptic crosstalk) between nerve fibers (e.g., afferents or the peripheral and sympathetic nervous systems) may occur (Galer, 1994).

Nociceptive cells are of two types: *specific* and *polymodal* (heat, pinch, cold). Nociceptive (reception and transmission) input to the thalamus, other limbic centers, and the cortex allows integration of pain with ongoing arousal and emotional state, prior pain experience, and other learned parameters. This renders determining the ultimate source of "pain" more difficult (Elliott, 1994; Katz, 1994; Perlman & Kroening, 1990). Segmental modulation of nociceptive stimuli occurs from one dorsal root to adjacent roots. Suprasegmental modulation occurs (1) via reciprocal pathways to the thalamic relay nuclei, (2) via the pyramidal tract to presynaptic spinal cord structures that mediate output from lamina IV, and (3) via descending feedback from periventricular and periaqueductal areas via serotonergic fibers in the dorsolateral funiculus of the spinal cord. The latter triggers enkaphalin-mediated presynaptic inhibition of nociceptive afferent input to spinal cord laminae I and IV (Perlman & Kroening, 1990). Nociceptive pain originates in deep somatic tissues, such as the meninges (ophthalmic division of vagus nerve [V]), cerebral and cranial vessels, muscles, fasciae, and temporomandibular and other joints. The posterior scalp and upper neck are innervated by C1–C3, including the greater occipital nerve (C2) and the lesser occipital nerve (C3). Central sensitization and expansion of receptive fields, and the overlap between receptive fields of secondary neurons of the trigeminal nerve, and the innervation of the occiput and neck may cause referral between somatic and trigeminal sources of pain.

Neuropathic Pain

Neuropathic pain is described as severe, persistent, refractory to current therapies (Pezet & McMahan, 2006), constant, burning, and related to the sensory distribution of a sensory nerve. It refers to a major injury to the pain pathways, which become hyperexcitable and the input of which to the brain is interpreted (perhaps inaccurately) as tissue injury (Oaklander, 2003). The physiological nature of pain can change over time (e.g., from nociceptive to neuropathic pain) (Elliott, 1994). Neuropathic pain is sustained by an aberrant somatosensory processing in the peripheral or CNS (peripheral sensitivity, the dorsal root or horn of the spinal cord, or the thalamus). Spinal inhibition and excitation can play a role. Central sensitization in the dorsal horn has been experimentally produced by persistent nociceptive or neuropathic stimuli and is the likely mechanism underlying neuropathic pain. It may be precipitated by direct injury to neural tissue, formation of a neuroma, dysfunctions at the spinal or thalamic levels that may involve gene activities, changes in the balance of spinal excitation and inhibition, and so on. Central sensitization (dorsal horn) is a manifestation of neuronal plasticity and seems driven by persistent signaling of either persistent nociceptive or neuropathic stimuli. In the peripheral nervous system, traumatic injury increases the excitability of nociceptors and involves components released from nerve terminals (neurogenic inflammation and immunological and vascular components from cells in or recruited into the affected area).

MISCELLANEOUS

Jump sign is a response to an unanticipated painful stimulus, with wincing, crying out, etc.

Latent myofascial trigger point: A focus of hyperirritability in a muscle or its fascia that is painful only when palpated. It may have all the other clinical characteristics of an active trigger point.

Active myofascial trigger point: A focus of hyperirritability in a muscle or its fascia, referring to a pattern of pain at rest and/or in motion. It is always tender, prevents the full lengthening of the muscle, weakens it, usually refers pain on direct compression, and often produces specific referred autonomic phenomena in its pain reference zone.

Paresthesis: Abnormal sensations in the affected area, usually difficult to describe, with the patient often using metaphorical expressions like "crawling" or "running water" (Margoles, 1999).

Allodynia: Pain resulting from an ordinarily non-noxious stimulus to normal skin. It may be explained by the death of C fibers ("second pain") and substitution by A fibers ("first pain"), which sprout onto second-order nociceptive neurons in the dorsal horn (Oaklander, 2003). There is amplification of the thalamic, insular, and somatosensory II areas, with paradoxical reduction of regional Cerebral Blood Flow (rCBF) on the anterior cingulate gyrus (Zasler et al., 2007).

Radiation: Experienced over wide regions. Radiation of pain is associated with the stimulation of the periosteum or tendinous attachments, and also with soreness and tenderness over bony prominences (Croft, 1995b).

Hyperalgesia is an increased pain to noxious stimuli.

Hyperpathia is an abnormally exaggerated, subjective response to painful stimuli.

PAIN DEVELOPMENT

The 1979 definition by the International Association for the Study of Pain recognized its subjective quality, its relationship to experience, and the fact that it may be reported in the absence of any apparent pathological cause (Wall, 1987). The *periaqueductal gray* matter (midbrain) contributes to analgesia, which is not mediated by endogenous opioids. Its stress responses also integrate cardiovascular behavioral responses to stress (Benarroch, 1997, p. 437). *Deep pain* (muscles, joints, viscera) elicits hypotension associated with quiescent inactivity and also tonic contractions of trunk muscles. Nociceptive stimulation of deep muscle or viscera usually produces sympathoinhibition. *Superficial pain* elicits hypertension and tachycardia, which are associated with protective motor responses. Nociceptive stimulation of the skin elicits *sympathoexcitation* (Benarroch, 1997a, pp. 433–434). Assessment involves tissue damage, anxiety, and depression. Pain's mobilizing qualities are related to input to the reticular system and other subcortical structures (Hamburger & Lohr, 1984, pp. 103–104, 110). The location, intensity, and frequency of pain should be determined through the interview.

CHRONIC PAIN

Chronic pain is defined as "pain that persists beyond the usual course of an acute disease or a reasonable time for an injury to heal." This period may be as short as 3–4 weeks. Chronic pain has been defined as pain persisting for more than 6 months, although there may not be any obvious tissue damage or pathological process. Objective findings need not explain the intensity of pain and the pervasiveness of symptoms (Kraft, 1996). It has been described as a disease in itself, not always diagnosable, inducing depression, the use of narcotics contraindicated by potential addiction, its pathological origin unclear, often with complex interactions, and whose cure may be impossible. Central sensitization not occurring in the acute phase seems to be present.

Chronic pain is associated with affective responses (anxiety, depression, and anger). There may be maladaptive protective responses, restriction of activity, avoidance behavior that enhances disability, problems stemming from prolonged use of medication or medical mistakes in prescribing pain medication, and cognitive impairment. Functional imaging studies have documented the interaction of neurobiological and psychological processes in chronic pain (Martelli et al., 2007).

Some assume that chronic pain is either faked or exaggerated to obtain secondary gain. Actually, a complex interaction of components determines *treatment outcome for pain*. The literature concerning chronic pain in patients receiving *workmen's compensation* is equivocal (i.e., some showing a poorer outcome and a larger number finding no relationship between compensation and treatment response). On the other hand, there do not seem to be differences in treatment response between patients with litigation pending and those without. Damage to soft tissue (e.g., the head) may cause chronic pain that may be described as an "*unhealed wound*" that continues stress long after the injury.

Pain level is maintained by mechanical stimulation, toxic effects from tissue breakdown, ischemia due to muscle tension, and sympathetic hyperactivity (Bonica & Chapman, 1986). Chronic pain may be experienced out of proportion to the physical findings. While damaged connective tissue may heal, nervous system signals may remain unchanged. New sites of symptomatic pain may be nowhere near the original site of injury; indeed they may be bilateral without evidence for injury. It is hypothesized that aberrant noxious impulses bombard and exhaust the autonomic nervous system (ANS). This stimulation creates headaches, sleep disturbance, gastrointestinal problems, agitation, anxiety, and so on.

The psychological component, not the biological, may lead to chronicity (Grzesiak, 2002). Failure to respond to appropriate medical treatment, slow healing time, and exaggerated clinical presentation falls into the category of the chronic pain syndrome (Monsein, 2002). It is estimated that over 50 million accidental injuries occur annually, in the United States, with more than one-third associated with moderate-to-severe pain.

Maintenance of Pain

Pain level is maintained by mechanical stimulation, toxic effects from tissue breakdown, ischemia due to muscle tension and sympathetic hyperactivity (Bonica & Chapman, 1986). Patients with chronic pain often experience pain in numerous scattered areas or one large area that may be lateralized; muscles that are abnormally tender; tingling, numbness, or burning sensations; tightness and/or stiffness of the muscles; soreness; fatigue; sleep disturbance; and depression (Brecher, 1999). Chronic pain may be experienced out of proportion to the physical findings. While damaged connective tissue may heal, nervous system signals may remain unchanged. Damage to soft tissues is a significant cause of chronic pain. New sites of symptomatic pain may be nowhere near the original site of injury, and indeed may be bilateral without evidence for injury. It is hypothesized that aberrant noxious impulses bombard and exhaust the ANS. This stimulation creates headache, sleep disturbance, gastrointestinal problems, agitation, anxiety, and so on (Barr, 1992).

Sensitization of Nerves: The Cause of Tenderness and Pain

Tenderness is defined as decrease in the pressure threshold (i.e., increased pressure pain sensitivity). The intensity at which pressure changes into pain is the pressure pain threshold. Normal tissues do not generate real pain on pressure. Tenderness is the leading diagnostic sign of inflammation. Sensitization is the process by which the nerve fibers react with pain to mechanical pressure. Cell damage causes the release of substances that induce a chemical cascade and inflammation, resulting in pain and sensitivity to pressure (Fischer, 2002a).

TRIGGER POINTS AND SOMATIC REFERRED PAIN

Diagrams illustrating different areas of the body susceptible to trigger points and targets of referred pain: Jay, 1999, pp. 51–57; Mackley, 1999; Nordhoff et al., 1996a, p. 53; Smolders, 1984; Travell & Simons Trigger Point Flip Charts, 1996 (Trigger Points I and II).

Clinical vignette: Trigger point: "I was driving, and all of a sudden I got hit by a huge trailer truck." He estimated his own vehicle's speed as 20 mph and was slowing down because of the traffic. His car was rear-ended by a truck whose speed he did not know. His arms were extended by holding the steering wheel. His body was shaking in all directions. His head hit the corner of the headrest, which was not totally padded. He was asked about bruises or cuts on his head and he pointed to the right frontal area above the eyebrow. He had pain all over his body, including his neck, back, and shoulders. At the hospital, the doctor asked him to stay for observation but he declined. He is afraid of hospitals.

Summary of the medical record: Orthopedic disorders (cervical, thoracic, sacral); brain scan (bilateral occipital lobe hypodensity, possibly due to hemorrhage; also described as probably encephalomalacia of the posterior parietal occipital lobes bilaterally but most prominently on the right side); disorders bilaterally of knees and shoulders. The issue was raised as to the implication of three previous motor vehicle accidents (MVAs).

Pain: Patient complained of a sharp pain in the right temporofrontal region.

Referred pain: Pain was referred from the lower trapezius region to the sacral area, and was referred from top of shoulder to the right side of the neck and occiput. When his neck was pressed, pain was referred to the entire occiput. Dorsiflection of his right hand radiated pain up his arm to his right side. Nine months later he was still complaining of generalized pain. Both acupuncture and progressive muscular relaxation (PMR) were very effective in temporarily reducing pain. Upon observation there was a raised area on the left side of his neck, of which he was unaware, and which was tender to the touch. Pressure on the right trapezius caused a sharp pain on the left trapezius.

Trigger Points

Trigger points are a focus of hyperirritability in a muscle or its fascia and refer to a pattern of pain at rest and/or on motion specific for the muscle. It is a tender area or a hard knot in a firm band of taut muscle that is easy to palpate (i.e., a tender node of degenerated muscle tissue that can cause local and radiating pain). They also occur in ligaments, tendons, joint capsule, skin, and periosteum (i.e., restricting joint mobility). An injury to a particular tissue can produce muscle spasm in another segmentally related although uninjured tissue. This secondary source of pain can be more intense than the original injury. It is perpetuated via a sensory reflex and segmental spinal cord connections, thus firing anterior horn cells that maintain the muscle spasm (Croft, 1995b).

Trigger points have a decrease of energy-producing chemicals, decreased vascularity from compression, and relative hypoxia. Dysfunctions include decreased range of motion, weakness, and increased fatigability. The course includes nociception, perhaps with affective contribution. It creates spasm, which if prolonged becomes tonic muscular contraction, creating hypoxia via the compression of small blood vessels, followed by ischemia and the production of pain-producing metabolites. This leads to an increase of muscle pain and spasm, which in turn may stimulate central mechanisms, inducing continued muscle contraction and maintenance of the myogenic nociceptive cycle (Gay, 1999, pp. 48–49, 128–129). When the trigger point is pressed, pain or (rarely) paresthesia is radiated in a myotomal distribution. When the trigger point is pressed, the pain is referred to the same place. Common sites for trigger points with nodular indurations are along the neck, shoulder, elbow, wrist, palms, lower back, or have a more generalized distribution (Jay, 1999; Younger, 2005d, citing Taut, 1968). An *active* point is always tender, prevents the full lengthening of the muscle, weakens it, and usually refers pain on direct stimulation.

Origin: Trigger points are created or activated due to an acute injury or chronic muscle pain (e.g., sustained muscle contraction due to tension or poor posture), nonrestorative sleep patterns, nerve root compression, and physical deconditioning secondary to immobility following acute trauma and/or lack of exercise (causes muscles to be more easily stressed). The acute injury or chronic microtrauma due to sustained muscle contraction damages the muscle membrane, inducing a cascade of chemical reactions that raise the sensitivity of nociceptors. Their output converges with other visceral and somatic input, creating local and referred pain. The pain stimulates motor units via the CNS, inducing muscle spasm and splinting. If untreated, a vicious cycle of muscle pain and spasm causes decreased blood flow to the muscle, resulting in other chemical reactions with increased contraction perpetuating the pain-spasm cycle. Over time, sustained noxious metabolites build up in the connective tissue creating *localized fibrosis* (Rachlin, 2002a). Headaches can evolve during prolonged skeletal muscle contraction, leading to the accumulation of waste products and the irritation of nerve fibers in the muscles. Contributing to prolonged contraction are psychosocial stress, muscle strain originating from the neck, eyes, or jaws, macrotrauma from a stretch injury (i.e., whiplash, impaired relaxation due to fatigue or lack of sleep), low thyroid function, and so on (Spierings, 1996).

Trigger points develop in areas with the greatest biomechanical stress. Located in myofascial tissue, they are self-sustaining, hyperirritable, tender, and extremely sensitive. They are usually about 1 cm in diameter, palpable (ropy, fibrotic, hard, nodular, or taut). Many show evidence of local vasoconstriction, pilomotor erection, or cutaneous erythema. Most develop 4–6 weeks after a neck injury, starting as tender points. With added trauma or other perpetuating factors they become latent trigger points. Initially under digital pressure, trigger points refer pain without producing pain at rest or with muscle activity. The pain is local. As they progress to more pathological and ischemic states, they become active, referring pain even at rest (Nordhoff et al., 1996a).

Trauma may activate a latent trigger point. Trigger points from muscles both outside and within the head may be referred to headaches. The clinician should be alert to reduced range of motion as a sign of a trigger point, although various somatic diseases also refer pain to muscles. Pain can be referred from the shoulder and neck to the head. Some headaches that are due to *referred pain* from *trigger points* in the muscles do not follow any known root or peripheral nerve pathway (Evans, 1992b). Referred pain also stems from lesions of fascia and tendon sheath, which produce sharply localized pain (Evans, 1992b). Pain in the supratentorial compartment is *referred* ipsilaterally to the orbital, retro-orbital, or frontal regions. Pain in the posterior fossa is referred to the occipital or suboccipital region or the upper cervical region (Gennarelli, 1986).

Referred pain: There are no neurological abnormalities, and since it does not follow segmental dermatomes, it is usually not helpful in localizing an injury. It has been attributed to convergence of inputs from deep sources (muscle, joint, visceral) and cutaneous on WDR lamina V cells of the dorsal horn. Nociceptive afferents from the viscera and afferents from specific somatic areas of the periphery converge on the same projection neurons in the dorsal horn. The brain mistakenly identifies the sensation from the viscera with the peripheral structure (Basbaum & Jessell, 2000, Figure 24–3; Craig, 2003). This may develop from visceral internal organs as well as myofascial trigger points in skeletal muscle. Trigger points may manifest focal or regional autonomic dysfunction (localized vasoconstriction, diaphoresis, hyperemia after palpation, pilomotor activity, etc.). There may be increased skin temperature over an active myofacial trigger point and an area of cooling or heating of the skin in the region of referred pain (Inbody, 1996).

There appears to be two distinct interpretations of referred pain (i.e., following or not following dermatomal spinal distributions).

(1) Referred pain has been described as a spinal segmental sensitization (SSS) (i.e., a hyperactive state of a spinal segment that develops in reaction to an irritative focus that constantly bombards the sensory ganglion with nociceptive stimuli). SSS is a component of radiculopathy, overuse, and myofascial pain, including trigger point injuries (Fischer, 2002b).Whereas a tender spot can be palpated where the patient indicates the most intense pain, the trigger point also refers to a small, exquisitely tender area, but causes pain in a distant part. Referred pain zones from trigger points and dermatomes overlap. Further, myotomal (muscle segment) is consistently a presenting complaint.

The myotome is a group of muscles innervated from a single spinal segment. It is manifested by a taut muscle band that is hard in consistency, tender on compression, and has a fibrotic core (Fischer, 2002a). It offers detailed contemporary diagrams of dermatomes, underlying muscles, and narrowly defined pain zones and their coincidence with dermatomes.

(2) The distribution of pain and/or altered sensation has been described as usually not coextensive with the trigger points (Bayer, 1999). An additional description of trigger points, more liberal than Fischer (2002a) is presented by Mackley (1999, citing Travell & Simons, 1983), with diagrams of trigger points, spillover areas, and referral sites in the head, neck, shoulders and other parts of the body. With reference to headaches, pain referrals are offered for the trapezius, sternocleidomastoid, masseter, temporalis, and scalene muscles. This is consistent with this writer's observations. Each skeletal muscle has a specific pattern of referred pain (i.e., a trigger area in a muscle will produce pain in another part of the body when activated).

Clinical Example (Referred pain): A middle-aged woman was in a MVA. Her car was going 50 mph. The oncoming car swerved left so that the passenger sides collided. The woman is completely amnestic. She arrived at the hospital via ambulance with cervical spine immobilization. She has not returned to work because of head pains, lower back pain, and having started to experience memory loss after the accident. She is still having problems. Her equilibrium is off; she loses balance and falls. When she came out of the hospital she could hardly stand, she was falling frequently and had a lot of nausea. She fell four to five times a day. The frequency has diminished to one to two times a day. *Pain study*: Pressure on the right trapezius muscle was referred to the neck. She was asked to squeeze her own shoulder and she told of pain referred into her arm. Pain was also referred to the right of the vertex from the hairline towards the occipital region. This illustrates the point that the area of experienced pain may be initiated by an unhealed injury elsewhere.

PAIN SYNDROMES

Fibromyalgia

Fibromyalgia is essentially a condition of widespread pain, occurring in 11 of 18 tender spots on digital palpation (American College of Rheumatology criteria, detailed in Romano, 1999) and as axial pain over the sternum and spine (Fischer, 2002a). FM may be expressed as a separate condition (2–3% incidence) or may be comorbid with the symptoms of *chronic fatigue syndrome* (*CFS*) (0.2–0.4%). The symptoms that are comorbid with CFS are arthralgias, myalgias, other pain complaints, fatigue, nonrestorative sleep, and psychological distress. It consists of nonpalpable, multiple tender points, often associated with depression, anxiety, and sleep disturbance. An alternate explanation offered by Oaklander (2003) is that these two groups may have occult small fiber myopathies. FM patients, including those with FM comorbid with CFS, experience abnormal pain to mild pressure at multiple sites (*allodynia*) at a rate about three times as that of those with CFS alone. These groups manifest dysregulations of the hypothalamic adrenal-axis (HPA-axis) (RH; GH), CRH production, locus ceruleus-norepinephrine, and the ANS; reduced serotonin input to antinociceptive pathways from the brain to the spinal dorsal horns and enhanced markers of *tissue injury* (substance P; calcitonin gene-related peptide) are a pattern of central pain sensitization. There may be diminished thalamic pain modulation. An overview is suggested of a *neuromatrix* that generates perceptions of pain and pain behavior and comprised of the thalamus, cortex, and limbic system (Bradley et al., 2000). The use of these tender points as diagnostic criteria has been criticized. Typical symptoms include fatigue, stiffness, nonrestorative sleep, headaches, mood disorders, weakness, paresthesias, and dizziness. Unless there is a distinct disorder, neurological examination is normal.

The pathogenic role of trauma and stress is controversial (Rosenbaum & Ciaverella, 2004). In the writer's practice, more than half of about a dozen and a half cases experienced occult mechanical brain trauma or neurotoxic exposure. It was characteristic that "accidents" were not inquired about or studied in prior medical treatment. Nevertheless, although there is no clear evidence of structural abnormalities, experienced pain is characteristic. It is possible that there is central summation of pain (i.e., abnormal processing of input to central nociceptive pathways since repetitive thermal stimulation created sensations greater in magnitude, lasted longer, and were more frequently painful) (Staud et al., 2001).

Myofascial Pain Syndrome

Myalgia is a muscle discomfort involving a single muscle group. Myofascial pain syndrome (MPS) is a form of chronic pain that may arise from trauma (Jay, 1999; Margoles, 1999a, 1999b; Nordhoff, 1996a; Parris, 2002). Myofascial pain is a localized syndrome with palpable tender modules called trigger points, and is associated with pain, stiffness, limitation of motion, weakness, and sometimes, autonomic dysfunction. It is also associated with acute muscle strain or chronic muscle overuse, and occurs in isolated or regional muscles.

An *active* myofascial trigger point is a focus of hyperirritability in a muscle or its fascia, referring to a pattern of pain at rest and/or in motion. It is always tender, prevents full lengthening of the muscle, weakens it, usually refers pain on direct compression, and often produces specific referred autonomic phenomena in its pain reference zone. A *latent* myofascial trigger point is a focus of hyperirritability in a muscle or its fascia that is painful only when palpated. It may have all the other clinical characteristics of an active trigger point.

Complaints and disorders associated with MPS include pain, stiffness, fatigue, deep tenderness, muscle weakness, restricted joint motion, referred pain or numbness, autonomic symptoms, dizziness, poor balance, emotional stress and tension, endocrine disorders (hypothyroidism; estrogen imbalance), sleep disturbance including waking up with pain or stiffness, dental problems (clenching of the teeth), joint dysfunction, bladder or bowel irregularities due to nerve entrapment, joint dysfunctions caused by muscles pulling a bone (e.g., a vertebra out of alignment), or referred pain patterns that are mistaken for nerve root compression by disc herniation (Rachlin, 2002b). Muscles in which trigger points are found may exhibit stiffness, weakness, fatigue, and decreased range of motion. Muscle shortening increases pain on stretching. Patients protect these muscles by adopting poor posture with sustained contraction, which is accompanied by poor standing and sitting posture (e.g., forward head tilt, rounded shoulders) (Gay, 1999, pp. 52, 52).

Trigger point spasticity may cause nerve root entrapment in vulnerable area (Rachlin, 2001a, 2002b). MPS generally affects a single muscle in the superficial regions of the musculoskeletal system, although it may affect an entire regional muscle group. It can be referred to distant locations. Its etiology includes major trauma, extremes of temperature, repetitive microtraumatic activities (occupational and recreational [e.g., typing, guitar playing, assembly-line work]). It can be associated with chronic debility, severe stiffness, and generalized fatigue. Chronic pain that is unresponsive to treatment invites duplication of services, misdiagnoses, and patient dissatisfaction (Parris, 2002).

Other major criteria for MPS include a regional distribution, the complaint or feeling of altered sensation being consistent with the expected distribution of referred pain from a myofascial trigger point, and a palpable taut band in the accessible muscles, with exquisite spot tenderness at one point along the length of the taut band in the muscle belly. There is some degree of restricted range of motion. Since the distribution of pain does not follow spinal root pathways (i.e., it is characterized by referred pain distributions, it leads to the misapprehension that the pain is faked or exaggerated) (Bayer, 1999). The severity of symptoms may range from painless restriction of motion (latent trigger points) to agonizing incapacitating pain caused by very active trigger points. Pain may be aggravated by emotional stress and change of weather. The prognosis is poor; a multidiscipline pain-management team may be needed. Normalizing sleep is important, along with reducing tension, stress, and anxiety.

MPS is associated with cervical spine stretching, acute muscle strain or chronic muscle overuse, and occurs in isolated or regional muscles. It is a localized syndrome with palpable tender modules or bands of muscles called trigger points, and is associated with pain, stiffness, limitation of motion, weakness, and sometimes autonomic dysfunction.

Chronic pain is characterized by trigger points with *referred pain*. What identifies referred pain is the pattern or target of specific muscles and the ultimate conversion of myofascial pain into a MPS (Travell & Rinzler cited by Gay, 1999). Referred pain might be confused with radicular pain of spinal disc origin. Focal muscle injury (i.e., muscle and tendinous sprains, strains, and tears) can lead to regional myalgia (Younger, 2005d).

Pain may occur at different intervals after the time of injury, and once initiated can become progressively worse over a period from days through years. Other major criteria for MPS include a regional distribution, the complaint or altered sensation is consistent with the expected distribution of referred pain from a myofascial trigger point, a taut band is palpable in muscles that are accessible, with an exquisite spot tenderness at one point along the length of the taut band in the muscle belly. There is some degree of restricted range of motion. Since the distribution of pain does not follow spinal root pathways (i.e., it is characterized by referred pain distributions, it leads to the misapprehension that the pain is faked or exaggerated, etc.) (Bayer, 1999).

Entrapment Neuropathies

These are sensory and motor syndromes caused at anatomically vulnerable positions by physical compression or irritation of named nerves, resulting in tissue injury and swelling. They are found at specific sites along the limbs, due to the presence of bony prominences, muscle insertions, fibrous and ligamentous attachments, and demarcated canals (neck, wrists, elbows, shoulders, pelvis, knees, and ankles) (Younger, 2005b).

Complex Regional Pain Syndrome

Complex regional pain syndrome (CRPS) is accompanied by disturbances of sensory, motor, and autonomic pathways. It may occur after minor trauma with a severe inflammatory response and possible neurogenic dysregulation. Motor symptoms are weakness, tremor, dystonia, and myoclonia, usually starting in the affected extremity, but also may spread to the other side. These symptoms are assumed to be generated in the CNS. There is also dysregulation of central thermoregulatory pathways. CRPS has been offered as an explanation for central sensitization; motor abnormalities; sympathetic dysfunction with neuropathic, nociceptive and inflammatory components (Zasler et al., 2007); or the result of incongruence between motor output and sensory feedback.

Varied: Other pathological pain disorders are *phantom limb pain disorders* and *motor disorders* such as dystonia.

ENDOGENOUS PAIN CONTROL

Endogenous Opiates: Analgesia

Pain control can be modified by nonpainful sensory input (*the "gate" theory*) and *brain opioids* (morphine, heroin, codeine) that bind to opioid receptors. The analgesic action of opioids is the blunting of the distressing affective component of the pain without dulling the sensation.

Beta-endorphin mediates the suppressive effects of CRH and stress upon the hypothalamic-gonadal axis, as well as the analgesic effects upon pain perception during stress (Hammer & Arana, 2000). Endogenous opiates serve as an analgesic in prolonged exercise and injury, although there are differences between trained and unfit (stressed) individuals (Cumming, 2001). There is evidence of HPA axis involvement in acute and chronic pain (Crofford, 2002). Musculoskeletal problems (common after accidents) are presented as nonspecific pain. Often it is due to a complex constellation of factors: ergonomic, physiological, motivational and emotional; particularly in the neck, shoulders, lower back and extremities (legs and forearms).

POSTTRAUMATIC HEADACHES

"The head hurts more often without harm having been done to it than any other part of the body" (Spierings, 1996). Curiously, it has been frequently observed that headaches are stronger in those with concussive level TBI than in those with more severe brain injuries. Posttraumatic headaches (PTH) incidence appears unrelated to the severity of the head trauma or evidence of TBI elicited by CT or MRI (Haas, 1993a, 1993b).

Lesions and Posttraumatic Headaches

The typical head injury involves comorbid multiple trauma: one or more impacts to the head and body; stretching exceeding the modulus of strain of the muscles, ligaments, and tendons of the torso and spinal column; formation of scar tissue; bone injury (spinal column, torso, skull); injury to nerve bodies; inflammation causing the release of pain-inducing substances; ischemia of varying periods, etc. The etiology of acute headache may have life-threatening consequences: hematoma; vascular dissection; venous sinus thrombosis; ischemic stroke; and cervical spine fracture (Ward & Levin, 2005). A CSF leak, which may result in meningitis, is suggested by otorrhea or rhinorrhea. Since headaches have multiple origins, diagnosis involves overlapping possibilities: trauma, emotional reactions, and medical etiology. Trauma-stimulating pain may have occurred from impact to the head or may be referred from damaged muscles within the neck or shoulders. An alternate headache explanation is the tension type, which may evolve from episodic migraine or rebound effect of medication.

Headaches have been classified as to whether they have vascular components (migraine and cluster), tension type, traction and inflammatory, or whether they are secondary to some underlying disease (intracranial mass; hemorrhage; infection; inflammation; structural abnormality) (Kunkel, 1996). PTH in part represents an unhealed injury consequent to head, neck, or body impact or movement (hyperextension and hyperflexion ["whiplash"]). There may be worsening of a preexisting headache or a new form (e.g., migraine, tension-type, or cluster). Some authors exclude headaches that have been consequent to scalp injury, intracranial hematoma, or posttraumatic hydrocephalus, but these should be considered. PTH type include orgasmic cephalgia, supraorbital, muscle contraction, occipital neuralgia (secondary to neck and temporomandibular joint injury), migraine, cluster, supraorbital and infraorbital neuralgia, and pain due to scalp lacerations or local trauma (Evans, 1992a). They have variable presentations: dull, throbbing, pressing, vascular in nature, intermittent and variable in intensity, burning, sharply localized, or polar. PTH may be precipitated by change of posture, fatigue, effort, or be unrelated to any known factor.

Classification of Headaches (Head and/or Neck Trauma)

Headaches are often associated with comorbid neck injury (i.e., the multiple injuries associated with whiplash). When headaches persist long after the brain injury and their presentation changes significantly, other unrelated etiologies should be considered (e.g., aneurysm and hydrocephalus). The three superior cervical roots (upper cervical area and posterior occiput) and the face (trigeminal nerve) offer input to the spinal nucleus of the trigeminal nerve *Safety belts* are associated with an increased risk of cervical injury and *shoulder harnesses* to trauma of the greater occipital nerve (right side in drivers; left side in passengers).

The following is based upon the *International Headache Society Classification*, 2nd Ed.: Acute PTH (with moderate or severe head injury; mild head injury); chronic PTH (moderate or severe head injury; mild head injury); acute post-whiplash injury headache; headache attributed to traumatic intracranial hematoma; chronic subdural hematoma; headache attributed to other head and neck trauma (Hecht, 2004).

Persistent headaches can disrupt a patient's life through effects upon school performance, work, sense of well being, and social/family relationships. Headaches may not fit a recognizable pattern and can be associated with dizziness, blurred vision, or memory loss. The statement that the great majority of common PTH "occur in the absence of structural intracranial abnormalities or notable damage to the skull or scalp" (Haas, 1993a) is controversial. Traction, a common consequence of accidents, is a recognized but infrequent source of headaches. Hypertensive headaches are typically bilateral and often present in the occipital region (Diamond & Urban, 2002). The writer has frequently observed that unilateral headaches are associated with trigger points of the pectoral and trapezius areas. The trauma can consist of edema, tendons and joints injury, hemorrhage, and direct trauma to the nerve roots. The occipitocervical junction may undergo strain. There may be injury to the intervertebral disk, with narrowing of the foramen, and possible fibrosis and abnormal motility of the vertebral joints. *Central mechanisms* include injury-related cerebral edema, axonal shearing, perturbation of pain-related structures such as the PAG, or damage to the meninges. Inflammation related to the injury may predispose one to central sensitivity and the development of chronic pain. Additional contributors can be genetic, comorbid disease, medications or other drugs, or psychological factors (Couch et al., 2007). *Trigger points* may radiate pain into the head or down the arm (Packard, 1999).

Incidence of Headaches

The most common problem in adult patients being discharged after minor head injury and damage to the torso appears to be headaches, with incidence estimated at 30–80% (Alves & Jane, 1990; Haas, 1993a; Packard et al., 1993). Another survey indicates that headache acutely follows head injury in 50–80% of patients and continues for 1–2 years in 20–30% of patients. In the injured group, 30–50% manifest *CDH*. Most of the evidence indicates that whiplash neck injury is also a risk factor for CDH. There is evidence that the lifetime risk of *CDH* increases with *head and neck injury* (*HANI*), although CDH may not be proximate to HANI. A telephone sample for clinical trials was selected to determine the effect of HANI to the incidence of chronic headaches. CDH were defined as lasting for 15 or more days per month. *Episodic headaches* (*EH*) were defined as 2 to 102 headaches per year. Migraine with aura was found in 41% of CDH subjects and in 19% of the EH control group subjects. The most common classification of the EH group was 40% tension-type headaches. The odds ratio for men with CDH of a HANI was 3.1 (estimated attribution was 36%). The odds ratio for HANI of women with CDH was not significant, but in a similar direction of 1.5 (with an estimated attribution of 11%). CDH cases were more likely than EH to be depressed (44% vs. 21%) (Couch et al., 2007).

Chronic Headaches

The following are causes and triggers of chronic PTH: whiplash or cervical spine injury; upper cervical root entrapment; TMJ injury; dysautonomous cephalalgia; dissection of the carotid and vertebral arteries; hematoma (subdural but rarely epidural); neuromas; neuralgias; CSF leak causing hypotension; hydrocephalus or intracranial hypertension; thrombosis (venous sinus or cerebral vein); posttraumatic seizures (Ward & Levin, 2005). Prolonged headache occurs in about one-quarter of patients, with an incidence range estimated as greater in minor head injury than in more severe head injury (i.e., 30–90%). Accident victims without head injury also experience persistent headaches (Haas, 1993b). Of the subset of head trauma patients who have headaches soon after the trauma, about one-half have headaches 1 week after discharge from the hospital, and one-third experience them 2 months later (Haas, 1993a). Women have a higher incidence of PTH than men (Haas, 1993b). Children are less likely to suffer headaches, but have a higher incidence of posttraumatic vomiting (see Migraine, below) (Haas, 1993a). Children's headaches may be intermittent. Late onset headaches (i.e., after 3 months) were more common than early onset, occurring in 31% of one sample (Haas, 1993b).

Conditions Comorbid with PTH

PTH is usually accompanied by other concussion symptoms: depression and anxiety, dizziness; memory problems; weakness; nausea; numbness; diplopia; tinnitus; hearing problems; sexual problems, and often posttraumatic stress disorder. Both acute and persistent PTH may have some of the characteristics of migraine. Head injury may be accompanied by injury to limbs, torso and neck (Evans, 1992a; Haas, 1993a; Lipton et al., 1994). Trauma may activate a migraine headache.

PTH has been attributed to damage to the contents of the head, including entrapment of the sensory nerves at the site of the injury, vasodilation, or excessive muscle contraction. Similar distress may stem from varied anatomical locations (King & Young, 1990). *Dysautonomic cephalgia* is a headache due to injury of the anterior triangle of the neck or the carotid sheath. It may occur a few times per month and last for hours or days, and is characterized by severe unilateral frontotemporal headache, ipsilateral facial sweating, dilation of the ipsilateral pupil, blurred vision, ipsilateral photophobia, and nausea (Evans, 1997). Headache or tinnitus may reflect injuries to scalp, inner ear, or other noncerebral structures (Gennarelli, 1986). Dysfunctions of neuropeptides may cause cerebral vasoconstriction and vasodilation and affect transmission of nociceptive stimuli to the CNS (Packard, 1994). Pain-sensitive structures in the supratentorial compartment are supplied by branches of cranial nerve V. The infratentorial surface, infratentorial basal dura, and the proximal major arteries and veins of the posterior are innervated by nerves IX, X, and primarily by cervical nerves C2 and C3.

Late development symptoms (after several months) that are persistent and perhaps progressive and localized may be consequent to *entrapment* of peripheral cutaneous nerve branches or the dural and arachnoid membranes after a basilar skull fracture. Soft-tissue injury of the scalp or face or a laceration and its suturing may entrap a sensory nerve, causing neuropathic pain (Jay, 1999). Diastatic linear fracture of the cranial vault (particularly in children) may incorporate the arachnoid layer and cerebral vessels in the line of fracture (King & Young, 1990).

Emotional and Psychiatric Components of PTH

There is a considerable emotional component to PTH, stemming from pain, difficulty in adjusting to a limited and impaired style of life, loss of money, problems of mobility, and so on. Persistent PTH has a higher incidence in patients with some preexisting emotional distress and trauma-related personality problems, particularly concerns about ability to work. Headaches may be initiated or enhanced by frustration, anger (victimization, restriction, pain), depression, stress, tension, or anxiety. Patients with PTH exhibit more psychopathology than those with other types of headaches, while those with chronic pain exhibit more psychopathology than control subjects (Ham et al., 1994). Some patients develop headaches as a symptom after hospital discharge; these are associated with anxiety and depression. Over half of these patients claim compensation (Cartlidge, 1991). Severe headaches are associated with higher levels of unemployment, comorbid depression, and not making academic or work progress. They contribute to downward economic drift (Stewart & Lipton, 1993). Not all patients with PTH have cognitive complaints. In one sample, these were expressed by 69% of females and 55% of males (Packard et al., 1993).

VARIETIES OF HEADACHE AND NECK PAIN

Occipital Neuralgia

This refers to a posterior cervical headache with complaints of suboccipital pain radiating up the back of the head or down the neck. The greater and lesser occipital nerves are derived from the C2–C3 nerves, which penetrate various muscles (trapezius, inferior oblique, semispinalis) Entrapment can occur at the C2–C2 face joint or compression on its route, muscle spasm, or neck flexion. Pain may be referred to the retro-orbital area via the spinal nucleus of the trigeminal nerve, and also to the temporal region. This condition is described as a frequent, lancinating pain occurring every second or two, or lasting for days. It may be confused with tension-type headache or trigger point radiation (Jay, 1999, pp. 129–130).

Musculoskeletal Headache

This varies with the particular musculature but is usually characterized by a cap-like discomfort. The sternocleidomastoid muscle may refer pain retro- or periorbitally. There may be autonomic components, myofascial pain, or it may be related to the TMJ syndrome.

Cervicogenic Headache

This is common posttraumatically. It is usually unilateral but may be bilateral as well. Most patients present unilateral or bilateral suboccipital pain and secondary oculofrontotemporal pain. Dysfunction of cervical zygapohyseal joints, particularly C2–C3, may refer pain to the head. Structures sensitive to pain include joints, ligaments, periosteum of the vertebrae, all parts of the vertebral column, musculature, nerve roots and nerves, and the vertebral arteries. In addition to identified structural disorder of pain-sensitive areas, involvement of the neck should include location in an area where the neural network permits referral of pain from the cervical locus to the head and the disease or dysfunction should be verifiable. Kovacs (cited by Hecht, 2004) attributed this to compromise of the vertebral artery and nerves by muscle spasm, due to subluxations of the cervical apophysial joints. Cervicogenic headaches may evolve from head or neck trauma, has a female predominance, and is accompanied by nausea, vomiting, and phono- and photophobia. Unilateral headache pain may originate in any cervical structure or soft tissue, is unilateral starting in the neck, and is referred or spread to the oculofrontotemporal regions, which may be the site of maximum pain. It may be precipitated by certain neck movements, poor posture, or pressure over the posterior neck or occiput. There may be nonradicular neck, shoulder, or arm pain. The headache is nonthrobbing and moderately intensive. It is identified by limitations of motion, palpable changes or tenderness in neck musculature, and radiological or pathological changes in bony structure (Hecht, 2004; Jay, 1999, pp. 134–138).

Neuritic and Neuralgic Head Pain

Neuritic scalp pain occurs after blunt or penetrating injures. It may vary from numbness to lancinating pain without apparent provocation. Occipital neuralgia occurs after acceleration/deceleration injuries or direct trauma to the craniocervical junction (perhaps C2, which may be tender to palpation that replicates the pain). Pain may be referred the ipsilateral frontotemporal scalp or retro-orbitally.

Migraine Headaches

Trauma may activate a migraine headache. There may be a genetic predisposition. Stress can lead to a migraine attack, perhaps associated with repression of emotions, increased perceptions of somatic symptoms, higher sympathetic activity, anxiety and depression (Passchier & Andrasik,

1993). The attack may occur from hours to 10 weeks later. It may be preceded by an aura (depressive mood, somatic symptoms, hallucinations of smell or taste, dysfunctions of hearing, speech, sensation, vision [scintillating scotoma, homonymous hemianopia; tunnel vision, hemiplegia], or vertigo). The headache itself is accompanied by a wide range of auras (psychic, somatic, rare, hallucinations) and symptoms (headache, psychic, gastric pain, nausea and vomiting, urinary frequency, hallucinations of smell and taste, numerous neurological phenomena, gastrointestinal phenomena, reactions of the heart and vessels, sympathetic inhibition or stimulation, and disorders of various cranial nerves) (Jenkner, 1995, pp. 214–215). The pain is throbbing, usually unilateral, when coughing or bending over. The pain involves the trigeminal afferent system that is sensitized by a complex group of neuroactive substances that produces vasodilation of extracerebral vessels (including the meningeal) and accentuates the inflammatory response (Zasler et al., 2007).

Posttraumatic Tension Headache

This is caused by trauma to the neck, cranium, cranial-attached structures, or the brain. Increased excitability of the brain from repetitive and sustained pericranial myfascial input may transform an episodic tension-type headache into a chronic form. There may be an interaction between descending control over second-order trigeminal brainstem nociceptors and interrelated peripheral changes (myofascial and pericranial muscular nociceptive afferent input). Stress and negative emotional states mediated through limbic circuitry may trigger such headaches centrally. The direct role of TBI in susceptible individuals is not clear. Over a period of time, there is sensitization of nociceptive neurons and decreased activity in the anti-nociceptive system, which gradually leads to a chronic condition.

INNERVATION OF TRAUMATIC HEADACHES

Sensitive Posttraumatic Structures of the Head, Face, and Neck

The bones and brain are insensitive to pain, but their encapsulating membranes are very sensitive (i.e., the *periost* and *arachnoid* respectively). *Skeletal muscles* cause pain when prolonged contraction leads to accumulation of waste products and irritation of nerve fibers. This is caused by increased arousal (psychosocial stress, excessive caffeine), muscle strain (neck, eyes, jaw), stretch injury (whiplash), impaired relaxation (fatigue, lack of sleep), low thyroid function, and reflex contraction of a muscle due to an irritative focus of the face or neck. Due to the CNS convergence of somatic and visceral pain information pain from visceral structures is referred to the corresponding somatic area (Spierings, 1996).

Arachnoid layer: The encapsulating membrane of the brain is the arachnoid, which is very sensitive to pain and which is derived from nerve fibers that travel along arteries and veins. Thus, it is most sensitive to pain in areas that are adjacent to the blood vessels. When the course of these vessels has passed through the arachnoid they have already given off their nerve fibers so that they are insensitive to pain within the brain itself.

Cranial Afferents

The descriptions of Jay (1999) and Spierings (1996) offer specific information concerning the location of a neck or head lesion, the afferent structure innervating that region, and the referral of pain sensation to a different structure. These structures are innervated by the superior cervical ganglion (C8–T4) and trigeminal and cervical nerves. The border is formed by the tentorium, which separates the anterior and middle fossa from the posterior fossa, and extracranially by an imaginary line running over the vertex from ear to ear. In the CNS there is convergence of somatic and visceral pain information so that pain is referred from visceral structures to the corresponding somatic areas.

There are separate afferents to the anterior and middle cranial fossa (mandibular nerve) and the posterior fossa (major and minor occipital nerves), respectively C2 and C2 and sometimes C3. Pain originating in the posterior fossa causes pain referred to the back of the head and upper neck. The lesser occipital nerve (a branch of the cervical plexus) supplies the scalp above and behind the ear, and the greater cervical nerve supplies the skin of the back of the scalp up to the vertex of the skull (Dorland, 2007, p. 1265, Plate 37; Standring, 2005, pp. 515, 532). Nociceptive information from the trigeminal and cervical innervation systems of the head, face, and neck converge in the dorsal horn of the upper cervical spinal cord (*cervicotrigeminal relay*). It is here that facial or head pain refers pain to the neck, and painful stimulation of the neck causes pain in the head or face.

- *Cerebral arteries*, particularly at the base of the brain;
- *Vertebral arteries*;
- *Venus vasculature, brain surface* (dural sinuses; cerebral veins);
- *Dura mater* (adjacent to meningeal arteries and venous sinuses);
- *Cranial cavities* (venous and other sinuses; eye socket, ear, nasal, and oral pharynx);
- *Cranial structures* (skin, nerves, muscles, periosteum, arteries, veins, and skin);
- *Face*: Innervated by the *trigeminal nerve* (V) and its branches: (1) the ophthalmic nerve innervates the forehead and anterior vertex; (2) the maxillary nerve innervates the cheek; (3) the mandibular nerve innervates the anterior jaw and temple;
- Cervical/cranial zygapophyseal joints;
- *Cervical facets* (plane surface of a hard bone, where it articulates with another);
- *Cervical sympathetic plexus*;
- *Mucosal membranes* (the nasal cavities and sinuses);
- *Intraspinous ligaments*;
- *Intervertebral disks*;
- *Head muscles*: Refers pain within the head (temporalis; sternocleidomastoid; lateral pterygoid; upper and lower trapezius; medial pterygoid; masseter; frontalis, occipitalis);
- *Neck muscles*: Refers pain within the head (platysma; splenius capitus; splenius cervicus, upper and lower tip; semispinalis cervicus; multifidi; suboccipital).

Innervation and Referred Head Pain

The trigeminal and cervical nerves innervate the structures involved in headaches . The border between these two innervating systems is formed intracranially by the tentorium, separating the anterior and middle from the posterior cranial fossa, and extracranially by an imaginary line running from ear to ear over the vertex (Spierings, 1996). In addition, the neck's skin, joints, and muscles (potentially trigger points for referred pain) are also served by the glossopharyngeal (IX), vagus (X), and spinal accessory (XI) nerves (XI), as well as the cervical sympathetic trunk (Standring, 2005, pp. 554–560).

Dura mater in the anterior or middle fossa: Innervated by the *mandibular nerve* traveling along the *meningeal artery*, which vascularizes most of the dura mater above the tentorium. The auriculotemporal branch innervates the temple, the anterior ear, and the anterior jaw.

Painful stimulation results in pain referred to the temple.

Posterior fossa : Innervated by the recurrent meningeal nerves of the first three *cervical spinal nerves*. These also innervate the back of the head and upper neck (Zasler et al., 2007, Figs. 38–10, 11).

The major occipital nerve (C2) innervates the medial side of the back of the head, and the minor occipital nerve (C3) innervates the lateral surface. Painful stimulation of structures in the posterior fossa refers pain to the back of the head and upper neck.

- *Posterior neck muscle*; *zygapophyseal joints*; *interspinous ligaments*: Posterior branches of the cervical spinal nerves.

- *Anterior neck muscles; intervertebral discs*: Innervated by *anterior branches of the cervical spinal nerves.*
- *Turbinates of the lateral side of the nasal cavity, the maxillary sinus, and the cheek* are innervated by the *maxillary nerve.*
- Pain is referred to the cheek.
- *Sphenoid sinus* pain (maxillary nerve) is referred to the vertex (Spierings, 1996).

Afferent spinal nerves project to particular sections (lamina) of the dorsal horn and ipsilateral and contralateral projections rostrally. Contralateral pain occurs secondarily when pain-sensitive structures get involved (e.g., craniocervical muscles).

CLINICAL STUDY

Headaches may be caused by *peripheral nerve injuries* from scalp laceration, local tissue injury, or skull fracture. Pain experienced in the head, face, and neck has complex innervation. One may separately consider intracranial headaches, superficial headaches, and myofascial pain experienced directly and referred from neck and torso muscles.

When the patient reports unilateral headache or somatic pain, the clinician may suspect a traumatic origin. Palpation of the head may reveal a residual tender spot or a previously unidentified elevation.

Frequently, unilateral PTH seem to be referred from soft-tissue trigger points.

Cerebral Arteries

However, the pattern of innervation is different for fibers accompanying the carotid artery to the *cerebral arteries.* The ophthalmic nerve (a branch of the trigeminal), gives off fibers to the carotid artery as it accompanies it through the carotid sinus, exiting via the superior orbital fissure to the orbit, forehead, and anterior vertex. The innervated arteries are the anterior, middle, and posterior arteries, all branches of the carotid artery. These fibers offer pain innervation to the arteries and to the arachnoid of the anterior and middle cranial fossae. Convergence of somatic and visceral information in the CNS results in referral of pain from stimulation of carotid and cerebral arteries, and from the adjacent arachnoid to the ipsilateral eye and forehead. The dura mater is innervated by nerve fibers accompanying the main meningeal artery, a branch of the maxillary artery. This artery's parietal and frontal branches vascularize most of the dura mater above the tentorium. It is accompanied by the mandibular nerve, innervating arteries, the dura, the temple, the anterior ear, and the anterior jaw. Thus, painful stimulation of the dura mater in the *anterior fossa* results in pain referred to the temple.

Cerebral Veins

The *cerebral veins* enter into the venous sinuses of the dura mater. They were innervated by a branch of the ophthalmic nerve known as the tentorium nerve, which originates close to the ganglion and travels along the tentorium to the sagittal sinus in the falx. Thus, painful stimulation of the cerebral sinuses and veins are referred to the ipsilateral anterior vertex, the area of somatic innervation of the ophthalmic nerve (Spierings, 1996).

Posterior Fossa

The recurrent meningeal nerves of the first three cervical spinal nerves innervate the pain-sensitive structures of the posterior fossa . They also innervate the dura mater, the basilar artery, its cerebellar branches, and the adjacent arachnoid. The first three cervical nerves innervate the medial side of the back of the head (occipital nerve, C2) and the lateral side (minor occipital nerve, C3). Painful stimulation of structures in the posterior fossa refers pain to the back of the head and upper neck.

TEMPOROMANDIBULAR JOINT SYNDROME

The TMJ is particularly vulnerable to rear-end collisions or direct blows. It is identified by pain of the jaw precipitated by movement and/or clenching, decreased range of movement, noise during joint movements, and tenderness of the joint capsule. It includes a loose joint capsule and nonlimiting bony shape. As the head hyperextends backwards in a rear-end collision, the jaw opens on its hinge joint, reaches maximal extension, and rapidly closes, causing localized joint soft-tissue tearing and anterior meniscus dislocation (Nordhoff et al., 1996a). A blow can cause the jaw joints to be forced out of alignment or can create spasms of the muscles that operate the jaw (which is of greater etiological significance, according to Pincus & Tucker, 1985, p. 295). When the head is not in the proper position, it does not rest comfortably on the neck and shoulders, which causes headaches, muscle tension, spasms and trigger points, jaw clicking and other noises, earaches, pains in various parts of the head, various somatic symptoms, and also ringing in the ears, difficulties in swallowing, migraine-like pain of the head or face, dizziness, and neck spasms (Mackley, 1999).

Characteristics of TMJ: Pain should relate directly to jaw movements and mastication. Also, there should be tenderness in the masticatory muscles or over the TMJ on palpation, its presence and location can be confirmed by anesthetic blocking.

TMJ dysfunction: Clicking, incoordination and crepitus with mandibular movement; restriction of mandibular movement; sudden change in the occlusion of the teeth (Boes et al., 2004).

TMJ-related symptoms may not be attributed to the injury. The examiner should be alert to asymmetry that is consequent to ipsilateral muscle spasm, general tenderness, trigger points, and deviation of the jaw when opened, and also to a decreased cervical range of motion.

JOINT PAIN

Whiplash trauma can injure the TMJ, causing jaw pain often associated with headache. The *zygapophyseal* (*facet*) *joints* between vertebrae determine the range of gliding movements. Injury affects spinal nerves that emerge from nearby intervertebral foramina. This causes pain along the distribution of the dermatomes and muscular spasms in the associated myotomes. Headache may be referred from the C2–C3 facet joint, innervated by the third occipital nerve (Evans & Wilberger, 2003).

HEADACHE EXAMINATION AFTER INJURY

Pain referrals are offered for the trapezius, sternocleidomastoid, masseter, temporalis, and scalene muscles. Sheftel (1996) notes that pain is a personal experience, that patients should be believed, and that "proper treatment focuses on the pain in the context of the entire patient." Parameters to be explored, such as age of onset, frequency, location, laterality, description, duration, prodrome, associated symptoms, psychosocial history (psychological and trigger factors). Patients may be unaware of how tight they keep themselves. Interviews with collaterals can reveal the impact of pain experience upon occupational, family and academic functioning; whether the family reaction is supportive or punitive; and whether or not patients reveal their pain because others do not understand it or are critical of it (Weeks, 1996). *Palpation* of the entire head and neck is recommended, being alert to excessive muscular tenderness and pericranial tenderness. Palpation of the trapezius, paracervical and cervical spine has been recommended. Range of neck motion with pain should be studied.

Study of the occiput may be associated with tenderness of the greater and lesser occipital nerves and referral of pain to other parts of the head.

Experiencing Pain

The experience of pain may depend upon the interval after injury and such personal characteristics as stoicism, histrionics, or faking. Once initiated, pain can become progressively worse over a period ranging from days to years. Anxiety and depression can interfere with treatment (Grzesiak, 2002). The experience of pain varies over time, interacting with psychological, social and economic factors, and subjective dynamics (mood, fears, expectancies, coping efforts, resources, social support and other responses, and the impact of the pain on the patient's life). The experience of pain may or may not have visible consequences. A broader experiential study explores the adaptive reality of the patient. Since descending impulses from the brain may close down nerve transmission from the periphery, pain perception is affected by sensation and also by evaluative and affective systems.

Pain as a "Distractor"

Patient self-description: "I do have mild headaches...I just can't think straight. I heard my mother say that I'm not 'the same any more.' That sort of shook me up."

Distractors are defined as uncomfortable symptoms that interfere with adaption (i.e., activities of daily living). "Distractors" include headache, pain, hyperarousal, depression and partial seizures that impair capacity to function, balance problems, restricted motion, long-lasting illness and biological dysfunctions, changes of identity after TBI, headaches, somatic pain, partial seizures, balance problems, restricted motion, long-lasting illness, and biological dysfunctions (Parker, 1995). Consequently, the experience of stress is continued as direct discomfort, as is the experience of impairment. This has been described for chronic job stress as the "burnout" syndrome of emotional exhaustion, depersonalization, and reduced personal accomplishment (Burke & Richardson, 1996), but seems appropriate for accident-related stress as well. CRH plays a major roll in anorexia nervosa, Alzheimer disease, and possibly other psychiatric and neurodegenerative diseases (Hartline et al., 1996).

Pain contributes to dysfunction of complex activities of daily living. It has a variety of neurological and somatic etiologies, including somatic injury; referral from one area of the body to another; disturbed neural functioning at peripheral, spinal, and thalamic levels; physiological disturbance (vegetative arousal; vascular); and dysphoria by stress because of opiate depletion. Psychodynamic factors interact with pain (i.e., conversion from depression and anger). The pain experience is enhanced by emotional problems such as anger and tension. Pain can preclude treatment participation, yet analgesics to treat pain can reduce mental ability. One's style of reacting to pain may have developed during childhood due to personal experience or viewing a family member. Pain or disability can be associated with over-self-concern, learning to obtain one's needs, or attention seeking (Devine, 1999). Pain behavior includes the role as patient and is composed of attention-seeking behavior, pessimistic outlook, and learned components (belief in nonrecovery, disturbed relationship with physicians and family; postconcussion syndrome). Pain anywhere in the body can reduce efficiency (Anderson et al., 1990; Craig, 1994; Haas, 1993a, 1993b; Loubser & Donovan, 1996; Parker, 1996; Parker & Rosenblum, 1996; Perlman & Kroening, 1990).

Emotional Interactions with Pain

Psychogenic pain is observed in a variety of symptoms, including somatization disorder, conversion disorder, depression, anxiety, addiction, and attention seeking disorder with secondary gain (Loubser & Donovan, 1996). The clinician's study of pain, as well as other disorders, benefits from the knowledge of the potential disorders consequent to a particular injury and the persistence on obtaining the patient's cooperation in self-revelation. Studies have associated *pain* with *PTSD* after

MVA (Chibnall & Duckro, 1994). Increased attention to memories of unpleasant events might account in part for indelible memories, (i.e., PTSD) (Craig, 1994). Posttraumatic stress lowers the *pain thresholds* of patients suffering from physical injury (Friedman, 1990).

Mood disorders alter the evaluative component of pain. Somatic preoccupation may alter the response pain-transmission neurones so that a non-noxious stimulus is increased into the noxious range where there was no prior pain (Fields, 1991). Posttraumatic stress lowers the *pain thresholds* of patients suffering from physical injury (Friedman, 1990). Pain is also enhanced by emotional factors (e.g., inability to express feelings normally, cultural patterns, depression, and muscular tension causing stress). Its expression can be socially reinforced (i.e., for secondary gain). Pain is enhanced by depression, and by emotional stress that causes muscular tension. Pain and PTSD are moderately related, and can exacerbate each other. There is evidence that their relationship is mediated by depression (Poundja et al., 2006).

Cognitive and Conditioning

Pain expression has a learned component from familial and cultural patterns. Early development can help predispose one to experiencing a psychological component to pain. Such expressions include physically or verbally abusive parents, who may overcompensate with rare displays of affect, cold or distant parents responsive only when the child was ill; a parent who suffered from chronic pain, etc. Chronic pain may reflect a failure to cope, irrational beliefs about pain, postural problems, hyperreactivity, somatic preoccupation, and denial of psychological difficulties. A patient who was seriously ill as a child, or who had been exposed to others with chronic pain, could have learned or assimilated a repertoire of sick role and illness behaviors (Grzesiak, 2002). Pain can be conditioned to nontissue damaging stimuli analogously to fear and other emotional responses. Its presence can reinforce aversive behavior (Hamburger & Lohr, 1984, pp. 111–114).

Some chronic pain risk factors are medical history (dissatisfaction with medical care and conflicting medical opinions, lifelong history of health problems); pain behavior ("doctor shopping" and dramatic display); chemical dependency; emotional reactions; vocational (job dissatisfaction, near retirement age, threatened loss of job, unstable work history, limited education), secondary benefits; emotional stress factors (retirement, financial problems, marital discord, illness or death in family); personality factors (passive, dependent, hypochondriacal, hysteroid, perfectionistic); childhood trauma (loss in early life); chemical dependency in family; and abuse (Monsein, 2002).

Peripheral stimuli are modulated by cortical variables ("gate control") (Jamison, 2003). Pain can reinforce aversive behavior (Hamburger & Lohr, 1984, pp. 111–114). Emotional problems interacting with pain (depression; anger; tension; frustration) affect its intensity and quality of experience. Pain's effect on cognitive function goes beyond its immediate sensory-discriminative features (intensity and location). Chronic pain leads to depression and interferes with cognitive functioning. One can develop a belief that one's lifestyle is being interfered with, in which case social reinforcement can develop for pain-related behavior where maladaptive beliefs and negative thoughts contribute to pain-related suffering. The cognitive impairment of chronic pain is associated with mood changes, emotional distress, increased somatic preoccupation, sleep disturbance, fatigue, and perceived interference with daily activities that can cause chronic stress. A somatic focus and associated emotional reactions may increase the disruptive influence of pain on cognition by facilitating awareness of pain. Anxiety also reports greater affective responses to pain. Affective-motivational and cognitive-evaluative components of the pain experience may be mediated by the ACC. There can be interference effects when there are two attention-demanding tasks with limited resources. Noting that there are cognitive and emotional divisions of the ACC, competitive demand on resources may cause interference effects or reciprocal suppression between these subdivisions. Anticipation of pain symptoms that are difficult to predict is a stressor that activates the HPA axis and ACC areas, resulting in cognitive impairment (Hart et al., 2003).

Pain may be exaggerated to obtain attention or may represent an emotional conversion of depression. It increases after chronic stress (i.e., opiate depletion). Expecting pain increases anxiety, and if it is chronic, leads to depression and problems with marriage, including reduced sexual activity, the feeling of being a burden to family and friends, and reduced self-esteem (Hendler, 1990). Pain feelings can be enhanced by psychological factors such as the inability to express feelings normally, cultural patterns, depression, anxiety, and anger, as well as muscular tension causing the experience of stress. Its expression can be socially reinforced (i.e., for secondary gain). Pain is enhanced by depression and emotional stress that causes muscular tension. It can be a gross mistake to attribute headache after TBI to emotional causes, since the head is susceptible to a variety of bone and soft-tissue damage to the head and the nerve roots, as well as trauma to the neck (Parker, 1990, pp. 102–103).

Pain's presence can be conditioned to nontissue damaging stimuli analogously to fear and other emotional responses. Its presence can reinforce aversive behavior (Hamburger & Lohr, 1984, pp. 111–114).

Anxiety and depression participate in the use of injury to exaggerate the extent of complaints. While chronic pain and depression are separate interacting phenomena, patients who present with pain claim less depression and anxiety than patients presenting without pain. While patients may be unwilling or unable to confide feelings of distress, they give nonverbal messages (facial and bodily activity) to which the clinician is sensitive (Craig, 1994). Perhaps pain substitutes for anxiety and depression (Hendler, 1990, citing Pilling et al.). Patients have learned that lesser distress does not gain attention. Interaction between pain and anxiety can lead to physical decompensation and psychophysiological disorders. Mental disorders can present as pain, such as *dementias* (senile and cerebrovascular) and *personality disorders* (compulsive, histrionic, narcissistic, dependent). While the setting has more of an influence on the perception of acute pain than the patient's personality, anticipation of pain produces anxiety, reducing the pain threshold. Among the differential diagnoses to be considered is *Briquet's syndrome* (i.e., hysterical personality, high somatic concern, sociopathy and delinquent behavior) (Hendler, 1990). One characteristic of the *malingerer* is the presence of a definable lesion with attribution of a prior difficulty to a current minor injury. *Conversion* and *depression* may be expressed as low back pain. *Depression and pain.*

Depression is often associated with chronic pain, and these conditions have similar *neurovegetative symptoms*, including hypochondriasis and somatic preoccupation. Pain and depression vary diurnally. Pain is speculatively attributed to disruption of *circadian rhythms* (Healy & Williams, 1988). It promotes physical deterioration due to its disturbing effects on sleep, appetite, libido, and activity level. Apparent depletion of endorphins and serotonin reduces pain tolerance, so that even minor injury may provoke a major response (Bonica & Chapman, 1986). Fifty-five percent of one group experienced mild to severe depression after severe brain injury (Garske & Thomas, 1992), and it may be presumed that a significant proportion of the dysphoria was more than reactive (i.e., there were lesion effects) (Garske & Thomas, 1992). *Magnesium deficiency* has been associated with *chronic pain and depression* (Cox et al., 1996) and such postconcussive symptoms as *weakness and irritability* (Pagana & Pagana, 2006, p. 367).

THE MEANING OF PAIN

A Clinical Approach to Pain

Integrating the adaptive consequences of chronic pain is needed to assess the patient's status with regards to regression, permanent impairment, capacity for rehabilitation and independence, and fair compensation for injury.

Subjective components of pain: What are its qualities in terms of intensity, frequency, location, and daily variability? What moods accompany the pain? How do ongoing events and moods enhance or detract from the pain?

Patient beliefs: What does the patient believe is the cause of the pain, and is it? Is it medically reasonable, or does it interfere with recovery and treatment?

Neurotrauma: Peripheral, varied levels of the CNS, complex bidirectional pathways, and the development of over time of various neurological processes of sensitization.

Somatic trauma: There are direct injuries to a tissue; slipped discs intruding upon nerve tracts, interfering with the normal functioning of peripheral nerves through inflammation; scar tissue; crushing; and entrapment.

Psychodynamics: This is the reaction to the injured lifestyle (e.g., changed identity or the view of one's body, mind, and social status as being impaired, undesirable, and incompetent); the meaning of pain (e.g., undeserved, victimization, or a retribution for sin).

Mood reactions: Moods and affects such as anger, depression, and anxiety may enhance the intensity of pain or be created as a reaction to it. Thus, pain is a component in quality of life, an interference with rehabilitation, and an impairment to adaptive functioning.

Chronic stress: Pain contributes to dysregulation of physiological systems, interferes with sleep, and consequentially creates loss of stamina, fatigue, reduced cognitive efficiency, and vulnerability to stress-related conditions.

Coping style: The patient's reactions to being pained influences treatment response, social life, and the quality of the pained person's lifestyle and independence.

Two main dimensions of coping strategies to stressful events and trauma have been identified (Littleton et al., 2007):

- *Problem focused or emotion focused*. Problem-focused coping strategies directly address the problem (seeking information about the stressor, making a plan of action, and then planning to manage or resolve the stressor). Emotion-focused strategies manage the emotional stress (disengaging from stress-related emotions, seeking emotional support, and venting emotions). This style is not emotional regulation since it concentrates conscious processes addressing a stressor. Problem-focused strategies are more effective in controllable situations, whereas emotion-focused processes are more effective in uncontrollable situations.
- *Approach focused or avoidance focused*. Approach strategies are focused on the stressor or one's reaction to it (seeking emotional support, planning to resolve the stressor, and seeking information about it). Avoidance strategies focus on avoiding the stressor or one's reaction to it (withdrawing from others, denying that the stressor exists, and disengaging one's thoughts and feelings from it). Reliance on avoidance strategies seems maladaptive, as it is associated with emotional distress. Distress is related to age, consistent with the view that some individuals are ineffective in developing coping strategies and thus over time experience diminished energy and cognitive changes.

Activities of daily living: Severe pain reduces social enjoyment and one's capacity for participation in various domestic, professional, and pleasurable activities interferes with safety.

Pain exaggeration: Deliberate malingering for material benefits; a plea for help to overcome collaterals denial of one's difficult condition; a cry for help to overcome organizational resistance to appropriate compensation. Exaggeration can be used to achieve secondary gain (i.e., social support or other domestic advantage, or, when there has been social neglect).

Identity and Pain

Pain is associated with changes in self-perception (i.e., "I'm not the same person as before."). Pain may particularly interfere with the role of provider and with sexual performance, both of which hamper self-esteem (Fields, 1991; Nogales, 1992). Chronic pain sufferers may develop a belief in nonrecovery, become somatically oriented, lose insight into nonsomatic factors influencing pain, and becoming depressed is the most common affect accompanying pain. Pain contributes to irritability, impatience, poor compliance with treatment, and deteriorating relationships with physicians, employers, and family (Perlman & Kroening, 1990). Pain changes a marital relationship due to resentment, insufficient communication skills to cope with the change, and side effects of medication (sedation and sexual problems) (Devine, 1999). Pain and disability become a social role. Failure to find relief, difficulty in accepting the change in one's body, and difficulty in seeing new possibilities can be expected to yield to *hopelessness and helplessness* (Craig, 1994). Eventually, the patient may convince themselves that they are unable to function, focus on the pain—which increases in intensity—and become more dysfunctional while being convinced of his or her incapacity. This is termed the *chronic pain syndrome*, in which there is a lifestyle of dysfunctioning. When the pain state is undiagnosed for a long time, the patients may desperately try to prove that they do suffer pain and offer dramatic pain behavior (i.e., nearly falling upon rising, swaying from one piece of office furniture to another, or "walking funny") (Bayer, 1999). Reduced communication of pain also occurs.

Pain Behavior

Pain behavior is a reaction to chronic pain and may be found after closed head injury. It is useful for the clinician to approach the patient as experiencing both a real discomfort and also having a subjective experience influencing daily living. Pain behavior is a reaction to chronic pain. How does the patient think about and cope with the pain? Cognitive style may shape pain perception and activate the ANS, thereby increasing pain.

Some qualities of pain behavior are impaired concentration; overt communication of distress (moaning, sighing, exaggerated gestures); numerous medical consultations; the uses of devices such as cervical collars and canes; increased irritability; self-imposed limitation of activity and application for disability payments; decreased attention span; perseveration; ego-centered; easily fatigued; impaired relationships; increased dependence; increased medical contacts; impaired vocational capability; loss of anticipation; loss of initiation; anxiety and depression; precluding treatment participation; fatigue; impaired relationships; increased dependence; increased medical contacts; impaired vocational capability; anxiety and depression (Anderson et al., 1990; Kraft, 1996).

Etscheidt et al. (1995) describe three profiles:

Dysfunctional: high pain severity; interference with life; lower activity.

Interpersonally distressed: assert that their families and/or significant others are not very supportive.

Adaptive copers: experience a lower level of pain severity, lower level of adaptive distress, higher level of life control, high level of daily activity, and lower interference in their lives. Dysfunctional and interpersonally distressed patients are described as immature, self-centered, disregardful of rules, feeling they have a raw deal from life, distrustful of others for being responsible for their problems, anxious, and emotionally isolated.

ASSESSMENT ISSUES FOR PAIN

Obtaining Information

The clinician is concerned with the preexisting condition, the nature of the injury, the extent of suffering, the disability, and the inability to enjoy usual activities, as well as whether behavior appears to be appropriate or whether there is evidence of symptom exaggeration. Diagnostic caution is indicated since pain patients frequently endorse somatic complaints on psychometric tests, which may inappropriately influence the assessment of psychopathology. Negative thoughts about a pain problem may contribute to increased pain, emotional distress, decreased functioning, and greater reliance upon medication. Major psychopathology indicates poor prognosis (Jamison, 2003).

PAIN AND LITIGATION

Pain is a complex issue to evaluate within the context of causation and the outcome of an accident. Patients with legitimate injuries that are uncomfortable but adapting can be turned into people focused upon the "case" and their problems when examiners refuse to acknowledge their pain, do not examine it thoroughly, do not refer for further examination, and then the insurance companies do not offer reasonable settlements in a timely fashion. When the suffering person is involved in administrative or legal action for compensation, the examiner needs to proceed cautiously. Excessive delay in resolution of claims or inappropriate denial interacts with concern for income, debts, and disturbance of self-esteem from inability to work to create further emotional distress. The issues of malingering, eligibility for treatment, standard of evidence, reaction to developing events, and intensity of suffering are presented in detail (Hall et al., 2008). Information concerning pain assessment procedures and treatments for pain (behavioral, desensitization) is found in Martelli et al. (2007).

MALINGERING ISSUES

Malingerers would like to present themselves as honest and normal except for the problems associated with their injury (Lees-Haley, 1991). One must balance a range of credibility between two polarities—chronic pain being either faked or exaggerated to obtain secondary gain versus the patient's reaction to inappropriate denial of treatment or the validity of a claim.

Precise evaluation of pain's presence, intensity, and behavioral effect is difficult. This invites caution in rendering an opinion about response bias and malingering. The following is based upon Martelli et al. (2007):

Pain is subjective and complex. Its qualities also change with chronicity (i.e., sensory-affective, cognitive-evaluative, and behavioral). This process evolves into illness behavior. Adaptation to chronic pain requires learning to overcome innate avoidance reactions. Poor psychological adaptation may involve catastrophization rather than exaggeration for monetary or secondary gain. This is a nonphysiological coping response not characterized by symptoms of exaggeration or secondary gain.

Pain creates avoidance behavior that decompensates efficient adaptation. Reaction to pain with anxiety, distress, and avoidance may be misinterpreted as "nonorganicity" or exaggeration.

The lack of close correspondence between apparent level of injury and reported pain is illustrated by the progression from acute peripheral mechanisms to dysregulation of central pain processing mechanisms. It varies from acute pain in terms of identifiable tissue damage or neuroanatomical pathways, as well as processing in the CNS. The absence of clear pain, and the presence of emotional activation, may lead to erroneous conclusions of exaggeration.

Pain experience and the reporting of it may be influenced by allodynia, hyperalgesia, and hyperesthesia. It is difficult to disentangle the contributors to the subjective pain experience and its associated functional disability.

An enhanced report of pain can come from the tendency of patients with injury-derived chronic pain to experience physical sensations they consider to be harmful and noxious with comorbid depression, anxiety, and anger. This increases anxiety and subjective pain, with activity avoidance and functional limitations, which in turn facilitates an increased risk for chronic and disabling pain.

Markers of significant exaggeration and illness behavior include failure to comply with reasonable treatment; report of severe pain without psychological effects; marked inconsistencies in the effect on activities; poor work record and history of persistent appeals concerning awards; and previous litigation. In fact, with increasing medical-legal referrals, it is anticipated that an inflation of false positive diagnoses of malingering is to be expected.

The writers acknowledge the existence of "strong gate keeping pressures." There are issues of the validity of malingering measurement as well as false positives.

Nonphysiological Issues in Pain Assessment

What are usually ignored in the assessment of accident victims are *expressive deficits*, which are (1) the inability of the examiner to obtain all information in order to correctly assess the effects of an injury; (2) the tendency of some patients to refuse to communicate their difficulties; or (3) the patients' unawareness of their deficits. One may think, for example, of the "Spartan" mentality (i.e., family or cultural training not to complain).

Litigation as a marker of potential malingering may be a response bias by researchers. It ignores the realities of the requirement to utilize the adversary legal system to obtain treatment and compensation for injury. Injured people find it usually necessary to go through the agony and grave delay of the adversary legal system to obtain treatment and compensation for injury.

Compensation Issues

The status concerning the outcome of chronic pain in patients receiving *workmen's compensation* is equivocal (e.g., the effect of awarding compensation upon response to treatment or the outcome as to whether or not there is litigation pending). Significantly fewer moderate-to-severe medical problems occur when patients receive compensation, have litigation pending, and are employed. Successful outcome of pain treatment is related to employment status at the beginning of treatment, a greater number of visits to the pain service, and fewer prior treatment modalities. Similar outcomes were obtained with short-term and long-term treatment in Chronic pain patients with or without litigation pending. Those with or without compensation benefits do not vary from each other in such characteristics as number of nonorganic signs, likelihood of having an organic basis for their pain complaint, level of psychological distress, description of their pain, and indicators of pain behavior. Employment status, not compensation or litigation, may be the key predictor of pain treatment outcome (Dworwkin et al., 1999). Backache is associated with *muscle tension associated with feelings of hostility* and *depression*, which is associated with poor response to treatment (Mendelson, 1984). When *chronic low back pain* patients with or without in personal injury litigation were compared, there was no difference in the ratings of pain severity, pain description, or level of psychological disturbance. A review of compensation for back injuries is associated with less favorable response to treatment. The authors note the social, economic, psychological, vocational, marital, and other problems affecting physical status (Beals & Hickman, 1972). A question may be raised about the probability that those who are being compensated probably incurred greater somatic damage.

The idea, based upon presumed false claims, that patients typically resume employment when litigation following accidents is finally settled has little evidence (Dworkin et al., 1999, citing Mendelson, 1984). Since there is evidence that the longer the time the patient has been unemployed means the lower the probability of return to work, it is recommended that return to work be part of a treatment program for chronic pain. Objective assessment of the injury and psychiatric status with rapid settlement are recommended (Adams & Victor, 1989, pp. 114, 1193, 1196).

A BEHAVIORAL APPROACH TO PAIN CONTROL

Prevention of chronicity is the most important accomplishment of pain management. (Zasler et al., 2005, 2007). The hyper-responsiveness of the pain system, characteristic of chronic pain, directs attention to desensitization through multiple approaches: relaxation; imagery; changing cognitive focus and strategy; emotional desensitization of catastrophic reaction to injury, pain, fear, and trauma; graduated exposure and activity; and biofeedback (Martelli et al., 2007).

Understanding the Injury

The clinician's initial orientation is to understand the nature of the trauma and its resulting lesions. The writer finds it useful to study the injury by reconstructing the accident. What happened to the person's body? Sometimes information is elicited that is not in the records concerning the injury (e.g., single or multiple impacts to the head, torso, and limbs and self-protective musculo-ligamentous injuries). Paradoxical experience of pain may suggest a lesion elsewhere than in the area of complaint; referred pain is one example (e.g., lateralized headaches, or pain radiating to distant and apparently unrelated anatomical regions).

Understanding the Patient: Adaptation

The clinician needs to consider the adaptive reality of the patient. The experience of pain may or may not have visible consequences. A patient with a spinal column injury may require a cane or walker at all times. Another might appear to be unimpaired but experience pain during uneven movements in an automobile or on public transportation. Pain and its control during activity are very exhausting; thus, secondary impairment should be studied. Fear of pain and pain-avoidant behaviors such as antalgic gait may represent an impediment to recovery through perpetuating the painful activity and stance, as well as preventing normal restoration of function. Different social backgrounds encourage or discourage the expression of pain and discomfort. Therefore, the level of social support or the rejection of it is somewhat contingent upon the patient's style of expressing distress, as well as the parental and cultural training of the collaterals.

Cognitive difficulties make pain harder to manage, requiring more time to complete a program, as well as special attention (Martelli et al., 2007). Some clinicians utilize psychometric instruments in their assessment: varied domains of coping; health; pain; disturbances of mood, anger, and anxiety; stress; and more comprehensive personality assessment (the Minnesota Multiphasic Personality Inventory/MMPI) (Zasler et al., 2005).

Treatment Outcome

A complex interaction of components determines treatment outcome for pain. The clinician's self-assessment concerning accuracy of judgment of the intensity of pain should be humble: one study of children's capacity to feign malingering through instructions to "fake bad" on a comprehensive neuropsychological examination (with minimal guidance on how to do so) revealed that 93% of the clinical neuropsychologists diagnosed abnormality. Eighty-seven percent of these attributed the results to cortical dysfunction (Faust et al., 1988).

The complexity of the pain experience invites multidisciplinary treatment. It is supplemented by physical therapy, exercise, biofeedback, physical modalities (heat and cold; hydrotherapy, manual therapy (e.g., chiropractic), ultrasound and phonophoresis; electrical stimulation transcutaneous electrical nerve stimulation, or TENS), and cranioelectrotherapy that targets the CNS with microcurrents across the scalp. The application of electrical currents to many areas of the body for the treatment of various types of pain is presented by Jenkner (1995). Jenkner asserts that pain of psychic origin, which is associated with reduced endorphin levels in the CSF, is not treated effectively electrically nor with pharmacological nerve blocks or surgical intervention.

BEHAVIORAL PROCEDURES

Behavioral procedures are a necessary component of pain management. A biopsychosocial model addresses health and illness as occurring in interdependent biological, psychological, and social subsystems. It targets pain, the patient's reaction to it, and the value of self-control (Zasler et al., 2005). These authors describe the following behavioral treatments: patient education, biofeedback, relaxation training, operant treatment, cognitive behavioral treatment, social and assertiveness skills training, imagery and hypnosis, and habit reversal.

Relaxation Procedures Pain Management

Relaxation procedures relieve pain. Under current administrative payments for injury, it is the writer's experience that most regrettably, it is only palliative. For example, a combination of initial deep breathing (three sets of five breaths), PMR, imagery for relaxation, and instructions after initiation of PMR to gently rotate the head laterally and in the vertical plane for cervical relaxation, have proven effective for temporary major orthopedic pain relief consequent to a whiplash from a speeding truck impacting a moving car. Yet, the relief occurs characteristically for 2–3 days. This may be understood as treatment being intermittent since there are other counseling issues, or that specialized medical pain management has not been authorized.

Progressive Muscular Relaxation

Chronic pain incurs muscular tension due to bracing, posturing, and emotional arousal. Reduction of proprioceptive input reduces muscle stress and its physiological consequences as well as mental hyperarousal. Such behavior exacerbates pain and muscular discomfort in injured areas. Low back pain and limb injuries can lead to neck stiffness and tension-type headaches. The possible involvement of referred pain should be studied. Relaxation training may also reduce anxiety and enhance self-efficacy. Other forms of relaxation, imagery-utilizing procedures, hypnosis, and biofeedback merit consideration (see Martelli et al., 2007). The *relaxation response* is considered particularly relevant in mild or early disease states for some immunological, cardiovascular, and neurodegenerative diseases or mental disorders (Esch et al., 2004). Since biofeedback is often directed at muscular retraining, the practitioner should consider whether or not to focus upon the trigger point (when superficial) or the referred pained muscle.

Although PMR can be highly effective in pain reduction in a particular session, since the injured patient may also need advocacy or other topics outside of a formal pain control facility, it is hard to work regularly on pain control. Thus, after a few days the pain may return to baseline without the benefits of continuous application. Limitation of funds precludes the multidisciplinary approach required for the recommended pattern of treatment.

The writer starts PMR with a preliminary relaxation procedure (i.e., deep breathing—cautiously—to avoid hyperventilation). The writer models a deep breath (forceful enough so that the patient can hear it), holds it for a moment, then has the patient exhale deeply (forceful enough to be heard). The clinician may start with one set of five breaths and then increase one set at a time to three sets for a session. The patient is then instructed to relax each definable muscle group, starting with the scalp and head; then the face; oral cavity; neck; shoulders; arms and hand; sections of the back, thorax, and abdomen; and sections of the leg. The writer finishes the procedure with a general relaxation instruction. Particular attention, perhaps with repetition, is given in instructing the patient to relax specific painful areas.

Towards the end of the session, the writer may use imagery: "All of your tension is going into your belly. I will count to three, then it will drift upward in the form of a colored cloud. What color is the cloud? The wind is now pushing it over the horizon, and as it disappears your tension (pain) is reducing"; "You are lying on the most comfortable meadow in the world. The sun is warm, the skies are blue, you smell the grass, you can hear birds. You will relax more and more and fall asleep."

Most recently (experimentally), the writer has instructed the patient to send a "heal" signal to a chronically painful injured area. It appeared to help in reducing the pain level. CAUTION AND DISCLOSURE: This is anecdotal evidence; the procedure has not been double-blinded with governmental approval.

Additional Treatment Considerations

Much of the following is abstracted from Jamison (2003) and Martelli et al. (2007).

Education: The patient is educated concerning the basis for the pain. While avoiding exaggeration of completeness of information available, the significance of inflammation, adhesions, scarring, and muscle tension is revealed. The role of these lesions is introduced. Education about the expected symptoms and their course reduces the anxiety, selective attention, and misattribution that prolong symptoms. Allied topics include coping with stress and hyperarousal, postural training, and sleep hygiene.

Exercise: The combination of posttraumatic fatigability and a wish to prevent further injury can lead to further loss of stamina and also flexibility. For patients with chronic pain, resuming usual activities as soon as possible helps to prevent disability. Collateral activities are stretching, cardiovascular activity, and weight control. Recommended is a structured setting with monitoring and encouragement, graduated exercise with activity pacing, and persistence going beyond 6 months.

Cognitive Therapy

Our discussion of this complex procedure is focused upon pain management. Emphasized is taking personal responsibility for identifying habits and beliefs requiring change. Patients should be convinced that the treatment is relevant to their problem and that improvement is enhanced by their being actively involved. Supplements include skills acquisition, education, cognitive and behavioral rehearsal, homework, generalization, and maintenance. One approach is the substitution of positive thoughts for self-perceptions of being overwhelmed and hopeless ("catastrophizing"). Symptom magnification and rumination are to be avoided since they maintain physiological arousal, avoidance, and sickness behaviors. One's condition can be conceptualized as difficult, controllable, and more manageable. Assuming that negative thoughts perpetuate pain behaviors and feelings of helplessness, they are taught to substitute positive thoughts and procedures for active management of pain. Support reinforces success.

Operant strategies include limiting the reinforcement of sickness behavior (verbal complaints, avoidance, inactivity) while rewarding "well" behavior (controlled increased exercise, activity level, participation). Some specialized procedures are homework assignments, giving examples, organizing a daily routine, recruiting support from family members, encouraging outside activities, obtaining treatment resources, and actively following patients after treatment. Maladaptive habits are deliberately reversed: particular limb posture, articular muscular tension, stressful thoughts and the feelings that provoke them.

14 Chronic Posttraumatic Disorders of Consciousness

OVERVIEW

Behavioral components of altered consciousness include changed arousal, attention, motivation, sensory input, motor output, and cognitive and affective function (Harris & Berger, 1991; Hayes & Ellison, 1989). Lesions of the cerebral cortex and thalamus can alter the content of consciousness without altering the state of consciousness (i.e., partial seizures) (Walton, 1985, p. 640). In a sample weighted for severe traumatic brain injury (TBI), disordered arousal was observed for up to 10 years, sleep disturbance occurred in 47% of the sample, and there was significant adverse effect upon activity in two-thirds of the sample. Psychiatric illness, but not epilepsy, was associated with arousal and sleep disorder (Worthington & Melia, 2006). Dissociative phenomena (e.g., altered sense of self and amnesia) should be differentiated from reduced arousal, which can cause alterations of consciousness (AOC) commencing with head trauma that sometimes persists for years.

It is useful to differentiate between impairment due to lesions, the effect of seizures, and interictal deficits (e.g., memory and language [word finding; verbal fluency; comprehension]). In patients with TBI, problems falling asleep or maintaining sleep can exacerbate pain, cognitive deficits, fatigue, or irritability (Quellet et al., 2004). Seizures are associated with attention deficits and the slowing of cognitive processes. Subclinical discharges interfere with cognitive processes even when seizures are well controlled. *Interictal memory*: Adverse cognitive effects are greatest for phenobarbital (IQ deficits measured after six months). Phenytoin affects memory and cognition. Carbazepine and valproic have fewer cognitive side effects. Other side effects are mood changes, level of alertness, oculomotor and gastrointestinal (Brown, 1991). Frequency is greatest in early life and may be an epiphenomena or preexisting brain damage. Closed fractures are associated with a risk of posttraumatic epilepsy (PTE) in 10%, and compound fracture with a risk of 20% . The risk rises with torn dura mater (Chiofalo et al., 1986). Early focal seizures are associated with 14% poor outcome, as well as later seizures. Early posttraumatic seizures are followed by epilepsy beyond the first week in 20–25% of the patients followed for a minimum of 4 years. Seizures occurring after a week increase the risk of a subsequent seizure in the next 4 years by nearly 75% (Rosman et al., 1979, citing Jennett, 1973). Generalized seizures have 27% poor outcome (Raimondi & Hirschauer, 1986). Seizures may cause brain damage depending upon age (e.g., ischemic cell necrosis). By altering the connectivity of neuronal pathways, they increase later seizure susceptibility and impair the ability to inquire subsequent conditioned avoidance responses. Over a period of time, harmful affects depend upon the outcome measure, although the immature brain seems more resistant to structural brain damage than the mature brain (Moshe, 1987). There are numerous neuropsychological deficits consequent to anticonvulsant medication, including learning, attention, memory, concentration, personality changes, psychomotor performance, and so on (Parker, 1990, p. 308).

CHRONIC ALTERATIONS OF CONSCIOUSNESS

Clinical Example: Extended AOC—From Worker to Street Person

A middle-aged skilled construction worker fell five floors to the ground, landing on his back with confusion but no loss of consciousness (LOC). "I was trying to gain access to my apartment. I was climbing down from the roof. Somehow I managed to fall from the fifth to the first floor [a free fall]. The last thing that I remembered was falling from the roof. I did not lose consciousness in that fall. The police and ambulance attendants did not recognize that I was injured. One of the ambulance attendants found muscle tissue on one of the pickets on which I had fallen. I fell, stood up, and fell into the picket fence. I was in the hospital for about one week, during which I felt a lot of pain, confusion, and a lot of fear. The social worker told me that I did not have medical coverage [A gross error that prevented treatment and led to his subsequent life living on the street!]. I was disoriented, disorganized, I thought that because I was told that I had no coverage I was being told to leave the hospital." When the man returned to his apartment, his belongings had been thrown on the street and the door was locked.

Description of his homeless experience: "When you live on the street everyone looks at you as though you brought it on yourself." While in the hospital, denial of certain facts (e.g., his initial injury and that an intruder attempted to enter his room) initiated his belief that he could do nothing right. "I began to believe that no matter what I did or said I would be refused or it would be denied… There is a 2–3 year period that I can't remember." Whatever he was doing was wrong, and he felt vague about what he was actually doing. "It seemed to me that people were denying what I was saying, contradicting me—I felt it could be paranoia… If I was on a line, it was the wrong line… If I spoke, I spoke too loudly… I would avoid speaking to people." Eventually he was arrested for intoxication and sleeping in public. "After working all these years I changed into someone who wanted to hide from people. I began to question did it really happen or did I imagine it."

Clinical Examples: Extended AOC (Adult)

A 62-year-old man was struck by a car. He suffered left frontal contusions and a skull fracture and was comatose for 5 days. He did not remember being struck but did not exhibit long retrograde amnesia. When he woke up, he didn't know who the people were around him or why he was in the hospital, although he knew where he was. Twelve days elapsed before he knew who he was and who all the people were. Two months later, he tried to make sense of his life. He thought he was supposed to go on a trip, and would say, "I'm leaving tomorrow." He mixed up his family name with his first name.

A woman was asked whether she sees herself as not being her real self: "I can't deal with people and feel uncomfortable in my own body." She forgets words and loses her train of thought: "I'm not myself. I can't speak to people. I can't say what I want to say. Words don't come out right… The fog is lifting. I tell myself this will go away but it might not."

Clinical example: Confusion: A man goes to a place and forgets the purpose for why he went there. Once, when he was driving, he had the visual illusion that the light turned green. Another time he was looking for his cap; a friend said that he had it on his head. "It happens a lot. Even my childrens' names, I am confusing them. I was speaking to my wife about my bigger son, and called him by the name of another child. Yesterday, I was speaking about my son-in-law I called him with the name of my son."

Clinical Example: Very Late Awareness

A man was disabled when his car was struck in the rear while it was stopped at a traffic light. His head struck the dashboard, damaging his nose to the point that he needed a rhinoplasty and presumably transmitting the impact to the tips of his frontal lobes. After the accident, the man wondered why he was more introverted than he had been before the TBI; why he was not more outgoing and dynamic. He went to a college reunion and became aware that he was out of touch with his classmates. "I had lots of friends, but there I was vaguely aware that I had no friends any more." Then he answered his own question: "I realize that by asking the question it is a step forward."

"At first I had many problems. I did not know that I was injured. When I found out that I was, it took me several months to accept that. After finally accepting that I had a deficit, you [the writer] offered me your ego. You said: 'I will let you use my ego as yours, since you don't seem to have one.' I was walking around very confused and I did not realize the gravity of my impairment until you pointed it out to me. People have told me what I did in the days after the accident. I continued on to the car dealership [which was his destination] for a routine servicing, although there was major damage to both me and the car. I didn't understand what happened. When I started having seizures I knew there was something wrong... Its not easy for a brain that's damaged to be aware that there's something wrong... How I navigated my car, I don't know. It was as though I were navigating on autopilot. I have no recollection of the day at all." This man's self-description illustrates (1) posttraumatic amnesia (PTA); (2) poor judgment; (3) lack of insight; (4) recovery of insight after an interval; and (5) that therapy is the advocate in the absence of self-direction.

Clinical example: Multiple AOC and post concussive syndrome (PCS) symptoms: A 19-year-old patient was struck 15 months earlier by a brick and suffered a laceration on the left parietal frontal area. At the scene he denied experiencing LOC, dizziness, and nausea, and was assessed as alert and oriented. During the interview, conflicting information was elicited: He stated that he fell to the street and that the next thing he remembered was people around him. When the ambulance came, he felt nervous and stupid. He blanked out 2–4 times a week. A partial seizures screen (see below) indicated visual phenomena on the left and auditory phenomena on the right. He appeared to blank out on one occasion.

An epileptologist added the following information: on the way to the hospital the patient felt tingling in both legs for 3 min; straightening of the cervical spine; episodes of vertigo for seconds to minutes; a tendency to veer to the right; confusion; flashing light in his left eye for seconds; a few seconds of ringing in his right ear that occurred twice a month; occasional bilateral temporal pounding headaches that lasted 30 min and for which he does not take medication; diminished memory function; intermittent neck pain. Routine Electroencephalogram (EEG) was within normal limits, with video-EEG monitoring planned.

Memory and Seizures

There is evidence for material-specific memory deficits based on the laterality of a seizure disorder. Deficits of nonverbal memory are associated with right temporal lobe epilepsy (TLE) and deficits of verbal memory are associated with left TLE. In addition to the effect of temporal lobe involvement, the self-evaluation of memory loss is influenced by a large spectrum of subjective factors, including the effects of polytherapy and longer duration of treatment (Giavagnoli et al., 1997).

SLEEP DISORDER (INSOMNIA, HYPERSOMNIA, PARASOMNIA)

Sleep is the only normal form of altered consciousness. It is a common symptom of the postconcussive syndrome. A review suggests these complaints: initiating sleep, sleep interruptions through spontaneous awakenings, difficulty returning to sleep, increased time to function efficiently upon final awakening, nonrestorative sleep, snoring, daytime fatigue, and excessive daytime sleepiness. A study of adolescent minor head injury (MHI) patients substantiated sleep complaints. There were changes in the power spectrum for each of the predominant frequency bands in sleep, which required differing amounts of time to return to their lowest levels (Parsons et al., 1997).

Parasomnias are unusual behaviors occurring during sleep: arousal disorders (confusional arousals, sleepwalking, sleep terrors); sleep-wake transition disorders (rhythmic movement disorders, sleep starts, sleep talking nocturnal leg cramps); disorders associated with rapid eye movement

(REM) sleep (nightmares, sleep paralysis, painful erections, behavior disorders); and other disorders (bruxism, enuresis) (Chokroverty, 2004).

Cycles

Sleep is promoted by cytokines interleukin-2 [IL-2], receptor associated factor [TNF], and interferon [INF-α] and other substances. It is divided into REM and non-REM (NREM) sleep (70–80% in adults). NREM sleep is further divided into four cycles of 90–100 min: state I alpha rhythms, characteristic of wakefulness, diminish; stages II-IV are NREM sleep. In the first two sleep cycles, slow waves are more common in stages II and IV, but are uncommon or missing in later cycles (Chokroverty, 2004). Sleep disturbance, which is common in head injury and emotional stress, effects endocrine secretion (see Parker, 2001, Chapter 7, for review of the pacemaker).

Characteristic Sleep Disorders

Difficulty in falling asleep (sleep onset insomnia); frequent or sustained awakening (sleep maintenance insomnia); staying awake; change of diurnal sleep patterns; bad dreams; nightmares; early morning awakening, daytime sleepiness (nonrestorative sleep). Interruption of circadian rhythms may disturb cortisol production (Merola et al., 1994). Obstructive sleep apnoea may result from a functional or structural alteration in the brainstem respiratory control system (Chokroverty, 2004). Hypersomnia (daytime sleepiness) is common in adults with brain injuries. It is associated with sleep apnea-hypopnea syndrome, periodic limb movement disorder, and posttraumatic hypersomnia. Some subjects seem unable to perceive their hypersomnolence and therefore do not report it (Masel et al., 2001). Sleep disturbance (difficulty falling asleep and waking up) and hypersomnia (increased sleep duration and daytime naps) occur less frequently. Differential diagnosis is needed for disorders such as narcolepsy, sleep apnea syndromes, periodic limb movement disorder, depression, and posttraumatic hypersomnia. REM sleep disturbance is reduced in patients with EEG changes after TBI (Granacher, 2003). Posttraumatic sleep disturbance is common and is exacerbated by pain.

Daytime drowsiness: Hypersomnia may occur post-head trauma. It is a contributor to motor vehicle accidents (MVAs) (particularly rear-end collisions) and to industrial accidents. Sleep attacks should be differentiated from altered awareness originating in hypoglycemia, cardiac problems, and hypotension (Adams & Victor, 1989, p. 273ff; Parkes, 1991). Sleep disorder is a component of depression and raises the issue of suicidal risk.

Terrifying dreams or nightmares: These occur during the period of REM sleep. When awakening occurs, the person, although anxious, has a clear sensorium and recalls the details of the dream. Thus, it is considered *state dependent*. Nightmares may occur acutely or chronically after trauma (Broughton, 1989).

Sleep terrors, night terrors, pavor nocturnus: Sleep terrors are associated with change in arousal state (i.e., from slow wave sleep [SWS] to awakening), the greatest change that a human can experience. Thus, they are *change of state dependent*. They occur during NREM sleep or SWS. The sleeper awakes with a scream and appearance of terror, cannot be consoled for some time, and recollects difficulties in breathing and moving (feels "locked in") but has none or minimal recollection of mental activity. Predisposition includes a history of sleepwalking, sleep deprivation, sedatives and hypnotics, mild head injury, and daytime stress (Broughton, 1989). Sleep terrors decline with age (along with delta sleep and associated delta sleep conditions such as enuresis and somnambulism) and can be suppressed by benzodiazepines, which suppress delta sleep (in contrast to REM and its associated dreams, which neither decline nor are suppressed by the same drugs) (Kelly, 1991b).

Hypnagogic Hallucinations: Normal individuals falling into NREM sleep may have sensations, illusions, and hallucinations only recalled if they awaken at that time. Hypnagogic hallucinations may be terrifying and require differentiation from narcoleptic phenomena (Broughton, 1989).

REM sleep behavioral disorder: Individuals who generally older awaken from REM sleep showing signs of aggression that are not present during waking hours (violent kicking, punching, diving out of bed, ambulatory collisions with furniture and walls, and other wild behaviors). The patients often have brainstem lesions (Broughton, 1989).

Narcolepsy: The primary symptom is excessive daytime sleepiness. There are unanticipated sleep episodes that last from seconds to minutes and occur during reduced environmental stimulation, including driving a motor vehicle, attending classes or meetings, etc. The attack may be accompanied by diplopia and blurred vision. While most cases are idiopathic, and there is a strong genetic basis. It has been found in abnormalities of the diencephalon, hypothalamus, and pons.

Associated symptoms (which can be frightening) include *sleep paralysis* (awakening from a dream with total body paralysis except breathing and eye movements), *hallucinations* on sleep onset or awakening (which may be associated with sleep paralysis, feelings of oppression and dread), and *automatic behavior* (driving distances without awareness, inappropriate statements, bizarre writing, shoplifting). One half of narcoleptics report having fallen asleep while driving, and one-third report sleep-related automobile accidents. Daytime sleepiness disturbs family relationships, results in irritability, and can be mistaken for laziness, depression or avoidance behavior (Mahowald & Schenck, 1989).

Persons complaining of fatigue from insomnia report 2.5 times as many MVAs as those with fatigue from other causes. This may be due to increased drug or alcohol use. Curiously, insomniacs are no sleepier than age-matched control subjects. This may be due to *hyperarousal* or impaired perception of sleep (overestimating sleep latency or underestimating total sleep time) (Chokroverty, 2004). A general quality of sleep disorders is observed in a majority of PCS patients, as well as those with moderate-to-severe TBI. Its determinants may vary from neurological etiology early on to psychosocial etiology later on. Insomnia has been attributed to a variety of lesions in addition to the familiar medullary projections: subthalamic; thalamic lesions that interrupt rostral inhibitory connections to the waking system; and occipitocervical anomalies (Autret et al., 2001).

Sleep disturbance is reported in one-third of the general population. It may be described as *delayed* (3–6 AM) or *advanced* (8–9 PM) (Roth & Roehrs, 2000). A study of sleep quality determined that 31% of the variance was accounted for by particular measures of cognitive performance (14%), injury severity, and gender (Mahmood et al., 2004). Insomnia often interferes with job performance and interpersonal relationships, with chronic symptoms creating a risk of major depression. Yet, formal cognitive and motor skill tests generally do not objectively indicate impairment. Sleep disturbance is associated with depression and old age. It is a predictor of mortality risk (Irwin & Cole, 2005). Some frequent disorders include difficulty in falling asleep (sleep onset insomnia); frequent or sustained awakening (sleep maintenance insomnia); staying awake; change of diurnal sleep patterns; bad dreams; nightmares; early morning awakening; daytime sleepiness (nonrestorative sleep); and an irregular sleep-wake pattern (ISWP). Differential diagnosis is needed for disorders such as narcolepsy, sleep apnea syndromes, periodic limb movement disorder, depression, and posttraumatic hypersomnia. TBI patients with insomnia tend to overestimate their sleep disturbance when subjective measures (sleep diary) are compared to objective ones (two nights of polysomnography). When patients using psychotropic medications are excluded and the results of normal sleepers were compared, the findings resembled those of patients with either primary insomnia or insomnia related to depression. There were more awakenings lasting longer than 5 min and a shorter REM latency (Quellet & Morin, 2006). In this study, it was not associated with gender, education, age, or interval since injury (Fichtenberg et al., 2000). Posttraumatic sleep disturbance is common and is exacerbated by pain. TBI patients have more insomnia and pain complaints than non-TBI neurological patients. Poor sleep maintenance was the most common sleep problem; those subjects without pain reported more sleep complaints than non-TBI patients. Mild TBI patients report significantly more pain than those with a moderate or severe injury. Patients with mild TBI also report more insomnia than those with moderate or severe injury (Beetar et al., 1997). In a sample of children (with a mean of 8.3 years and an outcome 1–2 years posttrauma), 6% reported sleep problems (Klonoff et al., 1995).

Sleep-Wake Disturbance of Immune Function

Sleep and sleep loss affect immune function. It appears that short-term sleep loss may enhance host defenses whereas long-term sleep loss is devastating. Since the immune system has an ongoing influence on the nervous system, disturbances of the circadian rhythm are of concern (Williams et al., 2001).

Symptoms: Disruption of social patterns (i.e., connection between events or between particular responses and outcomes) may remove the social cues to which circadian rhythms are attuned (Healy & Williams, 1988). In women, an approximately 90-min ultradian rhythm is present, with larger bursts of gonadotropin secretion occurring during sleep than during the day (Reichlin, 1998). There can be a *delayed sleep phase syndrome (DSPS)* with chronic inability to fall asleep.

Circadian Rhythm and Neurobehavioral Disorder

A variety of behavioral disturbances (e.g., sleep and appetite) may be attributable to disorganization of the circadian rhythm. When neural control of gonadotropin regulation is perturbed (e.g., after stress), the amplitude of pulses is initially reduced and, if severe, ultradian cyclic rhythm may be lost completely. If stress occurs prior to puberty, puberty may be delayed indefinitely. The fundamental change during puberty has long been thought to be due to a reduction in tonic hypothalamic inhibition of Lutein hormone releasing hormone (LHRH) release. Sleep and stress affect growth hormone (GH) secretion since maximum secretion occurs at night. Head trauma may cause isolated GH deficiency or multiple anterior pituitary deficiencies. Symptoms include disruption of social patterns (i.e., connection between events or between particular responses and outcomes) and may remove the social cues to which circadian rhythms are attuned (Healy & Williams, 1988). In women, an approximately 90-min ultradian rhythm is present, with larger bursts of gonadotropin secretion occurring during sleep than during the day (Reichlin, 1998). There can be a DSPS with chronic inability to fall asleep at a desired time (Quinto et al., 2000).

Insomnia

Insomnia refers to an inadequate amount of sleep or impaired sleep, to complaints of difficulty in initiating or maintaining sleep, to sleep that does not restore or refresh, and to the impairment of daytime functioning. Hypersomnia is excessive sleepiness or sleeping and has a variety of psychological or physiological causes. Insomnia can create mild deficits similar to TBI (Rees, 2003). In post-acute TBI patients, insomnia is associated with both depression and milder brain injury.

Insomnia, as defined here, is frequently found in TBI patients and manifests itself difficulty falling asleep, staying awake, early morning awakenings with an inability to resume sleep, and nonrestorative sleep. The patient cannot initiate sleep within 30 min or maintain sleep without nocturnal awakening that occurs at least three times weekly for a month, resulting in daytime impairment. This is manifested as fatigue, sleepiness, confusion, moodiness, or cognitive impairment. Individuals with TBI may complain of impairing sleep-related problems that do not meet the above definition of insomnia. *Treatment* of insomnia may be behavioral (instruction in sleep hygiene, cognitive-behavioral therapy) or pharmaceutical (benzodiazipine sedative hypnotics, nonbenzodiazipine sedative hypnotics, antidepressants, melatonin agonists). In prescribing medication, one considers:

- The specific symptom of the patient;
- The general characteristics of TBI (e.g., impaired cognition);
- Optimum treatment time;
- Side effects of a particular medication, such as long half-life, psychomotor, sedation, impaired memory, dizziness, falls, motor vehicle collisions, potential for abuse (physical dependence, tolerance, rebound insomnia), etc. (Flanagan et al., 2007).

Physiological Aspects

During sleep, respiratory and temperature control and sensitivity, as well as sensorimotor functions, are diminished. This contributes to insomnia and hypersomnia (Hobson & Pace-Schott, 2003). Physiological sleepiness is both homeostatic (prior wakefulness determines the propensity to sleepiness) and circadian (timing, duration, and characteristics of sleep). The sleep debt is aided by sleep, but an exact number of hours are not needed to repay it. Sleep is promoted by cytokines (IL-2, TNF, and INF-α) and other substances. It is divided into REM and NREM sleep (70–80% in adults). NREM sleep is further divided into four 90–100-min cycles. In Stage I, alpha rhythms are characteristic of wakefulness diminish; Stages II–IV are NREM sleep. In the first two sleep cycles, slow waves are more common in Stages II and IV, but are either uncommon or missing in later cycles (Chokroverty, 2004). REM sleep disturbance is reduced in patients with EEG changes after TBI (Granacher, 2003).

Health Disorders

A variety of dysfunctions are reported due to sleep disorders, even after short periods of disrupted sleep. Cognitive hyperarousal predicts poor sleep quality, regardless of pain severity: Racing thought, intrusive thoughts, depressive cognitions, and worry. Compromise of immune, metabolic, and neuroendocrine functions, with chronic activation of the HPA axis. There are increased levels of evening cortisol, glucose, and insulin, as well as increased insulin resistance. Sleep disturbance increases the level of proinflammatory cytokines and may be the partial cause of inflammation in depressed people. There is bidirectional enhancement between pain and sleep disturbance. Sleep problems may reduce ability to cope with chronic pain. The consequence is increased mortality risk (Kendall-Tackett, 2007).

SLEEP SYNDROMES

Obstructive Sleep Apnea

This may result from a functional or structural alteration in the brainstem respiratory control system (Chokroverty, 2004). In a sample of consecutive patients admitted for rehabilitation after TBI, sleep-related breathing disorders were found in 36% of cases (Webster et al., 2001). In a sample of ten TBI patients with hypersomnia, three had hypersomnia before the injury, two of whom had been driving a car at the time of injury. Also found were obstructive sleep apnoea and upper airway resistance syndrome (Castriotta & Lai, 2001).

Daytime Drowsiness

Hypersomnia (daytime drowsiness) is common in adults with brain injuries. It is associated with sleep apnea-hypopnea syndrome, periodic limb movement disorder, and posttraumatic hypersomnia. Some subjects seem unable to perceive their hypersomnolence and therefore do not report it (Masel et al., 2001).

Daytime impairment of sleep loss is difficult to quantify. Sleepiness may not be proportional to sleep deprivation; subjective descriptions may have a different meaning between patients; sleepiness may affect judgment analogous to ethanol so that subjective awareness of cognitive and motor loss is reduced; and patients may be reluctant to admit it since it subjects them to criticism for lacking motivation. Sleepiness is related to issues of safety, work, and school performance, as well as of social and family life. The issue of differential diagnosis between fatigue and sleepiness arises in such medical conditions as fibromyalgia, chronic fatigue syndrome, and endocrine disorders (hypothyroidism or Addison disease).

Circadian rhythm sleep-pattern disorder (CRSD) represents the condition where the sleep-wake rhythm mismatches the 24-h environmental and social schedule. The influence of TBI upon the circadian clock seems variable. Circadian rhythm and sleep-pattern disorders (*sleep phase syndrome*) are reported after minor TBI using actigraphic recording (Quinto et al., 2000). On the other hand, there was no significant change in dim light melatonin onset and habitual sleep time when TBI and control groups were compared (Steele et al., 2005). The midbrain and upper pons region contains the *ascending reticular activating system* (*ARAS*), whose damage might explain the postconcussion syndrome's sleep-wake, attention, and concentration disturbances (Ward & Levin, 2005). Subjective reports indicate that the sleep difficulties occurred only after head trauma. Subtypes are the *DSPS* and the *ISWP*. DSPS subtypes include delayed melatonin and body temperature rhythms and a non-24 h sleep-wake pattern with abnormal melatonin rhythm. Since 36% of one sample with "mild" TBI and insomnia had a CRSD, it might be relatively common in this type of patient. Misdiagnosis as insomnia might lead to the prescription of hypnotic medications. Since they actually exhibit normal sleep duration and efficiency with self-chosen schedules, hypnotic drugs would help patients to fall asleep but would not normalize the sleep-wake cycle. Melatonin or bright lights to synchronize the sleep–wake cycle with the environmental dark-light cycle were considered more appropriate (Ayalon et al., 2007). Trauma appears to interfere with circuits that control the circadian rhythms. Norepinephrine tracts (locus ceruleus) course anteriorly from the brainstem and curve around the hypothalamus, the basal ganglia, and the frontal cortex (Silver et al., 1991). A variety of behavioral disturbances (e.g., sleep, appetite, etc.) may be attributable to disorganization of the circadian rhythm. When neural control of gonadotropin regulation is perturbed (e.g., after stress), the amplitude of pulses is initially reduced and, if severe, the ultradian cyclic rhythm may be lost completely. If stress occurs prior to puberty, puberty may be delayed indefinitely. The fundamental change during puberty has long been thought to be due to a reduction in tonic hypothalamic inhibition of LHRH release. Sleep and stress affect GH secretion. Maximum secretion occurs at night. Head trauma may cause isolated GH deficiency or multiple anterior pituitary deficiencies. Symptoms include disruption of social patterns (i.e., connection between events or between particular responses and outcomes) and may remove the social cues to which circadian rhythms are attuned (Healy & Williams, 1988). In women, an approximately 90-min ultradian rhythm is present, with larger bursts of gonadotropin secretion occurring during sleep than during the day (Reichlin, 1998).

DSPS is a CRSD occurring after TBI, which may be underdiagnosed in the PCS. It refers to a chronic inability to fall asleep at a desired time in order to meet a required schedule, lasting for more than 6 months, and normal sleep patterns when on a normal schedule. Since the human internal clock is more than 24 h, a small daily phase advance is necessary to entrain to the earth's 2-h day. Sleep onset is increasingly delayed. It may be managed by delaying the patient's bedtime until the desired synchronization of times is obtained (*chromotherapy*) or by exposing the patient to bright light in the morning (*phototherapy*). If the practitioner ignores DSPS, other PCS symptoms may resist treatment until sleep disturbances are managed (Quinto et al., 2000) (e.g., headache, emotional distress, and cognitive impairment) (Ayalon et al., 2007). In one case, physiological markers of the sleep-wake rhythm were disturbed: plasma melatonin, body temperature, sleep architecture/EEG, wrist activity (Nagtegaal et al., 1997). In another study, all patients with DSPS exhibited a 24-h periodicity of oral temperature rhythm, but those with an *ISWP* lacked this daily rhythm (Ayalon et al., 2007).

DISSOCIATION

Dissociation is a psychiatric term (see American Psychiatric Association, 2000, pp. 519–533), with symptoms that overlap those of altered states of consciousness immediately following trauma. Dissociation is defined as a disruption in the usually integrated functions of consciousness, memory, identity, and perception of the environment. Dissociative disorders were originally considered as classic hysteria (Freud). Janet offered an alternative explanation (Kihlstrom, 2001). Such conditions as hysteria and multiple personality were described as "dynamic illnesses" caused by an idea or suggestion originating in a psychological trauma of an unknown nature to the victim. Consequently, certain experiences, thoughts, and actions became separated from the monitoring and controlling functions of the central executive ego. Janet has asserted that the elementary structures of the mental system were "psychological automatisms." These were complex acts, finely attuned to the external and internal environments, which were preceded by an idea and accompanied by an emotion. The repertory of automatisms was considered to be bound together by a single and united stream of consciousness that was accessible to awareness and voluntary control. After trauma, one or more automatisms would be split off and function outside of awareness, be independent of voluntary control, or be both (i.e., dissociation). Hilgard's conception was that integration and organization of the individual control structures can be disrupted, producing divided consciousness. They might not be linked with each other or the link with the executive might be cut. Alternatively, many mental and behavioral functions are performed automatically by specialized cognitive modules. Dissociation does not reflect the imposition of an amnestic barrier, but is rather the failure of these modules to be integrated by higher levels of the system (i.e., *executive control structures* associated with the frontal lobes).

Current concepts regard dissoccation as separate disorders (DSM-IV-TR) or conditions with similar underlying mechanisms (ICD-10, Lowenstein & Putnam, 2005, p. 1845). The emphasis for the PCS is an alteration of consciousness, probably as a result of severe stress and/or anxiety, although possibly as comorbid with TBI for brief or extended intervals. In my clinical experience, the various subtypes are not characteristic of persons experiencing PCS. I have observed patients whose clouded consciousness cleared years later without any apparent reason.

Etiology and comorbidity include combat stress, civilian stress, group traumatization, the issue of false memories with claims against psychotherapists for malpractice, physiological and EEG characteristics, side effects of medications or illicit drug use, stress release of glutamate, brain structural changes associated with PTSD, issues of the differential diagnosis between PTSD and borderline personality, discrete behavioral states, sexual molestation and brutality by family members during childhood, physical assault, and comorbid psychophysiological, somatoform, and conversion symptoms. The clinician is alerted to malingering, factitious amnesia, factitious disorders, and suggestive influences. In balance, it is established that there are unusually high rates of trauma in dissociative disorder patients, though weaker association for depersonalization disorders (Lowenstein & Putnam, 2005). Dissociation may occur as a component of immediate PTSD or as an aspect of partial seizures (Paraiso & Devinsky, 1997). There may be a sudden or gradual alteration in mental integration (i.e., identity, memory, or consciousness). Amnesia for the recall of events can occur during the dissociative state, as can depersonalization, derealization, autoscopy, and/or personality changes. Capacity for dissociation occurs along a continuum and is enhanced by childhood trauma. Dissociation is a defense that insulates the person from overwhelming experiences that may generalize and thus become maladaptive. People with multiple personality disorder, a dissociative phenomenon, may have episodes of "time loss" or blackouts for complex activities, analogous to absence seizures.

DISSOCIATION: CHRONIC PHASE

Dissociation: Rorschach Percepts

Reenactment of trauma (*derealization*): A woman who watched while a car struck her own stopped car: "[Like a] tidal wave coming over an island."

Depersonalization (*altered identity*): A woman who doesn't feel like the same person: "I feel like a caricature."

Clinical example: *Depersonalization* (*believing that one is dead*): A woman was threatened by an oncoming car, extended an arm to protect herself and suffered back injuries that created persistent pain. It took a long time after the accident for her to realize that she was still alive: "For many weeks I thought I was dead. I was confused. Am I alive or dead? 'Till now I still dream about the car. I have interacted with people; that's when I know I am not dead... They wanted me to go to the hospital.

I didn't want to go because I didn't know if I was alive or not. I was in shock. I saw myself flying." Perhaps contributing to the depth of her experience was an event that occurred when she was 7–8 years old. She had not gone to school in the usual car because her mother had just given birth. This car met with an accident, killing several people and causing a man whom she knew well to lose an arm.

Clinical example: Wide range of dissociative symptoms: The patient was a middle-aged man who, beginning from when he was 8 years old, had three head injuries and experienced a 37-point loss of mental ability from the estimated baseline when he was an officer in a financial company. He experienced derealization: "Like I'm in the scene of a movie, but I'm not. Everything moves, but not me. Like I'm not there. Its like its all around me, but I'm outside. I'm crazy?" He also experienced depersonalization: "Like I'm not inside me. I can see myself. Like I'm not real. The body is the same." He sees himself as calm: "Not feeling what I'm feeling." He feels "distressed." He also experiences deja vu: "Sometimes people are introduced to me and I know that person before, but that person tells me that we never met."

Overview of Dissociative Disorders

There are disparate explanations of the nature of this phenomenon, including a process, various symptoms, mental structure, psychological defense, and a deficit of integrative capacity. Integrative functions are hampered by the neurochemicals released by a severe threat. These are integrated in the brain regions implicated in integrative mental acts (e.g., the hippocampus and the prefrontal cortex). Two parts of one's personality may become dissociated. One is a feeling of numbness and the avoidance of traumatic memories and associations can remain functional in daily life. The other type is enmeshed in traumatic memories and intermittently intrudes or replaces the numb dissociative system. The dissociative phenomena of psychoform and somatoform symptoms are both manifested with negative and positive symptoms. Negative symptoms indicate a failure to integrate dissociative systems of ideas and functions that are available. Positive symptoms occur when these systems retrieved (e.g., flashbacks). Dissociative symptoms may be confused with other disorders. Hearing voices maybe confused with psychosis, while intrusive reexperiencing manifested by disorientation of time, place, and person can be confused with intoxication, delirium, delusions, hallucinations, and psychosis. While PTSD is classified as an anxiety disorder, it might be regarded as a dissociative disorder. Movement and sensations are defined in the ICD-10 as dissociative disorders but are defined in the DSM-IV as conversion disorders. Yet, there is agreement upon the contribution of trauma and its consequence as neurobiological alterations (Nijenhuis et al., 2002).

Symptoms

The following list is modified from Krystal et al. (1995):

- Social: withdrawal and detachment
- Identity, body schema; self-monitoring: depersonalization; derealization; conversion; fugue multiple personality; proprioception (body shape; position); analgesia
- Memory: flashbacks; amnesia; shifts in modes of memory encoding (pictorial/iconic vs. linguistic; dèja vu; jamais vu)
- Perceptual: time; visual (shape, color size); auditory; context (proximity; temporal relatedness); olfaction; taste; tactile
- Cognitive: constricted attention; neglect; confusion; information processing; altered depth of associative processing; holistic versus detail

A patient in PTA might appear to be normal (i.e., clinically considered to be uninjured or slightly injured), resulting in the entry of inaccurate non-TBI statistics after an accident. The patient may have apparent short-term memory and seem lucid in the period after the accident but subsequently be unable to recall either the accident or any current conversations the next day. There was failure to consolidate the events. Moreover, "memories" of the events around the accident may be inaccurate.

The association of dissociative phenomena with trauma requires two cautions: (1) verification (the events have not always been verified) and (2) patients with dissociative tendencies have a strong propensity to have a firm conviction of the reality of ideas suggested to them by others (Nemiah, 1995). The development of dissociative symptoms in a group of policemen after a traumatic event was associated with the absence or increasing complexity of the PTSD (partial or complete). The level of awareness of events after a frightening experience varies between patients. With lesser levels of dissociation, the individual can monitor his or her relationship to the environment, which is lost in more complete dissociation. Learning during dissociation may be state dependent (Krystal et al., 1995). Defenses fail when the victim is reminded of the event by internal or external stimuli. The consequences may be episodic rage attacks (Salley & Teiling, 1984) or fugue states (Spiegel, 1988). Traumatic events may be repressed or dissociated entirely or in part, or suppressed (Blank, 1994). Dissociation during combat trauma was associated with a greater risk for PTSD and dissociative symptomatology (Bremner & Brett, 1997). Even when dissociation is addressed as a psychological stress phenomenon, neurobiological phenomena include memory and hyperarousal so that the overlapping with TBI dysfunctions is considerable. For example, tunnel vision, which is sometimes considered to be a sign of hysterical sensory loss, has been obtained through administration of ketamine, an antagonist of one type of glutamate receptor. Dissociative states are also obtained through serotonergic hallucinogens, a neurotransmitter involved in stress-related functions (Krystal et al., 1995). Other examples of the difficulty of determining the etiology of dissociative-related phenomena are found. The association of phobic anxiety and panic attacks with depersonalization (Steinberg, 1991) is no doubt more common after head injury. Furthermore, the characteristics and examples offered for psychogenic amnesia (PA, including inability to recall certain important information) overlap with cerebral trauma (retroactive amnesia and PTA). Yet, the criterion for PA that the dysfunction of neurobiological systems is reversible (Loewenstein, 1991) would not be determined incorrect for some time.

Dissociative amnesia (DA) is recognized as a component of PTSD (American Psychiatric Association, 1994, p. 480), which is often comorbid with concussion. Correct identification of dissociation is significant since the presence and length of the PTA interval is considered to be a measure of the intensity of TBI. Therefore, confusion of a different phenomena for a symptom based upon neurotrauma would lead to incorrect diagnosis, assessment of performance, and possibly incorrect treatment. DA may occur in three forms: generalized (patients have no recall of any events in their lives); systematized (certain kinds of events are forgotten); and circumscribed (an island of amnesia creates a gap in the patient's historical memory). DA can be differentiated from complex partial seizures since the episode is triggered by a particular emotional stress and the memory may be enhanced as opposed to permanently lost.

Dissociative fugue is differentiated from complex partial seizures by the development of a new identity. Nevertheless, in the postictal stage, patients with cerebral personality symptom (CPS) may exhibit personality changes in the areas of prosody, humor, and expression of hostility (Moore, 1997, pp. 101–102). The memory loss is episodic. Other memories remain intact: semantic, skills, factual information, and social skills (Nemiah, 1995).

DEPERSONALIZATION

Clinical example: Momentary LOC, no head impact: A woman was holding open the door of a car struck by another vehicle. She was propelled forward, and suffered damage to her shoulder, neck, and trunk. There was perhaps momentary LOC, but no apparent direct impact to her head. There may have been a moment after the impact of brief LOC or reduced awareness. "I don't remember letting the handle go. Then I remember walking out of the car...I was a little dazed. I was walking, I wasn't myself...I still don't feel like myself. I was very active before. I feel like I am in slow motion. Everybody is doing 55 mph and I'm doing 5 mph. It still occurs that, even if I have no pain, I feel 'dopey'."

Depersonalization has been defined as "a feeling of detachment from and being an outside observer of one's mental processes or body, or feeling like an automaton or as if in a dream." The person feels detached or estranged from oneself (i.e., an automaton or observer of their own mental processes, life, or body). The literature infers both an organic and psychodynamic origin. Depersonalization can be a rare manifestation of complex partial seizures, which are more often found in schizophrenia (DeLorenzo, 1991). Features differentiating TLE and a phobic-anxiety depersonalization syndrome are found in Trimble (1991, citing Harper & Roth, 1962). Depersonalizaton is represented by a separate diagnostic category in the third edition of the *Diagnostic and Statistical Manual* (see 300.60, Depersonalization Disorder or Neurosis, pp. 275–276), and is a component of panic disorder (see 300.01, pp. 237–238). The patient feels unreal, sees the self outside of the self, and experiences emotional numbness and affective loss. Depersonalization is characterized by an absence of feelings, a reduction of vividness, a sense of detachment and of being real or an automaton, and possibly by a state of low arousal. Sometimes the experience is pleasurable and can be followed by an appreciation of life. It may involve heightened arousal that is dissociated from consciousness (Steinberg, 1991). Psychogenic fugues can cause a prolonged altered responsiveness to the environment that mimics convulsive status epilepticus (DeLorenzo, 1991). Depersonalization can be found in a variety of conditions (including other dissociative disorders) and may or may not be a separate clinical entity in a particular patient. There are numerous theories concerning its etiology. It is most frequently associated with panic, but has also been associated with anxiety, agorophobia, depression, seizures (preictal, ictal, and postictal), substance abuse, organic illness, depression, and side effects of a medication.

Clinical Examples of Depersonalization

Acute: "I've got the feeling I'm looking at myself. It's somebody else, a kind of movie. My mind is detached from my body."

Chronic: An adolescent girl was in an automobile accident and comatose for a week. After 1–2 months she tried to go to school, but it didn't seem real. "I stood in the hallway and it seemed like I wasn't there. It was all foggy and like a dream. Everyone was moving really fast."

PARTIAL SEIZURES (FOCAL)

In cases of concussion, exploration frequently elicits information about sensory phenomena that are (correctly) assumed by the patient to not be based upon object internal or external realistic stimuli. Frequently, they are ridiculed by collaterals or it is privately assumed that these patients are mentally deranged. The writer routinely screens for the sensations associated with partial seizures and often receives expressions of relief from the patient that finally somebody is aware of his or her distress.

Focal seizures begin in a relatively localized area, and the narrow range of sensations contrasts with generalized ones (convulsive or nonconvulsive). They elicit altered states of consciousness that resemble the normal stimulation of the cortex involved. Temporal lobe lesions are prone to create multimodal symptoms (i.e., noxious odors with complex visual formations) and stereotyped for a given patient.

Kindling

Kindling refers to repeated subthreshold stimulation that increases behavioral responsivity, resulting ultimately in paroxysmal behavior such as seizures, migraines, and affective disorders. Repetitive stress (e.g., persistent PTSD or the discomforts of recovery after impairment and/or somatic damage) may over a period of time evoke minor kindling, culminating in an increase of biochemical and physiological responses with a full-blown seizure episode. Eventually, a significant disorder may appear "out of the blue," perhaps due to a reduced threshold (Post & Silberstein, 1994). These are a recognized but infrequent consequence of mild head injury (Evans, 1992a; Haas, 1993a; Lipton et al., 1994). Focal seizures may become generalized. Repetitive stimulation of certain tracts (e.g., by an irritative focus) can facilitate conduction in the future ("kindling"), leading to an enhancement of activity after mild stimulation (Horvath et al., 1989).

Symptoms of Partial Seizures

Partial seizures are considered controversial from the viewpoint of classification (Trescher & Lesser, 1996). They refer to phenomena that arise from a one part of a cerebral hemisphere, may progress to more extensive activity, and have a relatively limited impairment of consciousness—at least initially. The associated phenomena affect consciousness, mood, motor behavior, thinking, sensation, sense of self. They may not be detected by an examiner unless sought for specifically, nor may they be associated with brain trauma or the cerebral dysfunction of a different etiology. Partial seizures may occur with or without impaired consciousness (complex partial).

Symptoms: Transient distortions of consciousness (confusion; deja vu); cognition (memory; forced thinking); speech, word-finding problems, and jargon; affect (fearfulness; temper outbursts); sensory (pain; illusions of taste, smell, vision, movement touch); and behavior (automatisms). The hyperventilation syndrome has been described as a borderland with partial seizures. Overlapping complaints include deja vu, strange feelings, feelings of confusion, left-sided paresthesias, and syncope. The association of strange feelings and syncope makes differentiating hyperventilation from epilepsy difficult (Evans, 1995).

The following symptoms may have a different etiology than partial seizures, and description in greater detail is offered elsewhere for those that have a higher association with concussive brain trauma (Trescher & Lesser, 1996; Varney et al., 1993).

Visual phenomena: Partial seizures are location related—frontal lobe; supplementary motor area; cingulate gyrus; anterior frontal pole; orbitofrontal; lateral dorsal motor cortex; temporal lobe (hippocampal; amygdala; lateral posterior temporal; operculum; parietal lobe; occipital lobe) (DeLorenzo, 1991, pp. 1462–1464). This type is discussed in greater detail below.

Abdominal epilepsy with paroxysmal abdominal pain: Primarily occurs in children, with sudden pain and other autonomic phenomena, as well as vomiting, incontinence, sweating, salivation, and audible bowel noise.

The writer routinely screens for the sensations associated with partial seizures and often receives expressions of relief from the patient that finally somebody is aware of his or her distress.

When environmental demands exceed a person's sensorimotor and cognitive capacities, the result may be inappropriate behavior with inadequate motor control or behavioral deterioration (Weber & Verbanets, 1986). While it is claimed that the cortex's influence via descending systems over spinal motor neurons allows for significant recovery in case of injury, an exception would be the projection from the motor cortex to the fingers and hands. Injury to these fibers results in permanent loss of skilled movements, such as the manipulation of small objects (Ghez & Krakauer, 2000).

POSTTRAUMATIC EPILEPSY

Seizures

These are defined as a single event that results in an altered state of brain function, with a distinct beginning and end. One cannot be certain that clinical behavior represents a seizure. For example, one youth exhibited eye rolling and facial grimacing that were considered absence seizures. Yet, when these occurred during an EEG examination they were not accompanied by any change in normal background electrical activity. The writer raised the question as to whether they may have been conversion reactions.

Focal (partial) seizures arise from a localized are of the cerebral cortex. Simple partial seizures rarely have bilateral hemispheric involvement. The complexity of visual phenomena depends upon the cortical locale; organized hallucinations are initiated in the temporal lobe where the complex visual integration occurs. These may become generalized, sometimes so quickly that their initial focal nature is concealed. *Complex partial seizures* are defined by an impairment of consciousness. The epileptic discharge occurs in a region where structures subserving memory, consciousness, emotion, and vision are affected. This creates a dreamy state and may include affective, autonomic, and automatic behavioral symptoms. Psychomotor automatisms, the most prominent aspect of complex partial seizures, are activities that occur during altered consciousness. There is frequently bilateral involvement. An apparently generalized seizure may in fact result from secondary synchrony of a focal localized hemispheric disorder (Dreifuss, 1999).

Epilepsy is a recurrent brain disorder with varied etiologies, characterized by recurrent seizures due to excessive discharge of cerebral neurons. For a patient to be considered epileptic, AOC must occur repetitively. Epilepsies can be classified as localization related (focal, local, partial), generalized, undetermined, and as having special syndromes (e.g., febrile convulsions). Epileptic seizures may be accompanied by severe alterations of affect. Trauma associated with seizure generation includes extravasation of *red blood cells* into the brain parenchyma with astrocyte prolification, iron filled macrophages, neuronal changes, focal edema, and generation of free radicals. *Kindling* is a concept developed from a model of repeated transient electrical stimulation of the amygdala or hippocampus that produces brief electrical afterdischarges, which later produces seizures. It is associated with neuronal loss of the hilar region of the hippocampus. Another model is *kainic acid* injection (an analogue of the neurotransmitter glutamate), which produces hippocampal seizures and reorganization (Granner, 1996).

Etiology

Seizures are defined as a single event that results in an altered state of brain function, with a distinct beginning and end. Partial seizures have a stereotyped sequence. Epilepsy is a recurrent brain disorder of multiple etiology and is characterized by recurrent seizures due to the excessive discharge of cerebral neurones. For a patient to be considered epileptic, AOC must occur repetitively (DeLorenzo, 1991). Risk factors for early PTE (within one week of an injury) include being less than 5 years of age, PTA more than 24 h, skull fractures (particularly depressed fractures), and intracranial hemorrhage. Late posttraumatic seizures are those that occur more than 1 week after a head injury. They often become chronic and develop into epileptic conditions (DeLorenzo, 1991). Late seizures are particularly associated with penetrating brain injuries, as well as operations that produce cortical damage both minor (insertion of a ventricular shunt or stereotaxic probes) and major (Fisch, 1991, p. 319). Seizures may be expressed unexpectedly with an initial seizure perhaps 15 years after a trauma.

There are two classifications in general use (DeLorenzo, 1991), the rubrics of which are synthesized below by the writer. Comprehensive summaries are offered by Dreifuss (1989).

Epilepsy is estimated to occur in 4% of the population and perhaps even more, since its existence may not be recognized. Some seizure disorders create little or no outward manifestations of epilepsy (Benson, 1986). *Etiology* includes genetic, prenatal and perinatal factors, metabolic, toxic, head trauma, infection, anoxia and hypoxia, vascular, tumors, degenerative diseases, and demyelinating diseases. *Seizure self-prediction* may utilize premonitory feelings. Patients report these seizure precipitants: emotional stress and stressful life events via HPA axis, sleep deprivation, depression, anxiety, menstruation, and alcohol use. Loss of sleep for even 1 h can have a significant effect. The effects of stress and anxiety on seizure risk are independent of the effects of sleep (Haut et al., 2007).

Posttraumatic Seizures

Early onset (usually within 24 h) is associated with simple partial seizures and motor signs. Late onset (usually within years of injury) has a less favorable outcome than early onset and is more frequently tonic-clonic than partial (Dreifuss, 1999). The overlap between dissociation and TLE (and other seizure conditions) includes blackouts, amnesia, fugues, depersonalization, deja vu, derealization, somatic sensations, and auditory, visual, and olfactory hallucinations (Putnam, 1991). Whether only mental symptoms are involved, or whether somatic phenomena such as conversion and hysteria should be included, is controversial. Psychodynamic phenomena such as childhood trauma, psychological needs and conflicts, or an inability to cope can be central to "splitting of the ego," although neurological and psychiatric pathology also is present. While dissociated affects, fantasies, memories, and so on are removed from consciousness, they may be recalled or may manifest ongoing affects such an ego-alien disturbance of sensorimotor function (Loewenstein, 1991; Nemiah, 1991). Under normal waking conditions, sensory-driven neuronal groupings inhibit internally activated patterns (Bisiach & Geminiani, 1991). In their absence, fantastic images occur or the person may neglect part of external space (see above).

Seizure Classification

Impact or immediate (within 24 h): these seizures are of no prognostic value for the development of epilepsy.

Early (within 1 week): Occurrence of early seizures is a risk factor for late PTE.

Late (day 8 and beyond 2 years).

Risk Factors

After the acute effects of head injury have resolved, there is a 3.6-fold increase in late seizures, the majority of which occur in the first year with an increased risk occurring for 4 years. A summary of antiepileptic medications is available (Granner, 1996; Hernandez et al., 2004). PTE abnormalities occur most frequently as a manifestation of partial complex seizures in patients with TLE: *gelastic epilepsy* (laughter); *dacrystic epilepsy* (crying or tearing) (Cummings & Trimble, 1995, pp. 40–41).

Risk factors for early PTE (within one week of the injury) include being less than 5 years of age, PTA for more than 24 h that is incurred during the first week after injury, dural laceration or intracranial hematoma, iron deposition (hemosiderin) skull fractures (particularly depressed fractures), bone fragments, intracranial hemorrhage, and medications (including high dose of tricyclic antidepressants). Late posttraumatic seizures often become chronic and develop into epileptic conditions (Chiofalo et al., 1986; DeLorenzo, 1991; Kirby et al., 2007). Seizure foci can be separated into acute (penetration of parenchyma; shearing forces; disruption of the blood-brain barrier [BBB]), late, or secondary (vascular disruption; cicatricial contraction; synaptic reorganization) (Hernandez et al., 2004). About half of men with temporal lobe seizures are either impotent or hyposexual. Up to 40% of women with this condition have menstrual irregularities or reproductive dysfunction. Amenorrhea associated with hypogonadotropic hypogonadism is associated with low levels of luteinizing hormone (LH). The frequency of seizures often increases in females during menses and pregnancy since estrogens lower the seizure threshold (Sandel et al., 2007). Frequency is greatest in early life and may be an epiphenomenon of preexisting brain damage. Closed fractures are associated with a 10% risk of PTE, and a 20% risk of compound fracture. The higher incidence of epilepsy in developing countries contributes to our understanding of risk factors apart from trauma: limited access to health care; poor sanitation (infectious disorders of the central nervous system [CNS]); alcohol and substance abuse. Additional risk factors for early seizures include the severity of the injury, frontoparietal location, depression of the fractured bone, intracranial hematoma, and young age. Risk for early seizures declines throughout adult life. Missile injuries increase the rate of late PTE to 28–53%, compared to the civilian rate after closed head injury of 3–14% (Trescher & Lesser, 2004). Early focal seizures are associated with 14% poor outcome, as well as later seizures. Early posttraumatic seizures are followed by epilepsy beyond the first week in 20–25% of patients followed for a minimum of 4 years. The risk of developing late posttraumatic seizures is strongly determined by the severity of the brain injury. Late PTE after more than 1 week increases the risk of a subsequent seizure in the next 4 years to nearly 75% (Rosman et al., 1979, citing Jennett, 1973).

Late Posttraumatic Seizures

About half of PTE cases experience generalized seizures, that is, grand mal characterized by sudden loss of motor control and consciousness, bilateral tonic-clonic spasms followed by a postictal phase with reduced alertness, confusion, lethargy, and fatigue. Preventing late seizures (after 1 year) may not lead to a better functional outcome (Haltiner et al., 1996).

Outcome

Generalized seizures have 27% rate of poor outcome (Raimondi & Hirschauer, 1986). Seizures may cause brain damage depending upon age (e.g., ischemic cell necrosis). By altering the connectivity of neuronal pathways, the susceptibility to later seizure increases and the ability to inquire subsequent conditioned avoidance responses is impaired. Over a period of time, harmful affects depends upon the outcome measure, although the immature brain seems more resistant to structural brain damage than the mature brain (Moshe, 1987).

Adverse effects: There are numerous neuropsychological deficits consequent to anticonvulsant medication. These are dose dependent and host independent: learning, attention, memory, concentration, personality changes, psychomotor performance, and so forth (Parker, 1990, p. 308); sedation, dizziness, ataxia, diplopia, and headaches (Trescher & Lesser, 2004). Cognitive disorders are most frequently described, although disorders of psychomotor function and somatosensory recovery also occur. Prophylactic administration of antiepileptic drugs is not effective in preventing later PTE. Thus, since TBI carries a 5% risk of PTE, the remaining 95% would needlessly receive anticonvulsant medication (Hernandez et al., 2004).

SIMPLE ABSENCE SEIZURES

Clinical Examples: "Absence": "Does it ever happen to you that you lose time, you don't know what happens for a period?"

"Yesterday, I after I made dinner, my children and I ate, watched TV. I was talking with my daughter, it was 5:30. Later, I got a phone call from a friend. I told her to call me back a little later. 'Are you sure?' she said. 'Its 11:30 now, and I always say 'don't call my home after 11." It was 11:30, and she thought it was 7 PM."

A man goes to work, forgets his cell phone, goes back, but doesn't know what he went to pick up. When somebody is talking to him, he loses attention. His mind is going out. On transportation he may not get off at his stop.

An absence seizure is an abrupt, brief episode of decreased awareness without any aura or postictal symptoms. Formerly called "*petit mal*," it usually occurs in childhood, rarely persists into adulthood, and consists of momentary apparent inattentiveness without loss of muscle tone and posture (Fisch, 1991, p. 294). Awareness is variable, but the interrupted activity may resume with no memory of the lapse of awareness. There is an interruption of activity. A simple absence seizure is characterized only by an alteration of consciousness, with no changes in breathing, color, or muscle tone. A complex absence seizure exhibits additional symptoms, such as motor automatisms, and visceral symptoms such as change in pulse rate, flushing or pallor, and so on. Minor seizures, or "absences," are occasionally noticeable during the examination.

One cannot be certain that clinical behavior represents a seizure. For example, one youth exhibited eye rolling and facial grimacing that were considered absence seizures. Yet, when these occurred during an EEG examination they were not accompanied by any change in normal background electrical activity. The writer raised the question as to whether they may have been conversion reactions.

The writer raises the question as to whether posttraumatic "absence" phenomena may have some characteristics that are different from those with more medical etiology. Absence seizures are usually characterized by an abrupt, brief episode of decreased awareness without warning, aura, or postictal symptoms. The patient is described as remaining immobile, with no loss of postural tone and no motor manifestations. However, ongoing activity (speech, chewing, walking) may cease abruptly. The patient does not respond to external stimuli and is unaware of having had the seizure. The present writer observes that after concussive level TBI, the event may differ insofar as there is continued activity with some loss of external orientation and of memory for the effected interval. Driving and other daily activities may continue but may not be completely oriented to the outside world. Atypical absence seizures are longer lasting, have motor phenomena, and present loss of postural tone (Bruni, 2004; Trescher & Lesser, 2004).

MISCELLANEOUS SEIZURE EXPERIENCES

The overlap between dissociation (Steinberg, 2000a) and TLE (and other seizure conditions) includes blackouts, amnesia, fugues, depersonalization, deja vu, derealization, somatic sensations, and auditory, visual, and olfactory hallucinations (Putnam, 1991, citing Devinsky). Whether only mental symptoms are involved or whether somatic phenomena such as conversion and hysteria should be included is controversial. Psychodynamic phenomena such as childhood trauma, psychological needs, conflicts, or inability to cope can be central to the "splitting of the ego," although neurological and psychiatric pathology also is present. While dissociated affects, fantasies, memories, and so on, are removed from consciousness, they may be recalled or manifest ongoing effects, such as an ego-alien disturbance of sensorimotor function (Loewenstein, 1991; Nemiah, 1991). Under normal waking conditions, sensory-driven neuronal groupings inhibit internally activated patterns (Bisiach & Geminiani, 1991). In their absence, fantastic images occur or the person neglects part of external space (see above).

FUGUE STATES

The differential diagnosis of *fugue states* is difficult and perhaps controversial. Although dissociative disturbances of memory are attributed to fugue (e.g., losing time, blackouts, and gaps in the continuity of personal experiences) (Putnam & Loewenstein, 2000), this pattern creates concern about absences and partial seizures. Although Coons (2000) notes the implication of overwhelming stress, he also asserts that the most common organic fugue is secondary to epilepsy, especially complex partial seizures. Depersonalization occurs in both psychiatric and nonpsychiatric persons (Steinberg, 2000b). In addition to being an adaptation to overwhelming trauma, it may result from temporal lobe dysfunction and metabolic and toxic states.

AOC COMMON TO CONCUSSION AND PTSD

Flashbacks (Fischer, 1986): A flashback is defined as the involuntary recurrence of some aspect of a hallucinatory experience, or a perceptual distortion some time after taking a hallucinogen. It is a common occurrence after trauma but it may also occur as a normal experience. In addition, flashbacks are common after accidents, perhaps being caused by the profound neurotransmitter effects. They are also state related (i.e., evoked by imagery, melodies, and symbols of the content of an experience). A woman who was struck by a car offered this description of a flashback: "I can see the car coming and myself flying."

Other Alterations of Consciousness

The following symptoms may have a different etiology than partial seizures (Trescher & Lesser, 1996; Varney et al., 1993):

Abdominal epilepsy: Characterized by paroxysmal abdominal pain. This primarily occurs in children, with sudden pain and other autonomic phenomena, as well as vomiting, incontinence, sweating, salivation, and audible bowel noise.

Release phenomena: Are consequent to reduction of normal visual input and may occur with changes of illumination. These have less localizing value than irritative hallucinations.

NONEPILEPTIC SEIZURES

An accident causing concussion often is comorbid with an acute fearful or chronic generalized stress response that is initiated either acutely or after recovery of consciousness. AOC often occur in the context of persistent concussive symptoms, which are often described as nonepileptic seizures (NES), pseudoseizures, conversion nonepileptic seizures (C-NES), and nonepileptic events (NEE). Anxiety is a common aftermath of these events, thus neurobehavioral symptoms may be elicited that mimic traumatic ones. The stress response may result from (1) actual memory of the event; (2) the patient's imaginal reconstruction of what happened (in the absence of memory); (3) what the patient was told about the accident; or (4) frightening post-accident clinical circumstances (waking up at the accident scene in a hospital under treatment, etc.). The prevalence of psychogenic nonepileptic seizures (PNES) has been estimated to be as high as 20–30% in an outpatient epilepsy population (Aboukasm et al., 1998; Berkhoff et al., 1998). An estimated 10–15% of patients seen for intractable epilepsy present behavior that has been described as psychogenic seizures or pseudoseizures, but true origin of which may or may not be psychodynamic . Unusual symptoms and bizarre behavior do not necessarily indicate psychogenic seizures. The patients need psychotherapy and support (DeLorenzo, 1991). On the other hand, 40% of patients with psychogenic seizures also suffer from true epileptic seizures. Therefore, the diagnosis should be made cautiously.

Psychopathology: The most common cause of NES has been considered to be conversion disorder. The classical sign of "*la belle indifference*" may be absent or may be misinterpreted for patients who are stoic, have right parietal lobe lesions, or have the cerebral personality symptom of aprosodia. Other conditions that can also be responsible for NES are malingering, anxiety disorder, adjustment disorder of adolescence, anxiety disorders, psychotic disorders, and impulse control problems in the setting of attention deficit disorder. Malingering is difficult to differentiate from conversion disorder when the potential for secondary gain exists (i.e., disability or litigation). In fact, intentional and unconscious mechanisms may coexist (Devinsky, 1998). A cautionary note is offered by Berkhoff et al. (1998): since seizure-like activity of undetermined origins (SLAUE) are sometimes attributed to emotional problems with reference to developmental emotional stress, there was no differentiation between patients with complex partial seizures epileptic seizures (ES) and NES for disturbances in childhood or adolescence. Perhaps ES can offer a negative influence on emotional development, as well as a stigma, the frightening attacks of which may cause ambivalent reactions in relatives. Since the two groups displayed autonomic symptoms, they were both susceptible to somatization or conversion.

Psychiatric diagnostic categories include somatoform disorders; anxiety disorders; disorders with psychotic symptoms; reinforced behavior patterns; dissociative disorders; and malingering (Gates & Mercer, 1995). Lovitt (1987) describes the pseudoseizure as symbolizing an underlying psychological conflict and serving as a discharge of tension. LOC is a dissociative component. The convulsions serve to release tension or anxiety through conversion. A Rorschach case study was offered: conflict and anxiety (some responses similar to signs of intrusive anxiety; pathological inner life; disrupted ideational resources; poorly controlled discharge; and lack of insight. Psychological assessment is recommended since these patients have a high incidence of emotional disturbance, including hysteria, depression, personality disturbance, and secondary gain. Some nonepileptic disorders in children and adolescents (Bleasel & Kotagal, 1995) include trauma-like symptoms; migraine; movement disorders; psychological symptoms (panic, hyperventilation, malingering); sleep disorders (night terrors, sleepwalking, confusional arousal); absence-like events resembling partial seizures; histrionic personality; and depression (Berkhoff et al., 1998).

Treatment concerns: The NES is considered to be a response to an emotional stressor or chronic maladaptive behavior. Conversion phenomena benefit from psychotherapy but a support system is also needed (Devinsky, 1998). Misinterpretation of early NES can encourage the development of more severe epilepsy-like symptoms (Barry et al., 1998; Pedley, 1995) or expose the patient to the serious side effects of antiepileptic drugs, the doses of which might be increased because of a perceived drug resistance (Berkhoff et al., 1998). When the nature of seizure-like behavior cannot be firmly grounded through seizure findings on EEG, then a wide-range medical examination is useful (Pleet, 1995). Since not every paroxysmal event is a seizure, routinely construing them to be related to head trauma can delay diagnosis and lead to ineffective, unnecessary, and potentially harmful treatment. A multidiscipline approach is endorsed. There are ethical considerations regarding the use of placebos to elicit NES, particularly in the context of a low level of trust, such as with victims of sexual of physical abuse. Concerning psychogenic NES, psychotherapy or nonjudgmental counseling resulted in a beneficial outcome in two-thirds of the patients. Understanding the nature of their condition was a key element in the reduction of their symptoms. This could be difficult to accept and might require multiple therapeutic sessions. Video recordings were regarded as a supportive confrontational tool (Aboukasm et al., 1998).

Some EEG Considerations for SLAUE

One method of study is simultaneously recording with a video camera, electrocardiogram, and continuous EEG. However, a structured interview may be better tolerated (Berkhoff et al., 1998). Seizure-like activity may occur without a proven focus, or where proof of the focus was based on intracranial records and the results of surgical excision (Swartz, 1995). Changes of consciousness can occur with no EEG changes, while responsiveness may be unchanged in the presence of EEG changes (Zappulla, 1997). There are AOC where the etiology cannot be determined even with extended EEG and videotaped monitoring of "events" of concern.

A group of polysymptomatic presentations of seizure-like activity without predictable symptom sequence have been described as epilepsy spectrum disorder. The EEGs are either within normal limits or not clearly epileptiform. The neurobehavioral syndrome resembles multiple partial seizure-like symptoms in the context of persistent dysphoria and emotional lability (i.e., subjective experiencing of multiple cognitive, affective, and psychosensory phenomena). Many of these patients have abrupt performance decrements and mental lapses. It is hypothesized that the symptoms are attributable to the partial *kindling* of limbic structures that produce the behavioral changes relatively independent of the mechanisms creating motor-convulsive epileptogenic phenomena (Roberts et al., 1992). A majority of patients are responsive to carbamazepine or valproic acid (Hines et al., 1995). There is bilateral expression of the epileptiform discharge involving both cerebral hemispheres. While amnesia is characteristic of seizures, some consolidation of memories and awareness of ictal responsiveness may occur (Zappulla, 1997).

LATE POSTTRAUMATIC EPILEPSY

The risk of developing late posttraumatic seizures is strongly determined by the severity of brain injury. About half of PTE cases experience generalized seizures, that is, grand mal characterized by sudden loss of motor control and consciousness, bilateral tonic-clonic spasms followed by a postictal phase with reduced alertness, confusion, lethargy, and fatigue. Preventing late seizures (after 1 year) may not lead to a better functional outcome (Halter et al., 1996).

A study of the psychopathology associated with NES (Barry et al., 1998) found that 43% of patients with nonepileptic PTS had a premorbid psychiatric disease. Many had a concomitant psychiatric disease: 49% depression; 32% somatization; 27% substance abuse; 24% personality disorder; 19% anxiety disorder; 11% dissociative disorder; and 6% factitious disorder. Other conversion symptoms ranged from chronic pain (30%) to hemiparesis/hemisensory loss (19%) (with lesser proportions for other symptoms). Other complaints were headaches, dizziness, cognitive and memory difficulties, and emotional lability. Head injury is considered to be a triggering factor for NES. Even mild head injury can be associated with infractible seizures, being nonresponsive to multiple antiepileptic drugs and yet classified as nonepileptic. This group had a prevalence of women and was characterized by unusual or peculiar symptoms that may be mistaken for complex partial seizures. These events may begin on the first day after a head injury or shortly thereafter. Other physical trauma and organic brain disease, as well as childhood sexual and physical abuse, have been recognized in more than 30% of patients with NES (Barry et al., 1998). Using the Dissociative Experiences Scale (DES) as a source of items, a C-NES group and a group of subjects with complex partial epilepsy (CPE) were compared. The depersonalization-derealization factor (1) was best differentiated between denying childhood abuse (0.06) and disclosing a history of childhood abuse (0.089). The trend favored the concept that a history of childhood abuse contributes to the potential for a C-NES diagnosis, although other determinants contribute to the depersonalization-derealization factor in C-NES as well. Absorption-imaginative involvement (factor 3) was elevated in both groups who reported sexual abuse and did not differentiate between the clinical groups. These two factors relate more strongly to the sequelae of childhood abuse than to a tendency toward developing conversion symptoms (Alper et al., 1997). The incidence of NEE in outpatient populations has an estimated rate of 5–20%, and 10–40% in specialized or comprehensive epilepsy programs. A patient may have both epileptic and NEE.

AOC often occur in the context of a persistent concussive symptoms, which are often described as NES, pseudoseizures, C-NES, NEE, and SLAUE. Patients with frontal lobe epilepsy may present bizarre sudden-onset, short-duration seizures with wild motor movement and little or no postictal confusion. A differential diagnosis should consider hypoglycemia; syncope and vasovagal attacks; migraine; vertebrobasilar ischemia; transient ischemic attacks; and varied sleep disorders (Cummings & Trimble, 1995). The prevalence of PNES has been estimated to be as high as 20–30% in an outpatient epilepsy population (Aboukasm et al., 1998; Berkhoff et al., 1998). An estimated 10–15% of patients seen for intractable epilepsy present behavior that has been described as psychogenic seizures or pseudoseizures, but the true origin of which may or may not be psychodynamic. Unusual symptoms and bizarre behavior do not necessarily indicate psychogenic seizures. These patients need psychotherapy and support (DeLorenzo, 1991). On the other hand, 40% of patients with psychogenic seizures also suffer from true epileptic seizures. Therefore, the diagnosis should be made cautiously.

ALTERATIONS OF COGNITION ASSOCIATED WITH EPILEPSY

Seizures are associated with attention deficits and a slowing of cognitive processes. Subclinical discharges interfere with cognitive processes even when seizures are well controlled. Adverse cognitive effects vary with medications. Other side effects are mood changes, level of alertness, oculomotor and gastrointestinal (Brown, 1991). Various qualities of thinking have been described as characteristic of persons with epilepsy. They are stated here to serve to alert the clinician without implying the proportion of patients who clearly exhibit them:

- *Overinclusiveness*: in verbal output (circumstantiality), action (stickiness), and in writing (hypergraphia).

- *Intensification of mental activities* (i.e., philosophical, religious or political concerns) sometimes over abstract topics, leading to behavioral excesses.

EMOTIONAL REACTIONS

It is controversial as to whether the higher incidence of emotional problems of seizure victims is a direct response to the lesion or an adaptation to it. Seizure control, mood, social function, and cognition influence quality of life. *Anxiety* may be hemisphere-specific since it correlates only with a measure of *quality of life* in patients with left TLE and not right TLE. There is evidence that preexisting personality disorders contribute to SLAUE (see above). Significant concerns of patients with severe epilepsy include further seizures, health, discouragement about work, driving, and social functioning, which all serve to reduce a patient's sense of security. Depression, paranoia, and hostility may develop to the point that therapeutic intervention is needed.

According to Naito and Matsui (1988), *interictal auras* are likely to be feelings of fear, terror, or anxiety, and rarely joy or ecstasy. They present a case of a woman in religious ecstasy with visual hallucinations: "A halo appeared around God...I experienced a revelation of God and all creation glittering under the sun [which] engulfed me...my whole being was pervaded by a feeling of delight." Her writing of her experiences many times was described as hypergraphia. Her EEG was characterized by a dominant hemisphere-localized ictal spike focus.

DEPRESSION

The incidence of depression in patients with epilepsy is considerably more common than in the general population. The hippocampus plays a role in both epilepsy and depression, but there seems to be a multiple etiology for depression. Its prevalence increases sharply in patients with temporal lobe and recurrent epilepsy. Since women are twice as likely as men to suffer for depression, gonadal steroids may contribute to the pathophysiology of depression, with ovarian hormones linked to the incidence of seizures (Hajszan & MacLuskey, 2006) and elevated androgen levels in major depression indicating adrenal hyperactivity (McEwen, 2004). It has been suggested that a recurrent seizures directly produce a mood disturbance as a consequence of altered neurochemistry, changes in neural pathways, and adverse social and psychological effects (Tracy, 2007).

REPRODUCTIVE DISORDERS OF WOMEN

Epilepsy is associated in women with an increased frequency in disorders of reproductive function. Both epilepsy and the use of anitiepileptic drugs are associated with higher frequencies of menstrual disorders, infertility, and premature menopause. Menstrual disorders are associated with anovulatory cycles that may increase the risk of infertility, migraine, emotional disorders, and female cancers. Epilepsy and antiepileptic drugs alter reproductive hormone levels and promote the development of reproductive endocrine disorders (e.g., *polycystic ovarian syndrome*). Consistent with concussive brain trauma, menstrual disorders may result from abnormal neuroendocrine regulation or the altered autonomic neurally mediated trophic effects of the temporolimbic system, hypothalamus, or autonomic nuclei on the gonads (Herzog, 2006). Abnormally increased ovarian steroid secretion in women with epilepsy occurs in part due to adverse effects of antiepileptic medication, and also possibly due to seizure-related derangement of hypothalamic circuitry controlling pituitary gonadotrophin release. Epilepsy in men is also frequently associated with subnormal testosterone levels. The predominant effect of recurrent seizures may be induction of deficits of hippocampal neuron connectivity. The clinical changes of epilepsy seem similar to idiopathic depression and hormone withdrawal. A unifying consideration may be the extent to which hippocampal circuitry retains the capacity for growth and repair. Diminished capacity increases the risk for developing clinical depression (Hajszan & MacLuskey, 2006).

PSYCHODYNAMIC CONTRIBUTIONS TO AOC

Phobic-anxiety depersonalization temporal lobe epilepsy
Family history of neurosis; history of TBI
Migraine; phobias in childhood; episodic anxiety
Automatic behavior; complete LOC

No change of consciousness; derealization; loss of familiarity; gradual termination; self-injury; incontinence; attacks followed by amnesia phobias; persistent anxiety; depressive episodes; feelings of unsteadiness; irrational fears; hypochondriasis; immaturity and dependence; one or more attacks per day.

Dissociation and depersonalization seem to be protective psychological mechanisms. Dissociative memory phenomena such as amnesia and flashbacks are differentiated from intrusive disturbances such as nightmares and traumatic memories. It is difficult to differentiate short traumatic impairment of consciousness from dissociative reactions or very short PTA (Radanov et al., 1992). AOC symptoms are physiological and psychological reactions, partial seizure phenomena and hyperarousal, and also variations in the intensity of experience that range from gaps in awareness to total absorption in activities (one is not aware of what is happening), as well as imaginative involvement (uncertainty as to whether something has occurred in reality as opposed to having dreamed it). The latter are items from DES. The frequency of items on DES varies in the population (cited in Alper et al., 1997). Contrary to expectation, in an experimental study, the induction of cold pain led to a small but significantly greater report of pain and not greater pain tolerance. It is possible that the pain reduced the potency of the dissociative manipulation, which was not applied at the magnitude comparable to what occurs during a traumatic event (Horowitz & Telch, 2007).

A patient in PTA might appear to be normal (i.e., clinically be considered to be uninjured or slightly injured) and enter inaccurate non-TBI statistics after an accident. The patient may have apparent short-term memory and seem lucid in the period after the accident, but is subsequently unable to recall either the accident or current conversations the next day. There was failure to consolidate the events. Moreover, "memories" of the events around the accident may be inaccurate.

In 1986, Grigsby stated the difficulty of differentiating functional from neural etiology of depersonalization in cases of head trauma. Depersonalization reflects a physiological disturbance of the brain's dynamic equilibrium, the mechanism of which is likely to be a deficit in the integration of perceptual, visceral, affective, and cognitive information. This is consistent with the evidence that depersonalization can result from the high level of arousal accompanying panic. He offered a case in which there was mild neuropsychological disturbance and a depersonalization syndrome: "Neuropsychological assessment was largely within normal limits, and the patient's complaints were in large part a function of feelings of depersonalization and derealization... Depersonalization... appears to have been of psychogenic origin. Initially is probably served as a defensive response to a potentially life-threatening danger...... Her life was disrupted and she began to question her previous lifestyle."

15 Cognition, Language, and Intelligence

OVERVIEW

This chapter offers an overview of the effect of concussive-level injuries upon the enormously complex process of gathering information from our world in order to react successfully and survive. The effects of blast injury (improvised explosive devices, i.e., IEDs) are explained in a separate chapter, 19. One study, relevant to what follows, offers evidence that this type of trauma seems not to be categorically different from other traumatic brain injury (TBI) mechanisms with regard to cognitive sequelae for particular procedures (Belange et al., 2009).

When studying cognition, let us use the principle that a person's responses to our procedures are not precise data leading to a "valid" measurement of a narrowly defined function. Rather, we receive data that is influenced by lifelong styles of responding, with performance affected by varied and perhaps contradictory processes.

- Some responses are in "awareness" and/or the conscious control of the patient.
- Most processing is out of awareness ("unconscious"), though not exclusively in the psychoanalytic sense of this term. Processing may be influenced by the "primary process," that is, without regard to logic or reality, and determined by hidden feelings, the pleasure principle, fear, and wish-fulfillment.
- Responses are largely automatic, for example, overlearned, holistic, and designed for immediate response in the immediate world.
- Analytic and sequential processing is best utilized for long-term planning but inappropriate for an immediate and direct response to some challenges.
- Cognitive response complexity is created by the interaction between cognition and language, communication, executive functioning, and self-regulation of behavior.

TOWARDS DEFINING COGNITION

Cognition is a generic term referring to a wide range of disparate functions used for assessing information, adaptation, and planning. Cognition has been defined (American Psychological Association, 2007) as follows: "All forms of knowing and awareness, such as perceiving, conceiving, remembering, reasoning, judging, imagining, and problem solving" [including] "an individual percept, idea, memory, or the like." A different definition is offered by O Shanick and O Shanick (2005): It is "the sum of all processes involved in the analysis and management of data based activity." It includes the sensory acquisition of data, use of a hierarchy of choice and nonchoice options involving a predefined set of comparisons, executing the option chosen, and further analysis and comparison utilizing executive functions (EFs). The end products (mental units) are the result, in part, of information processing.

It has been proposed that cognitive process should replace the term intelligence, since the concept would rely as little as possible on acquired knowledge (Strauss et al., 2006, p. 133). The present writer disagrees: Intelligence, or general mental ability, refers to the capacity to achieve practical consequences than would be less likely with weaker general ability. Learning environmental facts and considerations is contingent upon what we ordinarily consider to be intelligence: the ability to learn difficult and complex facts (declarative memory) and concepts, comprehension, appreciating and integrating complex situations, and the capacity to solve complex problems.

Among the characteristic traumatic disorders of cognition are slow reaction time and processing of information, poor performance after delay, imprecision of judgments and thinking, inability to use language to retrieve information and solve problems, sensorimotor dysfunctions, and loss of what has been learned up to the time of injury.

ATTENTION

Attention refers to a state of awareness in which the senses or other consciousness are focused selectively on aspects of the environment, body, or mental contents. It has been classified as to the source of the stimulus, although prefrontal and parietal cortices are also involved. Its complex role may be illustrated by visual attention. After one selects a target, it remains active for coordinated motor behavior, for example, looking at the object while turning the body and reaching. Shifting attention improves with time after brain trauma, more slowly in the more severely injured.

ATTENTION IN THE SOMATOSENSORY SYSTEM

- *"Bottom-up,"* ascending, mechanisms that occur after a particular stimulus. This requires minimal attentional resources. They are salient stimuli arising from the sensory cortex.
- *"Top-down,"* or descending, signals are those from a set or task, arising from the prefrontal cortex (PFC). It represents the ability to filter out irrelevant information and enhance the central representation of stimuli that are immediately relevant, for example, a pin prick (Hsiao & Vega-Bermudez, 2002). Signal strength is influenced by the quality of the encoded information and the top-down and bottom-up processes of signals. Competition at each level helps to eliminate distracting stimuli and to select the most salient stimulus.

SELECTIVE ATTENTION (DIRECTED)

Effective selective attention, in the presence of distractors, implies allocation of attentional resources to *target detection.* The effects of selective attention are observed at all levels of central nervous system (CNS) processing, becoming stronger as one moves more rostrally. Readiness to respond interacts with selective attention. This is the result of a choice (set) or an environmental demand, in which attention is concentrated on particular stimuli from among others. The sensory modalities do not function independently; there are limited attentional resources for processing information across sensory systems (Hsaio & Vega-Bermudez, 2002). Sensory processing may have a limited capacity to carry out more than one cognitive or perceptual task at a time. Attention mediates this selection. Saccade planning and visual attention activate centers of the frontal eye fields of the PFC, lateral area, and the superior colliculus. *Exogenous (stimulus-driven)* attention focuses upon salient, potentially important events. *Endogenous* attention focuses upon internally defined goals and against external distractions.

FUNCTIONING ATTENTION

Attention is required for efficient performance. It assists *information processing* by increasing neuronal responses to stimuli with different physical characteristics. Directed attention enhances information processing at attended locations by counteracting suppression caused by stimuli competing for limited processing resources. Attention components include sustained attention or vigilance (attending as long as necessary), shifting focus when required, flexibility, and divided attention.

Impairment

There is some overlap between brain regions particularly vulnerable to TBI and the structures believed to support attention. Brain activation is increased for some tasks after TBI. Adults with mild TBI demonstrate overactivation in the neural circuit mediating working memory (WM), sometimes activating the same areas as orthopedically injured, but recruiting areas bilaterally or in the opposite hemisphere. Task performance between the groups tended to be similar (Kramer, et al., 2008). Aspects of attention potentially impaired after TBI have been summarized (McCullagh & Feinstein, 2005): arousal/alertness receptivity to sensory information in preparation to make a response; selective attention, that is, selecting a target and inhibiting irrelevant stimuli; divided or shared attention when there are multiple sources of information or task demands; appropriate processing speed, which allows cognitive activities to occur; and supervisory control of top-down coordination from a particular set or task, which allocates limited attentional resources, particularly for nonroutine tasks.

Varying deficits contribute to attentional deficiency. Reduced information-processing speed is prominent. Reaction time is slowed proportional to task complexity. Some attention disorders are caused by disorders of the receptive field. Unless distraction is reduced by suppressing inappropriate stimuli, attention is ineffectual. Distraction reduces focused attention and prevents learning compensatory strategies or encoding information. This can mimic a memory disorder or offer the appearance of a problem-solving deficit. An attentional disorder may enhance impulsiveness, may prevent solution of a problem, or may cause a person to start many tasks without completing them. A social effect is shifting from a conversational topic without having dealt with it thoroughly (Pelham & Lovell, 2005). Changes of level of attention, information-processing speed, memory, and consistency of performance are characteristic of milder brain damage in the first week. However, these deficits can occur with accidents not involving the brain. While more characteristic of moderate TBI, and most spontaneous recovery after mild traumatic brain injury (mTBI) is complete by 3 months postinjury, there is a minority with persistent, significant cognitive defects. Poorer outcome follows a history of TBI or older age (McAllister, 2005; McCullagh & Feinstein, 2005). The effectiveness of sustained attention varies with both the severity of TBI and the complexity of the task. Three groups of children (mild; moderate; severe) were examined 24 months postinjury using the Continuous Performance Test. No significant differences were detected by simple reaction time or length of interstimulus interval. However, on the most complex task, which required speed, accuracy, and decision making, the procedure discriminated the mild and severe TBI groups (Catroppa & Anderson, 2003).

Spatial Attention

When visual fixation is changed (saccade), the same neural structures may be involved as those utilized for visual attention. Readiness to respond is also maintained by *set* (a temporary readiness to respond in a certain way to a specific situation or stimulus). It ensures that irrelevant data is not included in a task or does not overwhelm the activity. Cognitive efficiency will be influenced by the appropriateness of set. Attentional accuracy may be lost through distractibility, or maintained inappropriately causing *perseveration*. The elicited information is used to achieve either an *internal goal* (e.g., a complex cognitive activity such as writing, communicating one's needs, or relating to the demands of the social environment) or *to perform external actions* (household actions, employment, etc.).

Spatial attention improves perception of an object within an attended region. It is subserved by a highly interconnected *frontocingular-parietal network*. These are sites of heavy sensory and motor connection, receiving visual, somatosensory, auditory, and proprioceptive information. They contain representations of movements (eye, head, locomotion, reaching, and grasping). It is significant that the network has access to limbic system information concerning the motivational value or behavioral significance of particular objects (Reynolds et al., 2003).

SOME COMPONENTS OF ATTENTION

Alertness

General receptivity to stimulation, varying from very low in sleep to high in wakefulness. Alertness normally results in symmetrical alertness to the external world (Zomeren, 1992).

Neglect

This may be described as hemi-attention to a perceptual half of a perceptual field, possibly a defect in the orienting response, or a disturbance in the spatial distribution of spatial attention.

Focused Attention

This is a process of selection, since the individual cannot cope with all of the information registered by his senses, the bulk of which is irrelevant. Since there is a limited processing capacity, selection of the most relevant information is needed to reach the most satisfactory performance.

Distraction

This refers to the interference of selective attention by responses to stimuli which are irrelevant for task performance.

Divided Attention

Multiple simultaneous stimuli may be task-relevant. Nevertheless, due to limited information-processing capacity, multiple stimuli may lead to a decline in performance in one or more components. Processing of information may be relatively automatic, for example, the experienced reader, or controlled for nonroutine tasks and is slow and serial. A bottleneck occurs, since we can perform only one task at a time in WM. Thus, for dual processing tasks, efficiency is determined by the capacity of controlled processing and the strategy that is used to divide it among the subtasks or sources of information. With overload, negative effects are minimized by allocating processing capacity according to the relevance of the task or information.

Sustained Attention

Attention over longer periods involves both sustained focusing and sustained exclusion. *Time-on-task* involves variation of effectiveness due to practice effect, fatigue, boredom, or drowsiness. One may distinguish between *vigilance* (attention in a low-event-rate situation within an environment with a low level of stimulation) and *monitoring* (presentation of a continuous stream of stimuli requiring the detection of targets). *Lapses of attention* are short periods, perhaps seconds, with response omissions in a continuous task or extremely long reaction times. These occur during sleep deprivation. There is *intraindividual variation* even when the person is alert and active.

Supervisory Attentional Control

This is the effective allocation of processing capacity, including tasks involving focused or sustained attention. It involves flexibility, and is utilized to determine when attention will be shifted to other aspects of the stimulus situation, the sequencing of responses, and different processes according to whether the task or stimulus is routine or nonroutine.

COGNITION: FUNCTIONAL IMPLICATIONS

Cognition includes a highly varied group of integrated skills. It serves functions of daily living, from problem solving to coping with difficult or dangerous situations. It is a highly evolved function. Our species contrasts with those that respond to their environment with a relatively direct pathway between sensory input and motor output. The human species interpolates verbal intelligence and forethought into complex activities (Davachi et al., 2004). Therefore, cognition's definition affects the functions to which it can be applied. If it is defined narrowly as the highest level of thought and intellect, then cognition is independent of *emotion*. However, if conceived broadly as information processing, then emotion is dependent upon cognition, since there is cognitive input to emotional systems. The emotional processing itself is noncognitive. Whereas the goal of cognition is the elaboration of stimulus input and the elaboration of "good" stimulus representations, emotional processing has as its goal determination of the relevance of the stimulus for individual welfare (LeDoux, 1993). The association of TBI and depression with cognitive compromise creates complex assessment problems. Their relative contribution to cognitive deficit is controversial, perhaps additive or synergistic (Keiski et al., 2007). The cognitive contribution to emotional processing has been described as *emotional intelligence*.

COGNITION IS MORE THAN CLEAR THINKING

Reduced Efficiency Consequent to "Distracting Symptoms"

Not all cognitive loss is related directly to TBI. Symptoms reducing efficiency include pain, headache, reduced range of motion, and fears caused by dizziness and imbalance (Parker, 1995). Pain has an alerting function; anywhere in the body it can reduce performance efficiency. Pain interacts with cognition, depression, anxiety, and anger, affecting their nature and severity. One has to consider the effects of medication and reaction to chronicity (Grzesiak, 2002). Pain's concomitants, more than the pain itself, may produce cognitive impairment: emotional distress; somatic preoccupation and pain catastrophization; sleep disturbance; fatigue; and perception that daily activities could create stress. Sleep deprivation impairs cognitive and motor performance, and may produce hyperalgesic changes counteracting analgesic medications. Sleep disturbances associated with depression reduce the efficiency of executive functioning and information processing speed. This creates vulnerability to cognitive dysfunction (Martelli et al., 2007). Descending influences (cognitive, attentional, and emotional) affect the peripheral response. Beyond some level, its effects are disorganizing: attentional capacity, processing speed, memory, executive functioning, concentration, focus, and information processing (Anderson et al., 1990; Zasler et al., 2005). However, caution is needed since the capacity of a person with chronic pain to answer simple questions and perform simple tasks may encourage the examiner not to explore in depth, leading to nondetection of cognitive deficits.

Aware or Unaware?

Although cognition has been defined using such terms as information, planning, knowledge, reasoning, etc., it has been suggested that the human mind operates largely out of view of its owner, probably because it works best that way under many circumstances, that is, offering an efficient, powerful, and fast means of understanding and acting in the world. Consequently, people lack access to a broad range of mental processes (Wilson & Bar-Anan, 2008). This perspective is related to the *Freudian unconscious*, which describes a maladaptive system not up to the task of adaptive behavior in the real world. A different view of the unconscious is described as the *cognitive unconscious*, which is adaptive. Most information processing occurs automatically and effortlessly outside of awareness, because it is far more efficient than conscious deliberative thinking (Epstein, 1994).

Subjective or Objective

The intensity of feelings affects the nature of cognition processing and the components of planning. Information obtained from experience is more compelling than abstract knowledge, for example, textbooks and lectures. Information processing and decision making can be conceived as having a varied group of alternative processes. The following alternatives will alert the clinician to a range of possible cognitive styles. These might suggest psychopathology or a TBI based on the preponderance or lack of one of these cognitive processing styles. Precise assessment would be difficult; yet, taking these influences into consideration offers information about outcome and functioning in the community. Initially, one might consider the domains of emotion and intellect, but some higher power has determined that clinical practice should be difficult.

Information Processing: Experiential or Rational

- The experiential informational processing system is described as holistic; motivated by the pleasure-pain system; with associationistic connections; affected by past experiences; encodes reality in concrete images, metaphors, and narratives; more rapid processing to achieve rapid action; broad generalizations with stereotyped thinking; more crudely integrated and context-specific. We are seized by our emotions, that is, they are experienced passively and preconsciously. Nonrational thinking is highly prevalent and very satisfying. It is complex, with a long evolutionary history. At its lower levels, it processes information rapidly and effortlessly. Emotional arousal and experience shift processing to the experiential system. The Freudian primary process may explain dreams, drug-induced states, and psychotic reactions. However, more adaptive subconscious cognitions that automatically organize experiences and direct daily behavior can be better explained by an experiential, intuitive system.
- The rational system is analytical logical, or reason-oriented. Conscious appraisal mediates behavior and encodes reality as abstract symbols, words, and numbers. Processing varies between slowness for delayed action and rapid changes with the speed of thinking; more differentiated yet integrated. If we are in control of our thoughts, the experience is active and conscious. The rational system operates in the medium of language, is deliberative, and capable of long-term gratification. It is inefficient in responding to everyday events (Epstein, 1994).

Social Cognitive Neuroscience

Social disturbance is characteristic of concussed patients. We have emphasized trauma and its impairing physical, economic, and cosmetic effects. Yet, understanding and remediating social impairment requires that different and complex mental functions be considered. Three levels of performance have been suggested: *social* (motivational and social factors); *cognitive* (information-processing mechanisms give rise to social-level phenomena); and *neural* (brain processes identified with particular cognitive processes). Rather than utilizing cognitive neuroscience's analysis of emotions as though they were a unitary stimulus property, they may be a situation-dependent property of a person. The gestalt of cognition varies with mood, for example, emotion, mood, motivation, self- versus other-focused attention, culture, and analytic versus intuitive mental sets (Ochsner & Lieberman, 2001).

Social cognition has automatic and controlled processes, involving different neural regions and different foci, inviting further study: understanding others; understanding oneself; controlling oneself; and processes occurring at the interface of self and others (Lieberman, 2007).

COGNITIVE CONTROL

Although neurological control has been defined as the ability to take charge of one's actions and direct them toward future aims (Miller & Wallis, 2003), in the present writer's opinion, the attribution to "cognition" is grossly incomplete. It does not include all of the subjective, affective, historical, and unconscious bases for action, including the irrational or self-destructive components of actions in life (Parker, 1981, 1983).

The distinction is made between direct reactions to an environment that does not tax our attention and sudden demands for a very different response to increase our safety. The latter may require knowledge from previous experiences to weigh alternatives and consequences. In a new situation, behavior is not triggered by the environment ("bottom-up"); rather, knowledge shapes and controls it ("top-down"). "Top-down" signals are derived from knowledge about the current task (frontal cortex). The "bottom up" capture of attention is driven by salient properties of stimuli (parietal sensory cortex) (Buschman & Miller, 2007).

Characteristics of Controlled Processing

- Controlled behaviors are goal-directed and learned: Directing behavior towards unseen goals requires knowledge of what goals are available and what behavior has been successful in achieving them in the past.
- Cognitive control is multimodal: The external world is processed in separate systems; a system for cognitive control must have access to and encode relationships between a wide range of brain systems.
- Controlled processing depends on maintaining relevant information "online": WM is needed to maintain goal-related information while we wait for new information or wait for a more appropriate time to act.
- Controlled processing is limited in capacity: While some automatic functions can be performed with more skillful acts, with some actions, only one or a limited number of functions can coexist.
- Limited capacity requires selection: Attention is one way, in which some processes are favored over others in order to influence them.
- Controlled behaviors are flexible: They are altered according to variations in surroundings, and are temporarily interrupted and resumed, as opposed to operating the same way every time (Miller & Wallis, 2003).

NEUROLOGICAL SUPPORT OF COGNITION

Caution is needed when attempting to associate brain lesions with cognitive dysfunction. Since the neurological support for cognition is so anatomically widespread, it is highly susceptible to diffuse axonal injury (DAI) consequent to impact and acceleration/deceleration trauma. Reduced performance is only partially attributable to impaired destruction of particular neural centers and cranial nerves.

Wide Distribution

Broad regions contributing to semantic processes are widely distributed in the brain, that is, not merely restricted to areas associated with Wernicke's and Broca's conditions or with cognitive psychological models positing a central semantic store, etc. (Bookheimer, 2002). Cognitive loss is associated with white matter lesions. Thus, cognition is an extraordinarily complex domain involving the integration of many cortical and subcortical structures. This renders it vulnerable to diffuse TBI, and requires consideration of neural components not usually considered involved in verbal ability. There are different neurobehavioral outcomes from a diffuse TBI (e.g., the postconcussive symptoms) and focal lesions (e.g., aphasia, which may be superimposed on postconcussion syndrome [PCS] symptoms). An example of neurobehavioral association (cognitive function) is cerebellar circuits involving the prefrontal and parietal cortices. Impairment of language may conceal other cognitive disorders. The description of many cognitive functions will enable the practitioner to clinically recognize potential deficits, and then select appropriate examination procedures. Neuroanatomical associations are difficult after concussion; PCS symptoms usually reflect subtle impairment from diffuse injury.

Holistic, Localization, and Modular Representation

Holistic: Historically, there has been modification of Lashley's holistic concept of brain function and lesion effect. He stressed that every portion of a particular brain region carries out a particular function in a given person; the larger the neuronal pool, the more efficiently that function is performed, and lesions of a particular size in a region have equivalent effects, and the magnitude of a functional deficit is proportional to the size of the lesion in the appropriate area.

Localization: Evidence against the holistic concept comes from the familiar association of lesion localization with particular functional defects. A lack of complete description of lesion effects arises from the observation of *"graceful degradation"* of performance of specific domains of language after TBI. Disruption of processing of particular language functions (e.g., words) may be inversely proportional to preinjury exposure. Such patterns of performance may occur after lesions to systems in which information is represented and processed in massively parallel, distributed forms. This theory of language requires an integration of sensory and motor processes. This theory deals with words, not other levels of the language code. The syndromes are composed of many processing deficits, which differ among patients. Yet, it is considered reasonable that different neural areas support different language functions, and each of the areas works as a whole. Holism should be applied on the right scale (Caplan & Gould, 2003).

Modular representation: Knowledge is not stored as a complete representation. It is divided into small categories, which are stored separately. The example is offered of a lesion to the left temporal lobe's association area, which can obliterate knowledge of living things, especially people, while leaving knowledge of inanimate objects quite intact. Our sense of self ("I") is constructed through independent circuits, each with its own sense of awareness, carrying out separate operations in the two cerebral hemispheres. This is the lesson of the split-brain procedure of Sperry and Gazzaniga (Saper et al., 2000).

Neural Networks

The hypothalamic histaminergic system is involved in the regulation of physiological functions, and learning and memory. It also projects to the cerebral cortex. The sensorimotor cortex constitutes 10% of the 2.5 m^2 surface; the remaining association cortex subsumes unimodal, heteromodal, paralimbic, and limbic areas. The association cortex mediates integrative processes subserving cognition, emotion, and behavior.

Large-scale Networks

Rather than centers for particular functions, it is now believed that cognitive and other behavioral domains are integrated into clinically relevant *large-scale neural networks*: (1) the perisylvian network for *language*; (2) parietofrontal network for *spatial recognition*; (3) occipitotemporal for *face and object recognition*; (4) limbic for *retentive memory*; and (5) prefrontal for *attention and behavior.*

Clinical Implications of this Network Hypothesis

- A single domain (e.g., memory) can be disrupted by damage to any area belonging to the network.
- Damage to a single area can disrupt those functions participating in the network.
- Damage to a network component can give rise to temporary impairment if other network parts undergo compensatory reorganization.
- Different anatomical sites within the network may offer a relative specialization for different behavioral aspects of the particular function (Mesulam, 2005).

Co-morbid Influences upon Cognition

Cognitive effectiveness is influenced by altered consciousness, physiological disorders, fatigue and illness, faking, poor mental control, and so on. There is a significance to the high heritability of brain structure (corpus callosum [CC] and ventricles, but not gyral patterns or the volume of the hippocampus) and gray matter volume. Since the importance of environmental influence on IQ was four times stronger in the poorest families than higher-status ones, it was inferred that nature matters more on the high end of socioeconomic status, and nurture more on the low end (Toga & Thompson, 2005). Vestibular dysfunction and brain trauma are associated with cognitive complaints. Perceived disability in dizzy patients is higher in patients with, rather than without, brain trauma; cognitive complaints are likely due to concurrent affective disturbance (Gizzi et al., 2003).

BRAIN RESERVE CAPACITY

The concept of brain reserve capacity (BRC) contributes greatly to our understanding the outcome of a particular injury in a person of a certain age, and the frequent imprecision or errors of our estimates. It is conceived as an extra level of ability that permits some loss after a TBI so that the person can adapt with little or no apparent impairment. Repeated loss of the reserve, or a significant trauma, creates difficulty in coping or adaptation. A high degree of imprecision is introduced into prediction and assessment of outcome of a person with a TBI due to the fact that seemingly significant traumata may have no acute effect upon adaptive capacity. In fact, there may be no direct relationship between the degree of brain pathology and the clinical manifestation of that damage in terms of the cognitive deficit. Individuals with higher pretrauma cognitive skills may have a greater ability to optimize emotional functioning through alternative cognitive strategies, recruitment of alternative cognitive strategies, recruitment of alternative neural pathways, or more efficient use of brain networks. In addition, they may have better family support or access to external support. In contrast, in a subgroup of head injury survivors, depression is associated with a lower premorbid intelligence (Salmond et al., 2000). However, observable cognitive ability may decline prematurely, perhaps decades later. This writer will utilize the concepts of *performance demand* and *ceiling of performance* as contributions to understanding the paradox of neurotrauma with no apparent effect. In addition to head injury, the theory was explicated with Parkinson's disease, Alzheimer's disease, Down's syndrome, poliomyelitis, multi-infarct dementia, etc. A theory to explain the absence of an acute effect of trauma, and its apparent later expression as an early start of mental decline, is the concept of BRC (Satz, 1993). BRC may be a function of overall and regional brain size, dendritic branches, neuronal efficiency, and indirectly, adaptability, intelligence, and education. It provides a margin of safety, that is, the brain can lose a substantial amount of tissue without apparent consequences. BRC's explanation for continued adequate performance is the brain's neurological redundancy, capacity to compensate for functional loss, and the extended period of neuropathological development (e.g., the neurofibrillary tangles and plaques of Alzheimer's disease, and neurochemical processes), which only later creates observable symptoms.

Threshold Effect

The threshold effect describes the phenomenon, that is, late symptom onset. There are individual differences in BRC, which alter the threshold for clinical appearance of dysfunction. The cumulative effects of the neuropathological process progress (summate) until the characteristic clinical picture appears. By inference, there is a nonobserved, subclinical condition. The occult condition may appear earlier under appropriate challenge, which exceeds the BRC. The capacity for apparent neurobehavioral posttraumatic adequacy is the BRC. General intelligence and educational level are indirect, imprecise measures of it. Risk factors for earlier neurological disorders include low level of education, alcoholism, chronic drug abuse, prematurity, prior head injury, HIV-1, and learning disability. Lower education may also be a marker of positive serostatus for HIV-1 infection.

ENDOCRINE AND ENVIRONMENTAL GENDER DIFFERENCES

Sex Differences in Cognitive Ability

As if neuropsychology was insufficiently complicated, one has to consider gender trends in cognitive performance. Steroid hormones are involved in the sexual differentiation of the vertebrate brain during early development. The size of hormonal effects is variable, and does not diminish the role of environmental or social factors. Sex steroids affect cognitive and motor functions outside the hypothalamus and pituitary, perhaps establishing behavioral markers of sex. Sex steroids modulate a variety of cognitive and motor processes in the adult brain. The greatest detected effect is estrogen in females. The effects are restricted to, or are most pronounced, in functions manifesting sex differences. However, with regard to intellectual abilities, the roles of nature and nurture are not clearly defined. During the *menstrual cycle*, verbal output (speed and accuracy of articulation) may be improved during phases with high estrogen content. After a potential TBI, one would consider not only temperamental characteristics at baseline and potential posttraumatic change, but a prior gender-related cognitive advantage. Thus, a given score after trauma represents an occult injury since the baseline level had a gender advantage.

Due to the normative standards of intelligence test construction, there are no sex differences in IQ. There may be a prenatal hormonal effect upon spatial functions. The direction of the advantage varies for different abilities. There is usually considerable overlap in scores. Sex differences in verbal abilities are usually smaller than spatial abilities, but vary tremendously, depending on the exact verbal function assessed. A list of *sexually differentiated abilities* is offered (Hampson, 2002).

Female advantages: Verbal skills (fluency, rate of speech acquisition, spelling, grammar); computation (accuracy, procedural knowledge of arithmetic and mathematics); fine motor skills and finger dexterity; short-term memory, including object location.

Male advantages: Spatial ability (mental rotation, route learning, visualization of spatial relationships); mathematical reasoning and problem solving; gross motor skills involving strength.

NEUROCHEMICAL EFFECTS UPON MEMORY

Neurotransmitters

Memory and learning may be partially explained by the strengthening of some synapses and the growth of neural processes. The *N*-methyl-D-aspartate (NMDA) receptors bind to the neurotransmitter glutamate. The postsynaptic cells become modified and potentially more sensitive to neurotransmitter release. Then a retrograde signal (*neurotropin*) may be released from the postsynaptic neuron that influences presynaptic cell growth or the level of transmitter released by that axon (Wiencken-Barger & Casagrande, 2002, citing Donald Hebb). In addition, central *catecholaminergic* dysregulation seems to be involved (McAllister et al., 2006) in regions important in *WM*. These circuits overlap with regions commonly vulnerable to damage in TBI. The *serotonin* system participates in a wide range of systems affecting cognition. It plays a role in modulating the responsiveness of cortical neurons, physiological regulation via the hypothalamus, sleep–wake cycle, affective and sexual behaviors, food intake, and thermoregulation (Saper, 2000b). It is located primarily in the raphe nuclei, located along the midline of the brainstem from the midbrain to the medulla. Pre- and postsynaptic *serotonin (5HT)* receptor is involved in the acquisition and consolidation of learning (Meneses et al., 1998). *Monamines (dopamine and serotonin)* have been implicated in the reinforcing actions of stimulants, opiates, nicotine, alcohol, and tetrahydrocannabinol (THC) (Koob, 2003). Dopamine plays a central role in reward-related learning (Hyman et al., 2006). It marks the motivational significance and value of particular experiences, cues, or responses. Thus, it brings together information about the motivational state with sensory information (interoceptive or environmental) and stored memories. Natural rewards and addictive drugs increase synaptic dopamine in the *nucleus accumbens* (ventral striatum).

Cytokines

Another possible interfering effect upon memory is indicated by the finding that *cytokines* (released during enhanced immune response) inhibit long-term potentiation in hippocampal slices. The effects of the cytokines are so dire, that it is believed that they are normally modulated by cytokine antagonists. These become available during host challenge such as inflammation, that is, a consequence of an accident-related injury. One of these is the neuropeptide *á-melanotonin-stimulating hormone*, important in communication among neuroendocrine, immune, and the CNS (Catania et al., 1994).

Cellular Damage

Cellular damage affects memory, that is, shear forces affect the *second messenger* system. These are large molecules within the cell membrane that modify and amplify signals brought by neurotransmitters and mediators. These seem to play a role in such complex neurological processes as encoding memory; therefore, changes may account for memory changes occurring after neurotrauma (Bullock & Nathoo, 2005).

FRONTAL LOBE INTEGRATION

Although many disorders are described as consequent to frontal lobe injuries, this may not be the case. Consider its bidirectional interaction with all other areas of the brain. Since frontal lobe damage may be part of a diffuse injury, its functioning is affected by the lesions of other parts of the brain. Disorders will be complex, varied, and not always described by familiar syndromes. Frontal lobe interactions influence a wide range of sensory, motor, mood, and physiological functions. The dynamic and flexible modulation in prefrontal neurons, as well as between the PFC and other cortical and subcortical areas, participates in the PFC contribution to executive control (Funahashi, 2001). Traumatic prefrontal lesions are associated with psychiatric diseases (attention deficit hyperactivity disorder [ADHD], autism, obsessive-compulsive disorder) and neuropsychological impairment (planning, behavioral inhibition, personality changes, reduced creativity). The PFC has important reciprocal connections with monoaminergic systems whose cell bodies are in the brainstem. Brainstem lesions may lead to symptoms similar to prefrontal lesions. Alterations in basal ganglia catecholamine mechanisms are also implicated in the deficits of EF observed in ADHD (Salgado et al., 2007).

Prefrontal Cortex

Complex adaptation with respect to future consequences is associated with the evolution of a greatly enlarged PFC. It appears to be one of the last brain regions to mature. Associated memory and attention functions develop behaviorally and physiologically during childhood and adolescence. There are both common and discrete regions of brain activity in regions of the PFC. Increasing cognitive capacity during childhood may coincide with a gradual loss of new synapses, with a strengthening of remaining synaptic connections (Casey et al., 2000). The continued development of the frontal lobes subserves cognitive functions, executive control, behavioral inhibition, temporal ordering, WM, and allocation of attention to specific tasks and processes. Patients with frontal lobe defects have a deficit in the allocation of attention, for example, shifting attention in the process of moving information from one site to another (e.g., Wisconsin Card Sorting Test, planning, estimating).

The PFC contributes to *memory organization* at the times of learning and retrieval, including temporal order. Information processing is achieved by a gate function (filters irrelevant information) or inhibition of activity in posterior cortical regions. Patients with prefrontal lesions are less likely to engage in useful organizational strategies (subjective organization or category clustering) and temporal order. Structural information aids semantic organization that patients with prefrontal lesions cannot detect by themselves. Prefrontal injuries impair *metamemory*, that is, the belief that one knows or does not know the answer to a question, and knowledge of strategies that can aid memory.

The Dorsolateral Prefrontal Cortex

The dorsolateral prefrontal cortex (DLPFC) participates in *WM* by maintaining internal representations of the spatial locations of visual targets and manipulating them for planning adaptive behavior. It projects to the posterior cingulate cortex that is involved in attention and eye movements (Pavuluri et al., 2005). Brodmann's area 46 seems responsible for maintaining a *representation* across time to mediate the response when the stimulus has been withdrawn. The perceptual representation system also includes the extrastriate occipital cortex, the temporal lobe, and the frontal cortex. WM and long-term memory may substitute for stimulus representation (Holyoak & Kroger, 1995).

Orbitofrontal Cortex

Controlling impulses (self-regulation) is part of individual decision making and aids harmonious social interactions. The orbitofrontal cortex (OFC) receives information from all sensory modalities, including polysensory and secondary and tertiary sensory areas. Damage to the *OFC* results in intact cognitive abilities but impaired everyday decision making. Multisensory features of a reward are integrated to determine its value. The OFC also projects to limbic, hypothalamic (autonomic), and brainstem structures (periaqueductal gray). The OFC processes award by integrating multiple sources of information to derive a value signal, that is, how rewarding the reward will be. This value is held in WM to be used by the lateral PFC to plan and organize behavior to obtain the reward. Then, the medial PFC (MPFC) evaluates the action in terms of its success and the effort that was required (Wallis).

Lateral Prefrontal Cortex

Dysfunctioning of the PFC interferes with selecting and organizing neural activity (Shimamura, 1995). The lateral PFC enables the construction of plans and organization of behavior necessary to obtain goals and outcomes. It is possible that information about the value of outcomes is held in WM, enhancing formation of action plans and predicting and monitoring expected outcomes. Lesions to the PFC, particularly those that are right-sided, reduce the capacity to maintain self-interest, that is, *riskier behavior occurs, that is, "out of character" decision making and disregard for negative consequences* (Knoch & Fehr, 2007).

Yet, the PFC is not alone in modifying impulsivity. *Serotonin* has been linked to an inability to wait, specifically, choosing a smaller immediate reinforcer rather than a larger but delayed outcome. In short, delayed rewards do not motivate behavior (Schweighofer et al., 2007).

The *MPFC* uses information from the OFC concerning the value of the outcome. Positive and negative consequences have to be evaluated to determine how rewarding a reward actually is. Information is obtained from the lateral PFC concerning probability of success in order to perform a cost–benefit analysis and generate the overall value of an action. Lesions of the OFC can result in failure to perform stimulus–reward reversals, that is, an inability to learn new contingencies causes *perseveration* due to lack of inhibition of prior contingencies (Wallis, 2007).

The *DLPFC* is part of a circuit including the cingulate cortex, basal ganglia, and thalamic nuclei, projecting back to the DLPFC. This signals the value of expected reward and actual outcome, and EFs (DeLong, 2000; Lee & Seo, 2007). It serves the EF by *regulating choices*, both those that are impulsive and those based upon self-interest. Its state-based influence upon the effect of reward to influence action selection contrasts with the lateral PFC in which motives (basic desires) influence choice (Balleine et al., 2007).

The *striatum* is the main input unit of the basal ganglia. It is part of a circuit involved in evaluating current rewards to guide future actions. Rewards are desirable outcomes that influence behavior. The striatum receives projections from the motor cortex and dopaminergic input from the substantia nigra, other midbrain nuclei, and the amygdala, which is involved in emotional processing. It seems to integrate information regarding cognition, motor control, and motivation. The striatum seems to be involved in reward (appetitive processing), aversive processing, and fear conditioning. By processing reward-related information, it mediates goal-directed behavior (Delgado, 2007).

TEMPORAL LOBE

Medial Temporal Lobe

Injury to the frontal lobes creates far less *learning impairment* than do lesions to the medial temporal lobe (hippocampus) or diencephalic midline (thalamic nuclei). The latter injuries produces organic amnesia, that is, extreme difficulty remembering information and events encountered since the onset of amnesia (Shimamura, 1995). Recognition and its components, familiarity and recollection, function together. The hippocampus is critical for recollection. The parahippocampal cortex also contributes to recollection and the perirhinal cortex is necessary for familiarity-based recognition. Encoding of distinct items (people, objects, events) takes place in the perirhinal and lateral rhinal cortices. These representations, with projections to the neocortical ("what") areas, support familiarity. Item information is combined with contextual ("where") representations formed in the parahippocampal cortex and medial entorhinal area. The hippocampus associates items and their context during item and contextual association in support of recollection.

THE CORPUS CALLOSUM AND COGNITION

Shearing injury to the CC is expected to impair function. The CC mediates excitation of one area causing inhibition of the contralateral cerebral area with excitement in the surrounding area (Baynes, 2002). Most callosal fibers are excitatory. Attention is a limited resource that depends partially on the CC for allocation. However, inhibitory processes occur in the analogous area of the opposite hemisphere (*homotopic callosal inhibition*) with excitation of the immediately surrounding area. Greater task difficulty leads to participation of the CC in division of labor for bihemispheric processing. This is involved in interhemispheric transfer of information. Changes in the CC have been related to the lateralization of language, disposition of attention, mnemonic processing, conscious behavior, disorders of learning, and schizophrenia (Baynes, 2002).

CEREBELLUM AND COMPLEX NEUROBEHAVIORAL FUNCTIONS

Cerebral association areas subserving higher-order behaviors are linked with the lateral hemispheres of the cerebellar posterior lobe. This activation is modified by feed-forward (basis pontis) and feedback loops (deep cerebellar nuclei via the thalamus), and reciprocal connections between the cerebellum and hypothalamus (Schmahmann, 2006). Cortical fibers project to pontine nuclei of the rostral pons (basis pontis), primarily from the premotor, motor, somatosensory cortices, and adjacent parts of the parietal lobe. They are joined to a lesser extent by visual, limbic, and association area fibers. They then form the middle cerebellar peduncle and project almost completely to virtually all parts of the contralateral cerebellar cortex (primarily the lateral hemisphere) for visual, limbic, and association fibers. The lateral hemispheres may be involved in the planning of movements acting by influencing the output of the motor cortex. In anticipation of a movement, activity occurs in the cerebral association cortex, then dentate nucleus, and then motor cortex. The dentate nucleus projects via the superior cerebral peduncle to the contralateral midbrain red nucleus, thence to the ventral lateral and ventral anterior thalamic nuclei. Cerebellar damage can be associated with a variety of cognitive or behavioral disturbances. It seems that connections between the lateral cerebellum and association cortex are involved in *cognition*, while *affective and autonomic* functioning involves connections between the medial cerebellum and limbic cortex, and also cerebellum and hypothalamus (Nolte, 2002, pp. 265, 498–505, 507). The dentate nucleus serves as the anatomical substrate for cerebellar influence over cognitive and visual-spatial functions. It is activated during tasks involving short-term WM, rule-based learning, and higher EFs (Dum et al., 2002). The cerebellum projects via the thalamus to several prefrontal areas involved in a wide range of cognitive processing (Van Mier & Petersen, 2002).

The cerebellum has been implicated in nondeclarative (implicit) learning; classical conditioning (eyeblink) and procedural learning; judgment of explicit time intervals; implicit learning, including classical conditioning and visual-motor procedural learning; verb generation; and solving pegboard puzzles. Dysfunctions after lesions involving the hemispheric regions of the cerebellar posterior lobes include: EF disorders associated with lesions of the hemispheric regions of the posterior cerebellar lobes: (planning, set shifting, verbal fluency, abstract reasoning, WM), spatial cognition (visual-spatial organization and memory), and linguistic processing (agrammatism and dysprosodia). Cerebellar circuitry detects and recognizes event sequences, rather than planning and executing them. Neurobehavioral deficits may occur in the absence of the cerebellar motor syndrome. The more general learning function, as opposed to simple motor learning, is illustrated: Patients with focal cerebellar lesions manifest procedural deficits that are not correlated with the degree of motor impairment. They occur regardless of the side of cerebellar damage and the somatic side involved in the motor execution of the procedure (Molinari et al., 1997).

Since the cerebellum participates in language, decision making, and affect, independently of motor defects, cerebellar motor disorders are a marker for neuropsychological cognitive study. The cerebellum and its associated brainstem circuitry participates in *sequential motor learning*. The lateral cerebellum seems involved in central timing processes determining when a response will be made, while the medial area involves execution of the timed response. Other motor learning features include coordination of multijoint movements, eye–hand coordination, error detection and correction, and control of motor timing (Van Mier & Petersen, 2002). Learning new movement trajectories is relayed via the dentate nucleus to contralateral motor and premotor cortices (Pocock & Richards, 2004). Cerebellar functioning is *lateralized*.

The cerebellum may participate in an interactive cortical–cerebellar network that initiates and motors the conscious retrieval of *episodic memories*. Positron emission tomography (PET) study suggested activation of these structures: right lateral cerebellum, left medial dorsal thalamus, left medial and left orbital frontal cortex anterior cingulate, and a left parietal region. This network is consistent with known anterior projections of the cerebellum and multimodal integration circuitry

(Andreasen et al., 1999). The neocerebellum and dorsolateral cortex are co-activated. The PFC may also play a role in motor function. Hypometabolism in either the PFC or the lateral cerebellum occurs after damage to the other structure. Cognitive sequelae are similar (although not always identical) when either of these structures is damaged (Diamond, 2000).

Cerebellar systems may play a critical role in the early acquisition stages of motor and visuomotor skill learning, and the precise timing of nonmotor functions. The findings suggest that the neuronal representation (engram) of the learning is not stored in the cerebellum, but may be mediated by cerebral cortical–subcortical (basal ganglia) systems. More activity is found with more difficult tasks (Barinaga, 1996; Mostofsky et al., 2000). The theory that the cerebellum may subsume verbal WM through covert articulatory rehearsal has not been supported (Ravizza et al., 2006). It may contribute to verbal WM through initial phonological encoding, strengthening memory traces, or correcting degraded sensory information that would be less prone to decay or error.

Prosodic disturbance has been observed as part of the *cerebellar cognitive affective syndrome* (Schmahmann, 2006; Schmahmann & Sherman, 1997). This is a generally unrecognized syndrome with language, personality, executive, visual–spatial, and sequencing disorders, reminiscent of the frontal lobe syndrome and other conditions. *Dysregulation of affect* occurs when lesions encroach upon the vermis.

SUBCORTICAL

The thalamus and basal ganglia participate in cognitive processes (Crosson & Haaland, 2003). Thalamic mechanisms, under the guidance of the frontal cortex, engage those elements of neural nets necessary to perform a semantic task. The thalamus and pulvinar bind linguistic details into a unitary object representation (e.g., "desert" and "humps" represent the known object "camel"). The basal ganglia are involved in controlled linguistic processing (i.e., attention is deliberately focused on relevant stimuli, e.g., single words of grammatical judgments). The right caudate nucleus and putamen are active in word but not nonsense syllable generation. The left basal ganglia are involved in lexical, semantic, and syntactical processes.

COMPONENTS OF COGNITION

Cognitive Processing

Luria (1980, pp. 348–60) asserts that the first stage of intellectual activity is analysis of the conditions of the task and identification of its most important elements. Resultant hypotheses give subsequent activities a goal-directed and selective quality. This shapes the "connections" formed with comparison between the attempts made and the original plans ("error monitoring"). Luria suggests that frontal lobe lesions disturb this complex process of intellectual activity.

Frequent cognitive activity is associated with reduced incidence of mild cognitive impairment and less rapid decline in cognitive function. The capacity to be cognitively active in old age seems to reduce vulnerability to *Alzheimer's disease* and may represent earlier *socioeconomic advantage* (Wilson et al., 2007). Cognitive performance is composed of many integrative functions: *reception* (selection, acquisition, classification, and integration of information), *memory and learning* (storage and retrieval), *thinking* (organization and reorganization of information), and *expression* (the means by which information is communicated and acted upon). It is supported by stimulation, focused attention, concentration (attention maintained for a useful interval), and processing of stimuli and mental fragments into preliminary mental products. Under overall direction of numerous subsidiary processes described as executive control, these cognitive fragments remain available (WM) while they are processed into usable ideas. These are further developed by such processes as analysis, synthesis, imagination, comparison, etc., into a mental product.

Cognitive Ability

The complexity of general mental ability, for example, intelligence quotient or IQ, is poorly represented by a unitary IQ score. Although scores on varied cognitive procedures (e.g., verbal and nonverbal) converge around the same level, individual scores vary in differing degrees from the central tendency (Lezak et al., 2004, pp. 19–21).

Among the contributing cognitive processes are attention (focusing and changing), planning, simultaneous processing, successive processing, information processing, multitasking, control of lower cognitive processes. foresight. error monitoring. and maturity of problem solving (preplanning vs. trial and error). These are not ordinarily considered to be mental ability as such. Moreover, cognitive efficiency may vary substantially between various stimulus types (verbal, visual, symbolic) and environmental conditions (structured, e.g., IQ and academic tests), unstructured (Rorschach), and visuoconstructive (drawing). This influences mental ability without being identified with it. Nevertheless, the planning and expression of a complex activity requires performing a set of activities. This is a semantic process, with hierarchical, temporal, and syntactic components. The activity of mental units is unpredictable, varying with situational dynamics.

In a study of 33 patients with persistent concussive complaints, there was an estimated loss of Wechsler Adult Intelligence Scale-Revised (WAIS-R) Full Scale IQ of 14 points (Parker & Rosenblum, 1996).

Analysis, Synthesis, Abstraction, and Concept Formation

The undamaged brain sees relationships between certain kinds of objects or ideas, forms categories, and understands the implications of a situation beyond the immediate or obvious. A new, complicated, or vague situation may be divided into meaningful parts and then integrated into a new idea, or resynthesized for a better or different solution. Former experience is applied when events and things are seen as belonging to a certain category. Assignment to a category may yield a feeling of familiarity, so that prior experience (cognitive and emotional) is applied to a situation, and new events and stimuli are not experienced as totally different. Conceptual thinking requires holistic processing, that is, integrating information, concepts, or potential outcomes. Maintaining an abstract attitude (a high level of cognitive performance) involves a change from dealing with the obvious or specific ("*the concrete attitude*") to concern with the essentials of a particular matter, the ability to keep different aspects of a situation in mind and shift consideration from one possibility to another, to think symbolically or metaphorically, and to leave the present, immediate situation and to think ahead and plan for the future and its consequences.

INTELLIGENCE

Clinical example: Intellectual loss: A professional woman after a slight blow to the head: "I used to be able to think around the question or give the answer that people expected. Now I can't. I can think but I can't reach that level." Her supervisor queried her, since she had apparently retained verbal ability: I can't understand how you have this problem, because when I speak to you there doesn't seem to be a problem. Intelligence involves the ability to apply knowledge to solve problems and to comprehend one's environment. Performance level may be influenced by the nature and quality of training and life experience, developmental disorders, generalized or lateralized brain injury, motivation, sensory disorders during development or currently, etc.

General Intelligence

This is a unifying concept reflecting mental ability that is used for successfully engaging in complex adaptive, goal-directed behavior. Intelligence reflects the range, level of difficulty, speed of comprehension, and complexity of information that can be comprehended, and problems that can be understood, solved, and learned, through reasoning processes or relatively intuitively (Parker, 2001, p. 250). Recently, the capacity for self-understanding has become a significant component. Caution: Nevertheless, the level of ability within a general estimate is not uniform, usually varying, sometimes considerably, between defined components or subtests. Moreover, a function's level can vary according to the procedure, for example, complexity, whether fluid or crystallized, structured or unstructured. This description does not deviate greatly from *Boring's* infamous principle (cited by Sternberg & Kaufman, 2002) that intelligence is what intelligence tests measure.

The application of intelligence utilizes practical experience, formal learning, and many cognitive functions, including such EFs as foresight (anticipation of consequences) and reality testing (detecting errors before or after action). The clinician might consider that assessing the effective application of a person's intelligence, regardless of neurological integrity, requires study of the shaping influence of identity, emotional reactions, motivation and prejudices, and moods upon problem solving.

In contrast, the concept that intelligence is a relatively global phenomenon has been denied to the point of denying the utility of summing scores that vary "in any way" (Lezak et al., 2004, p. 335). Closer to the present writer's view is that it is difficult to define or estimate "basic" intelligence because competence actually reflects lifetime experience and training, academic accomplishment, and the consequences of injury, illness, and stress. Assessing intelligence also implies the including circumstances and necessities, followed by the person's capacity to find a practical solution to a problem.

Ecological Validity

A worldwide survey of how local communities defined intelligence included many foci that would be ignored or not given high value in traditional definitions offered by Western academic professionals or many clinicians. Included were metacognition (ability to understand and control oneself); depth or processing rather than speed; interpersonal intelligence (PI); intrapersonal intelligence; intellectual self-assertion; intellectual self-effacement; de-emphasis upon verbal skills; memory for facts; noticing, recognizing, and determining mental efforts, feelings, and opinions; practical skills and speed; speaking less; skills that facilitate and maintain harmonious and stable intergroup relations; social responsibility, cooperativeness; obedience; depth (listening rather than talking); seeing all aspects of an issue; placing the issue in its overall context; practical and adaptive rather than school intelligence; and social competence skills.

Biological Approach to Intelligence

When considering biological approach to intelligence the following needs to be considered speed of neuronal conduction; economy of glucose metabolism reflecting less requirement for effort; brain size (in men, a larger left hemisphere predicted verbal ability more than nonverbal ability, while in women it predicted nonverbal better than verbal ability); genetics (prediction of IQ scores; environment and heredity are hard to separate, and probably, within the family, work together to produce phenotype intelligence) (Sternberg & Kaufman, 2002).

Broad Theories of Intelligence

The examiner may find it useful to consider intelligence from different perspectives than the familiar psychometric IQ scores.

Multiple intelligences: Not a single unified intelligence; rather, a set of relatively distinct, independent, and modular multiple intelligences.

Successful intelligence: The ability to adapt to, shape, and select environments so as to accomplish one's goals and those of our society and culture. This requires capitalizing on our strengths and compensating for our weaknesses.

True intelligence: The basic aspects are neural, experiential (what has been learned from experience), and reflective (strategies and cognitive monitoring).

Bioecological: There are multiple cognitive potentials, which are developed and expressed differently in different contexts.

Emotional intelligence: The ability to understand, perceive accurately, appraise, and express emotion, and to regulate emotions to promote emotional growth.

SOME COMPONENTS OF INTELLIGENCE

Personal Intelligence

This capacity is conceptualized as a basis for the individual's well-being, shaped by personality, and utilized in decisions and choices (Mayer, 2009). PI involves knowing something about who one is and who one wants to be to guide one's choices. Goals are selected that are both consistent with one another, and involve objects that are realistic with one's talents and resources. Self-awareness and identity are discussed in detail in this volume, and earlier in Parker (1977, 1983). Personal life decisions depend upon a multicomponent capacity, that is, the intelligence the individual brings to bear on life choices. Personality draws upon an intelligence that supports self-understanding, contributes to decision making, being a good judge of other people, personal goal management, *psychological mindedness* (learning about oneself with the capacity for change), accurate emotional perception of one's internal states, recognizing others' views of oneself, and the ability to evaluate others' personality. The description of low PI individuals is as being out of touch with their inner states, how others view them, and failing to take into account others' personalities and preferences. That their self-concept is undifferentiated and inaccurate, that their goals seem scattered or in conflict, and that their life in part seems to lack coherence is consistent with the clinical observation of many persons with concussion, and brings to mind the discussions of such topics as body schema and reduced motivation.

Educational Contribution to Mental Ability

It is generally assumed that intelligence is more related to age or development than to formal education. An attempt was made to isolate the relative effects of age (within a grade) and length of schooling (same age, but different grade placement). It was concluded that the major factor underlying increase in intelligence test scores was schooling as a function of age, with greater effects upon verbal than nonverbal tests. Moreover, even tests of "fluid" ability showed a schooling effect, with four of six procedures manifesting a greater schooling than age effect (Cahan & Cohen, 1989). The authors believe that age-based norms "penalize" individuals with less schooling experience, that is, a bias correction would use school exposure variables in norming.

Verbal ability and intelligence: This is the skill to effectively comprehend and communicate with words. Receptive ability emphasizes language perceived and mentally processed by a person. Expressive ability emphasizes verbal fluency and communicating through access to words, associations, and meanings. Verbal intelligence extends the usage to include comprehension and problem solving of tasks conveyed as words.

Nonverbal intelligence: This component of mental ability merits assessment on its own merits, as well as being the primary mode accessible to the examiner when examining a patient for whom proper translation is not available. Nonverbal reasoning and information processing are overlapping functions that integrate holistic, nonverbal, and simultaneous mental processing. These are sensitive to trauma and neurotoxins. Examples include learning letters or combining them into words, drawing from imagination, copying materials from the blackboard into notes, the ability to read a map, understand geometry, prepare and decipher technical drawings, recognize faces, orient oneself in space, comprehending visually presented arithmetic problems and many components of higher algebra, geometry, and higher mathematics, and drawing tasks (e.g., Bender Gestalt, and representational drawings).

Graphomotor: The process of drawing involves such abilities as visual perception, visual imagery, and graphic production, which includes planning and action programming.

Coping with Different Kinds of Intellectual Situations

Environments, including employment, make different cognitive demands, which may be differentially impaired.

Structured/crystallized: The situation and demands are familiar and/or the requirements for success are relatively precise. An example is arithmetic, spelling, or a job with a detailed procedural manual. Crystallized intelligence is applying cultural knowledge to problems, that is, the body of factual knowledge accumulated during a lifetime.

Unstructured/fluid: The situation is unfamiliar, and the requirements are not precise or subjective. *Fluid intelligence* describes solving problems in novel situations. Its components include analytical reasoning ability, WM, and information-processing speed. An example of a task requiring fluid intelligence is responding to the Rorschach Inkblot Procedure, or entering an employment position in which little supervision is given. Fluid intelligence has been asserted to be poorly measured by the older intelligence tests.

It is believed that the older intelligence tests overemphasized crystallized intelligence at the expense of fluid intelligence. In fact, tests of fluid intelligence may reflect *g* (general intelligence). Intelligence tests seemed insensitive to damage to the prefrontal cortices. Patients with frontal damage (disrupted decision making, planning, or social conduct) seemed by IQ performance to be seemingly unimpaired, or of high intelligence. Nevertheless, when the *Matrix Reasoning Subtest*, believed to be a measure of fluid intelligence of WAIS III was used, it did not differentiate between patients with and lacking frontal lobe damage (Tranel et al., 2008).

Acalculia (Calculations Deficiency)

The more generic title calculations, as opposed to arithmetic, is used because it refers to higher levels of education and is used in the problems offered by several tests of academic achievement. Arithmetic is classified with the Wechsler IQ tests in the Verbal IQ portion. However, calculation ability reflects numerous domains. It is a multifactor skill that utilizes attention, verbal, spatial, memory, body language, and EF abilities; WM; and processing speed components.

Solving a numerical problem depends upon utilizing verbal, spatial, and conceptual abilities, which require the active participation of numerous cerebral structures. Lesions in different parts of the cerebral cortex originate different errors. Numerical operations may differ from that of recognition of the numbers. Coping with numbers is differentiated from understanding the products of calculation.

Left cerebral hemisphere: The posterior parietal lobe is in a region that in the right hemisphere is involved with spatial functions. Spatial concepts are mediated through language. Semantic aphasia and dyscalculia have dysfunctions in common. Left angular gyrus disorder is associated with conceptualizing words with a spatial meaning, that is, left, right, up, down, over, and below.

Right cerebral hemisphere: This refers to visual–spatial defects involving the spatial organization of numbers; inability to invoke and remember the appropriate use of mathematical facts; and inability to conceptualize quantities and process numbers.

Calculation disabilities may be based upon a narrow disorder. TBI may result in a relatively restricted difficulty in performing arithmetic operations (number permutability, e.g., adding and subtracting). An example of the components that may contribute to errors of numerical task is elicited from errors made by children with developmental disability:

- Spatial: Placement of numbers to follow procedural directions
- Visual: Reading arithmetic signs (e.g., the meaning of decimal points)

- Procedural: Eliminating or adding a step, or misapplying a learned rule to a different problem
- Substitution of numbers: Due to lexical or aphasic disorder
- Graphomotor: Difficulty in forming the numbers
- Judgment: Offering responses that offer impossible results, for example, the relative size of two numbers
- Memory: Recall of the multiplication tables or arithmetic procedures
- Perseveration: Difficulty in changing from one task to another, or repetition of the same number
- Right-left orientation: (Ardila & Rosseli, 2002).

The following discussion is based upon Greiffenstein and Baker (2002), noting that in their selection of cases for the study of adults with arithmetic deficiency (AD), TBI patients seem not to have been screened out or identified. Moreover, their WAIS-R Full Scaled, Verbal Scale, and Performance Scale patterns for Arithmetic and Reading Deficiency subjects resembled the WAIS-R TBI patterns referred to earlier, and were reversed for Reading Deficiency. Arithmetic learning disability refers to children and adults whose computational abilities are significantly lower than would be predicted from intelligence and reading levels. Difficulties can result from social, environmental, neurological, and developmental factors. Patterns observed in children persist into adulthood. From their findings, Greiffenstein and Baker (2002) concluded that substantial AD in adulthood retains an association with cognitive difficulties: the same pattern of weaker nonverbal intelligence, switching mental set, and slower motor speed.

While a nonverbal learning disability pattern has been described (AD; nonverbal cognitive weakness; socioemotional difficulties), their findings did not confirm a psychopathological association. However, some studies associate AD in adults and adolescents with depression. Although children with visuospatial deficits display them as adults, those with linguistic problems evolved into more globally impaired individuals. Perinatal right brain lesions may affect subsequent intellectual development than do left-brain lesions. A summary of childrens' IQ findings by age and lateralization was consistent with this statement: Early right lesions offered a mean IQ of 91.9 and late left lesions of 102.5. All groups had somewhat higher verbal than performance IQ, while later right lesions offered a V-P IQ of +14.1, suggesting that interhemispheric compensation from the right hemisphere to the left may take place earlier in life, but a late injury does not permit right hemispheric functions to be efficiently transferred to the left hemisphere (Parker, 1990, p. 312).

Visuospatial Processing (Temporoparietal)

Visual information begins at the primary occipital area. It proceeds in two separate, parallel streams: Dorsal and ventral projections of the posterior cortex are continuous with the dorsal and ventral pathways of the frontal lobe. This functional continuity is maintained by WM. The *ventral stream* leading downward to the temporal lobe is involved in the "what" of recognition of facial patterns, including faces. Noting that rapid eye movement (REM) sleep is most sensitive to brain damage, and is reduced in all patients with nonspecific epileptiform changes in the electroencephalogram (EEG), and that *reduction of REM sleep* is a sensitive marker of epileptiform EEG changes is noteworthy (Busek & Faber, 2000). Possible epileptiform activity may explain why REM sleep deprivation is more potent in reducing perceptual learning than slow wave sleep (SWS) deprivation (Chokroverty, 2004). Inferior temporal cortex lesions produced deficits in visual discrimination, for example, shape and patterns. The *dorsal stream*, representing the "where" of perception, projects toward the parietal lobe. It contributes to conscious spatial awareness and spatial guidance of actions such as reaching and grasping. Parietal neurons construct a representation of space by combining multiple sensory signal modalities with motor signals. The transition from purely visual functioning to spatial awareness is accomplished gradually and terminated in the association cortex of the posterior parietal lobe. Damage may cause deficit of judging distance. The superior parietal lobule serves *somesthetic or tactile* perception; the inferior parietal lobule serves visuospatial cognition. Separate parietal centers are involved in spatial attention, and the coalescence of visual and somatosensory representations of space (Buschman & Miller, 2007; Tucker et al., 1995).

Parietal deficits of spatial behavior: See Nunez (2002) for details. *Simultanagonsia*: Inability to see multiple objects simultaneously; *optic ataxia*: Impairment of visually guided reaching; *hemispatial neglect*: Unawareness of the contralesional half of space (see "Neglect") (Reynolds et al., 2003).

Reading

Long axonal connections between regions may account for individual differences in cognitive skill and the development of such skills as reading. Genetic differences contributing to development involve genetic loading for white matter microstructure. There is a genetic role in developmental dyslexia. Reading involves development of cortical reorganization of functional cortical regions involved in an association between cortical regions supporting perception of visual words (orthographic) and spoken language (phonological). There are different influences upon task efficiency associated with performance and developmental levels. Age-related regions are left frontal and parietal cortices. Activity decreases with increasing age. Performance-related regions are bilateral extrastriate cortex and the left parietal-occipital-temporal (POT) junction. Reading skill (without age effects) is associated with activation of left ventral occipitotemporal regions. The visual word-form area appears associated with the left occipitotemporal region (midfusiform gyrus). Letter-sound integration involves heteromodal input (visual and auditory) for integration of orthographic and phonological processing in the left superior temporal cortex. Changes will differ between beginning readers and skilled readers (Schlaggar & McCandliss, 2007).

PROBLEM SOLVING AND DECISION MAKING

Clinical example: Problem-solving deficiency: A business woman was asked about her problem-solving ability. Her estimated full scale IQ (FSIQ) loss was around 10 points: She said that usually when she has a chore, she can figure out how to do it. "Today it took me half an hour to figure out how to fold the clothes for my family of three and separate them without creating a mess. I decided to take three different baskets for everybody's clothes. When it was all through, I was out of energy. Previously, I could work 15 to 18 hours a day."

Problem solving is the application of general intelligence to a task or necessity. Efficiency can be judged as appropriate or regressed according to expectations of age, education, and occupation during such procedures as assembly tasks. The writer assesses problem-solving maturity by the level of task difficulty at which trial and error or foresight is utilized in solving the task. Individuals of a particular age and education, when confronted with a progressively more difficult task with varied solution (e.g., Wechsler Block Design or Object Assembly), are expected to have a range of problem-solving maturity, from foresight at the easier range (anticipating the correct response, i.e., efficient) to trial and error at the difficult range (an apparently random erroneous response not utilizing error monitoring, that is, regressed or inefficient). When the range from foresight to trial and error is significantly reduced from expectations, one may consider *regression*. Other markers of reduced problem-solving levels are *inflexibility* or *perseveration*.

Decision Making: Choice and Effect

Decision making, as a complex cognitive activity, is affected by the integrity of neurological functioning. Our choices are subject to a *framing effect*, that is, the manner in which the options are presented. Its association with amygdala activity indicates a role for emotion in decision making.

The orbital and medial PFC integrates input from the amygdala, allowing evaluation of the incentive value of various outcomes, which guides future behavior. Those with a better representation of their own emotional biases may be able to modify their behavior, thus avoiding suboptimal behavior. However, reduced susceptibility to the framing effect accompanies orbital and medial prefrontal cortical activity (De Martino et al., 2006). Effective decision making involves evaluating the expected benefits of an action, as well as the costs incurred to obtain them. Examples of costs are time, effort, the risk that a reward may not be forthcoming, or the requirement to endure pain. Yet the components of decision making may be unspecified in the mind of the decision maker. The mind may be made up at an unconscious level, even if the claim is that no decision has been made. *Automatic associations* may predict future choices. These are associations that come to mind unintentionally, are difficult to control after being activated, and may not be endorsed consciously. After a time, biased processing brings future choices into line with existing automatic associations: Minds are made up although the person does not yet know it (Galdi et al., 2008).

MENTAL WORKLOAD

Mental workload has been defined as a brain state(s) that mediates performance of perceptual, cognitive, and motor tasks. Mental workload refers to task demands, the mental effort expended to meet them, how well the person performs the task, and the subjective perception of the expended effort. It may be driven exogenously by environmental sources, that is, task load, and endogenously by the voluntary application of mental effort. In short, it reflects the interaction between the task, the demands, and the capability of the individual to meet these demands (Parasuraman & Caggiano, 2002). *Time pressure* is an exogenous driver. When complex tasks are performed, the endogenous workload reflects the *strategies used*. Regarding *motor vehicle driving* it has been estimated that the average driver is confronted with ten or more traffic situations/second, makes one to three decisions per second, and is confronted by at least one driver error every 2 min (Nordhoff, 1996c). Patients with mTBI and persistent postconcussive symptoms were likely to have abnormal medial temporal lobe findings on PET and single photon emission cardiac tomography (SPECT). Even with task performance compared with uninjured persons, in a high-processing load condition they manifested a different allocation of processing resources. Perhaps injury-related changes in the allocation of WM resources may underlie some of the memory complaints after mild head injury (Jorge, 2005).

Slow Waves

Magnetoencephalography (MEG) slow waves are associated with cognitive problems. EFs were coupled with frontal slow waves, although executive disorders were not correlated with temporal, parietal, or occipital slowing; All six patients with epileptiform activity (independent of the region) showed executive deficits, but only four also had frontal slowing. It was inferred that general epileptiform activity can compromise the neural network supporting executive skills. There was a significant association between temporal lobe findings and decreased processing speed (Lewine, 2007).

Resources

Performance of mental tasks increases metabolic activity in specific cortical regions. Additional procedures to measure mental workload include PET, functional magnetic resonance imaging (fMRI), EEG, and event-related potential (ERP). Among the neural centers to consider are the anterior cingulate cortex (ACC) and anteromedial frontal cortex.

Unitary: Within the lower portion of the range of resources, the allocation of more resources leads to faster and more accurate execution. Working harder improves performance (*resource limited*). Beyond a maximum point, performance is *data limited*, that is, efficiency is determined by the quality of the input source (memory or sensory). With sufficient distraction, no amount of effort will improve the product.

Modular: It is hypothesized that a central executive controls the process and allocates resources (see "Working Memory"). While electrophysiological and functional brain imaging support a modular theory of mental workload, the commonly observed characteristics of TBI patients have a different implication. Although modules might exist in the uninjured person, when injured, the commonly observed cumulative dysfunction suggests a unitary loss of performance. Exceptions do occur where an uneven profile indicates a pattern of loss and sparing.

ALERTNESS

Alertness is the state of being awake, aware, attentive, and prepared to react. It is within the continuum of *sleep-wakefulness*. The components of attention include alertness, WM, competitive selective attention (directed), top-down sensitivity control, and bottom-up salience filters (Knudsen, 2007). Alertness (Reynolds et al., 2003) is mediated by a variety of neuromodulatory pathways originating in the brainstem that sends widespread projections into cortical and subcortical regions. Therefore, the level of alertness occurs within a circadian contest, that is, core temperature minimum between 4 and 7 AM, melatonin secretion, the time awake influencing fatigue, and sleep. *Mental performance* is influenced by wakefulness and rising core temperature. In the latter half of the day, tasks continue to be done well if they require little central processing or memory. *Decision making* deteriorates before the evening fall in core temperature because of fatigue (Waterhouse, 2007). There is preliminary evidence for pharmacological agents to improve alertness, mental processing speed, and memory (McCullagh & Feinstein, 2005).

WORKING MEMORY

Clinical example: Loss of WM: "I have watched you about to do something, e.g., listen to me, and about to write something down, go to another completely different project, and then snap back to what you were doing as though there were no break at all. Stop the flow and then flow in the same direction. If I stop the flow I am adrift and I cannot start the flow again. That's why I write things down. I think of something, write it down , e.g., going to the bank today, then I think of it as though it were the first time. The same as when I order a videotape of a film, and look at it as though I had never seen it before." (Description of memory problem by a brain-damaged victim.)

WM is a tool in service of other activities, for example, attention, executive functioning, information processing, reasoning, problem solving, and comprehending. The purpose is to guide behavior during intermediate steps, and to modify mental products so that components are available for complex operations. Sensory cues are incorporated into larger cognitive entities. Their representation affects existing functional entities (Llinas, 2001, p. 8). Neurons in the PFC continue to discharge for many seconds after a target has disappeared. Different sets of neurones appear to serve WM, that is, not a single system. There are different activation patterns according to whether the material stored briefly is spatial, visual, or verbal. While the PFC is most frequently cited as the basis for WM, nevertheless, the left hemisphere posterior parietal cortex (PPC) is involved in rehearsal and storage of verbal material as well as anterior speech-related areas for rehearsal. However, tasks requiring only WM storage do not extend into the DLPFC, while those that require partial updating or some other executive processing, as well as storage, cause activations, including the DLPFC (Smith & Jonides, 2003).

Thus, WM bridges temporally separate elements, compares several pieces of information, and holds information ("online") in one's mind for seconds while manipulating it for a mental operation. WM holds information online after sensory information has ceased. With a plan in mind, it carries it out step by step. Information is rehearsed, evaluated, updated, and manipulated on the basis of the current operation and stored memories. Along with information processing, WM leads to the creation of mental units, which are further organized and altered by executive functioning to achieve intended cognitive outcomes. *Verbal information* appears to be retained through a phonological loop, a limited-capacity store holding the information for a short period and a rehearsal system recirculating it to prevent its rapid decay. Damage to the dorsolateral cortex seems to have the greatest impact upon WM, possibly through disruption of posterior regions more directly involved in memory encoding and storage (Damasio & Anderson, 2003).

Models of Working Memory

- A central executive that coordinates and maintains information from "slave systems," for example, verbal and visual spatial (Della Sala et al., 1995; Parasuraman & Caggiano, 2002). A rehearsal system, specialized for the kind of information, keeps active the information to be manipulated to prevent its rapid decay (Molinari, 2002).
- Participation as one component of an attention system, including top-down and bottom-up information control, and competitive selection of data for processing (Knudsen, 2007).

Aspects of Working Memory

Features: Mental units may be verbal, sensory (e.g., visual or auditory), or objects. These are stored in WM, which is an active state in which potentially useful, undeveloped mental contents remain available so that they can be altered and manipulated in the short term for use in some goal, for example, prospective actions, problem solving, creative production, etc. These mental contents become central, through a combination of attention and gating, to avoid distraction by irrelevant information and other contents. The PFC appears to exert an inhibitory influence upon posterior perceptual regions.

WM and attention: It has been suggested that WM, not attention, selects targets for evaluation and decision making. Available information is selected for that which is most relevant for our purpose or situation. WM analyzes it, decisions about the information are made, and plans for action are elaborated (Cottencin et al., 2008; Knudsen, 2007). This process assumes a state of *readiness (alertness)* to respond to stimuli. Neuronal responsiveness to sensory input is synergized by states of attention and motivation correlated with the activities of widely projecting neuromodulatory systems (Kaas, 2002a).

Spatial location: Plans may have a spatial component, which would involve the DLPFC, which participates in reciprocal circuits with the PPC. Also involved are the hippocampal formation and possibly the left temporal lobe (Colby & Olson, 2003; Knudsen, 2007).

CONCENTRATION

Concentration refers to maintaining performance and completing a task in a useful interval. *Selective attention* is required so that irrelevant data is not included in a task or does not overwhelm the activity. However, there is an attentional bias toward trauma-related stimuli after posttraumatic stress disorder (PTSD) (Bustamante et al., 2001). *Mental control* is the capacity to maintain focus upon one or more significant, relatively narrow selected functions or activities while directing, coordinating, and ultimately integrating them in order to accomplish a valued purpose. An independent function is *suppression* of other stimuli, that is, filtering supports concentration (Wegner, 1988). Since *identity* is the higher-level function that creates personal meaning, it can be the determinant of what we focus upon. Thus, the tasks we concentrate upon are often guided by the values and personal experiences of the person.

INFORMATION PROCESSING

Information processing refers to creating usable mental contents from inchoate and vague sensations and ideas. Performance level is affected by reduced speed characteristic of TBI patients and impairment executive efficiency. These are lacking mentally usable qualities, but may be organizable into meaningful units. The original stimuli may be visual, auditory, somesthetic, etc., or mental bits from memory, associations, images, the unconscious, and so forth. This may be defined as neural processes that take stimuli (internal and external) or relatively simple ideas or stimuli and convert them into ideas, concepts, images, and so forth. These are further organized into components of meaningful behavior. Complex examples include the transformation of different retinal images and perspectives into the recognition of a cat, prior experience (e.g., a previously seen orientation leading to a faster recognition of an object), and at the phase of recognition the object is identified as a member of that object's class. Using the ERP N170 response, bird experts showed a greater response to pictures of birds than dogs, and dog experts responded more to the dog pictures (Tanaka, 2004). Activities included are receiving stimuli, encoding them so that they can be transmitted to the CNS, transforming them into meaningful behavioral or cognitive units, and then storing them in memory or in an ongoing program to influence the direction of outcome.

Sequential Processing

The process may be *sequential*: initial stimuli (internal and environmental; sensory encoding that creates cognitive and perceptual particles that are organized into usable ideas, perceptions, and concepts). The organization of the mental particles may occur according to be the principles of *Gestalt psychology* (configurations or patterns). Eventually, even greater organization can develop, for example, a perception beyond that of an idea, concept, object, or person, but of a situation. This can motivate action.

The individual perceives and adjusts ideas into cause-and-effect relationships or arranges them in sequence to solve problems. Examples include placing words into a comprehensible and grammatically correct order for efficient thinking and communication, or receiving information in order: "How much is 3 × 3, and then subtract this from 15." A test example is Picture Arrangement (WAIS-R). Coding and decoding verbal information is related to sequential processing, although verbal ability has some unique qualities: reception (auditory and/or visual channels), coding into words and concepts, offering meaning to the components, and finally expression in a grammatical sequence.

Simultaneous Processing

Certain types of information must be dealt with all at once, although problem solving involves alternating between attention to more and less inclusive hierarchies of information. Examples of simultaneous processing include reading a map, chart, or diagram; forming complex concepts; or having a concept of a complete solution to a problem. Normally, concurrent auditory and visual stimuli are congruent. Incongruent information is reacted to as not ecological. Incongruent facial and vocal information is associated with changes in the ERP known as mismatch negativity (De Gelder, 2000). Simultaneous processing can be a component of monitoring, that is, alternating between an inner model of what the outcome should be, details of the solution as the problem is approached, overall view of the status of a solution (monitoring), and ultimately matching against the desired solution for suitability. Nonverbal, spatial, and simultaneous processing are somewhat overlapping concepts. In practical situations, internal speech and memory contribute to the solution of nonverbal problems.

Utility

Information processing is used for functions such as memory and problem solving. Examples are the alteration, encoding, and integration of incoming stimuli, of single or various sensory modalities, into perceptions (meaningful units), abstract, symbolic, representational, that are familiar due to prior actions, memories, semantic or personal meaning, similarities, concepts, other associations, etc., or new mental units to be absorbed into an already assembled apperceptive mass. Without efficient encoding the information does is useless and seems not to exist.

Encoded according to mode of input: Mental processing exchanges information between sensory input, WM, and long-term memory. Sensory stimuli are encoded peripherally and then centrally. Unless this information is retrievable for practical purposes, it is nonexistent or its utility is minimal. The ability to utilize information depends on recognizing that it belongs to a class of objects. With impaired retrieval or processing (e.g., visual or verbal), new events are not stored in long-term memory and prior learning is not available. When use of particular kinds of information is impaired, the person cannot recognize classes of stimuli or retrieve them from long-term storage. Familiarity will contribute to awareness of a mental unit and its personal meaning. Incompleteness or unfamiliarity with a mental unit may prevent the ability to retrieve it from memory or awareness of its presence.

TOP-DOWN SENSITIVITY CONTROL

WM generates signals that improve the quality of the information that it processes. Increases in neuronal sensitivity elicited by the task relevance of stimuli occur at levels from the thalamus, primary sensory cortex, to the PFC. Activity is enhanced in the PFC, neocortex, limbic cortex, basal ganglia, pulvinar nucleus, superior colliculus, and cerebellum, depending upon what domain of information is being attended. Top-down modulation improves the signal-to-noise ratio in all domains of information processing: sensory, motor, internal state, and memory. Modulation of the features upon which judgments will be made distinguishes attention from general arousal. Space-specific signals improve the localization and representation of stimuli ("*spatial attention*").

BOTTOM-UP SALIENCE FILTERS

Stimulus-driven access to WM ("bottom-up attention") refers to the effects of salience filters at many levels of the CNS that select for important properties of stimuli. Stimuli may be salient because they are uncommon, intense, instinctive, or learned. Among the filters is adaptation to sustained or repeated stimuli. Isolated stimuli arouse unusually strong activation, "popping out," and offering an advantage in the competition for access to WM although it is processing other information. When the information enters WM, its relative importance is evaluated, and it comes under the control of top-down sensitivity. Highly salient stimuli modulate the sensitivity of ascending circuits, and it has been established that they can trigger eye saccades mediated by the superior colliculus (not the motor cortex). An element of judgment is involved since this will not occur if the property of the stimulus determines the correct endpoint for the saccade (Knudsen, 2007).

NEUROLOGICAL ASPECTS

The cytoarchitecture, connections, and function of different prefrontal areas differ. Various neurotransmitters modulate, attenuate, or magnify the response of a "WM" neuron (Davachi et al., 2004). Prefrontal regions are involved in WM, with mechanisms determined by circuit relationships with posterior association cortices. Maintenance of information in WM is associated with activation of ventrolateral prefrontal cortex (VLPFC) and DLPFC cortices; Broca's area and parietal cortical regions. fMRI studies of brain activation of verbal WM tasks have revealed activity in the left Broca's area, the supplementary motor area, and the cerebellum, which may represent the rehearsal system (Molinari, 2002). Memory is vulnerable to prefrontal injury (Nunez, 2002). The volume of activated tissue increases with the demand for processing exogenous and probably endogenous tasks (Parasuraman & Caggiano, 2002).

Domain Specificity (Localization)

There is a segregation of functions according to structural location (Davachi et al., 2004). In the DLPFC cortex, retention of *spatial* information is the organizing principle. It has reciprocal projections with the PPC (dorsal visual stream). The PPC is organized into different areas representing different kinds of movements, and that integrate sensory spatial information with body or limb position information (Knudsen, 2007). Small lesions create specific WM deficits ("scotomas") for particular visual field locations. The VLPFC (ventral stream) appears to be the locale for processing *nonspatial* information (color, form, objects, faces) with the participation of projections from the inferotemporal cortex, a relay in the ventral visual pathway for *object vision. Damage to the right ventro occipital temporal cortex results in prosopagnosia, a selective loss in the ability to recognize faces.* Anterior temporal regions associate face information with stored information about individuals. The frontal cortex contributes to the analysis of frontal facial expression (Eriksson, 2002; Tanaka, 2004). There are separate auditory streams targeting the DLPFC (spatial) and VLPFC (nonspatial) areas (Davachi et al., 2004). *Numerosity* is a function that is found early in life and is the source of symbolic numerical thinking (Piazza & Dehaene, 2004). Numerosity is extracted in intraparietal sulcus and transmitted to prefrontal circuits to fulfill the task. Numerous associated parietal functions create vulnerability to dissociations of performance according to the site of injury: verbal, visual-spatial, symbolic, and nonsymbolic numerical functions.

EXECUTIVE FUNCTION (EF): CONTROL OF COGNITIVE ACTIVITY

Defining Activities

There have been many definitions of EF (American Psychological Association, 2007; Lezak et al., 2004, p. 611). EF has been defined as "a neuropsychological system that translates awareness into action and consists of those abilities that allow the person to engage in purposeful, autonomous, self-serving, and prosocial activities" (Hall, 1993). A prior presentation by Lezak included formulating goals; planning; carrying out plans to reach goals; and capacity for performing these activities effectively. Stuss and Benson appended attention, anticipation, modification, and oversight. In considering this type of performance, the examination should consider these limitations: deficits of definition, complexity of functioning of any gross neurological area (e.g., the frontal lobes), the structure of the tests and testing situation with a push for maximum performance, and focus upon outcome scores with a neglect of process. Further, there are differences between the clinical setting and everyday life, that is, less structure, little task focus, criticism for errors, individual must plan rather than be dependent upon the examiner, greater self-motivation and persistence required by the person, less protective environment including inadequacies publicly seen, and competition (Cripe, 1996).

Executive functioning is a generic description of autoregulation of varied functions that participate in cognition or other mental activities that accomplish some purpose. High-level cognitive functions modify and direct lower-level cognitive functions, that is, the coordinated operation of various processes to accomplish a particular goal in a flexible manner (Elliott, 2003). EF and information processing organize, alter, and integrate simpler mental content in a series of steps, making them useful for higher cognitive activity. Simpler functions include flexibility or response inhibition, mental speed, WM, reality testing, error detection, and judgment, which may modify ideas, memories, or concepts. Then, major mental contents contribute to volition, initiative, and motivation; purposive action (planning, carrying out, and terminating goal-directed behavior); metacognition; logic and reasoning; abstract thinking; coping with novel situations; problem solving; concept formation; and so on.

It is unclear how EF exerts its influence on functional ability (Lewis & Miller, 2007). EFs are distinguishable from specific adaptive cognitive functions, for example, memory, language and communication, and praxis (practical activities). They are controls and modifiers of lower-level functions that precede the ultimate cognitive function or goal, which is expressed in a flexible manner (Elliott, 2003; Funahashi, 2001): volitional behavior, planning, and purposeful actions. EF is differentiated from specific cognitive functions: memory, language, praxis (learned activities). Efficiency requires mental speed, error monitoring, and mature forms of problem solving, for example, foresight as opposed to trial and error.

Scope of Executive Function

The EF is an abstract concept concerning the regulation of cognitive and adaptive functions, but not the level of ability of these tasks (Aston Jones et al., 1999; Lezak, 1983, pp. 38–9, 1993, 1995, pp. 650–676; Osmon, 1999). What is involved are cognitive skills for carrying out solutions to problems whose resolutions are not immediately evident or overtly specified in the environment. The examiner should be alert to inefficiencies in these areas: formulation of efficient problem-solving strategies; ability to disregard nonessential strategies; modification of ongoing plans in response to static or dynamic task requirements; ability to learn from past and present experiences by using previously successful strategies; and avoiding future hindrances or present distractions (Dugbartey et al., 1999). Its components include volition (including awareness of self), surrounding and motivational state, drive, sequencing, deciding on the nature of the problem and the required processes for solution, allocating the mental resources, planning, creating alternatives, coping with and conceptualizing change, purposive action and self-regulation, and monitoring performance in order to ensure that it achieves an internal or external goal (Hall, 1993; Lezak, 1993; Woodcock, 1993).

EF supports those abilities that allow the person to engage in purposeful, autonomous, self-serving, and prosocial activities. It is dependent upon varied neuropsychological systems: consciousness (attention); physiological and health (health and stamina); cerebral personality functions (motivation level); information processing (generating alternatives; monitoring; foresight; reality testing); memory (short-term), and identity and other personality functions (experience of self, effect of values and experience upon behavior, freedom from disorganization by dysphoric moods, good morale, and a constructive sense of one's identity).

EF reflects the efficiency of cognitive tasks applied to problem solving, many employment situations, and in adaptability in the community (e.g., their planning and execution of goal-directed activities). It is an incomplete explanation of adaption, since effectiveness also relies upon considerably different functions (e.g., mood stability, interpersonal behavior, and insight) (Mateer, 1999). The EF functions at different levels of complexity. First, it is a group of mental abilities that are selected to enable a particular task: attention and target selection; concentration; establishing and changing sets; planning; WM; etc. These functions probably cannot be isolated and measured with current instruments. Second, a higher level of control is utilized for special situations (Norman & Shallice, cited by Aston Jones et al., 1999): planning or decision making; error correction; novel responses; situations judged to be difficult or dangerous; and situations requiring overcoming habitual responses.

Effective performance utilizes monitoring, self-correcting, and regulating the intensity, tempo, and qualitative aspects of actions. Mistakes may not be corrected because of nonperception or perseveration. Details of poor performance may be ignored although clearly perceived. This author attempts to observe the selection of regressive or mature problem-solving styles (e.g., trial and error or preplanning one's response). The Wechsler Object Assembly and Block Design subtests are useful procedures to obtain this information.

ANATOMICAL

Although the EF is usually attributed to the frontal lobes, assessment need not assume any anatomical localization. Deficits of attention and efficiency may have varied causes: lead poisoning or altered circadian rhythms associated with inappropriate neuropeptide release. Any dysfunction may be associated in different patients with more than a single lesion location. Frontal lobe function has been categorized as *regulatory* (initiation, modulation, inhibition of ongoing mental attention); *executive* (planning, goal setting, and controlling behavior by its intended result); *social discourse* (productive interactions through conversation); *purposeful self-serving behavior*; and *control over a stimulus situation*.

There are no firm links of aspects of EF to discrete prefrontal foci (Elliott, 2003; Pachalska et al., 2002; Royall et al., 2002). Rather, EF is mediated by dynamic and flexible networks that exchange with basal ganglia-thalamic, cerebellar, and brainstem centers. While EF is attributed to the frontal lobe (Catroppa & Anderson, 2006), this complex structure interacts with all other portions of the nervous system, and therefore its functioning is vulnerable to trauma elsewhere. Yet, in general, the PFC is a central integrating center: It is connected to more brain regions than any other cortical region; it is positioned to integrate cognitive and sensorimotor information with emotional valence and internal motivation; and it is the primary target of the basal ganglia-thalamocortical circuit. Thus, the frontal lobe integrates motivational, mnemonic, emotional, somatosensory, and external sensory information into unified goal-directed action (Royall et al., 2002). Insight need not be followed by appropriate action. The person may realistically self-describe his own condition and what should be done to improve it, yet be unable to perform according to his own desires or those on whom he is reliant. The patient cannot take significant action.

MAJOR EF FUNCTIONS

Flexibility

Flexibility is expressed through the capacity to generate multiple response alternatives or a new strategy. It is manifested by changing one's problem-solving strategies, perceptions, or specific actions when there are changes in the situation (environment or specific demands of some other person), or it is apparent that an activity is ineffective or inappropriate. This is termed *task set switching*, which requires the use of feedback and error monitoring. Flexibility depends on prompt reaction of the system to a change in the positive or negative reinforcement signals. Change of valence imposes a change in the internal activity of prefrontal neurons, and hence a switch to a new plan of behavior (Dehaene et al., 1999). Shifting attention improves with time after brain trauma; more slowly in the more severely injured.

Perseveration: Applying a previous response to new situations inappropriately. It may represent a failure of inhibition, perhaps associated with damage of the dorsolateral region of the frontal cortex.

Response Inhibition

This concerns inhibition of responses that ordinarily are part of established patterns of behavior, as well a response to unexpected demands or circumstances. EF is associated in young children with inhibition and with "theory of mind" (Carlson et al., 2004), that is, awareness that other people have their own reactions, which may differ from ours. This would enhance performance efficiency through effective error monitoring. Inhibitory efficiency seems largely independent of global cognitive impairment.

The frontal cortex, together with the basal ganglia, plays an integrating, initiating role in response inhibition, together with sensory information from posterior brain regions. It participates in tasks requiring inhibition (response tendencies that are distracting or prepotent; mental set). This appears to explain the fact that there are comparable deficits in patients with frontal and basal ganglia lesions. Damage to these structures may also impair response initiation. There is some overlap of the systems for response initiation and inhibition (Rieger et al., 2003).

Impulsivity occurs in a wide variety of neuropsychiatric disorders. It may be contrasted with *inhibition*, which suggests that some disorders of EF have a temperamental and developmental characteristic. Short-term gratification hampers assessment of the risk–benefit ratio, which I consider to be a cognitive inefficiency or executive dysfunction leading to poor decision making. The pervasiveness or episodic nature of impulsivity varies with the condition, but in TBI and other neurological diseases, it is subacute, chronic, and relatively nonepisodic (Chambers & Potenza, 2003).

Response inhibition has been conceptualized as part of a developmental pattern initiated by moderate stress inoculation. A major but not exclusive role is played by the frontal lobes. In squirrel monkeys, this stimulates the development of larger prefrontal cortical volume without affecting hippocampal volume. Prefrontal-dependent cognitive control plays a role in emotional regulation and resilience. Stress-inoculated animals inhibit incorrect responses compared to non-stress-inoculated, engage in more exploration (curiosity), and manifest diminished expression of the hypothalamic-pituitary axis (HPA) expressed over time and across situations (Lyons & Parker, 2007).

One of the characteristic disorders of immature individuals, and those who have impact or other injury to the front of the head, is *impulsivity*. This common in cerebral personality disorders, for example, the "*frontal lobe syndrome*." The orbitofrontal cortex has been implicated in personality changes after lesions: indifference to others, irritability, tactlessness, and impulsivity (Kemenoff et al., 2002).

Processing Speed

Reduced processing speed is a frequent marker of TBI. Processing speed loss has been ascribed to white matter microstructure damage (Levin et al., 2008; Lewine, 2007). There was some lateralization of function with a particular role for the right PFC and bilateral lesions creating deficits on tasks requiring complex cognitive resources (Anderson et al., 2005b). Its academic expression is described as slowness in responding to written or verbal directions, questions, and requests. The patient finds difficulty in coping with information that is presented at a rapid rate or is complex. The person is unable to formulate a response to a question in the expected time, although he may know the correct response or behavior (DePompei & Tylor, 2004). The age of the first memory with loss of consciousness (LOC) predicted processing speed, delayed memory, and abstraction (Corrigan & Bogner, 2007).

Volition, Initiative, and Motivation

This behavior is used to carry out purposeful actions, to determine what one needs or wants, and to plan the realization of that need or want. It involves intention or goal creation, and regulating one's own behavior. Initiative means the power to begin, or to follow through energetically, with a plan or task. It implies enterprise or determination. A person may indicate full knowledge of what is needed, yet do little except when under direct pressure and then rarely. Loss of initiative for daily activities is a common symptom after TBI, although it may be impaired by multiple effects of chronic stress, injury, and impairment.

Error Monitoring, Reality Testing, Monitoring, and Metacognition

Error monitoring refers to self-monitoring, to assess the accuracy or social value of one's mental productions and social activities. It differs from *reality testing*, which relates to psychiatric issues, awareness of personal limitations, and sense impressions that enable an individual to distinguish between the internal and external worlds, and between fantasy and reality. *Social sensitivity* is a significant component of error monitoring, that is, sensitivity to others. It includes social feedback, enhancing the likelihood for appropriate social behavior, and for enhancing the success of goal-directed behavior, monitoring detects when actions are erroneous and permitting appropriate remedial mechanisms. Event-related potentials can be used as markers for error detection. Monitoring involves a widespread network of brain structures, including a functional interconnection of the lateral PFC and the cingulate motor area. It is abolished with lesions of the lateral frontal cortex (Ullsperger et al., 2002). Monitoring is overseeing a process to see how it is functioning. Neuropsychologically significant for error monitoring are evaluating mental products, imagination, and sensory experience to determine whether they are accurate enough to be the basis for action.

Successful monitoring utilizes metacognition, the awareness of one's own thought processes. Metacognition describes one's judgment concerning major aspects of a project or assessment of the overall task. The person makes a judgment as to the degree that a complex product is in conformity to an internal or external model. Access to memory (visual and verbal) is required for the ability to create a mental image. Memory and imagery are then used for monitoring, that is, determining whether planning and accomplishment match in the sense of achieving a goal. Awareness of deviation from the requirements of a task, or from community standards of behavior, increases performance quality by improving the selection of actions and programs or correcting deviations.

There is a difference between error knowledge (posterior cerebral function) and error utilization or correction of inappropriate or erroneous responses (anterior functions) (Stuss, 1991, citing Konow & Prigram, 1970). Reduced reality monitoring has been attributed to diffusion of the boundary between externally and internally derived information as ideas, fantasies, and hopes, as well as breakdown in the systematic organization of memory (Johnson, 1991). A patient with cognitive impairment may have an inability to judge poor quality of performance, change of behavior, or maladaptive reactions and their implications. There is also inability to grasp the facts describing an accident. Reduced motivation for reality monitoring, or loss of social constraints for truth, increases errors (Johnson, 1991), such as an inability to detect that the order of the picture arrangement cards did not make a correct story (inability to correct an error), which was consistent with right frontal lobe damage (McFie & Thompson, 1972). Right-hemisphere-damaged patients averaged two scaled score points below left-hemisphere patients, which is consistent with picture arrangements placement on the Performance Scale, but paradoxical to the author's belief that it is a measure of sequential ability, a classically left-hemisphere task. However, the right temporal group had the lowest scores (relative to frontal and temporal lesions).

Rorschach example of breakdown of reality testing: This is a rare kind of response, that is, the perceived figure is experienced as attacking the viewer: "A monster hanging over me, trying to hit me with a ray."

Visual Images

Dreams appear to obey their own rules for image creation: Bizarre features have similarities to neuropsychological syndromes after brain damage, such as delusional misidentifications of faces and places (Maquette & Schwartz, 2002).

Hallucinations are images resembling real experiences: A prisoner of war who suffered death threats and head trauma, two months later heard previously conversations occurring spontaneously or induced by effortful concentration. SPECT scan indicated increase of blood flow in the left temporal lobe and in the brainstem (Stephane et al., 2004). Head injury–related psychosis is usually paranoid-hallucinatory, with a genetic predisposition to schizophrenia, and an acute or chronic course. Significant brain damage and cognitive impairment may create vulnerability (Sachdev et al, 2001).

Reality testing is also degraded by subjectivity of neurotic or psychotic origin and proportions. The examiner can make an effort to determine whether the patient could not detect an error, or was forced by impairment to respond in a dysfunctional way. For example, Hartlage (1990) suggested that, rather than assume that Bender-Gestalt Test (B-G) errors are perceptual or motor, one can give the cards back to the patient (after recall procedure, please) and inquire whether the reproduced figure is the same or different. The response will determine whether the error is perceptual or motor. The person decides whether the mental product should be rejected or not offered. One example is the patient's response to the *Rorschach inkblots* (the entire olate or a portion of it). A "percept" is offered and then rejected, or is privately considered and not revealed to the examiner. Form accuracy is judged by using group norms of acceptable or erroneous responses from the plates (e.g., Exner, 1993, including overall percentage expectations by age), augmented by the examiner's personal estimate or use of an authoritative picture of the object. Other useful assessment tools are examination procedures such as assembly subtests or house-figures-clock drawings. Performance originates in the person's mental model, which is compared to the product, and the process then is continued, modified, or rejected.

Foresight, Judgment, and Error Monitoring

Foresight has several meanings: *anticipation of future conditions* (actions are created so that the present status will be changed), *predicting the consequences* of a particular social or cognitive action. Prediction serves survival, moment-to-moment actions, and conserving time and energy. Skipping to the next moment's need for processing, without getting stuck on what it is doing, has been called the *look ahead function* , an inherent property of neuronal circuits (Llinas, 2001, pp. 23–24). Good judgment requires anticipatory error monitoring to avoid errors, and ongoing error monitoring to change direction or remove poor performance. Participating in judgment are comprehension and experience (episodic memory).

Programming

I will use the term programming for complex behavior, performed as a sequence or other unit, which is part of a larger action. One may think of a musical conductor, instrumentalist, or vocalist whose note patterns are so complex and rapidly performed that they cannot be consciously executed as one note or chord played consciously after the previous one. However, anticipation enhances the formation of automatic schemata for routine tasks (Healy & Williams, 1988), and also the next part-program, that is, modification according to the anticipated effect of the activity upon the environment or the remainder of the entire sequence. Thus, a plan of action (e.g., motor program, information acquisition, sequence of mental activities, etc.) organizes a (generic) performance, while utilizing ongoing error correction in its expression.

Goal Establishment and Planning

Imagining a future, planning for it, and then shaping it. This has been a prime determinant of human evolution. Foresight, more than an accurate memory of the past, offers a greater adaptive advantage. Clinically, amnesic patients who cannot answer simple questions about yesterday's events are unable to say what may happen tomorrow (Suddendorf, 2006). Correct anticipation of consequences utilizes adequate error monitoring. Success and utility will depend upon the level of motivation, and the level of intelligence when varied and difficult information and processes must be comprehended and integrated into a course of action. Planning involves reconsideration of preexisting habits, responses, and programs (whether or not successful). Error monitoring is part of the consideration of these alternatives: Preexisting behavior may be recognized as inappropriate, and therefore rejected, or selected as part of a new program for action.

A plan of action can be directed towards some specific goal, change of status, mood, or reduction of discomfort in the internal milieu or mental environment, physical location, or relationship to some external object. Modification may only be possible when an activity proceeds at a leisurely pace. Otherwise, it may be modified by feed-forward control, or when it is possible to stop and reconsider, and then be revised, corrected, or repeated if this is appropriate.

Planning and programming without action result in an increase of *regional cerebral blood flow (rCBF)* in the premotor and prefrontal areas, but not the motor cortex. Prefrontal metabolic rates and rCBF reflect the participation of the PFC in *sequencing* (temporal organization of serial behavior) and in the cognitive processes that make this organization possible (Fuster, 1997, pp. 197–198). In older patients there is a strong link between the planning ability and functional ability, with implications for risk of injury in independent living situations (Lewis & Miller, 2007).

EXECUTIVE DYSFUNCTIONING

This is defined as "impairment in the ability to think abstractly, and to plan, initiate, sequence, monitor, and stop complex behavior" (American Psychological Association, 2007). Loss of autoregulation (cognitive control) is central to executive dysfunctioning (ED). EF encompasses a variety of higher-level functions that may be affected differently in different patients.

Executive Dysfunctioning Syndromes

- *Goal-oriented complexities*: Loss of mental flexibility, manipulation of novel materials, sequential processing, multitasking, sequential processing, decision anticipation and foresight, searching, etc. (Eslinger & Geder, 2000).
- *Active* (continuing to make wrong or bad decisions, e.g., sequencing, incorrect choices, irrelevant action).
- *Passive* (despite being able to describe in detail what needs to be done, the patients never actually begin, i.e., they make no decisions rather than wrong decisions) (Prochalska, 2002). Inability to convert motivation into action has been variously attributed, including bilateral disruption of a circuit centering on the basal ganglia (Habib, 2000).

Novel situations are difficult for persons with executive dysfunction since they are harder to match with a "crystallized" or familiar model. The physical environment and the rules of behavior are obscure. Not understanding the ineffectiveness of one's own action or obvious self-destructiveness (Parker, 1981) leads to social actions that are disliked by others. Poor consequences could be avoided either by anticipating their results or by watching the social reactions to one's behavior. Occurring in a child, it interferes with social acceptance and participation in activities that are enjoyable, permit the developing ability to work with a team, make friends, and other practical learning based upon watching and participating with others.

MEASUREMENT OF EXECUTIVE DYSFUNCTIONING

There is no single measure of frontal-executive functioning. EF activities are so behaviorally diverse, that this domain can only be sampled by a practical and small group of measurements. Tests of EF are not specific. Consequently, its evaluation requires a multimodal approach involving standardized tests (structured situations), projective tests (for unstructured situations), and study of clinical evidence of daily functioning. They can be affected by breakdown in lower-order abilities. A variety of brain regions and systems are likely to be involved (Cerhan et al., 1996). There is no single measure of frontal-executive functioning. Familiar procedures are multifactorial. Age effects are small, but are associated with certain kinds of errors (Wecker et al., 2000). Many EF tests do not clearly delineate which of the many processes are being studied. It is important to factor out component parts. A frontal system battery is suggested (Royall et al., 2002). However, dissociations between an individual's data does not support a single executive foresight and junction within the frontal lobes (Della Sala et al., 1995). For one study, factors isolated included planning/initiation and reasoning; problem solving and mental flexibility;

awareness/comprehension; and verbal categorization (Bamdad et al., 2003). When healthy individuals between the ages of 20 and 35 are matched with those between the ages of 45 and 65, age effects are found (Royall et al., 2002).

It is hard to account for a significant discrepancy between level of performance on formal testing and assessment of efficiency in the patient's world. EF measures are associated more with functional status than syndrome-specific positive symptoms or nonexecutive cognitive domains. It has been asserted that performance on traditional tests of EF do not capture deficits in real-life decision making and interpersonal functioning following severe TBI. This is conceived as a deficit of self-regulation, which is correlated with TBI severity. Unstructured procedures are useful for this assessment. These functions are attributed to ventral frontal regions whose performance is not routinely tested. It is not certain whether formal psychological cognitive examinations measure real-world capacities, for example, operating a motor vehicle or readiness for a rehabilitation program (McCullagh & Feinstein, 2005).

Since EF tests tap different aspects of executive ability, more than one measure should be used in measuring cognitive performance (Bamdad et al., 2003). It remains difficult to develop a unified estimate of the level of functioning of complex cognitive performance. Moreover, due to the complexity of cortical connections, no single measurement estimating an integrated level of functioning will describe a person's EF, since so many different functions are considered to be members of this domain.

Illness and poor stamina: Reduced performance due to illness, depression, and loss of stamina should be differentiated from ED, which is associated with brain lesions.

Poor Morale due to Impaired Performance

The patient experiences reduced confidence in their ability to cope with adaptive problems such as work, often expressing discomfort directed at particular aspects of impairment, such as concentration, verbal ability and comprehension, and somatic complaints.

Lack of self-confidence or anxiety about success or competence: identity (self-image) is competent when it is directed primarily at a goal and incompetent when the person is reluctant to begin action, or there is reduced interference caused by self-doubts or destructive self-directed stimuli about the likelihood of success.

Test materials: Building a tower with children's blocks (provide 16 or more) using one hand. Is the tower visibly tilted so that the person should observe that it will fall? Rorschach form accuracy, that is, is the score at community norms, appropriate for the person's education, or very high, suggesting some obsessiveness or depression; or very low indicating failure of self-monitoring?

Block design and object assembly: Make a mental judgment whether the task is of low-, medium-, or high-level difficulty for the person's age, education, and occupation. At what level does preplanning of responses cease and trial and error begin? Is this the expected level for a person of particular characteristics in the examiner's judgment?

Interview: Ask about changes in planning, level of motivation, control of feelings, and creativity.

MEMORY AND LEARNING

Clinical vignette: Adult short-term memory: A middle-aged man suffered multiple whiplash injuries, without loss of consciousness. He complained of short-term memory loss, which interfered with daily activities, and which could be observed in his therapeutic sessions. He relates repeatedly the same incidents within the session and later sessions. If he is interrupted, he asks with some bewilderment, "What were we talking about?"

This is a very complex function comprising various types of performance (Barkley, 1988; Boyd, 1988; Hitch, 1990). No single mechanism can explain all forms of memory. Some component of self affects the mental efficiency system, integrating need for and retrieval of memory with ongoing activities. Memory matures in complexity, capacity, and strategies (Boyd, 1988). The frontal lobe's participation in formation and activation of stored representation is indicated by its bidirectional connection with sensory association cortices, thalamic nuclei, and the amygdalohippocampal regions.

Even between apparently closely related procedures (e.g., visual or verbal memory), there are differences of performance related to executive functioning tests with a related component or level of structure (verbal learning) (Temple et al., 2006). Memory can only be assessed by sampling a broad spectrum of procedures, for example, assessment after varying intervals, rote learning in various sensory inputs, levels of difficulty, degree of meaning, integration of disparate components into skills and knowledge, etc. Dissociation of memory into components is illustrated by retention of skills in the presence of amnesia (Ellis & Young, 1988, p. 296).The process of memory can be impaired by a wide variety of dysfunctions: Sensory input (receptors, e.g., visual and auditory); attention (sustained, alternating, or multiple targets); encoding (central pathways); WM (capacity, strategies, concentration, comprehension, knowledge base); storage (visual, verbal, spatial); integration into complex skills, programs, and cognitive/affective complexes in both right and left cerebral hemispheres; retrieval upon demand or adaptive necessity (meanings, experiences, precise recall, recognition, lexicon and grammatical usage, etc.); and expression (oral, written, nonverbal).

There are multiple memory systems (Beggs et al., 1999), and these involve changes in neural circuits, multiple cellular mechanisms within a particular cell, changes in the cell membrane, and new protein synthesis for long-term memory. *Synapses* are the physical site of many or all forms of memory storage. Synaptic strength, the mean amplitude of the postsynaptic response, depends upon its previous activity, which leads to a change in its future effectiveness, and is the basis for building memory into a neural circuit (Connors, 2005a). Particular types of memories permit the storage and retrieval of particular types or classes of information (Eichenbaum et al., 1999).

It is useful to differentiate between learning (the process of acquiring new and relatively enduring information, behavior patterns or abilities, characterized by a modification of behavior as a result of practice, study, or experience) and memory (the ability to retain information or a representation of past experience based upon encoding, retention, and retrieval). Memory is a process that results in a relatively permanent change in behavior (Kolb & Whishaw, 1990, p. 526) and the capacity of the nervous system to benefit from experience (Tulving, 2000).

Long-term storage and retrieval are especially impaired after severe injury. Even after a 10-year follow-up, 25% of children had persistent memory deficits, with verbal deficits more common than spatial (Lehr, 1990, p. 105). Intermodal learning is vulnerable to white matter lesions (Selzer et al., 1992). Yet, new learning can occur despite an amnesic or poor recollection of recent experiences (Eichenbaum, 2003). Degree of recovery after severe head injury was found to be better for the 13- to 15-year-old group than younger children (Levin & Eisenberg, 1991), perhaps this is a function of recovery as opposed to impairment of functional memory.

After stress, combat veterans manifested immediate and delayed verbal recall, verbal fluency, visual tracking, sensitivity to retroactive interference, sustained attention, mental manipulation, etc. In a different sample (both traumatized and injured patients), there were negative significant (.01) correlations between measures of PTSD and learning 6 weeks later: Cumulative learning (–.32); recall after interference (–.42); delayed recall (–.42); verbal fluency (–.37). Reduced functioning might not have been due to cognitive abnormalities, but possibly to pain analgesic, medication, or other factors related to medical or surgical status (Bustamante et al., 2001).

PARAMETERS OF MEMORY

There are numerous forms of memory, imprecisely defined, and overlapping with other types. Theoretical descriptions may not match the names of familiar memory procedures (Petersen & Weingartner, 1991). Clinically, memory can only be sampled due to its complexity. Anterograde and retrograde amnesia are considered under "Acute Alterations of Consciousness." A purpose of the description of different memory systems is to alert the examiner to identify particular memory disorders as currently or previously observed, and enhance the selection of techniques to document dysfunctions suggested by the baseline or clinical history (Petersen & Weingartner, 1991).

Motivation and Reward

The personal circumstances of the potential learning experience may determine both the target and the amount learned (Hyman et al., 2006). Reward is a general term referring to consequence, that is, some sense of improvement experienced by the person. Rewards create pleasure or "wanting," rapid learning of predicted cues, and efficient behavioral sequences to obtain the reward. The value of the reward will be determined by the nature and level of motivation, for example, *physiological* (drugs, food sex), *avoiding punishment* (physical or social), or *self-esteem*. *Reinforcement* is the event that increases the frequency of behaviors, and it may be routine or intermittent (gambling). The negative consequences of drug use may not be avoidable since withdrawal from some substances creates serious physical symptoms.

Recall

Free recall: Retrieving information from memory upon demand but without any external cues. It is considered to be the most difficult memory retrieval task.

Cued recall: Assessing memory by offering a cue to focus performance. The cue alerts the subject to the required information.

Recognition

This is the sense of awareness and familiarity experienced when one encounters people, events, or objects that have been encountered or learned before. Recollection supports high confidence recognition responses, whereas familiarity increases gradually over a wide range of recognition confidence (see "Neurological Support").

Familiarity: A belief of varying intensity in prior exposure to a situation.

Recollection: The recovery of associations prompted by a critical cue.

Types of Information

Visual-spatial: This is mediated by the PFC and the hippocampus, although nonspatial learning is also mediated by this structure. As is characteristic of any cognitive domain, despite its name, a variety of functions mediate each ability. Measured visual memory performance is supported by varied domains: verbal memory, WM, memory retrieval more than encoding, attention, processing speed, and executive functioning (planning, perceptual organization, mental flexibility). With reference to the Ray Complex Figure Test, higher intelligence scores contributed to better organization of the figure (Schwarz et al., 2009).

Perceptual representation system: This operates at a presemantic level. It is involved in the identification of words and objects based upon their structure and form (extrastriate occipital cortex) contrasted with their meaning (temporal lobe and frontal cortex).

Verbal Memory

The to-be-remembered items are presented along with a foil; the instruction is to remember the correct item(s). Multiple choice items belong in this category.

Prospective Memory

This is the realization of delayed intentions, that is, memory for activities to be carried out in the future. *Event-based tasks* are carried out when a specific external event occurs. *Time-based tasks* occur at a specific time. Remembering to perform a task at a particular point in the future is considered to be at the core of independent living. It is a multiphase process: storage of the future task, followed by a complex EF (monitoring for the appropriate moment to initiate the action, inhibition of ongoing activities, and flexible switching from the ongoing activities to the planned action). Time-based tasks place a greater demand on EFs. The self-awareness required for time-based tasks may be reduced after TBI, affecting willingness to participate in rehabilitation and thus outcome. Patients with longer PTA and EF impairment display poorer prospective memory, inviting them to be targeted for specific prospective memory rehabilitation (Altgassen et al., 2007; Fleming et al., 2008).

Retrieval Interval

Short-term memory: Data that is kept in storage for a relatively brief period and then retrieved without thinking about or rehearsing the items. The distinction between short-term and *recent memory* is not clear (perhaps hours to days). One criterion might be the retrieval of data after distraction, that is, several hours.

Clinical example: Teacher's loss of WM: A teacher returned to school a week after suffering slight trauma to her head. "I went back to teaching and my kids asked me about things we did the week before and I wasn't sure what they were talking about. 'Don't you remember such and such?' They had to keep on until I had vaguely remembered so I could agree with them. My principal could not understand it when after a day I wanted to take an extended leave of absence. I had to tell him that it was necessary for me to make notes on everything that was said to me."

Long-term memory: Retrieval after more than a few days to a lifetime. It refers to *personal experiences or autobiographical memory* and *academic learning. Calculations* and other skills utilize *procedural* memory.

Context

Procedural memory refers to long-term memory for the habits, skills, and sensorimotor adaptations that occur constantly in the background of intentional, planned behaviors of particular tasks. It includes stereotyped and unconscious behaviors, both simple refinements of often repeated motor patterns and the learning of long action sequences in response to highly complex stimuli (sports; playing musical instruments). It is demonstrated by skilled performance, and may be separate from ability to verbalize this knowledge. It includes the unique characteristics of personal style and tempo in the expression of skilled activities. It utilizes corticostriatal systems and possibly the cerebellum. For motor behavior, the system modifies cortical motor representation, rather than control behavior through direct motor output (Eichenbaum, 2003).

Declarative memory: Information is recalled in response to a request to remember. It is directly accessible to consciousness. It deals with the encoding, storage, and retrieval of facts, events, and data. Examples are the acquisition of new memory, for example, episodic memory for events and semantic memory of general facts. Its major anatomical components are cortical association areas that exchange with the hippocampal region and the medial temporal lobe (the cortical region adjacent to the hippocampal region, i.e., perirhinal, parahippocampal, and entorhinal cortices) (Eichenbaum, 2003).

Semantic memory: Refers to general knowledge, without depending upon the particular learning circumstances. Lateral and medial regions of the temporal lobe participate. Associations between semantically related visual objects are formed in neural representations in the temporal neocortex and limbic cortex. Automatic retrieval flows backwards from the limbic cortex; active retrieval involves a top-down signal from the PFC to the temporal cortex (Fujimichi et al., 2004).

Episodic memory: A type of *autobiographical memory* in which the event or information is recalled along with when and where it was learned, including a spatial component. There are hippocampal and PFC contributions. Executive processes are believed to play an important role in the intentional retrieval of specific autobiographical memories. These must be distinguished from other information, which may interfere with new information. The EF then makes a final selection in WM. *Intentional retrieval* utilizes shifting between mental sets, updating and monitoring WM representations, and inhibition of dominant and automatic responses (Cottencin et al., 2008).

Explicit memory: The conscious recollection of recent events. The person is aware and can process in consciousness. Examples include the instruction to learn a list and being asked to remember "what occurred."

Learning has taken place without consciousness or intention to learn or recollect. More is known than is initially apparent, and despite possibly forgetting, the material is retrieved through some implicit test of memory. The person is unaware of having learned it ("incidental"). It is inferred to have taken place when retrieval is improved or occurs, although there is no recollection of learning. Explicit memory is dependent on structures in the temporal lobe of the cerebral cortex. Implicit memory involves the cerebellum, the striatum, the amygdala, and in the simplest case, only the sensory and motor pathways recruited for particular perceptual or motor skills (Bailey & Kandel, 2004).

Declarative, explicit memory: Explicit (declarative) memory is a combination of episodic memory (our record of unique, personal experiences) and semantic memory (our general world knowledge) (Eichenbaum, 2004). *Episodic* memory permits us to travel back to a previous time and to reexperience a particular event. It is organized on a temporal dimension. *Semantic* memory involves an accumulation of time-independent factual information and is organized by logical and abstract dimensions. Declarative memory is considered flexible since the content of memory can guide a variety of behaviors outside the learning event. The hippocampus lacks both a sufficient number of cells and sufficient topographic organization to support representation of details of events and places, composing the information in declarative memories. It is likely that the details of information in memory are represented in numerous cortical areas. The neocortical mediates the representation of stimulus details, which converge as representation of figural items in the hippocampal region. Information is recorded and identified. Subsequently, hippocampal representations are used for directing the contents and timing of recovery of detailed cortical representation of episodic and semantic information. This accounts for the full phenomena of declarative memory (Eichenbaum, 2004). It involves information of which the person is aware and can process in consciousness. Explicit recollection is aided by retrieval of the context within which the event occurred, for example, temporal, spatial, and external sources (Schachter et al., 1991). When facts and lists are involved, the memory is considered "declarative." It is attributed to an interaction between medial and temporal cortices, hippocampus, and amygdala during the period that the event is being encoded and integrated within the current cognitive context. Conceptually, there is an *index or partial memory trace* to the various neocortical components of the memory (Bancaud et al., 1994).

Implicit memory: Implicit (nondeclarative) memory refers to learned material, of which the person is unaware of having learned it ("incidental"). It is inferred to have taken place when retrieval is improved or occurs, although there is no recollection of learning. Patients can demonstrate knowledge that they are not aware they possess (Schachter et al., 1991). Implicit memory is nonconscious and expressed through performance, and expresses cumulative changes in perceptions and the development of new skills and habits. Since it has no access to conscious recollections, it is inflexible (Squire & Zola-Morgan, 1991).

Emotional memory: A nondeclarative form of memory, perhaps an unconscious process, that mediates preferences and aversions independent of the declarative memory for the events in which the bias was acquired. Primary focus is the amygdala, which receives sensory input from the thalamus and cortex, and sends output to response systems mediating emotional expression (e.g., heart rate, blood pressure, sweating, and changes in the startle response, and hypothalamic structures mediating hormone release). The amygdala participates in the acquisition of positive and negative biases towards previously neutral stimuli. It is critical to the acquisition, consolidation, and expression of emotional memories. There is evidence for interference with emotional conditioning by amygdala damage, and for declarative memory for the learning situation by hippocampal damage (Eichenbaum, 2003).

SLEEP AND MEMORY

Sleep phases influence the consolidation of memories. The process has been described as memories being replayed, modified, stabilized, and even enhanced. Two effects are observed in the brain that only occur during sleep: alterations in the level of neuromodulators and distinctive oscillations of chemical activity. Induction of the latter improves the recall of word-pairs recalled the previous night. There are many different memory systems, and memory consolidation in different systems correlates with different states of sleep. Of the series of events of memory consolidation, it is unknown which of these are sleep-dependent (Stickgold, 2006). Of special concern for stress reactions are those that are emotionally charged. Animal evidence indicates that *REM sleep* is increased after training, and retention is decreased after REM sleep deprivation (Hobson & Pace-Schott, 2003). *Procedural memory* (skills not depending on the hippocampus) benefits particularly from REM sleep, which dominates the late part of nocturnal sleep. It is hypothesized that in wakefulness, information is encoded in the hypothalamus. During the subsequent slow-wave sleep, the encoded memory trace is replayed by the hippocampus. Information moves between the neocortex and hippocampus, repeated each sleep cycle. This reiteration may facilitate consolidation of memory traces (Ellenbogen, 2005). In fact, induction of slow oscillations (0.75 Hz) during early non-REM sleep enhances the retention of hippocampus-dependent declarative memories (Marshall et al., 2006).

Declarative memory for facts and episodes (hippocampus-dependent) depends upon SWS, which dominates the early part of nocturnal sleep. This theory was confirmed by the association of an odor during sleep, previously presented as a context, during the learning of a new, declarative memory. This improved the retention of hippocampus-dependent declarative memories but not hippocampus-independent procedural memories. It was hypothesized that what takes place is synchronized dialogue between thalamocortical and hippocampal circuitry. This leads to the transfer of representation to neocortical regions for long-term storage (Rasch et al., 2007). (The present writer observes that odor does not proceed to the cortex via the thalamus.)

It has been suggested that sleeplessness has at least two negative effects, using medical interns' long working hours as a model (Ellenbogen, 2005):

Detriment: Attentional errors from sleepiness cause cognitive errors.

Loss of cognitive benefits of sleep: Loss of memory and insight formation leads to deficits of learning, creativity, and scientific discovery.

Emotional memory, dependent upon the amygdala, is enhanced more by late REM sleep than by the SWS period. It is proposed that consolidation of emotional memories involves two memory systems: (1) Declarative memory (hippocampus-dependent) benefits from early SWS and is also enhanced by REM sleep; and (2) The emotional memory system is enhanced by REM sleep. These effects interact with the *circadian secretion of cortisol*: Early SWS sleep secretion is the low period for pituitary-adrenal activity, which is also actively inhibited by SWS. During late sleep, with REM prominent, the hormonal system is activated during the non-REM intervals. Plasma cortisol and adrenocorticotrophic hormone (ACTH) concentrations increase to a maximum at about the time of awakening. Cortisol infusion during SWS blocked declarative memory (and also hippocampal functioning), but not procedural memory. Inhibition of cortisol secretion is critical for formation of hippocampally mediated declarative memories. This also requires that glucocorticoid receptors (GR) be inactive. It has been experimentally determined that inhibition of cortisol secretion induces significant increase in emotional memory retention, while (normal) increase of cortisol may diminish emotionality of memories (Born & Wagner, 2004). Partial sleep deprivation has a variable effect upon individuals with regard to neuroactive steroids. In *depression*, commonly co-morbid with TBI, diurnal hypersecretion of both cortisol and dehydroepiandrosterone (DHEA) was found (Schûle et al., 2004).

TRAUMA EFFECTS UPON COGNITION

The cognitive processes occur in a widely distributed network whose integration and integrity are vulnerable to TBI. TBI appears to reduce the efficiency of the "gating process" that minimizes neural impulses that would disorganize the production of useful mental products. It is noteworthy that in a sample of TBI patients, those with sleep disturbances of posttraumatic hypersomnia (associated with sleep apnea) did not perform worse on measures of intellectual functioning, attention, memory, or executive functioning (Masel et al., 2001). However, inability to have restorative sleep would be expected to lead to some cognitive loss. One is interested in both the estimated preinjury level of a domain and performance as a measure of regression. A difficulty arises: Disparate results are often found in procedures which by name or face validity appear to measure the same function. Often, the discrepancy seems related to the complexity of the task. Therefore, I recommend procedures with a higher ceiling and greater complexity be used when the baseline or return to duty involves a high level of performance. A review of procedures to estimate preinjury ability is offered by Lezak et al. (2004, pp. 92–97).

WHIPLASH

Cognitive findings are stated to be similar to those reported following mild closed head injury, or with mixed results. Positive findings are reported for attentional deficits at 6 months, and also concentration, memory, and cognitive flexibility. The postinjury duration of cognitive defects varies with the function. There is some evidence for residual deficits in attention and concentration for up to 2 years. There are opposing views whether psychological variables are critical in symptom maintenance. In a retrospective study: the neurological examination was abnormal in only 1 of 32 patients. Six of 13 EEGs displayed abnormalities (excessive slow-wave activity in frontocentral and frontoanteriotemporal areas). Quantified EEG was more sensitive than routine clinical EEG in detecting abnormalities in all 11 examinations. The only participant offered functional neuroimaging (PET) produced frontocentral hypometabolism. Sleep studies were abnormal. Ninety-seven percent of patients complained of problems in attention and concentration. Executive functioning was impaired (reduced information-processing capacity; impaired divided attention). There was an advantage of trying to recall meaningful information: delayed story recall improved, but not delayed list learning.

Behavioral problems were prevalent and lasting. Seventy-two percent of patients offered complaints characterized as immaturity, irritability, and anger; 53% reported sleep disturbance, and 47% sexual problems. Depression and anxiety were not common (Henry et al., 2000).

FRONTAL LOBE

The effects of frontal lobe injuries upon cognition are not straightforward. Even bilateral frontal lobe lesions do not provide striking cognitive impairments, for example, IQ test reduction, except when patients present with confusion or dementia. However, such persons may behave "unintelligently." Defects do not result from deficient social knowledge or inability to reason regarding social situations. However, they may be socially impaired in assessing the effectiveness of solutions to social problems, with a tendency to social isolation and disruption of emotion impairing judgment and decision making (Dimasio & Anderson, 2003). Cognitive impairment does not appear as a loss of specific skill, information, reasoning, or problem-solving ability. Further, patients do not perform poorly on formal ability tests in which procedures are structured, and time-limited, and another person directs the examination (pacing, initiation, and cessation). On the other hand, scores are somewhat reduced by perseveration or carelessness. Cognitive defects are clearest in activities of daily living, that is, more observed by relatives and coworkers than health care professionals (Lezak et al., 2004, p. 82). Consistent with this, some focal lesions restricted to the frontal lobes leave routine behavior intact. Others see this lesion location differently. Luria's examples of breakdown of routine actions have been reattributed to massive frontal injury (Schwartz, 1995). However, capacity for higher-level reasoning is impaired when the task involves maintenance of a temporary representation in WM. The patient with a frontal lobe lesion does poorly on delay tasks that rely upon WM (Dubois et al., 1995; Fuster, 1997, pp. 4, 157). The attentional allocation system cannot update information, particularly when there is more than one chunk of information to be maintained in WM (Dunbar & Sussman, 1995). The frontal lobes also mediate behavioral responses to complex environmental situations in the absence of control by direct and immediate perceptual cues. This may be achieved by maintaining an association across time to match a response to a stimulus or monitoring a condition to permit an appropriate response.

Memory

There is an anatomical basis for the loss of memory in which the temporal lobe participates. In an acceleration/deceleration accident, the brain's momentum causes it to be propelled anteriorly. The lesser sphenoid wing is in a position to injure the uncinate funiculus projecting from the temporal tip to the frontal lobe as the frontal and temporal lobes move forward.

The amnestic syndrome: This is a severe and relatively pure impairment in new learning as a result of brain damage, that is, inability to remember recent events in the context of relatively intact cognitive, linguistic, and perceptual abilities, and retained fund of general information (Albert & Lafleche, 1991; Brandt, 1992; Schachter et al., 1991). For similar phenomena based upon a psychodynamic of anxiety-protective basis.

Seizures: There is evidence for material-specific memory deficits based on laterality of seizure disorder. Deficits of nonverbal memory are associated with right temporal lobe epilepoy (TLE) and deficits of verbal memory associated with left TLE. In addition to the effect of temporal lobe involvement, the self-evaluation of memory loss is influenced by a large spectrum of subjective factors, including the effects of polytherapy and longer duration of treatment (Giavagnoli et al., 1997).

TBI AND IQ

Research for seven studies of head trauma victims who were administered the WAIS (Farr et al., 1986) found that the average Full Scale IQ was 90.5, Verbal Scale IQ was 94.8, and Performance Scale IQ, 86.4.

Stability of IQ after 10 Years: TBI and Stress Compared

The writer addressed the question of whether there is measurable change in WAIS-R IQ scales over an extended period of up to 10 years (Parker, 1990, pp. 160–162) when two matched groups referred for head injury or stress are compared. All results are from single measurements. Also considered are deviations from the estimated IQ (Matarazzo & Herman, 1984). All patients were ambulatory, and almost all were engaged in litigation. Characteristic TBI trauma included vehicular accidents as passenger and pedestrian, falls in elevators, objects falling on the head (e.g., ceilings), and toxins. No one required neurosurgery or had extended periods of unconsciousness, etc. Stress victims suffered from incidents involving rape, false arrest, etc. Mean age: TBI, 35; Stress 36. Mean level of education for both groups was slightly more than 12 years. Estimated preinjury IQ for this demographic pattern (Matarazzo & Herman, 1984): VIQ: 97.4; PIQ: 98.6; FSIQ: 98.6. The patients were subdivided by the average length of interval since the accident or incident: I. 1–8 months; II. 9–23 months; III. 24 months to 10 years. Each group averaged 35 years of age, and slightly more than 12 years of education. The estimated preinjury IQ for 12 years of schooling, age 35, is Full Scale IQ 97.6, Verbal IQ 97.4, and Performance IQ 98.6.

Ten-year WAIS-R Stability

WAIS-R Data by Interval Since Injury: Brain Damage and Stress

	Brain Damaged				Stress			
Group	**I**	**II**	**III**	**Total**	**I**	**II**	**III**	**Mean P+**
N	75	39	47	161 13	9	13	35	
SCALE								
Verbal	91.3	95.6	89.7	91.7	94.7	90.2	97.5	94.8
Performance	86.4	87.3	84.6	85.9	90.6	93.0	92.5	92.4 .014
Full Scale	88.4	90.8	86.5	88.2	92.6	91.4	94.7	93.4
FACTOR (See WAIS-R, below)								
Verbal	8.6	9.7	7.8*	8.6	9.6	9.2	8.8	9.2
Perc-Orgn.	7.8	8.1	7.6	7.8	8.4	9.2	8.2	8.6
Frdm/Dist.	8.0	8.6	7.6	8.0	8.8	8.3	9.1	8.7

Interval: I. 1–8 months; II. 9–23 months; III. greater than 23 months (to 10 years).
* F = 4.13 (.018) 1-WAY ANOVA
+ (2-tailed t-test). Occasionally a bit of data was unavailable.

The deviations between the estimated preinjury IQ (Matarazzo & Herman, 1984) and current measurement were:

IQ	Brain damage	Stress
Full Scale	–9.1	–4.2
Verbal	–6.2	-2.8
Performance	–12.6	-6.2

There was a significant difference between Performance Scale IQ of 6.5 points in favor of the Stress Group. This is consistent with Farr et al. (1986) suggesting that Performance IQ deficit is characteristic. When PCS TBI Ss were compared to a group who incurred only stress, there was a greater loss of 5.2 FSIQ points, 3.1 VIQ points, and 6.5 PIQ points (P.014). These findings suggested TBI loss of fluid mental ability, and were consistent with Farr et al. (1986). They are also consistent with reduction from the estimated preinjury IQ: FSIQ, –9.1; VIQ, –6.2; PIQ, –12.6.

The findings indicate that when single measurements are used, Average IQ after TBI remains stable for about 10 years; a stressful experience alone affects IQ level, but to a lesser extent than

PCS; the greatest loss is in PIQ, suggesting a deficit of fluid intelligence, and perhaps mental speed and visuoperceptual processing. The Verbal–Performance Score advantage is consistent with apparently frequent reports of the effects of traumatic brain damage, that is, the vulnerability to TBI of unfamiliar, less well-learned holistic and nonverbal tasks when performed under timed conditions.

Analogous results indicating nonimprovement were also obtained for Rorschach variables based upon objective scoring categories. When TBI and Stress (ST) groups were compared, TBI had significantly fewer total responses, good form (F+) responses, and F+ original responses. The number of rejections was higher (plates for which no response was obtained). Although the overall form+ average was the same, this probably was made possible by the fact that the TBI group was offered fewer responses, that is, may have restricted the number of responses that were signaled as poor through error monitoring. There was a significant trend towards creating vaguer or poorly structured whole responses (the entire plate), a deterioration of cognitive effectiveness.

A Range of TBI

A group of patients was studied with MRI after injuries represented by Glasgow Coma Score (GCS) of 3–15, mean = 8.4). Scores on the Wechsler Memory Scale were reduced according to the severity of injury. However, the relationships were small between fornix size and memory ($r = 0.32$ and hippocampal volume and memory ($r = 0.32$). Perhaps memory disruptions after TBI reflect disseminated injury effects, caused by a combination of specific, yet independent injury effects at the hippocampal-fornix level, and nonspecific effects that disrupt cerebral connectivity and integration of neural networks responsible for memory (Tate & Bigler, 2008).

PHYSIOLOGICAL STRESS EFFECTS UPON COGNITION

The Hippocampus

The hippocampus contains GR (Lambert et al., 1999). Stress is associated with damage to the hippocampal CA3 region. This has been attributed to elevated levels of cortisol or excitatory amino acids, reduction of brain-derived neurotrophic factor, and inhibition of neurogenesis. With reference to potential permanent volume changes of the hippocampus, the process may begin with acute trauma. Major or sustained stress can impair hippocampal-dependent declarative memory. However, milder or transient stress enhances hippocampal-dependent cognition since it is stimulatory. Stressors, over their full range of severity, enhance implicit learning associated with vigilance and fear conditioning centered in the amygdala (Sapolsky, 2004). Although it is asserted that a high cortisol level interferes with (declarative) memory, acutely administered glucocorticoids or GR agonists enhance long-term memory consolidation on emotionally arousing tasks (Nathan et al., 2004) but impair delayed retrieval (Wolf et al., 2004). The ongoing chronic stress of PTSD symptoms, or increased vulnerability to retraumatization, contributes to hippocampal volume changes, which are seen primarily in chronic, severe PTSD. It has also been suggested that there is a genetic predisposition to smaller hippocampal volume and lower memory function, which leads to a vulnerability to PTSD (Bremner, 2005). Hippocampal participation in memory consolidation is probably enhanced by *long-term potentiation*, which occurs in excitatory synapses. It is a voltage change (*excitatory postsynaptic potential* or voltage change), which participates in *learning* by strengthening associations.

Endocrine Effects

Glucocorticoids: A glucocorticoid surge is characteristic of stress. The interaction between glucocorticoids and memory is complex, including extinction of fear responses (see earlier) and consolidation of fear experiences. Cognitive impairment is a frequent consequence of glucocorticoid treatment, that is, transiently impaired attention, concentration, episodic memory, and executive functioning. Excessive amounts of *cortisol* in the hippocampus (involved in memory formation) are modulated by hippocampal feedback to the hypothalamus where gamma-aminobutyric acid (GABA) inhibits corticotropin-releasing hormone (CRH) release (Kaye & Lightman, 2005).

An experimental study of the effect of dexamethasone (a synthetic glucocorticoid 25 times as potent as cortisol) upon *declarative memory* showed lowest performance occurring on day 4 followed by a return to baseline performance by day 11 (Newcomer et al., 1994). A case is reported with paresthesias, proximal muscle weakness, atrophy, pain, insomnia, irritability, and a questionable seizure. Potential impairment of hippocampal and prefrontal cortical functioning is posited, which rarely leads to a "*steroid dementia syndrome*" (Wolkowitz et al., 2004). Glucocorticoids interact upon multiple, interacting brain systems, including interaction with β-adrenergic activation in the *basolateral nucleus of the amygdala (BLA)*, to modulate memory stored in other parts of the brain, for example, the hippocampus. The BLA inhibits HPA responsiveness to novel stressors, but not chronic stress (Bhatnagar et al., 2004). It affects long-term memory consolidation, memory retrieval, and WM. It is part of an integrated network of cortical and subcortical brain regions that regulate the effects of stress hormone upon memory. Its integrity is required to enable glucocorticoid effects in other brain regions to modulate memory consolidation. Exposure to glucocorticoids or emotionally arousing training experiences evokes noradrenergic activation. This induces BLA activation, which functions in concert with inhibitory effects on the hippocampus and MPFC to create a brain state promoting long-term storage of these arousing events. However, high levels of glucocorticoids impair memory retrieval and WM (Nathan et al., 2004; Wolf et al., 2004).

The *age-related decline in short-term memory* (animals and humans) may be due to either loss of specific neurones or physiological processes, or compromise of neuroendocrine regulators of memory in the presence of adequate neurological substrate. High doses of experimentally injected epinephrine may produce *retrograde amnesia* (Mabry et al., 1995). In a nonstress situation, but with stimuli of varying neutral and emotional valences, *cortisol* has an impairing effect upon memory in animals and humans (Wolf et al., 2004). The direction of the effect varies with the task. Victims of a motor vehicle accident (MVA) were divided into amnestics, partial amnestics, and complete memory (Delahanty, 2004). Amnestics secreted the highest initial urinary cortisol levels. After 6 weeks they were less likely to meet PTSD diagnostic criteria, particularly for intrusive, reexperiencing phenomena. The difference between amnestic and partial/full memory persons may be due to differences in cortisol excretion.

COMMUNICATION AND LANGUAGE

Communication's original evolutionary goal in animals was reproduction. Human speech and language may have evolved from a repertory of two to three dozen innate calls typical of primates. This is consistent with babies' innate discrimination of a small number of sounds in human language, which function as acoustic sign stimuli. Animal communication serves the purpose of identifying members of a species, for example, a predator or a food source. There is evidence that language's function has been somewhat elaborated: safety, culture, thinking and reasoning, political structure, transmission of information and achievements, etc. We can designate items, actions, and properties to one another using varieties of meaning and relationships, including temporal and causation (Caplan & Gould, 2003), which explicate the details of processing the major components of language.

Elements of Language (Definitions)

- *Words*: An arbitrary association between a sound and a meaning.
- *Grammar*: The system specifying how vocabulary units are combined into words, phrases, sentences, and how the meaning of a combination can be determined by the meanings

of the units and they way that they are arranged. Grammar has three main components: *morphology* (the rules for combining words, prefixes, and suffixes); *syntax* (rules for combining words into phrases and sentences, and determining relations among words); and *phonology* (rules combining sounds into consistent patterns within a language) (Dronkers et al., 2000).

- *Lexicon*: The vocabulary of a language; the lexical knowledge of an individual.
- *Lexicology*: The study of the meaning of words and their idiomatic combinations.
- *Semantics*: The study of meaning in language, contrasted to the study of formal relationships (grammar) or sound systems (phonology).
- *Syntax*: The study of rules describing how words and phrases in a language are arranged into grammatical sentences. Syntax and morphology are considered to be the two subdivisions of grammar.
- *Morphology*: A branch of linguistics that investigates the form and structure of words.

Language: Communication, Comprehension, and Expression

Language is *supported* by intelligence, sensorimotor integrity, memory, emotional expression, speed of auditory processing, auditory resolution, sequencing, etc. It comprises expressive and receptor skills. The range of communications channels or levels has been described (Borod, 1993a, 1993b; DePompei & Tylor, 2004; Doupe & Kuhl, 1999): facial; prosodic; lexical; gestural; postural; with an emotional dimension, that is, pleasant–unpleasant and approach-avoidance, semantics; lexicon; phonology; expressive receptive and pragmatic language.

Nonverbal Language

Shrieks of fear are high in amplitude in order to be heard and rouse others to action. They are high in pitch since they are produced by muscles that are tensed for flight or fight. Llinas (2001, pp. 228–239) considers language to be generalized motor behavior by which one animal communicates with another. Biological prosody is purposeful, unspoken communication, for example, the dance of bees, posturing by wolves when attacked, or threats by animals with horns. Capacity to understand the other's fearful or threatening vocalization contributed to social understanding. Meaning is communicated by explanatory mimicry and gestures, that is, an outward gesturing of an internal state (centrally generated abstraction) that means something to another animal. Gestures, or *kinesics*, are limb, body, and facial movements that modulate the verbal message. *Pantomime* conveys specific semantic information, for example, the use of mutually agreed-upon symbols. Recognition of other animal's intent (e.g., growling) was essential for survival, ecologically and genetically. Mimicking is part of adaptive learning, that is, associating the action (one's own threat to the other) with the vocalized threat of the other beast. Animals that can mimic each other, that is, vocalization, engender a sense of familiarity or recognition, and the potential for living in groups. Vocalization is a sign of arousal from internal as well as external stimuli. As a form of defense it startles, distracts, and warns the attacker without actual threat or injury. Our complex laryngeal mechanism is capable of creating complex sounds and patterns, which have been selected as a mechanism for communication. With reference to prosody, the larynx, *velopharynx (soft palate and pharynx)*, and sensorimotor reactions and mimicking, associated with observation of the other beast's reactions, leads to social learning. One can anticipate threat or retreat by learning what noises mean when made by oneself or by the other.

Gestures: Expressive use of posture, face, and limbs is considered as an evolutionary precursor of speech (Caplan et al., 1999). These were the major means of communication prior to articulation, and are important in the pre-speech period of ontogenesis. In prehominids, gesture manifests emotional reactions. By imitating the object, they are a visual representation of it. Expressive movements and sound signals involve inseparable information about emotion and the situation. These are considered right-hemisphere language (Glezerman & Balkoski, 1999, p. 79). The kinesthetic modality eventually was accompanied by a voice gesture, that is, a sound similar to that of the object. Llinas (2001) refers to the phylogenetic modification of the fixed action pattern of breathing and swallowing, which slowly become modified to work together to produce vocalization.

Consistent with our evolution, gestures and speech have a common origin developmentally. Early in development, the inferior portion of the PFC that differentiates into Broca's area is equipotential for control of the articulatory apparatus and fine manual gestures of sign language (Caplan et al., 1999). From infancy, relationship with the caregiver is integrated with movement as part of communication (Cross, 2001). The infant follows, responds, and eventually initiates temporal regularities in vocalization and movement. It has been suggested (Lichtenberg, 2002) that prosody evolves from the infant's reaction to the mother's voice, thus, a vital learning experience with significant emotional associations.

Speech

Research on nonhuman acoustic signals creates doubt that human speech perception is a unique process (Fitch et al., 1997). Vervet monkeys innately offer three different alarm calls (eagle, leopard, or snake), which evoke appropriate evasive actions in the recipient. However, the appropriate expression is a matter of social learning, that is, the reactions of adults to correct and incorrect alarm calls reinforce the development of selective socially appropriate usage of the calls. Moreover, it has not been demonstrated that nonhuman primates alter the vocal tract shape to achieve distinguishable vowel-like configurations. However, monkeys can finely control airway constriction at one or more paces of articulation to produce calls distinguished by stopping or fricative control (Brown & Cannito, 1996). Human speech is built on preexisting brain structures, but an enormous evolutionary step occurred with the convergency of cognitive capacities and auditory and motor skills to create *language* (Doupe & Kuhl, 1997). The evolution of human prosody is a late-comer, perhaps the most recent addition to the psycholinguistic structure of speech and language acquisition by the human species. Early in the development of speech, its components were elements of singing: Changes of word meaning were accomplished through individual characteristics of sound change: intonation, rhythm, frequency, and so on. (Glezerman & Balkoski, 1999, pp. 79, 85, 87).

Language is associated with the development of consciousness and symbolic thinking. It is described by anthropologists as an *exaptation*, that is, a feature developed in an earlier context before being utilized in a different one. Language and consciousness may have developed through *emergence*, a mechanism of evolution suggested by Alfred Russell Wallace, distinctive from the more familiar *natural selection* proposed by Charles Darwin (Tattersall, 1998, 2000). The accretion of small changes per se does not express the later trait. The key anatomical change was not alone the development of increasingly larger brains. The immediate ancestor of modern humans possessed a brain that evolved to a point where a single genetic development or group of changes created a structure with an entirely new potential. Some enormous advantage is ascribed to a change in a gene (FOXP2) in the past 6 million years (postseparation from our chimpanzee cousins) (Wade, 2003). The present writer notes that the mechanism of speech is enormously complex, both somatically and neurologically. He reserves judgment whether a tweak in one gene created, on the one hand, Adolf Hitler's capacity to rouse the passions in crowds, and on the other hand, Franklin Roosevelt's capacity to soothe them.

Earlier in evolution, speech and language were contraindicated by the narrowness of the thoracic vertebral canal that carries the innervation of the thoracic musculature. Hominids are believed to have possessed modern vocal tracts for hundreds of thousands of years before they employed the articulate speech permitted by the peculiar form of our vocal tract. The anatomical development was flexion of the base of the skull, a structural quality that permits a vocal tract to produce the sounds associated with articulate speech. This refers to the extension of the oropharynx and nasopharynx into the base of the skull, then continuing into the nasal cavity (see Moore & Dalley, 1999, figures on pp. 1039, 1050). Nevertheless, a differentiating neurological characteristic is Broca's area, since not even nonhuman primates have such an area for social vocalization (Brown & Cannito, 1996). Language and symbolic thought seems to be represented around the time that Cro-Magnons (*Homo sapiens*) invaded Europe around 40,000 years ago, and replaced the Neanderthals (*H. neanderthalensis*) in about 10,000 years.

It has been suggested that the sequence between nonverbal primates and hominids evolved through an initial status using gestures and nonword sounds (e.g., alarm calls) without syntax (an order which determines meaning), to words (a combination of sounds) and syntax. Words and syntax are products of a brain combinatory system. Words represent symbols that can be used in a variety of contexts (e.g., for alarm or food). The proto-language may have been words without syntax, including click sounds, which are still in use. Complexity was driven by the need to communicate as they ventured out of the close environs of a forest in order to forage in the savanna (Wade, 2003).

Fitness: Darwin (1860, p. 70) described the preservation of favorable individual differences and variations and the destruction of injurious ones as natural selection, or the survival of the fittest. By this may be understood the connection between selection and capacity to reproduce. Darwin stated specifically (1871, vol. 2, p. 330) that "the vocal cords were primarily used and perfected in relation to the propagation of the species." Auditory signals (to other organisms) can deter predators, deter rivals, and attract mates.

Word Competence

Expressive language difficulty refers to poor retrieval of words, speaking off the topic, rambling, with written work tangential and disorganized. Narratives are long and unrelated, out of order, and disorganized. Substitute language is used, for example, "you know" and "thing," rather than the noun or verb.

Receptive language difficulty is poor comprehension of vocabulary, associated with an inability to communicate sequentially, follow complex directions, appearing not to hear what somebody says, and to ask for multiple repeats.

Pragmatic Communication Skills

Pragmatic communication skills are the skills underling competence in contextually determined, functional language use: the ability to meet the informational needs of the listener, verbalize ideas in a logical and coherent sequence, share the burden of conversation with a partner, and monitor the appropriateness of one's own spoken productions in a particular context. Impairments within the pragmatic domain may represent the most pervasive communication problem in the chronic stage after TBI.

These include daily living skills of language, for example, volume, pitch, emphasis, articulation; nonverbal behaviors (eye contact, physical proximity, body posture, gestures and facial expressions) (Murdoch et al., 1990); and social interaction (turn-taking, topic maintenance, social conventions) (Brookshire, 2003). Deficient individuals talk excessively; begin and end conversations abruptly; have difficult staying on topic; interject irrelevant, tangential, and inappropriate comments; and do not yield priority to others (turn-taking). Cognitively based linguistic and pragmatic disorders are particularly severe on adolescents, who are in a period of growth in semantic and syntactic abilities, metalinguistic and pragmatic language skills, knowledge base, and EFs that will permit them to communicate as adults (Turkestra et al., 1995). Pragmatic impairments have been attributed to right hemisphere brain injury (Brookshire, 2003). *Social deficits* lead to isolation, increased reliance upon the family for social support, and problems in returning to work, school, and premorbid avocations. These deficits may be a consequence of subtle deficits of pragmatic language during conversation. Even individuals who had recovered a high level of functional language did not initiate a great deal, depended upon the other person to maintain the conversation, provided more information than necessary, and by not encouraging disclosure on the part of the conversational partner, did not create an opportunity to talk about a subject of interest to strengthen the social interaction (2007).

Aprosodia is considered to be impairment of the *pragmatics of nonverbal communication* resulting in impoverished language and reduced communication proficiency (Jorge et al., 2000). This contrasts with cognitive and aphasic deficits that include incapacity for communicating inner states, inability to comprehend the emotional communications of others (Heilman et al., 1993), and deficit of symbolization and comprehension of the complexity of emotions (Lane et al., 1997).

SOCIAL COGNITION AND COMMUNICATIONS

Social signals and their reception are considered to have a genetic basis. Hearing and vision are part of the language process. Using specialized forebrain systems, speakers learn to control multiple craniomotor systems and correlate somatosensory feedback from the vocal tract, with auditory feedback generated by the vocal apparatus. There is almost seamless coordination of the complex vocal apparatus (Brauth et al., 1997). Voice-selective regions are found bilaterally long the upper bank of the superior temporal sulcus. It is hypothesized that these may be the counterpart of face-selective areas of the visual cortex (Belin et al., 2001). Communication of affect shapes behavior and forges identities. Effective vocalization requires concentration (staying on task). Learned vocal signals are acquired through *auditory-vocal learning* in which an individual learns to match self-produced sounds to an auditory memory. Communication also involves written language and symbolic communication. Peripheral hearing loss affects oral communication. It has been estimated at 56% immediately after brain damage, resolving to about 15% 6 months later. Also occurring are auditory distractibility and difficulty with competing passages. In this Canadian sample, persons with speaking difficulties were less frequently married (49%) than those with hearing difficulties (58%), while hearing difficulties was more associated with poverty (40%) than speaking disability (27%). Many individuals maintain hearing and speaking disorders for 5 years or longer. They have more difficulty being understood outside their family (Lubinski et al., 1997). Vision also participates in social understanding: Perception of someone's face to enables one to assess the other person's mood. Hearing also involves biofeedback. If you can't hear, it's hard to learn to speak. However, animal studies suggest that isolates, incapable of reacting appropriately to social signals, can develop a species-typical social structure when given social opportunities (Ginsburg & Buck, 1997).

SOCIAL COGNITION

This is an adaptive function referring to the processing of information, leading to the accurate perception of the dispositions and intentions of other individuals (*empathy*) (Brothers, 2001). Social cognition is mediated by a distributed network of interdependent brain regions subserving discrete cognitive and affective processes that are integrated during social behavior (Yeates et al., 2004). Social cognition involves skills such as turn-taking, politeness, negotiation, topic management, and sensitivity to the role and status of all participants (Coehlo, 2007). The syndrome of socially inappropriate behavior, compromised decision making, impaired social cognition (identifying facial expression of emotions), and poor self-regulation has been attributed to a right and bilateral ventromedial frontal and orbitofrontal lesions. Adults with primarily right ventromedial prefrontal lesions experienced postinjury deterioration of social and occupational functioning. These deficits were observed despite relatively preserved intelligence in children and adults. Dysfunctions following injury to these areas include social and interpersonal behavior, employment status, laboratory decision making, anticipatory skin conductance responses, linkage of somatic markers to events and situations, and behavioral reversal based upon social cues. This leads to the predictions that persons with orbitofrontal and ventromedial lesions will have an increased risk of *psychosocial sequelae* and difficulties in *inhibiting aggression* towards people who display fear or submission (Levin et al., 2004).

Discourse

This is the ability to convey a message, or tell a story, by communicating a series of ideas, usually in sentences. It is an interdependent process, relying upon abilities in various domains, including linguistic ability (lexicon and syntax), cognition (attention, memory, planning selection, organization of information, and social domains), using language to exchange information. It involves meaning, a genre (description, narration, conversation procedural, expository) serving different communication functions. Discourse measures are more accurate indices of communication competence than lexical or grammatical abilities (Chapman et al., 1995). Language dysfunctions after TBI are usually cognitively based, as opposed to aphasia.

Deficits of discourse (reduced content, cohesion, increased hesitation), detectable after apparent recovery on structured language measures, can limit the capacity for social relations and appropriate conversations (Dennis, 1991). With increasing brain injury there is loss of essential information that is required for maintaining structure and preserving core content (Chapman et al., 1995).

Hampered discourse has varied origins. The majority of those with cognitive communicative impairment show deficit of discourse and do not use single words or isolated sentences. There is a difficulty with *organization* of semantic information, probably caused by executive dyscontrol. Closely related is the flexible application of *social rules* (Coehlo, 2007). Speech is less productive when it is *fragmented*. The speaker cannot interpret direct communication, and finds it difficult to initiate and maintain a topic of conversation. Consequently, the needs of the other person are not met (McCullagh & Feinstein, 2005). *Nonverbal discourse* also is required for social communication. It has been estimated that more than 90% of a message's effect is communicated by nonverbal information, and the small remainder would then be verbal. It has been suggested that emotions may be conceptualized as emotional behavior (autonomic and overt behavior), emotional feelings and experiences that are subjectively reported, and the communication of emotions through facial expressions and gestures. Children learn to judge the emotional meaning in facial expression earlier than the ability to correctly identify vocal emotions. These functions are vulnerable to TBI and consequently may hamper social understanding (Van Lancker & Breitenstein, 2000).

Informativeness and coherence is described by the amount and quality of the information communicated and its efficiency. Discourse is considered to be the most natural unit of normal verbal communication, requiring linguistic, cognitive, and social abilities, all sensitive to disruption by TBI. A narrative's information content may be described as irrelevant, redundant, off-topic, overpersonalized, lengthier, and slower relative to the content produced. There are concerns about accuracy, implied meaning, amount of pertinent content, and number of concepts intruded. The examiner will observe cohesion and continuity, pragmatic success of the narrative, reduction in the number of components, failing to signal new episodes, and omitting essential action information. A procedure for detailed analysis of discourse is offered (Coelho, 2007).

Coherence refers to conceptual organization and thematic unity. It depends upon simultaneous attention and manipulation of information within the speaker's plan and the listener's perspective. Global coherence refers to a general goal, plan, or topic. It requires semantic memory (real-world knowledge) and conceptual integration necessary to maintain the plan and the organization of the discourse. Local coherence refers to maintaining meaningful conceptual links between sentences (Coehlo, 2007).

Aprosodia refers to inability to express moods nonverbally, that is, with a musical, expressive quality. Flat affect creates a false impression of indifference or lack of distress. This writer considers aprosodia to be a motor vocal disorder.

MEMORY

The Amnestic Syndrome

The amnestic syndrome refers to a severe and relatively pure impairment in new learning as a result of brain damage, that is, inability to remember recent events in the context of relatively intact cognitive, linguistic, perceptual abilities, and retained fund of general information (Albert & Lafleche, 1991; Brandt, 1992; Schachter et al., 1991). Amnestic patients perform poorly on tests of explicit memory, and almost normally on tests of skill acquisition, perceptual priming, etc. Recognition is more effective than recall (Brandt, 1992). Working (short-term) memory is preserved, although retention across delays is impaired. Amnestic patients have variable ability to retain information short-term, perhaps varying with the procedure, and whether interference occurs between exposure and retrieval (Schachter et al., 1991). There is a more rapid rate of forgetting than controls. Implicit memory retention can be demonstrated. Proactive interference causes intrusion of earlier exposed information into retrieval of lateral information.

APHASIA: NONMOTOR LANGUAGE DISORDER

Aphasia may be defined as a disorder of language acquired secondary to brain damage. Language refers to communication symbols and rules for their use. It requires facility for expression and reception. Aphasia is a disorder of language secondary to brain damage. It is differentiated from motor speech disorders as a disorder of language as a complex system of communication symbols and rules for their use, whereas speech is the articulation and phonation of language sounds (dysarthria, dysphonia, voice disorders), stuttering and speech dyspraxia. Aphasia is also distinguished from disorders of thought, that is, the mental processing of images, memories, and perceptions, usually involving language symbols. Psychiatric disorders derange thought and alter the content of speech without affecting its linguistic structure. While diffuse brain disorders (e.g., encephalitis, dementia) are aphasias, they involve other cognitive functions, which distinguish them from focal brain lesions. Involvement of the language cortex creates disorders of naming, reading, writing, memory, and visuospatial processes (Kirshner, 2004).

Communication is cognitively very complex: The expresser has a communication intent, organizes thoughts, decides what he wants to say, selects words and phrases to communicate meaning, places these words in the order specified by the grammatical rules of the language, and formulates the motor commands to program the speech mechanism. In additional to the familiar dominant, left-hemisphere processes, the right hemisphere has a language contribution: attentional deficits; neglect; visual perceptive deficits; affective and emotional (facial expressions, i.e., the use of and sensitivity to); prosody (use and sensitivity); cognitive and communication deficits (Coelho, 2007). Expressing words that communicate what was intended, and are prosodic to maintain credibility, is the product of a complicated neural and somatic mechanism representing motor programming, cognition, somatic structures, etc.

Communication can be compromised by lack of logic and cohesiveness, lack of clarity, insensitivity to the other person's needs and interests, insensitivity to the amount of explanation required to have the other person understand, confabulation, inconsistencies, and impaired pragmatics (knowledge and activities of socially appropriate communication) (Lezak et al., 2004).

The components of an observational study of language ("bedside") invite sensitivity to a range of language disorders (Coelho, 2007; Kirshner, 2004a). Their impairment ranges from subtle to impairing, depending upon the location and extent of neurotrauma: spontaneous speech; auditory comprehension and writing comprehension; circumlocutions; paraphasia; naming; use of grammatical rules; deficits of word retrieval; repetition; reading; writing; naming after confrontation; fluency in the generation of category-specific words; and associated signs.

16 Posttraumatic Personality Disorders

"I feel, therefore I am." A philosophically oriented reader may recognize this conclusion as deviant from the more cognitive concern of Rene Descartes ("I think, therefore I am"). Perhaps there is room for both views.

OVERVIEW

This chapter stresses the personality disturbances of lesser brain injuries, including the moderate and severe level only for clarification. Personality changes, affecting family life, are the most disturbing behavioral disorders up to 15 years after a traumatic brain injury (TBI). These vary from subtle disruptions of social life, not clearly obvious to the injured person or the family, to extreme departures from socially acceptable norms of behavior (O Shanick & O Shanick, 2005). Neuropsychiatric illness has been described as a highly prevalent complication of TBI. Prior psychiatric illness is a significant predictor of posttraumatic psychiatric morbidity, greater for the moderate-to-severe group than following mild TBI (mTBI). There is a risk factor for prior mood and anxiety disorders, as well as alcohol abuse.

Although the rate of psychiatric disorder at 1-year post-TBI in one sample was 18.3%, this did not differ much from the general population rate of 16.4%. The rates of depression and panic were increased, although the following were not elevated: generalized anxiety, phobias, obsessive-compulsive disorder, and schizophrenia. Increased depression and anxiety were associated with greater functional disability, but these patients were more likely to have mild injuries than the nondepressed anxious group and to report more subjective distress, neurological, and cognitive conditions (Moldover et al., 2004). When individuals with a prior history of psychiatric illness are compared to those without, the relative risk for a psychiatric illness for the 6 months postinjury was increased for moderate-to-severe TBI to 2.1 and for mTBI to 1.6. In the period of 1–5 years post-TBI, cognitive and communication functions tended to be stable or even improve, although there was a worsening of social interaction (Jorge, 2005). The etiology is complex: There is an exaggeration of premorbid traits, personality traits consequent to physiological dysregulation, direct consequence of cerebral trauma, psychodynamic reaction to being impaired, with physical injury and difficulties, being socially less desirable, and the struggle to have one's injury understood and treated.

Emotions help coordinate responses to such evolutionarily relevant scenarios as threat. They are associated with specific autonomic profiles and stereotyped, recognized facial expressions. Voluntary suppression of emotional displays affects the magnitude of behavioral and autonomic responses to the emotional stimulus. Disruption of the coordination of multisystem emotional responses may be the cause of the emotional dysregulation seen in bipolar disorder, schizophrenia, and frontotemporal dementia. Neural circuits in the lateral prefrontal cortex (PFC) are important for voluntary emotional regulation (Woolley et al., 2004). The emotional life and personality style of a TBI patient influence a wide range of functions, with implications for understanding behavior in the community and the inner emotional experience of the patient. Personality style and change has to be considered in assessing performance on formal assessment procedures. Adaptation and outcome are influenced by personality and affect in various domains: cognition; reaction to pain and impairment; motivation to recuperate; quality of life (QOL); outcome; diagnosis; impaired performance; recognition of emotions (personal or insight), and others (empathy and emotional intelligence). In the community, personality changes may adversely affect interpersonal relationships, psychosocial outcome, and qualification for services and compensation.

Emotional experience and expression depends upon an interaction with a wide range of other functions: the executive function (EF); comprehending and labeling verbal and prosodic stimuli; capacity for self-regulation (SR); insight, and so forth. This chapter focuses upon the effects of lesser levels of trauma (the postconcussion syndrome, or PCS). While lateralized lesions are not stressed, it is recognized that diffused injury, which usually does not appear on computed tomography (CT) or magnetic resonance imaging (MRI), may in fact be lateralized or differentially emphasized in particular neurological centers or circuits. Emotions express cortical, subcortical, and brainstem functions, and physiological and autonomic influences. Vulnerability to an affective disturbance following TBI is enhanced by genetic, developmental, and psychosocial factors (Jorge & Starkstein, 2005). The frontal lobes are most frequently the locus of interest in research reports, but it is worth repeating that the widespread bidirectional projections make their performance vulnerable to injuries elsewhere.

THE INTERACTION OF TBI AND EMOTIONAL CHANGES

The emotional distress and personality dysfunctions posthead trauma have varied sources. Some are directly consequent to brain trauma, some are reactions to impairment and fear, and in some instances, direct causation is unproven. Extremes of emotional expressiveness are characteristic, that is, *reduced affective expression* and *released affect*. Analysis involves *content* and *valence* (*positive*, i.e., experienced); *negative* (a function is lacking, and not directly experienced).

Double diagnoses are characteristic. The patient with brain trauma may have dual or multiple diagnoses, that is, TBI, and an additional diagnosis in the psychiatric or personality dysfunction areas (e.g., cerebral personality disorders [CPD], posttraumatic stress disorder [PTSD], depressive reaction, or a major psychosis). When posttraumatic psychoses develop, one may consider endocrine and metabolic disorders (Little & Sunderland, 1998). In one study, the findings of various brain scans (MRI, single photon emission cardiac tomography [SPECT], magnetoencephalography [MEG]) failed to correlate significantly with the presence of PCS symptoms. Perhaps such symptoms are secondary to TBI factors (e.g., cognition) or non-TBI factors (social or economic) (Lewine, 2007). Depressive symptoms and those of PTSD may emerge from or blend with those of PCS (Montgomery et al., 1991). Psychodynamic reactions are prominent, that is, changes of identity, reactive mood changes (e.g. anger, depression, social withdrawal) due to the distress of impairment, injury, loss of status, and so forth; reduced morale concerning recovery with a pessimistic view of life, that is, weltanschauung.

SOME PRINCIPLES OF TBI/PERSONALITY INTERACTION

- *Multiple neurobehavioral components*: Intelligence and coping skills; preexisting personality; level of support or opposition by social and community persons; neurological; somatic; autonomic; psychodynamic; neuromotor programs; emotional evaluation of the situation; capacity for foresight and control; autonomic reactions.
- *Multiple brain areas (diffuse brain injury)*: More than the frontal and temporal tips are vulnerable. The effect upon the brainstem, which is bent, compressed, and rotated, is often underestimated. Diffuse brain damage creates dysfunctioning of complex circuits processing the interaction of affect, personality, and cognitive input. Localization and lateralization are only weakly associated with particular emotional and personality dysfunctions. The centers and circuits for arousal, autonomic regulation, muscular action, and balance are interactive and interact with moods, drives, and motives.
- *Varied types of neurological dysfunctions*: Neuronal (incapacity, release, dysfunction, irritative, stimulation, arousal), control, stress effects (acute and chronic).

- *Psychological components*: Reactions; evaluative; comprehension. Cognitive loss consequent to TBI participates in the creation of CPDs, for example, "frontal lobe syndrome": poor judgment, inability to learn from experience, initiating and cessation problems contribute to social dysfunction, emotional distress, and interfering with the resolution of problems.
- *Internal milieu affects outcome*: Precipitating events and trauma are expressed through brain injury, autonomic functioning, somatic injury, central nervous systems, and somatic systems (autonomic; immune; circadian; inflammatory; hormonal); reexisting personality, social support, social opposition. Various "psychological" symptoms are considered either psychodynamic or consequent to neurological disturbances of consciousness, and are intimately associated with dysfunctions of autonomic functions. This leads to a possible interaction of panic, PTSD, autonomic dysfunction, and focal neurological dysfunction (vestibular apparatus), hyperventilation and other somatic symptoms (palpitations; sweating; paresthesias; feeling dizzy or faint); feelings of unreality (depersonalization; derealization); or fear of going crazy or of dying.

Consequently, in the assessment and treatment of post-TBI patients, emotional and personality disorders, endocrine, health, and vegetative functioning must be considered.

- *Emotional disorders affect outcome*: Motivation for recovery is determined by appropriate emotional regulation, lack of social rejection, support from the community and health care providers, and so forth.
- *The complete meaning of a symptom is complicated*: Preexisting condition; traumatic injury (neurological; somatic); reaction to stress; psychodynamic, neurochemical, endocrine, and brain dysfunctional causes of post-TBI behavior disorders.
- *Distress is dose-related*: It is common that affect disturbances and headaches are worse after so-called mTBI. There are greater distresses (and awarenesses) in so-called "mild" TBI.

MOODS, AFFECT, AND EMOTION

This section describes disturbances of the level and quality of mood after concussive brain trauma: variation in intensity, in the capacity for self-control, and reduced ability to express affect (apathy and endogenous depression). Affective release refers to disinhibition of behavioral controls and overt expression of feelings and moods: irritability; rage; violence dyscontrol; mood swings or lability (anger, anxiety, depression). There can be a discrepancy between experience and expression, for example, aprosodia and a reluctance to self-describe as illustrated in "Expressive Deficits." Intensive interviewing of the patient, and with the collaterals, may clarify the reasons for differences between self-described emotional pain and apparent blandness, or the reverse: claims of distress with lack of evidence for an attempt to cope with the distress. Clinical assessment requires interviewing the patient and collaterals, history, and observation, seeking expressions of mood disturbance such as reduced intensity of feelings (anhedonia; apathy; indifference; endogenous depression). It is important to differentiate between brain trauma–related depression (neurotransmitter dysfunction; reduced arousal; site-related) and ordinary unhappiness (psychogenic depression).

Clinical example: False impression of indifference: A 25-year-old man was examined several years after an auto accident in which he was rendered unconscious. A normal electroencephalogram (EEG) and CT scan of the brain were reported, although he was diagnosed as having a "concussion." He described his feelings as "weaker," could not find words to express himself, and does not remember what he does not remember. Clinically, his mood was in the pleasant-to-dull range. He expressed depression and wept when he was led to discuss his situation, but did not show anxiety. He asserted that others don't understand him. (His unhappy mood and subclinical aphasia would have been difficult to detect without a thorough interview and/or examination.)

Emotional States: Definitions

Mood is the pervasive and sustained emotion that colors the perception of the world. It is reactive to the situation, and may be transient or long-lasting. The symptoms of a mood disorder may vary with their occurrence during the course of the illness, induction by a medical disorder such as hypothyroidism, and their response to treatment. Abnormal moods change our perceptions, social interactions, and decisions. Common conditions include dysphoric mood (anger, anxiety, depression, hopelessness, guilt); suicidal fantasy and attempts; need for nurturance; and low morale.

Emotion is the experience and expression of feeling states, for example, sadness, elation, changes in motivation, reduced empathy or inability to experience another's feeling state, irritability, lability, fear, anxiety, sexual desire, persecution, threat of personal harm.

Affect is a pattern of observable behavior that objectively expresses the feeling state (Cummings & Bogousslavsky, 2002; Eslinger & Geder, 2000).

Weltanschauung (world outlook): After brain damage, the view of the world in which one lives becomes changed. One's security is precarious because of the life changes occurring as a consequence of the accident. The familiar support system may have disappeared, and others have become unfriendly. The environment is seen as having deteriorated because one's handicaps prevent obtaining the customary material and spiritual supplies. The patient wonders about the meaning of the event: Why did so great a catastrophic injury and impairment happen to me? The themes expressed in the person's Weltanschauung include:

- Search for the meaning of life.
- One's world or adaptation to it has been destroyed.
- The world is dangerous.
- Life has deteriorated.
- The world is unsupportive or unfriendly.
- To retreat is the best choice.

TEMPERAMENT

Temperament refers to persistent components of personality, which are styles of reacting in many different circumstances. Examples include energy level, emotional responsiveness, response tempo, and willingness to explore (American Psychological Association, 2007, p. 928). The identification of varied temperamental qualities is more than clinically useful personality traits. Since temperament participates in the baseline personality, changes are noteworthy. A temperamental trait is considered to be a biologically based, inherited personality characteristic. For example, early vulnerability to anxiety is accompanied by increased physiological reactivity. This is described as behavioral inhibition and withdrawal in the face of novel stimuli or challenging situations (Merikangas, 2005). Temperament is nonverbal; therefore, its expression is somewhat independent of the specific meaning of the response or situation.

Some Relevant Temperamental Traits

Temperament may represent behavioral qualities that in the patient's youth were vividly expressed, were minimized or controlled by experience, but were released as a disturbing quality after a TBI. The style of dealing with impairment and distress is vital in determining participation in treatment and thus shaping outcome. Thus, temperamental qualities are part of the clinician's assessment of capacity to cope with rehabilitation for particular occupations or roles within the community. A characteristic approach to handling human relations and situations in vital matters is the role in life. This may or may not be active, depending upon the life history and the consequences of the

TBI-related accident, but it could influence the patient's morale and attitude towards the rigors of rehabilitation.

Harm avoidance (fearful or daring): This describes the inhibition of behavior in response to signals of punishment, frustration, and nonreward. Relevant issues are fear of imbalance, coping with seizures, and resuming various means of transportation. A different approach stresses conditioned behavioral responses to sensations eliciting fear or anger independently of awareness, recognition of danger, observation, reflection, or reasoning. There are automatic responses to danger, novelty, and reward (Svrakic & Cloninger, 2005).

Novelty seeking (impulsive or deliberate): Contrasting a drive to approach rewards, avoidance of conditioned signals of punishment, escape from punishment after an injury, or valued intellectual or stirring activities, there is substituted general inactivity or passive attendance at undemanding TV or radio.

Reward dependence (affectionate or independent): Initial responses to cues of social reward are substituted with social indifference and retreat. These are common sequelae of concussive accidents.

Persistence (determined or spoiled): The maintenance of behavior despite frustration, fatigue, and intermittent reinforcement. Loss of this trait is a common finding, reflected in loss of stamina, apathy, and loss of motivation. Allostatic load can be a contributor insofar as chronic stress, and the systemic reaction may reduce the patient's capacity to persist in an activity.

TEMPERAMENTAL DISORDERS OF ENERGY EXPRESSION

Disinhibited

Orbitofrontal: dyscontrol, that is, inability to delay a response. Impulsivity and disinhibition cause inappropriate behavior; impulsive release of sex. This is described by Fuster (1989, p. 130) as hyperkinesis. Pseudopsychopathic behavior, that is, jocular, euphoric, emotionally labile, poor judgment and insight, distractibility, instinctual disinhibition, hypomanic, irritable, boastful and loud, inability to shift (e.g., from attack to escape). This syndrome has characteristics in common with the antisocial personality, and can be involved in violent behavior. The discrepancy between euphoric unconcerned attitude out of line with a devastating debility is attributed to a "global" error-monitoring deficit.

Apathetic

Frontal convexity: Reduced motivation and goal achievement, response delay; Inhibition: Inability to initiate responses causes reduced effectiveness in life, that is, lack of activity. Problems selecting a goal: Apathy alternates with occasional outbursts of anger, irritability, or euphoria; indifferent; psychomotor retardation, perseveration, and impersistence; motor programming deficits; loss of set yet stimulus boundedness; discrepant motor and verbal behavior. There is a question whether the "apathetic" reaction of the frontal lobe patient, which may alternate with emotional outbursts, really reflects the true definition of apathy, that is, absence of feelings of emotions, interests, or concern.

Akinetic

This may vary from loss of language and social behavior to inability to perform voluntary actions, as well as paucity of spontaneous movements; sparse verbal output; lower extremity weakness and loss of sensation; and incontinence. These symptoms may be summarized as motivational, affective, loss of energy.

Rage and Violence

Some issues are altered state of consciousness, level of comprehension, loss of control, appropriateness of affect, and anticipation of consequences (Hall, 1993). Explosive rage takes over, which is poorly controlled or easily provoked, with primitive expression such as gauging, biting, spitting, and use of a hammer or knife. Persistent or recurrent aggressive outbursts, either verbal or physical; outbursts are disproportionate to the stress or provocation; an organic factor is etiologically related to the disturbance; outbursts are not primarily related to other conditions.

SELF-REGULATION

Self-regulation is the conscious effort of a person to control affect, impulses, and personal activity so that it conforms simultaneously to the intended task and requirements of the field of action. SR is a biologically based attribute in which the PFC interacts with cortical, subcortical, and brainstem structures to integrate cognition and affect. Cognition, emotional experiences, and emotional stimuli influence each other. Successful SR enhances the capacity to manage one's own thoughts, feelings, and actions in adaptive and flexible ways. This contributes to appropriate goal-directed activity and recovery from emotional distress and trauma. Since SR affects cognitive processes, it overlaps *EF* (Genesalingam et al., 2007, citing Saarni, 1997). Yet, SR differs from the EF by involving active and self-aware behavioral control. In contrast, EF proceeds automatically, and its processes and intermediate products may never be known to the person. SR also differs from identity, which is an emotional or objective self-description. It, too, functions both in consciousness and in the unconscious. After identity has initiated a significant activity or determined a response to external activity, performance then utilizes SR and the EF.

Chronic inability to regulate affect is characteristic of both major mood disorders (depression; bipolar disorder) and TBI. The TBI-vulnerable PFC is integrated with the limbic brain: medial; orbitofrontal; cingulate gyrus (anterior and posterior), the dorsolateral prefrontal cortex (DLPFC), other areas of the neocortex (posterior perceptual and sensory systems), subcortical areas (basal ganglia), and the brainstem (thalamus; substantia nigra; hippocampus; hypothalamus; amygdala). Loss of SR of emotions may follow diffuse TBI, and be reflected in impaired conceptual and problem-solving behavior, and aspects of memory and learning.

Examples of Loss of Self-Regulation

- Erratic productivity: Initiation of an activity is slow, for example, stuttering sounds preparatory to speaking, or agitating a body part before its activation.
- Describing an intended action without performing it.
- Inflexibility: Responses are perseverative, stereotyped, and nonadaptive. Approaches to understanding and problem solving are concrete or rigid. Attention cannot be withdrawn from the perceptual field of current thoughts. In addition to perseveration, there is defective scanning and inability to change perceptual set easily.
- Not following instructions in a structured, uncomplicated task.
- Prefrontal language problems that are neither language nor cognitive problems (Lezak et al., 2004).

Self-regulatory Centers

Several brain systems are involved in affect regulation. The structures and the sequence of events are indicated by functional studies. Chronic inability to regulate affect is associated with major mood disorders like depression and bipolar disorder (Pavuluri et al., 2005).

- *Dorsolateral prefrontal cortex*: Strong positive emotions enhance task-related activation, while negative emotion reduces its responses, although activating the orbitofrontal cortex. It functions efficiently with working memory but less efficiently with negative emotions. The orbitofrontal cortex reacts during emotionally unpleasant situations.

- *Ventrolateral PFC*: Activated by sadness and participates in labeling tasks
- *Amygdala*: Determines the emotional significance of sensory stimuli and their representation in memory systems
- *Anterior cingulate*: An affective regulatory area, activated by emotional events, projecting to the amygdala and orbitofrontal cortex. It is activated during labeling tasks, mediates motivated behavior, and participates in EF.
- *Dorsal anterior and posterior cingulate*: Activated by cognitive tasks, it may be involved in attentional and motor responses through projections to prefrontal, premotor, and supplementary motor areas.
- *Lateral PFC*

Self-Regulation (Control)

When a subject has acquired control over a potentially dangerous situation and has been given clear feedback that this is the case, there are reduced responses in the psychoneuroendocrinological, psychoneuroimmunological, and psychophysiological systems (Ursin, 1998). *Locus of control* effects the capacity to habituate the cortisol response (Preussner et al., 2004). Low external locus of control (perceived controllability of the situation) reduces stress. The subject habituated quickly to a psychosocial stressor and did not show significant cortisol responses after the second stimulus. When controllability seems low (low self-esteem and high locus of control scores), then high levels of cortisol remain persistent to the same psychosocial stressor. It is hypothesized that persons with low self-esteem experience higher levels of stress, regardless of the specific type of stressor. The resulting stress load, if accompanied by a higher frequency of hypothalamic-pituitary axis (HPA) activation, accumulated over a lifetime, might have the result of *allostatic load*, that is, increased wear and tear on the organism.

Emotional Lability

Clinical Examples: Overexcitability, Temper, Irritability

- "Am I crazy? My sisters tell me I am crazy? I thought I was crazy because I burst out at them. I fight for no reason". This was the plea of a middle-aged mother of four children, who had suffered for many years after a heavy object fell on her head.
- A middle-aged woman who slipped and hit the right side of her head described herself as follows: She has become very temperamental, ordering people out of her office. She has lost tact. She gets irritated too easily if they bring up questions of taxes or price. "I always had a temper, but I have never been so easily set off as now. My manager told my psychiatrist that having me was like having a hand grenade with the pin out, you never know when it will go off.
- Here is how a middle-aged, educated woman described her condition after she was struck in the right temple by a batted hard baseball (with altered but not LOC): She feels like she is on a roller coaster, with high highs and low lows. Before, she was even. The changes of mood are not related to events to which she can attribute them. She becomes angry, knows that she looks angry, or speaks in an angry tone, and doesn't know it until she hears herself. The writer observed during counseling sessions that, on several occasions she was angry or irritable, could try to be objective, on another occasion stating that she doesn't know why she is angry.

Frequent unstimulated fluctuation of moods without an intense external event or deep experience ordinarily is a temperamental quality. It is also observed after TBI, and the clinician may on occasion consider it as a possible marker for an underlying partial seizure disorder. A variety of unpleasant reactions interfere with social relationships. Three kinds of emotional lability are described (Lezak et al., 2004, pp. 37–38).

Weakened controls with lowered frustration tolerance: Overreaction with fatigue and stress. Their emotional expression and feelings are congruent; sensitivity and capacity for emotional response are intact.

Loss of emotional sensitivity and capacity to modulate emotionally charged behavior: There is an overreaction to external stimulation. When alone and comfortable, they seem emotionless.

Pseudobulbar state: The patient's feelings are usually appropriate. Brief episodes of strong, affective expression (tearful crying; laughter) can be triggered by quite mild stimulation. This is associated with structural lesions involving the frontal cortex and connecting pathways to lower brain centers, usually left anterior. Feelings are not congruent with their appearance, but the patient is usually able to report the discrepancy. Nevertheless, observers may misinterpret the patient's actual experience. After determining the nature of the problem and whether the patient can recover efficiency, an examination should be addressed to the unique set of dysfunctional responses, with a formal, structured test not required. The range of tasks and the complexity of the procedure vary with the kind of documentation needed.

Although some mood disorders have the characteristic of a temperamental trait, for example, cyclothymic and dysthymic disorders, mania, and depression or mania (Akiskal, 2005a), temperament also refers to the strong expression or consequences of moods and strong dispositions, for example, hostility or pervasive anger.

STRESS RESISTANCE: PREEXISTING CHARACTERISTICS

The stress-resistant person has a sense of personal control over external events, a deep feeling of commitment and purpose, and flexibility in adapting to change. Kobassa (1979) noted that staying healthy under stress was critically dependent on "a strong commitment to self, ability to recognize one's distinctive values, goals, and priorities and an appreciation of one's capacity to have purpose and to make decisions." When life events create changes in the level of demand, it may create the experience of *loss of control* or *helplessness* to deal with the new lifestyle. A person's decision that a situation is uncontrollable there may be distress, with failure to gain control; or if there is failure to become committed to effort, it can result in passivity and being overwhelmed (Fisher, 1988). The intensity of the stress reaction, and resistance to regression and dysregulation, varies between people and from the acute to the chronic phase. There is an interaction between the trauma (injury and fear), level of physical injury (chronic and acute), and subsequent physiological and subjective reactions to the injured state. Preexisting individual differences in stress resistance (personal qualities) will affect initial reactions and outcome. The level of the stressful experience, including morale, and recovery from injury, will also be influenced by the availability of treatment and social support or hostility. The common assertion by insurance companies and collaterals that the injured person is malingering or could do better, where unjustified, adds to the ongoing trauma. The clinician will need to offer support to relieve the person's reaction to being injured, impaired, rejected, unable to work or participate in education, feeling unattractive, and so forth.

CONSTITUTION AND INDIVIDUAL DIFFERENCES

The stress-response system maintains an organism's functional integrity in the face of shifting environmental conditions (Raison et al., 2005). Health and strength contribute to resistance to stress. Stable, athletic, tough-fibered individuals take a concussive injury in stride, while the sensitive, nervous, complaining types may be so overwhelmed that they cannot expel the incident from their minds (Adams & Victor, 1985, p. 659). Factors associated with later physical and mental health risks for children include low birth weight, and other early trauma may have a lifelong influence over stress hormone responsiveness. These environments have long-term effects on adult health and well-being: lack of attachment security; lack of family warmth and support; parental overregulation or underregulation of children's behavior; and abuse (McEwen, 2003a). Genetic predisposition and early development, that is, abuse, neglect, and other forms of early life stress, predispose to overreact physiologically and behaviorally throughout life (McEwen & Seeman, 1999).

STRESS RESISTANCE: REACTION TO DIFFICULTY

PERCEPTION OF THE SITUATION

This author has observed that some physically adventuresome individuals have a relatively low level of posttraumatic stress after significant head trauma; perhaps they expect danger and injury. Information processing participates in stress resistance in the following ways: (1) Individuals vary in their tendency to attend to aspects of their self. They take action as a function of perception of the difference between the current and normal selves. Thus, people low in self-focus are more likely to report physical symptoms, perhaps not taking action to reduce stress and take action to successfully negotiate their response to a dangerous event; and (2) They differ in their reactions to threat-relevant cues, monitoring or blunting them. High internal monitors predict greater health care utilization and depression (Miller & Birnbaum, 1988).

RESILIENCE

This has been defined as an ability to maintain positive adaptation in the face of stress or trauma, and to recover from or adjust easily to misfortune or change, that is, to withstand shock without permanent deformation. Resilience is grounded in genetic, biological, psychological, and environmental factors (Campbell-Sills & Stein, 2007). The resilient person recovers from or adjusts easily to misfortune or change. Flexible adaptation is a component of bouncing back from a negative experience. Under extreme conditions, such as a *concentration camp*, a sense of moral commitment and accepting the reality of camp life increased the likelihood of survival (Totman, 1988). Individuals who report childhood maltreatment tend to also report psychiatric symptoms, but not those who describe themselves as resilient. A non-TBI study of abused inner-city women indicated that the *emotional numbing* symptoms of PTSD were more associated with future resource loss than reexperiencing, avoidance, or arousal symptoms. These resources would be needed to protect themselves from ongoing stress (Johnson, et al., 2007).

Stress inoculation: Early life experiences contribute to mood, anger, anxiety, and substance abuse disorders for some but not all individuals. For some, early life stress can contribute to the development of resilience. This has been described as inoculating or immunizing, and so forth. Early life stressors provide a challenge: When they are overcome, the adaptive learning enhances emotional processing, cognitive control, curiosity, and neuroendocrine regulation. Animal study (monkeys) indicates that information seeking and sensory curiosity plays a role in the maintenance of lifelong resilience. Exploration of novelty (stress inoculation) is more than diminished anxiety, since stress-inoculated monkeys are more curious than noninoculated monkeys. Postnatal stress inoculation stimulates the development of larger prefrontal cortical volumes without affecting hippocampal volume. The inoculated more readily exercise cognitive control and inhibit impulsive behavior compared to noninoculated monkeys. Separation experiences of juvenile monkeys led to diminished cortisol responses in subsequent environmental stress tests compared to noninoculated monkeys. Corticolimbic circuits play a role in cognitive control. Increased prefrontal activation corresponds with diminished amygdala activation and with increased cognitive control of emotions in humans (Lyons & Parker, 2007).

One must consider *genetic factors* that influence both vulnerability to trauma and resilience following trauma exposure. Depression and anxiety with a genetic component may influence the risk for PTSD (Koenen, 2007). Resilience has been conceptualized as the consequence of a sequence: postnatal stress inoculation; cognitive, emotional, and neurobiological adaptations that enhance SR; self-selection for novelty and arousal (information seeking and sensory curiosity); and finally, the *resilient phenotype*.

Developing PTSD does not preclude developing a constructive lifestyle. *Coping mechanisms* can be higher in the resilient group and low in the vulnerable group, that is, a countervailing factor overriding PTSD risk factors. Resilience is associated with left hippocampal volume. Recovery is associated with the neurotransmitter *neuropeptide Y (NPY) and the steroid dehydroepiandrosterone (DHEA)*. These are higher as a consequence of trauma exposure. Their role aids recovery rather than confronting PTSD directly (Yehuda & Flory, 2007).

Physiological and psychological systems enable survival for extended periods under increasingly demanding conditions. Developing a stress reaction might actually arise from adversity's damage to the adaptive capacity. There is evidence that most people do not become psychiatrically ill after natural or terroristic disasters. Arousal and intrusion were less likely to become pathological unless they were accompanied by avoidance and numbing (North, cited in Boschert, 2007).

I attribute resilience in part to lifestyle and constitutional considerations: health and stamina at the time of injury, a firm sense of identity as one copes with difficulty, social support enhancing the process of recovery while minimizing the consequences of injury, and so forth. Resilience is allied to hardiness; its absence implies vulnerability to stress-related illness. It is enhanced by psychotherapy, social support, and pharmacological intervention (antidepressants and adrenergic blockers). *Eustress* contributes to resilience. It is defined as multiple beneficial experiences that create a sense of fulfillment or achievement. It also refers to acute or circumscribed amounts of stress, which, by maintaining equilibrium, may result in immuno-preparatory or immunoenhancing conditions: increased leukocyte redeployment; innate immunity; effector cell function; cell-mediated immunity; humoral immunity; resistance to infections; and possible resistance to cancer (Southwick et al., 2003). Elsewhere (Sherman, 2006), Southwick is quoted concerning social and personal factors comprising resilience: A supportive social network bolsters useful neurobiologic responses; the influence of a mentor or role model enhances capacity for self-soothing under stress; finding meaning, purpose, and spiritual support; accepting a situation as unchangeable; cognitive reappraisal means finding opportunity, focusing on useful features remaining; an active coping style and dealing with fear.

Using maltreated children, resilience was conceptualized as the individual's capacity for adapting successfully and functioning competently despite chronic stress, adversity, or exposure to prolonged or severe trauma. Positive self-esteem, ego resiliency, and ego over-control predicted resilient adaptation. This contrasts with nonmaltreated children, where relationship features were more influential. Children play an active role in constructing their outcome and influencing their ultimate adaptation (Cicchetti, 2004).

Coping

Coping is a process by which the person attempts to master, reduce, or tolerate stressful internal or external demands. It involves regulation of emotion, or management of the problem, independently of its success (Knussen & Cunningham, 1988). At both conscious and unconscious levels, the patient tries to cope with his changed world. This includes new means of dealing with adaptive necessities, fantasied outcomes, spiritual reactions, neurotic defenses, substance abuse, and concealing true feelings of anger and despair. Coping involves managing demands (external and emotional) that are experienced as taxing or exceeding one's resources. Lazarus and Folkman (1984) discriminated between general techniques for adapting and nonroutine actions dealing with specific events. It may be more advantageous for the injured person to improve coping skills, in contrast to cognitive remediation, which focuses in brain damage (Taylor et al., 1996). Coping requires the belief in the possibility of a positive outcome, needing control, high self-esteem, hardiness, and affective stability). Thus, it contrasts with hopelessness, which is the expectation that the outcome will be negative.

Coping skills have been categorized as follows: *knowledge* (information about the world and how it operates; *inner resources* (techniques we have learned, and our perception of the world and how it works); *social support* (interpersonal systems that we participate in; and *spirituality* (beliefs about purpose, order, mortality, and a higher power) (Ray, 2004). *Psychostasis* refers to the mechanisms and processes that are essential to maintaining psychological equilibrium. The balance between coping skills and environmental demands has been termed the *psychostasis equilibrium*, which depends upon learned processes and may require conscious effort (Ray, 2004).

Available capacities: These are internal, for example, skills, muscles, memory, and external (friends, facilities). Positive feedback indicates the appropriateness of coping strategies and provides encouragement. When a strategy fails to produce benefits, discouragement is stressful and a costly strategy may develop, for example, repeating the activity (Schoenpflug & Battmann, 1988). A maladaptive learning pattern is characterized by maintenance of physiological stress responses when the person fails to recognize, respond appropriately to, or resolve stressors. This may create negative health outcomes. Internal factors also determine success, for example, a hyperaroused person for whom an appropriate stimulus or situation generates very little response. Thus, an overview considers both coping pattern and context (Temoshok, 2000).

Characteristic Coping Styles

Faced with a difficulty, individuals vary in their attempts to regain their adjustment. There are those whose basic hardiness leads them to attempt to reconstitute their lives, sometimes with success in returning to work and community. Stress becomes dangerous when the person fails to find a way of coping with altered circumstances (Totman, 1988). Others have characterological problems of dependency or isolation that interfere with overcoming difficulty or gaining the cooperation of family and friends, health care providers, insurance companies, attorneys, and so forth, for needed rehabilitation services. When a persistent behavioral style preexisted the injury, utilizing it may enhance or reduce the outcome.

These styles have implication for therapeutic focus and rehabilitation as useful or self-destructive: withdrawn/detached; conspicuous risk taking; style of engagement: fight, approach, disengagement (flight/peacemaking); brake: ambivalence/combined strategies of approach and anxious retreat; immobilization (reduced activity to avoid threat and conserve resources including slow heart rate); copes to extent possible; motivated to be a high achiever; minimizes one's expression of complaints.

Learned Resourcefulness

This is a set of behaviors and skills by which people self-regulate internal responses that interfere with the smooth execution of ongoing behavior. Involved are automatic reactions to a situation, appraisal of the event, and activities to reduce disruption of ongoing functioning. Appraisal is affected by self-confidence and hardiness. People high in learned resourcefulness cope better with cognitive reactions to such stressors as epileptic seizures. Upon being aware of a disruption, there is an appraisal of its importance. Since self-control behaviors are private events, the clinician must inquire into the process (Rosenbaum, 1988).

Hardiness

Hardiness has been defined as a pattern of courage and strategies that facilitate turning stressful circumstances from potential disasters into growth opportunities. (The author notes here than the strengths required by this trait are frequently grossly reduced by a TBI.) Hardiness components include willingness to respond to challenge (viewing stress as an opportunity to grow, rather than insistence on comfort and security), commitment (remaining involved), and control (influencing outcomes rather than passivity and powerlessness) (Maddi, 2008). It is considered to be a mediating variable that might counteract the adverse effects of stressful events (Howard et al., 1986; Kobassa, 1979; Kobassa et al., 1983). Hardy individuals have a reduced level of symptoms of physical illness and depression. An independent source of increased symptoms under stress was Type A personality (impatience, time urgency, competitiveness, hostility, together with a vulnerability toward coronary disease) in the low-hardy group (Kobassa et al., 1983; Nakano, 1990).

Stamina

Stamina is affected by depression-like reduction of activity based on stress-related endocrine changes (see Chapter 12 on CPD) (Williams, 1998). It is enhanced by self-esteem; a warm relationship with parents; an open, flexible approach to life; and minimal nervous tension, anxiety, depression, and anger under stress (Thomas, 1982). Dysfunctioning is characterized by morbid, frightening, and psychodynamic undertones of vulnerability and being traumatized in response to the Rorschach test. Such feeling tones are quite characteristic of stressed people, and it is inferred that they contribute to poor morale in facing problems.

STRESS RESISTANCE: POSTINJURY CHARACTERISTICS

The Will to Survive

Dangerous and degrading circumstances offer examples of individuals whose desire for self-preservation brought them through when others went under, for example, those incarcerated in concentration camps and prisoner-of-war camps.

Morale

This trait reveals the level of confidence concerning one's capacity to cope with impairment. Injured people vary in their self-confidence for overcoming problems and expectation of improvement. Morale is hampered by reduced motivation and low self-esteem. Morale is a function of the patient's reaction to impairment and an injured lifestyle. The clinician studies how the person copes with his or her condition? What is the patient's expectation for improvement or permanency? Is there motivation to recover, struggle, be dependent, to give up, to play the sick role, to recover, change of self-esteem, to cheat? Is there a desire for financial or secondary gain leading to exaggeration or pretense of symptoms?

Reduced QOL

A common result of TBI is an overall deterioration of QOL (Reichard et al., 2007; Tracy, 2007). The outsider experiences disabled persons as living in a different world than the nondisabled. There may be a barrier to free and intimate contact with the obviously impaired or injured. While structured activities, including support groups and community programs for people with disabilities, render skilled rehabilitation, these may be unavailable in the community or not authorized by insurance organizations.

Some components of loss of QOL:

- They are forced to depend upon others, thus suspending control over aspects of their life that formerly defined them as a person. They need to search for help and to cope with the injury. TBI persons cannot pay for quality counseling or health care.

- Behavioral changes: Cognitive, emotional, mental, physical, with difficulties in controlling temper and decreased memory.
- Adaptive loss: Friends, spouse, family.
- Loss of employment and/or mobility impairment impoverishment, embarrassment, or other interference with social participation and other community activities.
- Structured activities included support groups and community programs for people with disabilities.
- Concern with the social effect of the handicap: Secretive, or feel that others consider them retarded or stupid.
- Handicaps create defenselessness: There is difficulty in following standard safety precautions or in recognizing abuse. The impaired are perceived as vulnerable. Impaired people experience violence, abuse, and neglect. They experience a sense of isolation, or increased passivity or trust. They are vulnerable to sexual assault. Males tended to encounter a single perpetrator; females reported one-time and multiple incidents from a single perpetrator.

Ameliorating vulnerability: To ameliorate vulnerability, the following is recommended: community-based coalitions (victims must learn about their rights and available services): Law enforcement, health care, social service, mental health, and public health. Service providers need increased knowledge of the nature of TBI.

Self-efficacy

This is defined as an individual's capacity to bring about desired results, especially as perceived by the individual. Perceived self-efficacy is an individual's subjective perception of his or her performance in a given setting or ability to attain desired results. It was proposed by Albert Bandura as a primary determinant of emotional and motivational states and behavioral change. Efficacy has been defined as the power to produce an effect, that is, the injured person's belief in his or her capacity to influence events to rehabilitate oneself. Bandura (1982) describes psychological qualities affecting the injured person's effort to adapt to the impaired state. Inability to influence events and social conditions affecting one's life arouses feelings of futility, despondency, and anxiety. Self-efficacy, locus of control, perception of stigma, and psychological distress are components of QOL (Tracy, 2007).

Self-efficacy is a process intimately supportive of high-level adaptive processes. In uninjured people, a meta-analysis suggested that self-efficacy, self-esteem, locus of control, and absence of neuroticism are strongly intercorrelated and represent a set of self-evaluative processes that are strongly related to job satisfaction and overall life satisfaction. For TBI victims, perceived self-efficacy, particularly for the management of cognitive symptoms, shows a strong, consistent association with post-TBI life satisfaction (Cicerone & Azulay, 2007).

Self-efficacy derives from these sources of information (Gagnon et al., 2005):

- Previous performance of the behavior.
- Physical and emotional state when enacting the behavior.
- Observations of others.
- Encouragement of others.
- This writer adds encouragement early in life.

Sensation of physiological arousal in a taxing situation may signal vulnerability to dysfunction. Fatigue, aches, and pains may indicate physical inefficacy. Recognizing events that create anxiety and anger, with stress management and relaxation training, the patient can raise the threshold for emotional discharge. Desensitization from phobic activities, together with self-relaxation coping skill, has boosted perceived self-efficacy. Fear arousal declines as belief in one's capacity to control fear increases. Stress reaction in phobics (elevated blood pressure and cardiac acceleration) were reduced as a function of improved self-percept of efficacy.

Judgment of one's capability affects motivation, effort to achieve, and persistence in the face of obstacles and others' hostility and rejection. One's belief in what one can or can't do will affect return to work, effort to obtain rehabilitation, reentry into the community, and so forth. Thus, the child (or older person) whose self-image is athletic, is more likely to attempt activity earlier than others, but be confronted with lack of return to normal. While severe injury (disturbed self-efficacy) has been asserted to negatively affect performance even after the resolution of the initial injury, one must be certain that the examination was thorough enough to warrant a clean bill of health. A paradoxical conclusion has been found for children, suggesting *the generalizability of feelings of impairment, even when they seem unjustified* (Gagnon et al., 2005): Children diagnosed with mTBI were studied 1 day and 12 weeks postinjury. Although the injured children returned to their previous performance level, they lacked confidence in their ability to perform athletic activities compared to before their injury. It was as though the ongoing experience of postconcussive symptoms affected self-acceptance in the face of preserved ability.

Prior attitudes may reduce the person's capacity to improve belief in his self-efficacy, even with successful performance. Skills added to belief in coping efficiency mobilize the effort needed to succeed in high-risk situations. Disbelief in one's efficacy, leading to undeveloped and poorly used coping skills, contributes to substance abuse relapse. Self-efficacious persons reestablish control after a slip, contrasting with the less self-efficacious person's decrease in this characteristic with complete relapse. Passivity may be the outcome when a person believes that they are unskilled and should yield control to others to cope with a difficult environment.

EXPRESSIVE DEFICITS: PATIENTS' INABILITY TO SELF-DESCRIBE

"Expressive deficits" is a term the author has proposed to describe brain-damaged patients' inability or unwillingness to describe their personal experience of TBI. Expressive deficits may conceal symptoms from the examiner in every area of the neuropsychological taxonomy. When the patient cannot offer reasonably correct and complete information concerning difficulties, significant conditions may be ignored. To offer an accurate self-report requires complex cognitive functioning—indeed, an intact brain. Reduced comprehension interferes with insight, and consequently reduces the correctness of the report to the examiner. Without awareness or recollection that a change has taken place, the patient cannot offer a report of a problem. An accurate self-description may not be immediately presented. Self-understanding can come later in the interview, some time after the question was initially raised. Right hemisphere-damaged patients may state that they are now "okay" or "much better"—even with physical disability—when, in fact, they may behave childishly or without expression.

REDUCED PATIENT COMMUNICATION

There are numerous reasons why a patient may be unable or unwilling to communicate to the examiner or other party legitimate deficiencies consequent to an accident or other circumstance. It is a frequent finding that the patient is not aware of the degree of impairment. On occasion, this may be part of a dissociative reaction in which the integration of one's self-awareness with environmental realities and demands is incomplete.

The problem is compounded by the fact that as TBI is more severe, there is likely to be enhanced difficulties of communication. Certain types of information come directly from the patient so that loss of self-description may cause specific problems to be ignored: (1) sensory complaints: (A) primarily based on neurological damage; (2) partial seizures (see earlier) in which sensory phenomena may reflect not environmental stimuli but activity of the related cortex: (B) mood and personality disorders; (3) intrusive anxiety; (4) identity as an injured and impaired person with poor morale and feeling vulnerable to more injury.

The Examiner's Contribution to Inaccurate Information Gathering

Various cerebral and personality traits create inaccurate impressions or similar-appearing affects despite disparate etiologies.

Reduced expression of distress: Cerebral damage, leading to reduced intensity of affect, perhaps prevents the patient from discomfort, and so he or she is not motivated to relate problems to others. A *gross absence of complaints* in a seemingly highly impaired person may be a sign of brain damage per se. One differentiates between verbal explicit denial of illness (*anosognosia*) and indifference (*anosodiapheria*) (Heilman, 1991). Subtypes include anosognosia (indifference to or unawareness of a neurological deficit), indifference toward failure or events concerning the family, minimization of hemiplegia through attributing it to some less anxiety-provoking cause, or emotional indifference and unthinking resignation (Bisiach & Geminiani, 1991). Inability to reveal concerns may be due to avoidance of anxiety-related repression of particular experiences. Reduced expression or experience of anxiety prevents the patient from revealing problems to the examiner. Capacity to report any condition depends on the integrity of the language apparatus. Cultural differences between examiner and patient can also affect communication, for example, the described intensity of a complaint may vary according to background ("keep a stiff upper lip" or "let it all hang out"). Neurological impairment (e.g., commissurotomy) may render the contents of awareness inaccessible to the language expressive system (Rugg, 1992). This might be a "split brain" effect in which interference between the two hemispheres is attributable to injury of the corpus callosum. This has been described as two independent conscious selves. Lateralization would cause some information to be available primarily to one hemisphere (Dronkers et al., 2000). After a high-velocity, rapid acceleration/deceleration head injury, the edge of the dura mater extending down to the corpus callosum can create lacerations of the corpus callosum as the brain moves vertically and laterally (Rosenblum, 1989), and thus interferes with integration of the two cerebral hemispheres.

Flat affect and aprosodia: The brain-damaged person often expresses himself with a flat affect. This can be due to *aprosodia*, a motor disorder of vocalization, which causes the voice to appear flat or uninflected. The patient offers a misleading impression of lack of concern, even when describing the most painful kind of impairment, despair, or damage to lifestyle. Seeming emotional blandness conceals from the observer the deep distress experienced because of impairment, disorganization of life, and hopelessness. The pseudo-dull mood can be misunderstood as indifference, and the patient may not be taken seriously despite reports of suffering. Thus, a disorder can remain unrecognized because of the lack of overtly expressed distress. Depression, anxiety, and anger when described, although deeply experienced, may appear to the examiner as false because of the incorrect assessment of pretended feelings. Since prosody offers motivation, mood, and linguistic information, aprosodia reduces the accuracy and range of communication. The accident victim seemingly blandly describes a frightening accident, dysphoric moods, and bleak life. This raises the possibility of the speaker's level of distress being significantly underestimated. The aprosodic patient can have deep feelings, well described verbally, but be unconvincing concerning the level of distress unless the examiner considers this condition as a possibility and explores unexpressed affect. Assessment of depression is also difficult. Alternatively, endogenous depression, perhaps consequent to a high-secretion posttrauma of corticotropin-releasing hormone (CRH), or a thyroid deficiency, can be misperceived by the examiner as psychodynamic depression rather than a medical disorder. Paradoxically, impaired self-awareness seems to serve as a barrier to depression. Presumably if there is lack of knowledge of dysfunctioning, there is less cause for being depressed (Malec et al., 2007b). Impaired self-awareness after TBI is associated with lack of appreciation for the severity of deficits, and their impact on ability to work and live independently. It is a factor in long-term functional outcome after TBI because of its association with low motivation for treatment, longer stay in rehabilitation, and poor vocational outcome (Evans et al., 2005).

Alexythymia: Potential Misinterpretation of Affective State

This condition, not necessarily caused by brain damage, is one in which the person does not identify or label feelings, fantasies, or physiological reactions (Acklin & Bernat, 1987; Taylor, 1984). The patient is unable to formulate and express affect and psychological conflicts verbally (Yager & Gitlin, 1995), that is, does not verbally describe feeling states, impoverished fantasy life, reduced dreaming, and so on. It may not be a single personality characteristic (Norton, 1989). Rather, the patient expresses affective distress through somatic language. Alexythymics are prone to develop somatoform disorders and psychosomatic illnesses. It is claimed that they are seen in the ranks of patients with persistent PCS, chronic pain, and other traumatic disability.

Examiner's Data Gathering Is Affected by Prior Clinicians

The examiner's task is to enable the patient to understand his present condition and then to convey information to later clinicians. Benvenga et al. (2000) note that an examiner's use of too sophisticated language with a poorly educated man would not facilitate recollection of the details of a trauma suffered years previously. Thus, obtaining information concerning prior injuries is partially dependent upon the efficiency of prior examiners' conveying of findings to the patient. When I inquire whether this is the first time a patient has been queried concerning prior injuries, the reply is usually "Yes." Preexisting accident-related conditions are frequently ignored by health care providers. Thus, the patient is not alerted that some conditions experienced for some time may be attributed to a known accident. Therefore, the patient may not associate injury with the symptom. The event causing TBI, when not properly diagnosed at the time of trauma, may be forgotten, and therefore not associated with subsequent problems of personality and cognitive effectiveness. Pain that is caused by a trigger point in the torso or neck and then referred (usually unilaterally) to the head can be misunderstood. The injury may be in the area of the trigger point, and not in the head where the pain is experienced. Temporomandibular joint syndrome can also cause headaches that are not associated with a blow. If brain damage or concussion is not diagnosed at the time of trauma, then the potential nonskull origin of such symptoms may not be suspected, and thus can be misattributed to personality problems. Should the examiner become aware of the possible attribution of the pain paradoxically attributed to a body area away from the headache, a possible explanation may be offered to the patient, and a referral made for further examination by a physician.

Cognitive Deficits and Dysfunctions

Reduced comprehension: Patients tend to overrate their abilities compared to estimates by their families, and to report more physical than nonphysical impairment. Without awareness that a change has taken place, the patient cannot offer a report of a problem. Family members and the clinician are more likely to agree in their ratings than with patient self-ratings (Sherer et al., 1998). Reduced patient insight (self-awareness) is associated with lower verbal IQ and reduced temporal orientation, but association with the extent of lesion is not definitely known. A lack of foresight or judgment will cause patients not to realize that they have created acts damaging to themselves, or that they are likely to do so in the future.

Reduced judgment: Brain trauma induces maladaptive overt behavior: Inability to learn from experience; impaired error monitoring of ongoing social activity (inability to recognize or retrieve errors, due to deficits of matching performance against inner and/or external models of success);

defective foresight, that is, poor judgment (inability to anticipate social rejection for inappropriate behavior); reduced SR creating inability to cease rejected activity.

Poor memory: The examiner must make great efforts to elicit information about both daily activities and possible traumatic history. Poor memory reduces the details of the patient's self-description. In a long interview, more material may be spontaneously offered later. It may be necessary for the examiner to probe, although this leads to possible confabulation.

Avoidant coping style: This refers to PTSD criterion C: Persistent avoidance of stimuli associated with the trauma and numbing of general responsiveness: Thoughts, feelings, conversations, activities, places, people; inability to recall aspects of the trauma; reduced interest or participation; detachment or estrangement from others; restricted rage of affect; loss of loving feelings; sense of a foreshortened future (career, marriage, etc.).

Various kinds of agnosia and neglect imply lack of awareness of the nature of the injury. When there is lack of insight, patients may not realize the extent of their intellectual impairment. Professional-level individuals return to work unaware of IQ deficiencies of 20–30 points that produce gross deficits of ability. If compensation for deficits is utilized, the client may not be completely aware of their deficits. Successful use of coping strategies also conceals dysfunctioning from the examiner. Examples include memory problems that are solved through the use of lists, and someone with a poor sense of direction always using a compass and map. Sometimes, a patient is unaware of visual loss due to a combination of occipital and parietal injury (Anton's syndrome), which can lead to incorrect identification of their behavior as an inappropriate emotional reaction to injury, that is, denial of the visual defect and confabulation (Selhorst, 1989).

Reduced Awareness and Insight

Poor self-awareness may result in inability to appreciate performance errors, impaired judgement of the facts of one's own life, and poor understanding of the self as a continuous entity across time, in the context of failure to appreciate social and interpersonal norms (Kemenoff et al., 2002). Lack of self-awareness interferes with self-reporting of problems of everyday memory and is also a major barrier to rehabilitation. The patients may not understand the true requirements or standards for a given situation and thus may not know why the job performance or fulfillment of domestic responsibilities is considered poor. Both preexisting low intelligence and postinjury impaired intelligence or comprehension contribute to an inability to understand the deficit and its effects, and hamper the patient's effort to communicate the circumstances and result of an injury. The illiterate subject would have particular difficulty in judging and then providing a correct self-description (Eslinger et al., 2007; Lecours et al., 1987). Victims of *severe head injury* tend to underreport cognitive and behavioral change compared to observer ratings, whereas less than 10% of patients overreport symptoms. The informational distortion leading to nonself-report of symptoms is extremely complex (Gasquoine, 2000). These conditions contribute to lack of awareness (Flashman et al., 2005):

Agnosia: Impaired recognition in the absence of primary motor or sensory impairment.

Anosognosia: Lack of awareness of an illness, or frank denial of a neurological deficit, for example, left hemiplegia (Bisiach et al., 1986).

Denial of illness: A multidimensional construct including psychological defenses and involving lack of cognitive skills.

Anosodiaphoria: Indifference to an acknowledged deficit or illness.

Misattribution: Patients with PTSD may not attribute a range of symptoms to the precipitating trauma. This may be due to an avoidant coping style. Thus, accurate assessment requires (for all domains) intensive interviewing concerning the frequency, intensity, and nature of symptoms (Bryant, 2001).

Aphasic Problems

Concealment in an aphasic patient has varied origins. A minimum verbal capacity is needed to alert the examiner. Aphasic communication deficits imply an inability to understand others, find appropriate self-descriptive words, or express otherwise understood thoughts coherently. Aphasia contributes to voluntary concealment. There may be a combination of actual inability to express oneself correctly and clearly and embarrassment concerning this difficulty (see "Embarrassment"). Embarrassment that conceals information is consequent to aphasic difficulties that are observable in talking, for example, word-finding difficulties, or other receptive and expressive problems. The patient may remain silent, evasive, or stay away from any topic requiring detailed use of language. Perhaps they state that everything is fine, use nonverbal communication, express generalities, refuse to talk, or remain isolated.

Quality of Social Feedback

Everyone is dependent on feedback from family, friends, and employers to inform us whether our behavior is appropriate and satisfactory. Inability to monitor behavior, to learn from experience, and therefore to be aware of and report a problem may be an example of the frontal lobe syndrome.

Nonverbal and indirect expressions of dissatisfaction are common. Noncomprehension of cues of dissatisfaction, or withdrawal of comment by family and friends, eliminates social and emotional cues.

Concealment of Symptoms

There can be deliberate concealment of legitimate symptoms to the examiner due to embarrassment. One possibility is reluctance to describe a specific environmental or emotional condition. The patient does not want to discuss a situation in which they are unable to succeed, or that will create anxiety. The following are some clinical examples:

Clinical examples: Spartan mentality (trained to deny difficulty)

- One woman was reluctant to discuss the difficulties caused by an accident because of *religious beliefs*. She engaged in litigation only because of her husband's insistence. Since she believed that God had visited this affliction on her for His own reasons, there was no purpose in offering complaints.
- One woman was examined after a heavy object fell on her head. There was no doubt as to the distress, that is, she more than once broke into tears when describing the pain that she experienced at home. However, when I specifically asked her about her current status, she tended to minimize the effects. When she was asked whether the spot where the object struck her was tender, instead of affirming this, she replied: "It's getting better." I asked her whether at home she was trained not to complain, which she affirmed. Her training? "Count your blessings; since people got a lot less than you, you should be happy."

Reluctance to relive the trauma: The individual may be reluctant to express a complaint in order to avoid reexperiencing the pain. Perhaps crying would make the patient feel conspicuous when discussing the trauma and its after-effects. Moreover, repetitive or intrusive memories and reminders of impairment, pain, and loss of the QOL lead to active attempts to avoid discussion of the experience. One woman who was knocked down by a car said that she "pretended that there was no accident."

Spartan mentality: Social inhibitions against revelation of difficulties: One woman was reluctant to discuss the difficulties caused by an accident because of *religious beliefs*. She engaged in litigation only because of her husband's insistence. Since she believed that God had visited this affliction on her for His own reasons, there was no purpose in offering complaints. Some families offer *social training* to their children not to complain but to accept their lot (a "Spartan mentality"). Paradoxically, *genuine pain may not be revealed*, that is, the "good sport syndrome", contributing to perpetuation of the problem, that is, abuse of muscles with trigger points (Mackley, 1999). *Pride* can create a wish to overcome affliction by oneself, leading to concealment of the extent of impairment. This attitude is expressed by some people who are trained not to express emotional pain. They give themselves reasons not to ask for what is coming to them, or to assert their rights (Parker, 1972, 1981).

Employment anxiety: Inability to work may be concealed by an injured worker due to fear of loss of employment. Perhaps impairment may cause one to take a less-demanding job or avoid challenging situations. Active-duty troops, believing that revealing psychological symptoms will delay their return home or jeopardize their careers, do not report them. Service members may believe that it is unmilitary or a sign of weakness to display the symptoms of psychological distress (U.S. Government, 2007). The employee is afraid that if the employer knew the extent of the inability to function due to headaches, loss of concentration, problem-solving ability, and memory, then employment could be terminated (Parker, 1987). He or she may succeed temporarily in concealing reduced effectiveness since other workers cover up deficiencies. Unless the examiner asks about reduced effort or achievement, the individual may appear to be stable and uncomplaining.

Embarrassment: The patient whose accident was described in Chapter 1 sobbed in a psychotherapeutic session: "It's not mature to cry." I elicited that when he was young he was encouraged to be a brave man: "Don't cry for little things."

People fear loss of friendship due to reduced social acceptability should their limitations be known. One man did not want to discuss his loss of potency with his friends since bragging about sex with their wives was a source of prestige. In a psychometric test (true/false responses), one boy would not acknowledge any significant degree of anxiety. On the Rorschach procedure, which makes it more difficult to conceal basic feelings, he revealed a gross level of anxiety and feelings of bodily damage. One woman's seizures were concealed because she was reluctant to tell her neurologist that she lost bladder control. A child did not tell his parents of a serious fall resulting in unconsciousness.

The defenses of denial and avoidance: Denial is a psychological defense against admitting to oneself the existence of a condition that reduces the person's value in his own self-esteem or in the opinion of others. It is an anxiety-reducing mechanism in which the meaning of an event is repudiated. It can be a component of any defense (Lewis, 1991), and has meaningful, motivational, and adaptive aspects (Weinstein, 1991). To admit to a weakness is a blow to one's self-esteem. The loss of a better condition of life or higher capacity to achieve is so painful that self-concealment is common. Denial is to be distinguished from "agnosias" and inability to express emotional pain (Parker, 1972). Individuals avoid facing situations in which they are unable to succeed, or that will create anxiety.

Denial is a motivated response to impairment as opposed to neglect, that is, lack of representation of the body image or other self-monitoring, so that the person literally does not know that there is no afferent or efferent activity. Differentiating between neurologically based denial (*anosognosia*) and unconscious mechanisms is helped by these considerations: First, anosognosia may follow right anterior brain trauma, when the patient may complain that the paralyzed limb is not his. Second, when denial reflects disrupted brain functioning, it is not an isolated symptom. Third, the brain-damaged patient remains bland or perplexed when confronted, whereas the neurotic patient may become increasingly agitated (Lewis, 1991). Confronting the denial may involve therapeutic risk since awareness may involve guilt in having involved oneself or someone else in an accident, loss of support due to impairment, or reduced self-esteem due to loss of a valued quality.

There are different levels of denial, ranging from indifference to different styles of coping: delusions (a limb is someone else's); joking about a deficit's meaning; or detachment (giving a paralyzed limb a name). Denial can be enhanced by the sharp pattern of functioning, that is, apparent adequacy of such functions as sensorimotor mobility and performing activities of daily living, although higher functions are not put to the test.

EXAMINING THE PATIENT WITH EXPRESSIVE DEFICITS

One acknowledged value of the Rorschach is its ability to help the reluctant or unwilling person to express feelings and attitudes that are repressed or suppressed, that is, expressive deficits. *Expression Through Rorschach in a Case of Denial*: A woman did not express any problems on a four-page symptom questionnaire. She perceived: "Red rodents and orange witches." These are not only unreal, but reflect her denial of deficit and attempt to paint a rosy picture.

"At first I had many problems. I did not know that I was injured. When I found out that I was injured, it took me several months to accept that there was a deficit. After finally accepting that I had a deficit, you offered me your ego: Literally you said 'I will let you use my ego as yours, since you don't seem to have one.' I was walking around, very confused, and I did not realize the gravity of my impairment until you pointed it out to me. People have told me what I did in the days after the accident. I took the car to a routine servicing, although I was in pain, and the car was damaged, I continued on to the car dealership, although there was major damage to both me and the car. I didn't understand what happened (rear end collision). When I started having seizures, I knew there was something seriously wrong. (After neuropsychological and neurological testing) I became aware that there was something wrong. It's not easy for a brain that's damaged to be aware that there's something wrong.... How I navigated my car, I don't know. It was as though I were navigating on auto-pilot. I have no recollection of the day at all."

FLUCTUATING LEVELS OF EMOTIONAL EXPRESSION

Emotional expression at different times may be intense (intrusive) or inhibited (constricted), due to denial alternating with anxiety. Reexperiencing and denial have been considered to be distinct dimensions of PTSD. There may be a developmental pattern of moving from denial to reexperiencing, with the latency or denial-numbing phase, of a few days to decades, followed by an intrusive-repetitive phase, with experiencing of nightmares, frustration, and guilt. This theory is in contradistinction to the *DSM-III-R* requirement of simultaneous reexperiencing and numbing symptoms, which could hinder identification of PTSD victims (Miller-Perrin & Wurtele, 1990).

Ascertaining the balance between patient insight and denial of any deficits that impair adaptation is a goal of the examination. Varieties of lack of insight have been classified by Ben-Yishay and Prigatano (1990):

- Unrealistic expectations concerning recovery of functions and the possibility of resuming preinjury lifestyle
- Inability to assess severity and implications of deficits
- Poor compliance to treatment and resistance to guidance from rehabilitation professionals

Level of insight is estimated through:

- Discrepancy between self-description/plans for the future and examination findings
- Reports of attempts to maintain a lifestyle or particular activities that are grossly inappropriate

ILLNESS (SICKNESS) BEHAVIOR

The characteristics of illness behavior include: The perception of being ill, the expression of illness-related concerns to other, and utilization of health care services. While the usual description of this syndrome implies that there is a subjective quality, perhaps the hope of secondary gain, consideration of the following symptoms suggests that in some persons, part of its etiology may be medical, that is, trauma and allostatic load. Specific symptoms and complaints: Fever, increased metabolic rate, malaise, loss of appetite or anorexia, weakness, inactivity, fatigue, hyperesthesia, hyperalgia (with or without headache), apathy or listlessness, anhedonia, social withdrawal, depression, sleep disturbance (somnolence or hypersomnia; slow-wave sleep), diminished libido; and irritability (Berczi & Szevntivanhi, 1996; Dantzer, 2000; Kop & Cohen, 2001; Marshall & Rossio, 2000).

Sickness behavior is a complex stress reaction, whose emphasis varies between patients, expressing the physiological requirement for less expenditure of activity during illness and a coping reaction to chronic impairment with possible secondary gain. The Sickness Syndrome may have evolved as an adaptive response that reduces both effort when sick and the possibility of contagion. Clinical administration of cytokines (particularly pro-inflammatory cytokines IL-1, IL-2 IL-6, TNF, and IFN-α/β) creates development of behavioral symptoms called *sickness behavior* (Hansen-Grant et al., 1998).

The author approaches this topic assuming that its etiology is based upon an aberrant physiological reaction associated with chronic injury or stress. Stress is associated with increase of sickness behavior, but the association with verified objective disease is complex and less firm (Cohen & Williamson, 1991). One may differentiate between sickness behavior (e.g., sensitivity to physical sensations, labeling these as symptoms, or as a disease, and then seeking medical care) and some objective evidence of illness. Illness behavior does have *adaptive value*, in part from the organism-fighting function of fever, sleep hypomotility, anorexia, and reduced libido. Further, it may have the protective value of retreating from and escaping predators (Dunn, 1996).

Hormonal responses to stress have been studied in the context of three types of *nonassociative learning* (animal studies).

Habituation: Chronic intermittent exposure to a stressor of low or moderate intensity, within the context of familiarity, leads to minimal activation of cardiovascular and metabolic homeostasis, providing for significant conservation of energy expenditure.

Under stress, if adrenal steroid secretion does not increase appropriately, then secretion of inflammatory cytokines increases, which ordinarily would be counterregulated by these adrenal steroids. In rats it has been demonstrated that inadequate HPA response results in increased vulnerability to autoimmune and inflammatory disturbances. Analogous human examples of HPA hyporesponsiveness may be fibromyalgia and chronic fatigue syndrome. Further clinical examples of allostatic load are offered: (1) Elevated glucocorticoids may damage hippocampal neurons and perhaps affect the developing human brain (Lambert et al., 1999), cause insulin hypersecretion and then insulin resistance, leading to obesity and atherosclerotic plaque formation. The hippocampal damage interferes with its ability to suppress negative associations, and match expected outcome with reality. This further increases the likelihood of self-sustaining fears and anxiety that do not match with real events; (2) PTSD is associated with hypervigilance and elevated somatic nervous system (SNS) activity. While cortisol and insulin levels are not as high, chronically high sympathetic tone does have adverse consequences; and (3) Extreme physical exercise is associated with elevated cortisol and adrenal cortex and medulla size. Stress that occurs over a period is habituated in most individuals (McEwen, 1998). The remainder experience hypersecretion of cortisol of a cardiovascular stress response with high blood pressure.

Persons with performance efficiency of less than the estimated baseline or paradoxically low by other criteria should be considered from the viewpoints of trauma (mental and physical) creating dysregulation and the possible presence of an allostatic load, that is, the partial failure of the regulatory apparatus caused by trauma or persistent stress beyond a recovery period of a mental or physiological nature.

Cytokine release during response to injury and chronic stress creates responses frequently observed in TBI, but the etiology may be described as primarily the behavioral sequelae of the physiological reaction to primarily noncerebral injury, that is, sickness behavior: Malaise, social withdrawal, somnolence, hyperesthesia and depression (somnolence, anorexia, diminished libido, fatigue, slow-wave sleep, apathy, and irritability) (Kop & Cohen, 2001; Marshall & Rossio, 2000).

While PTSD sometimes creates similar alterations of consciousness (e.g., dissociation), concussion's more distinguishing alterations of consciousness (confusion, dizziness, "islands of memory") are not characteristic of nonhead injury stress. While the list of some symptoms is appropriate, specifically excluded are amnestic disorder due to head trauma, personality change due to head trauma (although "changes in personality, e.g., social or sexual inappropriateness is utilized as an identifying symptom), and dementia due to head trauma. Impairment in school, social, or occupational functioning is recognized. Other characteristic TBI conditions enter through the back door as mood or cognitive disorders.

Cytokines induce a pattern of behavioral changes that characterize the *sickness syndrome*. Immune molecules (*interleukins*) signal the brain through the bloodstream and nerve pathways, and result in feelings and reactions described as "sickness behavior" (Sternberg, 2000). Cytokines simultaneously activate or inhibit the release of hypothalamic peptides altering the pattern of pituitary hormone secretion. In addition, the cytokines alter the responses of the hypothalamic peptides and other transmitters that act directly on the pituitary gland, altering the response of the various pituitary cell types to the hypothalamic peptides and other transmitters acting directly on the gland. The pituitary hormones then modulate the responses of the immune cells, either directly or via secretions of their target glands (McCann, 2003). *IL-1* appears in the brain when stimuli activate astrocytes and microglia. It modifies EEG sleep patterns, activates the pituitary-adrenal axis, and elevates body temperature. It is produced by stimulation of astrocytes and microglia, with behavioral, neurochemical, and physiological effects and reduction of brain immune responses. Infusion of IL-1 into the brain suppresses the immune response with the contribution of adrenal factors. It also activates the pituitary-adrenal axis. IL-1's effect in the brain seems to activate the sympathetic ganglia, which activate peripheral cellular immune reactions (Weiss et al., 1994). IL-1β activates the HPA axis in many species, with systematic elevation of corticosteroids acting as a brake on the immune system (IS). Negative feedback limits immune hyperactivity that would then limit autoimmune hyperactivity and perhaps other physiological actions (Dunn, 2000).

Temperature: One of the frequent postconcussive symptoms is loss of control over body temperature. The patients frequently complain of feeling cold or warm in ambient temperatures that seem comfortable to others. In one study of patients hospitalized with severe head injury, only one of ten patients manifested circadian rhythms for body temperature and heart rate with the peaks occurring at appropriate times. The low (but significant) correlations between body temperature and heart rate suggest desynchronization of rhythms and may have clinical significance (Lanuza et al., 1989). *Dysthermia* (hypothermia and hyperthermia) after brain trauma and other causes is more common than is generally recognized and is considered to be an unregulated change in the hypothalamic set point. There are a variety of patterns: Flattened curves; reversed cycles; sleep phase delay; irregular unexplained fluctuations; and exaggerated reactions. There is some relationship between the type of dysthermia and the etiology, for example, hypothalamic damage, prenatal injury, abnormal circadian rhythms, various medical conditions, and so forth. (Chaney & Olmstead, 1994).

THE CATASTROPHIC REACTION

The catastrophic reaction (Goldstein, 1942) is the opposite of aprosodia, that is, it is an extreme reaction of distress when confronted with an ability to perform a task at the preinjury level of success. The catastrophic reaction is defined as extreme emotional distress aroused by awareness of impairment, that is, unable to perform familiar activities, particularly cognitive tasks: crying, giving up prematurely, self-criticism, anxiety, and so forth. It expresses the emotional pain and depression caused by inability to perform familiar tasks, that is, the loss of *self-efficacy* or the belief that one can perform tasks. Self-efficacy differs from self-esteem, which is related to the value that a culture values traits the person sees in himself.

Rorschach Signs

A Rorschach Inkblot Procedure description of patient discomfort with inadequacy was offered by pioneering neuropsychologist Zygmunt A. Piotrowski (1937):

Impotence (poor error monitoring): Giving a response in spite of recognition of its inadequacy, that is, recognizing the poor quality of the response, yet not withdrawing it. There is a paucity of ideas and an inability to think of a new and better response, but the patient goes through with the process once initiated.

Perplexity (need for support): This is distrust of one's ability, which is unpleasant, and associated with pressing the examiner to decide the whether their responses have been adequate. The patient is interested in the quality of the results, but needs someone else to tell him whether they are adequate.

Automatic phrases (perseveration): A phrase is used indiscriminately, without determining whether or not it is sensible. They are stereotyped reactions to new situations, which may be recognized as inappropriate.

The Catastrophic Reaction was named by Goldstein (1942, pp. 71–79), who attributed it in a clinical example to a frontal lobe lesion. The organism is described as attempting to return to a uniform condition. The sick person has a strong urge to meet all demands as well as possible; his existence is bound up with this to a greater degree than a healthy person. A totally blind man recovered some vision. As he became aware of his defect, he became upset with imperfect orientation. He changed from reasonable content to depression: "What's to become of me if I can't see?" Another patient was presented with a problem in simple arithmetic: He was unable to perform it, resulting in agitation, anxiety, autonomic responses, change from amiability to sullenness, and so forth. Since the patient was disturbed in "his whole behavior," the reaction was termed a catastrophic situation" (also Bruno, 1984; Gainotti, 1972; Heilman et al., 1985). Lezak et al., (2004, p. 62) describes the catastrophic reaction as an extreme, disruptive, and disorganizing emotional disturbance attributable to left hemisphere lesions.

Clinical example: Catastrophic Reaction: After the computerized continuous performance test (responding to varied targets), the patient wanted to discontinue. He said that he had a headache, was agitated, shaky, tense, and felt frustrated because it was hard to concentrate. After the Stroop Neuropsychological Screening Test (in which he had to report the color of a printed text representing the name of different colors), he stated: "I started getting emotional because I couldn't get it (weeping)."

Individuals with brain injury vary in their insight as to the level of deficit. They may be acutely aware of difficulty in solving problems that once were easy, or be slow in achieving insight (Klonoff & Lage, 1991). Loss of perceived self-efficacy has been considered to be the cause of the catastrophic reaction (Kihlstrom & Tobias, 1991). It can be defined as a painful reaction to the awareness that one is impaired, in particular, unable to perform the same type of familiar activities as before the neurological injury, particularly cognitive tasks.

The symptoms of the catastrophic reaction are crying, giving up prematurely, self-criticism, and so forth. While it is traditionally associated with left cerebral hemisphere damage (lesions in both Broca's and Wernicke's areas), it has also been observed with right hemisphere damage and attributed to release from limbic centers (Bruno, 1984). The basic cause of the emotional expression is controversial, that is, whether the more functional right hemisphere processes "negative" emotions, or merely expresses feelings in a nonpropositional, nonverbal manner.

There are lateralization influences upon the valence of the person's reaction, that is, left lesions associated with the catastrophic reaction and right lesions associated with minimization, anosognosia, indifference, and joking (Devinsky, 1992, p. 204). This author assumes that the catastrophic reaction requires simultaneous higher-level general comprehension, and also self-awareness of dysfunctioning. It appears more frequently among left hemisphere-damaged patients than right, and is associated with aphasia (Gainotti, 1972, 1991). It has been assigned to the class of organic mental disorders, that is, high levels of anxiety when environmental demands exceed cognitive or perceptual capacity (Horvath et al., 1989; see "Rorschach Signs").

The criteria demonstrated by Gainotti (1972) to indicate the probability of left- rather than right-sided lesions are anxiety; tears; restlessness; hyperemotionality; vegetative; swearing (curses, religious invocations); displacement of anxiety or aggressiveness to extraneous events; refusal to continue the examination; and renouncement (presumably refusal to follow directions).

Graded but Disinhibited Emotional Displays: Mood Changes

Ross (1993) differentiates between graded emotional displays (attributed to the right hemisphere) and extreme emotional display (laughing, crying, anger), including the catastrophic reaction. The latter are described as all or none, uncontrollable, and socially embarrassing. Emotional experience and graded affective behavior are considered dissociable, because patients with motor aprosodia lose the ability to encode affective behavior except for extreme emotional displays, although they may continue the entire range of emotional feeling states. Characteristic mood changes after head injury are excitability, mood swings, poorly controlled anger, and irritability. The patient's mood will fluctuate according to levels of arousal, the presence of PTSD, endocrine conditions, and so forth.

SEXUAL DISORDERS

Sexual difficulties arise from disorders at multiple levels and after all degrees of brain injury, and are consequent to complex etiology. Sexual difficulties vary according to the age, gender, nature of the injury, and support and rehabilitation availability. Maintaining existing relationships and forming new relationships can be difficult.

Neuroanatomical structures occur at cortical (both hemispheres), subcortical, brainstem, and peripheral nervous system levels. Some pharmacotherapeutic agents have beneficial or adverse effects upon neurotransmitters and neuromodulators with sexual function. Detailed information is available concerning the effects of classes of hormones and pharmaceuticals, rehabilitation clinical assessment, and approach to the sexual physical examination. Specific neuroendocrine evaluation includes for all ages: follicle-stimulating hormone (FSH), luteinizing hormone (LH), prolactin (PRL), free testosterone (males), estradiol and DHEA (females). See Chapter 17, "Developmental Considerations" and "Disorders of Physiological Development"). There are posttraumatic sexual changes in the direction of loss and enhanced function. The following disturbances should be considered: Reduced spousal affection; loss of libido, including decreased frequency of intercourse for males and females; sexual preoccupation; erectile, ejaculatory, and/or orgasmic dysfunction; alterations of sexual interest and cognitive (intellectual impairment); medication side effects; behavioral problems affecting family relationships (mood changes such as irritability); temporal lobe epilepsy; impulsivity; and inappropriate behavior. Time since injury and advancing age were correlated with sexual dissatisfaction. Counselors should consider issues of masturbation, appropriate behavior, sexually transmitted diseases, sexual surrogates (legal problem of prostitution), birth control, sexual abuse, sterilization, dating and marriage, child rearing, and changed sexual orientation due to the unavailability of heterosexual partners (Zasler & Martelli, 2005).

There is not a close relationship between injury to a given structure and the nature of sexual disorder. The physiological basis for the posttrauma sexual dysfunction involves many areas of the nervous system and numerous physiological and neurochemical effects. Neurochemical influences occur in different structures: Dopaminergic influence on desire (mesolimbic; mesocortical); serotonergic pathways (inhibiting sexual function); nitric oxide (genital level); hypothalamus receives (gonadal steroids and other hormones) and transmits chemical messages (releasing hormones for the anterior pituitary; oxytocin, a hormone involved in birthing, lactation, and orgasm). Neural structures mediating sexual functions include spinal, brainstem, and subcortical and cortical systems. *Temporolimbic* structures, which are highly epileptogenic, play an important role in relating emotions to behavior. They modulate hormonal secretion and mediate hormonal feedback. These structures influence emotion directly by neural pathways, and indirectly by altering hypothalamo-pituitary regulation of gonadal steroid secretion. This may lead to abnormal hormonal influences on emotional behavior. *Estrogen* is highly epileptogenic and exerts energizing and antidepressant effects. Excessive estrogen can reduce agitation, irritability, lability, anxiety, and the development of anxiety manifestations such as panic, phobias, and obsessive-compulsive disorder. *Progesterone* inhibits kindling and seizure activity, and has mood-stabilizing effects, possibly through GABA-ergic activity. In excess it produces sedation and depression (Herzog, 1997).

Primary: Damage to brain, pituitary gland, spinal cord, peripheral nerves, and autonomic nervous system (ANS).

Somatic: Loss of capacity consequent to anesthesia, hypesthesia, dysthesia, pain, amputations; muscular spasticity; deficits of mobility, strength or coordination; dysfunctions of bladder and bowel.

Psychosocial: Cosmetic defects; reduced interest and reduced self-esteem cause avoidance of sexual contact and rejection of the patient. The combination of cognitive, behavioral, and physical disabilities causes the patient to be experienced as a different person, interrupting the usual relationship. Being different and impaired prevents the formation of new relationships.

Physiological: Desire (aversion and hypoactive desire); arousal (poor lubrication in women; erectile disorders in both sexes); orgasmic disorders (anorgasmia in both sexes; premature ejaculation in men); and pain (dyspareunia; vaginismus).

Psychological: Deficits of cognition and communication, depression and other disturbing moods, and side effects of medication. The TBI patient may also experience co-morbidly many of the "distracting" conditions that hamper sexual initiation and enjoyment of both partners, for example, anxiety, scarring, mobility disorder, reduced social status, insufficient funds to participate in social activities, reduced self-esteem, and the identity of an injured and socially unattractive person (see "Rorschach; Identity").

The social aspect of sexuality is impaired after brain injury in a large proportion of the patient samples, though some injured people do not report difficulties. In fact, increased desire after TBI has been reported. Patients with more recent injuries, frontal lesions, and right hemisphere lesions reported higher sexual satisfaction and function. Measures of affect and severity of injury were not correlated with sexual dysfunction. Some studies did not find an association between the locus of the lesion and sexual dysfunction. Others report that medial basal-frontal or diencephalic injury was associated with hypersexuality and limbic injury with sexual orientation. After a TBI, there can be an adjustment involving detachment and the acceptance of the injury's permanence. However, the adjustment process may progress to separation or divorce.

Impairment of both the quality of the relationship and sexual functioning can result from these difficulties: Cognitive (not reported in all studies), emotional, behavioral, physical impairment, and substance abuse. Patients of both sexes with TBI have difficulties due to sexual positioning, sensation, and body image. Women with physical disabilities, and those with lesbian and gay sexual orientation, are poorly understood by practitioners, and may have difficulty in receiving good health care. Sexual drive and levels of sexual activity may be reduced. The wife may report disliking physical contact with her husband. While a worse relationship is common postinjury, some married people rated the relationship good or very good compared to the preinjury condition. The level of reduced sexuality is not related to the duration of posttraumatic amnesia (PTA) or the level of cognitive or physical impairment. While men with TBI were less sexually active, and fewer were involved in a relationship than disabled controls, there was no difference between women with and without TBI. A sample of premorbidly sexually active subjects presented some men with reduced erectile dysfunction and orgasm difficulties. Psychiatric levels of anxiety and depression are reported in both spouses. Already formed relationships, or the capacity to form new ones, may be hampered by deficits in cognitive and social skills and self-esteem, insight, judgment, self-awareness, and awareness of social cues and others' needs. For these reasons, persons with TBI are at increased risk for exposure to sexually transmitted diseases (review by Sandel et al., 2007).

MOTIVATION

Motivation refers to the characteristics and determinants of goal-directed behavior. Particular concerns after the level of trauma we are studying include apathy and hostility. Motivation includes the direction, subjective reaction, and persistence of action. Components include how the behavior starts, is energized, sustained, directed, and stopped. There is intact consciousness, attention, language, and sensorimotor capacity. Its disorders include increase, decrease, and dysregulation.

A core motivational system has been hypothesized in the forebrain: PFC and anterior cingulum; nucleus accumbens; ventral pallidum; and mediodorsal nucleus of the thalamus. The current motivational state is represented by the information in the core circuit. Before engaging the motivational system, it is necessary to appraise what is there and where it is. This information is represented in a highly processed form in the anterior temporal lobe and the insular cortex. This information is projected to limbic structures (amygdala, hippocampus, PFC), which continuously modulate the core circuit on the basis of the motivational significance of the internal and external environment.

The motivational process: (1) Represent the current motivational state. (2) Determine the environmental reward potential. (3) Modify the motivational state. (4) Select a new behavioral response. (5) Translate motivation into action.

Reduced Motivation

This is the common inability to initiate or sustain significant activities, and may have multiple origins: Reduced initiative directly based upon cerebral trauma; discouragement consequent to realistic inability to perform due to impairment and significant "distracting" symptoms; reduced community support and encouragement in obtaining employment, health care, and so forth.

Overt behavior: Attenuated social and occupational functioning
Goal-related cognition: Attenuated interests, plans and goals
Emotional responses: Attenuated responses to goal-related events; affect is shallow, labile, indifferent
Treatment modalities: Environmental interventions; psychological interventions; behavioral interventions; pharmacological treatment (Marin & Chakravorty, 2005)

For motivational disorders based upon concussion-level trauma, the present writer places less emphasis upon neurological mechanisms than suggested in the previous section. One may consider loss of motivation based upon doubt of success (pessimism with poor morale due to impairment), and giving up trying because of resistance and lack of support within the community (weltanschauung). Efforts for rehabilitation can fail due to unresponsiveness, negative bias, or rejection within the community. The need for further examination and treatment can be refused over a period of years, including the direct accusation of faking. One woman with a credible brain injury after an accident was accompanied to the examination by her husband. He overtly implied to the examiner that she was making up her difficulties. On a symptom validity test, her reduced performance seemed blatantly faked. This was her way of protesting against her husband's criticism.

CEREBRAL PERSONALITY DISORDERS

Personality Definition

The integrated description of a person's distinctive qualities of behaving and experiencing: temperament (energy; impulses); moods; motives (unconscious and conscious); fantasies; self-perception (identity); social roles and coping mechanisms; adaptive techniques; and lifestyle.

Cerebral personality symptoms and disorders are changes in behavior directly consequent to brain damage, However, it is not assumed that there is a constant association between a given dysfunction and a specific lesion site. Some syndromes and dysfunctions are caused by constitutional disorders or interference with achievement of personality maturity due to TBI or illness at critical periods: emotional control; motivational level; biological drives; moods; level of experience of emotions; overt expression of emotions; judgment; information-processing efficiency; behavior changes interfering with interpersonal relations; and maturational dysfunctions as a result of brain damage at critical periods. Secondary personality changes are psychodynamic reactions to impairment and stress: feelings of being impaired and less attractive due to the brain damage (identity), stress reactions, and overt and indirect expression of dysfunctions in school, work, independence, or other adaptive requirements.

CPD excludes:

- Distinctive neurological conditions (sensorimotor; dementia; memory, aphasia); illness; drug effects
- Stress reactions
- Psychodynamic reactions to injury and impairment (identity), and to changed lifestyle (dysphoria, reactive depression, anxiety, irritability and anger).

The Range of Cerebral Disorders

Emotional control, motivational level, biological drives, altered overt expression of emotions, altered social judgment and informational processing, which reduces the success of interpersonal relations.

- Mood changes: altered level of experience and expression of emotions, altered overt expression of emotions, cerebral damage.
- Emotional blunting (reduced ability to experience feelings): Lack of reaction to injury, appearance, impairment, failure, family problems, not motivated by embarrassment; indifference (dull mood and affect). Apathy, which may accompany psychomotor retardation and emotional blunting; interferes with rehabilitation.
- Outbursts: Euphoria; crying; laughing; rage. The expressed state may be incongruent with the reported experience.
- Reduced motivation and arousal.

Apathy and Blunted Affect

Apathy refers to indifference to the environment. It has the characteristics of diminished motivation not attributable to decreased level of consciousness, cognitive impairment, or emotional distress. It is different from depression, which is the experience of sadness and related feelings. Apathy has been observed in patients with impaired neural pathways, for example, neurodegenerative disorders (Levy et al., 1998). Stroke patients, with and without poststroke depression, may show similar impairments in emotional prosody, comprehension aprosody (sensory), and poststroke depression (Starkstein et al., 1994). While aprosodic patients may be apathetic (lost motivation, activity level), apathy is a different symptom. Blunt or flattened affect is found as a deficit system in chronic schizophrenia, some organic mental syndromes, and severe depression (Yager & Gitlin, 2000).

Mood Disorders: Reduced Motivation, Morale, and Stamina

Cytokine release during stress, in the control of immune and inflammatory responses, creates responses characteristic of TBI. These are sickness behavior (malaise, social withdrawal, somnolence, hyperesthesia) and depression (somnolence, anorexia, diminished libido) (Marshall & Rossio,

2000). Both PTSD and TBI elicit serious fears, and reduce self-confidence and a variety of adaptive abilities, and perhaps social attractiveness. The sense of foreshortened future with PTSD resembles the discouragement from persistent impairment and discomfort of the TBI victim. Depression is common in TBI and PTSD. Organic mood disorder (depressed type consequent to TBI) and/or organic anxiety disorder (Van Reekum et al., 2000) can be confused with the dysphoric moods of PTSD. Head injury predicts depression more than PTSD severity (Vasterling et al., 2000). Amnesia associated with depression is characterized by replies which are more frequently "I don't know" than near misses and confabulation.

Alexythymia

Alexythymia has these characteristics: deficient ability to express emotions caused by a lack of understanding of the emotions; concrete, bland, and tedious communication style; tends to be externally oriented in thinking; and poor ability to use imagery. One may speculate that a bland manner, confusable with aprosodia, could result from lack of appropriate emotional expression due to a combination of lack of understanding of one's emotions (Williams et al., 2001). Association has been demonstrated between self-reported head injury and levels of alexythymia (Williams et al., 2001). Alexythymia has been described as a complex condition not necessarily caused by brain damage in which bland affect or diminished facial expressions may occur (Lane et al., 1997). It refers to a relative constriction in emotional functioning, poverty of fantasy life, and inability to find appropriate words to describe emotions. Clinical implications are discussed in "Expressive Deficits." The original description (Sifneos, 1973) was based upon a group of psychosomatic patients, whose conditions are more understood contemporaneously to have a higher medical than psychological basis than originally understood (ulcerative colitis, asthma, peptic ulcer, rheumatoid arthritis). Sifneos considers the possibility of an etiology in neurophysiological, neuroanatomical, biochemical defects, or developmental difficulties involving the learning process. Cerebral disconnection, that is, nondominant limbic-neocortical disconnection, has been assumed to be related to alexythymia insofar as it includes flat affect, lack of spontaneous prosody or gesturing, and problems in expressive affective prosody. The basic affective prosodic set is attributed to an intact right superior posterior temporal posterior parietal area (Fricchione & Howanitz, 1985).

However, in one study of patients in a family practice, 49% reported a history of head injury (LOC; alteration of consciousness), and of the total, 18% were alexithymic (Toronto Alexithymia Scale-20). Those reporting head injury had significantly higher scores (Williams et al., 2001). Alexythymia has been found in individuals with cerebral commissurotomy (Bogen, 1993). The person may not identify or label or distinguish between feelings, fantasies, or physiological reactions (Acklin & Bernat, 1987); is unable to formulate and express affect and psychological conflicts verbally; or to symbolize emotion occurring in verbal behavior, fantasy, and dreams (Yager & Gitlin, 1995); or to distinguish between feelings and physical sensations (Yager & Gitlin, 2000). Bodily symptoms are the outlet for anxiety and depression, while the person is concerned with one's own body and the adequacy of its physiological functioning although unaware of their own emotional reactions (Lane et al., 1997), unable to verbally describe feeling states, impoverished fantasy life, reduced dreaming, and so forth. It may not be a single personality characteristic (Norton, 1989). Alexithymics are believed to express their feelings preferentially through physical channels, that is, are prone to develop somatoform disorders and psychosomatic illnesses. It is claimed that they are seen in the ranks of patients with persistent PCS, chronic pain, and other traumatic disability syndromes. Sifneos's (1973) therapeutic observation remains important: Alexithymic patients are not good candidates for dynamic therapy due to their inability to express their emotions verbally and diminution of fantasy. Supportive psychotherapy, behavior therapy, hypnosis, casework, and so forth, may be more appropriate. Musical ability has been identified in persons with nonfluent aphasia, transcortical aphasia, Wernicke's aphasia, aphasia with impaired speech comprehension and severe agraphia, and so forth (Brust, 2001). Aprosodia should be differentiated from amusia, which can be expressive, receptive, or both. Amusia stems from damage to the articulatory apparatus or primary receptor mechanisms.

DEPRESSION

Rorschach examples of depression: "Vampire flying away ... Symbol of death; two bodies leaning against a fence ... propped up: two dead monkeys ... mummified; a dead person; a dead nun flying on Pegasus."

Mood disorder symptoms overlap with those of TBI. They are frequent complications of TBI, and influence recovery, but are often overlooked. Post-TBI symptoms of major depression cluster in the domains of low mood, distorted self-representation, lack of motivation and anhedonia, subjective cognitive complaints, and hyperactive and disinhibited behavior (the latter activity findings seem to differ from the frequent fatigability and reduced activity of persons with lesser degrees of TBI). Other studies determined these characteristics of depression: depressed mood, lack of energy, feelings of worthlessness, suicidal ideas, fatigue, frustration, poor concentration, boredom (and in a group of moderate-to-severe injuries) distractibility, irritability, hopelessness, inability to enjoy activities, trouble falling asleep, restlessness, and weakness (Jorge & Starkstein, 2005). Depression may have evolved as a useful conservation strategy: The functions of a low mood would be the reassessment of major life strategies, replace lost resources, secure help from kin, stop pursuing unavailable goals, and so forth (Gardner, 2001). In patients with TBI, depression is associated with poorer social functioning, decline in activities of daily living in patients of varied age, failure to return to work, and financial stress, and long-term outcome. The clinician capability for appropriate treatment will follow assessment of depression's many possible etiologies.

HETEROGENEITY

Post-TBI depression has many different etiologies and clinical expressions. Yet, in one sample, depressed and nondepressed persons did not differ in age, sex, race, marital status, socioeconomic status, substance abuse, current medications, or neurological findings. Vegetative symptoms (autonomic, weight loss, delayed sleep, anergia, early awakening, and loss of libido) may be mistaken for depression, while "psychological" symptoms (worrying, hopelessness, suicidal ideation, irritability, social withdrawal) may be under-reported because of lack of insight. Depression may have a neurological basis, or be "reactive" and perhaps developing with insight into physical and neuropsychological deficits. An avoidance-based coping style is associated with poor outcome. Reactive depression is related to maladaptive belief systems, with negative events felt to be inevitable. TBI leads to a profound change of self and of development within the confines of lesser neurological functional capacity and the ability to cope with external reality (see "Identity"). Both patient and psychotherapist must accept their limitations and avoid pushing the patient into the experience of failure. From a cognitive viewpoint, depression may be consequent to a helpless attributional style and an external locus of control. Failure to reintegrate into prior social roles, with a destructive cognitive approach, leads to a self-perpetuating cycle of depression and withdrawal.

The strongest correlate of major depression was a left anterior lesion (left DLPFC and/or left basal ganglia). After 3 months, lesions location is not associated with depression, although the association between lesion location and affective status may be partially dependent on the measuring procedure. Social functioning maintains its association (Moldaver et al., 2004).

EPIDEMIOLOGY

Neurological disorders may be comorbid with depression, although this condition is not rampant. The reported frequency of depression after head injury has ranged from 10% to 77%, with a mode of one-third. A review of football players who incurred concussion offered the findings that retired players with five or more previous concussions had nearly a three-fold risk of incident depression, while those with three to four previous concussions had a two-fold risk of incident depression, compared to retired players with no history of concussion (Cajigal, 2007). The incidence varies with the intensity of the disorder and with the methodology. In one sample during a 1-year recovery, major depression occurred in 42% of survivors, while a minor depression ranged from 3% to 8% (Levin et al., 2007). Mood disorders, for example, major depression, are the most frequent complications of TBI. However, depression as a secondary condition after TBI is often undiagnosed, and may be the basis for somatoform disorders. While correct diagnosis is required for management, in a legal case, the actual etiology may be irrelevant if the disorder can be attributed to the accident (Restak, 1997). Vulnerability evolves from genetic, developmental, and psychosocial factors. Depression clusters with low mood and distorted self-attitude, low motivation and anhedonia, subjective cognitive complaints, and hyperactive and disinhibited behavior (Jorge & Starkstein, 2005). A significant minority of persons with mTBI have prolonged, complicated, or incomplete recoveries. Those members of one group of adults with problematic recoveries had psychiatric co-morbidity, contrasting with the good recoveries of those with only TBI. Psychiatric conditions responsible for poor recovery consisted of depression, anxiety disorders, or conversion disorder. Treatment is discussed by Robinson and Jorge (2005). Dissociative phenomena were common (Mooney & Speed, 2001).

One study did not find brain injury localization (CT scan) to be associated with depression incidence, nor was depression associated with damage to a particular anatomical region. Diffuse TBI may disrupt multiple systems and connections, leading to depression in a different mechanism than the psychiatric population (Salmond et al., 2006). Major depression and mood disturbance contribute to days lost at work and disability. Delayed depression may appear when the patient returns home and confronts the impact of the injury on daily functioning (Levin et al., 1997a). This author (Parker) notes the significance of the difference between a psychodynamic and endogenous depression, wherein psychological depression is an understandable and inevitable consequence of the loss of functioning relating to the injury. For example, loss of cognitive/behavioral efficiency will make life goals unattainable (Atteberry-Bennett et al., 1986). However, the capacity for insight into dysfunctions based upon lateralization or other brain trauma lays the basis for differences between reactions (e.g., right brain trauma associated with indifference and left hemisphere lesions with catastrophic reactions) (Heilman et al., 1983).

Diagnostic Considerations for Depression

Caution is needed when offering the diagnosis of depression, since a variety of essentially different conditions are included under this classification. Particular symptoms have been classified as *psychological* (worrying; brooding; loss interest; hopelessness; suicidal tendencies; social withdrawal; self-depreciation; ideas of reference) or *autonomic* (autonomic anxiety; loss of appetite; initial insomnia; middle insomnia; early morning awakening; anergic and retardation; loss of libido). There are multiple diagnostic codes and assessment procedures (interview, psychometric scales, and the Rorschach procedure—more in use with clinical psychologists than neuropsychologists). An important issue is whether mood disorders such as depression and anxiety found in patients with neurological illness are created by the disease itself. Particular symptoms are found in patients with TBI or stroke, as well as those with "primary" depression, that is, no known brain injury: insomnia, psychomotor retardation, loss of appetite, loss of libido, loss of energy. Further, patients with brain injury or stroke may have the syndromes of major depression, dysthymia, and minor depression. Variable depressive symptoms among nondepressive entities are illustrated by the finding that stroke patients have two different profiles: *endogenous* (suicide, anhedonia) and *reactive* (catastrophic reaction, hyperemotionality, anhedonia, diurnal mood variation). Most studies show a high concordance of autonomic and affective symptoms of depression and their consistency over time (Starkstein & Lischinsky, 2002).

Consistent with the fact that disparate conditions exhibit depression, the specificity of particular symptoms is low, that is, a symptom characteristic of one disorder is found in many others. In fact, being assigned a depressive diagnosis does not infer that there is a depressed mood. Even though meeting all the requirements for the diagnosis of major depression, some persons deny a depressed mood. Post-TBI major depression occurs with somatic symptoms that also may have a medical origin (sleep, appetite, libido). Patients with Parkinson's disease may or may not have a depressed mood, but those that do show a higher frequency of most of the autonomic symptoms and all of the psychological symptoms listed earlier. Starkstein and Lischinsky (2002) finally concluded that the diagnostic criteria for depressive disorders in various diagnostic entities (e.g., stroke, Parkinson's disease, Alzheimer's disease) should be determined separately for each condition.

Underdiagnosis of depression may follow denial or unawareness of a mood disorder. Depressive symptoms can be divided into "autonomic" and "psychological," whose frequency was about three times that in patients who deny a depressed mood. It is asserted that current requirements for three specific symptoms, as the criterion for a major depression, (DSM-IV-TR) have high sensitivity and specificity. The differential diagnosis of post-TBI major depression should consider adjustment disorder with depressed (and anxious mood), apathy, emotional lability, and PTSD. Thus, assessment of depression after TBI requires comprehensive study, for example, structured diagnostic interview, self-report, and caregiver-based measures. Clinical experience suggests that the Rorschach Inkblot Procedure also offers significant psychodynamic, affective, and cognitive information to contribute significantly to diagnostic issues (Robinson & Jorge, 2005).

Activation of the IS contributes to *mood dysregulation*. Cytokines in the brain influence the neurochemical systems involved in depression. Medical conditions involving activation of the IS are associated with psychological and neuroendocrine changes resembling depression. The effects of *infectious diseases* include fatigue, psychomotor retardation, anorexia, somnolence, lethargy, muscle aches, cognitive disturbances, and depressed mood. Many *noninfectious conditions* are associated with chronic activation of the IS and secretion of cytokines, including trauma, stroke, neurodegenerative diseases such as Alzheimer's, and multiple sclerosis. Even in rodents, immune activation induces a depressive-like syndrome, characterized by anhedonia, anorexia, body weight loss, and reduced locomotor, exploratory, and social behavior. Vaccination of vulnerable individuals (girls from low socioeconomic status) showed a virus-induced increase of depressed mood up to 10 weeks after vaccination (Yirmaya et al., 2000). There are high rates of depression in such disorders as multiple sclerosis, lupus erythematosus, and cardiovascular disease. In conditions manifesting episodic immune dysregulation, depression typically precedes disease exacerbation, suggesting that the depressive symptoms result from IS activity rather than being a psychological reaction to being sick (Raison et al., 2005).

Etiology

Endogenous opioids are involved in the mediation of mood with the contribution of other neurochemical systems and affect stress, appetite regulation, learning, memory, motor activity, and immune function (Harris et al., 2005). *Chronic numbing and blunting* of emotional responses may be due to stress-reduced analgesia or dysregulation of opioid systems in PTSD.

Cerebral lesions: TBI, which selectively affects prefrontal and anterior temporal structures, is associated with an increased prominence of mood disorders. The amygdala, hippocampus, basal ganglia, and thalamus are also involved. The right hemisphere is more commonly associated with depressive symptoms (Robinson & Jorge, 2005).

Subjective sense of impairment, rather than more objective indicators of injury severity and support, are the most important postinjury factors in depression. Patients' self-assessment of their acute level of impairment is most strongly related to early and late depression, regardless of injury type or severity (Malec et al., 2007b).

Relationships: Loss of a loved person may shape the expression of the illness, for example, by leading to earlier onset, more severe episodes, and an increased likelihood of personality disorder and suicide attempts. The depression may be reflected in impaired coping under stress (Akiskal, 2005a). Psychosocial and emotional repercussions of TBI are particularly disabling in young persons, who are in the process of establishing interpersonal and financial independence. Multiple injuries may have an indirect effect upon the development of depression through its effect on social support and functional disability (Levin et al., 1997b).

Temperament: While adjustment disorder with mixed anxiety and depressed mood, acute or chronic (309.28) is often observed after head injuries, the syndrome of a worrying temperament, associated with generalized anxiety and complicated by depressive episodes, may be prevalent in 5% of the general population (Akiskal, 2005b).

HPA: It has been suggested that a defect of negative feedback of the HPA has a role in the pathophysiology of depression (Posener et al., 1998), specifically a defect at the level of brain centers regulating hypothalamic secretion of CRH. It was suggested that normalization of the HPA system would prevent development of depression in chronically stressed subjects: Possible modalities would be pharmacological intervention or psychotherapeutic strategies (e.g., cognitive therapy), aimed at improving resilience and control in stressful situations (Tafet & Smolovich, 2004). Hormonal dysregulation may account for the finding that after severe TBI there is a pattern of sleep disturbances and changes in hormone secretion, which has some similarities to that of patients with remitted depression (Frieboes et al., 1999).

Major Depression

Since major depression is polygenic, it is inferred that posttraumatic vulnerability, augmented by nongenetic factors, varies among individuals (Kandel, 2000d). There is increased cytokine production in response to injury, illness, psychological, and physical stress (Slimmer et al., 2001). Melancholic depression is characterized by CRH hypersecretion, impaired cortisol feedback, and hypercortisolaemia. It contrasts with PTSD, which has a relative hypocortisolaemia (Kaye & Lightman, 2005).

Depression and Stress

Depression, in part, is a failure of normal reactive mechanisms to repeated or long stress. Stress is not a homogenous concept. Its components include the timing of the stress in terms of the subject's age, its severity and chronicity of exposure, the role of controllability, prior exposure to the same or related stressors, and so forth. Stress contributes to affective disorder, which is a complex interaction of genetic vulnerability, life events, and biology. The effect of life events in the occurrence of depressive episodes must not only be recognized, but there should be related adjustments of both mood-stabilizing medication and nonpharmacological treatment. It also increases morbidity and mortality of medical illnesses. In understanding depression, and one presumes other general disorders, overly fine discriminations actually interfere with basic understanding since depression may be a *final common pathway* resulting from diverse events, which in aggregation or interaction, result in a set of shared clinical features (McKinney & Tucker, 2001). Some mood disorders are associated with a hyperactive HPA (chronic anxiety, melancholic depression), and GH and IGF-1 (insulin-like growth factor 1) are significantly reduced (Chrousos, 1998). Chronic *psychosocial stress* is associated with depression, in which *increased cortisol* levels, decreased somatostatin level (Bremner et al., 1997), and altered *serotonic neurotransmission* been observed consequent to dysregulation of the HPA system. In addition to increases in circulating cortisol, there is increased cortisol excretion, or a failure to suppress cortisol release (*dexamethesone suppression test*). Immune and inflammatory cytokines are associated with depression. These factors may mediate the link between *depression, inflammatory disorders* (cardiovascular disease, rheumatoid arthritis), and the behavioral symptoms known as *sickness behavior.*

FRONTAL LOBE SYNDROMES: DYSREGULATION

Executive Dysfunction

It has been asserted that the term *"frontal lobe syndrome"* characterizes disorders of EF, and it comprises cognition, behavior, and mood disturbance, overlapping with psychiatric disturbances. Presumed frontal lobe symptoms may be consequent to the fact that the PFC is in reciprocal relations with the visual, auditory, and somatic sensory areas, the thalamus, the limbic area, and the medial frontal cingulate cortex. Therefore, EFs are not synonymous with "frontal functions." Executive dysfunction is found in diffuse brain damage, and correlated with metabolic alterations in specific prefrontal, premotor, and anterior cingulate regions (Boller et al., 1995).

One aspect of the behavioral consequence of executive dysfunction is difficulty in controlling behavior when there is an issue of choice, ambiguity, or complexity. The loss of control may be (1) a result of impaired emotional reactions or (2) the inability to predict the complex reward (and aversive) consequences of behavior (Alexander et al., 2007). An approach to understanding executive dysfunctioning offers these assumptions: At the core is disruption of problem-solving mechanisms; problem solving is supported or thwarted by problems of emotional regulation, and EF, emotional regulation, and learning are mediated by attention. Thus, remediation of attentional defects provides the foundation for more complex learning, and must be included as an essential element of rehabilitation (*executive plus model*) (Gordon et al., 2006).

Complex Psychopathology

Some cases described as a *frontal lobe syndrome* (dyscontrol) do not have obvious frontal lobe damage, for example, many neurological diseases, schizophrenia, attention deficit hyperactivity disorder, and obsessive-compulsive disorder (Lyketsos et al., 2004). Nevertheless, damage to the prefrontal region, particularly ventral and medial aspects, results in severe impairment of social decision making and emotional behavior, in the presence of intact language and other cognitive skills. These patients have difficulty planning their future at immediate, medium, and long ranges. Their plans, associates, and activities lead to unsuccessful outcomes. Nevertheless, IQ testing, retention of factual knowledge, and the learning and retention of skills remain "normal." *Psychosis* is a relatively infrequent consequence of TBI, and is characterized by paranoid delusions, auditory hallucinations, and less frequently, negative symptoms and other hallucinatory experiences. Head injury may contribute to a limited risk for schizophrenia and for bipolar affective illness. Posttraumatic psychosis is associated with cognitive disturbance and neuropathological changes in the frontal and temporal lobes (Jorge, 2005).

Somatic Markers of Consequences

While attributed to frontal lobe lesions, similar personality changes have been described in patients with temporal lobe and diencephalic lesions. These may be consequent to the rich reciprocal interconnections linking the temporal, limbic, and frontal regions (Price et al., 2002). These cortices receive projections from all sensory modalities, directly or indirectly. Their convergence zones hold a record of activity from varied neural structures (sensory cortices; limbic structures) that have received external and internal stimuli. They project to central autonomic control structures with a physiological influence on visceral control. In addition, they have bidirectional connections with the hippocampus and amygdala.

The ventromedial frontal cortex participates in a larger neural network, devoted to the integration of somatic tags with stored knowledge reflecting social conduct. This helps us to understand why some patients passively watch stimuli rather than actively respond (galvanic skin response). Patients with ventromedial frontal lobe lesions select appropriate interpersonal behavior on verbal tests but fail to perform in real situations (Grafman, 1995). Although ventromedial patients were intact in neuropsychological laboratory tests, they have a compromised ability to express emotion and experience feelings. Tucker's "*somatic marker hypothesis*" holds that these somatic changes are an essential aspect of the process of emotion, although rationality depends upon other systems. In assessing outcome or possible adaptive success of the injured person, a source of information not clinically available are *somatic markers*. They serve as an "*alarm signal*," which, if not activated, do not represent a negative or positive outcome for a response option. When earlier neural states are reprocessed, this signal is received by the ventromedial cortices, which then activates somatic effectors in the amygdala, hippocampus, and brainstem. The reenacted somatic state signals cortical and subcortical somatosensory processing structures, which modify appetitive or aversive behaviors. Different networks are used for processing unconditioned stimuli and those that require complex information processing. Ventromedial frontal damage precludes this process (Damasio, 1995; Damasio & Anderson, 2003; Tucker et al., 1995).

Syndromes of Frontal Lobe Damage

Prefrontal cortex: Damage to the PFC has paradoxical consequences. Patients can exhibit:

- Both prolonged deliberation about choices and subsequent irresponsible, risky behavior
- Apathy and indifference, and poor impulse control (Walton et al., 2007)

Disinhibited: The *orbitofrontal cortex* receives input from the basolateral component of the limbic system, and through it from parietal lobe and olfactory pathways. It is concerned with the sensory-receptive and interpretive cortex. This results in the so-called "positive" symptoms found in frontal lobe damage. This has been described as "pseudopsychopathic." Behavior is restless, explosive, distractible, hyperactive, and inappropriate. Mood is labile and jocular. Reflecting the prefrontal and limbic systems circuit, it appears that if the PFC does not control impulses from the limbic system, the result is impulsive or disorganized behavior. In contrast, if limbic impulses are not transmitted to the prefrontal motor strip, the result is adynamism (Prochalska et al., 2006). It appears "*Witzelsucht*" may occur, that is, inappropriate facetiousness and a tendency to pun. Judgment is poor. Paranoia may ensue. These patients are more impaired than others by drugs or alcohol, that is, they overreact to the disinhibiting effects of these substances (Golden, 1986).

Substance abuse: While there is a postinjury reduction of mood, anxiety, and *substance abuse* disorders over time (Axis I), the prevalence in TBI patients is significantly higher than in control groups many years after the injury. A group of people with concurrent TBI and posttraumatic alcohol dependence appears to use alcohol to self-treat affective and anxiety lability (Beresford et al., 2005). Alcohol misuse is significantly associated with TBI, for example, assault and motor vehicle accident (MVA). One study indicated a frequency of alcohol abuse/dependence of 34.8%, and an increase of mood disorders greater in this group than in normal drinkers or nondrinkers. Sixty percent of patients with a history of substance abuse relapsed into alcohol abuse in the year following the accident (Jorge & Starkstein, 2005). Increased brain atrophy has been observed in TBI patients with a history of moderate-to-heavy preinjury alcohol use. It has been speculated that the combined effects of substance abuse, particularly alcohol, produce greater disruption of the neural circuits involved in mood regulation, motivation, and reward processing. These changes would result in more severe psychopathology, uncontrolled addictive behavior, and poor psychosocial outcome (Jorge, 2005).

The dysexecutive syndrome: This may follow dorsolateral damage: Inability to integrate different sensory stimuli into a complete gestalt, a limited or stereotypical response pattern, easy loss of cognitive set, inflexible and/or perseverative responding, and lack of error monitoring (Bamdad et al., 2003).

Apathetic: The "pseudodepressed" state is associated with mesiofrontal damage. The medial limbic system is concerned with the reticular core. Angular acceleration injuries are associated with coma, followed by posttraumatic confusion and a pattern of "negative" symptoms. The results are inconspicuous, that is, apathy, lethargy, little spontaneity, inattention, slowness, long latency of response, unconcern, reduced sexual interest, little overt emotion, and inability to plan ahead. There may be little or no conspicuous sensorimotor or intellectual deficit.

Diffuse axonal injury can cause executive dysfunction following white matter damage involving widespread frontal systems. A *prefrontal lobotomy* severs fiber tracts between the frontal lobe and the dorsomedial nucleus of the thalamus. What was observed was difficulties in anticipating the future, lack of concern for pain and obsessions, and perseveration.

Akinetic

Medial frontal area: This involves a paucity of spontaneous movement, gestures, and verbal output. Somatic signs include lower extremity weakness and loss of sensation, and incontinence (Trimble, 1988, p. 164).

Symptoms

Witzelsucht describes the personality change: irresponsibility, childishness, facetiousness, disinhibition, indifference to others, and inappropriate sexual behavior.

Poor judgment: Error monitoring leading to correction of inappropriate behaviors or erroneous responses has been attributed to the frontal limbic area (Stuss, 1991). Poor error monitoring and inability to learn from experience gives the observer the impression of indifference or immaturity (frontal lobe syndrome). Lesions may permit repetitive maladaptive behavior, which seems unmodifiable by social criticism or repetitive unsatisfactory experiences. This can give the impression of indifference. The patient appears *psychopathic* since behavior that is socially rejected is expressed as if it were actually satisfactory. The apparent indifference by the patient to the self-destructive consequences of one's behavior is misunderstood; the basis is actually poor foresight and judgment.

PERSONALITY PSYCHODYNAMICS

In addition to the actual misfortune of the effects of a trauma, the patient is often burdened by inaccessibility of assessment, treatment, and compensation. There is often refusal to deal with the injured person in a matter-of-fact way, that is, provide treatment and compensation according to the laws of the jurisdiction. The situation has not improved since Lishman noted in 1987, resolution of the impaired condition involves litigation, that is, chronic conflict, doubts as to the person's integrity, and long drawn-out frustration, all of which reduce the capacity for rehabilitation.

Clinical example: Personality change: The patient was a middle-aged Chinese immigrant. In China he was a university graduate, and had worked as a sales manager. In the United States, he had modest English proficiency, for example, he could read a newspaper page in an hour with the help of a dictionary. At the time of the accident, he was struck by a car while on a bicycle delivering food for a restaurant. The left side of his head struck the ground. He asserts 30 min LOC, and was told he had a fracture, and he was hospitalized about 15 days. He was in a wheelchair for half a year, used crutches, but still does not walk as efficiently as before. His knee (where he was struck) requires an operation for which he is waiting authorization. Among his complaints are headaches, dizziness, double vision, dysesthesia of the right cheek, anxiety, and depression. He thinks that his condition is getting worse and that he will never recover. He conceals crying from his wife. He claims that memory problems would interfere with employment.

His wife states that he is like a different person. He becomes angry easily without a reason. When she prepares a meal, he complains and upsets everybody. He says that he doesn't like what she prepares, but this is food that he previously liked. He forgot the address of a very important appointment with an insurance company doctor and forgets to mail the telephone bill, even when he was told to do it. Before the accident he had many hobbies: drew pictures, painted, liked music (sang songs, had a good voice). Now he is very quiet and looks sad. He speaks about returning to work, but says that he feels worse. She believes he has become another person. "A very strange person to me. He's not a good husband. He has a sexual problem. He is always complaining about our family. The place is no good, the room is no good, the city is no good."

Clinical example: Conversion reactions and hysteria: Stress reactions may mimic illness (Herskowitz & Rosman, 1982, pp. 139–140; Schilder, 1964). These entities are a form of somatoform reactions, which are characterized by physical symptoms suggesting medical disease (or trauma), but without apparent organic pathology or pathophysiology. The examiner can be misled by a conversion reaction that resembles a neurological deficit. Additional somatoform reactions include somatization, undifferentiated somatoform disorder pain disorder, hypochondriasis, body dysmorphic disorder, and somatoform disorder NOS (American Psychiatric Association, *DSM-IV-TR*, 2000, pp. 485–511; Barsky 1989).

Hurwitz (1989) offers case evidence that conversion, mimicking such neurological symptoms as tremor, convulsions with urinary incontinence, hemiparesis, flattened affect, finger paralysis, and so forth, may be detected through narcoanalysis and enhanced symptoms when the examiner brought the schema of the affected part to the patient's attention. The pattern occurred within the context of depression. The proposed dynamic would be an inability to express themselves in psychological terms because of fear of stigmatization. Intrapsychic relief is obtained by believing that they are physically ill, which would explain dysphoria and dysfunction.

This writer has observed one case of unilateral muscular weakness and reduced ability to move the eyes to one side (causing double vision) that was relieved by an Amytal interview (Wada procedure).

Clinical example: Mumbling as a conversion symptom: An adolescent with an unquestioned traumatic brain damage was observed in school to have seizure-like activities (eyes rolling, loss of urinary control). (The EEG, which I received later, did not confirm cortical seizure-like activity.) In the examination, he was the worst mumbler I ever examined. With practically every statement I had to urge him to repeat or speak louder in order to be understood. Certainly, the question of damage to the frontal motor areas, tracts controlling peripheral speech areas, or somatic damage had to be considered.

His mother described his situation as follows: "The kids in the class are picking on him. He doesn't know why. He lost confidence in himself. He doesn't want to wash himself. He has to be told to change his clothing." I asked her the worst consequence of the accident: "He's very angry, tense. He wants nobody to touch his things. He has claustrophobia. He doesn't like to be in the car or in close places. In the car he was screaming to open up the window, he couldn't breathe, like somebody was attacking him. In your elevator, which is small, he was upset."

To obtain an objective level of enunciation difficulty, he was administered the Gray Oral Reading Test—3rd Ed. (GORT-3). Comprehension Standard Score was 85 (16th percentile). Passage Standard Score (rate plus accuracy) was 110, or 75th percentile)! His ability to read difficult passages was breathtaking. On the way out, I told him in the presence of his mother that he could speak very clearly.

In a feedback interview his mother told me of an improvement in his behavior. It was likely that mumbling represented an attempt to prevent communication and to give the impression of greater impairment. There was no doubt that there was some impairment, but he did not feel that he was obtaining sufficient support. I referred him to a psychologist for cognitive rehabilitation and psychotherapy.

SOMATOFORM DISORDERS

These are a pattern of physical symptoms suggesting a general medical condition but not fully explained by a general medical condition, the direct effects of a substance, or by another mental disorder (American Psychiatric Association, 2000, pp. 485–511).

SOMATIZATION DISORDER (HYSTERIA OR BRIQUET'S SYNDROME)

Somatization disorder is a polysymptomatic disorder, beginning before 30 years, occurring over a period of several years, and which cannot be explained by any known general medical condition or the direct effects of a substance. Its subtypes include undifferentiated; conversion disorder; pain disorder; hypochondriasis; body dysmorphic disorder; and somatoform disorder, not otherwise specified. It is characterized by a combination of pain, gastrointestinal, sexual, and pseudoneurological symptoms (American Psychiatric Association, 2000, p. 486). Additional somatoform reactions include somatization, hypochondriasis, and the somatoform pain disorder (American Psychiatric Association, 1987; Barsky 1989). The examiner should suspect a somatoform disorder in any patient claiming symptoms involving multiple organ systems (Bradley et al., 2004). Stress reactions may mimic illness (Herskowitz & Rosman, 1982, pp. 139–140; Schilder, 1964). Somatization disorders overlap with anxiety, depressive, dissociative, and personality disorders. Unexplained symptoms, with high distress, are a common medical problem. More than one-half of the most common primary care problems are not adequately explained (Hollifield, 2005).

Conversion reactions are a form of somatoform reactions, which are defined as conditions characterized by unexplained symptoms or deficits affecting voluntary motor or sensory function that suggest a neurological or other medical condition. Psychological factors appear to be involved. The examiner can be misled by a conversion reaction that resembles a neurological deficit. Medically naive persons present implausible symptoms. More sophisticated persons have more subtle symptoms and deficits that closely simulate medical or other neurological conditions. Thus, the diagnosis of conversion disorder should be made after thorough medical investigation, and should be reevaluated periodically. The rate of misdiagnosis of persons diagnosed with conversion disorder has been reduced from one-quarter to a half to an estimated 15% (Hurwitz & Pritchard, 2006).

Neurological-type symptoms listed in *DSM-4-TR* are described as "pseudoneurological." They include:

- Motor symptoms: Impaired coordination or balance; paralysis or localized weakness; aphonia; difficulty swallowing or the sensation of a lump in the throat; and urinary retention
- Sensory symptoms: Loss of touch or pain sensation; double vision; blindness; deafness; hallucinations
- Cognitive intellectual abilities
- Seizures or convulsions

Recent functional MRI (fMRI) studies suggest one mechanism, that is, active inhibition in the anterior cingular gyrus, and possibly the orbital frontal cortex. Attention and active inhibition are part of the conversion process; psychogenic deficits, which are more severe when patients attend to their neurological problem, become less severe when patients are distracted, particularly when mediated by narcoanalysis. The conversion process occurs when fixed beliefs concerning somatic function arise from psychological distress that controls cortical and subcortical pathways to produce patterns of loss or gain of function that are not organic, as ordinarily understood (Hurwitz & Pritchard, 2006).

CONVERSION REACTIONS

Conversion disorder is defined by the presence of symptoms or deficits that suggest a neurological or other general condition. Conversion symptoms do not conform to known anatomical patterns or physiological mechanisms, but may be shaped by the individual's conceptualization of some function. Misdiagnosis may occur with later identification of definite medical symptoms; the examiner should be cautious about overcertainty concerning current diagnostic techniques. While one cannot be certain, and secondary gain occurs, conversion disorder is differentiated from malingering or factitious disorder by the assumption that symptoms are not intentionally produced to gain benefits (American Psychiatric Association, 2000, pp. 492–498, DSM-TR 300.11). Motor symptoms include impaired coordination or balance, paralysis or localized weakness, aphonia, difficulty swallowing or the sensation of a lump in the throat, urinary detention, and seizures or convulsions. Sensory symptoms include loss of touch, anesthesia, pain, double vision, blindness, and hallucinations.

SELF AND IDENTITY

Clinical example (Injured identity): An active 71-year-old employed woman suffered a mild concussion when knocked down by a moving object, striking her right temple and suffering other somatic injuries. Prior to the accident, she was in treatment for depression, anxiety, and multiple phobias: "I feel vulnerable, nervous. I never thought of myself as fragile. Suddenly I felt like a fragile old woman. I aged several years in this one year." Her moods are "terrible depression, fearful, afraid to go out, I suddenly became afraid of falling."

From the viewpoint of evolutionary biology, the self is the individual, a coherent unit selected to nourish, protect, and reproduce itself (Queller & Strassmann, 2002). Psychologically, it is the personal experience that guides action and reaction to events (Parker, 1983). Its components include consciousness or self-awareness, body image (sensorimotor), and identity (stress and psychodynamic). Each of us has a preferred strategy preference (Rudel, 1978). In the uninjured or unimpaired person, the self is experienced as integrated. One can differentiate between the self as experiencer or observer and an objectified self, which is regarded as having feelings and qualities, and knowing, and includes the sense of "me" as subject or "I" as actor.

Any aspect of experience that comes into awareness may have a meaning attributed to it. Its effect upon our self (identity) and its feelings can determine our mood and select and direct our actions. Thus, our self, or who we think we are, is part of our consciousness, and a determinant of feelings and activities that interact with our attitude to our self or identity (Parker, 1983). In turn, our identity is a determinant of action. Mental representation of the self resides in working memory along with a coexisting representation of the current external environment. Disruptions of the sense of self as integrated within one's person, and with the environment, are considered to be dissociative disorders .

The self as a stable mental component represents a *memory system*. Reorganization may involve catastrophic destruction. A catastrophe (emotional stress) may create both disorganization of the earlier self and a different, highly cathected self. If the new experience is incompatible with the self, then the self must be reorganized.

Repression, understood as loss of access to the initial self, need not be defensive in the psychoanalytic concept. Repression occurs due to disorganization or to the existence of fragmentary experiences that are not fully encoded in the self. Information not consistent with the existing structure (memory of the self) cannot be processed (Tucker & Luu, 1998).

SENSE OF SELF

Our sense of self is essentially the totality of the individual, differentiated from the environment, which is described psychologically and physiologically as nonself. Damasio (cited by Nijenhuis et al., 2002) associates consciousness and sense of self. These include sensory feelings that involve several integrative levels. Self may be the target of appraisal, or the agent of action (William James in APA, p. 827). It includes states within ourselves, health, well-being, satisfaction, and identification of nonself. Self and identity are overlapping concepts involving both the unified perception of the individual experiencing one's self (self) and a self-definition and value of a range of characteristics, both aware and unconscious (American Psychological Association, 2007). Yet, self may extend far beyond the body wall, as when we identify with another person, a country, or a sports team.

Self has many dimensions. It refers to our *person* as differentiated from the environment (*nonself*). *Body schema* refers to neurological pathways projecting to numerous centers in the brain, that is, primarily somesthetic images of who we are, in motion and static. It is not known how the various *homunculi* are integrated so that motor activities that may be complex, unplanned, and highly skillful are performed moment-to-moment. *Identity* is largely a self-description, verbal and values-heavy–oriented to self-esteem or its lack, reflecting our emotional history influenced by our value to ourselves or to others. Are we attractive, complete, important, strong, or impaired, damaged, vulnerable to attack, and so forth? Our *psychosocial self* is the narrative of our life and its events (Booth, 2005). The physiological self shows the effect of illness, disturbed homeostatic functioning, momentary body disturbances, and the level of physical conditioning. Our adaptation involves a relationship between our sociocultural milieu and our biological and psychological selves. The latter communicates with our sociocultural context through language, meaning, and emotion. The integrated immune and nervous systems, as part of the "physiological self," integrate with the "psychosocial self" maintain a coherent, self-consistent relationship between the self and nonself (environmental context). This coherence is a strong predictor of life outcomes and satisfaction, and buffers the adverse immunological effects of stress in older adults (Booth, 2005).

Identity

This concept reflects our self-image, that is, the integrated picture that we have of ourselves. It is characterized by the varied labels with which we describe ourselves. These are learned from the reactions others express to us, as well as the conclusions that we draw from our own experiences. Among the identity themes influencing capacity for stress resistance are self-confidence, passivity, and self-image as weak or incompetent (Parker, 1981, pp. 10–13, 124). There are gross differences between people in their degree of self-awareness (at its lower extreme not knowing who you are) and the level of accuracy (including neurotic self-disparagement and narcissistic self-enhancement) (Parker, 1983, pp. 15–16). Black's (2006) discussion of self-protection explicates some aspects of identity through a concept of a "thrust to wholeness." We attempt to maintain our functional integrity of body and self, that is, preventing harm. This utilizes physiological systems and psychological defenses. A stable, nonfractured self-image enables both effective decision making and interactions with our environment. We defend against personal failure, social disparagement, and physical harm.

LACK OF INSIGHT

Clinical example: Lack of insight: One patient's forehead struck his car's windshield when it was rear-ended. He, a physician, described his condition: "I didn't see the change in myself. I didn't know that was wrong." His psychotherapist (this writer) pointed out that he was once better able to concentrate. "I thought that's the way it's always been. How else could it have been? I was unaware that my symptoms were related to the accident. I denied I was ill and acted as if nothing was wrong. I went to a conference, received 35 hours of credit, and I don't remember anything. I thought everybody else was strange. I was seeing them differently. I thought that you (the author) were asking strange questions."

Reduced patient insight (self-awareness) is associated with lower verbal IQ and reduced temporal orientation. Association with the extent of lesion is not definitely known. With lack of foresight or judgment, the patients will not realize that they have created acts damaging to themselves, or that their actions are likely to do so in the future. Reduced performance, foresight, and planning ability may not be recognized by the patient. With certain kinds of injury (e.g., dementia or the *frontal lobe syndrome*) (Parker, 2001, pp. 233–234 for cases), poor adaptation is not currently recognized by the patient.

It also reduces motivation to improve behavior. Socially inappropriate behavior and nonadaptive decision making may reflect a disconnection between the consequences of past actions and the current guidance of behavior (Damasio & Anderson, 2003; Garcia et al., 1998). Impaired self-awareness is associated with more severe injuries and report of fewer depressive symptoms. Those with better self-awareness are more likely to report symptoms of depression, to be less severely injured, and may be more likely to dwell on their perceived deficits (Malec et al., 2007b).

IDENTITY CHANGE AND WELTANSCHAUUNG AFTER TBI

Identity as a feature of the preinjury personality may be the most vulnerable to disorganization and regression after brain trauma. Among the unhappy qualities of identity that are frequently found after TBI are vulnerability, unattractive, damaged, reduced self-esteem and shame, incompetent with poor morale, helpless, discouraged, and depersonalized. The TBI victim has a different sense of posttraumatic identity, incompetent versus formerly competent, or damaged, or far less self-aware. As Dr. Ravella Levin described the identity change in a personal communication: "I am a stranger to myself. This person with aphasia, memory problems, seizures, etc., really isn't Me!"

Identity is in a two-directional exchange with Weltanschauung, that is, world outlook or philosophy of life: To the accident victim, the world is dangerous and ungiving, and one's future is bleak. It is experienced as dangerous, with forces too violent and overwhelming to be coped with. The victim feels trapped in an unhappy life. There is a search for the meaning of one's experience, that is, "Why me?" or "Why has my life been changed so rapidly and destructively, with no forewarning for no good reason?" (Parker, 1990, pp. 240–241).

Shame

It is necessary to differentiate between *guilt* (the belief that one has done something wrong) and shame. This is based upon the belief that one is a victim of others' contempt, for example, dishonor or ridicule. In addition to the thought that we do not meet the (social) standards of others, one does not meet the standards of one's own ego ideal (Piers & Singer, 1953, p. 16), or we believe that we do not meet the standards of other people. TBI may elicit feelings of shame in some ethnic groups. Since identity is, in part, the belief that we have certain qualities that are socially valuable (as well as socially disparaged), self-identification as having qualities that are contemptible becomes a painful reaction to our identity that is called "shame." The brain-damaged person may experience shame because of one's condition. This is part of the patient's *social distress*: shame, withdrawal, and isolation or loneliness. Since identity stems considerably from others' prior and current cumulative reactions, isolation can be experienced as a reduction in the value, intensity, or clarity of our self-experience. Shame and anger also affect PTSD developments, although they are not part of the definition.

Nevertheless, despite the facial and other disfigurements incurred after an accident, the extent of self-disparagement diagnosed as the *body dysmorphic disorder* (300.7; *DSM-IV-TR*) is not characteristic of injured people. It is an excessive preoccupation with a slight physical anomaly creating clinically significant impairment. While frontal serotonin pathways are believed to be involved, it is the somatoform disorder with the strongest sociocultural basis. Culture influences the individual experience of psychopathology, its clinical manifestations, and course (Hollifield, 2005).

ROLE IN LIFE

The role in life (Piotrowski, 1956, pp. 140–143, 155–163) has been described as reflecting the conception of life according to which the individual makes his adjustment to reality. The role is determined by health, heredity, intelligence, prior social experiences, and so forth, and is consistent with familiar preexisting and posttraumatic issues known to affect the outcome of a TBI. It is comparable to a steering mechanism. Yet, it is a deeply imbedded tendency, not easily modified, to assume the same attitude in dealing with matters that are personal and important. With higher reality testing (Rorschach Form + %), the more likely is the individual to use judgment and prudence. The main roles in life are self-assertion, compliance, and indecisiveness. Acting at variance with the prototypical role(s) in life causes anxiety and is not conducive to an efficient handling of vital life problems. Lack of Rorschach human movement responses may indicate either a traumatic or lifelong lack of insight or self-awareness. Actions at variance with the prototypical role cause anxiety. They may aid in the struggles of life, or be a severe handicap. The compliant role infers a quality often observed after a significant TBI: dependence upon somebody psychologically stronger to assume ultimate responsibilities. When failure is expected (e.g., depressive hopelessness), the person is convinced that nothing can be done to influence a future fraught with danger, and the desire to act is absent. This contrasts with elation and the expectation of success, when the person tackles anything that comes his way with undiminishing energy. Relevant psychological issues are the need for the strength of a protective atmosphere and whether social relationships are experienced as enervating or stimulating. Understanding the role in life is important to the clinician: It determines the individual's handling of interhuman relations only when the individual feels that he is vitally involved in the situation. Thus, the pattern of overt actions comprise such issues as the patient's role in life, identity (perhaps impaired, unattractive, withdrawn, dependent), and autoregulation (moods, impulses, motivation).

EMPATHY AND THEORY OF MIND

THEORY OF MIND

Theory of mind refers to primates' ability to understand the mental state of others and to predict behavior based on those states. It seems to depend upon the PFC (Kaas & Preuss, 2003). This process appears to develop in a consistent pattern during early childhood and perhaps is also consistent across cultures. Another component of social cognition is *autonoetic consciousness*, referring to an aspect of self-awareness allowing us to imagine our own experiences at different places and at different times. Two cases were presented where they experienced severe difficulties in recollecting any periods from their own lives, but had no measured difficulty in taking other persons' perspectives and inferring their thoughts, feelings, and intentions (Rosenbaum, 2007).

Children appear to possess *a "theory of mind,"* that is, assigning complex mental states such as beliefs to themselves and others (Zeman et al., 1997). Autistic children have difficulties in the area of theory of mind, which appears dissociable from overall mental deficiency. Deficits appear after acquired right hemisphere brain damage and focal frontal lesions. It was inferred that theory of mind may have a dedicated modular brain function. Although separate functions are inferred for theory of mind and executive functioning, they may have in common involvement of the PFC. The latter's participation in a larger distributed circuit is evidence against a localized region for theory-of-mind functioning (Saltzman et al., 2000).

Empathy

Empathy, described as a binding force of social cognition, refers to feeling what another person is feeling (*emotional empathy*), knowing what another person is feeling (*cognitive empathy*), or responding compassionately to another person's distress (*compassionate empathy*). Empathy involves an interaction between emotional processes and cognitive functions that are involved in social awareness and judgment. These are not lateralized, nor confined to prefrontal lesions, since right parietal lesions caused impairment similar to prefrontal ones. Emotional empathy, although a component of social cognition, may be relatively independent of cognitive ability per se. Since no relationship was determined between severity of TBI and emotional empathy, it suggests that relatively minor head injury in vulnerable individuals altered some functions important for social cognition. Moreover, since neither depression nor anxiety made a unique contribution to emotional empathy scores, these measures of affect may be independent of emotional empathy. Cautiousness in maximizing the effects of empathy is indicated by the possibility that some clinical effects appearing to be low empathy are actually alexythymia (Wood & Williams, 2008).

Empathy played an evolutionary role in consciousness. Awareness of likeness to another (empathy) contributes to imitative learning or consensual validation. Thus, combining consciousness and social process would contribute to the rapid dissemination of new behaviors. However, awareness of the other is insufficient for survival. The social advantages of consciousness infers prediction.

SOCIAL PARTICIPATION

Reduced social interest and participation are common postconcussive symptoms with multiple etiologies: impairment (physical and psychological); reduced emotional intelligence; economic loss; belief that one is unattractive; varied symptoms such as seizures, orthopedic, imbalance, and dizziness, and so forth; psychotherapy and rehabilitation also invite attention to functioning of specifically social qualities.

Intimacy

Social intimacy and success (predicting others' activity and influencing them to meet our own needs) involves a process of understanding others having:

- A similar experience
- A more detached cognitive process involving our own knowledge and experience to predict how people react in similar situations
- Inferences concerning their intentions from what they actually do

Social awareness involves perceptions similar to those experienced by others (Prigatano & Schachter, 1991). The clinician's inferences concerning someone else (in this instant, a patient being studied) assumes that we are ascribing to someone's behavior that can be ascribed to oneself (Bisiach, 1992).

Affiliation, Social Attachment, and Social Behavior

Affiliation and social attachment (*bonding*) arise in infancy. As a complex brain function (Panksepp et al., 1997), it is vulnerable to trauma. Complex social behavior has been attributed to the increased size of the neocortex compared with the rest of the forebrain, and interestingly, to the influence of the mother, including neural and hormonal mechanisms of care (Keverne et al., 1997). The injured person must adjust to *societal demands*: supportive structures, the availability of jobs of suitable skills and personal demands, and perhaps ability to learn academic and social skills. *Social skills* (incorporating communication) includes self-knowledge, attention to personal appearance, social cognition (perception, knowledge, decision making), communication skills, and the social environment) (Coelho, 2007). Return to work varies with many personal qualities, including motivation, anxiety, residual functions, presence of pain, scarring, muscular and orthopedic injuries, availability of social support, transportation, and so forth.

17 Children's Brain Trauma

OVERVIEW

IMPRECISION

The neurobehavioral characteristics of a traumatic brain injury (TBI) occurring during childhood are both more complex and obscure than a similar accident during adult years. The assessment of children is conceptually more complex than that of adults. Imprecision exists at the time of the initial examination, since the younger the child, the less information is available to assess a baseline. At the time of injury, an apparent area of impairment may reflect injury, a constitutional deficiency, or a reduced measurement of a function that might have matured later. Particularly with young children, the preinjury baseline is unknown; therefore, the level of injury cannot be assessed acutely or later. Since many disorders are "negative," they cannot be known until the child does not (in the distant future) or has not reached developmental markers of anatomical, physiological, or mental functions. There is a potential for profound disorders of physiological development, which may not appear until the age of expected puberty, and then will not be recognized as having an etiology consequent to trauma to the central nervous system (CNS) and/or pituitary gland. Therefore, the examiner has difficulty in determining whether the initial or later assessment reflects no loss, traumatic loss, or previous developmental deficits. There are also functions that are expressed precociously (such as physiological sexual development).

TBI is often not recognized immediately, or the parent is told that there is no disorder currently or to be anticipated. Frequently, there is no current examination, and if the accident occurs during a nonschool period, it is not known to the academic staff. Even if the traumatic effect of the accident is recognized, the issue of later expression of dysfunctioning is not known to the practitioner or conveyed to the caregiver. Moreover, older individuals who as a child had a TBI probably were treated by practitioners who believed that the pediatric TBI victim was more likely to make a better recovery than an adult. Thus, late-developing symptoms probably did not receive an acute-phase alert, and would not be attributed to the earlier injury. Children's accidents are less likely to be reported or described accurately by the child or by the family. If at some time there are cognitive, physiological, or behavioral disorders, accurate attribution is unlikely (see "Nonrecognition of Children's TBI").

Unfortunately, according to my experience, schools frequently do not take appropriate acute remedial action or prescribe remediation when an accident occurs during the active school season. Should an accident occur during the vacation period and the child arrives in an injured or impaired state, appropriate assessment with modified academic programming is rare. As years and decades elapse, the person frequently experiences neuropsychological impairment and other multiple impairments without recognizing their source or having compensatory coping skills. Frustration and a life of failure and social incompetence may ensue.

EPIDEMIOLOGY

Injuries are the largest cause of morbidity and mortality among children in the United States. One estimate is that more than 300,000 children ages up to age 15 are injured in motor vehicle accidents (MVAs) annually (Allen et al., 2008). TBI is the most frequent cause of children's and adolescents' disability and death in the United States. One-third of all pediatric injuries are brain injury. An estimated half million to 1 million children sustain TBI each year, and 300,000 have permanent disabilities as a result of the brain injury. Traffic crashes are the leading health threat to children in the United States, that is, nearly 1 million injuries annually (De Vries et al., 1999; Patrick & Hostler, 2002; Taylor et al., 2008).

Injuries are reported to result in more deaths in the United States and the United Kingdom than all diseases combined: 40% of deaths among children aged 1–4 years, and almost 70% of children and adolescents 5–19 years of age. Fifty percent of injury-related deaths are due to MVAs. Fall-related and assault-related injuries were higher for infants than 1-year-olds (Keppel-Bensen et al., 2002). In the first 2 years of life, falls and assaults account for the majority of hospitalizations. Falls are the leading case of TBI-related hospitalization and were classified as a fall between levels or from furniture. Short falls (4 ft or less) often result in simple skull fractures but not serious intracranial injury, which was reported in 40% of these cases. This may be an example of underestimating later outcome. Assault-related cases frequently cause intracranial injury. Intracranial injuries were diagnosed in approximately 38% of fall-related cases and 83% of assault-related cases (Eisele et al., 2006).

Risk Factors for TBI

Demographic: Poverty; single-parent households; congested living conditions; parental history of psychiatric disorder, drug and alcohol abuse, and physical illness; unsupervised play; cognitive problems; behavioral disorder; and, school learning disorder (Arffa, 1998).

Behavioral: Behavioral disturbance increases the risk of head injury. Therefore, in cases of mild TBI (mTBI), to make a secure attribution of behavioral disturbance, preinjury functioning must be taken into account (Yeates, 2000). In the immediate postinjury period, there is a greater risk for further head injury, which does not reflect a permanent accident-prone personality (Gronwall et al., 1997). The vulnerable child is the one who is prone to risk taking. Adolescents' riskier behavior compared to adults has been attributed to an immature prefrontal cortex (Knoch & Fehr, 2007). This may be exaggerated after a head injury in which foresight, judgment, and insight are impaired. Return of a vulnerable child who has incurred TBI to a disorganized family increases the risk of further head injury.

Contrasting Child and Adult TBI

Ordinary development, that is, uninjured or unimpaired, has been described as sequential and reflecting the somatosensory experiences of the first few years, that is, "use dependent." Capabilities are developed that are suitable to the environment of development but within the child's genetic potential. In contrast, TBI may grossly impair the development of a child. Whereas trauma in an adult modifies the state of organization or stability (called here homeostasis), (emotional) trauma in a child may be the original organizing experience. Up to later development, the child cannot fight or flee. Some posttraumatic events, whether comfortable or destructive to the young child, may have a different effect upon an adult (Perry & Pollard, 1998). The pattern of cognitive functioning following head injury provides no precise guide to the locus of the brain lesion in the individual child. The outcome of a child's TBI will be determined both by the brain's state of development and concurrent developmental and medical problems. For children, cerebral maturation, more than chronological age, is critical for recovery. Injury in early life has impairing consequences. The plastic potential of the younger brain need not prevent its compromise by concussive injury (Prins & Hovda, 2003). Differences exist in patterns of recovery and compensation between adults and the developing child. This review of children's TBI addresses primarily lesser brain injuries in which there is not conspicuous radiological or other evidence of lesions. Since the literature offers comparisons between lesser, mild, moderate, and severe injuries, some information from more serious injuries is included when it illustrates conceptual issues significant for lesser injuries.

Although one may generalize that injury severity is closely related to recovery, and this association may diminish with time (Anderson et al., 2005a), prediction of outcome is complex. Numerous parameters influence it. A trauma occurring before maturity may not be immediately apparent because undeveloped aspects of brain function are not challenged by tasks. Only when a developmental schedule is not met does it become apparent that permanent compromise of functioning has occurred. Such children need services reintegrating them into the school or community, since such levels of TBI may create significant impairment immediately or years later. There may be loss of school attendance due to somatic injuries. Lack of lateralization, or of gross sensorimotor deficits, does not exclude the possibility of brain trauma. The outcome of more significant brain trauma is discussed in Parker (1990, Chapter 19).

A normal child grows up with the ability to think, solve problems, learn, enjoy life, and get along with others in the family and community. Intellectual standing relative to others is maintained over the years. Nonintellectual functions such as feelings, motivation, and concentration maintain goal-directed behavior. Psychomotor functioning is well developed and contributes to skills and social acceptability. Self-esteem provides an impetus for learning and social integration. Development involves both losing irrelevant old functions to increase control over cognitive and sensory and motor functions (Dennis, 1988), and developing new learning.

The complexity of development must be considered in predicting and estimating outcome. In normal development, functions evolve and develop in an *integrated and/or synchronous* fashion, that is, there is an "orchestration" of development. Therefore, delayed development in one function can have a cumulative effect on many subsequently developing functions. For example, what appears to be a relatively unitary function, for example, verbal ability, becomes appreciably more complex between the ages of 5 and 15 (Crockett et al., 1981), that is, many aspects of verbal and nonverbal ability are utilized in language and communication.

Educability is a central issue in a child's development. Impaired components include general intellectual ability, memory, attention, problem solving, inconsistent learning profiles, and disorders of executive function (EF) (organizing, planning, monitoring behavior).

THE MULTIPLE CONSEQUENCES OF A CONCUSSIVE ACCIDENT

Clinical Example: Expressive deficits concealing a child's impairment: The informant was the mother of a patient, a physician experienced in trauma medicine. The focus is upon the patient's reluctance to complain, vulnerability to medical disorders after trauma, and practitioner and school's neglect of the association between current neurobehavioral disorders and an earlier trauma.

The injury occurred at age 10; the writer's neuropsychological exam was when she was 20. Her WAIS-III IQ was almost 145. She had a free fall, falling on her back at a velocity of 20.5 miles/h or 30 ft/sec. Prior to this event, she was healthy, socially active, and academically successful. Subsequently, she had exacerbation of a latent Lyme disease, other medical conditions including susceptibility to contagious diseases and endocrine disorders, gross loss of stamina preventing a very intelligent young woman from completing her high school courses in the normal time so that she could not receive a diploma by the date of examination (despite continuous effort that did not overcome the narrow definition by the school of attendance rules). She received these diagnoses: *cognitive or personality change of nonpsychotic severity* (310.1); *postconcussion syndrome (PCS)* (310.2); rule out *partial epilepsy* (345.45).

Current medical condition: lesion of posterior pituitary gland; neuroborreliosis; single photon emission cardiac tomography (SPECT) scan revealed hypoperfusion of brain; fatigue; polycystic ovary syndrome; recent doctor recommended avoidance of full-time schooling due to the systemic effects of multiple infections; permanent musculoskeletal affects of trauma (chiropractic and neurological examinations); postconcussive syndrome; TBI.

Her mother asserts that she lost an inch in height, and is shorter than either parent. Medical conditions have required numerous injections so that access to the veins of her arm is spoiled, and in case of emergency, it would be a hazard if an emergency doctor did not know her condition.

Expressive deficits interfered with assessment. Her mother described her as always having been "a tough kid" who had glossed over or concealed subtle symptoms and signs over the years since her accident. Avoidance of anxiety-related expression of accident symptoms complicated the diagnosis for most of her physicians. They simply don't ask and she doesn't tell. Whether certain physicians didn't ask for fear of involving themselves in litigation or others were simply ignorant in the subtle expression of concussive brain trauma remains a question. Her mother commented: "There is no question that every history my daughter or I gave in the previous ten years included the fact of her accident, a 14 ft. impact fall onto carpeted cement causing a momentary loss of consciousness."

Whether her reduced expression of her experience stems from effects of TBI or her "tough kid" personality, it is clear that the damage to her lifestyle from the accident is profound. Only by constructing timelines—physiologic and academic/social—from the neuropsychological evaluation, did those changes resonate. She endured many incorrect diagnoses. A pediatric endocrinologist dismissed her head injury as "inconsequential because she did not have diabetes insipidus" and her depression as "dysthymic personality."

Joan's depression was never psychodynamic; rather, it was an endogenous depression from elevated posttrauma cortisol-releasing hormone (CRH) in addition to laboratory evidence of thyroid deficiency, insulin resistance, and elevated testosterone levels. These abnormalities were substantial enough to already have elicited significant weight gain posttrauma and excessive hair growth in a male pattern. He dismissed those complaints as "bulimia" even though Joan was not throwing up or trying to combat the weight gain. Joan's mother shielded her from the self-esteem issues of male-pattern hirsutism by paying for laser hair removal, regardless of the nondiagnosis in endocrinology.

Joan was not able in the early years posttrauma to recognize her reduced performance in school, but she was good at adapting. Joan asked to have her seat moved to the front of the class because she had new-onset difficulty in seeing the blackboard. Foresight and planning abilities, previously stellar, were recognized as diminished only by Joan's mother. This led to a temporary withdrawal from school and home-schooling in an attempt to insulate Joan from failing where she had always succeeded. She made adaptations by making lists and drawing maps. Home-schooling shielded Joan from the heartbreak of not being able to make the team for outrigger canoe ocean paddling. Prior to trauma, she was competitive and excelled. Posttrauma she came in second to last at tryouts, did not make the team, and sobbed about it only at home with her mother. She would never again try out for sports at school.

Joan has honed her adaptive skills well over these past 10 years. She avoids discussing situations where she may be unable to succeed or experience anxiety. For example, preinjury, where she had lists of book projects, after a tough fifth grade teacher so highly praised her writing skills (and told Joan she could be published and paid at that age for her writings), she now says she doesn't like to write and avoids any reference to her previous writing skills or proposed projects. In order not to experience anxiety over any reduced focus and/or organizational skills, Joan maintains honor-grade work by taking one class at a time to make up missing credits to gain her high school diploma. Her absences from school due to postinjury medical problems prevented her from achieving the 75% attendance requirement to obtain the grade. The teachers and deans at her school knew of her injury but maintained the strict attendance requirement. Joan walked with her kindergarten mates at graduation but received a blank diploma because she finished high school with a year and a half of missing (A grade) credits. She could not get her hard-earned As, but instead earned an incomplete because of the attendance rule. The only accommodation Joan got was in being allowed to walk with her class. Joan was described by teachers and deans from kindergarten onward as "hard-working, self-motivated, an asset to the school" and received the only accommodation afforded to anyone in 100 years in being allowed to participate in graduation without earning her diploma.

Joan shows adaptation in all aspects of her life. Because she is able to keep up in an aquatic physical therapy class, she talks about being an exercise instructor as opposed to going to medical school, which had been her goal previously (she was unable to complete an assignment due to lack of stamina). She focuses on the positive at all times. Aside from the athletic disappointment, the most painful aspect of Joan's injury experience has been her reduced social acceptability and exclusion from her circle of friends. Joan was simply too ill too many times throughout high school to attend most of the social functions. Friends did not understand and gradually stopped calling and coming around. Joan's once friend-filled social life, in and out of school, became limited to Internet friendships made during the down times at home. The neuropsychologist's description concerning her quality of life is appropriate: "To admit to the loss of a better condition of life or higher capacity to achieve is so painful that self-concealment is an easier way." Joan had been so used to "fitting in" that for a time she ignored genuine pain in an attempt to continue to fit in and be a good sport. Embarrassment and pride contributed to her concealment of impairment. Perhaps worst of all, the combination of all of the above has contributed to Joan's inability to assert her rights or be able to ask for what should come to her.

OVERVIEW OF CHILDREN'S TBI

There are a variety of expressions of pediatric TBI, including the entire range of the (adult) taxonomy of neurobehavioral disorders, as well as the specifically developmental consequences of co-morbid brain and somatic injuries. As with adults, significant disorders may develop despite normal neurological and radiological examinations. All examining and treating clinicians should be informed concerning the trauma history.

The consequences of an accident causing TBI in children include neuromotor impairment, seizures, trauma-related orthopedic injuries, lowered cognitive and academic skills relative to age expectations or preinjury estimates, problems in school performance, behavior, socialization, and adaptive functioning. Especially pronounced impairments are memory, perceptual abilities, psychomotor speed, attention, EF, and discourse processing.

Two disparate markers of late pediatric TBI occur whose traumatic etiology may not be recognized.

- Positive: Late developing symptoms
- Negative: Missing age-sensitive developmental functions (neurobehavioral and physiological)

The adult PCS condition is described as multifactorial and interacting. In addition to the immediate traumatic consequences, developmental conditions interact with prior adaptational difficulties, with emotional reactions expressing different experiences at later ages over an entire life span. Brenner et al. (2007) offer a revealing 50-year follow-up of a man who when 5 years old had a TBI but whose parents were not educated concerning the nature of his brain injury. The factorial structure of PCS in children has not been studied (Ayr et al., 2009). Cognitive, somatic, and emotional factors were obtained at baseline and 3-month interval (rated by both parents and children), while a fourth dimension representing behavioral ratings did not emerge at the 3-month interval. The highest loadings for the factors were as follows:

- *Cognitive*: Trouble sustaining attention; easily distracted; difficulty concentrating; problems remembering what the child is told; difficulty following directions.
- *Somatic*: Feels dizzy; feels that the room is spinning; headaches; nausea; blurred vision; low energy level (or) gets tired a lot.
- *Emotional*: Unable to accept change and coping with change; withdrawn (and) lacks interest in coping with others; difficulty expressing feelings; difficulty showing emotions; anxious (or) depressed.
- *Behavioral (low loadings)*: Physically aggressive; sassy; high activity level (negative loading).

Further studies are needed concerning the relative importance of subjective symptom reports and standardized cognitive testing, and the relative sensitivity to mTBI of the various factors. Both injury- and noninjury-related variables explain the outcome, with noninjury-related variables often accounting for relatively more variance.

The rate at which concussion leads to increased rates of persistent symptoms likely varies with its definition and the sample that is studied. Persistent postconcussive symptoms may reflect premorbid difficulties, the effects of injury, the patient's specific fears and expectations associated with cerebral trauma, and posttraumatic stress. Many postconcussive symptoms are not specific to TBI (e.g., headache, alterations of consciousness such as appearing stunned or dazed) and might be attributable to psychological or medical factors. Comparison of PCS children with those with OI showed that concussive injuries were more likely to offer trajectories that involved high levels of acute symptoms that either resolved or persisted (24% vs. 7%). They were less likely than OI to show trajectories that involved no increase in postconcussive symptoms relative to premorbid levels (64% vs. 79%). The experience of PCS lasting more than 2–3 months is relatively infrequent. Identification of such individuals enhances management, speeds recovery, reduces morbidity, and improves public health outcome (Kirkwood et al., 2008).

THE CHILD'S SOMATIC ANATOMY: HEAD, MUSCLES, AND PROPORTION

The patterns of head and brain movement differ between children and adults. As an overview, let us view a young child's differences in anatomical maturation compared to an adult (Cheng et al., 2005): (1) The bony cranial vault is soft and pliable; (2) The fontanelles are open; (3) The water content is high; (4) The bone and dura are very vascular; (5) The dura is not firmly attached to the inner table; (6) The subarachnoid spaces are generous and the blood vessels abundant; and (7) The head is proportionally large compared to the body. Through the second decade myelinization continues. The decrease of water content of cortex and white matter alters the biophysical properties of the brain and its response to injury (Weiner & Weinberg, 2000).

PROPORTIONS

The infant's characteristic *somatic anatomy* renders it vulnerable to TBI. The child's muscles, skull, and brain are undeveloped (Parker, 2001, p. 101). The smaller body size, as well as the brain's smaller size and stiffer tissue, contributes to patterns of TBI and behavioral outcome differing from adults (Levchakov et al., 2006). The disproportionately large head is poorly supported by undeveloped neck muscles. This increases brain damage for several reasons: (1) The unmyelinated brain is softer, permitting excessive stretching of both brain and vessels. Since the head rotates more postaccident in various directions, there is shearing separation of brain layers and structures at different distances from the geometric center. (2) The relatively small head undergoes a greater velocity change on impact with more massive objects, due to the principle of conservation of momentum.

THE IMMATURE SKULL

The structure of the neonate's skull, maturing only after several years, renders it vulnerable to impact trauma. The brain, skull, eyes, and ears develop earlier than other parts of the body. The face is relatively small, that is, one-eighth of the cranium compared to one-half in the adult. After birth, the skull thickens and continues ossification towards the sutures, which are only unified by fibrous tissue or cartilage. At birth there are large gaps between some bones (fontanelles) and unossified sutures (Standing, 2005, pp. 495; 485–487; photographs of the neonate skull, p. 485). The immature brain is encased in a skull that is designed to allow the head to pass through the relatively restricted birth canal and to enable the maturing brain to grow. A child's skull is thinner, more pliant, and has unfused suture lines, which permits stretching of the brain and its blood vessels by external forces. The newborn's skull is soft, pliable, and thin. The bone becomes hard and brittle at 1 year, when suture fusion begins, reducing some skull pliability and vulnerability to penetration and impact indentation. The orbital roof and floors of the anterior fossa and middle fossa are smooth and offer little resistance to the shifting brain during the first years of life. At birth, the sutures between the bones are not completely formed, that is, the brain is protected only by soft membranes and skin. The bones forming the sides and roof of the skull are united by membranes. Some of those at the base of the skull are united by cartilage. Certain skull bones are in several pieces (Lewis, 1936, p. 141); of greatest relevance are the frontal, occipital, temporal, and sphenoid bones. Gaps between bone are called fontanelles. These close at different ages, ranging from posterior at 2 months after birth, to 26 months in late cases for the anterior fontanel (Peacock, 1986). There are outlets for intracranial pressure, common in the head-injured child, since the skull of the newborn and older infant consists of several membranous bones with fontanels and unfused bony structures. Once the cranial sutures have fused, the foramen magnum provides the only major outlet to accommodate increases in intracranial pressure (Rosman, 1999).

Vascular Structures

The arteries, bridging veins, and location of the superior sagittal sinus beneath the sagittal suture of the newborn are particularly susceptible to shearing forces (Raimondi, 1986). While the dura adheres more loosely to the cranium in the infant than older individuals (Rosman et al., 1979), the dural sinuses of the neonate are very large in relation to the size of the skull and brain (Gluhbegovic & Williams, 1980, p. 4). This design permits lesions such as interhemispheric subdural hematomas and tentorial-dural sinus tears rarely seen in adults. Greater cerebral perfusion may explain the higher proportion of acute diffuse cerebral swelling in the child than in the adult.

Vulnerability to Trauma

The child may be unrestrained in a MVA and be ejected from the vehicle, or fall behind the front seat, in a collision. The infant's characteristic anatomy renders it vulnerable to TBI: The disproportionately large head is poorly supported by undeveloped neck muscles. This increases brain damage since the head rotates more in various directions, causing separation of brain layers and structures at different distances from the geometric center (see "Child Abuse"). Since the immature skull is soft, thin, and pliant due to unfused suture lines, it is more easily deformed by external trauma (Shapiro, 1987). The lack of skull firmness contributes to potential depressed skull fracture in utero and during delivery (see Yamamoto & Sato, 1986, for a complete description; Hynd & Willis, 1988, p. 41 for a diagram). The subarachnoid space is smaller, and the growing brain is closer to the dura mater than in the adult. This increases vulnerability and affects the nature of traumatic injury. A somewhat different interpretation of the physical vulnerability of young children is offered by Lishman (1987, p. 171). The pressure of the blow is better absorbed, vessels are less readily ruptured, and transient rises of intracranial pressure are accommodated. The skull is more easily deformed by external trauma than the rigid skull of older children and adults. The pliable sutures and fontanelles stretch at the calvaria, inducing excessive tearing forces at the attachment of vessels to rigid, fixed soft tissues (e.g., falx cerebri). The thinness of the skull of infants and younger children, especially over the temporal areas, along suture lines, and at the orbital roofs, increases the risk of *penetrating injuries* to the brain. This includes self-inflicted injuries with sharp objects used as playthings. A careful search for puncture wounds is needed (Weiner & Weinberg, 2000).

BRAIN DEVELOPMENT

The brain grows and specializes according to a precise genetic program. Positive influences are stimulation and experience. The genetic program is modified by negative influences that impair or delay brain development (malnutrition, maternal drug abuse, viral infection) (Thompson et al., 2005). The developmental patterns vary between different cortical regions (Giedd et al., 1999; Thompson et al., 2005). After trauma, varied developmental patterns have outcome implications: Injury may occur to particular brain functions that are developed or not, or may be localized, or participate in distributed circuits.

NERVE CELL DEVELOPMENT

During development, cell loss is the rule, and is opposed by *trophic factors* that signal cell survival changes. Cells proliferate and influence each through an interplay of signals that commits them to specific locations early in development before they begin to migrate. Axon targets are determined by both chemoattractants and chemorepellents. Initially, molecular cues guide the axons to their target location; then, neural activity refines these connections. Concurrent cell death taking place at this stage helps to sculpt connections (Wiencken-Barger & Casagrande, 2002). This cell loss is related more to organ formation (embryogenesis) than to homeostatic regulation of cell number. Target-derived signals regulate the number of neurons that survive. Once a developing neuron has grown a process into its target, it competes with other neurones of the same type for a limited supply of the neurotrophic factor supplied by the target. These neurotrophins act on peripheral and CNS neurons to support their growth, differentiation, survival, and plasticity in the developing nervous system (Arvin & Holtzman, 2003). An example is the effect, described as retrograde, of limb (wing buds) loss or enhancement, upon the development of sensory or motor spinal neurons. During development as many as one half or more of all cells in a particular cell population will die (Oppenheim & Johnson, 2003). Neurons project to more targets than in the adult, with the adult pattern of connections resulting from selective axon or collateral neuron elimination: In the cortex, the limited adult distributions to the contralateral cortex via the corpus callosum are due to the loss of callosal axons, not neurons; layer 5 neocortical neurons projecting more broadly and collaterally to a larger set of targets than adult life; neocortex to specific neuronal subsets in the brainstem and spinal cord. Layer 5 neurons retain only projections that are functionally appropriate (Burden et al., 2003).

CEREBRAL DEVELOPMENT

At 3 months, the dendritic density is greater in the right hemisphere than the left, and the density is greater in the oral motor area of the cortex than in Broca's area. At 6 months, there is a peak in the development of the inner layers of the cortex in the language areas. By 15 months, the hippocampus is fully mature, with implications for memory. At 24 months, the dendritic density increases in Broca's area to catch up with the oral motor cortex. Left-hemisphere dendritic density catches up with the right hemisphere. At 72 months, the dendritic density of Broca's area is now greater than that in the oral motor cortex (Kent, 1999). Overall brain growth is not significant between childhood and adolescence. However, there is postadolescent brain growth primarily in the dorsal aspects of the frontal lobe bilaterally, and in the posterior temporo-occipital junction bilaterally. There is an inverse relationship between cortical gray matter density reduction and brain growth primarily in the superior frontal regions that control executive cognitive functioning. This inverse relationship is not as robust in the posterior temporo-occipital junction despite late brain growth between adolescence and adulthood (Sowell et al., 2001).

Cerebral Cortex: Volume

Development of volumes of gray matter are nonlinear and regionally specific. The primary visual cortex reaches adult thickness between the fourth and fifth months; that of the frontal cortex does not reach adult levels until 7 years. The frontal cortex does not reach adult thickness until age 7. The visual association areas do not reach adult thickness until 10 years of age. The volume of frontal and parietal gray matter increases until 12 years of age, followed by a decline. Temporal lobe gray matter reaches a maximum volume at 16, then declines slightly (Kaas & Preuss, 2003). Volume links of specific areas vary with age: children (anterior cingulate); adolescents (orbitofrontal and medial prefrontal cortex; and older adults (lateral prefrontal cortex).

Cerebral Cortex: Trajectory of Growth

The thickness of the cortex of the entire cerebrum is a sensitive index of normal brain development, although, paradoxically, it thins and thickens over time. The relationship between cortical thickness and IQ changes with age. There is a change in the valence of the correlations between IQ and cortical thickness in childhood, in the prefrontal cortex, and the left superior/middle temporal gyri. Intelligence level is more related to the trajectory of change of thickness rather than the thickness directly. Those of superior intelligence had the most rapid rate of initial increase of cortical thickness followed by a more rapid cortical thinning. The change was low for those of high intelligence, and almost flat for children of average intelligence. Using cumulative measurements of age, there is a correlation described as "modest" between psychometric measures of intelligence and total brain volume. There were positive correlations throughout most of the frontal, parietal, and occipital cortices, and similarly modest negative correlations in the anterior temporal cortex. Yet when different age groups were considered, there were age-related changes: a predominantly negative correlation between IQ and cortical thickness in the early childhood group. This contrasted with later positive correlations peaking in late childhood and still present but attenuated in adolescence and early adulthood (Shaw et al., 2006).

White Matter

The volume of white matter increases in children throughout childhood and early adulthood. However, the effect of TBI on this trajectory is unknown. Increased myelination and organization functional adequacy (FA) in school-aged children. FA is related to cognitive skills in children, while the apparent diffusion coefficient (ADC) for cortical gray matter has been reported not to differ significantly between school-aged children and young adults. This suggests to the present writer that cognitive maturity is related more to white matter than gray matter development. *Diffusion tensor imaging (DTI)* can be used to estimate the maturation of the brain's white matter tracts, including characterization of white matter injury in brain regions affected by TBI in children, including working memory, response inhibition, reading ability, and speed of reaction time (Levin et al., 2008). TBI results in cortical thinning with an impact on cognition and emotional regulation in children as they mature. (Merkley et al., 2008).

Brain Development and Myelinization

Functional adequacy is dependent on development of the myelin sheath (myelinization) around the neuronal axon. The development of the brain (volume percentage) proceeds in a caudocranial direction. At birth, the medulla, pons, and spinal cord are more developed than the cerebellum. The cerebellum, the least developed compared to other brain and CNS structures, grows rapidly from just before birth through the first year of life. At this stage, complex coordination is not needed. Cortical synaptic connections (responding to experience) increase during infancy, reach a maximum of about 50% above the adult mean at 12–24 months, and remain at that level until about age 16, when they begin to decrease (*pruning*) to adult levels (Augustyn & Zuckerman, 1998). Corticospinal tracts are not myelinated until around one and a half years. Myelinization is mostly complete by the end of the second year, but continues until after age 20. (See detailed chart, Adams & Victor, 1989, p. 461, for age and areas of myelinization.) Myelinization of the nervous system begins in the second trimester and may continue into adolescence. Myelin exists in posterior frontal and parietal lobes at birth. The occipital lobes (geniculocalcarine system) myelinate soon after. Myelinization of the frontal and temporal lobes proceeds during the first year of postnatal life. Most of the myelinization of the cerebrum is completed by the end of the second year. The prefrontal cortex is relatively late to mature. Myelinization continues to the end of the second decade (Katzman & Pappius, 1973). Simultaneously, synaptic and dendritic changes occur. There are also more subtle changes in neuronal development, and biochemical alterations linked to maturation (Norton, 1972).

Brain Vulnerability

The unmyelinated brain is softer, permitting excessive stretching of both brain and vessels. The relatively greater volume of cerebrospinal fluid (CSF) in the ventricles and subarachnoid spaces shifts farther and faster during *whiplash*, increasing their stretching effect on the more-resistant brain parenchyma and blood vessel attachments. There is a suspected greater vulnerability to an impaired blood-brain barrier, and a higher proportion of water content of cortical and white matter (87–89% at birth, versus adult values of 83–89%). Further, an animal study indicated that nonimpact inertial head rotation resulted in widespread axonal damage to a greater extent than in brain-injured adult animals (Raghupathi & Huh, 2007).

Brain Weight

The brain is the fastest growing organ, with a metabolic rate double that of an adult. The newborn's brain is one-fifth of its adult volume. Brain volume is 90% of its final adult volume by age 6. Girls' brains are, on the average, approximately 12% smaller than boys, which is explained largely by differences in height with large intersubject variability, and not by gender differences in cognitive performance. Amygdala volume increases more in males than females, while hippocampal volume is greater in females. Though body weight in the newborn is only 5% that of the adult, the brain comprises 15% of total body weight, compared to 3% in the adult. At birth, the brain weighs 350–400 g, 24% of the adult value. It reaches 1,000 g at 1 year and 75% of adult brain weight at the end of the second year of life. During the first 2 years of life there is approximately a 350% increase, followed by a 35% increase during the next 10 years; over 90% of adult brain weight is reached by the sixth year. The brain increases in size during the first year through increasing size and branching of neural processes, increasing glial cells, and growth of myelin. Although brain weight is 95% complete by age 10, bodily weight is only 50% complete (Peacock, 1986). By age 14 the brain has reached adult weight. Brain growth continues at a slow pace until age 12–15, when there is an average weight of 1,230–1,275 g in females and 1,350–1,410 g in males (Adams & Victor, 1985, pp. 420–422; Katzman & Pappius, 1973; Livingston, 1985, p. 1169; Martin, 1985; Wetherby, 1985; Willis & Widerstrom, 1986).

Ratio of Cortical to White Matter

The volume of white matter increases throughout childhood and extends to early adulthood. However, the effect of TBI on this trajectory is unknown (Levin et al., 2008). Growth of cerebral white matter lags behind cortical development and continues postnatally, long after the cerebral gray matter has reached its definitive volume. Myelinization proceeds rapidly to age 2, then slows markedly, although fibers to the association areas continue to myelinate until the third or fourth decade. Completion of myelinization is in the order of brainstem, cerebellum and basal ganglia, and then cerebral hemispheres. Sensory and motor systems display mature myelinization in the first 2 years of life; nonspecific thalamic radiations mature at 5 years; and intracortical fibers myelinate until the third decade (Rapp & Bachevelier, 2003). There are increases in white matter (axons or long neuronal connections between brain structures) across ages 4 through 20, and a preadolescent increase of cortical gray matter (masses of neurons) followed by a regionally specific postadolescent decrease. The corpus callosum seems not to complete its myelinization until puberty, reaching its maximum size at approximately age 25 (Baynes, 2002). The frontal and parietal lobes peak at about age 12, the temporal lobe at about age 16, and occipital lobe at age 20. The frontal and parietal gray matter peaks approximately 1 year earlier in females, corresponding with the earlier onset of puberty. The frontal and parietal thinning was associated with higher vocabulary scores, perhaps related to the higher verbal performance often found in females. This may indicate an influence of gonadal hormones. Further implications for neurobehavioral functioning are indicated by the parallel between the developmental pattern and cerebral glucose metabolism and slow-wave sleep amplitude.

Adult Neurogenesis

Brain development continues through life in various parts of the CNS. It represents structural plasticity in the mature CNS environment, and may participate in learning and memory. Ordinarily, it occurs in the subventricular zone (SVC) of the lateral ventricle, the subgranular zone (SGZ) of the dentate gyrus in the hippocampus, and the olfactory bulb. However, after trauma it has been experimentally demonstrated in the SVC and SGC, and sometimes causes migration of new neurons to injury sites. Neurogenesis may also occur posttraumatically in regions considered to be nonneurogenic. It is not yet known whether new neurons become functionally integrated (Ming & Song, 2005). There is an increasing complexity of fiber systems through late childhood and adolescence, and perhaps even into middle adult life. Until around age 20 (and perhaps later), the brain continues to mature anatomically and/or physiologically. The effect of trauma on these processes and the resumption of this orderly progression after physical injury are unknown.

A DESCRIPTION OF CHILDHOOD BRAIN TRAUMA

As with injuries occurring in other ages, it is important to try to reconstruct the particular trauma. Head injuries in children 24 months and younger differ from those of older children, that is, falls of a short distance may cause simple skull fractures, (supposedly) "not of great clinical consequence." *Linear skull fractures*, particularly temporoparietal, are especially frequent in children less than 2 years. They may be overly serious intracranial pathology such as epidural hemorrhage. When found in the infant or young child, these fractures suggest the possibility of neglect or inflicted injury. Depressed skull fractures may cause contusion or laceration by the fragments (see "Seizures"). Basal fractures (frontal and petrous) have similar symptoms to adults: hemorrhage around the mastoid bone or eye, nose, nasopharynx, middle ear, CSF otorrhea. Accompanying neurological trauma includes damage to cranial nerves I, VII, and VIII; conductive hearing loss; and facial palsy. Permanent sensorineural hearing loss and facial palsy that has poor recovery follows transverse rather than longitudinal petrous fractures (Rosman, 1999). The child's brain is more subject to traumatic distortion than an adult's. In children, the skull's higher elasticity, with open fontanelles and sutures, and poor myelinization contribute to greater plasticity (physical), and predisposition to cerebral tears (DiRocco & Velardi, 1986; Lindenberg & Freytag, 1969; Lockman, 1989). The trauma has been described as follows: The impact flattens or indents the skull, deforming the brain and shifting it along the line of impact depression; it occupies space and thus presses brain convolutions against the dura. Impact-caused brain deformation generates shearing and tearing forces in subcortical white matter causing splits in various directions. Tissue tears are more common in the poorly myelinated infant brain, most often in the cerebral white matter. Their presence primarily at temporal and orbital, and sometimes first and second frontal convolutions, at a distance from the impact, were likened to gliding or contracoup injuries, although coup contusions do occur.

Less myelinization also contributes to infants' greater plasticity of cerebral hemispheres (Zimmerman & Bilaniuk, 1981). Membrane damage and reduced Glasgow Coma Score (GCS) is associated with reduced cognitive skills in children (reading and working memory) (Levin et al., 2008). One year of age is considered the demarcation between both open and closed fontanelles and a higher incidence of skull fracture with greater morbidity and mortality. Whether causative factors are the primary or secondary effects of the impact are not known. The open fontanelles and sutures, and particularly the meningeal structure, predisposes the bridging cortical veins to tearing and for blood to accumulate along the vertex, or to form a significantly large clot volume. Skull malleability permits more severe compression and distortion injury to the brain (Raimondi & Hirschauer, 1986). While open sutures and a thin calvarium are stated to offer the advantage of a more flexible skull capable of absorbing a greater impact (Zimmerman & Bilaniuk, 1981), it may be that this merely changes the geometry of the force, that is, energy and distortion are absorbed directly into the brain, rather than the impact causing the head and enclosed brain to accelerate and decelerate, that is, whiplash or hyperextension followed by hyperreflexion). In one series of children with skull fracture, 69% had positive computed tomography (CT) findings, for example, subarachnoid hemorrhage (SAH), contusion, etc. In an infant, herniation of the arachnoid membrane into the margin of a fracture permits CSF pulsations to be transmitted to the fracture edges retarding osseous repair, with a growing fracture caused by the development of a leptomeningeal cyst.

A study of pediatric closed head injury (CHI) was made to assess the depth of brain lesion in relation to acute severity, and to assess long-term outcome. Late magnetic resonance imaging (MRI) revealed focal brain lesions in 55% of children and adolescents who sustained CHI of varying severity. It was found in 55% of moderate injuries and 81% of severe injuries. Head injury producing brief loss of consciousness (LOC) and a GCS of 13–15 resulted in normal MRI findings in 69% of the patients at 3 months or longer postinjury. The age of injury was not related to the MRI findings. High-velocity accidents were more likely to result in deep areas of abnormal signal. The severity of impaired consciousness was predictive of the depth of the abnormal signal. This was consistent with 3-month postinjury findings of gliosis, hemosiderin, and encephalomalacia. The depth of the lesion was inversely related to the outcome (Glasgow Outcome Scale; Vineland Adaptive Behavior Scale). Total lesion volume did not explain the depth of the lesion effect. There was a rostral–caudal gradient of the frequency of brain lesions. The frontal lobe white matter was the most frequent lesion site. The posterior region of the corpus callosum more frequently sustained focal lesions and atrophy than the anterior region. The central gray brainstem area sustained infrequent lesions relative to subcortical lesions (Levin et al., 1997b).

Anterior lesions are most likely to trigger epileptic attacks, and other predictable dysfunctions occur from cortical damage. Tears might extend to the cortex or ventricles. Bleeding into their cavity or borders is not significantly space-occupying, and there tends not to be secondary tissue necrosis. Contusion hemorrhages involve the cortex primarily, and might be within horizontal cortical tears at the crest of a hemispheral convexity. Children tend to have contralateral contusions after a blow and extensive tears beneath the site of a fall, a reversal of the adult trend (Lindenberg & Freytag, 1969). There are regional differences in the anterior brain for posttraumatic *metabolite changes*. A group of children with TBI between ages 3 and 6 had reduced metabolite levels (left frontal white matter; medial frontal gray matter) that correlated with GCS scores 1–3 years postinjury. More severe injury was related to lower metabolite concentrations 1–3 years postinjury. Some neurobehavioral measures are more correlated with brain metabolite concentrations than others, and these correlations differ between the TBI and orthopedic injury groups. For the TBI group, the level of particular metabolites was related to various academic outcomes and social competence (Walz et al., 2008).

Loss of Consciousness

Loss of consciousness of more than 1 h for infants and newborns indicates *severe head injury*, contrasted with 24 h for school-aged children. Outcome involves diminished intelligence and ability to succeed in school, disorders of visual motor function, emotion, and behavior. Late complications include posttraumatic epilepsy (PTE), hydrocephalus, and growing skull fracture. Immediate seizures are less frequent in infants than older children, but PTE is more frequent. PTE incidence is 5–10% in infants, and 25–30% in children. After severe head injury, it may require anywhere from 3 months to 3 years for a child to reach clear consciousness or a constant condition. Fatigue, reduced memory and thinking contribute to apathy and temper. Later, the child takes an interest in his surroundings and starts to play, but may offer unrestrained behavior (motor restlessness, affect, and uncontrolled action). Inability to handle environmental situations leads to regression of autonomic, motor, and behavioral functions (Kaiser et al., 1986). Length of coma is not established as a prognostic sign for language dysfunctions (Ozanne & Murdoch, 1990).

There is some association between cognitive impairment and coma for longer than 24 h, or posttraumatic amnesia (PTA) for 2 weeks or more. Severe CHI results in cognitive deficit, whereas mild CHI is not associated with a generalized cognitive deficit when examined at least l year postinjury. The findings with the Mental Processing Composite for the Kaufman Assessment Battery for Children were 71.4 versus 94.7 (Asarnow et al., 1991). It has been estimated that improvement may take place over 2–5 years, longer in the case of the more severely injured child. The examiner should avoid the mistake of attributing all deficits to a trauma (Lehr, 1990, p. 101). There is also a larger loss of mental ability in the interval from "moderate" to "severe" than "mild" to "moderate" brain damage, when the criterion is some combination of GCS, documented brain damage (contusion or laceration), and period of unconsciousness (<20 min; mean = 2.6 days; 11–77 days): Full Scale IQs of 95.7; 93.6; 82.1. There was a greater discrepancy of Performance Scale IQ between mild and severe cases (19.3) than for Verbal Scale IQ (5.7) (Bawden et al., 1985).

Seizures

Immediate seizures occurring within minutes of trauma are extremely common in children and infants after TBI. The incidence of early posttraumatic seizures is higher than expected for adults with comparable injury, and is estimated at 3–9%. The majority occurs in children less than 3 years of age (McLean et al., 1995).

PTE is more likely with depressed than linear skull fractures with a history of contusion or laceration. In this traumatic pattern, PTE is more likely when the LOC is more than 1 h and PTA lasts more than 24 h. Vulnerability is increased by damage to the mediotemporal, posterofrontal, and anteroparietal areas (Rosman, 1999). The incidence of early and late seizures in children is correlated with injury severity, for example, 35% with GCS of 8, 27% with GCS 9–12, and 6% with GCS 13. Children under 5 are more likely than adults to develop early seizures after mild brain damage, even after relatively mild brain injury (without LOC or depressed skull fracture). In contrast, children are less likely to develop late seizures. Ninety-five percent of one group developed seizures within 24 h, and 5% more than 1 week later (Dalmady-Israel & Zasler, 1992).

Children with epilepsy (older than 5) have a ten-fold heightened vulnerability to *submersion and drowning* in bathtubs, pools, or other bodies of water, even when additional handicaps are considered. This is related to greater activity and less supervision. Subtherapeutic anticonvulsant levels may be responsible for some submersion accidents, but children with poor seizure control probably are at greater risk (Diekema et al., 1993).

SPECIAL CHARACTERISTICS OF CHILDREN'S TBI

Biological Influences upon Outcome

The outcome of a child's and adolescent's TBI is substantially different from that of an adult, due to developmental differences of both anatomy and brain development. Brain injury implies some pattern of disturbed or impaired function, varying with the location and extent of the trauma. Developmental patterns vary over the adult lifespan. TBI is the leading cause of death in childhood (due to motor vehicle crashes, falls, and sports-related injuries). The rate of TBI death is stated to be to be five times the death rate of childhood leukemia (ages 1–15). Using an animal model, it has been established that impact, acceleration, and shaking are likely to produce higher levels of brain tissue stresses and strains in immature organisms due to the smallness of brain size and skull thickness across ages, and also brain tissue mechanical properties. The significantly stiffer brain tissue of the neonate creates elevated mechanical stress and strains when subjected to impact forces than more mature brains under the same loading conditions. Thus, there is greater vulnerability to TBI (Levchakov et al., 2006). Animal evidence indicates that during a crucial developmental phase until puberty, both sexes exhibit enhanced catecholaminergic and serotonergic activity (Knoll & Miklya, 1995).

The pattern of TBI is always influenced by the interaction of mechanical forces with the structure of both brain and soma. For young children, one considers that outcome after trauma is affected by weaker neck muscles; undeveloped skull and brain (Parker, 2001, p. 101); having less acquired information, skills, and personality resources to cope with adversity; smaller body size; and stiffer brain tissue (Levchakov et al., 2006). These contribute to different patterns of alterations of consciousness, primary TBI, and outcome. Subsequent developmental consequences are presented later: delayed or absent puberty (Parker, 1997); cognitive and behavioral milestones; premature achievement of developmental milestones; alterations in the level and pattern of cognitive development; and alterations in the level and pattern of emotional and personal development (maturity, identity, impulse control). Thus, it is difficult to assess the initial level of TBI in children and the prognosis. Only the passage of years may indicate the significance of the initial injury. Outcome may be different for a given level of injury. *Premorbid functioning* should be assessed to estimate more precisely the extent to which the TBI has produced neuropsychological deficits. Children's TBI may take years to be expressed, that is, as a dysfunction of development rather than observed as an immediate traumatic dysfunction. Progress should be tracked, since dysfunctions and deficits may be expressed later. Outcome is a balance between development, recovery, and late-developing neurological and stress-related dysfunctioning and deficits. The threshold for bringing a child to medical attention is lower, while head injury caused by abuse is probably underreported (Cerhan et al., 1996). Records are often incomplete and incorrect. Further, children can conceal or not report head injuries with LOC, and the same injury seems less likely to cause LOC in children than in adults. Brain injury has more significant consequences for younger children, and the sequelae of childhood brain injuries remain relatively constant over time or worsen (Taylor & Alden, 1997). Children's outcome deficiencies after head injury are multiple: physiological disorders of development; cognitive; personality and behavioral; sensorimotor, etc. These are conditioned by their developmental level, neurologically, osteologically, somatic structures, cognitive and personality level, etc.

Brain Trauma and Development

The effect of a lesion depends upon such factors as the level of development, extent, and whether it involves sites that are "hard-wired" maturing early (e.g., sensorimotor and physiological) or maturing slowly, and participation in complex functions integrating large brain areas, including the opposite hemisphere (e.g., cognition, memory, personality). Frontal lobe functions exhibit *multistage development.* Since TBI in young children impairs future development, it may not be detected until later maturational milestones are not expressed. More demanding tests may pick up developmental changes after that age (Llamas & Diamond, 1991). Children do not succeed on adult tests of frontal cortex function until ages 8–12. Measurement with adult procedures before maturity may tap different functions in the child. This is related to the principle that functions mature at different ages, for example, inhibitory control, goal direction, verbal and motor inhibition, conceptual flexibility, components of memory, external or internal mediation of self-control, and abstract thinking. Therefore, measurement at a given age may reflect different degrees of anticipated impairment of future development (Chelune & Baer, 1986; Passler et al., 1985; Welsh & Pennington, 1988).

The outcome of a given TBI will be influenced by the level of development of neurobehavioral and social functions, what has been learned, and the degree of consolidation of skills and functions at the time of injury. To the extent that a function is localized, and this author (Parker) believes that parallel processing and extended circuits should be assumed at all times, then a given lesion might prevent the development of skills that have not yet been expressed. The deficit is occult and postponed. A generalization has been offered that generalized brain injury is associated with more severe consequences for younger children, particularly for skills in a rapid state of development as opposed to well-established skills. The rate of development of more complex cognitive abilities may proceed at a slower rate (Ewing-Cobbs et al., 1997). Younger children had poorer outcome in a cognitive study, particularly in the domain of memory (Anderson et al., 1997).

Physiological Development

In normal development, *sex steroids* and the *glucocorticoids* (*GC*) are implicated. A burst of androgen secretion at birth programs the size, connection, and neurochemistry of specific hypothalamic nuclei, some sexual behaviors, and steroid-metabolizing enzyme expression in the liver. Estrogens also exert organizational effects on the developing CNS. GC accelerate organ maturation. Exposure in utero has acute effects upon neuronal structure and synapse formation, and may permanently alter brain structure. Experimental administration retards brain weight at birth. Antenatal dexamethasone causes degeneration of hippocampal neurones with persistent postnatal reduction of hippocampal volume. GC are important for normal maturation of most regions of the developing CNS. Used as medication, GC appear to reduce *birth weight*, which is associated with metabolic and cardiovascular disorders later in life (hypertension; insulin resistance; type II diabetes; cardiovascular disease deaths). There are differences in levels of cortisol depending upon sex and the length of exposure.

Maternal Stress During Pregnancy

Brain injury and maternal stress alter the brain and somatic internal milieu with potential consequences for sexual behavior. Exposure to gonadal hormones influences *sex-specific behavior* in adulthood, and *thyroid deficiency* in postnatal life severely impairs brain growth and development. Animal studies indicate that exposure to stressful circumstances (e.g., GC) reduces birth weight, long-lasting changes in immune response, altered neuroimmune responses and behavior (attentional defects, anxiety, social behavior, startle responses, increased alcohol preference, difficulties in spatial learning) cardiovascular disorders, etc. (Harris et al., 2005). *Hormonal cross-talk* between fetus and mother can affect sexual differentiation. Male fetuses of pregnant rats exposed to stress have lower-than-normal levels of fetal androgen and adult copulatory behavior, due to maternally secreted opiates damping fetal gonadotropin-releasing hormone (GnRH) release and therefore testosterone secretion. Whether sexual structural differences (dimorphism) arise from early steroid exposure and permanently alter the structure of the brain and behavior is somewhat controversial since late virilization can change female gender identity.

Prenatal exposure to *GC* (exogenous, maternal, or its own adrenal products) may experience CNS and cardiometabolic effects. Prenatal maternal stress seems associated with delayed motor development, reduced adaptive behavior, and impaired attention or learning ability, orientation, state regulation, robustness, and endurance. Stress during pregnancy is a predictor for gestational age, birth weight, neurological condition of the infant, neonatal behavior, delayed motor development, reduced adaptive behavior, and impaired attention or learning ability (Rieger et al., 2004).

Early Stress

One pole of the development of a flexible and adaptable stress response is a responsible and predictable caregiver. The combination of being able to explore the world and having a stable base to return to when overwhelmed contributes to resilience in coping with future trauma. In contrast, chaotic or threatening caregiving prepares for a sensitized stress-response system affecting arousal, emotional regulation, behavioral reactivity, and cardiovascular regulation (Perry & Pollard, 1998). Thus, a mother or family unit that has experienced behavioral and emotional disorganization due to developmental problems or trauma prepares the child for low stress-resistance. Early prenatal and environmental adversity alters the development of individual behavioral and endocrine responses to stress and threat. Variations in parent–offspring interaction seem to serve as a forecast of environmental quality for the child. If so, defensive responses to stress are neither innate nor invariant. Rather, they function as a preparation for environmental conditions ultimately faced after independence from the parent. Even subtle differences in parental behaviors, not necessarily gross stress, forecast the environmental conditions that the child is being prepared for. It is inferred that sustained parental influence, even within the normal range, influence behavioral expression (phenotype) apart from the basic genotype. It is assumed that enhanced stress reactivity has adaptive value. If environmental adversity influences parental fearfulness and stress reactivity, it alters parental care, commonly causing a decreased level of investment. The effect on the child has been termed a *pessimistic pattern* of development, that is, an enhanced stress reactivity that anticipates a high level of environmental adversity. The clinician may conclude that developmental preparation for stress may predispose to a larger-than-adaptive response, for example, posttraumatic stress disorder (PTSD). Research indicates that a cold, distant relationship with a parent leads to these stress reactions: increased GC and cardiovascular reactions and evidence of increased dopamine release in the ventral striatum. Victims of early abuse may experience adaptive and endocrine responses to stress. An additional health cost of persistent stress reactions may also be vulnerability to stress-related health disorders (Zhang et al., 2004).

Early experiences can influence the cognitive outcome. The developing brain's neural plasticity (nerve branches and synapses) permits stress to permanently affect systems such as learning (Kaye & Lightman, 2005). Maltreated children with PTSD have smaller intracranial volumes (the sum of brain tissue and CSF volume). The hypothalamic-pituitary axis (HPA) is stress-reactive from infancy with maternal stress, depression, and low *socioeconomic status (SES)* associated with basal cortisol elevations in young children. Elevated circulating GC hormones suppress bone formation and skeletal growth in infancy. Prenatal maternal ingestion of GC drugs and maternal psychological stress may predispose the child to vulnerability to PTSD. They are associated with smaller neonatal head circumference. This mechanism need not result in smaller adult stature or smaller cerebral tissue volume, because long bones and brain continue to grow into the second decade and may manifest "catch-up" growth denied the skull (Woodward et al. 2007).

Genetic/Environmental Interactions

The overt expression of genetic-environmental interaction (phenotype) is determined in part by the environment during pregnancy. The medical and emotional condition of the mother affects the developmental well-being of the child. The model of evolutionary biology is more comprehensive than the concept that adversity in early life alters the development of neural and endocrine systems to predispose to disease in adulthood. The relationship between the quality of the early environment and health in adulthood seems mediated by parental influences on the development of neural systems underlying the expression of behavioral and endocrine responses to stress. Not all individuals develop PTSD after significant trauma; vulnerability to stress seems influenced by early life events. Paradoxical to the usual assumption that a beneficial or benign early environment presumably has the optimal developmental outcome is the rigorous evolutionary concept that the only criterion worthy of consideration is *fitness*, that is, the extent to which conditions influence reproduction. Therefore, *traits that enhance survival* are favored.

Habituation and desensitization: Acute stress may require up to several hours for the HPA hormones to almost completely recover resting levels, depending upon the intensity of the stressors and on whether adrenocorticotrophic hormone (ACTH) or corticosterone (animals) is being measured. In contrast, a single exposure to a severe stressor can cause a long-term desensitization of the HPA to similar types of stresses (Amario et al., 2004). Chronic intermittent exposure to a stressor of low or moderate intensity, within the context of familiarity, leads to minimal activation of cardiovascular and metabolic homeostasis, providing for significant conservation of energy expenditure.

Dishabituation: This is the enhancement of a physiological response to a novel stressor in animals that had been exposed repeatedly or continuously to an unrelated stressor. If presented with an unfamiliar stressor, there is a much greater behavioral and physiological challenge than presentation of the same stressor to a naive (unstressed) control. The plasma catecholamine response is amplified compared to naive controls exposed to a similar stress.

Prior exposure to a stressor affects the subsequent secretory response (e.g., the adrenal medulla). After several weeks of exposure to the same stressor, the response is reduced significantly. If the same animals are exposed to a novel stressor, the adrenal medullary response is significantly greater than that of a first-time stressed control group. Stressed rats, at least, express this pattern: The adrenal medulla exhibits a significant, long-lasting enhancement of catecholamine synthetic capacity. However, habituation results in significantly less epinephrine (EPI) release with daily stress exposure. However, when a repeatedly stressed animal is exposed to a novel stressor, EPI release is higher than that of a first-time exposed stressed control. *Other locations* for stress-induced increases in catecholamine synthetic enzymes occur in various sympathetic ganglia, several hypothalamic nuclei, and noradrenergic cell beds of the N. locus ceruleus (Kvetnansky & McCarty, 2000). Several hormonal systems follow a pattern of sensitization and tolerance. Sensitization follows exposure to high-intensity stimuli, which create recurring interruptions in cardiovascular and metabolic processes. The organism has a higher plasma level of norepinephrine and epinephrine than animals exposed to the stress for the first time. In contrast is *densitization (tachyphylaxis)* or the diminished response to adrenergic agonists engendered by prior exposure to catecholamines (Landsberg & Young, 1992). GC also participate in adaptation to stress (Sorg & Kalivas, 1995). A transient sensitization response of GC (corticosterone in the rat) may initially occur. *Tolerance* appears after a few days, perhaps due to familiarity with the same stressful environment. Tolerance would not be observed if the animals are reexposed to a novel stimulus. Chronically stressed persons differ in their physiological patterns when not challenged. Adults with and without chronic life stress (assessed by questionnaires) were exposed to either 12-min mental arithmetic or a nature video. Although the groups were almost identical at baseline in psychological, sympathetic, neuroendocrine, and immunological domains, they reacted differently to acute stress: Greater subjective distress, higher peak levels of epinephrine, lower peak levels of β-endorphin, and differences in natural killer (NK) cells cytotoxicity and distribution, and delayed recovery. It was concluded that exaggerated psychological distress and sympathomedullary peak reactivity occurs in persons with antecedent life stress (Pike et al., 1997).

NONRECOGNITION OF CHILDREN'S TBI

Costs

Nonrecognition causes the educational, social, and medical needs of the child to be ignored, public health statistics underestimate the safety and service needs of the community, children's dysfunctions are misattributed to incorrect causes, and the child or adult feels rejected or is treated as a troublemaker, faker, or lazy. The social costs are higher than with adults, since the period of survival is longer than with adults. Children with TBI exhibit long-term behavior problems in spite of cognitive recovery. What is often termed "good recovery" may not be that at all.

Brain Characteristics

It has been difficult to detect axonal damage or ischemia in children or adolescents because of the relatively high water content of the developing white matter of the brain. The new imaging procedure more easily detects injured brain tissue of children since injured brain tissue has reduced intracellular water movement than normal children's brain tissue. It was found to predict outcome in 83.8% of a sample of children with TBI (Galloway, 2008).

Approach to Children's Assessment

The definition of the American Academy of Pediatrics (cited by Kirkwood et al., 2008) of "minor closed head injury" specifies: Normal mental status on initial examination; no abnormal or focal neurological findings; no skull fracture; LOC < 2 min; possible seizure, headache, or lethargy.

While guidelines can be offered for a child's assessment that reflects the viewpoints and technical findings reflected in this text, comprehensive examinations are discouraged wherever insurance and accident coverage is utilized. An examination is expected to reflect the obvious: Children vary, in age, language, education, health, injury, emotional reactions, attentiveness, cooperation, availability for examination, willingness or ability of some source to pay professional fees, etc. If one attempts to measure a relatively narrow function such as "intelligence" or "memory," it may be difficult to obtain an obviously suitable procedure for a child whose demographics are outside the norms of available assessment procedures.

Nonawareness

The potential TBI may be ignored, for several reasons: The child may not be considered to be badly injured and therefore not brought to an emergency room (ER) or admitted to a hospital. As a consequence of nonrecognition, children's and adult dysfunctions can be attributed to incorrect causes. Moreover, because of expressive deficits, or the lack of a thorough examination, the parent or school system may have an incomplete picture of the range of present, but obscure neuropsychological deficits. The recovery of motor-speech and simple auditory comprehension and linguistic structure can give the false impression of normal communication (Jaffe, 1986), creating the illusion that development is normal while academic achievement is very poor. Thus, a test battery based on complaints is likely to be incomplete and miss significant dysfunctioning (Berg, 1986). Subtle and delayed after-effects are ignored. The prior event of a TBI is frequently totally ignored in a child's care unless a thorough examiner takes a careful history. This may not occur until some event of adult life elicits concern.

Lack of Current Inquiry

At the ER, there is often nonrecognition of TBI due to lack of inquiry concerning the details of the injury, and to proper examination of the head and mental condition of the child. Thus, after a relatively severe accident that would cause LOC in an adult, the child appears to be shaken up but not severely injured. It occurs frequently that the parent is not advised to remember this event and to observe the child for some disorder and to report the prior accident to later clinicians. The consequence is inadequate examinations and information that is not considered in assessment by later health care providers. Therefore, symptoms are not attributed to a head injury since the possibility of an accident is not known. When the lack of development revealing TBI is manifested later, the postinjury interval may be so long, or the child be so undeveloped (verbal expression and memory) at the time of injury, that the connection between neurobehavioral problems and injury is lost.

Concealment of the Accident

Since the child does not have support systems available to the adult, revelation is less probable. If misbehavior caused the accident, it is concealed due to fear of punishment. The parent or caretaker may conceal injury due to child abuse or negligence. Even when the child is an innocent victim, the parent's attitude can create concealment. I remember a middle-aged man who was examined for the purpose of determining his fitness for child custody. During the administration of a Wechsler IQ test, it was clear that his Performance subtests were far inferior to the Verbal. When I inquired whether he had ever had an accident, he revealed that at age 10 while riding a bike, another boy who disliked him thrust a stick into a wheel causing him to fall with a brief LOC. He did not tell his parents.

Lack of Memory of the Event/Lesser Cognitive Development

A description of an injurious event may not be elicited when the child is very young. Apart from retrograde amnesia, due to inadequate cognitive development, the memory of the event may be permanently inadequate or reduced because of the length of the posttraumatic interval. Inadequate inquiry for prior accidents or current TBI by health care providers, or examination for current injuries, causes nonattribution of symptoms to the accident.

Lesser Expression of LOC

As mentioned, after a relatively severe accident that would cause LOC in an adult, the child appears to be shaken up but not severely injured. Babies can appear normal with massive brain damage or increased intracranial pressure. Bicycling movements may be misinterpreted as normal activity, but be a manifestation of a seizure in a young infant. The level of consciousness is extremely difficult to assess in very young infants, for example, stupor is difficult to distinguish from normal sleepiness (Arffa, 1998). Thus, potential TBI can be ignored. Fearnside and Simpson (2005) report that children's brain trauma is less likely to be associated with LOC than adults (1% vs. 5%) but is followed by lethargy, irritability, and vomiting, attributed to brainstem torsion. Older children who are frightened but fully conscious may withhold speech or cooperation (Simpson, 2005).

Duration of impaired consciousness has been defined as the time elapsing between injury until the child can follow commands, equaling the number of days in which the Motor Response Score of the GCS falls below 6 (Yeates, 2000). This writer believes that alteration of consciousness can actually exist (lack of alertness; confusion), with the child (or adult) seeming to perform normally (although not closely examined). Since it is sometimes difficult to assess the level of consciousness in infants and children, mistakes are often made: The severity of a head impact is overestimated; more frequently, the converse error is made. An injured infant cries or whimpers and is thought to be fully "conscious." causing serious brain damage to be overlooked. Preverbal infants are too undeveloped to offer motor and verbal responses of the GCS. Lethargy may be a sign of altered consciousness (DeLorenzo, 1991). Garvey et al. (1998) offer a case of a 6-year-old who had a generalized seizure within 1 h of minor head trauma not associated with LOC. Takahashi and Nakazawa

(1980) describe a pattern in which children under 10 years of age had no LOC after a "trivial" head injury, and then after a latent period manifested transient neurological disorders, with or without convulsion, with recovery. Convulsions were not associated with hematoma. The pattern included no initial LOC or skull fracture, headache, nausea or vomiting, pale complexion, disturbance of consciousness, hemiparesis or hemiplegia, motor aphasia, convulsion or no convulsion, with "complete recovery within 6–48 h.

The accident may be out of memory because it occurred many years previously. Older children have a higher incidence of amnesia and unconsciousness (Klonoff, 1979). The association between early head trauma and subsequent seizures is unsubstantiated, but dysfunctioning causing seizure activity could impair intellectual functioning and memory (Dreifuss, 1989). Early posttraumatic seizures occur in 5% of children hospitalized for head trauma. In one-third, the seizures occur between 1 and 7 days later. Early posttraumatic seizures occur most often in children less than 5 years of age; the trauma can be quite mild. Two-thirds have more than one seizure, and about 10% (especially young children) have status epilepticus. Children with early posttraumatic seizures are at risk for focal seizures, with or without secondary generalization (Rosman, 1999).

Late Development of Symptoms

There is an apparent disconnection between disturbed physiological development and TBI: Developmental disturbance: cognitive, personality, or physiological (inability to attain puberty or appropriate growth, or precocious puberty). The delayed onset of a disorder after apparent normal functioning. interferes with correct attribution of its etiology.

Posttraumatic Amnesia

Posttraumatic amnesia as an indicator of TBI may be less valuable than emotional disturbance, even without LOC. Older children have a higher incidence of amnesia and unconsciousness (Klonoff, 1979). Initial reaction to brain trauma, that is, recovering from coma can be experienced as "out of it" or not knowing what is going on (Bergland & Thomas, 1991). The association between early head trauma and subsequent seizures is unsubstantiated, but dysfunctioning causing seizure activity could impair intellectual functioning and memory (Dreifuss, 1989). Early posttraumatic seizures occur in 5% of children hospitalized for head trauma, and of these 25% will have subsequent seizures (Rosman, 1989, citing Jennett, 1975).

Insensitivity to Aphasia

Communication and comprehension deficits are easily missed by the professional. It requires very little mental ability to carry on a simple conversation, such as "How are you feeling?" or "What did you do today?" Although children's aphasic problems seem to show more rapid improvement than adults, one must be alert to other significant cognitive deficits. Even after apparent recovery, communications problems can be serious. Subtle problems in using words for problem solving or communication can lead to abandonment of verbal activities. In turn, development of normal social relationships is impaired at the cost of concealing the inability to speak properly. These problems can be detected on examination years later (Woods, 1987).

Incomplete Examination

Using parental reports of head injury with ambulatory care or overnight care, and comparing these children with a comparable group of children with limb fractures, lacerations, burns, and those without injury, Bijur et al. (1990) claimed that the only statistically reliable outcome were teachers' (not parents) reports of hyperactivity. Children with lacerations and burns scored as badly or worse on measures of intelligence, mathematics, reading, and aggression as the children with head injury. It was concluded that mild head injury (MHI) in school-age children does not have an adverse effect on global measures of cognition, achievement, and behavior 1–5 years after injury, and that the findings are consistent with the pediatric view of minor head injury as unfortunate but benign.

The conclusions are suspect for several reasons: The sampling is biased insofar as it ignores: (1) the many cases in which the parent does not know of a head trauma; (2) the personality factor consequences of accidents including posttraumatic stress reactions that were not studied in detail; (3) noncognitive effects of brain damage that are reflected in chronic immaturity, that is, the "frontal lobe syndrome(s)"; and (4) physiological dysfunctions consequent to damage to the HPA–cortex axis). Psychosocial effects upon performance that were detected are considered next.

Social Consequences of Nonrecognition

Nonrecognition of TBI causes the educational, social, and medical needs of the child to be ignored. The social costs of the injury are higher than with adults since the period of survival is longer than with adults. Children with TBI exhibit long-term behavior problems in spite of cognitive recovery. Thus, the assessment as "good recovery" might not be correct when assessed by a wide-range examination after several years. In practice, as a consequence of common nonrecognition of a brain injury, the forensic neuropsychologist's need for documentation of a child's claim of TBI is less likely to be satisfied than for an adult. In addition, public health statistics underestimate the safety and service needs of the community. Difficulties in assessing concussion add to nonrecognition of childrens' TBI. The child's lesser cognitive development is a contributing factor; a description of an injurious event may not be elicited. The child may not be considered to be badly injured and therefore not brought to an ER or admitted to a hospital. Since children appear to manifest LOC less frequently than adults, potential TBI is ignored. At the ER, there is lack of recognition of TBI due to lack of inquiry concerning the details of the injury, attention to proper examination of the head and mental condition of the child. Thus, after a relatively severe accident that would cause LOC in an adult, the child appears to be shaken up but not severely injured. There may be inadequate examination for TBI subsequently by health care providers subsequently; symptoms are not attributed to a head injury, or, the possibility of an accident is not explored. The accident may be occult: The child may not reveal an due to fear of punishment; the child may have retrograde amnesia for the events of the accident; the child may be too young to remember the event; the accident is out of memory because it occurred many years previously; the parent or caretaker may conceal injury due to child abuse or negligence. Since the child does not have support systems available to the adult, revelation is less probable. Late development of symptoms also hampers attribution to an accident. Correct assessment for the purposes of litigation are rendered difficult: Children's dysfunctions have been attributed to incorrect causes. Careful personality study is required to overcome statements from the school or previous examiners ignoring the accident history, that due to feeling rejected, the child's behavior results in assessment as a troublemaker, a faker, or lazy.

EPIDEMIOLOGY

Traumatic brain injury in children differs significantly in pathology and outcome from adults. Differences in outcome and diagnostic procedures make it necessary to discriminate between *inflicted* (physical child abuse, most frequently directed at the head and face) and accidental injuries. Although it has been asserted that most head injuries in children are "minor and require no medical or surgical intervention," the present author offers the same caution as for adult head injury, that is, without a wide range of examination, comparison with a baseline (if available), and follow-through after years, no robust conclusion about outcome is possible. As will be documented next, the brain tissue keeps maturing until at least the second decade, and consequently, the possibility of impaired development must be considered. The most common injuries are falls, pedestrian accidents, and bicycle and MVAs.

An infant's vulnerability to brain trauma, particularly shakelash, is summarized by Caffey (1971) and McLaurin and Towbin (1990). Children have a higher proportion of admission to hospitals after so-called "minor" head injury, with one report of 70% concussion as the primary diagnosis. There is evidence that head-injured children have a higher proportion of aggressive and premorbid personalities, come from lower-income families, and have younger mothers with higher depression scores (Weiner & Weinberg, 2000).The majority of pediatric head injuries are "minor," but the incidence is underestimated, as the victims do not present to the emergency department or require hospital admission. The trauma pattern differs from adults: ages 0–14 years: falls 39%; motor vehicle traffic 11%; assault 4%', other 41%; and unknown 5% (Langlois et al., 2004). The overall rates for TBI per 100,000 are age 0–4, 1,121; 5–9, 659; and 10–14, 629. Overall, falls are the most common cause of head injury in the age group 0–15 years. Rates of falling are highest for children 0–4 years and for adults 75 years and older. Note the rapid rise in deaths as childhood develops into adolescence calculated per 100,000 individuals: 0–4, 5.7; 5–9, 3.1; 10–14, 4.8; 15–19, 24.2. For serious injuries, pedestrian injuries from MVAs were most common, followed by falls, bicyclists, and then occupants of motor vehicles (Fearnside & Simpson, 2005).

The fiscal costs of pediatric brain injury have been described as "staggering," that is, nearly $16 billion a year on the acute and chronic management of children with head injuries. The emotional costs to the family and the loss of productive citizens was noted (Cheng et al., 2005). The educational, social, and medical needs of the child are ignored, public health statistics underestimate the safety and service needs of the community, children's dysfunctions are attributed to incorrect causes, and the child or adult feels rejected or is treated as a troublemaker, a faker, or as lazy. The social costs are higher than with adults because the period of survival is longer than with adults. With nonrecognition of TBI dysfunctions can be easily attributed to incorrect causes. Careful personality study is required to overcome statements from the school or previous examiners that incorrectly attribute disturbed behavior or cognitive dysfunction to inaccurate causes, for example, the child is a troublemaker, a faker, or lazy. Children with TBI exhibit long-term behavior problems in spite of cognitive recovery.

CHILDREN'S PCS

The symptoms of PCS in children are consistent with those of adults: cognitive dysfunctions (reduced IQ; language); lower academic achievement with increased likelihood of enrollment in special education; motor skills; psychosocial ability with troubled family relationships; health; neuropsychological functioning; maladaptive behavior with poor social competence that may cause rejection by peers and teachers; and increased likelihood of unemployment.

After so-called "minor" head injury (not associated with LOC), only 7% of children are likely to complain of headaches after 1 month. However, "there is a high incidence" of alterations in play, daily activities, and school absenteeism to the point that rates are twice that of population norms (Bruce, 1990). Even after "normal neurological examinations," some children display emotional disorders, probably a secondary reaction to diminished perceptual and cognitive ability associated with the brain injury (Chelune & Edwards, 1981). The question of incidence is somewhat clarified by considering the particular measurement of outcome. There was an increase in symptoms with increasing severity, that is, internalizing or externalizing (mild or severe CHI). However, if larger units of behavior are considered (adaptive), then only severe CHI cases may be affected. However, when preexisting conditions are controlled, mild CHI has a relatively small effect restricted to increased hyperactivity scores (Asarnow et al., 1991). Brown et al. (1981) also found that new psychiatric disturbance occurred in the severe head injury group: They were two to three times as frequent as in the controls of MHI group, and were associated with the upper extreme of PTA;

occurred more frequently with neurological disorders than without; and were associated with transient or persistent intellectual impairment.

Some common symptoms of PCS in children are *academic* (reading, spelling, and arithmetic, with no significant sensory component); *overt behavioral* (hyperactivity, impulsivity, short attention span, and emotional lability); and *socioemotional* (immaturity, withdrawal, and hyperactivity).

The number of symptoms is related to the intensity of the head injury, as measured by the GCS at hospital admission, CT, neurological examination, skull fracture, or a combination of these indicators. Intensity of anxiety seems not to be a reflection of the extent of neurological injury. Anxious children had a higher incidence of other symptoms after controlling for injury severity. After mild head trauma, adults reported a significantly larger number of symptoms (one), but in the moderate-to-severe head trauma range there was no difference in the reported number of symptoms. Ongoing stressors enhance symptom maintenance (Mittenberg et al., 1997). After 3 months, children with minor head injury (MHI) were characterized as having attention problems, headaches, low energy, and dizziness. Those who manifested more PCS symptoms were assessed as being worse adjusted preinjury than their siblings, and having at measurement lower scores on motivation. Enhanced PCS symptoms were associated with an initial and later smaller white matter volume as measured by MRI. An early deficit of mental ability (abbreviated WISC-3 short form) was attributed to executive dysfunction since Block Design performance depends on attention and planning skills (Yeates et al., 1999).

"DISTRACTING" SYMPTOMS

Outcome will be negatively affected by noncerebral "distracting" symptoms (Parker, 1995). Disturbed sleep patterns (altered circadian rhythms) would contribute to hormonal dysfunctions, ultimately attributable to chronic stress and direct brain injury. Posttraumatic outcome may be worse due to health disorders consequent to disturbance of hormonal secretion. Persistent insomnia can lead to impaired daytime functioning, injury due to accidents, depression, and variability of circadian rhythm associated with cardiac diseases and stroke (Creisler et al., 2005).

Sleep disorders: A study of adolescent MHI patients substantiated sleep complaints. There were changes in the power spectrum for each of the predominant frequency bands in sleep, requiring differing amounts of time to return to their lowest levels (Parsons et al., 1997). Disturbed sleep patterns (altered circadian rhythms) would contribute to hormonal dysfunctions, ultimately attributable to chronic stress and direct brain injury. Outcome may be worse due to health disorders consequent to disturbance of hormonal secretion during hours ordinarily assigned to sleep.

Anxiety contributes to a *pain-prone* emotional response to stress. Initially, the child's autonomic nervous system is under relatively poor control, in comparison to adult levels of function, so that anxiety may be transformed into colic. Subsequently, when regression occurs, pain is part of the affective response (Krystal, 1984). Depression can also follow from the child's awareness of deficiencies, social rejection, and subtle problems that are not understood by himself or others. As noted in Chapters 13 and 15, depression is extremely common after TBI.

Concussive Loss of Consciousness

Mistakes in assessing the severity of TBI follow from the difficulty of assessing the level of disturbed consciousness in infants and children. Most frequent is underestimating the injury. The writer asserts firmly: The only (reasonably) firm conclusion concerning the outcome of a credible mechanical accident, particularly young children (with decades of neurological development yet to occur) is observation after some rationally determined postinjury interval. There are no firm clinical guidelines for diagnostic or prognostic assessment in the acute phase. In the writer's experience, significant injuries to children are ignored for years, until later events cause a history to be obtained, and then suddenly into the picture comes falling on the head over the back seat, being dropped, or falling from a sofa.

Observational assessment cues are offered in the literature: It is asserted that children older than a few months who cry and open their eyes responsively have not suffered major diffuse TBI. Observation of the child is asserted to provide acute and prognostic information. An injured infant who cries or whimpers and is thought to be fully "conscious" can cause serious brain damage to be overlooked. Preverbal infants are too undeveloped to offer motor and verbal responses of the GCS. Lethargy may be a sign of altered consciousness (DeLorenzo, 1990; Ewing-Cobbs et al., 1997). Older children who are frightened but fully conscious may withhold speech or cooperation.

Childrens' brain trauma is less likely to be associated with LOC than adults' PTA but is followed by lethargy, irritability, and vomiting. Lethargy may be a sign of altered consciousness. A seizure may subsequently follow trauma without LOC. Garvey et al. (1998) offer a case of a 6-year-old who had a generalized seizure within 1 h of minor head trauma not associated with LOC.

There are different patterns of concussive alterations of consciousness and neurobehavioral outcome (Semrud-Clikeman, 2001). Childrens' brain trauma is less likely to be associated with LOC than adults (1% vs. 5%) (Fearnside & Simpson, 2005, citing Brookes et al., 1990), but is followed by lethargy, irritability, and vomiting, attributed to brainstem torsion (Rosman, 1989). Takahashi and Nakazawa (1980) describe a pattern in which children under 10 years of age had no LOC after a "trivial" head injury, and then after a latent period manifested transient neurological disorders, with or without convulsion, with recovery. Convulsions were not associated with hematoma. The pattern included no initial LOC or skull fracture; headache; nausea or vomiting; pale complexion; disturbance of consciousness; hemiparesis or hemiplegia; motor aphasia; convulsion or no convulsion, with "complete recovery within 6–48 h. Since it is not easy to assess the conscious level in infants and older children, mistakes are often made: An injured infant cries or whimpers and is thought to be fully 'conscious'; further examination to assess possible TBI is not performed. Preverbal infants are too undeveloped to offer motor and verbal responses of the GCS. Lethargy may be a sign of altered consciousness (DeLorenzo, 1991). Garvey et al. (1998) offer a case of a 6-year-old who had a generalized seizure within 1 h of a minor head trauma not associated with LOC. Takahashi and Nakazawa (1980) describe a pattern in which children under 10 years of age had no LOC after a "trivial" head injury, and then, after a latent period, manifested transient neurological disorders, with or without convulsion, with recovery. Children may not exhibit PTA but rather lethargy, irritability, and vomiting attributable to brainstem torsion (Rosman, 1989).

Prenatal to 2 Years

Prenatal auto accidents may cause fracture of the frontal, temporal, parietal, or occipital bones, with various hematoma locations, brain swelling, ischemia, etc. (Bowdler et al., 1987).

Neonatal trauma: The young child's proportions of head, neck, and trunk change the fulcrum of rotation, even when restrained, reducing the frequency of subdural hematoma and diffuse axonal injury (DAI). However, the unrestrained child becomes a missile, suffering multiple calvarial fractures. A school-age child, struck on the legs and knocked to the ground or upon the striking car, will suffer subdural hemorrhage and rotationally caused DAI. They may misleadingly appear normal, with eye opening and spontaneous motor activity in the presence of massive TBI. *Seizures* are manifested as "bicycling," that is, lower extremity stereotyped movements. Withdrawal of the limbs in response to noxious stimulation is not as reliable a marker for preserved level of consciousness as a vigorous, prompt cry with a facial grimace in response to a trapezius pinch (Ewing-Cobbs et al., 1995).

Brain Swelling

Brain swelling may occur after even minor head injuries in children: Since there is little space available for expansion within the skull of a child over 2 years, a minor degree of brain swelling causes a marked increase in intracerebral pressure (ICP); children's cerebral circulation may react more rapidly to trauma than adults (Mendelow & Crawford, 2005). There is evidence that the frequency of hematomas, contusions, and contracoup injury is less likely in children than adults (Weiner & Weinberg, 2000).

Diffuse brain swelling is 3.5 times as common in children as in an adult population (Zimmerman & Bilaniuk, 1981).

Acceleration Injury

The infant's vulnerability to brain trauma, particularly shakelash, is summarized by Caffey (1971) and McLaurin and Towbin (1990). The poorly developed neck musculature cannot support the relatively large head. The pliable sutures and fontanels are stretchable at the calvaria, inducing excessive tearing forces at the attachment of vessels to rigid fixed soft tissues (e.g., falx cerebri). The skull is thinner, more pliant, and has unfused suture lines, which permits stretching of the brain and its blood vessels by external forces. The floor of the anterior fossa and middle fossa is relatively smooth, offering little resistance to the shifting brain.

The unmyelinated brain is softer, permitting excessive stretching of both brain and vessels. The relatively greater volume of CSF in the ventricles and subarachnoid spaces shifts farther and faster during whiplash, increasing their stretching effect on the more-resistant brain parenchyma and blood vessel attachments.

Even carrying a child in a backpack carrier while *jogging* is sufficient movement of the brain in the skull to cause impact and/or rupture of the bridging veins between the static dura and the moving brain. Falls (tripping while running; falls from a mother's lap, bed, chair, or table); stairs), that is, "minor" head injury, has been associated with neurotrauma without LOC (Dharker et al., 1993). The mechanism is believed to be damage to perforating branches of the middle cerebral artery whose angle of origin is very acute, with stretching creating spasm and consequent decrease in local blood flow, that is, ischemic lesions in the *basal ganglia*. There may be immediate contralateral hemiparesis caused by ischemic changes in the adjacent internal capsule. Restoration of circulation can result in early and complete recovery although there is a persistent hypodense lesion.

Impact Injury

There appear to be fewer contracoup lesions in infants, increasing to a level in 4-year-old children slightly less than the proportion found in fatal injuries of adults (85–90%, McLaurin & Towbin, 1990, citing Courville and Pennington, 1988).

CHILD ABUSE ("SHAKEN BABY SYNDROME")

Head injury is the leading cause of death from *child abuse*, and half of the survivors are left with permanent neurological handicaps (Fenichel, 1988, p. 260). Inflicted injuries are associated with more severe TBI and less favorable outcome. Mortality and disability after inflicted injuries are greatest at the lower ages (Ewing-Cobbs et al., 1995). The following discussion is focused primarily upon nonsurgical cases. There is some doubt whether shaking alone suffices to explain subdural hemmorhage (SDH), SAH, cerebral contusions, edema, and infarction. Head impact during the whiplash injury may be necessary to cause significant intracranial injury, that is, being thrown or having the head forcibly banged on a hard surface (Britt & Heiserman, 2000). Subdural hemorrhage is common in children suffering from nonaccidental injury. Co-morbid injuries include retinal hemorrhage, skull, rib, and long bone fractures (Cheng et al., 2005). Caution is needed not to assume that an injury is inflicted when only the caregiver knows what took place, and objective signs of trauma may be absent. Further, the patterns of brain damage in infants under 1 year of age are markedly different from those seen in older children (Geddes & Whitwell, 2003). Shaking the baby is a common form of child abuse, not necessarily involving direct impact (which is more likely to cause cerebral contusion), but causing parenchymal damage (McLaurin & Towbin, 1990). The large angular excursion of the neuraxis injures the brainstem, causing hypoxi-ischemia with axonal injury (Anderson & McLean, 2005). Battering occurs in one-third to one-half of head injured children, and up to 85% of children less than 2 years old (Kaiser et al., 1986). Additional forms of trauma include direct blows, shaking, and abruptly jerking infants. It is recommended that all children with serious head injuries, regardless of cause, require long-term observation for impaired growth and hypopituitarism (Caffey, 1974; Miller et al., 1980).

Nonaccidental injury is a frequent question in children of age 2 and younger. However, it is questionable whether shaking alone, without impact, will cause subdural hemorrhage. This is attributed to sudden deceleration against a surface (tearing the bridging veins from or within the cortex into the subdural space). Fractures or bruises will be found if the surface is hard, while external injuries will be concealed if a soft surface is struck (Duhaime et al., 1992). Shaking the baby without direct impact causes parenchymal damage, while impact is more likely to cause cerebral contusion (McLaurin & Towbin, 1990). There can be a history of seizures, unconsciousness, bulging fontanelles, optic and retinal hemorrhages, and a swollen brain (Fenichel, 1993, p. 69). In addition, shaking injuries can create hypopituitary conditions.

Criteria for Abuse

(1) Inadequate, unlikely, or no explanation for injury; (2) Injury inconsistent with development; (3) Discrepancy between history given by various individuals; (4) Multiple injuries at various times; (5) Past history of suspicious injuries; (6) Delay in seeking medical treatment; (7) A passive attitude by the child to medical examinations; (8) Lack of parental concern; (9) Injury is excessive. Particularly suspect is when there is a claim that a child has fallen off a low sofa, or suddenly becomes limp and remains comatose; (10) There is a delay in seeking treatment (McLaurin & Towbin, 1990; Ward, 1989; Weiner et al., 1982, p. 197); (11) Retinal and subdural hemorrhage and intracerebral hematoma. Retinal hemorrhage is associated with seizures in inflicted injury (Duhaime et al., 1992; Shapiro, 1987) or a history of repeated head trauma, especially if accompanied by limb fractures or other injuries (Rosman, 1989). However, it is not pathognomonically associated with child abuse. However, in very young children, they are associated with inflicted injuries; and (12) Submersion in bathtub (Diekema et al., 1993); bruises of the ear associated with retinal bleeding; rib fractures. The blow may cause sufficient angular acceleration of the head to cause brain injury and seizure (Feldman, 1992); history of child abuse in the abuser (Aicardi, 1992, p. 732).

The setting that creates the suspicion of child abuse is the presence of retinal and subdural hemorrhage and intracerebral hematoma (Shapiro, 1987). Subdural hematoma is characterized by failure to thrive, pallor, irritability, jitteriness, hypertonia and hyperreflexia, etc. (Herskowitz & Rosman, 1982, p. 576). This has been called the "shaken baby syndrome." Abuse is to be suspected when there is a history of repeated head trauma, especially if accompanied by limb fractures or other injuries (Rosman, 1989). Of 13 infants with nonaccidental trauma, all presented with profound neurological impairment, seizures, retinal hemorrhages, and intracranial hemorrhage (Hadley et al., 1989). Autopsies on eight who died revealed that none had a skull fracture. The pathology was at the *cervicomedullary junction*, which impairs vegetative functions necessary for life. In addition, shaking injuries can create hypopituitary conditions.

Consequences: Physically abused children, compared to peers in similar SES levels, excluding those with severe TBI, are more likely to exhibit delays in intellectual functions, language development, social cognitive ability, and increased aggression, distractibility, negative emotionality, reduced self-control, and reduced frequency of positive emotional states (Ewing-Cobbs et al., 1995). Medical injuries include bilateral hemorrhages (subdural, subarachnoid, subpial, intraparenchymal, retinal) and retinal detachment, with concurrent absence of external signs of trauma to the head and neck. The infant's relatively large head and weak neck muscles prevents them from limiting head motion during shaking (Caffey, 1974; Christoffel & Zieserl, 1991). The consequences of subdural hematomas include meningoencephalitis, permanent brain damage, cerebral palsy, seizures, mental retardation, defects of vision and hearing, microcephaly, and death. Subdural hematoma is characterized by failure to thrive, pallor, irritability, jitteriness, hypertonia and hyperreflexia, etc. (Herskowitz & Rosman, 1982, p. 576).

DISORDERS OF PHYSIOLOGICAL DEVELOPMENT

It is recommended that all children with serious head injuries, regardless of cause, require long-term observation for impaired growth and hypopituitarism (Caffey, 1974; Dykes, 1986; Miller et al., 1980). Maturational disorders may be hypothalamic, pituitary, or gonadal. Alterations in puberty are related to reduced secretion or increased secretion of gonadal and other hormones. This may be consequent to changes in the level of control (stimulation or inhibition) by the brain or somatic feedback mechanisms. Homeostasis is controlled through feedback loops from the endocrine glands that maintain hormone levels rather precisely by feeding back to the hypothalamus and the pituitary. Damage to the *hypothalamic-pituitary-endocrine target organ axis* interferes with maintenance of body homeostasis. Hypothalamic influence is positive in all instances except secretion of prolactin (PRL), in which case, damage causes its release (Molitch, 1995; Treip, 1970). Growth is contingent upon thyroid secretions, growth hormone (GH), insulin, GC, catecholamines, CNS biogenic amines, inhibitory effects of the CNS independent of sex steroids, and reduced sensitivity of the hypothalamus to inhibitory effects of sex steroids and CNS inhibition (MacGillivray, 1995).

The examiner should recognize particular physiological developmental and health problems of children, consequent to trauma and child abuse, due to damage to the hypothalamus, anterior and posterior pituitary glands, and their influence upon endocrine organs and other vegetative functions. The clinician concerned with childhood injury will consider hypopituitarism (consequent to radiation therapy of the brain, head trauma, and child abuse, due to injury to the anterior pituitary, pituitary stalk, or hypothalamus) and endocrinological disorders directly consequent to trauma-related dysfunctions of particular glands. Postnatal growth is dependent upon a group of hormones (Styne, 1991a) that are implicated in brain trauma and emotional stress: GH and somatomedins (anterior pituitary hormones regulated by hypothalamic somatostatin and GRH); thyroid hormone; sex steroids; and GC (cortisol in excess can stop growth).

PUBERTY DEVELOPMENT

The CNS is the only major restraint on the onset of puberty. It inhibits the hypothalamic-pituitary-gonadal system during the prepubertal years, and is mediated by the hypothalamus acting on the neurosecretory neurons that synthesize and secrete luteinizing hormone-releasing hormone (LHRH). These act as an endogenous pulse (intermittent) secretion generator (Grumbach & Styne, 2002). Puberty represents disinhibition of the gonadal axis (i.e., its reactivation). Negative-feedback mechanisms of hypothalamic-pituitary gonadotropin-gonadal control are operative from fetal life onward. Episodic release is present in the neonatal period and childhood, although total gonadotropin secretion is diminished in childhood. At any age, the pituitary gland is capable of response to GnRH stimulation. Before puberty, the onset of pubertal hormonal changes is first evident in dramatic episodes of luteinizing hormone (LH) release of short duration that first occur during sleep. Sleep patterns normally change in puberty, and by inference, sleep disturbance may interfere with circadian control over hormonal function (Grumbach & Styne, 1998). The hormonal CNS changes responsible for the onset of puberty are not known. The primary determinants of the timing of puberty are probably genetic, but nutrition, physical health, and psychological factors can influence both the onset of and the rate of progression through puberty (Foster, 1996). With maturity, pituitary release occurs regularly throughout the day. Synchronized episodic increased gonadotrophin stimulation sets into motion all that is necessary for full development and ovulation or spermatogenesis. Gonadal steroids appear to be responsible for the rise in GH secretion characteristic of puberty. By the end of puberty, episodic release of gonadotropins is characteristic of the waking hours as well as those of sleep and, in the female, varies predictably during different phases of the menstrual cycle.

Traumatic *heterotopic ossification* (bone growth in an abnormal location in soft tissue, detectable by bone scan) is associated with loss of and range of motion pain. There is increased risk for fractures of long bones, prolonged coma, and spasticity. In association with various neurological processes, new bone formation may start at multiple sites within immature connective tissue (Ivanhoe & Hartman, 2004). The incidence in pediatric TBI is 14–23%, with the hip, elbow, knee, and shoulder most frequently involved (McLean et al., 1995). In cases with TBI, it is a marker of poorer functional outcome and slower progress through inpatient rehabilitation (Johns et al., 1999).

Dysfunctions of Growth

Growth is vulnerable to brain trauma. In infancy, growth is primarily dependent upon nutrition. In childhood, the major determinant is GH. The adolescent growth spurt is controlled by sex hormones and GH. GH deficiency can result from problems with delivery, child abuse, and accidental trauma. *Short stature* is associated with acquired hypothyroidism, precocious sexual maturation, diabetes insipidus, and lesions of the hypothalamus, pituitary, or adrenal glands. Thyroid function is essential to each of these growth stages (MacGillivray, 2001). Impairment of GH secretion in *children* leads to growth failure (Aron et al., 2004). GH secretion is enhanced by sleep, exercise, stress (trauma, physical, emotional, chemical), hormones (GHRH, ACTH, α-MSH, vasopressin, estrogen, neurotransmitters), and various pathological conditions. Somatostatin is a potent inhibitor of GH secretion.

Dysfunctions of Sexual Development

Disturbed sexual development can have a variety of causes: trauma, neurological disorders, developmental defects, medical, endocrine, tumors, genetic, and physiological (weight loss; anorexia nervosa; increased physical activity by female athletes).

GH and sex steroids contribute to the pubertal growth spurt. Acquired GH deficiency can result from perinatal disturbances, child abuse, accidental trauma, and perhaps GC excess (MacGillivary, 2001). *Precocious puberty* (sexual precocity) is defined as the appearance of secondary sexual development before 7 years in Caucasian girls and 6 years in African American girls, and 9 years in boys of either race (Styne, 2004). The upper and lower boundaries of assignment of outcome of the onset of puberty are ±2.5 SD from the age mean (mean age of *menarche* in the United States is 12.8 years (6 months later for Caucasian than African American girls). *Brain trauma* and *radiation therapy* may be followed by delayed puberty or precocious puberty. *Epilepsy* and *developmental delay* are associated with central precocious puberty. Growth outcome can be an interaction in a medical condition between factors that accelerate or decelerate growth (Styne, 2004). Juvenile *hypothyroidism* causes severe reduction of linear growth, with delayed puberty (sexual maturation). The child appears much younger than the chronological age. Precocious puberty may occur in the absence of imaging findings. It is considered rare, but may be have co-morbid galactorrhea, which is associated with elevated concentration of PRL and perhaps hypothyroidism (a traumatic potential) (Grumbach & Styne, 2003). Chronic hypothyroidism may be expressed as partial precosity of secondary sexual characteristics in girls, impaired growth, and galactorrhea, if there is an enhanced secretion of PRL (Aron et al., 2004). Hypogonadotropic hypogonadism, that is, absent or decreased ability of the hypothalamus to secrete GnRH or the pituitary gland to secrete LH and follicle-stimulating hormone (FSH), denotes an irreversible condition requiring replacement therapy. A rare cause is trauma (Styne, 2004).

Hemiatrophy (hemihyoplasia) is reduced limb growth without hemiplegia. This may be caused by an early acquired defect up to age 6 of one cerebral hemisphere (Harris & Carlson, 1988).

Symptoms of Endocrinological Developmental Disorder

Symptoms include growth retardation (Eichser et al., 1988; Yamanaka et al., 1993); lack of achievement of puberty; precocious puberty (defined as the onset of secondary sexual characteristics before age 8 in girls and 9 in boys) (Towbin et al., 1996), with growth acceleration and skeletal maturation, or dilation of the third ventricle (Woolf, 1992) consequent to a tear of the hypothalamus (Attie et al., 1990), which may commence within a few months of injury (Shaul et al., 1985); GnRH release in girls due to hypothalamic damage (Rosenfield, 1996); interference with inhibition of gonadotropin secretion by the mass effects of head trauma due to hypothalamus damage (Styne, 1996); GH deficiency (Attie et al., 1990) that may commence within a few months of injury (Shaul et al., 1985); absent secondary sexual development consequent to hypopituitary insufficiency (Miller et al., 1980; Pescovitz, 1992); gonadal failure with loss of libido, impotence, amenorrhea (Cytowic et al., 1986); amenorrhea and sexual infantilism consequent to hypothalamic insufficiency (Grossman & Sansfield, 1994).

Pituitary Gland Injury and Disorders of Sexual Development

Sexual development and growth disorders (MacGillivray, 1995) are commonly neglected as a consequence of a TBI. The writer has seen several late adolescents with undeveloped beards which were not considered to be clinically significant, and therefore were not attributed to juvenile head injuries. Yet, disturbed sexual development can have a variety of nontraumatic causes, so referrals should be made for further study. Since it is traditional for the examiner that is retained by a defendant (independent medical examination, i.e., IME) to be instructed to refrain from making clinical recommendations, in this writer's opinion, it is unethical to conceal this information if observation indicates a developmental or other medical disorder that is not already known to the parent or noted in a medical record. Since the retaining attorney has the right not to utilize the examiner's report, it cannot be assumed that information concerning a child's status will ultimately be conveyed to the claimant's attorney, and then to a parent. In any event, years might be go by. Since the mother's status during pregnancy may be a forensic issue, prenatal stress (exogenous and internal) is associated with adverse pregnancy outcomes, that is, fetal growth and maturation, and parturition (Wadhwa et al., 1997).

Since the pituitary gland is fixed in place in a cavity of the base of the skull (the sella turcica) and attached to the hypothalamus by the pituitary stalk, the secretory structure is vulnerable to trauma. A head impact, usually followed by movement of the brain in all directions (caused by its acceleration and deceleration), may stretch the pituitary stalk emerging from the hypothalamus with its end fixed to the enclosed pituitary gland. The hypothalamus and/or pituitary stalk may be stretched or torn. This is the etiology of traumatic hypopituitarism in the adult and developmental disorders in the child. Trauma may create neural injury, removing the inhibition of maturational secretion and resulting in premature puberty. Since this causes bone maturation before growth reaches its intended limit, there is a growth deficiency. An alternate consequence occurs when trauma injures various secretory cells in the hypothalamus or anterior pituitary, or blood vessels conveying hypothalamic-releasing hormones for the anterior pituitary's hormonal system. This delays or prevents puberty. The examiner's observations are crucial, since developmental deviations may be ignored by parents and by physicians. Neural pathways inhibit secretion of sexual development hormones until puberty. Later, sexual development is controlled by releasing hormones under some neural control, created in the hypothalamus and transported by blood vessels of the pituitary stalk to the secretory cells of the anterior pituitary gland. Hormones vital for development and normal physiological functioning are released: ACTH, GH, thyroid-stimulating hormone (TSH), LH, FSH, and PRL. Since these are entrained (linked in time) to the sleep-wake cycle, with maximum rate of secretion occurring at specific times after sleep onset (Molitch, 2001), the examiner may consider possible endocrine disturbance accompanying *sleep disorders*. Feedback mechanisms operate within the hypothalamo-pituitary-target gland axis to ensure fine control of endocrine function (Pocock & Richards, 2004, pp. 216–220).

Hypopituitarism: Anterior pituitary injury is less frequent than posterior pituitary injury, and has been reported after seemingly trivial trauma and the shake-impact injury of *child abuse*. It should be suspected when the following are observed: grossly reduced growth, regression of secondary sexual characteristics, and signs of hypothyroidism or Addison's disease (McLean et al., 1995). Acquired lesions of the pituitary, stalk, and suprahypophyseal hypothalamic zone are frequently associated with both GH and gonadotropic insufficiency, thus leading to short stature and hypogonadotropic hypogonadism to trauma (Foster, 1996). Growth retardation and absent secondary sexual development characterize hypopituitary insufficiency (Miller et al., 1980; Pescovitz, 1992). Osteoporosis is an endocrine disorder secondary to hypopituitarism with GH deficiency (Castels, 1996).

Delayed puberty: Delayed adolescence may be associated with lesions of the hypothalamus and pituitary gland causing varied hormonal deficiencies. Radiation therapy may be followed by delayed puberty or precocious puberty. Trauma can result in gonadal failure with loss of libido, impotence, amenorrhea, and sexual infantilism consequent to hypothalamic insufficiency (Cytowic et al., 1986; Grossman & Sansfield, 1994). Hypogonadotropic hypogonadism, that is, absent or decreased ability of the hypothalamus to secrete GnRH or the pituitary gland to secrete LH and FSH, denotes an irreversible condition requiring replacement therapy. A rare cause is trauma (Styne, 2004).

Hypogonadotrophism (reduced LH and FSH):

- *Prepubertal onset*: Reduced facial and body hair; eunuchoidal body proportions; reduced testicular volume and length; smooth scrotum; and small prostate.
- *Postpubertal onset*: Decreased libido; slow hair growth; deceased body hair; testes atrophic if long-standing; normal voice pitch; decreased muscle and bone mass; normal skeletal proportions, penis length, prostate size (Cone et al., 2002).

Late or reduced puberty can occur in both sexes. Hypothalamic-pituitary damage may not be detected until the characteristic developments of puberty are not observed as expected (Cooper, 1991; Jaffe et al. [Growth charts], 1990); Styne, 1991b). The examiner has frequently observed young men (late teens) who incurred brain trauma before the usual age of puberty, or around that time, who were beardless and otherwise lacking the musculature and other constitutional characteristics of the normally developed male. Epstein et al. (1987) point out that the temporal connection between an injury and its endocrine consequences may be missed due to the long period between an injury and the expected bodily expression of endocrine maturity. Boys are more insecure and vulnerable to peer pressure, especially in working-class and minority groups. Late-maturing girls are more comfortable receive more support of their families, and are less often brought to medical attention than early-maturing girls (Grumbach & Styne, 1998). Amenorrhea can be consequent to hypogonadotropic hypogonadism due to trauma. CNS pathology may also occur after the onset of menses resulting in secondary amenorrhea (Foster, 1996).

Growth deficiency: Growth outcome can be an interaction between factors that accelerate or decelerate growth (Styne, 2004), that is, between the results of injuries that reduce normal inhibition of puberty, that is, prematurely permitting it, and those that prevent normal development of puberty. GH deficiency with marked growth retardation and delay of puberty secondary to stress experiences has been reported (Harris et al., 2005).

GH and sex steroids contribute to the pubertal growth spurt. Acquired GH deficiency can result from perinatal disturbances, child abuse, accidental trauma, and perhaps GC excess (MacGillivary, 2001). Juvenile hypothyroidism causes severe reduction of linear growth, with delayed puberty (sexual maturation). The child appears much younger than the chronological age. Precocious puberty may occur in the absence of imaging findings.

Precocious puberty: Precocious puberty can be attributed to trauma, that is, a premature activation of the HPA (Lee, 1996). Trauma can prematurely remove hypothalamic inhibition of pituitary gonadotropin production (GnRH) resulting in premature puberty with short adult stature due to precocious sexual maturation. *Epilepsy* and *developmental delay* are associated with central precocious puberty. A pattern of precocious puberty is galactorrhea (an inappropriate secretion of milk), and hypothyroidism. Galactorrhea can be consequent to hyperprolactinemia, that is, reduced inhibition of PRL secretion, attributable to pituitary stalk section, hypothyroidism, and numerous medical disorders (Bulun & Adashi, 2003).

Hemiatrophy (hemihyoplasia) is reduced limb growth without hemiplegia. This may be caused by an early acquired defect up to age 6 of one cerebral hemisphere (Harris & Carlson, 1988).

Personality Consequences of Developmental Anomalies

The changes of puberty, that is, height, sexual characteristics, features of the head, are easily apparent to viewers. Consequently, deviations from the norm in terms of premature or delayed development arouse comment from others, and these children are subject to self-awareness. There are consequences in terms of self-esteem, identity, acceptance by same-sex peers and opposite-sex potential mates. Further, cognitive changes accompany puberty, that is, development of abstract thought and decision-making processes developing out of the concrete reasoning of childhood.

Growth charts, including adjustment for midparental stature, are offered by Styne (1991a). Jaffe et al. (1990) indicate the following physiological symptoms in children suggestive of anterior pituitary dysfunction (which in turn may be related to posterior pituitary dysfunction and damage to the hypothalamus): poor growth (height and weight); sexual immaturity; malaise; and anorexia.

SENSORIMOTOR DISORDERS

Motor

Normal motor development evolves from early infant mass synergistic reflex actions. Maturation combines two processes: (1) The development of mass movements and postural tone allows for the maintenance of body and limb position against gravity. It forms the postural background for voluntary sitting, standing, and walking. (2) For descriptions of the neonate and early childhood postural and righting reflexes (see Augustyn & Zuckerman, 1998; Campbell, 2005, pp. 500–501). Inhibition of mass responses enables selective movement, including manipulative skill. Motion disorder contributes to significant developmental problems: athletic skills, vocational training, and peer acceptance. Although "trivial" head injuries without LOC are considered common in toddlers, it is this writer's opinion that: (1) Significant disorders may not be expressed until years later; and (2) An equivalent mechanical impact results more frequently in children than adults with less alteration or LOC. Etiology includes child abuse in infants, and sports and play injuries and MVAs in adolescence (Fenichel, 1997, pp. 71–75).

Lateralized deficits are associated with different response patterns. RH damage tends to create unintegrated drawings with small details, while LH damage is associated with impoverished drawings, with satisfactory contours but lacking detail (Thal et al., 1991). While pure motor deficits often improve to a greater extent in children and adolescents, particularly after severe injuries, they are apparent on tasks requiring rapid motor responses. Performance deficits have been detected for trail-making tests, symbol-coding tasks, finger and foot tapping speed, form board assembly, and mazes. Deficits may be due to slowed reaction time, slowed information processing, or a combination of the two. Deficit increases with the complexity of the task. Speed problems can exist in the presence of relatively normal cognitive functioning.

Soft Signs

"Soft neurological signs" have been considered signs of developmental delay: overflow or mirror movements; immature grasp of pencil; clumsiness (poor performance on complex motor tasks for age); dyspraxia (poor motor planning); synkinesis (movement overflow); choreiform movements; reversed or incomplete manual dominance; lateness of developmental milestones such as standing, walking, and talking; motor impersistence; difficulties with gait, posture, and stance; speech articulation problems (Deuell & Robinson, 1987; Tupper, 1992). They are difficult to elicit and have poor reliability during the neurological examination.[1] Some consider them as not consequent to neurological conditions. Soft signs may occur and then disappear at a later age. One considers whether deviancy (e.g., in the lowest 5–10% of an age group is a normal pattern not precluding eventual maturity) and a soft motor sign has functional significance. Slower reaction time is a well-established consequence of TBI. Reaction speed of children with TBI is slow to the point that bicycling in traffic is discouraged. Children with poor reaction time and equilibrium are at greater risk of falling.

Soft signs are associated with a higher incidence of cognitive dysfunction and learning difficulties, attention deficit disorder, and psychiatric disturbance (Bigler, 1988). Tupper (1986) differentiates between developmental and abnormal soft signs. *Developmental* (persistence beyond a normal age, variously given as 7–9): Awkwardness, clumsiness (eye–hand activities); dyspraxia or poor motor planning; synkinesis (movement overflow) or mirror movements (Nass, 1985); impersistence; gait; posture; lateness in suppressing primitive reflexes; tactile extinction on double simultaneous stimulation. Clumsiness before school age is revealed in inability to perform adaptive functions, for example, tying a shoe or buttoning clothing. Subsequent dysfunctions include messy paper-and-pencil performance, poor games and sports skills, and difficulties in handling kitchen and writing utensils. *Abnormal* (appearance at any age is abnormal): Choreiform movements; visual-spatial confusion and errors; borderline intelligence; irregularities in the ability to learn; electroencephalogram (EEG) abnormalities without seizures; reversed, incomplete, or mixed manual dominance; right-left and spatial orientation difficulties, or body schema deficits hampering dressing; motor dysfunctions (tremors; tone; tremor; nystagmus, dysdiadochokinesis; astereognosis) (Brumback & Weinberg, 1990 [emphasize lateralization of dysfunction]; Deuel & Robinson, 1987; Hertzig & Shapiro, 1987; Nass, 1985; Tupper, 1986).

One may differentiate between "soft" and "focal" signs. Nonfocal neurological signs are expected to disappear by age 7–9. "Soft" motor signs are associated with a higher incidence of cognitive dysfunction and learning difficulties, attention deficit disorder, and psychiatric disturbance (Bigler, 1988). In short, if there is no known reason for inadequate sensorimotor development, the presence of soft signs points to the need for further examination and monitoring the child's progress over a wide range of functions. There are slight correlations between motor performance and IQ (Dennis, 1985; Ewing-Cobbs et al., 1989). Soft signs are not clearly defined, but may be considered as a nonfocal deviation from expected motor performance at a given age. Examples have been described as developmentally delayed, difficult to elicit and unreliable, behavioral, performance that is momentarily deviant, equivocal, nonlocalizing; in short, "subclinical" (Taylor, 1987).

Sensory and motor deficits are common among children with head injuries whose impairment may not be obvious. Soft signs can be aspects of normal development that become abnormal only if they persist beyond the age at which they are usually outgrown (Tupper, 1986), or borderline deficits in which the presence of impairment is not clear. Motor dysfunctions in children are expressed by deficits in development and regression to an earlier stage. This is observed for senses, motor function, body schema, and symptoms pathognomonic of brain trauma (Tupper, 1986). Examples of

[1] Tupper, D. (1987). Appendix A: Physical and neurological examination for soft signs. In D. E. Tupper (Ed.). *Soft Neurological Signs* (pp. 339–353). Orlando, FL: Grune and Stratton.

neurological soft signs are offered by Nichols (1987): poor coordination, abnormal gait, impaired position sense, nystagmus, strabismus, astereognosis, abnormal reflexes, mirror movements, other abnormal movements, and abnormal tactile finger recognition. Their incidence was from 0.8% to 14.2% of their sample. These findings are neuropsychologically reasonable, that is, reflecting cognitive functioning, hyperarousal and poor control, reduced affect, and brainstem deficits of motor control. The number of soft signs was associated with maternal illness, smoking, and pregnancy complications.

Speed: Fine Motor Skills

For psychomotor tasks requiring fine motor skills, the magnitude of the difference between groups with mild, moderate, and severe injuries was directly related to the demand for speeded output. The performance of children with severe injuries declined as the demand for speed increased, whereas the performance of children with mild and moderate injuries was less affected by speed demands.

Balance

Balance observations used with children include (Gagnon et al., 2001): duration of standing or makes a postural adjustment (removing hands from hips; moves feet from original position; falls or needs examiner's assistance), and stepping over an obstacle. One considers response to backward perturbations, anteroposterior and lateral sway, affect of the loss of visual cues, and balance measured by procedures such as the Bruininks-Oseretsky Test of Motor Performance and the Postural Stress Test (backward perturbations) indicating capacity to maintain balance. Children's balance has been tested by standing on one foot for 30 sec, or alternatively standing in place until the child makes a postural adjustment, such as removing hands from hips, moving feet from the original position, or falling.

Sensorimotor Development

Mature movement evolves from (1) mass movements and postural tone (maintains body and limb position against gravity as the postural background for sitting, standing, walking, etc.) and (2) mass responses. Development consists of differentiation of precise movements, in orderly temporal and spatial sequences, permitting selective movement or manipulation. Damage to the motor cortex, corpus callosum, or corticospinal spinal tracts directly impair motor function and maintain contralateral movement overflow due to undeveloped inhibition of the uncrossed corticospinal tracts. Sensorimotor deficits (soft signs and abnormalities) lead to low self-esteem, contribute to subsequent school failure, and may be the precursor for subsequent psychiatric problems even if overcome (Deuel & Robinson, 1987). Although the claim has been made (based largely upon animal experiments) that sensory systems recover well from perinatal damage (particularly when compared to adults), complex sensorimotor dysfunctions occur, for example, stereognosis, body schema, route finding, and clumsiness (Rudel, 1978).

Late-appearing sensorimotor signs include volitional dyskinesia due to injury to the superior cerebellar peduncles and sensory deficits. After TBI, motor performance recovery for mild and severe injury is less than for cognitive functions, reflecting topographically rather than associatively organized tissue (Ewing-Cobbs et al., 1989). When IQ < 70 is excluded, there are slight correlations between Full Scale IQ and motor signs such as hemiplegia (.29), ataxia (.14), and fine motor problems (.19). Mild motor deficits are associated with relatively late right-hemisphere lesions (Dennis, 1985), reflecting its substrate in topographically organized rather than associatively organized tissue (Ewing-Cobbs et al., 1989).

Motor dysfunctions in children are expressed by deficits in development and regression to an earlier stage. This is observed for senses, motor function, body schema, and symptoms pathognomonic of brain trauma (Tupper, 1986). A child with persistent weakness and hypotonicity in an upper extremity, especially in the presence of an ipsilateral clavicular fracture, may have suffered from traumatic brachial and lumbosacral plexopathies (McLean et al., 1995).

Handedness: During development (1–4 years) fine coordination is a better measure of hand preference than the relatively crude reaching (Harris & Carlson, 1988). While conventional descriptions of handedness implicate a dominance expressed contralaterally from the motor cortex, in actuality, control of the fingers, hand, and proximal muscles of the limb involves some ipsilateral and bilateral input involving the premotor and supplementary motor cortices, numerous brainstem nuclei, superior colliculus, the vestibular complex, etc. (Harris & Carlson, 1988; Kuypers, 1989). In the general population about 90% are natural right-handers. Deviations have numerous origins (Satz et al., 1988). *Natural left-handedness* springs from nonpathological causes, for example, genetic, accidental, and/or cultural. Sinistrals tend to have bilateralization of language and other functions lateralized in the dextral (Kinsbourne, 1988). *Mixed dominance* refers to differences between manual, ocular, and/or pedal preference. Ocular preference may be determined by intraocular not cerebral factors (Deuel & Robinson, 1987).

Premature highly consistent handedness is suspect under the assumption that mature handedness emerges in the range of 1–4 years. Therefore, hand preference from 4–6 to 9–12 months may reflect ipsilateral neuropathology. *Pathological right-handedness* is a rare event, occurring in what would otherwise be a left-handed (sinistral) person. Presumably the causation would be analogous to the timing and lesions causing pathological left-handedness. The proportion of left-handers increases in any population with unilateral brain damage, although the contribution of injury severity and critical age is unknown (Vargha-Khadem et al., 1985). *Ambiguous handedness*: This subset, found with raised incidence in autism and mental retardation, can be further subdivided on the basis of whether with repeated measurements with different-handedness tasks they vary in preferred hand for the same task (ambiguous) or vary between items for preferred hand but with a consistent choice for a given item (ambidextrous). It is attributed to bilateral brain damage preventing the establishment of manual dominance and cognitive development.

Pathological left-handedness is a controversial point. Not only is there a disproportionate number of left-handers among the intellectually gifted, but left-handed individuals as a group may be either gifted or brain damaged (Deuel & Robinson, 1987). They do not differ in intelligence, school performance, or other measures of intellectual achievement. Indeed, left-handedness can be caused by *in utero* or *postnatal* limb damage (Harris & Carlson, 1988). Kinsbourne (1988) asserts that direct evidence of early left-sided brain damage is weak except for explicit right-sided hemiplegic cerebral palsy. However, the proportion of left-handers increases in any population with unilateral brain damage, though the later the injury, the greater the proportion of right-hand dominance (Vargha-Khadam et al., 1985). Yet, in epilepsy the percentage of left-handedness is roughly twice that of control populations, and is found disproportionately in mental retardation, reading, speech and language disorders, and autism (Harris & Carlson, 1988).

Pathological left-handedness is ascribed to an early brain lesion, which may be manifested as mental retardation and epilepsy (Satz et al., 1985). Kinsbourne's review (1989) points out that right-hand preference yields to contralateral displacement only when early damage involves the locus of manual control or language areas. In addition to motor strip lesions, sensory damage, agenesis of the corpus callosum, subcortical sensory and motor deficits, and so on, should also be considered, including prenatal damage. There is an association between the likelihood of change of handedness and damage to the language areas, and also the basal ganglia. Left-handedness is increased in individuals who change manual dominance, with a low probability of familial sinistrality. Pathological left-handedness is found in a natural dextral population that incurred an early left-hemisphere injury. The examiner should ask for evidence for *change of handedness*. For example, one girl suffered a left cerebrovascular accident, accompanied by both aphasia and hemiparesis, requiring her to change dominance.

Children's trauma-related sensorimotor dysfunctions: motor speed; primitive reflexes, abnormal posture, or hypertonia; poor coordination; impaired position sense; strabismus; dyspraxia or poor motor planning; dystonia (uncontrolled and persistent movements of various speed and extent);

synkinesis or movement overflow; visual-spatial deficit and left–right confusion; body schema (which may hamper dressing, poor paper-and-pencil skills, utensil, and games performance; impersistence; gait; tactile extinction on double simultaneous stimulation; movement disorders (choreiform; tremors; tone; tremor; mirror movements; dysdiadochokinesis); nystagmus, astereognosis; handedness deviations (premature highly consistent handedness; change of handedness; mixed and ambiguous dominance); reduced and enhanced motor activity level (Parker, 1994; Sattler, 1990; Tupper, 1987). Procedures to assess movement are given by Sattler, 1988, p. 699; Tupper, 1987, pp. 339–353; Hertzig, 1987; and Peters, 1987.

Children's Vocalization Problems

In children with CHI, study is needed for motor-speech performance such as respiration, phonation, resonance, prosody, and articulation. However, a majority of children with CHI will not present with significant speech disorders (Jordan, 1990). Yet, brain-injured children display reduced speaking rate and impairment of articulatory speed and linguistic processing. The reduction in the speed of forming words and increased time between expressing these words places a burden upon the listener (Granacher, 2003, p. 33). Articulatory delay, rather than dyspraxia, is more likely with left hemispheric lesions than right hemispheric lesions. Language disorders in children with left-hemisphere vascular disorders are accompanied by dysarthria, oral dyspraxia, and a wide variety of additional linguistic disorders. There may be good recovery of linguistic skills if there are no seizures, but slow linguistic progress (Ozanne & Murdoch, 1990).

Psychomotor Disorder

The *DSM-IV-TR* diagnostic code is Developmental Coordination Disorder (315.4). The impairment must significantly interfere with academic achievement or activities of daily living, not be due to a general medical condition or pervasive developmental disorder, and when mental retardation is present, the motor difficulties exceed those usually present. Motor deficits persist throughout adulthood and can be associated with significant adjustment difficulties. Assessment is difficult: There is no stable baseline of performance, and the examiner has to judge whether the child is or is not reaching developmental milestones. Asymmetry of muscle tone suggests unilateral cortical or spinal cord damage. Developmental screening tasks include: drawing; imitative finger movements; hopping alternatively on left and right foot with sequence specified; rapid alternating movements (detecting dysdiadochokinesia); sustained motor stance (standing for 15 sec with arms extended, feet together); tandem balance (standing with one foot directly in front of the other, holding posture for 15 sec with eyes closed); maturity of pencil grip; ability to cut paper on a straight line. Concerning *psychomotor speed* (fine motor skills): The magnitude of the difference between groups with mild, moderate, and severe injuries was directly related to the demand for speeded output. The performance of children with severe injuries declined as the demand for speed increased, whereas the performance of children with mild and moderate injuries was less affected by speed demands (Yeates, 2000).

Careful history-taking is needed: Self-care skills such as buttoning clothes and tying shoelaces may be delayed. Older children may have difficulties with the motor aspects of assembling puzzles, building models, playing ball, and handwriting or printing. Clumsy children collide with people and furniture. They avoid competitive sports and may require repeated instructions for a new motor skill. Previous TBI may not have been recognized by a treating physician; children conceal injuries from their parents; caretakers may have been careless and deny the extent of a head injury or do not report them; an accident may have occurred years before at an age when the patient no longer remembers the details, suffered from LOC or PTA, or were too young to register and later describe bodily injury. Thus, there is a nebulous connection between dysfunction and incident. The question arises whether poor reaction time and equilibrium result from or were the cause of the accident.

Family study explores possible clumsiness, attention deficit hyperactivity disorder (ADHD), learning disabilities, or neurodevelopmental disorders.

One study of an 11-year-old girl measured before and after a mild trauma utilized a children's balance battery with measures of dynamic balance in various conditions, interaction of the various sensory systems required to maintain balance, and the ability to respond to backward perturbations while standing quietly. The child's preinjury baseline was developmentally accelerated. Consequently, achieving age expectations 12 weeks below injury (with the exception of an eyes closed item) was actually below baseline. A balance deficit was interpreted as reduced ability to react appropriately and therefore be a risk for reinjury. Such information could help parents to install activity restrictions. Restriction of the child's motor activity requires judgment, that is, balancing the consequences for self-esteem and self-efficacy, and avoidance of events that interfere with additional healing or adaption. Observation of single leg stance with eyes open or walking on lines may be insufficient to identify deficits. There may be an interaction between attentional difficulties and distractibility, balance, and obstructed vision in performance of particular tasks (Gagnon et al., 2004).

Children with mTBI display balance problems when confronted with a narrow base of support plus absence of vision. The difficulty has not been definitely attributed to attentional or cognitive problems. However, in adults, after mTBI, attention and information-processing deficits have been related to balance. Children with TBI are less effective than controls in a reaching task in both stable and unpredictable situations. This is an example of inability to adapt to changing tasks and environments (Zhang et al., 2002).

INFLUENCES UPON CHILDHOOD NEUROPSYCHOLOGICAL OUTCOME

Baseline: Preinjury Characteristics

Personal: Depression, anxiety, and stress may reduce both preinjury and posttraumatic performance. Injury and medical problems require consideration and treatment, not merely assumed cerebral and emotional factors. Family interviewing is vital. Family factors influence TBI recovery, including expectations and selective attentional biases (Kirkwood et al., 2008). While school records (particularly those supplied as standard scores or percentiles) are useful, the younger the child, the less documentation is available, dwindling to little use at age 6 and below, except for behavioral observations.

Emotional self-regulation is associated with better social and behavioral functioning; lability/negativity is related to poorer social and behavioral outcomes. Self-regulation mediates the relationship between childhood TBI and social and behavior functioning (Genesalingam et al., 2007). A preinjury psychiatric disturbance predicts one in the first 3 months, but not thereafter. Participation in litigation was not prognostic. Risk factors were family dysfunction, family psychiatric history, lower socioeconomic class, lower preinjury intellectual function, and lower preinjury behavior and adaptive function (Max et al., 1998).

Environmental: Preinjury personality, behavioral and cognitive level, and psychosocial circumstances, including family setting, affect the outcome (Lishman, 1987, pp. 172–173). Preinjury environmental factors can predict recovery from TBI in children (Yeates & Taylor, 1997). The preinjury family environment was a significant moderator of the effect of TBI, buffering its impact in high-functioning families and exacerbating it in low-functioning families. Shortly after an injury, the onset of postconcussive symptoms depends primarily on premorbid child and family factors, injury characteristics, and postinjury cognitive performance. The child's vulnerability to impairment is higher in families with poorer functioning because they cannot rely upon them to meet their stress (Yeates & Taylor, 2005).

Academic baseline: The school achievement scores on nationally standardized tests may be used as a cognitive baseline. Anticipating the outcome of a child's TBI is more uncertain than for adults for many reasons: (1) One does not know what proportion of development of the manifold functions has been achieved at the time of injury; (2) the plateau representing final development will be manifested years later; and (3) it is far more difficult to establish a baseline for the child because the opportunity for achievement is less and school grades can be significantly influenced by motivation, parental support, peer experiences, etc. The clinician cannot be certain when the ceiling of ability is reached, that is, when the increasing demands of school and community cannot be achieved by the injured brain that develops more slowly or to a lesser degree.

Injury Severity

Outcome varies with TBI severity: Behavior symptoms before and after injury correlate with injury severity. A group of children with TBI had more behavior symptoms than an orthopedic group. The lack of association between behavior symptoms and neuropsychological outcome was attributed to the greater recovery of cognitive functions, although it was possible that the test battery was insensitive to subtle residual deficits (Barry et al., 1996). Children with moderate TBI have more specific and less pronounced impairments than those with severe TBI. Severe TBI occurring during early childhood can result in generalized cognitive impairment and deficits of school readiness skills. More affected than language skills may be memory, spatial reasoning, and EF. Although cognitive deficits are documented in school-age children with moderate-to-severe TBI, children with mTBI offer inconsistent findings. More negative outcomes occur in less advantaged family environments. Children injured at age 2–7, contrasted with older ones, are more susceptible to deficits of expressive language, attention, and academic achievement (Taylor et al., 2008b).

Anatomical Complexity

The outcome of children's accidents is far more complex than that of adults. The child's musculoskeletal system is undeveloped and weaker, while the body is differently proportioned. Neuropsychological functioning, in contrast to the relative stability of the adult, varies at different levels of development in ways that significantly affect outcome. As in adults, we are concerned with the great differences between injuries in terms of anatomical distribution and severity of the lesion, and the fact that different neurobehavioral functions are anatomically widely distributed. Socioeconomic factors have a greater influence in children since they are narrowly experienced, compared to the greater exposure to a variety of experiences of most adults. Accuracy of assessment of outcome for children is worse than in the case of adults, where preinjury assessment is typically sparse or imprecise but may be completely lacking in children except for development information. With young children there is no record of even academic achievement.

Prediction of outcome is estimated on the interaction of many elements: maturation of the brain and the patient's age at the time of injury; the spatial extent, location, and intensity of the injury; whether it is focal, diffuse, develops secondary trauma, etc.; its progress through various phases (Eslinger et al., 1992; Parker, 1994); injury characteristics; environmental influences; and developmental variation (Yeates, 2000). Specifics of late expression of symptoms are noted in tertiary, quaternary, pentary phases and specifics of child development. In the immediate postinjury period there is a greater risk for further head injury, which does not reflect a permanent accident-prone personality (Gronwall et al., 1997).

Neurotraumatic Severity

Not all children with TBI manifest immediate or later behavioral disturbances, although LOC is prognostic at 1 year. Children with mTBI maintained their behavioral level postinjury, but those with moderate and severe TBI displayed significant increases in *problem behavior*. More severe injury lasted up to 30 months (Anderson et al., 2005a). *Epilepsy* is associated with mild underachievement (estimated at one-half a grade), particularly female gender, recent onset, and high seizure rate insufficiently controlled with medication (Aldenkamp et al., 1999). One study compared children with mild head injury according to whether they were hospitalized for fewer than 2 days.

Age at Injury

The age at injury can influence the outcome. Orbitofrontal and ventromedial frontal lesions (prior to age 5) create disruptive behavior, failure to follow rules, deficient empathy, and lack of moral reasoning (these were refractory to repeated instruction, treatment, and punishment); bilateral and left dorsolateral lead to socially inappropriate behavior and poor decision making. It was suggested that disruption of frontolimbic circuitry and other connections contributed to psychosocial problems, regardless of the site of the frontal cortical subregional lesion. The consequences of frontal lesions in children, with apparent loss of development of finely tuned social inhibitory control, may be magnified upon entering adolescence and adulthood (Levin et al., 2004).

Study of groups to associate outcome with age at injury is rendered somewhat ambiguous due to confounding sample characteristics: varying level of TBI, different ages described as earlier or later, co-morbidity of other injuries with TBI, varying age when estimating outcome, lower SES, and family dysfunction. To be considered are the preinjury history as well as the severity of the injury. Skills that are in a rapid stage of development, such as writing, may be more affected by cerebral injury than well-consolidated skills. One study of a wide range of functions did not elicit a model of prediction of a 30-month outcome. Different predictors were useful for physical and cognitive domains compared with psychosocial functions (Anderson et al., 2005a).

Language performance: In perinatal and postnatal damage, left-hemisphere language is achieved by sacrificing right-hemisphere functions. Even intrauterine injuries result in subtle, persistent dysphasia, with impairments most pronounced for injuries occurring at age 5 and later. While young children who have suffered left cerebral hemisphere lesions may overcome transient dysphasia, recovery is only apparent, and subtle but persistent language defects may persist (Vargha-Khadam et al., 1985). Early lesions in either hemisphere may cause aphasia, but speech is not transferred to the right hemisphere unless left-hemisphere damage occurs before age 5 (Rudel, 1978). By age 10, acquired aphasia is more permanent, consistent with the functional commitment of the left hemisphere to language (Shapiro, 1985). While considerable language competence may follow even major brain damage to the left hemisphere, with proper examination, deficits are detectable.

The critical feature appears to be the age at which the lesion occurred, not its severity. There are greater correlations between language test scores and FSIQ, VIQ, and age significant only for left-hemisphere damage (Vargha-Khadam et al., 1985). LH damage tends to decrease both verbal and nonverbal IQ, whereas RH damage hinders development of a narrower range of functions (Rudel, 1978). Linguistic disturbance is more common in the preschool years than in older children. Acquisition of complex skills (e.g., reading and writing) may be contingent upon the normal development of more basic skills (Crowley & Miles, 1991; Ewing-Cobbs et al., 1989). This conclusion contrasts strongly with the negative findings of Jordan et al. (1992).Using measures of receptive and expressive vocabulary, grammar, reading, and writing, and comparing matched individuals with (mild) CHI and with other injuries 10 years later, the CHI children did not demonstrate persistent speech and language deficits. They consider the possibilities that after such a protracted period, subtle linguistic deficits may have resolved, or that the subjects have developed compensatory strategies. Such deficits were demonstrable in children with severe CHI.

Caution Concerning Outcome

For children, the outcome of a possible TBI cannot be assessed acutely. The initial effects of early brain injury may worsen due to impaired developmental processes, including the acquisition of new skills and information, as well as psychosocial behavior. While some loss may be acutely observed, particular functions may require that years elapse before dysfunction is observed. Intellectual deficiencies are more immediately apparent in the older child, upon whom more demands are made. A child's outward appearance may appear "recovered" after an interval, but developmental lag or lack of achievement of maturity may be subsequently manifested (Berg, 1986). At all intervals since an accident, a wide range of study is indicated, first to establish current loss plus a baseline, and later an estimate of measured outcome. The outcome of younger versus older children's TBI is not clear, being confounded with the severity of the injury (Donders & Warschausky, 2007).

Considering the age at injury, the immediate results for different functions may be similar, but younger children manifested a decline of Full Scale IQ after 1 year, from 87.7 to 85.1, with a slight loss of Verbal IQ and a slight increase of Performance IQ. For language skills, the group with severe head injury had the greatest loss, with some recovery observed for all groups (Anderson et al., 1997). One study showed some recovery in the first 6 months, but no significant change 6–24 months after the injury. In the severe group, motor skills were most affected. In the mild-moderate group, Verbal IQ and expressive language were lower than Perceptual-Performance IQ and receptive language. The scores of the severe group did not catch up with the mild-moderate group. This suggests a model in which there is initial deficit, a variable recovery, and a stable persistent deficit (Ewing-Cobbs et al., 1985).

Time Since Injury

Children have a more truncated recovery trajectory, with few gains after the acute recovery period. In the psychosocial domain there may be a gradual deterioration or lack of development, in addition to increased parental stress. The greatest change was for physical dysfunction in the severe TBI sample, with 57.9% rated as having a good recovery by 30 months (Anderson et al., 2005a). Children with mild head injuries display more cognitive and somatic symptoms 3 months after injury than do children who are not injured, even after premorbid status was controlled for. They do not display more emotional or behavioral symptoms (Yeates, 2000).

Neuropsychological

Contributors to behavioral and cognitive sequelae in children include severity of the trauma; bilateral characteristic of the lesion; localization (laterality; cortical vs. subcortical); secondary complications such as seizures and subdural hematoma; level of intelligence at the time of trauma; history of psychiatric disturbance; family adversity; family reaction to the trauma; and parental mental disorder (Birmaher & Williams, 1994). Diffuse brain injury appears to cause a reduction of mental speed, efficiency, and integration, with deficits of information processing, attention, and reaction time among the most prominent effects of head injury of any severity (Beers, 1992). Reaction speed of TBI children is slow to the point that bicycling in traffic is discouraged. Children with poor reaction time and equilibrium are at greater risk of falling. The need for assistance in school 2 years postinjury was predicted by injury severity detectable in neuropsychological dysfunctions at 3 months (Kinsella et al., 1997). The discriminating profile was impairment of verbal learning, memory, and slowing in speed of information processing. This pattern still existed 2 years postinjury. Some criteria for academic impairment are change in academic rank (standardized scores), placement in special education programs, need for additional tutoring, and grade repetition. Balancing such criteria of the outcome of TBI are teacher reluctance to identify low academic achievement after TBI, lack of available educational information, community attitudes, parent coping strategies, and material resources (Kinsella et al., 1997).

Individual Psychosocial Qualities

Statistical evidence for decrements of cognitive performance becomes less when *social factors* are considered (Bijur et al., 1990). The following independently reduce effectiveness: social class and education of parents, housing quality, some personality features of the mother, number of siblings, and number of injuries prior to age 5. *Accident proneness* may affect performance. The accident group had higher aggression and hyperactivity scores at age 5; more injuries from birth to 5; and younger, more depressed mothers; and were more likely to be male. Thus, temperament and impoverished social conditions contribute to accident proneness.

COGNITIVE IMPAIRMENT

Age of Injury

Outcome within a single age group may depend upon the studied function interacting with such characteristics as age. An animal study indicated that the age of injury affected learning and memory, whose performance could be dissociated. For 17-day rats, learning was not impaired, but retention was. For brain-injured adult rats, there were deficits in both learning and retention tasks (Raghupathi & Huh, 2007). In contrast, and contrary to the theory of the developing brain's plasticity, there is greater impairment for younger individuals from preschool through adolescence (Granacher, 2003, p. 31). Moreover, the area of impairment may vary with age: In one study (age of injury 6–12 years vs. 16–20; mild-to-severe complicated TBI), while the groups did not vary in overall cognitive ability, education, or vocational accomplishments, the early-onset group is described as having worse outcome in attention and speed of information processing, social integration, likelihood of having a driver's license, and personal legal guardianship. It was considered evidence for the theory that early TBI interferes with development of immature or rapidly developing skills and may magnify deficits during later development. Further, in the early years after TBI, for both children and adults, lower SES and family dysfunction may affect outcome (Donders & Warschausky, 2007).

For children (and adults), PCS outcome must be considered to include family functioning and parental adjustment (Yeates et al., 1999). Outcome is affected by acute critical care and *premorbid status*: stressors, prior head injury, and premorbid family resources and coping strategy. For mTBI, length of PTA is not prognostic of PCS symptoms. In one sample, 17% of parents reported complaints. These children had an enhanced level of prior head injury with complaints of learning difficulties, premorbid stressors causing behavioral or emotional problems, or reported neurological or psychiatric problems (Ponsford et al., 1999; Woodward, 1995). It has been asserted that behavioral disturbance is rarely seen after mild head injury, but this point is controversial. The authors assert that parental overprotectiveness due to expectation of deficit may cause a child to be held out of school for an excessively long time or otherwise treated differently. Childhood head trauma seems to predispose for subsequent schizophrenia, that is, in comparison to depressives, manics, and surgical controls. Childhood head injury may play a role in the expression of schizophrenia in families with a strong genetic predisposition. Further, patients with a history of childhood head injury had a significantly younger median age of onset (Jorge, 2005; Wilcox & Nasrallah, 1987). With mild head injury, disturbances lasting longer than 3 months have been described as unlikely, with permanent changes exceedingly rare. One must consider premorbid behavioral and learning problems (Cerhan et al., 1996). Most academic and behavioral problems lasting more than 3 months after mTBI have been attributed to environmental factors. Another group (Gronwall et al., 1997) reviewed the literature, which led to an impression that the issue of pediatric mild head injury was equivocal in the domains of cognition or behavior. The latter authors acknowledge that when comparisons are available, children in the 0-to-6-years-old group may be more at risk than older children. When they conducted their own study of preschool children using appropriate controls, they determined that soon after injury, there were no significant differences on parental ratings of behavior or cognition. However, 6 months later, the MHI group showed deficits on a visual closure subtest, with a greater deficit 12 months later, and a deficit was still found 6.5 years later. Significantly, more MHI children needed special help with reading. It appeared that the MHI children were not affected in already-established skills, with no differences found in the first few weeks after the injury, but they did not develop other skills.

For children, outcome of TBI is partially determined by the level of development of the skill at the time of injury. Some definitions will clarify thinking: *emerging*, not yet functional, or in the preliminary stage of acquisition; *developing*, partially acquired and incompletely functional; and *Established*, fully acquired. The *order in which a skill is acquired* refers to its emergence relative to other skills. *Rate of acquisition* refers to the slope relating chronological age with skill development. Additional descriptors are useful for observational purposes of development or regression: *Mastery* refers to the level of final competence that may be normal or truncated (relative to estimated outcome). *Control* refers to the effectiveness of deployment. *Upkeep* refers to long-term maintenance of a skill (Dennis & Barnes, 1994).

Common traumatic deficits include performance speed, impaired memory for new information, social isolation, and impaired social skills. Impaired social skills in turn interfere with participating in rehabilitation programs as well as obtaining employment.

In one sample of *mild CHI* approximately 15% of the children manifested abnormalities on neuroimaging (mostly contusions and small hemorrhages or hematomas). Although there may be "resolution" of deficits on standardized cognitive testing, other postconcussive symptoms may persist for years (Yeates & Taylor, 2005). Deficits are more likely to be immediately detectable in adults since the child's inability to perform may take years to be observed. Assessment at a given age will not identify impairment of functions that are not developed at that age. For detection, one must wait until a developmental milestone is passed. Negative findings can be incorrect. It may require several years to detect measurable loss with reference to peers. The writer has observed, using standard scores for achievement tests, that several years may go by before there is a significant loss of rank relative to the child's peers. The combination of retained long-term memory, small requirements for new performance at the next grade, as well as later developmental milestones, creates the delay in impairment detection.

Brain injury may predominantly affect the acquisition of new skills. However, there is a tendency for all tests of scholastic attainment to show greater impairment with left-hemisphere lesions, somewhat more marked in the children who were under 5 years of age at the time of injury. Deficits in one area can impair learning and other achievements in another. Although IQ is a familiar measure of cognitive ability (though not the only component of effectiveness), below the age of 6, the unreliability of IQ measures makes it an uncertain indicator of long-term mental ability (Swisher, 1985). Deficits can be a reflection of both the severity of brain trauma and an interaction with personality and social variables (Beers, 1992). Although it is frequently stated that the child has a greater capacity for recovery, follow-up studies indicate that their deficits are as great as adults (Ewing-Cobbs et al., 1985). Thus, lifelong subtle deficits, which perplex the patient, originate in childhood.

Unsynchronized Maturation of Related Functions

Cognitive functioning depends on coordinated maturation of various functions (e.g., learning to read involves perception and sequencing). School tasks and other problem-solving situations become relatively more difficult for the brain-damaged child. Variations in the synchronous development of these skills lead to different patterns of linguistic, cognitive, and social development (Wetherby, 1985).

Later functions are dependent on the presence of earlier functions when academic and social demands are made. A delay in functioning impairs higher functions maturing later that would integrate or rely on these earlier functions. One dysfunction is added to another, causing secondary disruptive effects so that integration, synchrony, and sequencing of subsequent development do not occur. For example, learning calculations involves visual perception to acquire and organize information about digits and their visual presentation; numerical concepts, such as more or less, multiply, or divide; verbal ability to understand the teacher; memory; psychomotor skills to reproduce the assignment, etc.

Cognitive Symptoms

Symptoms observable in children after TBI include mutism; articulation disturbance; impaired repetition; decreased length and syntactic complexity of utterances; semantic errors; expressive aphasia (reduced initiation of speech); hesitations; dysprosodia; loss of narrative skills and pragmatic cues (reading faces, body language, and situational cues); loss of semantics (language content); written language; decrease of spontaneous speech, dysarthria; anomia; perseverative stock phrases; conversational filler; self-addressed questions; semantic paraphasias; anomia.

Rate of Recovery

The rate of recovery is most rapid in the first months after injury. Initial Performance IQ deficits tend to exceed Verbal IQ deficits, and may persist for 2 years, while Verbal IQ deficits tend to resolve (Bawden et al., 1985). There may be little difference in FSIQ between mild and severe head injury cases. However, with increased damage to the severe range, a sharp mean loss of 11 points is reported, with PIQ deficit > VIQ deficit (Bawden et al., 1985). Some trends exist for right-hemisphere lesions: Nonverbal, holistic intelligence develops laterality later than verbal. The right hemisphere can reorganize to subserve linguistic functions, although deficits are detected upon challenge (Feldman et al., 1992). Early right lesions are the most impairing, and late left the least impairing if Full Scale IQ is the criterion. Late right lesions have a larger effect on Performance than Verbal IQ (Aram & Ekelman, 1988).

IQ Pattern and Outcome

- Many children with brain damage are functioning in the average range.
- Early right lesions are the most impairing and late left the least impairing, if Full Scale IQ is the criterion.
- Early lesions have a general impairing effect, more so on nonverbal processing than verbal processing.
- Nonverbal processing is generally vulnerable, except for relatively late left lesions.
- Late right lesions have a larger effect on Performance than Verbal IQ.
- Nonverbal, holistic intelligence develops laterality later than Verbal.
- Verbal processing is more vulnerable earlier, regardless of laterality.
- There are great deviations within studied groups. Further, any given measurement or observation is subject to the warning that dysfunctions (positive or negative signs) may occur years later. Also, the range of initial examination is often incomplete.

Prognosis/Resolution

"Prognosis" cannot be defined without considering the definition of "recovery." A unitary statement of "resolution" or "recovery" is suspect. For example, it is asserted that there is "persuasive and consistent evidence" that the early cognitive deficits after mTBI are "largely resolved" after 3 months. Yet, there also are statements that confirm the suspicions of the clinician with a concern for a wide-range assessment of the injured child's condition that so-called mTBI is a major public health problem:

- The evidence suggests "few" short or long-term cognitive effects.
- "Most of the evidence" suggests that children with mTBI do not have higher rates of behavioral or school problems.
- "Only" one of four studies clearly attribute(d) the disability to the mTBI rather than to other associated injuries.
- "We cannot rule out the possibility that injury-related pain and distress play a role in cognitive deficits ... after TBI" (Carroll et al., 2004).

One may offer a different perspective than the symptoms described as "mild" TBI usually disappear.

Maturation of the brain and the patient's age at the time of injury; the spatial extent, location, intensity of the injury; whether it is focal, diffuse, develops secondary trauma, etc.; its progress through various phases (Eslinger et al., 1992; Parker, 1994); and the appearance of late expressed symptoms.

It has been asserted that the immature brain has a greater capacity for recovery after severe TBI than the mature brain. Optimum recovery may depend upon the normal functioning of the uninjured remainder of the brain (Fineman et al., 2000). This may be found in animal studies, but one review (Parker, 1990, pp. 309–312) offered contrary evidence for severely injured children.

Adaptive and Premorbid Functioning

Children with mild head injuries display more cognitive and somatic symptoms 3 months after injury than do children who are not injured, even after premorbid status was controlled for. They do not display more emotional or behavioral symptoms. Children with severe head injuries demonstrate more behavior problems and poorer adaptive functioning than do children with mild injuries or those with injuries not involving the head (using rating scales). Adaptive deficits and behavioral disturbance are also related to factors other than injury severity, including the children's premorbid functioning.

It may be that the presence of premorbid behavior problems actually increases the risk of head injury. A cited study of somatic, cognitive, and behavioral symptoms indicated that children with brain injuries displayed more symptoms in all three domains than did children with OI. The total number of symptoms correlated positively with injury severity.

Novel psychiatric disorders (those never before present) occurred in nearly 50% of all children (CHI). The most common novel diagnoses were organic personality syndrome, major depression, attention deficit/hyperactivity disorder, and oppositional defiant disorder. The onset of a novel psychiatric disorder was predicted by injury severity, preinjury intellectual functioning, SES, child psychiatric history, and global family functioning.

Cognitive outcomes were related more strongly to injury-related variables, whereas behavioral outcomes were related more strongly to measures of preinjury family functioning (Yeates, 2000).

Posttraumatic Cognitive Patterns

One criterion of traumatic dysfunction is progressively decreased academic and IQ standard scores, reflecting inability to keep up with one's age cohort. This may require years to be observed because school grades may reflect small demands and credit for long-term memory rather than expecting much new learning each year. Intellectual deficiencies are more apparent in the older child, upon whom more demands are made, although the child's outward appearance may appear "recovered," or subsequent developmental disorders (e.g., frontal lobes) may not have been manifested (Berg, 1986). Further, within limits, cognitive loss or reduced potential for learning and problem solving may be compensated for by the uninjured hemisphere. Late expression of developmental symptoms causes lack of attribution of the disorder to an earlier accident. The results may not show up until years later in lack of mental development or lack of physiological development (e.g., inability to attain puberty). Moreover, the connection between immaturity due to lack of endocrine development and the brain injury may not be recognized. Reduced stature and hypogonadotropic hypogonadism can be consequent to head trauma (Foster, 1996; Grumbach & Styne, 2003). Growth retardation and absent secondary sexual development characterize hypopituitary insufficiency (Miller et al., 1980; Pescovitz, 1992). Osteoporosis is an endocrine disorder secondary to hypopituitarism with GH deficiency (Castels, 1996).

Precocious puberty can be attributed to trauma, that is, a premature activation of the HPA (Lee, 1996). *Hypopituitarism*: Acquired lesions of the pituitary, stalk, and suprahypophyseal hypothalamic zone are frequently associated with both GH and gonadotropic insufficiency, thus leading to short stature.

I recommend examination for preschool children after completion of the first grade, which offers an opportunity for deficits of development to be displayed after a significant challenge. For older children, reexamination 3 years after the initial examination also creates substantially greater challenges to overcome.

There are numerous posttraumatic cognitive patterns, reported in the literature and observed in individual children.

An initial loss with the achieved plateau at a less than the expected level (Morris et al., 1992; Taylor & Alden 1997).

Reduced rate of development compared to preinjury development. The final plateau is lower than would have occurred had there been no TBI.

Premature plateau after initial baseline performance.

Immediate, permanent deficits. Cognitive symptoms remain permanent.

Subclinical deficits. After a recovery period, compensatory mechanisms are only partially effective in overcoming disability.

Late-appearing cognitive impairment.

A decline after 1 year followed by some recovery (Anderson et al., 1997).

Progressive Cognitive Deficiency

Disability in children may be defined as an interference with normal development, reducing ability to perform school and domestic responsibilities, and creating personality problems that affect social relations with peers, family, teachers, etc. Difficulties in school performance are created by hypersensitivity to noise, lower thresholds for stress, and lower levels of endurance (Begali, 1987, p. 71). It may be expressed immediately as cognitive or personality problems, but delayed development or unattained maturity of functions is common. Deficits can be a reflection of both the severity of brain trauma and an interaction with personality and social variables (Beers, 1992).

The following are characteristic dysfunctions in children: reduced IQ; delayed development of language; deficient behavioral adjustment and social competence that may cause rejection by peers and teachers; poor school performance or educational lag, with increased likelihood of enrollment in special education; delayed motor skills; increased likelihood of unemployment; troubled family relationships; poor health; and poor neuropsychological functioning.

Additional characteristic dysfunctions include:

- Processing inefficiency, slow speed.
- Processing inefficiency, amount of information: Deficits with increased demands for comprehension, length of assignment, and spoken language.
- Memory/learning problems: Teachers report a deficit in long-term memory for verbal material in severely head-injured students, more so than immediate recall.

- Attention-related problems: Children are easily distracted by peripheral auditory and visual stimuli. However, in a quiet environment (e.g., educational evaluation or the psychologist's office) they are better able to marshal their cognitive resources.

Cognitive Loss

The complexity of academic topics can hamper detection of the cause of a particular deficit. Using reading as an example, the basis is complex: phonological difficulties (the sound structure of speech), speed of auditory processing, speech perception, vocabulary, the relationship between graphemes (written letters) and phonemes (sound segments), visual processing, etc. Reduced comprehension of complex instructions or language and reading problems may interfere with responding to verbal directions, which may be misattributed to a disorder measured by the procedure being utilized. In contrast, this author has repeatedly observed that after TBI the child may initially achieve former IQ scores and rank on nationally standardized objective tests of academic achievement. Therefore, performance level immediately after an accident may actually be a baseline for studying future performance. There is so little difference in demands made from one grade to the next that the child coasts along on previous achievement and somewhat lesser mental abilities. It may require several years before there is a significant decline. Probably this is due to the use of prior learning, that is, relatively similar demands are made on the child from one grade to the next. When the child cannot keep up with learning demands, the parents and school may be unaware of the accident and potential TBI, and disorders are not related to it.

Although the outcome of TBI will be affected by the age at injury, the measured outcome depends upon the age, the postaccident interval, and the function studied. The immediate measurements for different functions may be similar, but younger children manifested a decline after 1 year of Full Scale IQ from 87.7 to 85.1 with a slight loss of Verbal IQ and a slight increase of Performance IQ. For language skills, the group with severe head injury had the greatest loss, with some recovery observed for all groups (Anderson et al., 1997). There is a model in which there is initial deficit, a variable recovery, and a stable persistent deficit (Ewing-Cobbs et al., 1997). This study showed some recovery in the first 6 months, but no significant change 6–24 months after the injury. In the severe group, motor skills were most affected. In the mild-moderate group verbal IQ and expressive language was lower than Perceptual-Performance IQ and receptive language. The scores of the severe group did not catch up with the mild-moderate group.

Anticipating the outcome of a child's TBI is more uncertain than for adults for many reasons: It is far more difficult to establish a baseline for the child because the opportunity for achievement was less and school grades can be significantly influenced by motivation, parental support, peer experiences, etc. One does not know the relative level of development that was achieved at the time of injury. The plateau representing final competence may be established years later. Even then, the clinician cannot be certain when the ceiling of ability is reached. How does one know that the increasing demands of school and community will not be achieved by an injured brain that develops more slowly or to a lesser degree? This uncertainty is particularly confusing the younger the age of the child when injured.

It may require several years to determine whether a child who may be functioning without measurable loss eventually loses ground with reference to his peers. Deficits can be a reflection of both the severity of brain trauma and an interaction with personality and social variables (Beers, 1992).

Plasticity and Compensation

It is presumed by some that since "*plasticity*" (pattern of recovery) of brain functioning aids in overcoming lesions, a child is less likely to manifest dysfunctions than an adult with comparable neurotrauma. The period of development in which the TBI occurs can determine the degree of performance at maturity (Stein & Hoffman, 2003). Plasticity of specific cortical areas may involve an inverse relationship, with myelinization of cortical neurons subserving these functions. Using language as an example, an injury to the dominant hemisphere is much less likely to result in language dysfunction than a similar injury occurring in an older child (Weiner & Weinberg, 2000). Some differences have been described based on presumed neuronal or functional neuroplasticity (Korkman, 1999). Examples include development of language in the right hemisphere, reserve capacity stemming from double representation of functions, and intrahemispheric functional compensation. Localized damage is supposed to be less prone to cause such adult characteristic location specific dysfunctions as aphasia and neglect to subside rapidly. Yet, data concerning brain-behavior relationships in adults may not be directly applicable to children, that is, different neural networks, different reorganization of the brain following brain damage, and possibly that children may have diffuse or multifocal rather than focal brain functioning. The belief in a good prognosis for *childhood* brain damage due to plasticity is doubtful (Levin et al., 1994). It evolves in part from utilizing inadequate criteria for recovery, for example, minimally adequate school performance, a "normal" IQ (whatever that means), not assessing a wide range of performance and behavioral assessment, and not waiting long enough to assess outcome. As with adults, there is underreporting of TBI: Children do not self-refer; lack awareness of difficulty; lesser level of assertiveness, which further reduces self-referral or self-identification; increased tendency to internalize symptomatology (Warschausky et al., 1999). Thus, dysfunction and slowed development may not be reported or detected. Reexamination of children who were 6 years or younger when first seen may result in considerable changes later on: reduced performance due to inability to keep up with peers or enhanced performance due to social and educational support, procedural differences, or some process of recovery, as well as an enhanced ability to concentrate and withstand frustration and anxiety.

While alternative sites may assume impaired contralateral functions (compensation), these may perform the function less efficiently (perhaps due to overcrowding of the alternative site, Rolland S. Parker (RSP)), and inefficiency is revealed later (Lazar & Menaldino, 1995). For example, until the age of 8 there may be no problems in such areas as expressive language since intact areas of the nondominant cerebral hemisphere are recruited to subserve this function. However, neuropsychological assessment may reveal impairment of the functions normally served by this area. Even in the estimated 15–25% of children who do not make spontaneous recovery, long-term follow-up reveals mild, residual neuropsychological deficits. Attention is called to mild blows to the head in *sports* (football, soccer, boxing) (Hartlage, 2002). A lack of strength of the skull and lack of development of the brain results in less acquired information and skills and less personality resources to cope with adversity, etc. Thus, it is even more difficult to assess the true level of mTBI in children than in adults. Outcome may be different for a given level of injury.

As children grow older, the intellectual demands on them, and the need for cognitive flexibility and for autonomy, are greater. There is an interaction between impaired development, organic personality damage, social consequences of failure and poor adaptation, deficient morale due to a damaged sense of identity, and emotional distress. Inability to learn leads to depression and further deficits in school achievement.

Baseline, Prognosis, and Recovery

Recovery is a complex phenomenon: It is superimposed over neural maturational/developmental changes in cognitive, behavioral, emotional, and social functioning (Lazar & Menaldino, 1995). Recovery and impaired developmental effects can operate in opposite directions. At the time of brain damage, children have less accumulated knowledge and established skills on which to rely, and suffer impairment in the acquisition of new skills. Expression of the effect of brain damage depends on the time since the injury, since evidence of impairment comes in part from inability to reach age-determined milestones. Different levels of performance in the same child have been described as dissociation of skills. These are attributable to neurological damage or congenital variations in development. A given deficit, for example, a reading disability, can be associated with dysfunctional abstract reasoning, visualmotor coordination, language, or memory (Taylor, 1984).

A preinjury baseline should be established. It may take years to be expressed, that is, as a dysfunction of development rather than observed as an immediate traumatic dysfunction. Progress should be tracked, since dysfunctions and deficits may be expressed later. Outcome is a balance between development, recovery, and late-developing neurological and stress-related dysfunctioning and deficits. The threshold for bringing a child to medical attention is lower, while head injuries caused by abuse are probably underreported (Cerhan et al., 1996). Records are often incomplete and incorrect. Further, children can conceal or not report head injuries with LOC, and the same injury seems less likely to cause LOC in children than in adults. Brain injury has more significant consequences for younger children, and the sequelae of childhood brain injuries remain relatively constant over time or worsen (Taylor & Alden, 1997). Children's outcome deficiencies after head injury are multiple: physiological disorders of development; cognitive; personality and behavioral; sensorimotor, etc. These are conditioned by their developmental level, neurologically, osteologically, somatic structures, cognitive and personality level, etc. The majority of pediatric head injuries are described as "minor," but this need not represent the actual outcome. The incidence may be underestimated, as the victims do not present to the emergency department or require hospital admission. Overall, falls are the most common cause of head injury in the age group 0–15 years. For serious injuries, pedestrian injuries from MVAs were most common, followed by falls, injuries to pedal cyclists, and then injuries to occupants of motor vehicles (Fearnside & Simpson, 2005).

Difficulties in Dealing with Complexity

Shapiro (1987) observes that although physical abilities seem normal, deficits still exist. Increasing difficulty in processing larger amounts of information creates problems as the complexity of learning required increases, so that the child may fall considerably behind his peers on intelligence and academic testing (Ylvisaker, 1989). Teachers estimate that head-injured children have the following rates of difficulties: 25%, reading vocabulary; 50%, rate of reading; 70%, higher levels of comprehension; 90%, comprehension of passages of substantial length. This is in accord with my suggestion that assessment should always use complex tasks with a high ceiling to detect possible deficits.

CHILDREN'S ATTENTION/CONCENTRATION PROBLEMS

Similar paradoxical findings as adults were determined for brain activation in a group of children age 5 when injured, with moderate or severe lesions, examined 1–3 years later, who had abnormal imaging at injury. They were compared to a group of children with orthopedic injury. Neural activation remained altered after a period of years. For a sensorimotor task (continuous performance test), the TBI group had significantly higher levels of activation in a variety of brain areas relative to the OI group. Similar findings were determined for a verbal task. The TBI group had significantly greater activation in language-related areas (right superior temporal gyrus and right middle temporal gyrus). It appears that children with TBI may require additional neural resources to achieve performance comparable to children without TBI (Kramer et al., 2008).

Learning deficits are partially attributed to distractibility, lack of persistence, and impulsivity (Ewing-Cobbs et al., 1985). This is an example of deficits of mental efficiency leading to ineffective performance when there may be a higher intellectual potential. Difficulties in selective attention are task-specific. Sustained attention is more likely to be impaired than selective attention (Lehr, 1990). A sample of children with prefrontal lesions were studied concerning the following facets of attention: selective, shifting, divided, and processing speed. Children assessed as inattentive-overactive were found to have deficient scores on such tasks as the Wisconsin Card Sorting Test, creating excess perseverative errors, and a memory task after a brief interval (Sequential Memory Task for Children). It was hypothesized that inattentive-overactive children (by inference suffering from a prefrontal-type deficit) do not sustain cognitive activity in the face of competing responses (Gorenstein et al. 1989, citing 1937). I relate this to deficient monitoring, that is, the inability to create a satisfactory inner model of the task or outcome, permitting repetition or selection of recent or nonrelevant responses.

A young woman reported that she fell on the floor onto her forehead at age 2 or 3, incurring an injury sufficient to impair aspects of subsequent development in this examiner's opinion. She described her condition as follows: "I have almost no trouble with what I hear, only with what I see, particularly if speed is involved. If I have to do it very fast, I can get confused about the order. I drop words, reverse words when writing, misspell words. It is harder to write by hand than to use a computer." She had recently completed a graduate professional degree with a Verbal IQ of 113; Performance IQ, 99; and Full Scale IQ, 107.

MENTAL EFFICIENCY AND EF

At different levels of maturity, the strategies available to solve particular tasks vary, for example, the early use of language to solve spatial tasks (Rudel, 1978). With increasing age, children inhibit motor movement, become less impulsive, increase attention, increase the influence of knowledge upon performance, and select relevant stimuli through ignoring distractors (Passler et al., 1985). TBI interferes with shifting to approaches that would be available with maturing tissues.

Processing Speed

Processing speed in formal psychological examinations increases during childhood and adolescence, and declines with age. It is an aspect of problem-solving capacity sensitive to brain damage. Since it is usually measured by paper-and-pencil tests, there are both mental and motor components in the psychometric performance. Processing speed loss has been measured in children with mild, moderate, and severe TBI (Processing Speed Index of the WISC-3 at posttraumatic intervals of 3, 12, and 24 months (Catroppa & Anderson, 2003, cited by Prigatano et al., 2008). Using the criterion of a deviation of 3 scaled score points between the lower of WISC-IV Vocabulary or Block Design subtest scores and Coding, about 5% of children without a history of TBI meet this criterion of reduced processing speed.

Late Developmental Effects of TBI

This discussion emphasizes lesser levels of TBI. The outcome is controversial, that is, age has sometimes been reported as unpredictive, and other studies indicated that injuries incurred at ages 0–6, are less favorable than in older children (Ewing-Cobbs et al., 1995). A 10–15 years' follow-up may be needed to chart the long-term effects of MHI in infancy. It is recommended that primary school teachers monitor the reading skills of infants who have sustained head injuries (Gronwall et al., 1997). The outcome of a given TBI will be influenced by the level of development of neurobehavioral and social functions, what has been learned, and the degree of consolidation of skills and functions at the time of injury. To the extent that a function is localized, and this author (Parker) believes that parallel processing and extended circuits should be assumed at all times, a given lesion might prevent the development of skills that have not yet been expressed. The deficit is *occult and postponed*. A generalization has been offered that generalized brain injury is associated with more severe consequences for younger children, particularly for skills in a rapid state of development as opposed to those that are more automatized and overlearned. The rate of development of more complex cognitive abilities may be impaired, that is, proceed at a slower rate (Ewing-Cobbs et al., 1989).

Patterns of cognitive and personality development (the school achievement scores on nationally standardized tests are a cognitive-level baseline): (1) immediate permanent deficits; (2) improvement through compensatory mechanisms, but with subclinical deficits; (3) initial progress with delayed expression of dysfunctions or lack of development according to schedule; (4) premature plateau after a variable period; (5) initial loss followed by apparently normal rate of growth, with inferred plateau at a less than optimal level (Morris et al., 1992; Taylor & Alden 1997); and (6) A "dramatic" increase in social difficulties at adolescence (Eslinger cited by in Taylor & Alden, 1997).

For children (and adults), PCS outcome must be considered to include family functioning and parental adjustment (Yeates et al., 1999). Outcome is affected by acute critical care and *premorbid status*: stressors, prior head injury, and premorbid family resources and coping strategy. For mTBI, length of PTA is not prognostic of PCS symptoms. In one sample, 17% of parents reported complaints. These children had an enhanced level of prior head injury with complaints of learning difficulties, premorbid stressors causing behavioral or emotional problems, or reported neurological or psychiatric problems (Ponsford et al., 1999; Woodward et al., 1999). It has been asserted that behavioral disturbance is rarely seen after mild head injury, but this point is controversial. The authors assert that parental overprotectiveness due to expectation of deficit may cause a child to be held out of school for an excessively long time period or otherwise treated differently. Childhood head trauma seems to predispose for subsequent schizophrenia (i.e., in comparison to depressives, manics, and surgical controls) (Wilcox & Nasrallah, 1987). With mild head injury, disturbances lasting longer than 3 months have been described as unlikely, with permanent changes exceedingly rare. One must consider premorbid behavioral and learning problems (Cerhan et al., 1996). Most academic and behavioral problems lasting more than 3 months after MTBI have been attributed to environmental factors Another group (Gronwall et al., 1997) reviewed the literature, which led to an impression that the issue of pediatric mild head injury was equivocal in the domains of cognition or behavior. The latter authors acknowledge that when comparisons are available, children in the 0–6 years old group may be more at risk than older children. When they conducted their own study of preschool children, using appropriate controls, they determined that soon after injury there was no significant differences on parental ratings of behavior or cognition. However, 6 months later, the MHI group showed deficits on a visual closure subtest, with a greater deficit 12 months later, and a deficit was still found 6.5 years later. Significantly, more MHI children needed special help with reading. It appeared that the MHI children were not affected in already-established skills, with no differences found in the first few weeks after the injury, but they did not develop other skills.

Children perform poorly on several cognitive tasks requiring working memory resources. Those with mild head, in contrast to moderate and severe head injury, received scores deemed to be normal (Dennis et al., 2000).

VERBAL AND COMMUNICATION DISORDERS

Verbal dysfunction and lack of development are characteristic signs of childhood brain damage. Deficits may reflect a combination of basic verbal dysfunction, loss of ability to memorize and learn, and concentration problems. Since language is involved with executive functioning and self-regulation, these functions may become dysfunctional. Language deficits contribute to later academic problems. Verbal IQ seems to be the best predictor of academic success, although IQ scores are usually not very reliable until age 6.

Before the age of 2 years language development can be delayed with damage to either the left or right hemisphere. The first symptom of an underlying neurological dysfunction can be a delay in language, or a failure in developing it. Communication difficulties can be a sensitive indicator of neurodevelopmental disorders. Children with preschool language disorder are at high risk for academic difficulties (Swisher, 1985).

Comprehension and Inferencing

Several deficits of verbal comprehension (including reading and social discourse) reduce communications capacity after TBI and can be detected in discourse and reading comprehension. Comprehension requires a representation of the interconnections within a text, and of the situation described by the text. Measuring children's comprehension by such questions as "what and "why" cues the formation of inferences by removing the requirement for monitoring when to make an inference.

The coherence of a story is based upon knowledge-based inferences that link events with outcomes in social situations (a deficiency in children with TBI) and text-based inference, that is, connecting ideas within a text.

Inferencing contributes to language comprehension. Some inferences maintain story coherence and others elaborate upon the story content. Inferencing draws upon a relevant knowledge base, speed of access to this knowledge or to the prior text, activation of different types of information in memory so that they can be integrated to form inferences as the text unfolds, metacognitive aspects of comprehension (knowing when and how often to make inferences), and causal reasoning that is required for some inferences. Inferencing problems are created by these deficits:

- Learning of general information.
- Slow information processing (interfering with the speed with which knowledge of text-based information is accessed).
- Working memory (deficits interfere with children's capacity to hold inference-relevant information in memory). Poor inferences by children with severe TBI are not accounted for by slow knowledge retrieval. However, slow word decoding does affect reading comprehension.
- Metacognition (monitoring one's understanding of ambiguous directions and also of the text itself). When the requirement for metacognition is reduced for severely brain-injured children, their inferencing deficit seems to disappear in comparison to normally developed of mild head-injured children.
- Causal reasoning (required to make an inference).

Social Cognition

Social cognition is a complex action that is vulnerable to neurological dysfunctions. Children with social impairment bear the burden of their disability in almost every environment. These children are at risk for poor relationships with peers, parents, and teachers. Understanding will improve treatment recommendations, and acceptance and understanding of these children. Social skills develop in spurts, believed to correspond to neurophysiological changes in synaptogenesis and myelination of the frontal cortex. It contributes to effective social interaction, which requires knowledge, interpretation, inhibition, and action. Misinterpretation may be at the root of *inappropriate behavior.* Social cognitive development proceeds through steps that are recognizably complex, that is, vulnerable to cognitive dysfunction. An example is recognizing the existence of false beliefs, that is, that people can hold beliefs and act upon them when they are objectively false. Children learn to manipulate the perceptions and experience of others by manipulating their beliefs, that is, deception and lying. *Theory of mind* participates in social cognition, and awareness of the mental states and intentions of others in their communications. More complex understanding develops in middle childhood: A *copy theory of mind* (beliefs are a copy of reality) develops into an *interpretive theory of mind* (people are able to derive different beliefs about the same event). This is a more difficult task, that is, identifying others' deliberate lies and sarcasm. Implicated brain regions include the orbitofrontal cortex, paracingulate cortex, and superior temporal sulcus. Social cognition seems related to theory of mind and conscious awareness (Saltzman-Benaiah & Lalonde, 2007).

Discourse

Stories of children with severe injuries were characterized as follows: deficits in interpreting ambiguous sentences, making inferences, formulating sentences from individual words, and explaining figurative expressions. Their stories contained less information, were more poorly organized, and less complete than stories of children with milder injuries or normal controls (Yeates, 2000). In children studied at 7 and 14 years (some with severe head injury), significant correlations were found between discourse ability and injury severity at 3- and 12-month intervals after the injury. There was a positive correlation between receptive vocabulary and discourse, indicating that a child with lexical comprehension deficits is likely to have difficulty comprehending and retrieving information from memory to retell a story. Since caregivers' ratings of communications correlated with discourse measures, adaptive deficits can be inferred. The parents may not fully acknowledge the deficits or may overestimate the impact of the injury (Chapman et al., 1995). Semrud-Clikeman (2001, pp. 47–49) emphasizes that helpful interventions for families include information about the child's health, available educational programs, and community agencies available for assistance. They need practical information about the child's progress, disabilities, and educational needs. Information should be provided in small bits, and answer the parents' questions. The caregiver or school official must take into account the parent's anxiety about the child or ongoing litigation. Officials of community agencies may be helpful in finding resources.

Linguistic and Pragmatic Disorders

These cognitively based dysfunctions are particularly severe in adolescents, who are in a period of growth in semantic and syntactic abilities, metalinguistic and pragmatic language skills, knowledge base, and EFs that will permit them to communicate as adults (Turkestra et al., 1995). In older children after head injury, initial *aphasic problems* are likely to be nonfluent (Ewing-Cobbs et al., 1985; Wetherby, 1985). When overall language performance and ability to name familiar objects were assessed a minimum of 1 year after CHI, increasing deficits appeared. The pattern of deficits was considered similar to that of adults (Jordan et al., 1988).

Postinjury *mutism* may have a motivational component. It can be followed by fluent aphasia with comprehension deficit and anomia (Klein et al., 1992) or by a nonfluent motor aphasia (Ozanne & Murdoch, 1990). Follow-up for 6 years is needed to determine the effect of TBI on fluent aphasia language and cognitive development (Klein et al., 1992). Classic aphasic symptoms after head injury" are less frequent long-term sequelae in children and adults than after vascular disease. After focal injury they have the same incidence as adults with similar focal injuries. Classic aphasic symptoms after head injury are less frequent in the long term, and subtle deficits of complex functions such as writing to dictation, hamper learning and academic performance, and have been detected up to 10 years after the injury (Lehr, 1990). Despite apparent rapid recovery, 25–50% of children will have aphasic symptoms after 1 year (Ozanne & Murdoch, 1990).

INTELLIGENCE LOSS

Intelligence and the Issue of Recovery

While "normal" IQ scores have been used to indicate recovery, this is doubtful since baseline studies are usually absent, and also impairment is manifested by emotional and activity problems, headaches, lack of control, etc. Many children with brain damage achieve IQs in the average range. Left late lesioned children in the Aram and Ekelman group had a mean Full Scale IQ of 117 in the High Average range. Deficits of concentration, memory, mental speed, and motivation can be impairing, even with seemingly retained IQ. IQ impairment is multiply determined, even within the parameters of laterality and localization. PIQ, for example, reflects motor status as well as cognitive level (Dennis, 1985). Deficits of speed represent a residual impairment after diffuse brain damage, which may not be picked up by the IQ, except perhaps in the severely injured (Bawden et al., 1985). Thus, a "normal" IQ may conceal an aberrant pattern indicating probable pathology (Dennis, 1985; Reynolds & Kaufman, 1985). Outcome after TBI is affected by the baseline principle that over-100

IQs tend to have a verbal advantage, and below-100 IQs have performance IQ advantage (Matarazzo & Herman, 1985).

Hemispheric Considerations in Language Performance

Verbal functioning of the right hemisphere after left-hemisphere damage differed in numerous parameters. Thal et al. (1991) offer the model of verbal comprehension being analogous to sensory analysis and production with motor abilities. However, one must consider the enormous contribution of memory and personal history in comprehension and pragmatic use of language. Full language competence (with variability) occurs around age 12–14, with subsequent increased vocabulary and more complex use of grammatical rules (Spreen & Risser, 1991). Here, as elsewhere, discrepancies in components require wide-range measurement to correctly assess adaptive status. For example, children with focal brain damage offer many discrepancies between language comprehension and expression. They are significantly delayed in vocabulary development (Thal et al., 1991). Severely injured young children exhibit greater persistent impairment of expressive language contrasted to receptive language (Levin & Eisenberg, 1991).

Anomalous language representation (ALR): This discussion of ALR is based upon (Kinsbourne, 1988). Seventy percent of left-handers who suffered early brain damage exhibited (as expected) left-hemisphere speech. When language dominance shifted to the right, it did not become bilateral (except for 15% of cases). Thus, handedness dominance shift (from left to right) does not require language representation shift at the cerebral level. Normally functioning left-handers, and individuals with developmental language and learning delays, tend to have bilateralized rather than right-hemisphere control. Even in the left hemisphere, nonright-handers may have a wider territorial representation for language, accounting for the fact that they run a greater risk of sustaining aphasia after LH damage (although primary processing is on the right). It is unsubstantiated that bilateral language representation is either immature or engenders inefficient language use. *Cortical areas can seemingly be used for different purposes, provided that this is done at different times.* Overlapping specialization of the cortex does not lead to significant psychometric deficits in left-handers. Dextrals performed better than matched sinestrals in simultaneous or competing verbal and spatial tasks, a finding attributed to interference across cognitive domains or inappropriate coding.

INTERVENTION AFTER SCHOOL REENTRY

One goal is to prepare for recovery and prevent the development of secondary psychiatric or stress-related disorders. The information provided, and the style of its delivery, depends upon the context, the child's developmental level, and the symptom severity. The child should be given an adequate understanding of what happened in the accident and what can be expected (Kirkwood et al., 2008).

Noncognitive Contributions to School Failure

Premature entry should be discouraged, since rest may be needed to prevent symptom exaggeration for cognitive or physical activities. School personnel should be made aware of the injury to monitor for neurological deterioration, and for common postconcussion symptoms (e.g., headache, fatigue, concentration). The ability to compete is hampered by noncognitive deficits such as poor social skills, anxiety, and inferiority feelings, such as being unattractive, incompetent, and rejected. Emotional regression, or immaturity, prevents the child from meeting social expectations, antagonizing teachers and peers. The children feel inadequate and are easily upset. Inattention, distractibility, and motivational problems can mask other deficits, for example, impaired reasoning and planning.

Disturbed Behavior

Academic problems are attributable to both the TBI and to *disturbed behavior*. Conduct problems, for example, oppositional defiance, may be manifested as passive aggressiveness and poor motivation (Teeter & Semrud-Clikeman, 1997, pp. 76–77). TBI can cause children to exhibit uncontrolled activity, tiredness, short temper, depression, agitation, confusion, and discouragement. TBI children are susceptible to stress, sensitive to noise, and easily fatigued. Therefore, they should be gradually reintroduced to a school program, and school personnel should coordinate with rehabilitation professionals. These behavioral problems may occur a number of years after the injury, and appear be permanent in some children (Begali, 1987, p. 71; McGuire & Rothenberg, 1986). School participation is more than cognitive efficiency: Behavioral disturbance prevents learning at the expected rate, that is, reduced problem-solving ability, problems of attention, memory, and organizing assignments (Cohen, 1986). The child slows down the class, being unable to do what teachers take for granted in terms of performance, creating irritation for the teacher and classmates. Severe, but not mTBI, yielded an increased occurrence of new psychiatric disorders. In noninflected injuries there was an exacerbation of preinjury behavioral tendencies in children experiencing psychosocial adversity (Ewing-Cobbs et al., 1995). It may be that these conditions were not newly elicited by the injury, but were detected by closer postinjury scrutiny. Initial expression of a psychiatric disorder occurred in 50% of children with TBI, most commonly organic personality syndrome, major depression, attention deficit/hyperactivity disorder, and oppositional defiant disorder. This is predicted by injury severity, preinjury intellectual functioning, SES, child and family psychiatric history, and global family functioning (Yeates, 2000).

Children's Return to Sports

This issue is controversial, but raises the issue as to when a child athlete is traumatically recovered to permit return to risky activities. *Neglected issues* in formal criteria include coordination, mental speed, motor speed, judgment, and alertness. Nevertheless, one author suggests that establishment of a baseline, with deviation measured by neuropsychological examination has only reached the "investigational" level of validity. Nevertheless, if a student athlete has not reached the baseline level after several seeks, a comprehensive neuropsychological examination is warranted. It has been suggested that the following criteria be met before pediatric athletes return to play: Be asymptomatic physically, cognitively, and behaviorally, both at rest and with exertion; the neurological examination should be unremarkable; and no neuroimaging findings should be apparent, if conducted. It is reasonable to restrict high-risk activities for 1–2 weeks in consideration of the child's developing brain. The child should usually not return to play in the acute period, and never before proper evaluation is complete. Particularly for symptomatic youth, a graduated transition includes an interval varying with the symptom level. One also takes into consideration the emotional toll caused by activity restrictions. Psychosocial problems include anxiety, depression, reduced self-esteem, and loss of contact with friends and coaches. An injury support group may be helpful. Evidence for vulnerability in later stages is manifested by repeated concussions, a lowered threshold for concussion, or cumulative effects from prior injuries (Kirkwood et al., 2008). However, the concept that rapid occurrence of concussive accidents on the same day in children could cause catastrophic edema has come into doubt (Ropper & Gorson, 2007).

Academic Difficulties

Learning deficits accompany both right- and left-sided cortical and subcortical lesions. It may require several years to determine whether a child who, initially functioning without measurable loss, eventually loses ground with reference to his peers. Deficits are more likely to be immediately detectable in adults since inability by the child to perform may take years to be manifested. Only after a considerable length of time does the demand to learn new material expose deficits in comprehension and learning. Therefore, the writer recommends both close monitoring and reexamination no more than 3 years after an accident in cases in which the child seems to have no measurable cognitive deficits. Although it is frequently stated that the child has a greater capacity for recovery, follow-up studies indicate that their deficits are as great as adults (Ewing-Cobbs et al., 1985). Thus, lifelong subtle deficits, which perplex the patient, originate in childhood.

Contributing Factors

- Problems of reasoning, perceptual speed, ability to read, and written language (Parker, 1990, pp. 298–299, citing Arem & Ekelman, 1988).
- Executive function: Organizing, planning, and monitoring behavior. The deficits described by Ylvisaker (1989) as "cognitive sequelae of moderate to severe head injuries" would be categorized as deficits of mental efficiency and control.
- Processing inefficiency: Slow performance speed; inability to process large amounts of information or the length of an assignment.
- Age-norm deficits for increased mental ability: Reduced IQ; demands for comprehension and reasoning.
- Reduced development of language: communication disorder (reading, written language, spoken language (Parker, 1990, pp. 298–299, citing Arem and Ekelman, 1988); receptive and expressive language skills (Semrud-Clikeman, 2001, chapters 7 and 8).
- Memory/learning problems: Short-term memory and new learning; deficit in long-term memory for verbal material in severely head-injured students, more so than immediate recall.
- Attention and concentration: Children are easily distracted by peripheral auditory and visual stimuli. However, in a quiet environment (e.g., educational evaluation, or the psychologist's office) they are better able to marshal their cognitive resources, thus concealing their deficiencies in a performing environment.
- Nonverbal performance: Fine and gross motor, spatial organizational, and visualmotor deficits.
- Impaired social skills: These interfere with participating in rehabilitation programs, and create social isolation (rejection by peers and teachers). Such disorders cause reduced school performance, educational lag, and/or increased likelihood of enrollment in special education. Ultimately there are problems in obtaining employment or social acceptance on the job.

APPROACHES TO COGNITIVE REHABILITATION

The range of disorders and stages of development are so wide that multiple treatment approaches should be considered. Children with TBI have multidimensional needs that vary with time, as late-developing symptoms are expressed and age-related norms create new and higher demands.

Transition Issues

Transition involves the change from high school to adulthood (Todis & Glang, 2008). It includes an eligibility category for children who sustain TBI in childhood under the *Individuals with Disabilities Education Act (DEA)*. Successful outcomes for youth with disabilities include such features as planning directed to the need of the individual; community support groups and psychological services directed to the individual and family; the student and family are actively involved in transition planning; development of self-determination; self-management skills; problem solving; decision making; and autonomy.

The services offered can vary in intensity and quality. Teachers and administrators may not offer accommodations so as to not appearing partial. Children may be passed along by teachers untrained in TBI so that they obtain a diploma. Life skills for employment and independent living may substitute for academic training. More satisfactorily, special education may be offered for those with learning problems. Transition to independent postsecondary educational programs for the student and family may be delayed until too late. Even with successful later education, employment problems can occur. There are both social and reality challenges: Living at home may create difficulties in making friends; untrustworthy associates; money management; immaturity in handling stress and alcohol; pressure to meet graduation needs in excess of the student's ability to progress; staff who are unaware of the special characteristics of the TBI-affected student.

Success is enhanced by accommodations; counseling directed at inefficiencies; guidance toward socially appropriate occupations, social skills, and confidence; independent functioning; and strategies for managing cognitive and other TBI-related problems. Pressure to maintain realistic achievements; intermediate goals such as an associate's degree or an achievable professional license.

Direct School Intervention

Although there is little empirical evidence for the effectiveness of educational intervention, despite the uniqueness of children with TBI, they have many commonalities with children with other disabilities. In this approach, subject content is analyzed and then presented in a way that reduces student confusion and maximizes mastery of skills and concepts in such areas as academics (reading, mathematics, and spelling), social behavior, and functional skills (shopping; job-related). Its principles include:

- The goal is to achieve success independently of teacher support.
- The approach addresses learning problems characteristic of the student with TBI: organizational impairment; inefficient learning and memory; failure to transfer; weak orientation to the task; attentional impairment. There is training to apply skills to appropriate stimuli; teaching examples build upon prior learning and teaching of the required discrimination.
- Communication should be logical, unambiguous, and clear: The order of presentation avoids confusion between procedures with similar features; memorization of reduced by teaching strategies that are explicit and generalizable over a range of examples.
- Regulation: Careful monitoring, avoidance of errors, and immediate nonjudgmental correction procedures.
- A mastery model ensuring a high rate of success: Carefully designed curriculum materials; efficient instructional delivery; students do not progress to new material without demonstrating mastery of what has already been taught (Glang et al., 2008).

Criteria for Return to School

Ability to attend to classroom instruction; ability to attend to simultaneous sensory modalities; ability to work unassisted for 30 min or longer; ability to understand and retain information, to reason and express ideas, to solve problems, and to plan and monitor one's own performance; and self-control. Other concerns in planning postinjury education include the extended time needed for learning and the interference caused by injury, which may require additional support; physical accessibility; faculty and staff awareness; academic support; flexibility in test administration opportunity for social interactions; and availability of personal and career counseling.

Special Education

Children with moderate-to-severe head injury require special education services because of residual disability. Even when children have multiple injuries, the majority of those who sustain TBI are unlikely to receive coordinated planning for school reentry. Frequently, rapid physical recovery causes the cognitive and behavioral effects of TBI to be unrecognized or minimized. Initial residential placement or special education may be required. However, it is estimated that only 9–38% of children with TBI are referred for special education programs. It is necessary to differentiate between TBI per se and the broader term *acquired brain injury*, which includes stroke, aneurysms, and anoxia (Dettmer et al., 2007). The school may not be notified of the injury even after hospitalization. On the other hand, school personnel may be the first to notice changes in a child's ability to follow directions, complete projects, remember information, and get along with other children. Dysfunctions such as memory and speed of information processing may require a reduced workload, tutoring, and more time to complete assignments. The school personnel should differentiate between trauma-related dysfunctions and ordinary cognitive immaturity (Shurtleff et al., 1995).

Executive Dysfunction

The interventions for *EF* difficulty are minimal, although it is possible that specific interventions could lead to positive outcomes (Catroppa & Anderson, 2006). Study of a sample of children with a range of brain trauma determined that children with severe TBI performed most poorly during the acute stage and exhibited the greatest recovery of EF over a 24-month period. Functional deficits remained most severe for this group. It was concluded that EF has a multidimensional nature with differential recovery of skills. Long-term deficits remain and may affect development (Anderson & Catroppa, 2005).

Innovative Rehabilitation Techniques

Laatsch et al. (2007) use the rubric of "acquired brain injury" (closed, open, and blast; infections; tumors; radiation; stroke) as the target of new procedures.

- Primary, early engagement of family members. Family-based procedures offer cognitive and social gains.
- Early distribution of an information booklet to minimize parental and child stress and misattribution of TBI symptoms to other causes.
- A cognitive reeducation program to assist children to adjust to, and compensate for their learning deficits, during the transition back to school. The academic curriculum was supplemented by facilitating motivation and ability to learn. The transfer of learning to the classroom utilized problem solving, tutoring and metacognitive strategies in light of one's own cognitive processes.
- Customized interventions to train academic skills, that is, while the student was engaged in the cognitive task, ongoing assessment was utilized to improve self-awareness, maturation, and error detection.
- Computer-assisted remediation to enhance language skills.
- Immediate feedback during skill acquisition training increases self-esteem and satisfaction.

POSTTRAUMATIC EMOTIONAL AND BEHAVIORAL PROBLEMS

Clinical example: Behavior disorder: An 11-year-old boy (JRcbd) was knocked down by a car, with altered state of consciousness (awake, but couldn't even recognize his mother) and subsequent

"PCS" and nightmares. This is how his teacher communicated his behavior on the report card: (Period 1) "Julio's behavior is terrible. Very little work gets done in the classroom and homework is never done! He is reported by every teacher because of his conduct...; (Period 2) Julio's behavior is impossible! Very little work gets done! He is "off the wall" almost all day long!. Please try to get some professional help for him. He should read an hour each day! He continues to talk "nonsense." (Period 3) Julio has been very disruptive these past few weeks. He needs to work on his self-control. His school work has improved, and I am pleased Julio will be attending summer school."

Incidence

The degree of increased incidence of postinjury behavioral disorders is controversial. On the one hand, Pelco et al. (1992) found that children with mild head injuries do not show a significant increase of behavioral problems in the first 12 months after injury. Yet, the incidence of psychiatric disorders among children crippled by brain injuries appears to be twice that of children crippled by other physical handicaps. Di Leo (1970, p. 242) observes that the more intelligent the child, the more likely he is to have emotional problems stemming from his inadequacies. A group of children with severe head injury (criterion is PTA ≥ 1 week) but without intellectual impairment had more new psychiatric disorders than the controls. More emphasis was placed on indirect rather than direct mechanisms. The presence of nonhandicapping physical disabilities did not increase the likelihood of a psychiatric disorder, but the presence of psychosocial adversity did.

The question of incidence is somewhat clarified by considering the trauma severity. There may be an increase in symptoms internalizing or externalizing style (mild or severe CHI). However, if larger units of behavior are considered (adaptive), then only severe CHI cases may be affected. However, when preexisting conditions are controlled, mild CHI has a relatively small effect, restricted to increased hyperactivity scores (Astern et al., 1991). Brown et al. (1981) also found that new psychiatric disturbance occurred in the severe head injury group. They were two to three times as frequent as in the controls of mild head injury group, and were associated with the upper extreme of PTA, occurred more frequently with neurological disorders, and were associated with transient or persistent intellectual impairment.

After so-called "minor" head injury (not associated with LOC), only 7% of children are likely to complain of headaches after 1 month. However, "there is a high incidence" of alterations in play, daily activities, and school absenteeism to the point that rates are twice population norms (Bruce, 1990). Even after "normal neurological examinations" some children display emotional disorders, probably a secondary reaction to diminished perceptual and cognitive ability associated with the brain injury (Chelune & Edwards, 1981).

Personality and Mood Disturbances

Children may not reveal their posttraumatic feelings, either by inten, or as a symptom of a motor vocalization disorder. After trauma, children express irritability, impulsiveness, disinhibition, decreased frustration tolerance, fatigue, poor anger control, hypoactivity with reduced motivation and initiative, aggressiveness and hyperactivity; reduced compliance; diminished behavior (reduced motivation, apathy, reduced initiative and fatigue, which can be perceived as laziness by outsiders or self-described as boredom; stress reactions; temperamental changes; loss of emotional control; feelings of rejection; pretended academic indifference; impaired sense of identity and self-acceptance; reduced development of autonomy; feeling vulnerability in a dangerous world; anxiety; withdrawal; emotional constriction; regression; propensity to pain; depression, social disinhibition or acting out inappropriately; irritability; reduced judgment and motivation; reduced frustration tolerance; reduced sensitivity to others; increased demanding behavior; feeling not like themselves, that is, out of control or crazy, anxiety, depression, impulsivity, distractibility, disinhibition, decreased frustration tolerance, apathy, reduced initiative, fatigue which can be perceived as laziness by outsiders or self-described as boredom, depression, social disinhibition or acting out inappropriately (Dean, 1986; Lehr, 1990, p. 160; Lehr & Lantz, 1990; Parker, 1990, pp. 303–307).

Disinhibited/Inappropriate/Disruptive

Behavior disturbances create a reaction by parents, school, and peers. This affects the child's perception of the accident and the personality consequences. Problems of control are conspicuous, for example, "disinhibition and socially inappropriate behavior reminiscent of the adult frontal lobe syndrome" (Lishman, 1987, p. 173). Co-morbid severe TBI and ADHD is a predictor of reduced response inhibition (Jorge, 2006). The most characteristic disorder attributed to severe brain injury was disinhibition or socially inappropriate behavior (outspokenness, nakedness, noise, kissing researchers, carelessness in personal hygiene, impulsiveness). Additional symptoms of the more severe group were speech abnormality, refusal to cooperate, and distractibility. Hyperactivity and stealing were more characteristic of preinjury than postinjury behavior (Brown et al., 1981). Persistent effects impede long-range adjustment (Chelune & Edwards, 1981; Tramontana & Hooper, 1989).

Temperamental Changes

Temperament has been defined as the nonverbal expression of energy and emotion (Parker, 1981, chapter 5) and behavioral traits that appear early and consistently (Lewis et al., 1988). "Difficult" patterns of temperamental characteristics play a significant role in the development of behavioral disorders (Thomas & Chess, 1977, p. 46). Temperamental problems can lead to adaptive deficits in school and problems with social acceptability. Brain damage enhances poor impulse control by reducing social inhibitions and impairing cognitive, adaptive, and defenses of the ego.

Temperamental problems frequently observed after brain damage include overactivity, restlessness, impulsive disobedience, explosive outbursts of anger and irritability, lying, stealing, destructiveness, aggression, increased tantrums, impulsiveness, socially uninhibited behavior, poor goal orientation (distractibility and lack of persistence), low frustration tolerance, and poor motivation. The *"difficult" child* further irritates the people around him, does not meet the values and expectations of parents and teachers, and does not fit in well with a peer group.

Risk Taking

Risk-taking behavior creates vulnerability to further injury. Thus, *premorbid behavior problems* may increase the risk of head injury. Risky behavior may be enhanced after a head injury since foresight, judgment, and insight are impaired. Return of a vulnerable child who has incurred TBI to a disorganized family increases the risk of further head injury. Behavior symptoms before and after injury correlate with *injury severity.* A group of children with TBI had more behavior symptoms than an orthopedic group. The lack of association between behavior symptoms and neuropsychological outcome was attributed to the greater recovery of cognitive functions, although it was possible that the test battery was insensitive to subtle residual deficits (Barry et al., 1996). Psychiatric disorders are more persistent than cognitive disorders, which may be compensated for.

PSYCHIATRIC OUTCOME

Childhood head trauma seems to predispose for subsequent schizophrenia, that is, in comparison to depressives, manics, and surgical controls (Wilcox & Nasrallah, 1987). One sample was characterized by these features (Patrick & Hostler, 2002).

Axis I disorders: Major depression, dysthymia, bipolar disorder, anxiety, panic disorder, obsessive-compulsive disorders, PTSD, phobia and substance use disorders.

Axis II disorders: Personality disorders predominated, for example, including borderline, avoidant, paranoid, obsessive-compulsive, and narcissistic features.

Functional Outcome

For children, there is little correlation of GCS score to functional outcome. This is contrary to the adult literature, which demonstrates a moderately high correlation between initial GCS score and functional outcome. In a study of children and adults in a comprehensive inpatient rehabilitation program, children recovered more completely and efficiently than adults, older children more so than younger children. The effects of adult modifying variables require careful consideration of the individual before being extrapolated to the pediatric TBI population (Niedzwecki et al., 2008). One significant difference between children and adults is the adult's retained material: facts, procedures, coping skills, and also more adult supporters. For the adult TBI, one assesses the loss or impairment of a previously acquired function. Assessment of the consequences of TBI in a child or adolescent is different. A guiding principle is that a child's capacity is composed of current level of performance plus expected undeveloped functioning. A significant cognitive loss is not immediately expressed; if it is, one may surmise a relatively serious TBI. Thus, the younger the child, the less has overtly been lost but the greater will be the extent of future psychological non-development. The *prefrontal cortex* is implicated because of its prolonged developmental course (McKinlay et al., 2002, 2008).

Behavior Problems

Children with severe head injuries demonstrate more behavior problems and poorer adaptive functioning than do children with mild injuries or those with injuries not involving the head. However, adaptive deficits and behavioral disturbance are also related to factors other than injury severity, including the children's premorbid functioning (Yeates, 2000). Disturbed behavior interferes with adaptive functions (e.g., communication, daily living skills, socialization, school success). Personality changes are greater in children than adults. Inability to succeed and to be accepted contributes to depression, which in turn hampers ability to function competently in school. Rutter et al. (1983) concluded that cognitive deficits may become much attenuated after 2¼ years postaccident, but psychiatric disorders continued as a persistent problem. Disturbances are related to the severity of injury, and are frequent when 1-year follow-up studies are performed. A higher proportion of children show emotional after-effects and increased psychiatric risk after moderate-to-severe brain trauma than mild injury (Rutter et al., 1983). Although childhood brain damage is not clearly associated with any particular diagnosis (with the probable exception of depression), having brain damage reduces one's adaptability and therefore capacity to deal with the ordinary demands of life. Associated fright, scars, etc., add to problems of adaptation.

Cerebral Personality Functions

Dysfunctions after frontal lobe damage are highly variable, but a familiar subtype includes impulsivity, lack of concern and foresight, absence of depression, boastfulness, and impairment of self-awareness (Benton, 1991). It is claimed that clinical problems are more associated with right cerebral dysfunction due to loss of socialization abilities, together with the expected motor and cognitive dysfunctions. It is vital to include medical conditions in the assessment, for example, panhypopituitarism can be mistaken for an amotivational syndrome.

PSYCHODYNAMIC POSTTRAUMATIC EMOTIONAL PROBLEMS

Maturity, Independence, and Autonomy

Children, more so than adults, experience being overwhelmed by *helplessness* after brain damage. They do not have the memory of, and therefore the hope of returning to, independence and achievement. The parents' protectiveness becomes increased, further impairing the child's ability to learn autonomy and to enjoy self-reliance.

After TBI, maturity and independence are more difficult. Brain insults in children affect family life and are associated with problems of functioning after brain damage. Difficulties are created by poor motor skills, increased likelihood of unemployment, troubled family relationships, poor health, and neuropsychological dysfunctions. Families have different resources to help the injured child (Taylor & Alden, 1997). Families also offer different degrees of initiative to mobilize community resources for the child. Good outcome is associated with family cohesion and support. On the other hand, there is an increased risk of psychiatric disorders in families experiencing psychosocial adversity, including mental disorder or marital problems. Nevertheless, socioeconomic status and family functioning were relatively unimportant predictors of outcome in a study of children's head injury over a wide range. In fact, the possible effects of social disadvantage in determining outcome were not determined in one study. Family problems may emerge later, or deficits enhanced, due to the restrictions and other problems associated with caring for an injured child (Anderson et al., 1997). This group seems to have changed its view with later data, since it was asserted that postinjury child and family function were less associated with injury factors and more dependent upon preinjury psychosocial functions, consistent with other studies (Anderson et al., 2005a).

Impaired Sense of Identity and Acceptance

The child's sense of identity, and awareness of his body, and development of a normal sense of self are grossly impaired by brain damage. Impaired bodily image results directly from brain damage and from the somatic and psychic effects of trauma. The sense of self remains primitive, although overt behavior may seem to be relatively mature (see Chapter 7 on parietal lobe and bodily image). They continue to experience themselves at the level of development, or even more regressed, than before. The injured child, particularly one who has undergone hospitalization and has scars from the accident or surgical procedures, experiences restraint due to casts, pain, and loss of mobility; is anxious about competing with peers; angry at the pain and loss of progress; may blame himself for the disaster; and fears disfigurement.

The kind of *ego distortion* will vary with the age at which the damage occurs (Pfeffer, 1985). During the first 2 years of life, the child has insufficient self-other differentiation. Illness is perceived in terms of terror and total dissolution of their world. The school-age child, who does not fully understand how the body functions, may perceive the after-effects of the trauma idiosyncratically, that is, their relationship to the parents and their own bodily functions are interrelated. In adolescents, body image and self-esteem are interwoven, and there are fantasies of bodily disintegration, loss of control, and shame.

Physical disability, illness, and varied kinds of incompetence create fears of rejection, abandonment, or loss of love. These are compounded by school failure, criticism, and sensory and motor deficits. Scorn and frustration result from being impaired, unable to keep up with one's peers, and then rejection. The brain-damaged child feels incompetent, scarred, unattractive, unliked, and damaged, and experiences pain, loss of mobility, loss of balance, etc. Further negative feedback actively adds to the already low self-esteem, social isolation, and belief in a bleak and unfriendly future. The lack of positive feedback, that is, absence of reinforcement, causes poor motivation to meet tomorrow's tasks.

Vulnerable in a Dangerous World

The author has observed that adults who were overwhelmed as children frequently have impaired *stimulus barriers*. The most casual events, for example, passing people on the street, or continued exposure to a noxious but not necessarily dangerous person, are experienced as though they were defenseless against some great threat. It can be inferred that the developing child who has been struck down by a major force will experience himself as vulnerable, that is, believing that he may be damaged more easily than other people.

Personality change was associated with severity of injury in children 5–14 years old, rather than psychosocial variables. It is considered to be an affective dysregulation of the dorsal prefrontal cortex system, specifically the superior frontal gyrus (Max et al., 2006).

Reduced judgment exposes the child or adolescent to risks of further injury, that is, including use of alcohol (with likelihood of greater effects of subsequent brain damage), reckless driving, other dangerous acts that are self-motivated due to poor judgment, or may occur because of the inability to withstand social pressure, dangerous actions in sports, combativeness, etc., to which children and youths are prone. The child's predisposition to further injury through impulsive risk-taking behavior may be poorly controlled by their family (Green et al., 1995).

After an injury, children experience themselves as different. The change of identity can be focused on the loss of specific abilities. They may be pessimistic concerning recovery, and perplexed as to why their behavior is extreme, unpredictable, or inappropriate. Disturbed behavior interferes with adaptive functions (e.g., communication, daily living skills, socialization, school success).

Frequency of Emotional Problems

The incidence of personality difficulties is influenced by several factors: severity of injury (to the point where self-awareness is reduced); preinjury adaptation (i.e., children with satisfactory preinjury functioning are less likely to develop new disorders); psychosocial adversity in the family and home environment (large family size and overcrowding; parental psychiatric disorder or criminal behavior; low social status; foster care; discordant family relationships) (Lehr, 1990, pp. 155–157). The degree of increased incidence of postinjury behavioral disorders is controversial. On the one hand, Pelco et al. (1992) found that children with mild head injuries do not show a significant increase of behavioral problems in the first 12 months after injury, yet, the incidence of psychiatric disorders among children crippled by brain injuries appears to be twice that of children crippled by other physical handicaps.

Parental Reactions

Lehr (1990, pp. 169–175) reviewed parental reactions to their child's impairment: persisting emotional disturbance; overprotectiveness; denial; personal disorganization; family overloading; paradoxical reaction (my term) (i.e., appreciation for hyporeactivity in a formerly poorly controlled child). Families of children with severe TBI suffer greater stress than families with OI. Anticipatory guidance concerning problems should be offered. Parental coping styles may be insufficient for ongoing stress (e.g., relations with other family members). Health care providers may have to probe in order to obtain emotional concerns and distress (analogous to the expressive deficits described for TBI patients (Wade et al., 1996). Moreover, there may be an adversarial relationship between medical professions and the school personnel. What the health care professionals recommend may be beyond the capacity of the schools to provide, while rehabilitation staff may suggest that medically based goals be incorporated into an education setting without considering the school context, resources, and academic objectives. A program has been described to improve parental communications skills to aid them as advocates (Glang et al., 2007).

FAMILY AND SOCIAL INFLUENCES UPON OUTCOME

Cross-cultural Issues

It is necessary to differentiate between injury-related and environmental variables in accounting for outcome. Precision of assessment is reduced by the unavailability of suitable norms, for example, dialect differences within presumably defined groups (e.g., Spanish and Arabic speakers). The parents of black and white children with TBI and orthopedic injury were studied. There were ethnic differences in preferred coping strategies, psychological distress, and perceived family burden, independent of SES. The initial negative consequences of TBI were less pronounced for parents of black children at baseline, but became more pronounced at follow-up (Yeates et al., 2002). Injury severity does not account for most of the variance in pediatric TBI; family and school influences play a role. There is evidence from studies of chronic childhood illness that behavioral adjustment is linked to SES, family stressors and resources, and parental and family adjustment. Environmental disadvantages predict lower scores on most tests for children with and without brain injury, and also amplify the effects of TBI on some tests (Taylor, 2008). Children with a wide range of TBI as measured by the GCS were studied to assess the relative significance of family and seriousness of injury upon outcome.

Psychosocial Level

Family variables include socioeconomic status, family stressors, family status, and demographics. The measures of family environment accounted for up to 25% of the variance in level of outcome, and 5% of rate of change. GCS score accounted for 20% of level of outcome, and 15% of the variance involving rate of change. The limitations of this study were acknowledged to be limited scope of outcome measures and not including volumetric measurement of the lesion (Yeates, 2000; Yeates et al., 1999).

Educational and cultural variables: Lack of familiarity with test taking; varying levels of linguistic competency; SES; lack of familiarity with writing implements; and level of formal education. A psychological procedure may emphasize particular cognitive abilities, which may be discrepant with a culture, or the stage of development at which particular abilities are learned. This may affect later neuropsychological performance. Further, the examiner is concerned whether a procedure normed on one segment of a population speaking a given language is generalizable to another without significant performance differences. A history of discrimination, or growing up in totalitarian regimes, may result in an attitude where "honest" behavior is rarely rewarded, therefore affecting the presentation of claims to obtain proper compensation.

Adverse psychosocial outcomes: Hyperactivity, inattention, and conduct disorder were found in children between the ages of 10 and 13, especially if the injury occurred before age 5. There were no definite cognitive deficits, irrespective of severity of mild injury or age of injury. There may be a level of mild injury for which there are adverse outcomes (suspected concussion and LOC for not more than 20 min) and one in which there is likely to be no detectable deficit. The outpatient and uninjured (reference) groups manifested one symptom. The inpatient group manifested two to three problem behaviors and large effect sizes for hyperactivity/inattention and conduct. It appeared as though teachers were more reactive to the child's hyperactivity were than the mothers. It was suggested that outcome depends upon the severity of the injury, the particular measurement (reliability and sensitivity), and whether the injury occurs before age 5 (McKinlay et al., 2008). Children who sustained deeper brain lesions had worse outcomes, as reflected by their Adaptive Behavior Composite Score on the Vineland Scale and GCS scores (Levin et al., 1997).

Social Rejection

Varied social disturbances lead to social rejection and disorganization. *Positive symptoms*: overbearing; intrusive, histrionic, threaten suicide and homicide, disruptive acts with little apparent affect, causing rejection because of unpredictable and volatile behavior (mood swings occur years later). *Negative symptoms*: hypoprosodia, hypoemotionality, inappropriate expression of and response to nonverbal social cues in others, difficulty in expressing and understanding prosodic components of language, giving the appearance of being disinterested, flat, or socially inept, sad or depressed (Brumback & Weinberg, 1990).

Reduced personality development (psychologically or physiologically) impairs social integration, and is an invitation to rejection. With increasing age children may not adapt to their peers' increasing bodily and social development. Their ability to get along, to learn social skills, and to participate are hampered by deficits of coordination and strength needed for games and other activities, reduced reinforcement from academic success, and reduced self-esteem due to scarring caused by impact and surgical interventions. Children experience loss of peer approval and become self-conscious due to the notoriety of the accident, being absent from school, changes in physical appearance, behavior, and adaptability.

After an initial period of support, friends drift away because of the inability of the youth to participate as adequately or fully as before. In addition to the usual concern about attaining independence and earning a living, the children worry about basic skills and attractiveness. Inability to develop and succeed at the same rate as one's peers, scars, and various disorders result in ridicule, peer impatience, etc. The child becomes fearful of rejection and is socially isolated. Children pretend indifference because of the pain of failure, leading to antagonism by teachers and peers. They are forgetful, experience "absence" attacks, and concentrate poorly. This, in turn, results in poor learning, social reinforcement as an inadequate person, identity problems, etc. Perhaps the child has encountered repeated failure and loss of self-esteem (Tramontana & Hooper, 1989, citing Dorman, 1982). During development, self-esteem and identity will be determined by the quality of exchange with other youth in school and neighborhood. Delays and change in vocational and academic progress cause isolation from friends, as well as assignment to classes of children that are younger. Poor neuropsychological performance is associated with externalizing behavior problems in boys 7–8 years old, but internalizing symptoms in 9–14-year-old boys.

Family Functioning

Brain insults in children both affect family life and are associated with problems of functioning after brain damage. Families have different resources to help the injured child (Taylor & Alden, 1997). Health care providers may have to probe in order to obtain emotional concerns and distress (analogous to the Expressive Deficits described for TBI patients (Wade et al., 1996). Families also offer different degrees of initiative to mobilize community resources for the child. Good outcome is associated with family cohesion and support. There is an increased risk of psychiatric disorders in families experiencing psychosocial adversity, including mental disorder or marital problems.

Family distress increases over time. Failure of the family to overcome problems of coping with the behavioral and cognitive consequences may jeopardize recovery and contribute to longer-term behavior problems. Additionally, there are realistic problems: the cost of care, the inability of the school to provide service. Nevertheless, SES and family functioning were relatively unimportant predictors of outcome in a study of childrens' head injury over a wide range. In fact, the possible effects of social disadvantage in determining outcome was not determined in one study. Single-parent households, those with less cohesiveness, and those with prior stress do not facilitate recovery. Higher behavior problems are associated with higher preinjury ratings, lower preinjury adaptive behavior scores, and poorer family functioning. Both pre- and postinjury family characteristics must be considered in assessing outcome (Taylor et al., 1995). Parents may experience anger, fearfulness, and/or guilt and self-blame. Disappointment can occur if initial progress is not maintained and they become aware of lasting deficits (Yeates, 1994). Family problems may emerge later, or deficits are enhanced, due to the restrictions and other problems associated with caring for an injured child (Anderson et al., 1997).

Referring to the burden of the parents of children studied 6 and 30 months after injury, the burden was high at both intervals, and was predicted by injury severity, functional impairment, and post injury child behavioral disturbance. Families of children with severe TBI report a higher level of burden and stress after 30 months (Anderson et al., 2005a).

Cultural Differences and SES

The rate of TBI in inflicted and accidental injury is related to lower *SES*. After MVAs, homicide is the most common cause of injury fatalities in children and adolescents (Ewing-Cobbs et al., 1995). The reported sequelae are unsurprisingly similar in different nationalities. Special concerns may arise because of cultural differences within a family in approaching such problems as illness. Some individuals do not understand the health care system, or may not want to challenge professionals. Brain injury causes shame in some ethnic groups; therefore, family members may not be informed of a relative's TBI. Cultural differences affect response to the health care system. Obtaining information may require the use of a translator, since friends and family may not be fluent in medical terminology; problems occur when they are not oriented as to the special characteristics of brain injury. Even these personnel may not be properly oriented to work with a TBI patient. To the extent that parental origin affects the caregiver style, the child's outcome is affected. African Americans and whites differ in health perceptions and methods of coping with negative life effects, which are not accounted for by differences in SES. African Americans assess their own health and their children's health as poor, reflecting lower access to adequate medical care. Coping strategies may differ: African American parents tend to rely more on religion, mental disengagement, and denial in coping with traumatic injuries. Avoidant strategies predict poorer family outcome than acceptance (Yeates et al., 2002). However, certain aspects of coping style are more related to SES than to race. Higher SES parents use active, problem-oriented strategies, which predict a higher level of distress than for parents of children with OI.

Since dementia is a potential outcome of TBI, and race plays a role in expression of its various manifestations, this may be a consideration in differential diagnosis (Gouvier et al., 2002). Examination of a child who is native to a culture using a language significantly distinct from that of the examiner requires care against mistaking characteristics of the language with familiar neuropsychological disorders. Using Japanese as an example, there are differences of syntax, lexicon, orthography, phonology, etc. The major European languages utilize the word order subject-verb-object, while Japanese utilizes subject-object-verb, which might suggest an aphasic disorder. Lateralization of verbal functions, which are dependent upon the visual structure of language, may also be reversed (to the left cerebral hemisphere as opposed to the right). Nevertheless, particular symptoms occur across languages, for example, aphasia (inability to understand or express oneself through language) and agrammatism (speech unconforming to grammatical rules). Acculturation can be categorized as traditional, marginal, bicultural, and assimilated. Particular procedures are normed on groups that may be socioeconomically or ethnically so disparate from the patient (rural and/or international) that they inaccurately indicate functional weaknesses and fewer strengths than they express in their environment. Further, those from rural environments are exposed to different conditions than those in urban environments: patterns of interpersonal violence (including intimate partners), types of neurotoxic exposure, stress, agricultural accidents, lack of medical coverage and treatment). These represent a potential for different preexisting conditions that must be considered in studying examination performance.

CHILDREN'S STRESS EXPERIENCE

Special Characteristics of Childhood Stress

Stress is not a unitary concept, and may be roughly categorized as (1) discrete, encapsulated traumatic events, and (2) chronic, pervasive trauma (Terr, 1991, cited by Perry & Pollard, 1998).

Analogous to childhood TBI, stress effects are influenced by current incomplete development and potential impairment of future development (emotional, behavioral, cognitive, social, and physical). Alternative diagnostic consequences to PTSD include dissociative disorders, major depression, attention deficit hyperactivity disorder, and various developmental disorders (following severe early life neglect and trauma).

Stress has been conceptualized as a fearful event, creating a set of memories that are generalized and then reactivated from associations, leading to sensitization of catecholamine systems, which leads to arousal, and dysregulation of numerous functions. The child is in a persistent fear state, that is, in a homeostasis of anxiety. The infant is not capable of activity such as fight or flight. The initial expression of distress is a precursor of hyperarousal. The limited behavioral repertoire is used to attract the attention of a caregiver: changes in facial expression, body movements, and vocalization. An inappropriate caregiver reaction may be stimulated by children who are fussy, difficult, or weepy.

Imprecision in Identifying Children's PTSD

There are both gaps and imprecision in our knowledge of children's TBI:

- Injury can cause varied developmental delays, which are commonly not attributed to the traumatic cause: disorders of physiological maturity; cognitive, personality, sensorimotor functioning; social adjustment; educability; and, capacity for employment.
- Children's head injury is commonly ignored as an event and as a potential cause of neurobehavioral dysfunction. For example, it is not even referred to in a review of PTSD in children (La Greca, 2000).
- There is no generally accepted classification of the levels of TBI, that is, mild, moderate, severe. To each category, different levels of the GCS, length of LOC, and of PTA have been assigned. Co-morbid trauma such as skull fracture or subdural hematoma may or may not be considered in the criteria for TBI severity (complicated vs uncomplicated) (Prigatano et al., 2008).
- Many symptoms in injured children, as in adults, may require further study to determine their precise etiology.
- It is more of an error for children than adults to assess permanent outcome in the acute condition than waiting a number of years to observe later dysfunctions and inability to achieve expected developmental markers.

Incidence of Children's PTSD

One report indicated that full PTSD symptomatology occurred 1 month postaccident in children who had been hospitalized for an accident (Allen et al., 2008).

In one study PTSD occurred in 25% of the children and 15% of the parents. After the child incurred a traffic injury, only 46% of the *parents* sought help of any kind for their child and only 20% of the parents sought help for themselves (De Vries et al., 1999). A literature review indicated that between 9% and 31% of parents develop PTSD-like symptoms after a child's injury. Their own PTSD symptoms affect their perceptions, leading to discrepancies between parent and youth symptom reports. Parental assessment of the child's condition might not agree with the child's self-report. Discrepancies seem related to greater levels of parental posttraumatic discourse. There is some correlation between parents' PTSD symptoms and their assessment of their child's PTSD symptoms (Ghesquiere et al., 2008). A significant minority of children manifest considerable PTSD symptoms up to 12 months after the accident. Sixty-eight percent of children in one study had at least one PTSD symptom. PTSD symptoms may be consequent to cognitive impairment secondary to TBI interacting with cognitive development (Bryant, 2001; Keppel-Benson et al., 2002). Those hospitalized after a crash, particularly for girls, have poorer adjustment (Curle & Williams, 1996).

Children's Resilience

Protective factors: Male gender; good cognitive abilities and ability to understand the stressor; good self-esteem and sense of self-efficacy; self-meaning in life, and talents valued by both self and society.

Family: Supportive and well structured; appropriate level of cohesion; open parent-child communication; caregiving adults involved in education; socioeconomic advantage.

Nonfamily characteristics: Supportive adults; peers who are supportive; effective educational setting; high level of public safety; available health care for the child; supporting institutions.

Children's Risk Factors

Child characteristics: Female gender; preaccident psychopathology or alcohol use; psychological reactions during or after the event; pain experience after the event; 12 years of age or older is predictive of PTSD or other psychiatric symptoms; younger age is predictive of acute stress symptoms.

Trauma characteristics: High level of exposure; severe, and/or prolonged event; injury to an extremity; parent involved in the accident.

Disturbed social environment: Disruption of social support system; a family with multiple problems, chaos, or disorganization.

Stress Outcome

Personal growth has been measured independently of the level of PTSD in an adolescent sample (grades 7 through 9). The theory that PTSD disturbs functioning and so is inversely related to personality growth was rejected. Rather, these are separate outcomes that coexist independently. Growth was represented as a general factor, with these components: (1) interpersonal relationships, and (2) interpersonal competence (self-perception and philosophy of life). The highest levels of growth were achieved by individuals with average PTSD levels, that is, moderate levels of PTSD facilitate the most growth (Levine et al., 2008). PTSD in children may present differently than in adults, including regression and higher levels of self-blame (Miller, 1999). Yet, the characteristics of children's stress are not specified in *DSM-IV* except for the substitution of disorganized or agitated behavior in place of fear, helplessness, or horror. The percentage of children who develop PTSD after accidents has been estimated between 4% and 50%, with the percentages being somewhat higher immediately following the accident. PTSD in children is accompanied by Rorschach signs of disordered thinking and inaccurate perception. Moreover, responses seem to represent violation of the child's naive belief that the world has predictable rules, people are trustworthy and fair; thus, punishment and pain are consequences of bad behavior (Holaday, 2000). Prediction of children at high risk for PTSD emphasizes reexperiencing, avoidance, and hyperarousal, but not dissociation (Bryant et al., 2007). There seems to be a linear relationship between *cumulative childhood trauma* that was experienced by university women, that is, total number of different types of childhood traumatic events, and symptom complexity. The practitioner should consider that child abuse and rape, in particular, can have severe, diverse psychological impacts. Accumulated trauma plays a role beyond specific trauma exposures (Briere et al., 2008). Children hospitalized after a crash, particularly girls, have poorer adjustment (Curle & Williams, 1996).

Complexity

One differentiates between preexisting conditions, the event-related trauma, posttraumatic stressors (medical conditions and treatment, feeling of impairment, rejection), the circumstances of recovery (support or frustration and denial of treatment), cerebral personality disorders, and exacerbation of a preexisting condition. Children, as do adults, express dysphoria in different ways: experience it inwardly (*internalize*); conceal it (*expressive deficits*); or act it out in social misbehavior (*externalize*). In the hospital the perception of life threat and the intensity of medical/surgical treatment is an important risk factor. The dose of morphine for burnt children diminished posttraumatic symptoms over 6 months. This may have been due to blocking fear conditioning in the locus ceruleus and amygdala. One should consider *preventative interventions* in injured children (Saxe et al., 2003). *Conditions co-morbid with PTSD*:

- *Diagnoses and symptoms*: Acute stress disorder; adjustment disorder; separation anxiety disorder; generalized anxiety disorder; depressive symptoms; attention deficit hyperactivity disorder; oppositional defiant disorder; conduct disorder; major depression; dysthymia; avoidant disorder; overanxious disorder; separation anxiety disorder; simple phobia not related to the accident; and PTSD unrelated to the accident (Allen et al., 2008; Keppel-Bensen et al., 2002; Yeates, 2000). The onset of a *"novel" psychiatric disorder*, that is, one never before present, occurred in nearly 50% of all children in a sample (CHI). The most common novel diagnoses were organic personality syndrome, major depression attention deficit/hyperactivity disorder, and oppositional defiant disorder. The onset of a novel psychiatric disorder was predicted by injury severity, preinjury intellectual functioning, SES, child psychiatric history, and global family functioning (Yeates, 2000).
- *Arousal/physical*: Motor hyperactivity, anxiety, impulsivity, sleep problems, tachycardia, hypertension, anxiety, impulsivity, stomachaches, headaches.
- *Psychological*: Diminished academic performance.

Styles of Retreat

- *Learned helplessness*: This is the opposite of use-dependent behavior, that is, a disuse-related behavior. It is a sense of defeat response and an overtly calm and accepting reaction should not be confused with resilience.
- *Dissociation*: This involves disengaging from the external world and attending to stimuli in the internal world (being distracted, avoidance, numbing, daydreaming, fugue, fantasy, derealization, depersonalization, fainting, or catatonia). There is a kind of depersonalization or derealization (going to a "different place," changing identity, a sense of floating, being in a movie, etc.).
- *Freezing*: In humans, it is considered analogous to the animal's loss of movement to prevent localization by a predator. By impairing thinking, freezing allows escalating anxiety to plateau and give the person a chance to mentally recover. Children rarely understand why they are anxious. During this state they are easily overwhelmed and less capable of processing complex information. They feel out of control and can cognitively and physically freeze. An adult's lack of awareness can lead to the instruction to do something, leading to refusal, with consequent persistence in the order with threats. This increases the child's loss of control and anxiety (Perry & Pollard, 1998).

The Acute Stress Disorder

The acute stress disorder has a low prevalence in children and adolescents involved in MVAs and assault (8–19%). It differs from PTSD by utilizing the criterion of dissociative symptoms. ASD has a moderate success in predicting PTSD, while ASD without the dissociation criterion (*subsyndromal ASD*) is an equally good or superior predictor. Thus, dissociation is one pathway to PTSD, while *separation anxiety* is another (Meiser-Stedman et al., 2007). When biological stress systems are dysregulated, there is an enhanced vulnerability for psychopathology (particularly PTSD and depression). This creates an increased risk for onset of adolescent or young adult onset of alcohol or substance abuse disorders (DeBellis, 2002).

Reexperiencing with Hyperarousal

One criterion for the PTSD is reexperiencing, although the *DSM-IV-TR* notes that in children there may be frightening dreams without recognizable content. Children under the age of 5 at the time when trauma ended did not report current replicative dreams at all (Schreuder, et al., 2000). This contrasts with adult findings in which a majority of nightmares were replicative of the event or mixed replicative and nonreplicative. The effect of the event depends upon the age of the child (Hornstein, 2000). Children manifest PTSD symptoms such as psychogenic amnesia, flashbacks, and infrequent dissociative phenomena. Children with dissociative disorders have difficulties regulating attention and concentration. If they manifest inattention, hypervigilance, and increased startle reactions, this can contribute to misdiagnosis of attention deficit hyperactivity disorders. Partial complex seizures occur, but should not be diagnosed in the absence of EEG evidence. *Avoidance* symptoms can evolve from confusion at the time of the accident or concussive loss of memory, according to a study of 50 children's accidents (age 7–17, mean = 11.6). Twenty-six percent reported moderate-to-severe levels of trauma. Fifty-six percent reported at least one reexperiencing symptom (34% were intrusive recollections). Fifty-four percent reported one avoidance symptom. Hyperarousal symptoms included irritability (18%) and sleep difficulties (16%). Other hyperarousal disturbances were slamming their feet on the car floorboard as though using the brakes, comments to their parents to slow down, and emphasizing seat belts and other safety measures. Providing social support (including emergency personnel) assists the child in dealing with the immediate traumatic effects and diminishes the likelihood of subsequent avoidance. For some of the participants, this was the second accident, and this seems to have reduced the traumatic response (Keppel-Benson et al., 2002).

Heart rate: Children with full or subsyndromal PTSD 3 or more months following an MVA had a significantly higher mean heart rate (HR) at triage than those who did not develop PTSD. This predicted persistent posttraumatic stress. Elevated HR at admission predicted PTSD symptoms 6 months later; elevated HR within 24 h of admission predicted traumatic stress symptoms 6 months later.

Emotional Numbing (EN)

Emotional numbing was studied in a sample of children who had incurred miscellaneous trauma, including vehicular/bicycle, sports, and assault. These findings concur that EN varies over time. Only 7% met full criteria for PTSD at 6-month follow-up and 1 of 57 (2%) at 6-month follow-up. Most experienced multiple traumas. The PTSD symptom level from prior trauma was low.

Cortisol level correlated positively with EN and hyperarousal after 6 weeks, and negatively after 6 months. It was suggested that children responding to trauma with increased acute biologic activity may become exhausted by their heightened arousal levels; consequently, they may respond with EN in the first weeks or months following trauma.

Initial HR and cortisol predicted EN after 6 weeks and 6 months posttrauma. Self-reported hyperarousal symptoms accounted for a significant amount of the EN variance. Initial hyperarousal predicted reexperiencing, avoidance, and EN. Pronounced hyperarousal symptoms were predictive of subsequent PTSD symptoms. EN did not simply reflect symptoms of depression (Nugent et al., 2006).

EN impairs adjustment in trauma-exposed children.

Rorschach Expression

Posttraumatic stress disorder in children is accompanied by Rorschach signs of disordered thinking and inaccurate perception. Moreover, responses seem to represent violation of the child's naive belief that the world has predictable rules, people are trustworthy and fair; thus, punishment and pain are consequences of bad behavior (Holaday, 2000). Children under the age of 5 at the time when trauma ended did not report current replicative dreams at all (Schreuder et al., 2000), contrasting with adult findings in which a majority of nightmares were replicative of the event or mixed replicative and nonreplicative. The effect of the event depends upon the age of the child (Hornstein, 2000). Children manifest PTSD symptoms such as psychogenic amnesia, flashbacks, and infrequent dissociative phenomena. Children with dissociative disorders have difficulties regulating attention and concentration. If they manifest inattention, hypervigilance, and increased startle reactions, this can contribute to misdiagnosis of attention deficit/hyperactivity disorders. Partial complex seizures occur, but should not be diagnosed in the absence of EEG evidence. Fantasy life resembled daydreams and fantasies, with primary process reminiscent or symbolic of the trauma (Benedek, 1985).

Stress reactions: The general features of posttraumatic stress applicable to adults also apply to children (see Chapters 8 and 9 on stress). After observing violent acts (involving the violent injury or death of a parent), children experience behavioral symptoms (intrusion of violent or mutilating imagery, problems of impulse control, need to account for the event) and changes in the state of arousal (sleep disturbances, startle reactions, and somatic symptoms) (Spiegel, 1985). When they are personally involved in a life-threatening situation, they are confused and regress (enuresis, clinging to parents, and phobias of such things as outdoor activities and open-air movies.) It was suggested that if a child patient denies distress after a serious experience, one must consider pathological denial. Anxiety is far more common than denial. Its level decreases a year later, with a high proportion of children claiming no recollection of the accident. Denial is more common in school-age than preschool children (Klonoff, 1979).

The identifying characteristics of TBI-related PTSD in *children* has been described (Miller, 1999): repetitive play (reenacting the event in play with dolls, toy cars, toy guns, etc.); sleep disturbance (checking for monsters or bogeymen, or rescue dreams); self-blame for what they have done to bring the trauma on themselves; foreshortened future (belief that they will never grow up); behavioral regression (wetting the bed; playing baby games; taste for previously abandoned foods; preference to play with younger children); atypical cognitive disturbance (impaired schoolwork; global psychogenic amnesia for blocks of time; amnesia for events contemporaneous with the trauma or all prior childhood experiences); somatization (any organ system). Emotional trauma (e.g., physical or sexual abuse) caused deleterious effect on brain development as measured by quantitative EEG measurements of lateralized coherence (Ito et al., 1998). It may be hypothesized that analogous effects will occur in children after the trauma of head injury and impairment. Psychophysiological reactions to stress are moderated by the social environment (Cacioppo et al., 1991). Deficits of coping created diminished interest and participation.

POSTTRAUMATIC SOCIAL, MOOD, AND BEHAVIORAL DISORDERS

Even after "normal neurological examinations" some children display emotional disorders, perhaps a TBI reaction to diminished perceptual and cognitive ability (Chelune & Edwards, 1981). Children with TBI exhibit long-term behavior problems in spite of cognitive recovery: motor skills; increased likelihood of unemployment; troubled family relationships; health; neuropsychological functioning.

Incidence

There is an increase in some symptoms with increasing CHI severity, that is, internalizing or externalizing. However, with larger units of behavior (adaptive) incidence severe CHI cases may be affected. When preexisting conditions are controlled, mild CHI has a relatively small effect restricted to increased hyperactivity scores (Asarnow et al., 1991). Brown et al. (1981) also found that new psychiatric disturbance occurred in the severe head injury group: They were two to three times as frequent as in the controls of mild head injury group, and were associated with the upper extreme of PTA, more with neurological disorders than without, and transient or persistent intellectual impairment.

Subsequent *delinquent* activity during adolescence may be considered to be an aspect of immaturity. However, when it is continued into adulthood, with connotations of the sociopathic personality (e.g., episodic dyscontrol with remorse, but lack of control during behavioral outbursts), the possibility of damage to the frontal lobes or to its pathways and interconnecting nuclei should be considered (Stuss & Benson, 1986, pp. 133–134).

Depression

Depression is the consequence of awareness of how brain damage and other results of an accident have impeded plans and dreams (Lehr, 1990, pp. 85–86). It is associated with reduced Performance Scale IQ, relative to Verbal Scale IQ (Brumback & Weinberg, 1990). Learned helpless is discussed by Winograd and Niquette (1988), who observe that while the criterion of success varies for each child, once a child is identified as a poor reader or learning disabled, instructional situations are even more ego-involving. Emphasis is placed on including the role of affect as well as skill in assessing reading difficulties.

Identity and Insight

Identity is our self-description and feelings about our qualities. It integrates our experiences, guides our responses and gives meaning to experience, and offers insight into motives and social role. After an injury, children experience themselves as different. The change of identity can focus on loss of specific abilities; loss of self-confidence in general; undesirability; and feeling different from others. They may be pessimistic concerning recovery, perplexed why their behavior is extreme, unpredictable, or inappropriate. Reduced behavioral inhibition is reflected in self-perception as out of control or crazy (Lehr, 1990, pp. 160–169). Conflicts with parents include lack of insight concerning their deficits (described as "living in the past"); on the other hand, some children experience their parents as having unrealistic and out-of-date expectations (Bergland & Thomas, 1991).

- *Representative dysfunctions*: Identity becomes vague or highly self-conscious; feels damaged, insecure, unattractive, victimized, vulnerable to further injury; loss of control; world is dangerous and bleak, meaningless, and requires a struggle to survive; loss of insight.
- *Adaptability: education; social relations;employment*: After brain trauma, available learned facts and procedures may be retained, relearned, or be too difficult for retraining.
- *Disability* in children may be defined as an interference with normal development, reducing ability to perform school and domestic responsibilities, and creating personality problems that affect social relations with peers, family, teachers, etc. Difficulties in school performance are created by hypersensitivity to noise, lower thresholds for stress, and lower levels of endurance (Begali, 1987, p. 71).
- *Social standards and disorders*: Children are required to attend school to learn academic skills. With increasing physical maturity, they also adapt to their own and peers' increasing bodily and social development. Learning and adaptation occurs within a changing social milieu. Urging development, self-esteem and identity will be determined by the quality of exchange with other youth in school and neighborhood. Ability to get along, to learn social skills, and to participate are hampered by: Social dysfunctions directly and indirectly

caused by TBI; Deficits of coordination and strength needed for games and other activities; Reduced academic ability in turn reduces the reinforcement of academic success; reduced self-esteem due to scarring caused by impact and surgical interventions.
- *Change of employment plans*: Seventy-five percent of a group of adolescents who experienced brain damage characterized by extended coma believed that they would have to restrict their career plans, while 92% with job experience had difficulties (memory; decreased planning and organizational abilities; slower response rate) Some were experienced as being unaware of deficits, unable to stay on schedule, accomplish goals, be consistent, or follow through (Bergland & Thomas, 1991).

PEER RELATIONSHIPS

RECOVERY

One study did not find substantial recovery in social functioning. Outcome was related to injury severity, and in some cases, the child's condition worsened. Prediction of social outcome was possible from the following functions, which were independent of intellectual ability: executive functioning, pragmatic language, and social problem solving, which predict a different set of social outcomes. There was little evidence of recovery after the first year. This is consistent with prior findings. The current findings suggest that poor social outcomes are persistent after childhood. Interventions to promote social competence in children with other disabilities have not been very successful (Yeates et al., 2004).

SELF-ESTEEM AND PEER APPROVAL

During development, self-esteem and identity will be determined by the quality of exchange with other youth in the school and neighborhood. Children learn that their beliefs may not be shared by all, that appearances may be deceiving, and intentions may differ from actions (*metarepresentations*) (Dennis et al., 1996). Ability to get along, to learn social skills, and to participate are hampered by social dysfunctions directly and indirectly caused by TBI; deficits of coordination and strength needed for games and other activities; reduced academic ability, which in turn reduces the reinforcement of academic success; reduced self-esteem due to scarring caused by impact and surgical interventions. Subnormal development (psychologically or physiologically) impairs social integration, and is an invitation to rejection. After an initial period of support from friends, they drift away because of the inability of the youth to participate as adequately or fully as before. They experience loss of peer approval , and becomes self-conscious, due to notoriety as having experienced an accident, being absent from school, changes in physical appearance, behavior, and adaptability. In addition to the usual concern ability to attain independence and earn a living, the children now worry about basic skills and attractiveness. Delays and change in vocational and academic progress cause isolation from friends, as well as assignment to classes of children that are younger. Poor neuropsychological performance is associated with externalizing behavior problems in boys 7–8 years old, but internalizing symptoms in 9–14 year-old-boys. Perhaps the child has encountered repeated failure and loss of self-esteem (Tramontana & Hooper, 1989, citing Dorman, 1982).

COUNSELING WITH CHILDREN AND ADOLESCENTS

The writer uses the term "counseling" because advocacy has a high priority and depth interpretations are usually inappropriate. The approach will vary according to the type and level of the neurotrauma, influence of stress upon mental life, socioeconomic influences, estimates of maturity and regression, social support, professional service availability (rehabilitation and medical), etc.

(Kirkwood et al., 2008; Miller, 1993). Consideration of the taxonomy of neurobehavioral dysfunctions will enable the psychotherapist to select a more precise area of dysfunction than overemphasis upon the psychodynamic, cerebral, etc. To be emphasized, as well, is the child's efforts towards independence and recovery of competence to achieve meaningful tasks.

Acknowledging the Injury

Many injured persons appear fine, that is, their complaints of suffering and impairment are cheerfully ignored by others. It is useful for the psychotherapist, when appropriate, to sympathetically and in a matter-of-fact way affirm performance difficulties and suffering. It is technically useless, and perhaps unethical, to exaggerate the likelihood of cure for particular dysfunctions if a good outcome is unexpected.

Work on Compensation and Recovery

Formal cognitive rehabilitation procedures are not yet firmly established for less severe injuries. Development of new approaches to impaired functions and demands is useful for postconcussive symptoms, for example, enhanced problem solving and coping skills.

Family Reactions

Various patterns may be observed (much of the following discussion is based upon Miller, 1993). Guilt feelings may follow the perceived responsibility of a family member for the accident, including parental rejection for not having taken care of the injured one. Change of status among older and young siblings may cause jealousy of the new role as more senior or more dependent. An accident increases dependency, regardless of the injury pattern. Alternatively, siblings may take the role of protector and educator. There could be an initial period of parental denial, with attention to preinjury patterns. An alternative is a response based upon exhaustion, distress, inability to cope with further injury about trauma, and lack of understanding of the consequences of TBI and other injury. Preexisting psychiatric behavior may remit temporarily in the acute phase, with the child being docile and tractable. When the displeasing behavior is expressed later, this may create family anger and feelings of betrayal. These may raise doubts in practitioners as to whether the family can cope with the child. What may be described as overprotection could really be a realistic effort to cope with a young person who is impaired intellectually and/or physically. Responsibilities to other family members may be neglected. Thus, the family becomes vulnerable to being considered uncooperative or overprotective.

Family Education

The practitioner will gather information from the family, including a description of preinjury functioning. The exchange of information may give the practitioner the opportunity to educate the family concerning the traumatic effects upon their child. Extreme behavior may lead to a reaction that "this is not my child," although psychotic attributions for change have been observed. Overprotection may lead to a perception of the child as unreasonably fragile or vulnerable. Thus, the therapeutic task is increasing independence and social and educational reintegration. If the child plays the sick role in the family one may expect self-sustaining secondary benefits (attention; avoidance of unpleasant activities as schoolwork; ego-protecting explanations for substandard performance).

Some Treatment Considerations

Learning and adaptation occur within a changing social milieu. Children's increasing physical maturity requires that they also adapt to their own and peers' increasing bodily and social development.

After TBI, maturity and independence are harder to attain because of reduced ability and anxiety. This creates *extended dependency* upon parents. The format may be both individual and family sessions. It may be useful, for example, when the parents are highly stressed or otherwise lacking strength, to have the therapist serve as a surrogate parent. Offering support in time of difficulty is an important therapeutic role. While mutual support and family cohesiveness are significant goals, the therapist also serves as the patient's advocate.

The full potential of psychotherapeutic success will probably not be achieved if the treatment is isolated. Significant adjuncts include pain management, an individualized education program, physiatric assessment, and physical therapy. While strategies developed specifically for brain-injured children have not been supported, educators should rely upon the best procedures for students with special needs.

OVERVIEW OF ASSESSMENT OF OUTCOME

The scope of a particular child's examination is influenced by the expected age of development of varied markers. The patient's age and the expected age of expression of a given behavior or physiological development offers a guideline as to whether a function can be reasonably examined for outcome or whether it is premature to determine whether the outcome indicates some impairment.

Good Recovery

Children with TBI exhibit long-term behavior problems in spite of cognitive recovery. What is often termed "good recovery" by clinicians, based upon incomplete examination or premature conclusions, might not be correct when assessed by a wide-range examination after several years. Thus, the assessment as "good recovery" might not be correct when assessed by a wide-range examination after several years. In practice, as a consequence of common nonrecognition of a brain injury, the forensic neuropsychologist's need for documentation of a child's claim of TBI is less likely to be satisfied than for an adult. In addition, public health statistics underestimate the safety and service needs of the community. Difficulties in assessing concussion add to nonrecognition of childrens' TBI.

Estimating Expected Ability

While the family intuitively seems to be a predictor of preinjury children's mental ability, it is recommended that family-child discrepancies be used cautiously with the use of corroborating evidence. Although intelligence is significantly correlated, its precision of IQ estimates is limited, with confidence limits indicating a high range of values. A soft indication of intellectual limitations of family members is insufficient evidence for a low expectation for a child's intelligence. The absence of a discrepancy could lead to the error of assessing another procedure's low score as representing normal variability. Generally, a significant discrepancy from estimated IQ is more significant than the absence of one. Correlations with the child's IQ have been determined to be for one parent .42, for two parents' mean IQ, .50; and for two parents plus three siblings plus demographics, .63.

The examiner takes into consideration premorbid conditions that affect neuropsychological outcome, for example, epilepsy, brain tumors, meningitis, and preinjury cognitive and behavioral problems. Consistent with the maturation of the brain, one observes the development of new problems or dysfunctions and declines in functioning over time (although immediate loss also occurs). Some limitation of the precision of preinjury functioning is overcome by taking a detailed developmental history, using tests sensitive to brain dysfunction, and analyzing performance style (Redfield, 2001).

Functional Domains

An approach to describing *participation* in home, school, and community life is the International Classification of Functioning, Disability, and Health (ICF) (Bedell et al., 2008). What is utilized here is its wide-range description of functions. The reader may find it useful in organizing an assessment of a child's baseline or posttraumatic development, and/of recovery in essential areas:

- *Learning/applying knowledge*: Basic learning and applying knowledge.
- *General tasks and demands*: Single and multiple tasks; daily routines; handling stress.
- *Communication*: Receiving and producing (spoken and nonverbal); conversation; communication devices and techniques.
- *Self-care*: Washing; grooming; toileting; dressing; eating and drinking; health.
- *Domestic life*: Acquisition of necessities; household tasks; caring for household objects; assisting others.
- *Interpersonal interactions and relationships*: Strangers; formal and informal relationships; family and intimate.
- *Major life areas*: Education; work; economics.
- *Community, social, civic*: Community life; recreation; religion; human rights; political; citizenship.

Emphases

Special emphasis is placed upon these issues:

- Different neurobehavioral functions initiate development at different ages and rates.
- Children differ in their extent of development at a specific age of injury, physiologically, and in crystallized learning.
- Functional loss varies according to the overall nature of the injury (neurological, somatic, psychological), the age at injury, the interval after injury, and the pattern of development at the time of injury.
- The ultimate neurobehavioral loss will be significantly determined by whether the function has not yet initiated development, has started to develop, is developing rapidly, or has substantially concluded its development.
- The interval postinjury should be precisely specified: the acute phase, a specific number of years later, or after presumed maturity.
- The baseline should be specified: pre- or postinjury.

Evidence for TBI

Evidence for TBI may be lack of development, rather than currently observable loss or dysfunction. In the physiological area, the examiner should be alert to both premature and delayed development. The examiner considers (using age norms and other clinical standards) whether development and performance in different domains proceed evenly until that age when maximum development is expected or whether the child falls behind the initial rank in the age cohort. In organizing the findings of a comprehensive examination, the neurobehavioral taxonomy is useful for children.

Measurement Imprecision

The extent of the impairment (somatic or neurobehavioral) need not be proportional to the estimated or observed trauma. As with adults, the trauma may affect different functions differently. As a result, comparison of one finding to another might lead to an inaccurate conclusion as to whether their relative level has been maintained or lost. It is necessary to monitor development and relative change between functions, to compare the child's overall rank in the age cohort, and to observe paradoxical differences of development. Does the child maintain the initial rank over a period of years or is there a substantial loss? As an examiner, the writer suggests that a loss of 20 percentile points is a reliable marker of loss.

Assessment of children is conceptually more complex than that of adults. The performance of an uninjured adult may be approximately described as having reached a plateau. There is usually an opportunity to obtain a range of information concerning preaccident style and level of activity. In contrast, with children, particularly younger ones, there may be no history of performance of a particular function from which to estimate a baseline. It is difficult to establish a loss since its expression may not have begun at the age of injury. Some generalizations differ between adults and children due to the much greater effect of developmental issues. The effect of preinjury cognitive and behavioral status, family functioning, and quality of the environment is less certain in children (Anderson et al., 2005a). Functions within approximately a uniform group (e.g., mild, moderate, severe) differ in their pattern of recovery, or when different levels of trauma are compared for the same function. When later functions are expressed, the examiner does not know whether it is earlier or later than what would have occurred without the accident. Subcomponents of a previously observed function may appear at an altered age and rate of development. Reduced performance level might not be possible to document. Lack of development may not be observed until many years pass and the child's cognitive, sensorimotor, or personality functions might not meet age expectations.

With all of these imprecisions of expression and expected development, the examination itself represents the baseline. Yet, the examiner may not know whether the current measurement is similar to preinjury performance; represents a loss; or what level is expected for the child to plateau for a variety of domains. This alerts us to an avoidable error: Particularly with children, outcome should not be prematurely stated. Since the brain's maturation continues into the third decade, the true outcome or establishment of a TBI may not be possible for decades!

Differential Diagnosis

One must differentiate between preexisting conditions and their exacerbation by trauma, stress reaction (acute and chronic), cerebral personality disorders, and reaction to being impaired. The clinician is concerned differentiating between psychodynamic reactions to being an injured person and cerebral or stress disorders. The reaction of parents, school, and peers affect the child's perception of the accident and its personality consequences. Lack of insight into deficits by parents creates conflict ("living in the past"). Thus, some children experience their parents as having unrealistic and out-of-date expectations (Bergland & Thomas, 1991). Problems in social functioning have numerous causes: injury severity, neuropsychological and information processing abilities vulnerable to TBI, and the family environment.

18 Outcome and Treatment
Concussive Trauma

OVERVIEW

This chapter emphasizes outcome issues for the patient with lesser levels of brain injury. Nevertheless, some may have been hospitalized for weeks or a month with co-morbid physical injuries. The appearance that they offer, that is, not appearing disabled, even to their intimates, creates a grave problem in obtaining rehabilitation and compensation. As documented earlier, it is widely asserted that lesser degrees of traumatic brain injury (TBI) "resolve" well. However, the examiner must also consider that "good recovery may include mild residual effects" (Carroll et al., 2004) and should consider the complexity of the injured person's condition:

- An accident causing a head injury is likely to be accompanied by somatic damage.
- The presence of chronic symptoms is a sign of unhealed wounds.
- Unhealed wounds cause ongoing functioning of various systems designed to repair injury and control stress: hormonal (stress), inflammatory (repair damage), and immune systems (detect damage).

To add a little difficulty to the process of assessing outcome:

- There is overlap between the symptoms of TBI and those reported by uninjured persons.
- Rehabilitating the concussed person does require a different psychotherapeutic approach than the one without brain injury.

The guidelines for managing mild TBI (mTBI) of the American Academy of Pediatrics focus upon severe TBI, with loss of consciousness (LOC) of less than 1 min, which excludes over 75% of children with mTBI. This contrasts with improvement of post-TBI functioning of individuals identified and evaluated in the emergency department who were provided with injury-specific information and coping strategies. A nonintervention group reported a greater number of post-TBI symptoms, behavioral maladjustment, and attention problems at 3 months (Gioia et al., 2008).

RETURN TO WORK: A VITAL OUTCOME

For adults, return to work is as important a marker of outcome as any. Its determinants and characteristics contribute to our understanding of the patient's well-being and activities of daily living (ADL). Among the dimensions of outcome that enable or disable return to work are sociodemographic, neurological, clinical, and neuropsychological variables. Seven percent to 15% of patients with mTBI still suffer from symptoms 1 year posttrauma, while 12–20% do not return to work. Some authors believe that women more than men suffer from persistent symptoms, including depression and somatization. Older age (age 40–50) influences the recovery process, increasing the duration and number of postconcussive symptoms while delaying the disappearance of cognitive defects. There is a negative correlation between return to work and a general list of variables and symptoms

($r = -.55$; $p = < .001$). One must consider that the number of symptoms present in members of a particular sample will be influenced by the rate of attrition during sample collection; participants hope to receive help, as opposed to a reluctance to take part by those who were no longer experiencing any symptoms. Nevertheless, individuals who return to work may have poorer performance than before the accident. Those who do not return to work may develop a postconcussive syndrome, or suffer from posttraumatic or work-related stress. These would contribute to work cessation or reduced performance quality (Nolin & Heroux, 2006).

THE LIFE OF SOME DISABLED PEOPLE

A study of the lifestyle of adults post injury with head or spinal cord injuries is tragic. Since the patient with a spinal cord injury (SCI) relies upon cognitive ability to learn new skills, a co-morbid TBI may hinder rehabilitation, since each form of injury has specific rehabilitation concerns, including cognitive and emotional (Ricker & Regan, 1999; Watanabe et al., 1999). Patients report that a TBI affected their lifestyle and culture. It forced them to depend upon others, that is, compelled them to surrender control of aspects of their lives that had previously defined them as a person. It was experienced as living with two different worlds, one for the disabled and one for people without disabilities. They had to face challenges from others and to conquer biases. It was assumed that they were retarded or couldn't comprehend, and were not actually capable, though seeing the world a little differently. This required concealment of their condition. To be handicapped means that you cannot expect cooperation to become the best that you can achieve. Vulnerability was created by physical disabilities, for example, being unable to yell or move, which prevented them from defending themselves. Further, since incurring a TBI they became trusting, passive, and less intuitive. Consequently, they became victims of violence, psychological and sexual abuse, and neglect. Perpetrators included family members and others (who abused their finances), strangers, health care and mental health workers, coworkers, personal care assistants, etc. In addition to economic advantages, perpetrators wanted to be in control, of emotional benefits from violence and abuse. Health care professionals were often unresponsive, did not follow up, were suspicious, or disbelieving. One personality change is becoming inappropriately compliant, thereby increasing susceptibility to abuse. It appears that there are few community efforts to understand and prevent abuse in this population or to provide supportive services (Reichard et al., 2007).

ORGANIZERS OF ADAPTIVE FUNCTIONING

There are units of adaptive behavior which organize major areas of our life. Some are physiological, not under our control, but are an automatic performance of body functioning. Others are within awareness. The following summary of major units of behavior that organize adaptive responses and enable enjoyment of our lives gives the practitioner the ability to organize information and establish the source and outcome of a psychological injury. The central organizers are presented according to domains detailed in the neurobehavioral taxonomy, and additional adaptive processes are presented elsewhere. The practitioner can focus clearly upon deficits, thus enabling assessment of required treatment and rehabilitation, prognosis, diagnosis, outcome, and compensation.

CNS/Systemic Interaction

I refer to the allostatic and directional exchange between the central nervous system (CNS) and stress-reactive functions: immune, inflammatory, hormonal, circadian, and autonomic nervous systems.

Emotional Regulation

Moods, impulse control, overt expression of affect, temperament, resilience.

Cognitive Regulation

Information processing (encoding and storage of different stimuli types); executive function (organization of cognitive details into larger units to enable participation in perception, comprehension, immediate reactions, planning, etc.); problem-solving strategy and efficiency; memory (retrieval); selective attention and concentration; set; selecting and alternating responses according to changing circumstances and feedback about appropriateness and success; setting priorities; etc.

Initiation of Action

The behavioral steering function (present or future) is based upon our identity, motivation (what we want), subjective reactions to internal and external events, feedback concerning the outcome of our activities, characteristic styles of social behavior, and experiential learning about how to cope with the world. While goal-directed behavior often disappears with problems of identity, motivation, fatigue, initiation, etc., its presence need not ensure an adaptive outcome. Personal goals play a role in adjustment disorders, and may be avoidant, grandiose, experienced as being externally imposed, or incongruent with the person's basic needs or personality (Karoly, 2006).

Body Schema and Programming

The neurological image of our body is represented in various structures at all levels of the CNS. Its integrity is required for integration of sensorimotor activities; the right hand needs to know what the left hand is doing. An athlete receives sensory information from the field, feeds it into multiple centers that organize body position and movement, and forwards an action schema to motor executive units consisting of a sequence of tiny actions. We cannot voluntarily expect a complex action to take place in three-dimensional space while integrating activities outside our body. Leg and limb movements are all occurring at the speed of an athletic action. Execute one segment of a movement while waiting for feedback concerning our activities and the world's reactions to them. Think of the sequence of a baseball player directing a bat towards the rapidly incoming ball, other players on the field, and integrating information from the body, the bat, and opposing players on the ground, and then running at a high speed and avoiding the awaiting first baseman.

TBI AND INTERFERENCE WITH ADAPTATION

Injury places a gross burden, sometimes not overcome, upon adaptive capacity. In this sense, adaptation means exerting one's (residual) capacities (resilience) to improve one's life while utilizing a changed brain and body, within a usually damaged environment.

TBI interferes with flexibility and capacity to deal with complex and difficult problems. It is useful to differentiate between pathology (i.e., neurotrauma), impairment (failure of a specified cognitive process), and disability (poor performance on an ecologically important task) (Whyte et al., 1996). Substantial inability to cope and maladaptive behavior create difficulties in school, employment, family and social life, and participation in the community. Chronic symptoms cause perplexity and despair concerning one's future.

The injured person's capacity to adapt to brain injury is influenced by preexisting personality qualities and self-destructive trends (Parker, 1981). Emotional experience and the personal style of being in this world create the experience of "being" and our quality of life. As practitioners involved in the consequences of accidents, we are concerned with the historic (pre and posttraumatic) emotions and lifestyle of the accident victim. Success depends on whether demands of the environment can be met by the spared and dysfunctional behaviors. The clinician is concerned with complex interaction: the lesion and its effect on the brain and other bodily functions; personal characteristics at the time of the injury (constitution, experiences, attitudes); interpersonal support and rejection; reaction of the patient to being injured; availability or denial of treatment by the patient's social and administrative milieu. After TBI, adjustment changes over time. Overall there is reduced confusion, helplessness, and social withdrawal, with improved emotional stability in the first year (Hanks et al., 1999). After a month, there can be a trend toward "acting out," disinhibition, and poor self-monitoring. Irritability and aggression are more than lesional in origin; they are a response to social rejection and denial of the validity of one's distress.

The discipline of behavioral ecology asks the question concerning the survival value of behavior or the conditions under which a behavior may serve an adaptive purpose. Maladaptive behavior, or generalized inability to cope, interacts with mental illness, ruins the quality of life, reduces the success of ADL, and threatens health, emotional well-being, and physical survival. Indicators of adaptive success are the level and pattern of achievement in school and employment, quality of social life and enjoyment of activities, and perhaps integration into one's community. TBI creates dysfunctions of complex adaptive functions, for example, combinations of injury, dysphoria, brain damage, and discouragement due to persistence of symptoms, can cause somatization and sexual dysfunctions.

A significant head injury creates complex consequences, somatic and cerebral trauma, and numerous mental and physiological reactions to being injured and attempting to heal the body and spirit. There are several guiding theses of this book: (1) There is no single stress reaction. The stress reaction following a momentary accident causing TBI (head injury) is an extremely complex phenomenon involving preexisting, subjective, and physiological responses: personal characteristics of the victim; acute and chronic physiological phenomena; the nature of the trauma; the duration of the trauma, etc; (2) Complications arise not only from the initial posttraumatic stress disorder (PTSD), but also the chronic effects of injury and impairment. Even a momentary accident can create a condition that is the source of a significant and chronic stress response; (3) An accident creating a head injury may create an initial and long-lasting fear, and also chronic bodily injury that creates a persistent somatic stress reaction as well as a chronic personality stress reaction (fear, impairment, rejection, and social withdrawal). Thus, assessment and treatment involve wounds, cerebral trauma, and varied psychological effects stemming from them; and (4) There is a subset of individuals who experience long-lasting and disabling dysfunctions and discomforts. Our professional task is to identify them, while reducing the possibility that the symptoms represent malingering, symptom exaggeration, symptom denial, or playing some type of sick role (factitious disorder).

The after-effects of an accident creating head injury are variable and controversial. They may include TBI (e.g., concussion); injury to the head, neck, torso, and extremities; a stress reaction (physiological and psychological) developed at the time of the accident or shortly thereafter; and a stress reaction consequent to the persistence of unhealed injuries, and the psychodynamic reaction to be impaired with the consequences of pain, disability, etc. The constellation of trauma and stress-related dysfunctions, including PTSD, are persistent, hard to treat, and create significant maladaptation. Forensic issues sometimes play a significant role, including blatant malingering and exaggeration in response to refusal by insurance companies to deal in a matter-of-fact way with compensation for an injured person.

The pattern is multidetermined: personal characteristics of the victim; the nature of the trauma; acute and chronic physiological phenomena; the duration of the trauma, etc. Study suggests that the stress reaction has many more components than the familiar PTSD, although this is complicated enough, involving subjective and physiological responses. We will consider more than the frightening effects of a head injury, but also the chronic effects of injury and impairment. The clinical focus presented is upon events that create a head injury, with the potential for injuries elsewhere in the body, as opposed to continuous trauma (concentration camps; combat).

Nevertheless, the point will be documented that even a momentary accident has to be regarded as a possible source of a significant and chronic stress response, which becomes itself a stressor with its chronicity during the passage of time. Moreover, in addition to fact that the experience and overt symptoms resulting from initial fear, impairment, rejection, and concussion and health effects are complex enough, the examination and review of outcome is rendered far more complex due to accompanying TBI.

SYMPTOM BASE RATES IN THE GENERAL POPULATION

Concussive symptoms occur in the general population and in patients with other conditions. While it has been stated that the symptoms of postconcussion syndrome (PCS) are found in both patients and noninjured persons, patients report significantly more severe symptoms compared with control persons (Bohnen et al., 1995). To avoid false attribution of findings, the patient is asked about the occurrence of other accidents and their after-effects. Clarification avoids errors of attribution, since a prior accident may cause a current one to have paradoxically grave consequences. Neurological, psychological, and environmental variables are related to PCS-like complaints and should be considered when PCS complaints are used as evidence for brain damage (Fox et al., 1995a). Many PCS symptoms rely upon self-report, implying independent verification before being associated with a brain injury. An affective component, for example, depression and neuroticism, can contribute to cognitive complaints (including memory and confusion). On the other hand, such symptoms as numbness, balance, vertigo, and localized sensitivity disorder may have a neurological etiology that can be more precisely associated with a trauma. Different clinical groups were selected to determine predictors of common PCS symptoms (psychotherapy, neurology, family practice, internal medicine, and underevaluation at a health maintenance organization (HMO). Etiological variables were being knocked unconscious, a bump on the head, and a lawsuit. Being knocked unconscious was a predictor for nine of the ten PCS symptoms, a bump on the head predicted eight of ten, and a lawsuit six of ten. The psychiatric group had high endorsement of PCS symptoms. In family practice, the predicted symptoms were concentration and impatience, a finding important but not surprising. The statement that certain symptoms not considered typical of PCS had a higher incidence in the knocked-out group is incorrect insofar as such symptoms certainly may accompany LOC (tremors 44%, loss of interest 68%, confusion 68%, and broken bones 32%). It was suggested that a head injury might increase the intensity of otherwise common symptoms. Several hypotheses were considered: Having cognitive complaints may be a reflection of psychological and emotional status; those who have suffered a bump to the head without LOC were less likely to have suffered a brain injury but the psychological trauma could have led to PCS complaints; PCS symptoms are more likely in psychiatric patients; being in litigation is associated with an enhanced level of symptoms. Concerning litigation, it is reasonable to assume an association between intensity of injury and the likelihood of claims for compensation. Also, psychiatric risk factors could increase the likelihood of having a head injury. That being a litigant was a predictor for six of ten PCS symptoms was noteworthy to the authors, but not to this writer. A credible reason might be that in the particular case the injury was more severe and not treated with a reasonable level of service and compensation merited by the clinical picture (Fox et al., 1995b).

Mittenberg et al. (1991) concluded that the anticipated cluster of syndromes by noninjured subjects, assuming that a head injury would occur, resembles that actually found in documented head trauma cases (outpatient practice). Head injury patients, compared with controls, underestimated the prevalence of benign symptoms. It is difficult to conclude, as the authors do, that this finding occurred despite the fact that the control group had "no opportunity to observe or experience postconcussive symptoms," since head injury is common and PCS persists in a subset of patients in the community. Mittenberg et al.'s (1991) further speculation is more reasonable: Arousal creates an expectancy that causes attentive bias for internal states, augmenting symptom perception and eliciting additional autonomic/emotional responses further reinforcing an expectation. Thus, the actual symptom level might be higher for a stressed than nonstressed group with the same physical condition.

ESTIMATING THE BASELINE

It is incorrect that the baseline cannot be estimated, for example, Faust's (1991) contention concerning the executive function. Prior psychometric measurement is augmented by interviewing the patient and collaterals, and reference to demographic tables provided for some intelligence and academic achievement procedures. Where possible, a baseline mental ability level is estimated from prior standard scores, work records, personal documents, crafts and preinjury percentiles, with a statement comparing current findings with the estimated baseline. The examiner compares current measured level and observed style with the estimated baseline.

Deviations from the estimated baseline provide diagnostic presence and information concerning the nature of any impairment. Not estimating a preinjury baseline contributes to significant generic and individual patient errors. Familiar research reports have created the general neuropsychological belief that the typical "outcome" of so-called "mild" head injury is benign, and by 3 months "resolution" or "recovery" is the norm. Another unhappy result is the conclusion that the patient has "no mTBI or PCS." The error of noncontrol of research, each by patient or with a control group, has been compounded by the restriction of study procedures to a narrow range of cognitive functions. Without a baseline, one invites the imprecise assessment that the person has a "normal" or "average" score on neuropsychological tests and thus nothing happened!

CONDITIONS EXISTING AT BASELINE

The examiner explores the developmental and pathological condition of the person when injured: constitutional and hereditary illness; acquired disease; prior accidents; social support and reactions of his peers. Premorbid traits interact with pathology to contribute to capacity for coping or vulnerability to trauma.

The examiner will explore constitutional and birth events, genetic, and developmental disorders, family, social and economic conditions, educational history (where it occurred, including technical and trade schools), employment history (including skills, and continuity), preexisting conditions, health and psychiatric disorders; activities in the community, social and personality functioning; overall social status, for example, marriage, children, and those with whom the patients live.

Note: It can be useless to use medical records to assess a past history of TBI since a significant proportion of head injuries receive no medical attention at all (Corrigan & Bogner, 2007). Further: In the writer's experience, examination of the head is often not a matter of record.

Estimating the preinjury behavioral (preexisting conditions and adaptive capacity) is compared to current functioning as a measure of traumatic regression. The examiner seeks dysfunctions not present before the accident and/or deviations from an estimated or known baseline. Some of the considerations of psychological tests include: *sensitivity* (the percentage of individuals with a disorder who are correctly classified); *specificity* (the percentage of individuals without a condition who are correctly classified) (Everitt & Wyks, 1999), and *predictive value* (a measure of confidence in the meaning of a test result, dependent upon sensitivity, specificity, and the base rate of the target condition in a given population) (Bianchini et al., 2005). Imprecision is invited when there is reliance upon test scores alone, or the excessive reliance upon the presence or lack of alterations or LOC in the lesser range. When the examiner considers noncerebral etiology of a symptom in a case of "concussion," correct attribution of a symptom or dysfunction after an accident becomes more likely, the possibility of appropriate treatment is enhanced, and the "outcome" of the accident may be less optimal than believed.

Preexisting Medical Conditions

Prenatal and developmental: The combination of the mother's stress during the prenatal period and the family ambience affects the level and nature of response to adversity and overt stress.

The stress experience of the mother may predispose the fetus to later dysregulation (Seckl & Meaney, 2004). The concept of "programming" or "imprinting" describes action during a sensitive period. In addition to effects upon prenatal development, there may evolve life-long structural and functional effects.

Prior psychometric measurements are typically sparse or absent, but qualitative functional description is useful: change in social activities, change of children's rate of development (emotional and physical), or level of performance in school, employment, or loss of leisure-time activities.

"Serious" medical events: I describe these to the patient or collaterals as those that were "life threatening, lasted for a long time, brought you to the hospital, or interfered with work or study, required surgery, seizures, etc." Since the patient's statements about prior diagnoses may not be supported by the medical records, outside verification is useful.

One group of patients with "mild" head injuries were divided into those with pretraumatic *psychosocial problems* and no such problems. The troubled group had 81% PCS at 1 month and 68% at 6 months. Those without such problems exhibited 66% PCS at 1 month and only 5% at 6 months. Employment for those with preexisting conditions increased from 5% at 1 month to 26% after 6 months. Employment for those without earlier problems increased from 33% at one month to 64% at 6 months. The incidence of specified disorders peaked around 1 month and then generally reduced somewhat from the initial incidence for affective, behavioral, and functional complaints (headache, dizziness, etc.), which contrasted for the much higher incidence for cognitive disorders, which then decline about three-quarters (Cohadon et al., 1991). Another study of mild head injury revealed that 39% became "cases" of depression or anxiety. There was a trend for women and persons over 25 to become a "case." Chronic social difficulties were four times more frequent among cases than noncases. The chronic PCS group was older. Chronic social difficulties were twice as common among those with persistent symptoms than with those whose symptoms had remitted. The presence of symptoms at 6 weeks was a strong predictor of chronicity. Chronic cases had, on the average, three social difficulties, twice as many as found among those whose symptoms had remitted. Compensation was not a factor in the great majority of cases in this and some other studies (Fenton et al., 1993).

- *Risk of developmental delay*: Premature birth, late birth, birth complications, walking or talking late, other developmental delays, muscle weakness, hearing impairments, speech problems, visual impairments, frequent ear infections, attention problems, hyperactivity, learning disability.
- *Childhood CNS compromise*: Seizure disorder, meningitis, lead toxicity.
- *Adult CNS compromise*: AIDS, brain tumor, Huntington's disease, meningitis, multiple sclerosis, psychosis, dementia, stroke, anoxia, seizure disorder.
- *Potentially confounding conditions*: Alcohol use, other drug use, past 30-day psychiatric history, and lifetime psychiatric history.

Preexisting Emotional or Personality Patterns

The clinician should inquire into previous psychiatric illness (hospitalization, psychotherapeutic or psychopharmacological treatment), drug use, psychosocial difficulties (domestic, financial, occupational), and constitutional (genetic vulnerability, personality, i.e., prone to accidents) (Lishman, 1988). Independent verification may be needed through record review and interviewing collaterals.

Caution is recommended concerning attribution of current personality dysfunctions (anxiety, psychodynamics, personality disorders, etc.) to a preexisting condition. Associations of the accident to prior experiences seem extremely rare. The overwhelming event, accompanied by fright, injuries, and reduced adaptability, and the meaning of the event to the victim, are usually separate contributing factors to posttraumatic emotional reactions, that is, of more importance than a preexisting personality disorder. A separate issue concerns coping and stress resistance. In the author's experience (intensive interviewing, projective testing, etc.), clinical personality studies of accident victims rarely offer any evidence that a preexisting neurosis shaped the symptoms. Rather, current impairment, depression, and anxiety contribute to regression. Symbolic interpretation of Rorschach responses frequently indicates the existence of intrusive anxiety and changes of identity (injury, victimization, vulnerable to further injury) that are reasonably related to the accident and its consequences. Preinjury depression did not seem to increase the incidence of postconcussive symptoms (Cicerone & Kalmar, 1997). Impulsive personalities and young males have a higher incidence of accidents causing TBI.

ASSESSING POSTCONCUSSIVE SYNDROME

The efficiency of the examination is enhanced by a precise referral, availability of medical and other provider records (e.g., history of drug or alcohol abuse, or emotional disturbance), and current information from collaterals (e.g., personality change, need for assistance, behavior creating danger).

There is a controversy concerning the optimal procedural content of a neuropsychological examination, that is, whether a *fixed battery* (e.g., the Halstead-Reitan-Battery [HRB]), standardized as a group, with the advantage of experience with particular diagnostic categories), and the *flexible battery*, whose components vary from a constant core to greater interpatient variability according to the referral question and the individual demographic qualities of the examinee (e.g., age, education, gender, ethnicity, occupation, health, senses, orthopedic condition, etc.) One study compared diagnosis of brain dysfunctioning of the HRB with an ability-focused neuropsychological battery (AFB). The latter emphasized language function, fine motor skill, working memory, processing speed, verbal and visual memory, and verbal and visual abstraction and problem solving. Processing speed was among the most sensitive measure of brain dysfunction. The data supported the current trend towards using a flexible battery (Larrabee et al., 2008).

The writer utilizes a flexible battery, since the demographic potentials from the baseline to be studied are dramatically large: nature of injury and apparent dysfunctions; gender; age; safety and social issues; medical condition, occupational and educational history; familiarity with the language of the community where the examination takes place, etc. Such procedures, as indicated earlier, are augmented with an intensive interview and the potential use of a range of sensorimotor procedures, intelligence tests, the Rorschach Inkblot Procedures (personality such as mood and identity, nonverbal cognitive performance, verbal loss possibly indicating an aphasic disturbance, stress reactions, etc.), achievement testing, inquiry concerning signs of stress and partial seizures, etc.

SOME EXAMINATION LIMITATIONS

The examiner should exercise caution concerning the patient's adaptive capacity from formal examination procedures. These are characterized by single, explicit problems; brief trials; and clear cues for task initiation and completion. The practical world requires integration of information from diverse sources over an extended time, an extended choice of response options, and perhaps an unstructured situation (no specific criteria for task or success).

The results of previous actions are not available among the data used for adaptive assessment. In particular, the examiner does not have available the presence of *somatic markers* that signal the prior consequences of particular actions.

Ecological Validity

This is defined as the degree to which results of a procedure are representative of conditions in the wider world. It is the functional and predictive relationship between the patient's performance on a set of neuropsychological tests and a patient's behavior in a variety of real-world settings, for example, home, work, school, community (Sbordone, 1996). Assessment may describe estimated performance in the narrowly defined personal preinjury niche, or the wider world in which the patient is judged against a general sample of competitors for jobs, for example, taxi drivers, accountants, professors of mathematics, etc. Thus, the test norms may be demographically related to the patient, that is, he is compared to a sample of individuals like herself or himself, or there can be a standard score representing a sample of the entire population, for example, an intelligence or academic achievement test, in which the comparison is made to an idealized community sample (Silverberg & Millis, 2009).

Differential Diagnosis

To diagnose TBI, assess its nature and severity, and to determine likely adaptive problems require a differential diagnosis between the (1) negative performance effects of TBI and (2) identification of somatic symptoms that reduce performance (e.g., musculoskeletal disorders), and (3) a mood disorder.

After a co-morbid brain and somatic injury, the issue of diagnosis is complex. Misdiagnosis may exacerbate emotional problems or offer an excuse for aggression and irresponsibility (Sbordone et al., 2007, pp. 111–116). Avoid describing impairment and the prognosis; the diagnosis classification is not clear. *Mood disorders* can be secondary to the location of the cerebral lesion, the dysregulation of hormonal and other physiological systems; the psychodynamic reaction to being injured, impaired, and socially rejected; and misattribution of symptoms, for example, fatigue to brain damage instead of the allostatic load. *Cognitive loss* that merits a psychiatric diagnosis may be posttraumatic (including both recent and prior accidents), constitutional (hydrocephalus, prenatal strokes), birth injuries, developmental, neurodegenerative, medical conditions (e.g., hormonal dysfunction, Lyme Disease) occult seizure disorder, side effects of medications (including antiepileptics), etc. Psychiatric disorders can mimic some PCS symptoms: somatoform disorder; anxiety disorder, panic disorder, dissociative symptoms, factitious disease, pain disorder, somatization and hypochondriasis. The overlap between PTSD and PCS has been described. In contrast is the concept that *panic attack* symptoms may be neurological in origin. After *hyperventilation* by patients with panic disorder, there was greater reduction in *basilar artery* flow rates and greater increases in dizziness than controls. Posttraumatic vertebral basilar insufficiency presents with headache, dizziness, diplopia, dysmetria, and drop attacks (Hecht, 2004). Noting that disturbances of basilar artery flow are associated with symptoms similar to panic attacks (visual disturbances, lightheadedness, unsteadiness), it is the primary supply to CNS regions implicated in panic, that is, the locus ceruleus and brainstem respirator and autonomic centers. It was suggested that panic attacks are a learned emotional response to catastrophically misinterpreted bodily symptoms (Ball & Shekhar, 1997).

This writer has observed one case of unilateral muscular weakness and reduced ability to move the eyes to one side (causing double vision) that was relieved by an Amytal interview (Wada procedure).

Credibility

The examiner relies upon credible evidence for an accident. Consideration is given to biomechanical factors such as acceleration and deceleration, the effects of different velocities, and direction of movement upon the interaction of the brain and its internal structures with the enclosing skull. These play a role in determining the credibility of claims of injury and the attribution of anatomical evidence of pathology to a particular mechanical accident. Reconstruction of the accident involves interviews, witnesses, photographs, etc.; the interaction of head and body position with the physical environment; physical forces (the speed and acceleration of the person and injuring objects or surfaces); the impacts of the head and body (number of blows, direction, and force involved); character of impacting surfaces (hardness, edges, points); speed of medical attention, etc.

Considering the Phase of the Injury

Primary phase: The immediate mechanical tissue damage; Secondary phase: Tissue pathology magnified after the initial trauma; Tertiary phase: Late-developing physiological disorders; Quaternary phase: Late-developing neurological conditions; Pentary phase: Chronic stress effects upon personality and health; developmental problems of children.

Gender

Unilateral brain damage seems to have less devastating consequences in women than men. Left hemisphere impairment of verbal abilities seems more characteristic of men (Hampson, 2002).

Chronic trauma: Complaints after Extended Interval

A gross or continuous injury, or environmental stressor, is considered to be potentially a part of a chronic stress disorder. It injures the spirit and body. A chronic stress is a physical or mental injury that is persistently unhealed. The consequence is dysregulation of hormonal, inflammatory, and neuroimmunological reactions affecting somatic or emotional systems, causing physical or mental distress.

Co-morbid Conditions

Somatic injury: The effect of extracranial injuries upon recovery and the intensity of PCS symptoms were studied in a group of mTBI patients admitted to a hospital with a Glasgow Coma Score (GCS) of 13–15. They were compared to a minor injury group (wrist or ankle). After 6 months, 44% of the patients were still in some form of treatment, compared to 14% of the patients with isolated mTBI and 5% of the controls. They resumed work less frequently and reported more limitations of physical functioning. Regardless of the presence of other injuries, patients still in treatment reported significantly more PCS, with the highest rates in patients with isolated mTBI. Thus, co-morbid injuries contribute to recovery delay (Stulemeijer et al., 2006b).

Substance abuse: About one-third of TBI victims are *intoxicated* at the time of injury. More than half of adolescents and adults served by brain injury rehabilitation programs have a history of *substance abuse* (alcohol or other drugs). Substance abuse following TBI increases for a sizable proportion of individuals. It is considered to be both a risk factor for injury and for rehabilitation. Preinjury substance abuse is associated with postinjury unemployment. For two or more years there is an increase of substance abuse in those with a history of this problem. A counseling program described as "skills-based" was effective in reducing substance abuse, coping skills, and the readiness for employment (Vungkhanching et al., 2007). After reduced use, preinjury levels of substance abuse may return (Walker et al., 2007). A history of substance abuse (also alcohol) was associated with acute complications, longer hospital stays, and poorer discharge status. Substance abuse additionally had a higher mortality rate, poorer neuropsychological outcome, greater likelihood of repeat injuries, and late deterioration (Corrigan, 1995). The co-morbidity of substance abuse and TBI created greater cortical atrophy changes in a group of adolescents and young adults than the presence of either condition alone (Barker et al., 1999).

Chronicity: The focus here is upon chronic symptoms that closely resemble TBI-related trauma: Bell-like palsy; chorea, encephalopathy (memory, mood, sleep, axonal polyneuropathy), facial palsy (bilateral); fatigue and lethargy, headache, meningitis, periventricular white matter lesions, photophobia, radiculoneuropathy, upper motor neuron bladder dysfunction (Kaye & Kaye, 2005; Malawista, 2004; Steere, 2005). An example is the instance of offenders whose performance may be reduced by substance abuse, learning disabilities, and less education (Diamond et al., 2007).

Forensic: A high proportion of persons with suspected or documented TBI are involved in forensic actions, whether a lawsuit for compensation of injuries, lost wages, reduced quality of life, etc.; workers compensation; disability; no-fault auto claims, insurance claims, Social Security disability and Social Security, etc. Thus, in such cases, the issue is broader than just whether a TBI can be documented: possible symptom exaggeration, impairing somatic injuries, prior injuries, substance abuse, prior epilepsy, medical conditions creating symptoms overlapping with TBI, for example, PTSD, substance abuse, and Lyme disease.

EXAMINATION OVERVIEW

This outline provides an approach to gathering information about an injured person, with the goal of a comprehensive report addressing such issues as baseline and preexisting conditions, treatment and discharge, living arrangements, assessment of current performance, diagnosis, need for treatment, anticipated outcome, issues of compensation, and educational and vocational planning.

DATABASE

The personal injury database includes behavioral dysfunctions, medical and emotional trauma, life sciences (physiology, gross anatomy, neuroscience, including neuroanatomy), chemistry (inorganic and biochemistry), trauma physiology (hormonal, immune, and inflammatory systems), the role of somatic injury in contributing to impairment and psychological disturbance, potential behavioral and health disorders after trauma, measuring impairment and recommending rehabilitation, psychological reactions to injury and rejection, developmental patterns, and personality reactions to being frightened and injured. Assessing an injured child requires a database that is in some ways significantly different from the adult patient.

A reasonably complete examination utilizes an extensive technical database (as mentioned earlier) to draw conclusions about diagnosis, performance, and outcome. Has the patient suffered dysfunction, deficits, or discomforts due to an accident? This is supported by direct examination or review of records.

The writer recommends extension of the knowledge database and role of the neuropsychological examination of a claim of TBI. This contrasts with one widely accepted representation of the neuropsychological examination that emphasizes the study of cognition in numerous conditions and aspects of behavior (Paulsen & Hoth, 2004).

THE STANDARD OF COMPETENCE

I recommend the board certification in neuropsychology. Training should include neuroscience, introduction to chemistry (inorganic and biochemistry), and life sciences (physiology, anatomy, and medical trauma), stress response, personality issues (mood, identity, defenses), neuropsychological principles (ability, cognition, concentration, sensorimotor, communication), and training in counseling. A related issue is the credibility of the evidence in court. There are varying standards recognized by federal and state courts concerning the admissibility of scientific evidence in relationship to legal issues and establishment of facts relating to the outcome of the case at hand. The issue is more than the qualifications of the "expert"; the scientific status and applicability to the particular case may be an issue (e.g., the Daubert and Frye "rules") (Granacher, 2003, p. 326). In addition, the examiner requires clinical skills to relate to adults and children who are injured, anxious, or suspicious, and their collaterals. Thus, empathic skills are required.

Quality of Procedures

The examiner is concerned with *ecological validity* (approximation of the actual ability the procedure attempts to predict) and *external validity* (prediction of performance regardless of the apparent similarity to the activity itself). *Verisimilitude* is important in selecting procedures, that is, the similarity between the procedure's data collection and skills required in the environment (Rabin, 2007).

Reasonably precise assessment of "outcome" requires consideration of a wide range of issues: estimation of the preinjury baseline; the injury phase; sampling a wide range of domains to render an opinion; and the kind of professional and social support available to the patient. These definitions are used: *impairment* (a departure from the body's typical functioning); *handicap* (inability to perform an educational, physical, or social task as a result of a physical or nonphysical obstacle; discrimination with regard to employment would be considered a *hindrance*); *disability* (a lasting physical or mental impairment that significantly interferes with the ability to function in a central life activity). Assessing real-world outcome requires the use of *ecologically valid* procedures, an examination description and group of procedures that are substantially different than "fixed" or "flexible" batteries. Issues addressed include rehabilitation potential, optimal living arrangements, functional independence, adaptation to everyday life, the impact of behavioral deficits on functioning in specific environments, and prediction of work behavior.

Study of a Wide Range of Functions

The examiner attempts to reduce the likelihood that particular dysfunctions will be ignored. An examination considers the nature of somatic injuries and physiological dysregulation. Then, misattribution of symptoms, or actually ignoring the consequences of trauma, may be avoided. Examples of misattribution of symptoms to head injury rather than somatic injury include motor aprosodia (involving both cerebral motor and programming networks, and injury to the vocalization apparatus; headaches consequent to referred pain from neck and shoulders; reduced activity and fatigue caused by dysregulation and exhaustion of physiological systems and mood disorders caused by physiological signals to the brain, perhaps enhanced by increased posttraumatic permeability of the various barriers between the brain and its liquid milieu.

Selection of Procedures

The examiner's choice will depend upon the referral, one's concept of the nature of the examination, the range of domains that one wishes to assess, what one has been trained to use, and the availability and value placed upon nonneuropsychological procedures such as records, interview, etc. The examiner may prefer one or another quality of psychological procedures:

- Ecological validity, for example, measuring a capacity to perform some realistic function ("spelling" or "math").
- Abstract function, for example, possessing an inferred function whose level affects other more conspicuous adaptive functions ("intelligence," "executive functioning," or "concentration."

The interview is a procedure second to none for gathering data trauma. As a clinical tool, it invites the examiner to form a relationship with a distressed and injured person. Goals include assessing the mood; effectiveness of consciousness and alertness; observing verbal, cognitive, and sensorimotor disturbances; studying the credibility of the claim of injury; alertness to indirectly or poorly expressed disorders requiring further interview, examination, or referral study, etc. (see "Expressive Deficits"). There are functional areas that can be studied best with an interview, taking into account the limitations of questionnaire and paper-and-pencil tests: screening for partial seizures; health disorders; focused emotional and social problems; the preexisting condition (family history and development; educational and employment history; education; prior accidents; etc.)

Prognosis

It has been asserted that resolution of cognitive defects occurs within 1–3 months after an uncomplicated TBI, and that after more severe injuries, residual deficits or global dysfunctions may be expected (McCullagh & Feinstein, 2005). This optimistic estimation of prognosis after TBI is controversial. The present writer's view is that with more thorough sampling of a neurobehavioral domain, a larger proportion of impaired individuals may be detected, some of whom may have substantial deficits. Due to the wide range of cognitive functions, no battery of tests or procedures can precisely and completely measure it. Subtle deficits may not be detected clinically or in a formal examination. The clinician's alertness, experience, and knowledge must be integrated with formal procedures to assess and counsel the patient. Even so, there is room for a bit of caution and humility.

Practically, due to the constraints of time and the potential range of disorders, cognitive functions must be sampled to attempt some generalization. Assessment of cognitive efficiency requires comparison of current adaptive efficiency with performance level and style in the preinjury or illness baseline. Impairment may be due to dysfunction of supportive functions such as *mental efficiency* (executive function) and information processing, or the impairing effect of "distracting" symptoms.

Some Errors

There are some common reporting statements concerning status and outcome that should be avoided as imprecise or possibly erroneous.

- A result in the average or higher range is not evidence for lack of dysfunction unless a baseline has been estimated.
- Premature conclusions implying permanency of outcome. This should be postponed until an appropriate postaccident interval has elapsed. The person's age should be considered in estimating outcome, with special consideration for the developmental status at injury and examination of the younger patients.

ORGANIZATION OF THE REPORT

Orientation of Reader

Name of patient:	**Name of referring source:**
Date(s) of examination:	Age at injury:
Date of injury:	Time since injury (years and months):
Present age:	Injury:

State the Reason for Referral

List the Specific Procedures Utilized

Background, Preexisting Conditions, and Baseline

Summary of the medical record: The section devoted to the actual examination, this is the place for summarizing the actual records provided as opposed to an overall summary. The format that this writer utilizes is to organize the record review chronologically by provider (individuals or organizations), with individual dated entries also chronologically arranged.

Major findings of this examination:

Patient's current status
Abstract of examiner's observations
Patient's complaints
Representativeness of the examination and motivation

Note that the examination took place in a protected environment, which may not be representative of work or home. Further, if appropriate, make a comment whether there is evidence for deliberate or unconscious misrepresentation of ability or distress.

Diagnoses
Attribution
Permanency, restitutive capacity; prognosis
Recommendations
Review of records: Organized perhaps as medical, psychological, educational, employment
Neuropsychological examination data
Interview with collaterals

This may be the only source of description of preinjury functioning, particularly with children or an unresponsive or uninsightful patient. Family, friends, and work associates can offer information about work, independence, responsibility, personality and temperament, social interest, frustration tolerance, and use of leisure time. Information about integrated and qualitative community functioning is useful. The examiner should seek preinjury behavior, personality, preferred activities, comprehension, employment or educational functioning, health and personal hygiene, social interest; walking and use of transportation, domestic duties, interests, and community participation. Pre- and posttraumatic status should be compared.

I begin an interview (or telephone conversation) with collaterals by asking the person to describe the patient before the injury, then asking about any changes that have occurred. I direct the conversation towards social participation; moods; sexuality; frustration tolerance; lifestyle changes (use of free time and interests); health, strength, and stamina; coordination; habits of work; and household participation.

Caution is indicated since veracity and memory may be inadequate. Response biases by patient and collaterals may create some inaccuracy in comparison of current and preinjury status. These include problems of memory, assigning meaning to current conditions, and expectations as to what is expected (Putnam et al., 1999). For children, such data as parent ratings, family demographics, and concurrent word reading skills obtained soon after an injury are not sufficiently accurate to be desirable for use for individual assessment. Yet, children with TBI were five to ten times more likely to show a large discrepancy between estimates and performance, although the question was raised whether this information after a longer interval would remain useful (Yeates & Taylor, 1997).

Interview with the Patient

Prior accidents or extended stressful experience: The patient is specifically queried concerning any history of mechanical injuries (vehicular, falling objects, slips or falls, assault, sports injury), electrical injuries, blast, toxins, and anoxia. If there are cues, the examiner may look into environmental stress (weather, temperature, clothing, nutrition); confinement (incarceration, criminal, concentration camp, prisoner of war); harassment (threats, sexual abuse).

Prior Accidents

For both adult and young patients, both direct inquiry and information elicited from collaterals will often provide material not offered by the patient:

- *Mechanical head injury:* Motor vehicle, falling objects, falls, assault

Inaccurate denials of injury are common. Statements made by patients or collaterals concerning lack of preinjury TBI must be approached with the love for precision of a tax examiner. Precise and persistent interviewing can bring out information about a credible injury that was never previously considered important by the patient, not carefully assessed by a clinician, or forgotten.

Different samples of college students offered a positive response when asked about TBI: 23–34% of males and 12–16% of females. The proportion of people having a prior head injury increases with age. Consequently, a larger proportion of people experience an unexpectedly grave level of impairment after apparently minor head injury (Crovitz et al., 1992).

- *Electrical*: Inquire about the route of the current, whether it created a fall with head and/or somatic injuries, what was the strength of the current, how long was the exposure, location of burns, unconsciousness, etc. (Chen et al., 1999).
- *Neurotoxins*: At an industrial workplace, hydrocarbon industrial site, farm, pesticides, business or home office equipment (e.g., printing), crafts, nearby leaking gasoline tanks, paint flaking off walls, paint circulated by air by careless workers scraping walls, household utensils (Derelanko & Hollinger, 2002)
- *Reduced oxygen*: Events include fire, stoves turning off while the gas is still emitted; poor ventilation from boilers, stoves, motor vehicles; carbon monoxide creation from incomplete burning or inadequate oxygen supply; confinement with lack of air; etc.

Educational History

The standard is academic, that is, the highest grade or college year completed. While the years credit for *general educational test (GED)* is controversial, since it is considered to be in some jurisdictions the equivalent of high school graduation, 12 (or 11 years) is appropriate. Obtain a description of any college or associate's degree obtained. Did the patient undergo technical or military training? How long was the training? What were the chief skills learned?

Questionnaire of Selected Symptoms

Questionnaire includes current medical treatment; somatic concerns (headache, pain elsewhere; sleep disturbance; oversensitivity to heat, cold, bright lights, loud sound; seizures; dizziness; balance; walking; strength change and lateralization); visual; hearing; movement (tics; tremors); autonomic (sweating; nausea; frequent urination; drinks too much but not thirsty; bowels; appetite change, weight change, activity level change; easily tired); women (do you have menstrual problems; unusual secretion of milk); do you get sick more often after the accident; military service (number of years, military occupation specialty, highest rank, skills and supervisory responsibility, combat, injuries); have others told you that you did something wrong and you did not know it; anxiety and hyperarousal (nightmares or bad dreams; flashbacks; avoidance due to anxiety or fear (objects, places, activities); rapid heartbeats; deep breathing; jumpiness; partial seizures screen: do you ever have strange experiences: Things happen and you can't explain it: vision, hearing, feel something move on your skin, taste something unpleasant that others do not taste; smell something unpleasant that others do not smell; do you ever think that somebody is standing behind you; sudden mood change for no reason; does it ever happen to you that you lose time, you don't know what happens for a period, you get to a place and don't know how you got there; is your memory normal or is there a change; change in level and control of anxiety, anger, depression (dull and empty OR sad OR both); do you have thoughts of injuring yourself; change in level and control of sexuality; dissociative experiences (does the outside world ever feel unreal (like being in a movie; can't explain); does your body feel changed or unreal (can't explain, not due to the current injury); do you ever feel that something has happened that never happened before; quality of life (what were you interested in or enjoyed doing before you were injured; what do you like to do now); fatigue (do you get tired more easily, what makes you tired); what is the worst effect of the accident; are you getting what you want out of life; what activities in your life are valuable (what access do you have to these activities, what people in your life are valuable, any change in your social life, have you had serious disagreements with your husband or wife, family members, or close friends, what access do you have to these people, how much does your condition interfere with your social life); will your condition get better, get worse, or remain the same; are you able to handle your problems, or they are too much for you, what do you do to make yourself feel better; is there anything else you want to tell me.

Reason for the Referral

Record summary: Documentation of brain and somatic injury; neurobehavioral performance (medical, psychological; vocational)

Current status: Employment or schooling; independence; mobility; safety and self-care

The examiner's observations: Motivation to do well; problem-solving maturity (preplanning vs trial-and-error; concentration); stamina; focused attention and concentration; comprehension; competent or impaired; English.

Expressive Deficits

To what extent do expressive deficits (inability or unwillingness of the patient to relate a true, complete self-report) contaminate the information?

Representativeness of the Examination and Motivation

The examination is a suitable sample of performance in a protected environment, but may not be typical of job or home. Assessment based upon observation or formal symptom validity testing whether there was evidence for exaggeration of difficulties.

Symptom organization (patient complaints, differentiated from exam findings). *Patient's complaints* (organized by category):

"Distracting" symptoms (e.g., headaches; pain; intrusive anxiety) (Parker, 1995); mental; somatic, health and stamina; potentially neurological; stress related; psychological; physiological (somatic; health; stamina); behavioral. Also specify: "distracting" symptoms; positive (intrusive anxiety, partial seizures).

Findings Organized by the Neurobehavioral Taxonomy

To assist in organizing an examination, an abbreviation of the neurobehavioral taxonomy is offered:

I. *Neurological*: Sensorimotor, arousal, attention, and cerebral personality disorder
II. *Consequent to somatic injury*: Performance and health disorders consequent to soft tissue damage, disturbed internal milieu (physiologically based symptoms); pain; developmental disorders of children
III. *Mental control:* Concentration (focused effort), information processing, and executive functioning
IV. General intelligence
V. Learning and memory
VI. Personality style: Psychodynamics, motivation, moods, defenses

VII. Identity, insight, morale, Weltanschauung (world view)
VIII. Empathy and social life
IX. Stress reactions: Acute/chronic; psychological; physiological; dissociative
X. Special problems of children

Assessment of Outcome

Impairment may be caused by relatively narrow disorders (e.g., the neurobehavioral taxonomy, or more global disorders: *Structure* (somatic and neurological); *supporting functions of adaptation* (executive function and information processing; *stamina*; *health*; *sleep efficiency*), and *adaptive performance* (goal-directed operations used for self-protection and affairs within the external environment).

Activities of daily living: Mental status; academic achievement; attention; language; memory; executive functions; and functional independence. Living skills may be assessed by direct observation, standard measures of independent behavior, and standard neurocognitive tests from which are extrapolated estimates of everyday living capacity (Lynch, 2008).

Record Summary

A brief statement about the inferences drawn from the records.
Current status: Employment or schooling; independence; mobility; safety and self-care.

Clinical Observation

The observation of the patient during the interview may offer insight into preinjury style: Appearance; verbal and affective expression; how the person expresses vital information and feelings during the interview and offers information as to personality development and education; the patient's style and emotional reactions while offering information and solving problems and approaching tasks, verbal style and vocabulary level; nonverbal aspects of behavior such as demeanor and expressed frustration at not performing particular activities. Problem-solving style can be observed with Wechsler Performance-type tasks such as Block Design and Object Assembly: observing whether mature problem-solving style is used (preplanning) or immature or regressive (trial-and-error).

Credibility

(Statement): The examination is a suitable sample of performance in a protected environment, but may not be typical of job or home; was there evidence for exaggeration of difficulties? When one compares the change in lifestyle, for example, loss of income, mobility, etc., during the period of alleged incapacity and estimates the likelihood of compensation leading to a life of affluence, is it credible that an uninjured person would sacrifice so much with the likelihood of gaining so little?

Extended Record Review

Organized by practitioner or institution, arranged in chronological order, that is, all entries in a category together, with the earliest one first (e.g., ambulance report).

If the record review involves numerous and varied documents, findings can be organized by types, for example, psychological, radiological, medical, physical therapy, school, vocational.

Educational records: Percentiles on nationally standard tests are a reasonable measure. If there is a range of subjects presented, an average for a given grade is useful. Observation suggests that years may go by before there is a clinically significant drop in average achievement scores. Early school records are more objective than grades, and offer percentiles from nationally standardized achievement tests that correlate well with IQ (Anastasi, 1988) and current psychometric findings. Caution is needed in interpretation of early low scores and few years of education: These can be reduced by lack of motivation, disinterest, leaving school because of poverty, change of residence, etc. The number of years of education is also imprecise, since advancement can occur due to social promotion, parental pressure, low educational standards, etc. Where possible, the completed years of education should be verified collaterally, due to memory deficits and sometimes deliberate false statements on the patient's part. The GED certificate is a 12-year educational equivalent (Dalton, 1990), although some clinicians consider it to represent a lesser academic achievement. The writer discourages use of procedures with long out-of-date standardizations. Formerly, this writer used WAIS-R Scales, with tables by age and education, to offer an estimated mean of preinjury IQ (Matarazzo & Herman, 1984; Sattler, 1988) or percentile ranks (Ryan et al., 1991). Current WAIS-III educational equivalents are available in the Manual of the Wechsler Test of Adult Reading—WTAR (Psychological Corporation, 2001). Norms are available by WTAR scores, sex, age, and ethnicity for Full Scale IQ and several indices, for United States and the United Kingdom. The lower means for African Americans should be considered in the use of these tables, noting that they probably reflect social conditions more than phenotype.

Current Psychological Test Performance

Current performance can be compared with earlier testing using standard scores or the examiner's estimates of relative status on such graphomotor procedures as the Bender Gestalt (2nd edition) and Drawing tests. Demographic procedures estimating preinjury mental abilities are not recommended by this writer. They use verbal and reading skills and demographic data. The examiner is cautioned concerning the narrow ability range of application and limited precision of estimate. The robustness of the scores after TBI from the viewpoint of estimating premorbid performance level is questionable. Many of the current measurements purporting to measure preinjury ability are verbal, and are thus vulnerable to dominant hemisphere TBI or aphasia (Granacher, 2003, pp. 177–179).

ESTIMATING THE BASELINE

Employment History/Level

Employment history is a significant component of the adult patient's baseline and is a reference point in determining whether current functioning permits return to work. Preinjury IQ and academic functioning can be estimated on the basis of vocational group characteristics (Hartlage 1990, pp. 8–10; *Wonderlic Personnel Test Manual* using percentiles of numerous vocational groups, 1992) and clinical experience with individuals with varying employment history.

Components useful for an estimate of preinjury functioning include the age of first employment; when full-time employment began; the most skilled position ever held; whether the patient was employed at the time of injury; employment characteristics (the number of people supervised, amount of training required, job requirements for planning, concentration, writing, calculations, personality, specific skills, etc.). Prior employment status seems not predict stress susceptibility (Rimel et al., 1981). The examiner should obtain work samples, employment reviews, the formal job description, training history, job stability, promotions, and temperament and personality qualities (Guilford, 1985).

Military History

This extreme environment demands sensorimotor capacity, physical strength and stamina, social ability, intelligence, stress resistance and health, leadership, responsibility, absorption and application of complex training demands, verbal and nonverbal skills and concepts, learning, and conceptual and administrative ability. The examiner should inquire about length of service, highest rank, duties and military occupational specialty, training, combat experience, etc. Military records may reveal conduct problems, psychiatric history, and medical disorders. Using weapons and maintaining vehicles requires nonverbal intelligence. Records maintenance and personnel actions involve sequencing, precise work habits, and verbal ability. A command position implies integrating considerable information (written, graphic, and oral), planning a sequence of events, error monitoring as the operation proceeds, anticipation of consequences, and a bit of tolerance for ambiguity.

Hobbies and Optional Time Activities

These reflect skills, interests, social capacity, conceptual and learning abilities, level of independence, social interests, available energy, initiative, and range of interests. Samples of writing, graphics, and handicrafts are useful as baseline representatives. Deviations may be conspicuous, for example, between preinjury art work and impaired house and figure drawings, or planning activities for friends, and regression to a solitary lifestyle. Sports reflect stamina, coordination, motor speed, strength, and social skills such as coaching.

Patient Characteristics

The examiner differentiates between symptoms (complaints) and descriptions of the patient's trauma (Corrigan & Bogner, 2007). Historical considerations are used in the determination of whether a new TBI has occurred: age, number, and severity of previous TBIs; ages of prior LOCs; and functional effects (prior; at time of injury; at current examination).

Current Test Patterns

Estimation of preinjury capacity through use of scores believed to be stable (most commonly utilized is vocabulary) called "hold" vs "non-hold" seems to be no longer accepted as a measure of before-injury ability: unreliability of scores for individual cases; reduction of a specific measurement by trauma; preinjury achievement reduced for noncognitive reasons, for example, poor education; traumatic lateralization effects reducing current score; sensory processing deficits; loss of long-term memory; aphasia; etc. Demographic formulae for predicting preinjury level (race, education, geographic region, and occupation) are relatively inefficient because they offer restricted range of scores, underestimate a bright person's ability, and use vague categories. It may be more useful to estimate single capacities rather than composite scores such as IQ (Larrabee et al., 1985; Phay et al., 1986; Putnam et al., 1999).

PHYSIOLOGICAL DETERMINANTS OF OUTCOME

Outcome may be defined as the patient's status after an extended interval when some stability is reached. It may be erroneous to state the outcome prematurely because of the possibility of late development disorders.

Genetic Issues

Long-term changes in brain function are elicited by both endogenous and environmental stimuli. These are likely to be mediated in large part by neural gene expression, normal physiological processes, experience, and pharmacologic agents. Metabolic pathways may differ due to biological differences affecting responses to injury and drug pharmacokinetics in drug trials. The risks and course of mental disorders are partly mediated by environmental effects that up- or downregulate the expression of certain genes. Mental disorders are partly based on differences in the sequences of genes at particular loci; psychotherapeutic drugs alter gene expression, and ultimately synaptic strength within relevant neural circuits (Molden & Hyman, 2005).

Functional outcome: Genetic factors significantly influence functional outcome after TBI (Diaz-Arrastia & Baxter, 2006). Some genetic types may be associated with poor outcome only in certain types of traumatic injuries. An example is the role played by those genes early in the course of trauma when inflammation is most prominent. It may be different from those playing a role during rehabilitation. Thus, determining such genetic influences is a step towards developing therapies associated with the type of injury and the phase of an illness. One focus has been rodent experiments demonstrating different susceptibilities to neurodegenerative insults.

Nevertheless, genetics is not destiny: Genetic influences upon common diseases result from a complex interplay of genetic and environmental influences. Prognostic factors include duration of posttraumatic amnesia (PTA), neuroimaging findings, etc. Estimation of the relative risk or odds ratio of a disease or neurotraumatic outcome requires complex statistics as well as a large number of subjects. Genetic pathways influencing TBI outcome include oxidative stress, excitotoxicity, inflammation, apoptosis, and aging. Nevertheless, incorrect conclusions can be drawn: There is a spurious association of an allele in a group that is genetically heterogeneous with a particular trait present at a high level; a particular ethnic group may be predisposed to suffer head injuries of greater severity (e.g., variations of musculature or skull structure) and more likely to obtain care at a particular institution.

Apoloprotein E (APOE) plays an important role in the CNS response to acute and chronic injury. Nevertheless, its eta4 allele, associated with Alzheimer's disease (AD) risk, exerts very little independent effect on normal neurocognitive performance. In assessing individual cases, age, education, and gender should be considered in determining the expected values for normal neurocognitive performance (Welsh-Bohmer et al., 2009). Its variant (*APO eta3*) protects against neuronal death. It is produced predominantly by astrocytes (also neurons and microglia), and it transports lipids among neural cells. It is upregulated after injury, and has a role in the modulation of learning and memory, structural plasticity during development and aging, and cell death after ischemia or convulsive brain injury. There is a complex epidemiological association between the EPO eta4 allele and late onset *AD*. One view holds that eta-4 increases the risk of AD twofold, and the risk of AD after TBI is increased tenfold. It is a risk factor for poor outcome after severe TBI (including risk of death), dementia in boxers, ischemic cerebral infarction, poor outcome after intracerebral hemorrhage, vasospasm after subarachnoid hemorrhage, and dementia in HIV-infected patients. Other findings are more precise: Cognitive status was studied in a TBI sample 3 decades after injury. Patients with the APOE4 allele (mean age 72.7) had an increased incidence of poorer overall cognitive presentation, but not poorer subjective memory, well-being, or ADL functioning. There was no association with physical or vocational outcome. However, cognitive decline was wholly accounted for by a subgroup of the APOE4 patients who developed subclinical or clinical dementia during the follow-up. Cognitive decline seems largely restricted to a subgroup of these patients (using the measures of the study). APOE 3 patients showed only a mild cognitive decline, while the *APOE2* allele was protective. The neuropathological cause of the decline is not certain, but TBI seems to contribute to dementia. Those patients who did not manifest decline were relatively young (56.2 years) and might develop decline in the ensuing 15 years. APOE4 seems to predispose to a disease-related decline, rather than predisposing to the disease itself. It appeared that the APO genotype affects some long-term brain functioning, but only certain functions are selectively vulnerable to the detrimental functions of APOE4, at least in TBI patients (Isonieml et al., 2006).

Brain Reserve Capacity

Ceiling of performance and environmental demand: I define ceiling of performance as the maximum level that can be achieved by a person for a task of a particular type when required or examined. One may consider that the effort to measure the current performance may be inefficient, regardless of the label placed by the publisher upon the procedure! In general, the ceiling is higher before an injury or illness. The environmental demand is the quality or amount of performance needed to succeed, on the job, in class, with avocations, etc. Thus, a person with sufficient talent may usually exceed the requirements; even with a good fit between the person and the job; occasionally, the demand may exceed the ceiling. In an examination, the goal is to assess the baseline, although usually it cannot be done precisely.

Assessment

When the accident victim returns to the job or school, the functional capacity may or may not be sufficient. If there is reasonable evidence for TBI, then success infers that at the moment the brain reserve capacity (BRC) was sufficient to support competent performance. Nevertheless, the reserve is depleted; after the next brain injury, or a series of smaller accidents, the BRC may be sufficiently depleted by lesions and neuronal loss through aging. The level of performance is uncomfortably difficult or overtly inadequate. Nevertheless, the patient, employer, or teacher may assess functioning as inadequate. An apparently high measured assessment may overestimate capacity to match the environmental demand. Is the person "faking bad"? Not necessarily. The writer believes that a more ecologically accurate measure of capacity is achieved from more complex materials. Assume that you are assessing verbal ability. The psychometrics of subtests (e.g., spelling, or the name of an object visually presented) are similar to those obtained from tasks of different difficulty. Nevertheless, the scores may differ for procedures that are more complex or have a higher ceiling. Examples include reading comprehension; responding to a request to explain a complex paragraph; or filling in the middle sentence of a trio of sentences. Moreover, using more difficult tasks may reduce the error of concluding that after an accident, there is no measurable deficit. In memory as well, the meaning of a measurement is influenced by underlying contributions, for example, visual tests relying upon verbal strategies to encode and recall; female advantage in visual memory (males exceed in spatial ability, but no transformation is needed here where the memory of the exact object is required).

Health and Stamina

Nature of the injury: Using the Glasgow Outcome Scale as the criterion for outcome after 3 or 6 months, in the moderate-to-severe TBI range, motor vehicle accidents (MVAs), assaults, and sports or recreational injuries, falls (associated with older age and a higher incidence of mass lesions) had the worst outcome.

"Distracting" symptoms: There is performance loss due to interference by "distracting" symptoms (seizures; pain; restricted range of motion; anxiety and other mood disorders, etc.) (Parker, 1995). Chronic pain reduces return to employment: After 6 months of unemployment, the probability of return to employment is 50%; after one year, the likelihood decreases to 10%.

Sleep disorders: Persistent insomnia can lead to impaired daytime function, injury due to accidents, depression, and variability of circadian rhythm to cardiac diseases and stroke (Creisler et al., 2005), and interference with hormonal function due to impaired circadian rhythms.

Allostatic load: Chronic stress causes "burnout" with health disorders and fatigability.

Older Age at Injury

The risk of TBI and death following TBI are increased for elderly individuals relative to nonelderly individuals. Their morbidity and mortality is related to different emphases than in younger accident victims (Tokutomi et al., 2008).

- Preexisting systemic disease, increasing intracranial hematomas associated with cerebral atrophy and limited intensive treatment resources).
- The proportion of individuals with multiple injuries is lower in the older group due to the lesser frequency of MVA with higher rates of falls (lower velocity). However, systemic complications had a higher adverse effect upon outcome, perhaps explaining the higher mortality.
- Secondary-phase mechanisms may progress more vigorously (increased expression of inflammatory molecules; vulnerability of cerebral vessels with age; indicators of cerebral endothelial injury).

Much of the increased death rate reflects differences in the distribution of TBI severity and age-associated increases in mortality. For elderly persons who sustained a TBI and survived for more than 6 months, subsequent mortality was no different from population expectations. However, older individuals do have a greater likelihood of becoming physically and financially dependent on others (Flaada et al., 2007). In addition, the consequences of TBI worsen with increasing age at each level of brain injury: Older adults showed delayed recovery compared to younger ones after 1 year. This concept is supported by some animal studies: for example, a study of a whiplash type of injury in rats that compared aged rats (20–23 months) to younger rats (2–3 months) in reflexes and motor learning (Maughan et al., 2000). Neuropathological findings were compared in rats 5–6 months and 13–24 months. Indicators of injury were generally worse in the older group: sensorimotor performance; acute edema resolved more slowly; more prolonged postacute opening of the blood–brain barrier (BBB). Secondary processes were implicated in age-related differences in outcome. Lesion cavity volumes did not differ. Prolonged and greater BBB opening in an aged brain may contribute to neurodegeneration. There was significantly greater neurodegeneration at 3 days postinjury and a trend at 7 days. The pattern of acute edema, increased BBB permeability, and neurodegeneration provide strong evidence for increased damage in the aged brain compared to adult brains following the same cortical impact injury. It is possible that the increased vulnerability of aged brains is consequent to *secondary processes* (Onyszchuk et al., 2008).

This issue is more complex than the simple assertion that increasing age is related to a poorer outcome. Age and outcome was studied with the Glasgow Outcome Scale (GOS) 3–6 months after injury. No gender differences were found. The outcome of Black patients was poorer relative to Caucasians. Higher levels of education (more than 12 years) were weakly related to a better outcome (Mushkudiani et al., 2007). The age of first injury with LOC contributed to the prediction of speed of processing, delayed memory loss, and abstraction. Those who sustained their first injury with LOC had greater impairment. After adolescence, the older a person with LOC, the worse the cognitive result. This is not due to recency, since time since injury did not relate to performance measures. However, since the sample was composed of sustained substance abusers, proximity to the injury or resumption of use after the injury may have reduced recovery (Corrigan & Bogner, 2007). Increasing age (birth to 80+) was strongly related to a poorer outcome (Mushkudiani et al., 2007).

Acute phase: Modifiers to a direct association of age and poorer outcome have been suggested for mTBI for at least the *acute phase*. In contrast to prior retrospective studies, including a broader range of severity, a sample of 60 year old patients and older seen at a TBI clinic within a month of a TBI were compared to 18–59 year old patients. The older group had a better outcome in the early period and offered no harbingers of a poor outcome. Possible reasons included returning to a less pressured environment, for example, the need to return to work, the need to care for a family, a higher level of musculoskeletal injuries occurring in MVAs (known to increase psychosocial distress), and more prior exposure to physical limitations (Rapoport & Feinstein, 2001).

Cortical atrophy may be documented after head injury. This condition may be described as "early Alzheimer's", or degeneration. Yet, the age of the patient and lack of pretraumatic impairment may preclude this diagnosis. One may speculate that diffuse cortical necrosis with white matter loss results in apparent degeneration of the cortical sulci after impact or acceleration brain injury.

ENDOCRINE/PITUITARY DETERMINANTS OF OUTCOME

Endocrine disorder can be consequent to *neurological disruptions*. *Cognitive disorders* are often found when TBI is accompanied by subarachnoid hemorrhages. It is not certain whether these are consequent to the brain injury or to associated hypopituitary defects (Leon-Carrion et al., 2005). Following sudden movement and head rotation, secondary injury mechanisms are initiated, that is, changes in neurotransmitters, ions, oxidative stress, blood flow, edema, and energy (Cernak et al., 1999; Molitch, 1995; Treip, 1970). mTBI interferes with endocrine function through diffuse axonal injury hampering homeostatic control and the initiation of delayed, multifactorial biochemical and physiological effects. A severe enough lesion of any part of the HPA leads to loss of endocrine homeostasis by disturbing excitatory and inhibitory stages of endocrine signaling to the hypothalamus and pituitary gland (Van Cauter et al., 1966). Hypothalamic damage and interrupted afferent supply disrupt delivery of releasing or inhibiting factors to the pituitary. This has been documented in fatal injuries, but lesser trauma can have a variety of endocrinological effects.

HYPOPITUITARISM

Pituitary hormone deficiencies are identified in a substantial proportion of patients with brain injury: Adrenalcorticotrophin releasing hormone (ACTH); growth hormone (GH); suppression of the gonadal axis (e.g., Lutenizin hormone-follicle stimulating hormone (LH-FSH) deficiency, gonadotroph); thyroid stimulating hormone (TSH); free T_4; basal morning cortisol. Hypogonadism is associated with poor outcome. GH deficiency is associated with depression, diminished quality of life, and cognitive disability. The latter concern would follow from its effects in adults: abnormalities in body composition, decreased bone mineral content, and reduced exercise capacity. Symptoms of anterior hypopituitarism may be overlooked or attributed to other causes. After TBI, endocrine dysfunction has been determined for all hypothalamic-pituitary axes (corticotropin, thyrotropin, GH, gonadotropin, prolactin, vasopressin). Hypopituitarism occurs in up to half of long-term survivors of moderate-to-severe head injury (6–36 months). With patients with minor-to-moderate injuries (GCS as the criterion), plasma cortisol levels increase during the early post-TBI period. Patients with severe trauma exhibit a significant decrease in cortisol. These abnormalities are transient in some patients, while the majority recovers by 6 months.

Appropriate procedure: It is recommended that all patients who undergo TBI, regardless of severity, should undergo baseline hormonal evaluation 3 and 12 months after the initial injury. Those with clinical signs of hypopituitarism (adrenal insufficiency, diabetes insipidus, or other clinical symptoms of hypopituitarism) should undergo immediate testing of the rest of the pituitary axis (D'Angelica et al., 1995; Ghigo et al., 2005). Thus, pituitary function testing is warranted in most patients with moderate or severe head injury, including children, who incur basilar skull fractures passing through the sella turcica (Bondanelli, 2007; Kelly et al., 2000).

Overlap with TBI: Signs of hypopituitarism may be misunderstood or erroneously ascribed to persistent neurological impairment. Hypopituitarism creates deficits and symptoms commonly observed in TBI, which may be ignored or misattributed: reduced strength, aerobic capacity, fatigue, erectile dysfunction, decreased libido, amenorrhea, cold intolerance, weight gain, cognitive impairment including memory disturbance, reduced sense of well-being, general health, vitality, and mental health, with commonly experienced depression and anxiety.

Delayed diagnosis: This is common. A survey of eight cases indicates that the average delay in diagnosis was 12.4 years, with a range of 0–44 years (D'Angelica et al., 1995).

Anterior Pituitary Symptoms

After head injury, the most important endocrine complication is anterior pituitary failure, which may take months or years to develop. Severe head injury with poor prognosis is associated with reduced levels of glucocorticoids, prolactin (PRL), and TSH response to thyrotropin releasing hormone (TRH) (Edwards & Clark, 1997). Earlier onset creates greater severity of thyroid, gonadal, adrenal hormone, GH, or water disturbances (Melmed & Kleinberg, 2003). Symptoms of hypopituitarism may be total or partial (D'Angelica et al., 1995; Grossman & Sanfield, 1994). Panhypopituitarism is rare, and is associated with alterations in arousal and awareness. Hypopituitarism is clinically evident after 70–75% destruction of the adenohypophysis. Total loss of pituitary secretion requires 90% glandular destruction and may accompany prolonged coma. Recovery of pituitary function usually parallels neurological improvement. Residual impairment is not uncommon (Frohman, 1995).

The developmental and neurobehavioral effects of posttraumatic endocrine dysfunctions vary between children and adults. They may occur from injuries that do not cause a LOC and may be unrecognized for many years (Hansen & Cook, 1993). The risk factors are unclear, with mechanical, physiological, and genetic factors participating. Hypopituitarism is clinically evident after 70–75% destruction of the adenohypophysis (80–90%) (Rosen & Cedars, 2004). It results in insufficient stimulation and hormonal output of target glands.

Pituitary Stalk

The integrity of hypothalamic-releasing factors, and of anterior and posterior pituitary secretion, is jeopardized by damage to the *infundibulum* (stalk of the pituitary) or to *portal (local) circulation* connecting the brain with the pituitary gland. Interference with delivery of hypothalamic-releasing hormones to the anterior pituitary creates *hypopituitarism*, and to the posterior pituitary may cause *diabetes insipidus* (lack of vasopressin, i.e., antidiuretic hormone [ADH]). This condition may develop within 24 h, and is resolved in only half of the patients (Bode et al., 1996; Schmidt & Wallace, 1998). Vulnerable anterior pituitary hormones include corticotropin (adrenal cortex: cortisol, androgens); GH (liver: insulin-like growth factor 1 [IGF-1]); TSH (thyroid: T_4, T_3); LH, and follicle stimulating hormone (gonads: estradiol; progesterone or testosterone); prolactin (breast: lactation).

NEUROCOGNITIVE DETERMINANTS OF OUTCOME

Brain maturation at the age of injury reflects the interaction of numerous conditions, including developmental, historical, and traumatic. Trauma hampers development and also creates characteristic impairments and symptoms. Both types of conditions may be concealed by compensation and reserve capacity. Outcome will be influenced by the type of neurological disorder, for example, focal, diffused, or combined. Impairment may be compensation for by the alternate hemisphere or other circuits. Impairment may be focal, cortically lateralized, occurring at different levels of the CNS, or functionally specialized (verbal, spatial, sensory, etc.).

Cognitive Reserve

This concept is used to explain the fact that often there is not a close relationship between the brain trauma and the outcome. When adaptive activities are considered, for example, academic or work performance, there may be no significant initial evidence of reduced performance. Resistance to reduced posttraumatic disability probably has different mechanisms, although what these have in common is *redundancy*. Built-in redundancy provides capacity to call upon a larger array of networks in the face of brain damage (Stern, 2002). Individuals with more reserve (skills, intelligences, occupational attainment) require neural recruitment at a higher task difficulty level. Thus, a potential marker for TBI is performance significantly reduced below the estimated premorbid baseline, indicating that the cognitive reserve has been exceeded (Levin, 1985).

The following study is evidence for enhanced cognitive loss following prior trauma. Two head injury groups were compared, with and without prior brain injury (alcoholism, drug abuse, psychiatric illness, and/or previous neurological insult), by the Wechsler Adult Intelligence Test-Rev., including an estimated preinjury IQ. The group with prior injury offered evidence of greater cognitive decline in excess of that expected from the head injury alone (Ropacki & Elias, 2003).

The *passive* reserve model is defined as the amount of damage that can be sustained before reaching a threshold of observable deficit. The concept of BRC or neural reserve infers mechanisms such as brain size or synapse count. Thus, whether there are impairing effects of a lesion of a particular size would be determined by the neurological capacity of the preinjured brain. It is asserted that each injury destroys neurons, thus diminishing the reserve available for performance. Later injury can bring a demand for a performance that is now above the patient's lowered mental level, that is, the patient's capacity to function is now below the *ecological demand.*

The *active* reserve model utilizes "software" rather than structure. Brain damage is assumed to increase task difficulty. Networks compensate for lesions by sustaining functioning through altered processing. Brain structures or networks are used that are not engaged by the undamaged brain. A common response is the more efficient use of brain networks:

- Recruiting areas that are utilized for an easier version of the task
- Recruiting additional brain areas

Ongoing Mental Activity

Prior cognitive activity influences outcome. Seven hundred elderly people were studied who underwent clinical evaluations for up to 5 years (Wilson et al., 2007). The level of late-life *cognitive activity* predicted later decline. Composite measures of amyloid, tangles, lewy bodies, and infarction were not associated with the level of cognitive activity at the study onset. Several possible explanations were offered: Earlier or cumulative experiences may be critical for subsequent performance level; environmental stimulation may enhance functional and structural changes in neural systems (see "Cognitive Reserve"); and there are experience-related differences in brain structure. These events can defer vulnerability to premature dementia that may occur in genetically vulnerable individuals. Neural systems underlying memory and information processing seem partially activity dependent. Thus, cognitive activity contributes to the cognitive reserve by maintaining efficiency in underlying neural systems, enhancing adaptation to age-related pathology, or a combination of these. This inference was supported by a study of the effect of *educational level* as a risk factor for AD (Wilson et al., 2008). Education is a well-established correlate of cognitive test performance at all ages, as well as a risk factor for AD. Persons with relatively more schooling are likely to begin old age at a relatively higher level of cognitive function and would need to express more cognitive decline before reaching a level of functioning describable as demented. However, educational level is not associated with cognitive decline. It does affect the risk of late-life dementia due to its association with the level of cognition, rather than predicting the rate of cognitive decline.

PSYCHOSOCIAL DETERMINANTS OF OUTCOME

Depression and Suicide

In TBI groups, a significant association was found between major depression, poor psychosocial outcome, and poor outcome for ADL. Depression may negatively affect participation in rehabilitation and social interaction during recovery, and these patients cannot recover early losses even when the depression is over. Traits not predicting poor psychosocial cognitive or ADL outcome were age, sex, education, socioeconomic status, premorbid level of social functioning, and previous history of alcohol and drug abuse (Robinson & Jorge, 2005).

The strongest predictors of suicide attempts among TBI survivors are strong feelings of hostility and aggression. With TBI there is more frequent suicidal ideation, suicide attempts, and completed suicide. Even mTBI is associated with a greater tendency to attempt suicide (Robinson & Jorge, 2005).

Personality Characteristics

In a sample of severe-to-very severe TBI patients, the interaction of capacity to appraise and to cope, as well as psychosocial variables affected every studied outcome except employment. Personality affected all predicted outcomes (Rutterford & Wood, 2006).

Victimization: Persons with disabilities are four to ten times as likely to become *victimized* (violence, abuse or neglect, that is, VAN) as a nondisabled person (Reichard et al., 2007).

The patient's anticipated outcome: Many PCS symptoms rely upon *self-report*, implying independent verification before being associated with a brain injury. Mittenberg et al. (1991) asked noninjured subjects to assume that a head injury would occur. It was concluded that the *anticipated cluster of syndromes* resembled that actually found in documented head trauma cases (outpatient practice). Head injury patients, compared with controls, underestimated the prevalence of benign symptoms. It is difficult to conclude, as the authors do, that this finding occurred despite the fact that the control group had "no opportunity to observe or experience post-concussive symptoms," since head injury is very common, and PCS persists in a subset of patients in the community. Their further speculation is more reasonable: Arousal creates an expectancy that causes attentive bias for internal states, augmenting symptom perception. Thus, eliciting additional autonomic/emotional responses further reinforces an expectation. And, the actual symptom level might be higher for a stressed than nonstressed group with the same physical condition.

Conditions Enhancing TBI Symptoms

Greater injury does not necessarily predict a greater number of symptoms; the opposite may occur (Corrigan & Bogner, 2007). This writer wonders whether a more intense TBI reduces the level of self-awareness, perhaps body schema, so that sensitivity to the body is impaired. An affective component, for example, depression and neuroticism, can contribute to cognitive complaints (including memory and confusion). On the other hand, such symptoms as numbness, balance, vertigo, and localized sensitivity disorder may have a neurological etiology that may be more precisely associated with a trauma.

OUTCOME OF CLINICAL GROUPS

There is no clear association between PCS status and category of group membership. Clinical groups were selected by Fox et al. (1995a) to determine predictors of common PCS symptoms (psychotherapy, neurology, family practice, internal medicine, and underevaluation at an HMO. Etiological variables were being knocked unconscious, a bump on the head, and a lawsuit. Being knocked unconscious was a predictor for 9 of the 10 PCS symptoms, a bump on the head predicted 8 of 10, and a lawsuit 6 of 10. The psychiatric group had high endorsement of PCS symptoms. In family practice the predicted symptoms were concentration and impatience, a finding important but not surprising. The statement that certain symptoms not considered typical of PCS had a higher incidence in the knocked-out group is incorrect insofar as such symptoms certainly may accompany LOC (tremors 44%, loss of interest 68%, confusion 68%, and broken bones 32%).

It was suggested that a head injury might increase the intensity of otherwise common symptoms. Several hypotheses were considered: Cognitive complaints may be a reflection of psychological and emotional status; those who have suffered a bump to the head without LOC were less likely to have suffered a brain injury but the psychological trauma could have led to PCS complaints; PCS symptoms are more likely in psychiatric patients; being in litigation is associated with an enhanced level of symptoms. Concerning litigation, it is reasonable to assume an association between intensity of injury and the likelihood of claims for compensation. Also, risk factors could increase the likelihood of having a head injury. Being a litigant was a predictor for 6 of 10 PCS symptoms was noteworthy to the authors, but this is no surprise. Litigation is sometimes the only means of obtaining compensation for an injury.

Nursing Home Residents

This is a varied sample: neurological, cardiac, metabolic/endocrine, musculoskeletal, psychiatric and pulmonary. The co-morbidity TBI and dementia decrease with age, that is, 44% (31–60), 27.5% (61–80), 28.4% (81 or older). TBI in this sample did not increase the likelihood of dementia. TBI alone, as a symptom of this disabled group, decreased from 63.6% to 9.8%, while dementia alone increased from 2.8% to 71.2%. Having neither of these conditions increased from 15.2% to 49.9%. These findings suggest that the likelihood of having a disabling condition based upon TBI is highest at a young age, and older individuals are more likely to be disabled by independently obtained dementia or a different medical condition (Gabella et al., 2007).

Ethnic (Cultural) Characteristics

Family and Cultural Influences: Persons of foreign origin require special services: the assistance of health interpreters to facilitate their communication with providers; overcoming problems of stigma and social isolation (Simpson, 2007). Ethnic and cultural differences can affect baseline performance, leading to misdiagnoses. It has been asserted that neuropsychological measures are so sensitive to demographic variables that without adjustments to the normative criteria, interpretations and diagnoses can be dramatically affected. Without this change, their specificity for a diagnosis would be reduced. (The present writer suggests demographic corrections have mixed utility: It is acceptable that some relatively low scores may be a cultural or educational artifact. However, the examiner should consider that measurements also have some ecological validity in estimating performance in the patient's environment.) On some procedures, reduced average cognitive scores are achieved by African Americans compared to Caucasians. The difference is found across age, gender, and educational levels. While such differences are described as normal variations, the meaning and significance of this term here is obscure. However, if cultural and contextual explanations are not accepted, the differences could create for African Americans a greater risk of misdiagnosis of neurocognitive impairment, dementia, and learning disabilities. Even when low-birth-weight children were compared, cognitive test differences were reduced by more than half when socioeconomic status and variations of the home environment were taken into account.

In a study of noninjured African Americans and Caucasian adults they differed significantly on their cognitive test performance, although the two groups had only slight but significant differences in their early environmental histories. The reduced effect of ethnic difference may have been due to an imprecise retrospective assessment method. Demographic variables to be considered are educational and occupational background of parents, family composition, availability at home of educationally oriented materials such as reading, early school experiences, and parental attitudes towards education, and quality of education. The findings suggest that the mere labeling of a person as coming from one category or another of background need not mean that stereotyped experiences have actually occurred. The patient's self-description could be expressed closer to some desired norm. Differences of attitude to the examination may have affected performance. African Americans were less concerned with the effectiveness of their performance, seemed to find the testing irrelevant to their lives and abilities, were less likely to guess, and more likely to interrupt to converse with the examiner.

Physiological aging: Should a person be examined after an injury and manifest impairment from the baseline, the examiner will attempt to assess the relative contribution of the injury, normal aging, and an accelerated aging process. The relationship of *early physical fitness* to late cognitive performance has been studied. Is there a relationship, and what is its direction, between components of physical fitness and cognitive decline from youth until old age? Is there a common aging factor accounting for age-related variance? This has been determined in simple procedures (reaction time) and complex cognitive tests. Cognitive domains age together. Yet, cognitive aging seems largely accounted for by sensory functions (balance, vision, hearing). Cognition in old age correlates with forced expiratory volume (FEV). Childhood and old age IQ are highly correlated; adolescent cognitive ability is correlated with FEV of middle age. Since IQ scores at age 79 correlate with three aspects of fitness, but fitness correlation with the same test at age 11 is close to zero, one infers that in older adults fitness in every domain is not the result of people with higher intelligence pursuing healthier lives. Nevertheless, survivors at age 79 averaged higher childhood cognitive ability (Deary et al., 2006).

RECOVERY IS THE RECIPROCAL OF THE PCS!

Recovery is defined as follows by *The Random House Compact Unabridged Dictionary*: "... the regaining of something lost or taken away; restoration or return to health from sickness; return to any former or better state or condition..." An alternate definition of recovery has been "no longer when higher receiving insurance benefits." While compensation factors may enhance the sick role and delay recovery, secondary losses (associated with chronic pain, financial loss, marital breakdown, loss of employment, anxiety, and depression) generally outweigh secondary gain available through third-party payers. The person's *appraisal of the situation* influences the outcome. Stress is a function of the appraisal of the injured person of the demanding environment situation and its demand in terms of coping resources (Knussen & Cunningham, 1988).

This author (Parker) offers as a definition of recovery: return to the estimated preinjury baseline without essential dysfunctions or discomforts. Anything less than that is not "recovery," although the patient may be functional to a considerable extent. Study of recovery is partially achieved by challenge with complex and difficult tasks. In addition, assessment of adaptive ability goes beyond performance measures to include general adaptive capacity, including functioning with people, employment, independence, and community activities. The description of a patient or research group as "recovered" is often inaccurate and incomplete. Among the functions to be considered are activities used in the community, family, employment or school, and quality of life. Using psychosocial reintegration as a measure of recovery, only a minority succeed to a major extent; its components include employment, schooling, interpersonal relationships, social contacts, and leisure interests (Levine, 1993).

Assessment that the patient has "recovered" or "returned to normal" requires the following points of reference:

1. Description of the preinjury pattern of functioning (baseline) as to level and quality over a wide range of representative activities; and
2. Comparison of current and estimated preinjury pattern over a wide range of neurobehavioral functions characteristic of preinjury neurobehavioral functioning.

There is a difference between stating that a patient has returned to work and stating that his efficiency, mood, self-esteem, and personality functioning are comparable to preinjury. This author does not believe that after any significant TBI, complete recovery occurs. Overestimation of returned capacity, apart from distorting the process of compensation for loss, frustrates the patient by creating unrealistic expectations of the employer and family. The patient may be fully aware of performance loss, but failure is more painful when it is attributed to weakness of spirit or desire to "cheat the system." Conclusions drawn concerning "recovery" will depend on the range of functions examined and the demands made on the patient. Individuals with mild injuries seek treatment for different reasons than those with more severe injuries: postconcussive syndrome, pain, depression, and litigation (!) versus medical and rehabilitation issues (Hanks et al., 1999). Recovery can be deemed complete or that no damage was determined, when the criterion was focal neurological sensory and motor functions. Simple tasks will not assess adaptive ability, only those that reflect environmental demands for quality, speed, and personality and relatedness.

When "higher" functions were assessed with neuropsychological tools, deficits accounted for the problems of employment and adaptation. For example, Ogden et al. (1990, citing Fortuny & Prieto-Valiente, 1989) illustrate that 100% of a sample of patients who had subarachnoid hemorrhages were rated as having a good neurological recovery, while only 37.5% were rated as having a good neuropsychological recovery (only 63% returned to leisure activities and only 50% had returned to premorbid work).

Realistic assessment whether a patient has "recovered" or "returned to normal" requires comparison of current and estimated preinjury pattern over a wide range of neurobehavioral functions. Assessing a wide range of functions reduces the likelihood that vague or unmentioned dysfunctions will be ignored. It also makes it more difficult for the conscious or unconscious exaggerator to exaggerate dysfunctions over a greater range of unfamiliar tasks. When a reliable psychometric baseline estimate of preinjury performance is not available, statistically significant deviations or gross qualitative deviations between disparate functions (e.g., verbal and spatial) may indicate brain trauma, although conclusions should not be drawn without considering the reliability of the indicated function, or the possibility of a chance downward deviation found in a large number of measurements.

The Rate of Recovery

There are numerous patterns of change after TBI. It has been suggested that there is a neurogenic etiology for postconcussional symptoms and cognitive sequelae during the early stage of recovery; whereas other factors probably account for the delayed onset and marked prolongation of these problems. It is necessary to wait to determine the extent of recovery. The interval after which most recovery has occurred is controversial (e.g., 3 months to 2 years). In any event, at the time of any assessment, a wide-range neuropsychological examination is needed to avoid missing dysfunctions that are not in the record. Levin (1985) asserts that MHI injuries are sufficient to produce cognitive deficit and postconcussional symptoms that have a characteristic time course of at least 2–6 weeks. The residual decrement in cognitive capacity would be evident only under stressful conditions. Level of recovery is influenced by the extent of injury and the function being measured, for there is improvement with regard to the number of distressing symptoms. For others, there may be worsening after some time. Still others may experience persistent, troubling somatic symptoms (e.g., headache, neck problems, or dizziness). Some persons are more symptomatic later than earlier after the injury, creating persistent postconcussion syndrome (PPCS) (Alexander, 1997). While patient understanding of the meaning of cognitive symptoms is troublesome, excessive attention by family, physicians, and other professionals may reify the symptoms into a diagnosis (Alexander, 1998). Most persons with head injury experience at least a transient phase of reduced cognitive efficiency and disturbances. This often arises because of premature resumption of stressful activities. It has been asserted that the recovery curve for Verbal IQ (WAIS) usually reaches a plateau by 6–12 months, whereas increments in Performance IQ may continue for a year or more after injury (Levin et al., 1982). This author's research indicates lack of evidence for recovery of baseline IQ up to 10 years when a practice effect is avoided by using only single measurements to determine posttraumatic status after an interval (Parker, 1990, p. 160; Parker & Rosenblum, 1996).

REHABILITATION

Since the neurobehavioral consequences of an accident causing TBI are so diverse, the skills and procedures of rehabilitation must also be sufficiently diverse to offer focused treatment. Assessment and treatment involve wounds, cerebral trauma, and the range of their physical and psychological effects. The goals for the rehabilitation of a disabled person are designed to improve function and independence. One may consider medication for cerebral or somatic disorders, physical therapy and prosthetics for broken bones and soft tissue, formal procedures to strengthen memory and concentration, internal medicine to cope with systemic dysregulation and the allostatic state of organ fatigue, pain management for headaches and radicular damage, and also personal face-to-face counseling and psychotherapy. The clinician gains information in order to develop a treatment plan and to make referrals for consultation and treatment: physical, intellectual, mental, physiologic, social, economic, and environmental. While the themes now described evolved from frontal lobe damage, they are generic and address self-awareness, self-regulation, and interactions with complex environments and social-emotional issues: environmental modification, behavioral management, and cognitive self-management (Eslinger & Geder, 2000).

UNMET NEEDS

Even when the injured person experiences a need for treatment, there are *institutional barriers*: not knowing that help is available or where to go for help (issues of advocacy and case management); transportation problems; lack of financial resources preclude accessing services; health problems (with lack of insurance for accident-related impairments) and functional limitations prevent accessing services; service inflexibility (eligibility issues); psychosocial (not wanting others to know they have a problem); and lack of motivation or complacency. One study determined that over half of the individuals discharged from an acute care facility had at least one unmet service need, and of which 35.2% had at least one unmet need that was critical to maintain ADL and avert secondary conditions (Pickelsimer et al., 2007).

APPROACH TO REHABILITATION

The patient must be trained and counseled to cope with a world changed for the worse. Some accident victims are seriously disabled or have dysfunctions of mobility, mood, language, and the counselor must serve as an advocate to obtain needed civil and medical services. As one considers the older person, domains of functional independence would include independence and survival, mobility, ADL, physical performance, use of health services, sphincter management, and costs. Physical conditioning plays a role in effective outcome (Cruise et al., 2006). Many injured people have such complex injuries and dysfunctions that rehabilitation must consider a matrix of unintegrated concepts and procedures: disparate understanding of the case by multiple practitioners; problems of drug interactions and polymedication resulting from multiple physicians who do not seek prescriptions offered by their colleagues; and the mistake of some practitioners for whom PTSD is diagnosed without considering possible TBI, or PTSD mistaken for developmental issues. I counsel my own patients that they should select a single physician to receive all medical information and collaborate with the others.

In the absence of long-term rehabilitation availability, family support is critical, but health and behavioral disorders represent a realistic burden upon them. The family should understand the injuries, treatments, and implications of the disorders. Caregivers should be alerted to the emotional and mental health needs of the injured person. They need help in finding a qualified provider. (I note the unwillingness of many providers to accept insurance and the insurance company's frequent delays and refusals to offer treatment and assessment for injured persons.) Appropriate treatment requires recognizing that with time, particular needs and their relative emphasis change (Rotondi et al., 2007).

The Patient's World

Please empathize with the experience of the TBI patient: Your new world is partially hostile and adversarial when you seek examination and treatment. You may not be a patient; rather, you are a "claimant" whose need for examination, treatment, and compensation is costly to organizations that are motivated to save money by reducing expenditures. When your health care provider fills out the familiar insurance claim form (HCFA 1500, Item 10) the question is asked: "Is the condition due to employment, auto accident, or other accident?" Since the answer is often "Yes," you have thereby lost your passport for treatment in the world of health insurance.

While the accident patient's learned and habitual reactions are not unlearned, their world is changed: unexpected mental and physical impairment; pain; dependency upon others with fear of impoverishment; new rules for obtaining support or perhaps none is easily available; etc. Life is changed: Previous modes of feeling and action no longer work.

Goals of TBI Psychotherapy

Different schools of psychotherapy offering service to uninjured or impaired persons express different missions, for example, "make the unconscious conscious" to paraphrase Sigmund Freud. I believe that the relatively universal goals of therapy for uninjured and unimpaired individuals who have not been emotionally crippled during development could be expressed as follows: improved coping with short-term and long-term adaptive problems; pleasant and mutually supportive family and social relationships; increasing the ratio of constructive to avoidable self-destructive actions (Parker, 1981, "Emotional Common Sense"); affect regulation in the sense of reducing where possible actions and contact with people that stir up toxic levels of anger, depression, and anxiety, while achieving mostly pleasant moods; achieving self-esteem and a personally valued identity (Parker, 1983); achieving a capacity to alternate between a constructive autonomy and satisfying social relationships as one's living status requires; and developing a sense of purpose and achieving valued aspirations (Parker, 1977).

Now that you have mastered the wisdom achieved by the writer in his less arduous life as a clinical psychologist practicing psychotherapy, kindly step off the elegant train that has offered scenic views of the Rocky Mountains, Alps, or Andes, and climb on board a truck carrying a heavy load, over unpaved roads, with inefficient shock absorbers, heading for a far destination over poorly charted roads: counseling the brain-injured.

Recognizing Different Treatment Requirements

Despite surface similarities with ordinary emotional distress, the actual etiology of a symptom may be posttraumatic. Psychotherapeutic success with TBI patients is difficult. It requires understanding and procedures for a different kind of person: brain and somatic trauma; pain; seizures; dizziness; reduced mental ability and motivation; behavioral and emotional dysregulation; etc. (Pollock, 1994). The knowledge database includes familiar principles of psychopathology, characteristic dysfunctions of different trauma sites and intensity of TBI, physical and physiological consequences of somatic trauma, and the psychodynamics of being injured, economically vulnerable, and socially troubled. The practitioner is also concerned with technical procedures, practitioner reimbursement, and referral to other specialists. The psychotherapist experienced in other areas of treatment utilizes these clinical skills (procedures and psychopathological understanding) but infrequently and cautiously. Knowledge of personality development remains valid and valuable.

However, although the TBI patient retains premorbid behavioral characteristics, there are also different symptoms that usually require equal or higher priority: psychopathology, ego structure, patterns of strengths and defects, impairing symptoms, financial resources, level of pain, etc. Consequently, the overall primary TBI therapeutic approach utilizes a substantially different emphasis and technique.

Concepts of Treatment

Traumatic brain injury application of psychotherapy utilizes concepts of treatment that involve substantially different issues than mental health and psychiatric disorders. Therefore, a clinician untrained in TBI may commit significant errors of assessment and treatment. The posttraumatic brain, the injured organ of adaptation, is less likely to support meaning functions, for example, work, social life, and personal satisfaction and pleasure. Thus, most TBI patients differ substantially from people who receive treatment for emotional, adjustment, or other psychiatric conditions. Among errors this writer has observed: not recognizing that a significant TBI has occurred in a new or ongoing therapy patient; mistaking TBI symptoms for those of other conditions; and focusing on stress conditions, while ignoring the possibility of brain trauma. Therefore, a psychotherapist without TBI training is advised to avoid treating a TBI patient without suitable supervision. An example of the different approaches between TBI and nontraumatic emotional disorders is the undesirability early in treatment of focusing upon familiar psychotherapeutic themes such as interpretation, confrontation, passivity by the psychotherapist, emphasizing earlier experiences, unconscious processes, symbols, etc. These outcomes would likely be time-wasting, confusing, might antagonize the patient, and often be technically incorrect.

Behavioral Approaches

The procedure now described differs substantially from the psychotherapeutic approach described later. While the latter is not primarily psychodynamic, it has learned from its principles. Its application, more than classical psychotherapy or behavioral approaches, utilizes transactions such as education and advocacy. The neurobehavioral approach, based upon learning theory, utilizes contingencies: Positive or aversive reinforcements are applied to modify maladaptive behavior or behavioral dysregulation. They reinforce adaptive behavior and reduce disruptive behavior. Their programs utilize staffs rather than individual therapists. Their transactions are time intervals, tokens, points, or privilege systems similar to those used for psychiatric and conduct-disordered populations. Interventions arise from the characteristics of the treatment environment, becoming the automatic and habitual ways in which individuals in a treatment culture interact. These are skill-building programs and cognitive therapy approaches designed to strengthen adaptive behaviors. The latter ameliorate or compensate for cognitive difficulties. However, emphasizing external reinforcement of desired behaviors can be counterproductive if behavioral change is attributed to the external reinforcer, which if withdrawn may lead clients to discontinue exhibiting the behavior. The procedure does not utilize "self-talk," thus ignoring a powerful commitment to behavior change (Giles & Manchester, 2006).

Operant neurobehavioral approach (ONA): The clients are supported to adapt to social norms within a structured environment. Clear contingencies strengthen desirable behaviors serving the same function as the disruptive behaviors. Undesirable behavior is extinguished by ignoring it or through or staff feedback that is direct, corrective, and authoritative. Immediate assistance is considered to reinforce the behavior difficulty. Increasing nonaversive approaches is more effective when unhelpful staff behavior (arguing, criticizing, blaming) is decreased simultaneously. The negative punishment principle is also used (deprivation of something they want).

Relationship neurobehavioral approach (RNA): Both ONA and RNA emphasize positive behavioral approaches. Extinction procedures are less central to the RNA than to the ONA approach. The therapeutic relationship is actively used. It would be helpful if behavioral treatments emphasize treatment factors and avoid confrontation and extinction methods. The staff utilizes a style that reduces aversive interventions and confrontation, but encourages clients to generate their own reasons for behavior change.

A PSYCHOTHERAPEUTIC APPROACH TO TBI

The complexity of the psychotherapeutic process merits consideration by the practitioner offering this service to TBI patients. "Verbal" therapy is a transaction mediated by expressive and receptive language capacity, verbal, nonverbal, and more! It is cerebrally bilateral, integrates with the remainder of the brain and spinal centers, torso, and spirit (whose localization has not been completely established). The patient's task is to understand the meaning of the therapist's interventions, and to express mood, feelings, desires, and intentions with clarity. Competent discourse (see this section) is vulnerable to trauma and stress: cognition (abstract reasoning to generalize to other situations; long- and short-term memory; comprehension) and nonsemantic language (understanding prosody, facial expressions, gestures).

The TBI patient considered frequently has a variety of personal qualities that render this treatment even more problematic than the crises and difficulties encountered by the treater of non-impaired persons, for example, serious injuries, lack of funds, impairments of comprehension, memory, self-awareness, cognitive processing, motivation, mood, etc. Miller (1991) notes: basic adaptive loss; emotional dysregulation; loss of motivation for varied reasons; inability to cope with different environments; need to redefine one's self-image; body disabled disproportionately to the brain's retained capacity; and flat affect and aprosodia misdiagnosed as depression. The co-morbid nature of the injury (cerebral and somatic) and the chronic nature of many symptoms invite rejection of an approach informing patients "about the benign effects of mild head trauma (and) reassurance that symptoms are not signs of brain damage and will resolve" (see "Litigation"). In fact, while a cognitive-behavioral approach was effective in reducing symptoms compared to a control group, even after 6 months of treatment the proportion of experiencing symptoms ranged from noise sensitivity (67%) to depression (27%) (Mittenberg et al., 1996). This writer is concerned that patients who remain significantly symptomatic will have an extra measure of distress if they have been incorrectly informed that their condition is essentially benign. Rehabilitation models offer procedures that differ from the approach usually described as psychotherapy and counseling, for example, behavioral, cognitive, compensatory, and holistic. However, awareness of various psychotherapeutic procedures, integrated into Orlinsky and Howard's Generic Model of Psychotherapy can help the clinician to select building blocks to plan psychotherapy for a given patient (Coetzer, 2007).

The therapeutic contract: Agreed-upon roles, session formats, duration of treatment, etc.

Therapeutic operations: The patient presents information to the psychotherapist to determine which intervention is useful.

Therapeutic bond: The relationship with its investment in their respective roles and the personal rapport that develops (aiding the resolution of interpersonal difficulties, Rolland S. Porker [RSP]).

Self-relatedness: An intrapersonal process, with openness or defensiveness, including response to the other person. (This has components of Freud's great concept of transference and counter-transference, describing the subjective qualities of the psychotherapeutic relationship and experience.)

In-session impacts: The therapeutic experiences of the patient, including emotional relief and insight.

Phases of treatment: Events in treatment that combine to form interventions into groups.

Therapeutic Reactions

Psychotherapy, regardless of the treater–patient duo, is an intimate experience that burdens the therapist with restraints usually not experienced or expressed in other civil and social relationships. The practitioner who has had conventional therapeutic experience and personal psychotherapy may find it easier to restrain rejecting and hostile attitudes. Pollock (1994) raises basic procedural and professional issues such as *transference* (patient's reactions based upon earlier experience and training) and *counter-transference* (the therapist's reaction to the patient on the basis of his or her own personal experiences). There is a strong ethical and professional need to aid the patient, while recognizing that one's personal experiences have created nontherapeutic reactions. The writer is in disagreement with Pollock's assertion that after TBI attitudes and responses brought into therapy stem from earlier interpersonal experiences. However, I am in accord with this concept: The therapist's expectation of success, with possible frustration and anger that interfere with the therapeutic alliance, could create a wish to abandon the patient, with feelings of guilt providing briefly a glue that prevents fracturing of the relationship. Thus, therapeutic success depends upon managing the transference and counter-transference. While positive components of the transference may be nurtured, the psychological boundary between patient and therapist must be respected, and reality testing is needed to preserve the therapeutic alliance.

Therapy Intake

For all practitioners, regardless of their clientele, careful screening should explore whether there has been a prior TBI. I have examined numerous patients in whom TBI was never considered in prior examination or treatment. Study may reveal that the patient is unaware that a trauma created TBI, or did not volunteer information about accidents in the interview (mechanical impact and whiplash; neurotoxic; electrical; anoxic, etc.).

Regardless of psychodynamic premorbid disorders, most likely, the initial emphasis should be on coping with the traumatic symptoms. Further, some psychiatric patients' neurological disorder would benefit from an approach overlapping with TBI. True!

Personality: Preexisting psychiatric disorders, temperament and behavior, commitment to independence, motivation to use strengths and tolerate their weaknesses, and motivation for psychotherapy or remediation. Such personality characteristics influence the level of behavioral disorder and its complexity that the therapist can address along with the resultant patient resistance. Both the pretrauma and current level of empathy or alienation will influence the firmness of the therapeutic alliance and the capacity or willingness to engage in social activities.

Cognition

Insight will help the patient create an agenda for the therapist's attention. Comprehension aids in planning tasks that are within the person's competence, that is, solving problems, error monitoring concerning their behavior, recognizing the level of success, and undertaking activities that are meaningful to the person's lifestyle. Loss of abstract thinking may cause attempts to utilize humor in the therapy to be misunderstood as deliberate humiliation. Cognitive rigidity can result in loss of empathy and egocentricity. When the patient is asked to reflect on a topic such as how another person reacted to a discussion or activity, there may be an attempt to empathize with the other, nonunderstanding of the topic, or the accusation that the therapist is focusing upon another person.

Specialized Databases: TBI and Psychodynamic

How does one resolve the requirement for two different informational databases and therapeutic procedures? The clinician must consider many possibilities, including some that are not in the familiar scope of neuropsychological or neurological practice. Clinical acumen, education, and experience derived from service to the "psychiatric" or "maladjusted" patient remains useful, although some components will rarely be expressed. Professional knowledge concerning nonneurological types of dysfunction enables the practitioner to increase diagnostic accuracy, and therefore render appropriate treatment or make useful referrals. Consider the differential diagnosis of "depression," a common problem. To be functionally impaired, injured, socially hampered, and, by the way, unemployed would seem to be good cause for "depression." It really is grounds for depression—honest! However, the careful practitioner will also consider whether the mood disorder may be psychodynamic (as noted earlier), left hemisphere trauma, fatigue caused by chronic physiological dysregulation (allostatic load), apathy due to a prefrontal lesion or endocrine dysfunction, or maybe something else! Consider irritability: Is this a result of chronic pain or sleeplessness, a brain lesion creating a low threshold for anger, or a reaction to a long delay in a request for treatment authorization?

The TBI database includes the consequences and interactions of both neurological and somatic trauma. This creates a priority of attention: While psychodynamic and psychiatric issues exist, treatment focus usually must be initially directed at rehabilitation, emotional support, and advocacy to obtain service from the community.

Alertness to a Range of Disorders

The psychotherapist or counselor must be alert to the possible existence of a wide range of medical and psychological disorders. Intake and record review should be screened for preexisting, traumatic, and current disorders, which may be psychological, physiological, neurological, medical, and behavioral. Any mental health practitioner has the same responsibility, but after injury, many additional possibilities occur creating vulnerability to additional classes of disturbance due to dysregulation and allostatic load. Therefore, while the practitioner may be "multimodal" in the sense of being widely trained in TBI treatments (Miller, 1993, p. 48) the patient's wider range of diseases and mental disorders requires awareness of a range of disorders and appropriate referrals. The period of altered consciousness, perhaps lack of clarity for mental processes rather than confusion, can extend for years, with clarity occurring suddenly for no apparent reason. Vague distress can now be experienced as depression.

When the therapeutic activity is primarily "psychological," the focus will alternate between cognitive rehabilitation, emotional dysregulation and stress, maladaptive behavioral activity, and occasional insight and interpretation. Self-directed activity is encouraged as being valuable for rehabilitation through disciplined mental activity, pride, and material for social contacts. Examples are teaching oneself computer skills and photography.

The Therapeutic Alliance

This is the personal bond between patient and therapist. It describes the collective mood and motivation, which enhance the effectiveness of treatment. Patients with a higher capacity for social relationships tend to develop a stronger alliance than those who are defensive or negative. Those who drop out of therapy prematurely liked the clinician less, felt less respected, viewed the therapist as more passive, and experienced a weaker therapeutic alliance However, the therapeutic relationship is more than the style of relating of the individuals:

- Trust that the therapist is committed to improvement of the patient's condition, and therefore will not be hostile or self-serving.
- The confidence experienced by the patient that the therapist will be reliable (continue therapy) despite the intense anger, anxiety, and depression expressed by the patient, and the unpleasant mood experienced by the therapist in this intense and intimate relationship.

When the therapeutic bond is reasonably strong, plans and goals can evolve. A positive outcome is more likely since disagreements will be coped with and the work will continue (Miller, 1993, pp. 47–69; Saunders & Lueger, 2005).

Style of Life and Support

After TBI the style of life and the social role changes considerably as a function of both the social response to the personality and mood changes of the patient (*cerebral personality disorder*), and the patient's behavioral change consequent to their psychodynamic response to the postbrain-trauma condition. Contrasting with pretrauma life, most people's self-determination now passes to others! The new social world may include two or more sets of attorneys (plaintiffs and defendants) and their assigned examiners, claims adjustors, bill collectors, dissatisfied family members, friends, and employers. Life has unexpected burdens in the community: difficulties consequent to mobility, seizures, anxiety, balance, and so forth. Spiritual life (morale) is assaulted by the traumatized identity and an unsupportive Weltanschauung.

The need for support has several bases:

- Disturbed lifestyle due to impaired work and family functioning
- Domestic and family immobility due to physical problems
- Denial of the reality or importance of dysfunctions by defendant's responsibility for providing examination and/or treatment
- Unwillingness or inability of the patient's personal attorney to properly represent him in such matters as prompt treatment and appropriate compensation for injury and lost wages
- *Isolation*: Loss of social contacts may mean a discontinuity between previous and current identity. There may be confinement due to anxiety, seizures, dizziness and imbalance; rejection by collaterals; the patient may live alone, without family or friends. Group therapy has been recommended, but in my practice, scheduling difficulties are gross
- Personal discomfort due to pain, fatigability, depression, etc.
- Economic problems due to employment loss
- Isolation
- *Rejection*: If the person's condition is not gross, doubts about its seriousness or reality may be expressed by those with a conflict of interest (i.e., family members and friends from whom there is an expectation of aid that would be experienced as inconvenient)

I believe that the expression of empathy with the condition, acknowledgement of difficulties and discomfort, and availability to perform appropriate advocacy assistance are often the most successful and useful therapeutic activities.

Successful actions should be explicitly acknowledged, for example, a patient's excellent work in representing himself in a marital financial dispute.

Advocacy

The TBI counselor's primary function with some patients may be indefinitely, or occasionally, only serving and facilitating! This transaction may be the greatest divergence from psychotherapy with nonbrain-injured people. The process of obtaining examination and varied psychological and medical treatment is more than adversarial. The patient needs referrals to members of the legal and health care professions who are specialists in TBI. Negotiation with governmental agencies, professional service organizations, landlords, schools, etc., may be too difficult for the patient due to deficits of language and community familiarity, loss of mental ability and communication skills, etc. Vocational counseling works on treatment goals and negotiation with former and potential employers. The Americans with Disabilities Act may offer protection against discrimination (Jamison, 2003).

The recommended approach emphasizes the principles of advocacy, the specialized (although highly varied) professional needs of the brain-injured person, and also some general principles of psychotherapeutic practice (development, psychodynamics, diagnostic issues, clinical experience in coping with the emotionally distressed patient and family). Some principles for counseling the TBI patient are offered.

Yet, the key theme in the practice of counseling with TBI victims is that a large proportion are cognitively, physically, and emotionally impaired. Many of them are unemployed, and had minimal savings prior to their injury. Yet, a common problem in treating accident victims is denial of examination procedures and treatment. The potential expenses of the accident victim can be extreme: Multiple specialists, medication, physical therapy, pain management, sleep medicine, counseling, cognitive and other rehabilitation, chronic stress, etc. Thus, the psychotherapist has responsibility for treatment of an injured person for whom the society is often indifferent or actively oppositional. Acknowledgement of liability means much cost. The procedure of obtaining service usually requires "preauthorization," which usually requires waiting for extended periods, perhaps years. The counselor becomes involved in writing detailed letters requesting authorization for varied, sometimes very costly, services. The officials responsible for decisions may be uneducated in the nature and outcomes of CNS and somatic trauma. Thus, the frequent negative decision due to "lack of medical necessity."

Intensive Multidisciplinary Study

Intake and other information will sometimes overlap during an intensive diagnostic or status evaluation. It includes demographics (place of birth, language of origin and culture), family structure before and after marriage, military history, technical and academic education, vocational history, avocations. The details of the injury, posttraumatic and retroactive amnesia, LOC, length of hospitalization, treatment and chronic injuires, current medical and psychological complaints, work, school, family, and social problems. The practitioner will make referrals for further study and treatment as indicated.

Misleading TBI Characteristics and Expressive Deficits

Some symptoms are ambiguous and therefore are more difficult to treat properly. Identifying them as TBI, psychodynamic, or medical enhances increased focus of treatment, and also patient education and capacity for self-monitoring and self-regulation.

The psychotherapist is concerned with expressive deficits interfering with the ability to reveal problems and maladaptation: The TBI patient may find it difficult to ask for help in a life full of unusually intense adversity. Adding to the patient's communication impairment, the unimpaired practitioner may find it difficult to change ingrained coping methods of patients who are needy, pained, immobile, anxious, depressed, resentful, incompetent, unreliable, demanding, motivated to deny or exaggerate their disorders, etc.

Some TBI symptoms are ambiguous:

Aprosodia: Lack of vocal emphasis could conceal a high level of distress.

Expressive deficits: Not reporting difficulties due to lack of insight, memory, or comprehension; social training not to complain; embarrassment; lack of orientation to TBI by previous practitioners, etc.

Poor self-monitoring: Apparently avoidable social and performance errors need not be due to reduced comprehension. Rather, neurological damage may have caused reduced foresight (judgment) and error monitoring of ongoing actions.

Cognitive loss: Persistent unadaptive behavioral actions may be a result of reduced comprehension of dysfunctions and performance errors rather than indifference or psychotherapeutic resistance.

Neurotraumatic personality change: Dysregulation may result in either reduced motivation or overt apathy, or impulsive behavior associated with lack of foresight and social sensitivity (e.g., anger and socially unacceptable actions).

Educating the Patient and Family about Trauma

Initially, the patient and/or spouse believes that the dysfunction is unimportant, easily curable, or not related to a trauma. One patient was injured when a heavy object fell on his head, striking his neck and back. There were multiple symptoms but no gross disorders such as seizures, orthopedic impairment, obvious cognitive loss, but rather subtle findings, for example, reduced mental ability, pain, and dissociative experiences.

Patients are often reassured when it is explained that a symptom is related to the injury, rather than an emotional problem, faked, or looking for sympathy. Miller (1993, p. 55) makes a major point: Since many practitioners do not understand that "minor" TBI can create significant impairment and distress, they may not recognize the actual etiology of the symptom, and may discourage the patient from relating distress to trauma. Education about trauma can be offered gently but directly. This is different than interpreting the experiential meaning of the disorder, a different topic that is not the initial therapeutic focus, and might not be a subject of attention in many patients. Brain lesions should not be confused with unresolved childhood trauma! Factual information can be highly useful if the documented CNS findings are revealed in a kind and matter-of-fact way. Information about significant injuries, expressed gently, is far more likely to be reassuring than disorganizing. Rather than the nature of the distress being a mystery, with implied self-criticism for weakness and failure, the symptom becomes experienced as an unfortunate event that is not blameworthy.

- *TBI findings*: It is useful to explain radiological reports and injured brain areas with neurological diagrams by plainly stating that particular symptoms are an expected consequence of the trauma, being realistic as to the likely slow healing or chronicity of a distressing symptom, explaining the basis for somatic symptoms, etc. Rather than be frightened, the patient may be relieved that dysfunctions are not imagined, willful, or crazy. For example, a man had negative magnetic resonance imaging (MRI) and computed tomography (CT) findings after a heavy object fell on his head. Thus, he was led to believe that there was no evidence for TBI. Contrary to his usual nature, he abused his family and struck his adolescent son. This upset him, and he wanted further study, even suggesting SPECT, for which I referred him. The findings were left hemisphere ventral frontal, and temporal-parietal hypoperfusion. He was relieved by the findings, since he interpreted them to mean that there was a natural basis for his work and family problems.
- *Medical problems*: The nonmedical psychotherapist helps the patient to understand and cope with somatic difficulties. This requires technical awareness and referrals to specialists who are familiar with trauma. A minimal trauma library would include general anatomy, neuroanatomy, physiology, general and psychotropic pharmacology, internal medicine, a medical dictionary, etc. Textbooks of immunology, endocrinology, and stress medicine are useful. Patients with chronic physical defects, pain, and sleeplessness are vulnerable to respiratory or other contagious diseases resulting from system "burnout." They are referred to their physician for study of possible reduced immune system function. One example of useful knowledge of trauma is the nature and radiation of referred pain from trigger points. This can lead to more precise treatment for chronic headaches. In one man, on whose head and shoulder a heavy object was dropped, there were two trigger points: a rostral trapezius area radiated up to the innervation of the lesser and greater occipital nerves (medial neck and occiput—C2/C3) and then anteriorly on the temple towards the forehead; A second trapezius area, below the first, radiated down his arm and to the contralateral shoulder. He was advised to alert his treating physicians to these findings.
- *Lack of awareness and denial*: This takes several forms. A psychodynamic form may be ego-protective, that is, an unwillingness to acknowledge disruption of an important component of one's self. These people are unlikely to modify their behavior, are resistant to treatment, and manifest reduced motivation and resistance to treatment directed at increased self-understanding. At the opposite pole is a neurological injury preventing self-understanding of a specific dysfunction. Patients with a cognitive deficit (attention, memory, reasoning) increase their awareness when provided with feedback and information concerning their disability, paralleling improvements in their cognitive domains. Various modalities are used, but they have limitations: education (difficulties with motivation if the

patient does not agree with the defined problem); group therapy (resistance to identifying with others as "similar"); education of the support system (provoking a catastrophic reaction by attempting to force awareness (Flashman et al., 2005).

While denial occurs over a range of domains, for example, cognitive, behavioral or emotional disability, particular disorders for an injured person may be either easy or difficult to admit. Social skills deficits may coexist with an inability to understand their ramifications in social situations. Among the dimensions describing lack of awareness: knowledge of a specific deficit or ability (*anosognosia*); the emotional response of indifference (*anosodiaphoria*) or bitter complaint; and comprehending the impact. Injured people and those who take care of them experience particular difficulties differently. Persons with TBI are less likely than their relatives to acknowledge changes in personality, such as irritability, impulsivity, and affective instability. They will acknowledge intellectual, memory, and speech deficits. Unawareness in the early recovery period creates implications for rehabilitation (Flashman et al., 2005; Miller, 1993). Loss of sexual responsiveness, or a skill needed for employment, or for social acceptance have different values according to personal history. Caution is recommended in approaching such a topic, or encouraging the person to acknowledge it or to actively cope with the disability through abandoning activities or compensating for the difficulty. The unprepared patient may undergo a catastrophic reaction, thus reinforcing the denial. Gradual insights and minimal interpretations may prepare the patient.

Catharsis

Expression of emotion regarding traumatic events, particularly repeatedly, can result in health benefits related to immune changes. The expression of varied moods, even by *"method" actors*, affected several immune parameters. In a study of emotional disclosure in patients positive for Epstein–Barr virus (EBV), the less disclosure, the higher the EBV antibody titre. Persons with *repressive interpersonal styles* generally had high antibody titres. The effectiveness of emotional disclosure may be determined by *cognitive processing* and restructuring, which alter understanding of the traumatic experience (Booth, 2005). Important immune system control stems from SNS input at spinal levels of T10 and lower (Maier et al., 2001).

Stress Avoidance

Premature work entry results in fatigue, irritation of the employer and colleagues, and potential failure. The patient is motivated to return to work and other obligations by self-esteem, employer demands, economic necessity, etc. Significant issues are reduction of insight, stamina, processing speed, work competence, physical mobility, etc. Psychotherapy plays a role in return to work by enhancing coping and avoidance of undue stress. The practitioner helps the patient balance ego and realistic demands with the actual functional level. Using military experience with counseling patients with PTSD as a guide, it is desirable to intervene with psychotherapy before work or relationships are compromised, before symptoms become chronically entrenched, or before co-morbid conditions develop. After initial reluctance to seek treatment, due to optimism or fear of the consequences, chronicity can motivate participation in psychotherapy (Milliken et al., 2007).

Coping Skills

The retraining effort depends upon the deficit. The counselor's treatment plan both estimates the complexity of tasks that are functionally deficient in ADL, as well as major daily necessities such as employment, family, recreation, and communication and negotiation. Limitations are respected to avoid humiliating the patient and thereby breaking the therapeutic alliance.

- Injury at the PCS level usually does not cause dyspraxia of overlearned skills of daily living. Where needed, the counselor may offer such tactics as avoiding fatigue; keeping a notebook; advising that a confusing message be repeated slowly and simply; following a sequence of predetermined steps; and setting up priorities.
- Aspects of the environment may require attention: Organizing systematically household equipment and utensils; practicing with an alarm clock or calendar watch (Pollock, 1994). Therapy appointments are often missed if a familiar schedule is changed; thus, a combination of a phone alert and insisting on the use of a formal calendar may be recommended.
- Individuals of achievement who cannot return to complex vocational assignments may require formal training from an extended cognitive rehabilitation program in such functions as memory, sequencing, and organization.

Achieving a Positive Attitude

The most important therapeutic task is often helping the patient to regain a belief in his or her self-efficacy, particularly for cognitive functions (see "Self-Efficacy").

Self-efficacy counseling: Rehabilitation should address self-efficacy and physical and cognitive limitations. Therapists and peers, when considered trustworthy and knowledgeable, can use interpersonal interactions to change unsatisfactory self-efficacy beliefs. Interventions address multiple aspects of performance:

- Specific task requirements (complexity, novelty, familiarity)
- The person's experience, recall, and evaluation of the task performance
- Self-assessment of resources (cognitive abilities, affective responses, effort)
- Situational influences (competing demands, time demands, distractions) (Cicerone & Azulay, 2007)

Counseling

- Patient agenda: It may be useful to propose during the first therapy session that the patient bring in an agenda for each subsequent session. I tell them that their concern or emotional pain should be the initial focus of our work, although it will evolve towards different topics. This is easier said than accomplished. The exchange of leadership roles with the practitioner is unexpected or even greeted with doubt. Yet, it is an expression of your confidence in the judgment of your patient. By increasing self-awareness, it makes more precise expression of disturbance and discomfort possible, which enhances the therapist's ability to treat and also to make referrals for other services.
- Encouraging assertion: Significant actions that may be resisted because of previous passivity, doubts, expected refusal, injury-related impediments, etc., are recommended.
- Success is maximized by emphasizing alternative or compensating strategies for deficits and avoiding known deficient functions.
- Coaching the patient to be more active as the process of recovery and therapeutic success proceeds.
- Hope: A positive attitude is needed in you, Dear Reader, as well as in your patient. There are champion scientists, athletes, politicians, musicians, poets with neurological disorders. Beethoven was deaf; composer Delius and writer John Milton were blind; a champion Olympic swimmer had an apparent childhood attention deficit disorder; F.D.R. has lower limb paralysis as an adult; an international classical concert star and vocal teacher has four undeveloped limbs (phocomelia), etc. I have made errors by underestimating the potential of my patients with deficits. Keep up your patients' activity and spirits.

Social: Group and Family Psychotherapy

One component is skills training. More effective communication increases needs fulfillment. This decreases stressful emotions, thus reducing arousal and lessening pain experience. Group therapies differ significantly from individual therapeutic procedures. Although issues of privacy may reduce the expression of some intimate concerns, the therapeutic efficiency may be balanced by the wider social stimuli. There is the opportunity for both expression of matters elicited by resemblance to important persons in the patient's history (transference) or identification with group members with pain and other family problems, with which the injured person can identify. Therapeutic contact with the family might be in the presence of the patient, without the patient (with consent), or in a group of families with injured family members who are not participating. A therapy group of injured persons utilizes mutual support, learning coping maneuvers from others, and if time-limited, benefiting from goals, rules, and endpoints. A family group with the injured person present offers an opportunity for the therapist to use group process to have the patient and family members offer their reactions to the style of the other party and to begin the process of negotiation in order to cope with the disturbing reactions of the other party.

FORENSIC ISSUES OF PCS

Forensic neuropsychology is a complex professional application. The neuropsychological examination studies the effects of an accident or medical problem upon the patient's adaptation to the daily requirements of life. A reasonably complete forensic neuropsychological examination utilizes an extensive technical database (see the earlier section) to draw conclusions about diagnosis, performance, and outcome. Has the patient suffered dysfunction, deficits, or discomforts due to an accident? This is supported by direct examination or review of records. In addition, it is important to assess whether the patient is deliberately or psychodynamically making a false presentation to obtain monetary or psychosocial gain.

A large proportion of accidents causing head injuries lead to lawsuits and claims of disability (impairment, loss of wages, suffering), involving numerous administrative and legal domains. The examination and report occur under the restraints of various laws and regulations. Since the economic stakes are large, the examiner must assess the credibility of the injury and its effects. This requires both a wide-range examination and an attempt to reconstruct the physical characteristics of the accident. High-quality forensic work requires extensive examination, thought, and written documentation. Yet should a case proceed to trial, since the stakes are high, the examiner can expect critical and sometimes abusive attack far beyond the usual clinical presentation.

Litigation

A segment of injured persons with head injury instigate litigation for compensation. The clinician or forensic examiner must consider the question whether claims of injury, impairment, and distress are merited or are motivated by a desire for undeserved compensation, a psychodynamically determined motivation to play the sick role, to achieve secondary gain, or otherwise exaggerate the disorder. This determination is difficult. The claimant's history is complex; the examiner should consider the range of neurological and somatic injuries that occur in an accident, the wide range of potential neurobehavioral and emotional disturbances that may occur after an accident, the necessity for defining a reduction from an estimated preinjury baseline, the possibility of later developing disorders, the imprecision regarding meaning and significance of scores for symptom validity tests, etc. Some findings:

- One hundred and twenty of 428 studies were considered adequate for a meta-analysis of children and adults with mTBI. For children, it was concluded that the prognosis was good, with quick resolution of symptoms and little evidence of residual cognitive, behavioral, or

academic deficits. For adults, cognitive deficits and symptoms are common in the acute stage. The majority of studies report recovery for most within 3–12 months. For patients with persistent symptoms, compensation and litigation were factors without consistent evidence for other predictors.
- A meta-analysis was performed of 39 selected studies, comparing patients with control groups. For unselected or prospective samples it was concluded that there was no residual neuropsychological impairment by 3 months postinjury. In contrast, clinic-based samples and participants in litigation were associated with cognitive sequelae of mTBI at 3 months or later (Belanger et al., 2005).

The present writer offers some cautions regarding these benign views of mTBI:

- Many current studies of mTBI utilized a restricted range of disorders for examination. Therefore, some current disorders may have been ignored.
- The potential for later-developing disorders, including physiological development of children, may not have been considered.
- Litigation may have an objectively higher level of brain injury and somatic injury, impairment, and distress than those who do not sue for damages.
- Co-morbid somatic injuries per se, that is, predominantly non-CNS, create reduced performance due to "distracting" symptoms.
- Disturbance of mood, motivation, and general mental efficiency, consequent to accident-related alteration of the brain's internal milieu may not have been directly or indirectly assessed.
- The range of documented TBI disorders indicates that firm conclusions concerning post-traumatic outcome must be reserved for studies that utilize a wider range of study than those confined to cognitive performance.
- Utilization of litigation to obtain compensation and exaggerating disorders on symptom validity tests are encouraged due to an inability to obtain prompt responses by insurance companies for specialized examinations and treatment, and compensation for lost wages. Some individuals with mTBI have injuries that, by their nature, are difficult to treat, and are therefore chronic. If, after a of time, treatment is cut off prematurely or is not offered, litigation may be required to obtain compensation.

Time to Settlement

A review of symptoms, return to sports, criteria for ordering imaging studies, symptomatic treatment, and controversies in these issues are available. Prompt resolution of litigation is encouraged. This is a controversial point. On the one hand, the delay of settlement of cases and unreasonable denial of treatment and assessment frequently worsen outcome and damage the mental condition of the accident victim. However, the possibility of delayed expression of symptoms should be reported by the examiner. These authors also state a return to baseline cognitive function within several weeks (Ropper & Gorson, 2007). A study of more severe concussion in athletics (Parker, 1990, pp. 161–162), during an interval after a TBI of 1 month to 10 years indicates that, on average, there is absolutely no cognitive improvement (Wechsler Adult Intelligence Scale, Revised), provided that only individual measurements are used, that is, there is no practice effect. The mean deviation from the estimated preinjury baseline (Matarazzo & Herman, 1984) was –9.1 points of Full Scale IQ.

Availability of Legal Counsel or Treatment

Should the forensic issue be liability, that is, lack of certainty who caused the injuries, or if there are no conspicuous bodily injuries convincing to a jury, or if it appears that the anticipated settlement will be small, there will be difficulty in obtaining an attorney to represent the accident victim. Thus, the process of obtaining examination and varied psychological and medical treatment is frequently time-consuming, delayed, or ultimately impossible. I know of a case in which a man in an MVA with extended altered consciousness who suffered for years from a posttraumatic sleep disorder was disqualified from obtaining treatment through the procedure of a neurologist disqualifying him for further neurological treatment.

Driving Competency

Examination or treatment may elicit information raising doubt concerning a driver's capacity to perform safely. Motor vehicle competency is an extremely complex activity: sensory, motor, and sensorimotor integrity; visuomotor organization ability needed for map reading and assessing moving traffic within the boundaries and structure of the road; motor skill and training for the particular vehicle; processing speed for motor and cognitive activities; driving experience and familiarity with the neighborhood; intelligence, foresight, and judgment; literacy; personality characteristics (cautiousness vs risk taking; control over anger ["road rage"], competitiveness, etc.); concentration versus distractibility; driving task (distances, road conditions, exchanges and access, type and amount of traffic; whether driving day or night, alone or accompanied); suspected or documented TBI or other impairment (mild cognitive impairment; neurodegenerative conditions; stroke; cardiovascular conditions; diagnosable emotional condition); etc.

The legal responsibility to report a seemingly impaired driver varies with the jurisdiction, and the examiner will always need to be credible in assessing the patient. There are too many neuropsychological and other assessment procedures relevant to driving ability to detail. The suspicions may be incidental to other issues, or a neuropsychological examination may be a reasonable screening procedure. Certain diagnoses indicating unquestioned loss of capacity may be detected (cognitive loss; psychosis; deviant personality characteristics). The examiner may decide to offer the patient and family an opinion that the patient is incompetent to drive, but probably cannot state that competence is assured. Desktop procedures are useful screening procedures, but some occupational therapists and specialists in other professions have far more credible driving simulating machines. These offer braking time, steering accuracy, and efficiency of processing through complex moving scenes. A driving school road test is also a credible criterion. However, the ultimate responsibility for deeming a driver fit or unfit rests with the legal authority.

Compensation Seeking

While some clinicians assume that the PCS is a psychological overreaction to a minor trauma, in the individual assessment, one must consider:

- The meaning of the injury to the person in terms of prior history;
- Objective evidence of injury, and the extent of somatic trauma (wide-range examination);
- The postaccident interval considered with a view to the balance between healing and continued distress ("the unhealed wound")
- The history of institutional and clinical response to the patient's claim of injury, impairment, and suffering. Resentment and a "cry for help" may create an experienced need to cheat the system. Infrequently considered is the relationship between litigation and denial of benefits and treatment, that is, objective measurement of whether litigants incurred a greater injury between those who did not seek court or administrative actions, and whether the claimants received prompt compensation for lost wages, appropriate diagnostic procedures, and treatment.

A review (Hall et al., 2005) suggests that patients involved in litigation demonstrate more prolonged and intense symptoms from PCS than do similar patients not involved in litigation. Proposed explanations include the possibility that litigating patients have read more about their condition, that is, they are influenced by "symptom knowledge"; fakers are coached by their lawyers who really know which symptoms to magnify; their symptoms are exacerbated by the stress of litigation; they are influenced by secondary gain; or malingering (they lie for money). No doubt all of these factors affecting claims and reduced performance exist. One might also consider the plight of injured people deprived of competent assessment, treatment, and timely settlement of legitimate claims by insurance companies and other defendants. Any injury health care practitioner is aware of the difficulty of offering service to injured people and then being paid in a prompt and fair-minded manner.

Litigation as a Marker for Malingering

The appellation is sometimes used to describe the patient with chronic complaints as "*the miserable minority*". This implies that most chronic complaints are subjective or exaggerated. For some developers of symptom validity procedures, litigation is a marker for suspicion of weak "effort," that is, unwillingness to perform to one's best ability in order to gain economic advantage. Patients involved in litigation are asserted to demonstrate more prolonged and intense PCS symptoms than similar patients not involved in litigation (Hall et al., 2005): Reading about their condition creates "symptom knowledge"; they are coached by their lawyers to magnify the "right" symptoms; litigation itself is a stressor; conscious or unconscious need for secondary gain; and, malingering for money. A review showed a relationship between those seeking financial compensation and delayed return to work, long-term symptoms, and greater symptom severity (Carroll et al., 2004). Evidence for the effect of potential compensation upon symptoms has been offered from two studies from Lithuania (Mickeviciene et al., 2002; Schrader et al., 1996) that received some attention. In that country, compensation is more infrequent than in the United States. The patients were identified from police records of rear-end collisions (estimated collision speed was 35 km/h, or about 21 mph) and hospital records of admissions for head trauma with LOC. Control groups were utilized. Information was gathered about 1–3 years after the accident. Mailed questionnaires were administered in both studies with participants entered into a raffle or receiving a gift of about $5. In one study it was asserted that these lesions causing persistent symptoms are rare: disc injury, fracture, cervical-disc herniation, ligament injury, etc. (Schrader et al., 1996). In the other paper (Mickeviciene et al., 2002), headaches disappeared in 96% of cases after 1 month. Response rates for the two concussion samples were 84% (n = 202) and 66% (n = 200). The outcome survey by postcard might not be accepted in some quarters as meeting a "gold standard" of professional research. Symptoms studied were familiar, that is, headache, dizziness, neck pain, subjective cognitive dysfunction, health, and other PCS symptoms. No CT, MRI, or neurological examination was performed. Mickeviciene et al. (2002) concluded that headache and dizziness did not differentiate between concussion patients and controls. Schrader et al. (1996) concluded that chronic symptoms were usually not caused by the car accident; rather, they were due to expectation of disability, family history, and attribution of preexisting symptoms to the trauma.

Your writer's impression: When one considers the very low rate of patients with documented cervical injuries, some of which cause headache by referral from trigger points and other injuries, one may wonder whether the similarity between accident cases and controls was due to selection bias. Further, questionnaire results without personal examination with robust and appropriate procedures invite skepticism.

A Cautious Attitude

While there is a need for malingering screening, please remember that the criteria for use of these assessment procedures are less stringent than those of the U.S. Food and Drug Administration.

In some SVT validity studies, litigation is considered a marker of suspicious motivation. This violates the civil right to sue for damages. Litigation need not be a marker of credibility for persistent postconcussion symptoms (Gasquoine, 1997). Using a litigation marker makes the frequently incorrect technical assumptions that litigation implies exaggeration of dysfunction. Since nonlitigants may not be able to afford a neuropsychological examination, the practitioner really does not know the baseline for noninjured people, injured nonlitigants, and injured litigants. Settlement of a case does not make all patients symptom-free, and persisting symptoms occur in the absence of possible financial gain. Claimants who engage in litigation are often regarded suspiciously, since mTBI is supposed to usually "resolve." Such an assertion leaves open the professional necessity for careful study of the remaining claimants: thorough study with a wide-range examination; estimate loss from an estimated preinjury baseline; documentation of both neural and co-morbid somatic injury; and follow-up to determine whether late-developing dysfunctions occur.

A frequent procedure in measuring the validity of symptom validity tests (SVT) ("effort") is the use of volunteers to *simulate* the assumed behavior of persons malingering on the procedure (faking bad). It is open to question whether college students and psychologists serving as subjects have the same attitudes and utilize the same faking procedures as dishonest persons seeking undeserved compensation. The domains SVTs utilize to assess honest performance (e.g., memory and sensorimotor), from which the utility of the entire battery is often assessed, are a narrow range of functions compared to the wide range of potential neurobehavioral disorders. Thus, evidence for false "effort" on SVTs may ignore important external evidence for injury and impairment. To the assessment of symptom distortion and exaggeration differential diagnoses must be included issues of somatization or conversion.

An objective examiner avoids both credulousness for exaggerated complaints and a premature assumption of exaggeration. Extremely important is considering that many patients cannot or will not present to the examiner the difficulties of their postaccident condition. The examiner should undertake an extensive range of study, and also obtain independent information from schools, collaterals, employers, etc. The writer believes that the accusation of making a false claim for damages requires a standard of proof as strong as that needed to convict a thief. Offering a reduced standard of proof that a person is malingering raises the possibility of depriving an injured person of compensation.

The meaning of a validity procedure should be assessed within the context of as much relevant data as possible, including the patient's personality and social support or pressure. Evidence of intent to malinger should consider that injury may have been separately empirically verified. However, even an injured person may believe that there is reason to exaggerate symptoms and minimize successful performance on a psychological procedure. Thus, the examiner should not assume that there is no impairment, even when convincing evidence for symptom exageration has been determined.

Clinical example: Faked performance from an injured woman: I examined a woman for whom there was objective evidence of a significant injury. Her scores on a symptom validity test were far below chance performance. She was accompanied by her husband, who in her presence firmly maintained that her complaints were exaggerations. She informed me of her husband's refusal to believe that she had been significantly injured. I inferred that her gross efforts to show grave impairment were "*a cry for help.*"

Response Styles for Deceiving

This data organizes several studies. It has been suggested that forensic distortion analysis utilize a combination of methods: interviewing, testing, observation, and base-rate comparison (Hall et al., 2008).

There are varieties of faking:

Intentional symptom exaggeration:

- Offering a nonexistent symptom
- Deliberate inaccurate performance
- Underperformance of a physical domain
- Inconsistencies (between systems within an examination and with what is known about normal variations of performance)
- Effort bias (intentional submaximal effort) (Bianchini et al., 2005)
- Deliberate, fraudulent malingering: Substyles include open fraud; exaggeration (faking bad to appear worse than one really is); faking good (denial of symptoms to maintain a job, concealment that drugs or improperly maintained safety equipment contributed to the accident; concealing symptoms such as impotence); perseveration (after genuine symptoms have ceased, they are alleged to continue); embarrassment (waxing and waning with the situation); invalidation (attempting to render the examination meaningless by irrelevant, random responses or not keeping appointments with an expert; switching response style after becoming aware of the testing procedures)
- Transference or decoy malingering: Genuine symptoms are fraudulently attributed to a cause other than the factual cause.

Psychopathological contribution:

- Fear of death, perhaps due to a family history of a disease, causes the real concern to be attributed incorrectly to an accident.
- *Hysteria*: A condition emerges following threatened or objective injury, in which being injured is part of a new identification, with heightened awareness of incapacity
- *Psychosis*: Bizarre or paranoid features
- *Organic*: Persons suffering from such conditions as Korsakoff invent pathology, enhanced by hypersuggestibility and shifting and uncertainty of symptoms.

Determining the Credibility of a Symptom or Pattern

- Have the symptoms followed an event with a credible head injury? Try to reconstruct the accident and specify the physical forces on the body within the accident's environment.
- What is the range of neurobehavioral domains that have been examined? What claims of discomfort and dysfunction have been made? Are these reasonable in terms of the accident?
- What disorders have been established by competent clinicians using robust procedures?
- Has the possibility of preexisting medical and psychological conditions and prior accidents been taken into account?
- What was the preinjury baseline? Is there a sufficient and reliable difference between the baseline and current behavior and performance to suggest a compensable injury or diagnostic conclusion?
- Directly estimate the presence of a deliberate desire to cheat, pretend, or exaggerate disorders. Is it reasonable that a previously productive person with significant income and satisfactory lifestyle would leave a job and pretend disability in order to obtain a hard-to-predict, possibly small, amount of money? Are there community situations that motivate the patient to exaggerate: lack of belief at home that a significant impairment or discomfort has occurred, and/or unprofessional deprivation of assessment, treatment and compensation by defendants or insurance companies?
- Within the validity limits of SVTs, and the range of the neurobehavioral domain(s) studied, is there evidence for symptom magnification? One must avoid overgeneralization from a limited procedure (e.g., purporting to measure memory or concentration) to the representativeness for indicating impairment to the entire test battery performance. Some clinicians

feel justified in discontinuing an examination observing poor performance on a SVT procedure, but in the writer's opinion, it is more appropriate to finish the examination than to offer some reasonable assessment of motivation. If the results suggest statistical evidence for unexpectedly poor performance (reduced "effort"), there may be alternative explanations besides attempt to deceive concerning performance ability.

Conclusion: If you believe that malingering for money or similar benefits is theft, then the assertion of malingering or fakery should be bound by high professional standards or that level of precision that would be required to offer testimony in a court of law in prosecution for theft (Parker, 1994).

19 Biomechanical Perspective on Blast Injury
Comorbid Brain and Somatic Trauma

Mariusz Ziejewski, Ph.D. Inż.
Ghodrat Karami, Ph.D.

INTRODUCTION

The nature of warfare is different today than in past conflicts because, instead of using ballistics, the enemy is now using high-order explosives (i.e., mines, grenades, improvised explosive devices (IEDs), etc.) (DePalma et al. 2005; Holden, 2005; Okie, 2005; Redhead et al., 2005; Ryan & Montgomery, 2005; Warden, 2006; Wessely, 2005).

Historically, most injuries have been caused by ballistics. However, due to the advanced nature of the armor used to protect today's soldiers in the field, it is believed that blast injuries to the head are now more common. Almost all modern soldiers wear body armor, without which blast injuries would most likely be fatal (Garner & Brett, 2007).

As a result, a large portion of wounded soldiers returning from Iraq and Afghanistan have suffered blast traumatic brain injuries (TBIs) (Drazen, 2005; Elsayed, 1997; Gawande, 2004; Lew, 2005; Lew et al., 2005; Okie, 2005; Peota, 2005; Sullivan, 2004). Additional statistics regarding blast injury include the following: (1) between 8% and 25% of those who suffer blast injuries die (Okie, 2005); (2) 155 injured soldiers who had returned from Iraq in 2003 were screened by the Walter Reed Medical Center's Defense and Veterans Brain Injury Center (DVBIC), and 62% of them were found to have a brain injury; and (3) according to Army data, 59% of all patients exposed to a blast have been diagnosed with TBI; of these, 56% of were considered moderate or severe, and 44% were considered mild (Okie, 2005). The number of veterans returning to the United States (U.S) from combat with a TBI, has created a clinical crisis for the U.S. Veterans Administration hospitals (Taber et al., 2006).

Clinical evidence of blast-induced, mild TBI can be difficult to identify. It is sometimes confused with post-traumatic stress disorder (PTSD) (Bhattacharjee, 2008; DePalma, et al., 2005; Guy, 2004; Lew et al., 2005), and in some cases, the symptoms do not develop until it is too late for treatment (Yilmaz & Pekdemir, 2007). The effects of a blast on a person's body may not be visible, even though vital organs may have been severely damaged by the blast (Abe & Takayama, 1990), and other complex clinical symptoms (i.e., cognitive abilities, etc.) may eventually occur (Cernak et al., 2001b). A delayed diagnosis can ultimately cause critical mental, social, and financial issues for returning military personnel. Even when a blast-induced brain injury is diagnosed, treatment is limited. This is partly because neither the conventional technology, nor medicine, can detect the total effects that a blast has had on a brain.

In order to better diagnose TBI, and to better understand how to treat it, there must be an increased knowledge base among researchers, physicians and soldiers in the field regarding the mechanisms of how blasts cause brain injuries. There is a critical need for the military to develop a conceptual understanding of blast-induced polytrauma, especially diagnostic biomarkers (see Biomarkers) of blast brain injury, which can be used to diagnose and triage brain-injured military personnel on the battlefield (Okie, 2005).

Blast-related forces can be grouped into four categories based on how they cause injury. The categories are: (i) primary, which is injury caused to the person by wave-induced changes in atmospheric pressure (i.e., velocity ≥ 300 m/sec, speed of sound in air); (ii) secondary, which occurs when flying debris hits the person; (iii) tertiary, which happens when the person hits a solid object; and (iv) quaternary, which is burns on the person and/or inhalation of gases from the blast (Cooper et al., 1983; DePalma, et al., 2005; Mayorga, 1997; Wightman & Gladish, 2001). A comprehensive evaluation of biomechanical aspects of injuries suffered by individuals (i.e., soldiers in combat) must include the cumulative effect of different force categories. Many components of blast-induced TBI are quite similar to what has been studied for decades, such as the effect of blunt impact and acceleration/deceleration. There is, however, the unique component, that is, the effect of overpressurization, with a subsequent negative pressure, that needs further research. (See Chapters 6 and 7 for further details of primary and later progression of trauma.)

This chapter will begin with a general description of brain injury mechanisms for blunt impact, inertia loading, and the less-studied mechanism of overpressurization.

BRAIN INJURY MECHANISMS DUE TO BLUNT IMPACT AND ACCELERATION/DECELERATION

General Concepts

Forces due to a blast, or a vehicular crash, can cause rapid acceleration to the head, causing neuronal, vascular, and cytoskeletal structural damage (Figure 19.1).

Factors that need to be considered in analyzing the mechanisms of brain injury include anatomical features (i.e., geometry of the brain and skull), mechanical properties of brain tissue and applied motion to the skull. Brain deformation is controlled by mechanical characteristics of the tissue. It has a high resistance to change in size (high bulk modulus, incompressible); low resistance to change in shape (low shear modulus, deformable); different properties within the brain (heterogeneous); different properties in different directions (anisotropic); and time-dependent properties. In

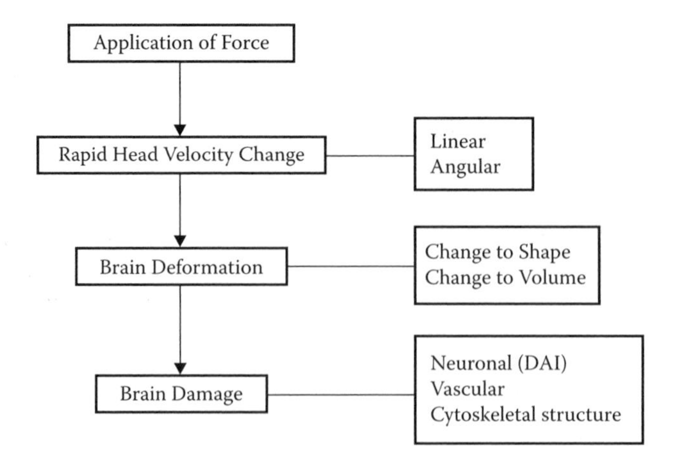

FIGURE 19.1 Flow chart of damage caused by force to the head.

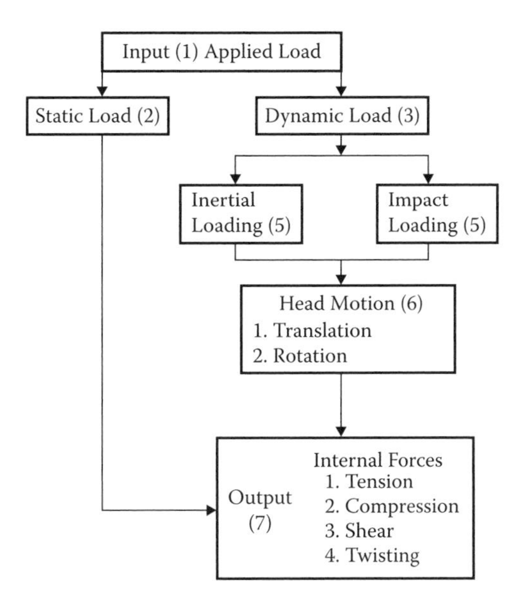

FIGURE 19.2 Flow chart for brain injury assessment.

other words, the level of deformation and the magnitude of the stress buildup in the tissue depend on the weight of the external load (viscoelastic). The head motion has to be carefully analyzed to determine the internal forces created in the form of tension, compression, shear, and twist, all of which are sources of brain damage. (See primary, secondary, tertiary, quaternary trauma.)

As shown in Figure 19.2, in a biomechanical analysis of a TBI, the motion of the head is best described in terms of linear and angular acceleration. At a constant magnitude of acceleration, the change in acceleration duration results in different types of loading and injury to the brain tissue. For short acceleration durations, the properties of the brain are such that much of the acceleration effects are usually damped, and consequently, the brain experiences little deformation. As the acceleration duration is increased, less damping occurs and, therefore, less acceleration is needed to produce injurious brain deformation. The deformation that occurs, however, is confined to the brain surface and may result in subdural hematoma (SDH). As the duration of acceleration increases, forces cannot be damped out, and the resulting deformations can propagate deeper into the brain. This can cause axonal injury.

Hirsch and Ommaya (1970) reported that rotational motion appeared to be more critical in causing brain injury than translational motion. Unterharnscheidt (1971) also studied the differences between brain injuries produced by translational and angular/rotational accelerations. The injury mechanism for pure translational motion appears to be pressure gradients, and the injury mechanism from pure angular/rotational acceleration appears to be shear stress as the skull rotates relative to the brain. Except for skull fracture and epidural hematoma, the majority of known types of brain injuries have been caused by angular/rotational acceleration (Adams et al., 1982; Gennarelli, 1983).

Impact loading moves the brain, which then impacts against the internal folds of the dura mater, and also shaves the corpus callosum against the lower edge of the falx on each side, supporting the cerebral hemispheres. Impact loading is caused by direct head contact with an object, or with a surface. Impact loading causes a group of complex mechanical events called contact phenomenon. This phenomenon can occur at the point of impact, or at points away from it. The significance of the contact phenomenon for brain injury depends on the shape and stiffness of the striking object, or of the surface impacted, as well as on the magnitude of the contact force. An impact load can cause focal (fairly localized), or diffuse (rather distributed), effects. Injury to the brain may result from a bone of the skull being pushed inward causing internal stresses to the brain (i.e., shear, tension, and compression). The presence of shear, tension, or compression, stresses can result in focal concussion, cerebral concussion, and primary brain lesions. If the impact load is large enough, the skull may fracture, causing a laceration at the site of impact. Additionally, shock waves due to impact traveling through the skull and brain tissue can potentially cause widespread damage. Noncontact inertial loading of the head is caused by forces transmitted through the neck. Rapid head motion itself is sufficient to create tension, compression, shear, and torsional forces within the brain tissue. Impact (compression) has a different spectrum of injury than acceleration-deceleration, which may cause internal impact against the meninges (e.g., falx and tentorium), as well as shear and axonal stretching.

The brain may be contused, or lacerated, as it moves against, or over, sections of the skull that are irregularly shaped, or that have relatively rough surfaces with some inward projections. Not surprisingly, less brain/skull interaction can be tolerated in these regions. It is important to note that injury may occur at these sites somewhat independently of the location and direction of inertial impact.

Investigations, ranging from human-volunteer experiments to animal model studies, confirm that inertial loading and its magnitude, duration, and direction relative to the anatomy can produce a broad spectrum of neuropathological injuries (Gennarelli & Thibault, 1972, 1982; Ommaya & Gennarelli, 1974; Ommaya et al., 1966).

The internal forces resulting from a dynamic load cause brain injury via different mechanisms. These mechanisms of brain injury can be grouped into three categories: (1) direct brain impact ongoing in the interior of the skull; (2) cavitation; and (3) brain deformation.

When the head is accelerated, or decelerated, due to direct or indirect loading, the relative motion of the brain, with respect to the skull, is controlled by inertia and centrifugal force (Figure 19.3). This brain motion causes the direct brain impact against the interior of the skull, as well as three-dimensional (3-D) brain deformations (shear, which separates brain layers, and pressure waves, which cause internal impacts, as well as shearing and stretching axons).

Cavitation is the creation and destruction of vapor bubbles in a fluid media. The destruction of the vapor bubbles may cause damage to the surrounding structure or, in the case of TBI, to brain tissue. Unterharnscheidt (1970) attributed the generation of lesions within the central brain to increases in cranial volume under certain conditions of deformation of the skull undergoing impact. The movement of the brain inside the skull may lead to a lowering of the internal pressure. This provides an environment for cavitation because the brain movement occurs faster than any possible inflow of fluid, particularly in the ventricular region.

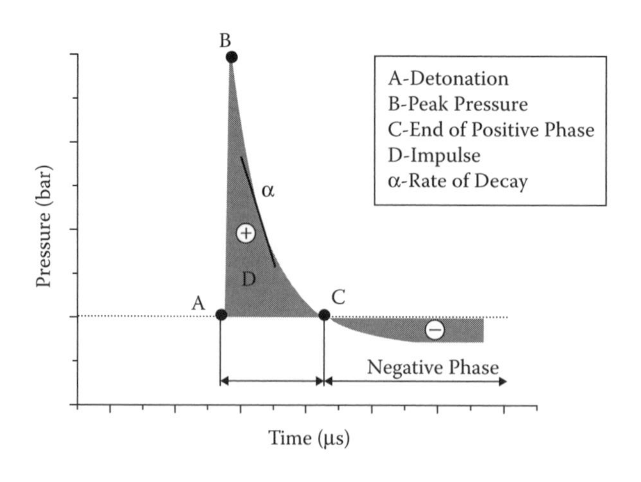

FIGURE 19.3 Brain motion.

When the head undergoes a rearward acceleration, the brain resists movement because of inertia, leaving a space at the back of the skull. The pressure in this space is low, allowing vapor-filled microbubbles to form. When the head stops and rebounds, the brain moves backward, collapsing the vapor microbubbles. Additionally, as the brain moves backward, it leaves a space at the front of the brain. The low pressure in this region allows a new set of microbubbles to be formed. These microbubbles collapse as soon as the pressure level returns to normal. The collapse, or destruction, of these vapor-filled microbubbles can cause forces that destroy tissue at the cellular level.

The effect of such an impact on the brain has been investigated theoretically and experimentally with certain idealizations (Benedict et al., 1970; Gennarelli, 1994; Luback & Goldsmith, 1988; McLean, 1996; Mertz & Nusholtz, 1996; Nusholtz & Ward, 1987; Nusholtz et al., 1984, 1986, 1994, 1995a, 1995b, 1997). Experimental and numerical analyses of simple models of the human brain under impact were used in these studies. Namely, a water-filled cylinder was struck by a free-flying mass. Rigid-body acceleration, time histories, and pressure at the fluid-cylinder interface were monitored during impact. Comparisons between the experimental results and the results of a computational model were made. As a conclusion of this work, the following view of head impact was developed: when the head receives a blow and a positive pressure develops under the point of impact, a small cavity is created between the skull and the dura, opposite the impact, due to the negative pressure (tension) that has developed there. Its subsequent collapse could be a mechanism of injury. Studies also indicated that under complex loading conditions, cavitation could occur in the brain material, not just at the boundary. It was found that if the head underwent pre-impact acceleration immediately before the head strike, internal cavitation was likely. Internal cavitation implies potential cellular tissue damage. It was admitted, therefore, that if the impact is severe enough, it may produce cavitation in the brain, and the associated violent cavity collapse could be a supplementary mechanism of brain injury.

Modeling of Injury

Biomechanics determine forces acting on the human body and the effects of those forces. Biomechanical engineering analysis is a tool that incorporates common-sense experience into the understanding of the human body's interaction with nature. Factored into the biomechanical analysis is the study of the environmental and human-body dynamics, as well as the human tolerance limits. There are currently two main approaches to biomechanical analysis: (1) experimental evaluation, which is most valuable, but not always practical; and (2) detailed computer simulations, using mathematically and numerically based biomechanical formulae. Computer simulations allow inclusion of all necessary parameters in the evaluation of specific scenarios. The results of computer simulations should be verified by clinical and laboratory studies.

As indicated earlier, in many biomechanical studies the head acceleration has been used to characterize the severity of an insult to the brain (Ziejewski, 2004). The complete global representation of the head motion, in terms of acceleration, can only be achieved if the complex input of linear and angular acceleration is known. This includes three components for linear acceleration and three for angular acceleration (Figure 19.4).

The head-acceleration data can be used to directly assess the probability of TBI by either extracting the resultant maximum values and the rate of change of acceleration, or by calculating head-injury assessment functions such as the Head Injury Criteria (HIC) (NHTSA, 2000), Head Impact Power (HIP), Power Index (PI) (Newman et al., 2000) and others (Ziejewski, 2008a).

Additional parameters dealing with local brain deformation have also been developed. It has been suggested that brain surface contusions, diffuse axonal injury (DAI), and acute SDH can be predicted using brain motion (Abel et al., 1978; Edberg et al., 1963; Gennarelli et al., 1982; Gurdjian et al., 1967; Meaney, 1991; Pudenz & Sheldon, 1946; Unterharnscheidt & Higgins, 1969); sudden change in the intracranial pressure; shear strain; stress/strain concentration (Holbourn, 1943; Ross et al., 1994); and the product of a stress and strain rate (Viano & Lovsund, 1999).

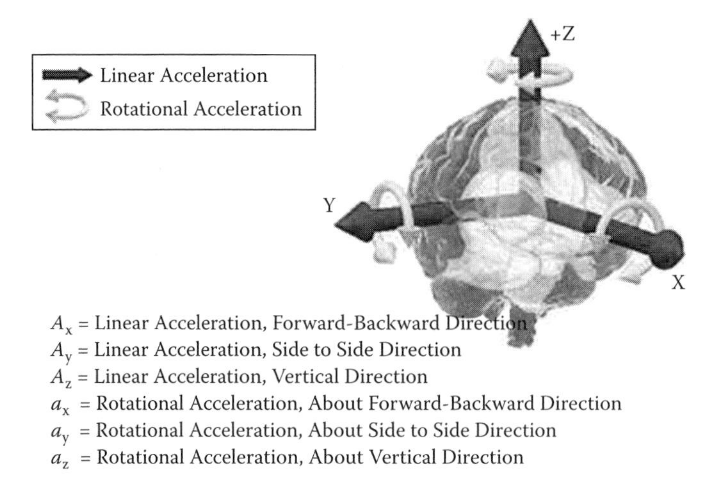

FIGURE 19.4 Types of acceleration.

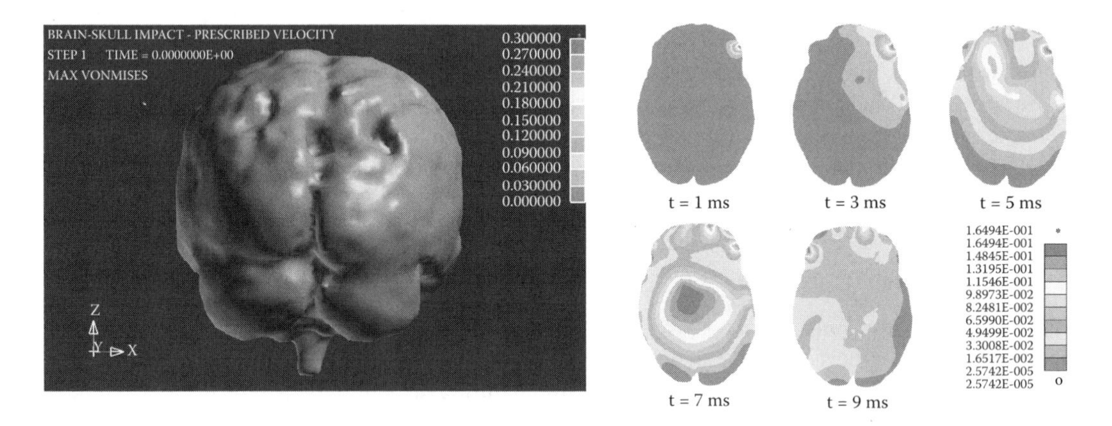

FIGURE 19.5 Macroscopic deformation of the brain under an impact.

In biomechanical simulations, sophisticated 3-D finite element (FE) analyses and rigid body bio-dynamics methods have been used to study impact injury events and the response of the human head (Figures 19.5 and 19.6). In general, TBI develops when the internal mechanical responses exceed tissue-tolerance levels. This injury process can be viewed as a load-injury scheme in equivalent biomechanical terms.

BRAIN INJURY MECHANISM DUE TO OVERPRESSURE

The issue of whether, or not, primary blast forces directly injure the brain is still unresolved. The vulnerability of the brain to a primary blast is supported by animal studies (Kaur et al., 1995). One of the outcomes has been identified as the formation of gas emboli, leading to infarction (Taber et al., 2006).

Physics of Blast Pressure Wave Analysis

The physics of blast waves are complex and often difficult to understand. The effect of blast waves on the body can be menacing. There are often no visual external injuries, but rather damaged

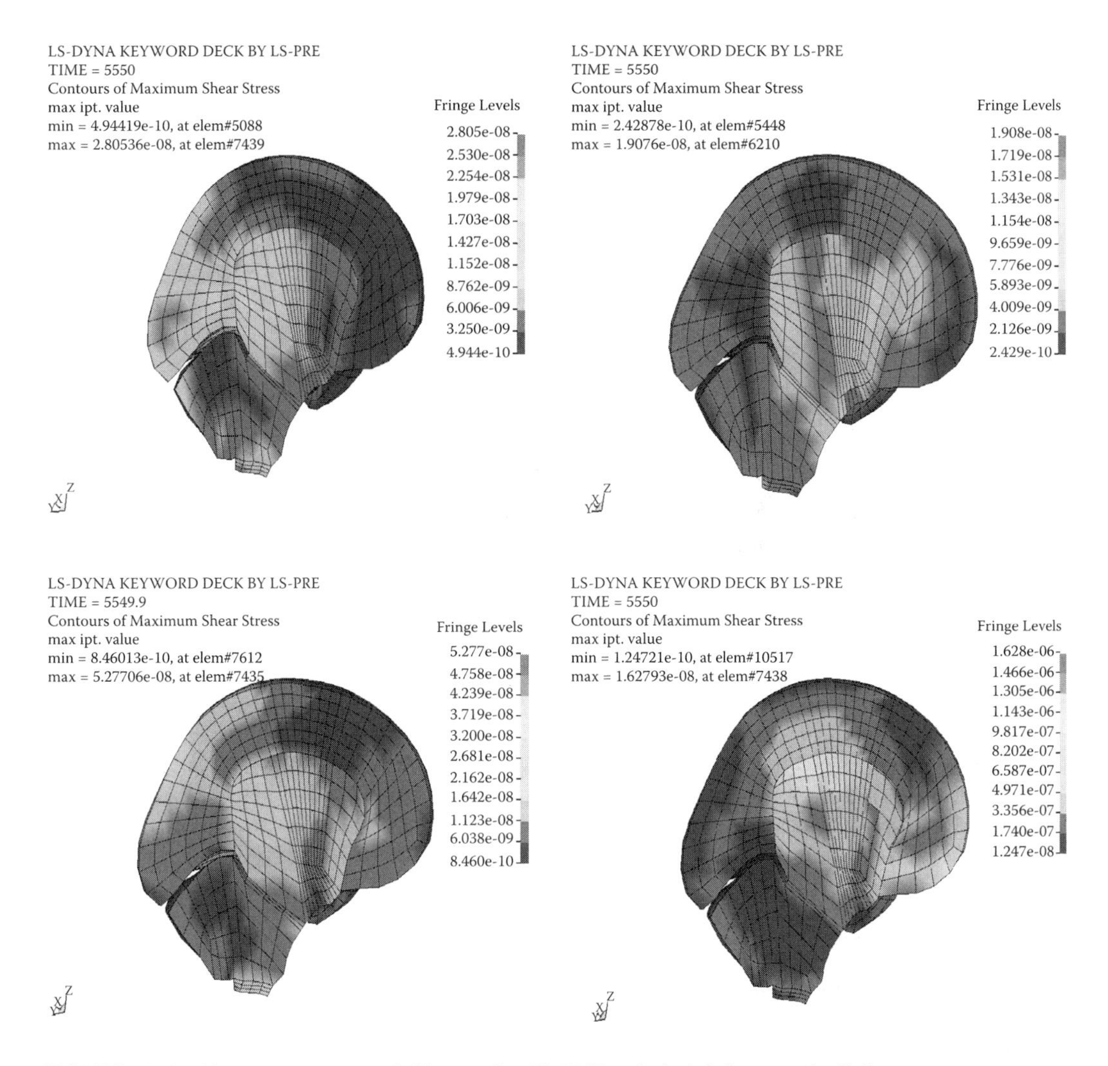

FIGURE 19.6 Shear stresses at t = 5.55 msec for 1lb TNT at 2, 3, 4 & 5 m stand-off distance.

internal organs. Blast waves are created when high-order explosives are detonated. Detonation of explosives creates a unique situation where, within a few microseconds, there is peak overpressure followed by negative pressure (Figure 19.7). The blast wave spherically propagates out from the detonation point, made up of both high-pressure shock waves and a blast wind. These two forces are capable of launching objects and causing blunt, or collision, injury. After the immediate explosion, the heated, detonated air chases the slower moving pressure areas and forms a discontinuity within the wave. This is known as the shock wave front.

Enhanced-blast explosive devices are even more damaging than standard high-order explosives. Enhanced-blast explosives form a primary blast followed by a secondary blast, which when combined, create a high-pressure zone much larger and lasting much longer than that of standard high-order explosives (DePalma et al., 2005).

It is challenging to understand how blast waves interact with soldiers in combat environments. Variables such as distance from the explosion, the amount and type of explosive used, and whether it was an open-air exposure, or a confined space, all factor into how blast waves have an effect. Some of the variables are never known. In addition, information such as whether the blast occurred above

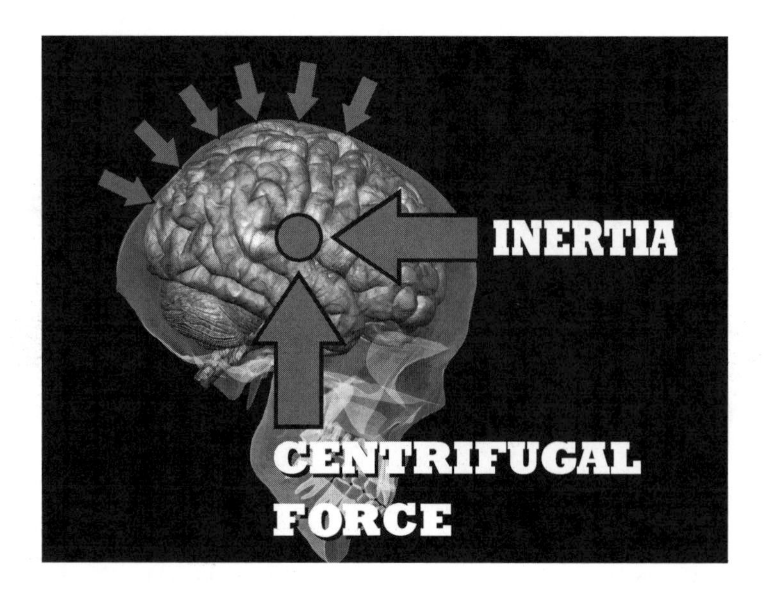

FIGURE 19.7 Overpressure generalization graph.

ground (therefore emitting reflected waves); was a ground explosion (which would not emit reflected waves); or whether it had a mach stem (a shock front formed by the joining of incident and reflected shock waves) may not be known. Waves can be reflected off buildings and other large objects within the environment. When this occurs, reflected sound waves double, and strong shock waves can increase up to eight times (Settles, 2006). This helps explain why someone standing further away from the detonation may be injured more than someone who is in closer proximity to it.

Pressure-Time Diagram

Blast waves are difficult to understand, due to the highly dynamic nature of the event. Figure 19.7 describes the pressure changes associated with a primary blast injury (PBI). From the viewpoint of a stationary observer, the arrival of a shockwave front is characterized by an abrupt acceleration and sudden rise in air pressure, density, and local temperature. As the time increases, the pressure decays to the point of being below the ambient atmospheric pressure (*negative phase*). In broad terms, waves may be characterized by the peak overpressure (pressure that is above atmospheric pressure, the *positive phase*) and duration of the blast event. Based on these two quantities, the intensity of the blast wave can be assessed and the exposure threshold limit can be determined. Figure 19.7 indicates a detailed characterization of the blast including peak overpressure, rate of decay, duration of shock loading, impulse (area under the positive pressure curve) and a negative pressure phase.

The amount of explosive and the distance from the explosion site control the severity of the blast. As simulated by the North Dakota State University (NDSU) Biomechanics Group, the blast parameters change as shown in Table 19.1 and in Figures 19.8 and 19.9. Several scenarios were run, varying the amount of explosive and the stand-off distances. As expected, with distance remaining constant, the data shows that overpressure and impulse are directly related to the amount of explosive used (Table 19.1; Figure 19.8). In Figure 19.9, the only change is in the distance of 50–250 cm. In both scenarios, the data shows the exponential decrease of peak pressure. While these graphs were created based on data gained from simulations conducted at NDSU, other researchers have obtained a similar pressure characteristic (e.g., Cullis, 2001; Guy et al., 2000; Kinney & Graham, 1985).

TABLE 19.1
Blast Parameters for Stand-Off Distance of 250 cm and Explosives of 0.05–5 kg

	Scenario I	Scenario II	Scenario III	Scenario IV	Scenario V	Scenario VI	Scenario VII
TNT Equivalent (kg)	0.05	0.125	0.25	0.375	0.5	1	5
Overpressure (kPa)	12.41	18.88	26.87	33.45	39.42	70.97	224.23
Positive-Phase Duration (msec)	1.25	1.32	1.45	1.45	1.52	1.65	1.55
Impulse/Unit Area (kPa-msec)	6	12	18	23	38	44	111

Note: The sensor location 250 cm distance from center of explosives.

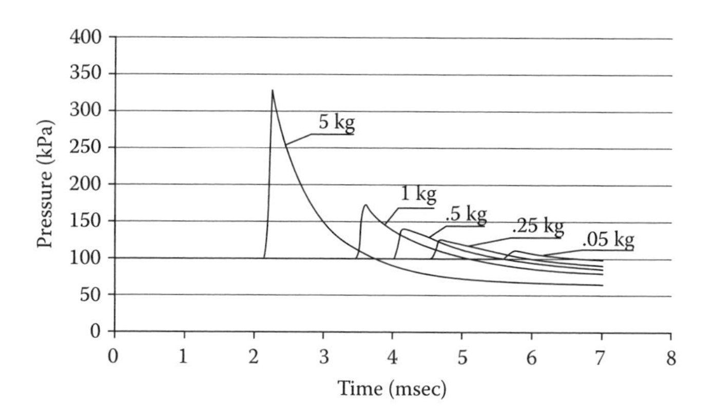

FIGURE 19.8 Pressure as a function of time from detonation point with varying magnitudes of 0.05–5 kg.

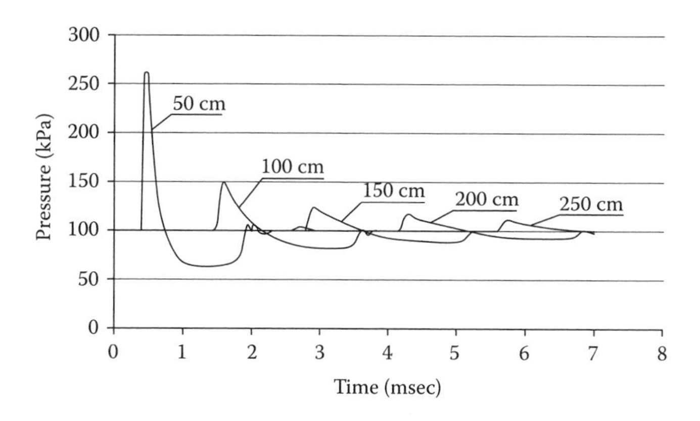

FIGURE 19.9 Pressure as a function of time with varying distances of 50–250 cm.

Mechanism of Brain Injury due to Blast

As stated earlier, there is more than one mechanism that can injure the brain. These mechanisms are complex, difficult to observe, and often cannot be detected by conventional medical procedures. There is a need to better understand their roles in TBI.

Kevlar helmets have been introduced to help reduce penetrating head injuries, but they have little effect on overpressure (Harcke et al., 2002; Lew, 2005). The pressure waves from a blast can be reflected many times before reaching the bystander. The peak overpressure following a blast occurs within the time span of only a few microseconds (μs) and, therefore, exposes the victim to pressure that is substantially greater than normal atmospheric pressure.

When the skull accelerates due to a force applied via blast or impact, there is a push and pull effect on the brain. Significant accelerations can take place in the time range of 2.5–20 msec (Viano et al., 2005). This causes the brain to stretch and also to change shape. These physical insults on the brain may be one of the leading causes of gray- and white-matter injury within the brain (Taber et al., 2006). This is what occurs during DAI when axons within the white matter are stretched beyond their physical limit (Graham et al., 1995).

Blunt impact injuries result from objects being propelled into the victim, or by the victim being thrown into an object. This can occur when, after an explosion, debris is propelled into a bystander, causing either penetrating, or closed, head injuries. Prior to the development of the new advanced combat helmet (ACH), no helmet that would stop, or reduce, the effects of blunt impact to the head had been designed. Now, Kevlar helmets, along with other antiballistic materials, have helped reduce the effects of blunt impact on today's soldiers in the field.

Impacts to the chest, or abdomen, have the ability to increase blood pressure to the point at which it can cause injury. Dr. Cernak's hypothesis involves a pressure increase that causes the blood vessels to fluctuate in size rapidly. The pulse ultimately ends up in the brain, where it still causes damage to brain structures (axonal fibers, neurons in the hippocampus, brain stem) (Cernak et al., 1999). Sharp increases in the vasculature pressure, or ultrasonic blast exposure, may also disrupt the blood-brain barrier (BBB) and, in doing so, other complications can occur within the central nervous system (CNS) (Chavko et al., 2007; Hynynen, 2007; McDannold et al., 2006).

Craig (2007) shed light on the fact that loading may occur through the chinstrap of a helmet after the projectile hits the helmet, or after a fall, thereby loading the skull and its contents. There are two possible means by which the loading travels: (1) the load travels through the mandible, eventually ending at the maxilla, or (2) the load travels through the mandible and into the TMJ, accessing the skull via the madibular fossa. The tympanum plate of the temporal bone, which forms the mandibular fossa, is small and fragile, and an impact fracturing it may result in TBI.

Experimental Research

Since blast physics is such a variable-dependent subject, careful consideration needs to go into the physics of the blast waves being used in the experiment. Different types of waves exist, and each has its own unique characteristics. It is incorrect to assume that since the overpressure is the same, the physical interactions of the wave must also be similar. Some researchers (e.g., Guy et al., 1998) placed specimen too far away from the blast tube exit for it to be subjected to an actual blast wave (Abe et al., 1990; Takayama & Sekiguchi, 1981). Some research has been conducted in which a negative pressure period did not exist as it would in the field (Ohinishi et al., 2001). Similarly, the manner in which specimen are restrained is controversial. Static restraint cannot offer the same comparison as a test run where the specimen may move freely (Boyer, 1960; Guy et al., 1998; Roberts et al., 2005; Taleb et al., 1999).

Animal models of blast injury have been conducted since World War II. In many studies, animals have been exposed to shock waves of a magnitude that has caused pathological indicators similar to those caused in humans. A few recent examples of blast-related research follow.

In a laboratory study using rat specimen, Moochhala et al. (2004) concluded that blast waves can harm neural tissue and cause cognitive, motor, and sensory deficits. The study involved subjecting the specimen to 20 kPa (kilopascal) peak overpressure. Within 24 h, the animals developed motor sensory deficits. Anatomical studies of the microscopic structure of tissue from the animals indicated that there had been an increase in the number of apoptotic cortical neural cells.

In another study, also involving rat specimen, Petras et al. (1997) noticed that with a peak overpressure of 104 kPa, axonal injury was prominent, along with visual impairment most likely due to retinal damage from the exposure.

Earle et al. (2007) ran laboratory studies with pigs in order to model blast trauma to multiple organ systems. The researchers discharged a nail gun over the brain and then over the lungs. Results showed that lung hemorrhage occurred, characteristic of that in a human injury.

Cernak et al. (2001a) ran laboratory studies with rats on the effect of whole body blast exposure (340 kPa peak overpressure). The study determined that the memory processes of the rats were affected by the blasts. The study also revealed that the blast caused a swelling of the neurons in the rats' brains, an increase in pinocytotic activity, debris of the myelin, and a reaction of the glial cells that supply oxygen and nutrients to the neurons.

Cernak et al. (2001b) simulated a blast exposure on rats using an air-driven shock tube. The duration of the shock wave was varied by changing the length of the high-pressure chamber. With the tube end open, a single blast wave could be simulated.

Experimental studies have shown that trauma to the head from a blast, and without direct head injury, can cause secondary alterations to the brain. This occurs by an increased release of neurotransmitters or by hyperexcitability to the CNS (Cernak et al., 2001a). Although there is not solid experimental information, it is relatively clear that ultrasound, combined with pressure on the vascular wall, presents a significant means of BBB disruption. This link between neurological diseases and BBB disruption has been the focus of the experiments of several researchers (Kanner et al., 2003; Kapural et al., 2002; Krizanac-Bengez et al., 2004, 2006a, 2006b Marchi et al., 2003).

Simulation of Blast Injury

Although there have been tremendous developments in neuroimaging, it still presents limitations, that is, not all the damage can be detected because some damage can be below detectable limits of neuroimaging technology. The image resolution is sufficient to delineate large white matter tracts, which consist mostly of neuroglia and axons that usually run parallel. A pixel of a diffusion tensor image (DTI) of the brain image contains bundles of axons and neuroglial cells (Mori & van Zijl, 2002). The size of the cell is on the order of μm (10^{-6} m) (Carpenter, 1976). Axons are filled with neuronal filaments running along the longitudinal axis. The size of the filament is on the order of nm (10^{-9} m) (Albert et al., 1989). A complementary technique that can provide a needed insight on the possible effect of the external force is a mathematical model of the human brain. A mathematical simulation is a noninvasive technique that can investigate injury mechanisms at the cellular level, including nanoscale cavitation (Ziejewski et al., 2007).

A more in-depth understanding of that outcome, including the location, size, and geometry of the damage site, would be of assistance to physicians in properly interpreting neurodiagnostic results. In an approach demonstrated briefly, the macro-/microscale solid mechanics modeling of brain tissues, cells, and head is used in conjunction with the micro- /nanoscale fluid mechanics modeling of cavitation created under sudden movement of the elements of brain constituents. The data that follows were gained through simulation conducted by the NDSU Biomechanics Group (see also Abolfathi et al., 2008a, 2008b, 2008c, Karami et al., in press, 2008; Li et al., 2006; Sotudeh et al., 2007a, 2007b, 2008, 2009; Ziejewski et al., 2004, 2008b.

The generation of blast injury models that define the causal mechanisms of TBI in sufficient detail needs to begin with the characterization of blast phenomenon. The general areas of relevant research are illustrated in Figure 19.10.

A more detailed description of the methodology for the analysis procedures of TBI is as follows:

1. *FEM blast analysis to measure the waves*: The blast of a mass of trinitroluene (TNT) at, or near, the ground surface has been modeled. The outcome is a profile of the pressure distribution.

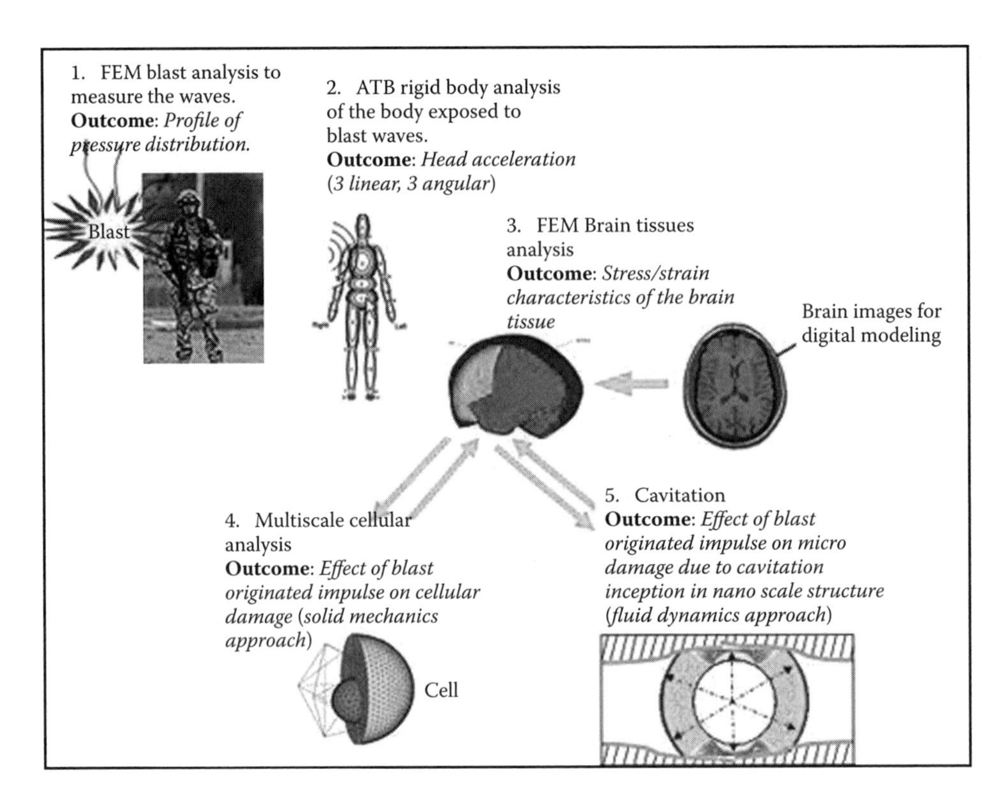

FIGURE 19.10 Flow chart of the analysis procedures for TBI under blast.

2. *Model Articulate Total Body (ATB) to determine the dynamics of the human body exposed to external loading, including impact and blast waves*: The outcome of the analysis of the rigid body exposed to wave fronts is a measurement of the motion, and acceleration of the body, particularly the head and neck regions.
3. *Development of a multiscale FEM of the brain and its cellular structure*: A 3-D model of individual cells, combined with the micro-infrastructural elements and components, is developed. The outcome is a characterization of the brain tissue under sudden stress and strain.
4. *Investigation of cell–matrix interactions under dynamics loading*: Following the development of the cell model, cell–cell interaction and cell–matrix adhesion are studied under dynamic loadings to understand the natural behavior of cells in motion, as well as dynamic loadings. The force transformation and the stress distribution between the interacting cells are examined. An augmented Lagrangian Eulerian (ALE) approach is introduced to study how external forces are transferred through the fluid matrix to the cell and vice versa.
5. *Cavitation modeling under blast conditions*: In order to have a complete understanding of the cumulative pattern of injury for a TBI caused by a blast, cavitation needs to be studied. In cases of a blast injury, the damage is ultimately caused by a global mechanical impact followed by high, transitory acceleration. If these cases are severe enough, they may lead to a complicated pressure pattern in the brain that may produce cavitation, causing a special type of brain injury.

Macro- and Microscale Solid Mechanics Modeling

The analysis is conducted at macroscale for head and brain tissues. LS-Dyna (Hallquist, 2006), a powerful FE software and tool for the analysis under impact loading, has the capacity to determine

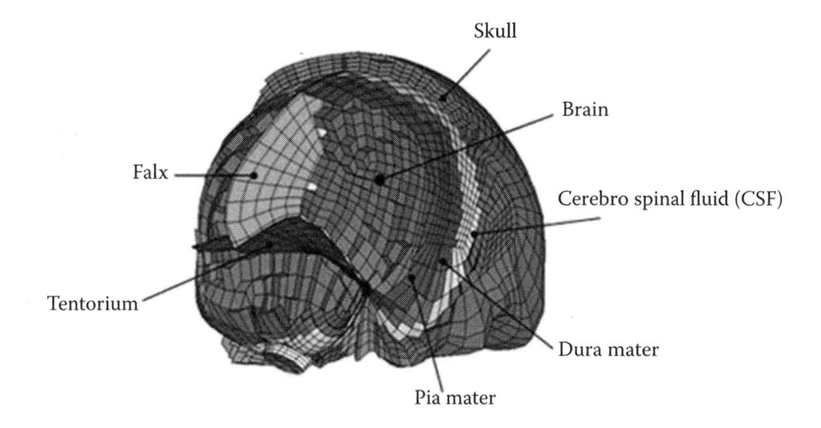

FIGURE 19.11 The right-half model of the brain, CSF, and skull bone.

the motion of the brain elements due to blast. Part of the input data to the LS-Dyna program can be determined using the ATB computer program (Ziejewski & Song, 1998). ATB is a rigid-body dynamic simulation program that is especially suited to measure head and neck motion parameters when the body is exposed to attacking waves of explosion. The program is used to measure linear and angular accelerations of the head. The output is forwarded to FEM for detailed stress and deformation of the skull, as well as for brain-tissue analysis.

The macroscale global FEM brain analysis takes into account the detailed structure of the human head anatomy, including the brain, CSF, dura mater (including the falx and tentorium), pia mater, skull bone, and scalp. The brain, CSF and skull bone are modeled as first-order brick elements with eight-nodes (Kleiven & von Holst, 2003; Li et al., 2006; Ruan, 1994; Willinger et al., 1999). The falx, tentorium, dura, pia and scalp are modeled as a four-node membrane, or shell elements, with uniform thickness. Figure 19.11 shows the 3-D FEM of these components (Li et al., 2006).

In an effort to examine the injury in a multiscaled approach (i.e., from cellular to the global head), the multiscale modeling of the brain and its components is advanced. The verifications are done through magnetic resonance imaging (MRI), which creates a link between the mechanism of brain injury at the cellular level and the mechanism of mechanical loading on the head at the macroscopic level. The work focuses on:

- blasts and creation of the destructive wave;
- determination of TBI at various scales;
- accurate modeling of brain material;
- damage and tolerance at the cellular level;
- correlation of damage at different levels;
- material characterization of the brain and its components; and
- data for a systematic design of injury-protection devices

Cellular analysis is important to the understanding of injuries. The cell, as a basic unit of life, is a biologically complex system. Understanding the behavior of cells requires a combination of various disciplines and approaches, including biomechanics. Cells must use genetic information; perform synthesis; sort, store and transport biomolecules; convert different forms of energy; transduce signals; maintain internal structures; and respond to external environments (Abolfathi et al., in press; Fabry et al., 2001; Guilak et al., 1999). Many of these processes involve mechanisms that should be addressed using the principles of mechanics.

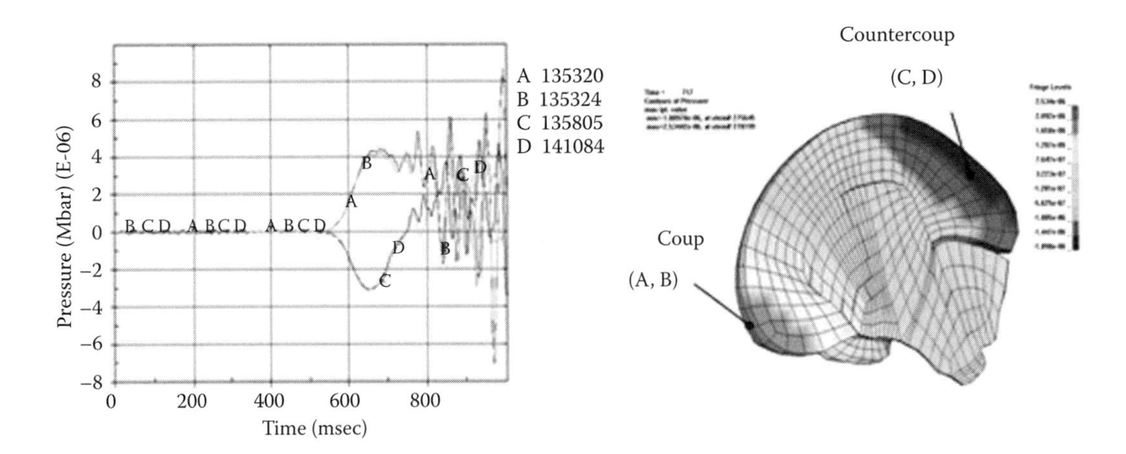

FIGURE 19.12 Pressure changes in the brain.

An example of simulation results for the pressure loading of an explosion is shown in Figure 19.12. The curve illustrates the smooth, uniform loading that initially takes place, followed by a very rapid alteration in pressure levels. This rapid shift in pressure levels could prove to be an important new mechanism of blast related TBI.

Micro- and Nanoscale Fluid Mechanics Modeling (Cavitation)

An additional effect of PBIs is based on the pressure differences that cause micro-/nanoscale cavitation, making cavitation a component of overall injury evaluation.

The complexity of the problem is mainly due to the fact that liquid in the brain is confined between membranes and cells at very different scales, from millimeters (10^{-3} m) to nanometers (10^{-9} m). In macroscale confinement, a brain injury may be caused by cavitation bubble collapse. In micro- and nanoscale confinement (neurons), cavitation inception may be a possible cause of brain injury (Akhatov et al., 2007; Ziejewski et al., 2007).

Possible mechanisms of injury include classically studied macroscale bubble collapse and the more recent nanoscale bubble inception. Within macroscale cavitation, there are two possible results after the inception of the bubble: (i) if the bubble is spherical, tensions will cause the sphere to destruct from the center outward in all directions and (ii) if the bubble is nonspherical (oval), a liquid jet forms near the center of the bubble. Nanoscale collapse is much different. Due to strong tensions, a high amount of energy is stored within the liquid. At inception, this energy is released, generating a "recoil pressure," which may be a factor in tissue damage (Akhatov et al., 2007).

Developing new theoretical approaches and experimental methods that allow an adequate prediction of liquid micro and nano film dynamics and cavitation in the brain, are crucial for improving early diagnoses and interpretations of TBIs. Physical concepts and mathematical tools for the dynamics of liquid micro and/or nano films confined between two solid elastic surfaces, or membranes, need to be developed.

CONCLUSION

Historically, in combat operations, the fragments from an explosive have been the primary cause of death from blasts. In current combat operations, the protective devices (helmets, armor systems for vehicles, etc.) have made it possible for more soldiers to survive blasts. Many of these soldiers, however, are experiencing delayed TBIs. With an increasing number of exposures to IEDs for soldiers in Iraq, the continued study of blast-induced TBI is critical for the physical and mental health of military personnel. With the conventional medical procedures (i.e., observation of external injuries, neuron-imaging, brief cognitive screening procedures, etc.), blast-induced brain injuries can be underdetected, and the treatment of those who have suffered such injuries can be delayed. Soldiers who are not diagnosed properly may even be returned to duty because they show no visible signs of injury. The health of any soldier who has suffered a blast-induced brain injury may deteriorate, and the outcome may be more-damaging permanent injury, or, in some cases, even sudden death.

Because of the current and anticipated future of geopolitical turmoil around the world, there is a strong possibility that conflicts will involve more cases of TBIs. As a result, TBI will become an increasingly costly health issue. A better understanding of blast-induced brain injuries will allow for the development of improved protective devices, as well as for earlier and more appropriate interventions for TBI patients. Early intervention has the potential to be more effective and to help alleviate the cognitive problems associated with TBI, along with job-related issues, relationship problems, and other functional difficulties. Understanding the mechanisms of the brain injury will lead to improved diagnoses of the severity of the injury, as well as to therapeutic interventions that target it.

REFERENCES

Abe, A., & Takayama, K. (1990). Shock wave diffraction from the open end of a shock tube. *Proceedings of the American Institute of Physics, Conference: Current Topics in Shock Waves, 17th International Symposium on Shock Waves and Shock Tubes.*

Bethlehem, PA, Jully 17–21, 1989 (A91-40576 17-34). New York, American Institute of Physics, 1990, p. 270–250.

Abel, J. M., Gennarelli, T. A., & Segawa, H. (1978). *Incidence and Severity of Cerebra Concussion in the Rhesus Monkey Following Sagittal Plane Angular Acceleration.* Society of Automotive Engineers, Inc. (SAE# 780886)

Abolfathi, N., Karami, G., & Ziejewski, M. (2008a). Biomechanical cell modeling under impact loading. *International Journal of Modeling and Simulation, 28(4)*, 470–476.

Abolfathi, N., Naik, A., Sotudeh, M., Karami, G., & Ziejewski, M. (2008b). A micromechanical procedure for the anisotropic modeling of mechanical properties of brain white matter. *Computer Methods in Biomechanics and Biomedical Engineering*, doi:10.1016/j.jmbbm.2008.08.003.

Abolfathi, N., Naik, A., Sotudeh, M., Karami, G., & Ziejewski, M. (2008c). Diffuse axonal injury and degradation in mechanical characteristics of brain white matter. *Proceedings of the ASME 2008 Summer Bioengineering Conference (SBC2008)*, June 25–29, Marriott Resort, Marco Island, Florida, USA, SBC2008–192251, 90%.

Adams, J. H., Graham, D. I., Murray, L. S., & Scott, G. (1982). Diffuse axonal injury due to nonmissile head injury in humans: an analysis of 45 cases. *Annals of Neurology, 12,* 557–563.

Akhatov, I., Wang, C., & Ziejewski, M. (2007). Cavitation in microscale confinement: New concept of mild brain injury. *60th Annual Meeting of the Division of Fluid Dynamics.*

Albert, B., Bray, D., Lewis, J., Raff, M., Roberts, K., & Watson, J. D. (1989). *Molecular Biology of the Cell.* New York: Garland Publishing.

Benedict, J. V., Harris, E. H., & von Rosenberg, D. U. (1970). An analytical investigation of the cavitation hypothesis of brain damage. *American Society of Mechanical Engineers Journal of Basic Engineering,* September, 597–603.

Bhattacharjee, Y. (2008). Shell shock revisited: Solving the puzzle of blast trauma. *Science*, 319, 406–408.

Boyer, D. W. (1960). An experimental study of the explosion generated by a pressurized sphere. *Journal of Fluid Mechanicals, 9,* 401–429.

Carpenter, M. (1976). *Human Neuroanatomy.* Baltimore, MD: Williams & Wilkins.

Cernak, I., Savic, J., Ignajatovic, D., & Jevtic, M. (1999). Blast injury from explosive munitions. *Journal of Trauma, 47(1)*, 96–103.

Cernak, I., Wang, Z., Jiang, J., Bian, X., & Savic, J. (2001a). Ultrastructure and functional characteristics of blast injury induced neurotrauma. *Journal of Trauma, 50,* 695–706.

Cernak, I., Wang, Z., Jiang, J., Bian, X., & Savic, J. (2001b). Cognitive deficits following blast injury-induced neurotrauma: Possible involvement of nitric oxide. *Brain Injury, 15,* 593–612.

Chavko, M., Koller, W. A., Prusackzyk, W. K., & McCarron, R. M. (2007, January 30). Measurement of blast wave by a miniature fiber optic pressure transducer in the rat brain. *Journal of Neuroscience Methods, 159(2)*, 277–281.

Cooper, G. J., Maynard, R. L., Cross, N. L., & Hill, J .F. (1983). Casualties from terrorist bombings. *Journal of Trauma, Injury, Infection and Critical Care, 23*, 955–967.

Cullis, I. G. (2001, February). Blast waves and how they interact with structures. *Journal of the Royal Army Medical Corps, 147(1)*, 16–26.

DePalma, R. G., Burris, D. G., Champion, H. R., & Hodgson, M. J. (2005). Blast injuries. *New England Journal of Medicine, 352*, 1335–1342.

Drazen, J. M. (2005). Using every resource to care for our casualties. *New England Journal of Medicine, 352*, 2121.

Earle, S. A., de Moya, M. A., Zuccarelli, J. E., Norenberg, M. D., & Proctor, K. G. (2007). Cerebrovascular resuscitation after polytrauma and fluid restriction. *Journal of the American College of Surgeons, 204(9)*, 261–275.

Edberg, S., Rieker, J., & Angrist, A. (1963). Study of impact pressure and acceleration in plastic skull models. *Laboratory Investigation, 12*, 1305–1311.

Elsayed, N. M. (1997). Toxicology of blast overpressure. *Toxicology, 121(1)*, 1–15.

Fabry, B., Maksym, G., Butler, J., Glogauer, M., Navajas, D., & Freedberg, J. (2001). Scaling the microrheology of living cells. *Physical Review Letters, 87(14)*, 1–4.

Garner, J., & Brett, S. J. (2007). Mechanisms of injury by explosive devices. *Anesthesiology Clinics, 25*, 147–160.

Gawande, A. (2004). Casualties of war-military care for the wounded from Iraq and Afghanistan. *New England Journal of Medicine, 351*, 2471–2475.

Gennarelli, T. A. (1994). Animate models of human head injury. *Journal of Neurotrauma, 11*, 357–368.

Gennarelli, T. A. (1983). Head injury in man and experimental animals: Clinical aspects. *Acta Neurochirurgica Supplement, 32*, 1–13.

Gennarelli, T. A., & Thibault, L. E. (1982). Biomechanics of acute subdural hematoma. *Journal of Trauma, 22(8)*, 680–686.

Gennarelli, T. A., & Thibault, L. E. (1972). Pathophysiologic responses of rotational and translational accelerations of the head. *Proceedings of the 16th Stapp Car Crash Conference*, Society of Automotive Engineers, 296–308.

Graham, D. I., Adams, J. H., Nicoll, J. A. R., Maxwell, W. L., & Gennarelli, T. A. (1995). The nature, distribution and causes of traumatic brain injury. *Brain Pathology, 5*, 397–406.

Guilak, F., Jones, W. R., Ting-Beall, H. P., & Lee, G. M. (1999). The deformation behavior and mechanical properties of chondrocytes in articular cartilage. *Osteoarthritis Cartilage, 7*, 59–70.

Gurdjian, E. S., Hodgson, V. R., & Thomas, L. M. (1967). High speed techniques in head injury research. *Medical Science, 18(11)*, 45–56.

Guy, R. J., Glover, M. A., & Cripps, N. P. (2000). Primary blast injury: Pathophysiology and implications for treatment. Part III: Injury to the central nervous system and the limbs. *Journal of the Royal Naval Medical Service, 86*, 27–31.

Guy, R. J., Glover, M. A., & Cripps, N. P. (1998). The pathophysiology of primary blast injury and its implications for treatment. Part 1: The thorax. *Journal of the Royal Naval Medical Service, 84*, 79–86.

Guy, R. J. (2004). Shell shock. *Journal of the Royal Society of Medicine, 97(5)*, 255–256.

Hallquist, J. O. (2006). LS-Dyna, Theory Manual, Livermore, California: Livermore Software Technology Corporation.

Harcke, H. T., Schauer, D. A., Harris, R. M., Campman, S. C., & Lonergan, G .J. (2002). Imaging body armor. *Military Medicine, 167*, 267–271.

Hirsch, A. E., & Ommaya, A. K. (1970). *Protection from Brain Injury: The Relative Significance of Translational and Rotational Motions of the Head after Impact*. Society of Automotive Engineers, Inc. (SAE# 700899).

Holbourn, A. H. S. (1943). Mechanics of head injuries. *The Lancet, 243(6293)*, 438–441.

Holden, P. J. (2005). The London attacks—a chronicle: Improvising in an emergency. *New England Journal of Medicine, 353*, 541–543.

Hynynen, K. (2007, January). Focused ultrasound for blood-brain disruption and delivery therapeutic molecules into the brain. *Expert Opinion on Drug Delivery, 4(1)*, 27–35.

Kanner, A. A., Marchi, N., Fazio, V., Mayberg, M. R., Koltz, M. T., Siomin, V., et al. (2003, June 1). Serum S100beta: A noninvasive marker of blood-brain barrier function and brain lesions. *Cancer, 97(11)*, 2806–2813.

Kapural, M., Krizanac-Bengez, L., Barnett, G., Perl, J., Masaryk, T., Apollo, D., et al. (2002, June 14). Serum S-100beta as a possible marker of blood-brain barrier disruption. *Brain Research, 940(1–2)*, 102–104.

Karami, G., Grundman, N., Abolfathi, N., Naik, A., and Ziejewski, M. A micromechanical hyperelastic modeling of brain white matter under large deformation. *Journal of the Mechanical Behavior of Biomedical Materials*. in press, 2: 243–254.

Kaur, C., Singh, J., Lim, M. K., Ng, B. L., Yap, E. P., & Ling, E. A. (1995). The response of neurons and microglia to blast injury in the rat brain. *Neuropathology and Applied Neurobiology*, *21*, 369–377.

Kinney, G. F., & Graham, K. J. (1985). *Explosive shocks in air*. Berlin: Springer-Verlag.

Kleiven, S., & von Holst, H. (2003). Review and evaluation of head injury criteria. *Proceedings RTO Specialist Meeting. NATO's Research and Technology Organization (RTO),* Koblenz, Germany, 19–23.

Krizanac-Bengez, L., Hossain, M., Fazio, V., Mayberg, M., & Janigro, D. (2006a, October). Loss of flow induces leukocyte mediated MMP/TIMP imbalance in dynamic in vitro blood-brain barrier model: Role of proinflammatory cytokines. *American Journal of Physiology: Cell Physiology, 291(4)*, C740–C749.

Krizanac-Bengez, L., Mayberg, M. R., Cunningham, E., Hossain, M., Ponnampalam, S., Parkinson, F. E., et al. (2006b, January). Loss of shear stress induces leukocyte-mediated cytokine release and blood-brain barrier failure in dynamic in vitro blood-brain barrier model. *Journal of Cell Physiology*, *206(1)*, 68–77.

Krizanac-Bengez, L., Mayberg, M. R., & Janigro, D. (2004, December). The cerebral vasculature as a therapeutic target for neurological disorders and the role of shear stress in vascular homeostatis and pathophysiology. *Neurological Research*, *26(8)*, 846–853.

Lew, H. L. (2005, July–August). Rehabilitation needs of an increasing population of patients: Traumatic brain injury, polytrauma, and blast-related injuries. Editorial. *Journal of Rehabilitation Research and Development*, *42(4)*, xiii–xvi.

Lew, H. L., Poole, J. H., Alvarez, S., & Moore, W. (2005). Soldiers with occult traumatic brain injury. *American Journal of Physical Medicine & Rehabilitation*, *84(6)*, 393–398.

Li, D., Ziejewski, M., & Karami, G. (2006). Parametric studies of brain materials in the analysis of head impact. *ASME Congress, Chicago*. (IMECE-2006–15596).

Luback, P., & Goldsmith, W. (1988). Experimental cavitation studies in a model head-neck system. *Journal of Biomechanics*, *13*, 1041–1052.

Marchi, N., Fazio, V., Cucullo, L., Kight, K., Masaryk, T. J., Barnett, G., et al. (2003). Serum transthyretin as a possible marker of blood-to-CSF barrier disruption. *Journal of Neuroscience*, *23(5)*, 1949–1955.

Mayorga, M. A. (1997). The pathology of primary blast overpressure injury. *Toxicology*, *121*, 17–28.

McDannold, N., Vykhodtsva, N., & Hynynen, K. (2006, February 21). Targeted disruption of the blood-brain barrier with focused ultrasound: Association with cavitation activity. *Physics in Medicine and Biology*, *51(4)*, 793–807.

McLean, A. J. (1996). Brain injury without head impact? In F. A. Bandak, R. H. Eppinger, & A. K. Ommaya (Eds.), *Traumatic brain injury: Bioscience and mechanics* (pp. 45–49). Mary Ann Liebert, New York: Larchmont.

Meaney, D. F. (1991). The biomechanics of acute subdural hematoma in the subhuman primate and man (Doctoral Dissertation, University of Pennsylvania, 1991).

Mertz, H. J., & Nusholtz, G. S. (1996). *Head Injury Risk Assessment for Forebrain Impacts*. Society of Automotive Engineers, Inc. (SAE # 960099).

Moochhala, S. M., Md, S., Lu, J., Teng, C. H., & Greengrass, C. (2004). Neuroprotective role of aminoguanidine in behavioral changes after blast injury. *Journal of Trauma*, *56*, 393–403.

Mori, S., & van Zijl, P. C. M. (2002). Fiber tracking: Principles and strategies—a technical review. *Nuclear Magnetic Resonance in Biomedicine*, *15*, 468–480.

National Highway Traffic Safety Administration (NHTSA). (2000). *Federal Motor Vehicle Safety Standards; Occupant Crash Protection* (49 CFR Parts 552, 571, 585, 595 Docket # NHTSA 00–7013, pp. 143–144).

Newman, J., Shewchenko, N., & Welbourne, E. (2000). A proposed new biomechanical head injury assessment function—the maximum power index. *44th Stapp Car Crash Conference*, Atlanta, GA.

Nusholtz, G. S., Kaiker, P. S., & Lehman, R. J. (1986). Critical limitations on significant factors in head injury research. *Proceedings of 30th Stapp Car Crash Conference Paper # 861890*, 237, 1986.

Nusholtz, G. S., Kaiker, P. S., Wylie, E. B., & Glascoe, L. G. (1994). The effects of the skull/dura interface and foramen magnum on the pressure response during head impact. *14th ESV Conference*, Munich, Paper # 94 S1 0 26, 1994.

Nusholtz, G. S., Lux, P., Kaiker, P. S., & Janicki, M. A. (1984). Head impact response-skull deformation and angular acceleration. *Proceedings of 28th Stapp Crash Conference*, Paper #841657, 41.

Nusholtz, G. S., & Ward, C. C. (1987). Comparison of epidural pressure in live anesthetized and postmortem primates. *Aviation, Space, and Environmental Medicine Journal*, *58(1)*, 9–17.

Nusholtz, G. S., Wylie, E. B., & Glascoe, L. G. (1995a). Cavitation/boundary effects in a simple head impact model. *Aviation, Space, and Environmental Medicine Journal, 66(7)*, 661–667.

Nusholtz, G. S., Wylie, E. B., & Glascoe, L. G. (1995b). Internal cavitation in simple head impact model. *Journal of Neurotrauma, 12(4)*, 707–714.

Ohinishi, M., Kirkman, E., Guy, R. J., & Watkins, P. E. (2001). Reflex nature of the cardiorespiratory response to primary thoracic blast injury in the anaesthetised rat. *Experimental Physiology, 86(3)*, 357–364.

Okie, S. (2005). Traumatic brain injury in the war zone. *New England Journal of Medicine, 352*, 2043–2047.

Ommaya, A., & Gennarelli, T. A. (1974). Cerebral concussion and traumatic unconconsciousness. *Brain, 97*, 633–654.

Ommaya, A. K., Hirsch, A. E., & Martinez, J. L. (1966). The role of whiplash in cerebral concussion. *Proceedings of the 10th Stapp Car Crash Conference*, November 1966.

Peota, C. (2005). Invisible wounds. *Minnesota Medicine, 88(1)*, 13–14.

Petras, J. M., Bauman, R. A., & Elsayed, N. M. (1997). Visual system degeneration induced by blast overpressure. *Toxicology, 121*, 41–49.

Pudenz, R. H., & Sheldon, C. H. (1946). The lucite calvarium—a method for direct observation of the brain, cranial trauma and brain movement. *Journal of Neurosurgery, 3*, 487–505.

Redhead, J., Ward, P., & Batrick, N. (2005). The London attacks—response: Prehospital and hospital care. *New England Journal of Medicine, 353*, 546–547.

Roberts, J. C., Biermann, P. J., O'Connor, J. V., Ward, E. E., Cain, R. P., Carkhuff, B. G., et al. (2005). Modeling nonpentrating ballistic impact on a human torso. *Johns Hopkins APL Technical Digest, 26(1)*, 84–92.

Ross, D. T., Meany, D. F., Sabol, M. K., Smith, D. H., & Genneralli, T. A. (1994). Distribution of forebrain diffuse axonal injury following inertial closed head injury in miniature swine. *Experimental Neurology, 126*, 291–299.

Ruan, J. S. (1994). Impact biomechanics of head injury by mathematical modeling (Doctoral Dissertation, Wayne State University, 1994).

Ryan, J., & Montgomery, H. (2005). The London attacks—preparedness: Terrorism and the medical response. *New England Journal of Medicine, 353*, 543–545.

Settles, G. S. (2006, January–February). High-speed imaging of shock waves, explosions and gunshots. *American Scientist Magazine, 94*(1), 22–31.

Sotudeh-Chafi, M., Abolfathi, N., Nick, A., Dirisala, V., Karami, G., & Ziejewski, M. A Multi-scale Finite Element Model for Shock Wave-induced , Axonal Brain Injury SBC2008–192342, Proceedings of the ASME 2008 Summer Bioengineering Conference (SBC2008), June 25–29, Marriott Resort, Marco Island, Florida, USA.

Sotudeh Chafi, M., Karami, G., & Ziejewski, M. Numerical analysis of blast-induced wave propagation using FSI and ALE multi-material formulations. *International Journal of Impact Engineering*, 2008 (under review).

Sotudeh Chafi, M., Dirisala, V., Karami, G., & Ziejewski M. A parametric study on the effects of the CSF constitutive model and material parameters on the dynamic response of human brain. *Journal of Biomechanics and Modeling in Mechanobiology*, 2008 (under review),

Sotudeh-Chafi, M., Karami, G., & Ziejewski, M. (2007b). An assessment of primary blast injury in human brain—a numerical simulation. SBC2007–176155.

Sullivan, M. G. (2004). Returning soldiers may have blast injury to brain. *Family Practice News, 34*, 24.

Taber, K. H., Warden, D. L., & Hurley, R. A. (2006, Spring). Blast-related traumatic brain injury: What is known? *Journal of Neuropsychiatry and Clinical Neurosciences, 18(2)*, 141–145.

Takayama, K., & Sekiguchi, H. (1981). Formation and diffraction of spherical shock waves in a shock tube. *Rep. Inst. High Speed Mech*, Tohoku University, *43(336)*, 89–119.

Taleb, L., Brown, M. J., & Sadeghi, M. M. (1999). Toward the development of a comprehensive HIC: Predicted injury—brain material dependency. *AMD-Vol. 237/BED-Vol. 45, Crashworthiness, Occupant Protection and Biomechanics in Transportation Systems—1999, ASME 1999.*

Unterharnscheidt, F. J. (1970). Discussion. In E. S. Gurdjian, W. A. Lange, L. M. Patrick, & L. M. Thomas (Eds.), *Impact injury and crash protection* (pp. 43–62). New York: Charles C. Thomas.

Unterharnscheidt, F. J. (1971). *Translational versus Rotational Acceleration—Animal Experiments with Measured Input.* Society of Automotive Engineers, Inc. (SAE# 710880).

Unterharnscheidt, F., & Higgins, L. S. (1969). Traumatic lesions of brain and spinal cord due to non-deforming angular acceleration of the head. *Texas Reports on Biology and Medicine*, 27, 127–166.

Viano, D. C., Casson, I. R., Pellman, E. J., Zhang, L., King, A. I., & Yang, K. H. (2005, November). Concussion in professional football: Brain responses by finite element analysis: Part 9. *Neurosurgery*, 57(5), 891–916.

Viano, D. C., & Lovsund, P. (1999). Biomechanics of brain and spinal cord injury: Analysis of neurophysiological experiments. *Crash Prevention and Injury Control*, 1, 35–43.

Warden, D. (2006, September–October). Military TBI during the Iraq and Afghanistan wars. *Journal of Head Trauma Rehabilitation*, 21(5), 398–402.

Wessely, S. (2005). The London attacks—aftermath: Victimhood and resilience. *New England Journal of Medicine*, 353, 548–550.

Wightman, J. M., & Gladish, S. L. (2001). Explosions and blasts injuries. *Annals of Emergency Medicine*, 37, 664–678.

Willinger, R., Kang, H. S., & Diaw, B. (1999). Three-dimensional human head finite element model validation against two experimental impacts. *Annals of Biomedical Engineering,* 27, 403–410.

Yilmaz, S., & Pekdemir, M. (2007). An unusual primary blast injury traumatic brain injury due to primary blast injury. *American Journal of Emergency Medicine*, 25(1), 97–98.

Ziejewski, M. (2004). The biomechanical assessment of traumatic brain injury. *Brain Injury Professional*, 1(1), 26–30.

Ziejewski, M. (2008a). Biomechanics: Use of evidence based physics as a foundation for understanding the nature of traumatically induced brain injury. *International Brain Injury Association, Seventh World Congress on Brain Injury, Lisbon, Portugal.*

Ziejewski, M. (2008b). Physics of blast related traumatic brain injury. *Traumatic Brain Injury Conference and Think Tank Session, College Station, TX.*

Ziejewski, M. (2004). State-of-the-art biomechanical issues in relation to mild traumatic brain injury. *North American Brain Injury Society Conference, Amelia Island, FL.*

Ziejewski, M., Karami, G., & Akhatov, I. (2007). Selected biomechanical issues of brain injury caused by blasts. *Brain Injury Professional*, 4(1), 10–15.

Ziejewski, M., & Song, J. (1998). Assessment of brain injury potential in design process of children's helmet using rigid body dynamics and finite element analysis. *ATB Users' Group Conference sponsored by US Air Force, Armstrong Laboratory, Dayton OH.*

Glossary

This information has been designed primarily to define technical terms and concepts of neurobehavioral significance. Some related terms are described together since they benefit from being combined. The clinical significance of these definitions will be found in the text.

Adrenal androgens: Moderately active male sex hormones (e.g., *dehydroepiandrosterone*) are continually secreted by the adrenal cortex, especially during fetal life. Ordinarily, adrenal androgens have a weak effect in humans; part of the early development of the male sex organs may result from a childhood secretion of adrenal androgens. Female sex hormones are secreted in minute quantities (*progesterone*; *estrogens*). Adrenal androgens exert mild effects in females before puberty and throughout life. Much of the growth of the female pubic and axillary hair results from the action of these hormones (see Developmental Considerations; Disorders of Physiological Development) (Guyton & Hall, 2006, p. 957).

Agonists: Bind to a receptor, leading to a characteristic chemical response (perhaps by a different compound).

Allele: One of a series of two or more genes that occupy the same locus on a specific chromosome.

Anabolism: A metabolic process by which substances are converted into other components of the body's chemical architecture.

Antagonists: bind to a receptor but prevent agonist actions by competing for the agonist that would bind to that receptor.

Antibody: These are immunoglobulins. The function of antibodies is adaptive immunity (i.e., to recognize a great variety of new antigens, presumably including necrotic tissue) (see below). This is in contrast to innate immunity (i.e., available prior to exposure) (see Innate Immunity, below). Antibodies bond to molecules of the pathogen that elicit immune response, neutralize them, and mark them for destruction by phagocytes and complement. They also recruit other cells and molecules to destroy the pathogen once the antibody is bound to it.

Antigen and immunogen: Any substance that can bind to a specific antibody (however, some antigens need to be attached to an immunogen in order to do so). These are able to generate antibodies. A complete antigen induces an immune response and reacts with the product of it. The incomplete antigen (actually immunogen) cannot induce an immune response but can react to its products. Immunogens induce cell-mediated immunity and antibody synthesis.

Astrocytes and gliosis: A form of neuroglia that is the most abundant cell in the brain. Astrocytes provide metabolic support, act as scavengers, and buffer potassium ions (Eriksson, 2002). They are responsible for neuronal metabolic and trophic support (Bambrick et al., 2004). Astrocytes are the principle repository of brain glycogen, as neurons have very little glycogen. They also provide physical and metabolic support for neurons. They play a role in reaction to a central nervous system (CNS) injury, and have been considered to be both impediments and promoters of regeneration: gliosis (proliferation of astroglia in both gray and white matter of damaged portions of the CNS); swelling (release of excitotoxic compounds, such as L-glutamate); reducing the vulnerability of dopaminergic neurons to Parkinson disease, protecting neurons by secreting neurotrophic factors, and (perhaps) control of glutamate-induced neurotoxicity (Baumann & Pham-Dimnh, 2002; Kimelberg & Norenberg, 1994). Footlike processes form a continuous layer around the blood vessels participating in the blood-brain barrier (BBB).

Ataxia: Lack of accuracy or coordination of movement that is not due to paresis, alteration in tone, sensory loss, or the presence of involuntary movements. It is considered a cardinal sign of cerebellar disease.

Balance: The condition in which all forces and torques acting on the body are in equilibrium so that the person's center of mass (COM) is within his or her limits of stability.

C-reactive protein (CRP): A serum protein induced by inflammation caused by necrosis or trauma.

Calvaria: The domelike superior portion of the cranium. It is thin, platelike, and has external layers of relatively thick compact bone with an intervening layer of spongy bone.

Catabolism: A destructive metabolic process by which substances are converted into excreted compounds.

Cell death (necrosis and apoptosis): Approximately 50% of initially generated neurons die (i.e., "programmed cell death"). This is a normal feature of the developing nervous system. Necrosis is the death of cells due to physical or chemical injury. Dead or necrotic tissue is taken up and degraded by phagocytes, which clear the damaged tissue and heal the wound. Spilling of cellular contents into the surrounding tissue attracts immune cells, which produce cytokines that generate an inflammatory response. Apoptosis is an ordered normal developmental function that causes remodeling through selective death of cells. It is characterized by cell death without inflammatory response. Necrosis (but not apoptosis) leaves extensive cellular debris that is removed by phagocytes. Cell death induced by free radicals (a result of trauma) may have the characteristics of both apoptosis and necrosis (Arvin & Holztman, 2003; Cruse & Lewis, 2003, pp. 61–64; Janeway et al., 2001, p. 352; Lee & Chao, 2005).

Chemokines: Macrophages secrete proteins (i.e., *chemokines*). These attract cells from the blood stream with chemokine receptors (i.e., neutrophils and monocytes), which initiate *inflammation* (heat, pain, redness, swelling) (see below). Chemokines direct specific white blood cells to the inflammatory focus (Rabin, 2005).

Cytokines are immune system proteins secreted by macrophages (blood cells with immune system functions) affecting the behavior of other cells with receptors for them (i.e., biological response modifiers). They are polypeptides or soluble proteins that serve as regulatory or effector molecules of the immune system and elsewhere (i.e., signals in cell-cell interactions). Subtypes include monokines and lymphokines (Cruse & Lewis, 2003, pp. 181–183). They release chemokines (proteins), which attract cells with chemokine receptors from the blood stream (neutrophils and monocytes). Cytokines are classified according to the type of immunity they mediate and whether they promote or retard inflammation (Giulian, 1994; Penedo & Dahn, 2005).

Cytokines are released by activated inflammatory cells that regulate bidirectional communication between cells to facilitate and perpetuate the inflammatory process, and interactions between immune cells. They function in both pathological and normal states, and have overlapping biological activities. Cytokines that are produced within the brain or penetrate it can promote peripheral inflammation (Lipton et al., 1994; Young & Ott, 1996). Since there are very few cytokines in healthy individuals it may be inferred that high levels are a marker of dysfunction or disease (Vedhara & Wang, 2005).

The cytokines and chemokines initiate inflammation (Janeway et al., 2005, p. 12) consist of polypeptides or soluble small protein molecules produced by various blood, tissue, and CNS cells. They serve as signals and influence the activity of other immune system cells (Ross et al., 2003, p. 366). They serve as regulatory or effector molecules of the immune (Giulian, 1994; see Vedhara & Wang, 2005, for functions of different types) and inflammatory systems (Gallowitsch-Puerta & Tracey, 2005).

Chemotaxis: Cell movement in response to chemicals, including leukocytes and macrophages involved in the immune and inflammatory systems.

Complement: A group of more than two dozen liver- and macrophage-derived proteins that pay a role in immunity as well as mediate aspects of inflammation. Although characterized as antimicrobial, they are significant in TBI response since they are chemoattractants for leukocytes, binding to receptors and triggering defensive reactions (Cunnion et al., 2001; Parslow & Bainton, 2001).

Connective tissue (fascia): Is formed of fibers and cells that maintain form in the body by providing a matrix that connects and binds cells and organs, giving support. These tissues vary in elasticity, flexibility, and resistance, thus offering variable vulnerability to different kinds of mechanical forces (Junqueira & Carneiro, 2005, pp. 91–92; 117–120).

Corticosteroids: Any of the 21-carbon steroids elaborated by the adrenal cortex (see Mineralocorticoids and Glucocorticoids) in response to pituitary corticotropin (ACTH) and angiotensin II. This excludes sex hormones of adrenal origin (androgenic hormones). Corticosteroids comprise the glucocortocoids and mineralocorticoids. They express anti-inflammatory functions by regulating the expression of many genes, with different genes regulated in different tissues. They reduce the production of inflammatory mediators, inhibit inflammatory cell migration to sites of inflammation, and promote the death of leukocytes and lymphocytes by apoptosis (Janeway et al., 2001, p. 555).

Dehydroepiandrosterone (DHEA): Secreted with cortisol and helps modulate its effects (i.e., under conditions of extreme stress DHEA may serve a protective role).

Dermatomes, myotomes, and somites: Somites are embryological structures, adjacent paired blocks of tissue along the neural tube. They form the vertebral column and segmental musculature. The dermatome is the area of skin innervated by a single dorsal root (corresponding to a somite). The different dermatome fields are not distinct; afferent innervation overlaps between segments. Thus, a lesion of a particular nerve root produces a more restricted clinical deficit than one based upon the apparent innervation of a structure of the dermatome, presumably (and incorrectly) served by a specific single nerve root. A myotome is all the muscles derived from a particular somite. Thus, it is innervated by one segmental spinal nerve (Campbell, 2005, p. 432).

Dexamethasone test: Dexamethasone is an artificial glucocorticoid that is twenty-five times as potent as cortisol. It is used as an anti-inflammatory and immunosuppressant. In the test, urinary levels of 17-hydrocorticosteroid are measured after administration. Cortisol secretion is suppressed in Cushing syndrome (hyperadrenocorticism), but not in adrenal tumors or ectopic ACTH syndrome.

Diploic and emissary veins; dural venous sinus: The diploic veins are channels located between the inner and outer tables of the calvaria. The emissary veins pass from the various venous sinuses on the surface of the brain into the calvaria (skullcap) through the two layers of the dura mater, exchange with the diploic veins of the spongy bone inside the skull, and link with the veins on the surface of the skull (Netter, 1983, pp. 54–55).

Dysmetria: Disturbance of the direction or placement of a body part during movement.

Electrolytes: A chemical compound that breaks into charged particles when dissolved (e.g., salt becoming $sodium^{+}$ and $chloride^{-}$).

Endothelial cells: A layer of flat cells lining blood and other vessels. It forms the blood-brain barrier (BBB).

Ependymal cells: Column-shaped cells lining the ventricles of the brain and the central canal of the spinal cord.

Epithelial cells: The covering of internal and external surfaces of the body, including the skin, as well as the lining of vessels and other small cavities.

Fascia (see also Connective Tissue; Tissue Injuries; Delayed Healing): A form of connective tissue immediately underneath the skin, which covers most of the body deep to the skin with extensions that cover deeper structures such as individual muscles and neurovascular bundles. Deep fascia may be the anatomical origin of muscles, but in most places muscles are free to glide deep to it. Thus, its location increases injury-related restriction of motility (Moore & Dalley, 2006, pp. 16–17).

Feed-forward control: Refers to signals sent to a system that prepares it for future motor action or anticipated sensory input. Feed-forward control is used when body movements are so rapid that there is no time for the sensory signals from moving parts to appraise the brain as to whether movements are performed correctly and return to the periphery to control the movement. If sensory signals indicate that a movement is incorrect, the brain corrects the feed-forward signal it sends to the muscles the next time the movement is required (see Control Systems, Guyton & Hall, 2006, pp. 6–9).

Free radicals: In chemistry, a molecule that contains at least one unpaired electron. Most molecules contain an even numbers of electrons, and the covalent chemical bonds holding the atoms together within a molecule normally consist of pairs of electrons jointly shared by the atoms linked by the bond. Although free radicals contain unpaired electrons, they may be electrically neutral. Because of their odd electrons, free radicals are usually highly reactive. They combine with one another or with single atoms that also carry free electrons to give ordinary molecules, all of whose electrons are paired. They may also react with intact molecules, abstracting parts of the molecules to complete their own electron pairs and generating new free radicals in the process. In all these reactions, each simple free radical, because of its single unpaired electron, is able to combine with one other radical or atom containing a single unpaired electron. Most free radicals are capable of only the most fleeting independent existence (Encyclopedia Britannica).

Galanin: Involved in pain control, cardiovascular regulation, food intake, neuroendocrine control, learning and memory, and anxiety. It is expressed by a high proportion of NE neurons in the locus coeruleus and enhances GH responses to GHRH (MacGillivray, 2001).

Genetic nomenclature (traumatic focus): (Note: this text emphasizes outcome of trauma. For further information concerning linkage of genetic components with a disease, see Diaz-Arrastia and Baxter, 2006.) An *allele* is one of the two alternative forms of a gene found at a particular chromosomal locus that determines inheritance. These may be identical or different (one from each parent). An individual may have up to two different alleles for each gene. *Locus* is the fragment of a genome that can be inherited. It can be a gene, a cluster of genes, or regulatory elements. The *genome* is the entirety of the genetic information of an organism, cell, or cell organelle (i.e., its DNA). A *genotype* is the alleles present at one or more specific loci. The *phenotype* is the observable morphological, physiological, or biochemical characteristics of an individual, either as a whole or with respect to individual traits. It is determined by the interaction of the genotype and the environment. *Genetic polymorphism*: The existence within a population of several different forms of individuals. It refers to a variant form of a gene that arises from a mutation.

Glucagon (glucose regulation): A hormone secreted by the islets of Langerhans (pancreas) in response to glucose deficiency. It stimulates the breakdown of glycogen to glucose and regulates plasma glucose levels (in direct opposition to insulin). Its effects on carbohydrates, fat, and protein metabolism raise the plasma levels of glucose. When plasma glucose falls below a set point, there is increased catecholamine (adrenal cortex) and norepinephrine (sympathetic nerve terminals) secretion. Norepinephrine and epinephrine work with glucocorticoids, GH, and glucagon in hypoglycemia to maintain plasma glucose levels and to reserve the available glucose for use by the brain (Pocock & Richards, 602–606).

Glucocorticoids (cortisol): Adrenal cortical steroid hormones secreted by a sequence of activations starting with hypothalamic CRH (corticotropin releasing hormone), which regulates ACTH, causing secretion of cortisol, all linked into a negative feedback loop. They increase blood glucose concentration and effect protein and fat metabolism. Glucocorticoids are secreted by the cortex of the adrenal gland and are part of the stress response (e.g., cortisol). They promote gluconeogenesis (formation of glucose by converting noncarbohydrate precursors) and the formation of glycogen, enhance the degradation of fat and protein, and inhibit the inflammatory and immune responses. Cortisol acts through intracellular receptors found in almost every cell of the body. When the glucocorticoids bind to the receptor, these receptors regulate the transcription of specific genes—perhaps 1% of the genome. The responsive genes are induced or suppressed (Janeway et al., 2005, p. 614).

Immunogens: Any substance that elicits an immune response (i.e., is immunogenic).

Jump sign: A response to an unanticipated painful stimulus with wincing, crying out, etc.

Leukocytes: Connective tissue cells that migrate as white blood cells from the blood vessels. Crossing the walls of venules and capillaries, these cells penetrate the tissues and display their defensive capabilities. This process increases greatly during inflammation (see Secondary Phase: Cell Pathology).

Lipids: A group of water-insoluble fats and fat-like substances. They are an important constituent of cell structures and also serve as a source of fuel.

T cells; B cells: T lymphocyte cells are derived from blood-forming cells that migrate first to the thymus and then to other lymphoid tissues. T cells are involved in the control of the immune response (i.e., helping or suppressing these responses). In the context of reduced cortisol, higher level of T cells is a marker of posttraumatic stress disorder (PTSD). This pattern suggests that chronic sufferers of PTSD may be at risk for autoimmune diseases (high levels of immune mediators with low levels of an anti-immune hormone). T cells are part of the adaptive immune system (i.e., a defense against invading or damaged matter). B lymphocyte cells mature in the bone marrow and carry antigen receptors that detect invaders.

Long-term potentiation (LTP): This is a persistent increase in synaptic strength, induced by a brief burst of spike activity in the presynaptic afferents. It is hypothesized that this is a form of synaptic plasticity that may participate in information storage in several brain regions (see Inhibition of Plasticity).

Macrophages: Derived from blood cells (monocytes) that migrate to the connective tissue and differentiate. Macrophages play a role in the regulation of wound healing by facilitating the removal of fragments of damaged tissue (debridement; phagocytosis) consequent to necrosis and apoptosis. They are the major source of cytokines, which are important for the healing process. They display *phagocytic* activity.

Magnetoencephalography (MEG): Measurement of the current that flows within the apical dendrites of pyramidal cells oriented parallel to the skull (Lewine et al., 2007).

Microglia (see Wound Healing): Inconspicuous cells that originate in the bone marrow and enter the CNS from the vascular system. Microglia possess phagocytotic properties.

Mineralocorticoids: Adrenal cortical steroid hormones affecting mineral electrolytes of the extracellular fluids, in particular sodium and potassium. Aldosterone acts on the kidney to retain Na+ (sodium) and to secrete K^+, H^+ and NH_4^+.

Modulation: (1) A relation between two systems (e.g., immune and nervous) and (2) regulation leading to stimulation or inhibition.

Myotome: The group of muscles innervated by a single anterior spinal root.

Natural killer cells (NK): Lymphoid cells of the natural immune system. NK cells do not require prior contact with an antigen. They destroy tumor cells and some virus-infected cells. They are a part of the stress response.

Neuroglia (glia): These are supporting cells (Ross et al., 2003). Glia and neurons are the primary cell types of the brain, with glia outnumbering the neurons an estimated 10–50 times. Functions: support the neurons, thus creating structure; separate and sometimes insulate neuronal groups and synaptic connections; produce myelin; promote efficient signaling between neurons by taking up neurotransmitters released during synaptic transmission; regulate the presynaptic terminal; help form the blood-brain barrier (BBB); release growth factors; and help nourish nerve cells (Kandel, 2000b). Glia secrete neurotrophic factors that stimulate sprouting and regeneration, contribute to axonal repair and neural transmission in response to brain injury, and give rise to new neurons in the hippocampus (Stein & Hoffman, 2003).

Neuropeptides (NP): A chain of two or more amino acids linked by peptide bonds and produced by the nervous system and somatic cells. Examples include adrenocorticotropic hormone (ACTH); melanocyte-stimulating hormone (MSH); vasopressin; enkephalins; and gut peptides produced in the brain (cholecystokinin; vasoactive intestinal peptide; and, somatostatin). Included in NP are hypothalamic hormones (oxytocin; vasopressin); hypothalamic-releasing and inhibiting hormones (somatostatin; growth hormone-releasing hormone; thyrotropin-releasing hormone; gonadotropin releasing hormone); tachykinins (substance P; neurokinins); and opioids (enkaphalin; endorphins; dynorphins).

Neurosecretory cells: These are specialized neurons that secrete substances that act as hormones directly into the bloodstream. All neurons in the CNS receive multiple synaptic largely onto the dendrites and cell bodies. They are able to respond to and integrate input through specific receptors. Their action potentials release neurotransmitters and neuromodulators into synapses that are formed with postsynaptic neurons. Since the vast majority of communication between neurons is accomplished by classical neurotransmitters, neurosecretion is a fundamentally new concept in understanding the mechanisms used by the CNS to control behavior and maintain homeostasis (Cone et al., 2003).

Neurosteroids: Steroids manufactured in the brain. Enzymes permit the brain to express steroids in the neurons and glia of the central and peripheral nervous systems. Neurosteroids modulate neurotransmitter receptors for GABA and NMDA and numerous other functions (Mellon, 2003).

Neurotransmitter: A substance released from the axon terminal of a presynaptic neuron, which travels across the synaptic cleft and excites or inhibits the target neuron. They are the active agent in widespread systems (Dunn, 1996). The same substances may be expressed and be physiologically active on a systemic basis. For example, the cholinergic and noradrenergic systems are not correlated with traditional functional systems. Agonists influence synapses in a variety of functional systems; a set of which could be defined as a *pharmacological system*. Thus, if a drug is targeted at a particular system (e.g., the cholinergic system of Alzheimer disease), it may produce "side effects" (Swanson, 2003).

Second messengers: Large molecules within the neuronal membrane or adjacent to its inner surface that modulate or amplify external signals from neurotransmitters and other mediators. Second messengers are vulnerable to the shear forces of neurotrauma, with second messenger systems amplified up to 200 times, downregulated, or deactivated (Bullock & Nathoo, 2005).

Neurotrophins: Polypeptide growth factors that influence the proliferation, differentiation, survival, and death of neuronal and nonneuronal cells (Lee & Chao, 2005).

Nicotinic and muscarinic receptors: These are two varieties of acetylcholine receptors. Each come in different subtypes. In both the sympathetic nervous system (SNS) and parasympathetic nervous system (PSNS), synaptic transmission between the preganglionic and postganglionic neurons is mediated by acetylcholine. Since the various types have heterogenous distribution among tissues, it is difficult to predict the effects of applying acetylcholine to a particular tissue (Richerson, 2005).

NMDA-Glutamate receptor: Nearly all neurons respond to glutamate stimulation, among the receptors of which are *N*-methyl-D-aspartate (NMDS) (Knapp, 2003). Glutamate is the major excitatory neurotransmitter. It binds to several glutamate receptors (GluR), of which NMDA is one. Glutamate is critical for normal nervous system function, but is toxic to neurons (*excitotoxicity*) at abnormally high concentrations if the GluRs are excessively activated (Martin, 2002).

Oligodendrocytes: Form and maintain myelin (i.e., the insulation covering most large axons in the CNS).

Osmolality; osmosis: The concentration of osmotically active particles in solution; diffusion across a membrane from a solution of lesser concentration to one of greater solute concentration.

Osmolarity, osmosis, osmostats: *Osmolarity* is the concentration of osmotically active particles in solution (e.g., the blood or cerebrospinal fluid). Alterations can occur due to posttraumatic endocrine dysfunction (see below). *Osmosis* involves the diffusion of substances through a cellular membrane from regions of lesser to greater concentration or of greater to lesser concentration. Thus, the process involves the adjustment of the concentration of solutes in such fluids as blood. The level of water volume affects concentration. Therefore, the authors of this book are concerned with the actions of adrenal cortical mineralocorticoids such as aldosterone. *Osmostats* are regulatory centers that control the osmolality of extracellular fluids. They are outside the blood-brain barrier (BBB). Diseases of this system after head injury affecting the hypothalamus and pituitary stalk and gland include inappropriate vasopressin secretion (i.e., syndrome of inappropriate antidiuretic hormone [*SIADH*]) and diabetes insipidus (see below) (Robertson, 2001; Robinson & Verbalis, 2003).

Pacemaker: Measures the passage of time independently of actual periodic stimulation from its environment and timed biological events.

Paracrine: A chemical influence when locally secreted hormones act on cells in the immediate vicinity of their release.

Parenchyma: The distinguishing cells of the brain, supported by a connective tissue framework.

Paresthesis: Abnormal sensations in the affected area, usually difficult to describe. Patients often use metaphorical expressions like "crawling" or "running water" (Margoles, 1999b).

Partial pressure: This refers to the fraction of total gas pressure caused by a particular molecular species. For example, in pulmonary capillary blood, the efficiency of lung function is measured by comparing the partial pressures of O_2 and CO_2. The symbol for partial pressure is P(gas).

pH: A measure of hydrogen ion (H^+) concentration or activity. A neutral solution has pH 7; above it is alkaline (accepts a hydrogen ion) and below it is acid (can donate a hydrogen ion). A tenfold change represents a unit change of 1. This function and its level affect biochemical reactions and a variety of processes critical for homeostasis of both the entire body and individual cells. Its value is maintained within a narrow range. pH changes (caused by metabolism, respiration, etc.) are compensated for by alterations of other functions (Boron, 2005).

Phagocytosis: The process of ingesting particles (cell fragments; bacteria) by invagination of the membrane, surrounding it as part of a vesicle and placing it within the cell structure.

Plexus(es): A network of nerves involving spinal nerve roots from several segmental levels.

Primed: A cell or organism that has been exposed to a specific antigen and that mounts a rapid and heightened response upon a second exposure to this same antigen.

Receptors, ligands: Physiological effects are a not solely a function of substances such as hormones and neurotransmitters, but of the interaction of these substances with a variety of receptors (with subtypes) to create varied bodily responses far more widespread than the classical hormonal target. When hormones arrive, a cell with an appropriate receptor (target cell) in the cell surface, body, or nucleus triggers a cascade of signals (*second messengers*) that release preformed products that change cell function by altering gene-expression patterns (Kaye & Lightman, 2005). Some signal molecules (estrogens and other steroid hormones) diffuse through the cell membrane and enter the cell. Most signal molecules are too large and too polar to pass through the membrane. Their information is transmitted across the cell membrane without the molecules themselves entering the cell. A *ligand* is a molecule that binds to another molecule, for example, an antigen binding to an antibody, a hormone or neurotransmitter binding to a receptor, antigen and antibody, hormone and receptor, neurotransmitter and receptor. The ligand-receptor complex initiates a process that modifies the expression of specific genes by binding to control elements in the DNA. This information is called the *primary messenger*, which creates changes in the concentration of small molecules known as *second messengers*. The effects of second messengers are multiple: they may amplify the signal; when there are multiple signaling pathways utilizing the same second messenger, there is *cross talk* that may finely regulate cell activity or inappropriate cross talk that can cause the second messenger to be misinterpreted (Berg et al., 2002, pp. 396–397, 880).

Neuroendocrine: An action on distant organs via the neurotransmitter mode. Neuroendocrine hormones act directly on the CNS to influence and modify behavior.

Releasing factors: Chemicals secreted in the hypothalamus that stimulate secretion of anterior pituitary (AP) hormones as well as hormones released by the anterior pituitary that directly stimulate hormonal glands (trophic hormones) that in turn stimulate target endocrines to secrete hormones: ACTH (corticotropin); growth hormone releasing hormone (GHRH); corticotropin-releasing hormone (CRH); thyroid stimulating hormone (TSH); gonadotropin-releasing hormone (GnRH); somatostatin; dopamine; and thyrotropin-releasing hormone (TRH).

Romberg test (eye closed-tandem and sign): A test of balance reflecting bilateral vestibular loss. Standing heel to toe with eyes closed for six seconds or more (inversely to age). The examiner also considers posterior column loss (proprioception) and cerebellar and basal ganglia dysfunction (Hain & Micco, 2003).

Signal transduction: The process by which a message carried by a hormone ultimately affects the rate that a regulatory enzyme in a cell enhances the passage of a substance from one state to another in a physiological pathway (e.g., turnover rate, level of synthesis, etc.) (Marks et al., 1996, pp. 381–382). Hormones initiate their actions on target cells by binding to specific receptors on the cell membrane (see above). Hormones reach their target in various ways. They may enter the cell and move directly into the cell nucleus. They may bind to the receptor to send a second messenger (see above) to intracellular enzymes. They may also pass through the "gated" ion channels into the cell. The variety of a particular receptor and the nature of the tissue determine the physiological response and side-effects of medications, and so forth. A particular hormone can react to different receptors in varied tissues, instigating a variety of effects.

Somatotopic: Neural organization organized by particular areas of the body (e.g., motor cortex).

Stance: The posture and general orientation of a person standing.

Steroid hormones: Have a similar core composed of four cyclical carbon-based rings united to each other. Steroid hormones are a family of hormone molecules derived from cholesterol, and can easily cross lipid membranes. Cholesterol is the precursor of bile acids, adrenocortical hormones, sex hormones, D vitamins and other biologically active substances (mineralocorticoids, such as aldosterone), androgens, and estrogens.

Strains and sprains: Both involve the muscles, ligaments, or tendons, and commonly occur after *whiplash* injury. *Strain* refers to overworking or overstretching due to excessive effort or undue force. There may be localized pain and swelling. *Sprain* refers to overworking with a partial tear. There is usually localized pain and swelling in the area of the tear. Sprain also refers to the tissues about a joint in which some of the fibers of a supporting ligament that attaches the bones are ruptured, that is, torn away from the bone, but the integrity of the ligament remains intact (Margoles, 1999a, 1999b).

Syncope: A relatively sudden nonepileptic loss of consciousness due to a global diminution of brain metabolism, causing decrease in brain perfusion and oxygen delivery. Presyncope and near syncope often include lightheadedness. It could be caused by rising to a standing position, the loss of sympathetic stimulation after urinating, and cutaneous vasodilation from a warm sleeping environment.

Syncopes were first described in the hypothalamus and pituitary gland in connection with endocrine functions, but have also been found in virtually all regions of the brain and spinal cord. Many of them, produced in the gut, are also produced by the brain. Neuropeptide transport systems can be either unidirectional or bidirectional (i.e., transporting peptides in and out of the brain). Systematically administered peptides have potent central effects and individual types can penetrate the BBB. They act on specific peptidergic receptors, producing specific physiological functions (e.g., regulation of reproduction, growth, food and water intake; electrolyte balance; control of respiratory, cardiovascular, and intestinal functions; and modulation of sensibility and emotions) (Neilan & Pasternak, 2003).

Torso organization: The body axis is formed from a series of *segments* or *somites* that develop into vertebrae and spinal nerves (sensory and motor).

Transcription and translation: DNA stores genetic information in cells concerning the sequence of proteins in cells that are involved in essentially all of life's processes. By the process of transcription, the genetic information is converted into ribonucleic acid (RNA). The subsequent conversion into protein is known as *translation* (Berg et al., 2002, pp. 4–6, 118).

Trophic factors: These are biologically active substances that support growth, differentiation, and survival of specific tissues after injury. Their production can be induced by inflammatory stimuli, ischemia, or mechanical damage.

Viscosity, viscoelasticity, elasticity: *Viscosity* is a characteristic of a liquid with resistance to flow or alteration of shape by a shearing force. Therefore, *viscoelasticity* refers to a viscous material that also shows elasticity (certain tissues). *Elasticity* is the property of returning to its original shape after being stretched, bent, compressed, or otherwise distorted.

References

Abbadie, C., & Pasternak, G. W. (2002). Endorphins and their receptors. In V. Ramachandran (Ed.), *Encyclopedia of the human brain* (Vol. 2, pp. 193–200). Boston: Academic Press.

Abelson, J. L., et al., (1996). Neuroendocrine responses to laboratory panic: Cognitive intervention in the Doxapram model. *Psychoneuroendocrinology, 21*, 375–390.

Aboukasm, A., et al. (1998). Retrospective analysis of the effects of psychotherapeutic interventions on outcomes of psychogenic nonepileptic seizures. *Epilepsia, 39*, 470–473.

Abu-Judeh, H. H., et al. (1999). SPECT brain perfusion imaging in mild traumatic brain injury without loss of consciousness and normal computed tomography. *Nuclear Medicine Communications, 20*, 505–510.

Ackerman, R. J., & Banks, M. E. (2002). Looking for the threads: Commonalities and differences. In Ferraro, F. R. (Ed.), *Minority and cross-cultural aspects of neuropsychological assessment* (pp. 387–415). Lisse, NL: Swets and Zeitlinger.

Acklin, M. W., & Bernat, E. (1987). Depression, alexithymia, and pain proneness disorder. Rorschach study. *Journal of Personality Assessment, 51*(3), 462–479.

Adair, J. C., et al. (2003). Anosognosia. In K. M. Heilman & E. Valenstein (Eds.), *Clinical neuropsychology, 4th Ed.* (pp. 185–213). New York: Oxford University Press.

Adams, J. H., et al. (1982a). Diffuse axonal injury due to nonmissile head injury in humans: An analysis of 45 cases. *Annals of Neurology, 12*, 557–563.

Adams, J. H., et al. (1982b). Neuropathology of acceleration-induced head injury in the subhuman primate. In R. G. Grossman & P. L. Goldenberg (Eds.), *Head injury* (pp. 141–149). New York: Raven Press.

Adams, J. H., et al. (1985). The contusion index: A reappraisal in human and experimental non-missile head injury. *Neuropathology and Applied Neurobiology, 11*(4), 299–308.

Adams, J. H., et al. (1986). Deep intracerebral (basal ganglia) haematomas in fatal non-missile head injury in man. *Journal of Neurology, Neurosurgery, and Psychiatry*, 49, 1039–1043.

Adams, J. H., et al. (1989). Diffuse axonal injury in head injury: Definition, diagnosis and grading. *Histopathology, 15*, 49–59.

Adams, R. D., & Victor, M. (1985). *Principles of neurology, 3rd Ed.* New York: McGraw Hill.

Adams, R. D., & Victor, M. (1989). *Principles of neurology, 4th Ed.* New York: McGraw Hill.

Ademec, R. E., et al. (1997). Blockade of CCC_B but not $CCCK_A$ receptors before and after the stress of predator exposure prevents lasting increases in anxiety-like behavior: Implications for anxiety associated with posttraumatic stress disorder. *Behavioral Neuroscience, 111*, 435–439.

Ader, R. (1996). Historical perspectives on psychoneuroimmunology. In H. Friedman, et al. (Eds.), *Psychoimmununology, stress, and infection* (pp. 1–24). Boca Raton, FL: CRC Press.

Ader, R. (2005). Integrative summary: On the clinical relevance of psychoneuroimmunology. In K. Vedhara & M. Irwin (Eds.), *Human psychoneuroimmunology* (pp. 344–349). New York: Oxford University Press.

Afdila, A., & Roselli, J. (2002). Acalculia and dyscalculia. *Neuropsychology Review, 12*, 179–231.

Afifi, A. K., & Bergman, R. A. (1998). *Functional neuroanatomy*. New York: McGraw-Hill.

Aicardi, J. (1992). *Diseases of the nervous system in childhood.* New York: Oxford University Press.

Akil, H., et al. (1999). Neuroendocrine systems I: Overview—thyroid and adrenal axes. In M. J. Zigmond et al. (Eds.), *Fundamental neuroscience* (pp. 1127–1150). San Diego: Academic Press.

Akiskal, H. S. (2005a). Mood disorders: Historical introduction and conceptual overview. In B. J. Sadock & V. A. Sadock (Eds.), *Comprehensive textbook of psychiatry, I, 8th Ed.* (pp. 1559–1574). Philadelphia: Lippincott Williams & Wilkins.

Akiskal, H. S. (2005b). Mood disorders: Clinical features. In B. J. Sadock & V. A. Sadock (Eds.), *Comprehensive textbook of psychiatry, I, 8th Ed.* (pp. 1611–1652). Philadelphia: Lippincott Williams & Wilkins.

Akiyama, K., et al. (2001). Bilateral cerebellar peduncle infarction caused by traumatic vertebral artery dissection. *Neurology, 56*, 693–694.

Alarcon, D., et al. (1997). Should there be a clinical typology of posttraumatic stress disorder? *Australian and New Zealand Journal of Psychiatry, 31*, 159–167.

Albert, M. S., & Lafleche, G. (1991). Neuropsychological testing of memory disorders. In T. Yanagihara & R. C. Petersen (Eds.), *Memory disorders* (pp. 165–193). New York: Marcel Dekker.

Aldenkamp, et al. (1999). Factors involved in learning problems and educational delay in children with epilepsy. *Child Neuropsychology,* 5, 130–136.

Alexander, M. P. (1997). Mild traumatic brain injury: A review of physiogenesis and psychogenesis. *Seminars in Clinical Neuropsychiatry, 2*(3), 177–187.

Alexander, M. P. (1998). In pursuit of proof of brain damage after whiplash injury. *Neurology, 51*, 336–340.

Alexander, M. P. (2003). Whiplash: Chronic pain and cognitive symptoms. *Neurology, 60*, 734.

Alexander, M. P., et al. (2007). Regional frontal injuries cause distinct impairments in cognitive control. *Neurology, 68*, 1515–1523.

Allan, S. M. (2000). The role of pro- and antiinflammatory cytokines in neurodegeneration. In A. Conti et al. (Eds.), *Neuroimmunomodulation: Perspectives at the new millennium* (Vol. 917, pp. 84–93). Annals of the NY Academy of Sciences.

Allen, S. F., et al. (2008). Early identification of children at risk for developing posttraumatic stress symptoms following traumatic injuries. *Journal of Psychological Trauma, 7*, 235–252.

Alper, K., et al. (1997). Dissociation in epilepsy and conversion nonepilepetic seizures. *Epilepsia,* 38, 991–997.

Altenmuller, E. O. (2001). How many music centers are in the brain? In R. Zatorre & I. Peretz (Eds.), *The biological foundations of music* (Vol. 930, pp. 273–280). Annals of the New York Academy of Sciences.

Altgassen, M., et al. (2007). Patients with Parkinson's disease can successfully remember to execute delayed intensions. *Journal of the International Neuropsychological Society, 13*, 888–892.

Alves, W. M., & Jane, K. A. (1990). Posttraumatic syndrome. In R. Youmans (Ed.) *Neurological Surgery,* vol. 3, pp. 2230–2242. New York: Saunders.

Amaral, D. G. (2000a). The anatomical organization of the central nervous system. In E. R. Kandel et al. (Eds.), *Principles of neural science, 4th Ed.* (pp. 317–336). New York: McGraw Hill.

Amaral, D. G. (2000b). The functional organization of perception and movement. In E. R. Kandel et al. (Eds.), *Principles of neural science, 4th Ed.* (pp. 347–348). New York: McGraw Hill.

Amaral, D. G., et al. (1992). Anatomical organization of the primate amygdaloid complex. In J. P. Aggleton (Ed.), *The amygdala: Neurobiological aspects emotion, memory, and mental dysfunction* (pp. 1–66). New York: Wiley-Liss.

Ambrosini, A. (2007). Occipital nerve stimulation for intractable cluster headache. *The Lancet, 369*, 1063–1064.

American Psychiatric Association (1985). *Diagnostic and statistical manual of mental disorders* (3rd Ed., Rev.). Washington, DC: Author.

American Psychiatric Association. (1987). *Diagnostic and statistical manual, 3rd Ed., Rev.* Washington, DC: Author.

American Psychiatric Association. (1994). *Diagnostic and Statistical Manual of Mental Disorders,* 4th ed. Washington DC.

American Psychiatric Association. (2000). *Diagnostic and statistical manual of mental disorders* (4th ed., Text Revision). Washington, DC: Author.

American Psychological Association. (2007). *APA Dictionary of Psychology*. Washington, D.C: Author.

Anastasi, A. (1988). *Psychological testing, 6th Ed.* New York: Macmillan.

Andersen, R. A., & Buneo, C.A. (2002). Intentional maps in posterior parietal cortex. *Annual Review of Neuroscience, 25*, 189–220.

Anderson, E. L., & Hall, E. D. (1994). Hydrolysis and free radical formation in central nervous system trauma. In S. K. Salzman & A. I. Faden (Eds.), *The neurobiology of central nervous system trauma* (pp. 131–138). New York: Oxford University Press.

Anderson, R., & McLean, J. (2005). Biomechanics of closed head injury. In P. L. Reilly & R. Bullock (Eds.), *Head injury, 2nd Ed.* (pp. 26–40). New York: Oxford University Press.

Anderson, V., & Catroppa, C. (2005). Recovery of executive skills following paediatric traumatic brain injury (TBI): A 2 year follow-up. *Brain Injury, 9*, 459–470.

Anderson, V. A., et al. (1997). Predicting recovery from head injury in young children. A prospective analysis. *Journal of the International Neuropsychological Society, 3*, 568–580.

Anderson, V. A., et al. (2005a). Identifying factors contributing to child and family outcome 3 months after traumatic brain injury in children. *Journal of Neurology, Neurosurgery, and Psychiatry*, *76*, 401–408.

Anderson, V., et al. (2005b). Prefrontal lesions and attentional skills in childhood. *Journal of the International Neuropsychological Society*, *11*, 817–831.

Andreasen, N. C., et al. (1999). The cerebellum plays a role in conscious episodic memory retrieval. *Human Brain Mapping, 8*, 226–234.

Andreassi, J. L. (2000). *Psychophysiology: Human behavior and physiological response, 4th Ed.* Mahwah, NJ: Lawrence Erlbaum.

Angevine, J. B. (2002). Nervous system, organization of. In V. S. Ramachandran (Ed.), *Encyclopedia of the human brain* (Vol. 3, pp. 313–371). Boston: Academic Press.

Ansbacher, H. L., & Ansbacher, R. R. (Eds.). (1956). *The individual psychology of Alfred Adler: A systematic presentation in selections from his writings*. New York: Basic Books.

Antoni, M. H. (2005). Behavioural interventions and psychoneuroimmunology. In K. Vedhara & M. Irwin (Eds.), *Human psychoneuroimmunology* (pp. 285–318). New York: Oxford University Press.

Appels, A. (2000). In G. Fink (Ed.), *Encyclopedia of stress* (Vol. 3, pp. 598–605). San Diego, CA: Academic Press.

Aram, D. M., & Ekelman, B. L. (1988). Scholastic aptitude and achievement among children with unilateral brain lesions. *Neuropsychologia, 26*, 903–916.

Arbour, N. (2005). In E. De Vries & A. Prat (Eds.), *The blood-brain barrier and its microenvironment* (pp. 481–504). New York: Taylor & Francis.

Ardila, A., & Rosselli, R. Acalculia and dyscalculia. *Neuropsychology Review,* 12, 4, 179–231.

Armario, A. et al., (1996). Acute stress markers in humans: Response of plasma glucose cortisol and prolactin to two examinations differing in the anxiety they provoke. *Psychoneuroendocrinology*, *21*, 17–24.

Armario, A., et al. (2004). Long term effects of a single exposure to immobilization on the hypothalamic-pituitary-adrenal axis: Neurobiologic mechanisms. In K. Pacak et al. *Stress: Current neuroendocrine and genetic approaches* (Vol. 1018, pp. 162–172). Annals of the New York Academy of Sciences.

Armstead, W. M., & Kurth, C. D. (1994). Different cerebral hemodynamic responses following fluid percussion brain injury in the newborn and juvenile pig. *Journal of Neurotrauma,* 14, 487–497.

Armstrong, J. G., & Loewenstein, R. J. (1990). Characteristics of patients with multiple personality and dissociative disorders on psychological testing. *The Journal of Nervous and Mental Disease, 178*, 454.

Arnold, S. E., et al. (1998). Cellular and molecular neuropathology of the olfactory epithelium and central olfactory pathways in Alzheimer's disease and schizophrenia. In C. Murphy (Ed.), *Olfaction and taste XII* (Vol. 855, pp. 762–775). Annals of the New York Academy of Sciences.

Aron, D. C., et al. (2004). Hypothalamus and pituitary gland. In F. S. Greenspan & D. G. Gardner (Eds.), *Basic and clinical endocrinology, 7th Ed.* (pp. 106–175). New York: Lange Medical Books.

Aron, D. C., et al. (2007a). Hypothalamus and pituitary gland. In F. S. Greenspan & D. G. Gardner (Eds.), *Basic and clinical endocrinology, 8th Ed.* (pp. 101–156). New York: Lange Medical Books.

Aron, D. C., et al. (2007b). Glucocorticoids and adrenal androgens. In F. S. Greenspan & D. G. Gardner (Eds.), *Basic and clinical endocrinology, 8th Ed.* (pp. 346–395). New York: Lange Medical Books.

Aronson, A. E. (1980). Organic voice disorders: Neurologic disease. In A. E. Aronson (Ed.), *Clinical voice disorders: An interdisciplinary approach* (pp. 77–122). New York: Thieme-Stratton.

Arvin, K. L., & Holtzman, D. M. (2003). Developmental neurotrophins. In M. J. Aminoff & R. B. Daroff (Eds.), *Encyclopedia of the neurological sciences* (Vol. 3, pp. 626–628). Boston: Academic Press.

Arzy, S., et al. (2006). Induction of an illusory shadowy person. *Nature*, *443*, 287.

Asarnow, R. B., et al. (1991). Behavior problems and adaptive functioning in children with mild and severe closed head injury. *Journal of Pediatric Neurology*, *5*, 543–555.

Ashman, T. A., et al. (2007). Objective measurement of fatigue following traumatic brain injury. *Journal of Head Trauma Rehabilitation, 23*, 33–40.

Astrocyte mitochondrial mechanisms of ischemic brain injury and neuroprotection.

Atteberry-Bennett, J., et al. (1986). The relationship between behavioral and cognitive deficits, demographics, and depression in patients with minor head injuries. *International Journal of Clinical Neuropsychology, 8*, 114–117.

Attie, K. M., et al. (1990). The pubertal growth spurt in eight patients with true precocious puberty and growth hormone deficiency: Evidence for a direct role of sex steroids. *Journal of Clinical Endocrinology & Metabolism, 71,* 975–983.

Auer, R. N., et al. (1994). Delayed symptoms and death after minor head trauma with occult vertebral artery injury. *Journal of Neurology, Neurosurgery, and Psychiatry, 57*, 500–502.

Augustyn, M., & Zuckerman, B. (1998). Normal behavioral development. In C. E. Coffey & R. A. Brumback (Eds.), *Textbook of neuropsychiatry* (pp. 117–139). Washington D.C.: American Psychiatric Press.

Autret, A., et al. (2001). Sleep and brain lesions: A critical review of the literature and additional new cases. *Neurophysiological Clinics, 31*, 356–375.

Avazov, V. N., et al. (2000). The syndrome of posttraumatic dysregulation of the neuroendocrine-immune system and the tasks of health-resort therapy (Russian; Abstract). *Vaprosy Kurortologii, Fizioterapii I Lechebnoi Fizicheskoi Kulltury, 3,* 22–24.

Ayalon, L., et al. (2007). Circadian rhythm sleep disorders following mild traumatic brain injury. *Neurology, 68*, 1136–1140.

Ayr, L. K., et al. (2009). Dimensions of postconcussive symptoms in children with mild traumatic brain injuries. *Journal of the International Neuropsychological Society, 15*, 19–30.

Bach-Y-Rita, P. (2002). Recovery from brain damage. In V. S. Ramachandran (Ed.), *Encyclopedia of the human brain* (Vol. 1, pp. 481–491). Boston: Academic Press.

Baggaley, M. R., & Rose, J. (1990). Case report. *British Journal of Psychiatry,* 156, 156.

Baguley, I., et al. (2000). Long-term mortality trends in patients with traumatic brain injury. *Brain Injury, 14*, 505–512.

Bailey, B. N., & Gudeman, S. K. (1989). Minor head injury. In D. P. Becker and S. K. Gudeman (Eds.) *Textbook of Head Injury,* pp. 308–318. Philadelphia: Saunders.

Bailey, C. H., & Kandel, E. R. (2004). Synaptic growth and the persistence of long-term memory: A molecular perspective. In M. S. Gazzaniga (Ed.), *The cognitive neurosciences, III* (pp. 647–663). Cambridge, MA: Massachusetts Institute of Technology.

Bakay, L., & Glasauer, F. E. (1980). *Head Injury.* Boston: Little, Brown.

Baker, G. A. (2006). Depression and suicide in adolescents with epilepsy. *Neurology*, *66*(6), S5–S12.

Ball, S., & Shekhar A. (1997). Basilar artery response to hyperventilation in panic disorder. *American Journal of Psychiatry, 154*, 1603–1604.

Balleine, B. W., et al. (2007). Introduction: Current trends in decision making. In B. W. Balleine et al. (Eds.), *Reward and decision making in corticobasal ganglia networks* (Vol. 1104, pp. xi–xiii). Annals of the New York Academy of Sciences.

Ballenger, J. C. (1995). Benzodiazepines. In A. F. Schatzberg & C. B. Nemeroff (Eds.), *Textbook of psychopharmacology* (pp. 215–230). Washington D.C.: American Psychiatric Press.

Baloh, R. W. (1998). *Dizziness, hearing loss, and tinnitus*. Philadelphia: F. A. Davis Co.

Baloh, W. W. (2003). Vestibular system. In M Aminoff & R. B. Daroff (Eds.), *Encyclopedia of the neurological science* (Vol. 4, pp. 642–646). Boston: Academic Press.

Bambrick, L., et al. (2004). Astrocyte mitochondrial mechanisms of ischemic brain injury and neuroprotection. *Neurochemistry Research, 29*(3), 601–608.

Bamdad, M. J., et al. (2003). Functional assessment of executive abilities following traumatic brain injury. *Brain Injury, 17,* 1011–1020.

Bancaud, J., et al. (1994). Anatomical origin of *déjà vu* and vivid memories in human temporal lobe epilepsy. *Brain,* 117, 71–91.

Bandak, F. A. (1996). On the mechanics of impact neurotrauma: A review and critical synthesis. *Journal of Neurotrauma,* 12, 635–649.

Bandura, A. (1982). Self-efficacy mechanism in human agency. *American Psychologist*, *37*, 122–147.

Banerjee, S., & Bhat, M. A. (2007). Neuron-glial interactions in blood-brain barrier formation. *Annual Review of Neuroscience, 30*, 235–258.

Banks, W. (2001). Cytokines, CVOs, and the blood-brain barrier. In Ader, R., Felten, D. L., & Cohen, N. *Psychoneuroimmunology,* 3rd Ed. (Vol. 1, pp. 483–497). San Diego: Academic Press.

Barabam, J. M. (2005). Intraneuronal signaling pathways. In B. J. Sadock & V. A. Sadock (Eds.), *Comprehensive textbook of psychiatry, I, 8th Ed.* (pp. 89–98). Philadelphia: Lippincott Williams & Wilkins.

Baram, T. Z., et al. (1997). Developmental neurobiology of the stress response: Multilevel regulation of corticotropin-releasing hormone function. In B. E. Beckwith et al. (Eds.), *Neuropeptides in development and aging* (Vol. 814, pp. 262–265). Annals of the New York Academy of Science.

Barinaga, M. (1996). *Science*, *272*, 482–483.

Barker, L. H., et al. (1999). Polysubstance abuse and traumatic brain injury: Quantitative magnetic resonance imaging and neuropsychological outcome in older adolescents and young adults. *Journal of the International Neuropsychological Society, 5*, 598–608.

Barkley, R. A. (1988). Attention. In M. G. Tramontana & S. R. Hooper (Eds.), *Assessment issues in child neuropsychology* (pp. 145–176). New York: Plenum Press.

Barnes, M. A., & Dennis, M. (2001). Knowledge-based inferencing after childhood head injury. *Brain and Language, 76*, 253–265.

Barnes, P. J. (2003). Autonomic control of the airways. In C. L. Bolis et al. (Eds.), *Handbook of the autonomic nervous system in health and disease* (pp. 439–461). New York: Marcel Dekker.

Barr, M. L., & Kiernan, J. A. (1988). *The human nervous system, 5th Ed.* New York: Harper and Row.

Barr, W. B. (2007). Recovering from mild traumatic brain injury: What psychology has learned from sports concussion research. *NYS Psychologist, 19*(5), 24–28.

Barrett, E. J. (2005a). Organization of endocrine control. In W. F. Boron & E. L. Boulpaep (Eds.), *Medical physiology, up-dated edition* (pp. 1005–1021). Philadelphia: Saunders.

Barrett, E. J. (2005b). The endocrine regulation of growth. In W. Boron & E. Boulpaep (Eds.), *Medical physiology, up-dated edition* (pp. 1023–1034). Philadelphia: Saunders.

Barrett, E. J. (2005c). The adrenal gland. In W. Boron & E. Boulpaep (Eds.), *Medical physiology, up-dated edition* (pp. 1049–1065). Philadelphia: Saunders.

Barrett, E. J., & Barrett, P. (2005). The parathyroid glands and vitamin D. In W. F. Boron & E. L. Boulpaep (Eds.), *Medical physiology, up-dated edition* (pp. 1086–1102). Philadelphia: Saunders.

Barry, C. T., et al. (1996). Validity of neurobehavioral symptoms reported in children with traumatic brain injury. *Child Neuropsychology,* 2, 213–226.

Barry, E., et al. (1998). Nonepileptic posttraumatic seizures. *Epilepsia, 39,* 427–431.

Barsky, A. J. (1989). Somatoform disorders. In H. I. Kaplan & B. J. Sadock (Eds.), *Comprehensive textbook of psychiatry, 5th Ed.* (Vol. I, pp. 1009–1027). Philadelphia: Lippincott, Williams & Wilkins.

Bartolo, A., et al. (2008). Cognitive approach to the assessment of limb ataxia. *The Clinical Neuropsychologist, 22,* 27–45.

Basbaum, A. I., & Jessell, T. M. (2000). The perception of pain. In E. R. Kandel et al. (Eds.), *Principles of neural science, 4th ed.* (pp. 472–491). New York: McGraw Hill.

Basso, A., et al. (1994). Differential influence of a thymic extract on α- and β-adrenoceptors of mouse brain cortex. In N. A. Fabris, et al. (Eds.), *Immunomodulation: The state of the art* (Vol. 741, pp. 124–128). Annals of the New York Academy of Sciences.

Bastian, A. J. (2002). Cerebellar limb ataxia: Abnormal control of self-generated and external forces. In S. M. Highstein & W. T. Thach (Eds.), *The cerebellum: Recent developments in cerebellar research* (Vol. 978, pp. 16–27). Annals of the New York Academy of Sciences.

Bastian, A. J., et al. (1999). Cerebellum. In M. J. Zigmond et al. (Eds.), *Fundamental neuroscience* (pp. 973–992). San Diego: Academic Press.

Batjer, H. H., et al. (1993). Intracranial and cervical vascular injuries. In P. R. Cooper (Ed.) *Head Injury,* 3rd ed., pp. 373–403. Baltimore: Williams and Wilkins.

Baumann, C. R., et al. (2005). Hypocretin-2 (orexin A) deficiency in acute traumatic brain injury. *Neurology, 65,* 147–149.

Baumann, N., & Pham-Dinh, D. (2002). Astrocytes. In V. S. Ramachandran (Ed.), *Encyclopedia of the human brain* (Vol. 1, pp. 251–268). Boston: Academic Press.

Bawden, H. N., et al. (1985). Speeded performance following head injury in children. *Journal of Clinical and Experimental Neuropsychology,* 7, 39–54.

Baxter, J. D., et al. (2004). Introduction to endocrinology. In F. S. Greenspan & D. G. Gardner (Eds.), *Basic and clinical endocrinology, 7th Ed* (pp. 1–37). New York: Lange Medical Books.

Baynes, K. (2002). Corpus callosum. *Encyclopedia of the neurological sciences* (Vol. 2, pp. 51–64). Boston: Academic Press.

Bazarian, J., et al. (2000). Minor head injury: Predicting follow-up after discharge from the emergency department. *Brain Injury, 14,* 285–294.

Bazarian, J. J., et al. (2007). Diffusion tensor imaging detects clinically important axonal damage after mild traumatic brain injury: A pilot study. *Journal of Neurotrauma, 24,* 1447–1450.

Beals, R. K., & Hickman, N.W. (1972). Industrial injuries of the back and extremities. *The Journal of Bone and Joint Surgery, 54*(A), 1593–1611.

Beato, M., & Klug, J. (2000). Steroid hormone receptors. In G. Fink (Ed.), *Encyclopedia of stress* (Vol. 3, pp. 495–504). San Diego, CA: Academic Press.

Becker, D. P. (1989). Common themes in head injury. In D. P. Becker and S. K. Gudeman (Eds.) *Textbook of Head Injury,* pp. 1–22. Philadelphia: Saunders.

Beckham, J. C., et al. (1998). Trail marking test performance in Vietnam combat veterans with and without posttraumatic stress disorder. *Journal of Traumatic Stress,* 11, 811–819.

Beckham, J. C., et al. (1998). Health status, somatization, and severity of posttraumatic stress disorder in Vietnam combat veterans with posttraumatic stress disorder. *American Journal of Psychiatry,* 155, 1565–1569.

Bedell, G., et al. (2008). Measuring participation of school-aged children with traumatic brain injuries: Considerations and approaches. *Journal of Head Trauma Rehabilitation, 23,* 220–229.

Beers, S. H. (1992). Cognitive effects of mild head injury in children and adolescents. *Neuropsychology Review, 3,* 281–320.

Beetar, J. T., et al. (1997). Sleep and pain complaints in symptomatic traumatic brain injury and neurologic populations. *Archives of Physical Medicine & Rehabilitation, 77,* 1298–1302.

Begali, V. (1987). *Head injury in children and adolescents.* Brandon, VT: Clinical Psychology Publishing Co.

Begaz, T., et al. (2006). Serum biochemical markers for post-concussion syndrome in patients with mild traumatic brain injury. *Journal of Neurotrauma, 23*, 1201–1210.

Beggs, J. M., et al. (1999). Learning and memory: Basic mechanisms. In M. J. Zigmond et al. (Eds.) *Fundamental Neuroscience,* pp. 1411–1454. San Diego: Academic Press.

Belanger, H. G., et al., (2005). Factors moderating neuropsychological outcomes following mild traumatic brain injury: A meta-analysis. *Journal of the International Neuropsychological Society, 11,* 215–227.

Belanger, H. G., et al. (2009). Cognitive sequelae of blast-related versus other mechanisms of brain trauma. *Journal of the International Neuropsychological Society, 15*, 1–8.

Bell, W. L., et al. (1990). Acquired aprosodia in children. *Journal of Child Neurology, 5*, 19–26.

Bellinger, D. L., et al. (1992). Neural-immune interactions. In A. Tasman & M. R. Riba (Eds.), *Annual review of psychiatry* (Vol. 11, pp. 127–144). Washington D.C.: American Psychiatric Press.

Bellinger, D. L., et al. (2001). Innervation of lymphoid organs: Association of nerves with cells of the immune system and their implications in disease. In R. Ader, et al. (Eds.), *Psychoneuroimmunology, 3rd Ed.* (Vol. I., pp. 55–113). San Diego: Academic Press.

Bellisle, F. (1992). Effects of monosodium glutamate on human food palatability. In C. Murphy (Ed.). *Olfaction and taste XII* (pp. 438–331). Annals of The New York Academy of Sciences, Vol. 855.

Ben-Yishay, Y., & Prigatano, G. (1990). Cognitive remediation. In M. Rosenthal et al. (Eds.), *Rehabilitation of the adult and child with traumatic brain injury, 2nd Ed.* (pp. 392–409). Philadelphia: F. A. Davis.

Benarroch, E. E. (1997a). *Central autonomic network: functional organization and clinical considerations.* Armonk, N.Y.: Future Publishing Co.

Benarroch, E. E. (1997b). The central autonomic network. In P. A. Low (Ed.), *Clinical autonomic disorders: Evaluation and management, 2nd Ed.* (pp. 17–23). Philadelphia: Lippincott-Raven.

Benarroch, E. E. (2007a). Sodium channels and pain. *Neurology, 68*, 152–154.

Benarroch, E. E. (2007b). Enteric nervous system. *Neurology, 69*, 1953–1957.

Benedek, E. P. (1985). Children and psychic trauma: A brief review of contemporary thinking. In S. Eth & R. S. Pynoss (Eds.), *Post-traumatic stress disorder in children* (pp. 1–16). Washington, DC: American Psychiatric Press.

Benson, D. F. (1985). Language in the left hemisphere. In D. F. Benson & E. Zaidel (Eds.), *The dual brain: Hemispheric specialization in humans* (pp. 193–203). New York: Guilford.

Benson, D. F. (1986). Interictal behavior disorders in epilepsy. In R. M. Restak (Ed.), *Psychiatric clinics of North America* (pp. 283–292). Philadelphia: W.B. Saunders.

Benson, D. F. (1993). Aphasia. In K. M. Heilman & E. Valenstein (Eds.), *Clinical Neuropsychology, 3rd Ed.* (pp. 17–36). New York: Oxford University Press.

Benton, A. (1985a). Body schema disturbances: Finger agnosia and right-left disorientation. In K. M. Heilman and E. Valenstein (Eds.) *Clinical Neuropsychology,* 2nd ed., pp. 115–129. New York: Oxford University Press.

Benton, A. (1991). Prefrontal injury and behavior in children. *Developmental Neuropsychology, 7*, 275–281.

Benton, A., & Sivan, A. B. (1993). Disturbances of the body schema. In K. M. Heilman and E. Valenstein (Eds.) *Clinical Neuropsychology,* 3rd ed., pp. 123–140. New York: Oxford University Press.

Benvenga, S., et al. (2000). Hypopituitarism secondary to head trauma. *The Journal of Clinical Endocrinology and Metabolism, 85*, 1353–1361.

Berczi, I., & Szentivanhi, A. (1996). The pituitary gland, psychoneuroimmunology and infection. In H. Friedman et al. (Eds.), *Psychoimmunology, stress, and infection* (pp. 79–107). Boca Raton: CRC Press.

Beresford, T. P., et al. (2005). Reduction of affective lability and alcohol use following traumatic brain injury: A clinical pilot study of anticonvulsant medications. *Brain Injury, 19*, 309–313.

Berg, J. M., et al. (2002). *Biochemistry, 5th Ed.* New York: W. H. Freeman.

Berg, K. A., et al. (1998). Interactions between effectors linked to serotonin receptors. In G. R. Martin et al. (Eds.), *Advances in serotonin receptor research* (Vol. 861, pp. 111–120). Annals of the New York Academy of Science.

Berg, R. (1986). Neuropsychological effects of closed-head injury in children. In J. E. Obrzut & G. W. Hynd (Eds.), *Child neuropsychology* (Vol. 2, pp. 113–135). New York: Academic Press.

Berger, R. P. (2006). The use of serum biomarkers to predict outcome after traumatic brain injury in adults and children. *Journal of Head Trauma Rehabilitation, 21*, 315–333.

Berger, R. P., et al. (2002). Neuron specific enolase and S100B in cerebrospinal fluid after severe traumatic brain injury in infants and children. *Pediatrics, 109*(2). Retrieved from http://www.pediatrics.org/cgi/content/full/10o9/2/e31.

Berger, R. P., et al. (2007). Serum biomarker concentrations and outcome after pediatric traumatic brain injury. *Journal of Neurotrauma, 24*, 1793–1801.

Bergland, M. M., & Thomas, K. R., (1991). Psychosocial issues following severe head injury in adolescence: Individual and family perceptions. *Rehabilitation Counseling Bulletin, 35*(1), 5–22.

Berkhoff, M., et al. (1998). Developmental background and outcome in patients with nonepileptic versus epileptic seizures: A controlled study. *Epiliepsia,* 39, 463–469.

Berne, R. M., & Levy, M. N. (1998). *Physiology, 4th Ed.* St. Louis: Mosby.

Besedovsky, H. O., & Del Rey, A. (2001). Cytokines as mediators of central and peripheral immune-neuroendocrine interactions. In Ader, R., Felten, D. L., & Cohen, N. *Psychoneuroimmunology*, 3rd ed. (Vol. 1, pp. 1–17). San Diego: Academic Press.

Beutler, L. E., et al. (1986). Inability to express intense affect: A common link between depression and pain? *Journal of Consulting and Clinical Psychology, 54*, 752–759.

Beversdorf, B. Q. (2007). Pharmacotherapy of aphasia. *Journal of Head Trauma Rehabilitation, 22*, 65–66.

Bhatia, K. P., & Marsden, C. D (1994). The behavioural and motor consequences of focal lesions of the bsal ganglia in man. *Brain,* 117, 859–876.

Bhatnagar, S., et al. (2004). Regulation of chronic stress induced changes in hypothalamic-pituitary-adrenal activity by the basolateral amygdala. In R. Yehuda & B. McEwen (Eds.), *Behavioral stress responses* (Vol. 1032, pp. 301–303). Annals of The New York Academy of Sciences.

Bhattacharyya, A., & Svendsen, C. (2003). Neurotrophins. In M. J. Aminoff & R. B. Daroff (Eds.), *Encyclopedia of the neurological sciences* (Vol. 3, pp. 621–623). Boston: Academic Press.

Bianchini, K. J., et al. (2005). On the diagnosis of malingered pain-related disability lessons from cognitive malingering research. *The Spine Journal, 5*, 404–417.

Biberthaler, P., et al. (2000). Influence of alcohol exposure on S1000B serum levels. *Acta Neurochirurgica-Supplement, 76*, 177–179.

Biberthaler, P., et al. (2001). Evaluation of S-100B as a specific marker for neuronal damage due to minor head trauma. *World Journal of Surgery, 25*, 93–97.

Biberthaler, P., et al. (2002). Rapid identification of high-risk patients after mild head trauma (MHT) by assessment of S-100B: Ascertainment of a cut-off level. *European Journal of Medical Research, 7*, 164–170.

Biberthaler, P., et al. (2004). Identification of high risk patients after minor craniocerbral trauma: Measurement of nerve tissue protein S 100. (German, Abstract). *Unfallchirurgie, 107*, 197–202.

Biernacki, K., et al. (2005). Immune functions of brain endothelial cells. In E. De Vries & A. Prat (Eds.), *The blood-brain barrier and its microenvironment* (pp. 167–191). New York: Taylor & Francis.

Bigler, E. D. (1988). The role of neuropsychological assessment in relation to other types of assessment with children. In M. G. Tramontana and S. R. Hooper (Eds.) *Assessment Issues in Child Neuropsychology,* pp. 67–91. New York: Plenum Press.

Bigler, E. D., et al. (1999). Head trauma and intellectual status: Relation to qualitative magnetic resonance imaging findings. *Applied Neuropsychology,* 6, 217–225.

Bijur, P. E., et al. (1990). Cognitive and behavioral sequelae of mild head injury in children. *Pediatrics, 86*, 337–344.

Biller, J., & Love, B. L. (2004). Ischemic cerebovascular disease. In W. G. Bradley et al. (Eds.), *Neurology in clinical practice, 4th Ed.* (Vol. 2, pp. 1197–1249). Boston: Butterworth-Heinemann.

Binder, H. J. (2005). Organization of the gastrointestinal system. Integration of salt and water balance. In W. Boron & E. Boulpaep (Eds.), *Medical physiology, up-dated edition* (pp. 879–890). Philadelphia: Saunders.

Biondi, M. (2001). Effects of stress on immune functions: An overview. In R. Ader, et al. (Eds.) *Psychoneuroimmunology, 3rd Ed.* (Vol. I, pp. 189–223). San Diego: Academic Press.

Birmaher, B, & Williams, D. T. (1994). Children and adolescents. In J. M. Silber et al. (Eds.) *Neuropsychiatry of Traumatic Brain Injury,* pp. 393–412. Washington DC: American Psychiatric Press.

Bisiach, D. E. (1992). Understanding consciousness: Clues from unilateral neglect and related disorders. In A. S. Milner & M. D. Rugg (Eds.), *The neuropsychology of consciousness* (pp. 114–137). San Diego: Academic Press.

Bisiach, E., & Geminiani, G. (1991). Anosognosia related to hemiplegia and hemianopia. In G. P. Prigatano & D. L. Schachter (Eds.), *Awareness of deficit after brain injury* (pp. 17–39). New York: Oxford University Press.

Bisiach, E., et al. (1986). Unawareness of disease following lesions of the right hemisphere: Anosognosia for hemiplegia and anosognosia for hemianopia. *Neuropsychologia, 24,* 471–482.

Black, P. (2006). Thrust to wholeness: The nature of self protection. *Review of General Psychology, 10*, 191–209.

Black, R. M. (1970). *The elements of paleontology*. New York: Cambridge University Press.
Blanchard, E. B., et al. (1997). Prediction of remission of acute posttraumatic stress disorder in motor vehicle accident victims. *Journal of Traumatic Stress*, *10*, 215–234.
Blanchard, E. B., & Hickling, E. J. (1997). *After the Crash.* Washington DC: American Psychological Association.
Blank, A. S. (1994). Clinical detection, diagnosis, and differential diagnosis of post-traumatic stress disorder. *Psychiatric Clinics of North America*, *17*(2), 351–381.
Blasiak, T. (2002). Eruption of neuronal activity in selected structures of the mammalian brain (Polish; Abstract). *Postepy Higieny I Medycyny Doswiadczalnej, 56*, 323–330.
Bleasel, A., & Kotagal, P. (1995). Paroxysmal nonepileptic disorders in children and adolescents. *Seminars in Neurology,* 15, 203–217.
Bleck, T. P. (2003). Levels of consciousness and attention. In C. G. Goetz (Ed.), *Textbook of clinical neurology, 2nd Ed.* (pp. 2–16). Philadelphia: Saunders.
Blessing, W. W. (2002). Brain stem. In V. S. Ramachandran (Ed.), *Encyclopedia of the human brain* (Vol. I, pp. 545–567). Boston: Academic Press.
Bloom, F. E. (2003). Fundamentals of neuroscience. In L. R. Squire et al. (Eds.), *Fundamental neuroscience, 2nd Ed.* (pp. 3–13). Boston: Academic Press.
Blumbergs, P. C. (2005). Pathology. In P. L. Reilly & R. Bullock (Eds.), *Head injury, 2nd Ed.* (pp. 41–72). New York: Hodder Arnold.
Blumstein, S. (1991). Phonological aspects of aphasia. In. M. T. Sarno (Ed.), *Acquired aphasia, 2nd Ed.* (pp. 151–180). San Diego: Academic Press.
Blumstein, S. E. (1988). Approaches to speech production deficits in aphasia. In F. Boller & J. Grafman (Eds.), *Handbook of neuropsychology* (Vol. I, pp. 349–365). New York: Elsevier Science.
Boake, C., et al. (2005). Diagnostic criteria for postconcussional syndrome after mild to moderate traumatic brain injury. *Journal of Neuropsychiatry and Clinical Neuroscience*, *17*, 350–356.
Bode, H. H., et al. (1996). Disorders of antidiuretic hormone homeostasis: Diabetes insipidus and SIADH. In F. Lifschitz (Ed.), *Pediatric endocrinology, 3rd Ed.* (pp. 731–751). New York: Dekker.
Boes, C. J., et al., (2004). Headache and other craniofacial pain. In W. G. Bradley et al. (Eds.), *Neurology in clinical practice, 4th Ed.* (Vol. 2, pp. 2055–2106). Boston: Butterworth-Heinemann.
Bogen, J. E. (1993). The callosal syndromes. In K. M. Heilman & E. Valenstein (Eds.), *Clinical neuropsychology, 3rd Ed.* (pp. 337–407). New York: Oxford University Press.
Bohnen, N. J., et al. (1992). Post-traumatic and emotional symptoms in different subgroups of patients with mild head injury. *Brain Injury, 6,* 481–487.
Bohnen, N. J., et al. (1995). The constellation of late post-traumatic symptoms of mild head injury patients. *Journal of NeuroRehabilitation*, *9*, 33–39.
Bohus, B., & Koolhaas, J. M. (1996). Psychoimmunology & psychobiology of parasitic infestation. In H. Friedman, T. W. Klein, & A. L. Friedman (Eds.), *Psychoimmununology, stress, and infection* (pp. 263–272). Boca Raton, FL: CRC Press.
Boll, T. J. (1982). Behavioral sequelae of head injury. In P. Cooper (Ed.), *Head injury* (pp. 363–375). Baltimore: William & Wilkins.
Boller, F., et al. (1995). Cognitive functioning in "diffuse" pathology: Role of prefrontal and limbic structures. In J. Grafman et al. (Eds.), *Structure and functions of the human prefrontal cortex* (Vol. 769, pp. 23–39). Annals of the New York Academy of Sciences.
Bondanelli, M., et al. (2007). Anterior pituitary function may predict functional and cognitive outcome in patients with traumatic brain injury undergoing rehabilitation. *Journal of Neurotrauma, 24*, 1687–1697.
Bonica, J. J., & Chapman, C. R. (1986). Biology, pathophysiology, and treatment of chronic pain. In P. A. Berger & H. K. H. Brodie (Eds.), *American handbook of psychiatry,* 2nd ed., Vol. VIII, *Biological psychiatry* (pp. 711–761). New York: Basic Books.
Bookheimer, S. (2002). Functional MRI of language: New approaches to understanding the cortical organization of semantic processing. *Annual Review of neuroscience*, *25*, 151–188.
Booth, R. (2005). Emotional disclosure and psychoneuroimmunology. In K. Vedhara & M. Irwin (Eds.), *Human psychoneuroimmunology* (pp. 319–341). New York: Oxford University Press.
Borgaro, S. R., et al. (2005). Subjective reports of fatigue during early recovery from traumatic brain injury. *Journal of Head Injury Rehabilitation, 20,* 416–425.
Born, J., & Wagner, U. (2004). Memory consolidation during sleep: Role of cortisol feedback. In R. Yehuda & B. McEwen (Eds.), *Behavioral stress responses* (Vol. 1032, pp. 198–201). Annals of The New York Academy of Sciences.

Bornstein, R. A. (1985). Normative data on intermanual differences on three tests of motor performance. *Journal of Clinical and Experimental Neuropsychology, 8*, 12–20.

Borod, J. C. (1993a). Emotion and the brain—anatomy and theory: An introduction to the special section. *Neuropsychology, 7*, 437–432.

Borod, J. C. (1993b). Central mechanisms underlying facial, prosodic, and lexical emotional expression: A review of neuropsychological studies and methodological issues. *Neuropsychology, 7*, 445–463.

Borodinsky, L. N. (2006). How fast can you go? *Nature, 440*, 158–159.

Boron, W. F. (2005). Acid-base physiology. In W. Boron & E. Boulpaep (Eds.), *Medical physiology, up-dated edition* (pp. 633–653). Philadelphia: Saunders.

Boscarino, J. A. (2004). Posttraumatic stress disorder and physical illness: Results from clinical and epidemiologic studies. In R. Yehuda & B. McEwen (Eds.), *Behavioral stress responses* (Vol. 1032, pp. 141–153). Annals of The New York Academy of Sciences.

Bosch, E. P., & Smith, B. E. (2004). Disorders of peripheral nerves. In W. G. Bradley et al. (Eds.), *Neurology in clinical practice, 4th Ed.* (Vol. 2, pp. 2299–2401). Boston: Butterworth-Heinemann.

Boschert, S. (2007). Assume resilience, not pathology, after disaster. *Clinical Psychiatry News, 35*(5), 1–2.

Bottjer, W. W., & Arnold, A. P. (1997). Developmental plasticity in neural circuits for a learned behavior. *Annual Review of Neuroscience, 20*, 459–481.

Bouloux, P. M. G., & Grossman, A. (1984). The involvement of opioid peptides in the catecholamine response to stress. In F. Fraioli, et al. (Eds.), *Opioid peptides in the periphery* (pp. 209–222). New York, Elsevier.

Boulpaep, E. L. (2005). Regulation of arterial pressure and cardiac output. In W. Boron & E. Boulpaep (Eds.), *Medical physiology, up-dated edition* (pp. 534–557). Philadelphia, Saunders.

Boulpaep, E. L., & Boron, W.F. (2003). Foundations of physiology. In W. F. Boron & E. L. Boulpaep (Eds.), *Medical physiology* (pp. 3–6). Philadelphia, Saunders.

Bovbjerg, D. H., & Stone, A. A. (1996). Psychological stress and upper respiratory illness. In H. Friedman, et al. (Eds.), *Psychoneuroimmunology, stress and infection* (pp. 195–213). Philadelphia: Lippincott-Raven.

Bowdler, N., et al. (1987). Fetal skull fracture and brain injury after a maternal automobile accident: A case report. *The Journal of Reproductive Medicine, 32*, 375–378.

Bowers, D., et al. (1993). The nonverbal affect lexicon: Theoretical perspectives from neuropsychological studies of affect perception. *Neuropsychology, 7*, 433–444.

Bowman, E. S. (1999). Nonepileptic seizures: Psychiatric framework, treatment, and outcome. *Neurology,* 53 (Suppl 2), S84-S88.

Boyd, T. A. (1988). Clinical assessment of memory in children. In M. G. Tramontana & S. R. Hooper (Eds.), *Assessment issues in child neuropsychology* (pp. 177–204). New York: Plenum Press.

Bradley, L. A. (2000). Pain complaints in patients with fibromyalgia versus chronic fatigue syndrome. *Current Review of Pain, 4*, 148–157.

Bradley, W. G., et al. (2004). The diagnosis of neurological disease. In W. G. Bradley et al. (Eds.), *Neurology in clinical practice, 4th Ed.* (Vol. 1, pp. 3–28). Boston: Butterworth-Heinemann.

Bradshaw, J. L. (1997). *Human evolution: A neuropsychological perspective*. East Sussex, UK: Psychology Press.

Bradvik, B., et al. (1991). Disturbances of speech prosody following right hemisphere infarcts. *Acta Neurologika Scandinavika, 84*, 114–126.

Braitenberg, V. (2002). In defense of the cerebellum. In S. M. Highstein & W. T. Thach (Eds.), *The cerebellum: Recent developments in cerebellar research* (Vol. 978, pp. 175–183). Annals of The New York Academy of Sciences.

Brandt, J. (1992). Detecting amnesia's imposters. In L. R. Squire & N. Butters (Eds.), *Neuropsychology of memory, 2nd Ed.* (pp. 151–157). New York: The Guilford Press.

Brauth, S. E., et al. (1997). Functional anatomy of forebrain vocal control pathways in the budgerigar (Melopsittacus undulatus). In C. S. Carter, et al. (Eds.), *The integrative neurobiology of affiliation* (Vol. 807, pp. 368–385). Annals of the New York Academy of Sciences.

Brazis, P. W., et al. (2001). *Localization in clinical neurology, 4th ed.* Philadelphia: Lippincott, Williams & Wilkins.

Breed, S. T., et al. (2004). The relationship between age and the self-report of health symptoms in persons with traumatic brain injury. *Archives of Physical Medicine & Rehabilitation, 85*(4 Suppl 2), 61–67.

Breedlove, S. M., & Hampson, E. (2002). Sexual differentiation of the brain and behavior. In J. B. Becker, et al. (Eds.), *Behavioral endocrinology, 2nd Ed.* (pp. 75–114). Cambridge, MA: MIT Press.

Bremner, J. D. (2005). Effects of traumatic stress on brain structure and function: Relevance to early responses to trauma. *Journal of Trauma ad Dissociation, 6*(2), 51–68.

Bremner, J. D., & Brett, E. (1997). Trauma-related dissociative states and long-term psychopathology in posttraumatic stress disorder. *Journal of Traumatic Stress, 10*, 37–49.

Bremner, J. D., & Vermetten, E. (2004). Neuroanatomical changes associated with pharmacotherapy in posttraumatic stress disorder. In R. Yehuda & B. McEwen (Eds.), *Behavioral stress responses* (Vol. 1032, pp. 154–155). Annals of The New York Academy of Sciences.

Bremner, J. D., et al. (1997). Elevated CSF corticotropin-releasing factor concentrations in posttraumatic stress disorder. *American Journal of Psychiatry, 154*, 624–629.

Brenner, L. A., et al. (2007). The long-term impact and rehabilitation of pediatric traumatic brain injury: A 50-year follow-up case study. *Journal of Head Trauma Rehabilitation, 22*, 56–64.

Breslau, N., et al. (1998). Trauma and posttraumatic stress disorder in the community: The 1996 Detroit Area Survey of Trauma. *Archives of General Psychiatry*, 557, 626–632.

Brewin, C. R., et al. (2000). Fear, helplessness, and horror in posttraumatic stress disorder: Investigating DSM-IV Criterion A2 in victims of violent crime. *Journal of Traumatic Stress, 13*, 499–509.

Briere, J., et al. (2008). Accumulated childhood trauma and symptom complexity. *Journal of Traumatic Stress, 21*, 223–226.

Brindley, D. N., & Rolland, Y. (1989). Possible connections between stress, diabetes, obesity, hypertension and altered lipoprotein metabolism that may result in atherosclerosis. *Clinical Science, 77*, 453–461.

Britt, P. M., & Heiserman, J. E. (2000). Imaging evaluation. In P. R. Cooper & J. G. Golfinos (Eds.), *Head injury, 4th Ed.* (pp. 63–131). New York: McGraw Hill.

Brodal, A. (1981). *Neurological anatomy, 3rd Ed.* New York: Oxford University Press.

Broman, S. H., & Michel, M. E. (Eds.). (1995). *Traumatic head injury in children.* New York: Oxford University Press.

Brontke, C. F., et al. (1996). Rehabilitation of the head-injured patient. In R. J. Narayan et al. (Eds.), *Neurotrauma* (pp. 841–858). New York: McGraw-Hill.

Brookes, M., et al. (1990). Head injuries in accident and emergency departments. How different are children from adults? *Journal of Epidemiology and Community Health, 44*(2), 147–151.

Brooks, W. M., et al. (2000). Metabolic and cognitive response to human traumatic brain injury: A quantitative proton magnetic resonance study. *Journal of Neurotrauma, 17*, 629–640.

Brookshire, R. H. (1997). *Introduction to neurogenic communication disorders, 5th Ed.* St. Louis, MO: Mosby.

Broughton, R. J. (1989). Nightmare, sleep terror. In G. Adelman (Ed.) *Neuroscience Year: Supplement 1 to the Encyclopedia of Neuroscience,* pp. 125–127. Boston: Birknauser.

Brown, A. C. (1989). Pain and itch. In H.D. Patton, et al. (Eds.), *Textbook of physiology* (Vol. I, pp. 346–364). Philadelphia: Saunders.

Brown, C. H., & Cannito, M. P. (1996). Articulated and inflected primate vocalizations. In A. Donald, et al. (Eds.), *Disorders of motor speech* (pp. 41–63). Baltimore: Paul H. Brooks.

Brown, D. H., & Zwilling, B. S. (1996). Neuroimmunology of host-microbial interactions. In H. Friedman, et al. (Eds.), *Psychoimmunology, stress, and infection* (pp. 153–172). Boca Raton, FL: CRC Press.

Brown, E. (1991). Interictal cognitive changes in epilepsy. *Seminars in Neurology, 11*(2), 167–174.

Brown, F. D., et al. (1982). Cardiac and cerebral changes following experimental head injury. In R. G. Grossman & P. L. Goldenberg (Eds.), *Head injury* (pp. 151–157). New York: Raven Press.

Brown, G., et al. (1981). A prospective study of children with head injuries: III. Psychiatric sequelae. *Psychological Medicine, 11*, 63–78.

Brown, R. H. (2001). Amytrophic lateral sclerosis and other motor neuron diseases. In E. Braunwald, et al. (Eds.), *Harrison's 15th edition: Principles of internal medicine* (pp. 2412–2416). New York: McGraw Hill.

Bruce, D. (1990). Scope of the problem (Child with traumatic brain injury)—Early assessment and management. In M. Rosenthal, E. R. Griffith, M. R. Bond, & J. D. Miller (Eds.) *Rehabilitation of the Adult and Child with Traumatic Brain Injury,* 2nd ed., pp. 521–538. Philadelphia: F. A. Davis.

Bruce, S. (2005). Introduction to immunology and immune-endocrine interactions. In K. Vedara & M. Irwin (Eds.), *Human psychoneuroimmunology* (pp. 1–24). Oxford: Oxford University Press.

Brumback, R. A., & Weinberg, W. A. (1990). Pediatric behavioral neurology: An update on the neurologic aspects of depression, hyperactivity, and learning disabilities. *Neurologic Clinics, 8*(3), 677–703.

Bruni, J. (2004). Episodic impairment of consciousness. In W. G. Bradley, et al. (Eds.), *Neurology in clinical practice, 4th Ed.* (Vol. 2, pp. 11–22). Boston: Butterworth-Heinemann.

Bruno, R. S. (1984). The catastrophic reaction: Release of cortical inhibition following cortical lesion. *Newsletter: The New York Neuropsychology Group, 3*(1), 1, 5, 6.

Brust, J. C. M. (2001). Music and the neurologist: A historical perspective. In R. Zatorre & I. Peretz (Eds.), *The biological foundations of music* (Vol. 930, pp. 143–152). Annals of the New York Academy of Sciences.

Bryant, R. A. (1996). Posttraumatic stress disorder, flashbacks, and pseudomemories in closed head injury. *Journal of Traumatic Stress,* 9, 621–629.
Bryant, R. A. (2001). Posttraumatic stress disorder and traumatic brain injury: Can they co-exist? *Clinical Psychology Review, 21*, 631–648.
Bryant, R. A., & Harvey, A. G. (1996). Visual imagery in posttraumatic stress disorder. *Journal of Traumatic Stress,* 9, 613–619.
Bryant, R. A., et al. (2000). Coping style and post-traumatic stress disorder following severe traumatic brain injury. *Brain Injury, 14,* 175–180.
Bryant, R. A., et al. (2007). The relationship between acute stress disorder and posttraumatic stress disorder in children. *Journal of Traumatic Stress, 20*, 1075–1779.
Buchan, A. M. J. (1989). Digestion and absorption. In H. D. Patton, et al. (Eds.), *Textbook of physiology* (Vol. II, pp. 1438–1460). Philadelphia: Saunders.
Buchanan, S. L., & Powell, D. A. (1993). Cingulothalamic and prefrontal control of autonomic function. In B. A. Vogt & M. Gabriel (Eds.), *Neurobiology of cingulate cortex and limbic thalamus* (pp. 381–414). Boston: Birkäuser.
Buckingham, H. Q. (1991). Explanations for the concept of apraxia of speech. In M. T. Sarno (Ed.), *Acquired aphasia, 2nd Ed.* (pp. 271–312). San Diego: Academic Press.
Bullock, R., & Nathoo, H. (2005). Injury and cell function. In P. L. Reilly & R. Bullock (Eds.), *Head injury, 2nd Ed.* (pp. 113–139). New York: Hodder Arnold.
Bulun, S. E., & Adashi, E. Y. (2003). The physiology and pathology of the female reproductive axis. In P. R. Larsen, et al. (Eds.), *Williams textbook of endocrinology, 10th Ed.* (pp. 587–664). Philadelphia: Saunders.
Burden, S. J., et al. (2003). Target selection, topographic maps, and synapse formation. In L. R. Squire, et al. (Eds.), *Fundamental neuroscience, 2nd Ed.* (pp. 469–498). Boston: Academic Press.
Burke, R. J., & Richardson, A. M. (1996). Stress, burnout, and health. In C. L. Cooper (Ed.), *Handbook of stress, medicine, and health* (pp. 101–117). Boca Raton, FL: CRC Press.
Burneo, J. G., et al. (2000). Amusement park stroke. *Neurology*, *55*, 564.
Burns, B., et al. (2007). Treatment of medically intractable cluster headache by occipital nerve stimulation: Long-term follow-up of eight patients. *The Lancet, 369*, 1099–1206.
Burt, A. M. (1993). *Textbook of neuroanatomy*. Philadelphia: Saunders.
Burton, N. (2002). Cerebral cortical regions devoted to the somatosensory system: Results from brain imaging studies in humans. In R. J. Nelson (Ed.), *The somatosensory system: Deciphering the brain's own body image* (pp. 27–60). Boca Raton, FL: CRC Press.
Busch, C. R., & Alpern, H. P. (1998). Depression after mild traumatic brain injury: A review of current research. *Neuropsychology Review,* 8, 43–78.
Buschman, T. J., & Miller, E. K. (2007). Top-down versus bottom-up control of attention in the prefrontal and posterior parietal cortices. *Science, 375*, 1860–1862.
Busek, P., & Faber, J. (2000). The influence of traumatic brain lesion on sleep architecture. *Sbornik Lekarsky, 101*, 233–239.
Bushnik, T., et al. (2007a). The experience of fatigue in the first 2 years after moderate-to-severe traumatic brain injury: A preliminary report. *Journal of Head Trauma Rehabilitation, 23*, 3–16.
Bushnik, T., et al. (2007b). Patterns of fatigue and its correlates over the first 2 years after traumatic brain injury. *Journal of Head Trauma Rehabilitation, 23*, 25–32.
Bushnik T., et al. (2007c). Objective measurement of fatigue following traumatic brain injury. *Journal of Head Trauma Rehabilitation, 23*, 33–40.
Buss, R. R., et al. (2006). Adaptive roles of programmed cell death during nervous system development. *Annual Review of Neuroscience, 29*, 1–35.
Bustamante, V., et al. (2001). Cognitive functioning and the early development of PTSD. *Journal of Traumatic Stress, 14*, 791–797.
Butcher, I., et al. (2007). Prognostic value of cause of injury in traumatic brain injury: Results from the IMPACT study. *Journal of Neurotrauma, 24*, 281–286.
Butler, A. B., & Hodos, W. (1966). *Vertebrate neuroanatomy: Evolution and adaptation.* New York: Wiley-Liss.
Buzdon, M. M., et al. (1999). Femur fracture induces site-specific changes in T-cell immunity. *Journal of Surgical Research, 82*, 201–208.
Byrd, D. A., et al. (2006). Early environmental factors, ethnicity, and adult cognitive test performance. *The Clinical Neuropsychologist, 20*, 243–260.
Byrne, J. H. (2003). Learning and memory: Basic mechanisms. In L. Squire, et al. (Eds.), *Fundamental neuroscience, 2nd Ed.* (pp. 1275–1298). London: Academic Press.

Byrne, T. N., & Waxman, S. G. (2004). Paraplegia and spinal cord syndromes. In W. G. Bradley, et al. (Eds.), *Neurology in clinical practice, 4th Ed.* (Vol. 1, pp. 351–365). Boston: Butterworth-Heinemann.

Cachetto, D., & Topolovec, J. C. (2002). Cerebral cortex. In V. S. Ramachandran (Ed.), *Encyclopedia of the human brain* (Vol. I, pp. 663–679). Boston: Academic Press.

Cacioppo, J. T., et al. (1991). Psychophysiological approaches to the evaluation of psychotherapeutic process and outcome, 1991: Contributions from social psychophysiology. *Psychological Assessment, 3*, 321–336.

Caffey, J. (1974). The whiplash shaken infant syndrome: Manual shaking by the extremities with whiplash-induced intracranial and intraocular bleedings, linked with residual permanent brain damage and mental retardation. *Pediatrics, 54*, 396–403.

Cahan, S., & Cohen, N. (1989). Age versus schooling effects on intelligence development. *Child Development, 60*, 1239–1249.

Cahill, L. (1999). A neurobiological perspective on emotionally influenced, long-term memory. *Seminars in Clinical Neuropsychiatry, 4*, 266–273.

Cahill, L. M., et al. (2003). Perceptual and instrumental analysis of laryngeal function after traumatic brain injury in childhood. *Journal of Head Trauma Rehabilitation, 18*, 268–283.

Cajigal, S. (2007). Football concussions linked depression, cognitive impairment-experts seek prospective studies. *Neurology Today, 7*(4), 1, 22, 23.

Calogero, A. E. (1995). Neurotransmitter regulation of the hypothalamic corticotropin-releasing hormone neuron. In G. P. Chrousis, et al. (Eds.), *Stress: Basic mechanisms and clinical implications* (Vol. 771, pp. 31–40). Annals of the New York Academy of Science.

Camicioli, R. E., & Nutt, J. K. (2003). Gait and balance. In C. G. Goetz, (Ed.), *Textbook of clinical neurology, 2nd Ed.* (pp. 317–331). Philadelphia: Saunders.

Campagnolo, D. E., et al. (1994). Alteration of immune system function in tetraplegics. *American Journal of Physical Medicine & Rehabilitation, 73*, 387–393.

Campagnolo, D. I., et al. (1997). Impaired phagocytosis of Staphylococcus aureas in complete tetraplegics. *American Journal of Physical Medicine, 76*, 276–280.

Campagnono, D. I., et al. (2000). Influence of neurological level on immune function following spinal cord injury: A review. *Journal of Spinal Cord Medicine, 23*, 121–128.

Campagnuolo, G., et al. (2003). IL-1β and TNF-α produce divergent acute inflammatory and skeletal lesions in the knees of Lewis rats. In N. Chiorazzi, et al. (Eds.), *Immune mechanisms and disease* (Vol. 987, pp. 295–298). Annals of the New York Academy of Science.

Campbell, T. W. (1999). Challenging the evidentiary reliability of DSM IV. *American Journal of Forensic Psychology, 17*(1), 47–68.

Campbell, W. W. (2005). *DeJong's: The neurologic examination, 6th Ed.* Philadelphia: Lippincott Williams & Wilkins.

Campbell-Sills, L. & Stein, M. B. (2007). Psychometric analysis and refinement of the Connor-Davidson Resilience Scale (CdpRISC): Validation of a 10-item measure of resilience. *Journal of Traumatic Stress, 20*, 1010–1028.

Cancelliere, A. E. B., & Kertesz, A. (1990) (1, 15, 34956.990). Lesion localization in acquired deficits of emotional expression and comprehension. *Brain and Cognition, 13*, 133–147.

Cantor, J. B., et al. (2007). Fatigue after traumatic brain injury and its impact on participation and quality of life. *Journal of Head Trauma Rehabilitation, 23*, 41–51.

Cantu, R. C. (1996). Head injuries in sport. *British Journal of Sports Medicine, 30,* 289–296.

Cantu, R. C. (1997). Athletic head injuries. *Clinics in Sports Medicine,* 16, 3, 351–541.

Cantu, R. C. (1998a). Second-impact syndrome. *Clinics in Sports Medicine,* 17, 1, 37–44.

Cantu, R. C. (1998b). Return to play guidelines after a head injury. *Clinics in Sports Medicine, 17*(1), 45–60.

Cantu, R. C. (2006). An overview of concussion consensus statements since 2000. *Neurosurgical Focus, 21*(4).

Caplan, D. (2003). Aphasic syndromes. In K. M. Heilman & E. Valenstein (Eds.), *Clinical neuropsychology, 4th Ed.* (pp. 14–34). New York: Oxford University Press.

Caplan, D. H., & Gould, J. L. (2003). Language and communications. In L. Squire et al. (Eds.), *Fundamental neuroscience, 2nd Ed.* (pp. 1329–1352). London: Academic Press.

Caplan, D., et al. (1999). Language and communication. In M. J. Zigmond, et al. (Eds.), *Fundamental neuroscience* (pp. 1487–1519). San Diego: Academic Press.

Carey, B. (2005, November 26). The struggle to gauge a war's psychological cost. *The New York Times,* A1, A4.

Carlier, V. E., et al. (1996). PTSD in relation to dissociation in traumatized police officers. *American Journal of Psychiatry, 153*, 1325–1328.

Carlson, S. M., et al. (2004). Individual differences in executive functioning and theory of mind: An investigation of inhibitory control and planning ability. *Journal of Experimental Child Psychology, 87*, 299–319.

Carroll, L. (November 2004). Antibody may explain cognitive decline in patients with systemic lupus erthyematosus. *Neurology Today*, 62, 65.

Carroll, L., et al. (2004). Prognosis for mild traumatic brain injury: results of the WHO Collaborating Centre Task Force on Mild Traumatic Brain Injury. *Journal of Rehabiltation Medicine, 43*(Suppl), 84–105.

Carter, C. R., et al. (1996). Effects of semantic and syntactical context on actual and estimated sentence intelligibility of dysarthric speakers. In A. Donald, et al. (Eds.), *Disorders of motor speech* (pp. 67–87). Baltimore: Paul H. Brooks.

Cartlidge, N. (1991). Head injury: Outcome and prognosis. In M. Swash and J. Oxbury (Eds.) *Clinical Neurology,* vol. 1, 699–707.

Caselli, R. J., et al. (1991). Thalamocortical diaschisis: Single-photon emission tomographic study of cortical blood flow changes after focal thalamic infarction. *Neuropsychiatry, Neuropsychology, and Behavioral Neurology,* 4, 193–214.

Casey, B. J., et al. (2000). Structural and functional brain development and its relation to cognitive development. *Biological Psychiatry, 54*, 241–225.

Casper, R. C. (1998). Serotonin, a major player in the regulation of feeding and affect. *Biological Psychiatry, 44*, 795–797.

Castels, S. (1996). Metabolic bone disease. In F. Lifschitz (Ed.), *Pediatric endocrinology, 3rd Ed.* (pp. 521–534). New York: Dekker.

Castriotta, R. J., & Lai, J. M. (2001). Sleep disorders associated with traumatic brain injury. *Archives of Physical Medicine & Rehabilitation, 82*, 1403–1406.

Catania, A., et al. (1994). Cytokine antagonists in infectious and inflammatory disorders. In N. A. Fabris, et al. (Eds.), *Immunomodulation: The state of the art* (Vol. 741, pp. 149–161). Annals of the New York Academy of Sciences.

Catroppa, C., & Anderson, V. (2003). Children's attentional skills 2 years post-traumatic brain injury. *Developmental Neuropsychology, 23*, 359–373.

Catroppa, C., & Anderson, V. (2006). Planning, problem-solving and organizational abilities in children following traumatic brain injury: Intervention techniques. *Pediatric Rehabilitation, 9*, 89–97.

Catt, K. J. (1995). Molecular mechanics of hormone action: Control of target cell function by peptide and catecholamine hormones. In P. Felig, J. D. Baxter, & L. A. Frohman (Eds.) *Endocrinologyand Metabolism,* 3rd ed., pp. 713–748. New York: McGraw-Hill.

Center for Disease Control. (2006). *Facts about traumatic brain injury.* Atlanta, GA: Author.

Cereda, C., et al. (2002). Strokes restricted to the insular cortex. *Neurology, 59*, 1950–1955.

Cerhan, J. H., et al. (1996). Neurologic and neuropsychological aspects of minor head injury in children. In M. Rizzo & D. Tranel (Eds.), *Head injury and postconcussive syndrome* (pp. 441–456). New York: Churchill Livingston.

Cernak, I., et al. (1999). Neuroenendocrine responses following graded traumatic brain injury in male adults. *Brain Injury, 13*, 1005–1015.

Cervero, F., & Laird, M. M. (2003). Visceral pain. In M. J. Aminoff & R. B. Daroff (Eds.). *Encyclopedia of the neurological sciences* (Vol. 3, pp. 773–778). Boston: Academic Press.

Chad, D. A. (2004). Disorders of nerve roots and plexuses. In W. G. Bradley, et al. (Eds.), *Neurology in clinical practice, 4th Ed.* (Vol. 1, pp. 2267–2297). Boston: Butterworth-Heinemann.

Chambers, R. A., & Potenza, M. N. (2003). Impulse control disorders. In M. J. Aminoff & R. B. Daroff (Eds.), *Encyclopedia of the neurological science* (Vol. 2, pp. 642–646). Boston: Academic Press.

Champ, P. C., et al. (2005). *Biochemistry, 3rd Ed.* Philadelphia: Lippincott Williams & Wilkins, p. 486.

Chan, H. S., & Liu, Y. K. (1974). The asymmetric response of a fluid-filled spherical shell-a mathematical simulation of a glancing blow to the head. *Journal of Biomechanics, 7,* 43–59.

Chandler, W. F. (1990). Trauma to the carotid artery and other cervical vessels. In R. Youmans (Ed.), *Neurological Surgery* (Vol. 4, pp. 2367–2376). New York: Saunders.

Chaney, R. H., & Olmstead, C. E. (1994). Hypothalamic dysthermia in persons with brain damage. *Brain Injury, 8*, 475–481.

Changeux, J. P. (1993, November). Chemical signaling in the brain. *Scientific American*, 58–62.

Chapman, S., et al. (1995). Discourse ability in children with brain injury: Correlations with psychosocial, linguistic, and cognitive factors. *Journal of Head Trauma Rehabilitation, 10*(5), 36–54.

Charney, D. S., et al. (1995). Neural circuits and mechanisms of post-traumatic stress disorder. In M. J. Friedman, et al. (Eds.), *Neurobiological and clinical consequences of stress* (pp. 271–287). Philadelphia: Lippincott-Raven.

Chelune, G. J., & Baer, R. A. (1986). Developmental norms for the Wisconsin Card Sorting Test. *Journal of Clinical and Experimental Neuropsychology, 8,* 219–228.

Chelune, G. J., & Edwards, P. (1981). Early brain lesions: Ontogenetic-environmental considerations. *Journal of Consulting and Clinical Psychology,* 49, 777–790.

Chen, C-T., et al. (1999). *Occupational electrical injury: An international symposium.* Annals of The New York, Academy of Sciences, Vol. 888.

Cheng, M. L., et al. (2005). Pediatric head injury. In P. L. Reilly & R. Bullock (Eds.), *Head injury, 2nd Ed.* (pp. 356–367). New York: Hodder Arnold.

Cherian, L., et al. (1996). Secondary insult increase injury after controlled cortical impact in rats. *Journal of Neurotrauma, 13,* 371–383.

Chestnut, R. M. (1996). Treating raised intracranial pressure in head injury. In R. K. Naryan, J. E. Wilberge, & J. T. Povlisheck (Eds.) *Neurotrauma,* pp. 445–469. New York: McGraw-Hill.

Chibnall, J. T., & Duckro, P. H. (1994). Post-traumatic stress disorder in chronic post-traumatic headache patients. *Headache, 34,* 357–361.

Chiofalo, N., et al. (1986). Perinatal and posttraumatic seizures. In A. J. Raimondi, et al. (Eds.), *Head injuries in the newborn and infant* (pp. 217–232). New York: Springer-Verlag.

Chiu, W. C., & Lilly, M. P. (2000). Acute trauma response. In G. Fink (Ed.), *Encyclopedia of stress* (Vol. 1, pp. 17–24). San Diego: Academic Press.

Chokroverty, S. (2004). Sleep and its disorders. In W. G. Bradley et al. (Eds.), *Neurology in clinical practice, 4th Ed.,* (Vol. 2, pp. 1993–2054). Boston: Butterworth-Heinemann.

Chow, T. W., & Cummings, J. L. (2000). Neuropsychiatry: Clinical assessment and approach to diagnosis. In B. J. Sadock & V. A. Sadock (Eds.), *Comprehensive textbook of psychiatry, 7th Ed.* (pp. 221–242). Philadelphia: Lippincott Williams & Wilkins.

Chows, T. K., et al. (1996). Spectrum of injuries from snowboarding. *Journal of Trauma: Injury, Infection, and Critical Care, 41,* 321–325.

Christoffel, K. K., & Zieserl, E. J. (1991). Jogging with infant in backback carrier. *Journal of the American Medical Association, 266,* 1146.

Chrousos, G. P. (1998). Stressors, stress, and neuroendocrine integration of the adaptive response. In P. Csermely (Ed.), *Stress of life: From molecules to man* (Vol. 851, pp. 311–335). Annals of The New York Academy of Sciences.

Chrousos, G. P. (2000). The stress response and immune function: Clinical implications. In A. Conti, et al. (Eds.), *Neuroimmunomodulation: Perspectives at the new millennium* (Vol. 917, pp. 38–67). Annals of the NY Academy of Sciences.

Churchland, P. S. (2002). Self-representation in nervous systems. *Science, 296,* 308–310.

Cicchetti, D. (2004). An Odyssey of discovery: Lessons learned through three decades of research on child maltreatment. *American Psychologist, 59,* 731–741.

Cicerone, K. D., & Azulay, J. D. (2007). Perceived self-efficacy and life satisfaction after traumatic brain injury. *Journal of Head Trauma Rehabilitation, 22,* 257–266.

Cicerone, K. D., & Kalmar, K. (1995). The persistent postconcussion syndrome: The structure of subjective complaints after mild traumatic brain injury. *The Journal of Head Trauma Rehabilitation, 10*(3), 1–17.

Cicerone, K. D., & Kalmar, K. (1997). Does premorbid depression influence symptoms and neuropsychological functioning? *Brain Injury, 11,* 643–648.

Cisek, P. E. (2002). Computational perspective on proprioception and movement guidance in parietal cortex. In R. J. Nelson (Ed.), *The somatosensory system: Deciphering the brain's own body image* (pp. 275–298). Boca Raton, FL: CRC Press.

Clauw, D. J., & Williams, D. A. (2002). Relationship between stress and pain in work related upper extremity disorders: the hidden role of chronic multisymptom illnesses. *American Journal of Industrial Medicine, 18,* 370–382.

Code, C. (1987). *Language, aphasia, and the right hemisphere.* New York: John Wiley.

Coelho, C. A. (2007). Cognitive-communication deficits following TBI. In N. D. Zasler, et al. (Eds.), *Brain injury medicine* (pp. 895–910). New York: Demos.

Coetzer, R. (2007). Psychotherapy following traumatic brain injury: Integrating theory and practice. *Journal of Head Trauma Rehabilitation, 22,* 39–47.

Coffey, S. F., et al. (2006). Screening for PTSD in motor vehicle accident survivors using the PSSSR & IES. *Journal of Traumatic Stress, 19,* 119–128.

Cohadon, R. F., et al. (1991). Head injuries: Incidence and outcome. *Journal of the Neurological sciences, 103,* S27–S31.

Cohen, H., & Zohar, J. (2004). An animal model of posttraumatic stress disorder: The use of cut-off behavioral criteria. In R. Yehuda & B. McEwen (Eds.), *Behavioral stress responses* (Vol. 1032, pp. 167–178). Annals of The New York Academy of Sciences.

Cohen, H., et al. (1998). Analysis of heart rate variability in posttraumatic stress disorder patients in response to a trauma-related reminder. *Biological Psychiatry, 44,* 1954–1959.

Cohen, N., & Kinney, K. S. (2001). Exploring the phylogenetic history of neural-immune system interactions. In R. Ader, et al. (Eds.), *Psychoneuroimmunology, 3rd Ed.* (Vol. 1, pp. 21–54). San Diego: Academic Press.

Cohen, S., & Williamson, G. M. (1991). Stress and infectious disease in humans. *Psychological Bulletin, 109,* 5–24.

Cohen, S.B. (1986). Educational reintegration and programming for children with head injuries. *Journal of Head Trauma Rehabilitation, 1*(4), 220–229.

Cohen, T. I., & Gudeman, S. K. (1996). Delayed traumatic intracranial hematoma. *Journal of Neurotrauma, 25,* 689–701.

Colby, E. L., & Olson, C. R. (2003). Spatial cognition. In L. R. Squire, et al. (Eds.), *Fundamental neuroscience, 2nd Ed.* (pp. 1229–1247). Boston: Academic Press.

Collins, M. W., et al. (1999). Relationship between concussion and neuropsychological performance in college football players. *Journal of the American Medical Association, 282,* 964–970.

Cone, R. D., et al. (2003). Neuroendocrinology. In P. R. Larsen et al. (Eds.), *Williams Textbook of Endocrinology, 10th Ed.* (pp. 81–176). Philadelphia: Saunders.

Confavreux, C., et al. (1992). Progressive amusia and aprosody. *Archives of Neurology, 49,* 971–976.

Connors, B. W. (2005a). Synaptic transmission in the nervous system. In W. Boron & E. Boulpaep (Eds.), *Medical physiology, up-dated edition* (pp. 295–324). Philadelphia, Saunders.

Connors, B. W. (2005b). Sensory transduction. In W. F. Boron & E. L. Boulpaep (Eds.), *Medical physiology, up-dated edition* (pp. 325–358). Philadelphia. Elsevier Saunders.

Connors, B. W. (2005c). Circuits of the central nervous system. In W. Boron & E. Boulpaep (Eds.), *Medical physiology, up-dated edition* (pp. 359–357). Philadelphia: Saunders.

Coons, P. M. (2000). Dissociative fugue. In B. J. Sadock & V. A. Sadock (Eds.), *Comprehensive textbook of psychiatry, 7th Ed.* (pp. 1549–1552). New York: McGraw-Hill.

Cooper, P. E. (2004). Neuroendocrinology. In W. G. Bradley, et al. (Eds.), *Neurology in clinical practice, 4th Ed.* (Vol. 1, pp. 849–968). Boston: Butterworth-Heinemann.

Cooper, P. R. (1987). Skull fracture and traumatic cerebrospinal fluid fistulas. In P. E. Cooper (Ed.), *Head injury, 2nd ed.* (pp. 89–107). Baltimore: Williams & Wilkins.

Cooper, P. R. (2000). Post-traumatic intracranial mass lesions. In P. R. Cooper & J. G. Golfinos (Eds.). *Head injury, 4th Ed.* (pp. 292–348). New York: McGraw Hill.

Cooper, P. W. (1991). Neuroendocrinology. In W. G. Bradley, R. B. Daroff, G. M. Fenichel, & C. D. Marsden (Eds.) *Neurology in Clinical Practice,* vol. 1, pp. 611–625. Boston: Butterworth-Heinemann.

Corkin, S., et al. (1985). In D. S. Olton, et al. (Eds.), *Memory dysfunctions: An integration of animal and human research from preclinical and clinical perspectives* (Vol. 444, pp. 10–40). Annals of the New York Academy of Science.

Corkin, S. H., et al. (1987). Consequences of nonpenetrating and penetrating head injury: Retrograde amnesia, posttraumatic amnesia, and lasting effects on cognition. In H. S. Levin, et al. (Eds.), *Neurobehavioral recovery from head injury* (pp. 318–329). New York: Oxford University Press.

Corrigan, J. D. (1995). Substance abuse as a mediating factor in outcome from traumatic brain injury. *Archives of Physical Medicine and Rehabilitation, 76,* 302–309.

Corrigan, J. D., & Bogner, J. (2007). Initial reliability and validity of the Ohio State University TBI Identification method. *Journal of Head Trauma Rehabilitation, 22,* 315–317.

Cottencin, O., et al. (2008). Directed forgetting in depression. *Journal of the International Neuropsychological Society, 14,* 895–899.

Cotterill, R. M. J. (2001). Cooperation of the basal ganglia, cerebellum, sensory cerebrum and hippocampus: Possible implications for cognition, consciousness, intelligence and creativity. *Progress in neurobiology, 64,* 1–33.

Couch, J. R., et al. (2007). Head or neck injury increases the risk of chronic daily headache. *Neurology, 69,* 1169–1177.

Cox, I. M., et al. (1996). Significant magnesium deficiency in depression. *Journal of Orthopedic Medicine and Surgery, 17,* 7–9.

Cox, T. (1988). Psychobiological factors in stress and health. In S. Fisher & J. Reason (Eds.), *Handbook of life stress, cognition and health* (pp. 603–628). New York: John Wiley and Sons.

Cozolino, L. J., & Siegel, D. J. (2005). Sensation, perception and cognition. In B. J. Sadock & V. A. Sadock (Eds.), *Comprehensive textbook of psychiatry, I, 8th Ed.* (pp. 512–528). Philadelphia: Lippincott Williams & Wilkins.

Craig, A. D. (2003). Pain mechanisms: Labeled lines versu convergence in central processing. *Annual Review of Neuroscience, 26,* 1–30.

Craig, K. D. (1994). Emotional aspects of pain. In P. D. Wells & R. Melzack (Eds.). *Textbook of pain* (pp. 261–274). New York: Churchill Livingston.

Crank, M., & Fox, P. T. (2002). Broca's area. In V. S. Ramachandran (Ed.), *Encyclopedia of the human brain* (pp. 569–586). Boston: Academic Press.

Crawford, T. O. (2004). The floppy infant. In W. G. Bradley et al. (Eds.), *Neurology in clinical practice, 4th Ed.* (Vol. 1, pp. 393–406). Boston: Butterworth-Heinemann.

Creisler, C. A., et al. (2005). In *Harrison's principles of internal medicine, 16th Ed.* (Vol. I, pp. 153–162). New York: McGraw-Hill.

Cripe, L. I. (1996). The ecological validity of executive function testing. In R. J. Sbordone & C. L. Long (Eds.), *Ecological validity of neuropsychological testing* (pp. 171–202). Delray Beach, FL: GR Press/St. Lucie Press.

Crockett, D., et al. (1981). An overview of neuropsychology. In S. B. Filskov & T. J. Boll (Eds.), *Handbook of neuropsychology* (pp. 1–37). New York: John Wiley.

Crofford, L. J. (2002). The hypothalamic-pituitary-adrenal axis in the pathogenesis of rheumatic diseases. *Endocrinology and metabolism clinics of North America, 31,* 1–13.

Croft, A. C. (1995a). Biomechanics. In S. M. Foreman & A. C. Croft (Eds.). *Whiplash injuries: The cervical acceleration/deceleration syndrome, 2nd ed.* (pp. 1–92). Baltimore: Williams & Wilkins.

Croft, A. C. (1995b). Soft tissue injury: Long-and short-term effects. In S. M. Foreman & A. C. Croft (Eds.), *Whiplash injuries: The cervical acceleration/deceleration syndrome, 2nd Ed.* (pp. 288–362). Baltimore: Williams & Wilkins.

Crompton, M. R. (1971). Hypothalamic lesions following closed head injury. *Brain, 94,* 165–172.

Cross, I. (2001). Music, cognition, culture, and evolution. In R. H. Zatorre & I. Peretz (Eds.), *The biological foundations of music* (Vol. 930, pp. 28–42). Annals of the New York Academy of Sciences.

Crosson, B., & Haaland, K. Y. (2003). Subcortical functions in cognition: Toward a consensus. *Journal of the International Neuropsychological Society, 9,* 1027–1030.

Crosson, B., et al. (2002). Basal Ganglia. In V. S. Ramachandran (Ed.), *Encyclopedia of the human brain* (Vol. I, pp. 367–379). Boston: Academic Press.

Crovitz, H. F., et al. (1992). Consistency in recalling features of former head injuries: Retrospective questionnaire vs. interview retest. *Cortex, 28,* 509–512.

Crowley, J. A., & Miles, M. A. (1991). Cognitive remediation in pediatric head injury: A case study. *Journal of Pediatric Psychology, 16,* 611–627.

Cruse, J. L., et al. (1996). Immune system-neuroendocrine dysregulation in spinal cord injury. *Immunologic Research, 15,* 306–314.

Cryer, P. E. (1995). Diseases of the sympathochromaffin system. In P. Felig, J. D. Baxter, & L. A. Frohman (Eds.) *Endocrinologyand Metabolism,* 3rd ed., pp. 713–748. New York: McGraw-Hill.

Cullen, D. K., & LaPlaca, M. C. (2006). Neuronal response to high rate shear deformation depends on heterogeneity of the local strain field. *Journal of Neurotrauma, 23,* 1304–1319.

Cullinan, W. E., et al. (1995). A neuroanatomy of stress. In M. J. Friedman et al. (Eds.), *Neurobiological and clinical consequences of stress: From normal adaptation to PTSD* (pp. 3–26). Philadelphia: Lippincott-Raven.

Cumming, D. C. (2001). Hormones and athletic performance. In P. Felig & L. A. Frohman (Eds.), *Endocrinology and metabolism, 4th Ed.* (pp. 1483–1519). New York: McGraw-Hill.

Cummings, J. L. (1985). Neurological syndromes associated with right hemisphere damage. In M. S. Burns, et al. (Eds.), *Clinical Management of right hemisphere dysfunction* Baltimore, MD: Aspen.

Cummings, J. L., & Bogousslavsky, J. (2002). Emotional consequences of focal brain lesions: An overview. In J. Bogousslavsky & J. L. Cummings (Eds.), *Behavior and mood disorders in focal brain lesions* (pp. 1–20). Cambridge, UK: Cambridge University Press.

Cummings, J. L., & Trimble, M. R. (1995). *Neuropsychiatry and behavioral neurology.* Washington, DC: American Psychiatric Press.

Cunnick, J., et al. (1992). In N. Fabris, et al. (Eds.), *Ontogenetic and phylogenetic mechanisms of neuroimmunomodulation* (Vol. 650, pp. 283–287). Annals of the NY Academy of Sciences.

Cunnion, K. M., et al. (2001). Complement & kinin. In T. G. Parslow, et al. (Eds.), *Medical immunology, 10th Ed.* (pp. 175–187). New York: Lange.

Cuny, E., et al. (2001). Dysautonomia syndrome in the acute recovery phase after traumatic brain injury: Relief with intrathecal Baclofen therapy. *Brain Injury, 15*, 917–925.

Curle, C. E., & Williams, C. (1996). Post-traumatic stress reactions in children: Gender differences in the incidence of trauma reactions at two years and examination of factors influencing adjustment. *British Journal of Clinical Psychology, 35*(2), 297–309.

Cytowic, R. E., et al. (1986). Transient amenorrhea after closed head trauma. *New England Journal of Medicine, 314*, 715.

Czeisler, C. A., et al. (2005). Sleep disorders. In D. L. Kasper, et al. (Eds.), *Harrison's principles of internal medicine, 16th Ed.* (Vol. 1, pp. 153–162). New York: McGraw-Hill.

Da Silva, F. H. L. (2002). Electrical potentials. In V. S. Ramachandran (Ed.), *Encyclopedia of the human brain* (Vol. II, pp. 147–167). Boston: Academic Press.

Dalmady-Israel, C., & Zasler, N. D. (1992). Post-traumatic seizures: A critical review. *Brain Injury, 7*, 263–273.

Dalton, J. E. (1990). Neuropsychological equivalence of the G.E.D. *International Journal of Clinical Neuropsychology, 12*, 138–139.

Damasio, A. R. (1995). On some functions of the human prefrontal cortex. In J. Grafman, et al. (Eds.), *Structure and functions of the human prefrontal cortex* (Vol. 769, pp. 213–239). Annals of the New York Academy of Sciences.

Damasio, A. R. (1999). *The feeling of what happens…* New York: Harcourt Brown & Co.

Damasio, A. R., & Anderson, S. W. (1993). The frontal lobes. In K. M. Heilman & E. Valenstein (Eds.), *Clinical neuropsychology, 3rd ed.* (pp. 409–460). New York: Oxford University Press.

Damasio, A. R., & Anderson, S. W. (2003). The frontal lobes. In K. M. Heilman & E. Valenstein (Eds.), *Clinical Neuropsychology, 4th Ed.* (pp. 404–446). New York: Oxford University Press.

Daniloff, R., et al. (1980). *The physiology of speech and hearing.* Englewood Cliffs, NJ: Prentice Hall.

Dantzer, R. (2000). Psychoneuroimmunology. In G. Fink (Ed.), *Encyclopedia of stress, 3* (pp. 294–298). San Diego, CA: Academic Press.

Darley, F. L., et al. (1975). *Motor speech disorders.* Philadelphia: W. B. Saunders.

Darwin, C. (1860). *The Origin of species by means of natural selection.* New York: Hurst & Co.

Darwin, C. (1871, reprinted 1881). *The descent of man and selection in relation to sex.* Princeton, NJ: Princeton University Press.

Davachi, L., et al. (2004). Domain specificity in cognitive systems. In M. S. Gazzaniga (Ed.), *The cognitive neurosciences, 3rd Ed.* (pp. 665–678). Cambridge, MA: A Bradford Book.

Davidovicz, H. M. (1996). Autistic spectrum disorders. In E. Frank (Ed.), *Pediatric behavioral neurology* (pp. 73–86). Boca Raton, FL: CRC Press.

Davidson, R. J. (1993). Cerebral asymmetry and emotion: Conceptual and methodological conundrums. *Cognition and Emotion, 7*, 115–138.

Davidson, T. M., et al. (1998). Rapid clinical evaluation of anosmia in children: The Alcohol Sniff Test. In C. Murphy (Ed.), *Olfaction and taste XII* (Vol. 855, pp. 787–792). Annals of the New York Academy of Sciences.

Davies, M. J., & King, T. T. (1997). Cranial nerve injuries (I, V and IX-XII). In R. Macfarlane & D. G. Hardy (Eds.), *Outcome after head, neck and spinal trauma* (pp. 55–63). Oxford: Butterworth-Heinemann.

Davis, G. W. (2006). Homeostatic control of neural activity: From phenomenology to molecular design. *Annual Review of Neuroscience, 29*, 307–323.

Davis, J. M., & Zimmerman, R. A. (1983). Injury to carotid and vertebral arteries. *Neuroradiology, 25*, 55–69.

Davis, M. (1992). The role of the amygdala in conditioned fear. In J. P. Aggleton (Ed.), *The amygdala: Neurobiological aspects emotion, memory, and mental dysfunction* (pp. 254–306). New York: Wiley-Liss.

Dayer, A., et al. (1998). Post-traumatic mutism in children: Clinical characteristics, pattern of recovery and clinicopathological correlations. *European Journal of Paediatric Neurology, 2*(3), 109–116.

De Gelder, B. (2000). Recognizing emotions by ear and by eye. In R. D. Lane & L. Nadel (Eds.), *Cognitive neuroscience of emotion* (pp. 84–105). New York: Oxford University Press.

De Kloet, E. R., & Deruk, R. (2004). Signaling pathways in brain involved in predisposition and pathogenesis of stress-related disease: Genetic and kinetic factors affecting the MR/GR balance. *Annals of the New York Academy of Sciences, 1032*, 14–34.

De Kruijk, J. R., et al. (2002). Prediction of post-traumatic complaints after mild traumatic brain injury: Early symptoms and biochemical markers. *Journal of Neurology, Neurosurgery, and Psychiatry,* 73 (6), 727–732.

De Krujik, J. R., et al. (2003). Olfactory function after traumatic brain injury. *Brain Injury, 17*, 73–78.

De Krujik, J. R., et al. (2006). Prediction of post-traumatic complaints after mild traumatic brain injury: Early symptoms and biochemical markers. *Journal of Neurology, Neurosurgery and Psychiatry*, *73*, 727–732.

De Martino, B., et al. (2006). Frames, biases, and rational decision-making in the human brain. *Science, 313*, 684–687.

De Vries, A., et al. (1999). Looking beyond the physical injury: Posttraumatic stress disorder in children and parents after pediatric traffic injury. *Pediatrics, 104*, 1293–1299.

De Vries, E., & Prat, A. (2005). Preface. In E. De Vries & A. Prat (Eds.), *The blood-brain barrier and its microenvironment* (v–vii). New York: Taylor & Francis.

De Young, A. C., et al. (2007). Elevated heart rate as a predictor of PTSD six months following accidental pediatric injury. *Journal of Traumatic Stress, 20*, 751–756.

DeAngelis, L. M., et al. (2001). Neurological complications of chemotherapy and radiation therapy. In M. J. Aminoff (Ed.), *Neurology and general medicine, 3rd Ed.* (pp. 437–458). New York: Churchill Livingstone.

Deary, I. J., et al. (2006). Physical fitness and lifetime cognitive change. *Neurology, 67*, 1195–1200.

DeBellis, M. D. (2002). Developmental traumatology: A contributory mechanism for alcohol and substance abuse mechanisms. *Psychoneuroendocrinology, 27*, 155–170.

Delahanty, D. (2004). Peritraumatic amnesia, PTSD and cortisol levels after trauma. In R. Yehuda & B. McEwen (Eds.), *Behavioral stress responses* (Vol. 1032, pp. 183–184). Annals of the New York Academy of Sciences.

Delgado, M. R. (2007). Reward-related responses in the human striatum. In B. W. Balleine et al. (Eds.), *Reward and decision making in corticobasal ganglia networks* (Vol. 1104, pp. 70–88). Annals of the New York Academy of Sciences.

Della Sala, S., et al. (1995). Dual-task paradigm: A means to examine the central executive. In J. Grafman et al. (Eds.), *Structure and functions of the human prefrontal cortex* (Vol. 769, pp. 161–171). Annals of the New York Academy of Sciences.

DeLong, M. R. (2000). The basal ganglia. In E. R. Kandel et al. (Eds.), *Principles of neural science, 4th ed.* (pp. 853–867). New York: McGraw Hill.

DeLorenzo, R. J. (1991). The epilepsies. In W. G. Bradley, et al. (Eds.), *Neurology in clinical practice, II.* (pp. 1443–1447). Boston: Butterworth-Heinemann.

Delye, H., et al. (2007). Biomechanics of frontal skull fracture. *Journal of Neurotrauma, 24*, 1576–1586.

Demellweek, C., et al. (2002). A prospective study and review of pre-morbid characteristics in children with traumatic brain injury. *Pediatric rehabilitation, 5*, 81–89.

Denes, G., et al. (1984). Discrimination and identification of emotions in human voice by brain-damaged subjects. *Acta Neurologica Scandinavica, 69*, 154–162.

Dennis, M. (1985). Intelligence after early brain injury: I. Predicting IQ scores from medical variables. *Journal of Clinical and Experimental Neuropsychology, 7*, 526–554.

Dennis, M. (1991). Frontal lobe function in childhood and adolescence: A heuristic for assessing attention regulation, executive control, and the intentional states important for social discourse. *Developmental Neuropsychology,* 7, 327–358.

Dennis, M., & Barnes, M. (1994). Developmental aspects of neuropsychology. In D . Zaidel (Ed.) *Handbook of Perception and Cognition,* vol. 15, pp. 219–246. New York: Academic Press.

Dennis, M., et al. (1996). Appraising and managing knowledge: Metacognitive skills after childhood head injury. *Developmental Neuropsychology, 12*, 77–103.

Dennis, M., et al. (2000). Working memory after mild, moderate, or severe childhood head injury. *Journal of the International Neuropsychological Society*, *6*, 132.

Dennis, M. M. (1988). Language and the young damaged brain. In T. Boll & B. K. Bryant (Eds.), *Clinical neuropsychology and brain function* (pp. 89–123). Washington, DC: American Psychological Association.

DePompei, R., & Tylor, J. (2004). Children and adolescents: Practical strategies for school participation and transition. In M. J. Ashley (Ed.), *Traumatic brain injury: rehabilitative treatment and case management* (pp. 559–580). Boca Raton, FL: CRC Press.

Derelanko, M. J., & Hollinger, M. A. (Eds.). (2002). *Handbook of toxicology, 2nd Ed.* Boca Raton: CRC Press.

Dettmer, J. L., et al. (2007). Putting brain injury on the radar: Exploratory reliability and validity analyses of the Screening Tool for Identification of Acquired Brain Injury in school-aged children. *Journal of Head Trauma Rehabilitation*, *22*, 339–349.

Deuel, R. K., & Robinson, D. J. (1987). Developmental motor signs. In D. E. Tupper (Ed.), *Soft neurological signs* (pp. 95–129). Orlando, FL: Grune & Stratton.

Deutch, A. Y., & Young, C. D. (1995). A model of the stress-induced activation of prefrontal cortical dopamine systems: Coping and the development of post-traumatic stress disorder. In M. J. Friedman, et al. (Eds.), *Neurobiological and clinical consequences of stress: From normal adaptation to PTSD* (pp. 163–175). Philadelphia: Lippincott-Raven.

Devinsky, O. (1992). *Behavioral Neurology.* St. Louis: Mosby.

DeMyer, W. (1988). *Neuroanatomy.* Malverne, PA: Harwall.

Dehaene, S., et al. (1999). Thinking and problem solving. In M. J. Zigmond et al. (Eds.) *Fundamental Neuroscience,* pp. 1543–1564. San Diego: Academic Press.

Devinsky, O. (1998). Nonepileptic psychogenic seizures: Quagmires of pathophysiology, diagnosis, and treatment. *Epilepsia, 39,* 458–462.

Dewitt, D. S., et al., (1997). L-arginine and superoxide dismutase prevent or reverse cerebral hypoperfusion after fluid-percussion traumatic brain injury. *Neurotrauma, 14,* 223–233.

Dhabar, F. S., & McEwen, B. S. (2001). Bidirectional effects of stress and glucocorticoid hormones on immune function: Possible explanations for paradoxical observations. In R. Ader et al. (Eds.), *Psychoimmunology, 3rd Ed.* (Vol. 1, pp. 301–338). San Diego: Academic Press.

Dharker, S. R., et al. (1993). Ischemic lesions in basal ganglia in children after minor head injury. *Neurosurgery, 33,* 863–865.

Di Leo, J. H. (1970). *Young Children and Their Drawings.* New York: Brunner/Mazel.

Di Stefano, G., et al. (1994). Correlation between estradiol serum levels and NK cell activity in endometriosis. In N. A. Fabris, et al. (Eds.), *Immunomodulation: The state of the art* (Vol. 741, pp. 197–203). Annals of the New York Academy of Sciences.

Diamond, A. (2000). Close interrelation of motor development and cognitive development of the cerebellum and prefrontal cortex. *Child Development, 71,* 44–56.

Diamond, M. E., et al. (2002). Sensory learning and the brain's body map. In R. J. Nelson (Ed.), *The somatosensory system: Deciphering the brain's own body image* (pp. 183–195). Boca Raton, FL: CRC Press.

Diamond, P. M., et al. (2007). Screening for traumatic brain injury in an offender sample: A first look at the reliability and validity of the Traumatic Brain Injury Questionnaire. *Journal of Head Trauma Rehabilitation, 22,* 330–338.

Diamond, S., & Urban, G. J. (2002). Headaches. In V. S. Ramachandran (Ed.), *Encyclopedia of the human brain* (Vol. 2, pp. 415–428). Boston: Academic Press.

Diaz-Arrastia, E., & Baxter, V. K. (2006). Genetic factors in outcome after traumatic brain injury: What the human genome project can teach us about brain trauma. *Journal of Head Trauma Rehabilitation, 21,* 361–374.

Diekema, D. S., et al. (1993). Epilepsy as a risk factor for submersion injury in children. *Pediatrics, 91,* 612–616.

Dietrich, W. D., & Bramlett, H. M. (2004). Basic neuroscience of neurotrauma. In W. G. Bradley, et al. (Eds.), *Neurology in clinical practice, 4th Ed.* (Vol. 2, pp. 1115–1125). Boston: Butterworth-Heinemann.

Dietrich, W. D., et al. (1994). Widespread metabolic depression and reduced somatosensory circuit activation following traumatic brain injury in rats. *Journal of Neurotrauma, 11,* 629–638.

Dijkers, P. J. M., & Bushnik, T. (2008). Assessing fatigue after traumatic brain injury: An evaluation of the Barroso Fatigue Scale. *Journal of Head Trauma Rehabilitation, 23,* 3–16.

DiRocco, C., & Velardi, F. (1986). Epidemiology and etiology of craniocerebral trauma in the first two ears of life. In A. J. Raimondi, et al. (Eds.), *Head injuries in the newborn and infant* (pp. 125–139). New York: Springer-Verlag.

Dixon, C. I., et al. (1991). A controlled cortical model of traumatic brain injury in the rats. *Journal of Neuroscience Methods, 39,* 253–262.

Dluhy, R., et al. (2002). Endocrine hypertension. In P. R. Larsen et al. *Williams textbook of endocrinology, 10th Ed.* (pp. 552–585). Philadelphia: Saunders.

Dobie, R. A., & Rubel, E. W. (1989). The Auditory System: Central auditory pathways. In H. D. Patton, et al., (Eds.), *Textbook of physiology* (Vol. I, pp. 386–409). Philadelphia: Saunders.

Dobkin, B. H. (2006). Short distance walking speed and timed walking distance: Redundant measures for clinical trials? *Neurology, 66,* 584–586.

Donders, J., & Warschausky, S. (2007). Neurobehavioral outcomes after early versus late childhood traumatic brain injury. *Journal of Head Trauma Rehabilitation, 22,* 296–302.

Dorland's illustrated medical dictionary, 31st Ed. (2007). Philadelphia, PA: Author.

Doty, R. L., et al. (1996). Development of the 12 item Cross Cultural Smell Identification Test (CC-SIT). *Laryngoscope, 196,* 353–356.

Doupe, A. J., & Kuhl, P. K. (1999). Birdsong and human speech: Common themes and mechanisms. *Annual Review of neuroscience, 22*, 597–631.

Downing, J. E., & Miyan, J. A. (2000). Neural immunoregulation: emerging roles for nerves in immune homeostasis and disease. *Immunology Today, 21*, 281–289.

Doyon, D., et al. (2002). *The cranial nerves*. Teterboro, NJ: Icon Learning Systems.

Drachman, D. B. (2001). Myesthenia gravis and other diseases of the neuromuscular junction. In E. Braunwald, et al. (Eds.), *Harrison's 15th edition: Principles of internal medicine* (pp. 2515–2520). New York: McGraw Hill.

Drake, R. E. (2007). Is comorbidity a psychological science? *Clinical Psychology: Science and Practice, 14*, 20–22.

Dreifuss, F. E. (1989). Focal and multifocal cortical seizures. In K. Swaiman (Ed.), *Pediatric neurology: Principles and practice* (Vol. I, pp. 393–411). St. Louis, MO: C. V. Mosby.

Drevets, W. C., et al. (1992). The motor cortex and the coding of force. *Science*, *256*, 1696.

Drislane, S. W. (1996). Transient events. In M. S. Samuels & S. Feske (Eds.), *Office practice of neurology* (pp. 111–1213). New York: Churchill Livingstone.

Drolet, G., et al. (2001). Role of endogenous opioid system in the regulation of the stress response. *Progress in Neuro-Psychopharmacology and Biological Psychiatry, 25*, 729–741.

Dronkers, N. F., et al. (2000a). Language and the aphasias. In E. R. Kandel, et al. (Eds.), *Principles of neural science, 4th ed.* (pp. 1169–1187). New York: McGraw Hill.

Dronkers, N. F., et al. (2000b). The neural architecture of language disorders. In M. S. Gazzaniga (Ed.), *The new cognitive neurosciences, 2nd Ed.* (pp. 949–958). Cambridge, MA: MIT Press.

Duane, D. (1996). Learning disabilities. In E. Frank (Ed.), *Pediatric behavioral neurology* (pp. 203–227). Boca Raton, FL: CRC Press.

Dubois, B., et al. (1995). Experimental approach to prefrontal functions in humans. In J. Grafman et al. (Eds.), *Structure and functions of the human prefrontal cortex* (Vol. 769, pp. 41–60). Annals of the New York Academy of Sciences.

Duffy, J. R., & Folger, W. N. (1996). Dysarthria associated with unilateral central nervous system lesions: A retrospective study. *Journal of Medical Speech-Language Pathology, 4*, 55–70.

Dugbartey, A. T., et al. (1999). Neuropsychological assessment of executive functions. *Seminars in Clinical Neuropsychiatry,* 4, 5–12.

Duhaime, A. C., et al. (1992). Head injury in very young children: Mechanisms, injury types, and ophthalmologic findings in 100 hospitalized patients younger than 2 years of age. *Pediatrics, 90*(2), 179–185.

Dum, R. P., et al. (2002). Motor and non-motor domains in the monkey dentate. In S. M. Highstein & W. T. Thach (Eds.), *The cerebellum: Recent developments in cerebellar research* (Vol. 978, pp. 289–301). Annals of the New York Academy of Sciences.

Dunbar, K., & Sussman, D. (1995). Toward a cognitive account of frontal lobe function: Simulating frontal lobe deficits in normal subjects. In J. Grafman, et al. (Eds.), *Structure and functions of the human prefrontal cortex* (Vol. 769, pp. 289–303). Annals of the New York Academy of Sciences.

Dunlap, J. C. (2006). Running a clock requires quality time together. *Science*, *311*, 184–186.

Dunn, A. J. (1993). The Role of cytokines in infection-induced stress. In Y. Taché & C. Rivier (Eds.), *Corticotropin-releasing factor and cytokines: Role in the stress response* (Vol. 691, pp. xiii–xiv). Annals of the New York Academy of Science.

Dunn, A. J. (1996). Psychoneuroimmunology, stress and infection. In H. Friedman, et al. (Eds.), *Psychoimmunology, stress, and infection* (pp. 25–46). Boca Raton, FL: CRC Press.

Dunn, A. J. (2000). Cytokine activation of the HPA axis. In Conti, A. et al. (Eds.), *Neuroimmunomodulation: Perspectives at the new millennium* (Vol. 917, pp. 608–617). Annals of the NY Academy of Sciences.

Dupont, S., et al. (2003). Functional anatomy of the human insula. *Surgical and Radiological Anatomy*, *25*, 113–119.

Dykes, L. J. (1986). The whiplash shaken infant syndrome: What has been learned? *Child Abuse & Neglect, 10*, 211–221.

D'Angelica, M., et al. (1995). Hypopituitarism secondary to transfacial gunshot wound. *The Journal of Trauma: Injury, Infection, and Critical Care, 39*, 768–771.

Eckelman, W. C. (2004). New tools to monitor stress using non-invasive PET imaging. In K. Pacak, et al. (Eds.), *Stress: Current neuroendocrine and genetic approaches* (Vol. 1018, pp. 487–494). Annals of the New York Academy of Sciences.

Edgerton, V. R., et al. (2004). Plasticity of the spinal neural circuitry after injury. *Annual Review of Neuroscience, 27*, 145–167.

Edvinsson, L. (2005). Neuronal influence on the local control of microcirculation. In E. De Vries & A. Prat (Eds.), *The blood-brain barrier and its microenvironment* (pp. 87–101). New York: Taylor & Francis.
Edwards, O. M., & Clark, J. D. A. (1997). Hypothalamic and pituitary dysfunction. In R. Macfarlane & D. G. Hardy (Eds.), *Outcome after head, neck and spinal trauma* (pp. 120–129). Oxford, UK: Butterworth Heinemann.
Efron, P. A., & Moldawer, L. L. (2004). Cytokines and wound healing: the role of cytokine and anticytokine therapy in the repair process. *Journal of Burn Care and Rehabilitation, 25*, 149–160.
Eichenbaum, A. P., et al. (2007). The medial temporal lobe and recognition memory. In S. E. Hyman, et al., (Eds.), *Annual Review of Neuroscience, 30*, 123–152.
Eichenbaum, H. (2004). An information processing framework for memory representation by the hippocampus. In M. S. Gazzaniga (Ed.), *The cognitive neurosciences, III* (pp. 679–690). Cambridge, MA: Massachusetts Institute of Technology.
Eichenbaum, H. B. (2003). Learning and memory: Brain Systems. In L. R. Squire et al. (Eds.), *Fundamental neuroscience, 2nd Ed.* (pp. 1299–1327). Boston: Academic Press.
Eichenbaum, H. B., et al. (1999). Learning and memory: Systems analysis. In M. J. Zigmond et al. (Eds.) *Fundamental Neuroscience,* pp. 1455–1486. San Diego: Academic Press.
Eisele, J. A., et al. (2006). Nonfatal traumatic brain injury-related hospitalization in very young children-15 states, 1999. *Journal of Head Trauma Rehabilitation, 21*, 537–553.
Ellenbogan, J. M. (2005). Cognitive benefits of sleep and their loss due to sleep deprivation. *Neurology, 64*, E25–E27.
Elliott, R. (2003). Executive functions and their disorders. *British Medical Bulletin, 65,* 49–59.
Ellis, A. W., & Young, A. W. (1988). *Human cognitive neuropsychology*. Hillsdale, NJ: Lawrence Erlbaum.
Elovic, E., & Kirschblum, S. (1999). Epidemiology of spinal cord injury and traumatic brain injury: The scope of the problem. *Topics in Spinal Cord Injury Rehabilitation, 5*, 1–20.
Elovic, E. P. (2003). Anterior pituitary dysfunction after traumatic brain injury. *Journal of Head Trauma Rehabilitation, 6*, 541–543.
Elovic, E. P., et al. (2004). Outcome assessment for spasticity management in the patient with traumatic brain injury. *Journal of Head Trauma Rehabilitation*, 19, 155–177.
Ely, D. L. (1995). Organization of cardiovascular and neurohumoral responses to stress. In G. P. Chrousis et al. (Eds.), *Stress: Basic mechanisms and clinical implications* (Vol. 771, pp. 594–608). Annals of the New York Academy of Science.
Emerson, R. G., & Pedley, T. A. (2004). Electroencephalography and evoked potentials. In W. Bradley et al. (Eds.), *Neurology in clinical practice, 4th Ed.* (Vol. 2, pp. 465–489). Boston: Butterworth-Heinemann.
Engelborghs, K., et al. (2000). Impaired autoregulation of cerebral blood flow in an experimental model of traumatic brain injury. *Journal of Neurotrauma, 17*, 667–677.
Epstein, E. M., et al. (1987). Medical complications of head injury. In P. R. Cooper (Ed.) *Head Injury,* pp. 390–421. Baltimore: Williams and Wilkins.
Epstein, S. (1994). Integration of the cognitive and the psychodynamic unconscious. *American Psychologist, 49*, 709–724.
Eriksson, P. S. (2002). Nerve cells and memory. In V. S. Ramachandran (Ed.), *Encyclopedia of the human brain* (Vol. 3, pp. 305–317). Boston: Academic Press.
Esch, T., et al. (2004). The therapeutic use of the relaxation response in stress-related diseases. *Medical Science Monitor 9*(2), RA23–RA34.
Eslinger, P. J., & Geder, L. (2000). Behavioral and emotional changes after focal frontal lobe damage. In J. Bogousslavsky & J. L. Cummings (Eds.), *Behavior and mood disorders in focal brain lesions* (pp. 217– 260). Cambridge, UK: Cambridge University Press.
Eslinger, P. J., et al. (1992). Developmental consequences of childhood frontal lobe damage. *Archives of Neurology, 49*, 764–769.
Eslinger, P. J., et al. (2007). Cognitive impairments after TBI. In N. D. Zasler, et al. (Eds.), *Brain injury medicine* (pp. 779–790). New York: Demos.
Etcoff, N. L. (1991). Asymmetries in recognition of emotion. In F. Boller & J. Grafman (Eds.), *Handbook of neuropsychology* (Vol. 3, pp. 363–382). New York: Elsevier.
Etscheidt, M. A., et al. (1995). Multidimensional pain inventory profile classifications and psychopathology. *Journal of Clinical Psychology, 51*, 29–66.
Evans, C. C., et al. (2005). Early impaired self-awareness, depression, and subjective well-being following traumatic brain injury. *Journal of Head Trauma Rehabilitation, 20,* 488–500.
Evans, R. S. (1992b). Some observations on whiplash injuries. *Neurological Clinics,* 10, 4, 974–997.

Evans, R. W. (1996). The postconcussion syndrome and the sequelae of mild head injury. In R. W. Evans (Ed.), *Neurology and trauma* (pp. 91–116). Philadelphia: Saunders.

Evans, R. W. (1997). Whiplash injuries. In R. Macfarlane & D. G. Hardy (Eds.), *Outcome after head, neck and spinal trauma* (pp. 359–372). Boston: Butterworth Heinemann.

Evans, R. W. (2001). The postconcussion syndrome. In M. M. Aminoff (Ed.), *Neurology and general medicine, 3rd Ed.* (pp. 499–507). New York: Churchill Livingston.

Evans, R. W., & Wilberger, J. E. (2003). Traumatic disorders. In C. G. Goetz, (Ed.), *Textbook of clinical neurology, 2nd Ed.* (pp. 1129–1153). Philadelphia: Saunders.

Everitt, B. S., & Wyks, T. (1999). *A dictionary of statistics for psychologists.* New York: Oxford University Press.

Everly, G. S., & Horton, A. M. (1988). Cognitive impairment and posttraumatic stress disorder. *Bulletin of the National Academy of Neuropsychologists,* 5, 1.

Ewing-Cobbs, L., et al. (1985). Language disorders after pediatric head injury. In J. Darby (Ed.), *Speech and language evaluation in neurology: Childhood disorders* (pp. 97–111). New York: Grune & Stratton.

Ewing-Cobbs, L., et al. (1989). Intellectual, motor, and language sequelae following closed head injury in infants and preschoolers. *Journal of Pediatric Psychology, 14,* 531–547.

Ewing-Cobbs, L., et al. (1995). Inflicted and non-inflicted traumatic brain injury in infants and preschoolers. *Journal of Head Trauma Rehabilitation, 10*(5), 13–24.

Ewing-Cobbs, L., et al. (1997). Longitudinal neuropsychological outcome in infants and preschoolers with traumatic brain injury. *Journal of the International Neuropsychological Society, 3*(6), 581–591.

Exner, J. E. (1993). *The Rorschach: A comprehensive system. Vol. 1: Basic Foundations (3rd Ed.).* New York: John Wiley.

Ezure, K. (1996). Respiratory control. In B. J. Yates & A. D. Miller (Eds.), *Vestibular autonomic regulation* (pp. 53–84). Boca Raton, FL: CRC Press.

Fahn, S. (2003). Hypokinesia and hyperkinesia. In C. G. Goetz (Ed.), *Textbook of Clinical Neurology, 2nd Ed.* (pp. 279–297). Philadelphia: Saunders.

Faist, E., et al. (2004). The immune response. In E. A. Moore, et al. (Eds.), *Trauma, 5th Ed.* (pp. 1383–1396). New York: McGraw-Hill.

Falaschi, P., et al. (1994). Immune system and the hypothalamus-pituitary-adrenal axis: Common words for a single language. In N. A. Fabris, et al. (Eds.), *Immunomodulation: The state of the art* (Vol. 741, pp. 137–148). Annals of the New York Academy of Sciences.

Farah, M. J., & Feinberg, T. E. (1997). Consciousness of perception after brain damage. *Seminars in Neurology, 17,* 145–152.

Farr, S. P., et al. (1986). Disease process, onset, and course and their relationship to neuropsychological performance. In S. B. Filskov & T. J. Boll (Eds.), *Handbook of Clinical Neuropsychology* (Vol. II, pp. 213–253). New York: Wiley and Sons.

Faust, D. (1991). Forensic Neuropsychology: The art of practicing a science that doesn't exist. *Neuropsychology Review, 2,* 205–231.

Faust, D., et al. (1988). Pediatric malingering: The capacity of children to fake believable deficits on neuropsychological testing. *Journal of Consulting and Clinical Psychology, 56,* 578–582.

Fearnside, M. R., & Simson, D. A. (2005). Epidemiology. In P. L. Reilly & Bullock, R. (Eds.), *Head injury, 2nd Ed.* (pp. 3–25). New York: Oxford University Press.

Feeny, N. C., et al. (2000). Exploring the roles of emotional numbing, depression and dissociation in PTSD. *Journal of Traumatic Stress, 13,* 489–498.

Fehm, H. L., & Born, J. (1989). Non-traditional aspects in the control of cortisol secretion. In H. Weiner, et al. (Eds.), *Neuronal control of bodily function: Basic and clinical aspects* (pp. 250–264). Toronto: Hans Huber.

Feldman, H. M., et al. (1992). Language development after unilateral brain injury. *Brain and Language, 42,* 89–102.

Feldman, J. L., et al. (2003). Breathing: Rhythmicity, plasticity, chemosensitivity. *Annual review of neuroscience, 26,* 239–266.

Feldman, K. W. (1992). Patterned abusive bruises of the buttocks and the pinnae. *Pediatrics, 90,* 633–636.

Felker, B., & Hubbard, J. R. (1998). The influence of mental stress on the endocrine system. In J. R. Hubbard & E. A. Workman (Eds.), *Handbook of stress medicine: An organ system approach* (pp. 69–85). Boca Raton, FL: CRC Press.

Felten, D. L., & Maida, M. E. (2000). Neuroimmunomodulation. In G. Fink (Ed.), *Encyclopedia of stress* (Vol. 1, pp. 37–48). San Diego: Academic Press.

Felten, D. L., & Maida, M.E. (2002). Psychoneuroimmunology. In V. Ramachandran (Ed.), *Encyclopedia of the human brain* (Vol. 4, pp. 103–127). Boston: Academic Press.

Felten, D. L., et al. (1991). Neurochemical links between the nervous and immune systems. In R. Ader, et al. (Eds.), *Psychoneuroimmunology, 2nd Ed.* (Vol. I, pp. 3–25). San Diego, CA: Academic Press.

Fenichel, G. (1993). *Clinical pediatric neurology.* Philadelphia: Saunders.

Fenichel, G. M. (1988). *Clinical pediatric neurology*. Philadelphia: Saunders.

Fenichel, G. M. (1997). *Clinical pediatric neurology, 3rd Ed.* Philadelphia, Saunders.

Fenton, G., et al. (1993). The postconcussional syndrome: Social antecedents and psychological sequelae. *British Journal of Psychiatry*, *162*, 493–497.

Ferin, M. (2001). Gonadotropin secretion, effects of stress on. In *Encyclopedia of stress* (Vol. 3, pp. 598–605). San Diego, CA: Academic Press.

Ferini-Strambi, L., et al. (1996). Cardiac autonomic regulation during sleep in panic disorder. *Journal of Neurology, Neurosurgery & Psychiatry, 61*, 921–922.

Ferrante, M. A., & Wilbourne, A. J. (2000). Plexopathies. In K. H. Levin & M. O. Lüders (Eds.), *Comprehensive clinical neurophysiology* (pp. 201–214). Philadelphia: Saunders.

Ferrari, E., et al. (2000). Variability of interactions between neuroendocrine and immunological functions in physiological aging and dementia of the Alzheimer's type. In A. Conti, et al. (Eds.), *Neuroimmunomodulation: Perspectives at the new millennium* (Vol. 917, pp. 582–586). Annals of the NY Academy of Sciences.

Ferrari, S., & Crosignani, P. G. (1986). Ovarian failure without gonatotropin elevation in a patient with post-traumatic isolated hypogonadotropic hypogonadism. *European Journal of Gynecology & Reproductive Biology, 21*, 241–244.

Ferraro, F. R. (Ed.). (2002). *Minority and cross-cultural aspects of neuropsychological assessment.* Lisse, NL: Swets and Zeitlinger.

Ferrero, M. E., et al. (1992). Conditioning of immune response by anesthetics. In N. Fabris et al. (Eds.), *Ontogenetic and phylogenetic mechanisms of neuroimmunomodulation* (Vol. 650, pp. 331–336). Annals of the New York Academy of Sciences.

Fichtenberg, N. L., et al. (2000). Factors associated with insomnia among post-acute traumatic brain injury survivors. *Brain Injury, 14*, 659–667.

Fichtenberg, N. L., et al. (2002). Insomnia in a post-acute brain injury sample. *Brain Injury, 3*, 197–206.

Fields, H. (1991). Depression and pain. *Neuropsychiatry, Neuropsychology, and Behavioral Neurology, 4,* 83–92.

Findling, J. W., & Tyrell, J.B. (1986). Anterior pituitary and somatomedins: I. Anterior pituitary. In F. G. Greenspan & P. H. Forsham (Eds.), *Basic and clinical endocrinology, 2nd Ed.* (pp. 43–94). Los Altos, CA: Lange Medical.

Fineman, I., et al. (2000). Inhibition of neocortical plasticity during development by a moderate concussive brain injury. *Journal of Neurotrauma,* 17, 739–749.

Finkelstein, J. S. (1996). Osteoporosis. In J. C. Bennett & F. Plum (Eds.), *Cecil textbook of medicine* (pp. 1379– 1384). Saunders: Philadelphia.

Fisch, B. J. (1991). *Spehlmann's EEG Primer,* 2nd ed. New York: Elsevier.

Fischer, R. (1986). On the remembrance of things present: The flashback. In B. B. Wolman & M. Ullman (Eds.), *Handbook of states of consciousness* (pp. 395–427). New York: Van Nostrand Reinhold.

Fisher, S. (1988). Life stress, control strategies and the risk of disease: A psychobiological model. In S. Fisher & J. Reason (Eds.), *Handbook of life stress, cognition and health* (pp. 581–602). New York: John Wiley and Sons.

Fisher, S. (1996). Life stress, personal control, and the risk of disease. In C. L. Cooper (Ed.), *Handbook of stress, medicine, and health* (pp. 123–136). Boca Raton, FL: CRC Press.

Fitch, R., et al. (1997). Neurobiology of speech perception. *Annual Review of Neuroscience*, *20*, 331–353.

Flaada, J. T., et al. (2007). Relative risk of mortality after traumatic brain injury: A population-based study of the role of age and injury severity. *Journal of Neurotrauma, 24*, 435–445.

Flanagan, S. R., et al. (2007). Pharmacological treatment of insomnia for individuals with brain injury. *Journal of Head Trauma Rehabilitation, 22,* 67–70.

Flashman, L. A., et al. (2005). Awareness of deficits. In J. M. Silver, et al. (Eds.), *Textbook of traumatic brain injury* (pp. 355–367). Washington DC: American Psychiatric Publishing.

Fleegal, M. A., et al. (2005). Molecular modulation of the blood-brain barrier during stroke. In E. De Vries & A. Prat (Eds.), *The blood-brain barrier and its microenvironment* (pp. 359–386). New York: Taylor & Francis.

Fleming, J., et al. (2008). Predictors of prospective memory in adults with traumatic brain injury. *Journal of the International Neuropsychological Society, 14*, 283–831.

Fletcher, M. A., & Hubbard, M. D. (1998). Influence of mental stress on the endocrine system. In J. R. Hubbard & E. A. Workman (Eds.), *Handbook of stress medicine: An organ system approach* (pp. 69–85). Boca Raton, FL: CRC Press.

Fletcher, M. A., et al. (1998). In J. R. Hubbard & E. A. Workman, (Eds.), *Handbook of stress medicine: An organ system approach* (pp. 69–85). Boca Raton, FL: CRC Press.

Flood, J. F., et al. (1987). Antagonism of sndogenus opioids modulates memory processing. *Brain Research,* 422, 218, 234.

Florence, S. L. (2002). The changeful mind: Plasticity in the somatosensory system. In R. J. Nelson (Ed.), *The somatosensory system: Deciphering the brain's own body image* (pp. 335–366). Boca Raton, FL: CRC Press.

Forrester, G., et al. (1994). Measuring post-traumatic amnesia (PTA): An historical review. *Brain Injury 8,* 175–184.

Foster, C. M. (1996). Adolescent menstrual abnormalities. In F. Lifschitz (Ed.), *Pediatric endocrinology, 3rd Ed.* (pp. 223–234). New York: Dekker.

Fowler, C. J. (2004). Neurological causes of bladder, bowel, and sexual dysfunction. *Neurology in clinical practice, 4th Ed.* (Vol. 1, pp. 419–431). Boston: Butterworth-Heinemann.

Fox, D. D., et al. (1995a). Base rates of postconcussive symptoms in health maintenance organization patients and controls. *Neuropsychology, 9,* 606–611.

Fox, D. D., et al. (1995b). Post-concussive symptoms: Base rates and etiology in psychiatric patients. *The Clinical Neuropsychologist, 9,* 89–92.

Fraser, D. M. (1994). Whiplash: A total body approach. *Journal of Neurological and Orthopaedic Medicine and Surgery, 15,* 10–12.

Freeman, B. D., et al. (2004). Genetic and genomic aspects of the immunoinflammatory response. In E. A. Moore, et al. (Eds.), *Trauma, 5th Ed.* (pp. 1197–1209). New York: McGraw-Hill.

Fricchione, G., & Howanitz, E., (1985). Aprosodia and alexithymia: A case report. *Psychotherapy and Psychosomatics, 43,* 156–160.

Frieboes, R. M., et al. (1999). Nocturnal hormone secretion and the sleep EEG in patients several months after traumatic brain injury. *Journal of Neuropsychiatry and Clinical Neurosciences, 11,* 354–360.

Friedman, E. M., & Irwin, M. R. (1995). A role for CRH and the sympathetic nervous system in stress-induced immunosuppression. In G. P. Chrousis, et al. (Eds.). *Stress: Basic mechanisms and clinical implications* (Vol. 771, pp. 396–418). Annals of the New York Academy of Science.

Friedman, M. J. (1990). Interrelationships between biological mechanisms and pharmacotherapy of posttraumatic stress disorder. In M. E. Wolf & A. D. Mosnaim (Eds.), *Posttraumatic stress disorder* (pp. 204–225). Washington DC: American Psychiatric Press.

Friedman, M. J. (1999). Progress in the psychobiology of post-traumatic stress disorder: An overview. *Seminars in clinical neuropsychiatry, 4,* 230–233.

Friedman, M. J., & McEwen, B. S. (2004). Posttraumatic stress disorder, allostatic load and medical illness. In P. P. Schnurr & B. L. Green (Eds.), *Trauma and health: Physical consequences of exposure to extreme stress* (pp. 157–188). Washington DC: American Psychological Association.

Frohman, L. A. (1995). Diseases of the anterior pituitary. In P. Felig, et al. (Eds.), *Endocrinology and metabolism, 3rd Ed.* (pp. 289–384). New York: McGraw Hill.

Frohman, L. A., & Felig P. (2001). Introduction to the endocrine system. In P. Felig & L. A. Frohman (Eds.), *Endocrinology and metabolism, 4th Ed.* (pp. 3–17). New York: McGraw Hill.

Frueh, B. C., et al. (2005). Documented combat exposure of US veterans seeking treatment for combat-related post-traumatic stress disorder. *British Journal of Psychiatry, 186,* 467–472.

Fuchs, A. F. (2002). The cerebellum as a device for the coordination of movement. In S. M. Highstein & W. T. Thach (Eds.), *The cerebellum: Recent developments in cerebellar research* (Vol. 978, pp. 46–49). Annals of the New York Academy of Sciences.

Fujiwara, R., & Yokoyama, M. M. (1994). Influence of pain stimulation on interleukin-2 production in mice. In N. A. Fabris, et al. (Eds.), *Immunomodulation: The state of the art* (Vol. 741, pp. 244–251). Annals of the New York Academy of Sciences.

Fukuda, K., et al. (1995). The blood-brain barrier disruption to circulating proteins in the early period after fluid percussion brain injury in rats. *Journal of neurotrauma, 12,* 315–324.

Fukutake, T., et al. (2000). Roller coaster headache and subdural hematoma. *Neurology, 54,* 264.

Fuller, C. A., & Fuller, C. M. (2002). Circadian rhythms. In V. S. Ramachandran (Ed.), *Encyclopedia of the human brain* (Vol. I, pp. 793–812). Boston: Academic Press.

Fuller, R. W. (1996). Mechanisms and functions of serotonin neuronal systems. In J. C. Crawley & S. McLean (Eds.), *Neuropeptides: Basic and clinical advances* (Vol. 780, pp. 176–184). Annals of the New York Academy of Sciences.

Fumamichi, R., et al. (2004). Associative memory: Representation, activation, and cognitive control. In M. S. Gazzaniga (Ed.), *The cognitive neurosciences, III* (pp. 905–913). Cambridge, MA: Massachusetts Institute of Technology.

Funahashi, S. (2001). Neuronal mechanisms of executive control by the prefrontal cortex. *Neuroscience Research, 39*(2), 147–165.

Furman, J. M. (2003). Vestibular loss. In M. J. Aminoff & R. B. Daroff (Eds.), *Encyclopedia of the neurological science* (Vol. 4, pp. 655–656). Boston: Academic Press.

Fuster, J. M. (1989). *The prefrontal cortex.* New York: Raven Press.

Fuster, J. M. (1995). Temporal processing. In J. Grafman, et al. (Eds.), *Structure and functions of the human prefrontal cortex* (Vol. 769, pp. 173–181). Annals of the New York Academy of Sciences.

Fuster, J. M. (1997). *The prefrontal cortex, 3rd Ed.* Philadelphia: Lippincott-Raven.

Gaab, J., et al. (2005). Reduced reactivity and enhanced negative feedback sensitivity of the hypothalamus-pituitary-adrenal axis in chronic whiplash-associated disorder. *Pain, 119*, 219–225.

Gabella, B. A., et al. (2007). Comparison of nursing home resident with and without traumatic brain injury: Use of the minimum data set. *Journal of Head Trauma Rehabilitation, 10*, 368–376.

Gaetz, M. (2004). The neurophysiology of brain injury. *Clinical Neurophysiology, 115*, 4–18.

Gagnon, I., et al. (2004). Children show decreased dynamic balance after mild traumatic brain injury. *Archives of Physical Medicine and Rehabilitation, 85,* 444–452.

Gagnon, I., et al. (2005). Exploring children's self-efficacy related to physical activity performance after a mild traumatic brain injury. *Journal of Head Trauma Rehabilitation, 20*, 436–449.

Gainotti, G. (1972). Emotional behavior and hemispheric side of the lesion. *Cortex, 8*, 41–55.

Gainotti, G. (1991). Frontal lobe damage and disorders of affect and personality. In M. Swash & J. Oxbury (Eds.) *Clinical Neurology,* vol. 1, 71–81. Edinburgh: Churchill Livingston.

Galdi, S., et al. (2008). Automatic mental associations predict future choices of undecided decision-makers. *Science, 321*, 1100–1102.

Galloway, et al. (2008). Diffusion-weighted imaging improves outcome prediction in pediatric traumatic brain injury. *Journal of Neurotrauma,* 25, 1153–1162.

Gallowitch-Puerta, M., & Tracey, K. J. (2005). Immunologic role of the cholinergic anti-inflammatory pathway and the nicotiic acetylcholine $\alpha 7$ receptor. In R. M. Steinman (Ed.), *Human immunology: Patient-based research* (Vol. 162, pp. 209–219). Annals of the New York Academy of Sciences.

Ganong, C. A., & Kappy, M. S. (1993). Cerebral salt wasting in children. *American Journal of Diseases of Children, 147*(2), 167–169.

Garada, B., et al. (1997). Neuroimaging in closed head injury. *Seminars in Clinical Neuropsychiatry,* 2, 3, 188–195.

Garcia, M. P., et al. (1998). Neuropsychological evaluation of everyday memory. *Neuropsychology Review, 8*, 203–227.

Gardner, E. P., & Martin, J. H. (2000). Coding of sensory information. In E. R. Kandel, et al. (Eds.), *Principles of neural science, 4th ed.* (pp. 411–429). New York: McGraw Hill.

Gardner, E. P., et al. (2000). The bodily senses. In E. R. Kandel, et al. (Eds.), *Principles of neural science, 4th Ed.* (pp. 430–450). New York: McGraw Hill.

Gardner, R. (2001). Evolutionary perspectives on stress and affective disorder. *Seminars in Clinical Psychiatry, 6,* 32–42.

Garrett, T. J., & Hubbard, S. A. (1992). Vertebral artery thrombosis due to motor vehicle accident. *Annals of Emergency Medicine*, *22*, 141–143.

Garske, G. G., & Thomas, K. R. (1992). Indexes of psychosocial adjustment following severe traumatic brain injury. *Rehabilitation Counseling Bulletin,* 36, 1, 44–52.

Garvey, M. A., et al. (1998). Emergency brain computed tomography in children with seizures: Who is most likely to benefit? *The Journal of Pediatrics*, *133*, 664–669.

Gasquoine, P. G. (1997). Postconcussion symptoms. *Neuropsychology Review*, *7*, 77–85.

Gasquoine, P. G. (2000). Postconcussional symptoms in chronic back pain. *Applied Neuropsychology, 7,* 83–89.

Gates, J. R., & Mercer, K. (1995). Nonepileptic events. *Seminars in Neurology,* 15, 167–174.

Gatti, S., et al. (2000). Thermotolerance, thermoresistance, and thermosensitivity. In G. Fink (Ed.), *Encyclopedia of stress* (Vol. 3, pp. 585–594). San Diego, CA: Academic Press.

Gay, G. W. (1999). *The headache handbook.* Boca Raton, FL: CRC Press.

Ge, Y., & Grossman, R. E. (2008, March). Thalamic involvement following mild brain injury: In vivo occult pathology by quantitative MRI. Paper presented at the 2008 Annual Convention of The New York Academy of Traumatic Brain Injury, New York.

Gean, A. D. (1994). *Imaging of head trauma.* New York: Raven Press.

Gebhart, A. L., et al. (2002). Role of the posterolateral cerebellum in language. In S. M. Highstein & W. T. Thach (Eds.), *The cerebellum: Recent developments in cerebellar research* (Vol. 978, pp. 318–333). Annals of the New York Academy of Sciences.

Geddes, J. F., & Whitwell, J. L. (2003). Neuropathology of fatal infant head injury. *Journal of Neurotrauma, 20,* 905.

Geenen, V., et al. (1994). Cytocrine signaling in the thymus network: Implications for central t-cell tolerance of neuroendocrine functions. In N. A. Fabris et al., (Eds.), *Immunomodulation: The state of the art* (Vol. 741, pp. 338–357). Annals of the New York Academy of Sciences.

Geeraerts, T., et al. (2006). Changes in cerebral energy metabolites induced by impact-acceleration brain trauma and hypoxic-hypotensive injury in rats. *Journal of Neurotrauma, 23,* 1059–1071.

Gelber, D. A., & Callahan, C. D. (2004). The neurological examination of the patient with traumatic brain injury. In M. J. Ashley (Ed.), *Traumatic brain injury: Rehabilitative treatment and case management* (pp. 3–26). Boca Raton, FL: CRC Press.

Genesalingam, K., et al. (2007). Self-regulation as a mediator of the effects of childhood traumatic brain injury on social and behavioral functioning. *Journal of the International Neuropsychological Society, 13,* 298–311.

Gennarelli, T. A. (1983). Head injury in man and experimental animals: Clinical aspects. *Acta Neurochirurgica, Suppl. 32,* 1–13.

Gennarelli, T. A. (1986). Mechanisms and pathophysiology of cerebral concussion. *Journal of Head Trauma Rehabilitation,* 1 (2), 23–29.

Gennarelli, T. A. (1987). Cerebral concussion and diffuse brain injuries. In P. R. Cooper (Ed.) *Head Injury,* 2nd ed., pp. 108–124. Baltimore: Williams and Wilkins.

Gennarelli, T. A. (1993). Cerebral concussion and diffuse brain injuries. In P. R. Cooper (Ed.) *Head Injury,* 3rd ed., pp. 137–158. Baltimore: Williams and Wilkins.

Gennarelli, T. A., & Graham, D. I. (1998). Neuropathology of head injuries. *Seminars in neuropsychiatry, 3,* 160–175.

Gennarelli, T. A., & Graham, D. I. (2005). Neuropathology. In J. M. Silver, et al. (Eds.), *Textbook of traumatic brain injury* (pp. 27–50). Washington, DC: American Psychiatric Publishing.

Gennarelli, T. A., et al. (1982). Physiological response to angular rotation of the head. In R. G. Grossman & P. L. Gildenberg (Eds.), *Brain injury: Basic and clinical aspects* (pp. 129–140). New York: Raven Press.

Gentry, L. R. (1989). Facial trauma and associated brain damage. *Radiologic Clinics of North America, 27,* 435–446.

Genuth, S. M. (1998a). General principles of endocrine physiology. In R. M. Berne & M. N. Levy (Eds.), *Physiology, 4th Ed.* (pp. 779–799). St. Louis, MO: Mosby.

Genuth, S. M. (1998b). The adrenal glands. In R. M. Berne & M. N. Levy (Eds.), *Physiology: 4th Ed.* (pp. 930–964). St. Louis: Mosby.

Genuth, S. M. (1998c). Whole body metabolism. In R. M. Berne & M. N. Levy (Eds.), *Physiology, 4th Ed.* (pp. 800–821). St. Louis: Mosby.

Georgiades, A., & Fredrikson, M. (2000). Hyperreactivity (cardiovascular). In G. Fink (Ed.), *Encyclopedia of stress* (Vol. 3, pp. 192–200). San Diego: Academic Press.

Georgopoulos, A. P. (1989). Visuomotor coordination in reaching and locomotion. *Science, 245,* 1209–1210.

Georgopoulos, A. P., et al. (1992). The motor cortex and the coding of force. *Science, 256,* 1692–1695.

Germaine, A., et al. (2004). Clinical correlates of poor sleep quality in posttraumatic stress disorder. *Journal of Traumatic Stress, 17,* 487–484.

Ghesquiere, A., et al. (2008). Adolescents' and parents' agreement on posttraumatic stress disorder symptoms and functioning after adolescent injury. *Journal of Traumatic Stress,* 21 (5), 487–491.

Ghez, C., & Krakauer, J. (2000). The organization of movement. In E. R. Kandel et al. (Eds.), *Principles of neural science, 4th ed.* (pp. 653–673). New York: McGraw Hill.

Ghez, C., & Thach, W. T. (2000). The cerebellum. In E. R. Kandel, et al. (Eds.), *Principles of neural science, 4th ed.* (pp. 832–852). New York: McGraw Hill.

Ghigo, E., et al. (2005). Consensus guidelines on screening for hypopituitarism following traumatic brain injury. *Brain Injury, 19,* 711–724.

Ghosh, P. K., & O'Dorisio, T. M. (2001). Gastrointestinal hormones and characinoid syndrome. In P. Felig & L. A. Frohman (Eds.), *Endocrinology and metabolism, 4th Ed.* (pp. 1317–1353). New York: McGraw-Hill.

Giebisch, G., & Windhager, E. (2005). Integration of salt and water balance. In W. Boron & E. Boulpaep (Eds.), *Medical physiology, up-dated edition* (pp. 861–876). Philadelphia: Saunders.

Giedd, J. N., et al. (1999). Brain development during childhood and adolescence: a longitudinal MRI study. *Nature Neuroscience, 2*(10), 861–863.

Giles, G. M., & Manchester, D. (2006). Two approaches to behavior disorder after traumatic brain injury. *Journal of Head Trauma Rehabilitation, 21*, 168–178.

Gill, G. N. (1985). The hypothalamic-pituitary control system. In J. B. West, *Best and Taylor's Physiological Basis of Medical Practice,* 11th ed., pp. 856–871. Baltimore: Williams and Wilkins.

Gill, G. N. (1996). Principles of endocrinology. In J. C. Bennett & F. Plum (Eds.), *Cecil Textbook of Medicine, 20th Ed.* (pp. 1176–1185). Saunders: Philadelphia.

Gilman, S., et al. (1981). *Disorders of the cerebellum.* Philadelphia: F. A. Davis.

Gioia, G. A., et al. (2008). Improving identification and diagnosis of mild traumatic brain injury with evidence: Psychometric support for acute concussion evaluation. *Journal of Head Trauma Rehabilitation, 23*, 230–242.

Giulian, D. (1994). The consequences of inflammation after injury to the central nervous system. In S. K. Salzman & A. I. Faden (Eds.), *The neurobiology of central nervous system trauma* (pp. 155–164). New York: Oxford University Press.

Giza, C. C. (2000). Ionic and metabolic consequences of concussion. In R. C. Cantu (Ed.), *Neurologic athletic head and spine injuries* (pp. 80–100). Philadelphia: Saunders.

Gizzi, M., et al. (2003). Vestibular disease and cognitive dysfunction. *Journal of Head Trauma Rehabilitation, 18*, 398–407.

Glaesser, J., et al. (2004). Posttraumatic stress disorder in patients with traumatic brain injury. *BMC Psychiatry, 4*(5), 1–6.

Glang, A., et al. (2007). Using interactive multimedia to teach parent advocacy skills: An exploratory study. *Journal of Head Trauma Rehabilitation, 12*, 198–205.

Glang, A., et al. (2008). Validated instructional practices: Application to students with traumatic brain injury. *Journal of Head Trauma Rehabilitation, 23*, 243–251.

Glezerman, T. B., & Balkoski, V. I. (1999). *Language, thought, and the brain.* New York: Kluwer Academic.

Gluhbegovic, N., & Williams, T. H. (1980). *The human brain: A photographic guide.* New York: Harper and Row.

Goebel, J. A. (2001). The ten-minute examination of the dizzy patient. *Seminars in Neurology, 21,* 391–398.

Goetz, C. G. (2003). Excitotoxins and excitotoxicity. In M. J. Aminoff & R. B. Daroff (Eds.), *Encyclopedia of the neurological sciences* (Vol. 2, pp. 316–317). Boston: Academic Press.

Goetz, C. G., & Pappert, E. J. (1996). Movement disorders: Post-traumatic syndromes. In R. W. Evans (Ed.), *Neurology and trauma* (pp. 569–580). Philadelphia: W. B. Saunders.

Gold, P. W., et al. (1987). Physiological, diagnostic and pathophysiological implications of corticotropin-releasing hormone. In C. B. Nemeroff & P. T. Loosen (Eds.), *Handbook of clinical psychoneuroendocrinology* (pp. 85–105). New York: Guilford Press.

Goldenberg, G. (2002). Body perception disorders. In V. S. Ramachandran (Ed.), *Encyclopedia of the human brain* (Vol. I, pp. 443–458). Boston: Academic Press.

Golden, C. J., et al. (1983). *Clinical Neuropsychology: Interface with Neurologic and Psychiatric Disorders.* New York: Grune and Stratton.

Goldstein, D. S. (2003). Imaging of the autonomic nervous system: Focus on cardiac sympathetic innervation. *Seminars in Neurology, 23*(4), 423–433.

Goldstein, F. C., et al. (1994). Neurobehavioural consequences of closed head injury in older adults. *Journal of Neurology, Neurosurgery and Psychiatry,* 57, 961–966.

Goldstein, J. (1991). Posttraumatic headache and the postconcussion syndrome. *Medical Clinics of North America, 75*, 641–651.

Goldstein, K. (1942). *Aftereffects of brain injuries in war.* New York: Grune & Stratton.

Goldstein, S., & Halbreich, U. (1987). Hormones and stress. In C. B. Nemeroff & P. T. Loosem (Eds.), *Handbook of clinical psychoneuroendocrinology* (pp. 460–469). New York: The Guilford Press.

Goozee, J. V., et al. (2000). Kinematic analysis of tongue movements in dysarthria following traumatic brain injury using electromagnetic articulography. *Brain Injury, 14*(2), 153–174.

Goozee, J. V., et al. (2001). Physiological assessment of tongue function in dysarthria following traumatic brain injury. *Logopedics, Phoniatrics, Vocology, 26*(2), 51–65.

Gorczynski, R. M. (1996). Conditioned immunity to *L. Major* in young and aged mice. In H. Friedman, et al. (Eds.), *Psychoneuroimmunology, stress and infection* (pp. 137–151). Philadelphia: Lippincott-Raven.

Gordon, M. R. (2006, July 12). A platoon's mission: Seeking and destroying explosives in disguise. *The New York Times.*

Gordon, W. A., et al. (2006). Treatment of post-TBI executive dysfunction. *Journal of Head Trauma Rehabilitation, 21*, 156–167.

Gore, A. C., & Roberts, J. L. (2003). Neuroendocrine systems. In L. Squire et al. (Eds.), *Fundamental neuroscience, 2nd Ed.* (pp. 1031–1065). London: Academic Press.

Gorenstein, E. E., et al. (1989). Performance of inattentive-overactive children on selected measures of prefrontal-type function. *Journal of Clinical Psychology, 45*, 619–632.

Gouvier, W. D., et al. (2002). Base rate analysis in cross-cultural clinical psychology: Diagnostic accuracy in the balance. *Minority and cross-cultural aspects of neuropsychological assessment* (pp. 375–386). Lisse, NL: Swets and Zeitlinger.

Goya, R. G., et al. (1994). In vitro studies on the thymus-pituitary axis in young and old rats. In N. A. Fabris, et al. (Eds.), *Immunomodulation: The state of the art* (Vol. 741, pp. 100–107). Annals of the New York Academy of Sciences.

Grafman, J. (1995). Similarities and distinctions among current models of prefrontal cortical functions. In J. Grafman, et al. (Eds.), *Structure and functions of the human prefrontal cortex* (Vol. 769, pp. 337–368). Annals of the New York Academy of Sciences.

Graham, D. I., et al. (1987). Pathology of brain damage in head injury. In P. R. Cooper (Ed.) *Head Injury,* 2nd ed., pp. 72–88. Baltimore: Williams and Wilkins.

Graham, D. I., et al. (1993). Pathology of brain damage in head injury. In P. R. Cooper (Ed.) *Head Injury,* 3rd ed., pp. 91–113. Baltimore: Williams and Wilkins.

Graham, D. I., & Gennarelli, T. A. (2000). Pathology of brain damage after head injury. In P. R. Cooper & J. G. Golfinos (Eds.), *Head injury, 4th Ed.* (pp. 133–153). New York: McGraw Hill.

Granacher, R. P. (2003). *Traumatic brain injury.* Boca Raton, FL: CRC Press.

Granier, M. (2006). Mild head trauma: Complications and acousticovestibular sequelae (French, Abstract). *Revue de Stomatologie et de Chirurgie Maxillo-Faciale, 107*, 253–263.

Granner, D. K., (1993). Hormones of the adrenal cortex. In R. K. Murray, et al. (Eds.), *Harpers Biochemistry* (pp. 523–535). New York: McGraw-Hill.

Granner, M. A. (1996). Post-traumatic epilepsy. In M. Rizzo and D. Tranel (Eds.) *Head Injury and Postconcussive Syndrome,* pp. 227–241. New York: Churchill-Livingston.

Graziano, M. (2006). The organization of behavioral repertoire in motor cortex. *Annual Review of Neuroscience, 29*, 105–134.

Green, A. I., et al. (1995). Mood disorders: Biochemical aspects. In H. I. Kaplan & B. J. Sadock (Eds.) *Comprehensive Textbook of Psychiatry,* 6th ed., pp. 1089–1102. Baltimore: Williams and Wilkins.

Greenspan, F. S. (2004). The thyroid gland. In F. S. Greenspan & D. G. Gardner (Eds.), *Basic and clinical endocrinology, 7th Ed.* (pp. 215–294). New York: Lange Medical Books.

Green, A. H. (1985). Children traumatized by physical abuse. In, Eth, S. and Pynoos, R.S. *Post-traumatic stress disorder in children* (pp. 133–154). Washington: American Psychiatric Association.

Greiffenstein, M. F., & Baker, W. J. (2002). Neuropsychological and psychosocial correlates of adult Arithmetic deficiency. *Neuropsychology, 16*, 451–458.

Griesemer, D., & Mautes, A. M. (2007). Closed head injury causes hyperexcitabity in rat hippocampal CA1 but not CA3 pyramidal cells. *Journal of Neurotrauma*, *24*, 1823–1832.

Grillner, S. (2003). Fundamentals of motor systems. In L. R. Squire, et al., (Eds.). *Fundamental neuroscience (2nd Ed.)* (pp. 753–766). Boston: Academic Press.

Gronwall, D., et al. (1997). Effect of mild head injury during the preschool years. *Journal of the International Neuropsychological Society, 3*, 592–597.

Grossman, W. F., & Sanfield, J. A. (1994). Hypothalamic atrophy presenting as amenorrhea and sexual infantilism in a female adolescent: A case report. *Journal of Reproductive Medicine, 39,* 738–740.

Grumbach, M. M., & Styne, D. M. (1998). Puberty: Ontogeny, neuroendocrinology, physiology, & disorders. In J. D. Wilson, et al. (Eds.), *Williams textbook of endocrinology, 9th Ed.* (pp. 1509–1625). Philadelphia: Saunders.

Grumbach, M. M., & Styne, D. M. (2003). Puberty: Ontogeny, neuroendocrinolgy, physiology, and disorders. In P. R. Larsen, et al. (Eds.), *Williams textbook of endocrinology, 10th Ed.* (pp. 1115–1286). Philadelphia: Saunders.

Gudeman, S. K., et al. (1989). Indications for operative treatment and operative technique in closed head injury. In D. P. Becker and S. K. Gudeman (Eds.) *Textbook of Head Injury,* pp. 138–181. Philadelphia: Saunders.

Guerrero, J. L., et al. (2000). Emergency department visits associated with traumatic brain injury: United States, 1995–1996. *Brain Injury*, *14*, 181–186.

Guesquiere, A., et al. (2008). Adolescents' and parents' agreement on posttraumatic stress disorder symptoms and functioning after adolescent injury. *Journal of Traumatic Stress, 21*, 487–491.

Guggenheim, F. G. (2000). Somatoform disorders. In B. J. Sadock & V. A. Sadock (Eds.), *Comprehensive textbook of psychiatry, 7th Ed.* (pp. 1504–1532). Philadelphia: Lippincott Williams & Wilkins.

Guilford, J. P. (1985). Structure-Intellect Model. In B. B. Wolman (Ed.), *Handbook of intelligence* (pp. 225–266). New York, NY: Wiley.

Gurdjian, E. S. (1975). Relative movements of scalp, skull and intracranial contents at the time of impact. *Impact Head Injury*, 163–180.

Gurdjian, E. S., & Gurdjian, E. S. (1975). Re-evaluation of the biomechanics of blunt impact injury of the head. *Surgery, Gynecology and Obstetrics, 140,* 845–850.

Gurdjian, E. S., Hodgson, V. R., Thomas, L. M., & Patrick, L. M. (1968). Significance of relative movements of scalp, skull, and intracranial contents during impact injury of the head. *Journal of Neurosurgery,* 29, 70–72.

Guskiewicz, K. M. (2001). Postural stability assessment following concussion: One piece of the puzzle. *Clinical Journal of Sport Medicine, 11*, 82–189.

Guthkelch, A. N. (1980). Posttraumatic amnesia, post-concussional symptoms and accident neurosis. *European Neurology, 19*, 91–102.

Guyton, A. G., & Hall, J. E. (1996). *Textbook of Medical Physiology,* 9th ed. Philadelphia: Saunders.

Guyton, A. C., & Hall, J. E. (2000). *Textbook of medical physiology, 10th Ed.* Philadelphia: W. B. Saunders.

Guyton, A. C., & Hall, J. E. (2006). *Textbook of medical physiology, 11th Ed.* Philadelphia: Elsevier Saunders.

Haas, D. C. (1993a). Acute posttraumatic headache. In J. Olesen, P. Tfelt-Hansen, & Welch, K. M. A. (Eds.) *The Headaches,* pp. 623–627. New York: Raven.

Haas, D. C. (1993b). Chronic posttraumatic headache. In J. Olesen, P. Tfelt-Hansen, & Welch, K. M. A. (Eds.) *The Headaches,* pp. 629–637. New York: Raven.

Habener, J. F. (2003). Genetic control of peptide hormone formation. In P. R. Larsen, et al. (Eds.), *Williams textbook of endocrinology, 10th Ed.* (pp. 17–34). Philadelphia: Saunders.

Hadani, M., et al. (1985). Unusual delayed onset of diabetes insipidus following closed head trauma. Case report. *Journal of Neurosurgery, 63,* 456–458.

Hadani, M., et al. (1997). Transiently increased basilar artery flow velocity following severe head injury: A time course Transcranial Doppler study. *Journal of Neurotrauma,* 14, 629–636.

Hadley, M. N., et al. (1989). The infant whiplash-shake injury syndrome: A clinical and pathological study. *Neurosurgery, 24*, 536–540.

Haerer, A. F. (1992). *DeJong's the neurological examination, 5th Ed.* Philadelphia: J. B. Lippincott.

Hain, T. C., & Helminski, J. O. (2003). Vestibular reflexes. In M. J. Aminoff & R. B. Daroff (Eds.), *Encyclopedia of the neurological science* (Vol. 4, pp. 655–656). Boston: Academic Press.

Hain, T. C., & Micco, A. G. (2003). Cranial nerve VIII: Vestibulocochlear system. In C. G. Goetz (Ed.), *Textbook of clinical neurology, 2nd Ed.* (pp. 195–210). Philadelphia: Saunders.

Hajszan, T., & MacLuskey, N. J. (2006). Neurologic links between epilepsy and depression in women: Is hippocampal neuroplasticity the key? *Neurology*, *66*(6), S13–S22.

Hales, R. E., & Yudofsky, S. C. (1987). *Textbook of neuropsychiatry.* Washington, DC: American Psychiatric Press.

Halgren, D. (1992). Emotional neurophysiology of the amygdala within the context of human cognition. In J. P. Aggleton (Ed.), *The amygdala: Neurobiological aspects emotion, memory, and mental dysfunction* (pp. 191–228). New York: Wiley-Liss.

Hall, D. A., et al. (2003). A French accent after corpus callosum infarct. *Neurology, 60*, 1551.

Hall, E. D. (1996). Free radicals and lipd peroxidation. In R. K. Narayan, J. E. Wilberger, & J. T. Povlishock (Eds.) *Neurotrauma,* pp. 1405–1419. New York: McGraw-Hill.

Hall, H. (1993). Criminal-forensic neuropsychology of disorders of executive functions. In H. V. Hall and R. J. Sbordone (Eds.) *Disorders of Executive Functions: Civil and Criminal Law Applications.* Winter Park, FL: PMD Publishers Group.

Hall, H. V., et al. (2008). Malingered pain and memory deficits in civil-forensic contexts. In H. Hall (Ed.), *Forensic psychology and neuropsychology for criminal and civil cases* (pp. 565–620). Boca Raton, FL: Taylor & Francis.

Hall, R. C. W., et al. (2005). Definition, diagnosis, and forensic implications of postconcussional syndrome. *Psychosomatics*, *46*, 195–202.

Halmaji, M. (2003). Vestibulocochlear nerve (cranial nerve VIII). In M Aminoff & R. B. Daroff (Eds.), *Encyclopedia of the neurological science* (Vol. 4, pp. 671–673). Boston: Academic Press.

Halpern, C. H., et al. (2008). Traumatic coagulopathy: The effect of brain injury. *Journal of Neurotrauma, 25*, 997–1001.

Halter, A. M., et al. (1996). The impact of posttraumatic seizures on 1-year neuropsychological and psychosocial outcome of head injury. *Journal of the International Neuropsychological Society,* 2, 494–504.

Ham, L. P., et al. (1994). Psychopathology in individuals with posttraumatic headaches and other pain types. *Cephalagia,* 14 (2), 118–126.

Hamburger, L. K., & Lohr, J. R. (1984). *Stress and Stress Management.* New York: Springer.

Hammen, V. L., & Yorkston, K. M. (1996). Respiratory patterning and variability in dysarthric speech. In A. Donald, et al. (Eds.), *Disorders of motor speech* (pp. 181–192). Baltimore: Paul H. Brooks.

Hammer, M. B., & Arana, G. W. (2000). Beta-adrenergic blockers. In G. Fink (Ed.), *Encyclopedia of stress* (Vol. 1, pp. 37–48). San Diego: Academic Press.

Hampson, E. (2002). Sex differences in human brain and cognition: The influence of sex steroids in early and adult life. In J. B. Becker, et al. (Eds.), *Behavioral neurology, 2nd Ed.* (pp. 579–628). Cambridge, MA: The MIT Press.

Hanks, R. A., et al. (1999). Emotional and behavioral adjustment after traumatic brain injury. *Archives of Physical Medicine and Rehabilitation, 80,* 991–999.

Hansen, J. R., & Cook, J. S. (1993). Post-traumatic neuroendocrine disorders. *Physical medicine and rehabilitation, 7,* 569–580.

Hansen, L. A., et al. (2003). The transfer of immunity from mother to child. In N. Chiorazzi, et al. (Eds.), *Immune mechanisms and disease* (Vol. 987, pp. 199–206). Annals of the New York Academy of Sciences.

Hansen-Grant, S. M., et al. (1998). Neuroendocrine and immune system pathology in psychiatric disease. In A. F. Schatzberg & C. B. Nemeroff (Eds.), *Textbook of psychopharmacology* (pp. 171–187). Washington DC: American Psychiatric Press.

Hanson, L. A., et al. (2003). The transfer of immunity from mother to child. In N. Chiorazzi, et al. (Eds.), *Immune mechanisms and disease* (Vol. 987, pp. 199–206). Annals of the New York Academy of Sciences.

Harden, C. L. (2006). The adolescent female with epilepsy: Mood, menstruation, and birth control. *Neurology, 66*(6), S3–S4.

Harris, D. S., et al. (2005). Psychoneuroendocrinology. In B. J. Sadock & V. A. Sadock (Eds.), *Comprehensive textbook of psychiatry, I, 8th Ed.* (pp. 126–137). Philadelphia: Lippincott Williams & Wilkins.

Harris, L. J., & Carlson, D. F. (1988). Pathological left-handedness: An analysis of theories and evidence. In D. L. Molfese & S. J. Segalowitz (Eds.), (1989). *Brain lateralization in children* (pp. 289–374). New York: The Guilford Press.

Harris, J. O., & Berger, J. R. (1991). Clinical approach to stupor and coma. In W. G. Bradley, R. B. Daroff, G. M. Fenichel, & C. D. Marsedn (Eds.) *Neurology in Clinical Practice,* vol. 1, pp. 43–63. Boston: Butterworth-Heinemann.

Hart, R. P., et al. (2003). Cognitive impairment in patients with chronic pain: The significance of stress. *Current Pain and Headache Reports,* 7, 116–127.

Harter, L., et al. (2001). Caspase-3 activity is present in cerebrospinal fluid from patients with traumatic brain injury. *Journal of Neuroimmunology, 1211,* 76–78.

Hartlage, L. G. (1990). *Neuropsychological evaluation of head injury.* Sarasota, Fl: Professional Resource Exchange.

Hartlage, L. C. (2002). Neuropsychological assessment, pediatric. In V. S. Ramachandran (Ed.), *Encyclopedia of the human brain* (Vol. 3, pp. 595–600). Boston: Academic Press.

Hartline, K. M., et al. (1996). Postmortem and cerebrospinal fluid studies of corticotroin-releasing factor in humans. In J. C. Crawley and S. McLean (Eds.) *Neuropeptides: Basic and Clinical Advances,* pp. 96–105. Annals of the New York Academy of Sciences, vol. 780.

Hartmann, E. L. (2003). Nightmares. In *Encyclopedia of the neurological sciences* (Vol. 3, pp. 628–631). Boston: Academic Press.

Harvard Medical School. (2009). Learning to walk: A graduate course. *Harvard Health Letter, 34*(4), 1–3.

Harvey, A. G., et al. (2003). Coexistence of posttraumatic stress disorder and traumatic brain injury: Toward a resolution of the paradox. *Journal of the International Neuropsychological Society, 9,* 663–676.

Harvey, B. H., et al. (2004). Serotonin and stress: Protective or malevolent actions in the behavioral response to repeated trauma? In R. Yehuda & B. McEwen (Eds.), *Behavioral stress responses* (Vol. 1032, pp. 267–272). Annals of the New York Academy of Sciences.

Hasinski, S. (1998). Assessment of adrenal glucocorticoid function: Which tests are appropriate for screening. *Postgraduate Medicine, 104,* 61–64, 69–72.

Hauser, K. F., et al. (2005). Pathobiology of dynorphins in trauma and disease. *Frontiers in Bioscience, 10,* 216–235.

Hauser, S. L., & Goodkin, D. E. (2001). Multiple sclerosis and other demyelinating diseases. In E. Braunwald, et al. (Eds.), *Harrison's 15th edition: Principles of internal medicine* (pp. 2152–2462). New York: McGraw Hill.

Haut, S. R., et al. (2007). Seizure occurrence: Precipitants and prediction. *Neurology, 69,* 1905–1910.

Haxby, J. V., et al. (2004). Spatial and temporal distribution of face and object representations in the human brain. *The cognitive neurosciences, 3rd Ed.* (pp. 889–904). Cambridge, MA: A Bradford Book.

Hayes, R. L., & Dixon, C. E. (1994). Neurochemical changes in mild head injury. *Seminars in Neurology, 14*, 25–31.

Hayes, R. L., & Ellison, M. D. (1989). Animal models of concussive head injury. In D. P. Becker and S. K. Gudeman (Eds.) *Textbook of Head Injury,* pp. 426–436. Philadelphia: Saunders.

Hayes, R. L., et al. (1984). Activation of pontine cholinergic sites implicated in unconsciousness following cerebral concussion in the cat. *Science, 283*, 301–303.

Hayes, R. L., et al. (1992). Neurochemical aspects of head injury: Role of excitatory neurotransmission. *Journal of Head Trauma Rehabilitation*, *7*(2), 16–28.

Hazeltine, E., & Ivry, R. (2000) Motor skill. In V. Ramachandran (Ed.), *Encyclopedia of the human brain* (Vol. 3, pp. 183–200). Boston: Academic Press.

Healy, D., & Williams, J. M. G. (1988). Dysrhythmia, dysphoria, and depression: The interaction of learned helplessness and circadian dysrhythmia in the pathogenesis of depression. *Psychological Bulletin, 103*, 163–178.

Hecht, J. S. (2004). Occipital nerve blocks in postconcussive headaches: A retrospective review and report of ten patients. *Journal of Head Trauma Rehabilitation, 19,* 58–71.

Heffner, H. E., & Heffner, R. S. (1984). Temporal lobe lesions and perception of species-specific vocalizations by Macaques. *Science*, *226*, 75–76.

Heid, L., et al. (2004). Vertigo, dizziness, and tinnitus after otobasal fractures. *The International Tinnitus Journal, 10*, 94–100.

Heikkila, H. V., & Wenngren, G-I., (1998). Cervicocephalic kinesthetic sensibility, active range of cervical motion, and oculomotor functions in patients with whiplash injury. *Archives of Physical Medicine and Rehabilitation, 79*, 1089–1094.

Heilbrun, M. P., & Ratcheson, R. A. (1972). Multiple extracranial vessel injuries following closed head and neck trauma. Case report. *Journal of Neurosurgery, 37*, 219–223.

Heilman, K. M. (1991). Anosognosia: Possible neuropsychological mechanisms. In G. P. Prigatano & D. L. Schachter (Eds.), *Awareness of deficit after brain injury* (pp. 53–62). New York: Oxford.

Heilman, K. M., & Rothi, L. J. G. (2003). Apraxia. In K. M. Heilman & E. Valenstein (Eds.), *Clinical neuropsychology, 4th Ed.* (pp. 215–235). New York: Oxford University Press.

Heilman, K. M., & Valenstein, E. (2003). Introduction. In K. M. Heilman & E. Valenstein (Eds.), *Clinical Neuropsychology, 4th Ed.* (pp. 1–13). New York: Oxford University Press.

Heilman, K. M., et al. (1983). Affective disorders associated with hemispheric disease. In K. M. Heilman & P. Satz (Eds.), *Neuropsychology of human emotions* (pp. 45–64). New York: Guilford Press.

Heilman, K. M., et al. (1993). Emotional disorders associated with neurological diseases. In K. M. Heilman & E. Valenstein (Eds.), *Clinical neuropsychology, 3rd Ed.* (pp. 461–497). New York: Oxford.

Heilman, K. M., et al. (1997). Apraxia. In S. C. Schachter & O. Devinsky (Eds.), *Behavioral neurology and the legacy of Norman Geschwind* (pp. 171–182). Philadelphia: Lippincott-Raven.

Heilman, K. M., et al. (2003). Emotional disorders associated with neurological diseases. In K. M. Heilman & E. Valenstein (Eds.), *Clinical neuropsychology, 4th Ed.* (pp. 447–475). New York: Oxford University Press.

Heinrich, P. C., et al. (2003). Principles of interleukin (IL)-6-type signaling and its regulation. *Biochemical Journal, 374*, 1–20.

Heller, W. (1993). Neuropsychological mechanisms of individual differences in emotion, personality, and arousal. *Neuropsychology*, *7*, 476–489.

Hellhammer, J., et al. (2004). Allostatic load, perceived stress, and health. *Annals of the New York Academy of Sciences, 1032*, 8–13.

Helm-Estabrooks, N., & Hotz, G. (1998). Sudden onset of "stuttering" in an adult: Neurogenic or psychogenic. *Seminars in Speech and Language*, *11*, 23–29.

Hendler, N. (1990). Psychiatric considerations of pain. In R. Youmans (Ed.) *Neurological Surgery,* pp. 3813–3855. New York: Saunders.

Heninger, G. R. (1995). Neuroimmunology of stress. In M. J. Friedman, et al. (Eds.), *Neuroimmunology of stress* (pp. 381–401). Philadelphia: Lippincott-Raven.

Henry, G. K., et al. (2000). Nonimpact brain injury: Neuropsychological and behavioral correlates with consideration of physiological findings. *Applied Neuropsychology*, *7*, 65–75.

Herdman, S. J., & Helminski, J. O. (1993). Vestibular deficits in the head-injured patient. *Physical Medicine and Rehabilitation,* 7 (3), 559–568.

Hermanowicz, N., & Truong, D. D. (1999). Cranial nerves IX (Glossopharyngeal) and X (Vagus). In C. G. Goetz & E. J. Pappert (Eds.), *Textbook of clinical neurology* (pp. 201–213). Philadelphia: W. B. Saunders.

Hernandez, T. D., et al. (2004). Posttraumatic epilepsy and neurorehabilitation. In M. J. Ashley (Ed.), *Traumatic brain injury* (pp. 27–55). Boca Raton, FL: CRC Press.

Herrmann, M., et al. (2000). Temporal profile of release of neurobiochemical markers of brain damage after traumatic brain injury is associated with intracranial pathology as demonstrated in cranial computerized tomography. *Journal of Neurotrauma, 17*, 113–122.

Herrmann, M., et al. (2001). Release of biochemical markers of damage to neuronal and glial brain tissue is associated with short and long term neuropsychological outcome after traumatic brain injury. *Journal of Neurology, Neurosurgery & Psychiatry, 70,* 95–100.

Herskowitz, J., & Rosman, N.P. (1982). *Pediatrics, neurology, and psychiatry-common ground.* New York: MacMillan.

Hertzig, M. E. (1987). Neurologic evaluation schedule. In D. E. Tupper (Ed.) *Soft Neurological Signs,* pp. 355–368. Orlando, FL: Grune and Stratton.

Hertzig, M. E., & Shapiro, T. (1987). The assessment of nonfocal neurological signs in school-aged children. In D. E. Tupper (Ed.), *Soft neurological signs* (pp. 71–93). Orlando, FL: Grune & Stratton.

Herzog, A. G. (1997). Neuroendocrinology of epilepsy. In S. C. Schachter & O. Devinsky (Eds.), *Behavioral neurology and the legacy of Norman Geschwind* (pp. 223–240). Philadelphia: Lippincott-Raven.

Herzog, A. G. (2006). Menstrual disorders in women with epilepsy. *Neurology, 66*(6), S23–S27.

Heurtult, B., & Benoit, J-P. (2005). Drug delivery through the BBB: Liposomes, nanoparticles, and other non-viral vectors. In E. De Vries & A. Prat (Eds.), *The blood-brain barrier and its microenvironment* (pp. 143–165). New York: Taylor & Francis.

Hi, J. H., et al. (2006). Early, transient increased in complexin I and complexin II in the cerebral cortex following traumatic brain injury is attenuated by N-acetylcysteine. *Journal of Neurotrauma, 23*, 86–96.

Hickling, E. J., et al. (1992). Motor vehicle accidents, headaches and post-traumatic stress disorder: Assessment findings in a consecutive series. *Headache,* 32, 147–151.

Hillbom, E. (1960). *After-Effects of Brain Injuries: Research on the Symptoms Causing Invalidism of Persons in Finland Having Sustained Brain Injuries during the Wars of 1939–1940 and 1941–1944.* Copenhagen: Ejnar Munksgaard.

Hilloowala, R. A., et al. (1998). Interrelationships of brain cranial base and mandible. *Journal of craniomandibular practice, 16*, 267–274.

Hines, M. E., et al. (1995). Characteristics and mechanisms of epilepsy spectrum disorder: An explanatory model. *Applied Neuropsychology, 2*(1), 1–6.

Hirmaher, B., & Williams, D. T. (1994). Children and adolescents. In J. M. Silver et al. (Eds.), *Neuropsychiatry of traumatic brain injury* (pp. 393–412). Washington, DC: American Psychiatric Press.

Hirshkowitz, M., & Moore, C. A. (2000). Nightmares. In G. Fink (Ed.), *Encyclopedia of Stress* (Vol. 3, pp. 181–186). San Diego, CA: Academic Press.

Hitch, G. J. (1990). Developmental fractionation of working memory. In G. Vallar & T. Shallice (Eds.), *Neuropsychological impairments of short-term memory* (pp. 221–246). New York: Cambridge University Press.

Hnisch, U-K. (2001). Effects of interleukin-2 and interferons on the nervous system. In R. Ader, et al. (Eds.), *Psychoneuroimmunology,* 3rd Ed. (Vol. 1, pp. 585–631). San Diego: Academic Press.

Hobson, J. A. (1999). Sleep and dreaming. In M. Z. Zigmond et al. (Eds.) *Fundamental Neuroscience,*pp. 1207–1227. San Diego: Academic Press.

Hobson, J. A., & Pace-Schott, E. (2003). Sleep, dreaming, and wakefulness. In L. Squire et al. (Eds.), *Fundamental neuroscience, 2nd Ed.* (pp. 1085–1008). London: Academic Press.

Hof, P. R., et al. (2003). Cellular components of nervous tissue. In L. R. Squire, et al. (Eds.), *Fundamental neuroscience, 2nd Ed.* (pp. 49–78). Boston: Academic Press.

Hohl, M. (1974). Soft-tissue injuries of the neck in automobile accidents. *Journal of Bone and Joint Surgery, 56-A*, 1675–1682.

Holaday, M. (2000). Rorschach protocols from children and adolescents diagnosed with posttraumatic stress disorder. *Journal of Personality Assessment, 75,* 143–157.

Hollifield, M. A. (2005). Somatoform disorders. In B. J. Sadock & V. A. Sadock (Eds.), *Comprehensive textbook of psychiatry, I, 8th Ed.* (pp. 1800–1828). Philadelphia: Lippincott Williams & Wilkins.

Holyoak, K. J., & Kroger, J. K. (1995). Forms of reasoning: Insight into prefrontal functions. In J. Grafman, et al. (Eds.), *Structure and functions of the human prefrontal cortex* (Vol. 769, pp. 253–263). Annals of the New York Academy of Sciences.

Holzer, R., et al. (2007). The relationship between specific cognitive functions and falls in aging. *Neuropsychology, 5*, 540–548.

Hooper, S. L. (1995). Crustacean stomatogastric system. In A. Arbib (Ed.), *The handbook of brain theory and neural networks* (pp. 275–278). Cambridge, MA: MIT Press.

Hoots, W. K. (1996). Coagulation disorders in the head-injured patient. *Journal of Neurotrauma, 25*, 677–668.

Hore, J., et al. (2002). Disorders in timing and force of finger opening in overarm throws made by cerebellar subjects. In S. M. Highstein & W. T. Thach (Eds.), *The cerebellum: Recent developments in cerebellar research* (Vol. 978, pp. 1–15). Annals of the New York Academy of Sciences.

Horger, B. A., & Roth, R. H. (1995). Stress and central amino acid systems. In M. J. Friedman, et al. (Eds.), *Neurobiological and clinical consequences of stress: From normal adaptation to PTSD* (pp. 61–81). Philadelphia: Lippincott-Raven.

Horn, L. J., & Zasler, N. D. (1990). Neuroanatomy and neurophysiology of sexual function. *Journal of Head Trauma Rehabilitation* 5, 2, 1–13.

Horner, J., & Massey, E. W. (1996). Dysarthria and dysphagia. In M. S. Samuels & S. Feske (Eds.), *Office practice of neurology* (pp. 104–111). New York: Churchill Livingstone.

Horner, M. D., et al. (2005). Patterns of alcohol abuse 1 year after traumatic brain injury: A population-based, epidemiological study. *Journal of the International Neuropsychological Society, 11*, 332–330.

Hornstein, N. L. (2000). Dissociative disorders in children and adolescents. In B. J. Sadock & V. A. Sadock (Eds.), *Comprehensive textbook of psychiatry, 7th Ed.* (pp. 2902–2917). New York: McGraw-Hill.

Horowitz, J. D., & Telch, M. J. (2007). Dissociation and pain perception. An experimental investigation. *Journal of Traumatic Stress, 20*, 597–609.

Horowitz, M., et al. (1996). Self-regard: A new measure. *American Journal of Psychiatry, 153*, 382–385.

Horton, J. C. (2005). Disorders of the eye. In *Harrison's principles of internal medicine, 16th Ed.* (I, pp. 162–176). New York: McGraw-Hill.

Horvath, T. B., et al. (1989). Organic mental syndromes and disorders. In H. I. Kaplan & B. J. Sadock (Eds.), *Comprehensive textbook of psychiatry, 5th Ed.* (Vol. I, pp. 599–641). Baltimore: Williams and Wilkins.

Howard, J. H., et al. (1986). Personality (Hardiness) as a moderator of job stress and coronary risk in Type A individuals: A longitudinal study. *Journal of Behavioral Medicine, 9*, 229–244.

Hsiao, S. S., & Vega-Bermudez, F. (2002). Attention in the somatosensory system. In R. J. Nelson (Ed.), *The somatosensory system: Deciphering the brain's own body image* (pp. 197–208). Boca Rqton, FL: CRC Press.

Hubbard, J. R., & Workman, E. A. (Eds.) (1988). *Handbook of Stress Medicine: An Organ System Approach.*

Hubbard, J. R., & Workman, E. A. (1998). On the nature of stress. In J. R. Hubbard & E. A. Workman (Eds.), *Handbook of stress medicine: An organ system approach* (pp. 3–13). Boca Raton, FL: CRC Press.

Hudson, L. J. (1990). Speech and language disorders in childhood brain tumours. In B. E. Murdoch (Ed.), *Acquired neurological speech/language disorders in childhood* (pp. 245–268). New York: Taylor and Francis.

Hudspeth, A. J. (2000). Hearing. In E. R. Kandel, et al. (Eds.), *Principles of neural science, 4th Ed.* (pp. 590–613). New York: McGraw Hill.

Hughes, J. T., & Brownell, B. (1968). Traumatic thrombosis of the internal carotid artery in the neck. *Journal of Neurology, Neurosurgery, and Psychiatry,* 31, 307–314.

Hults, K. N., et al. (2005). Outcome prediction after severe head injury. In P. L. Reilly & R. Bullock (Eds.), *Head injury, 2nd Ed.* (pp. 462–471). New York: Hodder Arnold.

Humphrey, N. K. (1987). The uses of consciousness. *The 57th James Arthur Lecture*. New York: American Museum of Natural History.

Huron, D. (2001). Is music an evolutionary adaptation? In R. H. Zatorre & I. Peretz (Eds.), *The biological foundations of music* (Vol. 930, pp. 43–61). Annals of the New York Academy of Sciences.

Hurwitz, T. A. (1989). Ideogenic neurological deficits: Conscious mechanisms in conversion symptoms. *Neuropsychiatry, Neuropsychology, and Behavioral Neurology, 1*, 301–308.

Hurwitz, T. A., & Prichard, J. W. (2006). Conversion disorder and fMRI. *Neurology, 67*, 1914–1915.

Hyman, S. E., et al. (2006). Neural mechanisms of addiction: The role of reward-related learning and memory. *Annual Review of Neuroscience, 29*, 565–598.

Hynd, G. W., & Willis, W. G. (1988). *Pediatric neuropsychology*. New York: Grune and Stratton.

Iadecola, C. (2004). Neurovascular regulation in the normal brain and in Alzheimer's disease. *Nature Reviews Neuroscience, 5*(5), 347–360.

Imboden, J., et al. (2001). Immunosuppressive, antiinflammatory, & immunomodulatory therapy. In T. G. Parslow, et al. (Eds.), *Medical immunology, 10th Ed.* (pp. 744–760). New York: Lange.

Indefrey, P., & Develt, W. J. M. (2000). The neural correlates of language production In M. S. Gazzaniga (Ed.), *The new cognitive neurosciences, 2nd Ed.* (pp. 845–865). Cambridge, MA: MIT Press.

Ingebrigtsen, T., & Romner, B. (2003). Biochemical serum markers for brain damage: A short review with emphasis on clinical utility in mild head injury. *Restorative Neurology and Neuroscience*, *21*, 171–176.

Inskip, P. D., et al. (1998). Incidence of intracranial tumors following hospitalization for head injuries (Denmark). *Cancer Causes and Control, 9*, 109–116.

Irwin, M. (1993). Stress-induced immune suppression: Role of the autonomic nervous system. In Y. Taché & C. Rivier (Eds.), *Corticotropin-releasing factor and cytokines: Role in the stress response* (Vol. 691, pp. 203–218). Annals of the New York Academy of Science.

Irwin, M. D., & Cole, J. C. (2005). Depression and psychoneuroimmunology. In K. Vedhara & M. Irwin (Eds.), *Human psychoneuroimmunology* (pp. 243–264). New York: Oxford University Press.

Isonieml, H., et al. (2006). Outcome of traumatic brain injury after three decades: Relationship to ApoE genotype. *Journal of Neurotrauma, 23*, 1600–1608.

Israellson, C., et al. (2008). Distinct cellular patterns of upregulated chemokine expression supporting a prominent inflammatory role in traumatic brain injury. *Journal of Neurotrauma, 25*, 959–974.

Ito, Y., et al. (1998). Preliminary evidence for aberrant cortical development in abused children: a quantitative EEG study. *The Journal of Neuropsychiatry and Clinical Neurosciences, 10*, 298–307.

Ivanhoe, C. R., & Hartman, E. T. (2004). Clinical caveats on medical assessment and treatment of pain after TBI. *Journal of Head Trauma Rehabilitation, 19*, 29–39.

Iverson, G. L. (2004). Relation between subjective fogginess and neuropsychological testing following concussion. *Journal of the International Neuropsychological Society, 10*, 904–906.

Iverson, G. L. (2005). Outcome from mild traumatic brain injury. *Current Opinion in Psychiatry,* 18, 301–317.

Iverson, G. L. (2006). Outcome from mild traumatic brain injury. *Current Opinion in Psychiatry, 18*, 301–317.

Iversen, S., et al. (2000). The autonomic nervous system and the hypothalamus. In E. R. Kandel, et al. (Eds.), *Principles of neural science, 4th ed.* (pp. 960–931). New York: McGraw Hill.

Jacob, R. G., et al. (1996). Panic, phobia, and vestibular dysfunction. In B. J. Yates & A. D. Miller (Eds.), *Vestibular autonomic regulation* (pp. 198–227). Boca Raton, FL: CRC Press.

Jacobson, R. I. & Abrams, G. M. (1999). Disorders of the hypothalamus and pituitary gland in adolescence and childhood. In K. F. Swaiman & S. Ashwal (Eds.), *Pediatric neurology: Principles and practice, 3rd Ed.* (pp. 1311–1351). St. Louis, MO: Mosby.

Jacome, D. E. (1986). Basilar artery migraine after uncomplicated whiplash injuries. *Headache, 26*, 515–516.

Jaeger, M., et al. (2000). Speech disorders following severe traumatic brain injury: Kinematic analysis of syllable repetitions using electromagnetic articulography. *Folia Phoniatrica et Logopaedica*, *52*, 187–196.

Jaffe, K. M. (1986). Preface: Pediatric head injury. *Journal of Head Trauma Rehabilitation*, *1*(4), ix.

Jaffe, K. M., et al. (1990). Specific problems associated with pediatric head injury. In M. Rosenthal, E. R. Griffith, M. R. Bond, & J. D. Miller (Eds.) *Rehabilitation of the Adult and Child with Traumatic Brain Injury,* 2nd ed., pp. 539–557. Philadelphia: F. A. Davis.

Jameson, J. L. (1996). In inherited disorders of the gonadotropin hormone. *Molecular and cellular endocrinology*, *125*(1,2), 143–149.

Jameson, J. L. (2001). Endocrinology and metabolism. In E. Braunwald, et al. (Eds.), *Harrison's principles of internal medicine, 15th Ed.* (pp. 2019–2029). New York: McGraw-Hill.

Jamison, R. N. (2003). Pain management, psychological strategies. In M. J. Aminoff & R. B. Daroff (Eds.), *Encyclopedia of the neurological sciences* (Vol. 3, pp. 753–758). Boston: Academic Press.

Jane, J. A., et al. (2002). Cerebral circulation. In V. S. Ramachandran (Ed.), *Encyclopedia of the human brain* (Vol. 1, pp. 629–661). Boston: Academic Press.

Janeway, C. A., et al. (2005). *Immunobiology, 6th Ed. The immune system in health and disease.* New York: Garland Science.

Jankovic, B. D. (1994). Neuroimmunomodulation: From phenomenology to molecular evidence. In N. A. Fabris, et al. (Eds.), *Immunomodulation: The state of the art* (Vol. 741, pp. 1–38). Annals of the New York Academy of Sciences.

Jay, G. W. (1999). *The headache handbook.* Boca Raton: CRC Press.

Jenkner, F. L. (1995). *Electric Pain Control.* New York: Springer-Verlag.

Jha, A., et al. (2008). A randomized trial of Modafinil for the treatment of fatigue and excessive daytime sleepiness in individuals with chronic traumatic brain injury. *Journal of Head Trauma Rehabilitation, 23*, 52–63.

Johns, J. S., et al. (1999). Impact of clinically significant hetertopic ossification on functional outcome after traumatic brain injury. *Journal of Head Trauma Rehabilitation, 14*, 269–276.

Johnson, D. M., et al. (2007). Emotional numbing weakens inner-city women's resilience resources. *Journal of Traumatic Stress, 20*, 197–206.

Johnson, J. L. (1991). Reality monitoring: Evidence from confabulation in organiz brain disease patients. In G. P. Prigatano and D. L. Schachter (Eds.) *Awareness of Deficit after Brain Injury,* pp. 176–197. New York: Oxford University Press.

Jones, C., et al. (2005). Traumatic brain injury, dissociation, and posttraumatic stress disorder in road traffic accident survivors. *Journal of Traumatic Stress, 18*, 181–191.

Jordan, F. M. (1990). Speech and language disorders following childhood closed head injury. In B. E. Murdoch (Ed.), *Acquired neurological speech/language disorders in childhood* (pp. 124–147). New York: Taylor and Francis.

Jordan, F. M., et al. (1990). Performance of closed head-injured children on a naming task. *Brain Injury, 1*, 27–32.

Jordan, F. M., (1992). Language abilities of mildly closed head injured (CHI) children 10 years post-injury. *Brain Injury,* 6, 39–44.

Jorge, R. E. (2005). Neuropsychiatric consequences of traumatic brain injury: A review of recent findings. *Current Opinion in Psychiatry, 18*, 289–299.

Jorge, R. E., & Starkstein, S. E. (2005). Pathophysiological aspects of major depression following traumatic brain injury. *Journal of Head Trauma Rehabilitation, 20*, 475–487.

Jorge, R. E., et al. (2000). Neuropsychiatric aspects of traumatic brain injury. In B. J. Sadock & V. A. Sadock (Eds.), *Comprehensive textbook of psychiatry, 7th Ed.* (pp. 273–285). Philadelphia: Lippincott Williams & Wilkins.

Joseph, R. (1988). The right cerebral hemisphere: Emotion, music, visual-spatial skills, body-image, dreams and awareness. *Journal of Clinical Psychology,* 44, 630–674.

Joseph, S., & Masterson, J. (1999). Posttraumatic stress disorder and traumatic brain injury: Are they mutually exclusive? *Journal of Traumatic Stress*, 12, 437–454.

Joëls, J., & van Riel, E. (2004). Mineralocorticoid and glucocorticoid receptor-mediated effects on serotonergic transmission in health and disease. In R. Yehuda & B. McEwen (Eds.), *Behavioral Stress Responses* (Vol. 1032, pp. 301–303). Annals of The New York Academy of Sciences.

Junger, C. E., et al. (1997). Cerebral autoregulation following minor head injury. *Journal of Neurosurgery, 87*, 425–432.

Junqueira, L. C., & Carneiro, J. (2005). *Basic histology, 11th Ed.* New York: McGraw Hill.

Kaale, R. K., et al. (2007). Active range of motion as an indicator for ligament and membrane lesions in the upper cervical spine after a whiplash trauma. *Journal of Neurotrauma, 24*, 713–721.

Kaas, J. H. (2002a). Functional implications of plasticity and reorganizations in the somatosensory and motor systems of developing and adult primates. In R. J. Nelson (Ed.), *The somatosensory system: Deciphering the brain's own body image* (pp. 367–381). Boca Raton, FL: CRC Press.

Kaas, J. H. (2002b). Neocortex. In V. S. Ramachandran (Ed.), *Encyclopedia of the human brain* (Vol. 3, pp. 291–303). Boston: Academic Press.

Kaas, J. H., et al. (2002). The organization of the somatosensory system in primates. In R. J. Nelson (Ed.), *The somatosensory system: Deciphering the brain's own body image* (pp. 1–25). Boca Raton, FL: CRC Press.

Kaas, J. H., & Preuss, T. H. (2003). Human brain evolution. In L. Squire et al. (Eds.), *Fundamental neuroscience, 2nd Ed.* (pp. 1147–1166). London: Academic Press.

Kagan, J. (1992). Behavior, biology, and the meanings of temperamental constructs. *Pediatrics, 90*, 510–513.

Kagan, J., & Snidman, N. (1991). Temperamental factors in human development. *American Psychologist, 46*, 856–862.

Kahn, C. R., et al. (1998). Mechanism of action of hormones that act at the cell surface. In J. D. Wilson, et al. (Eds.), *Williams textbook of endocrinology, 9th Ed.* (pp. 95–143). Philadelphia: Saunders.

Kaiser, G., et al. (1986). Rehabilitation medicine following severe head injury in infants and children. In A. J. Raimondi, et al. (Eds.), *Head Injuries in the Newborn and Infant* (pp. 263–279). New York: Springer-Verlag.

Kandel, E. R. (2000a). The brain and behavior. In E. R. Kandel, et al. (Eds.), *Principles of neural science, 4th ed.* (pp. 5–18). New York: McGraw Hill.

Kandel, E. R. (2000b). Nerve cells and behavior. In E. R. Kandel, et al. (Ed.), *Principles of neural science, 4th ed.* (pp. 19–35). New York: McGraw Hill.

Kandel, E. R. (2000c). From nerve cells to cognition. In E. R. Kandel, et al. (Eds.), *Principles of neural science, 4th ed.* (pp. 381–403). New York: McGraw Hill.

Kandel, E. R. (2000d). Disorders of mood: Depression, mania, and anxiety disorders. In E. R. Kandel, et al. (Eds.), *Principles of neural science, 4th ed.* (pp. 1209–1226). New York: McGraw Hill.

Kandel, E. R. (2000e). Disorders of thought and volition: Schizophrenia. In E. R. Kandel, et al. (Eds.), *Principles of neural science, 4th ed.* (pp. 1188–1208). New York: McGraw Hill.

Kandel, E. R., & Wurtz, R. H. (2000). Constructing the visual image. In E. R. Kandel, et al. (Eds.), *Principles of neural science, 4th ed.* (pp. 492–506). New York: McGraw Hill.

Kanter, E. D., et al. (1998). *Biological Psychiatry, 43*, 53S.

Kapoor, N., & Ciuffreda, K. J. (2005). Vision problems. In J. M. Silver, et al. (Eds.), *Textbook of traumatic brain injury* (pp. 405–415). Washington DC: American Psychiatric Publishing.

Karoly, P. (2006). Tracking the leading edge of self-regulatory failure: Commentary on "Where do we go from here? The goal perspective in psychotherapy." *Clinical Psychology: Science and Practice, 13*(4), 366–370.

Karon, S. L., et al. (2007). Challenges and approaches to the identification of traumatic brain injury among nursing home residents. *Journal of Head Trauma Rehabilitation, 22*, 350–359.

Kase, C. S. (2004). Intracerebral hemorrhage. In W. G. Bradley, et al. (Eds.), *Neurology in clinical practice, 4th Ed.* (Vol. 2, pp. 1251–1267). Boston: Butterworth-Heinemann.

Katayama, Y., et al. (1984). Behavioral evidence for a cholinoceptive pontine inhibitory area: Descending control of spinal motor output and sensory input. *Brain Research, 296*, 241–262.

Katayama, Y., et al. (1985). Dissociation of endogenous components of auditory evoked potentials following carbachol microinjection into the cholinoceptive pontine inhibitory area. *Brain Research, 334*, 366–371.

Katayama, Y., et al. (1988). Coma associated with flaccidity produced by fluid-percussion concussion in the cat: Contribution of activity in the pontine inhibitory system. *Brain Injury, 2*, 51–66.

Kato, T., et al. (2007). Statistical image analysis of cerebral glucose metabolism in patients with cognitive impairment following diffuse traumatic brain injury. *Journal of Neurotrauma, 24*, 919–926.

Katz, J., & Melzack, R. (2003). Assessment of pain. In M. J. Aminoff & R. B. Daroff (Eds.), *Encyclopedia of the neurological sciences* (Vol. 3, pp. 716–732). Boston: Academic Press.

Katzman, R., & Pappius, H. M. (1973). *Brain electrolytes and fluid metabolism*. Baltimore, MD: Williams and Wilkins.

Kavanagh, E. (2007). Letters: PTSD and Vietnam veterans. *Science, 315*, 184–187.

Kaye, E. T., & Kaye, K. M. (2005). Fever & rash. In *Harrison's principles of internal medicine, 16th Ed.* (Vol. I, pp. 108–116). New York: McGraw-Hill.

Kaye, J. M., & Lightman, S. L. (2005). Psychological stress and endocrine axes. In K. Vedhara & M. Irwin (Eds.), *Human psychoneuroimmunology* (pp. 25–52). New York: Oxford University Press.

Kazkrzak, H. A., et al. (2002). Enhanced lipid peroxidation processes in patients after brain contusion. *Journal of Neurotrauma, 8*, 793–797.

Keane, T. M., et al. (1998). Utility of psychophysiological measurement in the diagnosis of posttraumatic stress disorder: Results from a Department of Veterans Affairs cooperative study. *Journal of Consulting and Clinical Psychology, 66*, 914–923.

Keiski, M., et al. (2007). The role of depression in verbal memory following traumatic brain injury. *The Clinical Neuropsychologist*, 21, 744–761.

Kelley, D. B., & Brenowitz, E. (2002). Hormonal influences on courtship behaviors. In J. B. Becker, et al. (Eds.), *Behavioral endocrinology, 2nd Ed.* (pp. 289–329). Cambridge, MA: MIT Press.

Kelly, D. D. (1991b). Disorders of sleep and consciousness. In E. R. Kandel, et al. (Eds.), *Principles of Neuroscience*, 3rd Ed. (pp. 805–819). New York: Elsevier.

Kelly, D. D., et al. (1996). General principles of head injury management. In R. K. Narayan, et al. (Eds.), *Neurotrauma* (pp. 71–101). New York: McGraw-Hill.

Kelly, J. P. (1991). also referred to as Kelly, 1991a.

Kelly, J. P. (1991). The neural basis of perception and movement. In E. R. Kandel, J. H. Schwartz, & T. M. Jessell (Eds.) *Principles of Neural Sciences,* 3rd ed., pp. 283–295. New York: Elsevier.

Kelly, J. P., & Dodd, J. (1991). Anatomical organization of the nervous system. In E. R. Kandel, et al. (Eds.), *Principles of neural science, 3rd Ed.* (pp. 273–282). New York: Elsevier.

Kelly, J. P., et al. (1991). Concussion in sports: Guidelines for the prevention of catastrophic outcome. *Journal of the American Medical Association, 266*, 2867–2869.

Kelly, K. M., & Schramke, C. J. (2000). Epilepsy. In G. Fink (Ed.), *Encyclopedia of stress* (Vol. 2, pp. 66–70). San Diego, CA: Academic Press.

Kemenoff, L. A., et al. (2002). Frontal lobe. In V. S. Ramachandran (Ed.), *Encyclopedia of the human brain* (Vol. 2, pp. 317–325). Boston: Academic Press.

Kemeny, et. al. (1992). Psychoneuroimmunology. In C. B. Nemeroff (Ed.). *Neuroendocrinology* (pp. 563–591). Boca Raton, FL: CRC Press.

Kemp, P. M., et al. (1995). Cerebral perfusion and psychometric testing in military amateur boxers and controls. *Journal of Neurology, Neurosurgery, and Psychiatry, 59*, 368–374.

Kendall-Tackett, K. (2007). How sleep disorders impact health in trauma survivors. *Trauma Psychology Newsletter, 2*(2), 17–18.

Kendall-Tackett, K. (2008). Inflammation and traumatic stress: A likely mechanism for chronic illness in trauma survivors. *Trauma Psychology Newsletter, 3*(2), 12–14.

Kennedy, S. (1996). Herpes virus infection and psychoneuroimmunology. In H. Friedman, et al. (Eds.), *Psychoneuroimmunology, stress and infection* (pp. 195–213). Philadelphia: Lippincott-Raven.

Kennedy, M. R. T., et al. (2009). White matter and neurocognitive changes in adults with chronic traumatic brain injury. *Journal of the International Neuropsychological Society, 15*, 130–136.

Kent, G. C. (1978). *Comparative anatomy of the vertebrates*. St. Louis, MO: C. V. Mosby.

Kent, R. D. (1999). Motor control: Neurophysiology and functional development. In A. J. Caruso & E. A. Strand (Eds.), *Clinical management of motor speech disorders in children* (pp. 29–71). New York: Thieme.

Keppel-Benson, J. M., et al. (2002). Post-traumatic stress in children following motor vehicle accidents. *Journal of Child Psychology and Psychiatry, 43*, 203–212.

Kern, K. B., & Meislin, H. W. (1984). Diabetes insipidus: Occurrence after minor head trauma. *Journal of Trauma, 24*, 69–72.

Kerr, A. G. (1980). Trauma and the temporal bone: The effects of blast on the ear. *The Journal of Laryngology and Otology, 94*, 107–110.

Keverne, E. B., et al. (1997). Early learning and the social bond. In S. C. Carter, et al. (Eds.), *The integrative neurobiology of affiliation* (Vol. 807, pp. 329–339). Annals of the New York Academy of Sciences.

Kho, K. H., et al. (2006). Figuring out drawing-induced epilepsy. *Neurology, 66*, 723–726.

Khurana, R. K., & Nirankari, V. S. (1986). Bilateral sympathetic dysfunction in post-traumatic headaches. *Headaches,* 26, 183–188.

Kiecolt-Glaser, J., & Glaser, J. (1992). Stress and the immune system: Human studies. In A. Tasman & M. R. Riba (Eds.), *Annual review of psychiatry* (Vol. 11, pp. 169–180). Washington D.C.: American Psychiatric Press.

Kihlstrom, J. F. (2001). Dissociative disorders. In P. B. Sutker & H. E. Adams (Eds.). *Comprehensive handbook of psychopathology, 3rd Ed.* (pp. 259–276). New York: Kluwer Academic Publishers.

Kihlstrom, J. F., & Tobias, B. A. (1991). Anosognosia, consciousness, and the self. In G. P. Prigatano & D. L. Schachter (Eds.), *Awareness of deficit after brain injury* (pp. 198–219). New York: Oxford University Press.

Kimelberg, H. K., & Norenberg, M. D. (1994). Astrocytic responses to central nervous system trauma. In S. K. Salzman & A. I. Faden (Eds.), *The neurobiology of central nervous system trauma* (pp. 193–213). New York: Oxford University Press.

King, A. I. (2004). The biomechanics of brain injury: From historical to current perspectives. *Brain Injury/ Professional, 1*(1), 16–17.

King, N. (1997). Post-traumatic stress disorder and head injury as a dual diagnosis: Islands of memory as a mechanism. *Neurology, Neurosurgery & Psychiatry, 62*, 82–84.

King, N. D. (2002). Perseveration of traumatic re-experiencing in PTSD: A cautionary note regarding exposure based psychological treatments for PTSD when head injury and dysexecutive impairment are also present. *Brain Injury, 16,* 65–74.

King, N. S., et al. (1995). The Rivermead Post Concussion Symptoms Questionnaire: A measure of symptoms commonly experienced after head injury and its reliability. *Journal of Neurology, 242*, 587–592.

King, N. S., et al. (1997). Measurement of post-traumatic amnesia: How reliable is it? *Journal of Neurology, Neurosurgery and Psychiatry, 62*, 38–42.

King, R. B., & Young, R. F. (1990). Cephalic pain. In R. Youmans (Ed.) *Neurological Surgery,* pp. 3856–3879. New York: Saunders.

Kinsella, G. J., et al., (1997). Predictors and indicators of academic outcome in children 2 years following traumatic brain injury. *Journal of the International Neuropsychological Society, 3*(6), 608–616.

Kirby, D. F., et al. (2007). Gastrointestinal and nutritional issues. In N. D. Zasler, et al. (Eds.), *Brain injury medicine* (pp. 657–672). New York: Demos.

Kirkpatrick, J. B. (1983). Head-in-motion contusions in young adults. *Acta Neurochirurgica*, Suppl. 32, 115–117.

Kirkwood, M. W., et al. (2008). Management of pediatric mild traumatic brain injury: A neuropsychological review from injury through recovery. *The Clinical Neuropsychologist, 22,* 769–800.

Kirsch, P., et al. (2005). Oxytocin modulates neural circuitry for social cognition and fear in humans. *The Journal of Neuroscience, 25*(49), 11489–11493.

Kirshner, H. H. (1996). Speech and language disorders. In M. S. Cymbalist & S. Feske (Eds.), *Office practice of neurology* (pp. 718–722). New York: Churchill Livingstone.

Kirshner, H. S. (2000). Language disorders. In W. G. Bradley, et al. (Eds.), *Neurology in clinical practice* (Vol. 1, pp. 141–159). Boston: Butterworth-Heinemann.

Kirshner, H. S. (2004a). Language and speech disorders: A. Aphasia. In W. G. Bradley, et al. (Eds.), *Neurology in clinical practice, 4th Ed.* (Vol. 1, pp. 141–160). Boston: Butterworth-Heinemann.

Kirshner, H. S. (2004b) Language and speech disorders: B. Dysarthria and apraxia of speech. In W. G. Bradley, et al. (Eds.), *Neurology in clinical practice, 4th Ed.* (Vol. 1, pp. 161–164). Boston: Butterworth-Heinemann.

Kiss, I., & Ennis. T. (2001). Age-related decline in perception of prosodic affect. *Applied Neuropsychology, 8*, 251–254.

Kivioja, J., et al. (2001a). Systemic immune response in whiplash injury and ankle sprain: Elevated IL-6 and IL-10. *Clinical Immunology, 101*, 106–112.

Kivioja, J., et al. (2001b). Chemokines and their receptors in whiplash injury: Elevated RANTES and CCR-5. *Journal of Immunology, 21*, 272–277.

Klein, S. K., et al. (1992). Fluent aphasia in children: Definition and natural history. *Journal of Child Neurology, 7*, 50–59.

Kleiven, S. (2003). Influence of impact direction on the human head in prediction of subdural hematoma. *Journal of Neurotrauma, 20*, 365–379.

Klonoff, H. (1971). Head injuries in children: Predisposing factors, accident conditions, accident proneness, and sequelae. *American Journal of Public Health, 61*, 2405–2417.

Klonoff, H., et al. (1995). Outcome of head injuries from childhood to adulthood: A twenty-three year follow-up study. S. H. Broman & M. E. Michel (Eds.), *Traumatic head injury in children* (pp. 219–234). New York: Oxford University Press.

Klonoff, P. A., et al. Cognitive retraining in a milieu-oriented outpatient rehabilitation program. In J. A. Moses Jr. (Ed.) *Clinical Neuropsychology: Theoretical Foundations for Practitioners,* pp. 219–236. Mahwah, NJ: Erlbaum.

Klonoff, P. S., & Lage, G. A. (1991). Narcissistic injury in patients with traumatic brain injury. *The Journal of Head Trauma Rehabilitation, 6*(4), 11–21.

Knapp, T., et al. (2003). Neurotransmitters: Overview. In M. J. Aminoff & R. B. Daroff (Eds.), *Encyclopedia of the neurological sciences* (Vol. 3, pp. 614–623). Boston: Academic Press.

Knight, J. (1997). Neuropsychological assessment in posttraumatic stress disorder. In J. P. Wilson and T. M. Keane (Eds.) *Assessing Psychological Trauma and PTSD,* pp. 448–492. New York: Guilford Press.

Knightly, J. J., & Pulliam, M. W. (1996). Military head injuries. In R. K. Narayan et al. (Eds.), *Neurotrauma* (pp. 891–903). New York, McGraw-Hill.

Knoch, D., & Fehr, E. (2007). Resisting the power of Temptations: The right prefrontal cortex and self-control. In B. W. Balleine, et al. (Eds.), *Reward and decision making in corticobasal ganglia networks* (Vol. 1104, pp. 123–134). Annals of the New York Academy of Sciences.

Knoll, J., & Miklya, I. (1995). Enhanced catecholaminergic and serotoninergic activity in rat brain from weaning to sexual maturity: Rationale for prophyloactic (-)Deprenyl (Selegiline) medication. *Life Sciences, 56,* 611–620.

Knudsen, E. I. (2007). Fundamental components of attention. *Annual Review of Neuroscience, 30*, 57–78.

Knussen, C., & Cunningham, C. F. (1988) Stress, disability and handicap. In S. Fisher & J. Reason (Eds.), *Handbook of life stress, cognition and health* (pp. 335–350). New York: John Wiley and Sons.

Kobassa, S. C. (1979). Stressful life events, personality and health: An enquiry into hardiness. *Journal of Personality and Social Psychology, 37*, 1–11.

Kobassa, S. C., et al. (1983). Type A and hardiness. *Journal of Behavior Medicine, 67,* 41–51.

Kodama, T., et al. (1998). Enhanced glutamate release during REM sleep in the rostromedial medulla as measured by in vivo microdialysis. *Brain Research, 780*, 178–181.

Kodama, T., et al. (2004). Changes in inhibitory amino acid release linked to pontine-induced atonia: An in vivo microdialysis study. *Journal of Neuroscience, 3*, 1548–1554.

Koechlin, E., & Hyafil, A. (2007). Anterior prefrontal function and the limits of human decision-making. *Science, 318*, 594–598.

Koenen, K. D. (2007). Genetics of posttraumatic stress disorder: Review and recommendations for future studies. *Journal of Traumatic Stress, 20*, 737–750.

Koerding, K. (2007). Decision theory: What "should" the nervous system do? *Science, 318*, 606–610.

Koetke, C. & Doyle, R. M. (2005). *Handbook of Pathophysiology, 2nd Ed.* Philadelphia: Lippincott Williams & Wilkins.

Kohl, S. J. (1984). The process of psychological adaptation to traumatic limb loss. In D. W. Krueger (Ed.) *Emotional Rehabilitation of Physical Trauma and Disability,* pp. 113–148. New York: SP Medical and Scientific Books.

Koiv, L., et al. (1997) Changes of sympatho-adrenal and hypothalmo-pituitary-adrenocortical system in patients with head injury. *Acta Neurologica Scandinavica, 96*, 52–58.

Kokko, J. P. (1996). Osmolality disorders. In J. C. Bennett & F. Plum (Eds.), *Cecil textbook of medicine* (pp. 532–538). Philadelphia: W. B. Saunders.

Kokko, J. P. (2004). Fluids and electrolytes. In *Cecil textbook of medicine, 22nd Ed.* (pp. 669–687). Philadelphia: Saunders.

Kolb, B., & Taylor, L. (2000). Facial expression, emotion, and hemispheric organization. In R. D. Lane & L. Nadel (Eds.), *Cognitive neuroscience of emotion* (pp. 62–83). New York: Oxford University Press.

Kolb, B., & Whishaw, I. Q. (1990). *Fundamentals of Human Neuropsychology,* 3rd ed. New York: Freeman and Co.

Kolb, L. C. (1988). A critical survey of hypotheses regarding post-traumatic stress disorders in light of recent research findings. *Journal of Traumatic Stress,* 1, 291–304.

Koob, G. F. (2003). Drug reward and addiction. In L. R. Squire et al. (Eds.), *Fundamental neuroscience, 2nd Ed.* (pp. 1127–1143). London: Academic Press.

Kooij, G., et al. (2005). Tight junctions of the blood-brain barrier. In E. De Vries & A. Prat (Eds.), *The blood-brain barrier and its microenvironment* (pp. 47–69). New York: Taylor & Francis.

Kop, W. J., & Cohen, N. (2001). Psychological risk factors and immune system involvement in cardiovascular disease. In Ader, R. et al. (Eds.), *Psychoneuroimmunology,* 3rd Ed. (Vol. 2, pp. 525–544). San Diego: Academic Press.

Korkman, M. (1999). Applying Luria's diagnostic principles in the neuropsychological assessment of children. *Neuropsychology Review,* 9, 89–105.

Korn, A. K., et al. (2005). Focal cortical dysfunction and blood-brain barrier disruption in patients with post-concussion syndrome. *Journal of Clinical Neurophysiology, 22*, 1–9.

Kosteljanetz, M., et al. (1981). Sexual and hypothalamic dysfunction in postconcussional syndrome. *Acta neurol Scandin, 63*, 169–180.

Kraemer, G. W. (1997). Psychobiology of early social attachment in rhesus monkeys. In S. C. Carter, et al. (Eds.), *The integrative neurobiology of affiliation* (Vol. 807, pp. 401–418). Annals of the New York Academy of Sciences.

Kraft, K. (1996). Psychological evaluation of the chronic pain patient. In M. A. Samuels & S. Feske (Eds.), *Office practice of neurology* (pp. 1166–1170). New York: Churchill Livingston.

Krakauer, J., & Ghez, C. (2000). Voluntary movement. In E. R. Kandel et al. *Principles of neural science, 4th ed.* (pp. 756–781). New York: McGraw Hill.

Kramer, M. E., et al. (2008). Long-term neural processing of attention following early traumatic brain injury: fMRI and neurobehavioral outcomes. *Journal of the International Neuropsychological Society, 14*, 424–435.

Krassioukov, A. V. (2002). Peripheral nervous system. In V. S. Ramachandran (Ed.), *Encyclopedia of the human brain* (Vol. III, pp. 817–813). Boston: Academic Press.

Kraus, J., et al. (2005). Physical complaints, medical service use, and social and employment changes following mild traumatic brain injury: A 6-month longitudinal study. *Journal of Head Trauma Rehabilitation, 20*, 239–256.

Kretschmann, H. J., & Weinrich, W. (2004). *Cranial neuroimaging and clinical neuroanatomy, 3rd Ed.* New York: Thieme.

Kretschmann, H-L., & Weinrich, W. (1992). *Cranial neuroimaging and clinical neuroanatomy.* New York: Thieme Medical Publishers.

Krikorian, R., & Layton, B. S. (1998). Implicit memory in posttraumatic stress disorder with amnesia for the traumatic event. *Journal of Neuropsychiatry and Clinical Neurosciences,* 10, 359–362.

Kronenberg, H., et al. (2003). Principles of endocrinology. In P. R. Larsen, et al. (Eds.), *Williams textbook of endocrinology, 10th Ed.* (pp. 1–16). Philadelphia: Saunders.

Kropyvnytskyy, I. V., et al. (1999). Circadian rhythm of cerebral perfusion pressure and intracranial pressure in head injury. *Brain Injury, 13*, 45–52.

Krystal, J. H., et al. (1986). Assessment of alexithymia in posttraumatic stress disorder and somatic illness: Introduction of a reliable measure. *Psychosomatic Medicine, 48*, 84–89.

Krystal, J. H., et al. (1995). Toward a cognitive neuroscience of dissociation and altered memory functions in post-traumatic stress disorder. In M. J. Friedman, et al. (Eds.), *Neurobiological and clinical consequences of stress: From normal adaptation to PTSD* (pp. 239–269). Philadelphia: Lippincott-Raven.

Kuhn, H. G., et al. (2001). Adult neurogenesis: A compensatory mechanism for neuronal damage. *European Archives of Psychiatry and Clinical Neuroscience, 251*, 152–158.

Kuhl, P. K. (2000). Language, mind, and brain: Experience alters perception. In M. S. Gazzaniga (Ed.), *The new cognitive neurosciences, 2nd Ed.* (pp. 99–115). Cambridge, MA: MIT Press.

Kuijpers, A. H. W. M., et al. (1995). The influence of different boundary conditions on the response of the head to impact: A two-dimensional finite element study. *Journal of Neurotrauma, 12*, 715–724.

Kunkel, R. S. (1996). Classification and differential diagnosis of headache. In M. S. Samuels & S. Feske (Eds.), *Office practice of neurology* (pp. 1101–1104). New York: Churchill Livingstone.

Kupferman, I. (1991). Hypothalamus and limbic system: motivation. In E. R. Kandel, J. J. Schwartz, & T. M. Jessell (Eds.) *Principles of Neuroscience,* 3rd ed., pp. 750–760. New York: Elsevier.

Kvetnansky, R. (2004). Stressor specificity and effect of prior experience on catecholamine biosynthetic enzyme phenylethanolamine N-methyltransferase. *Annals of the New York Academy of Sciences, 1032*, 117–129.

Kvetnansky, R., & McCarty, R. (2000). Adrenal medulla. In G. Fink (Ed.), *Encyclopedia of stress* (Vol. 1, pp. 63–70). San Diego, CA: Academic Press.

La Greca, A. M. (2000). Posttraumatic stress disorder in children. In G. Fink (Ed.), *Encyclopedia of stress* (Vol. 3, pp. 17–24). San Diego: Academic Press.

Laatsch, L., et al. (2007). An evidence-based review of cognitive and behavioral rehabilitation treatment studies in children with acquired brain injury. *Journal of Head Injury Rehabilitation*, *22*, 248–256.

Labi, M. L. C., & Horn, L. J. (1990). Hypertension in traumatic brain injury. *Brain Injury,* 4, 365–370.

LaChapelle, D. L., & Finlayson, M. A. J. (1998). An evaluation of subjective and objective measures of fatigue in patients with brain injury and healthy controls. *Brain injury*, *12*, 649–659.

Lackner, J. R., & DiZio, P. (2002). Somatosensory and proprioceptive contributions to body orientation, sensory localization, and self-calibration. In R. J. Nelson (Ed.), *The somatosensory system: Deciphering the brain's own body image* (pp. 121–140). Boca Raton, FL: CRC Press.

Lacomis, D. (1996). Respiratory dysfunction. In M. S. Samuels & S. Feske (Eds.), *Office Practice of neurology* (pp. 121–129). New York: Churchill Livingstone.

Ladavas, E., et al. (1993). Emotional evaluation with and without conscious stimulus identification: Evidence from a split-brain patient. *Cognition and Emotion, 7*(1), 95–114.

Ladd, C. O., et al. (2000). Startle response. In G. Fink (Ed.), *Encyclopedia of stress* (Vol. 3, pp. 486–495). San Diego, CA: Academic Press.

Ladefoged, P. (1975). *A course in phonetics*. New York: Harcourt Brace Jovanovich.

Lai, Y. Y., & Siegel, J. M. (1992). Corticotropin-releasing factor mediated muscle atonia in pons and medulla. *Brain Research, 575*, 63–68.

Lake, A. E., et al. (1999). Headache level during neuropsychological testing and test performance in patients with chronic posttraumatic headache. *Journal of Head Trauma Rehabilitation,* 14, 1, 70–80.

Lalwani, A. K. (2008). Disorders of smell, taste, and hearing. In A. L Fauci, et al. (Eds.), *Harrison's principles of internal medicine, 17th Ed.* (pp. 196–204).

Lam, J. M. K., et al. (1997). Monitoring of autororegulation using laser Doppler flowmetry in patients with head injury. *Journal of Neurosurgery, 86*, 438–445.

Lambert, K. G., et al. (1999). Does chronic activity-stress produce hippocampal atrophy and basal forebrain lesions? In J. F. McGinty (Ed.), *Advancing from the ventral striatum to the extended amygdala* (Vol. 877, pp. 742–746). Annals of the New York Academy of Sciences.

Landsberg, L., & Young, J. B. (1992). Catecholamines and the adrenal medulla. In J. D. Wilson & D. W. Foster (Eds.), *Williams textbook of endocrinology, 8th Ed.* (pp. 621–705). Philadelphia: Saunders.

Lane, R., et al. (1997). Is alexithymia the emotional equivalent of blindsight? *Biological Psychiatry, 42*(9), 834–844.

Langlois, J. A., et al. (2004). *Emergency department visits, hospitalizations, and deaths*. Atlanta, GA: Centers for Disease Control and Prevention, National Center for Injury Prevention and Control.

Langlois, J. A., et al. (2006). The epidemiology and impact of traumatic brain injury. *Journal of Head Trauma Rehabilitation*, *5*, 375–378.

Lanuza, D. M., et al. (1989). Body temperature and heart rate rhythms in acutely head injured patients. *Applied Nursing Research, 2,* 135–138.

Larrabee, G. J., et al. (1985). Sensitivity age-decline resistant ("hold") WAIS subtests to Alzheimer's disease. *Journal of Clinical and Experimental Neuropsychology, 7*(5), 497–504.

Larrabee, G. J., et al. (2008). Sensitivity to brain dysfunction of the Halstad-Reitan vs. an ability-focused neuropsychological battery. *The Clinical Neuropsychologist, 22*, 813–825.

Larsen, P. R. & Davies, T. F. (2003). Hypothyroidism and thyroiditis. In P. R. Larsen, et al. (Eds.), *Williams Textbook of Endocrinology, 10th Ed.* (pp. 423–455). Philadelphia: Saunders.

Larsen, P. R., et al. (1998). The thyroid gland. In J. D. Wilson et al. (Eds.), *Williams textbook of endocrinology, 9th Ed.* (pp. 389–515). Philadelphia: Saunders.

Laterra, J., & Goldstein, G. W. (2000). Ventricular organization of cerebrospinal fluid: Blood-brain barrier, brain edema and hydrocephalus. In E. R. Kandel, et al. (Eds.) *Principles of neural science, 4th ed.* (pp. 1288–1300). New York: McGraw Hill.

Lazar, M. F., & Menaldino, S. (1995). Cognitive outcome and behavioral adjustment in children following traumatic brain injury: A developmental perspective. *Journal of Head Trauma Rehabilitation, 10*(5), 55–63.

Lazarus, R. S., & Folkman, S. (1984). Coping and adaptation. In W. D. Gentry (Ed.), *Handbook of behavioral medicine* (pp.282–325). New York: The Guilford Press.

Le Moal, M. L., et al. (1992). The behavioral neuroendocrinology of arginine vasopressin, adrenocorticoptropic hormone, and opioids. In C. B. Nemeroff (Ed.), *Neuroendocrinology* (pp. 365–396). Boca Raton, FL: CRC Press.

Le Roux, P. D., et al. (2000). Cerebral concussion and diffuse brain injury. In P. R. Cooper & J. G. Golfinos (Eds.), *Head injury, 4th Ed.* (pp. 175–199). New York: McGraw Hill.

LeBlanc, A. (1995). *The cranial nerves*. New York: Springer-Verlag.

Lecours, A. R., et al. (1987). Illiteracy and brain damage-I: Aphasia testing in culturally contrasted populations (control subjects). *Neuropsychologia, 25*, 231–245.

LeDoux, J. E. (1993). Cognition versus emotion, again—this time in the brain: A response to Parrott and Schulkin. *Cognition and Emotion, 7*(1), 61–64.

LeDoux, J. E. (1995). Setting "stress" into motion: Brain mechanisms of stimulus evaluation. In M. J. Friedman, et al. (Eds.), *Neurobiological and clinical consequences of stress: From normal adaptation to PTSD* (pp. 125–134). Philadelphia: Lippincott-Raven.

Lee, D., & Seo, H. (2007). Mechanisms of reinforcement learning and decision making in the primate dorsolateral prefrontal cortex. In B. W. Balleine, et al. (Eds.), *Reward and decision making in corticobasal ganglia networks* (Vol. 1104, pp. 108–122). Annals of the New York Academy of Sciences.

Lee, F. S., & Chao, M. V. (2005). Neurotrophic factors. In B. J. Sadock & V. A. Sadock (Eds.), *Comprehensive textbook of psychiatry, I, 8th Ed.* (pp. 84–89). Philadelphia: Lippincott Williams & Wilkins.

Lehr, E. (1990). *Psychological management of traumatic brain injuries in children and adolescents.* Rockville, MD: Aspen.

Lehr, E., & Lantz, J. A. (1990). Behavioral components. In E. Lehr (Ed.) *Psychological Management of Traumatic Brain Injuries in Children and Adolescents,* pp. 133–153. Rockville, MD: Aspen.

Leininger, B. E., et al. (1990). Neurological deficits in symptomatic head injury patients after concussion and mild concussion. *Journal of Neurology, Neurosurgery and Psychiatry, 53*, 293–296.

Lemack, G. E., et al. (1988). Effects of stress on male reproductive function. In J. R. Hubbard & E. A. Workman (Eds.), *Handbook of stress medicine: An organ system approach* (pp. 141–152). Boca Raton, FL: CRC Press.

Leon-Carrion, J., et al. (2005). Epidemiology of traumatic brain injury and subarachnoid hemorrhage. *Pituitary, 8*(3–4), 197–202.

Lerdahl, F. (2001). The sounds of poetry viewed as music. In R. Zatorre & I. Peretz (Eds.), *The biological foundations of music* (Vol. 930, pp. 337–354). Annals of the New York Academy of Sciences.

Leuschel, A., & Docherty, G. J. (1996). Prosodic assessment of dysarthria. In A. Donald, et al. (Eds.), *Disorders of motor speech* (pp. 155–177). Baltimore: Paul H. Brooks.

Levchakov, A., et al. (2006). Computational studies of strain exposures in neonate and mature rat brains during closed head impact. *Journal of Neurotrauma, 23*, 1570–1580.

Levin, H., et al. (1994). Tower of London performance in relation to magnetic resonance imaging following closed head injury in children. *Neuropsychology,* 8, 171–179.

Levin, H. S. (1985). Outcome after head injury: Part II. Neurobehavioral recovery. In D. P. Becker & J. T. Povlishock (Eds.), *Central nervous system status report-1985* (pp. 281–299). Washington, DC: National Institute of Neurological and Communicative Disorders and Stroke, National Institutes of Health.

Levin, H. S. (1991). Aphasia after head injury. In M. T. Sarno (Ed.), *Acquired aphasia, 2nd Ed.* (pp. 455–498). San Diego: Academic Press.

Levin, H. S., & Eisenberg, H. M. (1984). The relative durations of coma and posttraumatic amnesia after severe non-missile head injury: Findings from the pilot phase of the National Traumatic Coma Data Bank. In M. Miner & K. Wagner (Eds.), *Neural trauma: Treatment, monitoring and rehabilitation issues* (pp. 89–97). Stoneham, MA: Butterworth.

Levin, H. S., et al. (1982). *Neurobehavioral consequences of closed head injury.* New York: Oxford University Press.

Levin, H. S., et al. (1992). Posttraumatic and retrograde amnesia after closed head injury. In L. R. Squire & N. Butters (Eds.), *Neuropsychology of memory, 2nd Ed.* (pp. 290–308). New York: Guilford Press.

Levin, H. S., et al. (1997a). Depression as a secondary condition following mild and moderate traumatic brain injury. *Seminars in Clinical Neuropsychiatry, 2*(3), 207–215.

Levin, H. S., et al. (1997b). Magnetic resonance imaging in relation to functional outcome of pediatric closed head injury: A test of the Ommaya-Gennarelli model. *Neurosurgery Online, 40*(3), 432–441.

Levin, H. S., et al. (2004). Psychosocial outcome of TBI in children with unilateral frontal lesions. *Journal of the International Neuropsychological Society, 10*, 305–316.

Levin, H. S., et al. (2008). Diffusion tensor imaging in relation to cognitive and functional outcome of traumatic brain injury in children. *Journal of Head Trauma Rehabilitation, 23*, 197–208.

Levine, S. (1993). The psychoendocrinology of stress. In Y. Taché & C. Rivier (Eds.), *Corticotropin-releasing factor and cytokines: Role in the stress response* (Vol. 691, pp. 61–69). Annals of the New York Academy of Science.

Levine, S. C., et al. (2008). Posttraumatic growth in adolescence: Examining its components and relationship with PTSD. *Journal of Traumatic Stress, 21*, 492–496.

Levisohn, L., et al. (2000). Neurological consequences of cerebellar tumour resection in children: Cerebellar cognitive affective syndrome in a paediatric population. *Brain, 123*(5), 1041–1050.

Levite, M. (2000). Nerve-driven immunity: The direct effects of neurotransmitters on T-cell function. In A. Conti, et al. (Eds.), *Neuroimmunomodulation: Perspectives at the new millennium* (Vol. 917, pp. 307–321). Annals of the NY Academy of Sciences.

Levy, M. L., et al. (1998). Apathy is not depression. *The Journal of Neuropsychiatry and Clinical Neurosciences, 10*, 314–319.

Lewandowski, M. H., & Blasiak, T. (2004). Slow oscillation circuit of the intergeniculate leaflet. *Acta Neurobiologiae Experimentalis, 64*, 277–288.

Lewine, J. D. (2007). Objective documentation of traumatic brain injury subsequent to mild head trauma: Multimodal brain imaging with MEG, SPECT, and MRI. *Journal of Head Trauma Rehabilitation, 22*, 141–155.

Lewis, L. (1991). Role of psychological factors in disordered awareness. In G. P. Prigatano & D. L. Schachter (Eds.), *Awareness of deficit after brain injury* (pp. 223–239). New York: Oxford University Press.

Lewis, M. S., & Miller, L. S. (2007). Executive control functioning and functional ability in older adults. *The Clinical Neuropsychologist, 21*, 274–285.

Lewis, R. J., et al. (1988). Children at risk for emotional disorders. *Clinical Psychology Review, 8*, 417–448.

Lewis, W. H. (1936). *Gray's Anatomy*. Philadelphia: Lea and Febiger.

Lezak, M. D. (1983). *Neuropsychological Assessment,* 2nd ed. New York: Oxford University Press.

Lezak, M. D. (1989). Assessment of psychosocial dysfunctions resulting from head trauma. In M. D. Lezak (Ed.), *Assessment of the behavioral consequences of head trauma* (pp. 113–143). New York: Alan R. Liss.

Lezak, M. D. (1995). *Neuropsychological assessment, 3rd Ed.* New York: Oxford University Press.

Lezak, M. D., et al. (2004). *Neuropsychological assessment, 4th Ed.* New York: Oxford University Press.

Liau, L. M., et al. (1996). Pathology and pathophysiology of head injury. In J. R. Youmans (Ed.), *Neurological surgery, 4th Ed.* (pp. 1549–1594). Philadelphia: Saunders.

Lichtenberg, J. (2002). Core consciousness and expanded consciousness in the clinical setting. Presented at a conference on Consciousness, Self-Experience and the Brain. St. Vincent's Catholic Medical Center, New York, NY.

Lieberman, M. D. (2007). Social cognitive neuroscience: A review of core processes. *Annual Review of Psychology, 58*, 259–289.

Lieberman, S. A., et al. (2001). Prevalence of neuroendocrine dysfunction in patients recovering from traumatic brain injury. *Journal of Clinical Endocrinology and Metabolism, 86*, 2752–2756.

Lindenberg, R., & Freytag, E. (1969). Morphology of brain lesions from blunt trauma in early infancy. *Archives of Pathology, 87*, 298–305.

Lippincott Williams & Wilkins. (2005). *Handbook of pathophysiology, 2nd Ed.* Philadelphia: Author.

Lippincott Williams & Wilkins. (2006). *Stedman's medical dictionary, 28th Ed.* Philadelphia: Author.

Lipton, R. B., et al. (1994). Comorbidity of migraine: The connection between migraine and epilepsy. *Neurology,* 44, 10, Supplement 7, S38–S32.

Lishman, W. A. (1987). *Organic psychiatry*. Boston: Blackwell Scientific.

Lishman, W. A. (1988). Physiogenesis and psychogenesis in the 'post-concussional syndrome'. *British Journal of Psychiatry, 153*, 460–469.

Little, J. T., & Sunderland, T. (1998). Psychosis secondary to encephalitis and encephalopathies. *Seminars in Clinical Psychiatry,* 3, 4–11.

Littleton, H., et al. (2007). Trauma coping strategies and psychological stress: A metanalysis. *Journal of Traumatic Stress, 20*, 977–988.
Liu, G. T. (1996). Disorders of the eyes and eyelids. In M. S. Samuels & S. Feske (Eds.), *Office practice of neurology* (pp. 41–74). New York: Churchill Livingstone.
Livingston, R. B. (1985). Section IX. Neurophysiology. In J. B. West (Ed.), *Best & Taylor's physiological basis of medical practice, 11th Ed.* (pp. 970–1295). Baltimore: Williams & Wilkins.
Llamas, C., & Diamond, A. (1991). Development of frontal cortex abilities in children between 3–8 years of age. Presented at the Biennial Meeting of the Society for Research in Child Development.
Llinas, R. (2001). *I of the vortex. From neurons to self.* Cambridge, MA: MIT Press.
Lloyd, D. (2000). Virtual lesions and the not-so-modular brain. *Journal of the International Neuropsychological Society, 6*, 627–635.
Lo, Y., et al. (2007). The role of spinal inhibitory mechanisms in whiplash injuries. *Journal of Neurotrauma*, 1055–1067.
Lockman, L. A. (1989). Impairment of consciousness. In K. Swaiman (Ed.), *Pediatric neurology* (Vol. I, pp. 158–167). St. Louis, MO: C. V. Mosby.
Loewenstein, R. J. (1991). Psychogenic amnesia and psychogenic fugue: A comprehensive review. In A. Talisman & S. M. Goldfinger (Eds.), *Annual review of psychiatry, 10* (pp. 189–221). Washington, DC: American Psychiatric Press.
Lomas, J. P., & Dunning, J. (2005). Best evidence topic report. S100B protein levels as a predictor of long-term disability after head injury. *Emergency Medicine Journal*, *22*, 889–891.
Lombardi, G., et al. (1994). Neuroendocrine axis and behavioral stress. In N. A. Fabris, et al. (Eds.), *Immunomodulation: The state of the art* (Vol. 741, pp. 216–222). Annals of the New York Academy of Sciences.
Loubser, P. G., & Donovan, W. H. (1996). Chronic pain associated with spinal cord injury. In R. K. Narayan, J. E. Wilberger, & J. T. Povlishock (Eds.) *Neurotrauma,* pp. 1311–1322. New York: McGraw-Hill.
Lovell, M. R., & Collins, M. W. (1998). Neuropsychological assessment of the college football player. *Journal of Head Trauma Rehabilitation, 13*(2), 9–26.
Lovitt, R. (1987). A conceptual model and case study for the psychological assessment of hysterical pseudoseizures with the Rorschach. *Journal of Personality Assessment, 51*, 207–219.
Low, P. A. (2003). Testing the autonomic nervous system. *Seminars in Neurology*, *23*(4), 407–421.
Low, P. A., & Banniser, R. (1997). Multiple system atrophy and pure autonomic failure. In P. A. Low (Ed.), *Clinical autonomic disorders: Evaluation and management, 2nd Ed.* (pp. 555–573). Philadelphia: Lippincott-Raven.
Lowenstein, R. J., & Putnam, F. W. (2005). Dissociative disorders. In B. J. Sadock & V. A. Sadock (Eds.), *Comprehensive textbook of psychiatry, I, 8th Ed.* (pp. 1844–1901). Philadelphia: Lippincott Williams & Wilkins.
Luria, L. R. (1980). *Higher cortical functions in man*, 2nd Ed. Rev. Mew York: Basic Books.
Lyeth, B. G., et al. (1988). Effects of anticholinergic treatment on transient behavioral suppression and physiological responses following concussive brain injury to the rat. *Brain Research,* 488, 88–97.
Lyketsos, C. G., et al. (2004). Forgotten frontal lobe syndrome or "executive dysfunction syndrome". *Psychosomatics, 45*, 247–255.
Lynch, W. J. (2008). Everyday living assessment in cognitive evaluations. *Journal of Head Trauma Rehabilitation, 23,* 185–188.
Lyons, D. M., & Parker, K. J.(2007). Stress inoculation-induced indications of resilience in monkeys. *Journal of Traumatic Stress, 20*, 423–433.
Maas, A. I. R., et al. (2000). Prognosis and clinical trial design in traumatic brain injury. The IMPACT study. *Journal of Neurotrauma*, *24*, 232–238.
Mace, C. J., & Trimble, M. F. (1991). Psychogenic amnesias. In T. Yanigahara & R. C. Petersen (Eds.), *Memory disorders* (pp. 429–453). New York: Marcel Dekker.
MacGillivary, M. H. (1995). Disorders of growth and development. In P. Felig, et al. (Eds.), *Endocrinology and metabolism, 3rd Ed.* (pp. 1619–1673). New York: McGraw-Hill.
MacGillivary, M. H. (2001). Disorders of growth and development. In P. Felig & L. A. Frohman (Eds.), *Endocrinology and metabolism, 4th Ed.* (pp. 1265–1316). New York: McGraw-Hill.
Machulda, M. M., et al. (1998). Relationship between stress, coping and postconcussion symptoms in a healthy adult population. *Archives of Clinical Neuropsychology, 13*, 415–424.
Mackley, R. J. (1999). The role of trigger point in the management of head, neck, and face pain. In M. S. Margoles & R. Weinser (Eds.), *Chronic pain: Assessment, diagnosis and management* (pp. 153–166). Boca Raton, FL: CRC Press.

Maddi, S. R. (2008). The courage and strategies of hardiness as helpful in growing despite major disruptive stresses. *American Psychologist, 63*, 563.

Mahmood, O., et al. (2004). Neuropsychological performance and sleep disturbance following traumatic brain injury. *Journal of Head Trauma Rehabilitation, 19*, 378–390.

Mahowald, M. W., & Schenck, C. H. (1989). Narcolepsy. In G. Adelman (Ed.) *Neuroscience Year: Supplement I to the Encyclopedia of Neuroscience,* pp. 114–116. Boston: Birknauser.

Maier, S. F., et al. (1994). Psychoimmunology: The interface between behavior, brain, and immunity. *American Psychologist, 49*, 1004–1007.

Maier, S. F., et al. (2001). Multiple routes of action of interleukin-1 on the nervous system. In R. Ader et al. (Eds.), *Psychoneuroimmunology, 3rd Ed.* (Vol. I, pp. 563–581). San Diego: Academic Press.

Maikos, J. T., & Schreiber, D. L. (2007). Immediate damage to the blood-spinal cord barrier due to mechanical trauma. *Journal of Neurotrauma, 24*, 492–507.

Malawista, W. E. (2004). Lyme disease. In L. Goldman & D. Ausiello (Eds.), *Cecil textbook of medicine, 22nd Ed.* (pp. 1934–1939). Philadelphia: Saunders.

Malec, J. F., et al. (2007a). The Mayo Classification system for traumatic brain injury severity. *Journal of Neurotrauma, 24*, 1417–1424.

Malec, J. F., et al. (2007b). Self-assessment of impairment, impaired self-awareness, and depression after traumatic brain injury. *Journal of Head Trauma Rehabilitation, 122*, 156–166.

Maquet, P. (2004). Mechanism and function of dreams [French]. *Bulletin et Memoires de l'Academy Royale de Medecine de Belgique, 159*, 571–575.

Maquet, P., & Schwartz, M. S. (2002). Evaluation of dreams by the neuropsychological approach: Utility in the characterization in oneiric activity by functional neuroimaging. *Bulletin et Memoires de l'Academie Royale de Medicine de Belgique, 157,* 218–219.

Marchetti, B., et al. (2001). The hypothalamo-pituitary-gonadal axis and the immune system. In R. Ader, et al. (Eds.), *Psychoneuroimmunology,* 3rd Ed. (Vol. 1, pp. 363–385). San Diego: Academic Press.

Margoles, M. (1999). Soft tissue pain problems: Introduction. In M. S. Margoles & R. Weiner (Eds.), *Chronic pain: Assessment, diagnosis, and management* (pp. 91–92). Boca Raton, FL: CRC Press.

Marin, S., & Chakravorty, S. (2005). Disorders of diminished motivation. In J. M. Silver, et al. (Eds.), *Textbook of traumatic brain injury* (pp. 337–352). Washington DC: American Psychiatric Publishing.

Marion, D. W., et al. (2004). Craniocerebral trauma. In W. G. Bradley, et al. (Eds.), *Neurology in clinical practice, 4th Ed.* (Vol. 2, pp. 1127–1147). Boston: Butterworth-Heinemann.

Market, K. E., et al. (1990). Acquired stuttering: Descriptive data and treatment outcome. *Journal of Fluency Disorders, 15*, 21–33.

Marks, D. B., et al. (1996). *Basic medical biochemistry.* Baltimore: Williams & Wilkins.

Marrocco, R. T., & Field, B. A. (2002). Arousal. In V. S. Ramachandran (Ed.), *Encyclopedia of the human brain* (Vol. I, pp. 223–236). Boston: Academic Press.

Marshall, G. D., & Rossio, J. L. (2000). Cytokines. In G. Fink (Ed.), *Encyclopedia of stress, 1* (pp. 626–633). San Diego: Academic Press.

Marshall, L. F., & Marshall, S. B. (1985). Current clinical head injury research in the United States. In D. P. Becker and J. T. Povlishock (Eds.). *Central nervous system trauma status report: 1985,* pp. 45–51. National Institute of Neurological and Communicative Disorders and Stroke, National Institutes of Health. (No location cited.)

Marshall, L., et al. (2006). Boosting slow oscillations during sleep potentiates memory. *Nature, 444*, 610–613.

Martelli, M. F., et al. (2004). Psychological, neuropsychological, and medical considerations in assessment and management of pain. *Journal of Head Trauma Rehabilitation, 19,* 10–28.

Martelli, M. F., et al. (2007). Psychological approaches to comprehensive assessment and management following TBI. In N. D. Zasler, et al. (Eds.), *Brain injury medicine* (pp. 623–742). New York: Demos.

Martin, G., et al. (1990). Reticular formation of the pons and medulla. In G. Paxinos (Ed.), *The human nervous system* (pp. 203–230). San Diego: Academic Press.

Martin, J. B., & Reuchlin, S. (1987). *Clinical Neuroendocrinology,* 2nd ed. Philadelphia: F. A. Davis.

Martin, J. B., & Reichlin, S. (1989). *Clinical neuroendocrinology.* Philadelphia: F. A. Davis.

Martin, J. H. (1985). Anatomical substrates for somatic sensation. In E. R. Kandel & J. H. Schwartz (Eds.), *Principles of neural science, 2nd Ed.* (pp. 301–315). New York: Elsevier.

Martin, J. H. (2003). *Neuroanatomy: Text and atlas, 3rd Ed.* New York: McGraw-Hill.

Martin, L. J. (2002). Neurodegenerative disease. In V. S. Ramachandran (Ed.), *Encyclopedia of the human brain* (Vol. 2, pp. 441–465). Boston: Academic Press.

Martin, N. A., et al. (1992). Posttraumatic cerebral arterial spasm: Transcranial Doppler ultrasound, cerebral blood flow, and angiographic findings. *Journal of Neurosurgery, 77,* 575–583.

Martin, N. A., et al. (1995). Posttraumatic cerebral arterial spasm. *Journal of Neurotrauma, 12*, 897–906.

Marucha, P. T., et al. (2001). Stress and wound healing. In R. Ader, et al. (Eds.), *Psychoneuroimmunology, 3rd Ed.,* (Vol. 2, pp. 613–636). San Diego: Academic Press.

Marx, B. P., et al. (2008). Tonic immobility as an evolved predator defense: Implications for sexual assault survivors. *Clinical Psychology: Science and Practice, 15*, 74–90.

Masel, B. E., et al. (2001). Excessive daytime sleepiness in adults with brain injuries. *Archives of Physical Medicine and Rehabilitation, 82*, 1526–1532.

Matarazzo, J. D. (1990). Psychological assessment versus psychological testing: Validation from Binet to school, clinic and courtroom. *American Psychologist, 45*, 999–1017.

Matarazzo, J. D., & Herman, D. O. (1984). Relationship of education and IQ in WAIS-R standardization sample. *Journal of Consulting and Clinical Psychology, 52*, 631–634.

Matarazzo, J. D., & Herman, D. O. (1985). Clinical uses of the WAIS-R: Base rates of differences between VIQ and PIQ in the WAIS-R standardization sample. In B. B. Wolman (Ed.) *Handbook of Intelligence,* pp. 899–932. New York: Wiley.

Mateer, C. A. (1999). Executive function disorders: Rehabilitation challenges and strategies. *Seminars in Clinical Neuropsychiatry,* 4, 50–59.

Mathew, R. J. (1995). Sympathetic control of cerebral circulation: Relevance to psychiatry. *Biological Psychiatry, 37*, 283–285.

Matsumoto, A. M. (1996). The testis. In J. C. Bennett & F. Plum (Eds.), *Cecil textbook of medicine* (pp. 1325–1341). Philadelphia: W. B. Saunders.

Matsumoto, A. M. (2001). The testis. In P. Felig & L. A. Frohman (Eds.), *Endocrinology and metabolism, 4th Ed.* (pp. 625–705). New York: McGraw-Hill.

Maughan, P. H., et al. (2000). Recovery of water maze performance in aged versus young rats after brain injury with the impact acceleration model. *Journal of Neurotrauma, 17,* 1141–1153.

Max, J. E., et al. (1998). Traumatic brain injury in children and adolescents: Psychiatric disorders at one year. *Journal of Neuropsychiatry and Clinical Neurosciences,* 10, 290–297.

Max, J. E., et al. (2006). Predictors of personality change due to traumatic brain injury in children and adolescents in the first six months after injury. *Journal of the American Academy of Child and Adolescent Psychiatry, 44*, 434–432.

Max, W., et al. (1991). Head injuries: Costs and consequences. *The Journal of Head Trauma Rehabilitation, 6*(2), 76–87.

Maxwell, W. L., & Graham, D. I. (2003). Diffuse axonal injury (DAI). In M. J. Aminoff & R. B. Daroff (Eds.), *Encyclopedia of the neurological sciences* (Vol. 2, pp. 10–12). Boston: Academic Press.

Mayer, J. D. (2009). Personal intelligence expressed: injury (DAI). *Review of General Psychology, 13* 46–58.

Mayer, J. D., et al. (2008). Emotional intelligence: New ability or eclectic traits? *American Psychologist, 63*, 503–513.

Mayhofner, C., et al. (2004). Cortical reorganization during recovery from complex regional pain syndrome. *Neurology, 63,* 694–701.

Mayorga, M. A. (1997). The pathology of primary blast overpressure injury. *Toxicology, 121*, 17–28.

McAllister, T. W. (2005). Mild brain injury and the postconcussion syndrome. In J. M. Silver et al. (Eds.), *Textbook of traumatic brain injury* (pp. 279–308). Washington, DC: American Psychiatric Publishing.

McAllister, T. W, et al. (2006). Mechanisms of working memory dysfunction after mild and moderate TBI: Evidence from functional MRI and neurogenetics. *Journal of Neurotrauma, 23*, 1450–1467.

McCann, S. M. (2003). Introduction. In S. M. McCann (Ed.), *Neuroimmunomodulation* (Vol. 840, pp. xiii–xiv). Annals of the New York Academy of Sciences.

McCann, S. M. (2006). Chronology of advances in neuroendocrine immunomodulation. In G. P. Chrousos et al., (Eds.), *Neuroendocrine and immune crosstalk* (Vol. 1088, pp. 1–11). Annals of the New York Academy of Sciences.

McCarty, R., & Pacak, K. (2000). Alarm phase and general adaptation syndrome. In G. Fink (Ed.), *Encyclopedia of stress, 1* (pp. 126–130). San Diego, CA: Academic Press.

McCrea, M., et al. (2002). Immediate neurocognitive effects of concussion. *Neurosurgery, 50*, 1040–1042.

McCrory, P. (2000). Convulsions in contact and collision sports. In R. C. Cantu (Ed.), *Neurologic athletic head and spine injuries* (pp. 192–199). Philadelphia: Saunders.

McCrory, P. (2006). Revisiting chronic traumatic encephalopathy. *British Journal of Sports Medicine, 36*(1), 2.

McCrory, P. R., & Berkovic, S. F. (2001). Concussion. The history of clinical and pathophysiological concepts and misconceptions. *Neurology*, *57*, 2283–2289.

McCullagh, S., & Feinstein, S. (2005). Cognitive changes. In J. M. Silver et al. (Eds.), *Textbook of traumatic brain injury* (pp. 321–335). Washington, DC: American Psychiatric Publishing.

McElhaneny, et al. (1996). Mechanisms of basilar skull fracture. *Journal of Neurotrauma,* 12, 669–678.
McEwen, B. S. (1992). Effects of the steroid/thyroid hormone family on neural and behavioral plasticity. In C. B. Nemeroff (Ed.), *Neuroendocrinology* (pp. 333–351). Boca Raton: CRC Press.
McEwen, B. S. (1994a). Introduction: Stress and the nervous system. *Seminars in the Neurosciences, 6*, 195–196.
McEwen, B. S. (1994b). The plasticity of the hippocampus is the reason for its vulnerability. *Seminars in the Neurosciences, 6*, 239–246.
McEwen, B. S. (1995). Adrenal steroid actions on brain: Dissecting the fine line between protection and damage. In M. J. Friedman et al. (Eds.), *Neurobiological and clinical consequences of stress: From normal adaptation to PTSD* (pp. 124–147). Philadelphia: Lippincott-Raven.
McEwen, B. S. (1998). Protective and damaging effects of stress mediators. *New England Journal of Medicine, 338*, 171–179.
McEwen, B. S. (2000a). Allostasis and allostatic load. In G. Fink (Ed.), *Encyclopedia of stress* (Vol. 1, pp. 145–150). San Diego, CA: Academic Press.
McEwen, B. S. (2000b). The neurobiology of stress: from serendipity to clinical relevance. *Brain Research, 886*, 172–189.
McEwen, B. S. (2000c). Definitions and concepts of stress. In G. Fink (Ed.), *Encylopedia of stress* (Vol. 1, pp. 508–509). San Diego, CA: Academic Press.
McEwen, B. S. (2003a). Mood disorders and allostatic load. *Biological Psychiatry, 54*, 200–207.
McEwen, B. S. (2003b). Early life influences on life-long patterns of behavior and health. *Mental Retardation and Developmental Disabilities Research Reviews, 9*, 149–154.
McEwen, B. S. (2004). Protection and damage from acute and chronic stress: Allostasis and allostatic overload and relevance to the pathophysiolology of psychiatric disorders. In R. Yehuda & B. McEwen (Eds.), *Behavioral stress responses* (Vol. 1032, pp. 1–7). Annals of the New York Academy of Sciences.
McEwen, B. S., & Chattarji, S. (2004). Molecular mechanisms of neuroplasticity and pharmacological implications: The example of tianeptine. *European Neuropsychopharmacology, 14*, S497–S502.
McEwen, B. S., & Seeman, T. (1999). Protective and damaging effects of mediators of stress: Elaborating and testing the concepts of allostasis and allostatic load. *Annals of the New York Academy of Science,* 896, 30–47.
McEwen, B. S., & Seeman, T. (2004). Protective and damaging effects of mediators of stress: Elaborating and testing the concepts of allostasis and allostatic load. *Annals of New York Academy of Science, 896,* 30–47.
McFie, J., & Thompson, J. A. (1972). Picture arrangement: A measure of frontal lobe function? *British Journal of Psychiatry,* 121, 547–552.
McGehee, D. V. (1996). Head injury in motor vehicle crashes: Human factors, effects, and prevention. In M. Rizzo & D. Tranel (Eds.), *Head injury and postconcussive syndrome* (pp. 57–69). New York: Churchill Livingston.
McGuire, L. M., et al. (1998). Prevalence of traumatic brain injury in psychiatric and non-psychiatric subjects. *Brain Injury, 12,* 207–214.
McGuire, T. A., & Rothenberg, M. B. (1986). Behavioral and psychosocial sequelae of pediatric head injury. *The Journal of Head Trauma Rehabilitation, 1*(4), 1–6.
McHenry, M. A. (1996). Laryngeal airway resistance following traumatic brain injury. In A. Donald et al. (Eds.), *Disorders of motor speech* (pp. 229–240). Baltimore: Paul H. Brooks.
McHenry, M. A. (1998). Velopharyngeal airway resistance disorders after traumatic brain injury. *Archives of Physical Medicine and Rehabilitation, 79*, 545–549.
McHenry, M. A. (1999). Aerodynmic, acoustic, and perceptual measures of nasality following traumatic brain injury. *Brain Injury, 13*, 281–290.
McHenry, M. A. (2001). Vital capacity following traumatic brain injury. *Brain Injury,* 15, 8, 741–745.
McHenry, M. A., & Wilson, R. (1994). The challenge of unintelligible speech following traumatic brain injury. *Brain Injury, 4*, 363–375.
McIntyre, C. K., et al. (2003). Role of the basolateral amygdala in memory consolidation. In P. Shinnick-Gallagher, et al. (Eds.), *The amygdala in brain function* (Vol. 985, pp. 273–293). Annals of the New York Academy of Sciences.
McKinlay, A., et al. (2008). Long term psychosocial outcomes after mild head injury in early childhood. *Journal of Neurology, Neurosurgery and Psychiatry, 73*, 281–288.
McKinney, W. T., & Tucker, G. J. (2001). Introduction: Stress adaptation and affective disorders. *Seminars in Clinical Psychiatry, 6,* 1–3.
McLaurin, R. L., & Towbin, R. (1990). Diagnosis and treatment of head injury in infants and children. In R. Youmans, (Ed.), *Neurological Surgery* (Vol. 3, pp. 2148–2153). New York: Saunders.

McLean, D. E., et al. (1995). Medical and surgical complications of pediatric brain injury. *Journal of Head Trauma Rehabilitation, 10*(5), 1–12.
McLean, M. J. (2004). Principles of neuropharmacology and therapeutics. In W. G. Bradley, et al. (Eds.), *Neurology in clinical practice* (Vol. I, 877–920). Boston: Butterworth Heinemann.
Medzhitov, R., & Janeway, C. A. (2002). Decoding the patterns of self and nonself by the innate immune system. *Science* 296, 12 April, pp. 298–300.
Meiser-Stedman, R., et al. (2007). Dissociative symptoms and the acute stress disorder diagnosis in children and adolescents: A replication of the Harvey and Bryant (1999) study. *Journal of Traumatic Stress, 20*, 359–364.
Mellman, T. A., et al. (2007). Relationships between REM sleep findings and PTSD symptoms during the early aftermath of trauma. *Journal of Traumatic Stress, 20*, 893–904.
Mellon, S. H. (2003). Steroids: Overview. In M. J. Aminoff & R. B. Daroff (Eds.), *Encyclopedia of the neurological sciences* (Vol. 4, pp. 387–389). Boston: Academic Press.
Melmed, S., & Kleinberg, D. L. (2003). Anterior pituitary. In P. R. Larsen et al. (Eds.), *Williams textbook of endocrinology, 10th Ed.* (pp. 177–280). Philadelphia: Saunders.
Mendelow, A. D., & Crawford, P. J. (2005). Primary and secondary brain injury. In P. L. Reilly & R. Bullock (Eds.), *Head injury, 2nd Ed.* (pp. 73–92). New York: Hodder Arnold.
Mendez, M. F., & Cummings, J. L. (2002). Neuropsychiatric aspects of aphasia and related disorders. In S. C. Yudofsky & R. E. Hales (Eds.), *Textbook of neuropsychiatry and clinical neurosciences, 4th Ed.* (pp. 565–578). Washington, DC: American Psychiatric Publishing.
Meneses, A., et al. (1998). Involvement of 5-HT_{1a} receptors in the consolidation of learning in cognitively impaired rats. In G. R. Martin et al. (Eds.), *Advances in serotonin receptors research* (Vol. 861). Annals of the New York Academy of Sciences.
Menzaghi, F., et al. (1993). In Y. Taché & C. Rivier (Eds.), *Corticotropin-releasing factor and cytokines: Role in the stress response* (Vol. 691, pp. 142–154). Annals of the New York Academy of Science.
Merck Research Laboratories. (2006). *The merck manual.* White House Station, NJ: Author.
Merikangas, K. R. (2005). In B. J. Sadock & V. A. Sadock (Eds.), *Comprehensive textbook of psychiatry, I, 8th Ed.* (pp. 1720–1728). Philadelphia: Lippincott Williams & Wilkins.
Merisalu, E., et al. (1996). Simultaneous hormonal changes in the blood plasma of patients with traumatic isolated brain injury. *Acta Medica Baltica, 3*(2), 169–173.
Merkley, T. L., et al. (2008). Diffuse changes in cortical thickness in pediatric moderate-to severe traumatic brain injury. *Journal of Neurotrauma, 25*, 1343–1345.
Merola, B., et al. (1994). Hypothalamic-pituitary-adrenal axis in neuropsychiatric disorders. In N. A. Fabris et al. (Eds.), *Immunomodulation: The state of the art* (Vol. 741, pp. 263–270). Annals of the New York Academy of Sciences.
Merskey, H. (1993). Psychosomatic disorders. In R. Greenwood, M. P. Barnes, T. M. McMillan, & C. D. Ward (Eds.) *Neurological Rehabilitation,* pp. 413–421. New York: Churchill Livingstone.
Merson, R. M. (1967). Speech rehabilitation in congenital aglossia. *Journal of rehabilitation, 33*, 33–34.
Messam, C. A., et al. (2002). Glial cell types. In V. S. Ramachandran (Ed.), *Encyclopedia of the human brain, Vol. II* (pp. 369–387). Boston: Academic Press.
Messingham, K. A., et al. (2002). Alcohol, immunity, and cellular immunity. *Alcohol, 28*, 137–149.
Messner, M., & Messner, E. (1988). Mood disorder following stroke. *Comprehensive psychiatry, 29*(1), 22–27.
Mesulam, M. M. (2005). Aphasia, memory loss, and other focal cerebral disorders. In *Harrison's principles of internal medicine, 16th Ed.* (Vol. I, pp. 145–153). New York: McGraw-Hill.
Michelson, D., et al. (1995). Mediation of the stress response by the hypothalamic-pituitary-adrenal axis. In M. J. Friedman et al. (Eds.), *Neurobiological and clinical consequences of stress: From normal adaptation to PTSD* (pp. 225–238). Philadelphia: Lippincott-Raven.
Mickeviciene, D., et al. (2002). A controlled historical cohort study on the post-concussion syndrome. *European Journal of Neurology, 9*, 581–587.
Middleboe, T., et al. (1992). Minor head injury: Impact on general health after 1 year: A prospective follow-up study. *Acta Neurological Scandinavica, 85*, 5–9.
Migeon, C. J., & Lanes, R. L. (1996). Adrenal cortex: Hypo- and hyperfunction. In F. Lifschitz (Ed.), *Pediatric endocrinology, 3rd Ed.* (pp. 321–325). New York: Dekker.
Mileykovskiy, B. Y., et al. (2000). Activation of pontine and medulary motor inhibitory systems reduces discharge in neurons located in the locus coeruleus and the anatomical equivalent of the midbrain locomotor region. *Journal of Neuroscience, 20*, 8551–8558.
Miller, A. D., & Yates, B. J. (1996). Vestibular effects on respiratory activity. In B. J. Yates & A. D. Miller (Eds.), *Vestibular autonomic regulation* (pp. 113–125). Boca Raton, FL: CRC Press.

Miller, E. K., & Wallis, J. S. (2003). The prefrontal cortex and executive brain functions. In L. Squire et al. (Eds.), *Fundamental neuroscience, 2nd Ed.* (pp. 1353–1376). London: Academic Press.

Miller, F. (1977). *College Physics,* 4th ed. New York: Harcourt Brace.

Miller, G. (2000). Mental traits as fitness indicators: Expanding evolutionary psychology's adaptationism. *Evolutionary perspectives on human reproductive behavior* (Vol. 907, pp. 62–74). Annals of the New York Academy of Sciences.

Miller, J. D. (1989). Pathophysiology of human head injury. In D. P. Becker and S. K. Gudeman (Eds.) *Textbook of Head Injury,* pp. 507–524. Philadelphia: Saunders.

Miller, J. D., & Gudeman, S. K. (1986). Cerebral vasospasm after head injury. Vasospasm: Occurrence in conditions other than subarachnoic hemorrhage, In R, H, Wilkins (Ed.) *Cerebral Arterial Spasm* ,pp. 476–479. Baltimore: Williams and Wilkins.

Miller, L. (March/April 1991). Psychotherapy of the brain-injured patient: Principles and practices. *Cognitive Rehabilitation*, 24–30.

Miller, L. (1993). *Psychotherapy of the brain-injured patient.* New York: Norton.

Miller, L. (1999). Atypical psychological responses to traumatic brain injury: PTSD and beyond. *Neurorehabilitation, 13*, 79–90.

Miller, L., & Donders, J. (2001). Subjective symptomatology after traumatic head injury. *Brain Injury, 15*, 297–304.

Miller, N. (1989). Apraxia of speech. In C. Code (Ed.), *The characteristics of aphasia* (pp. 131–154). New York: Taylor & Francis.

Miller, N. R., & Newman, N. J. (Eds.) (1999). *The essentials: Walsh & Hoyt's clinical neuro-opthalmology.* Philadelphia: Lippincott Williams & Wilkins.

Miller, S. D., et al. (1999). Pre-natal stress-induced modifications of neuronal nitric acid synthase in amygdale and medial peptic area. In J. F. McGivney (Ed.), *Advancing from the ventral striatum to the extended amygdala* (Vol. 877, pp. 760–763). Annals of the New York Academy of Sciences.

Miller, S. M., & Birnbaum, A. (1988). Putting the life back into 'life events': Toward a cognitive social learning analysis of the coping process. In S. Fisher & J. Reason (Eds.), *Handbook of life stress, cognition and health* (pp. 497–509). New York: John Wiley and Sons.

Miller, W. L., & Chrousos, G. P. (2001). The adrenal cortex. In P. Felig & L. A. Frohman (Eds.). *Endocrinology and metabolism, 4th Ed.* (pp. 387–524). New York: McGraw-Hill.

Miller, W. L., & Tyrrell, J. B. (1995). The adrenal cortex. In P. Felig et al. (Eds.), *Endocrinology and Metabolism, 3rd Ed.* (pp. 555–711). New York: McGraw-Hill.

Miller, W. L., et al. (1980). Child abuse as a cause of post-traumatic hypopituitarism. *New England Journal of Medicine, 302*, 724–728.

Miller-Perrin, C. L., & Wurtele, S. K. (1990). Reactions to childhood sexual abuse: Implications for post-traumatic stress disorder. In C. Meek (Ed.), *Post-traumatic stress disorder: Assessment, differential diagnosis and forensic evaluation* (pp. 91–135). Sarasota, FL: Professional Resource Exchange.

Millesi, H. (1997). Brachial plexus injuries. In R. MacFarlane & D. J. Hardy (Eds.), *Outcome after head, neck and spinal trauma* (pp. 391–402). Oxford: Butterworth Heinemann.

Milliken, C. S., et al. (2007). Longitudinal assessment of mental health problems among active duty and reserve component soldiers returning from the Iraq war. *Journal of the American Medical Association, 298*(18), 2141–2148.

Ming, G-L., & Song, H. (2005). Adult neurogenesis in the mammalian central nervous system. *Annual Review of Neuroscience, 28*, 223–250.

Mink, J. W. (1999). Basal ganglia. Basal ganglia. In M. J. Zigmond et al. (Eds.), *Fundamental neuroscience* (pp. 951–972). San Diego: Academic Press.

Mink, J. W. (2003). The basal ganglia. In L. R. Squire et al. (Eds.), *Fundamental neuroscience, 2nd Ed.* (pp. 815–839). Boston: Academic Press.

Mirza, K. A., et al., (1998). *British Journal of Psychiatry,* 172, 443–447.

Mishkin, M., & Appenzeller, T. (1987). The anatomy of memory. *Scientific American,* 226, 6, 6/87. 80–89.

Mitrushina, M., et al. (2005). *Handbook of normative data for neuropsychological assessment, 2nd Ed.* New York: Oxford University Press.

Mitsumoto, H. (2000). Disorders of upper and lower motor neurons. In W. G. Bradley et al. (Eds.), *Neurology in clinical practice* (Vol. II, 1985–2018). Boston: Butterworth Heinemann.

Mittenberg, W., & Strauman, S. W. (2000). Diagnosis of mild head injury and the postconcussion syndrome. *Journal of Head Trauma Rehabilitation, 15*, 783–791.

Mittenberg, W., et al. (1991). Symptoms following mild head injury: Expectation as aetiology. *Journal of Neurology, Neurosurgery and Psychiatry, 51*, 200–204.

Mittenberg, W., et al. (1996). Cognitive-behavioral prevention of postconcussion syndrome. *Archives of Clinical Neuropsychology, 11*, 139–145.
Mittenberg, W., et al. (1997). Postconcussion syndrome occurs in children. *Neuropsychology, 11*, 447–452.
Mocchegiani, E., et al. (1994). Thymic endocrine function in neuroendocrine human diseases. In N. A. Fabris et al. (Eds.), *Immunomodulation: The state of the art* (Vol. 741, pp. 115–123). Annals of the New York Academy of Sciences.
Mokri, B. (2003). Recurrent events after a cervical artery dissection. *Neurology, 60*, 1321.
Molden, S. O., & Hyman, S. E. (2005). Genome, transcriptome, and proteome. In B. J. Sadock & V. A. Sadock (Eds.), *Comprehensive textbook of psychiatry, I, 8th Ed.* (pp. 115–125). Philadelphia: Lippincott Williams & Wilkins.
Moldover, E., et al. (2004). Depression after traumatic brain injury: A review of evidence for clinical heterogeneity. *Neuropsychology Review, 15*, 143–154.
Molinari, M. (2002). In V. S. Ramachandran (Ed.), *Encyclopedia of the human brain* (Vol. 1, pp. 611–627). Boston: Academic Press.
Molinari, M., et al. (1997). Cerebellum and procedural learning: evidence from focal cerebellar lesions, *Brain, 120*, 1753–1762.
Molitch, M. E. (1995). Neuroendocrinology. In P. Felig et al. (Eds.), *Endocrinology and Metabolism, 3rd Ed.* (pp. 221–288). New York: McGraw-Hill.
Molitch, M. E. (1995). Neuroendocrinology. In P. Felig, J. D. Baxter, & L. A. Frohman (Eds.) In *Endocrinology and Metabolism,* 3rd ed., pp. 221–288. New York: McGraw-Hill.
Molitch, M. E. (2001). Neuroendocrinology. In P. Felig & L. A. Frohman (Eds.), *Endocrinology and metabolism, 4th Ed.* (pp. 111–171). New York: McGraw-Hill.
Money, K. E., et al. (1996). The autonomic nervous system and motion sickness. In B. J. Yates & A. D. Miller (Eds.), *Vestibular autonomic regulation* (pp. 147–173). Boca Raton: CRC Press.
Monnot, M., et al. (2002). Neurological basis of deficits in affective prosody comprehension among alcoholics and fetal alcohol-exposed adults. *Journal of Neuropsychiatry and Clinical Neurosciences, 14*, 321–329.
Monrad-Krohn, G. B. (1947). The prosodic quality of speech and its disorders. *Acta Psycologia Scandanavia, 11*, 255–265.
Montgomery, E. A., et al. (1991). The psychobiology of minor head injury. *Psychological Medicine, 21*, 375–384.
Montgomery, K., et al. (1977). Some comparisons of the memory and visualperceptive deficits of chronic alcoholics and patients with Korsakoff's disease. *Alcoholism: Clinical and Experimental Research, 1*(1), 73–80.
Mooney, G., & Speed, J. (2001). The association between mild traumatic brain injury and psychiatric conditions. *Brain Injury, 15*, 865–877.
Moore, D. P. (1997). *Partial seizures and interictal disorders: The neuropsychiatric elements.* Boston: Butterworth-Heinemann.
Moore, E. E., et al. (Eds.). (2004). *Trauma, 5th Ed.* New York: McGraw Hill.
Moore, K. L., & Dalley, A. F. (1999). *Clinically oriented anatomy, 4th Ed.* Philadelphia: Lippincott Williams & Wilkins.
Moore, K. L., & Dalley, A. F. (2006). *Clinically oriented anatomy, 5th Ed.* Philadelphia: Lippincott Williams & Wilkins.
Moore, R. Y. (1999). Circadian timing. In M. J. Zigmond et al. (Eds.) *Fundamental Neuroscience,* pp. 1189–1206. San Diego: Academic Press
Moore, R. Y. (2003). Circadian timing. In L. R. Squire et al. (Eds.), *Fundamental neuroscience, 2nd Ed.* (pp. 1067–1084). London: Academic Press.
Moore, S. A. (2009). Cognitive abnormalities in posttraumatic stress disorder. *Current Opinion in Psychiatry, 21*, 19–24.
Morganti-Kossman, M. C., et al. (2005). Influence of brain trauma on blood-brain barrier properties. In E. De Vries & A. Part (Eds.), *The blood-brain barrier and its microenvironment* (pp. 457–479). New York: Taylor & Francis.
Morin, L. P., & Allen, C. N. (2005). The circadian visual system, 2005. *Brain Research Reviews, 51,* 1–60.
Morris, et al. (1992). Conceptual and psychometric issues in the neuropsychologic assessment of children: measurement of ability discrepancy and change. In I. Rapin & S. J. Segalowitz (Eds.). *Handbook of neuropsychology, 8, Child neuropsychology* (pp. 341–352). Amsterdam, Netherlands: Elsevier Science Publishing Co.
Moseley, G. L. (2005). Distorted body image in complex regional pain syndrome. *Neurology, 65*, 773.
Moss, N. E., et al. (1994). Post-concussion symptoms: Is stress a mediating factor? *Clinical Rehabilitation, 8,* 149–156.

Mostofsky, S. H., et al. (2000). Evidence for a deficit in procedural learning in children and adolescents with autism: Implications for cerebellar contribution. *Journal of the International Neuropsychological Society, 6*, 752–759.

Muir, J. K., et al. (1992). Continuous monitoring of posttraumatic cerebral blood flow using Laser-Doppler flowmetry. *Journal of Neurotrauma, 9*, 355–362.

Mulak, A., & Bonar, B. (2004). Irritable bowel syndrome: A model of the brain-gut interactions. *Medical Science Monitor, 10*(4), RA55–RA62.

Mulcahy, R., et al. (2003). Circadian and orthostatic blood pressure is abnormal in the carotid sinus syndrome. *American Journal of Geriatric Cardiology, 12*, 288–292, 301.

Murai, T., et al. (2002). Current issues in neuropsychological assessment in Japan. In Ferraro, F. R. (Ed.), *Minority and cross-cultural aspects of neuropsychological assessment* (pp. 99–107). Lisse, NL: Swets and Zeitlinger.

Murali, R., & Rovit, R. L. (2000). Injuries of the cranial nerves. In P. R. Cooper & J. G. Golfinos (Eds.), *Head injury, 4th Ed.* (pp. 201–219). New York: McGraw-Hill.

Murdoch, B. E., & Theodoros, D. G. (2001). Speech breathing impairments following traumatic brain injury. In B. E. Murdoch & D. G. Theodoros (Eds.), *Traumatic brain injury: Associated speech, language, and swallowing disorders* (pp. 109–119). San Diego: Singular Publishing Group.

Murdoch, B. E., et al. (1990). Acquired childhood speech disorders: Dysarthria and dyspraxia. In B. E. Murdoch (Ed.). *Acquired neurological speech/language disorders in childhood* (pp. 308–341). New York: Taylor and Francis.

Murison, R. (2000). Gastrointestinal effects. In G. Fink (Ed.), *Encylopedia of stress* (Vol. 1, pp. 191–196). San Diego, CA: Academic Press.

Murray, B. (2004). Peripheral nerve trauma. In W. G. Bradley et al. (Eds.), *Neurology in clinical practice, 4th Ed.* (Vol. 2, pp. 1176–1195). Boston: Butterworth-Heinemann.

Murray, B. E. (2003). Nerve injury. In M. J. Aminoff & R. B. Daroff (Eds.), *Encyclopedia of the neurological science* (Vol. 3, pp. 418–421). Boston: Academic Press.

Mushkudiani, N. A., et al. (2007). Prognostic value of demographic characteristics in traumatic brain injury: Results from the IMPACT study. *Journal of Neurotrauma, 24*, 259–269.

Mussack, T., et al. (2000). S100B as a screening marker of the severity of minor head trauma (MHT)—a pilot study. *Acta neurochirurgica – Supplement, 76*, 393–396.

Mussack, T., et al. (2002). Immediate S-100B and neuron-specific enolase plasma measurements for rapid evaluation of primary brain damage in alcohol-intoxicated, minor head-injured patients. *Shock, 18*, 481–482.

Mussack, T., et al. (2003). Serum S-100B protein levels in young soccer players after controlled heading and normal exercise. *European Journal of Medical Research, 8*, 754–764.

Naarding, P., et al. (2003). Aprosodia in major depression. *Journal of Neurolinguistics, 16*, 37–41.

Nadeau, S. E. (2003). Phonologic aspects of language disorders. Phonological aspects of language disorders. In K. M. Heilman & E. Valenstein (Eds.), *Clinical neuropsychology, 4th Ed.* (pp. 35–60). New York: Oxford University Press.

Nagtegaal, J. E., et al. (1997) Traumatic brain-injury associated delayed sleep phase syndrome. *Functional Neurology, 12,* 345–348.

Nakano, K. (1990). Hardiness, type A behavior, and physical symptoms in a Japanese sample. *The Journal of Nervous and Mental Disease, 178*, 52–56.

Nanda, A. (1996). Neurovascular trauma. In R. W. Evans (Ed.), *Neurology and trauma* (pp. 151–165). Philadelphia: Saunders.

Narayan, R. K. (1989). Emergency room management of the head-injured patient. In D. P. Becker and S. K. Gudeman (Eds.) *Textbook of Head Injury,* pp. 23–66. Philadelphia: Saunders.

Nasrallah, J. B. (2002). Recognition and rejection of self in plant reproduction. *Science, 296*, 305–308.

Nass, R. (1985). Mirror movement asymmetries in congenital hemiparesis: The inhibition hypothesis revisited. *Neurology, 35*, 1059–1062.

Nass, R. (1996). Disorders of speech and language development. In B. O. Berg (Ed.), *Principles of child neurology* (pp. 397–409). New York: McGraw-Hill.

Nass, R., & Stiles, J. (1996). Neurobehavioral consequences of congenital focal lesions. In E. Frank (Ed.), *Pediatric behavioral neurology* (pp. 149–178). Boca Raton, FL: CRC Press.

Nathan, S. V., et al. (2004). Basolateral amygdala interacts with other brain regions in regulating glucocorticoid effects on different memory functions. In R. Yehuda & B. McEwen (Eds.), *Behavioral stress responses* (Vol. 1032, pp. 179–182). Annals of the New York Academy of Sciences.

National Center for Injury Prevention and Control (2003). *Report to Congress on mild traumatic brain injury in the United States: Steps to prevent a serious public health problem.* Atlanta, GA: Author.

Neafsey, E. J., et al. (1993). Anterior cingulate cortex in rodents: Connections, visceral control functions, and implications for emotion. In B. A. Vogt & M. Gabriel (Eds.), *Neurobiology of cingulate cortex and limbic thalamus* (pp. 206–219). Boston: Birkäuser.

Neilan, C., & Pasternak, G. W. (2003). In M. J. Aminoff & R. B. Daroff (Eds.), *Encyclopedia of the neurological sciences* (Vol. 3, pp. 574–577). Boston: Academic Press.

Nelson, R. J., & Drazen, D. L. (2000). Melatonin mediates seasonal changes in immune function. In A. Conti, et al. (Eds.), *Neuroimmunomodulation: Perspectives at the new millennium* (Vol. 917, pp. 404–415). Annals of the NY Academy of Sciences.

Nemeth, K., et al. (1997). Reperfusion tissue injury during coronary bypass surgery. Poster. International Congress of Stress, Budapest.

Nemiah, J. D. (1991). Dissociation, conversion, and somatization. In A. Talisman & S. M. Goldfinger (Eds.), *Annual review of psychiatry, 10* (pp. 248–260). Washington DC: American Psychiatric Press.

Nemiah, J. C. (1995). Dissociative disorders. In H. I. Kaplan and B. J. Sadock (Eds.) *Comprehensive Textbook of Psychiatry,* 6th ed. Vol. II, pp. 1281–1293. Baltimore: Williams and Wilkins.

Neppe, V. M., & Goodwin, G. T. (1999). Neuropsychiatric evaluation of closed head injury of transient type (CHIT). In N. R. Varney & R. J. Roberts (Eds.), *The evaluation and treatment of mild traumatic brain injury* (pp. 149–208). Mahwah, NJ: Lawrence Erlbaum.

Netsell, R. (2001). Speech aeromechanics and the dysarthrias: Implications for children with traumatic brain injury. *Journal of Head Trauma Rehabilitation, 16*(5), 415–425.

Netter, F. H. (1983). *The Ciba collection of medical illustrations.* West Caldwell, NJ: Ciba.

New, M. I., & Rapaport, R. (1996). The adrenal cortex. In M. A. Sperling (Ed.), *Pediatric endocrinology* (pp. 281–314). Philadelphia: Saunders.

New York Academy of Science (2006, September–October). Resilience in children. *Update,* 5.

Newcomer, J. W., et al. (1994). Glucocorticoid-induced impairment in declarative memory performance in adult humans. *The Journal of Neuroscience, 14*, 2047–2053.

Newell, K. L., & Hedley-Whyte, E. T. (2003). Glia. In M. J. Aminoff & R. B. Daroff (Eds.), *Encyclopedia of the neurological sciences* (Vol. 2, pp. 253–255). Boston: Academic Press.

Ni, C. N., & Redmond, H. P. (2006). The immunological consequences of injury. *Surgeon Journal of the Royal Colleges of Surgeons of Edinburgh & Ireland, 4*, 23–31.

Nichols, J. S., et al. (1996). Detection of impaired cerebral autoregulation using spectral analysis of intracranial pressure waves. *Journal of Neurotrauma, 13*, 439–436.

Nichols, P. L. (1987). Minimal brain dysfunction and soft signs: The collaborative perinatal project. In D. E. Tupper (Ed.), *Soft neurological signs* (pp. 179–199). New York: Grune and Stratton.

Nicholson, K., & Martelli, M. F. (2004). *Journal of Head Trauma Rehabilitation, 19,* 2–9.

Nicolelis, M. A. L., et al. (2002). A critique of the pure feedforward model of touch. In R. J. Nelson (Ed.), *The somatosensory system: Deciphering the brain's own body image* (pp. 299–334). Boca Raton, FL: CRC Press.

Niedzwecki, C., et al. (2008). Traumatic brain injury: A comparison of inpatient functional outcomes between children and adults. *Journal of Head Trauma Rehabilitation, 23*, 209–217.

Nieuwenhuys, et al. (1988). *The Human Central Nervous System,* 3rd rev. ed. New York: Springer-Verlag.

Nijenhuis, E. R., et al. (2002). The emerging psychobiology of trauma-related dissociation and dissociative disorders. In H. D'haenen & P. Willner (Eds.), *Biological psychiatry* (pp. 1079–1097). New York: Wiley.

Nogales, A. Hispanic injured workers. *American Journal of Forensic Psychology,* 10, 3, 67–78.

Nogueiras, R., & Tschöp, M. (2005). Separation of conjoined hormones yields appetite rivals. *Science, 310*, 985–986.

Nolan, R. A., & Nordhoff, L. S. (1996). Soft tissue injury repair. In L. S. Nordhoff (Ed.) *Motor Vehicle Collision Injuries,* pp. 131–148. Gaithersburg, MD: Aspen.

Nolin, P., & Heroux, L. (2006). Relations among sociodemographic, neurologic, clinical, and neuropsychologic variables, and vocational status following mild traumatic brain injury: A follow-up study. *Journal of Head Trauma Rehabilitation, 21*, 514–526.

Nolle, C., et al. (2004). Pathophysiological changes of the central auditory pathway after blunt trauma of the head. *Journal of Neurotrauma, 21*, 251–258.

Nolte, J. (2002). *The human brain, 5th Ed.* St. Louis, MO: Mosby.

Noppeney, U., & Nacimiento, W. (1999). Bilateral hypoglossal nerve paralysis as isolated neurological symptom after craniocerebral trauma [German]. *Nervenarzt, 70*, 357–358.

Nordhoff, L. S. (1996a). Disability after car crashes. In L. S. Nordhoff (Ed.). *Motor vehicle collision injuries* (pp. 205–225). Gaithersburg, MD: Aspen.

Nordhoff, L. S. (1996b). Motor vehicle collision facts. In L. S. Nordhoff, Jr., (Ed.). *Motor vehicle collision injuries: Mechanisms, diagnosis, and management* (pp. 266–277). Gaithersburg, MD: Aspen.

Nordhoff, L. S. (1996c). Injury tolerance and injury factors. *Motor vehicle collision injuries: Mechanisms, diagnosis, and management* (pp. 328–342). Gaithersburg, MD: Aspen.

Nordhoff, L. S., & Emori, E. (1996). *Collision Dynamics of Vehicles and Occupants.* Gaithersburg, MD: Aspen.

Nordhoff, L. S., et al. (1996a). Diagnosis of common crash injuries. In L. S. Nordhoff, Jr., (Ed.), *Motor vehicle collision injuries* (pp. 1–69). Gaithersburg, MD: Aspen.

Nordhoff, L. S., et al. (1996b). Management of minor injuries. In L. S. Nordhoff (Ed.), *Motor vehicle collision injuries* (pp. 149–186). Gaithersburg, MD: Aspen.

Norris, D., & Wise, R. (2000). The study of prelexical and lexical processes in comprehension: Psycholinguistics and functional neuroimaging. In M. S. Gazzaniga (Ed.), *The new cognitive neurosciences, 2nd Ed.* (pp. 867–880). Cambridge, MA: MIT Press.

Norton, N. C. (1989). Three scales of alexythymia: Do they measure the same thing? *Journal of Personality Assessment, 53*, 621–637.

Norton, W. T. (1972). Formation, structure and biochemistry of myelin. In G. J. Siegel et al. (Eds.), *Basic neurochemistry* (pp. 74–99). Boston: Little Brown and Company.

Nosrat, C. A. (1998). Neurotrophic factors in the tongue: Expression patterns, biological activity, relation to innervation and studies of neurotrophin knockout mice. In C. Murphy (Ed.), *Olfaction and taste XII* (Vol. 855, pp. 28–49). Annals of the New York Academy of Sciences.

Nugent, N. R., et al. (2006). Initial physiological responses and perceived hyperarousal predict subsequent emotional numbing in pediatric injury patients. *Journal of Traumatic Stress, 19*, 349–359.

Nunez, P. L. (2002). Electrocephalography (EEG). In V. S. Ramachandran (Ed.), *Encyclopedia of the human brain* (Vol. II, pp. 169–179). Boston: Academic Press.

Nusholtz et al., (1996). Internal cavitation in simple head impact mode. *Journal of Neurotrauma, 12*, 707–714.

O'Brien, D. O., et al. (1997). Delayed traumatic cerebral aneurysm after brain injury. *Archives of Physical Medicine and Rehabilitation, 78*, 883–885.

O'Brien, S. L. (1998). *Traumatic events and mental health.* Cambridge, UK: Cambridge University Press.

O'Leary, A. (1990). Stress, emotion, and human immune function. *Psychological Bulletin,* 108, 363–382.

O'Neill, L. A. J. (2005). Immunity's early-warning system. *Scientific American, 12*, 38–45.

O Shanick, G. J., & O Shanick, A. M. (2005). Personality disorders. In J. M. Silver et al. (Eds.). *Textbook of traumatic brain injury* (pp. 245–258). Washington DC: American Psychiatric Publishing.

Oaklander, A. L. (2003). Basic neurobiology of pain. In M. J. Aminoff & R. B. Daroff (Eds.), *Encyclopedia of the neurological sciences* (Vol. 3, pp. 723–733). Boston: Academic Press.

Ochsner, K. N., & Lieberman, M. D. (2001). The emergence of social cognitive neuroscience. *The American Psychologist, 56*, 717–734.

Ogden, J. A., et al. (1990). Long-term neuropsychological and psychosocial effects of subarachnoid hemorrhage. *Neuropsychiatry, Neuropsychology, and Behavioral Neurology, 3*, 260–274.

Ohhashi, G., et al. (2002). Problems of health management of professional boxers in Japan. *British Journal of Sports Medicine, 36*, 346–352.

Okie, S. (2005). Traumatic brain injury in the war zone. *New England Journal of Medicine, 352*, 2043–2047.

Olden, K. W. (1998). Stress and the gastrointestinal tract. In J. R. Hubbard & E. A. Workman (Eds.), *Handbook of stress medicine: An organ system approach* (pp. 87–113). Boca Raton, FL: CRC Press.

Ommaya, A. K. (1990). Mechanisms of cerebral concussion, contusions, and other effects of head injury. In R. Youmans (Ed.) *Neurological Surgery,* Vol. 3, pp. 1877–1895. New York: Saunders.

Ommaya, A. K. (1996). Head injury mechanisms and the concept of preventive management; A review and critical synthesis. *Journal of Neurotrauma,* 12, 527–546.

Ommaya, A. K., & Gennarelli, T. A. (1974). Cerebral concussion and traumatic unconsciousness: Correlation of experimental and clinical observations on blunt head injuries. *Brain, 97*, 633–654.

Ommaya, A. K., & Hirsch, A. E. (1971). Tolerances for cerebral concussion from head impact and whiplash in primates. *Journal of Biomechanics,* 4, 13–21.

Ommaya, A. K., & Ommaya, A. K. (1997). Why neurobehavioral sequelae do not correlate with head injury severity: A biomechanical explanation for the traumatic disturbances of consciousness. *Seminars in Clinical Neuropsychiatry,* 2, 163–176.

Ommaya, A. K., et al. (1968). Whiplash injury and brain damage. *Journal of the American Medical Association,* 204 (4), 285–289.

Ommaya, A. K., et al. (1971). Coup and contre-coup injury: Observations on the mechanics of visible brain injuries in the rhesus monkey. *Journal of Neurosurgery, 35*, 503–516.

Ommaya, A. K., et al. (1996). Causation, incidence, and costs of traumatic brain injury in the U.S. military medical system. *The Journal of Trauma Injury, Infection and Critical Care, 40*, 211–217.

Onyszchuk, G., et al. (2008). Detrimental effects of aging on outcome from traumatic brain injury: A behavioral, magnetic resonance imaging, and histological study in mice. *Journal of Neurotrauma, 25*, 153–171.

Oppenheim, R. S., & Johnson, J. E. (2003). Programmed cell death and neurotrophic factors. In L. R. Squire et al. (Eds.), *Fundamental neuroscience, 2nd Ed.* (pp. 488–532). Boston: Academic Press.

Orr, S. P., & Kalouupek, D. G. (1997). Psychophysiological assessment of posttraumatic stress disorder. In J. P. Wilson & T. M. Keane (Eds.), *Assessing psychological trauma and PTSD* (pp. 69–97). New York: The Guilford Press.

Orth, D. N., et al. (1992). The adrenal cortex. In J. D. Wilson & D. W. Foster (Eds.), *Williams textbook of endocrinology, 8th Ed.* (pp. 489–619). Philadelphia: Saunders.

Osman, D. C. (1999). Complexities in the evaluation of executive functions. In J. J. Sweet (Ed.) *Forensic Neuropsychology,* pp. 185–226. Exton, PA: Swets and Zeitlinger.

Otto, M., et al. (2000). Boxing and running lead to a rise in serum levels of S-100B protein. *International Journal of Sports Medicine, 21*, 551–555.

Ovsiew, F. (2000). Bedside neuropsychiatry: Eliciting the clinical phenomena of neuropsychiatric illness. In S. C. Yudofsky & R. E. Hales (Eds.), *Textbook of neuropsychiatry and clinical neurosciences, 4th Ed.* (pp. 153–198). Washington, DC: American Psychiatric Publishing.

Ozanne, A. E., & Murdoch, B. E. (1990). Acquired childhood aphasia: Neuropathology, linguistic characteristics and prognosis. In B. E. Murdoch (Ed.), *Acquired neurological speech/language disorders in childhood* (pp. 1–65). New York: Taylor and Francis.

Pachalska, M., et al. (2002). Active and passive executive function disorder subsequent closed head injury. *Medical Science Monitor, 8*, CS1–CS9.

Packard, R. C., & Ham, L. P. (1999). Epidemiology and pathogenesis of posttraumatic headache. *Journal of Head Trauma Rehabilitation,* 14 (1), 9–21.

Packard, R. C., et al. (1993). Cognitive symptoms in patients with posttraumatic headache. *Headache,* 33, 365–368.

Pagana, K. D., & Pagana, T. J. (2006). *Mosby's manual of diagnostic ad aboratory tests.* St. Louis, MO: Mosby.

Pang, D. (1989). Physics and pathophysiology of closed head injury. In M. D. Lezak (Ed.), *Assessment of the behavioral consequences of head trauma* (pp. 1–17). New York: Alan R. Liss.

Panksepp, J., et al. (1997). Brain systems for the mediation of social separation-distress and social-reward. In S. C. Carter et al. (Eds.), *The integrative neurobiology of affiliation* (Vol. 807, pp. 78–100). Annals of the New York Academy of Sciences.

Pansky, B., et al. (1988). *Review of neuroscience*, 2nd Ed. New York: Macmillan.

Parasuraman, R., & Caggiano, D. (2002). Mental workload. In V. S. Ramachandran (Ed.), *Encyclopedia of the human brain* (Vol. 3, pp. 17–27). Boston: Academic Press.

Parent, A. (1996). *Carpenter's human neuroanatomy, 9th Ed.* Baltimore: Williams & Wilkins.

Paris, B. (2000). The practical application of trigger point work in physical therapy. In E. S. Rachlin & I. S. Rachlin (Eds.), *Myofascial pain and fibromyalgia: Trigger point management* (pp. 525–543). St Louis: Mosby.

Parker, R. S. (1972). The patient who cannot express pain. In R. S. Parker (Ed.), *The emotional stress of war, violence and peace* (pp. 71–85). Pittsburgh: Stanwix House.

Parker, R. S. (1977). *Effective decisions and emotional fulfillment.* Chicago, IL: Nelson-Hall.

Parker, R. S. (1978). *Living single successfully.* New York: Franklin Watts.

Parker, R. S. (1981). *Emotional common sense, 2nd Ed.* New York: Harper & Row.

Parker, R. S. (1983). *Self-image psychodynamics: Rewriting your life script.* Englewood Cliffs, NJ: Prentice Hall.

Parker, R. S. (1987). Recognizing the brain damaged employee. *EAP Digest, March–April*, 55–60.

Parker, R. S. (1990). *Traumatic brain injury and neuropsychological impairment.* New York, NY: Springer-Verlag.

Parker, R. S. (1994). Malingering and exaggerated claims after head injury. In C. Simkins (Ed.), *Analysis understanding and presentation of cases involving traumatic brain injury.* Washington, DC: National Head Injury Foundation.

Parker, R. S. (1995). Distracting effects of pain, headaches, and hyper-arousal upon employment after "minor" head injury. *Journal of Cognitive Rehabilitation, 13*(3), 14–23.

Parker, R. S. (1996). The spectrum of emotional distress and personality changes after minor head injury: The consequences of motor vehicle accidents. *Brain Injury, 10*, 287–302.

Parker, R. S. (1997). A taxonomy of neurobehavioral functions applied to neuropsychological assessment after head injury. *Neuropsychological Review*, *6*, 135–170.

Parker, R. S. (2001). *Concussive brain trauma.* Boca Raton, FL: CRC Press.

Parker, R. S. (2002). Recommendations for the revision of DSM-IV diagnostic categories for co-morbid post-traumatic stress disorder and traumatic brain injury. *NeuroRehabilitation*, *17*, 131–143.

Parker, R. S. (2005a). Dysregulation of the internal milieu after an accident causing head injury. In T. Corales (Ed.), *Trends in posttraumatic stress disorder research* (pp. 67–101). Hauppage, NY: Nova Science Publishers.

Parker, R. S. (2005b). Traumatic brain injury. In D. S. Younger (Ed.), *Motor disorders, 2nd Ed.* (pp. 707–710). Philadelphia: Lippincott Williams and Williams.

Parker, R. S. (2008). Forensic neuropsychological examination after accidents with emphasis upon the post-concussive syndrome. In H. Hall (Ed.), *Forensic psychology and neuropsychology for criminal and civil cases* (pp. 565–620). Boca Raton, FL: Taylor & Francis.

Parker, R. S., & Rosenblum, A. (1996). Intelligence and emotional dysfunctions after mild head injury incurred in a car accident. *Journal of Clinical Psychology*, *52*, 32–43.

Parkes, J. D. (1991). Excessive daytime drowsiness. In W. G. Bradley, R. B. Daroff, G. M. Fenichel, & C. D. Marsden (Eds.) *Neurology in Clinical Practice,* vol. 1, pp. 65–71. Boston: Butterworth-Heinemann.

Parslow, T. G. & Bainton, D. F. (2001). Innate immunity. In T. G. Parslow, et al., (Eds.), *Medical immunology, 10th Ed.* (pp. 19–39). New York: Lange.

Parsons, L. C., et al. (1997). Longitudinal sleep EEG power spectral analysis studies in adolescents with minor head injury. *Journal of Neurotrauma, 14*, 549–559.

Parsons, L. M. (2001). Exploring the functional neuroanatomy of music performance, perception, and comprehension. *The biological foundations of music* (Vol. 930, pp. 211–230). Annals of the New York Academy of Sciences.

Pary, L. F., & Rodnitzky, R. L. (2003). Traumatic internal carotid artery dissection associated with taekwondo. *Neurology*, *60*, 1392–1393.

Pascual, J. M., et al. (2007). Time course of early metabolic changes following diffuse traumatic brain injury in rats as detected by ^{1}H NMR spectroscopy. *Journal of Neurotrauma, 24,* 944–959.

Passani, M. B., et al. (2000). Central histaminergic system and cognition. *Neuroscience and Behavioral Reviews, 24*, 107–113.

Passler, M. A., et al. (1985). Neuropsychological development of behavior attributed to frontal lobe functioning in children. *Developmental Neuropsychology, 1*(4), 349–370.

Pasternak, G. W. (2003). Opioids and their receptors. In M. Aminoff & R. Daroff (Eds.), *Encyclopedia of the neurological sciences* (Vol. 3, pp. 675–679). Boston: Academic Press.

Patel, R. (2002). Prosodic control in severe dysarthria: Preserved ability to make the question-statement contrast. *Journal of Speech, Language, & Hearing Research*, *45*, 858–870.

Patrick, P. D., et al. (2002). DSM-IV: Diagnosis of children with traumatic brain injury. *NeuroRehabilitation, 17*(2), 123–129.

Patten, J. (1996). *Neurological differential diagnosis*. London: Springer-Verlag.

Paulis, M. P. (2000). Decision-making dysfunctions in psychiatry—altered homeostatic processing? *Science, 318*, 601–606.

Paulsen, J. S., & Hoth, K. F. (2004). Neuropsychology. In W. G. Bradley et al. (Eds.), *Neurology in clinical practice, 4th Ed.* (Vol. 1, pp. 675–700). Boston: Butterworth-Heinemann.

Pavuluri, M. N., et al. (2005). Affect regulation: a systems neuroscience perspective. *Neuropsychiatric Disease and Treatment, 1*, 9–15.

Peacock, W. J. (1986). The postnatal development of the brain and its coverings. In A. J. Raimondi et al. (Eds.), *Head Injuries in the Newborn and Infant* (pp. 53–66). New York: Springer-Verlag.

Pearce, J. M. S. (2001). Headache after head injury. *Headache, 12*, 101–107.

Pearson, K., & Gordon, J. (2000a). Spinal reflexes. In E. R. Kandel et al. (Eds.), *Principles of neural science, 4th Ed.* (pp. 713–736). New York: McGraw Hill.

Pearson, K., & Gordon, J. (2000b). Locomotion. In E. R. Kandel et al. (Eds.), *Principles of neural science, 4th Ed.* (pp. 737–755). New York: McGraw Hill.

Pedley, T. A., et al. (1995). Epilepsy. In *Merritt's Textbook of Neurology,* 9th ed., pp. 845–870. Baltimore: Williams and Wilkins.

Pekary, A. E., & Hershman, J. M. (2001). Hormone assays. In P. Felig & L. A. Frohman (Eds.), *Endocrinology and metabolism, 4th Ed.* (pp. 91–107). New York: McGraw-Hill.

Pelco, L. Sawyer, M., Duffield, G., Prior, M., & Kinsella, G. (1992). Premorbid emotional and beharioural adjustment in children with mild head injuries. *Brain Injury,* 6, 29–38.

Pelham, M. F., & Lovell, M. R. (2005). Issues in neuropsychological assessment. In J. M. Silver et al. (Eds.), *Textbook of traumatic brain injury* (pp. 159–172). Washington, DC: American Psychiatric Publishing.

Penedo, F. J., & Dahn, J. R. (2005). Psychoneuroimmunology and ageing. In K. Vedhara & M. Irwin (Eds.), *Human psychoneuroimmunology* (pp. 80–106). New York: Oxford University Press.

Penfield, W., & Roberts, L. (1959). *Speech and Brain-Mechanism.* Princeton, NJ: Princeton University Press.

Peretz, I. (2001). Brain specialization for music: New evidence from congenital amusia. *The biological foundations of music* (Vol. 930, pp. 153–165). Annals of the New York Academy of Sciences.

Perlman, S., & Kroening, R. J. (1990). General considerations of pain and its treatment. In R. Youmans (Ed.) *Neurological Surgery,* 3rd ed., pp. 3803–3812. New York: Saunders.

Perry, B. D., & Pollard, R. (1998). Homeostasis, stress, trauma, and adaptation. *Child and Adolescent Psychiatric Clinics of North America, 7*(1), 33–51.

Pescovitz, O. H. (1992). Pediatric neuroendocrinology: Growth and puberty. In C. B. Nemeroff (Ed.) *Neuroendocrinology,*pp. 473–500. Boca Raton, FL: CRC.

Peters, J. E. (1987). A special or soft neurological examination for school age children. In D. E. Tupper (Ed.) *Soft Neurological Signs,* pp. 369–377. Orlando, FL: Grune and Stratton.

Petersen, R. C., & Weingartner, H. (1991). Memory nomenclature. In T. Yanagihara & R. C. Petersen (Eds.), *Memory disorders* (pp. 9–20). New York: Marcel Dekker.

Pezet, S., & McMahon, S. B. (2006). Neurotrophins: Mediators and modulators of pain. *Annual Review of Neuroscience, 29*, 507–538.

Pfeffer, C. (1985). Children's reactions to illness, hospitalization and surgery. In H. I. Kaplan & B. J. Sadock (Eds.), *Comprehensive Textbook of Psychiatry, 4th Ed.* (pp. 1836–1842). Baltimore: Williams and Wilkins.

Phay, A., et al. (1986). Clinical interviewing of the patient and history in neuropsychological assessment. In T. Incagnoli et al. (Eds.), *Clinical application of neuropsychological test batteries* (pp. 45–73). New York: Plenum.

Piazza, M., & Dehaene, S. (2004). From number neurons to mental arithmetic: The cognitive neuroscience of number sense. *The cognitive neurosciences, 3rd Ed.* (pp. 865–875). Cambridge, MA: A Bradford Book.

Pickelsimer, E. E., et al. (2007). Unmet service needs of persons with traumatic brain injury. *Journal of Head Trauma Rehabilitation, 22,* 1–13.

Piers, G., & Singer, M. B. (1953). *Shame and guilt.* Springfield, IL: Charles C. Thomas.

Pike, J. L. et al. (1997). Chronic life stress alters sympathetic, neuroendocrine, and immune responsivity to an acute psychological stressor in humans. *Psychosomatic Medicine, 59*, 447–457.

Pincus, J. H., & Tucker, G. J. (1985). *Behavioral neurology, 3rd Ed.* New York: Oxford University Press.

Piotrowski, Z. A. (1937). The Rorschach Inkblot Method in organic disturbances of the central nervous system. *Journal of Nervous and Mental Diseases, 86*, 525–531.

Piotrowski, Z. A. (1956). *Perceptanalysis.* Privately republished. Philadelphia: Author.

Pleet, A. B. (1995). Funny smells in neuroendocrine disorders. *Seminars in Neurology,* 15, 133–150.

Pollmacher, T., et al. (2000). Experimental immunomodulation, sleep, and sleepiness in humans. In A. Conti, et al. (Eds.), *Neuroimmunomodulation: Perspectives at the new millennium* (Vol. 917, pp. 488–499). Annals of the NY Academy of Sciences.

Ponsford, J. et al., (1999). Cognitive and behavioral outcome following mild traumatic head injury in children. *Journal of Head Injury Rehabilitation, 14*(4), 360–372.

Poreh, A. (2002). Neuropsychological and psychological issues associated with cross-cultural and minority assessment. In F. R. Ferraro (Ed.), *Minority and cross-cultural aspects of neuropsychological assessment* (pp. 329–343). Lisse, NL: Swets and Zeitlinger.

Porges, S. W. (1993). Vagal tone: A physiologic marker of stress vulnerability. *Pediatrics*, *90*, 498–504.

Posener, J. A., et al. (1998). Late feedback effects of hypothalamic-pituitary-adrenal axis hormones in healthy subjects. *Psychoneuroendocrinology, 23*, 371–383.

Post, R. M., et al. (1991). Stress, conditioning, and the temporal aspects of affective disorders. In *Psychoneuroimmunology, 2nd ed.* (pp. 3–25).

Post, R. M., et al., (1995). A Sensitization and kindling: Implications for the evolving neural substrates of post-traumatic stress disorder. In M. J. Friedman, et al., (Eds.). *Neurobiological and clinical consequences of stress: From normal adaptation to PTSD* (pp. 203–224). Philadelphia: Lippincott-Raven.

Poundja, J., et al. (2006). The co-occurrence of posttraumatic stress disorder symptoms and pain: Is depression a mediator? *Journal of Traumatic Stress, 19,* 747–757.

Povlishock, J. T. (1989). Experimental studies of head injury. In D. P. Becker and S. K. Gudeman (Eds.) *Textbook of Head Injury,* pp. 437–450. Philadelphia: Saunders.

Povlishock, J. T. (2005). Update of neuropathology and neurological recovery after traumatic brain injury. *Journal of Head Trauma Rehabilitation*, *20*, 76–94.

Povlishock, J. T., & Christman, C. (1994). The pathobiology of traumatic brain injury. In S. K. Salzman & A. I. Faden (Eds.), *The neurobiology of central nervous system trauma* (pp. 108–120). New York: Oxford University Press.

Powell, J. W., & Farber-Ross, K. D. (1999) Traumatic brain injury in high school athletes. *JAMA, 282*, 958–963.

Powers, A. C. (2001). Diabetes mellitus. In E. Braunwald et al. (Eds.), *Harrison's 15th edition: Principles of internal medicine* (pp. 2109–2136). New York: McGraw Hill.

Powley, T. L. (2003). Central control of autonomic functions: Organization of the autonomic nervous system. In *Fundamental neuroscience, 2nd Ed.* (pp. 911–933). Boston: Academic Press.

Pretre, R., et al. (1995). Blunt carotid artery injury: Devastating consequences of undetected pseudoaneurysm. *Journal of Trauma: Injury, Infection and Critical Care,* 39, 1012–1014.

Preussner, J. C., et al. (2004). Effects of self-esteem on age-related changes in cognition and the regulation of the hypothalamic-pituitary-adrenal axis. In R. Yehuda & B. McEwen (Eds.), *Behavioral stress responses* (Vol. 1032, pp. 186–190). Annals of the New York Academy of Sciences.

Price, T. R. P., et al. (2002). Neuropsychiatric aspects of brain tumors. In S. C. Yudofsky & R. E. Hales (Eds.), *Textbook of neuropsychiatry and clinical neurosciences, 4th Ed.* (pp. 753–781). Washington DC: American Psychiatric.

Prigatano, G. P. (1987). Psychiatric aspects of head injury: Problem areas and suggested guidelines for research. In H. S. Levin et al. (Eds.), *Neurobehavioral recovery from head injury* (pp. 215–231). New York: Oxford University Press.

Prigatano, G. P. (1991). Disturbance of self-awareness of deficit after traumatic brain injury. In G. P. Prigatano & D. L. Schacher (Eds.), *Awareness of deficit after brain injury* (pp. 111–126). New York: Oxford University Press.

Prigatano, G. P., & Schachter, D. L. (Eds.) (1991). *Awareness of Deficit after Brain Injury.* New York: Oxford University.

Prigatano, G. P., et al. (1984). Neuropsychological rehabilitation after closed head injury in young adults. *Journal of Neurology, Neurosurgery, and Psychiatry, 47*, 505–513.

Prigatano, G. P., et al. (2008). Individual case analysis of processing speed difficulties in children with and without traumatic brain injury. *The Clinical Neuropsychologist, 22*, 603–619.

Prins, M. L., & Hovda, D. A. (2003). Developing experimental models to address traumatic brain injury in children. *Journal of Neurotrauma, 20*, 123–137.

Prochazka, A., & Yakovenko, S (2002). Locomotor control: from spring-like reactions of muscles to neural prediction. In R. J. Nelson (Ed.), *The somatosensory system: Deciphering the brain's own body image* (pp. 141–181). Boca Raton, FL: CRC Press.

Provencio, I. (2005). Chronobiology. In B. J. Sadock & V. A. Sadock (Eds.), *Comprehensive textbook of psychiatry, I. 8th Ed.* (pp. 161–171). Philadelphia: Lippincott Williams & Wilkins.

Provencio, J. J., & Vora, N. (2005). Subarachnoid hemorrhage and inflammation: Bench to bedside and back. *Seminars in Neurology*, *25*, 435–444.

Provini, L., et al. (1998). Somatotopic nucleocortical projections to the multiple somatosensory cerebellar maps. *Neuroscience*, *83*, 1085–1104.

Psychological Corporation (2001). *Manual: Wechsler test of adult reading.* San Antonio TX: Author.

Puder, J. J., & Wardlaw, S. L. (2000). Beta-endorphin. In G. Fink (Ed.), *Encyclopedia of stress* (Vol. 1, pp. 37–48). San Diego: Academic Press.

Puelles, L., & Rubenstein, J. (2002). Forebrain. In V. S. Ramachandran (Ed.), *Encyclopedia of the human brain* (Vol. 2, pp. 299–315). Boston: Academic Press.

Putnam, F. W. (1991). Dissociative phenomena. In A. Talisman & S. M. Goldfinger (Eds.), *Annual Review of Psychiatry, 10* (pp. 145–160). Washington DC: American Psychiatric Press.

Putnam, F. W., & Loewenstein, R. J. (2000). Dissociative identity disorder. In B. J. Sadock & V. A. Sadock (Eds.), *Comprehensive textbook of psychiatry, 7th Ed.* (pp. 1552–1564). Philadelphia: Lippincott Williams & Wilkins.

Putnam, S. H., et al. (1999). Considering premorbid functioning: Beyond cognition to a conceptualization of personality in postinjury functioning. In J. J. Sweet (Ed.), *Forensic neuropsychology* (pp. 39–81). Exton, PA: Swets & Zeitlinger.

Queller, D. C., & Strassman, J. E. (2002). The many selves of social insects. *Science*, *12*, 311–313.

Quellet, M-C., & Morin C. M. (2006). Subjective and objective measures of insomnia in the context of traumatic brain injury: a preliminary study. *Sleep Medicine, 7,* 486–497.

Quinto, C., et al. (2000). Posttraumatic delayed sleep phase syndrome. *Neurology, 54,* 250–252.

Quellet, M-C., et al. (2004). Insomnia following traumatic brain injury: a review. *Neurorehabilitation and Neural Repair, 18*, 187–198.

Rabin, B. S. (2005). Introduction to immunology and immune-endocrine interactions. In K. Vedhara & M. Irwin (Eds.), *Human psychoneuroimmunology* (pp. 1–24). New York: Oxford University Press.

Rabin, L. A. (2007). Utilization rates of ecologically oriented instruments among clinical neuropsychologists. *The Clinical Neuropsychologist, 21*, 727–743.

Rachlin, E. S. (2002a). Trigger points. In E. S. Rachlin & I. S. Rachlin (Eds.), *Myofascial pain and fibromyalgia: Trigger point management* (pp. 202–216). St Louis: Mosby.

Rachlin, E. S. (2002b). History and physical examination for myofascial pain syndrome. In E. S. Rachlin & I. S. Rachlin (Eds.), *Myofascial pain and fibromyalgia: Trigger point management* (pp. 217–230). St Louis: Mosby.

Radanov, et al. (1991). Role of psychosocial stress in recovery from common whiplash. *Lancet,* 338, 712–715.

Radanov, B. P., et al. (1992). Cognitive deficits in patients after soft tissue injury of the cervical spine. *Spine, 17*, 127–131.

Raghupathi, R., & Huh, J. W. (2007). Diffuse brain injury in the immature rat: Evidence for an age-at-injury effect on cognitive function and histopathologic damage. *Journal of Neurotrauma, 24*, 1596–1608.

Raichle, M. E., & Mintun, M. A. (2006). Brain work and brain imaging. *Annual Review of Neuroscience, 29*, 449–476.

Raimondi, A. J. (1986). Posttraumatic cerebral vascular injuries. In A. J. Raimondi et al. (Eds.), *Head injuries in the newborn and infant* (pp. 233–239). New York: Springer-Verlag.

Raimondi, J. R., & Hirschauer, J. (1986). Clinical criteria-children's coma score and outcome scale-for decision making in managing head-injured infants and toddlers. In A. J. Raimondi et al. (Eds.), *Head injuries in the newborn and infant* (pp. 141–150). New York: Springer-Verlag.

Raison, C. L., et al. (2005). Immune system and central nervous system interactions. In B. J. Sadock & V. A. Sadock (Eds.), *Comprehensive textbook of psychiatry, I, 8th Ed.* (pp. 137–161). Philadelphia: Lippincott Williams & Wilkins.

Ramasubbu, R., & Kennedy, S. H. (1994). Factors complicating the diagnosis of depression in cerebrovascular disease, Part II—Neurological deficits and various assessment methods. *Canadian Journal of Psychiatry, 39*, 601–607.

Ramsey, D. J. (1986). Posterior pituitary gland. In F. G. Greenspan & P. H. Forsham (Eds.), *Basic and clinical endocrinology, 2nd Ed.* (pp. 132–142) Los Altos, CA: Lange Medical.

Ransom, B. R. (2005). The neuronal microenvironment. In W. F. Boron & E. L. Boulpaep (Eds.), *Medical physiology, up-dated edition* (pp. 399–419). Philadelphia: Saunders.

Rapaport, M. J., & Feinstein, A. (2001). Age and functioning after mild traumatic brain injury: The acute picture. *Brain Injury, 15*, 857–864.

Rapp, P. R., & Bachevelier, J. (2003). Cognitive development and aging. In L. Squire et al. (Eds.), *Fundamental neuroscience, 2nd Ed.* (pp. 1167–1200). London: Academic Press.

Rasch, B., et al. (2007). Odor cues during slow-wave sleep prompt declarative memory consolidation. *Science, 315,* 1426–1428.

Rasmusson, A. M., & Charney, D. S. (2000). Posttraumatic therapy. In G. Fink (Ed.), *Encyclopedia of stress* (Vol. 3, pp. 421–425). San Diego: Academic Press.

Ravizza, S. M., et al. (2006). Cerebellar damage produces selective deficits in verbal working memory. *Brain, 129*, 306–320.

Ray, O. (2004). The revolutionary health science of psychoendoneuroimmunology. *Annals of the New York Academy of Sciences, 1032*, 35–51.

Razak, M. S., et al. (1983). Total glossectomy. *American Journal of Surgery*, *146*, 509–511.

Rebar, R. W. (1996). The ovaries. In J. C. Bennett & F. Plum (Eds.), *Cecil textbook of medicine* (pp. 1293–1313). Philadelphia: W. B. Saunders.

Redfield, J. (2001). Familial intelligence as an estimate of expected ability children. *The Clinical Neuropsychologist, 15*, 446–460.

Redmond, C., & Lipp, J. (2006). Traumatic brain injury in the pediatric population. *Nutrition in Clinical Practice, 21*, 459–461.

Rees, P. M. (2003). Contemporary issues in mild traumatic brain injury. *Archives of Physical Medicine and Rehabilitation*, *84*, 1884–1894.

Reeves, W. B., et al. (1998). Posterior pituitary and water metabolism. In J. D. Wilson et al. (Eds.), *Williams textbook of endocrinology, 9th Ed.* (pp. 341–387). Philadelphia: Saunders.

Reichard, A. A., et al. (2007). Violence, abuse, and neglect among people with traumatic brain injuries. *Journal of Head Trauma Rehabilitation*, *22*, 390–402.

Reichlin, S. (1992). Neuroendocrinology. In J. D. Wilson & D. W. Foster (Eds.), *Williams textbook of endocrinology, 8th Ed.* (pp. 135–219). Philadelphia: Saunders.

Reichlin, S. (1998). Neuroendocrinology. In G. D. Wilson et al. (Eds.), *Williams textbook of endocrinology, 9th ed.* (pp. 165–248). Philadelphia: Saunders.

Reilly, P. L. (2005). Management of intracranial pressure and cerebral perfusion pressure. In P. L. Reilly & R. Bullock (Eds.), *Head injury, 2nd Ed.* (pp. 331–355). New York: Hodder Arnold.

Reilly, P. L., & Bullock, R. (Eds.). (2005). *Head injury*. London: Hodder Arnold.

Reis, D. J., & Golanov, E. V. (1996). Cerebral circulation. In D. Robertson, P. A. Low, & R. J. Pollinsky (Eds.) *Primer in the Autonomic Nervous System,* pp. 56–58. San Diego: Academic Press.

Reitan, R. M., & Wolfson, D. (1994). A selective and critical review of neuropsychological deficits and the frontal lobes. *Neuropsychology Review,* 4, 161–198.

Reiter, E. O., & Rosenfeld, R. G. (2003). Normal and aberrant growth. In P. R. Larsen et al. (Eds.), *Williams textbook of endocrinology, 10th Ed.* (pp. 1003–1114). Philadelphia: Saunders.

Reiter, R. J. (2000). Melatonin and its relation to the immune system and inflammation. In A. Conti et al. (Eds.), *Neuroimmunomodulation: Perspectives at the new millennium* (Vol. 917, pp. 376–386). Annals of the New York Academy of Sciences.

Remler, B. F., & Daroff, R. B. (1991). Falls and drop attacks. In W. G. Bradley, R. B. Daroff, G. M. Fenichel, & C. D. Marsden (Eds.) *Neurology in Clinical Practice,* vol. 1, pp. 25–29. Boston: Butterworth-Heinemann.

Restak, R. M. (1997). Neuropsychiatry of minor head injury: An overview. *Seminars in Clinical Neuropsychiatry, 2*, 160–162.

Reynolds, J. H., et al. (2003). Attention. In L. R. Squire et al. (Eds.), *Fundamental neuroscience, 2nd Ed.* (pp. 1249–1273). Boston: Academic Press.

Rhodes, R. (2002). *Masters of death: The SS-Einsatzgruppen and the invention of the holocaust.* New York: Vintage Books.

Richard, D. (1993). Involvement of corticotropin-releasing factor in the control of food intake and energy expenditure. In Y. Taché & C. Rivier (Eds.), *Corticotropin-releasing factor and cytokines: Role in the stress response* (Vol. 691, pp. 155–172). Annals of the New York Academy of Science.

Richerson, G. B. (2005). The autonomic nervous system. In W. Boron & E. Boulpaep (Eds.), *Medical physiology, up-dated edition* (pp. 378–398). Philadelphia: Saunders.

Richerson, G. B., & Boron, W. F. (2003). Control of ventilation. In W. F. Boron & E. L. Boulpaep (Eds.), *Medical physiology* (pp. 712–734). Philadelphia, PA: Saunders.

Richter, E. F. (2005). Balance problems and dizziness. In J. M. Silver et al. (Eds.), *Textbook of traumatic brain injury* (pp. 393–404). Washington DC: American Psychiatric Publishing.

Ricker, J. H., & Regan, T. M. (1999). Neuropsychological and psychological factors in acute rehabilitation of individuals with both spinal cord injury and traumatic brain injury. *Topics in Spinal Cord Injury Rehabilitation, 5*, 76–82.

Rieder, M. J., et al. (1986). Patterns of walker use and walker injury. *Pediatrics, 78*, 488–493.

Rieger, M., et al. (2003). Inhibition of ongoing responses following frontal, nonfrontal, and basal ganglia lesions. *Neuropsychology, 17*, 272–282.

Rieger, M., et al. (2004). Influence of stress during pregnancy on HPA activity and neonatal behavior. *Annals of the New York Academy of Sciences, 1032*, 228–230.

Rimel, R.W., et al. (1981). Disability caused by minor head injury. *Neurosurgery, 9*, 221–228.

Rinetti, G., & Wong, M-L. (2003). Major depression and the autonomic nervous system. In C. Liana Bolis et al. (Eds.), *Handbook of the autonomic nervous system in health and disease* (pp. 666–658). New York: Marcel Dekker.

Riva, D., & Giorgi, C. (2000). The cerebellum contributes to higher functions during development: Evidence from a series of children surgically treated for posterior fossa tumours. *Brain, 123*, 1051–1061.

Rivera, F. P., et al. (2008). Prevalence of pain in patients 1 year after major trauma. *Archives of Surgery, 143,* 282–287.

Rivest, S. (2003). Immune function and regulation of the autonomic nervous system. In C. Liana Bolis et al. (Eds.), *Handbook of the autonomic nervous system in health and disease* (pp. 55–132). New York: Marcel Dekker.

Rivest, S., & Rivier, C. (1993). Central mechanisms and sites of action involved in the inhibitory effects of CRF and cytokines in LHRH neuronal activity. In Y. Taché & C. Rivier (Eds.), *Corticotropin-releasing factor and cytokines: Role in the stress response* (Vol. 691, pp. 117–141). Annals of the New York Academy of Science.

Robertson, C. S. (1996). Nitrous oxide saturation technique for CBF measurement. In R. J. Narayan et al. (Eds.), *Neurotrauma* (pp. 487–501). New York: McGraw-Hill.

Robertson, G. L. (2001). Posterior pituitary. In P. Felig & L. A. Frohman (Eds.), *Endocrinology and metabolism, 4th Ed.* (pp. 217–258). New York: McGraw-Hill.

Robertson, G. L. (2005). Disorders of the neurohypophysis. In *Harrison's principles of internal medicine, 16th Ed.* (Vol. II, pp. 2097–2104). New York: McGraw-Hill.

Robertson, I. H., & Garavan, H. (2004). Vigilant attention. In M. S. Gazzaniga (Ed.), *The cognitive neurosciences, III* (pp. 631–640). Cambridge, MA: Massachusetts Institute of Technology.

Robinson, A. G. (1996). Posterior pituitary. In G. C. Bennett & F. Plum (Eds.), *Cecil textbook of medicine, 9th Ed.* (pp. 1221–1226). Philadelphia: Saunders.

Robinson, A. G. (2004). Posterior pituitary. In L. Goldman et al. (Eds.), *Cecil textbook of medicine, 22nd Ed.* (pp. 1385–1391). Philadelphia: Saunders.

Robinson, A. G. (2007). Posterior pituitary (Neurohypophysis). In D. G. Gardner & D. Shoback (Eds.), *Greenspan's basic & clinical endocrinology* (pp. 157–170). New York: McGraw-Hill.

Robinson, A. G., & Verbalis, J. G. (2003). Posterior pituitary gland. In P. R. Larsen et al. (Eds.), *Williams Textbook of Endocrinology, 10th Ed.* (pp. 281–329). Philadelphia: Saunders.

Robinson, R. (2007, July 3). Amateur boxers suffer short-term brain damage. *Neurology Today*, 5, 8.

Robinson, R. G., & Jorge, R. E. (2005). Mood disorders. In J. M. Silver et al. (Eds.), *Textbook of traumatic brain injury* (pp. 201–212). Washington DC: American Psychiatric Publishing.

Robinson, R. G., & Starkstein, S. E. (2002). Neuropsychiatric aspects of cerebrovascular disorders. In S. C. Yudofsky & R. E. Hales (Eds.), *Textbook of neuropsychiatry and clinical neurosciences, 4th Ed.* (pp. 723–752). Washington, DC: American Psychiatric Publishing.

Roca, C., et al. (1998). Effect of reproductive hormones on the hypothalamic-pituitary axis response to stress. *Biological Psychiatry, 44,* 6S.

Rohen, J. W., & Yokochi, C. (1993). *Color atlas of anatomy, 3rd Ed.* New York: Iaku-Shoin.

Rolak, L. A. (1988). Psychogenic sensory loss. *Journal of Nervous and Mental Disease,* 176, 686–687.

Role, L. W., & Kelly, J. P. (1991). The brain stem: Cranial nerve nuclei and the monaminergic systems. In E. R. Kandel et al. (Eds.), *Principles of neuroscience, 3rd Ed.* (pp. 683–699). New York: Elsevier.

Rolls, E. T., et al. (1998). The neurophysiology of taste and olfaction in primates, and umami flavor. In C. Murphy (Ed.), *Olfaction and taste XII* (Vol. 855, pp. 426–437). Annals of the New York Academy of Sciences.

Romner, B., et al. (2000). Serum S-100 protein measurements related to neuroradiological findings. *Journal of Neurotrauma, 17,* 641–647.

Ronken, E., & van Scharrenburg, G. J. M. (2005). In E. De Vries & A. Prat (Eds.), *The blood-brain barrier and its microenvironment* (pp. 71–85). New York: Taylor & Francis.

Ropacki, M. T., & Elias, J. W. (2003). Preliminary examination of cognitive reserve theory in closed head injury. *Archives of Clinical Neuropsychology, 18,* 643–654.

Ropper, A. H. (1997). Management of the autonomic storm. In P. Low (Ed.), *Clinical autonomic disorders: Evaluation and management, 2nd Ed.* (pp. 791–801). Philadelphia: Lippincott-Raven.

Ropper, A. H., & Gorson, K. C. (2007). Concussion. *New England Journal of Medicine*, *356*, 166–172.

Rosen, M., & Cedars, M. I. (2004). Female reproductive endocrinology and infertility. In F. S. Greenspan & D. G. Gardner (Eds.), *Basic and clinical endocrinology, 7th Ed.* (pp. 511–593). New York: Lange Medical Books.

Rosen, S. A., et al. (2001). The integration of visual and vestibular systems in balance disorders: A clinical perspective. In I. B. Suchoff et al. (Eds.), *Visual and vestibular consequences of acquired brain injury* (pp. 174–200). Santa Ana: CA: Optometric Extension Program Foundation.

Rosenbaum, M. (1988). Learned resourcefulness, stress and self-regulation. In S. Fisher & J. Reason (Eds.), *Handbook of life stress, cognition and health* (pp. 483–496). New York: John Wiley and Sons.

Rosenbaum, R. B., & Ciaverella, D. P. (2004). Disorders of bones, joints, ligaments, and meninges. In W. G. Bradley et al. (Eds.), *Neurology in clinical practice, 4th Ed.* (Vol. 2, pp. 2189–2222). Boston: Butterworth-Heinemann.

Rosenbaum, R. S. (2007). Theory of mind is independent of episodic memory. *Science, 318,* 1257.

Rosenbek, J., et al. (1978). Stuttering following brain damage. *Brain and Language*, *6*, 82–96.

Rosenblum, W. I. (1989). Pathology of human head injury. In D. P. Becker & S. K. Gudeman (Eds.), *Textbook of head injury* (pp. 525–537). Philadelphia: Saunders.

Rosenfield, D. B. (1997). Stuttering. In S. C. Schachter & O. Devinsky (Eds.), *Behavioral neurology and the legacy of Norman Geschwind* (pp. 101–111). Philadelphia: Lippincott-Raven.

Rosenfield, D. B., & Barroso, A. O. (2000). Difficulties with speech and swallowing. In W. G. Bradley et al. (Eds.), *Neurology in clinical practice* (Vol. 1, pp. 171–185).

Rosman, N. P. (1999). Traumatic brain injury in children. In K. F. Swaiman & S. Ashwal (Eds.), *Pediatric neurology: Principles and practice, 3rd Ed.* (pp. 873–897). St. Louis, MO: Mosby.

Rosman, N. P., et al. (1979). Acute head trauma in infancy and childhood. *Pediatric Clinics of North America, 26,* 707–736.

Ross, E. D. (1993). Nonverbal aspects of language. *Neurological Clinics, 11*(1), 9–23.

Ross, E. D., & Rush, A. J. (1981). Diagnosis and neuroanatomical correlates of depression in brain-damaged patients. *Archives of General Psychiatry, 38*, 1344–1354.

Ross, E. D., et al. (1981) How the brain integrates affective and propositional language into a unified behavioral function: Hypothesis based on clinicoanatomic evidence. *Archives of Neurology, 38*, 745–748.

Ross, E. D., et al. (2001). Affective-prosodic deficits in Schizophrenia: profiles of patients with brain damage and comparison with relation to schizophrenic symptoms. *Journal of Neurology, Neurosurgery, and Psychiatry*, *70*, 597–604.

Ross, E. L., & Michna, E. (2003). Invasive procedures for pain. In M. J. Aminoff & R. B. Daroff (Eds.), *Encyclopedia of the neurological sciences* (Vol. 3, pp. 740–750). Boston: Academic Press.

Ross, M. H., et al. (2003). *Histology, 4th Ed.* Philadelphia: Lippincott Williams & Wilkins.

Rothoerl, R. D., et al. (2000). S-100 serum levels and outcome after severe head injury. *Acta Neurochirurgica-Supplement, 76*, 97–100.

Rotondi, A. J., et al. (2007). A qualitative needs assessment of persons who have experienced traumatic brain injury and their primary family caregivers. *Journal of Head Trauma Rehabilitation, 22*, 14–25.

Royall, D. R., et al. (2002). Executive control function: A review of its promise and challenges for clinical research. *Journal of Neuropsychiatry and Clinical Neuroscience, 14*, 377–405.

Ruan, J. S., & Prasad, P. (1995). Coupling of a finite element human head model with a lumped parameter hybrid III dummy model: Preliminary results. *Journal of Neurotrauma, 12*, 725–734.

Rudel, R. (1978). Neuroplasticity: Implications for development and education. In J.S. Chall & A. F. Mirsky (Eds.), *Education and the brain* (pp. 269–307). Chicago: University of Chicago Press.

Ruff, R. (2005). Two decades of advances in understanding of mild traumatic brain injury. *Journal of Head Trauma Rehabilitation*, *20*, 5–18.

Ruff, R. M. (1999). Discipline-specific approach versus individual care. In N. R. Varney & R. J. Roberts (Eds.), *The evaluation and treatment of mild traumatic brain injury* (pp. 99–113). Mahwah, NJ: Lawrence Erlbaum.

Rugg, M. D. (1992). Conscious and unconscious processors in language and memory: Commentary. In A. S. Milner & M. D. Rugg (Eds.), *The neuropsychology of consciousness* (pp. 263–278). San Diego: Academic Press.

Rutherford, W. H., et al. (1977). Sequelae of concussion caused by minor head injuries. *Lancet*, 1 January 1977, pp. 1–4.

Rutland-Brown, W., et al. (2003). Incidence of traumatic brain injury in the United States, 2002. *Journal of Head Trauma Rehabilitation*, *21*, 544–548.

Rutland-Brown, W., (2006). Incidence of traumatic brain injury in the United States, 2003. *Journal of Head Trauma Rehabilitation,* 21 (6), 544–548.

Rutterford, N. A., & Wood, R. L. (2006). Evaluating a theory of stress and adjustment when predicting psychosocial outcome after brain injury. *Journal of the International Neuropsychological Society, 12,* 359–367.

Ryan, J. R., et al. (1991). Percentile rank conversion tables for WAIS-R IQs at six educational levels. *Journal of Clinical Psychology*, *47*, 104–107.

Sabban, E. L., et al. (2004). Differential effects of stress on gene transcription factors in catecholamine systems. *Annals of the New York Academy of Sciences, 1032,* 117–129.

Sabbatini, M. E. (1998). Claude Bernard: A brief biography. *Brain & Mind Magazine*, *2*(6). Retrieved from http://www.cerebromente.org.br/n06/historia/bernard_i.htm

Sachdev, P., et al. (2001). Schizophrenia-like psychosis following traumatic brain injury: A chart-based descriptive and case-control study. *Psychological Medicine, 31,* 231–239.

Salgado, J. V., et al. (2007). Prefrontal cognitive dysfunction following brainstem lesion. *Clinical Neurology and Neurosurgery, 109*, 379–382.

Salley, R. D., & Teiling, P. A. (1984). Dissociated rage attacks in a Viet Nam Veteran: A Rorschach study. *Journal of Personality Assessment*, *48*, 98–104.

Salmond, D. H., et al. (2006). Cognitive reserve as a resilience factor against depression after moderate/severe head injury. *Journal of Neurotrauma, 23*, 1049–1058.

Salmon, D. P., et al. (2002). Neuropsychological assessment of dementia on Guam. In F. R. Ferraro (Ed.), *Minority and cross-cultural aspects of neuropsychological assessment* (pp. 129–144). Lisse, NL: Swets and Zeitlinger.

Saltzman, J., et al. (2000). Theory of mind and executive functions in normal human aging and Parkinson's disease. *Journal of the International Neuropsychological Society, 6*, 781–788.

Saltzman-Benaiah, J., & Lalonde, C. E. (2007). Developing clinically suitable measures of social cognition for children: Initial findings from a normal sample. *The Clinical Neuropsychologist, 21*, 294–317.

Samson, K. (2005). Analysis of criteria for using CT for mild head injury. *Neurology Today, 5*(1w), 22–23.

Samson, K. (2006). VA reinforces stateside rehab units for Iraq blast injuries. *Neurology Today, 6*(8), 18, 24.

Samuel, C., et al. (1998). Dysprosody after severe close head injury: An acoustic analysis. *Journal of Neurology, Neurosurgery & Psychiatry, 64*, 482–485.

Sandel, M. E., et al. (2007). Sexuality, reproduction, and neuroendocrine disorders following TBI. In N. D. Zasler et al. (Eds.), Brain injury medicine (pp. 673–695). New York: Demos.

Sanders, V. M., & Straub, R. H. (2002). Norepinephrine, the β-adrenergic receptor, and immunity. *Brain, Behavior, and Immunity, 16*(4), 290–332.

Sanfey, A. G. (2007). Social decision-making: Insights from game theory and neuroscience. *Science, 318*, 598–602.

Santiago, P., & Fessler, F. G. (2004). Spinal cord trauma. In W. G. Bradley et al. (Eds.), *Neurology in clinical practice, 4th Ed.* (Vol. 2, pp. 1149–1178). Boston: Butterworth-Heinemann.

Saper, C. B. (2000a). Brainstem, reflexive behavior and cranial nerves. In E. R. Kandel et al. (Eds.), *Principles of neural science, 4th ed.* (pp. 873–888). New York: McGraw Hill.

Saper, C. B. (2000b). Brain stem modulation of sensation, movement, and consciousness. The autonomic nervous system and the hypothalamus. In E. R. Kandel et al. (Eds.), *Principles of neural science, 4th ed.* (pp. 889–908). New York: McGraw Hill.

Saper, C. B. (2002). The central autonomic nervous system: Conscious visceral perception and autonomic pattern generation. *Annual Review of Neuroscience, 25*, 433–490.

Saper, C. B., et al. (2000). Integration of sensory and motor functions: The association areas of the cerebral cortex and the cognitive capabilities of the brain. In E. R. Kandel et al. (Eds.), *Principles of neural science, 4th ed.* (pp. 348–380). New York: McGraw Hill.

Sapolsky, R. M. (2004). Stress and cognition. *The cognitive neurosciences, 3rd Ed.* (pp. 1031–1042). Cambridge, MA: A Bradford Book.

Sapolsky, R. M., & Pulsinelli, W. A. (1985). Glucocorticoids potentiate ischemic injury to neurons: Therapeutic implications. *Science, 229*, 1397–1400.

Sapolsky, R. M., et al. (1984). Glucocorticoid-sensitive hippocampal neurons are involved in terminating and adrenocortical stress response. *Proceedings of the National Academy of Sciences,* 81, 6174–6177.

Sarnat, H. B., & Flores-Sarnet, L. (2004). Developmental disorders of the nervous system. In W. G. Bradley et al. (Eds.), *Neurology in clinical practice, 4th Ed.* (Vol. 2, pp. 1763–1789). Boston: Butterworth-Heinemann.

Sasaki, M., et al. (1987). The influences of neurotransmitters on the traumatic unconsciousness, immediate convulsion and mortality in the experimental mice model (Japanese). *Brain and Nerve, 39*, 983–990.

Sattler, J. M. (1988). *Assessment of children.* San Diego, CA: Author.

Satz, P. (1993). Brain reserve capacity on symptom onset after brain injury: A formulation and review of evidence for threshold theory. *Neuropsychology, 7*, 275–295.

Saunders Elsevier. (2007). *Dorland's illustrated medical dictionary, 31st Ed.* Philadelphia: Saunders Elsevier.

Saunders, S. M., & Lueger, R. J. (2005). Evaluation of psychotherapy. In B. J. Sadock & V. A. Sadock (Eds.), *Comprehensive textbook of psychiatry, 8th Ed.* (Vol. 2, pp. 963–1003). Philadelphia: Lippincott, Williams & Wilkins.

Savola, O., & Hilborn, M. (2003). Early predictors of post-concussion symptoms in patients with mild head injury. *European Journal of Neurology, 10*, 175–181.

Sawauchi, S., et al. (2005). Serum S-100B protein and neuro-specific enolase after traumatic brain injury. *No Shinkei Geka—Neurological Surgery, 3*, 1073–1080.

Saxe, G., et al. (2003). Traumatic stress in injured and ill children. *PTSD Research Quarterly, Spring*, 1–3.

Sbordone, R. J. (1996). Neuropsychological tests: A look at our past and the impact that ecological issues may have on our future. In R. J. Sbordone & C. J. Long (Eds.), *Ecological validity of neuropsychological testing* (pp. 1–14). Delray Beach, FL: GR Press/St. Lucie Press.

Sbordone, R. J. (1999). Post-traumatic stress disorder: An overview and its relationship to closed head injuries, *NeuroRehabilitation, 13*, 69–78.

Sbordone, R. J., et al. (2007). *Neuropsychology for psychologists, health care professionals, and attorneys.* Boca Raton: CRC Press.

Scaer, R. C. (2001). The neurophysiology of dissociation and chronic disease. *Applied Psychophysiology and Biofeedback, 26*, 73–91.

Schachter, D. L., & Prigatano, G. P. (1991). Forms of unawareness. In G. P. Prigatano & D. L. Schachter (Eds.), *Awareness of deficit after brain injury* (pp. 259–262). New York: Oxford University Press.

Schachter, D. L., et al. (1991). Models of memory and the understanding of memory disorders. In T. Yanagihara & R. C. Peterson (Eds.), *Memory disorders* (pp. 111–134). New York: Marcel Dekker.

Schachter, D. L., et al. (1993). Implicit memory: A selective review. In *Annual Review of Neuroscience, 16* (pp. 159–182). Palo Alto, CA: Annual Reviews.

Schauenstein, K., et al. (2006). Immunomodulation by peripheral adrenergic and cholinergic agonists/antagonists in rat and mouse models. In A. Conti et al. (Eds.), *Neuroimmunomodulation: Perspectives at the new millennium* (Vol. 917, pp. 618–627). Annals of the NY Academy of Sciences.

Scheid, R. (2007). Comparative magnetic resonance imaging at 1.5 and 3 Tesla for the evaluation of traumatic microbleeds. *Journal of Neurotrauma, 24*, 1811–1816.

Schelling, G., et al. (2004). Can posttraumatic stress disorder be prevented with glucocorticoids? In R. Yehuda & B. McEwen (Eds.), *Behavioral stress responses* (Vol. 1032, pp. 158–166). Annals of the New York Academy of Sciences.

Schieber, M. H., & Baker, J. F. (2003). The descending control of motion. In L. R. Squire et al. (Eds.), *Fundamental neuroscience, 2nd Ed.* (pp. 791–814). Boston: Academic Press.

Schiff, N. D. (2006). Measurements and models of cerebral function in the severely injured brain. *Journal of Neurotrauma, 23*, 1436–1449.

Schilder, P. (1950). *The image and appearance of the human body*. New York: Wiley.

Schilder, P. (1964) *Contributions to developmental neuropsychiatry*. New York: International Universities Press.

Schinkel, C., et al. (2006). Inflammatory mediators are altered in the acite phase of posttraumatic complex regional pain syndrome. *Clinical Journal of Pain, 22*, 235–239.

Schlaggar, B. L., & McCandliss, B. D. (2007). Development of neural systems for reading. *Annual Review of Neuroscience, 30*, 475–503.

Schlesinger, M., & Yodfat, Y. (1996). Psychoimmunology, stress, and disease. In H. Friedman et al. (Eds.), *Psychoimmunology, stress, and infection* (pp. 127–136). Boca Raton, FL: CRC Press.

Schmahmann, J. D. (1991). An emerging concept: The cerebellar contribution to higher function. *Archives of Neurology, 48*, 1178–1187.

Schmahmann, J. D. (2006). Cognition, emotion and the cerebellum. *Brain, 129*, 290–291.

Schmahmann, J. D., & Sherman, J. C. (1997). Cerebellar cognitive affective syndrome. *International Review of Neurobiology, 41*, 433–440.

Schmidt, D. N., & Wallace, K. (1998). How to diagnose hypopititarism. *Postgraduate Medicine, 104*, 77–87.

Schneider, H. J., et al. (2007). Hypothalamopituitary dysfunction following traumatic brain injury and aneurysmal subarachnoid hemorrhage: A systematic review. *JAMA, 298*, 1429–1438.

Schneider, J., et al. (2005). Anterior pituitary hormone abnormalities following traumatic brain injury. *Journal of Neurotrauma, 22*, 937–946.

Schnurr, P. P., & Green, B. L. (2004). *Trauma and health: Physical health consequences of exposure to extreme stress*. Washington, DC: American Psychological Association.

Schnurr, P. P., et al. (2007). Cognitive behavioral therapy for posttraumatic stress disorder in woman. *Journal of the American Medical Association, 27*(8), 820–830.

Schoenpflug, W., & Battmann, W. (1988). The costs and benefits of coping. In S. Fisher & J. Reason (Eds.), *Handbook of life stress, cognition and health* (pp. 699–713). New York: John Wiley and Sons.

Schrader, H., et al. (1996). Natural evolution of late whiplash syndrome outside the medicolegal context. *The Lancet, 347*, 207–1211.

Schreuder, B. J. N., et al. (2000). Nocturnal re-experiencing more than forty years after war trauma. *Journal of Traumatic Stress, 13*, 453–463.

Schwartz, M. (2000). Beneficial autoimmune T cells and posttraumatic neuroprotection. In Conti, A. et al. (Eds.), *Neuroimmunomodulation: Perspectives at the new millennium* (Vol. 917, pp. 341–347). Annals of the New York Academy of Sciences.

Schulkin, J. (2003). *Rethinking homeostasis*. Cambridge, MA: MIT Press.

Schutz, L. E. (2007). Models of exceptional adaptation in recovery after traumatic brain injury: A case series. *Journal of Head Trauma Rehabilitation, 22*, 48–55.

Schwartz, M. F. (1995). Re-examining the role of executive functions in routine action production. In J. Grafman et al. (Eds.), *Structure and functions of the human prefrontal cortex* (Vol. 769, pp. 321–335). Annals of the New York Academy of Sciences.

Schwarz, A. (2007a, June 16). Lineman, dead at 36, sheds light on brain injuries. *New York Times*, D1.

Schwarz, A. (2007b, August 2). New advice by N.F.L. in handling concussions. *New York Times*, D1.

Schwarz, L., et al. (2009). Factors contributing to performance on the Rey complex figure test in individuals with traumatic brain injury. *The Clinical Neuropsychologist, 23*, 255–267.

Schweighofer, N., et al. (2007). Serotonin and the evaluation of future rewards: Theory, experiments, and possible neural mechanisms. In B. W. Balleine et al. (Eds.), *Reward and decision making in corticobasal ganglia networks* (Vol. 1104, pp. 289–300). Annals of the New York Academy of Sciences.

Schwoebel, J., et al. (2002). Pain and the body schema. *Neurology, 59,* 775–777.

Schűle, C., et al. (2004). Neuroative steroids in responders and nonresponders to sleep deprivation. In R. Yehuda & B. McEwen (Eds.), *Behavioral stress responses* (Vol. 1032, pp. 216–223). Annals of the New York Academy of Sciences.

Scott, S. G., et al. (2006). Mechanism-of-injury approach to evaluating patients with blast-related polytrauma. *Journal of the American Osteopathic Society*, *106*, 265–270.

Scremin, O. U., et al. (1997). Cholinergic modulation of cerebral cortical blood flow changes induced by trauma. *Journal of Neurotrauma, 14*, 573–586.

Seckl, J. R., & Meaney, M. J. (2004). Glucocorticoid programming. *Annals of the New York Academy of Sciences, 1032*, 63–84.

Seif-Naraghi, A. H., & Herman, R. M. (1999). A novel method for locomotion training. *Journal of Head Trauma Rehabilitation*, 14, 146–162.

Selhorst, J. B. (1989). Neurological examination of head-injured patients. In D. P. Becker & S. K. Gudeman (Eds.), *Textbook of head injury* (pp. 82–101). Philadelphia: Saunders.

Semrud-Clikeman, M. (2001). *Traumatic brain injury in children and adolescents.* New York: The Guilford Press.

Servadel, F., et al. (2001). Defining acute mild head injury in adults: A proposal based on prognostic factors, diagnosis, and management. *Journal of Neurotrauma, 18*, 657–664.

Seymour, L., et al. (2008). Connecting kids! Effective recruitment for resource facilitation in the pediatric population. *Journal of Head Trauma Rehabilitation, 23*, 264–270.

Shapiro, B. E., & Danly, M. (1985). The role of the right hemisphere in the control of speech prosody in propositional and affective contexts. *Brain and language, 25*, 19–36.

Shapiro, K. (1985). Head injury in children. In D. B. Becker and J. T. Povlishock (Eds.) *Central Nervous System Trauma Status Report—1985,* pp. 243–253. National Institute of Neurological and Communicative Disorders and Stroke, National Institutes of Health.

Sharo, K. (1987). Special considerations for the pediatric age group. In Coope P. R. (Ed.), *Head injury* (pp. 367–389). Baltimore: Williams and Wilkins.

Shaul, P. W., et al. (1985). Precocious puberty following severe head trauma. *American Journal of Diseases of Children, 139*, 467–469.

Shaw, P., et al. (2006). Intellectual ability and cortical development in children and adolescents. *Nature, 440*, 676–679.

Sheftel, F. D. (1996). Approach to the patient with headache. In M. S. Samuels & S. Feske (Eds.), *Office practice of neurology* (pp. 1086–1096). New York: Churchill Livingstone.

Shekhar, A., et al. (1999). Role of the basolateral amygdala in panic disorder. In J. F. McGinty (Ed.), *Advancing from the ventral striatum to the extended amygdala* (Vol. 877, pp. 747–750). Annals of the New York Academy of Sciences.

Sherer, M., et al., (1998). Characteristics of impaired awareness after traumatic brain injury. *Journal of the International Neuropsychological Society, 4*, 380–387.

Sherman, C. (2006). Personal resilience staves off PTSD. *Clinical Psychiatry News,* 34 (4), 1, 8.

Sherwin, R. S. (1996). Diabetes mellitus. In J. C. Bennett & F. Plum (Eds.), *Cecil textbook of medicine* (pp. 1258–1277). Saunders: Philadelphia.

Sherwood, A., & Carels, R. A. (2000). Blood Pressure. In G. Fink (Ed.), *Encyclopedia of stress* (Vol. 1, pp. 283–289). San Diego, CA: Academic Press.

Shetter, A. G., & Demakas, J. J. (1979). The pathophysiology of concussion: A review. In, R. A. Thompson & J. R. Green (Eds.), *Advances in neurology* (Vol. 22, pp. 5–14). New York: Raven Press.

Shimamura, A. P. (1995). Memory and the prefrontal cortex. In J. Grafman et al. (Eds.), *Structure and functions of the human prefrontal cortex* (Vol. 769, pp. 151–159). Annals of the New York Academy of Sciences.

Shoback, D., & Funk, J. L. (2007). Humoral manifestations of malignancy. In D. G. Gardner & D. Shoback (Eds.), *Basic and clinical endocrinology, 8th Ed.* (pp. 817–830). New York: Lange Medical Books.

Shohami, E., et al. (1995). The effect of the adrenocortical axis upon recovery from closed head injury. *Journal of Neurotrauma, 12*, 1069–1077.

Shorter, N. A., et al. (1996). Skiing injuries in children and adolescents. Skiing injuries in children and adolescents. *The Journal of Trauma, 40,* 997–1001.

Shurtleff, H. A., et al. (1995). Screening children and adolescents with mild or moderate traumatic brain injury to assist school reentry. *Journal of Head Trauma Rehabilitation, 10*, 64–79.

Sifneos, P. (1973). The prevalence of "alexithymic" characteristics in psychosomatic patients. *Psychotherapy and Psychosomatics, 22*, 255–262.

Silver, J. M., et al. (1991). Depression in traumatic brain injury. *Neuropsychiatry, Neuropsychology, and Behavioral Neurology, 4*, 12–23.

Silver, S. M., & Salamone-Genovese, L. (1991). A study of the MMPI clinical and research scales for post-traumatic stress disorder diagnostic utility. *Journal of Traumatic Stress, 4*, 533–548.

Silverberg, N. D., & Millis, S. R. (2009). Impairment versus deficiency in neuropsychological assessment. *Journal of the International Neuropsychological Society, 15*, 94–102.

Silverman, E. T. (1953). Is speech a biologic function? M.A. Thesis, Brooklyn College, NY.

Silverman, E. T. (1986). Profile of speech and language disorders relevant to dental therapeutics. Second International Congress of Gerontology, Singapore.

Silverman, S. I. (2002). Osteogingival atrophy: A new descriptor of periodontal disease highly correlated with functional dental occlussion trauma. (In preparation for publication).

Simon, R. P. (2001). Breathing and the nervous system. In M. J. Aminoff (Ed.), *Neurology and general medicine, 3rd Ed.* (pp. 1–22). New York: Churchill Livingstone.

Simpson, D. A. (2005). Clinical examination and grading. In P. L. Reilly & R. Bullock (Eds.), *Head injury, 2nd Ed.* (pp. 143–163). New York: Hodder Arnold.

Simpson, G., et al. (2000). Cultural variations in the understanding of traumatic brain injury and brain injury rehabilitation. *Brain Injury, 14*, 125–140.

Sinatra, R. S. (2003). Postoperative pain. In M. J. Aminoff & R. B. Daroff (Eds.), *Encyclopedia of the neurological sciences* (Vol. 3, pp. 770–773). Boston: Academic Press.

Singer, B. J., et al. (2003). Velocity dependent passive plantar flexor resistive torque in patients with acquired brain injury. *Clinical Biomechanics, 18*, 167–165.

Sinha, R., et al. (2004). Neural circuits underlying emotional distress in humans. In R. Yehuda & B. McEwen (Eds.), *Behavioral stress responses* (Vol. 1032, pp. 254–257). Annals of the New York Academy of Sciences.

Slagle, D. A. (1990). Psychiatric disorders following closed head injury: An overviews of biopsychosocial factors in their etiology and management. *International Journal of Psychiatry in Medicine, 20*, 1–35.

Slemmer, E. J., et al. (2002). Repeated mild injury causes cumulative damage to hippocampal cells. *Brain, 125*(12), 2699–2709.

Slimmer, L. M., et al. (2001). Stress, medical illness, and depression. *Seminars in Clinical Psychiatry, 6*, 12–26.

Smith, G. A. (1998). Injuries to children in the United States related to trampolines, 1990–1995: A national epidemic. *Pediatrics, 101*, 406–412.

Smith, E. E., & Jonides, J. (2003). Executive control and thought. In L. Squire et al. (Eds.), *Fundamental neuroscience, 2nd Ed.* (pp. 1377–1394). London: Academic Press.

Smith, S. A. (1989). Disorders of the autonomic nervous system. In K. Swaiman (Ed.), *Pediatric neurology, principles and practice* (Vol. II, pp. 845–856). St. Louis: C. V. Mosby.

Smith, Y., & Sidbe, M. (2003). Basal ganglia. In M. J. Aminoff & R. B. Daroff (Eds.), *Encyclopedia of the neurological sciences* (Vol. 1, pp. 345–355). Boston: Academic Press.

Smolders, J. (1984). Charts of trigger points, I & II. Toronto, Ontario: Sjef Enterprises. Distributed by Lippincott, Williams & Wilkins, Hagerstown, MD.

Snape, W. J. (1996). Disorders of gastrointestinal motility. In J. C. Bennett & F. Plum (Eds.), *Cecil textbook of medicine* (pp. 680–688). Philadelphia: Saunders.

Snyder, B. D., & Daroff, R. B. (2004). Hypoxic/anoxic and ischemic encephalolopathies. In W. G. Bradley et al. (Eds.), *Neurology in clinical practice, 4th Ed.* (Vol. 2, pp. 1665–1672). Boston: Butterworth-Heinemann.

Snyder, P. J. (2001). Diseases of the anterior pituitary. In P. Felig & L. A. Frohman (Eds.), *Endocrinology and metabolism, 4th Ed.* (pp. 173–216). New York: McGraw-Hill.

Soechting, J. F., & Flanders, M. (2002). Movement regulation. In V. Ramachandran (Ed.), *Encyclopedia of the human brain* (Vol. 3, pp. 201–210). Boston: Academic Press.

Somjen, G. G. (2004). *Ions in the brain.* New York: Oxford University Press.

Song, C., et al. (1995). Behavioral, neurochemical, and immunological responses to CRF administration. In G. P. Chrousos et al. (Eds.), *Stress: Basic mechanisms and clinical implications* (Vol. 771, pp. 55–72). Annals of The New York Academy of Sciences.

Soo, C., et al. (2007). Reliability of the Care and Needs Scale for assessing support needs after traumatic brain injury. *Journal of Head Trauma Rehabilitation, 22*, 288–295.

Sorg, B. A., & Kalivas, P. W. (1995). Stress and neuronal sensitization. In M. J. Friedman et al. (Eds.), *Neurobiological and clinical consequences of stress* (pp. 83–102). Philadelphia: Lippincott-Raven.

Southwick, S. M., et al. (1992). Neurobiology of posttraumatic stress disorder. In A. Tasman & M. R. Riba (Eds.), *Annual Review of Psychiatry* (Vol. 11, pp. 347–367). Washington, DC: American Psychiatric Press.

Southwick, S. M., et al. (1994). Psychobiologic research in post-traumatic stress disorder. *Psychiatric Clinics of North America, 17*(2), 251–264.

Southwick, S. M., et al. (1995). Clinical studies of neurotransmitter alterations in post-traumatic stress disorder. In M. J. Friedman et al. (Eds.), *Neurobiological and clinical consequences of stress: From normal adaptation to PTSD* (pp. 345–349). Philadelphia: Lippincott-Raven.

Southwick, S. M., et al. (2003). Emerging neurobiological factors in stress resilience. *PTSD Research Quarterly, 14*(4), 1–3.

Sowell, E. R., et al. (2001). Mapping continued brain growth and gray matter density reduction in dorsal frontal cortex: Inverse relationships during postadolescent brain maturation. *The Journal of Neuroscience, 21*(22), 8819–8829.

Spencer, R. C., et al. (2003). Disrupted timing of discontinuous but not continuous movements by cerebellar lesions. *Science*, *300*, 1437–1439.

Spiegel, D. (1985). Introduction. In S. Eth & R. S. Pynoss (Eds.), *Post-traumatic stress disorder in children* (pp xi–xvi). Washington, DC: American Psychiatric Press.

Spiegel, D. (1988). Dissociation and hypnosis in post-traumatic stress disorders. *Journal of Traumatic Stress, 1*, 17–33.

Spiegel, D. (1991). Dissociation and trauma. In A. Talisman & S. M. Goldfinger (Eds.), *Review of psychiatry* (Vol. 10, pp. 261–275). Washington, DC: American Psychiatric Press.

Spierings, E. L. H. (1996). Anatomy and physiology of headache. In M. S. Samuels & S. Feske (Eds.), *Office practice of neurology* (pp. 1082–1086). New York: Churchill Livingstone.

Squire, L. R., & Zola-Morgan, S. (1991). The medial temporal lobe memory system. *Science, 252*, 1380–1386.

Sriram, T. G., & Silverman, J. J. (1998). The effects of stress on the respiratory system. In J. R. Hubbard & E. A. Workman (Eds.), *Handbook of stress: An organ system approach* (pp. 45–68). Boca Raton, FL: CRC Press.

Stalnacke, B. M., et al. (2003). Playing ice hockey and basketball increases serum levels of S-100B in elite players: A pilot study. *Clinical Journal of Sports Medicine, 13*, 292–302.

Stalnacke, B. M., et al. (2004). Playing soccer increases serum concentrations of the biochemical markers of brain damage S-100B and neuron-specific enolase in elite players: a pilot study. *Brain Injury, 18*, 899–909.

Stalnacke, B. M., et al. (2005). One-year follow-up of mild traumatic brain injury: Post-concussion symptoms, disabilities and life satisfaction in relation to serum levels of S-100B and neurone-specific enolase in acute phase. *Journal of Rehabilitation Medicine*, *37*, 300–305.

Stalnacke, B. M. et al. (2006). Serum concentrations of two biochemical markers of brain tissue damage S-100B and neuron specific enolase are increased in elite female soccer players after a competitive game. *British Journal of Sports Medicine*, *40*, 313–316.

Standing, S. (2005a). Nervous system: Principles of hormone production and secretion. In *Gray's Anatomy, 39th Ed.* (pp. 43–67). Edinburgh: Elsevier.

Standing, S. (2005b). Endocrine system: Principles of hormone production and secretion. In *Gray's Anatomy, 39th Ed.* (pp. 179–183). Edinburgh: Elsevier.

Stapert, S., et al. (2005). S-100B concentration is not related to neurocognitive performance in the first month after mild traumatic brain injury. *European Neurology*, 53, 22–26.

Starkstein, S. D., & Lischinsky, A. (2002). The phenomenology of depression after brain injury. *Neurorehabilition, 17*, 105–113.

Starkstein, S. E., et al. (1994). Neuropsychological and neuroradiologic correlates of emotional prosody comprehension. *Neurology, 44*, 515–522.

Staub, N. C. (1998a). Structure and function of the respiratory system. In R. M. Berne & M. N. Levy (Eds.), *Physiology* (pp. 517–533). St. Louis, MO: Mosby.

Staub, N. C. (1998b). Control of breathing. In R. M. Berne & M. N. Levy (Eds.), *Physiology* (pp. 572–585). St. Louis, MO: Mosby.

Staud, R., et al. (2001). Abnormal sensitization and temporal summation of second pain (wind up) in patients with fibromyalgia syndrome. *Pain, 91*(1–2), 165–175.
Steele, D. L., et al. (2005). The effect of traumatic brain injury on the timing of sleep. *Chronobiology International, 22*, 89–105.
Steere, A. C. (2005). Lyme borreliosis. In *Harrison's principles of internal medicine, 16th Ed.* (Vol. I, pp. 995–999). New York: McGraw-Hill.
Stein, D. G., & Hoffman, S. W. (2003). Concepts of CNS plasticity in the context of brain damage and repair. *Journal of Head Trauma Rehabilitation, 18*, 317–341.
Stein, M. B., et al. (2008). Pharmacotherapy to prevent PTSD: Results from a randomized controlled proof-of-concept trial in physically injured patients. *Journal of Traumatic Stress, 20*, 623–632.
Steinberg, M. (1991). The spectrum of depersonalization: Assessment and treatment. In A. Talisman & S. M. Goldfinger (Eds.), *Annual review of psychiatry, 10* (pp. 223–247). Washington, DC: American Psychiatric Press.
Steinberg, M. (2000a). Dissociative disorders. In B. J. Sadock & V. A. Sadock (Eds.), *Comprehensive textbook of psychiatry, 7th Ed.* (pp. 1544–1564). Philadelphia: Lippincott Williams & Wilkins.
Steinberg, M. (2000b). Depersonalization disorders. In B. J. Sadock & V. A. Sadock (Eds.), *Comprehensive textbook of psychiatry, 7th Ed.* (pp. 1564–1570). Philadelphia: Lippincott Williams & Wilkins.
Steinman, R. M., et al. (2003). Dendridic cell function in vivo during the steady state: A role in peripheral tolerance. In N. Chiorazzi et al. (Eds.), *Immune mechanisms and disease* (Vol. 987, pp. 15–25). Annals of the New York Academy of Sciences.
Stephane, M., et al. (2004). New phenomenon of abnormal auditory perception associated with emotional and head trauma: Pathological confirmation by SPECT scan. *Brain and Language, 89*, 503–507.
Steptoe, A. (2000). Hypertension. In G. Fink (Ed.), *Encyclopedia of stress,* (Vol. 3, pp. 425–431). San Diego: Academic Press.
Steptoe, A., & Brydon, L. (2005). Psychoneuroimmunology and coronary heart disease. In K. Vedhara & M. Irwin (Eds.), *Human psychoneuroimmunology* (pp. 107–1352). New York: Oxford University Press.
Sterling, P., & Eyer, J. (1988). Allostasis: A new paradigm to explain arousal pathology. In S. Fisher & J. Reason (Eds.), *Handbook of life stress, cognition and health* (pp. 629–649). New York: John Wiley and Sons.
Stern, P. (2007). Decisions, decisions… *Science, 318*, 393.
Stern, Y. (2002). What is cognitive reserve? Theory and research application of the reserve concept. *Journal of the International Neuropsychological Society, 8*, 448–460.
Sternberg, E. M. (2000). Does stress make you sick and belief make you well? The science connecting body and mind. In A. Conti et al. (Eds.), *Neuroimmunomodulation: Perspectives at the new millennium* (Vol. 917, pp. 1–3). Annals of the NY Academy of Sciences.
Sternberg, E. M. (2003). Introduction: Overview of the conference and the field. In *Neuroimmunomodulation* (Vol. 840, pp. 1–8). Annals of the New York Academy of Sciences.
Sternberg, R. J., & Kaufman, R. J. (2002). Intelligence. In V. S. Ramachandran (Ed.), *Encyclopedia of the human brain* (Vol. 2, pp. 587–597). Boston: Academic Press.
Stewart, C. F., et al. (1999). Evaluation and management of swallowing and voice disorders. In D. S. Younger (Ed.), *Motor disorders* (pp. 477–491). Philadelphia: Lippincott Williams & Wilkins.
Stewart, C. F., et al. (2005). Evaluation and management of swallowing and voice disorders. In D. S. Younger (Ed.), *Motor disorders, 2nd Ed.* (pp. 663–676). Philadelphia: Lippincott Williams & Wilkins.
Stickgold, R. (2006). A memory boost while you sleep. *Nature, 444*, 559.
Stierwalt, J. A. G., et al. (1996). Tongue strength and endurance: Relation to the speaking ability of children and adolescents following traumatic brain injury. In A. Donald et al. (Eds.), *Disorders of motor speech* (pp. 241–256). Baltimore: Paul H. Brooks.
Stigbrand, T., et al. (2000). A new specific method for measuring S-100B in serum. *International Journal of Biological Markers, 15*, 33–40.
Stoel-Gammon, C., & Dunn, C. (1985). *Normal and disordered phonology in children.* Austin, TX: Pro-Ed.
Stoll, G., et al. (2002). Microglia. In V. S. Ramachandran (Ed.), *Encyclopedia of the human brain* (Vol. 3, pp. 29–41). Boston: Academic Press.
Stollberger, C., et al. (2001). Dysarthria as the leading symptom of hypothyroidism. *American Journal of Otolaryngology, 22*, 70–72.
Stone, J. L., et al. (1996). Civilian cases of tangential gunshot wounds to the head. *Journal of Trauma: Injury, Infection and Critical Care,* 49, 57–60.
Stout, S. C., et al. (1995). Neuropeptides and stress: Preclinical findings and implications for pathophysiology. In M. J. Friedman et al. (Eds.), *Neurobiological and clinical consequences of stress: From normal adaptation to PTSD* (pp. 103–123). Philadelphia: Lippincott-Raven.

Stranjalis, G., et al. (2004). Elevated serum S-1000B protein as a predictor of failure to short-term return to work or activities after mild head injury. *Journal of Neurotrauma, 21*, 1070–1075.
Stratakis, C. A., & Chrousos, G. P. (1995). Neuroendocrinology and pathophysiology of the stress system. In G. P. Chrousis et al. (Eds.), *Stress: Basic mechanisms and clinical implications* (Vol. 771, pp. 1–18). Annals of the New York Academy of Science.
Strauss, E., et al. (2006). *A compendium of neuropsychological tests. Administration, norms, and commentary.* New York: Oxford University Press.
Strebel, S., et al. (1997). Impaired cerebral autoregulation after mild head injury. *Surgical Neurology,* 47, 128–131.
Stringer, A. Y. (1996). Treatment of motor aprosodia with pitch biofeedback and expression modeling. *Brain Injury, 10*, 583–590.
Strub, R. L., & Black, F. W. (1985). *The mental status examination in neurology, 2nd Ed.* Philadelphia: F. A. Davis.
Stryke, J., et al. (2007). Traumatic brain injuries in a well-defined population: Epidemiological aspects and severity. *Journal of Neurotrauma, 24*, 1425–1436.
Stulemeijer, M., et al. (2006a). Recovery from mild traumatic brain injury: A focus on fatigue. *Journal of Neurology, 253*, 1041–1047.
Stulemeijer, M., et al. (2006b). Impact of additional extracranial injuries on outcome after mild traumatic brain injury. *Journal of Neurotrauma, 23*, 1561–1569.
Stuss, D. T. (1991). Disturbance of self-awareness after frontal system damage. In G. P. Prigatano & D. L. Schachter (Eds.), *Awareness of deficit after brain injury* (pp. 63–83). New York: Oxford.
Stuss, D. T., & Benson, D. F. (1986). *The frontal lobes*. New York: Raven Press.
Stuss, D. T., et al. (1995). A multidisciplinary approach to anterior attentional functions. In J. Grafman et al. (Eds.), *Structure and functions of the human prefrontal cortex* (Vol. 769, pp. 191–211). Annals of the New York Academy of Sciences.
Styne, D. (2004). Puberty. In F. S. Greenspan & D. G. Gardner (Eds.), *Basic and clinical endocrinology, 7th Ed.* (pp. 608–636). New York: Lange Medical Books.
Styne, D. M. (1991a). Growth. In F. S. Greenspan (Ed.), *Basic and clinical endocrinology* (pp. 147–176). Norwalk, CT: Appleton and Lange.
Styne, D. M. (1991b). Puberty. In F. S. Greenspan (Ed.), *Basic and clinical endocrinology* (pp. 519–542). Norwalk, CT: Appleton and Lange.
Suddendorf, T. (2006). Foresight and evolution of the human mind. *Science, 312*, 1006–1007.
Sukhotinsky, I., et al. (2007). Neural pathways associated with loss of consciousness caused by intracerebral microinjection of GABAA-active anesthetics. *European Journal of Neuroscience, 25*, 1417–1436.
Sumitani, M., et al. (2007a). Pathologic pain distorts visuospatial perception. *Neurology, 68*, 42–49.
Sumitani, M., et al. (2007b). Prism adaptation to optical deviation alleviates pathologic pain. *Neurology, 68*, 42–49.
Suter, P. S. (2004). Rehabilitation and management of visual dysfunction following traumatic brain injury. In M. J. Ashley (Ed.), *Traumatic brain injury: Rehabilitative treatment and case management* (pp. 209–249). Boca Raton, FL: CRC Press.
Svenningsson, P., et al. (2006). Alterations in 5-HT_{1b} receptor function by Pll in depression-like states. *Science, 311*, 77–80.
Svrakic, D. M., & Cloninger, C. R. (2005). Personality disorders. In B. J. Sadock & V. A. Sadock (Eds.), *Comprehensive textbook of psychiatry, II, 8th Ed.* (pp. 2063–2104). Philadelphia: Lippincott Williams & Wilkins.
Swanberg, M. M., et al. (2003). Speech and language. In C. G. Goetz (Ed.), *Textbook of Clinical Neurology, 2nd Ed.* (pp. 77–97). Philadelphia: Saunders.
Swanson, L. W. (2003). The architecture of nervous systems. In L. R. Squire et al. (Eds.), *Fundamental neuroscience, 2nd Ed.* (pp. 15–45). Boston: Academic Press.
Swartz, B. E. (1995). Unusual seizure types. *Seminars in Neurology,* 15, 126–132.
Swift, P. (1999). Validation of a Spanish language test of verbal learning and memory: "The Perri Test de Aprendizaje & Memoria" (Doctoral Dissertation, The University of Connecticut). *Dissertation Abstracts.*
Swisher, L. (1985). Language disorders in children. In J. Darby (Ed.), *Speech and language evaluation in neurology: Childhood disorders* (pp. 33–96). New York: Grune and Stratton.
Tabbal, S. D., & Pullman, S. L. (2005). Neurophysiology of clinical motor control. In D. Younger (Ed.), *Motor disorders, 2nd Ed.* (pp. 55–64). Philadelphia, PA: Lippincott, Williams & Wilkins.
Taber, K. H., et al. (2006). Blast-related traumatic injury: What is known. *Journal of Neuropsychiatry and Clinical Neuroscience, 1,* 141–145.

Taber, M. T., & Fibiger, N. C. (1997). Activation of the mesocortical dopamine system by feeding: Lack of a selective response to stress. *Neuroscience, 77*, 295–298.

Tabuchi, M., et al. (2000). Rinsho Shinkeigaku. *Clinical Neurology*, *40*, 464–470.

Taché, Y. (2003). The parasympathetic nervous system in the pathophysiology of the gastrointestinal tract. In C. Liana Bolis et al. (Eds.), *Handbook of the autonomic nervous system in health and disease* (pp. 634–502). New York: Marcel Dekker.

Taché Y., et al. (1993). Role of CRF in stress-related alterations of gastric and colonic motor function. In Y. Taché & C. Rivier (Eds.), *Corticotropin-releasing factor and cytokines: Role in the stress response* (Vol. 691, pp. 233–243). Annals of the New York Academy of Science.

Tafet, G. E., & Smolovich, J. (2004). Psychoneuroendocrinological studies of chronic stress and depression. In R. Yehuda & B. McEwen (Eds.), *Behavioral stress responses* (Vol. 1032, pp. 276–282). Annals of The New York Academy of Sciences.

Taft, C. T., et al. (2007). Aggression among combat veterans: Relationships with combat exposure and symptoms of posttraumatic stress disorder, dysphoria, and anxiety. *Journal of Traumatic Stress, 2*, 135–145.

Taheri, S., et al. (2002). The role of hypocretins (orexins) in sleep regulation and narcolepsy. *Annual Review of Neuroscience, 25*, 283–313.

Takahashi, H., & Nakazawa, S. (1980). Specific type of head injury in children. *Child's Brain, 7*, 124–131.

Takamiya, M., et al. (2007). Simultaneous detections of 27 cytokines during cerebral wound healing by multiplexed bead-based immunoassay for wound age estimation. *Journal of Neurotrauma*, *24*, 1833–1845.

Talan, J. (2008, October 2). New report links sports concussion to chronic traumatic encephalopathy. *Neurology Today, 8*(19), 12–13.

Talman, W. T. (1997). The central nervous system and cardiovascular control in health and disease. In P. Felig et al. (Eds.), *Endocrinology and metabolism, 3rd Ed.* (pp. 47–59). New York: McGraw-Hill.

Tanaka, J. W. (2004). Object categorization, expertise, and neural plasticity. In M. S. Gazzaniga (Ed.), *The cognitive neurosciences, 3rd Ed.* (pp. 877–877). Cambridge, MA: A Bradford Book.

Tarnriverdi, R., et al. (2008). Poloprotein E/3/E/3 genotype decreases the risk of pituitary dysfunction after traumatic brain injury due to various cases: Preliminary data. *Journal of Neurotrauma, 10*, 1071–1077.

Tate, D. F., & Bigler, E. D. (2008). Fornix and hippocampal atrophy in traumatic brain injury. *Learning and Memory, 7*, 742–746.

Tattersall, I. (1998). The origin of the human capacity. The *68th James Arthur Lecture*. New York: American Museum of Natural History.

Taylor, A. E., et al. (1996). Persistent neuropsychological deficits following whiplash: Evidence for chronic mild traumatic brain injury? *Archives of Physical Medicine and Rehabilitation, 77*, 529–535.

Taylor, A. N., et al. (2008a). Injury severity differentially affects short and long-term neuroendocrine outcomes of traumatic brain injury. *Journal of Neurotrauma*, *25*, 311–323.

Taylor, A. R., & Bell, T. K. (1966). Slowing of cerebral circulation after concessional head injury: A controlled trial. *Lancet,* July 23, 178–180.

Taylor, H. G. (1987). The meaning and value of soft signs in the behavioral sciences. In D. E. Tupper (Ed.) *Soft Neurological Signs,* pp. 297–335. Orlando, FL: Grune and Stratton.

Taylor, H. G., & Alden, J. (1997). Age-related differences in outcomes following childhood brain insults: An introduction and overview. *Journal of the International Neuropsychological Society*, *3*, 555–567.

Taylor, H. G., et al. (1995). Recovery from traumatic brain injury in children: The importance of the family. In S. H. Broman & M. E. Michel (Eds.), *Traumatic head injury in children* (pp. 188–216). New York: Oxford University Press.

Taylor, H. G., et al. (2008b). Traumatic brain injury in young children: Postacute effects on cognitive and school readiness skills. *Journal of the International Neuropsychological Society, 14*, 734–745.

Taylor, J. G. (1984). Alexythymia: Concept, measurement, and implications for treatment. *American Journal of Psychiatry*, *145*, 725–732.

Teasdale, G., & Mathew, P. (1996). Mechanisms of cerebral concussion, contusion, and other effects of head injury. In J. R. Youmans (Ed.), *Neurological Surgery, 4th Ed.* (pp. 1533–1548). Philadelphia: Saunders.

Teasell, R. W., & Shapiro, A. P. (1998). Whiplash injuries: An update. *Pain Research Management,* 3 (2), 81–90.

Teeter, P. A., & Semrud-Clikeman, M. (1997). *Child neuropsychology*. Boston: Allyn & Bacon.

Teman, A. J., et al. (1991). Traumatic thrombosis of a vertebral artery in a hypoplastic vertebrobasilar system: Protective effect of collateral circulation. *New York State Journal of Medicine, 9*, 314–315.

Temoshok, L. R. (2000). Complex coping patterns and their role in adaptation and neuroimmunomodulation. In A. Conti et al. (Eds.), *Neuroimmunomodulation: Perspectives at the new millennium* (Vol. 917, pp. 446–465). Annals of the NY Academy of Sciences.

Temple, R. O., et al. (2006). Differential impact of executive function on visual memory tasks. *The Clinical Neuropsychologist, 20*, 480–490.
Tenedieva, T. D., et al. (2001). Involvement of thyroid hormones in mental recovery following severe craniocerebral trauma (Russian, Abstract). *Zhurnal Voprosy Neirokhirugii Imeni N-N-Burdenko, 15*, 10–15.
Testa, J. A., et al. (2001). Impaired affective prosody in AD: Relationship to aphasic deficits and emotional behaviors. *Neurology, 57*(8), 1474–1481.
Thal, D. J., et al. (1991). Early lexical development in children with focal brain injury. *Brain and Language, 40*, 491–527.
Thamel, P. (2007, November 12). Clarity lacking in discipline for hits to head. *New York Times*, D1.
Theodoros, D. G. & Murdoch, B. E. (2001a). Velopharyngeal dysfunction following traumatic brain injury. In B. E. Murdoch & D. G. Theodoros (Eds.), *Traumatic brain injury: Associated speech, language, and swallowing disorders* (pp. 53–73, 75–78). San Diego: Singular Publishing Group.
Theodoros, D. G., & Murdoch, B. E. (1996). Differential patterns of hyperfunctional laryngeal impairment in dysarthric speakers following severe closed head injury. In A. Donald et al. (Eds.), *Disorders of motor speech* (pp. 205–227). Baltimore: Paul H. Brooks.
Theodoros, D. G., & Murdoch, B. E. (2001b). Laryngeal dysfunction following traumatic brain injury. In B. E. Murdoch & D. G. Theodoros (Eds.), *Traumatic brain injury: Associated speech, language, and swallowing disorders.* (pp. 89–108). San Diego: Singular Publishing Group.
Theodoros, D. G., et al. (1995). Variability in the perceptual and physiological features of dysarthria following severe closed head injury: An examination of five cases. *Brain Injury, 9*, 671–696.
Theodoros, D. G., et al. (2001a). Dysarthria following traumatic brain injury: Incidence, recovery, and perceptual features. In B. E. Murdoch & D. G. Theodoros (Eds.), *Traumatic brain injury: Associated speech, language, and swallowing disorder* (pp. 27–51). San Diego: Singular Publishing Group.
Theodoros, D. G., et al. (2001b). Articulatory dysfunction following traumatic brain injury. In B. E. Murdoch & D. G. Theodoros (Eds.), *Traumatic brain injury: Associated speech, language, and swallowing disorders* (pp. 53–73). San Diego: Singular Publishing Group.
Thibault, L. E. & Gennarelli, T. A. (1985). Biomechanics and craniocerebral trauma. In Becker, D.P., & Povlishock, J.R. (Eds.), *Central nervous system trauma status report-1985* (pp. 379–389). Prepared for the National Institute of Neurological and Communicative Disorders and Stroke, National Institutes of Health.
Thomas, A., & Chess, S. (1977). *Temperament and development*. New York: Brunner Mazel.
Thomas, C. B. (1982). Stamina: The thread of human life. *Psychotherapy and Psychosomatics, 38*, 74–80.
Thomas, D. H., & Meaney, D. F. (2000). Axonal damage in traumatic brain injury. *The Neuroscientist, 6*, 483–495.
Thomas, S., et al. (2000). Cerebral metabolic response to traumatic brain injury sustained early in development: A 2-deoxy-D-glucose autoradiographic study. *Journal of Neurotrauma, 17*, 649–665.
Thomp, P. M., et al. (2005). Structural MRI and brain development. *International Review of Neurobiology, 67*, 285–323.
Thompson, J., et al. (2005). EEG and postural correlates of mild traumatic brain injury in athletes. *Neuroscience Letters, 377*, 158–163.
Thompson, P. D. (2004). Gait disorders. In W. G. Bradley et al. (Eds.), *Neurology in clinical practice, 4th Ed.* (Vol. 1, pp. 323–336). Boston: Butterworth-Heinemann.
Thompson, W. G. (2006). EEG changes and balance deficits following concussion: One piece of the puzzle. In S. Slobounov & W. Sebastianelli (Eds.), *Foundations of sport related brain injuries* (pp. 341–374). New York: Springer.
Timmann, D., & Diener, H. C. (2003). Coordination and ataxia. In C. G. Goetz (Ed.) *Textbook of clinical neurology, 2nd Ed.* (pp. 299–315). Philadelphia: Saunders.
Todis, B., & Glang, A. (2008). Redefining success: Results of a qualitative study of postsecondary transition outcomes for youth with traumatic brain injury. *Journal of Head Trauma Rehabilitation, 23*, 252–263.
Toga, A. W., & Thompson, P. M. (2005). Genetics of brain structure and intelligence. *Annual Review of Neuroscience, 28*, 1–24.
Tokutomi, T., et al. (2008). Age-associated increase in poor outcomes after traumatic brain injury. *Journal of Neurotrauma, 25*, 1407–1414.
Tomas, R. L. (2004). Vision loss. In W. G. Bradley et al. (Eds.), *Neurology in clinical practice, 4th Ed.* (Vol. 1, pp. 177–183). Boston: Butterworth-Heinemann.
Tong, K. A., et al. (2003). Hemorrhagic shearing lesions in children and adolescents with diffuse axonal injury: Improved detection and initial results. *Radiology, 227*, 332–339.
Topka, H. J., & Massaquoi, S. G. (2002). Pathophysiology of clinical cerebellar signs. In M-U. Manto & M. Pandolfo (Eds.), *The cerebellum and its disorders* (pp. 121–135). Cambridge University Press.

Topka, H., et al. (1993). Deficits in classical conditioning in patients with cerebellar degeneration. *Brain,* 116, 961–969.
Totman, R. (1988). Stress, language and illness. In S. Fisher & J. Reason (Eds.), *Handbook of life stress, cognition and health* (pp. 531–542). New York: John Wiley and Sons.
Townend, W. J., et al. (2002). Head injury outcome prediction in the emergency department: A role for protein S-100B? *Journal of Neurology, Neurosurgery, and Psychiatry, 73*, 542–546.
Tracy, J. I. (2007). The association of mood with quality of life ratings in epilepsy. *Neurology, 68*, 1101–1107.
Trainin, N., et al. (1996). Thymic hormones, viral infections and Psychoneuroimmunology. In H. Friedman et al. (Eds.), *Psychoimmunology, stress, and infection* (pp. 215–229). Boca Raton, FL: CRC Press.
Tramontana, M. G., & Hooper, S. R. (1989). Neuropsychology of child psychopathology. In C. R. Reynolds & E. Fletcher-Janzen (Eds.), *Handbook of clinical child neuropsychology* (pp. 87–106). New York: Plenum.
Tranel, D., et al. (2008). Is the prefrontal cortex important for fluid intelligence? A neuropsychological study using matrix reasoning. *The Clinical Neuropsychologist, 22*, 242–261.
Tranel, S. (2002). Functional neuroanatomy: Neuropsychological correlates of cortical and subcortical damage. In S. C. Yudofsky & R. E. Hales (Eds.), *Textbook of neuropsychiatry and clinical neurosciences, 4th Ed.* (pp. 71–113). Washington, DC: American Psychiatric Publishing.
Travell, J. G., & Simons, D. G. (1996). *Trigger point flip charts.* Baltimore: Williams & Wilkins.
Treip, C. S. (1970). Hypothalamic and pituitary injury. *Journal of Clinical Pathology, 32*(4), 178–186.
Tremblay, S., et al. (2003). Somatosensory basis of speech production. *Nature*, 423, 866–869.
Trescher, W. H., & Lesser, R. P. (2000). The epilepsies. In W. G. Bradley et al. (Eds.) *Neurology in Clinical Practice,* 3rd ed., pp. 1745–1779. Boston: Butterworth-Heinemann.
Trescher, W. H., & Lesser, R. P (2004). The epilepsies. In W. G. Bradley et al. (Eds.), *Neurology in clinical practice, 4th Ed.* (Vol. 2, pp. 1953–1952). Boston: Butterworth-Heinemann.
Trevarthen, C. (1990). Integrative functions of the cerebral commissures. In F. Boller & J. Grafman (Eds.), *Handbook of neuropsychology* (Vol. IV, pp. 115–150). New York: Elsevier Science.
Trimble, M. R. (1991). Behavior and personality disturbances. In W. G. Bradley et al. (Eds.), *Neurology in clinical practice* (Vol. I, pp. 81–100). Boston: Butterworth-Heinemann.
Troost, P. T. (2004). Dizziness and vertigo. In W. G. Bradley et al. (Eds.), *Neurology in clinical practice, 4th Ed.* (Vol. 1, pp. 233–246). Boston: Butterworth-Heinemann.
Trudeau, D. L., et al. (1998). Findings of mild traumatic brain injury in combat veterans with PTSD and a history of blast concussion. *The Journal of Neuropsychiatry and Clinical Neurosciences, 10*, 308–313.
Truettner, J. S., et al. (2007). Subcellular stress response after traumatic brain injury. *Journal of Neurotrauma, 24*, 599–612.
Tsigos, C., & Chrousos, G. P. (1996). Stress, endocrine manifestations, and diseases. In C. L. Cooper (Ed.), *Handbook of stress, medicine, and health* (pp. 61–85). Boca Raton, FL: CRC Press.
Tucker, D. M., & Luu, P. (1998). Cathexis revisited: Corticolimbic resonance and the adaptive control of memory. In R. M. Bilder & F. F. LeFever (Eds.), *Neuroscience of the mind on the centennial of Freud's Project for a Scientific Psychology* (Vol. 643, pp. 134–152). Annals of the New York Academy of Science.
Tucker, D. M., et al. (1995). Social and emotional self-regulation. In J. Grafman et al. (Eds.), *Structure and functions of the human prefrontal cortex* (Vol. 769, pp. 213–239). Annals of the New York Academy of Sciences, Vol. 769.
Tulving, E. (2000). Memory: Introduction. In M. S. Gazzaniga (Ed.) *The New Cognitive Neurosciences,* 2nd ed., 721–732. Cambridge, MA: MIT Press.
Tupper, D. (1986). Neuropsychological screening and soft signs. In J. E. Obrzut and G. W. Hynd (Eds.) *Child Neuropsychology,* vol. 2, pp. 139–186. Orlando, FL: Academic Press.
Tupper, D. (1987). Appendix A: Physical and neurological examination for soft signs. In D. E. Tupper (Ed.), *Soft neurological signs* (pp. 339–353). Orlando, FL: Grune & Stratton.
Tupper, D. E. (1992). The issues with "soft signs". In D. E. Tupper (Ed.), *Soft neurological signs* (pp. 339–353). Orlando, FL: Grune & Stratton.
Turk, D. C., & Marcus, D. A. (1994). Assessment of chronic pain patients. *Seminars in Neurology, 14*, 206–212.
Turkestra, L. S., et al. (1995). Assessment of pragmatic communication skills in adolescents after traumatic brain injury. *Brain Injury, 10*, 329–345.
Turnbull, S. J., et al. (2001). Post-traumatic stress disorder symptoms following a head injury. *Brain-Injury, 15*, 775–785.
Tusa, R. J., & Brown, S. B. (1996). Neuro-otologic trauma and dizziness. In M. Rizzo & D. Tranel (Eds.), *Head injury and postconcussive syndrome* (pp. 177–200). New York: Churchill Livingston.

U.S. Government. (2007, July). Serve, support, simplify. Report of the President's Commission on Care for America's Returning Wounded Warriors. Washington, DC: Author.

Uceyler, N. (2007). Differential expression of cytokines in painful and painless neuropathies. *Neurology, 68*, 152–154.

Ullsperger, M., et al. (2002). Interactions of focal cortical lesions with error processing: Evidence from event-related brain potentials. *Neuropsychology, 16*, 548–561.

Unger, R. H., & Foster, D. W. (1998). Diabetes mellitus. In J. D. Wilson et al. (Eds.), *Williams textbook of endocrinology, 9th Ed.*(pp. 973–1059). Philadelphia: Saunders.

Uno, K., et al. (2006). Neuronal pathway from the liver modulates energy expenditure and systemic insulin sensitivity. *Science, 312*, 1656–1659.

Unterharnscheidt, F. J. (1972). Translational versus rotational acceleration: Animal experiments with measured input. *Scandinavian Journal of Rehabilitation Medicine, 4*, 24–26.

Ursin, H. (1998). The psychology in psychoneuroendocrinology. *Psychoneuroendocrinology, 23*, 555–570.

Uryu, K., et al. (2002). Repetitive mild train trauma accelerates Aβ deposition, lipid peroxidation and cognitive impairment in a trangenetic mouse model of Alzheimer amylodosis. *Journal of Neuroscience, 22*(2), 446–454.

Utiger, R. D. (2001). The thyroid: Physiology, thyrotoxicosis, hypothyroidism, and the painful thyroid. In P. Felig & L. A. Frohman (Eds.), *Endocrinology and metabolism, 4th Ed.* (pp. 261–347). New York: McGraw-Hill.

Vaccaro, A. R., et al. (1998). Long-term evaluation of vertebral artery injuries following cervical spine trauma using magnetic resonance angiography. *Spine, 237*, 789–795.

Vacchio, M. S. (2000). Thymus. In G. Fink (Ed.), *Encyclopedia of stress* (Vol. 3, pp. 598–605). San Diego, CA: Academic Press.

Vakalopoulos, C. (2006). Neuropharmacology of cognition and memory: A unifying theory of neuromodulator imbalance in psychiatry and amnesia. *Medical Hypotheses, 66*, 394–431.

Valentino, J. J., et al. (1993). The locus coeruleus as a site for integrating corticotropin-releasing factor and noradrenergic mediation of stress responses. In Y. Taché & C. Rivier (Eds.), *Corticotropin-releasing factor and cytokines: Role in the stress response* (Vol. 691, pp. 173–188). Annals of the New York Academy of Science.

Van Cauter, E., et al. (1966). Effects of gender and age on the levels and circadian rhythmicity of plasma cortisol. *Journal of Clinical Endocrinology and Metabolism, 81*, 2468–2473.

van der Kolk, B. A. (1988). The trauma spectrum: The interaction of biological and social events in the genesis of the trauma response. *Journal of Traumatic Stress,* 1, 273–290.

Van Lancker, D., & Breitenstein, C. (2000). Emotional dysprosody and similar dysfunctions. In J. Bogousslavsky & J. L. Cummings (Eds.), *Behavior and mood disorders in focal brain lesions* (pp. 327–368). New York: Cambridge University Press.

Van Leden, H. (1961). The mechanism of phonation: Search for a rational theory of voice production. *Archives of Otolaryngology, 74*, 72–88.

Van Loey, N. E. E., et al. (2003). Predictors of chronic posttraumatic stress symptoms following burn injury: Results of a longitudinal study. *Journal of Traumatic Stress, 16*, 361–369.

Van Mier, H. I., & Petersen, S. E. (2002). Role of the cerebellum in motor cognition. In S. M. Highstein & W. T. Thach (Eds.), *The cerebellum: Recent developments in cerebellar research* (Vol. 978, pp. 334–353). Annals of The New York Academy of Sciences.

Van Reekum, R., et al. (2000). Can traumatic brain injury cause psychiatric disorders? *Journal of Neuropsychiatry and Clinical Neurosciences, 12*, 316–327.

Van Zomeren, A. H., & Brouer, W. H. (1992). Assessment of attention. In J. R. Crawford, et al. (Eds.). *A handbook of neuropsychological assessment.*

Vanderling, J. J., et al. (2006). Neuropsychological outcomes of Army personnel following deployment to the Iraq War. *Journal of the American Medical Association, 296*, 519–529.

Vanderploeg, R. D., et al. (2005). Long-term neuropsychological outcomes following mild traumatic brain injury. *Journal of the International Neuropsychological Society, 15*, 227–236.

Vargha-Khadam, F., et al. (1985). Aphasia and handedness in relation to hemispheric side, age at injury and severity of cerebral lesion during childhood. *Brain,* 108, 677–696.

Vargas, M. E., & Barres, B. A. (2007). Why is Wallerian degeneration in the CNS so slow? *Annual Review of Neuroscience, 30*, 153–179.

Varney, N. R., & Menefee, L. (1993). Psychosocial and executive deficits following closed head injury: Implications for orbital frontal cortex. *Journal of Head Trauma Rehabilitation,* 8 (1), 32–44.

Vasterling, J. J., et al. (2000). Head injury as a predictor of psychological outcome in combat veterans. *Journal of Traumatic Stress, 13*, 441–451.

Vecchiet, L., & Giamberardino, M. A. (2003). Referred pain. In M. J. Aminoff & R. B. Daroff (Eds.), *Encyclopedia of the neurological sciences* (Vol. 3, pp. 764–769). Boston: Academic Press.

Vedantam, S. (2006, May 11). Returning troops may not be getting help they need. *The Honolulu Advertiser*. Retrieved from http://www.honoluluadvertiser.com/apps/pbcs.dll/article?AID-20.1.

Vedhara, K., & Irwin, M. (Eds.). (2005). *Human psychoimmunology*. New York: Oxford University Press.

Vedhara, K., & Wang, E. C. Y. (2005). Assessment of the immune system in human psychoneuroimmunology. In K. Vedhara & M. Irwin (Eds.), *Human psychoneuroimmunology* (pp. 53–80). New York: Oxford University Press.

Veldhuis, J. D., et al. (1998). The effects of mental and metabolic stress on the female reproductive system and female reproductive hormones. In J. R. Hubbard & E. A. Workman (Eds.), *Handbook of stress medicine: An organ system approach* (pp. 115–140). Boca Raton, FL: CRC Press.

Verberne, A. J. M. (2003). Medulla oblongata. In M. J. Aminoff & R. B. Daroff (Eds.), *Encyclopedia of the neurological sciences* (Vol. 3, pp. 54–63). Boston: Academic Press.

Vernick, D. M. (1996). Hearing loss and tinnitus. In M. S. Samuels & S. Feske (Eds.), *Office practice of neurology* (pp. 91–98). New York: Churchill Livingstone.

Villarreal, G., & King, C. Y. (2001). Brain imaging in posttraumatic stress disorder. *Seminars in Clinical Neuropsychiatry, 6*, 131–145.

Voeller, K. K. S. (1998). Nonverbal learning disabilities and motor skills disorders. In C. E. Coffey & R. A. Brumback (Eds.), *Textbook of pediatric neuropsychiatry* (pp. 719–767). Washington, DC: American Psychiatric Press.

Voshall, L. B., & Stocker, R. F. (2007). Molecular architecture of smell and taste in *Drosophila. Annual Review of Neuroscience, 30*, 505–533.

Vungkhanching, M., et al. (2007). Feasability of a skills-based substance abuse prevention program following traumatic brain injury. *Journal of Head Trauma Rehabilitation, 122*, 167–176.

Wade, D. T., et al. (1998). Routine follow up after head injury: A second randomised controlled trial. *Journal of Neurology, Neurosurgery, and Psychiatry, 65*, 177–183.

Wade, N. (2003, July 15). Early Voices: the leap to language. *The New York Times*, F1, 4.

Wade, S. L., et al. (1996). Childhood traumatic brain injury: Initial impact on the family. *Journal of Learning Disabilities,* 29, 652–661.

Wadhwa, P. D., et al. (1997). Placental CRH modulates maternal pituitary-adrenal function in human pregnancy. *Neuropeptides in development and aging* (Vol. 814, pp. 276–281). Annals of the New York Academy of Sciences.

Wagner, A. W., et al. (2000). An investigation of the impact of posttraumatic stress disorder on physical health. *Journal of Traumatic Stress, 13*, 41–55.

Waheed, W., & Pestronk, A. (2004). Muscle pain and cramps. In W. G. Bradley et al. (Eds.), *Neurology in clinical practice, 4th Ed.* (Vol. 1, pp. 387–392). Boston: Butterworth-Heinemann.

Walker, L. G., et al. (2005). Psychoneuroimmunology and chronic malignant disease: Cancer. In K. Vedhara & M. Irwin (Eds.), *Human psychoneuroimmunology* (pp. 137–163). New York: Oxford University Press.

Walker, R., et al. (2007). Screening substance abuse treatment clients for traumatic brain injury: Prevalence and characteristics. *Journal of Head Trauma Rehabilitation, 22*, 360–367.

Walker, W. C. (2004). Pain pathoetiology after TBI: Neural and nonneural mechanisms. *Journal of Head Trauma Rehabilitation, 19,* 72–81.

Wallesch, C. W., et al. (2001). The neuropsychology of blunt head injury in the early acute stage: Effects of focal lesions and diffuse axonal injury. *Journal of Neurotrauma, 18*, 11–19.

Wallis, J. D. (2007). Orbitofrontal cortex and its contribution to decision-making. *Annual Review of Neuroscience, 30*, 31–56.

Walton, J. (1985). *Brain's disease of the nervous system*, 9th Ed. New York: Oxford University Press.

Walton, M. E. (2007). Calculating the cost of acting in frontal cortex. In B. W. Balleine et al. (Eds.), *Reward and decision making in corticobasal ganglia networks* (Vol. 1104, pp. 340–356). Annals of the New York Academy of Sciences.

Walton, M. E., et al. (2007). In B. W. Balleine et al. (Eds.), *Reward and decision making in corticobasal ganglia networks* (Vol. 1104, pp. 340–356). Annals of the New York Academy of Sciences.

Walz, N. C., et al. (2008). Late proton magnetic resonance spectroscopy following traumatic brain injury during early childhood: Relationship with neurobehavioral outcomes. *Journal of Neurotrauma, 25*, 94–103.

Ward, J. D. (1989). Pediatric head injuries: Special considerations. In D. P. Becker & S. K. Gudeman (Eds.), *Textbook of head injury* (pp. 319–325). Philadelphia: Saunders.

Ward, T. N., & Levin, M. (2005). Headaches. In J. M. Silver et al. (Eds.), *Textbook of traumatic brain injury* (pp. 385–391). Washington, DC: American Psychiatric Publishing.

Warden, D. L., et al. (1997). Posttraumatic stress disorder in patients with traumatic brain injury and amnesia for the event? *Neuropsychiatry, 9,* 18–22.
Warren, W. L., & Bailes, J. E. (1998). On the field evaluation of athletic head injuries. *Clinics in Sports Medicine, 17*(1), 13–26.
Warschausky, S., et al. (1999). Empirically supported psychological and behavioral therapies in pediatric rehabilitation of TBI. *Journal of Head Trauma Rehabilitation,* 14, 373–383.
Watanabe, T. K., et al. (1999). Traumatic brain injury associated with acute spinal cord injury: Risk factors, evaluation, and outcomes. *Topics in Spinal Cord Injury Rehabilitation, 5,* 83–90.
Waterhouse, J. (2007). Jet lag: Trends and coping strategies. *Lancet, 369,* 1117–1129.
Watson, S. J., & Akil, H. (1991). The brain's stress axis: An update. A. Tasman & S. M. Goldfinger (Eds.), *Annual review of psychiatry* (Vol. 10, pp. 499–514). Washington DC: American Psychiatric Press.
Watts, A. G. (2000). Brain and brain regions. In G. Fink (Ed.), *Encyclopedia of stress* (Vol. 3, pp. 342–348). San Diego, CA: Academic Press.
Weathers, F. W., & Keane, T. M. (2007). The criterion A problem revisited: Controversies and challenges in defining and measuring psychological trauma. *Journal of Traumatic Stress, 20,* 107–121.
Webb, P., & Baxter J. D. (2007). Introduction to endocrinology. In D. G. Gardner & D. Shoback (Eds.), *Greenspan's basic & clinical endocrinology* (pp. 1–34). New York: McGraw-Hill.
Weber, L. M., & Verbanets, J. (1986). Assessing balance performance in moderate brain injury. *Topics in Acute Care Trauma Rehabilitation, 1,* 84–94.
Webster, J. B., & Bell, K. R. (1997). Adrenal insufficiency following traumatic brain injury: A case report and review of the literature. *Archives of Physical Medicine and Rehabilitation, 78,* 314–318.
Webster, J. B., et al. (2001). Sleep apnea in adults with traumatic brain injury: a preliminary investigation. *Archives of Physical Medicine and Rehabilitation, 82,* 316–321.
Wecker, N. S., et al. (2000). Age effects on executive ability. *Neuropsychology, 14,* 409–414.
Weeks, R. E. (1996). Psychological assessment of the headache patient. In M. S. Samuels & S. Feske (Eds.), *Office practice of neurology* (pp. 1096–1101). New York: Churchill Livingstone.
Wegner, D. M. (1988). Stress and mental control. In S. Fisher & J. Reason (Eds.), *Handbook of life stress, cognition and health* (pp. 683–697). New York: John Wiley and Sons.
Weiner, H. L., & Weinberg, J. S. (2000). Head injury in the pediatric age group. In P. R. Cooper & J. G. Golfinos (Eds.), *Head injury, 4th Ed.* (pp. 418–456). New York: McGraw Hill.
Weiner, H. L., et al. (1982). *Pediatric neurology for the house officer, 2nd Ed.* Baltimore: Williams and Wilkins.
Weinstein, E. A. (1991). Anosognosia and denial of illness. In G. P. Prigatano & D. L. Schachter (Eds.), *Awareness of deficit after brain injury* (pp. 240–257). New York: Oxford University Press.
Weintraub, S. (2000). Neuropsychological assessment of mental state. In M. M. Mesulam (Ed.), *Principles of behavioral and cognitive neurology, 2nd Ed.* (pp. 121–173). New York: Oxford.
Weisinger, R. S., & Denton, D. A. (2000). Salt appetite. In G. Fink (Ed.), *Encyclopedia of stress, 3* (pp. 384–392). San Diego, CA: Academic Press.
Weiskrantz, L. (1992). Introduction: Dissociated issues. In A. S. Milner & M. D. Rugg (Eds.), *The neuropsychology of consciousness* (pp. 1–10). San Diego: Academic Press.
Weiss, J. M., & Sundar, S. (1992). Effects of stress on cellular immune responses in animals. In A. Tasman & M. R. Riba (Eds.). *Annual review of psychiatry* (Vol. 11, pp. 145–168). Washington D.C.: American Psychiatric Press.
Weiss, J. M., et al. (1985). Neurochemical basis of stress-induced depression. *Psychopharmacology Bulletin, 21*(3), 447–457.
Weiss, J. M., et al. (1994). Widespread activation and consequences of interleukin-1 in the brain. In N. A. Fabris et al. (Eds.), *Immunomodulation: The state of the art* (Vol. 741, pp. 338–357). Annals of the New York Academy of Sciences.
Welsh, M. D., & Pennington, B. F. (1988). Assessing frontal lobe functioning n children: Views from developmental psychology. *Developmental Psychology, 4*(3), 199–230.
Welsh-Bohmer, K. A., et al. (2009). Neuropsychological performance in advanced age: Influences of demographic factors and Apolipoprotein E: Findings from the Cache County Memory Study. *The Clinical Neuropsychologist, 23,* 77–99.
Wertz, R. T., et al. (1998). Affective prosodic disturbance subsequent to right hemisphere stroke: A clinical application. *Journal of Linguistics, 11*(1–2), 89–102.
Wetherby, A. M. (1985). Speech and language disorders in children: An overview. In J. K. Darby (Ed.), *Speech and language evaluation in neurology: Childhood disorders* (pp. 3–32). New York: Grune and Stratton.
Wetsel, W. C., et al. (1998). Functional Morphology of the immortalized hypothalamic LHRH neurons. In L. D. Van de Kar (Ed.), *Methods in neuroendocrinology* (pp. 1–16). Boca Raton, FL: CRC Press.

Whelan, B. M., et al. (2007). Delineating communication impairments associated with mild traumatic brain injury: A case report. *Journal of Head Trauma Rehabilitation, 122*, 192–197.
Whitehouse, B. J. (2000). Adrenal cortex. In G. Fink (Ed.), *Encyclopedia of stress* (Vol. 1, pp. 42–52). San Diego, CA: Academic Press.
Whyte, J. et al. (1996). Inattentive behavior after traumatic brain injury. *Journal of the International Neuropsychological Society, 2,* 274–281.
Wiencken-Barger, A. E., & Casagrande, V. A. (2002). Visual system development and neural activity. In V. S. Ramachandran (Ed.), *Encyclopedia of the human brain* (Vol. 4, pp. 751–804). Boston: Academic Press.
Wiercisiewski, D. R., & McDeavitt, J. T. (1998). Pulmonary complications in traumatic brain injury. *Journal of Head Trauma Rehabilitation, 13*, 28–35.
Wilberger, J. E. (2000). Emergency care and initial evaluation. In P. R. Cooper & J. G. Golfinos (Eds.), *Head injury, 4th Ed.* (pp. 27–40). New York: McGraw Hill.
Wilbourne, A. J. (2003). Brachial plexopathies. In M. J. Aminoff & R. B. Daroff (Eds.), *Encyclopedia of the neurological sciences* (Vol. 1, pp. 411–417). Boston: Academic Press.
Wilcox, J. A. & Nasrallah, H. A. (1987). Childhood head trauma and psychosis. *Psychiatry Research, 12*, 303–306.
Willenberg, H. S., et al. (2000). Adrenal insufficiency. In G. Fink (Ed.), *Encyclopedia of stress* (Vol. 1, pp. 58–63). San Diego, CA: Academic Press.
Williams, J. A., et al. (2001). A circadian output in Drosophila mediated by Neurofibromatosis-1 and Ras/MARK. *Science, 293*, 2251–2256.
Williams, J. H. G. (1998). Using behavioural ecology to understand depression. *British Journal of Psychiatry, 173*, 453–454.
Williams, P. L. (Ed., 1995). *Gray's anatomy, 38th ed.* New York: Churchill Livingston.
Willinger, R., et al. (1995). Modal and temporal analysis of head mathematical models. *Journal of Neurotrauma, 12*, 743–754.
Willingham, D. B. (1992). Systems of motor skill. In L. R. Squire & N. Butters (Eds.), *Neuropsychology of memory, 2nd Ed.* (pp. 166–178). New York: The Guilford Press.
Willis, W. G., & Widerstrom, A. H. (1986). Neuropsychological development. In J. E. Obrzut & G. W. Hynd (Eds.), *Neuropsychology* (Vol. 1, pp. 13–53). Orlando, FL: Academic Press.
Wilson, B. (1996). Ecological validity of neuropsychological assessment after severe brain injury. In R. J. Sbordone & C. J. Long (Eds.), *Ecological validity of neuropsychological testing* (pp. 414–428). DelRay Beach, FL: GR Press/St. Lucie Press.
Wilson, J. T. L. (1991). Significance of MRI in clarifying whether neuropsychological deficits after head injury are organically based. *Neuropsychology, 4*, 261–269.
Wilson, J. T. L., & Wyper, D. (1992). Neuroimaging and neuropsychological functioning following closed head injury: CT, MRI, and SPECT. *Journal of Head Trauma Rehabilitation* 7, 2, 29–39.
Wilson, R. S., et al. (2007). Relation of cognitive activity to risk of developing Alzheimer disease. *Neurology, 69*, 1911–1920.
Wilson, R. S., et al. (2008). Educational attainment and cognitive decline in old age. *Neurology, 72*, 460–465.
Wilson, T. D., & Bar-Anan, Y. (2008). The unseen mind. *Science, 321*, 1046–1047.
Wilson-Pauwels, L., et al. (1988). *Cranial nerves*. Philadelphia: B.C. Decker.
Wingfield, J. C., et al. (1997). Ecological constraints and the evolution of hormone-behavior interrelationships. In C. S. Carter et al. (Eds.), *The integrative neurobiology of affiliation* (Vol. 807, pp. 22–41). Annals of the New York Academy of Sciences.
Winograd, P., & Niquette, G. (1988). Assessing learned helplessness in poor readers. *Topics in Language Disorders, 8*(3), 38–55.
Wittmann, L., et al. (2006). Low predictive power of peritraumatic dissociation for PTSD symptoms in accident survivors. *Journal of Traumatic Stress, 19,* 639–651.
Woertgen, D., et al. (2002). Does bungee jumping release S-100B protein? *Journal of Clinical Neuroscience, 9*, 51–52.
Wolburg, H., & Warth, A. (2005). Endothelial cells, extracellular matrix, and astrocytes: Interplay for managing the blood-brain barrier. The transport systems of the blood-brain barrier. In E. De Vries & A. Prat (Eds.), *The blood-brain barrier and its microenvironment* (pp. 359–386). New York: Taylor & Francis.
Wolf, O. T., et al. (2004). Cortisol and memory retrieval in humans: Influence of emotional valence. In R. Yehuda & B. McEwen (Eds.), *Behavioral stress responses* (Vol. 1032, pp. 195–197). Annals of the New York Academy of Sciences.
Wolkowitz, O. M., et al. (2004). The "steroid dementia syndrome": An unrecognized complication of glucocorticoid treatment. In R. Yehuda & B. McEwen (Eds.), *Behavioral stress responses* (Vol. 1032, pp. 191–194). Annals of the New York Academy of Sciences.

Wonderlic Personnel Test & Scholastic Level Exam: User's Manual (1992). Libertyville, IL: Wonderlic Personnel Test, Inc.

Wood, R. L., & Williams, C. (2008). Inability to empathize following traumatic brain injury. *Journal of the International Neuropsychological Society, 14*, 289–296.

Woodcock, R. (1993). An information processing view of Gf-Gc theory. *Journal of Psychoeducational Assessment,* Monograph Series: Woodcock-Johnson Psycho-Educational Battery, Revised.

Woods, B. T. (1987). Impaired speech following early lesions of either hemisphere. *Neuropsychologia, 25*, 519–525.

Woodward, S. H. (1995). Neurobiological perspectives on sleep in post-traumatic stress disorder. In M. J. Friedman et al. (Eds.), *Neurobiological and clinical consequences of stress: From normal adaptation to PTSD* (pp. 315–333). Philadelphia: Lippincott-Raven.

Woodward, S. H., et al. (2007). Brain, skull, and cerebrospinal fluid volumes in adult posttraumatic stress disorder. *Journal of Traumatic Stress, 20*, 763–774.

Woolf, P. D. (1992). Hormonal responses to trauma. *Critical Care Medicine, 20,* 216–226.

Woolf, P. D., et al. (1986). Transient hypogonadotrophic hypogonadism after head trauma: Effects on steroid precursors and correlation with sympatic nervous system activity. *Clinical Endocrinology, 25,* 265–274.

Woolf, P. D., et al. (1988). Thyroid test abnormalities in traumatic brain injury: Correlation with neurologic impairment and sympathetic nervous system activation. *The American Journal of Medicine, 14*, 201–208.

Woolley, J. D., et al. (2004). The autonomic and behavioral profile of emotional dysregulation. *Neurology,* 64, 1740–1743.

Worthington, A. D., & Melia, Y. (2006). Rehabilitation is compromised by arousal and sleep disorders: Results of a survey of rehabilitation centers. *Brain Injury, 20*, 327–332.

Wurtz, R. H., & Kandel, E. R. (2000). Perception of motion, depth and form. In E. R. Kandel et al. (Eds.), *Principles of neural science, 4th ed.* (pp. 548–571). New York: McGraw Hill.

Wüst, S., et al. (2004). A psychobiological perspective on genetic determinants of hypothalamus-pituitary-adrenal axis activity. *Annals of the New York Academy of Sciences, 1032*, 52–62.

Yaffe, K., et al. (2003). Inflammatory markers and cognition in well-functioning African-American and white elders. *Neurology, 61*, 76–80.

Yager, J., & Gitlin, M. J. (1995). Clinical manifestations of psychiatric disorders. In H. I. Kaplan & B. J. Sacock (Eds.), *Comprehensive textbook of psychiatry, 6th Ed.* (Vol. I, p. 637–669). Baltimore: Williams & Wilkins.

Yager, J., & Gitlin, M. J. (2000). Clinical manifestations of psychiatric disorders. In B. J. Sadock & V. A. Sadock (Eds.), *Comprehensive textbook of psychiatry, 7th Ed.* (pp. 789–823). Philadelphia: Lippincott, Williams & Wilkins.

Yager, J., & Gitlin, M. J. (2005). Clinical manifestations of psychiatric disorders. In B. J. Sadock & V. A. Sadock (Eds.), *Comprehensive textbook of psychiatry, 8th Ed.* (pp. 963–1003). Philadelphia: Lippincott, Williams & Wilkins.

Yamamoto, I., & Sato, O. (1986). Intrauterine development of the skull. In A. J. Raimondi et al. (Eds.), *Head injuries in the newborn and infant* (pp. 1–18). New York: Springer Verlag.

Yamamoto, Y., & Friedman, H. (1996). Steroids and infection. Stress, burnout, and health. In C. L. Cooper (Ed.), *Handbook of stress, medicine, and health* (pp. 173–194). Boca Raton, FL: CRC Press.

Yang, E. V., & Glaser, R. (2005). Wound healing and psychoneuroimmunology. In K. Vedhara & M. R. Irwin (Eds.), *Human psychoimmunology* (pp. 265–284). New York: Oxford University Press.

Yeates, K. O. (1994). Head injuries: Psychological issues. In R. A. Olson et al. (Eds.), *The sourcebook of pediatric psychology* (pp. 262–275). Boston: Allyn & Baron.

Yeates, K. O. (2000). Closed head injury. In K. O. Yeates et al. (Eds.), *Pediatric neuropsychology* (pp. 92–116). New York: Guilford Press.

Yeates, K. O., & Taylor, H. G. (1997). Predicting premorbid neuropsychological functioning following pediatric traumatic brain injury. *Journal of Clinical and Experimental Neuropsychology, 19*, 825–837.

Yeates, K. O., & Taylor, H. G. (2005). Neurobehavioral outcomes of mild head injury in children and adolescents. *Pediatric Rehabilitation, 8*(12), 5–16.

Yeates, K. O., et al. (1999). Postconcussive symptoms in children with mild closed head injuries. *Journal of Head Trauma Rehabilitation, 14*, 337–450.

Yeates, K. O., et al. (2002). Race as a moderator of parent and family outcomes following pediatric traumatic brain injury. *Journal of Pediatric Psychology, 27,* 393–403.

Yeates, K. O., et al. (2004). Short- and long-term social outcomes following pediatric traumatic brain injury. *Journal of the International Neuropsychological Society, 10*, 412–426.

Yehuda, R., & Flory, J. D. (2007). Differentiating biological correlates of risk, PTSD, and resilience following trauma exposure. *Journal of Traumatic Stress, 20*, 435–447.

Yehuda, R., & McEwen, B. (2004). Introduction: Neurobehavioral stress responses. *Annals of the New York Academy of Sciences, 1032*, xi–xvi.

Yehuda, R., & Wong, C. M. (2000). Acute stress disorder and posttraumatic stress disorder. In G. Fink (Ed.), *Encyclopedia of stress* (Vol. 1, pp. 1–7). San Diego, CA: Academic Press.

Yehuda, R., et al. (1995). Hypothalamic-pituitary-adrenal functioning in post-traumatic stress disorder. In M. J. Friedman et al. (Eds.), *Neurobiological and clinical consequences of stress: From normal adaptation to PTSD* (pp. 351–365). Philadelphia: Lippincott-Raven.

Yehuda, R., et al. (2007). Ten-year follow-up study of cortisol levels in aging Holocaust survivors with and without PTSD. *Journal of Traumatic Stress, 20*, 757–761.

Yi, B. A., & Jan L. Y. (2002). Ion channels. In V. S. Ramachandran (Ed.), *Encyclopedia of the human brain* (Vol. 2, pp. 599–615). Boston: Academic Press.

Yirmaya, R., et al. (2000). Illness, cytokines, and depression. In A. Conti et al. (Eds.), *Neuroimmunomodulation: Perspectives at the new millennium* (Vol. 917, pp. 478–487). Annals of the NY Academy of Sciences.

Ylvisaker, M. (1989). Cognitive and psychosocial outcome following head injury in children. In J. T. Hoff et al. (Eds.), *Mild to moderate head injury* (pp. 203–216). Boston: Blackwell Scientific Publications.

Young, L. J., et al. (2005). Neuropeptides: Biology, regulation, and role in neuropsychiatric disorders. In B. J. Sadock & V. A. Sadock (Eds.), *Comprehensive textbook of psychiatry, I, 8th Ed.* (pp. 161–171). Philadelphia: Lippincott Williams & Wilkins.

Younger, D. S. (1999). Overview of motor disorders. In D. S. Younger (Ed.), *Motor disorders* (pp. 3–17). Philadelphia: Lippincott, Williams & Wilkins.

Younger, D. S. (2005a). Plexus disorders. In D. Younger (Ed.), *Motor disorders, 2nd Ed.* (pp. 247–263). Philadelphia, PA: Lippincott, Williams & Wilkins.

Younger, D. S. (2005b). Entrapment neuropathies. In D. Younger (Ed.), *Motor disorders, 2nd Ed.* (pp. 265–279). Philadelphia, PA: Lippincott, Williams & Wilkins.

Younger, D. S. (2005c). Laryngeal motor disorders. In D. Younger (Ed.), *Motor disorders, 2nd Ed.* (pp. 121–134). Philadelphia, PA: Lippincott, Williams & Wilkins.

Younger, D. S. (2005d). Myalgia, fibromyalgia, fasciculation, and cramps. In D. Younger (Ed.), *Motor disorders, 2nd Ed.* (pp. 219–230). Philadelphia, PA: Lippincott, Williams & Wilkins.

Younger, D. S. (2005e). Vertigo and vestibular disorders. In D. Younger (Ed.), *Motor disorders, 2nd Ed.* (pp. 99–119). Philadelphia, PA: Lippincott, Williams & Wilkins.

Zafonte, R. D., & Horn, L. J. (1999). Clinical assessment of posttraumatic headaches. *Journal of Head Trauma Rehabilitation, 14*(1), 22–33.

Zappulla, R. A. (1997). Epilepsy and consciousness. *Seminars in Neurology,* 17, 113–119.

Zasler, N. D., & Martelli, M. F. (2005). Sexual dysfunction. In J. M. Silver et al. (Eds.), *Textbook of traumatic brain injury* (pp. 437–450). Washington, DC: American Psychiatric Publishing.

Zasler, N. D., et al. (2005). Chronic pain. In J. M. Silver et al. (Eds.), *Textbook of traumatic brain injury* (pp. 419–433). Washington, DC: American Psychiatric Publishing.

Zasler, N. D., et al. (2007). Post-traumatic pain disorders: Medical assessment and management. In N. D. Zasler et al. (Eds.), *Brain injury medicin*e (pp. 697–721). New York: Demos.

Zattore, R. J. (2001). Neural specializations for tonal processing. In *The biological foundations of music* (Vol. 930, pp. 193–210). Annals of the New York Academy of Sciences.

Zee, P. C., & Manthena, P. (2007). The brain's master circadian clock: Implications and opportunities for therapy of sleep disorders. *Sleep Medicine Reviews, 11*, 59–70.

Zellweger, R., et al. (2001). Immunologic sequelae of surgery and trauma. In R. Ader et al. (Eds.), *Psychoneuroimmunology, 3rd Ed.* (Vol. 2, pp. 291–315). San Diego: Academic Press.

Zeman, A. Z., et al. (1997). Contemporary theories of consciousness. *Neurology, Neurosurgery, & Psychiatry, 62*, 549–552.

Zeplin, J., & Kent, R. D. (1996). Reliability of auditory-perceptual scaling of dysarthria. In V. L. Hammen & K. M. Yorkston (Eds.), *Disorders of motor speech* (pp. 145–154). Baltimore: Paul H. Brooks.

Zhang, J. V., et al. (2005). Obestatin, a peptide encoded by the ghrelin gene, opposes ghrelin's effects on food intake. *Science, 310*, 996–999.

Zhang, L., et al. (2002). The effect of predictable and unpredictable motor tasks on postural control after traumatic brain injury. *NeuroRehabilitation, 17,* 225–230.

Zhang, L., et al. (2004a). A proposed injury threshold for mild traumatic brain injury. *Transactions of the ASME, 126*, 226–236.

Zhang, T-Y., et al. (2004b). Maternal programming of individual differences in defensive responses in the rat. *Annals of The New York Academy of Sciences, 1032*, 85–101.

Zhang, W., & Stanimirovic, D. B. (2005). The transport systems of the blood-brain barrier. In E. De Vries & A. Prat (Eds.), *The blood-brain barrier and its microenvironment* (pp. 103–142). New York: Taylor & Francis.

Ziaja, M., et al. (2007). Nitric oxide spin-trapping and NADPH-Diaphorase activity in mature rat brain after injury. *Journal of Neurotrauma*, *24*, 1845–1854.

Zigmond, M. J., et al. (1995). Neurochemical studies of central noradrenergic responses to acute and chronic stress. In M. J. Friedman et al. (Eds.), *Neurobiological and clinical consequences of stress* (pp. 45–60). Philadelphia: Lippincott-Raven.

Zilles, K. (1990). Cortex. In G. Paxinos (Ed.), *The Human Nervous System* (pp. 757–802). San Diego: Academic Press.

Zimmerman, R. T., & Bilaniuk, L. T. (1981). Computed tomography in pediatric head trauma. *Journal of Neuroradiology, 8,* 257–271.

Zingler, V. C., & Pohlmann-Eden, B. (2005). Diagnostic pitfalls in patients with hypoxic brain damage: Three case reports. *Resuscitation, 65*, 107–110.

Zinser, L. (2007, March 21). After concussion, Shanahan is back for playoff chase. *New York Times*, D1.

Zoellner, L. A. (2008). Translational challenges with tonic immobility. *Psychology: Science and Practice, 15*, 98–101.

Zorumski, C. F., et al. (2005). Basic electrophysiology. In B. J. Sadock & V. A. Sadock (Eds.), *Comprehensive textbook of psychiatry, I, 8th Ed.* (pp. 99–115). Philadelphia: Lippincott Williams & Wilkins.

Zubkov, A. Y., et al. (1999). Posttraumatic cerebral vasospasm: Clinical and morphological presentations. *Journal of Neurotrauma, 16*, 763–770.

Index

A

B

C

D

E

F

G

H

I

Q

R

S

T

W